Kelly L. Murdock's 3ds Max 2020 Complete Reference Guide

Kelly L. Murdock's 3ds Max 2020 Complete Reference Guide

Kelly L. Murdock

SDC Publications

P.O. Box 1334
Mission, KS 66222
913-262-2664
www.SDCpublications.com
Publisher: Stephen Schroff

ISBN-13: 978-1-63057-253-2
ISBN-10: 1-63057-253-5

Printed and bound in the United States of America.

Dedication

When my life ends, I hope not too soon.

I will be in life's late afternoon.

I'll ascend to heaven on the wings of a dove, To paradise, a place that is filled with Godly love.

But, heaven is unfamiliar and everything will be new.

I will be like a lost child searching without a clue.

But, you'll be there to take me by the hand, and show me the beauty of this divine land.

You'll show all of Heaven's very best parts to me, and I'll marvel at the majestic sites there are to see.

The eternal choirs will be a wonder to behold, and glory of the streets that will be paved with gold.

We'll meet God and Jesus and mansions up on high, and singing praising angels flying through the sky.

There will be gatherings with loved ones who have gone on before, and knowledge a plenty, I'm sure my heart will soar.

Heaven is wonderful and more than I could believe.

There is joy being in a place where I never have to grieve.

It will be great to be free from sin and pain, but the best part of heaven will be seeing you again.

To my beloved cousin, Tina, who will be missed everyday until we're together again.

About the Author

Kelly Murdock has been authoring computer books for many years now and still gets immense enjoyment from the completed work. His book credits include various 3D, graphics, multimedia, and Web titles, including many editions of the book *3ds Max Bible.* Other major accomplishments include *Maya Basics Guide, Google SketchUp Bible, Edgeloop Character Modeling for 3D Professionals Only, Maya 6 and 7 Revealed, LightWave 3D 8 Revealed, The Official Guide to Anime Studio, Poser 6, 7, and 8 Revealed, 3D Game Animation For Dummies, gmax Bible, Adobe Atmosphere Bible, Master VISUALLY HTML and XHTML, JavaScript Visual Blueprint,* and co-authoring duties on two editions of the *Illustrator Bible* (for versions 9 and 10) and five editions of the *Adobe Creative Suite Bible.*

With a background in engineering and computer graphics, Kelly has been all over the 3D industry and still finds it fascinating. He's used high-level CAD workstations for product design and analysis, completed several large-scale visualization projects, created 3D models for feature films and games, worked as a freelance 3D artist, and even did some 3D programming. Kelly's been using 3D Studio since version 3 for DOS. Kelly has also branched into training others in 3D technologies. He currently works as a freelance graphic artist and video game producer.

In his spare time, Kelly enjoys playing basketball and collecting video games.

Preface

Every time I enter the computer room (which my wife calls the dungeon), my wife still says that I am off to my "fun and games." I, as always, flatly deny this accusation, saying that it is serious work that I am involved in. But later, when I emerge with a twinkle in my eye and excitedly ask her to take a look at my latest rendering, I know that she is right. Working with the Autodesk® 3ds Max® 2020 software is pure "fun and games."

My goal in writing this book was to take all my fun years of playing and working in 3D and boil them down into something that's worthwhile for you, the reader. This goal was compounded by the fact that all you 3ds Max-heads out there are at different levels. Luckily, this book is thick enough to include a little something for everyone.

The audience level for the book focuses on the beginner, with a smattering of intermediate and advanced topics for the seasoned user. If you're new to 3ds Max, then you'll want to start at the beginning and move methodically through the book. If you're relatively comfortable making your way around 3ds Max, then review the Table of Contents for sections that can enhance your fundamental base. If you're a seasoned pro, then you'll want to watch for coverage of the features new to Release 2020.

As this book has come together, I've tried to write the type of book that I'd like to read. I've tried to include a variety of scenes that are infused with creativity. It is my hope that these examples will not only teach you how to use the software but also provide a creative springboard for you in your own projects. After all, that's what turns 3D graphics from work into "fun and games."

The Growth of 3ds Max?

One way we humans develop our personalities is to incorporate desirable personality traits from those around us. The personality of the 3ds Max software is developing as well: every new release has incorporated a plethora of desirable new features. Many of these features come from the many additional plug-ins being developed to enhance 3ds Max. With each new release, 3ds Max has adopted many features that were available as plug-ins for previous releases. Several new features have been magically assimilated into the core product, such as the Character Animation Toolkit (CAT) and the Hair and Fur system. These additions make the software's personality much more likable, like a human developing a sense of humor.

Other personality traits are gained by stretching in new directions. 3ds Max and its developers have accomplished this feat as well. Many of the new features are completely new, not only to 3ds Max, but also to the industry. As 3ds Max grows up, it will continue to mature by adopting new features and inventing others. I just hope 3ds Max doesn't experience a midlife crisis in the next version.

Along with adopted features and new developments, the development teams at Autodesk have sought feedback from 3ds Max users. This feedback has resulted in many small tweaks to the package that enable scenes to be created more quickly and easily.

Some additional factors have appeared in the software's house that certainly affect its development. First is the appearance of the software's adopted brother, Maya. There are other siblings in the Autodesk household (including MotionBuilder, Softimage, and AutoCAD), but Maya is closest in age to 3ds Max, and its personality likely will rub off in different ways.

About This Book

Let me paint a picture of the writing process. It starts with years of experience, which are followed by months of painstaking research. There were system crashes and personal catastrophes and the always-present, ever-looming deadlines. I wrote into the early hours of the morning and during the late hours of the night—burning the candle at both ends and in the middle, all at the same time. It was grueling and difficult, and spending all this time staring at the 3ds Max interface made me feel like . . . well . . . like a 3d artist.

Sound familiar? This process actually isn't much different from what 3D artists, modelers, and animators do on a daily basis, and, like you, I find satisfaction in the finished product.

Tutorials aplenty

I've always been a very visual learner—the easiest way for me to gain knowledge is by doing things for myself while exploring at the same time. Other people learn by reading and comprehending ideas. In this book, I've tried to present information in a number of ways to make the information usable for all types of learners. That is why you see detailed discussions of the various features along with tutorials that show these concepts in action.

The tutorials appear throughout the book and are clearly marked with the "Tutorial" label in front of the title. They always include a series of logical steps, typically ending with a figure for you to study and compare. These tutorial examples are provided in the downloadable content for this book to give you a firsthand look and a chance to get some hands-on experience.

I've attempted to "laser focus" all the tutorials down to one or two key concepts. All tutorials are designed to be completed in 10 steps or less. This means that you probably will not want to place the results in your portfolio. For example, many of the early tutorials don't have any materials applied because I felt that using materials before they've been explained would only confuse you.

I've attempted to think of and use examples that are diverse, unique, and interesting, while striving to make them simple, light, and easy to follow. I'm happy to report that every example in the book is included in the downloadable content for this book along with the models and textures required to complete the tutorial.

The tutorials often don't start from scratch but instead give you a starting point. This approach lets me "laser focus" the tutorials even more, and with fewer, more relevant steps, you can learn and experience the concepts without the complexity. On the book's downloaded content set, you will find the 3ds Max files that are referenced in Step 1 of most tutorials.

In addition to the starting-point files, every tutorial has been saved at the completion of the tutorial steps. These files are marked with the word *final* at the end of the filename. If you get stuck in a tutorial, simply open the final example and compare the settings.

I've put lots of effort into this book, and I hope it helps you in your efforts. I present this book as a starting point. In each tutorial, I've purposely left out most of the creative spice, leaving room for you to put it in—you're the one with the vision.

How this book is organized

Many different aspects of 3D graphics exist, and in some larger production houses, you might be focused on only one specific area. However, for smaller organizations or the general hobbyist, you end up wearing all the hats—from modeler and lighting director to animator and post-production compositor. This book is organized to cover all the various aspects of 3D graphics, regardless of the hat on your head.

The book is divided into the following parts:

* **Part I: Getting Started with Autodesk 3ds Max 2020**—Whether it's understanding the interface, working with the viewports, or dealing with files, the chapters in this part get you comfortable with the interface so you won't get lost moving about this mammoth package.

* **Part II: Manipulating Objects**—3ds Max objects can include meshes, cameras, lights, Space Warps, and anything that can be viewed in a viewport. This part starts by introducing the various primitive objects and also includes chapters on how to reference, select, clone, group, link, and transform these various objects.

* **Part III: Modeling 3D Assets**—3ds Max includes several different ways to model objects. This part includes chapters covering the basic modeling methods and constructs including working with spline shapes, meshes, and polys. It also introduces modifiers and the Modifier Stack.

* **Part IV: Applying Materials and Textures**—This part shows how to apply basic materials and textures to objects including maps using the Slate Material Editor.

* **Part V: Working with Cameras and Lights**—This part delves into using cameras and lights.

* **Part VI: Rendering a Scene--** This part shows how to render out images using Quicksilver and Arnold, how to use atmospheric and render effects, and ends with some coverage of compositing.

* **Part VII: Animating Objects and Scenes**—The simplest animation features include keyframing, constraints, and controllers. With these topics, you'll be able to animate scenes. It also covers some advanced techniques, including animation layers and modifiers, wiring parameters, and the Track View.

* **Part VIII: Working with Characters**—This part covers creating and working with bone systems, rigging, skinning, and the CAT system.

* **Part IX: Adding Special Effects--** Advanced effects include volume lights, lens effects, and using the Particle Flow interface. It includes coverage of particles and Space Warps.

* **Part X: Using Dynamic Animation Systems**—This part covers creating animation sequences using physics calculations with MassFX. It also covers hair and cloth.

* **Part XI: Extending 3ds Max**—This part revisits the interface and shows how you can customize it to your liking. It also covers Max Creation Graphs and extending 3ds Max using scripting and plug-ins.

* **Appendixes**—At the very end of this book, you'll find three appendixes that cover the new features of 3ds Max 2020, the keyboard shortcuts, and installing 3ds Max.

Using the book's icons

The following margin icons are used to help you get the most out of this book:

Note
Notes highlight useful information that you should take into consideration.

Tip
Tips provide additional bits of advice that make particular features quicker or easier to use.

Caution
Cautions warn you of potential problems before you make a mistake.

New Feature in 2020
The New Feature icon highlights features that are new to the 2020 release.

Watch for the Cross-Reference icon to learn where in another chapter you can go to find more information on a particular feature.

Acknowledgments

I have a host of people to thank for their involvement in this major work. The order in which they are mentioned doesn't necessarily represent the amount of work they did.

Thanks as always to my dear wife, Angela, and my sons, Eric and Thomas, without whose support I wouldn't get very far. They are my QA team and my brainstorming team who always provide honest feedback on my latest example. We have had many family sessions to think of good tutorial examples, and I'm always amazed with what they come up with. One of my favorites that hasn't been implemented yet is a tutorial of a group of bicycles chasing an ice cream truck.

Huge kudos to Eric Murdock, my son, who tackled the technical editing for this edition. Eric painstakingly read the entire book cover to cover and found hundreds of errors that have been propagated across several editions. His attention to detail makes this the most accurate edition yet. Thanks, Eric.

In the first edition, the task at hand was too big for just me, so I shared the pain with two co-authors—Dave Brueck and Sanford Kennedy (both of whom have gone on to write books of their own). I still thank them for their work, which, although overhauled, retains their spirits. In a later edition, I again asked for help, a request that was answered by Sue Blackman. Sue provided several excellent examples that show off the power of the Track View interface. Thanks for your help, Sue.

Major thanks to the editors and personnel at SDC Publications. I'd like to specifically thank Stephen Schroff who picked up this title and gave it a new life. Thanks also to Karla and Zach Werner who did the managing, editing, and website. Special thanks to Tyler Bryant for his cover work and Megan Kemper for additional editing.

The various people who work in the graphics industry are amazing in their willingness to help and support. I'd like to thank first of all the entire Autodesk team for their timely support and help. I'd also like to thank the talented people at Curious Labs for many of their models, which make the examples much more interesting. (You can only do so much with the teapot after all.) Additional thanks go out to David Mathis, Sue Blackman, and Chris Murdock for completing models used in some of the tutorials.

Table of Contents

Part I

Getting Started with Autodesk 3ds Max 2020

IN THIS PART

Chapter 1

Exploring the Interface

IN THIS CHAPTER

Learning the interface elements

Previewing the menu commands

Becoming familiar with the toolbars

Using the Command Panel

Examining the Lower Interface Bar

Interacting with the interface

Defining workspaces

Getting help

Well, welcome to the latest version of the Autodesk® 3ds Max® 2020 software, and the first question on the minds of existing users is, "Did the interface change?" The answer is a happy "very little." Most serious users would rather go through root canal surgery than have their user interface (UI) change, and Autodesk has learned and respected this valued opinion by keeping the interface changes to a minimum.

As you look around the new interface, you'll see that everything is still there but that 3ds Max has a few new additions. You may find yourself saying, as you navigate the interface, "Where did that come from?" But just like encountering a new house in your neighborhood, over time you'll become accustomed to the addition and may even meet some new friends.

Why is the software interface so important? Well, consider this: The interface is the set of controls that enable you to access the program's features. Without a good interface, you may never use many of the best features of the software, or you may spend a frustrating bit of time locating those features. A piece of software can have all the greatest features, but if the user can't find or access them, the software won't be used to its full potential. 3ds Max is a powerful piece of software with some amazing features, and luckily, the interface makes these amazing features easy to find and use, but the interface can be a little daunting to new users.

The interface's purpose is to make the software features accessible, and in 3ds Max you have many different ways to access the features. Some of these access methods are faster than others. This design is intentional because it gives beginning users an intuitive command and advanced users direct access. For example, to open the Material Editor, you can choose Rendering→Material Editor→Slate Material Editor (requiring two mouse clicks), but as you gain more experience, you can simply click the Material Editor icon on the Main toolbar (only one click); an expert with his hands on the keyboard can press M without having to reach for the mouse at all. All three of these methods have the same result, but you can use the one that is easiest for you.

Has the 3ds Max interface succeeded? Yes, to a degree, but like most interfaces, it always has room for improvement, and we hope that each new version takes us closer to the perfect interface (but I'm still looking for the "read my thoughts" feature). Autodesk has built a loophole into the program to cover anyone who complains about the interface—customization. If you don't like the current interface, you can change it to be exactly what you want.

Customizing the 3ds Max interface is covered in Chapter 49, "Customizing the Interface."

This chapter examines the latest incarnation of the 3ds Max interface and presents some tips that make the interface feel comfortable, not cumbersome.

Note

When 3ds Max starts, the default color scheme uses dark gray colors with white text. Although this scheme works great for artists who stare at a computer monitor for long periods of time with little or no background light, it isn't the ideal setting for printing. All the figures in this book use the alternate lighter gray color scheme. You can easily switch between the different color schemes using the Customize→Custom UI and Defaults Switcher menu command.

Learning the Interface Elements

If you're new to the 3ds Max interface, the first order of business is to take a stroll around the block and meet the neighbors. The interface has a number of elements that neatly group all the similar commands together. For example, all the commands for controlling the viewports are grouped together in the Viewport Navigation Controls found in the lower-right corner of the interface.

Note

If all the details of every interface command were covered in this chapter, it would be an awfully long chapter. So for those commands that are covered in more detail elsewhere, I include a cross-reference to the chapter where you can find their coverage.

The entire interface can be divided into six easy elements. Each of these interface elements, in turn, has groupings of sub-elements. The six main interface elements are listed here and shown separated in Figure 1.1:

* **Title bar and menus:** This is the default source for most commands, but also one of the most time-consuming interface methods. The title bar and menus are found along the top edge of the 3ds Max window.

* **Toolbars:** 3ds Max includes several toolbars of icon buttons that provide single-click access to features. These toolbars can float independently or be docked to an interface edge. The main toolbar and the Viewport Layout toolbar are visible by default.

* **Modeling ribbon:** Located under the main toolbar, the Modeling ribbon provides quick access to the polygon modeling features, including the Graphite Modeling Tools. It is populated with panels, buttons, and menus when a modeling object is selected.

* **Viewports:** Four separate views into the scene show the scene from different points of view, including Top, Front, Left, and Perspective.

* **Command Panel:** The major control panel, located to the right of the four viewports, it has six tabbed icons at its top that you can click to open the various panels. Each panel includes rollouts containing parameters and settings. These rollouts change, depending on the object and tab that is selected.

* **Lower interface bar:** Along the bottom edge of the interface window is a collection of miscellaneous controls for working with animations and navigating the viewports.

FIGURE 1.1

3ds Max includes six main interface elements.

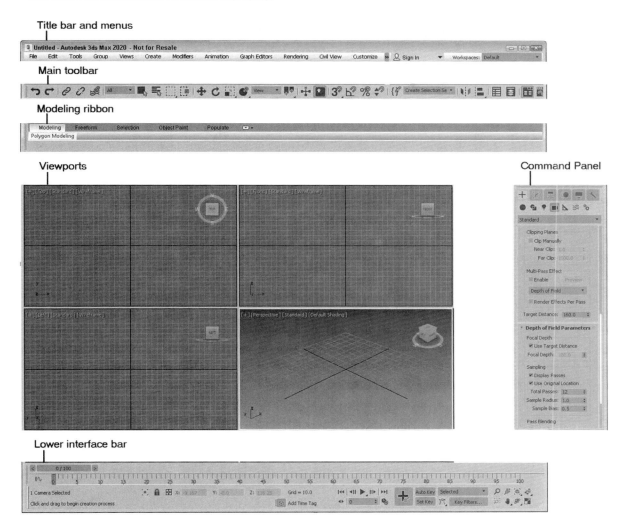

In addition to these default elements are several interface elements that aren't initially visible when 3ds Max is first loaded. These additional interface elements include the following:

* **Floating toolbars:** Several additional toolbars are available as floating toolbars. You access them by choosing Customize→Show UI→Show Floating Toolbars or by selecting them from the main toolbar's right-click pop-up menu.

* **Quad menus:** Right-clicking the active viewport reveals a pop-up menu with up to four panes, referred to as a quad menu. *Quad menus* offer context-sensitive commands based on the object or location being clicked and provide one of the quickest ways to access commands.

* **Caddy settings:** When modeling, you can open Caddy settings. This group of settings floats in a simple dialog box near the current selection and offers several settings that are immediately updated in the viewport.

* **Dialog boxes and editors:** Some commands open a separate window of controls, such as the Array dialog box or the Material Editor. These dialog boxes may contain their own menus, toolbars, and interface elements.

Using the Menus

The pull-down menus at the top of the 3ds Max interface include most of the features available in 3ds Max and are a great place for beginners to start. Several of the menu commands have corresponding toolbar buttons and keyboard shortcuts. To execute a menu command, you can choose it from the menu with the mouse cursor, click its corresponding toolbar button if it has one, or press its keyboard shortcut.

The main menu includes the following options: File, Edit, Tools, Group, Views, Create, Modifiers, Animation, Graph Editors, Rendering, Civil View, Customize, Scripting, Interactive, Content, Arnold and Help. Unlike some other programs, these menu options do not disappear if not needed. The list is set, and they are always there when you need them, but you can hide the toolbar that contains these tools if you wish.

Note

If you select the Alt Menu and Toolbar workspace, then the menus will change to be File, Edit, Objects, Modifiers, Animation, Simulate, Materials, Lighting/Cameras, Rendering, Scene, Civil View, Content, Publish, Scripting, Interactive, Customize, and Help. This is a different organization of the same menus.

If the width of the software window isn't wide enough to show all the menus, then the menu options at the end of the list are placed under an extend menu icon located at the end of the menu, as shown in Figure 1.2.

At the right end of the menu is a control for logging into your Autodesk account. From this login, you can manage your license and if you are using a trial version, it will show you the number of days remaining. The Workspace selector lets you quickly switch to a different Workspace by selecting the desired option from a drop-down list.

FIGURE 1.2

The main menu is located beneath the title bar.

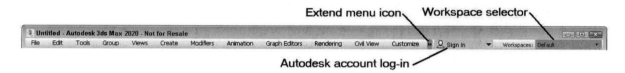

If a keyboard command is available for a menu command, it is shown to the right of the menu item. If a dialog icon appears after a menu item, that menu command causes a separate dialog box to open. A small black arrow to the right of a menu item indicates that a submenu exists. Clicking the menu item or holding the mouse over the top of a menu item makes the submenu appear. Toggle menu options (such as Views→Show Ghosting) change state each time they are selected. If a toggle menu option is enabled, a small check mark appears to its left; if disabled, no check mark appears.

A complete list of keyboard shortcuts can be found in Appendix C, "Keyboard Shortcuts."

You also can navigate the menus using the keyboard by pressing the Alt key by itself. Doing so selects the File menu, and then you can use the arrow keys to move up and down and between menus. With a menu selected, you can press the keyboard letter that is underlined to select and execute a menu command. For example, pressing and holding down Alt and then E (for Edit) and then use the down arrow to select a command, and press the Enter key.

Tip

By learning the underlined letters in the menu, you can use the keyboard to quickly access menu commands, even if the menu command doesn't have an assigned keyboard shortcut. And because you don't need to stretch for the Y key while holding down the Ctrl key, underlined menu letters can be faster. For example, by pressing and holding Alt, then pressing the G, and U keys successively, you can access the Group→Ungroup menu command. The keyboard buffer remembers the order of the letters you type regardless of how fast you key them, making it possible to quickly access menu commands using the keyboard.

Not all menu commands are available at all times. If a menu command is unavailable, it is grayed out, as shown in Figure 1.3, and you cannot select it. For example, the Clone command is available only when an object is selected, so if no objects are selected, the Clone command is grayed out and unavailable. After you select an object, this command becomes available.

FIGURE 1.3

All menus feature visual clues.

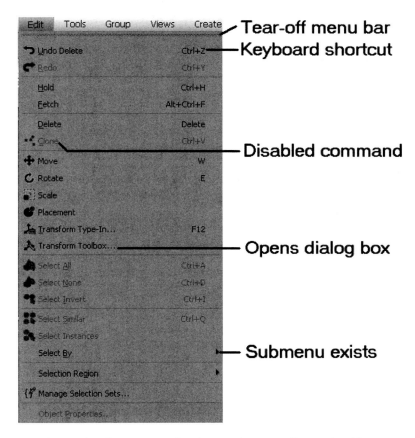

Located at the very top of each menu is a tear-off menu bar (they are two double dashed lines). If you click on these double dashed lines, the menu will become a floating dialog box, as shown for the Graph Editors menu in Figure 1.4 that can be moved independent of the interface. This makes the menu commands immediately accessible. Tearing off a menu doesn't remove it from the main menu and closing the menu dialog box simply removes it without changing the main menu.

FIGURE 1.4

All menus can be made into a separate dialog box using the tear-off menu feature.

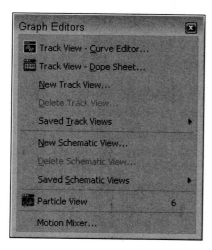

Using the Toolbars

Now that you've learned the menu two-step, it is time for the toolbar one-step. The main toolbar appears by default directly under the menus at the top of the 3ds Max window. Using toolbars is one of the most convenient ways to execute commands because most commands require only a single click.

Docking and floating toolbars

By default, the main toolbar is docked along the top edge of the interface just below the menus, but you can make any docked toolbar (including the main toolbar) a floating toolbar by clicking and dragging the two vertical lines on the left (or top) end of the toolbar away from the interface edge. After you separate it from the window, you can then drag and dock it to any of the window edges. Figure 1.5 shows the main toolbar as a floating panel. If you right click on the double vertical lines, you can access a pop-up menu that gives you options to Dock the toolbar to the Top, Bottom, Left or Right.

This pop-up menu also includes options to float the current toolbar, access the Customize User Interface window, or show or hide any of the toolbars. You can also hide and/or show the Command Panel, the modeling Ribbon, and the Timeline. The main toolbar can be hidden and made visible again with the Alt+6 keyboard shortcut toggle.

FIGURE 1.5

The main toolbar includes buttons and drop-down lists for controlling many of the most popular 3ds Max functions.

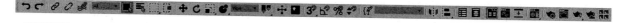

You can customize the buttons that appear on any of the toolbars. See Chapter 49, "Customizing the Interface."

If you select the Customize→Show UI→Show Floating Toolbars menu command, several additional toolbars appear. These are floating toolbars. You also can make them appear by selecting them individually from the toolbar right-click pop-up menu. These floating toolbars include Viewport Layout Tabs, Animation Layers, Axis Constraints, Brush Presets, Containers, Extras, Layers, MassFX Toolbar, Projects, Render Shortcuts, Snaps, and State Sets.

Using tooltips and flyouts

All icon buttons (including those found in toolbars, the Command Panel, and other dialog boxes and windows) include tooltips, which are identifying text labels. If you hold the mouse cursor over an icon button, the tooltip label appears. This feature is useful for identifying buttons. If you can't remember what a specific button does, hold the cursor over the top of it, and the tooltip gives you its name.

All toolbar buttons with a small triangle in the lower-right corner, such as the Rectangular Selection Region button, are flyouts. A *flyout* is a single toolbar button that expands to reveal additional buttons. Click and hold on the flyout to reveal the additional icons, and drag to select one. Figure 1.6 shows the flyout for the Selection Region button on just a portion of the main toolbar.

FIGURE 1.6

Flyout menus bundle several toolbar buttons together.

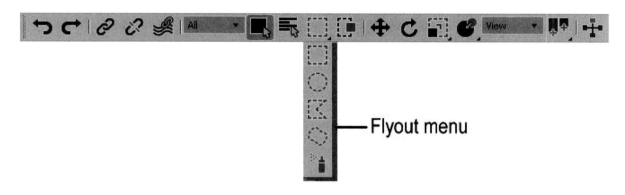

Note

The General panel of the Preference Settings dialog box contains an option for setting the number of milliseconds to wait before the flyout appears.

Learning the main toolbar

On smaller-resolution screens, the main toolbar is too long to be entirely visible. To see the entire main toolbar, you need to set your monitor resolution to be at least 1280 pixels wide. To scroll the toolbar to see the end, position the cursor on the toolbar away from the buttons, such as below one of the drop-down lists (the cursor changes to a hand); then click and drag the toolbar in either direction. Using the hand cursor to scroll also works in the Command Panel, Material Editor, and any other place where the panel exceeds the given space.

Tip

The easiest way to scroll the main toolbar is to drag with the middle mouse button because you can click anywhere on the toolbar and drag.

Toolbar buttons that open dialog boxes such as the Layer Manager, Material Editor, and Render Setup buttons are toggle buttons. When the dialog box is open, the button is highlighted, indicating that the dialog box is open. Clicking a highlighted toggle button closes the dialog box. Corresponding menus (and keyboard shortcuts) work the same way, with a small check mark appearing to the left of the menu command when a dialog box is opened.

Table 1.1 lists the controls found in the main toolbar. Buttons with flyouts are separated with commas.

TABLE 1.1 Main Toolbar Buttons

Toolbar Button	Name	Description
	Undo	Undo the last action.
	Redo	Redoes the last undo action.
	Select and Link	Establishes links between objects.
	Unlink Selection	Breaks links between objects.
	Bind to Space Warp	Assigns objects to be modified by a space warp.
All	Selection Filter drop-down list	Limits the type of objects that can be selected.
	Select Object (Q)	Chooses an object.
	Select by Name (H)	Opens a dialog box for selecting objects by name.
	Rectangular Selection Region, Circular Selection Region, Fence Selection Region, Lasso Selection Region, Paint Selection Region (Ctrl+F to cycle)	Determines the shape used for selecting objects in the viewport.
	Window/Crossing Toggle	Specifies whether an object must be crossed or windowed to be selected.
	Select and Move (W)	Selects an object and allows positional translations.
	Select and Rotate (E)	Selects an object and allows rotational transforms.
	Select and Uniform Scale, Select and Non-uniform Scale, Select and Squash (R to cycle)	Selects an object and allows scaling transforms using different methods.
	Select and Place, Select and Rotate	Moves and rotates the current selection across the face of the other scene objects.
View	Reference Coordinate System drop-down list	Specifies the coordinate system used for transforms.
	Use Pivot Point Center, Use Selection Center, Use Transform Coordinate Center	Specifies the center about which rotations and scaling are completed.
	Select and Manipulate	Selects an object and allows parameter manipulation via a manipulator.
	Keyboard Shortcut Override Toggle	Allows keyboard shortcuts for the main interface and the active dialog box or feature set to be used when enabled. Only

		main interface shortcuts are available when disabled.
	Snap Toggle 2D, Snap Toggle 2.5D, Snap Toggle 3D (S)	Specifies the snap mode. 2D snaps only to the active construction grid, 2.5D snaps to the construction grid or to geometry projected from the grid, and 3D snaps to anywhere in 3D space.
	Angle Snap Toggle (A)	Causes rotations to snap to specified angles.
	Percent Snap (Shift+Ctrl+P)	Causes scaling to snap to specified percentages.
	Spinner Snap Toggle	Determines the amount a spinner value changes with each click.
	Edit Named Selection Sets	Opens a dialog box for creating and managing selection sets.
Create Selection Se ▾	Named Selection Sets drop-down list	Lists and allows you to select a set of named objects.
	Mirror	Creates a mirrored copy of the selected object.
	Align (Alt+A), Quick Align, Normal Align (Alt+N), Place Highlight, Align Camera, Align to View	Opens the alignment dialog box for positioning objects, allows objects to be aligned by their normals, determines the location of highlights, and aligns objects to a camera or view.
	Toggle Scene Explorer	Opens the Layer Manager interface where you can work with layers.
	Toggle Layer Explorer	Opens the Layer Manager interface where you can work with layers.
	Toggle Ribbon	Opens the Modeling Ribbon panel.
	Curve Editor (Open)	Opens the Function Curves Editor.
	Schematic View (Open)	Opens the Schematic View window.
	Compact Material Editor (M), Slate Material Editor (M)	Opens either the Compact Material Editor window or the Slate Material Editor window.
	Render Setup (F10)	Opens the Render Setup dialog box for setting rendering options.
	Rendered Frame Window	Opens the Rendered Frame Window.
	Render Production (Shift+Q), Render Iterative, ActiveShade	Produces a quick test rendering of the current viewport without opening the Render Setup dialog box using the production settings, the iterative render mode, or the ActiveShade window.

	Render in the Cloud	Sends the scene to be rendered in Autodesk's Cloud Rendering Service.
	Open Autodesk A360 Gallery	Opens a gallery of images available in the Autodesk Cloud Service.

Using the Modeling Ribbon

The Modeling Ribbon interface is a deluxe toolbar with many different tool sections. It currently is populated with a variety of modeling tools that are collectively called the Graphite Modeling Tools. You can turn the Modeling Ribbon on and off using the Toggle Ribbon button on the main toolbar. When enabled, tabs for the Modeling, Freeform, Selection, Object Paint, and Populate are displayed.

Most Modeling Ribbon buttons are visible only when an Editable Poly object is selected. You can learn more about Editable Poly objects and the Graphite Modeling Tools in Chapter 13, "Modeling with Polygons," and Chapter 14, "Using the Graphite Modeling Tools and Painting with Objects."

Using the Minimize Ribbon button at the right end of the Ribbon, you can switch the display mode to minimize to only the tabs, only the panel titles, or only the panel buttons, or to enable the Minimize button to cycle through each of the modes. You also can double-click the Ribbon tabs to minimize the Ribbon or to cycle through the minimized modes.

Right-click the Ribbon title bar to access the standard pop-up menu of available floating toolbars. There is also an option to Dock the ribbon and to access the Customize menu. If you double click on the title bar, the floating Ribbon toolbar will return to its last location. If you right click on the toolbar itself, a different menu of options appears. These menus let you show or hide specific tabs or panels, customize the ribbon, save or load a custom ribbon configuration, reset the ribbon to its default, or enable tooltips. Figure 1.7 shows the different Ribbon display modes.

The Ribbon customization features are covered in Chapter 49, "Customizing the Interface."

FIGURE 1.7

The Ribbon can be set to be displayed using several different modes.

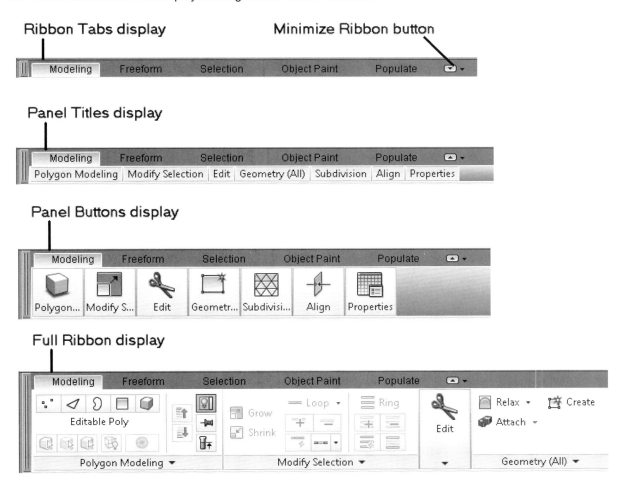

The entire Ribbon, as well as each individual panel of buttons, can be made into a floating control by dragging the Ribbon title bar or the lower panel bar away from the rest of the buttons. When a panel is made into a floating panel, like the one in Figure 1.8, the icons in the upper right of the floating panel let you return the panel to the Ribbon or toggle the orientation between vertical and horizontal. You also can move the floating panel about by dragging on the gray bar on either side of the panel.

FIGURE 1.8

Ribbon panels can float independently of one another.

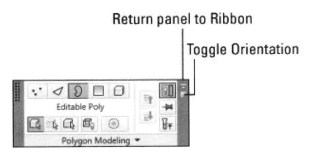

Using the Viewports

The four viewports make up the largest area of the entire interface and provide a way of viewing the objects within the scene. Each of the viewports is configurable and can be unique from the others.

Understanding how to work with the viewports is vital to accomplishing tasks with 3ds Max, so viewports have an entire chapter dedicated just to them—Chapter 2, "Controlling and Configuring the Viewports."

Using the Command Panel

If there is one place in 3ds Max, besides the viewports, where you'll spend all your time, it's the Command Panel (at least until you're comfortable enough with the quad menus). The Command Panel is located to the right of the viewports along the right edge of the interface. This is where all the specific parameters, settings, and controls are located. The Command Panel is split into six panels, each accessed via a tab icon located at its top. These six tabs are Create, Modify, Hierarchy, Motion, Display, and Utilities.

You can pull away the Command Panel from the right window edge as a floating dialog box, as shown in Figure 1.9, by dragging the double lines at the top of the Command Panel away from the interface edge. You also can dock it to the left window edge, which is really handy if you're left-handed. While it's a floating panel, you can resize the Command Panel by dragging on its edges or corners.

After you've pulled the Command Panel away from the interface, you can redock it to its last position by double-clicking its title bar. You also can right-click the title bar to access the pop-up menu of floating toolbars, but the pop-up menu options to Dock is limited to either Left or Right for the Command Panel.

FIGURE 1.9

The Command Panel includes six separate panels accessed via tab icons.

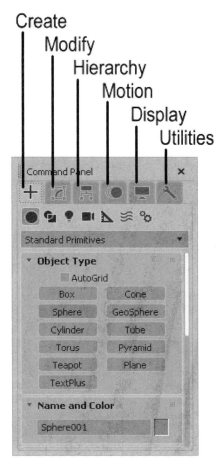

Working with rollouts

Most of the controls, buttons, and parameters in the Command Panel are contained within sections called rollouts. A *rollout* is a grouping of controls positioned under a gray, boxed title, as shown in Figure 1.10. Each rollout title bar includes a small triangle. If the triangle is pointing down, then the rollout it open, but if it points to the right, then the rollout is collapsed. Clicking the rollout title opens or closes the rollout.

You also can reposition the order of the rollouts by dragging the reposition icon located in the upper right corner of each rollout (it looks like a 3x3 set of dots) and dropping it above or below the other rollouts.

FIGURE 1.10

Open and close rollouts by clicking the rollout title.

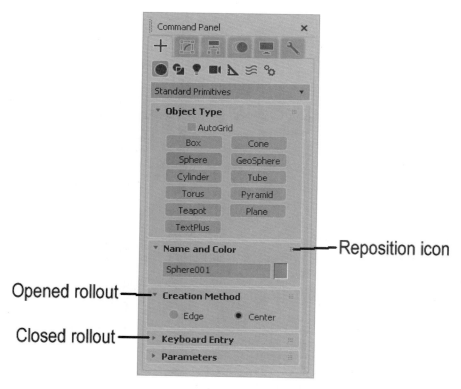

Right-clicking away from the buttons in a rollout presents a pop-up menu where you can select options to close the rollout you've clicked in, Close All, Open All, or Reset Rollout Order. The pop-up menu also lists all available rollouts within the current panel with a check mark next to the ones that are open.

Expanding all the rollouts often exceeds the screen space allotted to the Command Panel. If the rollouts exceed the given space, a small vertical scroll bar appears at the right edge of the Command Panel. You can drag this scroll bar to access the rollouts at the bottom of the Command Panel, or you can click away from the controls when a hand cursor appears. With the hand cursor, click and drag in either direction to scroll the Command Panel. You also can scroll the Command Panel with the scroll wheel on the mouse or by dragging with the middle mouse button.

Increasing the Command Panel's width

The Command Panel can be doubled or tripled (or any multiple, as long as you have room) in width by dragging its left edge toward the center of the interface. The width of the Command Panel is increased at the expense of the viewports. Figure 1.11 shows the Command Panel double its normal size.

FIGURE 1.11

Increase the width of the Command Panel by dragging its left edge.

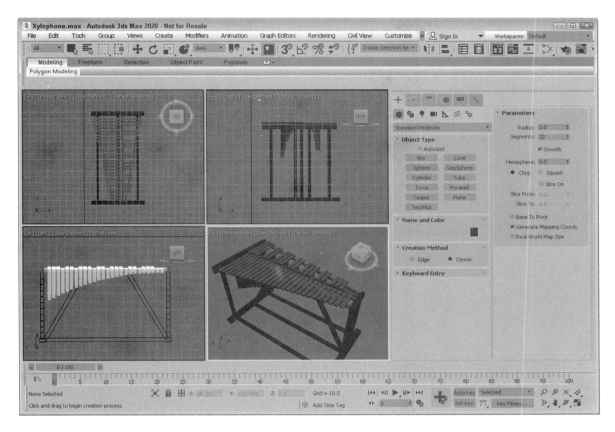

Tutorial: Rearranging the interface for lefties

I used to work for a company that required that all computers have the mouse to the left of the keyboard. We swapped computers often, and the boss hated having to move the mouse to the other side of the keyboard (and you thought your work environment was weird). The reality is that some people like it on the left and others prefer it on the right, and 3ds Max can accommodate both.

With the Command Panel on the right side of the interface, the default interface obviously favors right-handers, but with the docking panels, you can quickly change it to be friendly to lefties.

To rearrange the interface for lefties, follow these steps:

1. Click the double line at the top of the Command Panel and drag toward the center of the interface. As you drag the Command Panel away from the right edge, the cursor changes.

2. Continue to drag the Command Panel to the left edge, and the cursor changes again to indicate that it will be docked when released. Release the mouse button, and the Command Panel docks to the left side.

3. For an even easier method, you can right-click the Command Panel's title bar and select Dock→Left from the pop-up menu.

Figure 1.12 shows the rearranged interface ready for all you southpaws.

Tip

If you've made changes to the interface that you want to keep, try saving a workspace using the main toolbar. Workspaces can be immediately recalled using the drop-down list. Workspaces are covered later in this chapter.

FIGURE 1.12

Left-handed users can move the Command Panel to the left side.

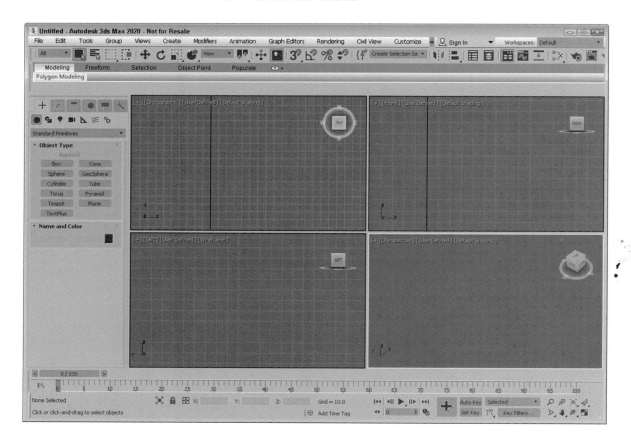

Using the Lower Interface Bar Controls

The last major interface element isn't really an interface element but just a collection of several sets of controls located along the bottom edge of the interface window. These controls cannot be pulled away from the interface like the main toolbar, but you can hide them using Expert Mode (Ctrl+X). These controls, shown in Figure 1.13, include the following, from left to right:

* **Time Slider:** The Time Slider, located under the viewports, enables you to quickly locate a specific animation frame. It spans the number of frames included in the current animation. Dragging the Time Slider moves you quickly between frames. Clicking the arrow buttons on either side of the Time Slider moves to the previous or next frame (or key).

* **Track Bar:** The Track Bar displays animation keys as color-coded rectangles with red for positional keys, green for rotational keys, and blue for scale keys. Parameter change keys are denoted by gray rectangles. Using the Track Bar, you can select, move, and delete animation keys. The button at the left end of the Track Bar is the Open Mini Curve Editor button. It provides access to the animation function curves.

* **Status Bar:** The Status Bar is below the Track Bar. It provides valuable information, such as the number and type of objects selected, transformation values, and grid size. It also includes the Selection Lock Toggle, Transform Type-In fields, and the value of the current grid size.

* **Prompt Line:** The Prompt Line is the line of text directly under the Status Bar. It is located at the bottom of the window. If you're stuck as to what to do next, look at the Prompt Line for information on what 3ds Max expects. The Prompt Line also includes buttons for enabling Progressive Display and adding and editing Time Tags, which are used to name specific animation frames.

* **Key Controls:** These controls are for creating animation keys and include two different modes—Auto Key (keyboard shortcut, N) and Set Key (keyboard shortcut, apostrophe). Auto Key mode sets keys for any changes made to the scene objects. Set Key mode gives you more precise control and sets keys for the selected filters only when you click the Set Keys button (keyboard shortcut, K).

* **Time Controls:** Resembling the controls on an audio or video device, the Time Controls offer an easy way to move through the various animation frames and keys. Based on the selected mode (keys or frames), the Time Controls can move among the first, previous, next, and last frames or keys.

* **Viewport Navigation Controls:** In the lower-right corner of the interface are the controls for manipulating the viewports. They enable you to zoom, pan, and rotate the active viewport's view.

Note
You may notice two text fields located to the left of the Status Bar and Prompt Line. This is the MAXScript Listener control, used to access scripts. You can hide it by dragging its right edge to the left. More on MAXScript and using this control are covered in Chapter 52, "Automating with MAXScript."

Most of the controls on the lower interface bar—including the Time Slider, the Track Bar, and the Key and Time Controls—deal with animation. You can learn more about these controls in Chapter 32, "Understanding Animation and Keyframes." The Viewport Navigation Controls are covered in Chapter 2, "Controlling and Configuring the Viewports."

FIGURE 1.13

The lower interface bar includes several sets of controls.

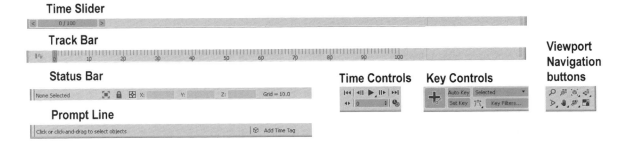

Interacting with the Interface

Knowing where all the interface elements are located is only the start. 3ds Max includes several interactive features that make the interface work. Learning these features makes the difference between an interface that works for you and one that doesn't.

Gaining quick access with the right-click quad menus

Quad menus are pop-up menus with up to four separate sections that surround the cursor, as shown in Figure 1.14. Right-clicking in the active viewport opens these quad menus. The contents of the menus depend on the object selected.

Tip

Many of the real pros use quad menus extensively. One reason is that they can access the commands from the mouse's current location using a couple of clicks without having to go all the way to the Command Panel to click a button.

FIGURE 1.14

Quad menus contain a host of commands in an easily accessible location.

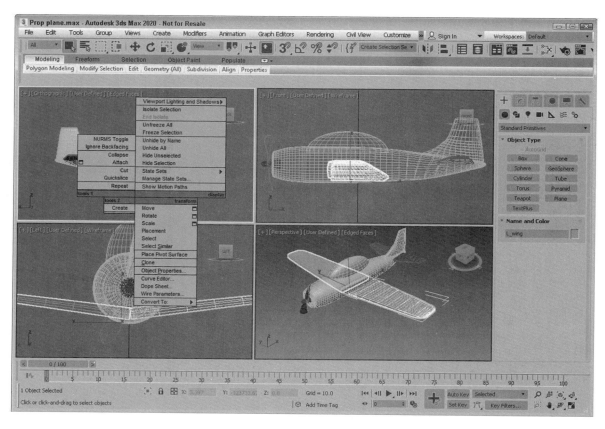

Clicking with the left mouse button away from the quad menu closes it. For each menu, the text of the last menu item selected is displayed in blue. To quickly access the blue menu item again, simply click the gray-shaded bar for the quadrant that contains the blue menu item. Using Customize→Customize User Interface, you can specify which commands appear on the quad menus, but the default options have just about everything you need.

You can learn more about customizing the 3ds Max interface in Chapter 49, "Customizing the Interface."

If you press and hold the Alt, Ctrl, and Shift keys while right-clicking in the active viewport, you can access specific sets of commands. Shift+right-click opens the Snap options, Alt+right-click opens Animation commands, Ctrl+right-click opens a menu of primitives, Shift+Alt+right-click opens a menu of MassFX commands, and Ctrl+Alt+right-click opens a menu of rendering commands.

Using Caddy controls

Quad menus are great for accessing specific commands, but changing the settings for the various features still requires that you visit the Command Panel. This is where the Caddy controls help. Certain modeling features, such as Bevel and Extrude, let you open a select set of controls, known as a Caddy, overlaid over the selected object, as shown in Figure 1.15. Changing any of these settings updates the selection and lets you see if the change is what you want. If you're happy with the setting, you can accept the change and dismiss the Caddy control.

A key benefit of the Caddy controls is that they stay near the selected subobject even if you change the viewport. In addition to several settings that are updated immediately, there are buttons to accept and commit the current change, to apply the change and continue to work with the tool, or to cancel. Using the Apply and Continue button keeps the tool around for more work.

FIGURE 1.15

Caddy controls appear above the selection and let you try several different settings.

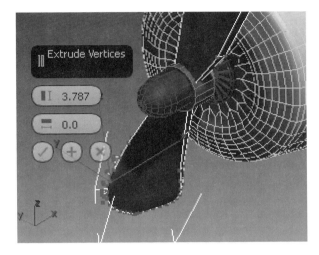

Understanding the button color cues

The interface uses color cues to help remind you of the current mode. When a button is yellow, it warns that it has control of the interface. For example, if one of the subobject buttons in the Command Panel is selected, it turns yellow, and the ability to select another object is disabled until this subobject mode is turned off. Knowing what the current mode is at all times can keep you out of trouble.

Another common button color is red. When either the Auto Key or Set Key button is active, it turns red. The edge of the active viewport being animated along with the Time Slider also turns red. This reminds you that any modifications will be saved as a key.

Toggle buttons can be turned on and off. Example toggle buttons include the Snap buttons. When a toggle button is enabled, it also turns yellow (or light gray, depending on the color scheme). Toggle buttons highlighted in blue are nonexclusive, but they notify you of a mode that is enabled, such as the Key Mode Toggle or the Affect Pivot Only button.

All interface colors can be customized using the Customize User Interface dialog box, which is discussed in Chapter 49, "Customizing the Interface."

Using drag-and-drop features

Dialog boxes that work with files benefit greatly from the software's drag-and-drop features. The Material Editor, Background Image, View File, and Environmental Settings dialog boxes all use drag and drop. These dialog boxes let you select a file or a material and drag it on top of where you want to apply it. For example, with the Maps rollout in the Material Editor open, you can drag a texture image filename from Windows Explorer or the Asset Manager and drop it on the Map button. You can even drag and drop 3ds Max files from Windows Explorer into the interface to open them.

Controlling spinners

Spinners are those little controls throughout the interface with a value field and two small arrows to its right. As you would expect, clicking the up arrow increases the value, and clicking the down arrow decreases the value. The amount of the increase or decrease depends on the setting in the General tab of the Preference Settings dialog box. Right-clicking the spinner resets the value to its lowest acceptable value. Another way to control the spinner value is to click the arrows and drag with the mouse. Dragging up increases the value, and dragging down decreases it.

The effect of the spinner drag is shown in the viewport if the Update During Spinner Drag menu option is enabled in the Views menu. If the cursor is located within a spinner, you can press Ctrl+N to open the Numerical Expression Evaluator, which lets you set the value using an expression. For example, you can set a spinner value by adding numbers together as you would if using a calculator. An expression of 30+40+35 sets the value to 105.

Understanding modeless and persistent dialog boxes

Many dialog boxes in 3ds Max are *modeless,* which means that the dialog box doesn't need to be closed before you can work with objects in the background viewports. The Material Editor is an example of a modeless dialog box. With the Material Editor open, you can create, select, and transform objects in the background. Other modeless dialog boxes include the Material/Map Browser, the Render Scene dialog box, the Caddy controls, the Video Post dialog box, the Transform Type-In dialog box, the Display and Selection Floaters, the Array dialog box, and the various graph editors. Pressing the Ctrl+~ keyboard shortcut closes all open dialog boxes. Pressing the same keyboard shortcut again reopens the dialog boxes that were previously closed.

Another feature of many, but not all, dialog boxes is *persistence,* which means that values added to a dialog box remain set when the dialog box is reopened. This feature applies only within a given session. Choosing the File→Reset command button or exiting and restarting 3ds Max resets all the dialog boxes.

Using Workspaces

If you've rearranged the various interface elements or customized different aspects of the interface, the Workspaces drop-down list, located at the right end of the main toolbar, lets you save the changes for quick recall. Workspaces let you create a unique interface environment for several different tasks. For example, if you are modeling, you can create an interface with a double-wide control panel and another with the control panel hidden for animating. If you use 3ds Max for architectural design, then the Design Standard workspace places all the main icons you need in the Ribbon interface, as shown in Figure 1.16.

FIGURE 1.16

The Design Standard workspace places all the icons you need on the Ribbon interface.

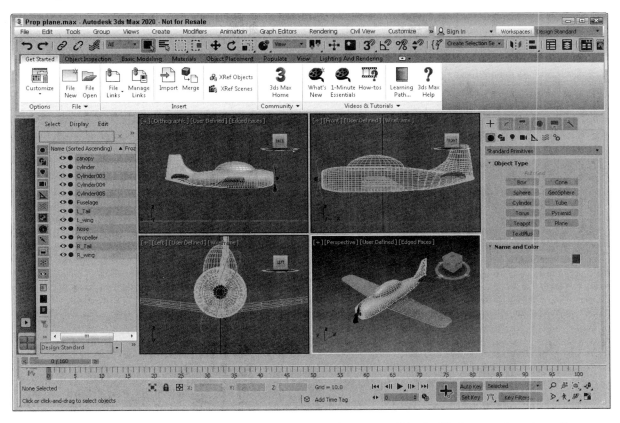

All the custom workspaces that are saved are available for instant recall in the Workspaces drop-down list. When 3ds Max is exited and reloaded, the last workspace that was open is automatically loaded. If you change a saved workspace, you can return it to its saved setting using the Reset to Default State menu, also found in the drop-down list.

To save a custom interface setup, simply select the Manage Workspaces from the Workspace drop-down list, and the dialog box shown in Figure 1.17 opens.

FIGURE 1.17

The Manage Workspaces dialog box lets you save custom interfaces for immediate recall.

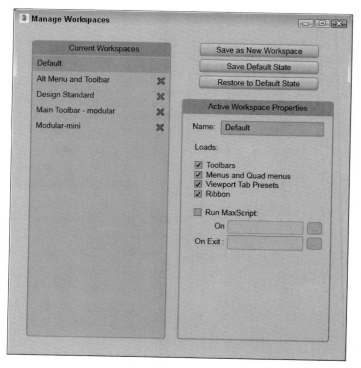

All the current defined workspaces are shown in a list to the left of the Manage Workspaces dialog box. You can delete any defined workspace by clicking the red X next to it. Clicking the Save as New Workspace button opens another dialog box where you can enter the workspace's name. For the selected workspace, you also can select which interface elements to include and even specify a MaxScript to run when the workspace loads or is exited.

Getting Help

If you get stuck, 3ds Max won't leave you stranded. You can turn to several places in 3ds Max to get help. The Help menu is a valuable resource that provides access to references and tutorials. The 3ds Max Help and MAXScript Help are comprehensive help systems. They are HTML pages that are accessible through the Autodesk.com website. Selecting Autodesk 3ds Max Help from the Help menu opens a web browser and loads the help files. This ensures that the latest and most up-to-date help files are available. Additional Help presents help systems for any external plug-ins that are loaded. The Tutorials command loads the tutorials, which offer a chance to gain valuable experience.

If you are working with 3ds Max offline, you can download and access a local copy of the help files. The Help panel in the Preference Settings dialog box lets you specify whether to use the online or the local help files.

Tip

If you choose to use the online help files, you may find yourself waiting for the pages to load, which can be annoying if you are anxious to get an answer. I find that downloading and using a local copy is much quicker than the online version.

Using the Search Command feature

If you know what you want to do, but can't remember where to find a specific command, you can use the Search Command text field to type the command and all matching commands are presented in a list. Simply choose the command you want from the list and it is executed. You can find the Search Command in the Help→Search 3ds Max Command menu or you can simply press the X key and it will immediately appear.

Tip

You can use wildcards when searching for keywords. The asterisk (*) replaces one or more characters, the question mark (?) replaces a single character, and the tilde (~) looks for prefixes and/or suffixes added to the word. For example, con* finds controller, construct, and contour; sta? finds star and stat; ~lit finds prelit and relit; and limit~ finds limited and limitless.

Viewing the Essential Skills Movies

When the software first loads, users are greeted with a Welcome to 3ds Max dialog box, shown in Figure 1.18, which includes several slides to help you get started. The last page has a link to several 1-Minute Startup Movies. These simple movies explain the basics of working with 3ds Max.

The Welcome to 3ds Max dialog box also includes links to What's New and to the Learning Channel, which are pages where you can find more tutorials and resources. If you get tired of this appearing every time the software starts, you can disable the Show this dialog at startup option to prevent this dialog box from appearing the next time you start 3ds Max. You can access the Learning Channel and tutorials at any time using the Help menu.

Note

The Essential Skills Movies require an installation of Flash.

FIGURE 1.18

The Welcome to 3ds Max dialog box includes video clips showing the basic skills you need for working with 3ds Max.

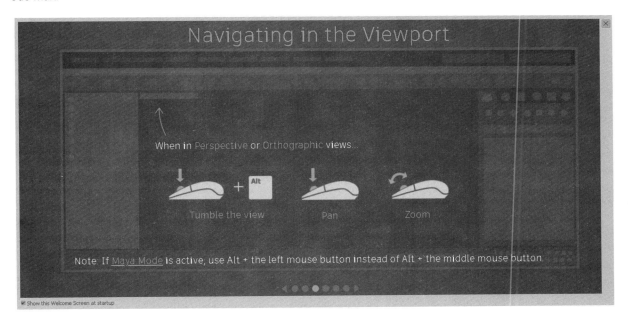

Using the online reference guides

Within the Help menu, the Autodesk 3ds Max Help, What's New, MAXScript Help, and Tutorials are all loaded within a Web browser. An organized list of topics is available in the left navigation pane, as shown in Figure 1.19, and the right side includes a pane where the details on the selected topic are displayed. The Help Home button returns to the first page of the Help file. There is also a Search control for entering a keyword search. The Share button lets you e-mail the selected page. The Back, Up, and Forward buttons move to the last, above, or previous page in the navigation.

Tip

You also can use the web browser buttons to move back and forth between the last-visited and next-visited pages.

FIGURE 1.19

The 3ds Max Reference includes panels for viewing the index of commands and searching the reference.

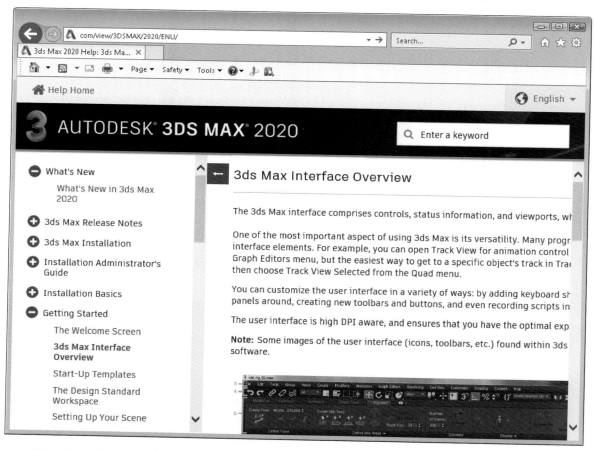

Throughout the textual descriptions, keywords linked to other related topics are highlighted in blue and underlined.

Using the rest of the Help menu

The Help→3ds Max Resources and Tools→Keyboard Shortcut Map menu command displays an interactive interface for learning all the keyboard shortcuts. The Help→3ds Max Services and Support→Data Exchange Solutions menu command opens a web page that explains how to use the FBX format to exchange files with other software packages. The Help→Feedback menu command provides an interface where you can send feedback to Autodesk regarding 3ds Max. The program lets you send feedback anonymously, or you can include your e-mail address. If you notice a problem with the software, you can report it with the Report a Problem feature.

The Help→3ds Max Communities options (The AREA, Student Community, and Facebook) automatically open a web browser and load the various web pages. The AREA website is another excellent resource for help. It is the community site for 3ds Max users.

Tip

If you need help from something more personable than a Help file, the AREA website is a 3ds Max community sponsored by Autodesk. It has some awesome help worth looking into.

The Help→Autodesk Product Information menu includes the About 3ds Max command, which opens the About 3ds Max dialog box, which displays the serial number, the version number, and information about your license. This menu also includes the Check for Updates command to make sure you have the latest updates to the software.

Summary

You should now be familiar with the interface elements for 3ds Max. Understanding the interface is one of the keys to success in using the software. 3ds Max includes a variety of different interface elements. Among the menus, toolbars, and keyboard shortcuts, you have several ways to perform the same command. Discover the method that works best for you.

This chapter covered the following topics:

* Learning the interface elements
* Viewing and using the pull-down menus
* Working with toolbars
* Accessing the Command Panel
* Learning the lower interface controls
* Interacting with the 3ds Max interface
* Managing workspaces
* Getting additional help

In this chapter, I've skirted about the viewports, covering all the other interface elements, but in Chapter 2, you're going to hit the viewports head-on.

Chapter 2

Controlling and Configuring the Viewports

IN THIS CHAPTER

Understanding 3D space

Using the ViewCube and the SteeringWheels

Navigating with the mouse and the Viewport Navigation Control buttons

Changing the visual style of the viewports

Changing viewport layouts and adding information to the viewports

Displaying materials, lighting, and shadows

Loading a viewport background image

Controlling the display settings with the Viewport Setting and Preference dialog box

Altering the viewport layout

Displaying safe frames and setting display performance

Defining regions and displaying statistics

Configuring the ViewCube and SteeringWheel controls

Although the Autodesk® 3ds Max® 2020 software consists of many different interface elements, such as panels, dialog boxes, and menus, the viewports are the main areas that will catch your attention. The four main viewports make up the bulk of the interface. You can think of the viewports as looking at the television screen instead of the remote. Learning to control and use the viewports can make a huge difference in your comfort level with 3ds Max. Nothing is more frustrating than not being able to rotate, pan, and zoom the view.

The viewports have numerous settings and controls that you can use to provide thousands of different ways to look at your scene, and beginners can feel frustrated at not being able to control what they see. 3ds Max includes several handy little gizmos that make navigating the viewports much easier. This chapter includes all the details you need to make the viewports reveal their secrets.

If the Viewport Navigation Controls help define what you see, the Viewport Configuration dialog box helps define how you see objects in the viewports. You can configure each viewport using this dialog box. To open this dialog box, choose the Views→Viewport Configuration menu command. You also can open this dialog box by right-clicking the viewport's General label located in the upper-left corner of each viewport and choosing Configure Viewports from the pop-up menu.

The viewport labels are divided into a General menu (shown as a plus sign), a point of view label, a shading quality label and a shading method label. The various pop-up menus for these labels also include many of the settings found in the Viewport Configuration dialog box, but the dialog box lets you alter several settings at once. You also can make this dialog box appear for the active viewport by right-clicking any of the Viewport Navigation Control buttons in the lower-right corner of the interface.

The Viewport Configuration dialog box contains several panels, including Display Performance, Background, Layout, Safe Frames, Regions, Statistics, ViewCube, and SteeringWheels. The Preference Settings dialog box also includes many settings for controlling the behavior and look of the viewports.

See Chapter 4, "Setting Preferences," for more on the Preference Settings dialog box and all its options.

Understanding 3D Space

It seems silly to be talking about 3D space because we live and move in 3D space. If we stop and think about it, 3D space is natural to us. For example, consider trying to locate your kids at the swimming pool. If you're standing poolside, the kids could be to your left or right, in front of you or behind you, or in the water below you or on the high dive above you. Each of these sets of directions represents a dimension in 3D space.

Now imagine that you're drawing a map that pinpoints a kid's location at the swimming pool. Using the drawing (which is 2D), you can describe the kid's position on the map as left, right, top, or bottom, but the descriptions of above and below have been lost. By moving from a 3D reference to a 2D one, the number of dimensions has decreased.

The conundrum that 3D computer artists face is this: How do you represent 3D objects on a 2D device such as a computer screen? The answer that 3ds Max provides is to present several views, called *viewports,* of the scene. A viewport is a small window that displays the scene from one perspective. These viewports are the windows into the software's 3D world. Each viewport has numerous settings and viewing options.

Learning Axonometric versus Perspective

When it comes to views in the 3D world, two different types exist: axonometric and perspective. Axonometric views are common in the CAD world, where the viewer is set at an infinite distance from the object such that all parallel lines remain parallel. A perspective view simulates how our eyes actually work and converges all points to a single location off in the distance.

You can see the difference between these two types of views clearly if you look at a long line of objects. For example, if you were to look down a long row of trees lining a road, the trees would eventually merge on the horizon. In axonometric views, lines stay parallel as they recede into the distance. Figure 2.1 shows this example with the axonometric view on the left and the perspective view on the right.

FIGURE 2.1

Axonometric and perspective views

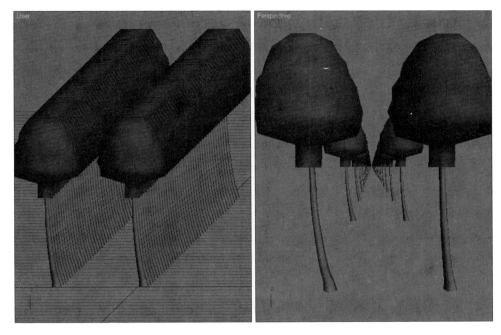

Learning orthographic and isometric views

If you dig a little deeper into axonometric views, you find two different types: orthographic and isometric. Orthographic views are displayed from the perspective of looking straight down an axis at an object. This reveals a view in only one plane. Because orthographic viewports are constrained to one plane, they show the actual height and width of the object, which is why the CAD world uses orthographic views extensively. Isometric views are not constrained to a single axis and can view the scene from any location, but all dimensions are still maintained. The words *isometric* and *orthographic* are often interchanged, but they refer to a view that has consistent dimensions.

Discovering the viewports in 3ds Max

Available orthographic viewports in 3ds Max include Front, Back, Top, Bottom, Left, and Right. 3ds Max starts up with the Top, Front, and Left orthographic viewports visible. The center label in the top-left corner of the viewport displays the viewport name. This label is called the Point-of-View viewport label. The fourth default viewport is a perspective view. Only one viewport, known as the *active viewport*, is enabled at a time. A yellow border highlights the active viewport.

Figure 2.2 shows the viewports with a cartoon chicken. You can see the model from a different direction in each viewport. If you want to measure the chicken's beak, you could get an accurate measurement using the Top or Right viewport, whereas you can use the Front and Right viewports to measure its precise height. Using these different viewports, you can accurately work with all object dimensions.

FIGURE 2.2

The 3ds Max interface includes four viewports, each with a different view.

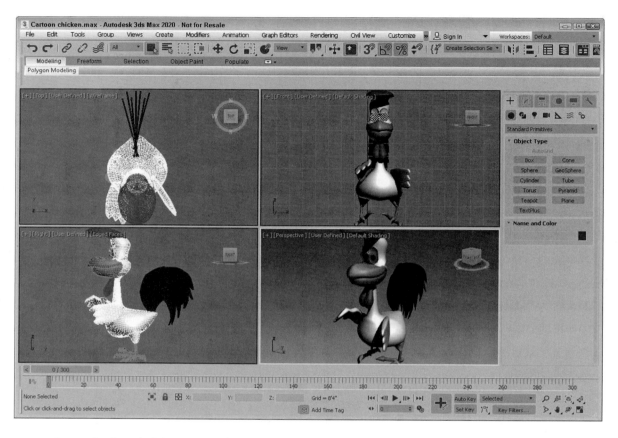

Isometric views in 3ds Max are called orthographic viewports. You can create an orthographic viewport by rotating any of the default non-perspective views.

Tip

3ds Max includes several keyboard shortcuts for quickly changing the view in the active viewport, including T (Top view), B (Bottom view), F (Front view), L (Left view), C (Camera view), $ (Spotlight view), P (Perspective view), and U (Orthographic User view). Pressing the V key opens a quad menu that lets you select a new view.

Using the Navigation Gizmos

One of the key advantages of working in 3D is that you can view your models from an endless number of viewpoints, but you won't be able to switch to these endless viewpoints until you learn to navigate the viewports. Being able to quickly navigate the viewports is essential to working in 3ds Max and one of the first skills you should master.

To make the process of navigating within the viewports and switching among the various views easier, 3ds Max has some navigation gizmos that make this chore easy. These semitransparent gizmos hover in the upper-right corner of each viewport and provide a way to change the view without having to access a tool, select a menu, or even use a keyboard shortcut.

Working with the ViewCube

The ViewCube consists of a 3D cube that is labeled on each side and centered in a ring aligned with the ground plane. Its purpose is to show the current orientation of the viewport, but it is also interactive and provides a way to quickly move among the different views.

If you drag the cursor over the top of the ViewCube, shown in Figure 2.3, you'll notice that the cube's faces, corners, and edges are highlighted as the cursor moves over them. If you click when any of the cube's parts are highlighted, the viewport is animated and moves to the new view so it's positioned as if it's pointing at the selected part. By slowly animating the transition to the new view, you get a better idea of the size and shape of the model. It also makes it easy to reorient the model if it gets twisted around to an odd angle. For example, if you click the cube's face labeled Top, the view moves from its current view to the same orientation as the top view.

FIGURE 2.3

The ViewCube lets you quickly change the current view.

Go to home view

The ViewCube also lets you click and drag on the cube to rotate the view around. The scene rotates along with the ViewCube. You also can click and drag on the base ring to spin the model about its current orientation. When you hover over the ViewCube, a small house icon appears. Clicking this icon changes the view to the defined home view. You can set the Home view by right-clicking the ViewCube and selecting the Set Current View as Home option from the pop-up menu. These same menu options also are available in the ViewCube menu and as an option in the General Viewport label (the small plus sign located in the upper-left corner of each viewport). If the ViewCube isn't visible, you can enable it using the ViewCube→Show For Active View. Another option is to Show For All Views.

Other pop-up menu options let you switch the view between orthographic and perspective views. You also can set the current view as Front, reset the Front view, and open the ViewCube panel that is located in the Viewport Configuration dialog box with the Configure option.

The ViewCube panel in the Viewport Configuration dialog box includes settings for turning the ViewCube on and off In All Views in Active Layout or In Only the Active View. You also can set the ViewCube size and its inactive opacity.

Tip

If you like the ViewCube but feel that it takes up too much of the viewport, you can change its size to Small or Tiny, or you can set its inactive opacity to 0. When its inactive opacity is set to 0, the ViewCube isn't visible at all until you move the cursor over its location, causing it to appear.

The ViewCube panel in the Viewport Configuration dialog box also has options to control what happens when you click or drag the ViewCube. You can snap to the closest view when dragging the ViewCube, and you have options to automatically make the models fit to the view when the view changes, to use animated transitions, and to keep the scene upright. If you find that the view keeps ending up at odd angles when you drag the ViewCube, try enabling the Keep Scene Upright option. Finally, you have an option to display the compass under the ViewCube and a setting for the Angle of North so you can change the compass' orientation. The compass is helpful in being able to spin the model around, but if your model is something like a planet that doesn't have a top or bottom, disabling the compass makes sense.

Using the SteeringWheels

The ViewCube is great for switching between the default views and for rotating the current view, but there are many additional navigation tools that aren't covered with the ViewCube. To handle many of these other navigation tools, such as zooming and panning, 3ds Max includes the SteeringWheels, another gizmo for navigating the viewports.

When 3ds Max is first started, the SteeringWheels are turned off, but you can enable this gizmo with the SteeringWheels→Toggle SteeringWheels in the General Viewport menu or by pressing the Shift+W shortcut. The SteeringWheels menu also is found in the General Viewport label, located in the upper-left corner of the viewport. Once enabled, this gizmo follows your mouse cursor around as it moves about the viewport, and different parts of the wheel are highlighted when you move over them. If you click and drag while a section of the SteeringWheels is highlighted, you can access that control. Right-clicking at any time will open the SteeringWheels menu and pressing the Escape key will toggle the SteeringWheels off again. There are several different modes for the SteeringWheels in both full-size and mini versions, including the Full Navigation Wheel, with all commands; the View Object Wheel, with commands for navigating about objects; and the Tour Building Wheel, for walking through buildings. The Full Navigation Wheel, shown in Figure 2.4, includes the following modes:

FIGURE 2.4

The SteeringWheel gizmo includes several different ways to navigate the viewports.

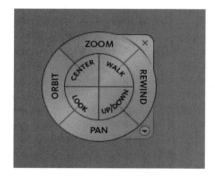

* **Zoom:** This causes the view to zoom in and out of the scene about the pivot. The Zoom pivot is set at the position of the mouse cursor in the scene when you click the Zoom button.
* **Orbit:** This causes the view to orbit about the pivot. The pivot is set by clicking and dragging on the Center button.
* **Pan:** This causes the view to pan in the direction that you drag the cursor.
* **Rewind:** As you change the scene, 3ds Max remembers each view where you stop and keeps these views in a buffer. The Rewind mode displays these views as small thumbnails, as shown in Figure 2.5, and lets you move through them by dragging the mouse. This allows you to rewind and move forward through the buffered views.

Tip
Moving between the buffered thumbnail views with the Rewind feature gradually animates the transition between adjacent thumbnails and allows you to click to change the view to one that is between two buffered views.

FIGURE 2.5

Rewind mode lets you move back and forth through recent views.

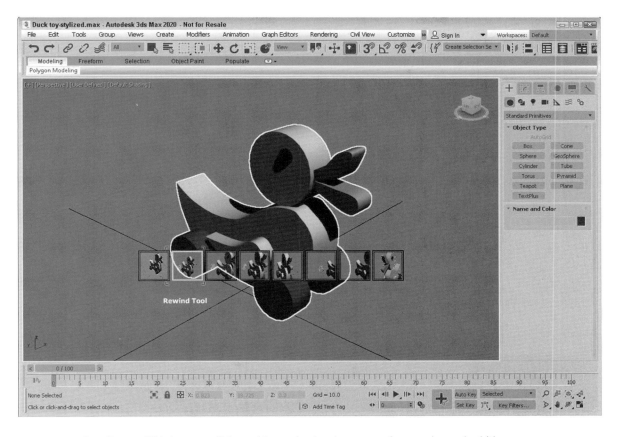

* **Center:** This lets you click an object to be the pivot center for zooming and orbiting.

* **Walk:** This moves you forward through the scene as if you were walking through it by holding down the left mouse button and moving the mouse.

* **Look:** This causes the camera to rotate side to side as if looking to the side.

* **Up/Down:** This moves the view up and down from the current location.

Note
It is covered later in the chapter, but you can maximize the active viewport by clicking the Maximize Viewport Toggle button (the button in the lower-right corner of the interface) or by pressing Alt+W.

In the upper-right corner of the wheel is a small X icon. This icon is used to close the SteeringWheels gizmo. In the lower-right corner of the wheel is a small down arrow. This icon opens a pop-up menu. Using the pop-up menu, you can select a different wheel type, go to the Home view as defined by the ViewCube, increase or decrease the walk speed, restore the original center, or open the SteeringWheels panel in the Viewport Configuration dialog box. These same options are also available in the SteeringWheels menu in the General viewport label, along with an option to Toggle the SteeringWheels on and off (Shift+W).

Using the SteeringWheels panel in the Viewport Configuration dialog box, you can set the size and opacity of the SteeringWheels. Settings for controlling many of the different modes are available as well.

Tutorial: Navigating the active viewport

Over time, navigating the viewports becomes second nature to you, but you need to practice to get to that point. In this tutorial, you get a chance to take the viewports for a spin—literally.

To practice navigating a viewport, follow these steps:

1. Open the Little Dragon.max file from the Chap 02 directory in the downloaded content set.

   This file includes a model of a cartoon dragon. It provides a reference as you navigate the viewport. The active viewport is the Perspective viewport.

2. Click the Maximize Viewport Toggle button (or press Alt+W) to make the Perspective viewport fill the space of all four viewports.

3. Click the Front face in the ViewCube to transition the view to the front view. Then move the cursor over the upper-right corner of the ViewCube, and click to return the view to an angled view.

4. Select the SteeringWheels→Toggle SteeringWheels from the General viewport menu in the upper left corner of the active viewport; then hold down the Ctrl key and click Bruce's head to set the pivot. Then move the cursor over the Zoom button, and drag until Bruce's head fills the viewport.

5. With the SteeringWheels still active, move the cursor over the Pan button, and drag the window until Bruce's head is centered evenly in the viewport, as shown in Figure 2.6. Right-click in the viewport to toggle the SteeringWheels gizmo off.

FIGURE 2.6

The Perspective viewport is zoomed in on the dragon's head using the Zoom and Pan controls.

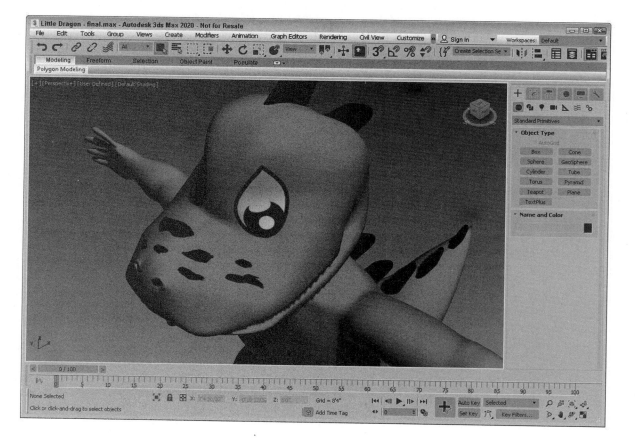

Controlling Viewports with a Scroll Wheel Mouse

Now that I've explained the viewport navigation gizmos, I'll explain another easy way to control the viewports. Often, the quickest way to control the viewports is with the mouse. To really get the benefit of the mouse, you need to use a mouse with a scroll wheel (which also acts as a middle mouse button).

Rolling the scroll wheel in the active viewport zooms in to and out of the viewport by steps just like the bracket keys ([and]). You can zoom gradually by holding down Ctrl+Alt while dragging the scroll wheel.

Clicking and dragging the scroll wheel button pans the active viewport. Clicking and dragging with the Alt button held down rotates the active viewport.

Caution

Be careful when zooming in with the scroll wheel. If you zoom in too far, the zooming becomes unstable. If this happens, you can select the Undo View Change (Shift+Z) command to undo the zoom or use the Zoom Extents button.

Strokes are covered in Chapter 49, "Customizing the Interface."

Using the Viewport Navigation Controls

Although the ViewCube, the SteeringWheels, and the scroll wheel make navigating the viewports easy, you can still use the standard navigation tools located in the bottom-right corner of the interface. The standard viewports show you several different views of your current project, but within each viewport, you can zoom in on certain objects, pan the view, or rotate about the center of the viewport. Clicking a viewport with any of the Viewport Navigation Controls automatically makes the selected viewport the active viewport. In Table 2.1, the keyboard shortcut for each button is listed in parentheses next to its name.

TABLE 2.1 Viewport Navigation Controls

Toolbar Button	Name	Description
	Zoom (Alt+Z or [or])	Moves closer to or farther from the objects in the active viewport by dragging the mouse or zooming by steps with the bracket keys.
	Zoom All	Zooms in to or out of all the viewports simultaneously by dragging the mouse.
	Zoom Extents (Ctrl+Alt+Z), Zoom Extents Selected (Z)	Zooms in on all objects or just the selected object until it fills the active viewport.
	Zoom Extents All (Ctrl+Shift+Z), Zoom Extents All Selected	Zooms in on all objects or just the selected object until it fills all the viewports.
	Field of View, Zoom Region (Ctrl+W)	The Field of View button (available only in the perspective view) controls the width of the view. The Zoom Region button zooms in to the region selected by dragging the mouse.
	Pan View (Ctrl+P or I), 2D Pan Zoom, Walk Through (Up arrow)	Moves the view to the left, to the right, up, or down by dragging the mouse or by moving the mouse while holding down the I key. The Walk Through feature moves through the scene using the arrow keys or a mouse like a first-person video game.

	Orbit (Ctrl+R), Orbit Selected, Orbit SubObject, Orbit Point of Interest	Rotates the view around the global axis, selected object, subobject, or point of interest by dragging the mouse.
	Maximize Viewport Toggle (Alt+W)	Makes the active viewport fill the screen, replacing the four separate viewports. Clicking this button a second time shows all four viewports again.

Caution

When one of the Viewport Navigation buttons is selected, it is highlighted yellow. You cannot select, create, or transform objects while one of these buttons is highlighted. Right-clicking in the active viewpoint or clicking the Select Objects tool reverts to select object mode.

Zooming a view

You can zoom into and out of the scene in several ways. Clicking the Zoom (Alt+Z) button enters zoom mode, where you can zoom into and out of a viewport by dragging the mouse. This works in whichever viewport you drag in. To the right of the Zoom button is the Zoom All button, which does the same thing as the Zoom button, only to all four viewports at once. If you hold down the Ctrl key while dragging in Zoom mode, the zoom action happens more quickly, requiring only a small mouse movement to get a large zoom amount. Holding down the Alt key while dragging in Zoom mode has the opposite effect; the zoom happens much more slowly, and a large mouse move is required for a small zoom amount. This is helpful for fine-tuning the zoom.

The Zoom Extents (Ctrl+Alt+Z) button zooms the active viewport so that all objects (or the selected objects with the Zoom Extents Selected button) are visible in the viewport. A Zoom Extents All (Ctrl+Shift+Z) button is available for zooming in all viewports to all objects' extents; the most popular zoom command (and the easiest to remember) is Zoom Extents Selected (Z), which is for zooming in to the extents of the selected objects in the active viewport.

You can use the brackets keys to zoom in ([) and out (]) by steps. Each key press zooms in (or out) another step. The Zoom Region (Ctrl+W) button lets you drag over the region that you want to zoom in on. If you select a non-orthogonal view, such as the Perspective view, the Zoom Region button has a flyout called the Field of View. Using this button, you can control how wide or narrow the view is. This is like using a wide angle or telephoto lens on your camera. This feature is different from zoom in that the perspective is distorted as the Field of View is increased.

Field of View is covered in more detail in Chapter 24, "Configuring and Aiming Cameras."

Panning a view

The Viewport Navigation Controls also offer two ways to pan in a viewport. In Pan View mode (Ctrl+P), dragging in a viewport pans the view. Note that this doesn't move the objects, only the view. In addition, the Ctrl and Alt keys can be held down to speed or slow the panning motions. The second way to pan is to hold down the I key while moving the mouse. This is known as an *interactive* pan.

Walking through a view

The Walk Through button (instantly accessed with the Up arrow button), found as a flyout button under the Pan View button, allows you to move through the scene in the Perspective or Camera viewport using the arrow keys or the mouse just as you would if you were playing a first-person computer game. When this button is active, the cursor changes to a small circle with an arrow inside it that points in the direction you are moving. You need to click in the viewport before you can use the arrow keys.

Caution

The Pan View button is a flyout only if the Perspective view or a Camera view is selected.

The Walk Through feature includes several keystrokes for controlling the camera's movement. The arrow keys move the camera forward, left, back, and right (or you can use the W, A, S, and D keys). You can toggle the speed of the motion with the Q (speed up) and Z (speed down) keys or with the [(decrease step size) and] (increase step size) keys. The E and C keys (or the Shift+up and Shift+down arrows) are used to move up and down in the scene. The Shift+spacebar key causes the camera to be set level. Dragging the mouse while the camera is moving changes the direction in which the camera points.

Caution

To use the W, A, S, and D keys for walk-through mode, make sure that the Keyboard Shortcut Override Toggle on the main toolbar is enabled.

A handy alternative to Walk Through mode is the Walkthrough Assistant, which is found on the Animation menu. This utility opens a dialog box that includes buttons for creating and adding a camera to a path. It also has controls for turning the view side to side as the camera moves along the path.

The Walkthrough Assistant is covered in more detail in Chapter 33, "Animating with Constraints and Simple Controllers."

Rotating a view

Rotating the view can be the most revealing of all the view changes. When the Orbit (Ctrl+R) button is selected, a rotation guide appears in the active viewport, as shown in Figure 2.7. This rotation guide is a circle with a square located at each quadrant. Clicking and dragging the left or right squares rotates the view side to side; the same action with the top and bottom squares rotates the view up and down. Clicking within the circle and dragging rotates within a single plane, and clicking and dragging outside the circle rotates the view about the circle's center either clockwise or counterclockwise. If you get confused, look at the cursor, which changes depending on the type of rotation. The Ctrl and Alt keys also can speed and slow the rotating view.

Within the Viewport Navigation Controls are several flyout options for rotating the scene. Using the Orbit Selection mode, you can orbit about the selected object. There are also options for orbiting about the selected subobject and about the cursor's location (known as the point of interest).

Note

The Orbit keyboard shortcut (Ctrl+R) selects whichever Orbit tool was the last to be used.

FIGURE 2.7

The rotation guide appears whenever the Orbit tool is selected.

Note

If you rotate one of the default non-perspective views, it automatically becomes an orthographic view, but you can undo the change using the Undo View Change command in the Views Viewport menu or the Shift+Z shortcut.

Maximizing the active viewport

Sooner or later, the viewports will feel too small. When this happens, you have several ways to increase the size of your viewports. The first trick to try is to change the viewport sizes by clicking and dragging any of the viewport borders. Dragging on the intersection of the viewports resizes all the viewports. Figure 2.8 shows the viewports after being dynamically resized.

Tip

You can return to the original layout by right-clicking any of the viewport borders and selecting Reset Layout from the pop-up menu.

FIGURE 2.8

You can dynamically resize viewports by dragging their borders.

The second trick is to use the Maximize Viewport Toggle (Alt+W) to expand the active viewport to fill the space reserved for all four viewports. Clicking the Maximize Viewport Toggle (or pressing Alt+W) a second time returns to the defined layout.

Maximizing the viewport helps temporarily, but you can take another step before convincing your boss that you need a larger monitor. You can enter Expert Mode by choosing Views→Expert Mode (Ctrl+X). It maximizes the viewport space by removing the main toolbar, the Command Panel, and most of the lower interface bar.

With most of the interface elements gone, you'll need to rely on the menus, keyboard shortcuts, and quad menus to execute commands. To re-enable the default interface, select the Views→Expert Mode menu again (or press Ctrl+X again). Figure 2.9 shows the interface in Expert Mode.

FIGURE 2.9

Expert Mode maximizes the viewports by eliminating most of the interface elements.

Controlling camera and spotlight views

You can set any viewport to be a camera view (C) or a spotlight view ($) if a camera or a spotlight exists in the scene. When either of these views is active, the Viewport Navigation Control buttons change. In camera view, controls for dolly, roll, truck, pan, orbit, and field of view become active. A light view includes controls for the falloff and hotspots.

Chapter 24, "Configuring and Aiming Cameras," and Chapter 25, "Using Lights and Basic Lighting Techniques," cover these changes in more detail.

Setting the navigation controls to match Maya

In many studios, you'll find both 3ds Max and Maya. These software packages have different navigation controls, but you can set 3ds Max to mimic the Maya controls using the Interaction Mode panel in the Preference Settings dialog box. This panel includes a simple drop-down box for switching the viewport navigation controls between 3ds Max and Maya. If you make the change, the designated controls will be used automatically the next time you start up the software.

Table 2.2 shows the basic keyboard and mouse navigation controls for both 3ds Max and Maya.

TABLE 2.2 Viewport Navigation Controls for 3ds Max and Maya

Viewport Navigation Control	3ds Max	Maya
Rotate View	Alt+middle mouse button	Alt+left mouse button

Pan View	middle mouse button	Alt+middle mouse button
Zoom View	scrub scroll wheel	Alt+right mouse button
Zoom Extents Selected	Z	F
Zoom Extents All	Shift+Ctrl+Z	A
Maximize Viewport	Alt+W	spacebar
Undo/Redo Viewport Change	Shift+Z/Shift+Y	Alt+Z/Alt+Y

Changing the Viewport Display

Although the Viewport Navigation Controls are focused on controlling what is visible in the viewports, a number of useful commands are available in the Views menu and in the viewport labels at the top-left corner of each viewport that directly affect the viewports. The four viewport labels, shown in Figure 2.10, include the General viewport label (which is a simple plus sign), the Point-of-View viewport label, the Shading Quality viewport label and the Shading viewport label. The last two labels show the current setting.

Within the Views menu and the viewport labels are options to configure the viewports using the Viewport Configuration dialog box.

FIGURE 2.10

Viewport labels in the upper-left corner of each viewport hold settings for that particular viewport.

General viewport label
Point-of-View viewport label
Shading Quality viewport label
Shading viewport label

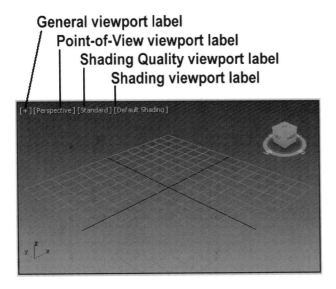

Undoing and saving viewport changes

If you get lost in your view, you can undo and redo viewport changes with Undo View Change (Shift+Z) and Redo View Change (Shift+Y). These same commands also are available at the bottom of the Point-of-View viewport label. These commands are different from the Edit→Undo and Edit→Redo commands, which can undo or redo geometry changes.

43

You can save changes made to a viewport by using the Save Active Viewport menu command in the Views viewport menu. This command saves the Viewport Navigation settings for recall. To restore these settings, use Restore Active Viewport.

Note

The Save and Restore Active Viewport commands do not save any viewport configuration settings, just the navigated view. Saving an active view uses a buffer, so it remembers only one view for each viewport.

Disabling and refreshing viewports

If your scene gets too complicated, you can experience some slowdown waiting for each viewport to be updated with changes, but fear not, because several options will come to your rescue. The first option to try is to disable a viewport.

You can disable a viewport by clicking the General viewport label and selecting the Disable View menu command from the pop-up menu, or you can press the keyboard shortcut, Shift+Ctrl+D. When a disabled viewport is active, it is updated as normal; when it is inactive, the viewport is not updated at all until it becomes active again. Disabled viewports are identified by the word *Disabled*, which appears next to the viewport's labels in the upper-left corner.

Another trick to increase the viewport update speed is to disable the Views→Update During Spinner Drag menu option. Changing parameter spinners can cause a slowdown by requiring every viewport to update as the spinner changes. If the spinner is changing rapidly, it can really slow even a powerful system. Disabling this option causes the viewport to wait for the spinner to stop changing before updating.

Sometimes when changes are made, the viewports aren't completely refreshed. This typically happens when dialog boxes from other programs are moved in front of the viewports or as objects get moved around, because they often mask one another and lines disappear. If this happens, you can force 3ds Max to refresh all the viewports with the Views→Redraw All Views (keyboard shortcut, `) menu command. The Redraw All Views command refreshes each viewport and makes everything visible again.

Floating viewports

At the bottom of the viewport's General label menu are three options to make the current viewport floating in the Float Viewport menu. This option doesn't remove the current docked viewport, but simply makes a copy of the current viewport and makes it floating. Floating viewports can be moved as needed and still has the same configuration menu options.

New Feature in 2020

Floating viewports are new to 3ds Max 2020.

Setting the viewport visual style

Complex scenes take longer to display and render. The renderer used for the viewports is highly optimized to be very quick, but if you're working on a huge model with lots of complex textures and every viewport is set to display the highest-quality view, updating each viewport can slow the program to a crawl.

By default, all the orthographic views are set to wireframe, and only the perspective view is set to show shading, but the Shading viewport label includes several options that let you set the visual style settings for the current viewport.

Tip

If you ever get stuck waiting for 3ds Max to complete a task, such as redrawing the viewports, you can always press the Escape key to suspend any task immediately and return control to the interface.

Note

These settings have no effect on the final rendering specified using the Rendering menu. They affect only the display in the viewport.

The Rendering Level options include the following:

*   **Default Shading:** Shows smooth surfaces with lighting highlights

*   **Facets:** Shows each face as a flat surface without smoothing the edges

*   **Bounding Box:** Shows a box that would enclose the object

*   **Flat Color:** Shows the entire object with minimal lighting

*   **Hidden Line:** Shows only polygon edges facing the camera

*   **Clay:** Shows the surface as red modeling clay, which is helpful for seeing deformations

*   **Stylized:** Includes artistic options such as Graphite, Color Pencil, Ink, Color Ink, Acrylic, Pastel, and Tech

*   **Edged Faces:** An option applied in addition to the shading method that shows all edge segments

Although it really isn't a rendering method, the Edged Faces option shows the edges for each face when a shaded rendering method is selected. You can enable and disable this option with the F4 keyboard shortcut. Figure 2.11 shows, side by side, all the various viewport rendering methods applied to a simple sphere.

FIGURE 2.11

The viewport rendering methods are shown from left to right. First row: Default, Shaded, Flat Color, Facets, and Hidden Line. Second row: Wireframe, Bounding Box, Clay, and Edged Faces applied to Shaded.

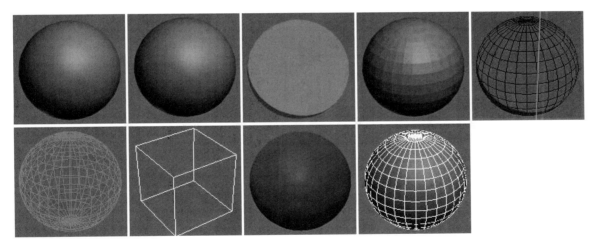

The default shading method uses any applied textures and high-quality lighting and shadows. The Shaded method is similar, but it uses a quick shading method called Phong shading. The Consistent Color method displays the entire object using a single color without any shading. The simplest rendering setting that represents the shape of the object is Wireframe. It gives a good representation of the object while redrawing very quickly. By default, the Top, Front, and Left viewports are set to Wireframe, and the Perspective viewport is set to Realistic. The Bounding Box method shows only the limits of the object as a rectangular box.

Note

Many material effects, such as bump and displacement maps, cannot be seen in the viewport and show up only in the final render.

Viewing stylized scenes

Stylized non-photorealistic effects often are not as computationally complex as realistic renderings, and as such, they can be enabled and displayed within the viewports as well as rendered images. Although you probably would not want to work in one of these stylized display modes, you can use the Shading viewport label to access a menu of available stylized display options.

The options include Graphite, Color Pencil, Ink, Color Ink, Acrylic, Pastel, and Tech. Figure 2.12 shows the clock tower model with the Color Ink option.

FIGURE 2.12

The clock tower scene is displayed using the Colored Ink display style.

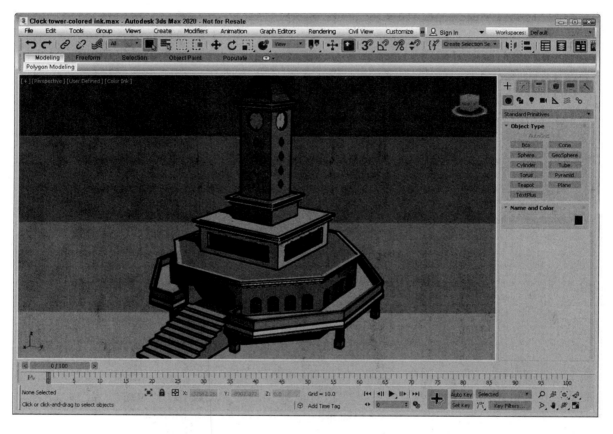

The various stylized display options also can be set in the Per-View Preferences panel in the Viewport Setting and Preference dialog box.

Enhancing the Viewport

In addition to changing the viewport display, several options allow for changing the viewport layout or adding information that will help during modeling and animating, such as xView, clipping view, and safe frames. These enhancements help the viewports be even more helpful by presenting information that you need.

Changing the viewport layout

So far, we've used only the default four-pane layout and the maximized viewport, but you can change the viewport layout to have one, two, three, or four panes orientated on top of one another or side by side. The easiest way to switch between the different layouts is with the Viewport Layout Tabs toolbar.

This toolbar is opened by right-clicking any toolbar and selecting it from the pop-up menu. The toolbar appears by default vertically at the left edge of the viewports, as shown in Figure 2.13. Clicking the Create a New Viewport Layout Tab button opens a panel of several layout options. Selecting one switches the current layout and places the selected layout in the toolbar, where it can quickly be selected again.

FIGURE 2.13

The Viewport Layout Tabs toolbar lets you quickly change between different layouts.

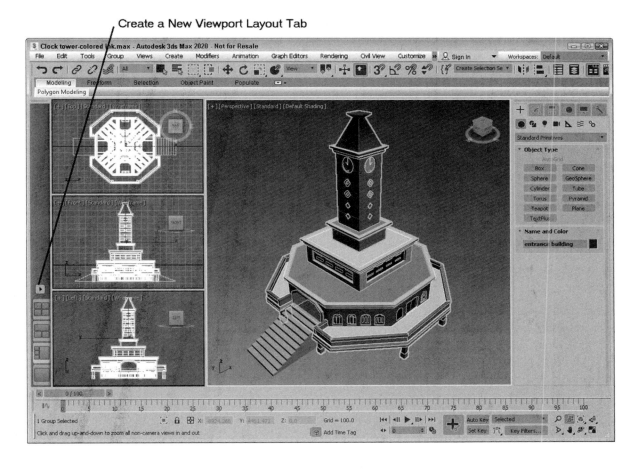

For more options, including the ability to change the view type for each pane, open the Viewport Configuration dialog box using the Views menu, and select the Layout panel.

Using clipping planes

Clipping planes define an invisible barrier beyond which all objects are invisible. For example, if you have a scene with many detailed mountain objects in the background, working with an object in front of the scene can be difficult. By setting the clipping plane between the two, you can work on the front objects without having to redraw the mountain objects every time you update the scene. This affects only the viewport, not the rendered output.

Enabling the Viewport Clipping option in the viewport Point-of-View label menu places a yellow line with two arrows on the right side of the viewport, as shown in Figure 2.14. The top arrow represents the back clipping plane, and the bottom arrow is the front clipping plane. Drag the arrows up and down to set the clipping planes.

FIGURE 2.14

The clipping planes can be used to show the interior of this plane model.

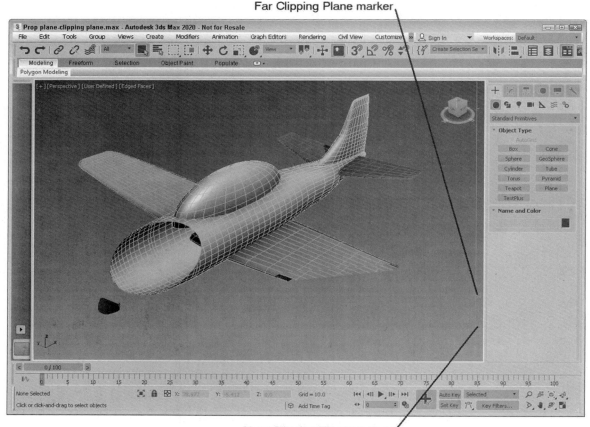

Far Clipping Plane marker

Near Clipping Plane marker

Tutorial: Viewing the interior of a tooth with clipping planes

You can use the Viewport Clipping option in the Point-of-View viewport label to view the interior of a model, such as this tooth model.

To view the interior of a tooth model, follow these steps:

1. Open the Tooth.max file from the Chap 02 directory in the downloaded content set.

2. In the Point-of-View viewport label in the upper-left corner of the viewport, enable the Viewport Clipping option.

   The clipping plane markers appear along the right edge of the viewport. The top marker controls the back clipping plane, and the bottom marker controls the front clipping plane.

3. Drag the bottom clipping plane marker upward to slice through the tooth model to reveal its interior, as shown in Figure 2.15.

FIGURE 2.15

By using clipping planes, you can reveal the interior of a model.

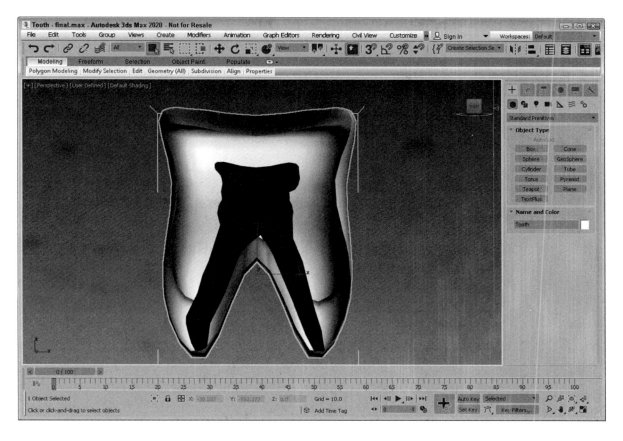

Locating mesh problems with xView

When modeling or importing mesh objects, a number of problems with the geometry can cause rendering artifacts such as flipped normals, overlapping faces, and open edges. Locating these problem areas can be tricky, requiring multiple renders to get it right. Within the viewports is a powerful analysis feature for locating a number of specific problem areas. This feature is called xView, and you can access it from the Views menu and the General viewport label.

The xView analysis tool can locate and highlight the following anomalies:

* **Show Statistics (7):** Displays the number of Polys, Tris, Edges, Verts, and Frames per Second (FPS) for the entire scene and for the selected object.

* **Face Orientation:** Highlights the back side of the faces in the current selection to quickly identify faces with flipped normals.

* **Overlapping Faces:** Highlights any faces that are stacked on top of each other, which can cause render problems.

* **Open Edges:** Identifies unwanted holes in the geometry.

* **Multiple Edges:** Checks for edges that are stacked on each other. Each edge should be connected to only two faces.

* **Isolated Vertices:** Highlights vertices that aren't connected to anything. These vertices just take up space.

* **Overlapping Vertices:** Flags vertices that are within a given tolerance.

* **T-Vertices:** Highlights vertices where three edges meet. This can terminate an edge loop.
* **Missing UVW Coordinates:** Shows any faces that have no UVW coordinates for applying textures.
* **Flipped UVW Faces:** Highlights any faces that are flipped with opposite-pointing normals.
* **Overlapping UVW Faces:** Displays any faces where the textures are overlapping.

Whichever option is selected is listed at the bottom of the viewport in green, along with the number of offending subobjects, such as Isolated Vertices: 12 Vertices. The menu also includes options to Select the Results, which provides a way to quickly select and delete problem subobjects like isolated vertices. If the selected option has a setting such as the Tolerance of overlapping edges, you can select the Configure option to set this setting or click the Click Here to Configure text at the bottom of the viewport. Additional menu options allow you to See Through the model, Auto Update the results, and display the results at the top of the viewport.

Tip

If any xView data is displayed in the viewport, you can click it to select another data option. This lets you quickly view all the potential problems with the current object.

By enabling the Show Statistics option, the selected statistics are overlaid on the active viewport, as shown in Figure 2.16. This is helpful for knowing how complex your model is and how many polygons it includes.

FIGURE 2.16

The active viewport can be set to display the selected statistics.

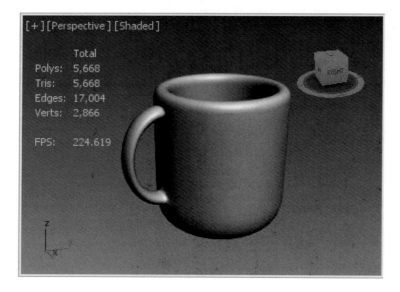

Displaying Materials, Lighting, and Shadows in the Viewport

The difference between what is seen in the viewport and the final render that is output is often the details like materials, lighting, and shadows. Over time, the viewport has gotten stronger and better at being able to display these details without slowing down the system, especially with the Nitrous display drivers. If you are working on a complex scene, you can turn down or turn off these details in order to work faster with the scene.

Viewing materials in the viewports

The Views menu and the Shading Quality viewport label also include several commands for making scene details such as materials, lighting, and shadows visible in the viewports. Each of these options can slow down the refresh rate, but they provide immediate feedback, which is often helpful.

Texture maps also can take up lots of memory. The Materials→Shaded Materials with Maps command in the Shading Quality viewport label shows all applied texture maps in the viewports. If you don't need to see the texture maps, switching to Materials→Shaded Materials without Maps will speed up the display. The option to use Realistic Materials without Maps and Realistic Materials with Maps uses the video card's memory to display the applied textures. The Materials menu also includes a toggle to Enable Transparency in the viewport.

Tip

The options for enabling materials in the viewports also are available as a submenu under the Shading Quality viewport label menu.

More on applying texture maps is covered in Chapter 18, "Adding Material Details with Maps."

Displaying lighting and shadows in the viewports

Options for enabling lighting and shadow effects within the viewports are located in the Shading Quality viewport label menu. By default, both shadows and ambient occlusion are enabled when the High Quality preset mode is selected, but you can disable them using the toggle menu options under the Lighting and Shadows menu. You also have an option to Illuminate the scene using Scene Lights or Default Lights. Default Lights are simply one or two lights added to the scene if no other lights are present to make sure the scene objects are visible. Figure 2.17 shows an example of viewport shadows and Ambient Occlusion.

FIGURE 2.17

Viewport Shading shown with shadows enabled

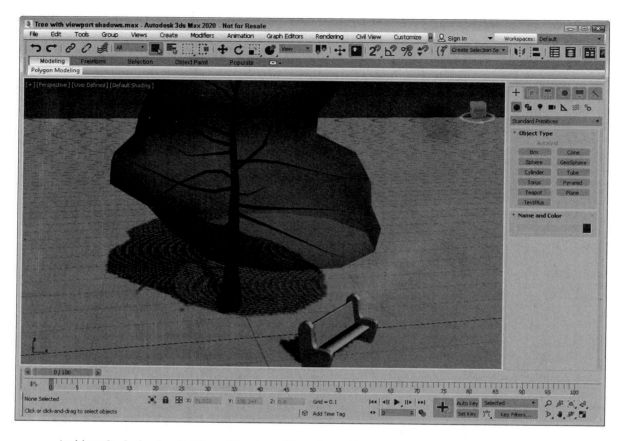

Ambient Occlusion is a lighting effect that adds to the realism of a scene by making objects cast shadows on surrounding objects based on how they block the light. Objects that are close to one another spread a soft shadow onto nearby objects. Figure 2.18 gives a good example. The columns in the left viewport have ambient occlusion turned off, but the right columns have it turned on. Notice how the columns on the right cast a light shadow onto the walls and onto the areas around the column caps.

FIGURE 2.18

Ambient Occlusion can often be used as an alternative to full shadows.

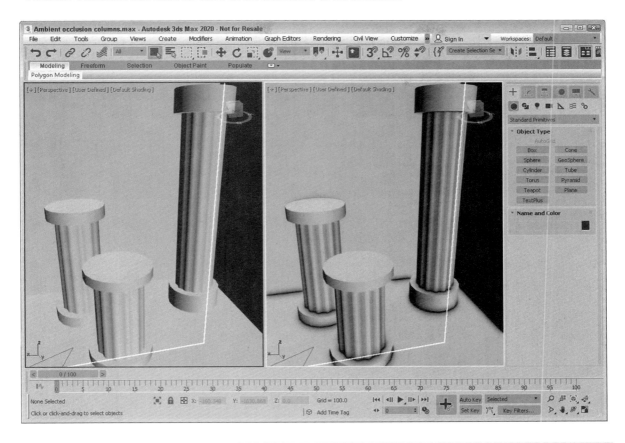

Caution

Displaying shadows in the viewport requires Direct3D 9.0 to be installed on your system, and your video card must support Shader Model 2.0 or Shader Model 3.0. If your computer doesn't support either of these Shader models, the options will be disabled in the viewport shading label menu and in the Viewport Configuration dialog box. You can check the capabilities of your video card using the Help→3ds Max Resources and Tools→Diagnose Video Hardware command.

To enable and configure lighting and shadows in the viewports, follow these steps:

1. Open the Duck toy.max file from the Chap 02 directory in the downloaded content set.

   This file includes a duck pull toy. An Omni light is added to the scene as well.

2. Click in the Shading Quality viewport label, and check the Lighting and Shadows→Shadows option to make sure it is enabled. Then select the Per-View Presets option from the same menu to access the Per-View Presets panel in the Viewport Setting and Preference dialog box.

Caution

If you don't see the Visual Style & Appearance panel in the Viewport Configuration dialog box, you probably have a different display driver enabled.

3. Enable the Ambient Occlusion setting with a Radius of 10 and an Intensity/Fade value of 1. Select the Point Lights/Soft-Edged Shadows option in the Lighting and Shadows Quality setting. Then click the OK button.

4. Click the Select and Move toolbar button, and drag the light object about the scene.

The shadows under the duck toy are automatically updated as the light is moved, as shown in Figure 2.19. Notice that the shadows are soft.

FIGURE 2.19

Lights and shadows are updated in real time when the light is moved about the scene.

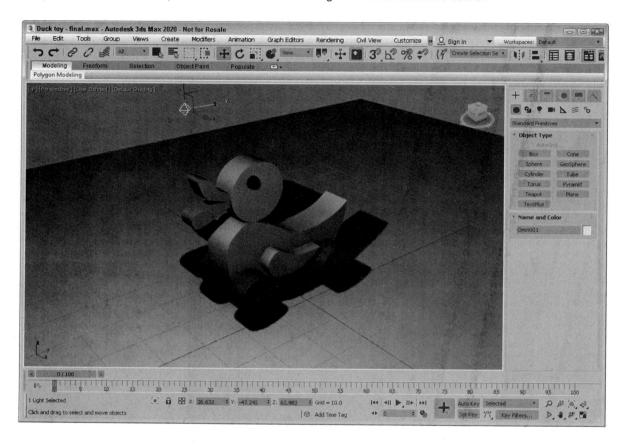

Working with Viewport Backgrounds

Remember in grade school when you realized that you could immediately draw really well using tracing paper (where all you needed to do was follow the lines)? Well, it's not quite tracing paper, but you can load background reference images into a viewport that can help as you create and position your objects.

Changing the viewport background

The Views→Viewport Background menu command includes submenus for switching the viewport background between a gradient and solid color. You also have options to display the Environment Background or a custom image file. These same commands also are available in the Shading viewport label.

Tip

Colors for the background gradient and solid color are set using the Colors panel in the Customize User Interface dialog box. You can open this dialog box using the Customize menu. The colors are found by selecting the Viewports option from the top drop-down list.

The default gradient is set to display dark gray at the top of the viewport and gradually changes to a lighter gray at the bottom. You can alter the displayed colors using the Colors panel in the Customize User

Interface dialog box opened using the Customize menu. Look for the Viewport Background, Viewport Gradient Background Bottom, and Viewport Gradient Background Top entries.

Loading viewport background images

To create a background image to be rendered along with the scene, you need to choose the Environment Background option and specify the background in the Environment dialog box, opened using the Rendering→Environment (keyboard shortcut, 8) menu command. If an environment map is already loaded into the Environment dialog box, you can simply click the Use Environment Background option. Keep in mind that the background image will not be rendered unless it is used as an Environment map.

Environment maps are covered in Chapter 27, "Rendering a Scene and Enabling Quicksilver."

To load images into the viewport background, select the Custom Image File option in the Views→Viewport Background menu or in the Shading viewport label, and then use the Views→Viewport Background→Configure Viewport Background menu command (Alt+B) to open a dialog box, shown in Figure 2.20, in which you can select an image or animation to appear behind a viewport. Each viewport can have a different background image. The displayed background image is helpful for aligning objects in a scene, but it is for display purposes only and will not be rendered.

FIGURE 2.20

The Background panel in the Viewport Configuration dialog box lets you select a background source image or animation.

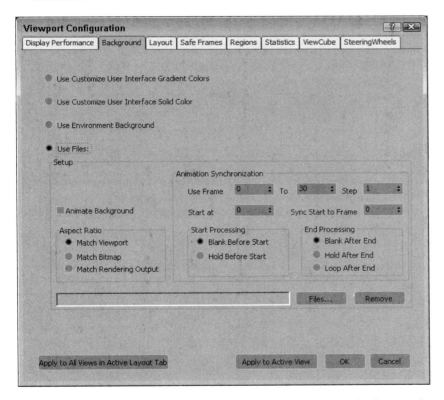

The Environment Background is set using the Background color swatch and Environment Map button located in the Environment and Effects dialog box. You can open this panel with the Rendering→Environment menu or by pressing the 8 key. The benefit of using an environment background is that it renders with the scene objects where background images are not.

The Files button in the Background panel of the Viewport Configuration dialog box opens the Select Background Image dialog box, where you can select the image to load. Once an image is selected, its path is added to the Background panel of the Viewport Configuration dialog box. You can remove the designated file path by clicking the Remove button. The Aspect Ratio section offers options for setting the size of the background image. You can select to Match Viewport, Match Bitmap, or Match Rendering Output.

Loading viewport background animations

The Animation Synchronization section of the Background panel lets you set which frames of a background animation sequence are displayed. The Use Frame and To values determine which frames of the loaded animation are used. The Step value trims the number of frames that are to be used by selecting every Nth frame. For example, a Step value of 4 would use every fourth frame.

Tip

Loading an animation sequence as a viewport background can really help as you begin to animate complex motions, like a running horse. By stepping through the frames of the animation, you can line up your model with the background image for realistic animations.

The Start At value is the frame in the current scene where this background animation would first appear. The Sync Start to Frame value is the frame of the background animation that should appear first. The Start and End Processing options let you determine what appears before the Start and End frames. Options include displaying a blank, holding the current frame, and looping. Make sure to enable the Animate Background option in order to play the background animation when the animation plays.

You can set the Apply to All Views in the Active Layout Tab or the Apply to Active View option to display the background in All Views or in the active viewport only.

Tutorial: Loading reference images for modeling

When modeling a physical object, you can get a jump on the project by taking pictures with a digital camera of the front, top, and left views of the object and then loading them as background images in the respective viewports. The background images can then be a reference for your work. This is especially helpful with models that need to be precise. You can even work from CAD drawings.

To load the background images of a brass swan, follow these steps:

1. Choose File→New (or press Ctrl+N) to open a blank scene file. Select New All in the New Scene dialog box.

2. Right-click the Front viewport to make it the active viewport, and choose Views→Viewport Background→Configure Viewport Background (or press Alt+B).

   The Background panel of the Viewport Configuration dialog box opens.

3. Enable the Use Files option, click the Files button, and in the Select Background Image dialog box that opens, select the Brass swan-front view.jpg image from the Chap 02 directory in the downloaded content set. Then click Open.

4. Select the Match Bitmap option and click the Apply to Active View button, then click OK to close the dialog box.

   The image now appears in the background of the Front viewport.

5. Repeat Steps 2 through 4 for the Top and Left viewports.

6. Zoom and pan each view until the image is centered in the background.

Figure 2.21 shows the 3ds Max interface with background images loaded in the Front, Top, and Left viewports.

FIGURE 2.21

Adding a background image to a viewport can help as you begin to model objects.

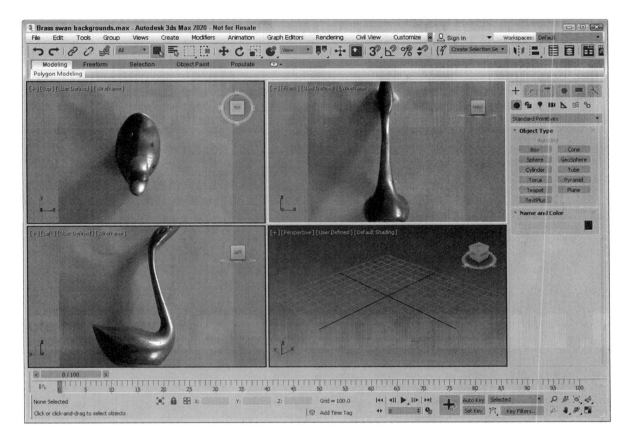

Setting Viewport Visual Style and Appearance

The Viewport Setting and Preferences dialog box, shown in Figure 2.22, holds all the visual settings of the viewports. It also holds several presets that you can quickly select. You can access it using the Views→Viewport Per-View Settings menu. It is also accessible from the Shading viewport label menu. Using this dialog box, you can configure how the scene objects are rendered within the viewports.

The Viewport Setting and Preferences dialog box is divided into two panels--Per-View Presets and Per-View Preferences. There is an option at the bottom of the dialog box that lets you apply the changes to all views. This lets you make changes to many viewports at once.

FIGURE 2.22

The Viewport Setting and Preference dialog box sets how the scene objects are displayed.

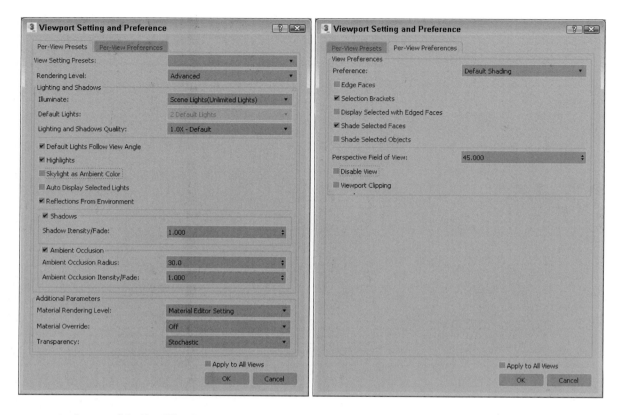

At the top of the Per-View Presets panel is a drop-down list of several presets including High-Quality, Standard, Performance and DX Mode. The High-Quality setting uses scene lights, shadows, ambient occlusion, and reflections. The Standard preset disables shadows and ambient occlusion, but keeps reflections enabled and uses the default lights. The Performance preset also uses default lights, but disables reflections and uses basic materials on all objects. The DX Mode preset uses DirectX shader materials. The Rendering Level setting can be set to Advanced, Basic or DX.

Using the Rendering levels

The Preference drop-down list in the Per-View Preferences panel includes all the shading options found in the Shading viewport label, including all the Stylized non-photorealistic options.

Tip

If you ever get stuck waiting for 3ds Max to complete a task, such as redrawing the viewports, you can always press the Escape key to suspend any task immediately and return control to the interface.

Note

These settings have no effect on the final rendering specified using the Render Setup dialog box. They affect only the display in the viewport.

Figure 2.23 shows some stylized display options.

FIGURE 2.23

Stylized renders here include Pastel and Color Pencil along the top and Tech and Color Ink along the bottom.

Viewing transparency

In addition to these shading types, you also can set the viewport to display objects that contain transparency (which is set in the Material Editor). The Enable Transparency option is located under the viewport Preset label in the Materials submenu. The Transparency option is located in the Per-View Presets panel of the Viewport Setting and Preference dialog box. Figure 2.24 shows the transparency option disabled and enabled with the help of a hungry little animated creature and his ghostly rival.

FIGURE 2.24

Transparency in the viewport can be enabled using the Materials menu in the Shading Quality viewport label.

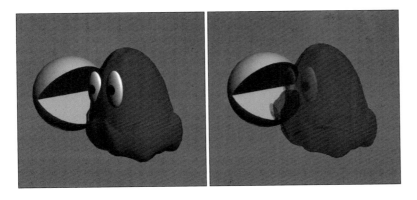

Other rendering options

The other rendering options, such as Disable View (D), Viewport Clipping, and an option to turn Textures on and off also are available. These options can help speed up viewport updates or increase the visual detail of the objects in the viewport.

Tip

At any time during a viewport update, you can click the mouse or press a key to cancel the redraw. 3ds Max doesn't make you wait for a screen redraw to be able to execute commands with the mouse or keyboard shortcuts.

Within the Preference section of the Per-View Preferences panel are several options that make it easier to see the selected object and/or subobjects. The Selection Brackets option displays white corners around the current selection. Selection brackets are useful for helping you see the entire size of a grouped object, but can be annoying if left on with many objects selected. Uncheck this option (or press the J key) to make these brackets disappear. The option to Display Selected with Edged Faces helps to highlight the selected object. If this option is enabled, the edges of the current selection are displayed regardless of whether the Edged Faces check box is enabled. Figure 2.25 shows the beak of a cartoon chicken selected with the Display Selected with Edged Faces and the Use Selection Brackets options enabled. These options make the current selection easy to see.

FIGURE 2.25

The Display Selected with Edged Faces and Use Selection Brackets options make identifying the current selection easy.

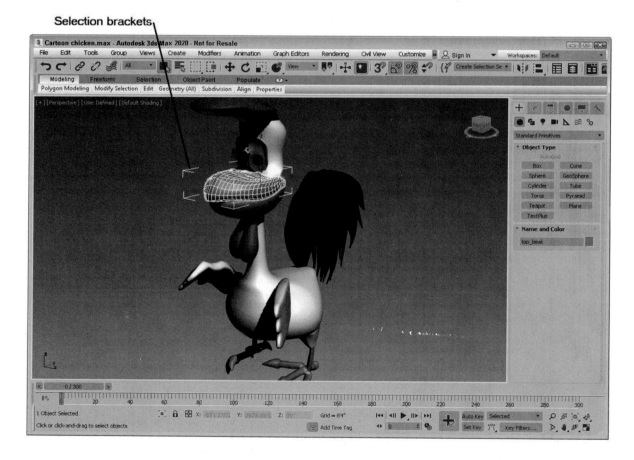

You use Shade Selected Faces (F2) to shade selected subobject faces in red, making them easy to see. The Shade Selected Objects option causes the selected object to be shaded. This is noticeable only if the render level is set to Wireframe or Hidden Line. It causes the selected object to be shaded.

Note

The Shade Selected Faces (F2) option, which shades selected subobject faces, is different from the Views→Shade Selected menu command, which turns on shading for the selected object in all viewports.

Setting the Field of View

You also can alter the Field of View (FOV) for the Perspective viewport in the Per-View Preferences panel of the Viewport Setting and Preference dialog box. To create a fish-eye view, decrease the FOV setting to 10 or less. The maximum FOV value is 180, and the default value is 45. You also can change the Field of View using the Field of View button in the Viewport Navigation Controls. The setting in the Per-View Preferences panel, however, lets you enter precise values.

See Chapter 24, "Configuring and Aiming Cameras," for more coverage on Field of View.

Grabbing a viewport image

It's not rendering, but you can grab an image of the active viewport using Tools→Preview - Grab Viewport→Capture Still Image. This option is also available in the Create Preview menu in the General Viewport label. Before grabbing the image, a simple dialog box appears asking you to add a label to the grabbed image. The image is loaded into the Rendered Frame Window, and its label appears in the lower-right corner of the image. The Tools→Preview - Grab Viewport menu also includes options for creating, viewing, and renaming animated sequence files, which provides a quick way to create animation previews.

Tip

If you want to save the image loaded in the Rendered Frame Window, simply click the Save Image button. You also can copy, clone, and print from this same window.

More on working with animation previews is covered in Chapter 32, "Understanding Animation and Keyframes."

Configuring viewport lighting and shadows

The Per-View Presets panel of the Viewport Setting and Preference dialog box includes a number of settings for the viewport lighting and shadows. This panel includes options to illuminate the scene with Default lights (either one or two) or with Scene lights, just like the viewport Shading label menu. The one-light option creates a single light positioned behind the viewer and at an angle to the scene. Scenes with one light update more quickly than scenes with two lights.

The Default Lights option deactivates your current scene lights and uses the default lights. This option can be helpful when you're trying to view objects in a dark setting because the default lighting illuminates the entire scene without requiring you to remove or turn off lights. This dialog box also includes options for enabling Highlights, Skylight as Ambient Color, Auto Display Selected Lights and Reflections from Environment.

Note

If you are not using the Nitrous display drivers, the configuration options for lighting and shadows are different.

The Lighting and Shadows Quality drop-down list lets you select from several different quality settings. The options range from Point Lights/Hard Shadows and Point Lights/Soft-Edged Shadows on the left through 0.125X-Very Low to 16X-Very High Quality on the right. The higher quality options can take longer to render and use area lights to improve the quality.

Beneath the Lighting and Shadows Quality setting is an option to have the Skylights add to the Ambient Color and settings for Shadows and Ambient Occlusion. You also have options to adjust the Intensity/Fade value for shadows and the Radius and Intensity/Fade amount of the Ambient Occlusion effect. These values are used to override the Viewport Shadow Intensity to dim the shadows if they are too dark. The Radius value sets how close objects need to be in order to be included in the ambient occlusion solution. You can have the environment reflections appear in the viewport as well.

Another option available in the Visual Style & Appearance panel is the Auto Display Selected Lights. This option also is available in the Lighting and Shadows menu in the Shading Quality viewport label. It is helpful when you're placing and aiming lights in the scene. It causes the selected light to be displayed in the shaded viewport automatically. The Viewport Lighting and Shadows menu also includes options for locking and unlocking selected lights.

Altering the Viewport Layout

Now that you've started to figure out the viewports, you may want to change the number and size of viewports displayed. The easiest way to work with viewport layouts is with the Viewport Layout Tabs toolbar, but the Layout panel in the Viewport Configuration dialog box offers more options and lets you configure each viewport.

The Layout panel, shown in Figure 2.26, in the Viewpoint Configuration dialog box, offers several layouts as alternatives to the default layout (not that there is anything wrong with the default and its four equally sized viewports).

FIGURE 2.26

The Layout panel offers many layout options.

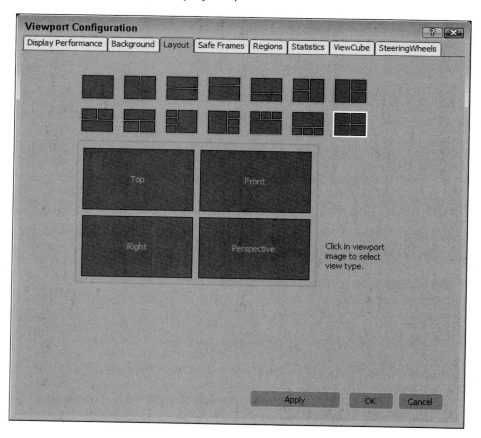

After selecting a layout from the options at the top of the panel, you can assign each individual viewport a different view by clicking each viewport in the Layout panel and choosing a view from the pop-up menu. The view options include Perspective, Orthographic, Front, Back, Top, Bottom, Left, Right, Track, Grid (Front, Back, Top, Bottom, Left, Right, Display Planes), Scene Explorer, Extended (Motion Mixer, Biped Animation WorkBench, State Sets Main Window, Boolean Explorer, Crease Explorer, Material Explorer, MAXScript Listener, Base Object Parameters, Materials, Modifiers, and Custom Attributes), and Shape. These view options are also available by clicking the viewport Point-of-View label in the upper-left corner of the viewport. Figure 2.27 shows a viewport layout with the Track view, Schematic view, Asset Browser, and Perspective views open.

FIGURE 2.27

Other interfaces such as the Track View, Schematic View, and Scene Explorer can be opened within a viewport.

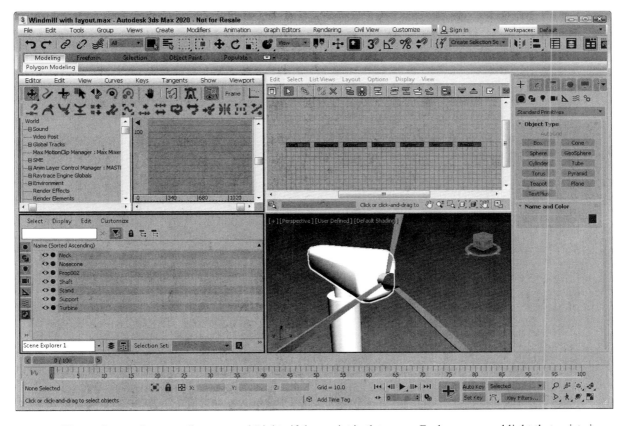

Views also can be set to Cameras and Lights if they exist in the scene. Each camera and light that exists is listed by name at the top of the pop-up menu.

Using Safe Frames

Completing an animation and converting it to some broadcast medium, only to see that the whole left side of the animation is being cut off in the final screening, can be discouraging. If you rely on the size of the active viewport to show the edges of the final output, you could be way off. Using the Safe Frames feature, you can display some guides within the viewport that show where the content must be to avoid such problems.

The Safe Frames panel of the Viewport Configuration dialog box lets you define several safe frame options, as shown in Figure 2.28, including the following:

* **Live Area:** Marks the area that will be rendered, shown as yellow lines. If a background image is added to the viewport and the Match Rendering Output option is selected, the background image will fit within the Live Area.

* **Action Safe:** The area ensured to be visible in the final rendered file, marked with light blue lines; objects outside this area will be at the edge of the monitor and could be distorted.

* **Title Safe:** The area where the title can safely appear without distortion or bleeding, marked with orange lines.

* **User Safe:** The output area defined by the user, marked with magenta lines.

* **12-Field Grid:** Displays a grid in the viewport, marked with a pink grid.

FIGURE 2.28

The Safe Frames panel lets you specify areas to render.

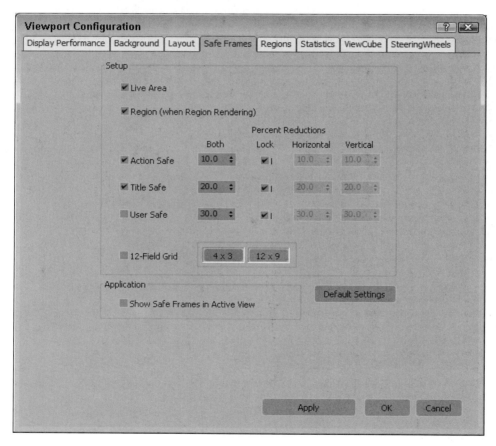

For each type of safe frame, you can set the percent reduction by entering values in the Horizontal, Vertical, or Both fields. The 12-Field Grid option offers 4 x 3 and 12 x 9 layout grids.

The Show Safe Frames in Active View option displays the Safe Frame borders in the active viewport. You can quickly enable or disable Safe Frames by right-clicking the viewport Point-of-View label and choosing Show Safe Frame in the pop-up menu (or you can use the Shift+F keyboard shortcut).

Figure 2.29 shows an elongated Perspective viewport with all the safe frame guides enabled. The Safe Frames show that the top and bottom of my clock tower will be cut off when rendered.

FIGURE 2.29

Safe frames provide guides that help you see when the scene objects are out of bounds.

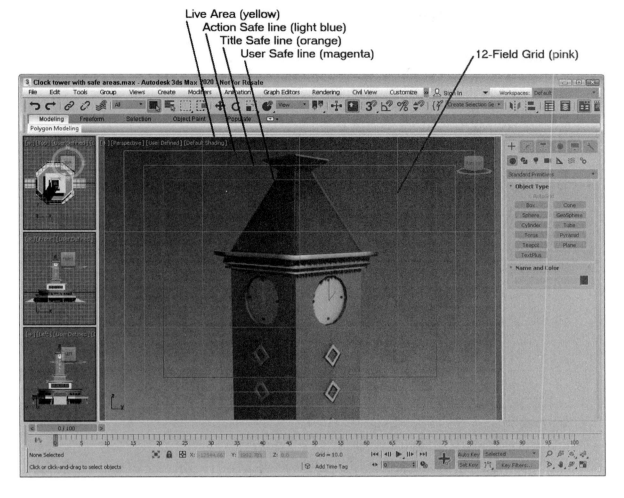

Setting Display Performance

Within the Display Performance panel, shown in Figure 2.30, are a couple of simple settings. The Improve Quality Progressively setting causes a rough approximation of the viewport update to be rendered immediately, and gradually the details are added to improve the displayed results as time progresses. This progressive update doesn't slow you down but provides more detail if there is time to add it before the next update.

Note

If the Nitrous display driver is not enabled, the Adaptive Degradation panel is available in the Viewport Configuration dialog box.

You also can set the display resolution used on procedural maps. Higher resolution values have more detail, and lower resolution maps render and are updated more quickly.

FIGURE 2.30

The Display Performance panel offers only a few simple settings.

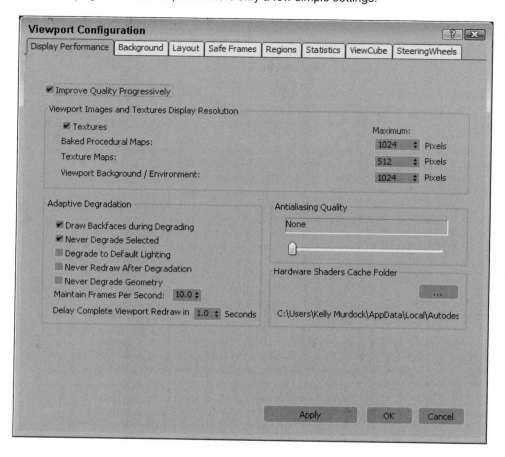

 You can enable Progressive Display option using the Views→Progressive Display menu command. You also can turn Adaptive Degradation on and off with a button located at the bottom of the interface between the Prompt Bar and the Add Time Tag (or by pressing the O key). The Adaptive Degradation button looks like a simple cube.

Tip

Right-clicking the Progressive Display button opens the Display Performance panel in the Viewport Configuration dialog box.

Defining Regions

The Regions panel enables you to define regions and focus your rendering energies on a smaller area. Complex scenes can take considerable time and machine power to render. Sometimes, you want to test render only a portion of a viewport to check material assignment, texture map placement, or lighting.

You can define the size of the various regions in the Regions panel of the Viewport Configuration dialog box, shown in Figure 2.31.

FIGURE 2.31

The Regions panel enables you to work with smaller regions within your scene.

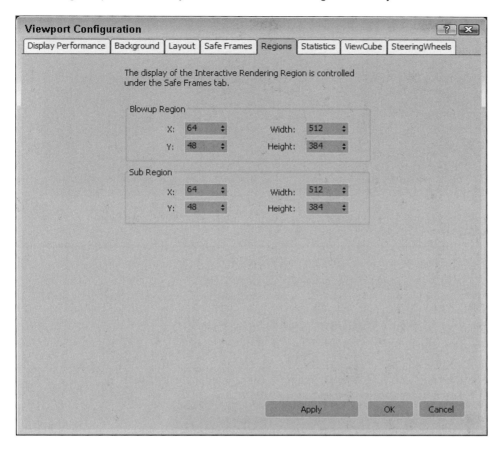

After you've specified a Blowup Region or a Sub Region, you can select to render using these regions by selecting Region or Blowup from the Render Frame Window and clicking the Render button. If you click the Edit Region button, the specified region is displayed as an outline in the viewport. You can move this outline to reposition it, or drag its edge or corner handles to resize the region. The new position and dimension values are updated in the Regions panel for next time. Click the Render button to begin the rendering process.

The difference between these two regions is that the Sub Region displays the Rendered Frame Window in black, except for the specified sub-region. The Blowup Region fills the entire Rendered Frame Window, as shown in Figure 2.32.

You can learn more about Render Types and the Rendered Frame Window in Chapter 27, "Rendering a Scene and Enabling Quicksilver."

FIGURE 2.32

The image on the left was rendered using the Sub Region option; the right image used the Blowup Region.

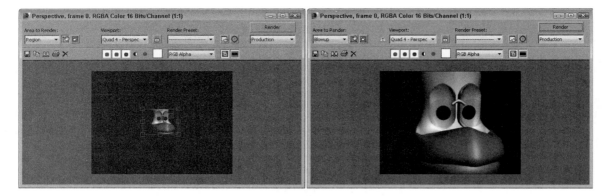

Viewing Statistics

The Statistics panel, shown in Figure 2.33, lets you display valuable statistics in the viewport window. These statistics can include Polygon Count, Triangle Count, Edge Count, Vertex Count, and Frames Per Second. You also can select to view these statistics for all the objects in the scene (Total), for just the selected object, or for both. You can toggle statistics on and off for the active viewport using the xView→Show Statistics menu in the General viewport label or the 7 key.

FIGURE 2.33

The Statistics panel lets you display polygon count and frames per second in the viewport.

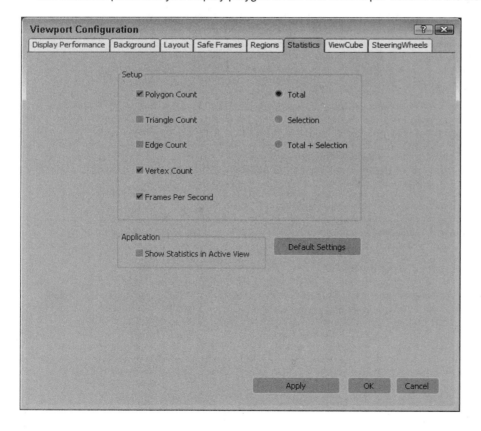

By enabling the Show Statistics in Active View option, the selected statistics are overlaid on the active viewport.

Configuring the ViewCube and SteeringWheels

The final two panels in the Viewport Configuration dialog box contain configuration settings for the ViewCube and SteeringWheels. The ViewCube panel includes settings for which viewports the ViewCube appears within, either All Views in the Active Layout or only in the Active View. You also can set the ViewCube's Size and its Inactive Opacity. An Inactive Opacity value of 0 makes the ViewCube invisible until you move over it when it appears again.

You also have an option for the ViewCube to snap to its closest view when dragging it. And you can define what happens when you click it. This provides an easy way to make the scene upright. Finally, you can define the compass orientation, which is the ring that surrounds the ViewCube.

The SteeringWheels panel lets you set the size and opacity for the Big Wheels or Mini Wheels modes. Several other options enable tool tips, pin the wheel, and configure the individual navigation tools.

Summary

Viewports are the windows into the 3ds Max world. Remember that if you can't see it, you can't work with it, so you need to learn to use the viewports. You also can configure viewports to display just the way you desire. Most of the commands for working with the viewports are located in the Views menu and in the various viewport labels, but all the configuration settings are located in the Viewport Configuration dialog box. Using these settings, you can configure viewports to display just the way you desire.

This chapter covered the following topics:

* Understanding 3D space and the various viewport points-of-view
* Navigating with the ViewCube, the SteeringWheels, and the scroll wheel
* Using the various Viewport Navigation Control buttons
* Changing the Visual Style of the viewport
* Changing the viewport layout, clipping planes, and xView information
* Turning materials, lighting, and shadows on and off in the viewport
* Working with viewport background colors, gradients, and images
* Changing the shading and display options in the Viewport Setting and Preference dialog box
* Discovering the other panels of the Viewport Configuration dialog box that allow you to change the layout, safe frames, regions, and statistics
* Finding out how to use Progressive Display to speed up viewport updates
* Exploring the configuration settings for the ViewCube and SteeringWheels

In the next chapter, you find out all the details about working with files, including loading, saving, and merging scene files. The next chapter also covers the import and export formats for interfacing with other software packages.

Chapter 3
Working with Files, Importing and Exporting

IN THIS CHAPTER

Saving, opening, merging, and archiving files

Converting scenes

Exporting objects and scenes

Importing objects from external packages

Working with file utilities

Accessing scene files' information

Importing objects from external packages

Exporting objects and scenes

Working with Asset Tracking

Complex scenes can end up being a collection of hundreds of files, and misplacing any of them will affect the final output, so learning to work with files is critical. This chapter focuses on working with files, whether they are object files, texture images, or background images. Files enable you to move scene pieces into and out of the Autodesk® 3ds Max® 2020 software. You also can export and import files to and from other packages.

This chapter also includes perhaps the most important feature in 3ds Max: the Save feature, which I suggest you use often. Remember the mantra: Save Early, Save Often.

Files enable you to move scene pieces into and out of the Autodesk® 3ds Max® 2020 software. You also can export and import files to and from other packages. This chapter looks at the more common formats used to import and export data.

Working with 3ds Max Scene Files

Of all the different file types and formats, you probably will work with one type of file more than any other—the 3ds Max format. 3ds Max has its own proprietary format for its scene files. These files have the .max extension and allow you to save your work as a file and return to it at a later time. 3ds Max also supports files saved with the .chr extension used for character files.

Starting new

When 3ds Max starts, a new scene opens. You can start a new scene at any time with the File→New→New All (Ctrl+N) command. Although each instance of 3ds Max can have only one scene open at a time, you can open multiple copies of 3ds Max, each with its own scene instance, if you have enough memory.

Starting a new scene deletes the current scene, but if you've made changes to the current scene, a dialog box appears asking if want to save your changes.

Starting a new scene maintains all the current interface settings, including the viewport configurations, any interface changes, viewport backgrounds, and any changes to the Command Panel. To reset the interface, choose File→Reset. When reset, all interface settings return to their default states, but interface changes aren't affected. For example, if you have the Modify panel open and the Command Panel made into a floating panel, then resetting the interface will close the Modify panel and reopen the default Create panel, but the Command Panel will remain floating.

Using a Start-Up Template

If you select the File→New→New from Template menu command, then the Create New Scene dialog box appears. This dialog box includes several thumbnails of templates that you can start from. These templates have all the settings for a specific type of project such as outdoor rendering or an underwater scene. Templates can include rendering configurations and also objects and backgrounds. The Create New Scene panel also includes a link to open the Template Manager, shown in Figure 3.1.

FIGURE 3.1

The Template Manager panel lets you create and save new templates.

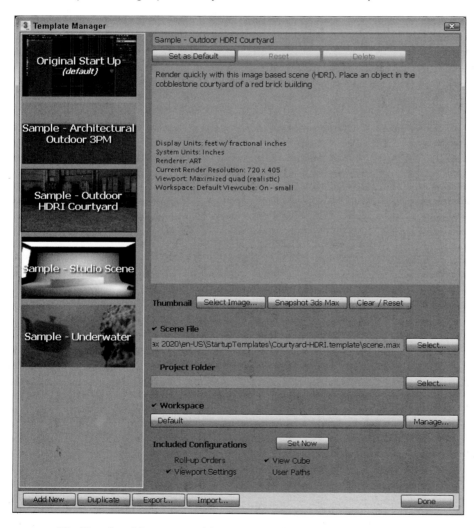

The Template Manager panel lets you save the current scene settings, give the template a name and you can also load a scene file, a project folder and a workspace. Saved templates can also be imported and exported.

Saving files

After you start up 3ds Max, the first thing you should learn is how to save your work. After a scene has changed, you can save it as a file. Before a file is saved, the word *Untitled* appears in the title bar; after you save the file, its filename appears in the title bar. Choose the File→Save (Ctrl+S) to save the scene. If the scene hasn't been saved yet, a Save File As dialog box appears, as shown in Figure 3.2. You also can make this dialog box appear using the File→Save As command. After a file has been saved, using the Save command saves the file without opening the Save File As dialog box. Pretty simple—just don't forget to do it often.

Within the Save File As dialog box is an option in the Save as Type field to save the file as a 3ds Max file, a 3ds Max 2017 file, a 3ds Max 2018, a 3ds Max 2019 file, or a 3ds Max Characters file Files saved using a format for a previous version of 3ds Max can be opened only within the designated version or any version newer than that version. Be aware that any new features included in 3ds Max 2020 are not included in the saved file using an older format. For example, if the current file uses a newer feature, and you save the file to an older format, support for the new feature is lost.

Caution

Be aware that 3ds Max files are not backward-compatible. A .max file saved using 3ds Max 2020 cannot be opened in an earlier version of 3ds Max. The solution to compatibility issues is to export the file using the FBX format and then import it in the older version of 3ds Max.

FIGURE 3.2

Use the Save File As dialog box to save a scene as a file.

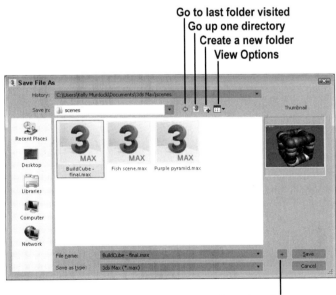

The Save File As dialog box keeps a history list of the last several directories that you've opened. You can select these directories from the History drop-down list at the top of the dialog box. The buttons in this dialog box are the standard Windows file dialog box buttons used to go to the last folder visited, go up one folder, create a new folder, and view a pop-up menu of file view options.

Note

If you try to save a scene over an existing scene, 3ds Max presents a dialog box confirming this action.

Clicking the button with a plus sign to the right of the Save button automatically appends a number onto the end of the current filename and saves the file. For example, if you select the myScene.max file and click the plus button, a file named myScene01.max is saved.

Tip

Use the auto increment file number and Save button to save progressive versions of a scene. This is an easy version control system. If you need to backtrack to an earlier version, you can.

The File menu includes options to Save, Save As, Save Copy As, Save Selected, and Archive. The File→Save Copy As menu command lets you save the current scene to a different name without changing its current name. The File→Save Selected option saves the current selected objects to a separate scene file. If you create a single object that you might use again, select the object and use the Save Selected option to save it to a directory of models.

Tip

Another useful feature for saving files is to enable the Auto Backup feature in the Files panel of the Preference Settings dialog box. This dialog box can be accessed with the Customize→Preferences menu command or by clicking the Options button at the bottom of the File menu.

Archiving files

By archiving a 3ds Max scene along with its reference bitmaps, you can ensure that the archived file includes all the necessary files. This is especially useful if you need to send the project to your cousin to show off or to your boss, and you don't want to miss any ancillary files. Choose File→Archive to save all scene files as a compressed archive. The default archive format is .zip (but you can change it to use whichever compression program you want in the Files panel of the Preference Settings dialog box).

Saving an archive as a ZIP file compiles all external files, such as bitmaps, into a single compressed file. Along with all the scene files, a text file is automatically created that lists all the files and their paths.

Opening files

When you want to open a file you've saved, you may do so by choosing File→Open (Ctrl+O), which opens a file dialog box that is similar to the one used to save files. 3ds Max can open files saved with the .max and .chr extensions. 3ds Max also can open VIZ Render files that have the .drf extension. Selecting a file and clicking the plus button opens a copy of the selected file with a new version number appended to its name.

The File→Open Recent command menu displays an extensive list of recently opened 3ds Max files. This list holds 10 recently opened filenames, but it can hold up to 50 filenames, as set in the Files panel of the Preference Settings dialog box. The most recently accessed files are organized by last date opened.

The File→Open menu also includes commands for opening files from Vault if the Vault plug-in is installed. Vault is a version control system for 3ds Max resources.

If 3ds Max cannot locate resources used within a scene (such as texture maps) when you open a 3ds Max file, the Missing External Files dialog box, shown in Figure 3.3, appears, enabling you to Continue without the file or to Browse for the missing files. If you click the Browse button, the Configure External File Paths dialog box opens, where you can add a path to the missing files. There is also a Remove All button that simply removes the missing maps from the scene objects. This is convenient if you want to load and render the scene without having to find the files.

Note

If 3ds Max cannot locate missing files, a similar warning dialog box also appears when you try to render the scene with missing files.

FIGURE 3.3

The Missing External Files dialog box identifies files for the current scene that is missing.

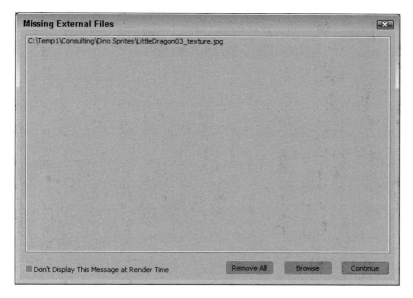

If you open a file saved using a previous version of 3ds Max that includes features that have changed since the previous version, 3ds Max presents an obsolete-data-format warning statement. Resaving the file can fix this problem. However, if you save a file created with a previous version of 3ds Max as a 3ds Max 2020 scene file, you won't be able to open the file again in the previous versions of 3ds Max.

Tip
You can disable the obsolete-file message in the Files panel of the Preference Settings dialog box.

Note
You also can open files from the command line by placing the filename after the executable name, as in 3dsmax.exe myFile.max. You also can use the –L switch after the executable name to open the last file that was opened.

Using the Scene Converter

In each new version of 3ds Max, there are changes to the available plug-in tools that cause trouble when legacy files try to load. These older files rely on a feature that has been replaced by a newer feature. A good example of this is when the mental ray rendering engine was replaced with the Arnold rendering system. This change causes a warning dialog box, shown in Figure 3.4, to appear. It also lets you access the Scene Converter, which is a tool for migrating old files to the newer features.

FIGURE 3.4

This warning dialog box reminds you that the current file references some out-of-date features.

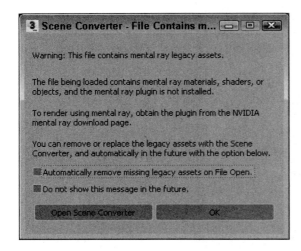

The Scene Converter dialog box, shown in Figure 3.5, lets you define and save presets for migrating older files to the new features. This can be done using scripts or by defining conversion rules that are shown in the right panel. To access this dialog box, click the Open Scene Converter button in the warning dialog box when the file is first opened or you can select it at anytime using the Rendering→Scene Converter menu command.

FIGURE 3.5

The Scene Converter dialog box lets you define rules to swap the old features for new features.

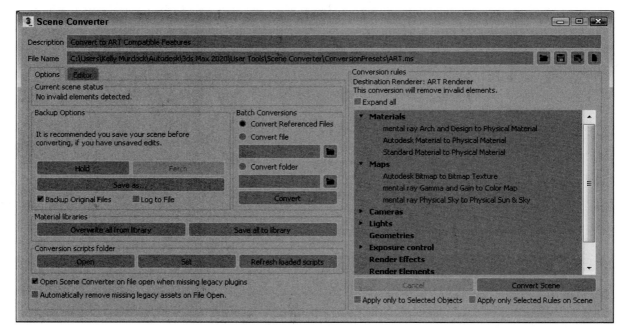

Within the Conversion Rules section is a tree of available changes. These changes include options such as changing mental ray Arch and Design materials to Physical Material. For each rule, you can select it and remove it from the list or choose to Apply Only the Selected rule.

To define new rules, select the Editor panel. This panel has two panes for Source features and Destination features. If you select a Source item such as a specific material like Autodesk Masonry or the Skylight, then the Destination pane shows the available conversion options such as Physical Material or Physical Sky. When all the rules you want are included, click the Convert Scene button to convert and save the file.

Setting a Project Folder

By default, the software's Open File dialog box opens to the Scenes folder in the Documents\3dsMax directory, but you can set a project folder that may be located anywhere on your local hard drive or on the network. All file dialog boxes will then open to the new project folder automatically. The File→Project→Set Active Project menu opens a dialog box where you can select a project folder to be the current active project. A list of recently opened projects is displayed at the top of the Project menu.

The File→Project→Create Empty menu creates a new project folder without the default resource files. The Create Default menu creates a new project folder with all the resource folders included and the Create from Current menu option creates a new folder that mimics the structure of the current active project. If the File→Project→Automatic Switch option is enabled, then the scene's project is automatically loaded when the scene file is opened.

Within the project folder's root is a file with the .mxp extension named the same as the project folder. This file is a simple text file that can be opened within a text editor. Editing this file lets you define which subfolders are created within the project folder. The defined project folder also is visible within the title bar if the interface is wide enough to display it.

Merging and replacing objects

If you happen to create the perfect prop in one scene and want to integrate the prop into another scene, you can use the Merge menu command. Choose File→Import→Merge to load objects from another scene into the current scene. Using this menu command opens a file dialog box that is exactly like the Open File dialog box, but after you select a scene and click the Open button, the Merge dialog box, shown in Figure 3.6, appears. This dialog box displays all the objects found in the selected scene file. It also has options for sorting the objects and filtering certain types of objects. Selecting an object and clicking OK loads the object into the current scene.

Note
Using the Merge command places a copy of the object into the current scene. The object maintains no links to the previous scene and can be edited. XRefs are different from merged objects in that they do maintain a link to the previous scene, which allows them to be updated when the original is changed.

You can learn about XRefs in Chapter 10, "Organizing Scenes with Layers, Containers, XRefs and the Schematic View."

FIGURE 3.6

The Merge dialog box lists all the objects from a merging scene.

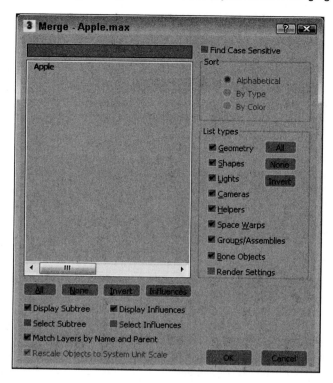

If you ever get involved in a modeling duel, you'll probably be using the File→Import→Replace menu command at some time. A modeling duel is when two modelers work on the same rough model of named objects and the animator (or boss) gets to choose which object to use. With the Replace command, you can replace a named object with an object of the same name in a different scene. The objects are selected using the Replace dialog box, which is identical to the Merge dialog box, but only the objects with identical names in both scene files display. If no objects with the same name appear in both scene files, a warning box is displayed.

Tip

When working with a team, one person, such as an environment modeler, can add a dummy object to the scene that shares the name of a more detailed model, such as "furniture." When the detailed model is completed, the Replace command adds the detailed model to the scene. This lets the environment modeler work, even though the detailed models aren't completed yet.

Getting out

As you can probably guess, you use the File→Exit menu to exit the program, but only after it gives you a chance to save your work. Clicking the window icon with an X on it in the upper right has the same effect (but I'm sure you knew that).

Importing and Exporting

If you haven't noticed, 3ds Max isn't the only game in town. A number of different 3D packages exist, and exchanging files between them is where the importing and exporting menu commands come in. You can find both of these commands in the File menu.

Each import and export format has its own dialog boxes of settings that affect how the transferred data appears.

Importing supported formats

Choose File→Import→Import to open the Select File To Import dialog box. This dialog box looks like a typical Windows file dialog box. The real power comes with the various Import Settings dialog boxes that are available for each format. These dialog boxes appear after you select a file to import. The settings in the Import Settings dialog box are different for the various format types.

3ds Max can import several different formats. All acceptable files are automatically displayed in the file dialog box, or you can filter for a specific format using the Files of Type drop-down list at the bottom of the file dialog box. The available import formats include the following:

* Autodesk (FBX)
* 3D Studio Mesh, Projects, and Shapes (3DS, PRJ, SHP)
* Alembic (ABC)
* Adobe Illustrator (AI)
* Pro/ENGINEER ASM (ASM)
* Pro/ENGINEER (PRT)
* CATIA V5 (CARPART, CGR, CATPRODUCT)
* CATIA V4 (MODEL, DLV4, DLV3, DLV, EXP, SESSION, MDL)
* Collada (DAE)
* LandXML/DEM/DDF (DEM, XML, DDF)
* AutoCAD and Legacy AutoCAD (DWG, DXF)
* Flight Studio OpenFlight (FLT)
* Motion Analysis (HTR, TRC)
* Initial Graphics Exchange Standard (IGE, IGS, IGES)
* Autodesk Inventor (IPT, IAM)
* JT Open (JT)
* OBJ Material and Object (OBJ)
* Unigraphics-NX (PRT)
* ACIS SAT (SAT)
* Google SketchUp (SKP)
* Revit (3DS, PRJ, SHP)
* SOLIDWORKS (RVT)
* StereoLitho (STL)
* STEP (STP, STEP)
* Autodesk Alias (WIRE)
* VRML (WRL, WRZ)
* VIZ Material XML Import (XML)

Note

Be aware that these formats are used for different types of data. For example, Adobe Illustrator files typically hold only 2D data, and Motion Analysis files hold motion capture data for animations.

New Feature in 2020

The Revit Importer has added several new features in 3ds Max 2020.

Import preference

The Files panel of the Preference Settings dialog box has a single option dealing with importing—Zoom Extents on Import. When this option is enabled, it automatically zooms all viewports to the extent of the imported objects. Imported objects can often be scaled so small that they aren't even visible. This option helps you locate an object when imported. This option also helps if the imported objects aren't located near the other scene objects.

Exporting supported formats

In addition to importing, you'll sometimes want to export 3ds Max objects for use in other programs. You access the Export command by choosing File→Export→Export. You also have the option to Export Selected (available only if an object is selected) and Export to DWF. DWF files use the Design Web Format, which enables them to be viewed by others that don't have access to 3ds Max. The files are compressed and can be viewed and navigated using the Autodesk Design Review program.

3ds Max can export to several different formats, including the following:

* Autodesk (FBX)
* 3D Studio (3DS)
* Alembic (ABC)
* Adobe Illustrator (AI)
* ASCII Scene Export (ASE)
* Arnold Scene Source (ASS)
* AutoCAD (DWG, DXF)
* Collada (DAE)
* Initial Graphics Exchange Standard (IGS)
* Flight Studio OpenFlight (FLT)
* Motion Analysis (HTR)
* Publish to DWF (DWF)
* OBJ Material and Object (OBJ)
* PhysX and APEX (PXPROJ)
* ACIS SAT (SAT)
* StereoLithography (STL)
* LMV SVF (SVF)
* Autodesk Alias (WIRE)
* VRML97 (WRL)

Moving files to other Suite packages

3ds Max is available as a stand-alone product, but it also ships within a Creative Suite of applications offered by Autodesk. These suites can include Maya, Softimage, MotionBuilder, and Mudbox, and you can easily move the current 3ds Max scene file to one of these other applications using the File→Send To menu. For each application, you can send the scene as a New Scene, Update the Current Scene, Add to the Current Scene, or Select the Previously Sent Objects.

Manually moving files to and from Maya

Maya is Autodesk's sister to 3ds Max, so you may find yourself having to move scene files between 3ds Max and Maya at some time. If you are not using the Send To command, the best format to transport files between 3ds Max and Maya is the FBX format. Autodesk controls this format and has endowed it with the ability to seamlessly transport files between these packages.

Tip

The FBX format also is the format to choose when transferring files back and forth with 3ds Max files in older versions of 3ds Max.

The FBX format includes support for all the scene constructs, including animation, bone systems, morph targets, and animation cache files. It has an option to embed textures with the export file or to convert them to the TIF format. Other import and export settings deal with the system units and world-coordinate orientation. You also have the ability to filter specific objects.

Tip

When exporting a file for use in Maya, be sure to set the Up Axis to Y-up, or the models will show up rotated.

The FBX format is being continually updated, but the FBX Export dialog box lets you select which FBX version to use. If you need to export a 3ds Max file for use in an older version of 3ds Max, be sure to select an older FBX version in the Version drop-down list.

Using the OBJ format

The OBJ format is a text-based format that has been around since the early days of Wavefront, an early 3D package. It is a common format and used to exchange 3D data with a variety of programs, including Poser and ZBrush.

One aspect of the OBJ format is that it separates the model data and the texture data into two different files. The OBJ file holds the geometry data, and the MTL file holds the texture data. Previous versions of 3ds Max required that you import each of these data files separately, but the latest OBJ workflow imports these two data files together. The new OBJ workflow is a plug-in developed by GuruWare.

The new OBJ import and export workflow is much smoother and automatically gets the right materials and textures for the object. The dialog box for importing OBJ files is shown in Figure 3.7. Notice how each of the individual objects is recognized and displayed in a list. This gives you the option of importing only specific objects. You also have control over how normals are handled, the ability to convert units, and several options for dealing with materials. The small green and red lights to the left of some options indicate whether the option is in the OBJ file. Green indicates that it exists, and red means it doesn't exist.

FIGURE 3.7

The OBJ Import Options dialog box provides another excellent choice for transporting files from other packages.

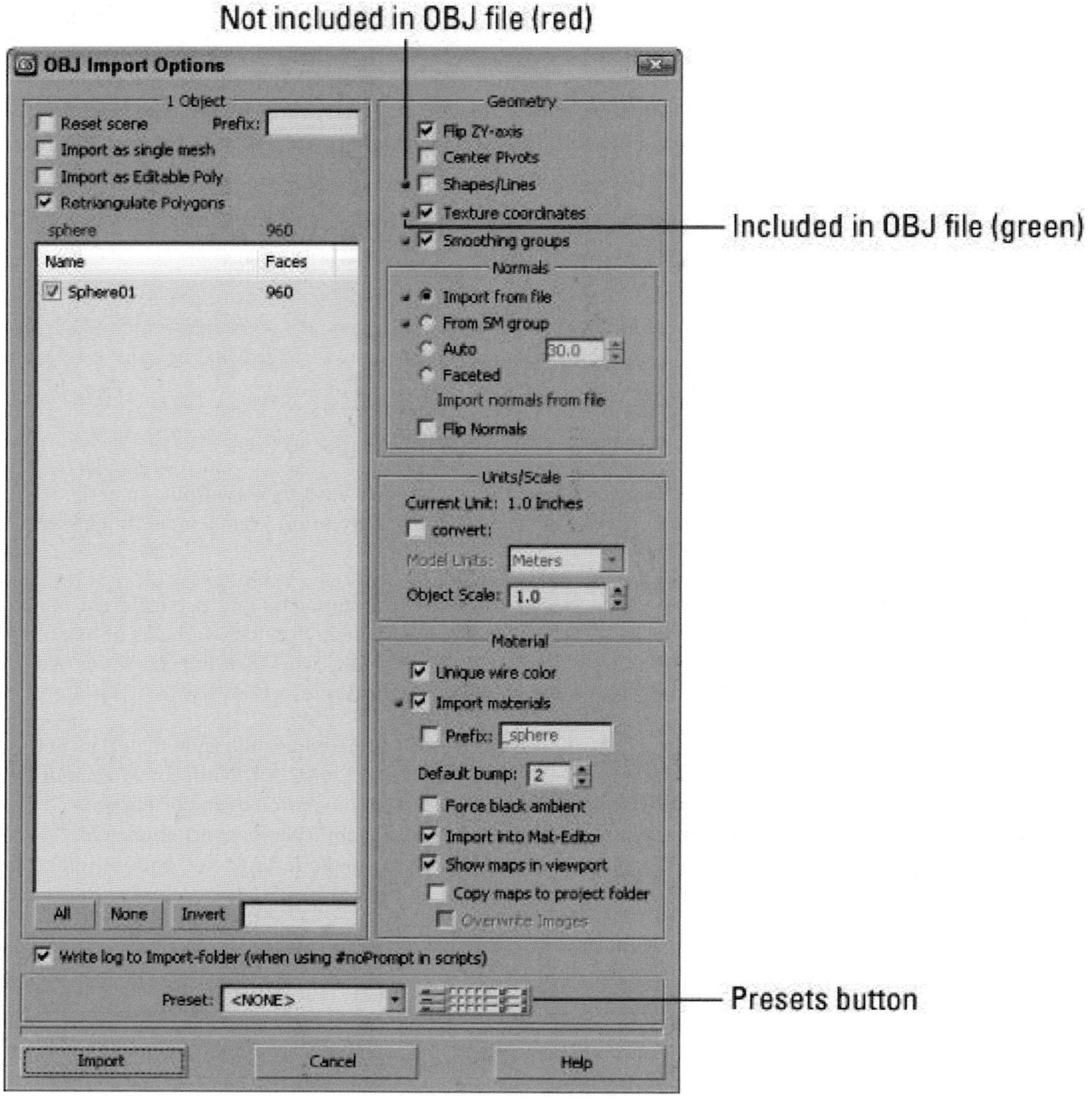

The Presets button opens a dialog box, shown in Figure 3.8, which includes the import settings for several different 3D packages. Using these settings, you can control how the files are imported and keep them consistent.

FIGURE 3.8

The Edit OBJ-Import Presets dialog box gives an overview of the available import settings for multiple packages.

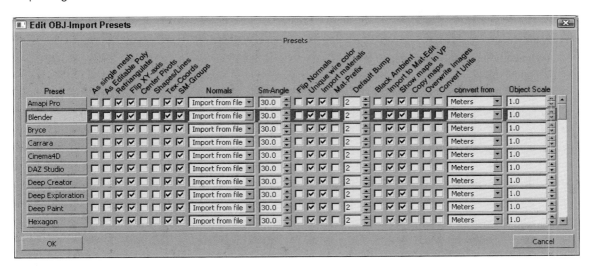

A similar dialog box of settings appears when exporting a scene to the OBJ format, as shown in Figure 3.9.

FIGURE 3.9

The OBJ Export Options dialog box lets you export 3ds Max scenes to other packages.

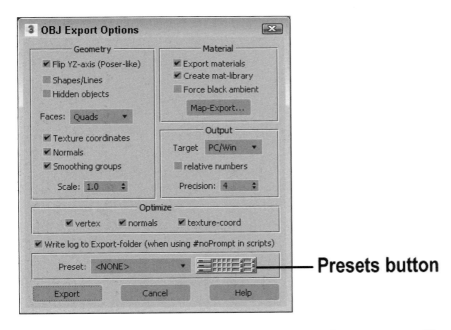

The OBJ Export Options dialog box includes presets for most common 3D apps, including Amapi Pro, Blender, Bryce, Carrara, Cinema-4D, DAZ Studio, Deep Paint, Hexagon, Lightwave, Maya, Modo, Motion Builder, Mudbox, Poser, Realflow, Rhino, Silo, Softimage XSI, UV Mapper, VUE, Worldbuilder, and ZBrush.

Clicking the Map Export button lets you specify the export map path where the textures for the scene are saved. You also can automatically convert the maps to a specific size or format. For each map format, you can configure the bits per pixel and any compression settings.

If you click the Presets button in the OBJ Export Options dialog box, the export options for each format are shown in a table, similar to the same one for the import options. Each of these settings can be quickly altered using this dialog box.

Using the FBX format

Autodesk has a host of other products that deal with 3D, including Maya and MotionBuilder, so you may find yourself having to move scene files between Max and Maya at some time. The best format to transport files among Max, Maya, and MotionBuilder is the FBX format. Autodesk controls this format and has endowed it with the ability to seamlessly transport files among these packages.

Tip

The FBX format also is the format to choose when transferring files back and forth with Softimage XSI and for opening 3ds Max files in older versions of 3ds Max.

The FBX format includes support for all the scene constructs, including animation, bone systems, morph targets, and animation cache files. It has an option to embed textures with the export file or to convert them to the TIF format. Other import and export settings deal with the system units and world coordinate orientation. You also have the ability to filter specific objects. The FBX Import dialog box, shown in Figure 3.10, lets you choose a preset as either the Autodesk Media & Entertainment or as Autodesk Revit for CAD drawings.

Note

The date on the FBX Import dialog box is behind by a year because this plug-in is updated separately from 3ds Max.

FIGURE 3.10

The FBX Import dialog box lets you choose which features to import.

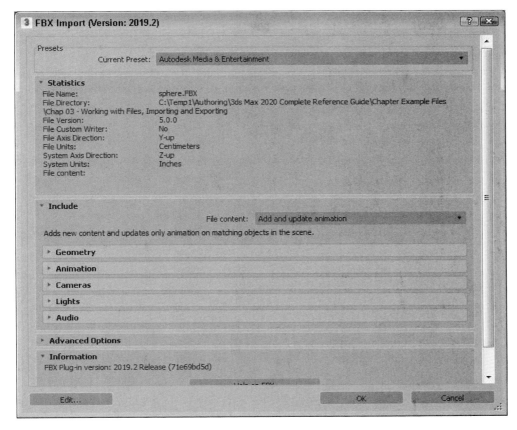

The FBX Import dialog box also reads and presents various statistics for the file being imported, including its file version, axis direction, and units.

Many of the same settings are included for exporting as well. The FBX Export dialog box is shown in Figure 3.11.

FIGURE 3.11

The FBX Export dialog box provides the best way of transferring among 3ds Max, Maya, and MotionBuilder.

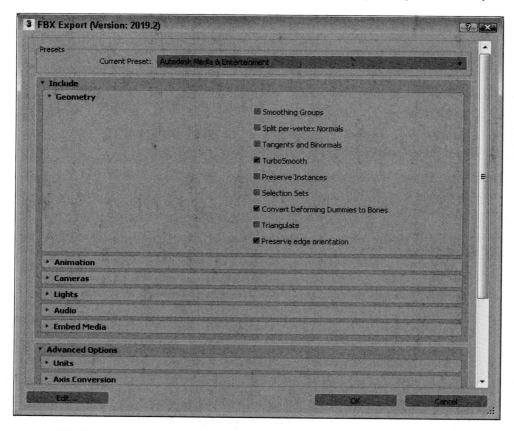

The FBX Import and Export dialog boxes include the ability to save and load configuration presets. This is helpful because, after you figure out the correct settings to get models in and out of 3ds Max, you can save the preset and instantly select it the next time you need to move a file. The Web updates button lets you check for updates to the FBX format online.

Tip
When exporting a file for use in Maya, be sure to set the Up Axis to Y-up, or the models will show up rotated.

When exporting the file to FBX format, the Type option lets you export the file as an ASCII text file or as a Binary file. Binary files are typically smaller, but ASCII files can be edited in a text editor.

The FBX format is being continually updated, but the FBX Export dialog box lets you select which FBX version to use. If you need to export a 3ds Max file for use on an older version of 3ds Max, be sure to select an older FBX version in the FBX Version drop-down list.

Using the Game Exporter utility

If you are working with game assets that need to be exported to a game engine, you can use the File→Export→Game Exporter menu command. The Game Exporter link is also available from the Utilities panel. This menu accesses the Game Exporter dialog box, shown in Figure 3.12. This dialog box is essentially just another FBX export method with only those features needed by the game engine.

New Feature in 2020
The Game Exporter utility isn't new, but it is new to the File→Export menu.

FIGURE 3.12

The Game Exporter dialog box is a subset of the FBX Export dialog box specific for game assets.

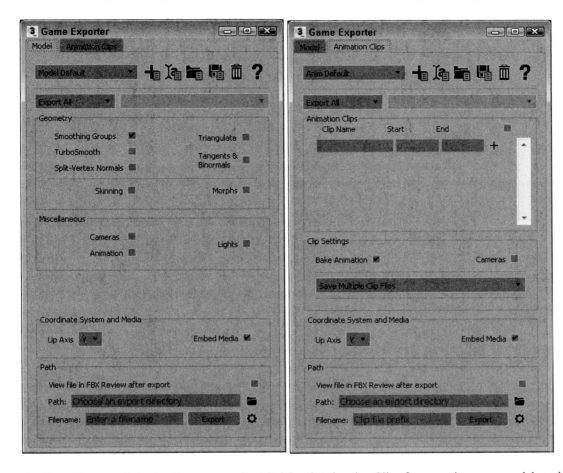

The Game Exporter dialog box has two panels--Model and Animation Clips for exporting game models and animation clips respectively. At the top of each panel are buttons for creating, loading and saving presets. Click the Export button at the bottom of the dialog box to export each after the settings are set.

Exporting to the DWF format

The Design Web Format (DWF) is an ideal format for displaying your textured models to others via the Web. It creates relatively small files that can be attached easily to an e-mail. You can use the File→Export→Publish to DWF menu command to export the current scene to this format. This command opens a dialog box of options, shown in Figure 3.13, that specify to Group by Object or Group by Layer. You also can choose to publish the Object Properties, Materials, Selected Objects Only, or Hidden Objects. Another option is to Rescale Bitmaps to a size entered in pixels.

FIGURE 3.13

The DWF Publish Options dialog box lets you set the options for the exported DWF file.

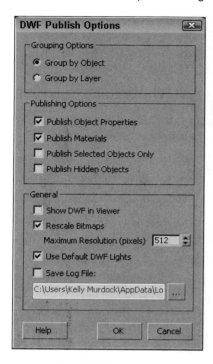

Saved files can be viewed in the Autodesk DWF Viewer. The Autodesk DWF Viewer can be downloaded for free from the Autodesk website. This provides a way for users without 3ds Max installed to view models.

If you want to view the exported files in the viewer, simply enable the Show DWF in Viewer option in the DWF Publish Options dialog box. The viewer includes controls for transforming the model, changing its shading and view, and printing the current view.

Exporting utilities

In addition to the menu commands found in the File menu, 3ds Max includes a couple of utilities that export specific information: the Panorama Exporter Utility, and the Lighting Data Export Utility. You can access these utilities from the Utilities panel in the Command Panel by clicking the More button and selecting them from the pop-up list that appears.

Lighting Data Export Utility

The Lighting Data Export Utility exports exposure control data for a scene's Illuminance and Luminance values. These files can be saved as PIC or TIF files, which you can select in the 2D Lighting Data Exporter rollout by clicking on the File Name button. You also can set an image's Width and Height dimensions.

Caution

Exposure Control must be enabled for this utility to be enabled. You can learn about exposure control in Chapter 44, "Using Atmospheric and Render Effects."

Panorama Exporter Utility

The Panorama Exporter Utility exports a scene into a format that allows all 360 degrees of the scene to be viewed. The scene must include a camera that marks the starting view location. Using this utility, you can open the Render Setup dialog box for rendering a panoramic scene or accessing a viewer for viewing rendered panoramic scenes.

Tutorial: Importing vector drawings from Illustrator

Before leaving this section, let's look at an example of importing a file for use in 3ds Max.

In most companies, a professional creative team uses an advanced vector drawing tool such as Illustrator to design the company logo. If you need to work with such a logo, learning how to import the externally created file gives you a jumpstart on your project.

Note

When importing vector-based files into 3ds Max, only the lines are imported. 3ds Max cannot import fills, blends, or other specialized vector effects. All imported lines are automatically converted to Bézier splines in 3ds Max.

Although 3ds Max can draw and work with splines, the software's spline features take a backseat to the vector functions available in Adobe Illustrator. If you have an Illustrator (AI) file, you can import it directly into 3ds Max.

To import Adobe Illustrator files into 3ds Max, follow these steps:

1. Within Illustrator, save your file as **Box It Up Co logo** using the .ai file format by choosing File→Save As. This file is also included with the downloadable content if you don't have Illustrator.

Note

When saving the Illustrator file, don't use the latest file format. For this example, I've saved the file using the Illustrator 8 format instead of one of the latest Illustrator CS formats.

Figure 3.14 shows a logo created using Illustrator.

FIGURE 3.14

A company logo created in Illustrator and ready to save and import into 3ds Max

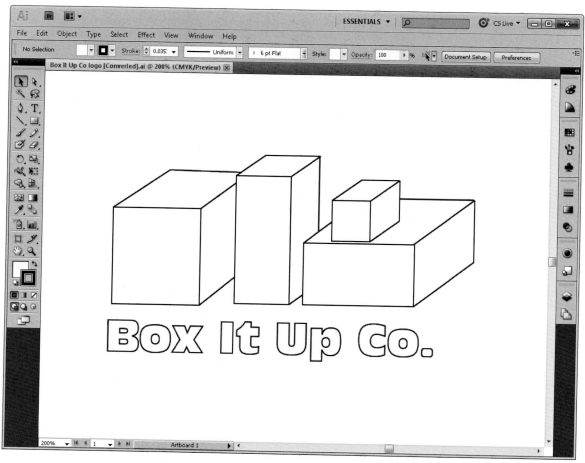

2. Open 3ds Max, and choose File→Import→Import.

   A file dialog box opens.

3. Select Adobe Illustrator (AI) as the File Type. Locate the file to import, and click Open.

   The AI Import dialog box asks whether you want to merge the objects with the current scene or replace the current scene.

4. For your purposes, select the replace the current scene option and click OK.

5. The Shape Import dialog box asks whether you want to import the shapes as single or multiple objects. Select multiple, and click OK.

Figure 3.15 shows the logo after it has been imported into 3ds Max.

FIGURE 3.15

A company logo created in Illustrator and imported into 3ds Max

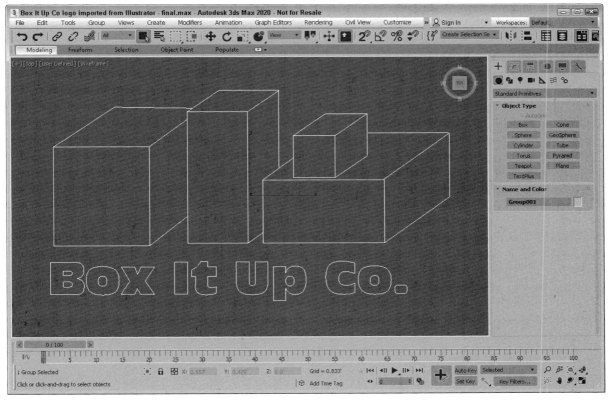

Spline objects that are imported from Illustrator appear in 3ds Max as Editable Spline objects. You can learn more about Editable Splines in Chapter 12, "Drawing and Editing 2D Splines and Shapes."

Using the File Utilities

A single scene can include multiple files, including model files, textures, environment lighting, script files, and so on. With all these various files floating around, 3ds Max has included several utilities that make working with them easier. The Utilities panel of the Command Panel includes several useful utilities for working with files. You can access these utilities by opening the Utilities panel and clicking the More button to see a list of available utilities.

Finding files with the Max File Finder utility

Another useful utility for locating files is the Max File Finder utility, which you get to by using the More button in the Utilities panel of the Command Panel. When you select this utility, a rollout with a Start button appears in the Utilities panel. Clicking this button opens the MAXFinder dialog box. Using MAXFinder, you can search for scene files by any of the information listed in the File Properties dialog box.

Tip

You also can access the MAXFinder dialog box using the MaxFind icon located in the same folder where 3ds Max is installed.

You can use the Browse button to specify the root directory to search. You can select to have the search also examine any subfolders. Figure 3.16 shows the MAXFinder dialog box locating all the scene files that include the word *blue.*

FIGURE 3.16

You can use the MAX File Finder utility to search for scene files by property.

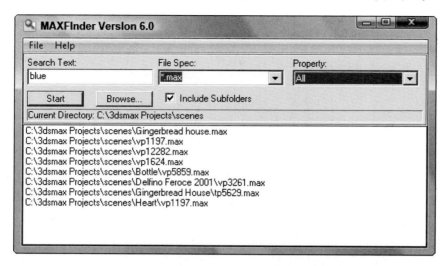

Collecting files with the Resource Collector utility

When a scene is created, image and object files can be pulled from several different locations. The Resource Collector utility helps you consolidate all these files into one location. The settings for this utility appear in the Parameters rollout in the Utilities panel of the Command Panel once the Resource Collector utility is selected, as shown in Figure 3.17. The Output Path is the location where the files are collected. You can change this location using the Browse button.

FIGURE 3.17

The Resource Collector utility can compile all referenced files into a single location.

The utility includes options to Collect Bitmaps/Photometric Files, to include the 3ds Max scene file, and to compress the files into a compressed MaxZip/WinZip file. The Copy option makes copies of the files, and the Move option moves the actual file into the directory specified in the Output Path field. The Update Materials option updates all material paths in the Material Editor. When you're comfortable with the settings, click the Begin button to start the collecting.

Accessing File Information and Sharing Views

As you work with files, several dialog boxes in 3ds Max supply you with extra information about your scene. You can use this information to keep track of files and record valuable statistics about a scene.

Autodesk also provides a way to share views of your scene with clients and collaborators. Using the Share View menu, you can email your current scene to others and they can use the freely available Autodesk Viewer to view and comment on the scene.

Displaying scene information

If you like to keep statistics on your files (to see whether you've broken the company record for the model with the greatest number of faces), you'll find the Summary Info dialog box useful. Use the File→Summary Info menu command to open a dialog box that displays all the relevant details about the current scene, such as the number of objects, lights, and cameras; the total number of vertices and faces; and various model settings, as well as a Description field where you can describe the scene. Figure 3.18 shows the Summary Info dialog box.

FIGURE 3.18

The Summary Info dialog box shows all the basic information about the current scene.

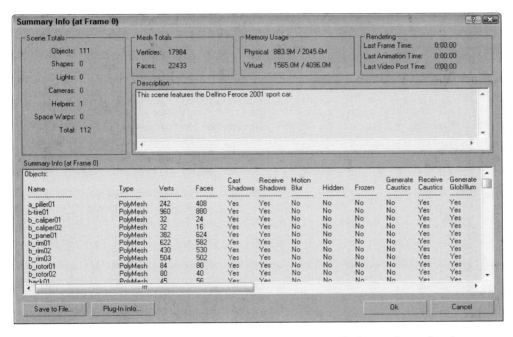

The Summary Info dialog box also includes a Save to File button for saving the scene summary information as a text file.

Viewing file properties

As the number of files in your system increases, you'll be wishing you had a card catalog to keep track of them all. 3ds Max has an interface that you can use to attach keywords and other descriptive information about the scene to the file. The File→File Properties menu command opens the File Properties dialog box. This dialog box, shown in Figure 3.19, includes three panels: Summary, Contents, and Custom. The Summary panel holds information such as the Title, Subject, and Author of the 3ds Max file and can be useful for managing a collaborative project. The Contents panel holds information about the scene, such as the total number of objects and much more. Much of this information also is found in the Summary Info dialog box. The Custom panel includes a way to enter a custom list of properties, such as client information, language, and so on.

FIGURE 3.19

The File Properties dialog box contains workflow information such as the scene author, comments, and revision dates.

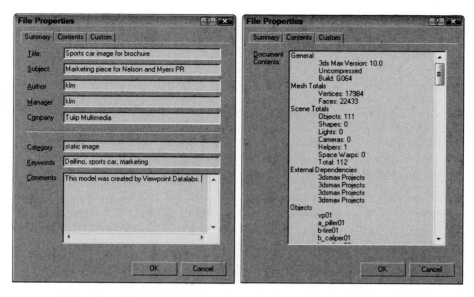

Viewing files

Sometimes, looking at the thumbnail of an image isn't enough to help you decide whether you have the right image. For these cases, you can quickly load the image in question into a viewer to look at it closely. The File→View Image File menu command opens the View File dialog box, shown in Figure 3.20. This dialog box lets you load and view graphic and animation files using the Rendered Frame Window or the default Media Player for your system.

FIGURE 3.20

The View File dialog box can open an assortment of image and animation formats.

The Rendered Frame Window is discussed in more detail in Chapter 27, "Rendering a Scene and Enabling Quicksilver."

The View File dialog box includes several controls for viewing files. The Devices and Setup buttons let you set up and view a file using external devices such as video recorders. The Info button lets you view detailed information about the selected file. The View button opens the file for viewing while leaving the View File dialog box open. The Open button opens the selected file and closes the dialog box. At the bottom of the View File dialog box, the statistics and path of the current file are displayed.

The View File dialog box can open many types of files, including Microsoft videos (AVI), MPEG files, bitmap images (BMP), Kodak Cineon images (CIN), Combustion files (CWS), Graphics Interchange Format images (GIF), Radiance HDRI image files (HDR), Image File List files (IFL), JPEG images (JPG), OpenEXR image files (EXR, FXR), Portable Network Graphics images (PNG), Adobe Photoshop images (PSD), QuickTime movies (MOV), SGI images (RGB), RLA images, RPF images, Targa images (TGA, VST), Tagged Image File Format images (TIF), Abekas Digital Disk images (YUV), and DirectDraw Surface (DDS) images.

You use the Gamma area on the View File dialog box to specify whether an image uses its own gamma settings or the system's default setting, or whether an override value should be used.

Sharing views

The File→Share View menu command opens a dialog box where you specify the email addresses of the individuals that you want to see your current scene. You will need to log into your Autodesk account before you can access this dialog box. Once shared, the recipient will receive a link to the view. Clicking on this link will open the shared view in the Autodesk Viewer. The Autodesk Viewer runs within a web browser, as shown in Figure 3.21, and includes tools for rotating, panning and zooming the scene. There are also tools to Measure, Section, Explode and Markup the view.

New Feature in 2020
The ability to share views in the Autodesk Viewer is new to 3ds Max 2020.

If the recipient is also logged into their Autodesk account, then they can comment on the view and those comments are returned to the sender. This provides a convenient way to collaborate with others without requiring that they have 3ds Max installed. From within the Autodesk Viewer, others can capture snapshots and print views for presentations. Note also that the Autodesk Viewer can also be used to view other Autodesk files including DWG, STEP and Revit views.

Note

Shared views will expire automatically after 30 days.

FIGURE 3.21

The Autodesk Viewer opens a shared view within a web browser and lets others comment and markup the view.

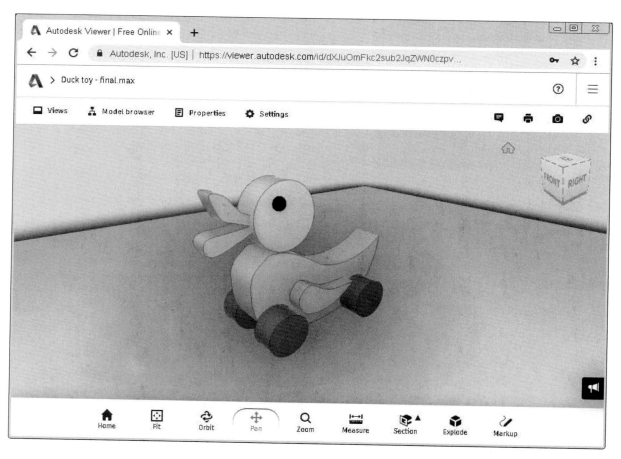

Working with Asset Tracking

How often do you go into your garage during the winter months? Unless you have a pair of warm overalls or a space heater, you probably just throw in any tools you use, the Christmas lights, spare quarts of oil, and the kids' bikes wherever they fit as you make a mad dash back into the warm house, and then when spring comes, you have a real mess on your hands.

Max projects can be the same. During an aggressive project, Max files, textures, and render passes get saved all over the network. By the time the project is finished, you may have a hard time reusing things simply because you can't find anything. This dilemma is compounded when you work on a team with several individuals throwing stuff all over the place.

An asset tracking system can help with this problem by introducing a system that acts as a little secretary, logging every object thrown to the network. This little secretary also is smart enough to keep track of the latest updates of all files, making sure that each accessed file includes them. It also locks any file that is being used to keep unwanted fingers out of the pie.

Max's asset tracking dialog box is opened with the File→Reference→Asset Tracking Toggle menu command (Shift+T). This dialog box supports several different systems, including Autodesk's Vault, Microsoft's Visual SourceSafe, CVS, and Perforce. This interface Each of these systems can be accessed using Server→Launch Provider menu command in the Asset Tracking interface.

Note

In order for an asset tracking system to work with Max, it must conform to the MSSCC standard. [b]

For an asset tracking system, assets are defined as any file that is used as part of a project. This could include Max scene files, XRefs, bitmap textures, MAXScript files, and so on. One of the key benefits to an asset management system is that it stores data in a hierarchical structure including files such as textures as dependents of the Max scene file.

Checking in and checking out

To request an asset from the asset tracking system, you "check out" the file. This command locks the file so you can edit it without worrying about others making changes at the same time. If another user tries to check out a file that is already checked out, he receives a polite message stating that the requested file is available for read-only access and lists you as the person who has the file checked out. The requested file can be loaded and viewed, but it cannot be edited.

When you're finished editing the file, you can "check in" the file, making it available for other users. As a file is checked in, a comment dialog box appears where you can enter a message about the latest changes. Over time, these comments are compiled into a historical list that marks the changes over the life of the file. The asset tracking system also can be used to recall the file at any point in its history.

Note

Asset tracking systems such as Vault are very good at reminding you to check out and check in files so changes aren't lost. For example, if you make changes to a checked out Vault file and try to open another file, a dialog box appears reminding you to check the file back in before opening another file. [b]

Using the Asset Tracking interface

The Asset Tracking interface, shown in Figure 3.22, is opened using the File→Reference→Asset Tracking Toggle menu command. It also may be opened using the Shift+T keyboard shortcut. This interface also allows you to access commands in the main Max interface while it is open.

Vault files that are opened appear in this interface including all dependent files such as texture bitmaps, XRefs, photometric files, render files, and so on. Using the buttons at the top of the interface, you can Refresh the current list to see the latest updates, view the Status Log (which is a running list of all Vault commands), and change the view between a Tree view and a Table view.

To the left of each filename is a small icon that marks the status of the file. These icons are handy when the Tree view is enabled, or you can read the status as text in the Table view. The Table view includes separate columns for displaying the Name, Full Path, Status, Proxy Resolution, and Proxy Status of the various files.

On the left side of the interface are two buttons for filtering the assets. The first button highlights the material and map assets. The second button highlights the assets for the selected object.

FIGURE 3.22

The Asset Tracking interface shows all the checked-out files and the status of each.

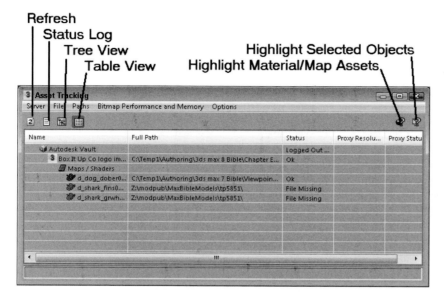

Logging in

The first step in using an asset tracking system is to log into the system. This lets the system know who you are, the access rights you have, and which name to place beside the assets you edit. You can log into the designated asset tracking system using the Server→Log in menu command in the Asset Tracking dialog box.

Note

If Vault or another asset tracking software isn't installed, then the Log In command is unavailable. [b]

Selecting a working folder

When files are checked out from the asset tracking system, they are copied to a local directory on your current machine where the changes are saved until the file is checked back in. The first time you try to check out a file from the asset tracking system, the system asks you to select a working folder. This folder is a temporary folder where the checked out files are saved while being worked on. You can manually set the working folder by selecting the Server→Options menu command in the Asset Tracking interface.

You can set which warning dialog boxes appear and which are handled automatically using the Options→Prompts menu command in the Asset Tracking interface. This command presents a list of all the available prompt dialog boxes and lets you right-click each to have it appear or not appear.

The Asset Tracking dialog box has commands for checking files in and out under the File menu. Each time a file is checked in, a confirmation dialog box appears, as shown in Figure 3.23. In this dialog box, you can enter a comment about the latest changes. The file is then saved to the Vault server with an incremented version number. The ability to save and keep track of different versions of a file is known as "version control" and is one of the key features of an asset management system.

FIGURE 3.23

The dialog box where comments on the latest changes are entered appears every time a Vault file is checked in.

An Undo Checkout command is available in the File menu that throws away the current changes and restores the current Vault file.

Getting and adding Vault files

After a database is created for a project and you've logged in, you can add open files to the Vault server using File→Add Files menu command. This command adds the current file to the Vault project folder, but the file must be saved to the local working folder before it can be added to the Vault folder. You can select multiple files such as bitmap textures by holding down the Ctrl key while clicking them; you also can add multiple files at the same time using this command.

The File→Get from Provider menu command does the opposite. It downloads any selected files from the Vault folder to your local working folder.

If you're missing files, such as a texture that has been moved to a different local folder, you can use the File→Browse menu command to locate the missing file.

Loading older file versions

Each time a file is checked in, it is given a new version number. This makes it possible to load older versions of a file. For example, if you want to reuse a character made for a previous game, but you want to access the character before any textures were applied, you can locate and load the version saved just before textures were applied.

Older file versions can be found in the History dialog box, which is also found in the File menu. This dialog box, shown in Figure 3.24, lists all the different versions of a file along with their creation dates, their creator, and any comments entered when the file was checked in. To load an older version, simply select the file from the list and click the Get Version button.

FIGURE 3.24

The History dialog box lets you access older versions of a file.

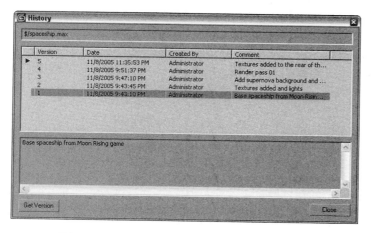

Changing asset paths

The Path menu includes options for setting and changing the paths for the various assets. The Highlight Editable Assets command selects all assets in the Asset Tracking interface that have paths that can change. Typically, only the base Max file loaded from the Vault cannot be edited. The Set Path command opens a simple dialog box where you can browse to a new path for an asset. This is helpful if the asset has moved and is marked as missing. The Path menu also includes commands for retargeting the root path, which is the path that all assets have in common, stripping the path from an asset so that only the filenames are visible, making the path absolute or relative to the project folder, and converting paths to the Universal Naming Convention (UNC).

Working with proxies

The Proxy System lets you use proxy texture maps in place of high-resolution maps across all objects in the scene. Using the Bitmap Performance and Memory menu, you can enable the use of proxies, enable bitmap paging, set the global settings for the proxy system, and set the proxy resolution to use. The table view also displays the current proxy resolution for each asset and its status.

Bitmap Proxies can be used outside of Vault. For information on using them, see Chapter 27, "Rendering a Scene and Enabling Quicksilver." [b]

Summary

If you look into the different commands located in the File menu, you'll find commands that let you save your work, share it with others, and reload it for more work. There are also commands to import and export files, and several utilities for viewing, accessing, and getting information about the file's contents.

Working with files lets you save your work, share it with others, and reload it for more work. This chapter covered the following topics:

This chapter covered the following topics:

* Creating, saving, opening, merging, and archiving files
* Converting scenes to use new features
* Understanding the various import and export types
* Importing models from other programs, such as Illustrator, Maya, and MotionBuilder

* Exporting data

* Working with the file utilities, such as the Resource Collector

* Using the Summary Info and File Properties dialog boxes to keep track of scene files

* Sharing views with the Autodesk Viewer

* Understanding what an asset management system is

* Setting up an asset tracking system to work with Max by logging in and selecting a working folder

* Using the Asset Tracking interface to work with Vault assets

The next chapter looks at all the extensive preferences that can be set in 3ds Max. Understanding and using these preferences correctly will help as you begin to use the interface.

Chapter 4

Setting Preferences

IN THIS CHAPTER

Learning the general preferences

Setting file preferences and configuring paths

Altering viewport preferences

Changing gamma preferences

Setting rendering and radiosity preferences

Setting animation, IK, and gizmos preferences

Setting other preferences

The Autodesk® 3ds Max® 2020 software also has a rather bulky set of preferences that you can use to set almost every aspect of the program. All the various preference settings are divided into several different panels. The advantage of having all these various settings in a single location is that many can be set at once. Understanding the various preferences can streamline your experience and make the interface even more helpful.

The Preference Settings dialog box lets you configure 3ds Max so it works in a way that is most comfortable for you. You open it by choosing Customize→Preferences or you also can open it using the File menu, (File→Preferences). The dialog box includes several panels: General, Files, Viewports, Interaction Mode, Gamma and LUT, Rendering, Radiosity, Animation, Inverse Kinematics, Gizmos, MAXScript, Containers, and Help.

Tip
The quickest way I've found to open the Preference Settings dialog box is to right-click the Spinner Snap Toggle button on the main toolbar.

Setting General Preferences

The first panel in the Preference Settings dialog box is for General settings, as shown in Figure 4.1. The General panel includes many global settings that affect the entire interface.

FIGURE 4.1

The General panel lets you change many UI settings.

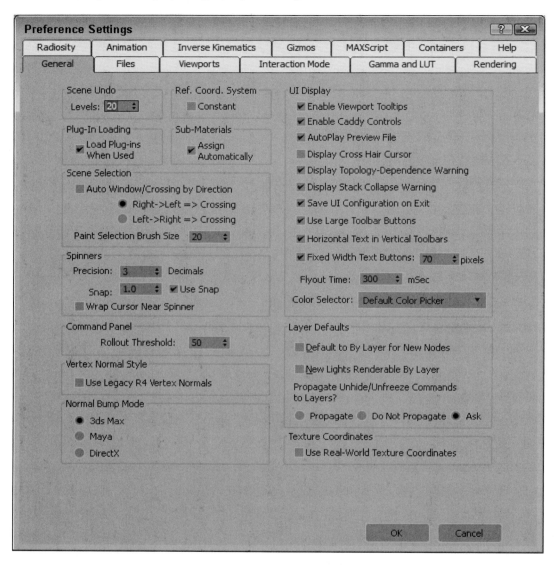

Undo Levels and the Reference Coordinate System

The Scene Undo spinner sets the number of commands that can be kept in a buffer for undoing. A smaller number frees up memory but does not let you backtrack as far through your work. The default Undo Levels is 20.

Tip

Although it takes up some valuable memory, I've found that increasing the number of Undos is very helpful. When working on a model, it takes almost no time to do 20 commands.

The Reference Coordinate System setting makes all transform tools use the same coordinate system and transform center when the Constant option is enabled. If disabled, each transform (move, scale, and rotate) uses the coordinate system last selected.

Loading Plug-Ins and Sub-Material settings

The Load Plug-Ins When Used option keeps plug-ins out of memory until they are accessed. This saves valuable memory and still makes the plug-ins accessible.

The Automatic Sub-Material Assignment option, when checked, enables materials to be dragged and dropped directly onto a subobject selection. This applies the Multi/Sub-Object material to the object with the dropped material corresponding with the subobject selection's Material ID. If you regularly use the Multi/Sub-Object material, enabling this option can be a great timesaver, but if you aren't familiar with the Multi/Sub-Object material, this option can lead to confusion, making it difficult to locate applied materials.

Scene Selection settings

The Auto Window/Crossing by Direction option lets you select scene objects using the windowing method (the entire object must be within the selected windowed area to be selected) and the crossing method (which selects objects if they are contained within or if their borders are crossed with the mouse) at the same time, depending on the direction that the mouse is dragged. If you select the first option, the Crossing method is used when the mouse is dragged from right to left, and the Window method is used when the mouse is dragged from left to right.

Tip

I like to keep the Auto Window/Crossing by Direction option disabled. I use the Crossing selection method and find that I don't always start my selection from the same side.

The Paint Selection Brush Size value sets the default size of the Paint Selection Brush. In the default interface, this size is set to 20. If you find yourself changing the brush size every time you use this tool, you can alter its default size with this setting.

Spinner, Rollout, and Vertex Normal settings

Spinners are interface controls that enable you to enter values or interactively increase or decrease the value by clicking the arrows on the right. The Preference Settings dialog box includes settings for changing the number of decimals displayed in spinners and the increment or decrement value for clicking an arrow. The Use Snap option enables the snap mode.

 You also can enable the snap mode using the Spinner Snap Toggle button on the main toolbar.

Tip

Right-click a spinner to automatically set its value to 0 or its lowest positive threshold.

You also can change the values in the spinner by clicking the spinner and dragging up to increase the value or down to decrease it. The Wrap Cursor Near Spinner option keeps the cursor close to the spinner when you change values by dragging with the mouse, so you can drag the mouse continuously without worrying about hitting the top or bottom of the screen.

The Rollout Threshold value sets how many pixels can be scrolled before the rollup shifts to another column. This is used only if you've made the Command Panel wider or floating.

The Use Legacy R4 Vertex Normals option computes vertex normals based on the 3ds Max version 4 instead of the newer method. The newer method is more accurate but may affect smoothing groups. Enable this setting only if you plan on using any models created using 3ds Max R4 or earlier.

When normal maps are rendered in the viewport, the results can depend on these different settings. 3ds Max and Maya each have a unique way to bake normal maps and the method that is used can be set here. The DirectX method is more optimized and is a good choice for game assets.

User Interface Display settings

The options in the UI Display section control additional aspects of the interface. The Enable Viewport Tooltips option can toggle tooltips on or off. Tooltips are helpful when you're first learning the 3ds Max interface, but they quickly become annoying, and you'll want to turn them off.

When editing Editable Poly objects, caddy controls appear near the selected object and let you see the changes before applying them. Caddy controls are only available for specific tools. If disabled, standard dialog box are used.

The AutoPlay Preview File setting automatically plays Preview Files in the default media player when they are finished rendering. If this option is disabled, you need to play the previews with the Tools→Preview - Grab Viewport→Play Preview Animation menu command.

The Display Cross Hair Cursor option changes the cursor from the Windows default arrow to a crosshair cursor similar to the one used in AutoCAD.

For some actions, such as non-uniform scaling, 3ds Max displays a warning dialog box asking whether you are sure of the action. To disable these warnings, uncheck this option (or you could check the Disable this Warning box in the dialog box). Actions with warnings include topology-dependence and collapsing the Modifier Stack.

The Save UI Configuration on Exit switch automatically saves any interface configuration changes when you exit 3ds Max. You can deselect the Use Large Toolbar Buttons option, enabling the use of smaller toolbar buttons and icons, which reclaims valuable screen real estate.

The Horizontal Text in Vertical Toolbars option fixes the problem of text buttons that take up too much space, especially when printed horizontally on a vertical toolbar. You also can specify a width for text buttons. Any text larger than this value is clipped off at the edges of the button.

The Flyout Time spinner adjusts the time the system waits before displaying flyout buttons. The Color Selection drop-down list lets you choose which color selector interface 3ds Max uses.

Layer settings

If you select an object and open its Object Properties dialog box, the Display Properties, Rendering Control, and Motion Blur sections each have a button that can toggle between ByLayer and ByObject. If ByObject is selected, the options are enabled and you can set them for the object in the Object Properties dialog box, but if the ByLayer option is selected, the settings are determined by the setting defined for all objects in the layer in the Layer Manager.

The settings in the Preference Settings dialog box set the ByLayer option as the default for new objects and new lights. You also have an option to propagate all unhide and unfreeze commands to the layer. You can select Propagate, Do Not Propagate, or Ask.

Real-World Texture Coordinates setting

The Use Real-World Texture Coordinates setting causes the Real-World Scale or the Real-World Map Size option in the Coordinates rollout to be enabled. This setting is off by default, but it can be enabled to be the default by using this setting.

Real-World Texture Coordinates is a mapping method explained in more detail in Chapter 18, "Adding Material Details with Maps."

Setting File Preferences

The Files panel of the Preference Settings dialog box holds the controls for backing up, archiving, and logging 3ds Max files. Figure 4.2 shows this panel.

FIGURE 4.2

The Files panel includes an Auto Backup feature.

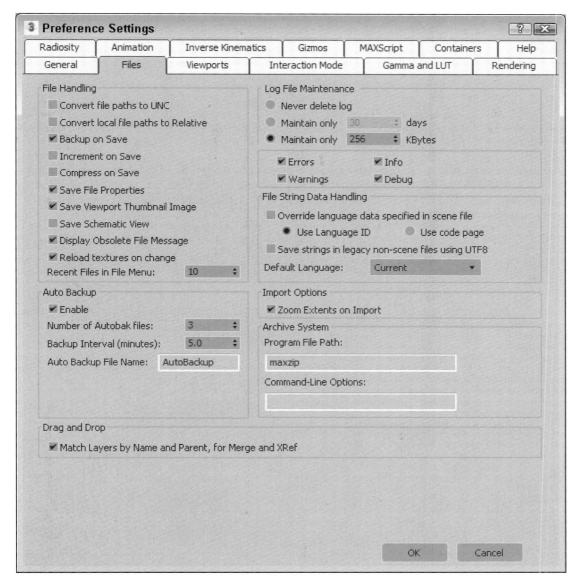

Handling files

The Files panel includes several options that define how to handle files. The first option is to Convert file paths to UNC (Universal Naming Convention). This option displays file paths using the UNC for any files accessed over a mapped drive. The Convert local file paths to Relative option causes all paths to be saved internally as relative paths to the project folder. This is useful if all files you access are in the same folder, but if you use files such as bitmaps from a different folder, be sure to disable this option.

The next option is Backup on Save. When you save a file using the File→Save (Ctrl+S) menu command, the existing file is overwritten. The Backup on Save option causes the current scene file to be saved as a backup (with the name 3ds MaxBack.bak in the 3dsmax\autobak directory) before saving the new file. If the changes you made were a mistake, you can recover the file before the last changes by renaming the 3ds MaxBack.bak file to 3ds MaxBack.max and reopening it in 3ds Max.

Another option to prevent overwriting your changes is the Increment on Save option. This option adds an incremented number to the end of the existing filename every time it is saved (the same as when clicking the plus button in the Save File As dialog box). This retains multiple copies of the file and is an easy version-control method for your scene files. This way, you can always go back to an earlier file when the client changes his mind. With this option enabled, the 3ds MaxBack.bak file isn't used.

The Compress on Save option compresses the file automatically when it is saved. Compressed files require less file space but take longer to load. If you're running low on hard drive space, you'll want to enable this option.

Tip

Another reason to enable the Compress on Save option is that large files (100MB or greater) load into the Network Queue Manager much more quickly when compressed for network rendering.

The Save Viewport Thumbnail Image option saves a 64 x 64-pixel thumbnail of the active viewport along with the file. This thumbnail is displayed in the Open File dialog box on Windows 10. This thumbnail is also used in the Open File dialog box when a file is selected, as shown in Figure 4.3. Saving a thumbnail with a scene adds about 9K to the file size.

Caution

Although thumbnails appear in Windows Explorer when viewed from within Windows 10, they do not appear in Windows 7.

Tip

The Save Viewport Thumbnail Image option is another good option to keep enabled. Thumbnails help you to find scene files later, and nothing is more frustrating than seeing a scene's filename without a thumbnail.

FIGURE 4.3

3ds Max files with thumbnails show up in Windows Explorer.

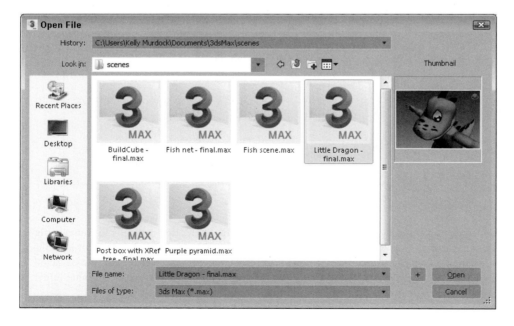

In addition to a thumbnail, 3ds Max offers an option to save the Schematic View with the file. Although 3ds Max can generate a new Schematic View from an existing file, saving the Schematic View with the file is quicker if you work with this view often. Saving File Properties with the file is also helpful, but be warned that saving this extra info with the file increases its file size slightly. Still, doing so is worth the effort because you can easily locate and understand the scene file later on.

More details on using Schematic View are covered in Chapter 10, "Organizing Scenes with Layers, Containers, XRefs and the Schematic View."

When a 3ds Max file created in a previous version of 3ds Max is opened, a warning dialog box appears that says, "Obsolete data format found—Please resave file." To eliminate this warning, disable the Display Obsolete File Message option. The warning dialog box also includes an option to Don't Display Again that enables this option when selected.

When textures are updated, the Reload textures on change option forces the textures to be reloaded when they are altered. This slows your system while 3ds Max waits for the textures to reload but offers the latest look immediately.

The Recent Files in File Menu option determines the number of recently opened files that appear in the File→Open Recent menu. The maximum value is 50.

Backing up files

The Auto Backup feature in 3ds Max can save you from the nightmare of losing all your work because of a system crash. With Auto Backup enabled, you can select the number of Autobak files to keep around and how often the files are backed up. The backup files are saved to the directory specified by the Configure User Paths dialog box. The default is to save these backups to the 3dsmax\autoback directory. You also can select a name for the backup files.

Note

Even if you have this feature enabled, you should still save your file often.

This is how it works: If you've set the number of backup files to two, the interval to five minutes, and the backup name to MyBackup, after five minutes the current file is saved as MyBackup1.max. After another five minutes, another file named MyBackup2.max is saved, and then after another five minutes, the MyBackup1.max file is overwritten with the latest changes.

If you lose your work as a result of a power failure or by having your toddler accidentally pull out the plug, you can recover your work by locating the Autobak file with the latest date and reloading it into 3ds Max. This file won't include all the latest changes; it updates only to the last backup save.

Tip

I highly recommend that you keep the Auto Backup option enabled. This feature has saved my bacon more than once. Also, if you enter a different Backup File Name for different projects, you won't accidentally overwrite a backed-up project.

Tutorial: Setting up Auto Backup

Now that I have stressed that setting up Auto Backup is an important step to do, here's exactly how to set it up.

To set up the Auto Backup feature, follow these steps:

1. Open the Preference Settings dialog box by choosing Customize→Preferences and click the Files panel.

2. Turn on Auto Backup by selecting the Enable option in the Auto Backup section.

3. Set the number of Autobak files to **5**.

Note

To maintain version control of your 3ds Max scenes, use the Increment on Save feature instead of increasing the Number of Autobak Files.

4. Set the Backup Interval to the amount of minutes to wait between backups.

   The Backup Interval should be set to the maximum amount of work that you are willing to redo. (I keep my settings at 15 minutes.) You also can give the Auto Backup file a name.

5. Auto Backup saves the files in the directory specified by the Auto Backup path. To view where this path is located, choose Customize→Configure Project Paths.

Maintaining log files

You also can use the Files panel to control log files. Log files keep track of any errors and warnings, general command info, and any debugging information. You can set log files to never be deleted, expire after so many days, or keep a specified file size with the latest information. If your system is having trouble, checking the error log gives you some idea as to what the problem is. Logs are essential if you plan on developing any custom scripts or plug-ins. You can select that the log contain all Errors, Warnings, Info, and Debug statements.

Each entry in the log file includes a date-time stamp and a three-letter designation of the type of message with DBG for debug, INF for info, WRN for warning, and ERR for error messages, followed by the message. The name of the log file is 3ds Max.log.

Configuring Paths

When strolling through a park, chances are good that you'll see several different paths. One might take you to the lake and another to the playground. Knowing where the various paths lead can help you as you navigate around the park. Paths in 3ds Max lead, or point, to various resources, either locally or across the network.

All paths can be configured using two distinct Configure Paths dialog boxes found in the Customize menu: Configure Project Paths and Configure User and System Paths. The Configure Project Paths dialog box is used to specify where to look for scene resource files such as scenes, animations, and textures. The Configure User and System Paths dialog box is used to specify where the system looks to load files that 3ds Max uses, such as fonts, scripts, and plug-ins.

Configuring project paths

The Configure Project Paths dialog box, shown in Figure 4.4, holds the path definitions to all the various resource folders. The dialog box includes three panels: File I/O, External Files, and XRefs.

FIGURE 4.4

The Configure Project Paths dialog box specifies where to look for various resources.

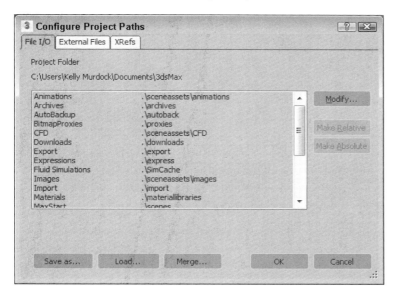

The main panel in the Configure Project Paths dialog box is the File I/O panel. The Project Folder is listed at the top of the dialog box and can be changed in this dialog box or with the File→Project→Set Active Project menu command. This panel includes entries for Animations, Archives, Auto Backup, Bitmap Proxies, CFD, Downloads, Export, Expressions, Fluid Simulations, Images, Import, Materials, Max Start, Photometric, Previews, Render Assets, Render Output, Render Presets, Scenes, Sounds, and Video Post. If you select any of these entries, you can click the Modify button to change its path. All paths are set by default to folders contained within the designated Project Folder, but you can change them to whatever you want. The Make Relative and Make Absolute buttons cause the selected entry to be displayed as a relative path based on the Project Folder or an absolute path.

Tip

Personally, I like to keep all my content in a separate directory from where the application is installed. That way, new installs or upgrades don't risk overwriting my files. To do this, simply change the Project Folder to a location separate from the 3ds Max installation directory.

Under the External Files and XRefs panels, you can add and delete paths that specify where 3ds Max looks to find specific files. All paths specified in both these panels are searched in the order they are listed when you're looking for resources such as plug-ins, but file dialog boxes open only to the first path. Use the Move Up and Move Down buttons to realign path entries.

Caution

Using the Customize→Revert to Startup Layout command does not reset path configuration changes.

At the bottom of the Configure Project Paths dialog box are buttons for saving, loading, and merging the defined configuration paths into a separate file. These files are saved using the .mxp format. This file can be found in the root of the Project Folder.

Tip

Setting up a Project Folder on the network gives every team member access to all the project files and synchronizes all the paths for a project.

Configuring user and system paths

3ds Max default paths are listed in the Configure User and System Paths dialog box, shown in Figure 4.5. When 3ds Max is installed, all the paths are set to point to the default subdirectories where 3ds Max was installed. To modify a path, select the path and click the Modify button. A file dialog box lets you locate the new directory.

The Configure User and System Paths dialog box also includes the 3rd Party Plug-Ins panel where you can add directories for 3ds Max to search when looking for plug-ins.

FIGURE 4.5

The Configure System Paths dialog box specifies additional paths.

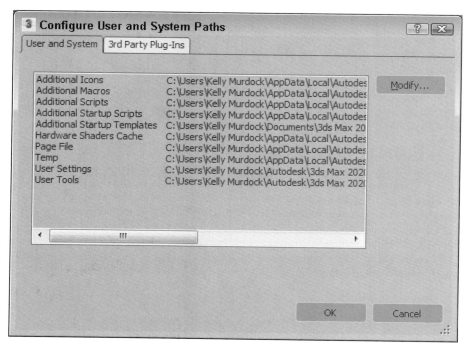

Setting viewport preferences

The viewports are your window into the scene. The Viewports panel, shown in Figure 4.6, contains many options for controlling these viewports.

FIGURE 4.6

The Viewports panel contains several viewport parameter settings.

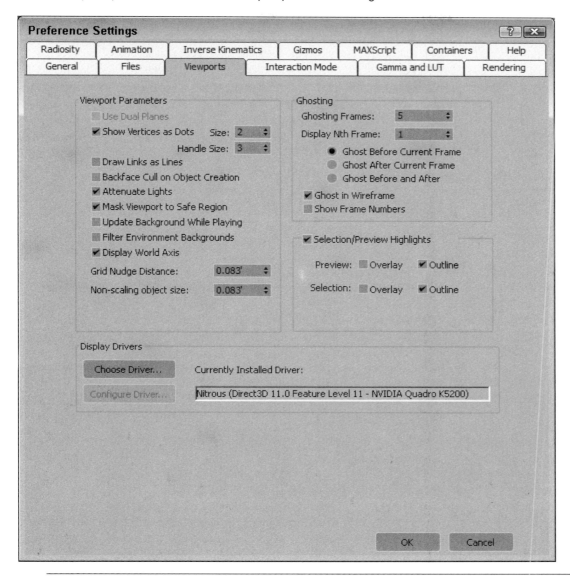

Although the viewports are the major topic in Chapter 2, "Controlling and Configuring the Viewports," the viewport preference settings are covered here.

Viewport parameter options

The Use Dual Planes option enables a method designed to speed up viewport redraws. Objects close to the scene are included in a front plane, and objects farther back are included in a back plane. When this option is enabled, only the objects on the front plane are redrawn.

In subobject mode, the default is to display vertices as small plus signs. The Show Vertices as Dots option displays vertices as either Small or Large dots. The Draw Links as Lines option shows all displayed links as lines that connect the two linked objects.

Caution

I've found that keeping the Draw Links as Lines option turned on can make it confusing to see objects clearly, so I tend to keep it turned off, but it is occasionally useful when trying to determine which objects are linked and to which other object.

When the Backface Cull on Object Creation option is enabled, the backside of an object in wireframe mode is not displayed. If disabled, you can see the wireframe lines that make up the backside of the object. The Backface Cull option setting is determined when the object is created, so some objects in your scene may be backface culled and others may not be. Figure 4.7 includes a sphere and a cube on the left that are not backface culled and a sphere and cube on the right that are.

Note

The Object Properties dialog box also contains a Backface Cull option.

FIGURE 4.7

Backface culling simplifies objects by hiding their backsides.

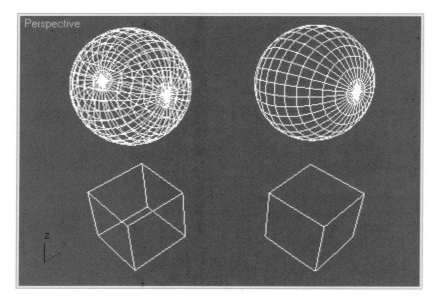

The Attenuate Lights option causes objects farther back in a viewport to appear darker. Attenuation is the property that causes lights to diminish over distance.

In the Viewport Configuration dialog box, you can set Safe Regions, which are borders that the renderer includes. The Mask Viewport to Safe Region option causes the objects beyond the Safe Region border to be invisible.

The Update Background While Playing option causes viewport background bitmaps to be updated while an animation sequence plays. Viewport backgrounds can be filtered if the Filter Environment Background option is enabled, but this slows the update time. If this option is disabled, the background image appears aliased and pixelated.

The Display World Axis option displays the axes in the lower-left corner of each viewport. This can be helpful as you learn to navigate in 3D space. The Grid Nudge Distance is the distance that an object moves when Grid Nudge (+ and − on the numeric keypad) keys are used. Objects without scale, such as lights and cameras, appear in the scene according to the Non-Scaling Object Size value. Making this value large makes lights and camera objects very obvious.

Enabling ghosting

Ghosting is similar to the use of "onion skins" in traditional animation, causing an object's prior position and next position to be displayed. When producing animation, knowing where you're going and where you've come from is helpful.

3ds Max offers several ghosting options. You can set whether a ghost appears before the current frame, after the current frame, or both before and after the current frame. You can set the total number of ghosting frames and how often they should appear. You also can set an option to show the frame numbers.

For a more detailed discussion of ghosting, see Chapter 32, "Understanding Animation and Keyframes."

Highlighting Selections

The Selection/Preview Highlights section lets you turn on and off selection highlighting. When enabled, scene objects are highlighted in yellow using an overlay or an outline or both when the cursor is over the top of the various objects. The Selection Overlay and Outline options mark the selected object with blue highlighting. These options make it easier to locate and select the exact object that you want before clicking to select it.

Choosing and configuring display drivers

When 3ds Max is installed, it loads the latest custom driver called Nitrous and sets the display to use that driver, but you can change the display driver to Direct3D, OpenGL, or to Software if your video card doesn't support the needed drivers.

The Display Drivers section in the Viewports panel of the Preference Settings dialog box lists the currently installed driver. Clicking the Configure Driver button opens a dialog box of settings for the current driver. Clicking the Choose Driver button opens the Display Driver Selection dialog box, shown in Figure 4.8. This dialog box lets you change the display driver to Direct3D, OpenGL, or some custom driver, but unless you have a reason to change it, keep it set to Nitrous or you'll disable some features. If you change the display driver, you need to restart 3ds Max.

Caution

The Display Driver Selection dialog box displays the options only for the drivers that it finds on your system, but just because an option exists doesn't mean it works correctly. If a driver hangs your system, you can restart it from a command line with the –h flag after 3dsmax.exe to force 3ds Max to present the Graphics Driver Setup dialog box again or use the Start→Programs→Autodesk→3ds Max 2020→Change Graphics Mode program icon to restart the program.

FIGURE 4.8

You use the Display Driver Selection dialog box to select a different display driver.

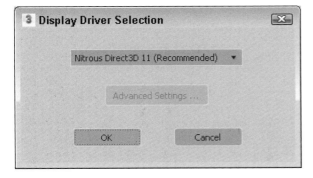

The Configure Driver option opens a dialog box of configurations for the driver that is currently installed. The various configuration dialog boxes include options such as specifying the Texture Size, which is the size of the bitmap used to texture map an object. Larger maps have better image quality but can slow down your display.

Note

If the Nitrous display drivers are enabled, the Configure Driver button is grayed out and unavailable because these drivers have no configuration options.

All the display driver configuration settings present tradeoffs between image quality and speed of display. By tweaking the configuration settings, you can optimize these settings to suit your needs. In general, the more memory available on your video card, the better the results.

You can learn more about the various display drivers in Appendix C, "Installing and Configuring Autodesk 3ds Max 2020."

Setting Interaction Mode

The Interaction Mode panel, shown in Figure 4.9, includes a simple set of features for switching the navigation commands used to control the keyboard and mouse between 3ds Max and Maya. 3ds Max uses the middle mouse button to pan and zoom the view and the Alt+middle mouse button to rotate. Maya uses the Alt key along with the left, middle, and right mouse buttons to rotate, pan and zoom the view.

FIGURE 4.9

The Interaction Mode panel lets you switch viewport navigation between 3ds Max and Maya.

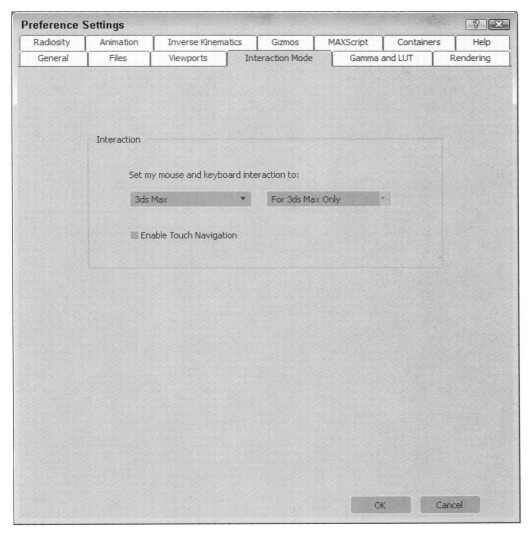

If you plan on using 3ds Max and Maya interchangeably or if you are a Maya user using 3ds Max, switching the mode to Maya will help make navigation easier for you.

Note

A similar feature is available in Maya 2020 to use the 3ds Max navigation controls.

Setting Gamma and Look-Up Table (LUT) Preferences

The Gamma and LUT panel, shown in Figure 4.10, controls the gamma correction for the display and for bitmap files. It also includes a Browse button for loading an Autodesk Look-up Table (LUT) file. A Look-up Table is a file that holds all the color calibration settings that can be shared across different types of software and hardware within a studio to maintain consistency.

FIGURE 4.10

Enabling gamma correction makes colors consistent regardless of the monitor.

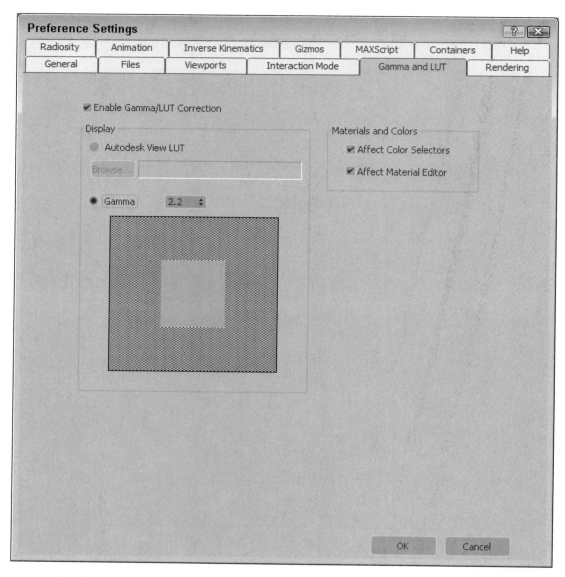

Setting screen gamma

Have you ever noticed in an electronics store that television-screen displays vary in color? Colors on monitor screens may be fairly consistent for related models but may vary across brands. *Gamma settings* are a means by which colors can be consistently represented regardless of the monitor that is being used.

Gamma value regulates the contrast of an image. It is a numerical offset required by an individual monitor in order to be consistent with a standard. To enable gamma correction for 3ds Max, open the Gamma and LUT panel in the Preference Settings dialog box and click the Enable Gamma/LUT Correction option. To determine the gamma value, use the spinner or adjust the Gamma value until the gray square blends in unnoticeably with the background.

Note

3ds Max cannot create LUT files, but it can use existing LUT files created in other software packages, such as Combustion.

Propagating gamma settings

Although gamma settings have a direct impact on the viewports, they do not affect the colors found in the Color Selector or in the Material Editor. Using the Affect Color Selectors and Affect Material Editor options, you can propagate the gamma settings to these other interfaces also.

Setting bitmap gamma

Many bitmap formats, such as TGA, contain their own gamma settings. The Input Gamma setting for Bitmap Files sets the gamma for bitmaps that don't have a gamma setting. The Output Gamma setting is the value set for bitmaps being output from 3ds Max.

Note

Match the Input Gamma value to the Display Gamma value so that bitmaps loaded for textures are displayed correctly.

Setting Rendering and Radiosity Preferences

The next two panels deal with rendering and radiosity settings. These settings can be used to speed up rendering and lighting solutions.

Changing rendering quality

In addition to the settings available in the Render Setup dialog box, the Rendering panel in the Preference Settings dialog box includes many global rendering settings. The Preference Settings dialog box can be opened using the Customize→Preferences menu command. Figure 4.11 shows this panel.

FIGURE 4.11

The Rendering panel in the Preference Settings dialog box lets you set global rendering settings.

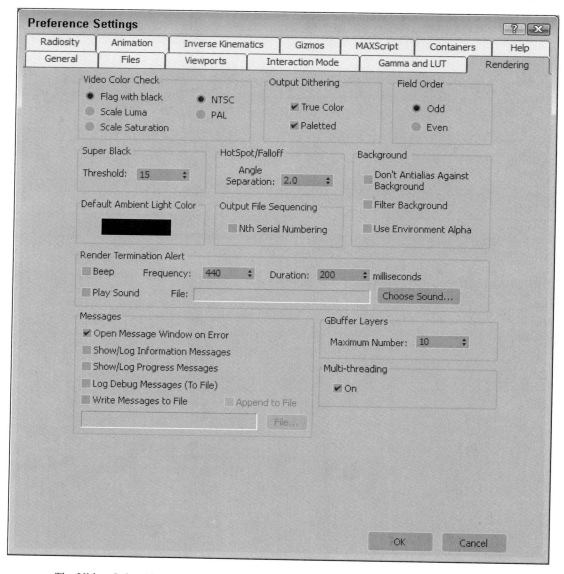

The Video Color Check options specify how unsafe video colors are flagged or corrected. The Flag with black option shows the unsafe colors, and the Scale Luma and Scale Saturation options correct them by scaling either the luminance or the saturation until they are in range. You also can choose to check NTSC or PAL format.

Caution

Be aware that the Scale options can discolor some objects.

Output Dithering options can enable or disable dithering of colors. The options include True Color for 24-bit images and Paletted for 8-bit images.

The Field Order options let you select which field is rendered first. Some video devices use even first, and others use odd first. Check your specific device to see which setting is correct.

The Super Black Threshold setting is the level below which black is displayed as Super Black.

The Angle Separation value sets the angle between the Hotspot and Falloff cones of a light. If the Hotspot angle equals the Falloff angle, alias artifacts will appear.

The Don't Antialias Against Background option should be enabled if you plan on using a rendered object as part of a composite image. The Filter Background option includes the background image in the anti-aliasing calculations. The Use Environment Alpha option combines the background image's alpha channel with the scene object's alpha channel.

The Default Ambient Light Color is the darkest color for rendered shadows in the scene. Selecting a color other than black brightens the shadows.

You can set the Output File Sequencing option to list the frames in order if the Nth Serial Numbering option is enabled. If the Nth Serial Numbering option is disabled, the sequence uses the actual frame numbers.

In the Render Termination Alert section, you can elect to have a beep triggered when a rendering job is finished. The Frequency value changes the pitch of the sound, and the Duration value changes its length. You also can choose to load and play a different sound. The Choose Sound button opens a file dialog box where you can select the sound file to play.

You also can specify whether error messages are displayed and which messages are added to the log file. Log files by default are written to the renderassets folder, but you can select a different path.

The GBuffer Layers value is the maximum number of graphics buffers to allow during rendering. This value can range between 1 and 1000. The value you can use depends on the memory of your system.

The Multi-threading option enables the renderer to complete different rendering tasks as separate threads. Threads use the available processor cycles more efficiently by subdividing tasks. This option should be enabled, especially if you're rendering on a multiprocessor computer.

Using Local and Global Advanced Lighting Settings

You can set advanced lighting settings locally for specific objects using the Object Properties dialog box, shown in Figure 4.12, opened with the Edit→Object Properties menu command. This dialog box includes an Advanced Lighting panel, and several of the settings are specific to radiosity.

More on Advanced Lighting and Radiosity is covered in Chapter 26, "Working with Advanced Lighting, Light Tracing, and Radiosity."

FIGURE 4.12

Local radiosity settings are set using the Object Properties dialog box.

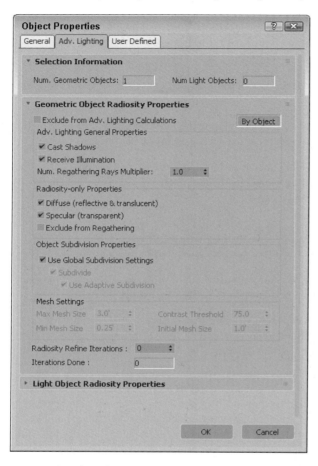

For the selected objects, you can specify whether they Cast Shadows and Receive Illumination. For a radiosity solution, you also can select to enable or disable Diffuse, Specular, Exclude from Regathering, and Subdividing. The Radiosity Refine Iterations value lets you set the number of iterations for the current selection. Think of Refine Iterations as the quality setting. You can set it for all objects or just for the selected object. It works by comparing the variance between adjacent faces.

If any light objects are selected, you can select to exclude them from radiosity processing or to store the illumination values with the mesh.

The Preference Settings dialog box also includes a Radiosity panel. Using this panel, shown in Figure 4.13, you can set the advanced lighting settings that apply to all objects globally.

FIGURE 4.13

Use the Radiosity panel of the Preference Settings dialog box to set global parameters.

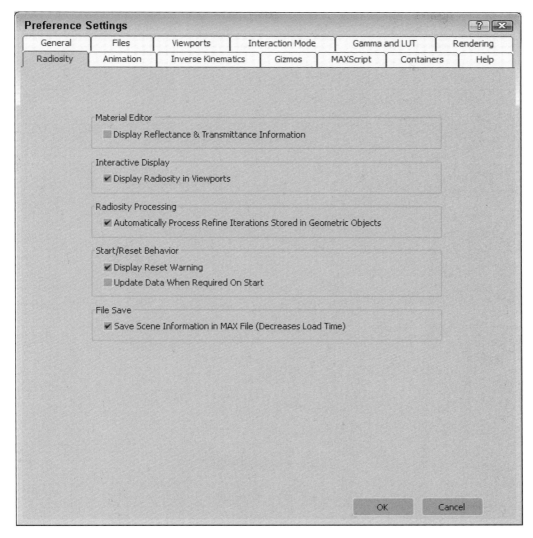

In the Radiosity panel of the Preference Settings dialog box is an option to Display Reflectance & Transmittance Information. If this option is enabled, this information (average and maximum percent values) appears directly below the sample slots in the Compact Material Editor. You also can select to have the radiosity solution displayed in the viewports and to automatically process any refine iterations noted in the Object Properties dialog box for a given object. Radiosity processing includes a couple of warning dialog boxes that appear: one when the current solution is reset, and another to update the solution when the Start button is clicked. You can disable both of these warnings. The final option is to save the radiosity solution with the 3ds Max file. This slightly increases the file size, but it doesn't require that you recalculate the radiosity solution when the file is opened again.

Setting Animation Preferences

The Animation panel of the Preference Settings dialog box, shown in Figure 4.14, contains several preference options dealing with animations. When a specific frame is selected, all objects with keys for that frame are surrounded with white brackets. The Animation panel offers options that specify which objects get these brackets. Options include All Objects, Selected Objects, and None. You also can limit the brackets to only those objects with certain transform keys.

Tip

The Key Bracket Display options are helpful when you need to locate specific keys. When the selected object for the given frame has a key, the object is surrounded with brackets.

FIGURE 4.14

The Animation panel includes settings for displaying Key Brackets.

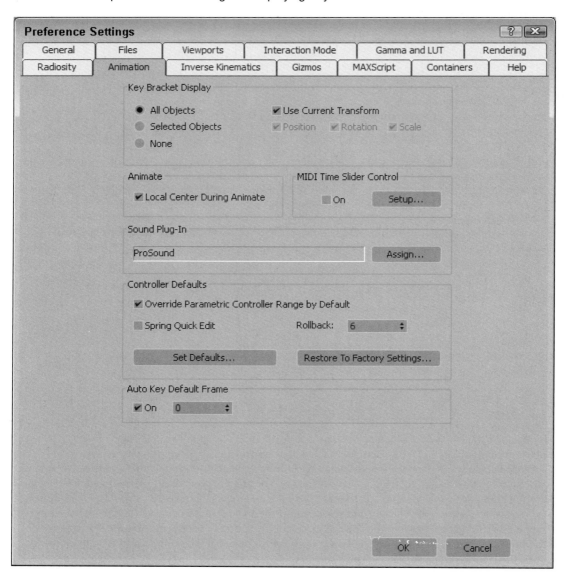

The Local Center During Animate option causes all objects to be animated about their local centers. Turning this option off enables animations about other centers (such as screen and world).

The MIDI Time Slider Controls include an On option and a Setup button. The Setup button opens the MIDI Time Slider Control Setup dialog box shown in Figure 4.15. After this control is set up, you can control an animation using a MIDI device.

FIGURE 4.15

The MIDI Time Slider Control Setup dialog box lets you set up specific notes to start, stop, and step through an animation.

You can use the Animation panel to assign a new Sound Plug-In to use, as well as to set the default values of all animation controllers. The Override Parametric Controller Range by Default option causes controllers to be active for the entire animation sequence instead of just their designated range. The Spring Quick Edit option lets you change the accuracy of all Spring controllers in the entire scene in one place. The Rollback setting is the number of frames that the Spring controller uses to return to its original position.

Clicking the Set Defaults button opens the Set Controller Defaults dialog box. This dialog box includes a list of all the controllers and a Set button. When you select a controller and click the Set button, another dialog box appears with all the values for that controller.

When you first start up 3ds Max, the default first frame on the Timeline is frame 0, but if you enable the Auto Key Default Frame option, you can set the first frame to be any frame you want. This is convenient if you like to use some frames to set up a shot or if the starting frame of the shot is not at frame 0.

You can learn more about specific controllers in Chapter 35, "Animating with Constraints and Simple Controllers."

Setting Inverse Kinematics Preferences

The required accuracy of the IK solution can be set using the Inverse Kinematics panel in the Preference Settings dialog box, shown in Figure 4.16. You can open this dialog box by choosing Customize→Preferences. For the Interactive and Applied IK methods, you can set Position and Rotation Thresholds. These Threshold values determine how close the moving object must be to the defined position for the solution to be valid.

FIGURE 4.16

The Inverse Kinematics panel of the Preference Settings dialog box lets you set the global Threshold values.

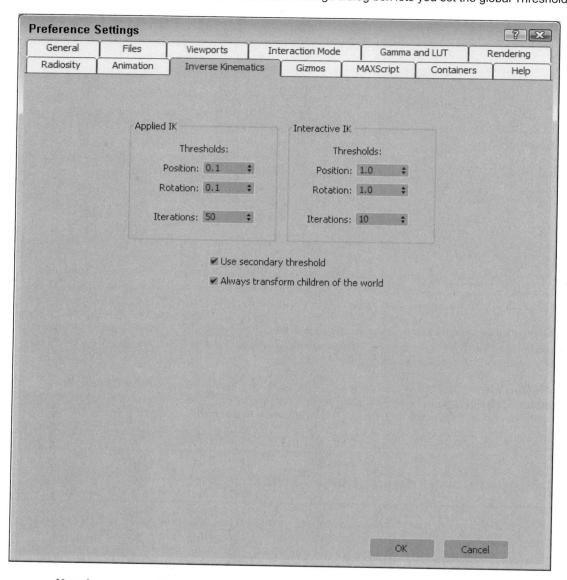

You also can set an Iterations limit for both methods. The Iterations value is the maximum number of times the calculations are performed. This value limits the time that Max spends looking for a valid solution. The Iterations settings control the speed and accuracy of each IK solution.

Note

If the Iterations value is reached without a valid solution, 3ds Max uses the last calculated iteration.

The Use Secondary Threshold option provides a backup method for determining whether 3ds Max should continue to look for a valid solution. This method should be used if you want 3ds Max to bail out of a particularly difficult situation rather than to continue to look for a solution. If you are working with very small thresholds, you want to enable this option.

The Always Transform Children of the World option enables you to move the root object when it is selected by itself, but constrains its movement when any of its children are moved.

Setting Gizmo Preferences

For each of these gizmos, you can set the preferences using the Gizmos panel in the Preference Settings dialog box, shown in Figure 4.17, which is accessed from the Customize menu. In this panel for all gizmos, you can turn the gizmos on or off, set to Show Axis Labels, Allow Multiple Gizmos, and set the Size of the gizmo's axes. The Allow Multiple Gizmos option enables a separate gizmo for each selection set object. The Labels option labels each axis with an X, Y, or Z.

FIGURE 4.17

The Gizmos panel in the Preference Settings dialog box lets you control how the Transform Gizmos look.

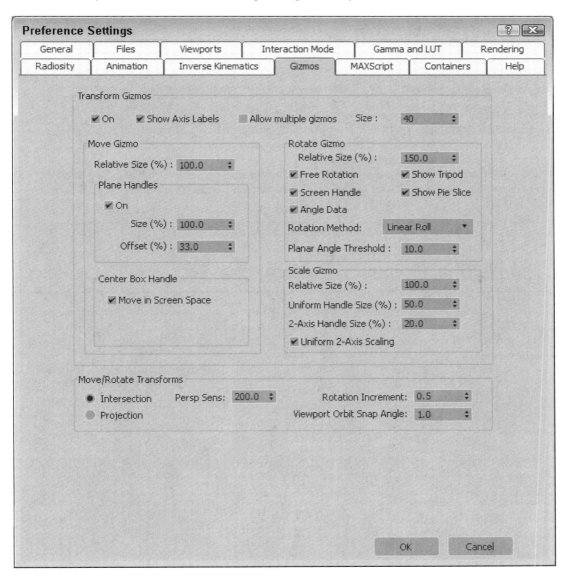

For the Move Gizmo section, you can set the Relative Size of the gizmo, which is relative to the top Size value, so a setting of 100 percent makes the size of the gizmo the full value, and a setting of 50 percent makes it half the full Size value. You also can select to turn the plane handles on or off and set their Size and Offset values, which determine how large the highlighted planes are and where they are located relative to the center of the gizmo. A Size value of 100 percent extends the plane handles to be as long as the axis handles. You also can enable the Center Box Handle for moving in all three axes.

The Rotate Gizmo preferences also include a Relative Size value. The Free Rotation option lets you click and drag between the axes to rotate the object freely along all axes. The Show Tripod option displays the axes tripod at the center of the object. The Screen Handle option displays an additional gray circle that surrounds all the axes. Dragging on this handle spins the object about the viewport's center. The Show Pie Slice option highlights a slice along the selected axis that is as big as the offset distance. The Angle Data option displays the rotation values above the gizmo as it is being rotated.

The Gizmos panel offers three Rotation Methods: Linear Roll, Circular Crank, and Legacy R4. The Linear Roll method displays a tangent line at the source point where the rotation starts. The Circular Crank method rotates using the gizmo axes that surround the object. The Legacy R4 method uses a gizmo that looks just like the Move Gizmo that was available in the previous 3ds Max version. The Planar Angle Threshold value determines the minimum value to rotate within a plane.

The Scale Gizmo section also can set a Relative Size of the gizmo. The Uniform Handle Size value sets the size of the inner triangle, and the 2-Axis Handle Size value sets the size of the outer triangle. The Uniform 2-Axis Scaling option makes scaling with the outer triangle uniform along both axes.

The Move/Rotate Transforms section has some additional settings that control how objects move in the Perspective viewport. The Intersection and Projection options are for two different modes. The Intersection mode moves objects faster the farther they get from the center. In Projection mode, the Perspective Sensitivity value is used to set the mouse movements to the distance of the transformation. Small values result in small transformations for large mouse drags. The Rotation Increment value sets the amount of rotation that occurs for a given mouse drag distance, and the Viewport Arc Rotate Snap Angle sets where the arc snaps to.

The MAXScript preferences are covered in Chapter 52, "Automating with MAXScript."

Setting Global Container Preferences

Within the Preference Settings dialog box is a panel for setting the global container preferences, shown in Figure 4.18. Using these settings you can define which rules are used when saving to the Max 2010 container format. You also can set how often containers are refreshed and whether to display the status of the scene containers.

FIGURE 4.18

The Containers panel of the Preference Settings dialog box holds the global container settings.

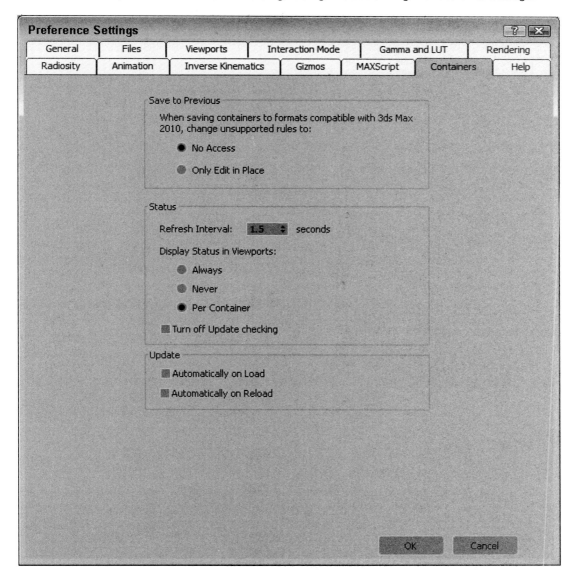

Setting Help Preferences

If you are working with 3ds Max offline, you can download and access a local copy of the help files. The Help panel in the Preference Settings dialog box, shown in Figure 4.19, lets you specify whether to use the online or the local help files.

FIGURE 4.19

The Help panel includes buttons for downloading the latest help files.

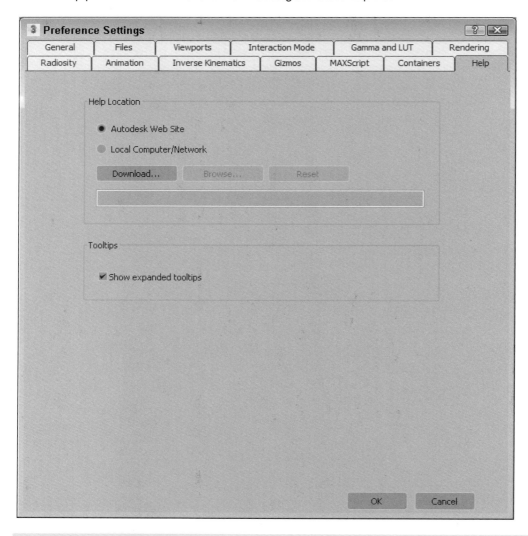

New Feature in 2020

The Show Expanded Tooltips option is new to 3ds Max 2020.

If you hold the mouse cursor over an interface button, the tool name will appear as a tooltip. Some tooltips are expanded with more information on using the tool. Enabling the Show Expanded Tooltips makes these expanded tooltips visible if they exist.

Summary

The Preference Settings dialog box is a one-stop shop where you can find all the various preferences. You may set some preferences once and never touch them again, and you may change others all the time.

This chapter covered the following topics:

* Setting general preferences

* Changing file preferences and configuring paths

* Changing viewport, interaction mode, and gamma preferences

* Setting rendering and radiosity preferences

* Setting animation, IK, gizmos, and other preferences

The next chapter jumps right into adding objects to a scene starting with the various primitive objects, which are the easiest objects to create.

Part II

Manipulating Objects

IN THIS PART

Chapter 5

Creating and Editing Primitive Objects

IN THIS CHAPTER

Setting system units

Creating primitive objects

Naming objects and setting object colors

Using creation methods

Setting object parameters

Using helper objects and utilities

So what exactly did the Romans use to build their civilization? The answer is lots and lots of basic blocks. The basic building blocks in the Autodesk® 3ds Max 2020® software are called *primitives*. You can use these primitives to start any modeling job. After you create a primitive, you can bend it, stretch it, smash it, or cut it to create new objects, but for now, you'll focus on using primitives in their default shape.

This chapter covers the basics of using primitive object types and introduces the various primitive objects, including how to accurately create and edit them by changing some simple parameters. You also use these base objects in the coming chapters to learn about selecting, cloning, grouping, and transforming.

Modeling is covered in depth in Part III, but first you need to learn how to create some basic blocks and move them around. Later, you can work on building a civilization. I'm sure workers in Rome would be jealous.

Selecting System Units

One of the first things you'll want to set before beginning a project is the scene units. Units can be as small as millimeters or as large as kilometers, or they can be generic, which means they have meaning only relative to the other parts of the scene. 3ds Max offers a large array of available units, and you can even define your own.

The system units have a direct impact on modeling and define the units that are represented by the coordinate values. Units directly relate to parameters entered with the keyboard. For example, with the units set to meters, a sphere created with the radius parameter of 2 would be 4 meters across.

3ds Max supports several different measurement systems, including Metric and U.S. Standard units. You also can define a Custom units system. (I suggest parsecs if you're working on a space scene.) Working with a units system enables you to work with precision and accuracy using realistic values.

Tip

Most game engines work with meters, so if you're building assets for a game, set the units to meters.

To specify a units system, choose Customize→Units Setup to display the Units Setup dialog box, shown in Figure 5.1. For the Metric system, options include Millimeters, Centimeters, Meters, and Kilometers. The

U.S. Standard units system can be set to the default units of Feet or Inches displayed as decimals or fractional units. You also can select to display feet with fractional inches or feet with decimal inches. Fractional values can be divided from 1/1 to 1/100 increments.

FIGURE 5.1

The Units Setup dialog box lets you choose which units system to use. Options include Metric, U.S. Standard, Custom, and Generic.

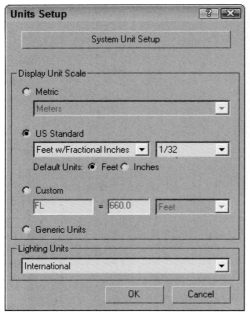

Using Custom and Generic units

To define a Custom units system, modify the fields under the Custom option, including a units label and its equivalence to known units. The final option is to use the default Generic units. Generic units relate distances to each other, but the numbers themselves are irrelevant. You also can set lighting units to use American or International standards. Lighting units are used to define Photometric lights.

At the top of the Units Setup dialog box is the System Unit Setup button. This button opens the System Unit Setup dialog box, also shown in Figure 5.1. This dialog box enables you to define the measurement system used by 3ds Max. Options include Inches, Feet, Miles, Millimeters, Centimeters, Meters, and Kilometers.

For example, when using 3ds Max to create models that are to be used in the Unreal game editor, you can use the Custom option to define a unit called the Unreal Foot unit that sets 1 Uft equal to 16 units, which matches the units in the Unreal editor just fine.

A multiplier field allows you to alter the value of each unit. The Respect System Units in Files toggle presents a dialog box whenever a file with a different system units setting is encountered. If this option is disabled, all new objects are automatically converted to the current units system.

The Origin control helps you determine the accuracy of an object as it is moved away from the scene origin. If you know how far objects will be located from the origin, entering that value tells you the Resulting Accuracy. You can use this feature to determine the accuracy of your parameters. Objects farther from the origin have a lower accuracy.

Caution

Be cautious when working with objects that are positioned a long way from the scene origin. The farther an object is from the origin, the lower its accuracy and the less precisely you can move it. If you are having trouble precisely positioning an object (in particular, an object that has been imported from an external file), check the object's distance from the origin. Moving it closer to the origin should help resolve the problem.

Handling mismatched units

Imagine designing a new ski-resort layout. For such a project, you'd want to probably use kilometers as the file units. If your next project is to design a custom body design for a race car, you'll want to use meters as the new units. If you need to reopen the ski-resort project while your units are set to meters, you'll get a File Load: Units Mismatch dialog box, shown in Figure 5.2.

This dialog box reminds you that the units specified in the file that you're opening don't match the current units setting. This also can happen when you're trying to merge in an object with a different units setting. The dialog box lists the units used in both the file and the system and offers two options. The Rescale the File Objects to the System Unit Scale option changes the units in the file to match the current system units setting. The second option changes the system units to match the file unit settings.

Tip

If you rescale the file object to match the system file units setting, the objects will either appear tiny or huge in the current scene. Use the Zoom Extents All button to see the rescaled objects in the viewport.

FIGURE 5.2

The File Load: Units Mismatch dialog box lets you synch up units between the current file and the system settings.

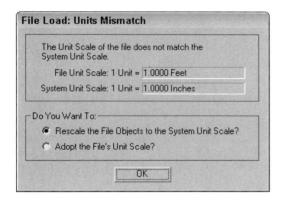

Rescaling world units

If you discover halfway through your scene that you're working with the wrong units, you can use the Rescale World Units utility to scale up the entire scene or just selected objects. To access this utility, click the Utilities panel and then the More button. In the utilities list, select the Rescale World Units utility, and click OK. Then click the loaded Rescale button in the Command Panel to open the Rescale World Units dialog box.

The Rescale World Units dialog box has a Scale Factor value, which is the value by which the scene or objects are increased or decreased. If your world was created using millimeter units, and you need to work in meters, increasing by a Scale Factor of 1000 will set the world right. You also can select to scale the entire Scene or just the Selection.

Creating Primitive Objects

3ds Max is all about creating objects and scenes, so it's appropriate that one of the first things to learn is how to create objects. Although you can create complex models and objects, 3ds Max includes many simple, default geometric objects—called *primitives*—that you can use as starting points. Creating these primitive objects can be as easy as clicking and dragging in a viewport.

Using the Create menu

The Create menu offers quick access to the buttons in the Create panel. All the objects that you can create using the Create panel, you can access using the Create menu. Selecting an object from the Create menu automatically opens the Create panel in the Command Panel and selects the correct category, subcategory, and button needed to create the object. After selecting the menu option, you simply need to click in one of the viewports to create the object.

Using the Create panel

The creation of all basic 3ds Max objects, such as primitive spheres, shapes, lights, and cameras, starts with the Create panel (or the Create menu, which leads to the Create panel). This panel is the first in the Command Panel, indicated by a big + sign.

Of all the panels in the Command Panel, only the Create panel—shown in Figure 5.3—includes both categories and subcategories. After you click the Create tab, seven category icons are displayed. From left to right, they are Geometry, Shapes, Lights, Cameras, Helpers, Space Warps, and Systems.

The Create panel is the place you go to create objects for the scene. These objects could be geometric objects such as spheres, cones, and boxes or other objects such as lights, cameras, and Space Warps. The Create panel contains a huge variety of objects. To create an object, you simply need to find the button for the object that you want to create, click it, and click in one of the viewports, and voilà—instant object.

After you select the Geometry category icon (which has an icon of a sphere on it), a drop-down list with several subcategories appears directly below the category icons. The first available subcategory is Standard Primitives. After you select this subcategory, several text buttons appear that enable you to create some simple primitive objects.

Note

The second subcategory is called Extended Primitives. It also includes primitive objects. The Extended Primitives are more specialized and aren't used as often.

FIGURE 5.3

The Create panel includes categories and subcategories.

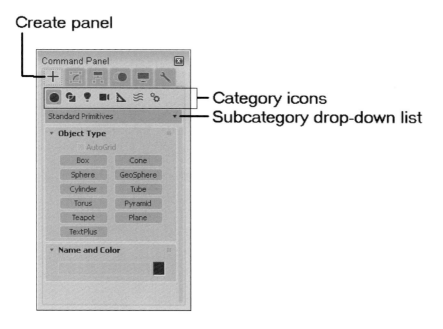

As an example, click the button labeled Sphere (not to be confused with the Geometry category, which has a sphere icon). Several rollouts appear at the bottom of the Command Panel. These rollouts for the Sphere primitive object include Name and Color, Creation Method, Keyboard Entry, and Parameters. The rollouts for each primitive are slightly different, as well as the parameters within each rollout.

If you want to ignore these rollouts and just create a sphere, simply click and drag within one of the viewports, and a sphere object appears. The size of the sphere is determined by how far you drag the mouse before releasing the mouse button. Figure 5.4 shows the new sphere and its parameters.

FIGURE 5.4

You can create primitive spheres easily by dragging in a viewport.

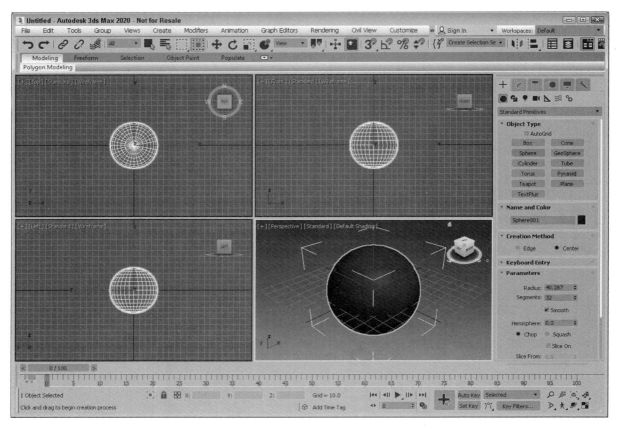

When an object button, such as the Sphere button, is selected, it is highlighted (in yellow or gray, depending on your color scheme). This color change reminds you that you are in creation mode. Clicking and dragging within any viewport creates an additional sphere. While in creation mode, you can create many spheres by clicking and dragging several times in one of the viewports. To get out of creation mode, right-click in the active viewport, or click the Select Object button or one of the transform buttons on the main toolbar.

After you select a primitive button, several additional rollouts magically appear. These new rollouts hold the parameters for the selected object and are displayed in the Create panel below the Name and Color rollout. Altering these parameters changes the most recently created object.

Naming and renaming objects

Every object in the scene can have both a name and a color assigned to it. Each object is given a default name and random color when first created. The default name is the type of object followed by a number. For example, when you create a sphere object, 3ds Max labels it "Sphere001." These default names aren't very exciting and can be confusing if you have many objects. You can change the object's name at any time by modifying the Name field in the Name and Color rollout of the Command Panel.

Note

3ds Max gives each newly created object a unique name. 3ds Max is smart enough to give each new object a different name by adding a sequential number to the end of the name.

Caution

Be aware that 3ds Max allows you to give two different objects the same name.

5. Drag in the Top viewport to create another object. The Cube/Octa option is still selected. Enter a value of **1.0** in the Parameters Q field this time, and set the Radius to **5**. Name this object **Cube**.

6. Drag in the Top viewport again to create the fourth Hedra object. In the Parameters rollout, select the Dodec/Icos option, enter a value of **1.0** in the P field, and set the Radius value to **5**. Name the object **Icosahedron**.

7. Drag in the Top viewport to create the final object. With the Dodec/Icos option set, enter **1.0** for the Q value, and set the Radius to **5**. Name this object **Dodecahedron**.

8. To get a good look at the objects, click the Perspective viewport, press the Zoom Extents button, and maximize the viewport by clicking the Maximize Viewport Toggle (or press Alt+W) in the lower-right corner of the window.

Figure 5.10 shows the five perfect solid primitive objects. Using the Modify panel, you can return to these objects and change their parameters to learn the relationships among them.

FIGURE 5.10

The octahedron, cube, tetrahedron, icosahedron, and dodecahedron objects; Plato would be amazed.

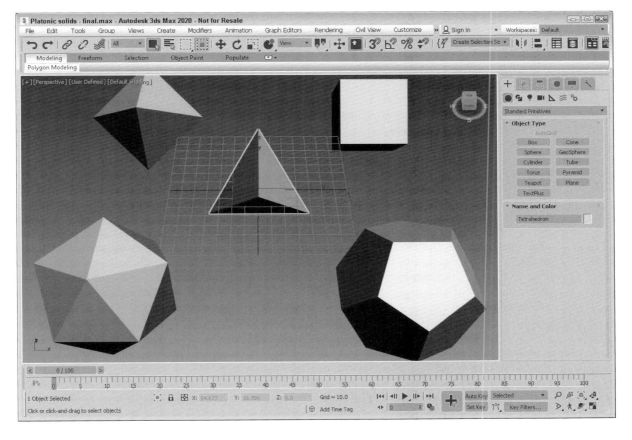

Using Modeling Helpers

Before leaving the discussion on primitive objects, let's take a look at some modeling helpers. In the Create panel (and the Create menu) is a category of miscellaneous objects called *helpers* (the icon looks like a triangle tool). These objects are useful in positioning objects and measuring dimensions.

Using Dummy and Point objects

The Dummy object is a useful object for marking specific locations in the scene with an object that isn't included in the final render. A Dummy object appears in the viewports as a simple cube with a pivot point at its center, but the object will not be rendered and has no parameters. It is used only as an object about which to transform objects. For example, you could create a Dummy object that the camera could follow through an animation sequence. Dummy objects are used in many examples throughout the remainder of the book.

The Point object is very similar to the Dummy object in that it also is not rendered and has minimal parameters. A Point object defines a point in space and is identified as a Cross, an Axis Tripod, or a simple Box. The Center Marker option places an X at the center of the Point object (so X really does mark the spot). The Axis Tripod option displays the X-, Y-, and Z-axes; the Cross option extends the length of the marker along each axis; and the Box option displays the Point object as a Box. The Size value determines how big the Point object is. You can display several of these options at once.

Tip
The Size parameter actually makes Point helpers preferable over Dummy helpers because you can parametrically change their size.

The Constant Screen Size option keeps the size of the Point object constant, regardless of how much you zoom in or out of the scene. The Draw On Top option draws the Point object above all other scene objects, making it easy to locate. The main purpose for the Point object is to mark positions within the scene.

Caution
Point objects are difficult to see and easy to lose. If you use a point object, be sure to name it so you can find it easily in the Select From Scene dialog box.

Measuring coordinate distances

The Helpers category also includes several handy utilities for measuring dimensions and directions. These are the Tape, Protractor, and Compass objects. The units are all based on the current selected system units.

Using the Measure Distance tool

In the Tools menu is a command to Measure Distance. This tool is easy to use. Just select it and click at the starting point and again at the ending point; the distance between the two clicks is shown in the Status Bar at the bottom of the interface. Measure Distance also reports the Delta values (the amount of change for each dimension) in the X, Y, and Z directions. You can use this tool with the Snap feature enabled for accurate measurements.

Using the Tape helper

You use the Tape helper object to measure distances. To use it, simply drag the distance that you would like to measure, and view the resulting dimension in the Parameters rollout. You also can set the length of the Tape object using the Specify Length option. You can move and reposition the end points of the Tape object with the Select and Move tool, but the Rotate and Scale transforms have no effect.

Using the Protractor helper

The Protractor helper object works in a manner similar to the Tape object, but it measures the angle between two objects. To use the Protractor object, click in a viewport to position the Protractor object. (The Protractor object looks like two pyramids aligned point to point and represents the origin of the angle.) Then click the Pick Object 1 button, and select an object in the scene. A line is drawn from the Protractor object to the pivot point of the selected object. Next, click the Pick Object 2 button, and select another object. A second line is drawn from the Protractor object and the pivot point of the second object. The Protractor object and the two lines form an angle and its value is displayed in the Parameters rollout. The value changes when either of the selected objects or the Protractor is moved.

Note

All measurement values are presented in gray fields within the Parameters rollout. This gray field indicates that the value cannot be modified.

Using the Compass helper

The Compass helper object identifies North, East, West, and South positions on a planar star-shaped object called a Compass Rose. You can drag the Compass object to increase its size or change its Radius value. The Show Compass Rose option lets you turn the compass on and off.

The Compass object is mainly used in conjunction with the Sunlight and Daylight Systems, which you can learn about in Chapter 25, "Using Lights and Basic Lighting Techniques."

Using the Measure utility

In the Utilities panel is another useful tool for getting the scoop on the current selected object: the Measure utility. You can open the Measure utility as a floater dialog box, shown in Figure 5.11. This dialog box displays the object's name along with its Surface Area, Volume, Center of Mass, Length (for shapes), and Dimensions. It also includes an option to lock the current selection.

FIGURE 5.11

The Measure dialog box displays some useful information.

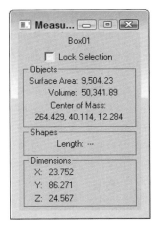

Tutorial: Testing the Pythagorean Theorem

I always trusted my teachers in school to tell me the truth, but maybe they were making it all up, especially my math teacher. (He did have shifty eyes, after all.) For my peace of mind, I want to test one of the mathematical principles he taught us: the Pythagorean Theorem. (What kind of name is that, anyway?)

If I remember the theorem correctly, it says that the sum of the squares of the sides of a right triangle equals the hypotenuse squared. So according to my calculations, a right triangle with a side of 3 and a side of 4 has a hypotenuse of 5. Because 3ds Max is proficient at drawing shapes such as this one, we test the theorem by creating a box with a width of 4 and a height of 3 and then measuring the diagonal.

To test the Pythagorean Theorem, follow these steps:

1. Start by setting the scene units. Select the Customize→→Units Setup to open the Units Setup dialog box. Select the Metric option with the Meters selection from the drop-down list, and click OK.

2. Select the Create→Standard Primitives→Box menu command, and drag and click in the Top viewport to create a Box object. Change its parameters to 40 for the Length, 30 for the Width, and 10 for the Height values. Then right-click in the active viewport to exit Box creation mode, and press the Z key to zoom in on the Box object.

3. Right-click the Snaps Toggle on the main toolbar to open the Snap and Grid Settings dialog box, select the Snaps panel, and set the Snap feature to snap to vertices by clicking the Clear All button and then selecting the Vertex option. Close the Grid and Snap Settings dialog box, and enable the 3D Snap feature by clicking the Snaps Toggle button in the main toolbar (or by pressing S).

4. Select the Create→Helpers→Tape Measure menu command.

5. In the Top viewport, move the cursor over the upper-left corner of the object, and click the blue vertex that appears. Then drag down to the lower-right corner, and click the next blue vertex that appears. Note the Length value in the Parameters rollout.

Well, I guess my math teacher didn't lie about this theorem, but I wonder whether he was correct about all those multiplication tables. Figure 5.12 shows the resulting box and measurement value.

FIGURE 5.12

I guess old Pythagoras was right. (Good thing I have 3ds Max to help me check.)

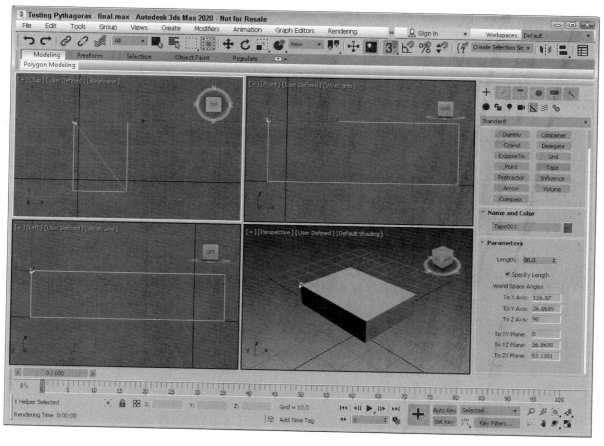

Summary

Primitives are the most basic objects and often provide starting points for more ambitious modeling projects. The two classes of primitives—Standard and Extended—provide a host of possible objects. This chapter covered the following topics:

* Handling system units

* Creating primitives by both dragging and entering keyboard values

* Naming objects and setting and changing object color

    *   Using the various creation methods for all the primitive objects

    *   Using helper objects and measuring distances

Now that you know how to create objects, you can focus on selecting them after they're created, which is what the next chapter covers. You can select objects in numerous ways. Setting object properties is also covered.

Chapter 6

Selecting Objects and Setting Object Properties

IN THIS CHAPTER

Selecting objects using toolbars and menus

Using named selection sets

Viewing object information

Setting display properties

Setting render properties

Enabling object motion blur

Hiding and freezing objects

Exploring the Scene Explorer

Now that you've learned how to create objects and had some practice, you've probably created more than you really need. To eliminate, move, or change the look of any objects, you first have to know how to select the object. Doing so can be tricky if the viewports are all full of objects lying on top of one another. Luckily, the Autodesk® 3ds Max® 2020 software offers several selection features that make looking for a needle in a haystack easier.

3ds Max offers many different ways to select objects. You can select by name, color, type, and even material. You also can use selection filters to make only certain types of objects selectable. And after you've found all the objects you need, you can make a selection set, which will allow you to quickly select a set of objects by name.

All objects have properties that define their physical characteristics, such as shape, radius, and smoothness, but objects also have properties that control where they are located in the scene, how they are displayed and rendered, and what their parent object is. These properties have a major impact on how you work with objects; understanding them can make objects in a scene easier to work with. These later properties are set in the Object Properties dialog box.

Sometimes, you'll find it easier to select a named object from a list rather than locating its object in the viewport. For these occasions, the Scene Explorer is really handy. Another huge benefit of the Scene Explorer is that you can quickly set properties for multiple objects at once. Now, where is that needle?

Selecting Objects

3ds Max includes several methods for selecting objects, the easiest being simply clicking the object or dragging over it in one of the viewports. Selected objects turn white and are enclosed in brackets called *selection brackets.*

In addition to turning white and displaying selection brackets, several options allow you to mark selected objects. You can find these options in the Per-View Preferences panel of the Viewport Setting and Preference dialog box (which you access with the Views→Viewport Per-View Settings menu command); they include selection brackets (keyboard shortcut J) and edged faces. Either or both of these options can be enabled, as shown in Figure 6.1. Another way to detect the selected object is that the object's axes appear at the object's pivot point. The Views→Shade Selected command turns on shading for the selected object in all viewports, including Wireframe viewports.

Tip

If clicking on an object isn't working, it may be that you have another tool selected. Try right-clicking in the viewport to release the current tool, or click the Select Object button in the main toolbar.

Caution

The Viewport Setting and Preference dialog box also includes an option to Shade Selected Faces (F2), but this option shades only selected subobject faces.

FIGURE 6.1

Selected objects can be highlighted with selection brackets (left), edged faces (middle), or both (right).

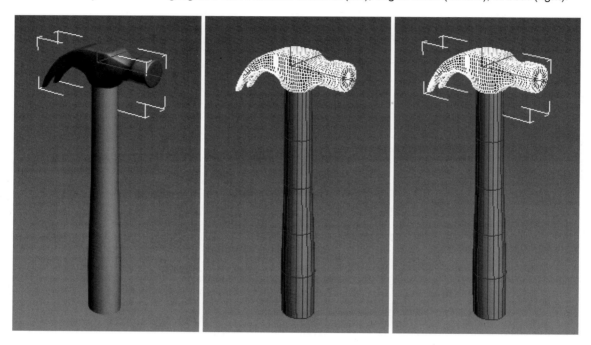

With many objects in a scene, clicking directly on a single object free from the others can be difficult, but persistence can pay off. If you continue to click an object that is already selected, the object directly behind the object you clicked is selected. For example, if you have a row of spheres lined up, you can select the third sphere by clicking three times on the first object.

Tip

In complicated scenes, finding an object is often much easier if it has a relevant name. Be sure to name your new objects using the Name and Color rollout. If a single object is selected, its name appears in the Name and Color rollout.

Selection filters

Before examining the selection commands in the Edit menu, I need to tell you about selection filters. With a complex scene that includes geometry, lights, cameras, shapes, and so on, selecting the exact object that you want can be difficult. Selection filters can simplify this task.

A selection filter specifies which types of objects can be selected. The Selection Filter drop-down list is located on the main toolbar to the left of the Select Object button. Selecting Shapes, for example, makes only shape objects available for selection. Clicking a geometry object with the Shapes Selection Filter enabled does nothing.

The available filters include All, Geometry, Shapes, Lights, Cameras, Helpers, and Warps. If you're using Inverse Kinematics, you also can filter by Bone, IK Chain Object, Point, and CAT Bone.

The Combos option opens the Filter Combinations dialog box, shown in Figure 6.2. From this dialog box, you can select combinations of objects to filter. These new filter combinations are added to the drop-down list. For example, to create a filter combination for lights and cameras, open the Filter Combinations dialog box, select Lights and Cameras, and click Add. The combination is listed as LC in the Current Combinations section, and the LC option is added to the drop-down list.

FIGURE 6.2

The Filter Combinations dialog box enables you to create a custom selection filter.

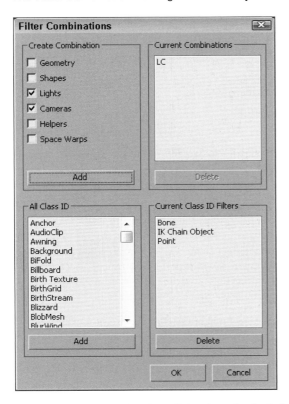

The Filter Combinations dialog box also includes a list of additional objects. Using this list, you can filter very specific object types, such as a Boolean object or a Box primitive. In fact, the Bone, IK Chain Object, Point, and CAT Bone filters that appear in the default main toolbar drop-down list all come from this additional list.

Select buttons

On the main toolbar are several buttons used to select objects, shown in Table 6.1. The Select Object button looks like the arrow cursor over a box. The other three buttons select and transform objects. They are Select and Move (W), Select and Rotate (E), and Select and Scale (R). There is also a Select and Place button that moves the selected object over the surface of the other scene objects. These commands also are available on the quad menu. The final selection button is the Select and Manipulate button. With this button, you can select and use special helpers such as sliders.

TABLE 6.1 Select Buttons

Button	Description
■	Select Object (Q)
✛	Select and Move (W)
↻	Select and Rotate (E)
■ ■ ■	Select and Scale (R), Select and Non-Uniform Scale, Select and Squash
● ●	Select and Place, Select and Rotate
✛	Select and Manipulate

See Chapter 7, "Transforming Objects, Pivoting, Aligning, and Snapping," for more details on the Select and Transform buttons.

Selecting with the Edit menu

The Edit menu includes several convenient selection commands. The Edit→Select All (Ctrl+A) menu command does just what you would think it does: it selects all unfrozen and unhidden objects in the current scene of the type defined by the selection filter. The Edit→Select None (Ctrl+D) menu command deselects all objects. You also can simulate this command by clicking in any viewport away from all objects. The Edit→Select Invert (Ctrl+I) menu command selects all objects defined by the selection filter that are currently not selected and deselects all currently selected objects.

The Edit→Select Similar (Ctrl+Q) command selects all objects that are similar to the current selection. Select Instances selects all instances of the current object. Instances are copies of an object that have a link to the original so that changing one automatically changes the other. If multiple objects are selected, the Select Similar command selects the objects that meet the criteria for being similar to each of the selected objects. Objects are similar if they meet one of the following criteria:

* Same object type, such as lights, helpers, or Space Warps
* Same primitive object, such as Sphere, Box, or Hedra
* Same modeling type, such as Editable Spline, Editable Poly, or Editable Patch
* Imported objects from an AutoCAD DWG file that have the same style applied
* Same applied material
* Objects existing on the same layer

Figure 6.3 shows a treasure chest of Hedra gems. With a single object selected, choosing Edit→Select Similar (Ctrl+Q) causes all Hedra primitive objects to be selected.

FIGURE 6.3

The Select Similar command selects all Hedra objects.

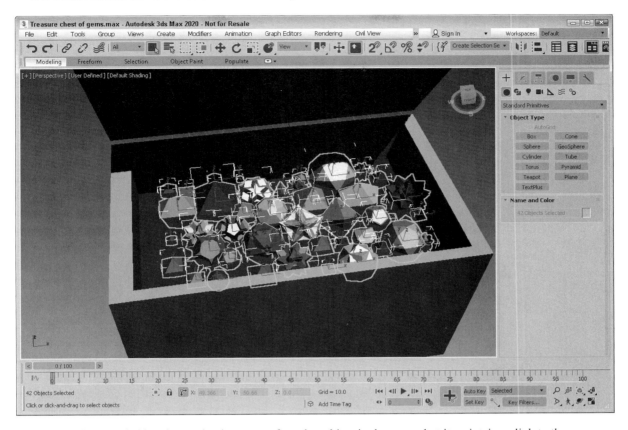

An instanced object is one that is a copy of another object in the scene, but it maintains a link to the previous object so that changing one automatically changes its instance. If an instanced object is selected, you can use the Edit→Select Instances menu to quickly select all other instances of the current object.

Select by Name

Choosing Edit→Select By→Name opens the Select From Scene dialog box, which is a version of the Scene Explorer dialog box, except that you can't change any parameters. Clicking the Select by Name button on the main toolbar, positioned to the right of the Select Object button, or pressing the keyboard shortcut, H, also opens this dialog box. The Scene Explorer dialog box is covered in detail later in this chapter.

You select objects by clicking their names in the list and then clicking OK or by simply double-clicking a single item. To pick and choose several objects, hold down the Ctrl key while selecting. Holding down the Shift key selects a range of objects.

Select by Layer

The Scene Explorer lets you separate all scene objects into layers for easy selection. The Edit→Select By→Layer command opens a simple dialog box listing the defined layers and lets you select a layer. All objects in the selected layer are then selected. The Scene Explorer is covered in detail later in this chapter.

Select by Color

Choosing Edit→Select By→Color lets you click a single object in any of the viewports. All objects with the same color as the one you selected are selected. Even if you already have an object of that color selected, you still must select an object of the desired color. Be aware that this is the object color, not the applied material color. This command, of course, does not work on any objects without an associated color, such as Space Warps. This same command is also accessible as a button in the Object Color dialog box.

157

Selection method

The Window/Crossing Selection method buttons on the main toolbar lets you select one of two different methods for selecting objects in the viewport using the mouse. First, make sure that you're in select mode, and then click away from any of the objects and drag over the objects to select. The first method for selecting objects is Window Selection. This method selects all objects that are contained completely within the dragged outline. The Crossing Selection method selects any objects that are inside or overlapping the dragged outline. You also can access these two selection methods via the Window/Crossing toggle button, found on the main toolbar and shown in Table 6.2.

Tip

If you can't decide whether to use the Crossing or Window selection method, you can select to use both. The General panel of the Preference Settings dialog box provides an option to enable Auto Window/Crossing by Direction. When this option is enabled, you can select a direction, and the Crossing selection method is used for all selections that move from that direction. The Window selection method is used for all selections that move from the opposite direction. For example, if you select Left to Right for the Crossing selection method, moving from Left to Right uses the Crossing selection method, and selecting from Right to Left uses the Window selection method.

Table 6.2 Window Selection Buttons

Button	Description
	Window
	Crossing

You also can change the shape of the selection outline. The Selection Region button on the main toolbar to the left of the Selection Filter drop-down list includes flyout buttons for Rectangular, Circular, Fence, Lasso, and Paint Selection Regions, shown in Table 6.3.

Table 6.3 Shape-Shifting Selection Region Buttons

Button	Description
	Rectangular
	Circular
	Fence
	Lasso
	Paint

The Rectangular selection method lets you select objects by dragging a rectangular section (from corner to corner) over a viewport. The Circular selection method selects objects within a circle that grows from the center outward. The Fence method lets you draw a polygon-shaped selection area by clicking at each corner. Simply double-click to finish the fenced selection. The Lasso method lets you draw the selection area by freehand. The Paint method lets you choose objects by painting an area. All objects covered by the paintbrush area are selected.

Pressing the Q keyboard shortcut selects the Select Object mode in the main toolbar, but repeated pressing of the Q keyboard shortcut cycles through the selection methods. Figure 6.4 shows the first four selection methods.

FIGURE 6.4

The mug is selected using the Rectangular, Circular, Fence, and Lasso selection methods.

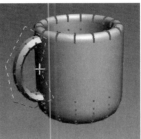

Selecting multiple objects

As you work with objects in 3ds Max, you'll sometimes want to apply a modification or transform to several objects at once. You can select multiple objects in several ways. By using the Edit→Select By→Name command, clicking the Select by Name main toolbar button, or pressing the H key, you can open the Select From Scene dialog box. With the Select From Scene dialog box open, you can choose several objects from the list, using the standard Ctrl and Shift keys. Holding down the Ctrl key selects or deselects multiple list items, but holding down the Shift key selects all consecutive list items between the first and second selected items.

The Ctrl key also works when selecting objects in the viewport using one of the main toolbar Select buttons. You can tell whether you're in select mode by looking for a button that's highlighted. If you hold down the Ctrl key and click an object, the object is added to the current selection set. If you drag over multiple objects while holding down the Ctrl key, all items in the dragged selection are added to the current selection set.

The Alt key deselects objects from the current selection set, which is opposite of what the Ctrl key does.

If you drag over several objects while holding down the Shift key, the selection set is inverted. Each item that was selected is deselected, and vice versa.

Object hierarchies are established using the Select and Link button on the main toolbar. You can select an entire hierarchy of objects by double-clicking its parent object. You also can select multiple objects within the hierarchy. When you double-click an object, any children of that object also are selected. When an object with a hierarchy is selected, the Page Up and Page Down keys select the next object up or down the hierarchy.

Hierarchies and linking objects are covered in Chapter 9, "Grouping, Linking and Parenting Objects."

Another way to select multiple objects is by dragging within the viewport using the Window and Crossing selection methods, discussed previously in the "Select by Region" section.

Caution

Although the Move, Rotate, and Scale buttons also may be used to select objects, they can cause problems when selecting multiple objects. If you are selecting multiple objects with the Select and Move tool, and you accidentally drag the mouse while moving to the next item, the entire selection is moved out of place. You can use the Undo feature to return it to its original position. To prevent this from happening, use the Select Object tool when selecting multiple objects.

Using the Paint Selection Region tool

The Paint Selection Region tool is the last flyout button under the Selection Region button. Using this tool, you can drag a circular paintbrush area over the viewports, and all objects or subobjects underneath the brush are selected.

The size of the Paint Selection brush is shown as a circle when the tool is selected and may be changed using the Paint Selection Brush Size field in the General panel of the Preference Settings dialog box. Right-clicking the Paint Selection Region button on the main toolbar automatically opens the Preference Settings dialog box. Figure 6.5 shows how the Paint Selection Region may be used to select several spheres by dragging over them.

FIGURE 6.5

The Paint Selection Region tool makes it easy to select spheres by dragging.

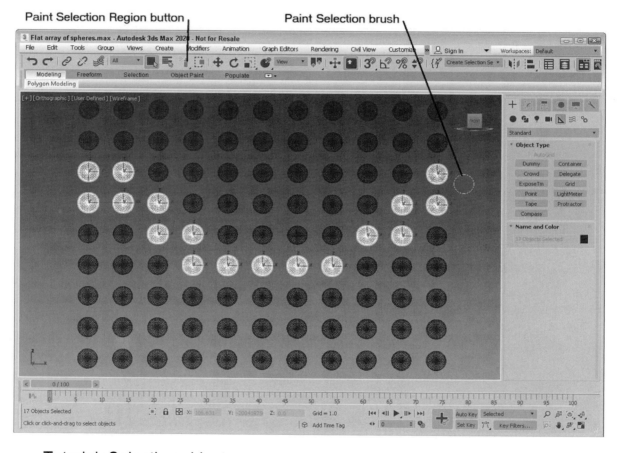

Tutorial: Selecting objects

To practice selecting objects, you'll work with a simple model of Marvin Moose.

To select objects, follow these steps:

1. Open the Marvin Moose.max scene, which you can find in the Chap 06 directory in the downloaded content set.
2. Click the Select Object button (or press the Q key), and click Marvin's body in one of the viewports.

   In the Command Panel, the name for this object, Shirt, is displayed in the Name and Color rollout, and shirt object is selected and highlighted in white.

3. Click the Select and Move button (or press the W key), click the shirt object, and drag in the Perspective viewport to the right.

Moving the selected body separates it from the rest of the model's objects.

4. Choose Undo Scene Operation from the left end of the main toolbar (or press Ctrl+Z) to undo the move and to piece Marvin back together.

5. With the Select and Move tool still selected, drag an outline around the entire character in the Top view to select all the Character parts, and then click and drag the entire Character.

This time, the entire character moves as one entity, and the name field displays 11 Objects Selected.

6. Open the Select From Scene dialog box by clicking the Select by Name button on the main toolbar (or by pressing the H key).

All the individual parts that make up this model are listed.

7. Select the Glasses object listed in the dialog box, and click OK.

The Select From Scene dialog box automatically closes, and the glasses object becomes selected in the viewports.

Figure 6.6 shows Marvin Moose with just his glasses object selected. Notice that the name of the selected object in the Name and Color rollout says "Glasses."

FIGURE 6.6

A moose cartoon character with its selected glasses

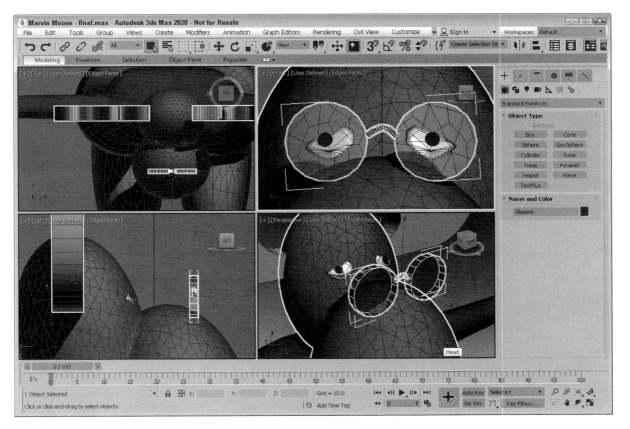

Using named selection sets

With a group of selected objects, you can establish a named selection set. After it's established as a named selection set, you can recall this group of selected objects at any time by selecting its name from the Named

Selection Sets drop-down list on the main toolbar or by opening the Named Selection Sets dialog box, shown in Figure 6.7.

 You can access this dialog box using the Edit Named Selection Sets button on the main toolbar or by selecting the Edit→Manage Selection Sets menu command. To establish a selection set, type a name in the Named Selection Set drop-down list toward the right end of the main toolbar or use the dialog box.

FIGURE 6.7

The Named Selection Sets dialog box lets you view and manage selection sets.

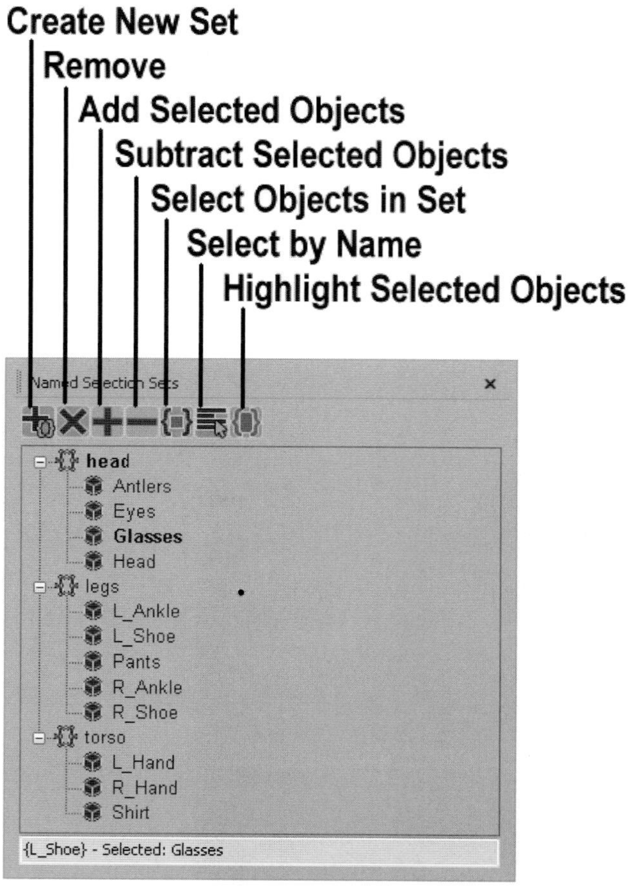

Create New Set
 Remove
   Add Selected Objects
     Subtract Selected Objects
       Select Objects in Set
         Select by Name
           Highlight Selected Objects

```
Named Selection Sets                              ×
⊞ ✕ ✚ ━ {} ≡ {}
  ⊟ {} head
       Antlers
       Eyes
       Glasses
       Head
  ⊟ {} legs
       L_Ankle
       L_Shoe
       Pants
       R_Ankle
       R_Shoe
  ⊟ {} torso
       L_Hand
       R_Hand
       Shirt
{L_Shoe} - Selected: Glasses
```

You also can create named selection sets for subobject selections. Be aware that these subobject selection sets are available only when you're in subobject edit mode and only for the currently selected object.

Tip

Anytime you spend a lot of time selecting a bunch of objects like trees in the landscape scene, you should create a selection set.

Editing named selection sets

After you've created several named selection sets, you can use the Named Selections Sets dialog box to manage the selection sets. The buttons at the top let you create and delete sets, add objects to or remove objects from a set, and select and highlight set objects. You also can move an object between sets by dragging its name to the set name to which you want to add it. Dragging one set name onto another set name combines all the objects from both sets under the second set name. Double-clicking a set name expands or contracts the set. If you right-click on the dialog box, you can access a pop-up menu of options including the ability to Rename the current selection and options to Cut, Copy and Paste objects between sets.

Locking selection sets

Another alternative to creating a selection set is to lock the current selection. If you've finally selected the exact objects you want to work with, you can disable any other selections using the Selection Lock Toggle button on the status bar. (It looks like a lock.) When this button is enabled, it is highlighted, and clicking objects in the viewports won't have any effect on the current selection. The keyboard shortcut toggle for this command is the spacebar.

Caution

In Photoshop and Illustrator, the spacebar is the keyboard shortcut to pan, but in 3ds Max it locks the current selection. If you accidentally lock the current selection, you can't select any other objects until the lock is removed.

Isolating the current selection

The Tools→Isolate Selection (Alt+Q) menu command hides all objects except for the selected object. This command is also available by clicking the Isolate Selection Toggle button located on the status bar at the bottom of the interface. The Isolate Selection Toggle button is highlighted when enabled. Clicking this button or selecting the End Isolate command exits isolation mode and displays all the objects again.

Caution

Although the Alt+Q keyboard shortcut isolates a selection, it will not end the isolation mode, you'll need to click on the Isolate Selection button or use the menu to leave isolation mode.

Isolate Selection mode is very convenient for working on a certain area. Figure 6.8 shows the Isolate Selection mode for just the glasses object.

FIGURE 6.8

Isolated Selection mode lets you focus on the details of the selected object.

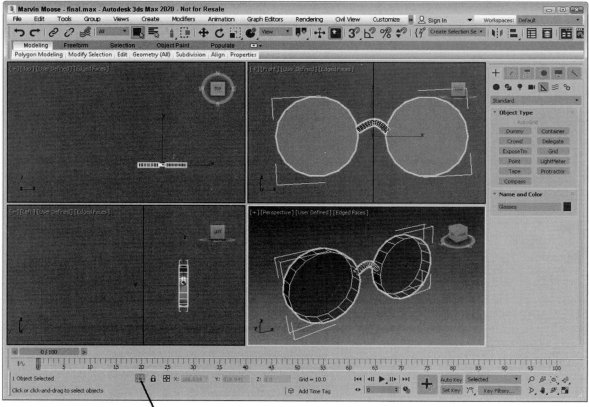

Isolate Selection button

Selecting objects in other interfaces

In addition to selecting objects in the viewports, you can use many of the other interfaces and dialog boxes to select objects. For example, the Material Editor includes a button that selects all objects in a scene with the same material applied.

 The Select by Material button opens the Select Object dialog box with all objects that use the selected material highlighted.

Another interface that you can use to select objects is the Schematic View, which is opened using the Graph Editors→New Schematic View menu command. It offers a hierarchical look at your scene and displays all links and relationships between objects. Each object in the Schematic View is displayed as a rectangular node.

To select an object in the viewport, find its rectangular representation in the Schematic View and simply click it. To select multiple objects in the Schematic View, you need to enable Sync Selection mode with the Select→Sync Selection command in the Schematic View menu and then drag an outline over all the rectangular nodes that you want to select.

The Schematic View also includes the Select by Name text field at the bottom of the interface for selecting an object by typing its name.

The Material Editor is covered in detail in Chapter 17, "Creating and Applying Standard Materials with the Slate Material Editor."

Setting Object Properties

After you select an object or multiple objects, you can view their object properties by choosing Edit→Object Properties. Alternatively, you can right-click the object and select Properties from the pop-up quad menu. Figure 6.9 shows the Object Properties dialog box. The settings located in this dialog box are different from the parameters found in the Command Panel. Parameters are used to change the resulting look and shape of the object, and object properties are used only to set how the object is displayed and rendered. This dialog box includes three panels—General, Advanced Lighting, and User Defined.

FIGURE 6.9

The Object Properties dialog box provides information about the selected object and sets how it is displayed and rendered.

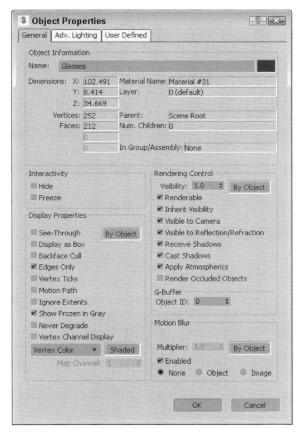

Viewing object information

For a single object, the General panel of the Object Properties dialog box lists details about the object in the Object Information section. These details include the object's name; color; bounding box measurements along the X-, Y-, and Z-axes; number of vertices and faces; the object's parent; the object's Material Name; the number of children attached to the object; the object's group name if it's part of a group; and the layer on which the object can be found. All this information (except for its name and color) is for display only and cannot be changed.

Tip

Knowing how many faces an object has is an important piece of information, especially when working with games where you need to know how much memory the object will take. The Object Properties dialog box is an easy place to find this information.

Note

The two fields under the Vertices and Faces are used only when the properties for a Shape are being displayed. These fields show the number of Shape Vertices and Shape Curves.

If the properties for multiple objects are to be displayed, the Object Properties dialog box places the text "Multiple Selected" in the Name field. The properties that are in common between all these objects are displayed. With multiple objects selected, you can set their display and rendering properties all at once.

The Object Properties dialog box can be displayed for all geometric objects and shapes, as well as for lights, cameras, helpers, and Space Warps. Not all properties are available for all objects.

The Hide and Freeze options are covered later in the next section.

Setting Display Properties

Display properties don't affect how an object is rendered, only how it is displayed in the viewports. In this section, along with the Rendering Control and Motion Blur sections, are three By Object/By Layer toggle buttons. If the By Object button is displayed, options can be set for the selected object, but if the By Layer option is enabled, all options become disabled and the object gets its display properties from the layer settings found in the Scene Explorer.

Note

You also can find and set the same Display Properties that are listed in the Object Properties dialog box in the Display Properties rollout of the Display panel in the Command Panel and in the Display Floater.

The See-Through option causes shaded objects to appear semi-transparent, which is helpful for seeing how it lines up with other objects. This option is similar to the Visibility setting in the Rendering Control section, except that it doesn't affect the rendered image. It is only for displaying objects in the viewports. This option really doesn't help in wireframe viewports. Figure 6.10 shows the duck toy model with spheres behind it with and without this option selected.

FIGURE 6.10

The See-Through display property can make objects transparent in the viewports.

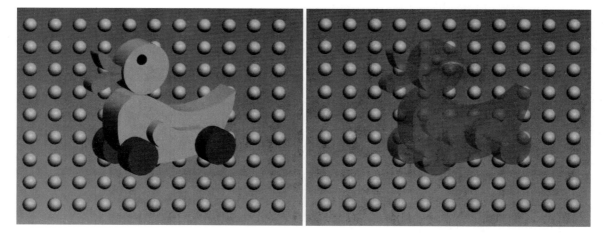

Many of these display properties can speed up or slow down the viewport refresh rates. For example, Bounding Box increases the viewport update rate dramatically for complex scenes, but at the expense of any detail. This setting can be useful to see how the objects generally fit in comparison to one another. This option also can be accessed from the Viewport Setting and Preference dialog box or from the Shading viewport label menu, but the Object Properties dialog box lets you set this option for a single object instead of for the entire viewport.

When the Backface Cull option is enabled, it causes the faces on the backside of the object to not be displayed. The Autodesk® 3ds Max® 2020 software considers the direction that each normal is pointing and doesn't display a face if its normal points away from the view. A *normal* is a vector that extends perpendicular to the face and is used to determine the orientation of individual faces. This option produces the same result as the Force 2-Sided option in the Render Setup dialog box, except that it can be applied to a single object and not the entire viewport. This display option works only in wireframe viewports.

The Edges Only option displays only the edges of each object when the viewport is set to Wireframe mode. When Edges Only is not selected, a dashed line indicates the junction of individual faces.

When the Vertex subobject mode is selected for an object, all vertices for the selected object appear as blue + signs. The Vertex Ticks option displays all object vertices in this same way without requiring the Vertex subobject mode. Figure 6.11 shows the duck toy mesh with this option enabled. The Motion Path option displays the animation path that the object follows. You also can make the motion path of the selected object appear without enabling the Motion Path option by selecting the Motion Paths button in the Motion panel.

FIGURE 6.11

The Vertex Ticks option displays all vertices as small, blue tick marks.

The Motion Path option displays any animated motions as a spline path.

To learn more about using animated motion paths, see Chapter 32, "Understanding Animation and Keyframes."

The Ignore Extents option causes an object to be ignored when you are using the Zoom Extents button in the Viewport Navigation controls. For example, if you have a camera or light positioned at a distance from the objects in the scene, then anytime you use the Zoom Extents All button, the center objects are so small that you cannot see them because the Zoom Extents needs to include the distance light. If you set the Ignore Extent option for the camera or light, the Zoom Extents All button zooms in on just the geometry objects.

When objects are frozen, they appear dark gray, but if the Show Frozen in Gray option is disabled, the object appears as it normally does in the viewport. The Never Degrade option causes the object to be removed from the Adaptive Degradation settings used to maintain a given frame rate to get the animation timing right.

The Vertex Channel Display option displays the colors of any object vertices that have been assigned colors. You can select to use Vertex Color, Vertex Illumination, Vertex Alpha, Map Channel Color, or Soft Selection Color. The Shaded button causes the meshes to be shaded by the vertex colors. If the Shaded button is disabled, the object is unshaded. You can assign vertex colors only to editable meshes, editable polys, and editable patches. If the Map Channel Color option is selected, you can specify the Map Channel.

For more information about vertex colors, check out Chapter 22, "Painting in the Viewport Canvas and Rendering Surface Maps."

Setting rendering controls

In the Object Properties dialog box, the Rendering Controls section includes options that affect how an object is rendered.

The Visibility spinner defines a value for how opaque (nontransparent) an object is. A value of 1 makes the object completely visible. A setting of 0.1 makes the object almost transparent. The Inherit Visibility option causes an object to adopt the same visibility setting as its parent.

Tip

The Visibility option also can be animated for making objects slowly disappear.

The Renderable option determines whether the object is rendered. If this option isn't selected, the rest of the options are disabled because they don't have any effect if the object isn't rendered. The Renderable option is useful if you have a complex object that takes a while to render. You can disable the renderability of the single object to quickly render the other objects in the scene.

You can use the Visible to Camera and Visible to Reflection/Refraction options to make objects invisible to the camera or to any reflections or refractions. This feature can be useful when you are test-rendering scene elements and raytraced objects.

Tip

If an object has the Visible to Camera option disabled and the Cast Shadows option enabled, the object isn't rendered, but its shadows are.

The Receive Shadows and Cast Shadows options control how shadows are rendered for the selected object. The Apply Atmospherics options enable or disable rendering atmospherics. Atmospheric effects can increase the rendering time by a factor of 10, in some cases.

Atmospheric and render effects are covered in Chapter 44, "Using Atmospheric and Render Effects."

The Render Occluded Objects option causes the rendering engine to render all objects that are hidden behind the selected object. The hidden or occluded objects can have glows or other effects applied to them that would show up if rendered.

You use the G-Buffer Object ID value to apply Render or Video Post effects to an object. By matching the Object Channel value to an effect ID, you can make an object receive an effect. A *g-buffer* is a temporary bit of memory used to process an image that isn't interrupted by transferring the data to the hard disk.

The Video Post interface is covered in Chapter 31, "Compositing with Render Elements and the Video Post Interface."

Enabling Motion Blur

You also can set Motion Blur from within the Object Properties dialog box. The Motion Blur effect causes objects that move fast (such as the Road Runner) to be blurred (which is useful in portraying speed). The render engine accomplishes this effect by rendering multiple copies of the object or image.

More information on these blur options is in Chapter 27, "Rendering a Scene and Enabling Quicksilver."

The Object Properties dialog box can set two different types of Motion Blur: Object and Image. Object motion blur affects only the object and is not affected by the camera movement. Image motion blur applies the effect to the entire image and is applied after rendering.

A third type of Motion Blur is called Scene Motion Blur and is available in the Video Post interface. See Chapter 31, "Compositing with Render Elements and the Video Post Interface," for information on using Scene Motion Blur.

You can turn the Enabled option on and off as an animation progresses, allowing you to motion blur select sections of your animation sequence. The Multiplier value is enabled only for the Image Motion Blur type. It is used to set the length of the blur effect. The higher the Multiplier value, the longer the blurring streaks. The Motion Blur settings found in the Object Properties dialog box can be overridden by the settings in the Render Scene dialog box.

Caution

If the Motion Blur option in the Object Properties dialog box is enabled but the Motion Blur option in the Renderer panel of the Render Setup dialog box is disabled, motion blur will not be included in the final rendered image.

Using the Advanced Lighting panel

The second panel in the Object Properties dialog box contain object settings for working with Advanced Lighting. Using the settings in the Advanced Lighting panel, you can exclude an object from any Advanced Lighting calculations, set an object to cast shadows and receive illumination, and set the number of refine iterations to complete.

Advanced Lighting is covered in Chapter 26, "Working with Advanced Lighting, Light Tracing, and Radiosity."

Using the User-Defined panel

The User-Defined panel contains a simple text window. In this window, you can type any sort of information. This information is saved with the scene and can be referred to as notes about an object.

Hiding and Freezing Objects

Hidden and frozen objects cannot be selected, and as such, they cannot be moved from their existing positions. This becomes convenient when you move objects around in the scene. If you have an object in a correct position, you can freeze it to prevent it from being moved accidentally, or you can hide it from the viewports completely. A key difference between these modes is that frozen objects are still rendered, but hidden objects are not.

You can hide and freeze objects in several ways. You can hide or freeze objects in a scene by selecting the Hide or Freeze options in the Object Properties dialog box. You also can hide and freeze objects using the Display Floater dialog box, which you access by choosing Tools→Display Floater. You also can hide and freeze objects using the quad menu commands.

Tip

Several keyboard shortcuts can be used to hide specific objects. These shortcuts are toggles, so one press makes the objects disappear and another press makes them reappear. Object types that can be hidden with these shortcuts include cameras (Shift+C), geometry (Shift+G), grids (G), helpers (Shift+H), lights (Shift+L), particle systems (Shift+P), shapes (Shift+S), and Space Warps (Shift+W).

The Hide option makes the selected object in the scene invisible, and the Freeze option turns the selected object dark gray (if the Show Frozen in Gray option in the Object Properties dialog box is enabled) and doesn't allow it to be transformed or selected. You cannot select hidden objects by clicking in the viewport.

Note

When you use the Zoom Extents button to resize the viewports around the current objects, hidden objects aren't included.

Using the Display Floater dialog box

The Display Floater dialog box includes two tabs: Hide/Freeze and Object Level. The Hide/Freeze tab splits the dialog box into two columns, one for Hide and one for Freeze. Both columns have similar buttons that let you hide or freeze Selected or Unselected objects, By Name or By Hit. The By Name button opens the Select Objects dialog box (which is labeled Hide or Freeze Objects). The By Hit option lets you click in one of the viewports to select an object to hide or freeze. Each column also has additional buttons to unhide or unfreeze All objects By Name or (in the case of Freeze) By Hit. You also can select an option to Hide Frozen Objects.

Note

Other places to find the same buttons found in the Display Floater are the Hide and Freeze rollouts of the Display panel of the Command Panel and in the right-click quad menu.

The Object Level panel of the Display Floater lets you hide objects by category such as All Lights or All Cameras. You also can view and change many of the Display Properties that are listed in the Object Properties dialog box.

Figure 6.12 shows the Hide/Freeze and Object Level panels of the Display Floater dialog box.

FIGURE 6.12

The Display Floater dialog box includes two panels: Hide/Freeze and Object Level.

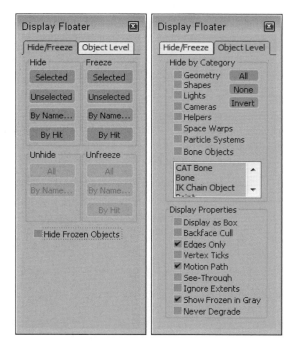

Using the Display panel

If you took many of the features of the Display Floater and the Object Properties dialog box and mixed them together, the result would be the Display panel. You access this panel by clicking the fifth icon from the left in the Command Panel (the icon that looks like a monitor screen).

The first rollout in the Display panel, shown in Figure 6.13, is the Display Color rollout. This rollout includes options for setting whether Wireframe and Shaded objects in the viewports are displayed using the Object Color or the Material Color.

FIGURE 6.13

The Display panel includes many of the same features as the Display Floater and the Object Properties dialog box.

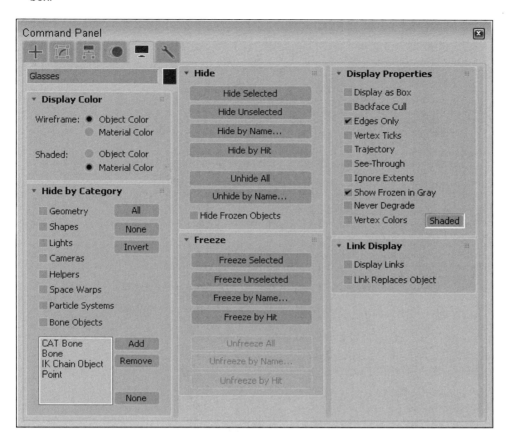

The panel also includes a Hide by Category rollout. Using this rollout, you can add new categories that will appear in the Object Level panel of the Display Floater. To add a new category, click the Add button of the Hide by Category rollout. The Add Display Filter list appears, as shown in Figure 6.14. From this list, you can choose specific object categories to add to the Hide by Category list.

FIGURE 6.14

From this dialog box, you can add new categories to the Hide by Category list.

The Display panel also includes Hide and Freeze rollouts that include the same buttons and features as the Hide/Freeze panel of the Display Floater. You also find a Display Properties rollout that is the same as the list found in the Display Floater's Object Level panel and the Object Properties dialog box.

The Link Display rollout at the bottom of the Display panel includes options for displaying links in the viewports. Links are displayed as lines that extend from the child to its parent object. Using the Link Replaces Object option, you can hide the objects in the viewport and see only the links.

Tutorial: Hidden toothbrushes

In this example, I've hidden several toothbrushes in the scene, and your task is to find them. To find the hidden objects, follow these steps:

1. Open the Toothbrushes.max scene file.

   This file appears to contain only a single toothbrush, but it really contains more. Can you find them? You can find this file in the Chap 06 directory in the downloaded content set.

2. Locate the hidden object in the scene by opening the Display Floater (choose Tools→Display Floater).

3. In the Display Floater, select the Hide/Freeze tab. In the Unhide section, click the By Name button.

   The Unhide Objects dialog box appears, which lists all the hidden objects in the scene.

4. Select the green toothbrush object from the list, and click the Unhide button.

   The Unhide Objects dialog box closes, and the hidden object become visible again.

Note

Notice that the Display Floater is still open. That's because it's modeless. You don't need to close it to keep working.

5. To see all the remaining objects, click the All button in the Unhide section of the Display Floater.

Figure 6.15 shows the finished scene with all toothbrushes visible.

FIGURE 6.15

Here are toothbrushes for the whole family; just remember which color is yours.

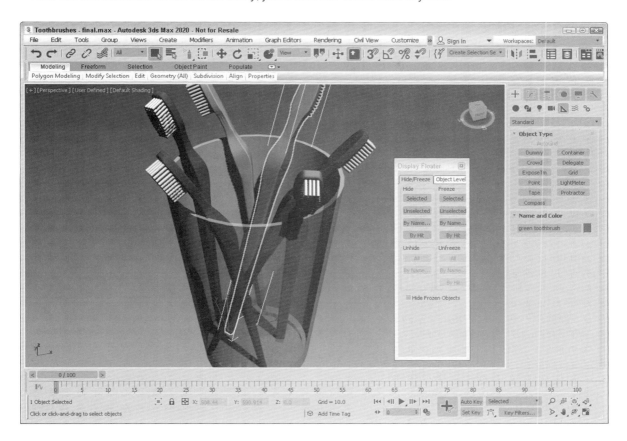

Using the Scene Explorer

The Scene Explorer is a one-stop shop for all scene objects and display properties. It displays all the objects in the scene in a hierarchical list or by layers, along with various display properties. It allows you to filter the display so you can see just what you want and customize the display so only those properties you want to see are visible. The Scene Explorer also lets you select, rename, hide, sort, freeze, link, and delete objects and change the object color and visibility.

A default Scene Explorer is open by default for the current workspace docked to the left side of the interface. You can undock the Scene Explorer or open a new Scene Explorer dialog box, shown in Figure 6.16, using the Tools→Scene Explorer menu command (Alt+Ctrl+O).

FIGURE 6.16

The Scene Explorer dialog box displays all scene objects and their display properties.

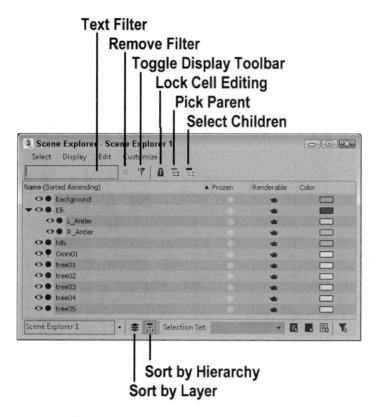

All scene objects in the Scene Explorer are listed in hierarchical order, with children objects indented under their parent objects. You can expand or contract children objects by clicking the arrow icons to the left of the parent object.

The Scene Explorer has two different sorting options--by Layer or by Hierarchy. If you've divided the scene objects into layers, then the Sort by Layer option would be best, but if there are no layer assignments, then the Sort by Hierarchy option will show all scene objects. You can switch between these modes using the buttons at the bottom of the interface.

Dividing the scene into layers is covered in Chapter 10, "Organizing Scenes with Layers, Containers, XRefs and the Schematic View."

Global versus local

The Scene Explorer window can be customized by changing its display settings, filtering specific object types and adding or reducing the displayed properties. There are also several variants already established that you can select from the Tools menu including Layer Explorer, Crease Explorer and within the All Global Explorers menu the following: Container Explorer, MassFX Explorer, Light Explorer, Missing Plugin Objects Explorer, Property Explorer and Revit Property Explorer.

Each of these variants show a limited set of global properties. For example, the Light Explorer shows only the lights in the scene and all the light properties. This makes it easy to access and change the properties for the scene lights without having to worry about selecting them in the scene. You can also select each of these global variants using the drop-down list at the lower left corner of the Scene Explorer window.

Each of the global views have a preset display, but you can convert any global view to a local view with the Make Active Explorer Local menu option in the lower left corner drop-down menu. Once a view is local, you can make changes to it and name it by typing a name in the text field in the lower left corner. Each named local view will be saved with the scene and can be quickly reopened using the Tools→Local Scene Explorers menu, and the Tools→Manage Local Explorers opens a simple dialog box where you can Load, Save, Delete, and Rename the saved views.

Note

Scene Explorer views are automatically saved and reloaded with the 3ds Max file.

Selecting objects in the Scene Explorer

If you click an object name in the Scene Explorer dialog box, the object row is highlighted. Holding down the Ctrl key lets you click to select multiple objects, or you can use the Shift key to select a range of adjacent objects. Selected objects also can be removed from the current selection with the Ctrl key held down.

The Select menu also includes options for selecting objects. The Select All (Ctrl+A), Select None (Ctrl+D), and Select Invert (Ctrl+I) menu commands work as expected, selecting all objects, deselecting all objects, and selecting the inverse of the current selection. You also can access these commands using the buttons at the bottom right corner of the dialog box.

The Select→Select Children (Ctrl+C) causes all children objects to automatically be selected when the parent is selected. The Select→Select Influences option selects all influence objects that are attached to the selected objects. An influence object is an object that controls or shapes another object. For example, when a sphere is constrained to follow an animation path, the path is an influence object to the sphere. Another example is a skin mesh being influenced by a character rig. The Select→Select Dependants option selects any objects that are dependent on the selected object, such as an instance and reference.

The Scene Explorer recognizes any defined Selection Sets and lets you select these sets from the drop-down list at the bottom of the interface.

Setting Scene Explorer display options

The Display toolbar includes several object type icons. This toolbar, shown in Figure 6.17, can be turned on and off using the Toggle Display Toolbar button on the toolbar. You can switch this toolbar to be horizontal using the Customize→Layout→Horizontal menu command. The shaded icons are selected and allowed to be viewed in the Scene Explorer. To filter out a specific object type, disable its icon, and then all objects of that type are no longer displayed in the list. These same commands are available in the Display→Object Types menu. This provides a way to quickly see only those objects you want.

FIGURE 6.17

The Display toolbar includes toggle buttons for Geometry, Shapes, Lights, Cameras, Helpers, Space Warps, Groups, Object XRefs, Bones, Containers, Frozen Objects, and Hidden Objects.

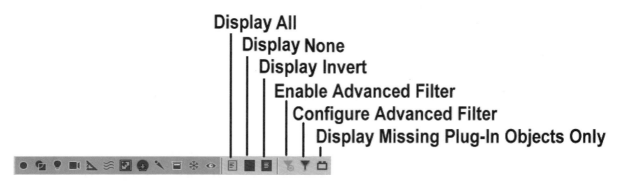

There are also buttons to turn all filter buttons on, off and invert. The Configure Advanced Filter button (which is the same as the Advanced Search dialog box covered below) opens a dialog box where you can customize the search.

The Display menu includes some additional commands for displaying children, influences, and dependents. You also have an option to Display in Track View. This option opens the Track View with the selected object's tracks visible.

The Track View interface is covered in more detail in Chapter 36, "Editing Animation Curves in the Track View."

Finding objects

You also can use the Text Field to search the hierarchy for a specific object by name. All objects that match the typed characters are selected. If you enable the Select→Find Case Sensitive option, uppercase characters are distinguished from lowercase characters.

If the Select→Find Using Wildcards option is selected, you can use wildcards to locate objects. Acceptable wildcards include an asterisk (*) for multiple characters in a row and a question mark (?) for single characters. For example, an entry of **hedra*** selects all objects beginning with "hedra," regardless of the ending, and **hedra?1** finds "hedra01" and "hedra11" but not "hedra02" or "hedra0001."

The Select→Find Using Regular Expressions option provides yet another way to search for specific objects. Regular expressions are commonly used in various scripting languages and require specific syntax in order to locate objects. Table 6.4 lists some common regular expression characters.

TABLE 6.4 Common Regular Expression Syntax

Character	Description	Example
[htk]	Used to define a group of search characters	Matches all objects beginning with the letters *h*, *t*, and *k*
eye|light|key	Used to separate words to search for	Matches all objects beginning with eye, light, or key
\w	Used to identify any letter or number, just like the ? wildcard	Matches any number or letter
\s	Used to identify any white space	Matches any space between words, no matter the length
\d	Used to identify any single-digit number	Matches any single-digit number, 0 through 9
[^geft]	Used to match all objects except for the	Matches all objects except for those that

	ones inside the brackets	begin with *g*, *e*, *f*, or *t*
t.*1	Used to match multiple letters between two specified characters	Matches all objects that begin with the letter *t* and end with the number 1

If regular expressions seem confusing, you also can search using the Advanced Search dialog box, shown in Figure 6.18. This dialog box is opened using the Select→Search menu command. In the Property field, you can search by Name, Type, Color, Faces, or any of the other available columns. In the Condition, the options for the Name property include Starts With, Does Not Start With, Contains String, Does Not Contain String, Regular Expression Matches, and Inverse Regular Expression Matches. The available Condition options will change if the Property is changed. Multiple criteria can be added to the search list.

FIGURE 6.18

The Advanced Search dialog box lets you select search criteria using drop-down lists.

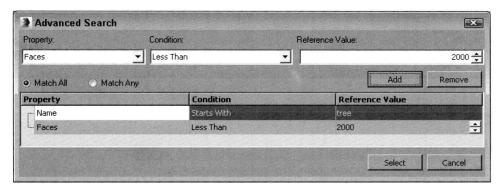

Editing in the Scene Explorer

You can edit exactly which properties are available using the Customize→Configure Columns menu, but the property for the visibility of objects is always available. You can change the visibility of any object by simply clicking on the eye icon to the left of its name. This will hide the object in the scene, but will not delete it from the scene list.

Any of the display properties listed in the Scene Explorer (such as Frozen and Renderable) can be changed by simply clicking on its icon to enable (or disable) the property. If multiple objects are selected when a property is enabled or disabled, the same property is enabled or disabled for all the selected objects at the same time.

Note

If the Lock Cell Editing button is enabled, none of the properties can be changed.

You also can rearrange the columns by dragging and dropping them to a new location. Selecting the Customize→Configure Columns menu opens the Configure Columns dialog box, shown in Figure 6.19.

This dialog box lists all the available remaining display property columns. To add one to the Scene Explorer, simply select it from the Configure Columns dialog box and drop it where you want it. You can remove any property column by simply dragging it away from its current location until an X appears.

The Configure Columns dialog box includes a large number of properties that can be added as columns to the Scene Explorer.

Tip

The width of each column can be altered by dragging on either side. To reset all column widths, right-click a column name and choose Best Fit (all columns) from the pop-up menu.

FIGURE 6.19

The Configure Columns dialog box holds all the display properties not currently available in the Scene Explorer.

If you click the column name, you can sort all the listed objects either in descending or ascending order. Click the column name once to sort in ascending order and again to sort in descending order. You also can right-click and select the sorting order from the pop-up menu. For example, if you click the Faces column, all the objects are sorted so the objects with the smallest number of faces are listed at the top of the interface and the objects with the most faces are listed at the bottom.

Using the Edit menu, you also can cut, copy, and paste selected objects, called nodes. Pasting objects opens the Clone Options dialog box. The Customize menu also includes options to hide various toolbars and a choice to lay out the window using horizontal or vertical icons.

Summary

Selecting objects enables you to work with them, and 3ds Max includes many different ways to select objects. In this chapter you learned how to work with selection sets, and hide and freeze objects. The Object Properties dialog box includes information about the selected object and settings for controlling how the object is displayed in the viewports and how it is rendered. The Scene Explorer dialog box also was covered. In this chapter, you've done the following:

* Learned how to use selection filters

* Selected objects with the Edit menu by Name, Layer, Color, and Region

* Selected multiple objects and used a named selection set to find the set easily

* Selected objects using other interfaces

* Accessed the Object Properties dialog box

* Learned valuable information about the selected object

* Set the Display and Rendering settings for an object

* Discovered Motion Blur options

* Learned how to hide and freeze objects
* Used the Scene Explorer dialog box

Now that you've learned how to select objects, you're ready to move them about using the transform tools, which are covered in the next chapter.

Chapter 7

Transforming Objects, Pivoting, Aligning, and Snapping

IN THIS CHAPTER

Transforming objects

Controlling transformations with the Transform Gizmos

Using the Transform Type-Ins and the Transform Managers

Working with pivot points and axis constraints

Aligning objects with the align tools

Using grids and snapping objects to common points

Although a *transformation* sounds like something that would happen during the climax of a superhero film, transformation is simply the process of repositioning or changing an object's position, rotation, or scale. So moving an object from here to there is a transformation. Superman would be so envious.

The Autodesk® 3ds Max® 2020 software includes several tools to help in the transformation of objects, including the Transform Gizmos, the Transform Type-In dialog box, and the Transform Managers.

This chapter covers each of these tools and several others that make transformations more automatic, such as the alignment, grid, and snap features.

Translating, Rotating, and Scaling Objects

So you have an object created, and it's just sitting there—sitting and waiting. Waiting for what? Waiting to be transformed. To be moved a little to the left or rotated around to show its good side or scaled down a little smaller. These actions are called transformations because they transform the object to a different state. Transformations are different from modifications. Modifications change the object's geometry, but transformations do not affect the object's geometry at all.

Using the transform buttons

The four transform buttons located on the main toolbar are Select and Move, Select and Rotate, Select and Uniform Scale, and Select and Place, as shown in Table 7.1. Using these buttons, you can select objects and transform them by dragging in one of the viewports with the mouse. You can access these buttons using three of the big four keyboard shortcuts: Q for Select Objects, W for Select and Move, E for Select and Rotate, and R for Select and Uniform Scale.

TABLE 7.1 Transform Buttons

Toolbar Button	Name	Description
✛	Select and Move (W)	Enters move mode where clicking and dragging an object moves it

↻	Select and Rotate (E)	Enters rotate mode where clicking and dragging an object rotates it
⬜ ⬜ ⬜	Select and Uniform Scale (R), Select and Non-Uniform Scale, Select and Squash	Enters scale mode where clicking and dragging an object scales it
⬤ ⬤	Select and Place, Select and Rotate	Moves or rotates across the surface of all scene objects mode

The three different types of transformations are translation (which is a fancy word for moving objects), rotation, and scaling.

Translating objects

The first transformation type is *translation,* or moving objects. This is identified in the various transform interfaces as the object's Position. You can move objects along any of the three axes or within the three planes. You can move objects to an absolute coordinate location or move them to a certain offset distance from their current location.

 To move objects, click the Select and Move button on the main toolbar (or press the W key), select the object to move, and drag the object in the viewport to the desired location. Translations are measured in the defined system units for the scene, which may be inches, centimeters, meters, and so on.

Rotating objects

 Rotation is the process of spinning the object about its Transform Center point. To rotate objects, click the Select and Rotate button on the main toolbar (or press the E key), select an object to rotate, and drag it in a viewport. Rotations are measured in degrees, where 360 degrees is a full rotation.

Scaling objects

 Scaling increases or decreases the overall size of an object. Most scaling operations are uniform, or equal in all directions. All Scaling is done about the Transform Center point.

To scale objects uniformly, click the Select and Uniform Scale button on the main toolbar (or press the R key), select an object to scale, and drag it in a viewport. Scalings are measured as a percentage of the original. For example, a cube scaled to a value of 200 percent is twice as big as the original.

Non-uniform scaling

 The Select and Scale button includes two additional flyout buttons for scaling objects non-uniformly, allowing objects to be scaled unequally in different dimensions. The two additional tools are Select and Non-Uniform Scale, and Select and Squash, shown in Table 7.1. Resizing a basketball with the Select and Non-Uniform Scale tool could result in a ball that is oblong and taller than it is wide. Scaling is done about whatever axes have been constrained (or limited) using the Axis Constraint buttons on the Axis Constraints toolbar.

Squashing objects

 The Squash option is a specialized type of non-uniform scaling. This scaling causes the constrained axis to be scaled at the same time that the opposite axes are scaled in the opposite direction. For example, if you push down on the top of a basketball by scaling the Z-axis, the sides, or the X- and Y-axes, it bulges outward. This simulates the actual results of such materials as rubber and plastic.

Tip

You can cycle through the different Scaling tools by repeatedly pressing the R key.

Figure 7.1 shows a basketball that has been scaled using uniform scaling, non-uniform scaling, and squash modes.

FIGURE 7.1

These basketballs have been scaled using uniform, non-uniform, and squash modes.

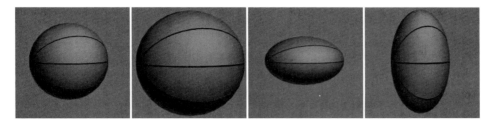

Note

It is also important to be aware of the order of things. Transformations typically happen after all object properties and modifiers are applied. So if you scale an object, it happens after the deforming modifier is applied.

Using the Select and Place tool

The first three transformation buttons are fairly common, but the Select and Place tool is a unique beast. Using this tool lets you move the current selection, but it is automatically constrained to move across the surface of the other scene objects so that the object's pivot is on the surface of the other objects. This makes it easy to place objects relative to each other.

Right clicking on the Select and Place tool opens the Placement Settings dialog box, shown in Figure 7.2. Using these settings, you can switch to Rotate mode which leaves the object in its current place and changes its orientation as you drag with the mouse. You can also use the Use Base as Pivot option to have the tool ignore the pivot and use the object's base to align to the other object's surfaces. The Pillow Mode option positions object so there are no intersections between them. The Autoparent button automatically links the moved object to the object that it is positioned over. You can also change the Object Up Axis as you move the object.

Using the Select and Rotate tool

The Select and Rotate tool lets you spin any object that you drag over about its y-axis regardless of its orientation. This lets you quickly realign objects without changing their placement. This tool is available as a flyout under the Select and Place tool.

FIGURE 7.2

The Placement Settings dialog box lets you change how the Select and Place tool works.

Working with the Transformation Tools

To help you in your transformations, you can use several tools to transform. These tools include the Transform Gizmos, the Transform Type-In dialog box (F12), Status Bar Transform Type-In fields, and the Transform Managers.

Working with the Transform Gizmos

The Transform Gizmos appear at the center of the selected object (actually, at the object's pivot point) when you click one of the transform buttons. The type of gizmo that appears depends on the transformation mode that is selected. You can choose from three different gizmos, one for each transformation type. Each gizmo includes three color-coded arrows, circles, or lines representing the X-, Y-, and Z-axes. The X-axis is colored red, the Y-axis is colored green, and the Z-axis is colored blue. Figure 7.3 shows the gizmos for each of the transformation types—move, rotate, and scale.

FIGURE 7.3

The Transform Gizmos let you constrain a transformation to a single axis or a plane.

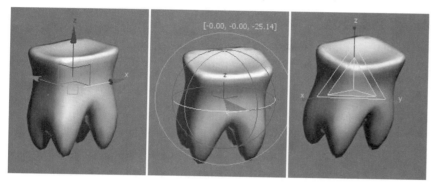

If the Transform Gizmo is not visible, you can enable it by choosing Views→Show Transform Gizmo. You can use the – (minus) and = (equal) keys to decrease or increase the gizmo's size.

Using the interactive gizmos

Moving the cursor over the top of one of the Transform Gizmos' axes in the active viewport selects the axis, which changes to yellow. Dragging the selected axis restricts the transformation to that axis only. For example, selecting the red X-axis on the Move Gizmo and dragging moves the selected object along only the X-axis.

Note

The transformation gizmos provide an alternate (and visual) method for constraining transformations along an axis or plane. This reduces the need for the Axis Constraint buttons, which are found on a separate floating toolbar. Learning to use these gizmos is well worth the time.

The Move Gizmo

In addition to the arrows for each axis, in each corner of the Move Gizmo are two perpendicular lines for each plane. These lines let you transform along two axes simultaneously. The colors of these lines match the various colors used for the axes. For example, in the Perspective view, dragging on a red and blue corner would constrain the movement to the XZ plane. Moving the cursor over the top of one of these lines highlights it. At the center of the Move Gizmo is a Center Box that marks the pivot point's origin.

Caution

If you find that the Move Gizmo isn't working as expected, check to make sure none of the Snap toggles are enabled.

The Rotate Gizmo

The Rotate Gizmo surrounds the selected object in a sphere that is made up of colored lines, one for each axis, and each circles the surrounding sphere. As you select an axis and drag, an arc is highlighted that shows the distance of the rotation along that axis, and the offset value is displayed in text above the object. Clicking within the sphere away from the axes lets you rotate the selected object in all directions. Dragging on the outer gray circle causes the selected object to spin perpendicular to the viewport.

The Scale Gizmo

The Scale Gizmo consists of two triangles and a line for each axis. Selecting and dragging the center triangle uniformly scales the entire object. Selecting a slice of the outer triangle scales the object along the adjacent two axes, and dragging on the axis lines scales the object in a non-uniform manner along a single axis.

Tip

To keep the various gizmo colors straight, simply remember that RGB = XYZ.

Additional settings that control how the transformation gizmos work are located in the Gizmo panel of the Preference Settings dialog box, opened using the Customize menu. These settings control the size and thresholds of the various gizmos. If you find that the gizmos are cumbersome, try tweaking these settings.

The preferences for the transformation gizmos are covered in Chapter 4, "Setting Preferences."

Using the Transform Toolbox

The Transform Toolbox, shown in Figure 7.4, is a pop-up panel that offers quick access to the most common transformation operations. You can open this panel using the Edit→Transform Toolbox menu command. The panel can be docked to the side of the interface by dragging it near the window border.

FIGURE 7.4

The Transform Toolbox provides quick access to the most common transformation operations.

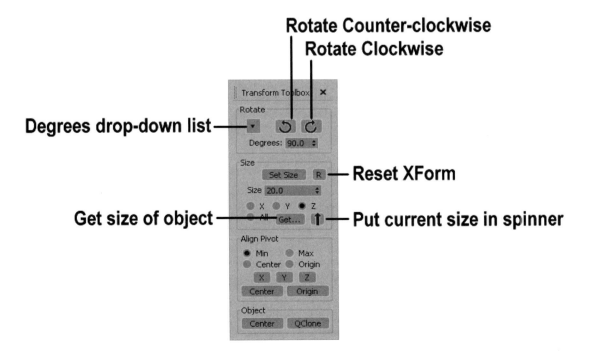

The Transform Toolbox is divided into four sections—Rotate, Size, Align Pivot, and Object. The Rotate section includes buttons for rotating the current selection by a set number of degrees in a clockwise or counterclockwise direction based on the current view. The drop-down list includes rotation values ranging from 1, 5, and 10 up to 180 and 240.

The Size section includes controls for scaling objects. The Set Size button scales the current object to the designated Size value along the specified axis or uniformly if the All option is selected. The R button resets the object transform by automatically applying the XForm modifier and then collapsing the stack to its base object. This sets the scaling values back to 100 percent for all axes. The Get button opens a small pop-up panel that lists the scale values for each of the axes, and the Put Current Size In Spinner button places the scale value for the selected object in the Size field for the specified axis.

The Align Pivot section changes the location of the selected object's pivot without your having to open the Hierarchy panel. Using the Min, Max, Center, and Origin options, you can move the pivot's origin for the X, Y, or Z axes, or you can use the Center and Origin buttons to move it for all three axes. Center moves the pivot to the object's center, and Origin moves the pivot to the center of the current scene.

The Object section includes only two buttons. The Center button moves the entire object to the world's origin. The QClone button, which stands for Quick Clone, creates a duplicate object and moves it to the side of the original object.

More information on the Quick Clone feature is available in Chapter 8, "Cloning Objects and Creating Object Arrays."

Using the Transform Type-In dialog box

The Transform Type-In dialog box (F12) lets you input precise values for moving, rotating, and scaling objects. This command provides more exact control over the placement of objects than dragging with the mouse.

The Transform Type-In dialog box allows you to enter numerical coordinates or offsets that can be used for precise transformations. Open this dialog box by choosing Edit→Transform Type-In or by pressing the F12 key.

Tip

Right-clicking any of the transform buttons opens the Transform Type-In dialog box for the transform button that is clicked.

The Transform Type-In dialog box is modeless and allows you to select new objects as needed or to switch between the various transforms. When the dialog box appears, it displays the coordinate values for the pivot point of the current selection if the Move tool is selected, rotation values in degrees if the Rotate tool is selected, or Scale percentages in the Absolute: World column.

Within the Transform Type-In dialog box are two columns. The first column displays the current Absolute coordinates. Updating these coordinates transforms the selected object in the viewport. The second column displays the Offset values. These values are all set to 0.0 when the dialog box is first opened, but changing these values transforms the object along the designated axis by the entered value. Figure 7.5 shows the Transform Type-In dialog box for the Move Transform.

Note

The name of this dialog box changes depending on the type of transformation taking place and the coordinate system. If the Select and Move button is selected along with the world coordinate system, the Transform Type-In dialog box is labeled Move Transform Type-In, and the column titles indicate the coordinate system.

FIGURE 7.5

The Transform Type-In dialog box displays the current Absolute coordinates and Offset values.

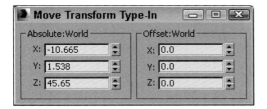

Using the Status Bar Type-In fields

The Status Bar includes three fields labeled X, Y, and Z for displaying transformation coordinates. When you move, rotate, or scale an object, the X, Y, and Z offset values appear in these fields. The values depend on the type of transformation taking place. Translation shows the unit distances, rotation displays the angle in degrees, and scaling shows a percentage value of the original size.

When you click the Select Object button, these fields show the absolute position of the cursor in world coordinates based on the active viewport.

You also can use these fields to enter values, as with the Transform Type-In dialog box. The type of transform depends on which transform button you select. The values that you enter can be either absolute coordinates or offset values, depending on the setting of the Transform Type-In toggle button that appears to the left of the transform fields. This toggle button lets you switch between Absolute and Offset modes, as shown in Table 7.2.

Tip

If you right-click any of these fields, a pop-up menu appears where you can cut, copy, or paste the current value.

TABLE 7.2 Absolute/Offset Buttons

Button	Description
⊞	Absolute mode
☝	Offset mode

Understanding the Transform Managers

To keep track of the position of every object in a scene, 3ds Max internally records the position of the object's vertices in reference to a Universal Coordinate System (UCS). This coordinate system defines the vertex position using the X, Y, and Z coordinates from the scene's origin.

However, even though 3ds Max uses the UCS to internally keep track of all the points, this isn't always the easiest way to reference the position of an object. Imagine a train with several cars. For each individual train car, it is often easier to describe its position as an offset from the car in front of it.

The Transform Managers are three types of controls that help you define the system about which objects are transformed. These controls, found on the main toolbar and on the Axis Constraints toolbar, directly affect your transformations. They include the following:

* **Reference Coordinate System:** This defines the coordinate system about which the transformations take place.

* **Transform Center settings:** The Use Pivot Point Center, Use Selection Center, and Use Transform Coordinate Center settings specify the center about which the transformations take place.

* **Axis Constraint settings:** These allow the transformation to happen using only one axis or plane. These buttons are on the Axis Constraints toolbar.

Note

The Axis Constraints are available on the Axis Constraints floating toolbar, but the transform gizmos make this toolbar mostly redundant.

Understanding reference coordinate systems

3ds Max supports several reference coordinate systems based on the UCS, and knowing which reference coordinate system you are working with as you transform an object is important. Using the wrong reference coordinate system can produce unexpected transformations.

Within the viewports, the UCS coordinates are displayed as a set of axes in the lower-left corner of the viewport, and the Transform Gizmo is oriented with respect to the reference coordinate system.

To understand the concept of reference coordinate systems, imagine that you're visiting the Grand Canyon and standing precariously at the edge of a lookout. To nervous onlookers calling the park rangers, the description of your position varies from viewpoint to viewpoint. A person standing by you would say you are next to him. A person on the other side of the canyon would say that you're across from her. A person at the floor of the canyon would say you're above him. And a person in an airplane would describe you as being on the east side of the canyon. Each person has a different viewpoint of you (the object), even though you have not moved.

3ds Max recognizes the following reference coordinate systems:

* **View Coordinate System:** A reference coordinate system based on the viewports; X points right, Y points up, and Z points out of the screen (toward you). The views are fixed, making this perhaps the most intuitive coordinate system to work with.

* **Screen Coordinate System:** Identical to the View Coordinate System, except the active viewport determines the coordinate system axes, whereas the inactive viewports show the axes as defined by the active viewport.

* **World Coordinate System:** Specifies X pointing to the right, Z pointing up, and Y pointing into the screen (away from you). The coordinate axes remain fixed regardless of any transformations applied to an object. For 3ds Max, this system matches the UCS.

* **Parent Coordinate System:** Uses the reference coordinate system applied to a linked object's parent and maintains consistency between hierarchical transformations. If an object doesn't have a parent, the world is its parent, and the system works the same as the World Coordinate System.

* **Local Coordinate System:** Sets the coordinate system based on the selected object. The axes are located at the pivot point for the object. You can reorient and move the pivot point using the Pivot button in the Hierarchy panel.

* **Gimbal Coordinate System:** Provides interactive feedback for objects using the Euler XYZ controller. If the object doesn't use the Euler XYZ controller, this coordinate system works just like the World Coordinate System.

* **Grid Coordinate System:** Uses the coordinate system for the active grid.

* **Working Coordinate System:** Lets you transform the selected object about the scene's Working Pivot, as defined in the Hierarchy panel.

* **Local Aligned Coordinate System:** When moving sub-objects, all axes for the sub-object are aligned to the selected object.

* **Pick Coordinate System:** Lets you select an object about which to transform. The Coordinate System list keeps the last four picked objects as coordinate system options.

Tip

All transforms occur relative to the current reference coordinate system as selected in the Referenced Coordinate System drop-down list, found on the main toolbar.

Each of the three basic transforms can have a different coordinate system specified, or you can set it to change uniformly when a new coordinate system is selected. To do this, open the General panel in the Preference Settings dialog box, and select the Constant option in the Reference Coordinate System section.

Using a transform center

All transforms are done about a center point. When transforming an object, you must understand what the object's current center point is, as well as the coordinate system in which you're working.

The Transform Center flyout consists of three buttons: Use Pivot Point Center, Use Selection Center, and Use Transform Coordinate Center, which are shown in Table 7.3. Each of these buttons alters how the transformations are done. The origin of the Transform Gizmo is always positioned at the center point specified by these buttons.

TABLE 7.3 Transform Center Buttons

Button	Description
	Use Pivot Point Center
	Use Selection Center
	Use Transform Coordinate Center

Pivot Point Center

Pivot points are typically set to the center of spherical objects and at the base of box-shaped and cylinder-shaped objects when the object is first created, but they can be relocated anywhere within the scene, including outside the object. Relocating the pivot point allows you to change the point about which objects are rotated. For example, if you have a car model that you want to position along an incline, moving the pivot point to the bottom of one of the tires allows you to easily line up the car with the incline.

If you select the Use Pivot Point Center button, the Select and Rotate tool rotates about the pivot point for the selected object, which can be located anywhere in the scene.

Note
Pivot points are discussed in detail in the next section.

Selection Center

The Use Selection Center button sets the transform center to the center of the selected object or objects regardless of the individual object's pivot point. If multiple objects are selected, the center is computed to be in the middle of a bounding box that surrounds all the objects.

Transform Coordinate Center

The Use Transform Coordinate Center button uses the center of the Local Coordinate System. If the View Coordinate System is selected, all objects are transformed about the center of the viewport grid. If an object is selected as the coordinate system using the Pick option, all transformations are transformed about that object's center.

When you select the Local Coordinate System, the Use Transform Coordinate Center button is ignored, and objects are transformed about their local axes. If you select multiple objects, they all transform individually about their local axes. Grouped objects transform about the group axes.

For example, the default pivot point for a cylinder object is in the middle of the cylinder's base, so if the Transform Center is set to the Use Pivot Point option, the cylinder rotates about its base pivot point. If the Use Selection Center option is selected, the cylinder rotates about its center point. If the Use Transform Coordinate Center option is selected for the View Coordinate System, the cylinder is rotated about the grid origin.

Figure 7.6 shows a simple cylinder object in the Left viewport using the different transform center modes. The left image shows the Pivot Point Center mode, the middle image shows the Selection Center mode with both objects selected, and the right image shows the Transform Coordinate Center mode. For each mode, notice that the Rotate Gizmo is located at different locations.

FIGURE 7.6

The Rotate Gizmo is located in different places, depending on the selected Transform Center mode.

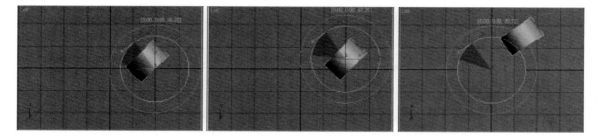

Selecting Axis Constraints

Three-dimensional space consists of three basic directions defined by three axes: X, Y, and Z. If you were to stand on each axis and look at a scene, you would see three separate planes: the XY plane, the YZ plane, and the ZX plane. These planes show only two dimensions at a time and restrict any transformations to the two axes. These planes are visible from the Top, Left, and Front viewports.

Tip

If you use the transformation gizmos, these constraints are used automatically without having to mess with the Axis Constraints toolbar.

By default, the Top, Left, and Front viewports show only a single plane and thereby restrict transformations to that single plane. The Top view constrains movement to the XY plane, the Left or Right side view constrains movement to the YZ plane, and the Front view constrains movement to the ZX plane. This setting is adequate for most modeling purposes, but sometimes you might need to limit the transformations in all the viewports to a single plane. In 3ds Max, you can restrict movement to specific transform axes using the Constrain Axis buttons in the Axis Constraints toolbar. You access this toolbar, shown in Figure 7.7, by right-clicking the main toolbar (away from the buttons) and selecting Axis Constraints options from the pop-up menu.

FIGURE 7.7

The Axis Constraints toolbar includes buttons for restricting transformations to a single axis or plane.

The first four buttons on this toolbar are Constrain axes buttons: Constrain to X (F5); Constrain to Y (F6); Constrain to Z (F7); and the flyout buttons, Constrain to XY, YZ, and ZX Plane (F8). The last button is the

Snaps Use Axis Constraints Toggle button. The effect of selecting one of the Constrain axes buttons is based on the selected coordinate system. For example, if you click the Constrain to X button and the reference coordinate system is set to View, the object always transforms to the right because, in the View Coordinate System, the X-axis is always to the right. If you click the Constrain to X button and the coordinate system is set to Local, the axes are attached to the object, so transformations along the X-axis are consistent in all viewports. (With this setting, the object does not move in the Left view because it shows only the YZ plane.)

Caution

If the axis constraints don't seem to be working, check the Preference Settings dialog box, and look at the General panel to make sure that the Reference Coordinate System option is set to Constant.

Additionally, you can restrict movement to a single plane with the Constrain to Plane flyouts consisting of Constrain to XY, Constrain to YZ, and Constrain to ZX. (Use the F8 key to cycle quickly through the various planes.)

Note

If the Transform Gizmo is enabled, the axis or plane that is selected in the Axis Constraints toolbar initially is displayed in yellow. If you transform an object using a Transform Gizmo, the respective Axis Constraints toolbar button is selected after you complete the transform.

Locking axes transformations

To lock an object's transformation axes on a more permanent basis, go to the Command Panel, and select the Hierarchy tab. Click the Link Info button to open the Locks rollout, shown in Figure 7.8. The rollout displays each axis for the three types of transformations: Move, Rotate, and Scale. Make sure that the object is selected and then click the transformation axes that you want to lock. Be aware that if all Move axes are selected, you won't be able to move the object until you deselect the axes.

Note

Another option is to use the Display floater to freeze the object.

Locking axes is helpful if you want to prevent accidental scaling of an object or restrict a vehicle's movement to a plane that makes up a road.

FIGURE 7.8

The Locks rollout can prevent any transforms along an axis.

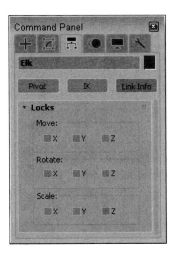

The Locks rollout displays unselected X, Y, and Z check boxes for the Move, Rotate, and Scale transformations. By selecting the check boxes, you limit the axes about which the object can be transformed.

191

For example, if you check the X and Y boxes under the Move transformation, the object can move only in the Z direction of the Local Coordinate System.

Note

These locks work regardless of the axis constraint settings.

Tutorial: Landing a spaceship in port

Transformations are the most basic object manipulations that you will do and probably the most common. This tutorial includes a spaceship object and a spaceport. The goal is to position the spaceship on the landing pad of the spaceport, but it is too big and in the wrong spot. With a few clever transformations, you'll be set.

To transform a spaceship to land in a spaceport, follow these steps:

1. Open the Transforming spaceship.max file from the Chap 07 directory in the downloaded content set.

2. To prevent any extraneous movements of the spaceport, select the spaceport by clicking it. Open the Hierarchy panel, and click the Link Info button. Then, in the Locks rollout, select all nine boxes to restrict all transformations so that the spaceport won't be accidentally moved.

3. To position the spaceship over the landing platform, select the Spaceship object, and click the Select and Move button in the main toolbar (or press the W key). The Move Gizmo appears in the center of the Spaceship object. Make sure that the Reference Coordinate System is set to View and that the Use Selection Center option is enabled. Right-click the Left viewport to make it active, select the red X-axis line of the gizmo, and drag to the right until the center of the spaceship is over the landing pad. Zoom out of the view if you can't see the spaceship fully.

4. Right-click the Front viewport, and drag the red X-axis gizmo line to the left to line up the spaceship with the center of the landing pad.

5. Click the Select and Uniform Scale button (or press the R key until it appears). Place the cursor over the center gizmo triangle, and drag downward until the spaceship fits within the landing pad.

6. Click the Select and Move button again (or press the W key), and drag the green Y-axis gizmo line downward in the Front viewport to move the spaceship toward the landing pad.

7. Click the Select and Rotate button (or press the E key). Right-click the Top viewport, and drag the blue Z-axis gizmo circle downward to rotate the spaceship clockwise so that its front end points towards the viewer.

Figure 7.9 shows the spaceship correctly positioned.

FIGURE 7.9

Transformation buttons and the Transform Gizmos were used to position this spaceship.

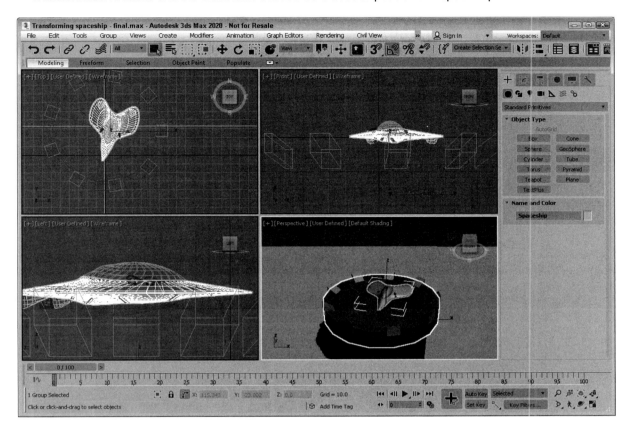

Using Pivot Points

An object's pivot point is the center about which the object is rotated and scaled and about which most modifiers are applied. Pivot points are created by default when an object is created and are usually created at the center or base of an object. You can move and orient a pivot point in any direction, but repositioning the pivot cannot be animated. Pivot points exist for all objects, whether or not they are part of a hierarchy.

Caution

Try to set your pivot points before animating any objects in your scene. If you relocate the pivot point after animation keys have been placed, all transformations are modified to use the new pivot point.

Positioning pivot points

To move and orient a pivot point, open the Hierarchy panel in the Command Panel, and click the Pivot button. At the top of the Adjust Pivot rollout are three buttons; each button represents a different mode. The Affect Pivot Only mode makes the transformation buttons affect only the pivot point of the current selection. The object does not move. The Affect Object Only mode causes the object to be transformed but not the pivot point. The Affect Hierarchy Only mode allows an object's links to be moved.

The pivot point is easily identified as the place where the Transform Gizmo is located when the object is selected, as shown in Figure 7.10.

FIGURE 7.10

The Transform Gizmo is located at the object's pivot point.

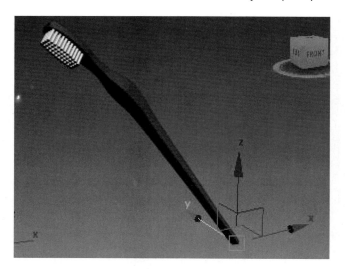

Note
Using the Scale transformation while one of these modes is selected alters the selected object but has no effect on the pivot point or the link.

Aligning pivot points

Below the mode buttons are three more buttons that are used to align the pivot points. These buttons are active only when a mode is selected. The buttons are Center to Object/Pivot, Align to Object/Pivot, and Align to World. The first two buttons switch between Object and Pivot, depending on the mode selected. You may select only one mode at a time. The button turns light blue when selected.

The Center to Object button moves the pivot point so it is aligned with the object center, and the Center to Pivot button moves the object so it is centered on its own pivot point. The Align to Object/Pivot button rotates the object or pivot point until the object's Local Coordinate System and the pivot point are aligned. The Align to World button rotates either the object or the pivot to the World Coordinate System. For example, if the Affect Object Only mode is selected and the object is separated from the pivot point, clicking the Center to Pivot button moves the object so its center is on the pivot point.

Under these three alignment buttons is another button labeled Reset Pivot, which you use to reset the pivot point to its original location.

Using the Working Pivot

Below the Adjust Pivot rollout is the Working Pivot rollout. Working Pivots are handy if you want to position an object using a temporary pivot without having to change the default object pivot. To position a Working Pivot, click the Edit Working Pivot button. This enters a mode just like the Affect Pivot Only button, described previously, except it works with the working pivot.

After the Working Pivot is in place, you can select to use it instead of the object pivot by clicking the Use Working Pivot button. The Working Pivot stays active until you disable it in the Hierarchy panel. The Working Pivot works for all objects in the scene. When the Working Pivot is active, reminder text "USE WP" appears in all the viewports under the viewport name; when the Edit Working Pivot mode is enabled, this text reads, "EDIT WP."

Tip

You can quickly enable the Working Pivot by selecting the Working option from the Reference Coordinate System drop-down list in the main toolbar.

The Working Pivot rollout includes several buttons to help position the Working Pivot. The Align To View button reorients the Working Pivot to the current view. The Reset button moves the Working Pivot to the object pivot location of the selected object or to the view center if no object is selected.

Note

There is only one working pivot that you can use, but it can be used by any selected object.

The View button enters a mode identified by "PLACE WP VIEW" in the viewports that lets you place the Working Pivot anywhere in the current viewport by simply clicking where it should be. This is great for eyeballing the Working Pivots location. The Surface button (identified by the PLACE WP SURFACE text in the viewport) enters a mode where you can position the Working Pivot on the surface of an object by interactively dragging the cursor over the object surface. The cursor automatically reorients itself to be aligned with the surface normal. If the Align to View option is selected, the Working Pivot is automatically aligned to the current view.

Transform adjustments

The Hierarchy panel of the Command Panel includes another useful rollout labeled Adjust Transform. This rollout includes another mode that you can use with hierarchies of objects. Clicking the Don't Affect Children button places you in a mode where any transformations of a linked hierarchy don't affect the children. Typically, transformations are applied to all linked children of a hierarchy, but this mode disables that.

The Adjust Transform rollout also includes two buttons that allow you to reset the Local Coordinate System and scale percentage. These buttons set the current orientation of an object as the World coordinate or as the 100 percent standard. For example, if you select an object, move it 30 units to the left, and scale it to 200 percent; these values are displayed in the coordinate fields on the status bar. Clicking the Reset Transform and Reset Scale buttons resets these values to 0 and 100 percent.

You use the Reset Scale button to reset the scale values for an object that has been scaled using non-uniform scaling. Non-uniform scaling can cause problems for child objects that inherit this type of scaling, such as shortening the links. The Reset Scale button can remedy these problems by resetting the parent's scaling values. When the scale is reset, you won't see a visible change to the object, but if you open the Scale Transform Type-In dialog box while the scale is being reset, you see the absolute local values being set back to 100 each.

Tip

If you are using an object that has been non-uniformly scaled, using Reset Scale before the item is linked saves you some headaches if you plan on using modifiers.

Using the Reset XForm utility

You also can reset transform values using the Reset XForm utility. To use this utility, open the Utilities panel, and click the Reset XForm button, which is one of the default buttons. The benefit of this utility is that you can reset the transform values for multiple objects simultaneously. This happens by applying the XForm modifier to the objects. The rollout for this utility includes only a single button labeled Reset Selected.

Tutorial: A bee buzzing about a flower

By adjusting an object's pivot point, you can control how the object is transformed about the scene. In this example, you position the pivot point for the bee's wings and then reposition the pivot point for the entire bee so it can rotate about the flower object.

To control how a bee rotates about a flower, follow these steps:

1. Open the Buzzing bee.max file from the Chap 07 directory in the downloaded content set.

    This file includes a bee created from primitives and a flower model.

2. Click the bee's body object to select it, and press Z to zoom in on it.

3. Select the right wing, and click the Select and Rotate button on the main toolbar. Notice how the gizmo axes are centered on the wing. Then open the Hierarchy panel, and click the Affect Pivot Only button. Select the Local coordinate system from the list in the main toolbar. This orients the Transform Gizmo to match the pivot's orientation. Drag the wing's pivot point along its X-axis with the Select and Move tool to place the pivot where the wing contacts the body object. Then disable the Affect Pivot Only button. Then select and repeat this step for the left wing. Select each wing and rotate it about with the Select and Rotate tool to see how it moves relative to the body.

Caution

Be sure you select the Rotate tool before changing the pivot. Because each transform tool can have its own pivot and coordinate system, if you don't select the Rotate tool, you'll change the pivot for the Move tool instead.

Working with the Track View is beyond the scope of this chapter, but you can find more information on the Track View in Chapter 36, "Editing Animation Curves in the Track View."

---

4. Select all parts that make up the bee in the Top viewport, select Group→Group, and name the object **bee**. Then select the bee and the flower in the Left viewport, and press Z to zoom in on them.

5. With the bee group selected and the Select and Rotate tool enabled, click the Affect Pivot Only button in the Hierarchy panel, and move the pivot point to the center of the flower, using the Select and Move tool in the Top and Front viewports. Click the Affect Pivot Only button again to disable it.

Note

If you don't want to move the object pivot in Step 5, you can use the Working Pivot to rotate the bee around the flower, or you could select the Pick coordinate system and select the flower.

6. Enable the Auto Key button (N) at the bottom of the interface, and drag to frame 35. With the Select and Rotate button (E), rotate the bee in the Top viewport a third of the way around the flower. Drag the Time Slider to frame 70, and rotate the bee another third of the way. With the Time Slider at frame 100, complete the rotation. Click the Auto Key button again to display key mode.

7. Click the Play Animation button (/) to see the final rotating bee.

Figure 7.11 shows the bee as it moves around the flower where its pivot point is located.

FIGURE 7.11

By moving the pivot point of the bee, you can control how it spins about the flower.

Using the Align Commands

The Align commands are an easy way to automatically transform objects. You can use these commands to line up object centers or edges, align normals and highlights, align to views and grids, and even line up cameras.

Aligning objects

Any object that you can transform, you can align, including lights, cameras, and Space Warps. After selecting the object to be aligned, click the Align flyout button on the main toolbar, or choose Tools→Align→Align (or press Alt+A). The cursor changes to the Align icon. Now click a target object with which you want to align all the selected objects. Clicking the target object opens the Align Selection dialog box with the target object's name displayed in the dialog box's title, as shown in Figure 7.12.

FIGURE 7.12

The Align Selection dialog box can align objects along any axes by their Minimum, Center, Pivot, or Maximum points.

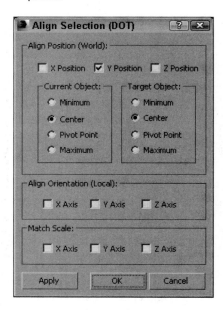

The Align Selection dialog box includes settings for the X, Y, and Z Positions to line up the Minimum, Center, Pivot Point, or Maximum dimensions for the selected or target object's bounding box. As you change the settings in the dialog box, the objects reposition themselves, but the actual transformations don't take place until you click Apply or OK.

Another way to align objects is with the Clone and Align tool, which is covered in Chapter 8, "Cloning Objects and Creating Object Arrays."

Using the Quick Align tool

The first flyout tool under the Align tool in the main toolbar (and in the Tools menu) is the Quick Align tool (Shift+A). This tool aligns the pivot points of the selected object with the object that you click without opening a separate dialog box. This is much quicker than the Align tool, which causes a separate dialog box to open.

Aligning normals

You can use the Normal Align command to line up points of the surface of two objects. A Normal vector is a projected line that extends from the center of a polygon face exactly perpendicular to the surface. When two Normal vectors are aligned, the objects are perfectly adjacent to each other. If the two objects are spheres, they touch at only one point.

To align normals, you need to first select the object to move (this is the source object). Then choose Tools→Align→Normal Align, or click the Normal Align flyout button under the Align button on the main toolbar (or press Alt+N). The cursor changes to the Normal Align icon. Drag the cursor across the surface of the source object, and a blue arrow pointing out from the face center appears. Release the mouse when you've correctly pinpointed the position to align.

Next, click the target object, and drag the mouse to locate the target object's align point. This is displayed as a green arrow. When you release the mouse, the source object moves to align the two points, and the Normal Align dialog box appears, as shown in Figure 7.13.

FIGURE 7.13

The Normal Align dialog box allows you to define offset values when aligning normals.

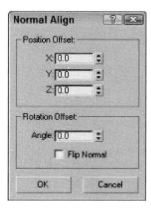

When the objects are aligned, the two points match up exactly. The Normal Align dialog box lets you specify offset values that you can use to keep a distance between the two objects. You also can specify an Angle Offset, which is used to deviate the parallelism of the normals. The Flip Normal option aligns the objects so their selected normals point in the same direction.

Objects without any faces, like Point Helper objects and Space Warps, use a vector between the origin and the Z-axis for normal alignment.

Tutorial: Aligning a kissing couple

Aligning normals positions two object faces directly opposite each other, so what better way to practice this tool than to align two faces?

To connect the kissing couple using the Normal Align command, follow these steps:

1. Open the Kissing couple.max file from the Chap 07 directory in the downloaded content set.

   This file includes two extruded shapes of a boy and a girl. The extruded shapes give you flat faces that are easy to align.

2. Select the girl shape, and choose the Tools→Align→Normal Align menu command (or press Alt+N). Then click and drag the cursor over the extruded girl shape until the blue vector points out from the front of the lips, as shown in Figure 7.13.

3. Click and drag the cursor over the boy shape until the green vector points out from the front of the lips. This vector pointing out from the face is the surface normal. Then release the mouse, and the Normal Align dialog box appears. Enter a value of **5** in the Z Position Offset field, and click OK.

Figure 7.14 shows the resulting couple with normal aligned faces.

FIGURE 7.14

Using the Normal Align feature, you can align object faces.

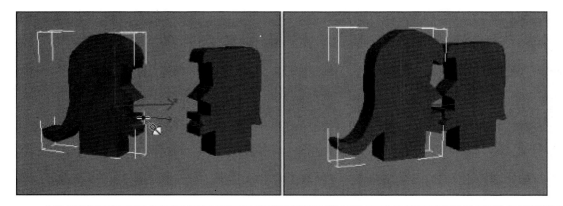

In the Align button flyout are two other common ways to align objects: Align Camera and Place Highlight. To learn about these features, see Chapter 24, "Configuring and Aiming Cameras," and Chapter 25, "Using Lights and Basic Lighting Techniques," respectively.

Aligning to a view

 The Align to View command provides an easy and quick way to reposition objects to one of the axes. To use this command, select an object, and choose Tools→Align→Align to View. The Align to View dialog box appears, as shown in Figure 7.15. Changing the settings in this dialog box displays the results in the viewports. You can use the Flip command for altering the direction of the object points. If no object is selected, the Align to View command cannot be used.

FIGURE 7.15

The Align to View dialog box is a quick way to line up objects with the axes.

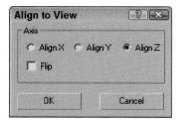

The Align to View command is especially useful for fixing the orientation of objects when you create them in the wrong view. All alignments are completed relative to the object's Local Coordinate System. If several objects are selected, each object is reoriented according to its Local Coordinate System.

Note

Using the Align to View command on symmetrical objects like spheres doesn't produce any noticeable difference in the viewports.

Using Grids

When 3ds Max is started, the one element that is visible is the Home Grid. This grid is there to give you a reference point for creating objects in 3D space. At the center of each grid are two darker lines. These lines meet at the origin point for the World Coordinate System where the coordinates for X, Y, and Z are all 0.0. This point is where all objects are placed by default.

In addition to the Home Grid, you can create and place new grids in the scene. These grids are not rendered, but you can use them to help you locate and align objects in 3D space.

The Home Grid

You can turn the Home Grid on and off by choosing Tools→Grids and Snaps→Show Home Grid. (You also can turn the Home Grid on and off for the active viewport using the G key.) If the Home Grid is the only grid in the scene, by default, it is also the construction grid where new objects are positioned when created.

You can access the Home Grid parameters (shown in Figure 7.16) by choosing Tools→Grids and Snaps→Grid and Snap Settings. You also can access this dialog box by right-clicking the Snap, Angle Snap, or Percent Snap toggle button, located on the main toolbar.

Tip
Right-clicking the Spinner Snap Toggle opens the Preference Settings dialog box.

In the Home Grid panel of the Grid and Snap Settings dialog box, you can set how often Major Lines appear, as well as Grid Spacing. (The Spacing value for the active grid is displayed on the Status Bar.) You also can specify to dynamically update the grid view in all viewports or just in the active one.

The User Grids panel lets you activate any new grids when created.

FIGURE 7.16

The Home Grid and User Grids panels of the Grid and Snap Settings dialog box let you define the grid spacing.

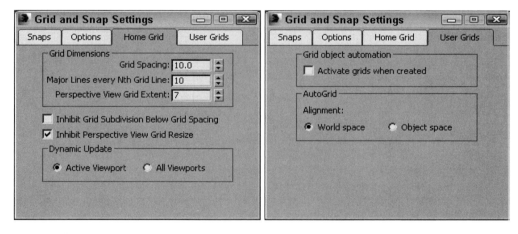

Creating and activating new grids

In addition to the Home Grid, you can create new grids. To create a new Grid object, select the Create→Helpers→Grid menu command, or open the Create panel, select the Helpers category, and click the Grid button. In the Parameters rollout are settings for specifying the new grid object's dimensions, spacing, and color, as well as which coordinate plane to display (XY, YZ, or ZX).

You can designate any newly created grid as the default active grid. To activate a grid, make sure it is selected, and choose Tools→Grids and Snaps→Activate Grid Object. Keep in mind that only one grid may be active at a time and that the default Home Grid cannot be selected. You also can activate a grid by right-clicking the grid object and selecting Activate Grid from the pop-up menu. To deactivate the new grid and reactivate the Home Grid, choose Tools→Grids and Snaps→Activate Home Grid, or right-click the grid object and choose Activate Grid→Home Grid from the pop-up quad menu.

You can find further grid settings for new grids in the Grid and Snap Settings dialog box on the User Grids panel. The settings include automatically activating the grid when created and an option for aligning an AutoGrid using World space or Object space coordinates.

Using AutoGrid

You can use the AutoGrid feature to create a new construction plane perpendicular to a face normal. This feature provides an easy way to create and align objects directly next to one another without lining them up manually or using the Align features.

The AutoGrid feature shows up as a check box at the top of the Object Type rollout for every category in the Create panel. It becomes active only when you're in Create Object mode.

To use AutoGrid, click the AutoGrid option after selecting an object type to create. If no objects are in the scene, the object is created as usual. If an object is in the scene, the cursor moves around on the surface of the object with its coordinate axes perpendicular to the surface that the cursor is over. Clicking and dragging creates the new object based on the precise location of the object under the mouse.

The AutoGrid option stays active for all new objects that you create until you turn it off by unchecking the box.

Tip

Holding down the Alt key before creating the object makes the new construction grid visible, disables the AutoGrid option, and causes all new objects to use the new active construction grid. You can disable this active construction grid by enabling the AutoGrid option again.

Tutorial: Creating a spyglass

As you begin to build objects for an existing scene, you find that working away from the scene origin is much easier if you enable the AutoGrid feature for the new objects you create. This feature enables you to position the new objects on (or close to) the surfaces of the nearby objects. It works best with objects that have pivot points located at their edges, such as Box and Cylinder objects.

In this example, you quickly create a spyglass object using the AutoGrid without needing to perform additional moves.

To create a spyglass using the AutoGrid and Snap features, follow these steps:

1. Before starting, click the Left viewport, and zoom way out so you can see the height of the spyglass pieces.
2. Select Create→Standard Primitives→Cylinder, and drag from the origin in the Top viewport to create a Cylinder object. Set the Radius value to **40** and the Height value to **200**. Then enable the AutoGrid option in the Object Type rollout.
3. Drag from the origin again in the Top viewport to create another Cylinder object. Set its Radius to **35** and its Height to **200**. Repeat this step three times, reducing the Radius by 5 each time.

Figure 7.17 shows the resulting spyglass object.

FIGURE 7.17

This spyglass object was created quickly and easily using the AutoGrid option.

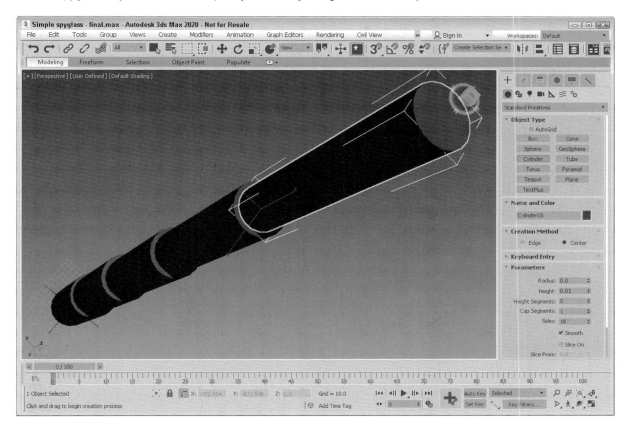

Using Snap Options

Often, when an object is being transformed, you know exactly where you want to put it. The Snap feature can be the means whereby objects get to the precise place they should be. For example, if you are constructing a set of stairs from box primitives, you can enable the Edge Snap feature to make each adjacent step align precisely along the edge of the previous step. With the Snap feature enabled, an object automatically moves (or snaps) to the specified snap position when you place it close enough. If you enable the Snap features, they affect any transformations that you make in a scene.

Snap points are defined in the Grid and Snap Settings dialog box, which you can open by choosing Tools→Grids and Snaps→Grid and Snap Settings or by right-clicking any of the first three Snap buttons on the main toolbar. (These Snap buttons have a small magnet icon in them.) Figure 7.18 shows the Snaps panel of the Grid and Snap Settings dialog box for Standard and NURBS objects. Options for configuring Body and Point Cloud snapping also are available. NURBS stands for Non-Uniform Rational B-Splines. They are a special type of object created from spline curves.

FIGURE 7.18

The Snaps panel includes many different points to snap to, depending on the object type.

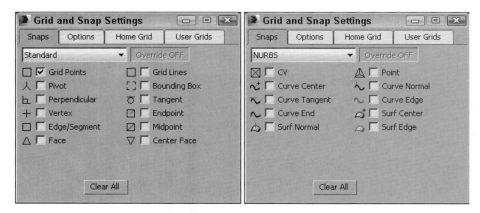

After snap points have been defined, the Snap buttons on the main toolbar activate the Snap feature, or you can press the S key. The first Snaps button consists of a flyout with three buttons: 3D Snap toggle, 2.5D Snap toggle, and 2D Snap toggle. The 2D Snap toggle button limits all snaps to the active construction grid. The 2.5D Snap toggle button snaps to points on the construction grid as well as projected points from objects in the scene. The 3D Snap toggle button can snap to any points in 3D space.

When snapping is enabled, a green square appears at the center of the pivot point. This square is a visual reminder that snapping is enabled. When you move an object while snapping is enabled, you can either drag the square icon to move the object freely between snapping points or drag the Move tool's controls to constrain the object's movement.

As you drag an object, the starting position is marked, and a line is drawn between this starting point and the destination point. Available snapping points are marked with a set of cross-hairs. If you release the object, it snaps to the highlighted set of cross-hairs. The line connecting the start and end points and the snapping point cross-hairs are colored green when the object is over an available snapping point and yellow when it is not.

In addition to the small circle icon and the Move tool controls, you can move the mouse over the object and any available snapping points on the object are highlighted. For example, if the Vertex option in the Grid and Snap Settings dialog box is enabled, moving the mouse over one of a box object's corners highlights the corner with a set of yellow cross-hairs. Dragging while a vertex's cross-hair is highlighted lets you snap the selected corner to another position.

Tutorial: Creating a 2D outline of an object

The 2.5D snap can be confusing. It limits snapping to the active construction grid, but within the active grid, it can snap to 3D points that are projected onto the active grid. You can create a 2D representation of a 3D object by snapping to the vertices of the suspended object.

To create a 2D outline of a cylinder object, follow these steps:

1. Select the Create→Standard Primitives→Cylinder menu to create a simple cylinder object.
2. Select and rotate the cylinder so it is suspended and rotated at an angle above the construction grid.
3. Click and hold the Snap toggle button, and select the 2.5D Snap flyout option. Right-click the Snap toggle, and select only the Vertex option in the Snaps panel. Then close the Grid and Snap Settings dialog box.
4. Choose the Create→Shapes→Line menu, and create a line in the Top viewport by snapping to the points that make the outline of the cylinder.

Figure 7.19 shows the projected outline. Using this method, you can quickly create 2D projections of 3D objects.

FIGURE 7.19

The 2.5D snap feature snaps to vertices of 3D objects projected onto the active grid.

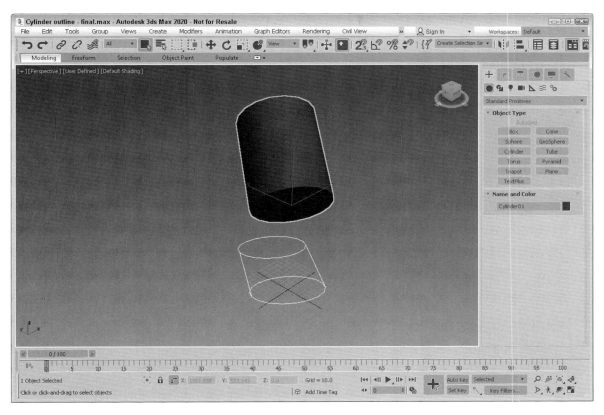

These Snap buttons control the snapping for translations. To the right are two other buttons: Angle Snap toggle and Percent Snap. These buttons control the snapping of rotations and scalings.

Note
The keyboard shortcut for turning the Snap feature on and off is the S key.

Setting snap points

The Snaps tab in the Grid and Snap Settings dialog box has many points that can be snapped to in several categories: Standard, Body Snaps, NURBS, and Point Cloud Objects. The Standard snap points (previously shown in Figure 7.18) include the following:

* **Grid Points:** Snaps to the Grid intersection points

* **Grid Lines:** Snaps only to positions located on the Grid lines

* **Pivot:** Snaps to an object's pivot point

* **Bounding Box:** Snaps to one of the corners of a bounding box

* **Perpendicular:** Snaps to a spline's next perpendicular point

* **Tangent:** Snaps to a spline's next tangent point

* **Vertex:** Snaps to polygon vertices

* **Endpoint:** Snaps to a spline's end point or the end of a polygon edge

> * **Edge/Segment:** Snaps to positions only on an edge
> * **Midpoint:** Snaps to a spline's midpoint or the middle of a polygon edge
> * **Face:** Snaps to any point on the surface of a face
> * **Center Face:** Snaps to the center of a face

The Body Snaps category includes a subset of the preceding list, including Vertex, Edge, Face, End Edge, and Edge Midpoint.

Several snap points specific to NURBS objects, such as NURBS points and curves, are also shown in Figure 7.18. These points include:

> * **CV:** Snaps to any NURBS Control Vertex subobject
> * **Point:** Snaps to a NURBS point
> * **Curve Center:** Snaps to the center of the NURBS curve
> * **Curve Normal:** Snaps to a point that is normal to a NURBS curve
> * **Curve Tangent:** Snaps to a point that is tangent to a NURBS curve
> * **Curve Edge:** Snaps to the edge of a NURBS curve
> * **Curve End:** Snaps to the end of a NURBS curve
> * **Surf Center:** Snaps to the center of a NURBS surface
> * **Surf Normal:** Snaps to a point that is normal to a NURBS surface
> * **Surf Edge:** Snaps to the edge of a NURBS surface

For Point Cloud Objects, you can enable snapping to a Point Cloud Vertex.

Setting snap options

The Grid and Snap Settings dialog box holds a panel of Options, shown in Figure 7.20, in which you can set the marker size and whether they display. The Snap Preview Radius defines the radial distance from the snap point required before the object that is being moved is displayed at the target snap point as a preview. This value can be larger than the actual Snap Radius and is meant to provide visual feedback on the snap operation. The Snap Radius setting determines how close the cursor must be to a snap point before it snaps to it.

The Angle and Percent values are the strengths for any Rotate and Scale transformations, respectively. The Snap to Frozen Objects lets you control whether frozen items can be snapped to. You also can cause translations to be affected by the designated axis constraints with the Use Axis Constraints option. The Display Rubber Band option draws a line from the object's starting location to its snapping location.

FIGURE 7.20

The Options panel includes settings for marker size and color and the Snap Strength value.

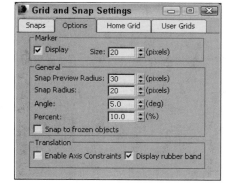

Within any viewpoint, holding down the Shift key and right-clicking in the viewport can access a pop-up menu of grid points and options. This pop-up quad menu lets you quickly add or reset all the current snap points and change snap options, such as Transformed Constraints and Snaps to Frozen Objects.

Using the Snaps toolbar

As a shortcut to enabling the various snapping categories, you can access the Snaps toolbar by right-clicking the main toolbar away from the buttons and selecting Snaps from the pop-up menu. The Snaps toolbar, shown in Figure 7.21, can have several toggle buttons enabled at a time. Each enabled button is highlighted in blue.

FIGURE 7.21

The Snaps toolbar provides a quick way to access several snap settings.

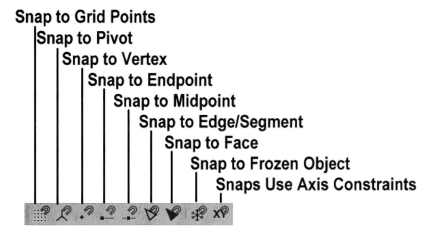

Snap to Grid Points
Snap to Pivot
Snap to Vertex
Snap to Endpoint
Snap to Midpoint
Snap to Edge/Segment
Snap to Face
Snap to Frozen Object
Snaps Use Axis Constraints

Tutorial: Creating a lattice for a methane molecule

Many molecules are represented by a lattice of spheres. Trying to line up the exact positions of the spheres by hand could be extremely frustrating, but using the Snap feature makes this challenge . . . well . . . a snap.

One of the simpler molecules is methane, which is composed of one carbon atom surrounded by four smaller hydrogen atoms. To reproduce this molecule as a lattice, you first need to create a tetrahedron primitive and snap spheres to each of its corners. The hedra isn't shown as part of the molecule but only used to place the spheres where they need to be.

To create a lattice of the methane molecule, follow these steps:

1. Right-click the Snap toggle button in the main toolbar to open the Grid and Snap Settings, and enable the Vertex option.

2. Select the Create→Extended Primitives→Hedra menu command, set the P Family Parameter to **1.0**, and drag in the Top viewport to create a Tetrahedron shape, then set the Radius value in the Command Panel to 100.

3. Click and hold the Snap toggle button, and select the 3D Snap flyout option. Select the Create→Standard Primitives→Sphere menu command. Right-click in the Left viewport, and drag from the top-left vertex to create a sphere. Set the sphere's Radius to **25**.

4. Create three more sphere objects with Radius values of 25 that are snapped to the vertices of the Tetrahedron object.

5. Finally, create a sphere in the Top viewport using the same snap point as the initial tetrahedron. Set its Radius to **80**.

Figure 7.22 shows the finished methane molecule.

FIGURE 7.22

A methane molecule lattice drawn with the help of the Snap feature

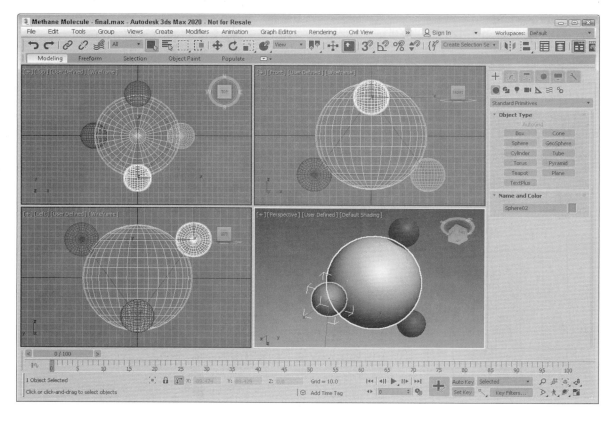

Summary

Transforming objects in 3ds Max is one of the fundamental actions you can perform. The three basic ways to transform objects are moving, rotating, and scaling. 3ds Max includes many helpful features to enable these transformations to take place quickly and easily. In this chapter, you learned these features:

* Using the Move, Rotate, and Scale buttons and the Transform Gizmos
* Transforming objects precisely with the Transform Type-In dialog box and Status Bar fields
* Using Transform Managers to change coordinate systems and lock axes
* Aligning objects with the align tool, aligning normals, and aligning to views
* Manipulating pivot points and using a Working Pivot
* Working with grids
* Setting up snap points
* Snapping objects to snap points

In the next chapter, you work more with multiple objects by learning how to clone objects. Using these techniques, you could very quickly have too many objects (and you were worried that there weren't enough objects).

Chapter 8

Cloning Objects and Creating Object Arrays

IN THIS CHAPTER

Cloning objects

Understanding copies, instances, and references

Using the Mirror and Snapshot tools

Spacing clones along a path with the Spacing tool

Using the Clone and Align tool

Creating object arrays

Using the Ring Array system

The only thing better than one perfect object is two perfect objects. Cloning objects is the process of creating copies of objects. These copies can maintain an internal connection (called an instance or a reference) to the original object that allows them to be modified along with the original object. For example, if you create a school desk and clone it multiple times as an instance to fill a school room, changing the parameter of one of the desks automatically changes it for all the other desks also. This is a huge timesaver and helps keep multiple scene objects up to date.

Another common way to create copies is with the Array dialog box. An *array* is a discrete set of regularly ordered objects. So creating an array of objects involves cloning several copies of an object in a pattern, such as in rows and columns or in a circle.

I'm sure you have the concept for that perfect object in your little bag of tricks, and this chapter lets you copy it over and over after you get it out.

Cloning Objects

You can clone objects in the Autodesk® 3ds Max® 2020 software in a couple of ways (and cloning luckily has nothing to do with DNA or gene splices). One method is to use the Edit→Clone (Ctrl+V) menu command, and another method is to transform an object while holding down the Shift key. You won't need to worry about these clones attacking anyone (unlike in *Star Wars: Episode II)*.

Using the Clone command

You can create a duplicate object by choosing the Edit→Clone (Ctrl+V) menu command. You must select an object before the Clone command becomes active, and you must not be in a Create mode. Selecting this command opens the Clone Options dialog box, shown in Figure 8.1, where you can give the clone a name and specify it as a Copy, Instance, or Reference. You also can copy any animation controllers associated with the object as a Copy or an Instance.

Caution

The Edit menu doesn't include the common Windows cut, copy, and paste commands because many objects and subobjects cannot be easily pasted into a different place. However, you will find a Clone (Ctrl+V) command that can duplicate a selected object.

FIGURE 8.1

The Clone Options dialog box defines the new object as a Copy, Instance, or Reference.

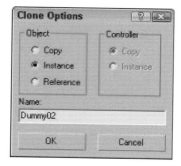

Note

The differences among Copy, Instance, and Reference are discussed in the "Understanding Cloning Options" section in this chapter.

When a clone is created with the Clone menu command, it is positioned directly on top of the original, which makes distinguishing it from the original difficult. To verify that a clone has been created, open the Select From Scene dialog box by pressing H, and look for the cloned object (it has the same name, but an incremented number has been added). To see both objects, click the Select and Move button on the main toolbar, and move one of the objects away from the other.

Using the Shift-clone method

An easier way to create clones is with the Shift key. You can use the Shift key when objects are transformed using the Select and Move, Select and Rotate, and Select and Scale commands. Holding down the Shift key while you use any of these commands on an object clones the object and opens the Clone Options dialog box. This Clone Options dialog box is identical to the dialog box previously shown, except it includes a spinner to specify the number of copies.

Performing a transformation with the Shift key held down defines an offset that is applied repeatedly to each copy. For example, holding down the Shift key while moving an object 5 units to the left (with the Number of Copies set to 5) places the first cloned object 5 units away from the original, the second cloned object 10 units away from the original object, and so on.

Tutorial: Cloning ducks

If you are babysitting a set of twins, then only one toy duck isn't enough. The clone features will really come in handy in this situation. To investigate cloning objects, follow these steps:

1. Open the Cloning ducks.max file, found in the Chap 08 directory in the downloaded content set.
2. Select the duck toy object by clicking it in one of the viewports.
3. With the toy duck model selected, choose Edit→Clone (or press Ctrl+V).
   The Clone Options dialog box appears.
4. Name the clone **First clone**, select the Copy option, and click OK.
5. Click the Select and Move button (or press the W key) on the main toolbar. Then, in the Top viewport, click and drag the duck toy model to the right.

As you move the model, the original model beneath it is revealed.

6. Select each model in turn, and notice the name change in the Create panel's Name field.

7. With the Select and Move button still active, hold down the Shift key, click the cloned duck toy in the Top viewport, and move it to the right again. In the Clone Options dialog box that appears, select the Copy option, set the Number of Copies to **3**, and click OK.

8. Click the Zoom Extents All button (or press Shift+Ctrl+Z) in the lower-right corner to view all the new ducks.

   Three additional duck toys have appeared, equally spaced from one another. The spacing was determined by the distance that you moved the second clone before releasing the mouse. Figure 8.2 shows the results of our duck cloning experiment. (Now you'll need to build a toy box to put the toys away.)

FIGURE 8.2

Cloning multiple objects is easy with the Shift-clone feature.

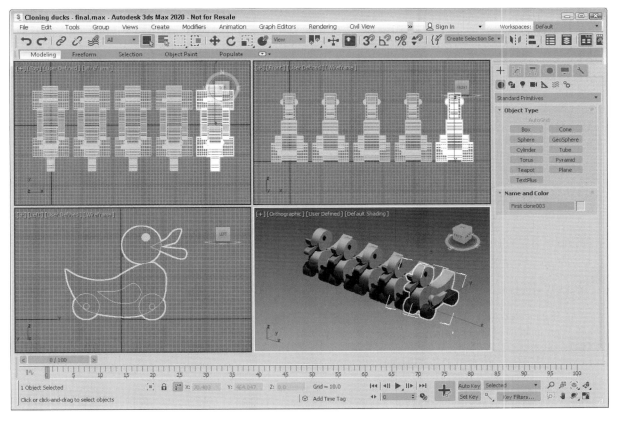

Using Quick Clone

Within the Transform Toolbox (opened with the Edit→Transform Toolbox menu) is a QClone button. This Quick Clone button creates a clone of the selected object and places it to the side of the selected object. The placement is exactly half the width of the selected object so the cloned object just touches the original. Placement also depends on the active viewport. If the Front or Left viewport is active, the clone is placed to the original's right, and if the Top viewport is active, the object is placed above the original object. Holding down the Shift key creates an Instanced copy, and holding down the Alt key creates two copies. This provides a quick and easy way to clone and move the object at the same time.

Understanding Cloning Options

When cloning in 3ds Max, you're offered the option to create the clone as a copy, an instance, or a reference. This is true not only for objects, but for materials, modifiers, and controllers as well.

Working with copies, instances, and references

When an object is cloned, the Clone Options dialog box appears. This dialog box enables you to select to make a copy, an instance, or a reference of the original object. Each of these clone types is unique and offers different capabilities.

A copy is just what it sounds like—an exact replica of the original object. The new copy maintains no ties to the original object and is a unique object in its own right. Any changes to the copy do not affect the original object, and vice versa.

Instances are different from copies in that they maintain strong ties to the original object. All instances of an object are interconnected, so any geometry modifications (done with modifiers or object parameters) to any single instance changes all instances. For example, if you create several instances of a mailbox and then use a modifier on one of them, all instances also are modified.

Note

Instances and references can have different object colors, materials, transformations (moving, rotating, or scaling), and object properties.

References are objects that inherit modifier changes from their parent objects but do not affect the parent when modified. Referenced objects get all the modifiers applied to the parent and can have their own modifiers as well. For example, suppose you have an apple object and a whole bunch of references to that apple. Applying a modifier to the base apple changes all the remaining apples, but you also can apply a modifier to any of the references without affecting the rest of the bunch.

Instances and references are tied to the applied object modifiers, which are covered in more detail in Chapter 11, "Accessing Subobjects and Modifiers and Using the Modifier Stack."

At any time, you can break the tie between instanced and referenced objects with the Make Unique button in the Modifier Stack. The Views→Show Dependencies command shows in magenta any objects that are instanced or referenced when the Modify panel is opened. This means that you can easily see which objects are instanced or referenced from the current selection.

Tutorial: Creating instanced doughnuts

Learning how the different clone options work will save you lots of future modifications. To investigate these options, you'll take a quick trip to the local doughnut shop.

To clone some doughnuts, follow these steps:

1. Create a doughnut using the Torus primitive by selecting Create→Standard Primitives→Torus, dragging and clicking to set the doughnut's radius, and then dragging and clicking a second time to set the cross-section radius in the Top viewport to create a torus object.

2. With the doughnut model selected, click the Select and Move button (or press the W key). Hold down the Shift key, and in the Top viewport, move the doughnut upward. In the Clone Options dialog box, select the Instance option, set the Number of Copies to **5**, and click OK. Click the Zoom Extents All (or press the Shift+Ctrl+Z key) button to widen your view.

3. Select all objects with the Edit→Select All (Ctrl+A) command and then Shift+drag the doughnuts in the Top viewport to the right. In the Clone Options dialog box, select the Instance option again, set **3**

for the Number of Copies, and click OK. This creates a nice array of two dozen doughnuts. Click the Zoom Extents All button (or press the Ctrl+Shift+Z key) to see all the doughnuts.

4. Select a single doughnut, and in the Parameters rollout of the Modify panel, set Radius1 to **20** and Radius2 to **10**.

    This makes a nice doughnut and changes all doughnuts at once.

5. Select the Modifiers→Parametric Deformers→Bend command. Then, in the Parameters rollout of the Command Panel, enter **25** in the Angle field, and select the X Bend Axis.

    This adds a slight bend to the doughnuts.

You can use modifiers to alter geometry. You can learn about using modifiers in Chapter 11, "Accessing Subobjects and Modifiers and Using the Modifier Stack."

Figure 8.3 shows the doughnuts all changed exactly the same way. You can imagine the amount of time it would take to change each doughnut individually. Using instances made these changes easy.

FIGURE 8.3

Two dozen doughnut instances ready for glaze.

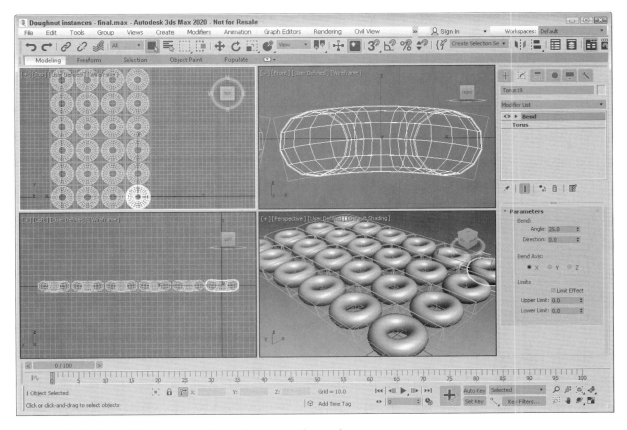

Tutorial: Working with referenced apples

Now that you have filled your belly with doughnuts, you need some healthful food for balance. What better way to add balance than to have an apple or two to keep the doctor away?

To create some apples using referenced clones, follow these steps:

1. Open the Referenced Apples.max file from the Chap 08 directory in the downloaded content set.

2. Select the apple, and Shift+drag with the Select and Move (W) tool in the Top viewport to create a cloned reference. Select the Reference option in the Clone Options dialog box. Then click OK to close the Clone Options dialog box.

3. Select the original apple again, and repeat Step 2 until several referenced apples surround the original apple.

4. Select the original apple in the middle again, and choose the Modifiers→Subdivision Surfaces→MeshSmooth command. In the Subdivision Amount rollout, set the number of Iterations to **2**.

   This smoothes all the apples.

5. Select one of the surrounding apples, and apply the Modifiers→Parametric Deformers→Taper command. Set the Amount value to **0.5** about the Z-axis.

6. Select another of the surrounding apples, and apply the Modifiers→Parametric Deformers→Squeeze command. Set the Axial Bulge Amount value to **0.3**.

7. Select another of the surrounding apples, and apply the Modifiers→Parametric Deformers→Squeeze command. Set the Radial Squeeze Amount value to **0.2**.

8. Select another of the surrounding apples, and apply the Modifiers→Parametric Deformers→Bend command. Set the Angle value to **20** about the Z axis.

Note

As you apply modifiers to a referenced object, notice the thick gray bar in the Modifier Stack. This bar, called the Derived Object Line, separates which modifiers get applied to all referenced objects (below the line) and which modifiers get applied to only the selected object (above the line). If you drag a modifier from above the gray bar to below the gray bar, that modifier is applied to all references.

Using referenced objects, you can apply the major changes to similar objects but still make minor changes to objects to make them a little different. Figure 8.4 shows the apples. Notice that they are not all exactly the same.

FIGURE 8.4

Even apples from the same tree should be slightly different.

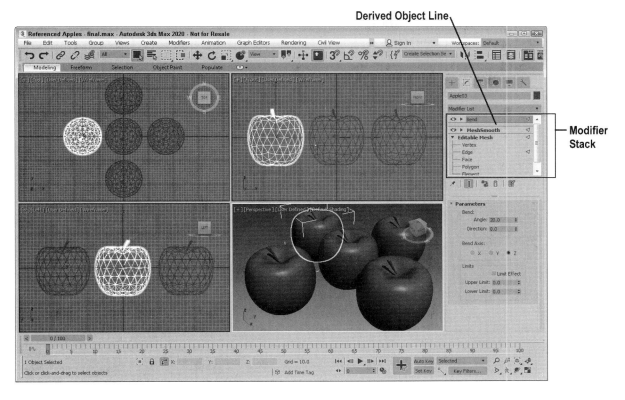

Mirroring Objects

Have you ever held the edge of a mirror up to your face to see half of your head in the mirror? Many objects have a natural symmetry that you can exploit to require that only half an object be modeled. The human face is a good example. You can clone symmetrical parts using the Mirror command.

Tip

The Symmetry modifier lets you model only half of an object. It automatically applies the modeling changes to the opposite side of the model.

Using the Mirror command

 The Mirror command creates a clone (or No Clone, if you so choose) of the selected object about the current coordinate system. To open the Mirror dialog box, shown in Figure 8.5, choose Tools→Mirror or click the Mirror button located on the main toolbar. You can access the Mirror dialog box only if an object is selected.

New Feature in 2020

The Transform and Geometry options are new to 3ds Max 2020.

FIGURE 8.5

The Mirror dialog box can create an inverted clone of an object.

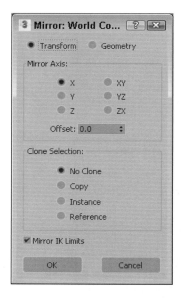

If the Transform option is selected, then all modifiers are also mirrored, but if the Geometry option is selected, then the mirror operation is applied as a modifier that appears in the Modifier Stack. This lets you revisit the mirror options as needed without having to reopen the Mirror dialog box.

Within the Mirror dialog box, you can specify an axis or plane about which to mirror the selected object. You also can define an Offset value. As with the other clone commands, you can specify whether the clone is to be a Copy, an Instance, or a Reference, or you can choose No Clone, which flips the object around the axis you specify. The dialog box also lets you mirror IK (inverse kinematics) Limits, which reduces the number of IK parameters that need to be set.

Learn more about inverse kinematics in Chapter 38, "Understanding Rigging, Kinematics, and Working with Bones."

Tutorial: Mirroring a robot's leg

Many characters have symmetry that you can use to your advantage, but to use this symmetry, you can't just clone one half; the cloned object would be oriented just like the original. Consider the position of a character's right ear relative to its right eye. If you clone the ear, the position of each ear will be identical, with the ear to the right of the eye, which would make for a strange-looking creature. What you need to use is the Mirror command, which clones the object and rotates it about a selected axis.

In this example, you have a complex mechanical robot with one of its legs created. Using Mirror, you can quickly clone and position its second leg.

To mirror a robot's leg, follow these steps:

1. Open the Robot mech.max file from the Chap 08 directory in the downloaded content set.

   This file includes a robot with only one leg.

2. Select all objects that make up the robot's leg in the Left viewport (which is easy because they are all grouped together), and open the Mirror dialog box with the Tools→Mirror menu command.

3. In the Mirror dialog box, select X as the Mirror Axis and Instance as the Clone Selection. Change the Offset value until the cloned leg is in position, which should be at around –**2.55**.

Note

The mirror axis depends on the viewport, so make sure the Left viewport is selected.

Any changes made to the dialog box are immediately shown in the viewports.

4. Click OK to close the dialog box.

Note

By making the clone selection an instance, you can ensure that any future modifications to the right half of the figure are automatically applied to the left half.

Figure 8.6 shows the resulting robot, which won't be falling over now.

FIGURE 8.6

A perfectly symmetrical robot, compliments of the Mirror tool.

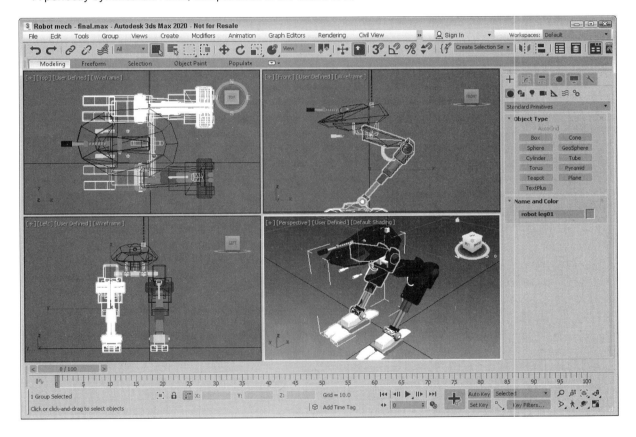

Cloning over Time

Another useful way to create multiple copies of an object is to have an object be created based on its position during a specific frame of an animation. This cloning at specific times is accomplished with the Snapshot feature.

Using the Snapshot command

The Snapshot command creates static copies, instances, references, or even meshes of a selected object as it moves along an animation path. For example, you could create a series of footprints by animating a set of footprints moving across the screen from frame 1 to frame 100, choose Tools→Snapshot, and enter the number of steps to appear over this range of frames in the Snapshot dialog box. The designated number of steps is created at regular intervals for the animation range. Be aware that the Snapshot command works only with objects that have an animation path defined.

You can open the Snapshot dialog box by choosing Tools→Snapshot or by clicking the Snapshot button (under the Array flyout on the Extras toolbar). Snapshot is the second button in the flyout. In the Snapshot dialog box, shown in Figure 8.7, you can choose to produce a single clone or a range of clones over a given number of frames. Selecting Single creates a single clone at the current frame.

Note

When you enter the number of Copies in the Snapshot dialog box, a copy is placed at both the beginning and end of the specified range, so if your path is closed, two objects are stacked on top of each other. For example, if you have a square path and you want to place a copy at each corner, you need to enter a value of **5**.

FIGURE 8.7

The Snapshot dialog box lets you create clones as a Copy, Instance, Reference, or Mesh.

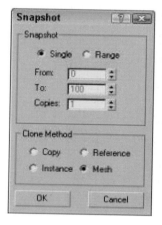

Tip

The Snapshot tool also can be used with particle systems.

Tutorial: Creating a path through a maze

The Snapshot tool can be used to create objects as a model is moved along an animated path. In this example, you create a series of footsteps through a maze.

To create a set of footprints through a maze with the Snapshot tool, follow these steps:

1. Open the Path through a maze.max file from the Chap 08 directory in the downloaded content set. This file includes a set of animated footprints that travel to the exit of a maze.
2. Select both footprint objects at the entrance to the maze.
3. Choose the Tools→Snapshot menu to open the Snapshot dialog box. Select the Range option, set the number of Copies to **20**, and select the Instance option. Then click OK.

Figure 8.8 shows the path of footsteps leading the way through the maze, which are easier to follow than breadcrumbs.

FIGURE 8.8

The Snapshot tool helps to build a set of footprints through a maze.

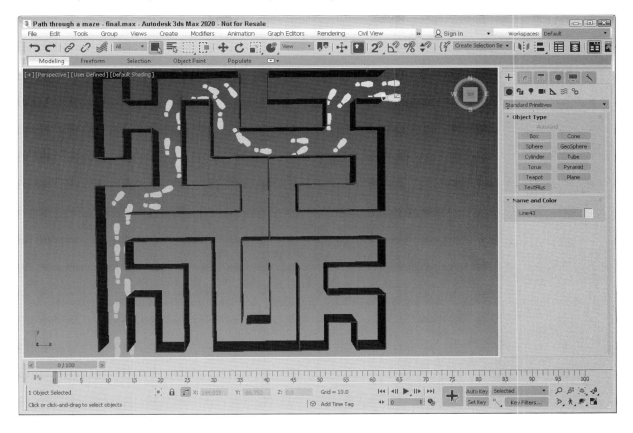

Spacing Cloned Objects

The Snapshot tool offers a convenient way to clone objects along an animation path, but what if you want to clone objects along a path that isn't animated? The answer is the Spacing tool. The Spacing tool can position clones at regular intervals along a path by either selecting a path and the number of cloned objects or by picking two points in the viewport.

Using the Spacing tool

You access the Spacing tool by clicking a button in the flyout under the Array button on the Extras toolbar. (The Extras toolbar can be made visible by right-clicking the main toolbar away from the buttons.) You also can access it using the Tools→Align→Spacing Tool (Shift+I) menu command. When accessed, it opens the Spacing Tool dialog box, shown in Figure 8.9. At the top of this dialog box are two buttons: Pick Path and Pick Points. If a path is selected, its name appears on the Pick Path button.

FIGURE 8.9

The Spacing Tool dialog box lets you select how to position clones along a path.

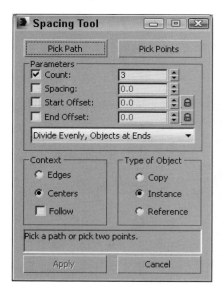

You also can specify Count, Spacing, Start Offset, and End Offset values. The drop-down list offers several preset options, including Divide Evenly, Free Center, End Offset, and more. These values and preset options are used to define the number and spacing of the objects. The spacing and position of the objects depend on the values that are included. For example, if you include only a Count value, the objects are evenly spaced along the path, including an object at each end. If an offset value is included, the first or last item is moved away from the end by the offset value. If a Spacing value is included, the number of objects required to meet this value is included automatically.

The Lock icons next to the Start and End Offset values force the Start or End Offset values to be the same as the Spacing value. This has the effect of pushing the objects away from their end points.

Before you can use either the Pick Path or Pick Points buttons, you must select the object to be cloned. Using the Pick Path button, you can select a spline path in the scene, and cloned objects are regularly spaced according to the values you selected. The Pick Points method lets you click to select the Start point and click again to select an end point. These points don't have to be objects; they are just locations selected on the construction grid in the viewport where you click. The cloned objects are spaced in a straight line between the two points.

The two options for determining the spacing width are Edges and Centers. The Edges option spaces objects from the edge of its bounding box to the edge of the adjacent bounding box, and the Centers option spaces objects based on their centers. The Follow option aligns the object with the path so the object's orientation follows the path. Each object can be a copy, instance, or reference of the original. The text field at the bottom of the dialog box displays for your information the number of objects and the spacing value between objects.

Tip

Lining up objects to correctly follow the path can be tricky. If the objects are misaligned, you can change the object's pivot point so it matches the viewport coordinates. This makes the object follow the path with the correct position.

You can continue to modify the Spacing Tool dialog box's values while the dialog box is open, but the objects are not added to the scene until you click the Apply button. The Close button closes the dialog box.

Tutorial: Stacking a row of dominoes

A good example of using the Spacing tool to accomplish something that is difficult in real life is to stack a row of dominoes. It is really a snap in 3ds Max, regardless of the complexity of the path.

To stack a row of dominoes using the Spacing tool, follow these steps:

1. Open the Row of dominoes.max file from the Chap 08 directory in the downloaded content set.

   This file includes a single domino and a wavy spline path.

2. Select the domino object, and open the Spacing tool by selecting the Tools→Align→Spacing Tool menu or with the flyout button under the Array button on the Extras toolbar (or by pressing Shift+I).

3. In the Spacing Tool dialog box, click the Pick Path button, and select the wavy path.

   The path name appears on the Pick Path button.

4. From the drop-down list in the Parameters section of the Spacing Tool dialog box, enable the Count option with a value of **35**. Then, select the Divide Evenly, Objects at Ends option in the drop-down list.

5. Select the Edges Context option, check the Follow check box, and make all clones Instances. Click Apply when the result looks right, and close the Spacing Tool dialog box.

Figure 8.10 shows the simple results. The Spacing Tool dialog box remains open until you click the Cancel button.

FIGURE 8.10

These virtual dominoes were much easier to stack than the set in my living room.

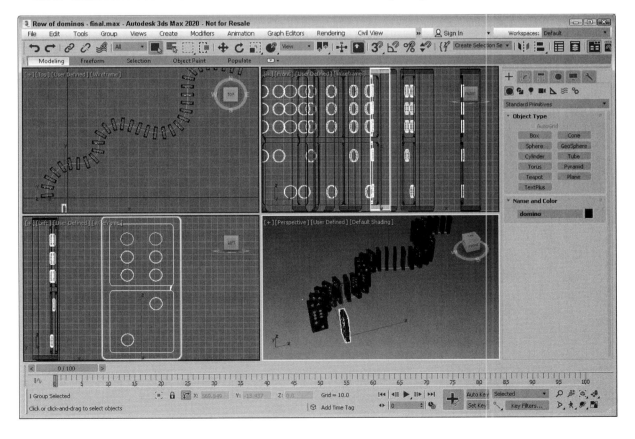

Using the Clone and Align Tool

Imagine you're working on a production team, and the modeler assigned to the project says he needs some more time to make the building columns "something special." Just as you prepare to give him the "deadlines don't die" speech, you remember the Clone and Align tool. Using this tool, you can place proxy objects where the detailed ones are supposed to go. Then, when the detailed object is ready, the Clone and Align tool lets you clone the detailed object and place it where all the proxies are positioned. This, of course, makes the modeler happy and doesn't disrupt your workflow. It's another production-team victory.

Aligning source objects to destination objects

Before selecting the Tools→Align→Clone and Align tool, you need to select the detailed object that you want to place. This object is referred to as the *source object*. Selecting the Clone and Align tool opens a dialog box, shown in Figure 8.11. From this dialog box, you can pick the proxy objects that are positioned where the source objects are supposed to go. These proxy objects are referred to as *destination objects*. The dialog box shows the number of source and destination objects that are selected.

The Align tool is covered in Chapter 7, "Transforming Objects, Pivoting, Aligning, and Snapping."

FIGURE 8.11

The Clone and Align dialog box lets you choose which objects mark the place where the source object should go.

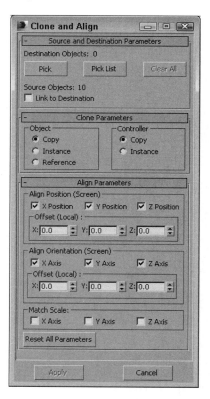

The Clone and Align dialog box also lets you select whether source objects are cloned as copies, instances, or references. In the Align Parameters rollout, you can specify the object's position and orientation using the same controls that are used to align objects, including any Offset values.

As you make changes in the Clone and Align dialog box, the objects are updated in the viewports, but these changes don't become permanent until you click the Apply button. The Clone and Align dialog box is *persistent,* meaning that after being applied, the settings remain until they are changed.

Tutorial: Cloning and aligning trees on a beach

To practice using the Clone and Align tool, you'll open a beach scene with a single set of grouped trees. Several other box objects have been positioned and rotated about the scene. The trees will be the source object, and the box objects will be the destinations.

To position and orient several high-res trees using the Clone and Align tool, follow these steps:

1. Open the Trees on beach.max file from the Chap 08 directory in the downloaded content set.
2. Select the tree object, and open the Clone and Align dialog box by selecting the Tools→Align→Clone and Align menu command.

Note

The Clone and Align dialog box remembers the last settings used, which may be different from what you want. You can reset all the settings with the Reset All Parameters button at the bottom of the dialog box.

3. In the Clone and Align dialog box, click the Pick button, and select each of the box objects in the scene.
4. In the Align Parameters rollout, enable the X and Y axes for the Positions and the X, Y, and Z axes for the Orientation. Then click the Apply button.

Figure 8.12 shows the simple results. Notice that the destination objects have not been replaced and are still there, but if you use dummy objects as the destination objects, you won't need to remove them because they won't be rendered.

FIGURE 8.12

Using the Clone and Align dialog box, you can place these trees to match the stand-in objects' position and orientation.

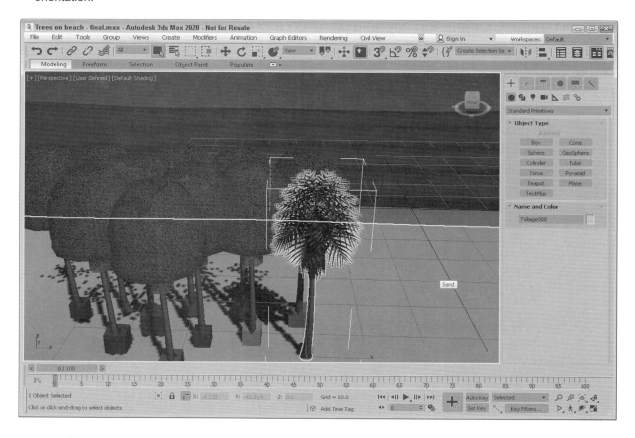

Creating Arrays of Objects

Now that you've probably figured out how to create arrays of objects by hand with the Shift-clone method, the Array command multiplies the fun by making it easy to create many copies instantaneously. The Array dialog box lets you specify the array dimensions, offsets, and transformation values. These parameters enable you to create an array of objects easily.

Access the Array dialog box by selecting an object and choosing Tools→Array or by clicking the Array button on the Extras toolbar. Figure 8.13 shows the Array dialog box. The top of the Array dialog box displays the coordinate system and the center about which the transformations are performed.

The Array dialog box is *modeless,* which means that you can still access and change the selected object or view in the viewports while the dialog box is open. The Array dialog box is also *persistent,* which means that it remembers the values you enter for the next time you open it. You can reset all the values at once by clicking the Reset All Parameters button within the Array dialog box. You also can preview the current array settings without actually creating an array of objects by using the Preview button. The Display as Box option lets you see the array as a bounding box to give you an idea of how large the array will be.

FIGURE 8.13

The Array dialog box defines the number of elements and transformation offsets in an array.

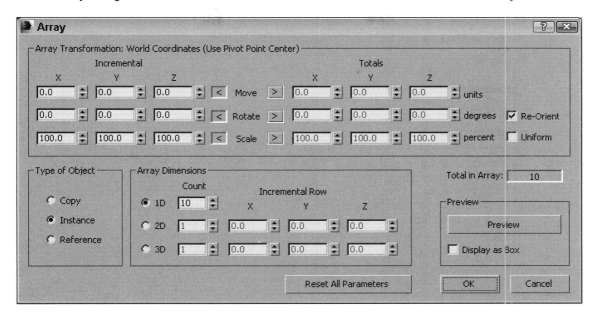

Linear arrays

Linear arrays are arrays in which the objects form straight lines, such as rows and columns. Using the Array dialog box, you can specify a positional offset along the X-, Y-, and Z-axes at the top of the dialog box and define this offset as an incremental amount or as a total amount. To change between incremental values and total values, click the arrows to the left and right of the Move, Rotate, and Scale labels. For example, an array with 10 elements and an incremental value of 5 will position each successive object a distance of 5 units from the previous one. An array with 10 elements and a total value of 100 will position each element a distance of 10 units apart by dividing the total value by the number of clones.

Note

Keep in mind that the original object is included in the total array count, which is different from cloning with the Shift key, which counts the copies.

The Move row values represent units as specified in the Units Setup dialog box. The Rotate row values represent degrees, and the Scale row values are a percentage of the selected object. All values can be either positive or negative values.

Clicking the Re-Orient check box causes the coordinate system to be reoriented after each rotation is made. If this check box isn't enabled, the objects in the array do not successively rotate. Clicking the Uniform check box to the right of the Scale row values disables the Y and Z Scale value columns and forces the scaling transformations to be uniform. To perform non-uniform scaling, simply deselect the Uniform check box.

The Type of Object section lets you define whether the new objects are copies, instances, or references, but the Array tool defaults to Instance. If you plan on modeling all the objects in a similar manner, you will want to select the Instance or Reference options.

In the Array Dimensions section, you can specify the number of objects to copy along three different dimensions. You also can define incremental offsets for each individual row.

Caution

You can use the Array dialog box to create a large number of objects. If your array of objects is too large, your system may crash.

Tutorial: Building a white picket fence

To start with a simple example, you'll create a white picket fence. Because a fence repeats, you need to create only a single slat; then you'll use the Array command to duplicate it consistently.

To create a picket fence, follow these steps:

1. Open the White picket fence.max file from the Chap 08 directory in the downloaded content set.

2. With the single fence board selected, choose Tools→Array or click the Array button on the Extras toolbar to open the Array dialog box.

3. In the Array dialog box, click the Reset All Parameters button to start with a clean slate. Then enter a value of **50** in the X column's Move row under the Incremental section. (This is the incremental value for spacing each successive picket.) Next, enter **20** in the Array Dimensions section next to the 1D radio button. (This is the number of objects to include in the array.) Click OK to create the objects.

Tip

The Preview button lets you see the resulting array before it is created. Don't worry if you don't get the values right the first time. The most recent values you entered into the Array dialog box stay around until you exit 3ds Max.

4. Click the Zoom Extents All button (or press Shift+Ctrl+Z) in the lower-right corner of the 3ds Max window to see the entire fence in the viewports.

Figure 8.14 shows the completed fence.

FIGURE 8.14

Tom Sawyer would be pleased to see this white picket fence, created easily with the Array dialog box.

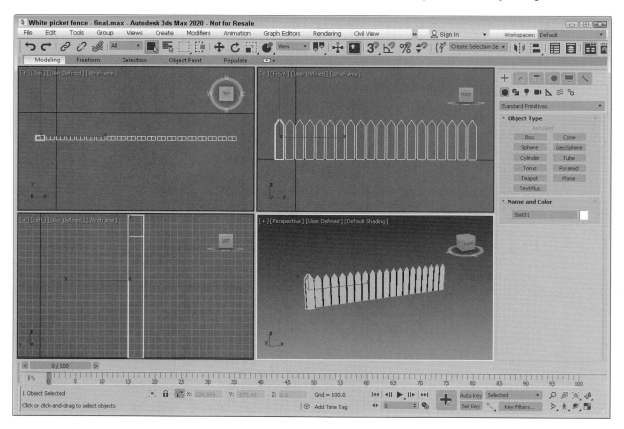

Circular arrays

 You can use the Array dialog box for creating more than just linear arrays. All transformations are done relative to a center point. You can change the center point about which transformations are performed by using the Use Selection Center button on the main toolbar. The three flyout options are Use Pivot Point Center, Use Selection Center, and Use Transform Coordinate Center.

For more about how these settings affect transformations, see Chapter 7, "Transforming Objects, Pivoting, Aligning, and Snapping."

Tutorial: Building a Ferris wheel

 Ferris wheels, like most of the rides at the fair, entertain by going around and around, with the riders seated in chairs spaced around the Ferris wheel's central point. The Array dialog box also can create objects around a central point.

In this example, you use the Rotate transformation along with the Use Transform Coordinate Center button to create a circular array.

To create a circular array, follow these steps:

1. Open the Ferris wheel.max file from the Chap 08 directory in the downloaded content set.

   This file has the Front viewport maximized to show the profile of the Ferris wheel.

2. Click the Use Pivot Point Center button on the main toolbar, and drag down to the last icon, which is the Use Transform Coordinate Center button.

   The Use Transform Coordinate Center button becomes active. This button causes all transformations to take place about the axis in the center of the screen. Navigate the scene view so the center of the Ferris wheel is in the center of the viewport.

3. Select the light-blue seat object, and open the Array dialog box by choosing Tools→Array or by clicking the Array button on the Extras toolbar. Before entering any values into the Array dialog box, click the Reset All Parameters button.

4. Between the Incremental and Totals sections are the labels Move, Rotate, and Scale. Click the arrow button to the right of the Rotate label. Set the Z column value of the Rotate row to **360** degrees, and make sure that the Re-Orient option is disabled.

   A value of 360 degrees defines one complete revolution. Disabling the Re-Orient option keeps each chair object from gradually turning upside down.

5. In the Array Dimensions section, set the 1D spinner Count value to **8**, and click OK to create the array.

6. Next, select the green strut, and open the Array dialog box again with the Tools→Array command. Select the Re-Orient option, and leave the rest of the settings as they are. Click OK to create the array.

Figure 8.15 shows the resulting Ferris wheel. You can click the Min/Max toggle in the lower-right corner to view all four viewports again.

FIGURE 8.15

A circular array created by rotating objects about the Transform Coordinate Center

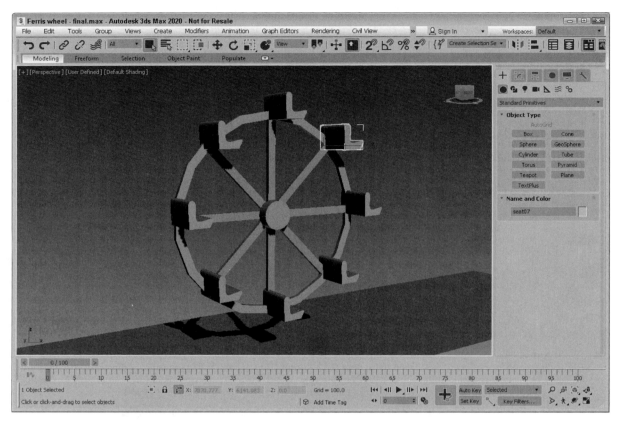

Working with a Ring Array

You can find the Ring Array system by opening the Create panel and selecting the Systems category. Clicking the Ring Array button opens a Parameters rollout. In this rollout are parameters for the ring's Radius, Amplitude, Cycles, Phase, and the Number of elements to include.

Caution

The Ring Array button can be found in the Systems category of the Create panel, but is not in the Create→Systems menu.

You create the actual array by clicking and dragging in one of the viewports. Initially, all elements are simple box objects surrounding a green dummy object.

The Amplitude, Cycles, and Phase values define the sinusoidal nature of the circle. The Amplitude is the maximum distance that you can position the objects from the horizontal plane. If the Amplitude is set to 0, all objects lie in the same horizontal plane. The Cycles value is the number of waves that occur around the entire circle. The Phase determines which position along the circle starts in the up position.

Tutorial: Using Ring Array to create a carousel

Continuing with the theme-park-attractions motif, this example creates a carousel. The horse model comes from Poser but was simplified using the MultiRes modifier.

To use a Ring Array system to create a carousel, follow these steps:

1. Open the Carousel.max file from the Chap 08 directory in the downloaded content set.

This file includes a carousel structure made from primitives along with a single carousel horse.

2. Open the Create panel, select the Systems category, and click the Ring Array button. Drag in the Top viewport from the center of the carousel to create a ring array. Enter a Radius value of **250**, an Amplitude of **20**, a Cycles value of **3**, and a Number value of **6**. Then right-click in the active viewport to deselect the Ring Array tool.

Note

If the Ring Array object gets deselected, you can access its parameters in the Motion panel, not in the Modify panel.

3. Select the Ring Array's center Dummy object in the Left viewport, select the Tools→Align→Align menu command, and then click the center cylinder. The Align Selection dialog box opens. Enable the X, Y, and Z Position options, choose the Center options for both the Current and Target objects, and click the OK button. This aligns the ring array to the center of the carousel.

4. Select the horse object, and choose the Tools→Align→Clone and Align menu command. In the Clone and Align dialog box that opens, select the Instance option along with the X, Y, and Z Position and Orientation options. Then click the Pick button, and click each of the boxes in the ring array. Set the Z-axis Orientation Offset to 90. Then click the Apply button, and close the Clone and Align dialog box.

Figure 8.16 shows the finished carousel. Notice that each horse is at a different height. After the horses are placed, you can delete or hide the Ring Array object.

FIGURE 8.16

The horses in the carousel were created using a Ring Array system.

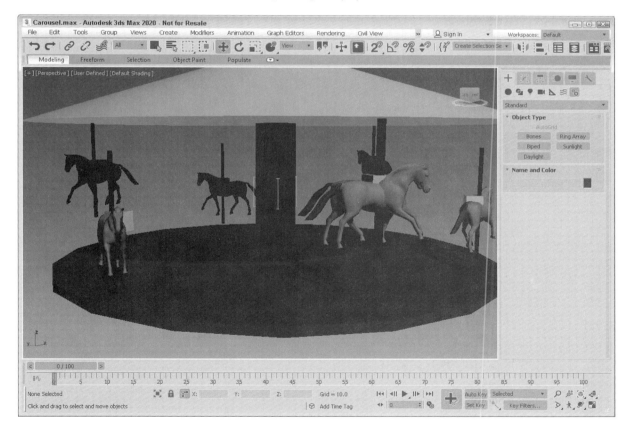

Summary

Many ways to clone an object are available. You can use the Clone command under the Edit menu or the Shift-clone feature for quickly creating numerous clones. Clones can be copies, instances, or references. These methods differ in how they retain links to the original object. You also can clone using the Mirror, Snapshot, and Spacing tools.

Arrays are another means of cloning. You can use the Array dialog box to produce clones in three different dimensions, and you can specify the offset transformations.

This chapter covered the following cloning topics:

* Cloning objects and Shift-cloning
* Understanding copies, instances, and references
* Using the Mirror, Snapshot, Spacing, and Clone and Align tools
* Building linear and circular arrays of objects
* Using the Ring Array system

In the next chapter, you learn to group objects and link them into hierarchies. Then you'll be able to organize into structures all the objects that you've learned to create. These hierarchies become important as you prepare to animate the scene.

Chapter 9

Grouping, Linking, and Parenting Objects

IN THIS CHAPTER

Grouping objects

Understanding root, parent, and child relationships

Linking and unlinking objects

Now that you've learned how to select and clone objects, you'll want to learn how to group objects in an easily accessible form, especially as a scene becomes more complex. The grouping features in the Autodesk® 3ds Max® 2020 software enable you to organize all the objects that you're dealing with, thereby making your workflow more efficient.

Another way of organizing objects is to build a linked hierarchy. A *linked hierarchy* attaches, or links, one object to another and makes it possible to transform the attached object by moving the object to which it is linked. The arm is a classic example of a linked hierarchy: when the shoulder rotates, so do the elbow, wrist, and fingers. Establishing linked hierarchies can make moving, positioning, and animating many objects much easier.

Working with Groups

Grouping objects organizes them and makes them easier to select and transform. Groups are different from selection sets in that groups exist like one object. Selecting any object in the group selects the entire group, whereas selecting an object in a selection set selects only that object and not the selection set. You can open groups to add, delete, or reposition objects within the group. Groups also can contain other groups. This is called *nesting groups.*

Creating groups

The Group command enables you to create a group. To do so, simply select the desired objects and choose Group→Group. A simple Group dialog box opens and enables you to give the group a name. The newly created group displays a new bounding box that encompasses all the objects in the group.

Tip

You can easily identify groups in the Select From Scene dialog box by using the Groups display toggle. Groups appear in bold in the Name and Color rollout of the Command Panel.

Ungrouping objects

The Ungroup command enables you to break up a group (kind of like a poor music album). To do so, simply select the desired group and choose Group→Ungroup. This menu command dissolves the group, and all the objects within the group revert to separate objects. The Ungroup command breaks up only the currently selected group. All nested groups within a group stay intact.

Caution

If you animate a group and later use the Ungroup command, all the keys created for the whole group are lost when you ungroup.

The easiest way to dissolve an entire group, including any nested groups, is with the Explode command. This command eliminates the group and the groups within the group, and makes each object separate.

Opening and closing groups

The Open command enables you to access the objects within a group. Grouped objects move, scale, and rotate as a unit when transformed, but individual objects within a group can be transformed independently after you open a group with the Open command.

To move an individual object in a group, select the group, and choose Group→Open. The white bounding box changes to a pink box. Then select an object within the group, and move it with the Select and Move button (keyboard shortcut, W). Choose Group→Close to reinstate the group.

Attaching and detaching objects

The Attach and Detach commands enable you to insert or remove objects from a group without dissolving and recreating the group. To attach objects to an existing group, you select an object, select the Attach menu command, and click the group to which you want to add the object. To detach an object from a group, you need to open the group and select the Detach menu command. Remember to close the group when finished.

Tutorial: Grouping a plane's parts together

Positioning objects relative to one another takes careful and precise work. After spending the time to place the wings, tail, and prop on a plane exactly where they need to be, transforming each object by itself can misalign all the parts. By grouping all the objects together, you can move all the objects at once.

For this tutorial, you can get some practice grouping all the parts of an airplane together. Follow these steps:

1. Open the Prop plane.max file from the Chap 09 directory in the downloaded content set.
2. Click the Select by Name button on the main toolbar (or press the H key) to open the Select From Scene dialog box. In this dialog box, notice all the different plane parts. Click the Select All button to select all the separate objects, and click OK to close the dialog box.
3. With all the objects selected, choose Group→Group to open the Group dialog box. Give the group the name **Plane**, and click OK.
4. Click the Select and Move button (or press W), and click and drag the plane. The entire group now moves together.

Figure 9.1 shows the plane grouped as one unit. Notice how only one set of brackets surrounds the plane in the Perspective viewport. The group name is displayed in the Name field of the Command Panel instead of listing the number of objects selected.

FIGURE 9.1

The plane moves as one unit after its objects are grouped.

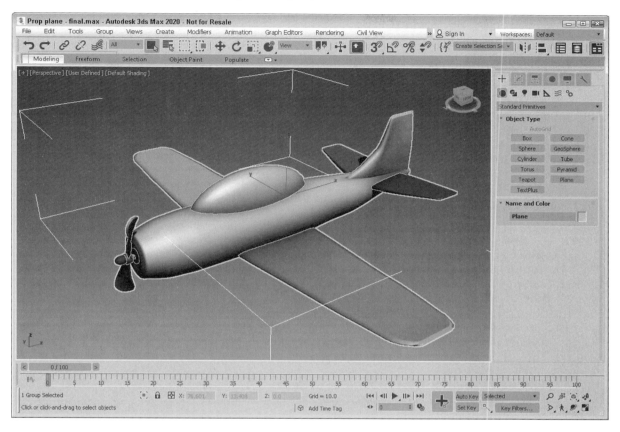

Alternatives to grouping

Grouping objects together makes them easier to transform about the scene, but they are limited in their functionality and can cause some headaches when animating. One alternative to grouping objects is to combine them together into a single object using the Attach command for editable objects. Editable objects, like the Editable Poly, also can make use of an Attach feature, but attaching objects to an editable object permanently combines the objects. Another alternative is to use the Container feature, which allows the grouped assets to be saved as an external file, where they can be shared with other team members.

You can learn more about the Editable Poly objects in Chapter 13, "Modeling with Polygons." Containers are discussed in Chapter 10, "Organizing Scenes with Layers, Containers, XRefs and the Schematic View."

Understanding Parent, Child, and Root Relationships

3ds Max uses several terms to describe the relationships between objects. A *parent object* is an object that controls any secondary, or child, objects linked to it. A *child object* is an object that is linked to and controlled by a parent. A parent object can have many children, but a child can have only one parent. Additionally, an object can be both a parent and a child at the same time. Another way to say this is:

* Child objects are linked to parent objects.

* Moving a parent object moves its children with it.

* Child objects can move independently of their parents.

A hierarchy is the complete set of linked objects that includes these types of relationships. Ancestors are all the parents above a child object. Descendants are all the children below a parent object. The root object is the top parent object that has no parent and controls the entire hierarchy.

Each hierarchy can have several branches or subtrees. Any parent with two or more children represents the start of a new branch.

The default hierarchies established using the Select and Link tool are referred to as forward kinematics systems, in which control moves forward down the hierarchy from parent to child. In forward kinematics systems, the child has no control over the parent. An inverse kinematics system (covered in Chapter 38, "Understanding Rigging, Kinematics, and Working with Bones") enables child objects to control their parents.

All objects in a scene, whether linked or not, belong to a hierarchy. Objects that aren't linked to any other objects are, by default, children of the *world object,* which is an imaginary object that holds all objects.

Note

You can view the world object, labeled Objects, in the Track View. Individual objects are listed under the Objects track by their object name.

You have several ways to establish and edit hierarchies using 3ds Max. The simplest method is to use the Select and Link and Unlink Selection buttons, found on the main toolbar. You can find these buttons in the Schematic View window as well. The Hierarchy panel in the Command Panel provides access to valuable controls and information about established hierarchies.

The Schematic View window is covered in Chapter 10, "Organizing Scenes with Layers, Containers, XRefs and the Schematic View."

Building Links between Objects

The main toolbar includes two buttons that you can use to build and edit a hierarchy: Select and Link and Unlink Selection. The order of selection defines which object becomes the parent and which becomes the child.

Linking objects

The Select and Link button always links children to the parents. To remind you of this order, remember that a parent can have many children, but a child can have only one parent.

To link two objects, click the Select and Link button. This places you in Link mode, which continues until you turn it off by selecting another button, such as the Select Object button or one of the Transform buttons. When you're in Link mode, the Link button is highlighted.

With the Link button highlighted, click an object, which will be the child, and drag a line to the target parent object. The cursor arrow changes to the link icon when it is over a potential parent. When you release the mouse button, the parent object flashes once, and the link is established. If you drag the same child object to a different parent, the link to the previous parent is replaced by the link to the new parent.

Once linked, all transformations applied to the parent are applied equally to its children about the parent's pivot point. A *pivot point* is the center about which the object rotates and scales.

Unlinking objects

The Unlink Selection button is used to destroy links, but only to the parent. For example, if a selected object has both children and a parent, clicking the Unlink button destroys the link to the parent of the selected object but not the links to its children.

To eliminate all links for an entire hierarchy, double-click an object to select its entire hierarchy, and click the Unlink Selection button.

Tutorial: Linking a family of ducks

What better way to show off parent–child relationships than with a family? I could have modeled my own family, but for some reason, my little ducks don't always like to follow me around.

To create a linked family of ducks, follow these steps:

1. Open the Linked duck family.max file from the Chap 09 directory in the downloaded content set. This file includes several simple ducks lined up in a row.

2. Click the Select and Link button in the main toolbar, and drag a line from the last duck to the one just in front of it.

Tip
You can link several objects at once by highlighting all the objects you want to link and dragging the selected objects to the parent object. This procedure creates a link between the parent object and each selected object.

3. Continue to connect each duck to the one in front of it.

4. Click the Select and Move button (or press the W key), and move the Mommy duck object. Notice how all the children move with her.

Figure 9.2 shows the duck family as they move forward in a line. The Select and Link button made it possible to move all the ducks simply by moving the parent duck.

FIGURE 9.2

Linked child ducks inherit transformations from their parent duck.

Displaying Links and Hierarchies

The Display panel includes a rollout that lets you display all the links in the viewports.

After links have been established, you can see linked objects listed as a hierarchy in several places. The Select From Scene dialog box, opened with the Select by Name button (or with the H key), can display objects in this manner, as well as the Schematic and Track Views.

Displaying links in the viewport

You can choose to see the links between the selected objects in the viewports by selecting the Display Links option in the Link Display rollout of the Display panel. The Display Links option shows links as lines that run between the pivot points of the objects with a diamond-shaped marker at the end of each line; these lines and markers are the same color as the object.

Note

The Display Links option can be enabled or disabled for each object in the scene. To display the links for all objects, use the Edit→Select All (Ctrl+A) command and then enable the Display Links option.

The Link Display rollout also offers the Link Replaces Object option, which removes the objects and displays only the link structure. This feature removes the complexity of the objects from the viewports and lets you work with the links directly. Although the objects disappear, you can still transform the objects using the link markers.

Viewing hierarchies

The Select From Scene dialog box, the Scene Explorer and the Schematic and Track Views can display the hierarchy of objects in a scene as an ordered list, with child objects indented under parent objects.

Clicking the Select by Name button (H) on the main toolbar opens the Select From Scene dialog box; select the Display→Display Children menu to see all the children under the selected object. Click the small arrow to the left of the parent to see its children. Figure 9.3 shows the Select From Scene dialog box with the Display Children menu enabled.

You can learn more about the Scene Explorer dialog box in Chapter 6, "Selecting Objects and Setting Object Properties."

FIGURE 9.3

The Select From Scene dialog box indents all child objects under their parent.

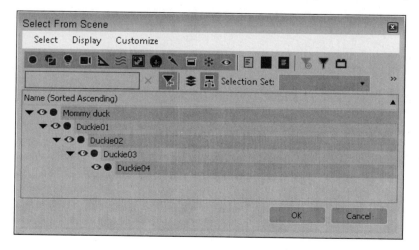

The Schematic View (opened with the Graph Editors→New Schematic View menu command) presents a graph in which objects are represented by rectangle nodes with their hierarchical links drawn as lines running between them.

For more information on using the Schematic View, see Chapter 10, "Organizing Scenes with Layers, Containers, XRefs and the Schematic View."

The Track View (opened with the Graph Editors→New Track View menu command) displays lots of scene details in addition to the object hierarchy. In the Track View, you can easily expand and contract the hierarchy to focus on just the section you want to see or select.

For more information on using the Track View, see Chapter 36, "Editing Animation Curves in the Track View."

Working with Linked Objects

If you link some objects together and set some animation keys, and the magical Play Animation button starts sending objects hurtling off into space, chances are good that you have a linked object that you didn't know about. Understanding object hierarchies and being able to transform those hierarchies are the keys to efficient animation sequences.

All transformations are done about an object's pivot point. You can move and reorient these pivot points as needed by clicking the Pivot button under the Hierarchy panel.

Several additional settings for controlling links are available under the Hierarchy panel of the Command Panel. (The Hierarchy panel tab looks like a mini organizational chart.) Just click the Link Info button. This button opens two rollouts if a linked object is selected. You can use the Locks and Inherit rollouts to limit an object's transformations and specify the transformations that it inherits.

I present more information on object transformations in Chapter 7, "Transforming Objects, Pivoting, Aligning, and Snapping."

Locking inheriting transformations

The Inherit rollout, like the Locks rollout, includes check boxes for each axis and each transformation, except that here, all the transformations are selected by default. By deselecting a check box, you specify which transformations an object does not inherit from its parent. The Inherit rollout appears only if the selected object is part of a hierarchy.

For example, suppose that a child object is created and linked to a parent, and the X Move Inherit check box is deselected. As the parent is moved in the Y or Z directions, the child follows, but if the parent is moved in the X direction, the child does not follow. If a parent doesn't inherit a transformation, its children don't either.

Using the Link Inheritance utility

The Link Inheritance utility works in the same way as the Inherit rollout of the Hierarchy panel, except that you can apply it to multiple objects at the same time. To use this utility, open the Utilities panel, and click the More button. In the Utilities dialog box, select the Link Inheritance (Selected) utility, and click OK. The rollout for this utility is identical to the Inherit rollout, discussed in the previous section.

Selecting hierarchies

You need to select a hierarchy before you can transform it, and you have several ways to do so. The easiest method is to simply double-click an object. Double-clicking the root object selects the entire hierarchy, and double-clicking an object within the hierarchy selects it and all of its children.

After you select an object in a hierarchy, pressing the Page Up or Page Down keyboard shortcut selects its parent or child objects. For example, if you select the Mommy duck object and press Page Down, the first baby duck object is selected, and the Mommy duck object is deselected. Selecting any of the baby duck objects and pressing Page Up selects the duck object in front of it.

Linking to dummies

Dummy objects are useful as root objects for controlling the motion of hierarchies. By linking the parent object of a hierarchy to a dummy object, you can control all the objects by moving the dummy.

To create a dummy object, select Create→Helpers→Dummy, or open the Create panel, click the Helpers category button (this button looks like a small tape measure), and select the Standard category. Within the Object Type rollout is the Dummy button; click it, click in the viewport where you want the dummy object to be positioned, and drag to set its size. Dummy objects look like wireframe box objects in the viewports, but dummy objects are not rendered.

Tutorial: Circling the globe

When you work with complex models with lots of parts, you can control the object more easily if you link it to a dummy object and then animate the dummy object instead of the entire model. To practice doing this, you'll create a simple animation of an airplane flying around the globe. To perform this feat, you create a dummy object in the center of a sphere, link the airplane model to it, and rotate the dummy object. This alternative doesn't require that you move any pivot points. This tutorial involves animating objects, which are covered in other chapters.

The basics of animation are covered in Chapter 32, "Understanding Animation and Keyframes."

To link and rotate objects using a dummy object, follow these steps:

1. Open the Circling the globe.max file, found in the Chap 09 directory in the downloaded content set.

   This file includes a texture mapped sphere with an airplane model positioned to the side of it.

2. Select Create→Helpers→Dummy and then drag in the center of the Sphere to create a dummy object. With the dummy object selected, choose the Tools→Align→Align menu command (or press the Alt+A shortcut), and click the globe. In the Align Selection dialog box, enable the X, Y, and Z Position options, along with the Center options, and click OK to align the centers of the dummy and globe objects.

3. Because the dummy object is inside the sphere, creating the link between the airplane and the dummy object can be difficult. To simplify this process, select and right-click the sphere object and then select Hide Selection from the pop-up quad menu.

   This hides the sphere so you can create a link between the airplane and the dummy object.

4. Click the Select and Link button on the main toolbar, and drag a line from the airplane to the dummy object.

5. Click the Auto Key button (or press N) to enable animation key mode, and drag the Time Slider to frame 100. Click the Select and Rotate button on the main toolbar (or press E), and select the dummy object. Then rotate the dummy object about its X-axis one full revolution, and notice how the linked airplane also rotates over the surface of the sphere. Click the Auto Key button again to disable Auto Key mode.

6. Right-click to access the pop-up quad menu. Select the Unhide All menu command to make the sphere visible again. Then click the Play button to see the plane fly about the globe.

By linking the airplane to a dummy object, you don't have to worry about moving the airplane's pivot point to get the correct motion. Figure 9.4 shows a frame from the final scene.

FIGURE 9.4

With a link to a dummy object, making the airplane circle the globe is easy.

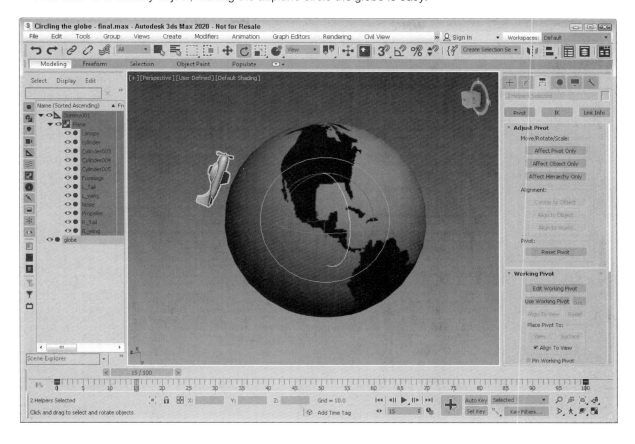

Summary

As scenes become more complex, the name of the game is organization. You can organize objects within the scene in several ways, including grouping, linking, and building hierarchies.

In this chapter, you've done the following:

* Grouped objects using the Group menu and learned to work with groups

* Learned about parent, child, and root relationships

* Created a hierarchy of objects using the Select and Link and Unlink Selection buttons

* Viewed links in the viewport

* Learned how to create a link to a dummy object

In the next chapter, you will look at several other ways to organize and transport objects between several files using Layers, Containers and XRefs. We'll also look at the Schematic View, which makes linking complex object sets together easy.

Chapter 10

Organizing Scenes with Layers, Containers, XRefs and the Schematic View

IN THIS CHAPTER

Working with layers

Working with containers

Externally referencing objects and scenes

Using the File Link Manager

Working with the Schematic View window

Working with hierarchies

Setting Schematic View preferences

Using List Views

One way to organize objects so they are easier to locate and select is to use layers. The Object Layer features in 3ds Max allow you to collect several objects together. Once established, a layer can quickly hide or freeze many scene objects at once with a single command.

Using containers and external references (XRefs), you can pull multiple scenes, objects, materials, and controllers together into a single scene from a set of external files. Both of these features allow a diverse team to work on separate parts of a scene at the same time. They also provide a great way to reuse existing resources.

A valuable tool for selecting, linking, and organizing scene objects is the Schematic View window. This window offers a 1,000-foot view of the objects in your scene. From this whole scene perspective, you can find the exact item you seek.

The Schematic View window shows all objects as simple nodes and uses arrows to show relationships between objects. This structure makes the Schematic View window the easiest place to establish links and to wire parameters. You also can use this view to quickly see all the instances of an object.

Using Layers

So what does 3ds Max have in common with a wedding cake? The answer is layers. Layers provide a way to separate scene objects into easy-to-select and easy-to-work-with groupings. These individual layers have properties that can then be turned on and off.

The Scene Explorer can also be used to create and divide your scene objects into layers. To switch the Scene Explorer to layers mode simply click on the Sort by Layers button at the bottom of the Scene Explorer interface. When layers are enabled, the Create New Layers button on the top toolbar becomes active. Using this button, you can create new layers. These layers appear in the Scene Explorer and are identified by a layer icon to its left.

You can rename a layer by clicking on its name and typing a new name. You can also easily turn layers on and off using the eye icon to the left of the layer name. Selected objects are added to layers by dragging and dropping them in the Scene Explorer. Figure 10.1 shows several layers in the Scene Explorer.

These layers are different from Animation Layers, which are used to break an animation sequence into several different parts that can be blended together. Animation Layers are covered in Chapter 33, "Using Animation Layers and Animation Modifiers."

FIGURE 10.1

The Scene Explorer dialog box is also used to manage layers.

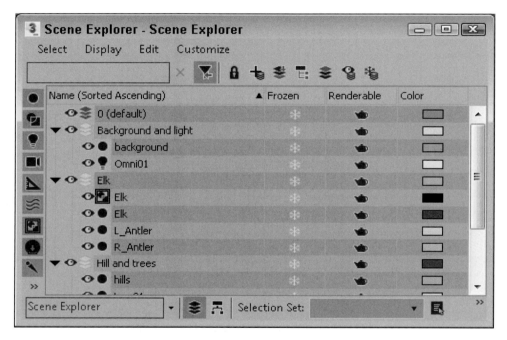

After you've set up your layers, you can control them using the Layers toolbar, shown in Figure 10.2. You can access the Layers toolbar by right-clicking the main toolbar away from the buttons and selecting the Layers toolbar from the pop-up menu or by selecting the Customize→Show UI→Show Floating Toolbars menu command.

FIGURE 10.2

Use the Layers toolbar to set the active layer.

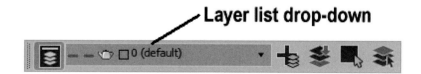

Table 6.1 lists the buttons found in the Layers toolbar.

TABLE 6.1 Layers Toolbar Buttons

Button Icon	Name	Description
	Toggle Layer Explorer	Opens special Scene Explorer named Layer Explorer
	Create New Layer (Containing Selected Objects)	Creates a new layer that includes the selected objects
	Add Selection to Current Layer	Adds any selected objects to the current layer
	Select Objects in Current Layer	Selects in the viewports all objects for the current layer
	Set Current Layer to Selection's Layer	Sets the selected object's layer as the current layer

With the Scene Explorer open, you can create new layers by clicking the Create New Layer (Containing Selected Objects) button. This adds a new layer to the manager, names it "Layer001," and includes any selected objects as part of the layer. If you click the layer's name, you can rename it. Layer 0 is the default layer to which all objects are added, if other layers don't exist. Layer 0 cannot be renamed.

Creating a new layer automatically makes the new layer the current layer, as denoted by the highlighting of the layer icon. All new objects that are created are automatically added to the current layer. Layers can be nested one within another. Only one layer can be current at a time, but several layers or objects can be highlighted. To highlight a layer, click its icon in the Scene Explorer. Highlighted layers are highlighted in blue.

A highlighted layer can be deleted, but only if it isn't the current layer and doesn't contain any objects.

Newly created objects are added to the current layer. If you forget to select the correct layer for the new objects, you can select the objects in the viewports, highlight the correct layer, and use the Add Selected Objects to Highlighted Layer button to add the objects to the correct layer.

Note

Every object can be added only to a single layer. You cannot add the same object to multiple layers.

If you expand the layer name in the Scene Explorer, you see a list of all the objects contained within the layer. If you right click the Layer icon (to the left of the layer's name) and select Properties, the Layer Properties dialog box, shown in Figure 10.3, opens. Right clicking the Object icon and selecting Properties opens the Object Properties dialog box. You also can open either of these dialog boxes by right-clicking the layer name and selecting either from the pop-up menu.

FIGURE 10.3

The Layer Properties dialog box is similar to the Object Properties dialog box, but it applies to the entire layer.

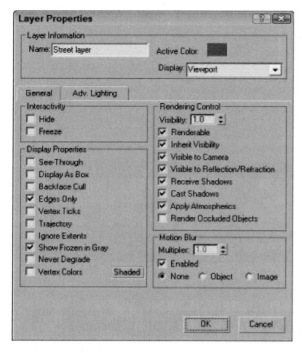

Using the layer list

In the Layers toolbar is the layer list (the drop-down list to the right of the Toggle Layer Explorer button) and its columns, which allow you to turn certain properties on and off. The properties in the columns include Hide, Freeze, Render, Color, and Radiosity. If a property is enabled, a simple icon is displayed; if disabled, a dash is displayed. If an object is set to get its property from the layer (by clicking the By Layer button in the Object Properties dialog box), a dot icon is displayed. Individual objects within a layer can have different properties. You can sort the column properties by clicking the column head.

You can toggle these properties on and off by clicking them. You also can set these properties in the Layers toolbar. The Hide toggle determines whether the layer's objects are visible in the viewports. The Freeze toggle makes objects on a layer unselectable. The Render toggle enables the layer's objects to be rendered. The Color toggle sets the layer color. Layer 0 is set to assign random colors and cannot be changed. The Radiosity toggle includes the layer's objects in the radiosity calculations.

Tutorial: Dividing a scene into layers

As a scene begins to come together, you'll start to find that it is difficult to keep track of all the different pieces. This is where the layers interface can really help. In this example, you take a simple scene and divide it into several layers.

To divide a scene into layers, follow these steps:

1. Open the Elk on hill layers.max scene file.

   You can find it in the Chap 10 directory in the downloaded content set. This file includes an Elk model.

2. Select Tools→Layer Explorer to open the Layer Explorer.

3. With the default layer selected, select the Sort by Layers button at the bottom or the Scene Explorer interface, then click the Create New Layer button and name the layer **Environment**. Click the Create New Layer button again and name this layer **Elk**. Click the Create New Layer button again, and

create a layer named **Background and light**. The Scene Explorer now includes four layers, including Layer 0.

4. Drag and drop the various objects into their respective layers with the trees and hills going into the Environment layer.

5. Expand the Elk layer by clicking the drop-down triangle icon to the left of its name.

 This displays all the objects within this layer.

You can now switch among the layers, depending on which one you want to add objects to or work on, and you can change properties as needed. For example, to focus on the elk object, you can quickly hide the other layers using the Scene Explorer. Figure 10.4 shows the various layers and the objects in each layer.

FIGURE 10.4

All objects assigned to a layer can be viewed in the Scene Explorer dialog box.

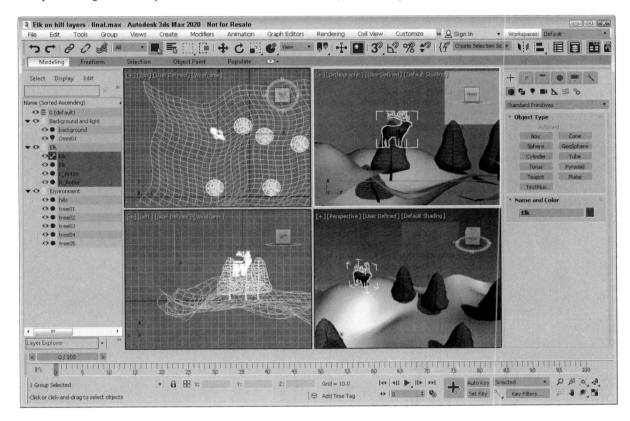

Working with Containers

Containers provide a great way to group several objects together, but containers have several additional features that extend beyond simply grouping objects together. For example, a container can be saved as an external file, which makes them easy to reuse in other scenes. Containers also can be unloaded from a scene to make the rest of the scene load quicker and to improve performance. Containers also can be locked, and the creator can set edit permissions for others.

Creating and filling containers

Containers are created using the Create→Helpers→Container menu and dragging in the viewport to create the Container icon, which looks like an open box, as shown in Figure 10.5. When a container is created, it is initially empty, but you can easily add objects to the container by clicking the Add button in the Local Content rollout of the Modify panel. This opens an Object Selection list where you can choose the objects to add to the container.

FIGURE 10.5

Containers are displayed as simple open boxes.

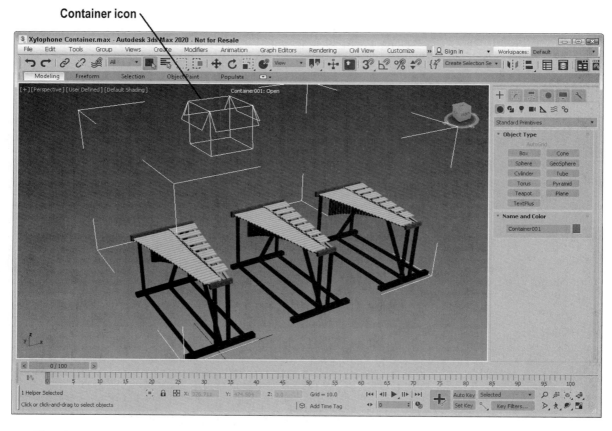

You also can create a container and automatically add the selected objects using the Tools→Containers→Create Container from Selection menu command. Container objects can be removed using the Remove button, which also opens an Object Selection list.

Another way to add and remove contents from a container is with the Containers toolbar, shown in Figure 10.6. You can access this toolbar by selecting it from the pop-up menu that appears by right-clicking the main toolbar away from the buttons. The Containers toolbar buttons are described in Table 10.2.

Note

These same commands for working with containers are available in the Tools→Containers menu.

FIGURE 10.6

The Containers toolbar

TABLE 10.2 Containers Toolbar Buttons

Toolbar Button	Name	Description
	Inherit Container	Opens a file dialog box where a source container is loaded into the current scene
	Create Container from Selection	Creates a new container and adds the current selection to the container
	Add Selected to Container	Opens an object dialog box where you can select which objects to add to the current container
	Remove Selected from Container	Removes the selected object from the container it is part of
	Load Container	Loads and displays the current container's objects in the scene
	Unload Container	Saves the current container and removes the display of its objects
	Open Container	Allows the container's objects to be edited
	Close Container	Saves the container and makes its objects so they can't be edited
	Save Container	Saves the current container
	Update Container	Reloads the saved container with any new edits
	Reload Container	Throws away any recent edits and reloads the saved container
	Make All Content Unique	Converts all displayed container objects including nested containers to a unique container
	Merge Container Source	Loads the most recent saved container without opening its nested containers
	Edit Container	Allows the current container to be opened and edited
	Override Object Properties	Uses the display settings for the container object instead of the display settings for the individual objects
	Override All Locks	Temporarily overrides the locks for the current local container

Note

The various container commands can be accessed from within the Scene Explorer by enabling the Containers toolbar.

When a container is created, you can move, rotate, scale, and change the display properties of all the objects contained within the container by simply selecting the container icon first.

Closing and saving containers

When a container is first created, it is open by default. The icon for an open container displays an open box. When you select the command to close a container, the icon changes to display a closed box. You can also select to display the Container's Name and Status and whether to expand the container's bounding box using the options in the Display rollout. The first time you close a Container, a file dialog box opens where you can save and name the container. 3ds Max container files are saved using the .maxc file extension.

Closed containers cannot be edited by anyone except their creator. Closing a container also causes its objects to be removed from the scene, but their display is still visible because the objects are referenced from the saved file. This provides a way to keep unwanted hands from changing assets that are finished.

Updating and reloading containers

When a container is loaded into a scene from a saved container file using the Inherit Content button, it is inherited into the current scene. This is different from the Inherit Container button, which gets the settings from the container definition. The inherited file maintains a link to the original container, and if the original container is edited, you can use the Update button to get the most recently saved changes.

If the container is open and allows edits, the Edit in Place button may be clicked to allow the current user to make edits to the container contents. This locks the container from all other users while the Edit in Place button is enabled. Once released, the edits are saved and made available for other users.

If you make edits that you'd like to ignore, you can use the Reload command to throw away any changes that have occurred since the last save. You also can break the link to the container with the Make All Content Unique button.

Setting container rules

Before a container is closed, you can set rules for defining exactly what content can be edited by other users. These rules are found in the Rules rollout in the Modify panel of the Command Panel, shown in Figure 10.7. The options include No Access, which is applied when a container is closed; Only Edit in Place, which gives anyone that opens the container full rights to edit the container's contents; Only Add New Objects, which gives the user that opens the container the right to add new objects to the container, but not to edit any of the existing content; and Anything Unlocked, which gives the user that opens the container access to edit anything that is unlocked. Using the four toggle buttons at the bottom of the Rules rollout, you can select to lock all modifiers, materials, transforms, and objects. Clicking the Edit button opens the Track View where you can choose tracks to lock.

FIGURE 10.7

The Rules rollout lets you define precisely what in the container can be edited.

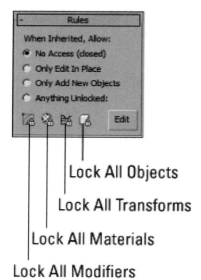

Lock All Objects

Lock All Transforms

Lock All Materials

Lock All Modifiers

Using container proxies

As an alternative to unloading a container from memory, you can specify a proxy container to take the place of the current container with the Proxies rollout of the Modify panel, shown in Figure 10.8. The drop-down list can hold multiple proxy containers, allowing you to quickly change between different resolutions. Click the Modify List button to add proxy containers to the list.

FIGURE 10.8

The Proxies rollout lets you substitute proxy containers for the current one.

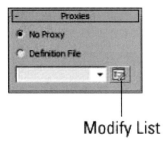

Modify List

Using the Container Explorer

The Container Explorer, shown in Figure 10.9, is another version of the Scene Explorer, except it provides a way to look at the various container objects and their content. It is accessed using the Container Explorer option in the Scene Explorer View drop-down list or using the Tools→All Global Explorers→Container Explorer. The Container Explorer also includes all the buttons from the Container toolbar and it lets you set several settings quickly all at once. As with the Scene Explorer, you can save multiple views and access them in the Tools→Local Scene Explorers menu.

FIGURE 10.9

The Container Explorer lets you work with all the container commands within a single interface.

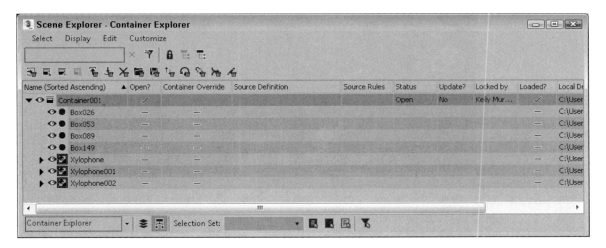

The Container Explorer also differs from the normal Scene Explorer by including column heads that are unique to containers. These column heads include information such as whether to open or closed, who it was locked by, its proxy and whether it has been saved or not.

Referencing External Objects

No man is an island, and if Autodesk has its way, no 3ds Max user will be an island, either. XRefs (which stands for eXternal References) make it easy for creative teams to collaborate on a project without having to wait for another group member to finish his or her respective production task. External references are objects and scenes contained in separate 3ds Max files and made available for reference during a 3ds Max session. This arrangement enables several artists on a team to work on separate sections of a project without interfering with one another or altering each other's work.

3ds Max includes two different types of XRefs: XRef scenes and XRef objects. You also can use XRefs for materials, modifiers, and controllers.

Note

Although XRefs are helpful and maintained for backward compatibility, in many ways using containers is the preferred method for loading in external files.

Using XRef scenes

An externally referenced scene is one that appears in the current 3ds Max session, but that is not accessible for editing or changing. The scene can be positioned and transformed when linked to a parent object and can be set to update automatically as changes are made to the source file.

As an example of how XRef scenes facilitate a project, let's say that a design team is in the midst of creating an environment for a project while the animator is animating a character model. The animator can access the in-production environment as an XRef scene in order to help him move the character correctly about the environment. The design team members are happy because the animator didn't modify any of their lights, terrain models, maps, and props. The animator is happy because he won't have to wait for the design team members to finish all their tweaking before he can get started. The end result is one large, happy production team (if they can meet their deadlines).

Choose File→Reference→XRef Scene to open the XRef Scenes dialog box, shown in Figure 10.10, which you use to load XRef scenes into a file.

FIGURE 10.10

The XRef Scenes dialog box lets you specify which scenes to load as external references.

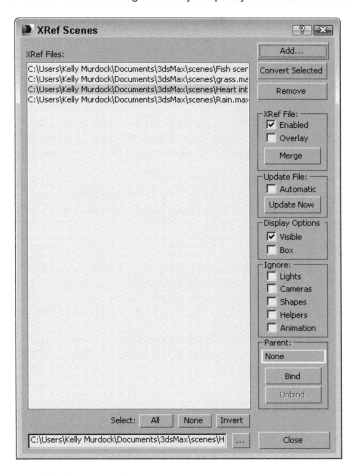

XRef scene options

In the XRef Scenes dialog box are several options for controlling the appearance of the scene objects, how often the scene is updated, and to which object the scene is bound. This dialog box is modeless, and you can open and change the options in this dialog box at any time.

The pane on the left lists all XRef scenes in the current scene. These scenes are displayed using their full path unless the Convert Local File Paths to Relative option in the Files panel of the Preference Settings dialog box is enabled. To the right are the settings, which can be different for each XRef scene in the list. To view or apply a setting, you first need to select the scene from the list. You can remove any scene by selecting it from the list and clicking the Remove button.

Caution

If an XRef scene in the list is displayed in red, the scene could not be loaded. If the path or name is incorrect, you can change it in the Path field at the bottom of the list.

The Convert Selected button converts any selected objects in the current scene to XRef objects by saving them as a separate file. This button opens a dialog box to let you name and save the new file. If no objects are selected in the current scene, this option is disabled.

Use the Enabled option to enable or disable the selected XRef scenes. Disabled scenes are displayed in gray. The Merge button lets you insert the current XRef scene into the current scene. This button removes the scene from the list and acts the same way as the File→Import→Merge command.

Updating an external scene

Automatic is a key option that can set any XRef scene to be automatically updated. Enable this option by selecting a scene from the list and checking the Automatic option box; thereafter, the scene is updated anytime the source file is updated. This option can slow the system if the external scene is updated frequently, but the benefit is that you can work with the latest update.

The Update Now button is for manually updating the XRef scene. Click this button to update the external scene to the latest saved version.

External scene appearance

Other options let you decide how the scene is displayed in the viewports. You can choose to make the external scene invisible or to display it as a box. Making an external scene invisible removes it from the viewports, but the scene is still included in the rendered output. To remove a scene from the rendered output, deselect the Enabled option.

The Ignore section lists objects such as lights, cameras, shapes, helpers, and animation; selecting them causes them to be ignored and to have no effect in the scene. If an external scene's animation is ignored, the scene appears as it does in frame 0.

Positioning an external scene

Positioning an external scene is accomplished by binding the scene to an object in the current scene (a dummy object, for example). The XRef Scenes dialog box is modeless, so you can select the object to bind to without closing the dialog box. After a binding object is selected, the external scene transforms to the binding object's pivot point. The name of the parent object is also displayed in the XRef Scenes dialog box.

Transforming the object to which the scene is bound can control how the external scene is repositioned. To unbind an object, click the Unbind button in the XRef Scenes dialog box. Unbound scenes are positioned at the World origin for the current scene.

Specifying an XRef as an overlay

The Overlay option in the XRef Scenes dialog box makes the XRef visible to the current scene but not to any other scenes that XRef the scene including the overlay. This provides a way to hide XRef content from more than one level. Overlay XRefs also make it possible to avoid circular dependencies. For example, in previous 3ds Max versions, 3ds Max wouldn't allow two designers to XRef one another's scenes, but if one of the scenes is an overlay, this can be done.

Working with XRef scenes

You can't edit XRef scenes in the current scene. Their objects are not visible in the Select from Scene dialog box or in the Track and Schematic Views. You also cannot access the Modifier Stack of external scenes' objects. However, you can make use of external scene objects in other ways. For example, you can change a viewport to show the view from any camera or light in the external scene. External scene objects are included in the Summary Info dialog box.

Tip

Another way to use XRef scenes is to create a scene with lights and/or cameras positioned at regular intervals around the scene. You can then use the XRef Scenes dialog box to turn these lights on and off or to select from a number of different views without creating new cameras.

You also can nest XRef scenes within each other, so you can have one XRef scene for the distant mountains that includes another XRef for a castle.

Note

If a 3ds Max file is loaded with XRef files that cannot be located, a warning dialog box appears, enabling you to browse to the file's new location. If you click OK or Cancel, the scene still loads, but the external scenes are missing.

Tutorial: Adding an XRef scene

As an example of a project that would benefit from XRefs, I've created a maze environment. I open a new 3ds Max file and animate a simple mouse moving through this maze that is opened as an XRef scene.

To set up an XRef scene, follow these steps:

1. Create a new 3ds Max file by choosing File→New.
2. Choose File→Reference→XRef Scenes to open the XRef Scene dialog box.
3. Click the Add button, locate the Maze.max file from the Chap 10 directory in the download content set, and click Open to add it to the XRef Scenes dialog box list, but don't close the dialog box just yet.

Tip

You can add several XRef scenes by clicking the Add button again. You also can add a scene to the XRef Scenes dialog box by dragging a .max file from Windows Explorer or from the Asset Manager window.

4. Select Create→Helpers→Dummy, and drag in the Perspective viewport to create a new Dummy object.
5. In the XRef Scenes dialog box, click the Bind button and select the dummy object. This enables you to reposition the XRef scene as needed.
6. Select the Automatic update option, and then click Close to exit the XRef Scenes dialog box.
7. Now animate objects moving through the maze, such as the mouse.

Figure 10.11 shows the Maze.max scene included in the current 3ds Max file as an XRef.

Tip

With the simple mouse animated, you can replace it at a later time with a detailed model of a furry mouse using the File→Import→Replace command.

FIGURE 10.11

The maze.max file loaded into the current file as an XRef scene

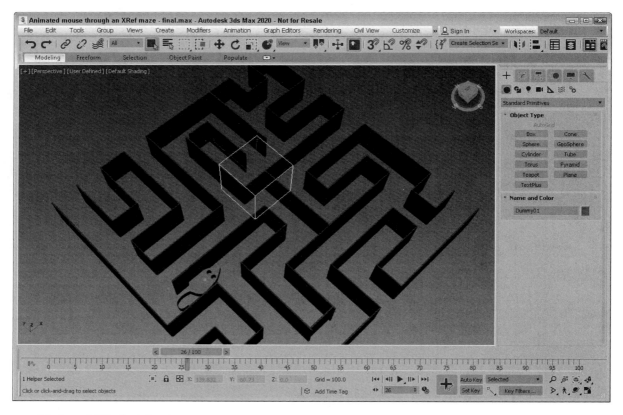

Using XRef objects

XRef objects are slightly different from XRef scenes. XRef objects appear in a scene and can be transformed and animated, but the original object's structure and Modifier Stack cannot be changed.

An innovative way to use this feature would be to create a library of objects that you could load on the fly as needed. For example, if you had a furniture library, you could load several different styles until you got just the look you wanted.

You also can use XRef objects to load low-resolution proxies of complex models in order to lighten the system load during a 3ds Max session. This method increases the viewport refresh rate.

The XRef Objects dialog box also includes several presets that you can select. For each XRef object, you can select to use the animation tracks from the XRef, from the Local object or to Merge the tracks. For Modifiers, you can select to Merge, Ignore or use the XRef modifiers. This lets you keep the XRef separate from the master copy if you choose. You can also choose to merge the materials and/or manipulators.

Many of the options in the XRef Objects dialog box, shown in Figure 10.12, are the same as in the XRef Scenes dialog box.

FIGURE 10.12

The XRef Objects dialog box lets you choose which files to look in for external objects.

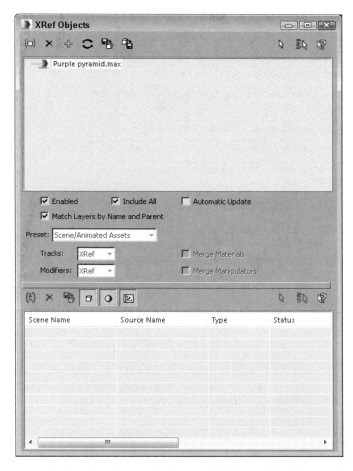

The interface buttons for the XRef Objects dialog box are listed in Table 10.3.

TABLE 10.3 XRef Objects Dialog Box

Button	Name	Description
{□}	Create XRef Record from File	Opens the XRef Merge dialog box where you can select the XRef source file.
×	Remove XRef Record	Deletes the selected XRef record.
✛	Combine XRef Record	Allows two or more selected XRef records pointing to the same file to be combined into a single record.
↻	Update	Updates the content of all XRefs.
🗗	Merge in Scene	Makes all XRefs for the selected record part of the scene file. This also removes the XRef record from the dialog box.
🗗	Convert Selected Objects to XRefs	Opens a save dialog box and saves the selected scene objects as a separate scene file, which is an XRef in the current scene.
↳	Select	Selects the objects that are part of the selected XRef record.

255

	Select by Name	Opens the Select Objects dialog box listing all objects that are part of the selected XRef record.
	Highlight Selected Object's XRef Records	Highlights the XRef record that contains the objects selected in the viewport.
	Add Objects	Opens the XRef Merge dialog box where additional objects from the selected XRef record can be loaded.
	Delete XRef Entity	Deletes the current object from the XRef record.
	Merge in Scene	Merges the selected object to the current scene and removes its XRef.
	List Objects	Filters the display to show the XRef objects.
	List Materials	Filters the display to show the XRef materials.
	List Controllers	Filters the display to show the XRef controllers.

The XRef Objects dialog box is divided into two sections. The top section displays the externally referenced source files (or records), and the lower section displays the objects, materials, or controllers selected from the source file. If multiple files are referenced, a file needs to be selected in the top pane in order for its objects and materials to be displayed in the lower pane.

The Convert Selected button works the same as in the XRef Scenes dialog box. It enables you to save the selected objects in the current scene to a separate file just like the File→Save As→Save Selected command and to have them instantly made into XRefs.

In the XRef Objects dialog box, you can choose to automatically update the external referenced objects or use the Update button or you can enable the Automatic Update option. You also can enable or disable all objects in a file with the Enabled option. The Include All option skips the XRef Merge dialog box and automatically includes all objects in the source file.

If the Merge Tracks, Modifiers, Materials, and Manipulators options are enabled before an XRef file is added, all animation tracks, modifiers, materials, and manipulators are automatically combined with the current scene instead of being referenced. When merged, the link between the source file is broken so that changes to the source file aren't propagated.

Using Material XRefs

When a source file is loaded into the XRef Objects dialog box, both its objects and materials are loaded and included. If the Merge Materials option is selected before the source file is loaded, the materials are included with the objects, but if the Merge Materials option isn't enabled, the objects and the materials appear as separate entities. You can use the List Objects and the List Materials buttons to list just one type of entity.

Materials also can be referenced from directly within the Material Editor. If you used a material in a previous scene that would be perfect in your current scene, you can just select the XRef Material from the Material/Map Browser. This material type includes fields where you can browse to an external scene file and select a specific object. The selected material is added automatically to the XRef Objects dialog box.

You can learn more about applying materials and using the Material Editor in Chapter 17, "Creating and Applying Standard Materials with the Slate Material Editor."

Merging animation tracks and modifiers

If the Merge Tracks option is enabled before loading the XRef file, any animation track that is part of the XRef object is merged and loaded along with the object or you can select to use the local animation tracks. You also can specify how modifiers are included with XRef objects, but you must select the option from the

Modifiers drop-down list before the file is selected. The XRef option loads all modifiers with the XRef object, but hides them from being edited. New modifiers can be added to the object. The Merge option adds all modifiers to the XRef object and makes these modifiers accessible via the Modifier Stack. The Ignore option strips all modifiers from the XRef object.

XRef objects appear and act like any other object in the scene. You may see a slight difference if you open the Modifier Stack. The Stack displays "XRef Object" as its only entry.

Using proxies

When an XRef object is selected in the viewport, all details concerning the XRef object—including its source filename, Object Name, and status—are listed in the Modify panel. The Modify panel also includes a Proxy Object rollout, where you can select a separate object in a separate file as a proxy object. The File or Object Name buttons open a file dialog box where you can select a low-resolution proxy object in place of a more complex object. This feature saves memory by not requiring the more complex object to be kept in memory. You also can select to enable or disable the proxy or use the proxy in rendering.

Tip

The real benefit of using proxies is to replace complex referenced objects with simpler objects that update quickly. When creating a complex object, remember to also create a low-resolution version to use as a proxy.

Configuring XRef paths

The Configure Paths dialog box includes an XRefs panel for setting the paths for XRef scenes and objects, shown in Figure 10.13. Choose Customize→Configure Project Paths to open the XRefs panel.

FIGURE 10.13

The XRefs panel in the Configure Project Paths dialog box lets you specify paths to be searched when an XRef cannot be located.

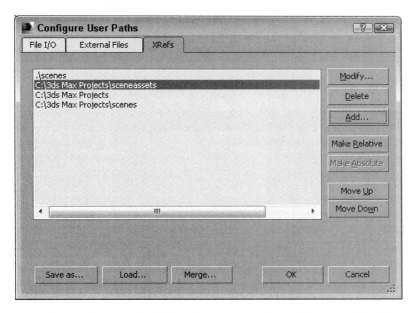

3ds Max keeps track of the path of any XRefs used in a scene, but if it cannot find them, it looks at the paths designated in the XRefs panel of the Configure Paths dialog box. For projects that use lots of XRefs, populating this list with potential paths is a good idea. Paths are scanned in the order they are listed, so place the most likely paths at the top of the list.

To add a new path to the panel, click the Add button. You also can modify or delete paths in this panel with the Modify and Delete buttons.

Tutorial: Using an XRef proxy

To set up an XRef proxy, follow these steps:

1. Open the Post box with XRef tree.max file from the Chap 10 directory in the downloaded content set.

2. Open the XRef Objects dialog box by choosing File→Reference→XRef Objects.

3. Click the Create XRef Record from File button, and locate the Park bench under a tree.max file from the Chap 10 directory in the downloaded content set.

Caution
The XRef Merge dialog box will not open if the Include All option in the XRef Objects dialog box is selected.

This file includes the old tree and park bench models. The XRef Merge dialog box, shown in Figure 10.14, automatically opens and displays a list of all the objects in the file just added.

FIGURE 10.14

The XRef Merge dialog box lets you choose specific objects from a scene.

4. Select the Tree object to add to the current scene, and click OK.

Note
If an object you've selected has the same name as an object that is currently in the scene, the Duplicate Name dialog box appears and lets you rename the object, merge it anyway, skip the new object, or delete the old version.

5. Select the Tree object in the lower pane of the XRef Objects dialog box, and click the Select button. The tree object is selected in the viewport. Close the XRef Objects dialog box.

6. With the tree object selected, open the Modify panel, in the Proxy rollout, check the Enable option, and click the button under the File Name field to open a file dialog box. Select the Tree Lo-Res.max file from the Chap 10 directory in the downloaded content set, and click the Open button.

The Merge dialog box opens.

7. Select the Cylinder01 object, and click OK.

Caution

If the proxy object has a different offset than the original object, a warning dialog box appears, instructing you to use the Reset XForm utility to reset the transform of the objects.

8. With the Tree object still selected, disable the Enable option in the Proxy Object rollout to see the actual object.

XRef objects that you add to a scene instantly appear in the current scene as you add them. Figure 10.15 shows the post box with the actual tree object. The Modify panel lets you switch to the proxy object at any time.

FIGURE 10.15

The tree object is an XRef from another scene. Its proxy is a simple cylinder.

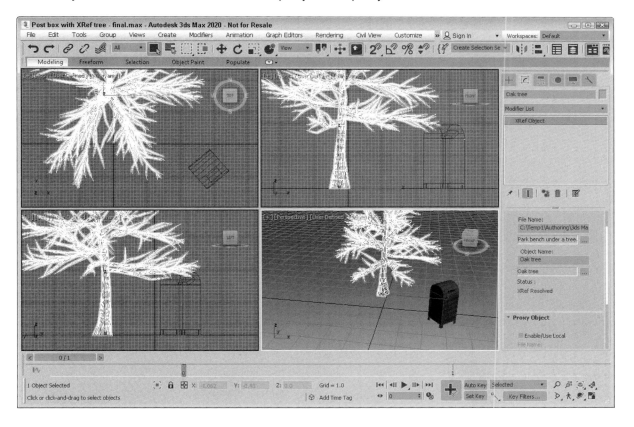

Using the File Link Manager

The File Link Manager (which also can be accessed using the File→Reference→Manage Links menu) lets you use external AutoCAD, FBX and Revit files in the same way that you use the software's XRef features.

By creating links between the current 3ds Max scene and an external AutoCAD, FBX or Revit file, you can reload the linked file when the external file has been updated and see the updates within 3ds Max.

The Manage Links dialog box, shown in Figure 10.16, includes several different presets for linking to the different external packages. These presets options are updated as the file type is selected. Within the File→Import menu are options to Link Revit, Link FBX, and Link AutoCAD. Each of these commands opens the Manage Links dialog box and automatically selects the correct preset.

FIGURE 10.16

The Manage Links dialog box lets you choose an external Revit, FBX, or AutoCAD file to link to the current 3ds Max scene.

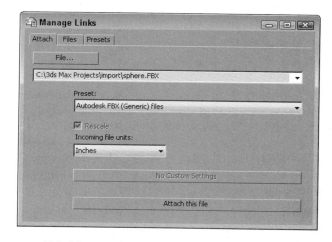

This Manage Links dialog box is divided into three panels: Attach, Files, and Presets. The Attach panel includes a File button to select and open a DWG, DXF, RVT, or FBX file. The Attach panel also includes options to choose a preset, rescale the file units and a button to attach the file.

The Files panel, shown in Figure 10.17, displays each linked file along with icons to show if the linked file has changed. A Reload button allows you to click to reload the linked file within 3ds Max. The Detach button removes the link and the Bind button removes the link but keeps the geometry within the current scene. Since binding breaks the link, any changes to the linked geometry is no longer updated when changes are made.

FIGURE 10.17

The Files panel of the Manage Links dialog box displays the path and filename for each linked file.

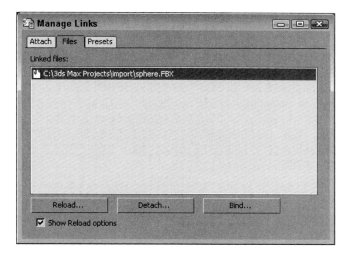

The Presets panel, shown in Figure 10.18, lets you see the current presets and define new file linking presets.

FIGURE 10.18

The Presets panel of the Manage Links dialog box displays all the currently defined presets.

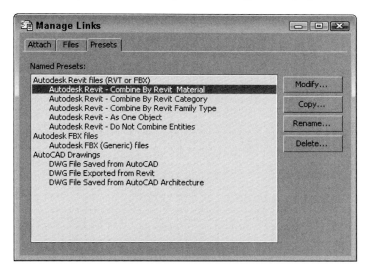

Using i-drop

To make accessing needed files from the web even easier, Autodesk has created a technology known as i-drop that lets you drag files from i-drop–supported web pages and drop them directly into 3ds Max. With i-drop, you can drag and drop 3ds Max-created light fixture models, textures, or any other 3ds Max-supported file from a light manufacturer's website into your scene without importing and positioning a file. This format allows you to add geometry, photometric data, and materials.

Using the Schematic View Window

A great way to organize and select objects is to use the Schematic View window. Every object in the Schematic View is displayed as a labeled rectangular box. These boxes (or nodes) are connected to show the relationships among them. You can rearrange them and save the customized views for later access.

You access the Schematic View window via the Graph Editors→New Schematic View menu command or by clicking its button on the main toolbar in the Autodesk® 3ds Max® 2020 software. When the window opens, it floats on top of the interface and can be moved by dragging its title bar. You also can resize the window by dragging on its borders. The window is modeless and lets you access the viewports and buttons in the interface beneath it.

The Schematic View menu options

The Schematic View menu options enable you to manage several different views. The Graph Editors→New Schematic View command opens the Schematic View window, shown in Figure 10.19. If you enter a name in the View Name field at the top of the window, you can name and save the current view. This name then appears in the Graph Editors→Saved Schematic Views submenu and also in the title bar when the saved view is open.

Every time the Graph Editors→New Schematic View menu command is used, a new view name is created and another view is added to the Saved Schematic Views submenu. The Graph Editors→Delete Schematic View command opens a dialog box in which you can select the view you want to delete.

Tip

You can open any saved Schematic View window (or a new Schematic View window) within a viewport by clicking the Point-of-View viewport label, choosing Extended Viewports→Schematic, and clicking the view name in the pop-up menu.

FIGURE 10.19

The Schematic View window displays all objects as nodes.

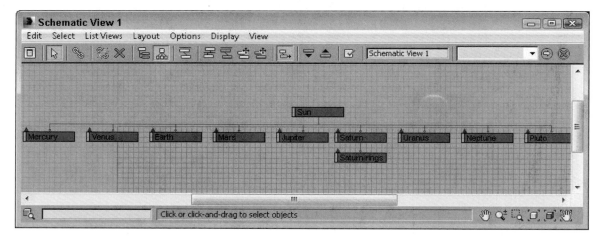

The Schematic View interface

The Schematic View window includes several common interface elements, including menus, toolbar buttons, and a right-click quadmenu. Just like the main interface, you can access the commands in many ways.

Using the Schematic View menus

The Schematic View window includes menus at the top of its interface, including Edit, Select, List Views, Layout, Options, Display, and View.

The Edit menu includes commands to Connect (C) and Unlink Selected object nodes. It also includes a Delete command, which deletes an object from the viewports as well as from the object node. The Edit menu includes commands to Assign Controllers, Wire Parameters, and open the Object Properties dialog box.

Note

Many of the keyboard shortcuts for the Schematic View window are the same as those in the main interface. If you enable the Keyboard Shortcut Override Toggle, you can use the Schematic View keyboard shortcuts.

The Select menu includes commands for accessing the Select tool (S or Q); selecting All (Ctrl+A), None (Ctrl+D), and Invert (Ctrl+I); selecting (Ctrl+C) and deselecting children; and commands to sync the selected nodes in the Schematic View with the scene (Select from Scene) and vice versa (Select to Scene).

The List Views menu determines what is shown in the Schematic View. Options include All Relationships, Selected Relationships, All Instances, Selected Instances, Show Occurrences, and All Animated Controllers. Many of these options are available in the Display Floater as well.

The Layout menu includes various options for controlling how the nodes are arranged. The Align submenu lets you align selected nodes to the Left, Right, Top, Bottom, Center Horizontal, or Center Vertical. You also can Arrange Children or Arrange Selected. The Free Selected (Alt+S) and Free All (Alt+F) commands keep nodes from being auto arranged. With the Layout menu, you also can Shrink Selected, UnShrink Selected, UnShrink All, and Toggle Shrink (Ctrl+S).

The Options menu lets you select the Always Arrange option and view mode (either Hierarchy or Reference Mode). You also can select the Move Children (Alt+C) option and open the Schematic View Preferences (P) dialog box.

The Display menu provides access to the Display Floater. The Display Floater (D) command opens a floating panel, which can be used to select the types of nodes to display. You also can hide and unhide nodes and expand or collapse the selected node.

The View menu includes commands for selecting the Pan (Ctrl+P), Zoom (Alt+Z), and Zoom Region (Ctrl+W) tools. You also can access the Zoom Extents (Alt+Ctrl+Z), Zoom Extents Selected (Z), and Pan to Selected commands. The View menu also includes options to Show/Hide Grid (G), Show/Hide Background, and Refresh View (Ctrl+U).

Learning the toolbar buttons

You can select most of these commands from the toolbar. Many of the toolbar buttons are toggle switches that enable and disable certain viewing modes. The background of these toggle buttons is highlighted blue when selected. You'll also find some buttons along the bottom of the window. All Schematic View icon buttons are shown in Table 10.4 and are described in the following sections.

Note

The Schematic View toolbar buttons are permanently docked to the interface and cannot be removed.

TABLE 10.4 Schematic View Toolbar Buttons

Toolbar Button	Name	Description
	Display Floater	Opens the Display Floater, where you can toggle which items are displayed or hidden.
	Select (S)	Toggles selection mode on, where nodes can be selected by clicking.

	Connect (C)	Enables you to create links between objects in the Schematic View window; also used to copy modifiers and materials between objects.
	Unlink Selected	Destroys the link between the selected object and its parent.
	Delete Objects	Deletes the selected object in both the Schematic View and the viewports.
	Hierarchy Mode	Displays all child objects indented under their parent objects.
	References Mode	Displays all object references and instances and all materials and modifiers associated with the objects.
	Always Arrange	Causes all nodes to be automatically arranged in a hierarchy or in references mode, and disables moving of individual nodes.
	Arrange Children	Automatically rearranges the children of the selected object nodes.
	Arrange Selected	Automatically rearranges the selected object nodes.
	Free All	Allows all objects to be freely moved without being automatically arranged.
	Free Selected	Allows selected objects to be freely moved without being automatically arranged.
	Move Children	Causes children to move along with their parent node.
	Expand Selected	Reveals all nodes below the selected node.
	Collapse Selected	Rolls up all nodes below the selected node.
	Preferences	Opens the Schematic View Preferences dialog box.
Schematic View 1	View Name field	Allows you to name the current display; named displays show up under the Graph Editors→Saved Schematic View submenu.
bookmark	Bookmark Name	Marks a selection of nodes to which you can return later.
	Go to Bookmark	Zooms and pans to the selected bookmarked objects.
	Delete Bookmark	Removes the bookmark from the Bookmark selection list.

As you navigate the Schematic View window, you can save specific views as bookmarks by typing an identifying name in the Bookmark drop-down list. To recall these views later, select them from the drop-down list and click the Go to Bookmark icon in the Schematic View toolbar. Bookmarks can be deleted with the Delete Bookmark button.

Note

Most of the menu commands and toolbar buttons are available in a pop-up menu that you can access by right-clicking in the Schematic View window.

Navigating the Schematic View window

As the number of nodes increases, it can become tricky to locate and see the correct node to work with. Along the bottom edge of the Schematic View window are several navigation buttons that work similarly to the Viewport Navigation Control buttons. Using these buttons, you can pan, zoom, and zoom to the extents of all nodes. These buttons are described in Table 10.5.

The Schematic View navigation buttons also can be accessed from within the View menu. These menu commands include Pan Tool (Ctrl+P), Zoom Tool (Alt+Z), Zoom Region tool (Ctrl+W), Zoom Extents (Alt+Ctrl+Z), Zoom Extents Selected (Z), and Pan to Selected.

Tip

You also can navigate the Schematic View window using the mouse and its scroll wheel. Scrubbing the mouse wheel zooms in and out of the window in steps. Holding down the Ctrl key and dragging with the scroll wheel button zooms smoothly in and out of the window. Dragging the scroll wheel pans within the window.

TABLE 10.5 Schematic View Navigation Buttons

Toolbar Button	Name	Description
	Zoom Selected Viewport Object	Zooms in on the nodes that correspond to the selected viewport objects.
	Search Name field	Locates an object node when you type its name.
	Pan	Moves the node view when you drag in the window.
	Zoom	Zooms when you drag the mouse in the window.
	Region Zoom	Zooms to an area selected when you drag an outline.
	Zoom Extents	Increases the window view until all nodes are visible.
	Zoom Extents Selected	Increases the window view until all selected nodes are visible.
	Pan to Selected	Moves the node view at the current zoom level to the selected objects.

Working with Schematic View nodes

Every object displayed in the scene has a *node*—a simple rectangular box that represents the object or attribute. Each node contains a label, and the color of the node depends on the node type.

Node colors

Nodes have a color scheme to help identify them. The colors of various nodes are listed in Table 10.6.

Table 10.6 Schematic View Node Colors

Color	Name
White	Selected node
Blue	Geometry Object node
Cyan	Shape Object node
Yellow	Light Object node
Dark Blue	Camera Object node
Green	Helper Object node
Purple	Space Warp Object node
Goldenrod	Modifier node
Dark Yellow	Base Object node
Brown	Material node
Dark Green	Map node

Salmon	Controller node
Magenta	Parameter Wires

Note

If you don't like any of these colors, you can set the colors used in the Schematic View using the Colors panel of the Customize→Customize User Interface dialog box.

Selecting nodes

When you click the Select (S) button, you enter select mode, which lets you select nodes within the Schematic View window by clicking the object node. You can select multiple objects by dragging an outline over them. Holding down the Ctrl key while clicking an object node selects or deselects it. Selected nodes are shown in white.

The Select menu includes several selection commands that enable you to quickly select (or deselect) many nodes, including Select All (Ctrl+A), Select None (Ctrl+D), Select Invert (Ctlr+I), Select Children (Ctrl+C), and Deselect Children.

If the Select→Sync Selection option in the Select menu is enabled, the node of any object that is selected in the viewports is also selected in the Schematic View window, and vice versa. If you disable the Sync Selection option, you can select different objects in the viewports and in the Schematic View at the same time. The node of the object selected in the viewports is outlined in white, and the interior of selected nodes is white. To select all the objects in the viewports that match the selected nodes without the Sync Selection option enabled, just use Select→Select to Scene.

Tip

All animated objects have their node border drawn in red.

Rearranging nodes

The Schematic View includes several options for arranging nodes. In the Options menu, you can toggle between Hierarchy and Reference modes. Hierarchy mode displays the nodes vertically with child objects indented under their parent. Reference mode displays the nodes horizontally, allowing for plenty of room to display all the various reference nodes under each parent node. Figure 10.20 shows these modes side by side.

FIGURE 10.20

The Schematic View window can automatically arrange nodes in two different modes: Hierarchy and Reference.

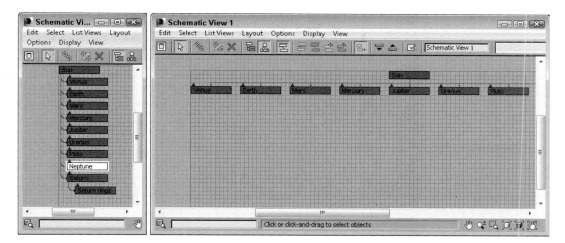

You can move nodes and rearrange them in any order. To move a node, simply click and drag it to a new location. When a node is dragged, all selected nodes move together, and any links follow the node movement. If a child node is moved, all remaining child nodes collapse together to maintain the specified arrangement mode. The moved node then becomes free, which is designated by an open rectangle on the left edge of the node. Figure 10.21 shows two nodes that were moved and thereby became free. The other children automatically moved closer together to close the gaps made by the moving nodes.

FIGURE 10.21

Free nodes are moved independent of the arranging mode.

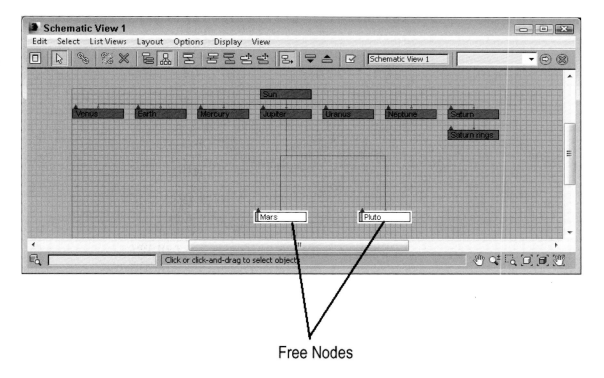

Free Nodes

Using the Layout→Free Selected (Alt+S) and Free All (Alt+F) menu commands, you can free the selected nodes or all nodes. You also can auto arrange all the children of a node with the Layout→Arrange Children menu command or arrange just the selected nodes (Layout→Arrange Selected). The Options→Move Children (Alt+C) command causes all children to be moved along with their parent when the parent is moved. This causes free and non-free nodes to move with their parent.

If the Options→Always Arrange option is enabled, 3ds Max automatically arranges all the nodes using either the Hierarchy mode or Reference mode, but you cannot move any of the nodes while this option is enabled. If the Always Arrange option is enabled, the Arrange Children, Free All (Alt+F), Free Selected (Alt+S), Move Children (Alt+C), and all the Align options are all disabled. If two or more nodes are selected, you can align them using the Layout→Align menu. The options include Left, Right, Top, Bottom, Center Horizontal, and Center Vertical.

Hiding, shrinking, and deleting nodes

If your Schematic View window starts to get cluttered, you can always hide nodes to simplify the view. To hide a node, select the nodes to hide and use the Display→Hide Selected menu command. The Display→Unhide All menu command can be used to make the hidden nodes visible again.

Note

If you hide a parent object, its children nodes are hidden also.

Another useful way to reduce clutter in the Schematic View window is with the Layout→Shrink Selected command. This command replaces the rectangular node with a simple dot, but all hierarchical lines to the node are kept intact. Figure 10.22 shows a Schematic View with several shrunk nodes. Shrunk nodes can be unshrunk with the Layout→UnShrink Selected and UnShrink All menu commands.

Note

The Shrink commands work only when Layout→Toggle Shrink (Ctrl+S) is enabled. With this command, you can turn on and turn off the visibility of shrunken nodes.

FIGURE 10.22

Shrunken nodes appear as simple dots in the Schematic View.

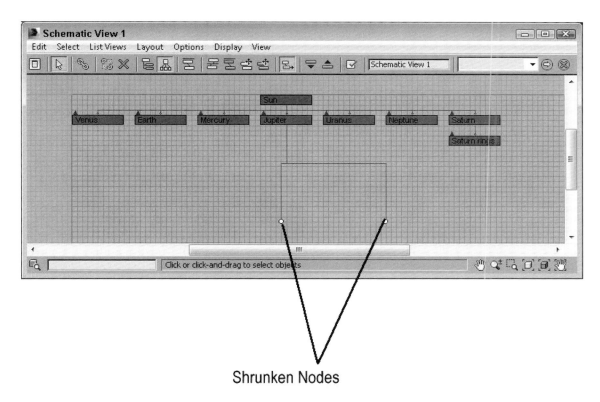

Shrunken Nodes

To delete a node, select the node and click the Delete Objects button on the Schematic View toolbar or press the Delete key. If several nodes are selected, they are all deleted. This deletes the object in the viewports also.

Renaming objects

In the Schematic View window, you can rename objects quickly and conveniently. To rename an object, click a selected node and click again to highlight the text. When the text is highlighted, you can type the new name for the object. This works only for nodes that have a name, which includes materials.

Tutorial: Rearranging the solar system

To practice moving nodes around, you'll order the solar system model. When 3ds Max places nodes in the Schematic View, it really doesn't follow any specific order, but you can move them as needed by hand.

To rearrange the solar system nodes, follow these steps:

1. Open the Ordered solar system.max file from the Chap 10 directory in the downloaded content set.

   This file includes several named spheres representing the solar system.

2. Select Graph Editors→New Schematic View to open the Schematic View window.

   All planets are displayed as blue nodes under the Sun object.

3. Select Options→Reference Mode (if it is not already selected) to position all the nodes horizontally. Click the Select tool on the Schematic View toolbar, or press the S key.

4. Make sure the Options→Always Arrange option is disabled. Then click and drag the Mercury node to the left, and place it in front of the Venus node.

5. Select the Options→Move Children (Alt+C) menu command to enable it, and drag and drop the Saturn node between the Jupiter and Uranus nodes.

With the Move Children option enabled, the Saturn rings node moves with its parent.

6. Drag and drop the Pluto node beyond the Neptune node.

7. Select all the planet nodes, and choose Layout→Align→Top to align all the nodes together.

Note

Although astronomers no longer classify Pluto as a planet, I doubt that this book is required reading for astronomers. I think we can get away with calling Pluto a planet.

Figure 10.23 shows the rearranged hierarchy with all the planets lined up in order.

FIGURE 10.23

After rearranging nodes to the correct order, the planets are easy to locate.

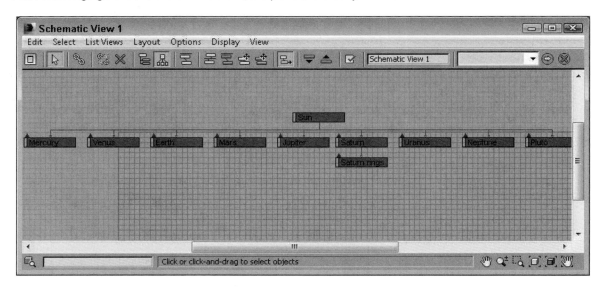

Working with Hierarchies

Another key benefit of the Schematic View is to see the relationships between different objects. With the Schematic View open, you can quickly tell which objects are children and which are parents. You also can see which objects have modifiers and which have materials applied. You can get a wealth of knowledge from the Schematic View.

Using the Display floater

With all relationships enabled, the Schematic View becomes a mess. Luckily, you can control which Relationships and which Entities are displayed using the Display floater, shown in Figure 10.24.

FIGURE 10.24

The Display floater can turn nodes and lines on and off in the Schematic View.

The top section of the Display floater shows or hides relationships between nodes, which are displayed as lines. The relationships that you can control include Constraints, Controllers, Parameter Wires, Light Inclusion, and Modifiers. If you hold the mouse over these relationship lines, the details of the relationship are shown in the tooltip that appears.

Tip

For some relationships, you can double-click the relationship line to open a dialog box where you can edit the relationship. For example, double-clicking a Parameter Wire relationship line opens the Parameter Wiring dialog box.

The lower section of the Display floater lets you show or hide entities that are displayed as nodes, including Base Objects, Modifier Stack, Materials, and Controllers. The P, R, and S buttons let you turn on Positional, Rotational, and Scale controllers. When a node has a relationship with another node, the right end of the node displays an arrow. Clicking this arrow toggles the relationship lines on and off.

The Expand button shows the actual nodes when enabled but only an arrow that can be clicked to access the nodes if disabled. The Focus button shows all related objects as colored nodes, and all other nodes are unshaded.

Figure 10.25 shows a Schematic View with the Base Objects and Controllers Entities selected in the Display floater. The Expand button also is disabled. This makes up and down arrows appear above each node. Clicking the up arrow collapses the node, rolling it up into its parent. Clicking the down arrow expands the node and displays the Base Object and Controller nodes for the node that you clicked, such as the Earth node in Figure 10.23. You also can expand and collapse nodes with the Display→Expand Selected and Collapse Selected menu commands.

FIGURE 10.25

Schematic View nodes can be collapsed or expanded by clicking the up and down arrows.

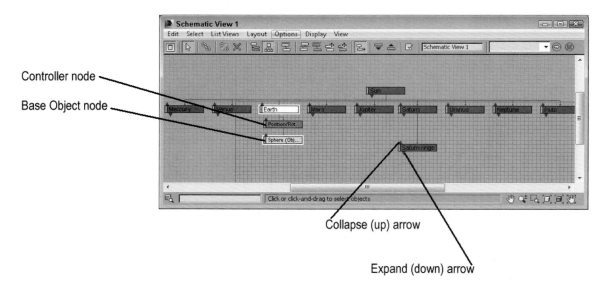

Hierarchical relationships are shown as lines that connect the nodes. Even if the nodes are moved, the lines follow as needed to show the relationship between the nodes.

Connecting nodes

To create a hierarchy, use the Edit→Connect menu command, or press the C shortcut, or click the Connect button on the Schematic View toolbar. This enters Connect mode, which lets you link objects together; copy modifiers, materials, or controllers between nodes; or even wire parameters.

For linking nodes, the Connect button works the same way here as it does on the main toolbar—selecting the child node and dragging a line from the child node to its parent. You can even select multiple nodes and link them all at once.

The Edit→Unlink Selected menu command (and toolbar button) destroys the link between any object and its immediate parent. Remember that every child object can have only one parent.

Copying modifiers and materials between nodes

Before you can copy materials or modifiers between nodes, you need to make sure they are visible. Material nodes and modifier nodes show up only if they are enabled in the Display floater. You can access this floater by clicking the Display floater button (or by pressing the D key).

To copy a material or modifier, select the material node for one object, click the Connect (C) button, and drag the material to another object node.

Note

In the Schematic View, materials can be copied only between objects; you cannot apply new materials from the Material Editor to Schematic View nodes.

When modifiers are copied between nodes, a dialog box appears, giving you the chance to Copy, Move, or Instance the modifier.

You can learn more about applying modifiers and the Modifier Stack in Chapter 11, "Accessing Subobjects and Modifiers and Using the Modifier Stack."

Assigning controllers and wiring parameters

If controller nodes are visible, you can copy them to another node using the same technique used for materials and modifiers using the Connect (C) tool. You also can assign a controller to an object node that doesn't have a controller using the Edit→Assign Controller menu command. This opens the Assign Controller dialog box, shown in Figure 10.26, where you can select the controller to apply.

FIGURE 10.26

Controllers can be assigned using the Schematic View window.

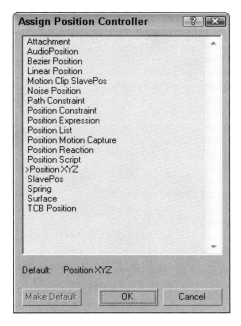

Nodes can be wired using the Schematic View window. To wire parameters, select the node you want to wire and select Edit→Wire Parameters. A pop-up menu of wire parameters appears that works the same as in the viewports. All parameter wiring relationships are shown in magenta.

You can learn more about parameter wiring in Chapter 32, "Understanding Animation and Keyframes."

Tutorial: Linking a character with the Schematic View

Perhaps one of the greatest benefits of the Schematic View is its ability to link objects. This can be tricky in the viewports because some objects are small and hidden behind other items. The Schematic View with its nodes that are all the same size makes it easy, but only if the objects are named correctly.

To link a character model using the Schematic View, follow these steps:

1. Open the Linked Marvin Moose.max file from the Chap 10 directory in the downloaded content set.

   This file includes a model of Marvin Moose with no links between the various parts.

2. Select Graph Editors→New Schematic View to open a Schematic View window, and name the view **Linked character**. Click the Zoom Region button in the lower-right corner of the Schematic View interface, and drag over all the nodes at the left end of the Schematic View.

   For this model, you want the shirt to be the parent node.

3. Click the Connect button on the toolbar (or press the C key), and drag from the L_Hand node to the Shirt node to link the two nodes. Continue linking by connecting the following nodes: R_Hand to Shirt, Head to Shirt, Glasses to Head, Antlers to Head, R_Shoe to R_Ankle, L_Shoe to L_Ankle, R_Ankle to Pants, L_Ankle to Pants, and Pants to Shirt.

This completes the hierarchy.

Note

Typically, when rigging characters, you want the pelvis to be the parent object because it is the center of most of the character movement.

Figure 10.27 shows the final geometry object nodes of the linked character. If you move the Shirt part in the viewports, all the parts move together.

FIGURE 10.27

All character parts are now linked to the Moose's Shirt part.

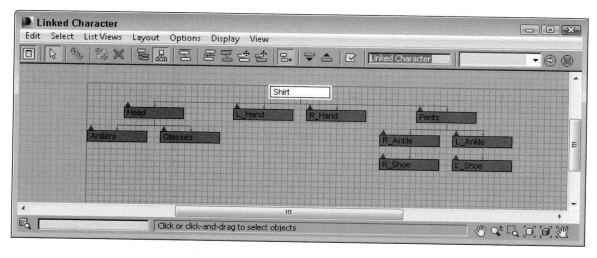

Setting Schematic View Preferences

The Preferences button (or the Options→Preferences command) opens the Schematic View Preferences dialog box, shown in Figure 10.28, where you can set which items are displayed or hidden, set up grids and background images, and specify how the Schematic View window looks.

FIGURE 10.28

The Schematic View Preferences dialog box lets you customize many aspects of the Schematic View window.

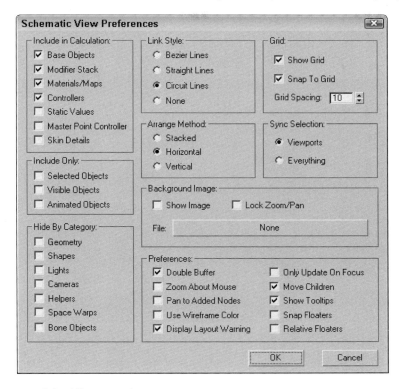

Limiting nodes

When the Schematic View window is opened, 3ds Max traverses the entire hierarchy looking for objects and features that can be presented as nodes. If you have a complex scene and don't intend to use the Schematic View to see materials or modifiers, you can disable them in the Include in Calculation section of the Schematic View Preferences dialog box. This provides a way to simplify the data presented. With less data, locating and manipulating what you are looking for becomes easier.

The Include in Calculation section includes options for limiting the following:

* **Base Objects:** The geometry type that makes up a node. The node is the named object, such as Earth; the Base Object is its primitive, such as Sphere (Object).

* **Modifier Stack:** Identifies all nodes with modifiers applied.

* **Materials/Maps:** Identifies all nodes with materials and maps applied.

* **Controllers:** Identifies all nodes that have controllers applied.

* **Static Values:** Displays unanimated parameter values.

* **Master Point Controller:** Displays nodes for any subobject selections that include controllers.

* **Skin Details:** Displays nodes for the modifiers and controllers that are used when the Skin modifier is applied to a bones system.

You also can limit the number of nodes by using the Include Only options. The Selected Objects option shows only the objects selected in the viewports. The nodes change as new objects are selected in the viewports. The Visible Objects option displays only the nodes for those objects that are not hidden in the viewports, and the Animated Objects option displays only the nodes of the objects that are animated.

Object categories that can be hidden include Geometry, Shapes, Lights, Cameras, Helpers, Space Warps, and Bone Objects. Figure 10.29 shows a single sphere object in the Schematic View window with all the Include in Calculation options selected.

FIGURE 10.29

Without limiting nodes, the Schematic View window can get very busy.

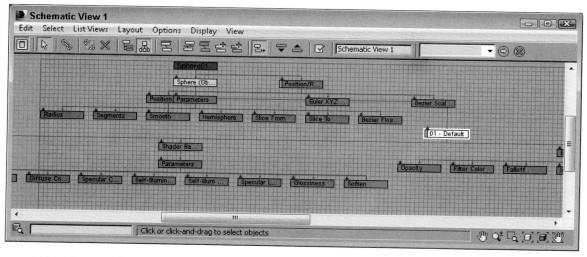

Working with grids and backgrounds

The Schematic View Preferences dialog box includes settings to Show Grid, Snap to Grid, and set Grid Spacing. The keyboard shortcut for toggling the grid on and off is G. Enabling the Snap to Grid option makes the nodes snap to the closest grid intersection. This helps keep the nodes aligned and looking neat.

The Background Image section of the Schematic View Preferences dialog box includes a File button that opens a file dialog box when clicked. Selecting an image file opens and displays the image as a background image. This is helpful as you arrange nodes. You need to select the Show Image option to see the background image; the Lock Zoom/Pan option locks the nodes to the background image so zooming in on a set of nodes also zooms in on the background image.

Tip

One of the easiest ways to get a background image of a model to use in the Schematic View is to render a single frame and save it from the Rendered Frame Window to a location where you can reopen it as the Schematic View background. If you want to print the hierarchy, you can do a screen capture of the Schematic View window, but it would be nice to have a print feature added to the window.

Display preferences

In the Schematic View Preferences dialog box, you can select the style to use for relationship lines. The options include Bezier, Straight, Circuit, and None. When the Always Arrange, Arrange Children, or Arrange Selected options are used, you can select to have the nodes arranged Stacked, Horizontal, or Vertical. The Sync Selection options enable you to sync the selection between the Schematic View and the Viewports or between Everything. If the Everything option is selected, not only are geometry objects in the viewports selected, but if a material is selected in the Schematic View, the material is selected in the Material Editor also. Sync Selection Everything also affects the Modifier Stack, the Controller pane in the Display panel, and the Wiring Parameters dialog box.

The Schematic View Preferences dialog box also includes a Preferences section. These preference settings include Double Buffer, which enables a double-buffer display and helps improve the viewport update performance. The Zoom About Mouse preference enables zooming by using the scroll wheel on your mouse or by pressing the middle mouse button while holding down the Ctrl key. The Move Children option causes children nodes to move along with their parent. The Pan to Added Nodes preference automatically resizes and moves the nodes to enable you to view any additional nodes that have been added.

The Use Wireframe Color option changes the node colors to be the same as the viewport object color. The Display Layout Warning preference lets you disable the warning that appears every time you use the Always Arrange feature. The Only Update On Focus option causes the Schematic View to update only when the window is selected. Until then, any changes are not propagated to the window. This can be a timesaver when complex scenes require redraws.

The Show Tooltips option allows you to disable tooltips if you desire. Tooltips show in the Schematic View window when you hover the cursor over the top of a node. Tooltips can be handy if you've zoomed out so far that you can't read the node labels; just move the cursor over a node, and its label appears. The Snap Floaters option allows the Display and List floaters to be snapped to the edge of the window for easy access, and the Relative Floaters option moves and resizes the floaters along with the Schematic View window.

Tutorial: Adding a background image to the Schematic View

You can position nodes anywhere within the Schematic View window. For example, you can position the nodes to look something like the shape of the model that you're linking. When positioning the different objects, having a background image is really handy.

To add a background image for the Schematic View, follow these steps:

1. Open the Marvin Moose with background.max file in the Chap 10 directory in the downloaded content set.

   This file uses the same Marvin Moose model used in the preceding example.

2. With the Perspective viewport maximized, select Rendering→Render. The resulting image opens in the Rendered Frame Window.

3. Click the Save Image button in the upper-left corner. Save the image as **Marvin Moose-front view**. Then close the Rendered Frame Window.

4. Select Graph Editors→New Schematic View to open a Schematic View window, and name the view **Background**. Click the Preferences button on the Schematic View toolbar, and click the File button in the Background Image section.

5. Locate the saved image, and click the Open button. Select the Show Image option in the Schematic View Preferences dialog box, and click OK.

   You can perform this step using the image file saved in the Chap 10 directory in the downloaded content set, if you so choose.

6. Select the View→Show Grid menu command (or press the G key) to turn off the grid. Drag the corner of the Schematic View interface to increase the size of the window so the whole background image is visible.

7. Before moving any of the nodes, enable the Lock Zoom/Pan option in the Schematic View Preferences dialog box so the image resizes with the nodes. Then select each of the nodes, and drag them so they are roughly positioned on top of the part they represent. Start by moving the parent objects first, and then work to their children.

Figure 10.30 shows all the nodes aligned over their respective parts. From this arrangement, you can clearly see how the links are organized.

FIGURE 10.30

Using a background image, you can see how the links relate to the model.

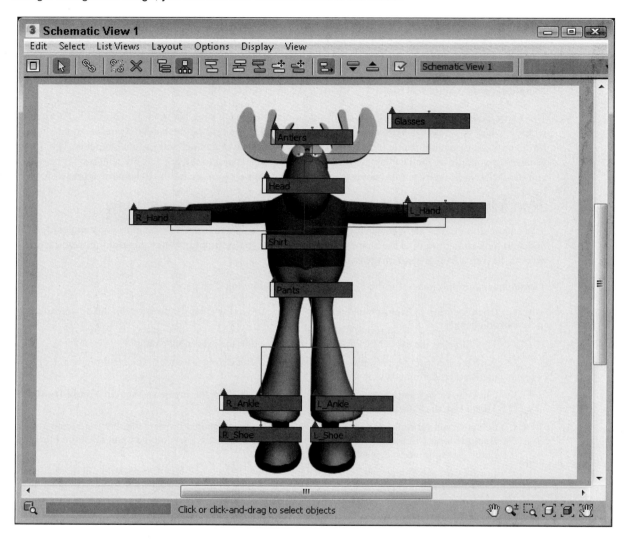

Using List Views

One of the last uses of the Schematic View is to list all nodes that have things in common. Using the List Views menu, you can select to see All Relationships, Selected Relationships, All Instances, Selected Instances, Show Occurrences, and All Animated Controllers.

The List Views→All Relationships menu command displays a separate dialog box, shown in Figure 10.31, containing a list of nodes and their relationships. The Selected Relationships menu command limits the list to only selected objects with relationships. The List Views dialog box also includes a Detach button to remove the relationships if desired. Double-clicking a relationship in the list opens its dialog box, where you can edit the relationship.

Tip
You can click each column head to sort the entries.

FIGURE 10.31

The List Views dialog box includes a list of nodes with relationships.

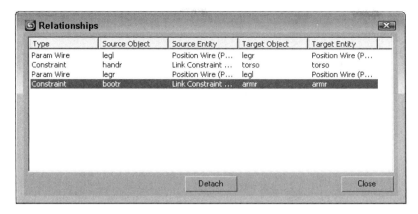

The List Views→All Instances menu command displays all the instances found in the scene. This includes all types of instances, including geometry, modifiers, controllers, and so on. For the Instances list view, the Detach button is replaced with a Make Unique button.

Note
Another way to identify instances is to look for bold text in the node. All label text for all instanced nodes is displayed in bold.

If a node is selected and you want to see all other nodes that share the same type of relationship or share a property, the List Views→Show Occurrences menu command displays them. The final list view shows All Animated Controllers.

Summary

Working with layers, containers and XRefs lets you combine the work of several users and creatively collaborate across teams.

Some tasks in the viewport, such as linking objects into a hierarchy, can be difficult. The Schematic View represents all data as simple rectangular nodes. These nodes make easy work of accomplishing a variety of tasks.

This chapter covered the following topics:

* Separated objects using layers

* Using containers and externally referenced scenes and objects to work on the same project at the same time as your fellow team members without interfering with their work (or they with yours)

* Configuring XRef paths to help 3ds Max track your XRef Scenes and Objects

* Linking external Revit, FBX and AutoCAD files

* Viewing all objects as nodes using the Schematic View window

* Learning the Schematic View interface

* Using the Schematic View window to select, delete, and copy objects, materials, and modifiers

* Using the Schematic View to assign controllers and wire parameters

* Setting preferences for the Schematic View window

* Listing views of nodes with common properties

In the next chapter, you jump headfirst into modeling by covering the basics of modeling and working with subobjects, modeling modifiers, and using the Modifier Stack.

Part III

Modeling 3D Assets

IN THIS PART

Chapter 11

Accessing Subobjects and Modifiers and Using the Modifier Stack

IN THIS CHAPTER

Understanding the modeling types

Using normals

Working with subobjects

Using Soft Selection

Introducing Modifiers

Using the Modifier Stack to manage modifiers

Learning to work with modifier gizmos

Modeling is the process of pure creation. Whether you're sculpting, building with blocks, construction work, carving, architecture, or advanced injection molding, you can create objects in many different ways. The Autodesk® 3ds Max® 2020 software includes many different model types and even more ways to work with them.

This chapter introduces the various modeling methods in 3ds Max. It also explains the common modeling components, including normals and subobjects. The purpose of this chapter is to whet your appetite for modeling and to cover some of the general concepts that apply to all models. More specific details on the various modeling types are presented in the subsequent chapters, so onward into the realm of creation.

Think for a moment of a woodshop with all its various (and expensive) tools and machines. Some tools, like a screwdriver or a sander, are simple, and others, like a lathe or router, are more complex, but they all change the wood (or models) in different ways. In some ways, you can think of modifiers as the tools and machines that work on 3D objects.

Each woodshop tool has different parameters that control how it works, such as how hard you turn the screwdriver or the coarseness of the sandpaper. Likewise, each modifier has parameters you can set that determine how it affects the 3D object.

Modifiers can be used in a number of different ways to reshape objects, apply material mappings, deform an object's surface, and perform many other actions. Many different types of modifiers exist. This chapter introduces you to the concept of modifiers and explains the basics on how to use them. The chapter concludes by exploring some of the common modifiers that are used to deform geometry objects.

Exploring the Model Types

You can climb a mountain in many ways, and you can model one in many ways too. You can make a mountain model out of primitive objects like blocks, cubes, and spheres, or you can create one as a polygon mesh. As your experience grows, you'll discover that some objects are easier to model using one method, and some are easier using another. 3ds Max offers several different modeling types to handle various modeling situations.

Parametric objects versus editable objects

All geometric objects in 3ds Max can be divided into two general categories: parametric objects and editable objects. *Parametric* means that the geometry of the object is controlled by variables called parameters. Modifying these parameters modifies the geometry of the object. This powerful concept gives parametric objects lots of flexibility. For example, the sphere object has a parameter called Radius. Changing this parameter changes the size of the sphere. Parametric objects in 3ds Max include the primitive objects found in the Create menu.

Editable objects do not have this flexibility of parameters, but they deal with subobjects and editing functions. The editable objects include Editable Spline, Mesh, Poly, Patch, and NURBS (Non-Uniform Rational B-Splines). Editable objects are listed in the Modifier Stack with the word *Editable* in front of their base object (except for NURBS objects, which are simply called NURBS Surfaces). For example, an editable mesh object is listed as Editable Mesh in the Modifier Stack. This identifies it as an object that is edited by adjusting its subobjects.

Note

Actually, NURBS objects are different beasts altogether. When created using the Create menu, they are parametric objects, but after you select the Modify panel, they are editable objects with a host of subobject modes and editing functions.

Editable objects aren't created; instead, they are converted or modified from another object. When a primitive object is converted to a different object type like an Editable Mesh, it loses its parametric nature and can no longer be changed by altering its base parameters. Editable objects do have their advantages, though. You can edit subobjects such as vertices, edges, and faces of meshes—all things that you cannot edit for a parametric object. Each editable object type has a host of functions that are specific to its type. These functions are discussed in the coming chapters.

Note

Several modifiers enable you to edit subobjects while maintaining the parametric nature of an object. These include Edit Patch, Edit Mesh, Edit Poly, and Edit Spline.

3ds Max includes the following model types:

* **Primitives:** Basic parametric objects such as cubes, spheres, and pyramids. The primitives are divided into two groups consisting of Standard and Extended Primitives. The AEC Objects also are considered primitive objects. A complete list of primitives is covered in Chapter 5, "Creating and Editing Primitive Objects."

* **Shapes and splines:** Simple vector shapes such as circles, stars, arcs, and text, and splines such as the Helix. These objects are fully renderable. The Create menu includes many parametric shapes and splines. These parametric objects can be converted to Editable Spline objects for more editing. These are covered in Chapter 12, "Drawing and Editing 2D Splines and Shapes."

* **Meshes:** Complex models created from many polygon faces that are smoothed together when the object is rendered. These objects are available only as Editable Mesh objects. Meshes are covered in Chapter 13, "Modeling with Polygons."

* **Polys:** Objects composed of polygon faces, similar to mesh objects, but with unique features. These objects also are available only as Editable Poly objects. Poly objects are covered in Chapter 13,

"Modeling with Polygons." The Graphite Modeling tools are designed to work on Editable Poly objects. These tools are covered in Chapter 14, "Using the Graphite Modeling Tools and Painting with Objects."

* **OpenSubDiv:** Applied as a modifier, this modeling type adds more polygons as needed to make the model smooth.

* **Patches:** Based on spline curves; patches can be modified using control points. The Create menu includes two parametric Patch objects, but most objects also can be converted to Editable Patch objects.

* **NURBS:** Stands for Non-Uniform Rational B-Splines. NURBS are similar to patches in that they also have control points. These control points define how a surface spreads over curves.

* **Compound objects:** A miscellaneous group of model types, including Booleans, loft objects, and scatter objects. Other compound objects are good at modeling one specialized type of object, such as Terrain or BlobMesh objects. The Compound objects are presented in Chapter 15, "Working with Compound Objects."

* **Body objects:** Solid objects that are imported from an SAT file produced by a solid modeling application like Revit have the concept of volume. 3ds Max mesh objects typically deal only with surfaces but can be converted to Body objects.

* **Particle systems:** Systems of small objects that work together as a single group. They are useful for creating effects such as rain, snow, and sparks. Particles are covered in Chapter 42, "Creating Particles and Particle Flow."

* **Point cloud:** Landscape scanners are used to scan large-scale real-world objects producing a point cloud dataset where the position and colors of thousands of points are collected. These point clouds can be loaded and displayed in 3ds Max using the Point Cloud object type.

* **Hair and fur:** Modeling hundreds of thousands of cylinder objects to create believable hair would quickly bog down any system, so hair is modeled using a separate system that represents each hair as a spline. The Hair and Fur modifiers are covered in Chapter 47, "Working with Hair and Cloth."

* **Cloth systems:** Cloth—with its waving, free-flowing nature—behaves like water in some cases and like a solid in others. 3ds Max includes a specialized set of modifiers for handling cloth systems. Creating and using a cloth system is discussed in Chapter 47, "Working with Hair and Cloth."

* **Fluid systems:** Fluids and liquids can be simulated using the Fluid dynamic system. Fluid can be made to respond to forces and gravity. Creating and using a fluid system is discussed in Chapter 48, "Creating Fluid Simulations."

Note

Hair, fur, cloth, and fluids are often considered effects or dynamic simulations instead of modeling constructs, so their inclusion in this list should be considered a stretch.

With all these options, modeling in 3ds Max can be intimidating, but you learn how to use each of these types the more you work with 3ds Max. For starters, begin with primitive or imported objects and then branch out by converting to editable objects. A single 3ds Max scene can include multiple different object types.

Converting to editable objects

Of all the commands found in the Create menu and in the Create panel, you won't find any menus or subcategories for creating editable objects.

To create an editable object, you need to import it or convert it from another object type. You can convert objects by right-clicking the object in the viewport and selecting the Convert To submenu from the pop-up quad menu, or by right-clicking the base object in the Modifier Stack and selecting the object type to convert to in the pop-up menu. The Modifier Stack is located at the top of the Modify panel in the Command Panel.

Once converted, all the editing features of the selected type are available in the Modify panel, but the object is no longer parametric and loses access to its common parameters such as Radius and Segments. However,

3ds Max also includes specialized modifiers such as the Edit Poly modifier that maintain the parametric nature of primitive objects while giving you access to the editing features of the Editable object. More on these modifiers is presented in the later modeling chapters.

Caution

If a modifier has been applied to an object, the Convert To menu option in the Modifier Stack pop-up menu is not available until you use the Collapse All command.

The Modifier Stack pop-up menu includes options to convert to Editable Mesh, Editable Poly, Editable Patch, and NURBS. If a shape or spline object is selected, the object also can be converted to an editable spline. Using any of the Convert To menu options collapses the Modifier Stack.

Note

Objects can be converted between the different types several times, but each conversion may subdivide the object. Therefore, multiple conversions are not recommended.

Converting between object types is done automatically using the software's best guess, but if you apply one of the Conversion modifiers to an object (located in the Modifiers→Conversion menu or by selecting them from the Modifier List at the top of the Modifier Stack), several parameters are displayed that let you define how the object is converted. For example, the Turn to Mesh modifier includes an option to Use Invisible Edges, which divides polygons using invisible edges. If this option is disabled, the entire object is triangulated. The Turn to Patch modifier includes an option to make quads into quad patches. If this option is disabled, all quads are triangulated.

The Turn to Poly modifier includes options to Keep Polygons Convex, Limit Polygon Size, Require Planar Polygons, and Remove Mid-Edge Vertices. The Keep Polygons Convex option divides any polygon that is concave, if enabled. The Limit Polygon Size option lets you specify the maximum allowable polygon size. This can be used to eliminate any pentagons and hexagons from the mesh. The Require Planar Polygons option keeps adjacent polygons as triangles if the angle between them is greater than the specified Threshold value. The Remove Mid-Edge Vertices option removes any vertices caused by intersections with invisible edges.

All Conversion modifiers also include options to preserve the current subobject selection (including any soft selection) and to specify the Selection Level. The From Pipeline option uses the current subobject selection on the given object. After a Conversion modifier is applied to an object, you must collapse the Modifier Stack to complete the conversion.

Understanding Normals

Before moving on to the various subobjects, you need to understand what a normal is and how it is used to tell which way the surface is facing. *Normals* are vectors that extend outward perpendicular to the surface of an object. These vectors aren't rendered and are used only to tell which way the surface face is pointing. If the normal vector points toward the camera, the polygon is visible, but if it points away from the camera, you are looking at the polygon's backside, which is visible only if the Backface Cull option in the Object Properties dialog box is disabled.

Several other properties also use the normal vector to determine how the polygon face is shaded, smoothed, and lighted. Normals also are used in dynamic simulations to determine collisions between objects.

Viewing normals

In all mesh subobject modes for the Editable Mesh object, except for Edge, you can select the Show Normals option in the Selection rollout to see any object's normals for the selected subobject and set a Scale value. Figure 11.1 shows a Plane, a Box, and a Sphere object. Each object has been converted to an Editable Mesh with all faces selected in Face subobject mode and with the Show Normals option selected.

Tip

For the other object types like Editable Poly, just apply the Edit Normals modifier to see and edit the object's normals.

FIGURE 11.1

The Show Normals option shows the normal vectors for each face in a Plane, a Box, and a Sphere.

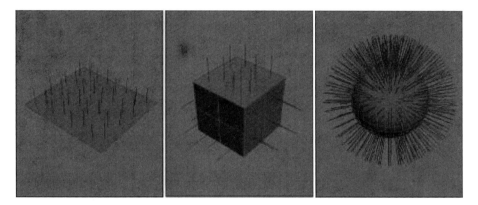

Tutorial: Cleaning up imported meshes

Many 3D formats are mesh-based, and importing mesh objects sometimes can create problems. By collapsing an imported model to an Editable Mesh, you can take advantage of several of the editable mesh features to clean up these problems.

The Modifier menu includes two modifiers that you can use to work with normals--Normal Modifier and Edit Normals modifier.

Figure 11.2 shows a model that was exported from Poser using the 3ds format. Notice that the model's waist is black when viewed using the Hidden Line viewport shading method. It appears this way because I've turned off the Backface Cull option in the Object Properties dialog box. If it were turned on, his waist would be invisible. The problem here is that the normals for this object are pointing in the wrong direction. This problem is common for imported meshes, and you'll fix it in this tutorial.

To fix the normals on an imported mesh model, follow these steps:

1. Open the Hailing taxi man with incorrect normals.max file from the Chap 11 directory in the downloaded content set.

2. Select the problem object—the waist on the right mesh. Open the object hierarchy by clicking the drop-down arrow icon to the left of the Editable Mesh object in the Modifier Stack if it isn't already expanded; then select Element subobject mode, and click the waist area.

3. In the Selection rollout, select the Show Normals option, and set the Scale value to a small number such as **0.1**.

   The normals are now visible. Notice that some of them point outward, and some of them point inward.

4. With the element subobject still selected, click the Unify button in the Surface Properties rollout and then click the Flip button until all normals are pointing outward.

This problem is fixed on the right model, and the waist object is now a visible part of the mesh. The fixed mesh on the right looks just like the original mesh on the left without being invisible (or, more correctly, inside-out), as shown in Figure 11.2.

FIGURE 11.2

This mesh suffers from objects with flipped normals, which makes them invisible.

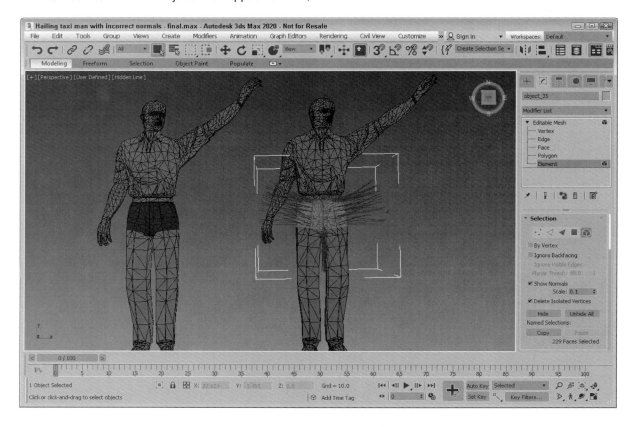

Working with Subobjects

All the editable modeling types offer the ability to work with subobjects. Subobjects are the elements that make up the model and can include vertices, edges, borders, faces, polygons, and elements, as shown in Figure 11.3. These individual subobjects can be selected and transformed just like normal objects using the transformation tools located on the main toolbar. But before you can transform these subobjects, you need to select them. You can select subobjects only on the object that is selected and only when you're in a particular subobject mode. Each editable object type has a different set of subobjects.

FIGURE 11.3

Expanding an editable object in the Modifier Stack reveals its subobjects.

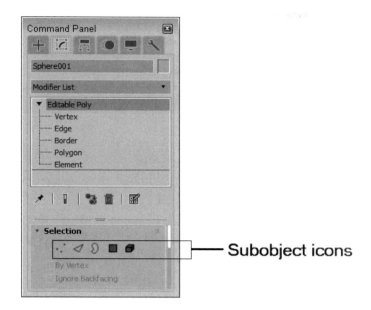

— Subobject icons

To select multiple subobjects, simply drag over the subobjects that you want to select. All subobjects within the dragged marquee are selected and highlighted in red. Just like with standard objects, you can choose a different selection tool from the main toolbar and change the Window/Crossing toggle. To add to the current subobject selection, hold down the Ctrl key and click on more subobjects. Holding down the Alt key lets you remove subobjects from the current selection.

If you hold down the Shift key and click on the subobject directly adjacent to the current selection, then all the subobjects within the same row or column are automatically selected. You can also hold down the Shift key before selecting a subobject to enable point-to-point selection. This lets you click on a single subobject and then drag the cursor to another part of the current object and all the subobjects in a direct line between the first and second selections are selected. This provides a quick way to select multiple subobjects at once.

If you expand the object's hierarchy in the Modifier Stack (by clicking the small drop-down arrow icon to the left of the object's name), all subobjects for an object are displayed. Selecting a subobject in the Modifier Stack places you in subobject mode for that subobject type. You also can enter subobject mode by clicking the subobject icons located at the top of the Selection rollout or by pressing the 1 through 5 keys on the keyboard. When you're in subobject mode, the subobject title and the icon in the Selection rollout are highlighted. You can work with the selected subobjects only while in subobject mode. To transform the entire object again, you need to exit subobject mode, which you can do by clicking either the subobject title or the subobject icon or by pressing one of the keyboard shortcuts, 1-5.

Tip

You also can access the subobject modes using the right-click quad menu. To exit a subobject mode, select Top Level in the quad menu.

Subobject selections can be locked with the Selection Lock Toggle (spacebar) located along the bottom edge of the interface to the left of the transform type-in fields. Subobject selections also can be made into a Selection Set by typing a name into the Named Selection Set drop-down list on the main toolbar. After a Selection Set is created, you can recall it any time you are in that same subobject mode. Named Selection Sets can then be copied and pasted between objects by using the Copy and Paste buttons found in the Selection rollout for most editable objects.

Using Soft Selection

When working with editable mesh, poly, patches, or splines, the Soft Selection rollout, shown in Figure 11.4, becomes available in subobject mode. Soft Selection selects all the subobjects surrounding the current selection and applies transformations to them to a lesser extent. For example, if a face is selected and moved a distance of 2, with linear Soft Selection, the neighboring faces within the soft selection range move a distance of 1. The overall effect is a smoother transition.

Note

The Soft Selection options are different for the various modeling types. For example, the Editable Mesh includes a standard set of options, but the Editable Poly object has more options, including a Paint Soft Selection mode.

FIGURE 11.4

The Soft Selection rollout is available only in subobject mode.

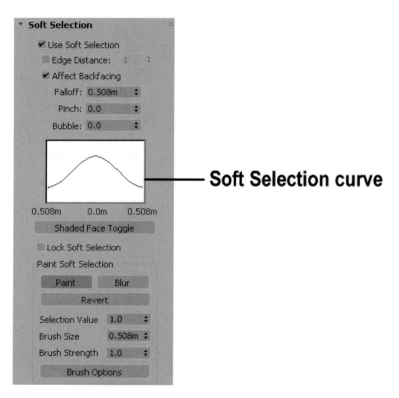

The Use Soft Selection parameter enables or disables the Soft Selection feature. The Edge Distance option sets the range (the number of edges from the current selection) that the Soft Selection will affect. If disabled, the distance is determined by the Falloff amount. The Affect Backfacing option applies the Soft Selection to selected subobjects on the backside of an object. For example, if you are selecting vertices on the front of a sphere object and the Affect Backfacing option is enabled, vertices on the opposite side of the sphere also are selected. This provides a way to work with subobjects on both sides of an object such as a wheel.

The Soft Selection curve shows a graphical representation of how the Soft Selection is applied. The Falloff value defines the spherical region where the Soft Selection has an effect. The Pinch button sharpens the point at the top of the curve. The Bubble button has an opposite effect and widens the curve. Changing the Falloff, Pinch, and Bubble values also changes the shape of the curve. Figure 11.5 shows several sample values and the resulting curve.

If you open the Customize User Interface dialog box, found in the Customize menu, you'll find an Action called Edit Soft Selection in the Keyboard panel. If you assign a hotkey to this action, you can use that keyboard shortcut to access Edit Soft Selection mode in the viewport. This mode lets you interactively change the Soft Selection Falloff, Pinch, and Bubble values by dragging in the viewports.

Once the Edit Soft Selection mode is enabled by using its assigned shortcut, the cursor changes to a custom cursor that looks like two circles. When this cursor appears, you can drag to change the Soft Selection's Falloff value. If you click, the cursor changes (to an upside-down V shape) and lets you drag to change the Pinch value. One more click, and you can edit the Bubble value (the cursor looks like an upside-down U for this mode), and another click returns you to the falloff edit mode. Pressing the keyboard shortcut again exits Edit Soft Selection mode.

FIGURE 11.5

The Soft Selection curve is affected by the Falloff, Pinch, and Bubble values.

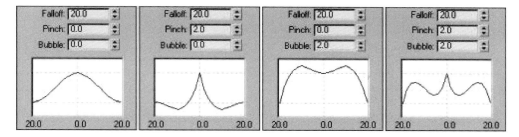

Tutorial: Soft selecting a heart shape from a plane

Soft Selection enables a smooth transition between subobjects, but sometimes you want the abrupt edge. This tutorial looks at moving some subobject vertices in a plane object with and without Soft Selection enabled.

To move subobject vertices with and without Soft Selection, follow these steps:

1. Open the Soft selection heart.max file from the Chap 11 directory in the downloaded content set.

   This file contains two simple plane objects that have been converted to Editable Mesh objects. Several vertices in the shape of a heart are selected.

2. Select the left plane, and you'll see that the vertices are already selected. In Vertex subobject mode, click the Select and Move button (or press the W key), move the cursor over the selected vertices, and drag upward on the Y-axis in the Left viewport away from the plane.

3. Exit subobject mode, select the right plane object, and enter Vertex subobject mode. The same vertices are again selected. Open the Soft Selection rollout, enable the Use Soft Selection option, and set the Falloff value to **40**.

4. Click the Select and Move button (or press the W key), and move the selected vertices upward. Notice the difference that Soft Selection makes.

Figure 11.6 shows the two resulting plane objects with the heart selections.

FIGURE 11.6

Soft Selection makes a smooth transition between the subobjects that are moved and those that are not.

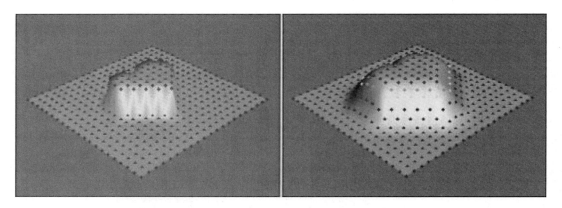

When you select subobjects, they turn red. Non-selected subobjects are blue, and soft selected subobjects are a gradient from orange to yellow, depending on their distance from the selected subobjects. This visual clue provides valuable feedback on how the Soft Selection affects the subobjects. Figure 11.7 shows the selected vertices from the preceding tutorial with Falloff values of 0, 20, 40, 60, and 80.

FIGURE 11.7

A gradient of colors shows the transition zone for soft-selected subobjects.

For the Editable Poly and Editable Patch objects, the Soft Selection rollout includes a Shaded Face Toggle button below its curve. This button shades the surface using the soft selection gradient colors, as shown in Figure 11.8. This shaded surface is displayed in any shaded viewports. The cooler colors have less of an impact on the transform.

FIGURE 11.8

The Shaded Face Toggle shades the surface using the soft selection gradient colors.

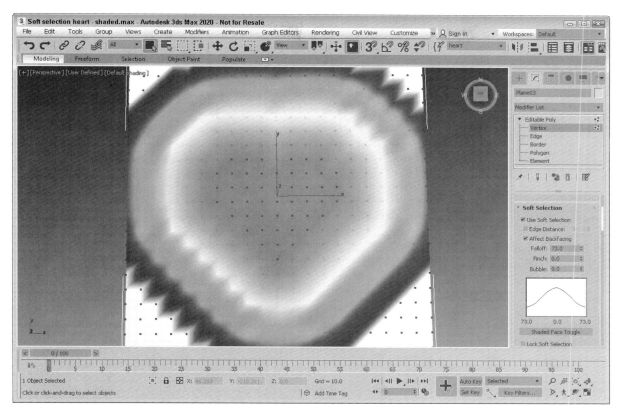

For Editable Poly objects, the bottom of the Soft Selection rollout includes a Paint Soft Selection section. Clicking the Paint button presents a brush icon in the viewports. You can change the Brush Size to increase or decrease the paintbrush. The Brush Strength sets the weight of the painted area. A Brush Strength of 1 sets the painted subobjects to their maximum value (which is only to orange when painting). Smaller Brush Strength values make more subtle changes, and a setting of 0 makes no change when painting.

In addition to using Paint mode, you can Blur the selections, which evens out the subobjects where you paint removing the high and low values. The Revert mode lets you return the selected subobjects to their previous values.

Introducing Modifiers

Modifiers are functions that you can apply to an object that allow you to change its structure without altering its base nature. For example, if you apply a Twist modifier to a cylinder, you'll still be able to change its base parameters, such as the cylinder radius and height. The Twist modifier adds additional parameters, such as the center of the twist and the amount to twist the model. Modifiers are stored separate from the object in the Modifier Stack. The Modifier Stack lets you remove, reorder, copy, and paste modifiers between objects.

Because modifiers provide an easy way to add more parameters to objects, they are used for all the different modeling types. Some specific modifiers are used on spline objects, and others are for polygon-based models. Another set of modifiers helps you animate. To keep all the various modifiers organized, 3ds Max has grouped them into several distinct modifier sets. The modifier sets, as they appear in the Modifier menu, include those listed in Table 11.1.

TABLE 11.1 Modifiers Menu Items

Menu	Submenu Items
Selection Modifiers	FFD Select, Mesh Select, Patch Select, Poly Select, Select by Channel, Spline Select, Volume Select
Patch/Spline Editing	Cross Section, Delete Patch, Delete Spline, Edit Patch, Edit Spline, Fillet/Chamfer, Lathe, Normalize Spline, Renderable Spline Modifier, Surface, Sweep, Trim/Extend, Optimize Spline, Spline Mirror, Spline Relax
Mesh Editing	Cap Holes, Chamfer, Delete Mesh, Edit Mesh, Edit Normals, Edit Poly, Extrude, Face Extrude, MultiRes, Normal Modifier, Optimize, ProOptimizer, Quadify Mesh, Smooth, STL Check, Symmetry, Tessellate, Vertex Paint, Vertex Weld
Conversion	Turn to Mesh, Turn to Patch, Turn to Poly
Animation	Attribute Holder, Flex, Linked XForm, Melt, Morpher, PatchDeform, PatchDeform (WSM), PathDeform, Spline Influence, Spline Morph, Spline Overlap, PathDeform (WSM), Skin, Skin Morph, Skin Wrap, Skin Wrap Patch, SplineIk Control, SurfDeform, SurfDeform (WSM)
Cloth	Cloth, Garment Maker, Welder
Hair and Fur	Hair and Fur (WSM)
UV Coordinates	Camera Map, Camera Map (WSM), MapScaler (WSM), Projection, Unwrap UVW, UVW Map, UVW Mapping Add, UVW Mapping Clear, UVW XForm
Cache Tools	Point Cache, Point Cache (WSM)
Subdivision Surfaces	Crease, Crease Set, HSDS Modifier, MeshSmooth, OpenSubDiv, TurboSmooth
Free Form Deformers	FFD 2x2x2, FFD 3x3x3, FFD 4x4x4, FFD Box, FFD Cylinder
Parametric Deformers	Affect Region, Bend, Data Channel, Displace, Lattice, Mirror, Noise, Physique, Push, Preserve, Relax, Ripple, Shell, Slice, Skew, Stretch, Spherify, Squeeze, Twist, Taper, Substitute, XForm, Wave
Surface	Disp Approx, Displace Mesh (WSM), Material, Material By Element
NURBS Editing	Disp Approx, Surf Deform, Surface Select
Radiosity	Subdivide, Subdivide (WSM)
Cameras	Camera Correction

You can find roughly these same sets if you click the Configure Modifier Sets button in the Modifier Stack. Within this list is a single selected set. The selected set is marked with an arrow to the left of its name. The modifiers contained within the selected set appear at the very top of the Modifier List.

Covering all the modifiers together would result in a very long chapter. Instead, I decided to cover most of the modifiers in their respective chapters—for example, animation modifiers in Chapter 33, "Using Animation Layers and Animation Modifiers"; the UV Coordinates modifiers in Chapter 21, "Unwrapping UVs and Mapping Textures"; and so on.

Exploring the Modifier Stack

All modifiers applied to an object are listed together in a single location known as the *Modifier Stack*. This Stack is the manager for all modifiers applied to an object and can be found at the top of the Modify panel in the Command Panel. You also can use the Stack to apply and delete modifiers; cut, copy, and paste modifiers between objects; and reorder them.

Understanding Base Objects

The first entry in the Modifier Stack isn't a modifier at all; it is the Base Object. The Base Object is the original object type. The Base Object for a primitive is listed as its object type, such as Sphere or Torus. Editable meshes, polys, patches, and splines also can be Base Objects. NURBS Surfaces and NURBS Curves also are Base Objects.

You also can see the Base Objects using the Schematic View window if you enable the Base Objects option in the Display floater.

Applying modifiers

An object can have several modifiers applied to it. Modifiers can be applied using the Modifiers menu or by selecting the modifier from the Modifier List drop-down list, located at the top of the Modify panel directly under the object name. Selecting a modifier in the Modifiers menu or from the Modifier List applies the modifier to the current selected object. Modifiers can be applied to multiple objects if several objects are selected.

Tip

You can quickly jump to a specific modifier in the Modifier List by pressing the first letter of the modifier that you want to select. For example, pressing the T key when the Modifier List is open immediately selects the Taper modifier.

Note

Some modifiers aren't available for some types of objects. For example, the Extrude and Lathe modifiers are enabled only when a spline or shape is selected.

Tutorial: Bending a tree

If you have a tree model that you want to bend as if the wind were blowing, you can apply the Bend modifier. The tree then bends about its Pivot Point. Luckily, all the trees and plants found in the AEC Objects category have their Pivot Points set about their base, so bending a tree is really easy.

To bend a tree using the Bend modifier, follow these steps:

1. Select the Create→AEC Objects→Foliage menu command to access the available trees. Select a long, thin tree like the Yucca, and click in the Top viewport to add it to the scene.

2. With the tree selected, select the Modifiers→Parametric Deformers→Bend menu command to apply the Bend modifier to the tree.

3. In the Parameters rollout, found in the Modify panel, set the Bend Axis to **Z** and the Bend Angle to **60**.

   The tree bends as desired.

Figure 11.9 shows the bending Yucca plant and a non-bent one behind for reference. To animate this tree bending back and forth, just set keys for the Angle parameter.

FIGURE 11.9

The Bend modifier can be used to bend trees.

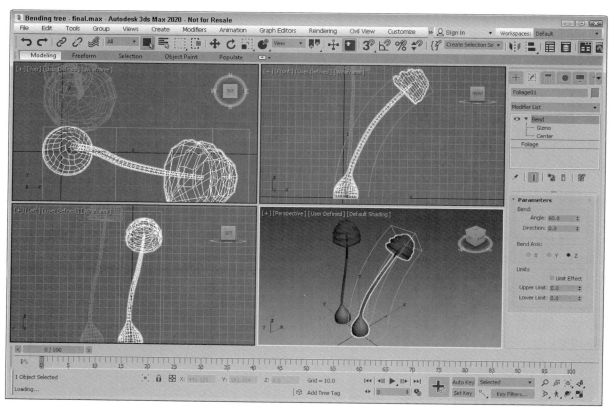

Other Modifier Stack entities

Most modifiers are Object-Space modifiers, but another category called World-Space modifiers also exists. World-Space modifiers are similar to Object-Space modifiers, except they are applied using a global coordinate system instead of a coordinate system that is local to the object. More on World-Space modifiers is presented later in this chapter, but you should be aware that World-Space modifiers (identified with the initials WSM) appear at the top of the Modifier Stack and are applied to the object after all Object-Space modifiers.

In addition to World-Space modifiers, Space Warp bindings appear at the top of the Modifier Stack.

Space Warps are covered in Chapter 43, "Using Space Warps."

Using the Modifier Stack

After a modifier is applied, its parameters appear in rollouts within the Command Panel. The Modifier Stack rollout, shown in Figure 11.10, lists the base object and all the modifiers that have been applied to an object. Any new modifiers applied to an object are placed at the top of the stack. By selecting a modifier from the list in the Modifier Stack, all the parameters for that specific modifier are displayed in rollouts.

FIGURE 11.10

The Modifier Stack displays all modifiers applied to an object.

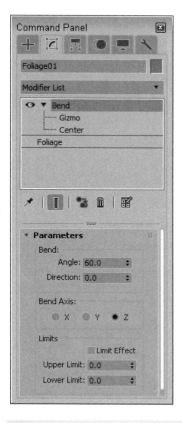

Tip

You can increase or decrease the size of the Modifier Stack by dragging the horizontal bar that appears beneath the Modifier Stack buttons.

Beneath the Modifier Stack are five buttons that affect the selected modifier. They are described in Table 11.2.

TABLE 11.2 Modifier Stack Buttons

Button	Name	Description
	Pin Stack	Makes the parameters for the selected modifier available for editing even if another object is selected (like taking a physical pin and sticking it into the screen so it won't move).
	Show End Result On/Off Toggle	Shows the end results of all the modifiers in the entire Stack if enabled and only the modifiers up to the current selected modifier if disabled.
	Make Unique	Used to break any instance or reference links to the selected object. After you click this button, an object will no longer be modified along with the other objects for which it was an instance or reference. Works for Base Object and modifiers.
	Remove Modifier from the Stack	Used to delete a modifier from the Stack or unbind a Space Warp if one is selected. Deleting a modifier restores it to the same state it was in before the modifier was applied.

	Configure Modifier Sets	Opens a pop-up menu where you can select to show a set of modifiers as buttons above the Modifier Stack. You also can select which modifier set appears at the top of the list of modifiers. The pop-up menu also includes an option to configure and define the various sets of modifiers.

For more information on configuring modifier sets, see Chapter 49, "Customizing the Interface."

If you right-click a modifier, a pop-up menu appears. This pop-up menu includes commands to rename the selected modifier, which you might want to do if the same modifier is applied to the same object multiple times. This pop-up menu also includes an option to delete the selected modifier, among other commands.

Copying and pasting modifiers

The pop-up menu also includes options to cut, copy, paste, and paste instance modifiers. The Cut command deletes the modifier from the current object but makes it available for pasting onto other objects. The Copy command retains the modifier for the current object and makes it available to paste onto another object. After you use the Cut or Copy command, you can use the Paste command to apply the modifier to another object. The Paste Instanced command retains a link between the original modifier and the instanced modifier, so any changes to either modifier affect the other instances.

You also can apply modifiers for the current object onto other objects by dragging the modifier from the Modifier Stack and dropping it on the other object in a viewport. Holding down the Ctrl key while dropping a modifier onto an object in a viewport applies the modifier as an instance (like the Paste Instanced command). Holding down the Shift key while dragging and dropping a modifier on an object in the viewport removes the modifier from the current object and applies it to the object on which it is dropped (like the Cut and Paste commands).

You also can cut, copy, and paste modifiers using the Schematic View window. See Chapter 10, "Organizing Scenes with Layers, Containers, XRefs and the Schematic View," for more details.

Using instanced modifiers

When you apply a single modifier to several objects at the same time, the modifier shows up in the Modifier Stack for each object. These are *instanced modifiers* that maintain a connection to one another. If one of these instanced modifiers is changed, the change is propagated to all other instances. This feature is very helpful for modifying large groups of objects.

When a modifier is copied between different objects, you can select to make the copy an instance.

To see all the objects that are linked to a particular modifier, select an object in the viewport and choose Views→Show Dependencies. All objects with instanced modifiers that are connected to the current selection appear in bright pink. At any time, you can break the link between a particular instanced modifier and the rest of the objects by using the Make Unique button below the Modifier Stack.

Identifying instances and references in the Modifier Stack

If you look closely at the Modifier Stack, you notice that it includes some visual clues that help you identify instances and references. Regular object and modifier copies appear in normal text, but instances appear in bold. This applies to both objects and modifiers. If a modifier is applied to two or more objects, it appears in italic.

Referenced objects and modifiers can be identified by a Reference Object Bar that splits the Modifier Stack into two categories—ones that are unique to the referenced object (above the bar) and ones that are shared with the other references (below the bar).

Figure 11.11 shows each of these cases in the Modifier Stack.

FIGURE 11.11

The Modifier Stack changes the text style to identify instances and references.

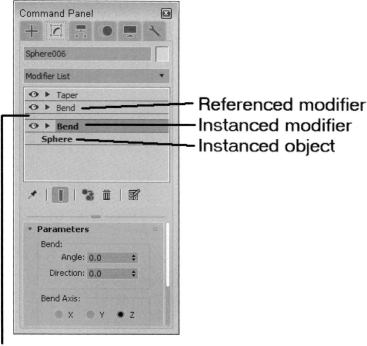

Disabling and removing modifiers

Clicking the eye icon to the left of the modifier name toggles the modifier on and off. The right-click pop-up menu also offers options to turn the modifier off in the viewport or off for the renderer.

To remove a modifier from the Modifier Stack, just select the modifier and press the Remove Modifier from the Stack button below the stack. This button removes the selected modifier only. You can select multiple modifiers at once by holding down the Ctrl key while clicking the modifiers individually or by holding down the Shift key and clicking the first and last modifiers in a range. This feature lets you apply and experiment with modifiers. If you try one that doesn't work, you can simply remove it without altering the object.

Reordering the Stack

Modifiers are listed in the Modifier Stack with the first applied ones on the bottom and the newest applied ones on the top. The Stack order is important and can change the appearance of the object. 3ds Max applies the modifiers starting with the lowest one in the Stack first and the topmost modifier last. You can change the order of the modifiers in the Stack by selecting a modifier and dragging it above or below the other modifiers. You cannot drag it below the object type or above any World-Space modifiers or Space Warp bindings.

Tutorial: Creating a molecular chain

Whether you're working with DNA splices or creating an animation to show how molecular chains are formed, you can use the Lattice and Twist modifiers to quickly create a molecular chain. Using these chains shows how reordering the Modifier Stack can change the outcome.

To create a molecular chain using modifiers, follow these steps:

1. Select Create→Standard Primitives→Plane, and drag in the Top viewport to create a Plane object. Set its Length to **300**, its Width to **60**, its Length Segments to **11**, and its Width Segments to **1**.

2. With the Plane object selected, select Modifiers→Parametric Deformers→Lattice to apply the Lattice modifier. Enable the Apply to Entire Object option. Then set the Struts Radius value to **1.0** with **12** sides and the Joints Base Type to **Icosa** with a Radius of **6.0** and a Segments value of **6**.

3. Select Modifiers→Parametric Deformers→Twist, and set the Twist Angle to **360** about the Y-axis.

4. Notice that the Sphere objects have been twisted along with the Plane object creating non-spherical shaped objects. You can fix this by switching the modifier order in the Modifier Stack. Select the Lattice modifier, and drag and drop it above the Twist modifier in the Modifier Stack.

   This step corrects the elongated spheres.

Figure 11.12 shows both the original and corrected molecular chains and is a good example of how the order of the modifiers can affect the final outcome.

FIGURE 11.12

Changing the order of the modifiers in the Modifier Stack can affect the end result.

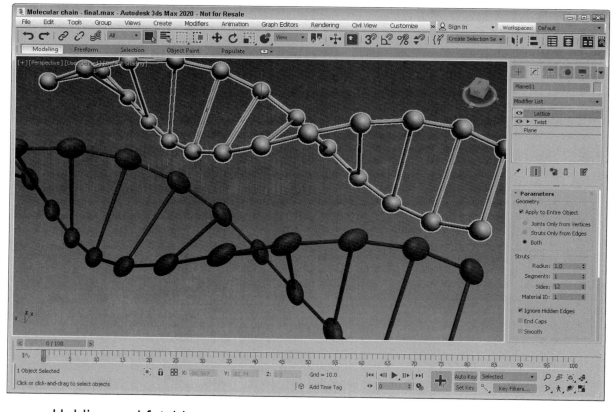

Holding and fetching a scene

Before going any further, you need to know about an important feature in 3ds Max that allows you to set a stopping point for the current scene. The Edit→Hold menu command saves the scene into a temporary buffer for easy recovery. After a scene is set with the Hold command (Ctrl+H), you can bring it back instantly with the Edit→Fetch menu command (Alt+Ctrl+F). These commands provide a quick way to backtrack on modifications to a scene or project without having to save and reload the project. If you use these commands before applying or deleting modifiers, you can avoid some potential headaches.

Tip

Along with saving your file often, using the Hold command before applying any complex modifier to an object is a good idea.

Collapsing the Stack

Collapsing the Modifier Stack removes all its modifiers by permanently applying them to the object. It also resets the modification history to a baseline. All the individual modifiers in the Modifier Stack are combined into the base object. This feature eliminates the ability to change any modifier parameters, but it simplifies the object. The right-click pop-up menu offers options to Collapse To and Collapse All. You can collapse the entire Stack with the Collapse All command, or you can collapse to the current selected modifier with the Collapse To command. Collapsed objects typically become Editable Mesh objects.

Tip

Another huge advantage of collapsing the Modifier Stack is that it conserves memory and results in smaller file sizes, which makes larger scenes load much quicker. Collapsing the Modifier Stack also speeds up rendering because 3ds Max doesn't need to calculate the stack results before rendering.

When you apply a Collapse command, a warning dialog box appears, shown in Figure 11.13, notifying you that this action will delete all the creation parameters. Click Yes to continue with the collapse.

Note

In addition to the Yes and No buttons, the warning dialog box includes a Hold/Yes button. This button saves the current state of the object to the Hold buffer and then applies the Collapse All function. If you have any problems, you can retrieve the object's previous state before the collapse was applied by choosing Edit→Fetch (Alt+Ctrl+F).

FIGURE 11.13

This warning dialog box offers a chance to Hold the scene.

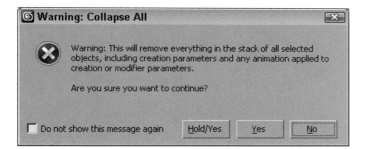

Using the Collapse utility

You also can use the Collapse utility found on the Utility panel to collapse the Modifier Stack. This utility enables you to collapse an object or several objects to a Modifier Stack Result or to a Mesh object. Collapsing to a Modifier Stack Result doesn't necessarily produce a mesh but collapses the object to its base object state, which is displayed at the bottom of the Stack hierarchy. Depending on the Stack, this could result in a mesh, patch, spline, or other object type. You also can collapse to a Single Object or to Multiple Objects.

If the Mesh and Single Object options are selected, you also can select to perform a Boolean operation. The Boolean operations are available if you are collapsing several overlapping objects into one. The options are Union (which combines geometries together), Intersection (which combines only the overlapping geometries), and Subtraction (which subtracts one geometry from another).

Boolean operations also can be performed using the Boolean compound object. See Chapter 15, "Working with Compound Objects," for details on this object type.

If multiple objects are selected, then a Boolean Intersection results in only the sections of the objects that are intersected by all objects; if no objects overlap, all objects disappear.

If you use the Boolean Subtraction option, you can specify which object is the base object from which the other objects are subtracted. To do so, select that object first and then select the other objects by holding down the Ctrl key and clicking them. Figure 11.14 shows an example of each of the Boolean operations.

FIGURE 11.14

Using the Collapse utility, you can select the following Boolean operations (shown from left to right): Union, Intersection, and Subtraction.

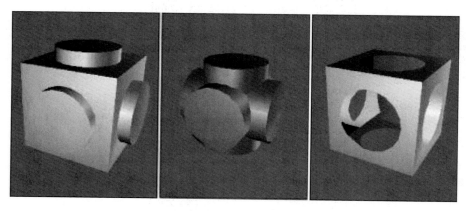

Using gizmo subobjects

As you've worked with modifiers, you've probably noticed the orange wireframe box that surrounds the object in the viewports when you apply the modifier. These boxes are called *modifier gizmos,* and they provide visual controls for how the modifier changes the geometry. If you want, you can work directly with these gizmos to affect the modifier.

Clicking the drop-down arrow icon to the left of the modifier name reveals any subobjects associated with the modifier. To select the modifier subobjects, simply click the subobject name. The subobject name is highlighted when selected. Many modifiers create *gizmo subobjects.* Gizmos have an icon, usually in the shape of a box that can be transformed and controlled like regular objects by using the transformation buttons on the main toolbar. Another common modifier subobject is Center, which controls the point about which the gizmo is transformed.

Tutorial: Squeezing a plastic bottle

To get a feel for how the modifier gizmo and its center affect an object, this tutorial applies the Squeeze modifier to a plastic bottle; by moving its center, you can change the shape of the object.

To change a modifier's characteristics by moving its center, follow these steps:

1. Open the Plastic bottle.max file from the Chap 11 directory in the downloaded content set.

   This file includes a plastic squirt bottle with all the parts attached into a single mesh object.

2. With the bottle selected, choose the Modifiers→Parametric Deformers→Squeeze menu command to apply the Squeeze modifier to the bottle. Set the Radial Squeeze Amount value to **1**.

3. In the Modifier Stack, click the drop-down arrow icon to the left of the Squeeze modifier to see the modifier's subobjects. Select the Center subobject.

   The selected subobject is highlighted.

4. Click the Select and Move (W) button on the main toolbar, and drag the center point in the Perspective viewport upward.

   Notice how the bottle's shape changes.

Figure 11.15 shows several different bottle shapes created by moving the modifier's center point.

FIGURE 11.15

By changing the modifier's center point, the bottle's shape changes.

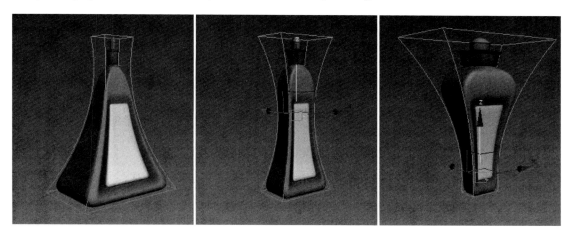

Applying modifiers to subobject selections

Modifiers are typically applied to entire objects, but you also can apply modifiers to subobjects. If the modifier isn't available for subobjects, it is excluded from the Modifier List or disabled in the Modifiers menu.

To work in subobject selection mode, click the drop-down arrow icon to the left of the object name to see the subobjects. Several modifiers, including Mesh Select, Spline Select, and Volume Select, can select subobject areas for passing these selections up to the next modifier in the Stack. For example, you can use the Mesh Select modifier to select several faces on the front of a sphere and then apply the Face Extrude modifier to extrude just those faces.

If the selected object isn't an editable object with available subobjects, you can still apply a modifier using one of the specialized Select modifiers. These modifiers let you select a subobject and apply a modifier to it without having to convert it to a non-parametric object. These Select modifiers include Mesh Select, Poly Select, Patch Select, Spline Select, Volume Select, FFD (Free Form Deformers) Select, and Select by Channel. You can find all these modifiers in the Modifiers→Selection Modifiers submenu.

After you apply a Select modifier to an object, you can select subobjects in the normal manner using the hierarchy in the Modifier Stack or the subobject icons in the Parameters rollout. Any modifiers that you apply after the Select modifier (which appear above the Select modifier in the Modifier Stack) affect only the subobject selection.

Tutorial: Applying damage to a plane wing

In this tutorial, you use the Volume Select modifier to select one wing of a plane and then apply Noise and XForm modifiers to make it look like it's been damaged in a collision. With the Volume Select modifier, only the selected portion of the plane wing gets the damage instead of the whole plane.

To use modifiers to make a section of a plane's wing appear damaged, follow these steps:

1. Open the Damaged plane.max file from the Chap 11 directory in the downloaded content set.

   This file includes a plane model.

2. With the left-wing of the plane selected, choose the Modifiers→Selection Modifiers→Volume Select menu command.

   This command applies the Volume Select modifier to the group.

3. In the Modifier Stack, click the drop-down arrow icon to the left of the modifier name, and select the Gizmo subobject. Move the gizmo in the Top viewport so only the front corner of the plane's wing is selected. In the Parameters rollout, select the Vertex option as the Stack Selection Level.

4. Choose the Modifiers→Parametric Deformers→Noise menu command to apply the Noise modifier to the selected volume. In the Parameters rollout, enable the Fractal option, and set the X, Y, and Z Strength values to **30**.

5. Choose Modifiers→Parametric Deformers→XForm to apply the XForm modifier, and use its gizmo to push the selected area up and to the left in the Top viewport to make this section look dented.

Figure 11.16 shows the resulting damaged plane. Notice that the rest of the object is fine; only the selected volume area is damaged.

FIGURE 11.16

The Noise and XForm modifiers are applied to just the subobject selection.

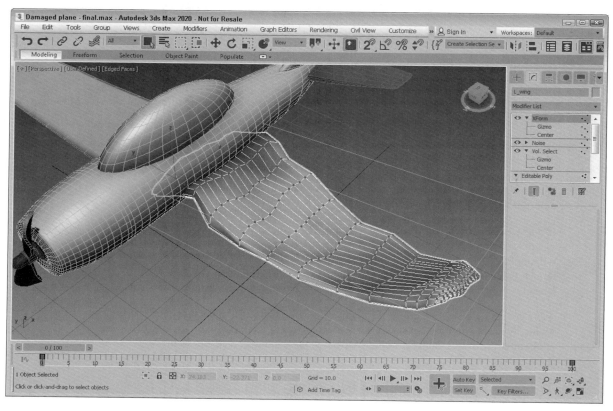

Topology dependency

When you attempt to modify the parameters of a Base Object that has a modifier applied, you sometimes get a warning dialog box that tells you that the modifier depends on topology that may change. 3ds Max is telling you that the surface of the object with that particular modifier is dependent on the subobjects that are selected, and if you change the underlying subobjects, you may change the resulting topology. For example, the CrossSection and Surface modifiers build the surface using a set of splines, but if you change the original spline, you can destroy the resulting surfaced object. You can eliminate this problem by collapsing the Modifier Stack.

You can disable the warning by selecting the "Do not show this message again" option on the dialog box or by opening the Preference Settings dialog box and turning off the Display Topology-Dependence Warning option in the General panel of the Preference Settings dialog box. Disabling the warning does not make the potential problem go away; it only prevents the warning dialog box from appearing.

Summary

Understanding the basics of modeling helps you as you build scenes. In this chapter, you've seen several different object types that are available in 3ds Max, including parametric and editable object types. Editable objects have subobjects that you can move to change the object.

With the modifiers contained in the Modify panel, you can alter objects in a vast number of ways. Modifiers can work with every aspect of an object, including geometric deformations, materials, and general object maintenance. In this chapter, you looked at the Modifier Stack and how modifiers are applied. This chapter covered the following topics:

* Understanding parametric objects and the various modeling types
* Viewing normals
* Using subobjects and soft selections
* Introducing and applying modifiers
* Working with the Modifier Stack to apply, reorder, and collapse modifiers

Now that you have the basics covered, you're ready to dive into the various modeling types. The first modeling types on the list are splines and shapes, which are covered in the next chapter.

Chapter 12

Drawing and Editing 2D Splines and Shapes

IN THIS CHAPTER

Working with shape primitives

Editing splines and shapes

Working with spline subobjects

Using Shape Booleans

Many modeling projects start from the ground up, and you can't get much lower to the ground than 2D. But this book is on 3D, you say? What place is there for 2D shapes? Within the 3D world, you frequently encounter flat surfaces—the side of a building, the top of a table, a billboard, and so on. All these objects have flat 2D surfaces. Understanding how objects are composed of 2D surfaces will help as you start to build objects in 3D. This chapter examines the 2D elements of 3D objects and covers the tools needed to work with them.

Working in 2D in the Autodesk® 3ds Max® 2020 software, you use two general objects: splines and shapes. A *spline* is a special type of line that curves according to mathematical principles. In 3ds Max, splines are used to create all sorts of shapes such as circles, ellipses, and rectangles.

You can create splines and shapes using the Create→Shapes menu, which opens the Shapes category on the Create panel. Just as with the other categories, several spline-based shape primitives are available. Spline shapes can be rendered, but they are normally used to create more advanced 3D geometric objects by extruding or lathing the spline. You can even find a whole group of modifiers that apply to splines. You can use splines to create animation paths as well as Loft and NURBS (Non-Uniform Rational B-Splines) objects, and you will find that splines and shapes, although they are only 2D, are used frequently in 3ds Max.

Shape Booleans let you combine, subtract, merge or find the intersection of two or more overlapping shapes. This provides a quick way to quickly create many new and interesting shapes.

Drawing in 2D

Shapes in 3ds Max are unique from other objects because they are drawn in 2D, which confines them to a single plane. That plane is defined by the viewport used to create the shape. For example, drawing a shape in the Top view constrains the shape to the XY plane, whereas drawing the shape in the Front view constrains it to the ZX plane. Even shapes drawn in the Perspective view are constrained to a plane such as the Home Grid.

You usually produce 2D shapes in a drawing package such as Adobe Illustrator (AI) or CorelDRAW. 3ds Max supports importing line drawings using the AI format.

See Chapter 2, "Working with Files, Importing and Exporting," to learn about importing AI files.

Whereas newly created or imported shapes are 2D and are confined to a single plane, splines can exist in 3D space. The Helix spline, for example, exists in 3D, having height as well as width values. Animation paths in particular typically move into 3D space.

Working with shape primitives

The shape primitive buttons are displayed in the Object Type rollout of the Create panel when either the Create→Shapes or the Create→Extended Shapes menu is selected. The Shapes category includes many basic shapes, including Line, Circle, Arc, NGon (a polygon where you can set the number of sides), Text, Egg, Rectangle, Ellipse, Donut, Star, Helix, and Section, as shown in Figure 12.1. The Extended Shapes category includes several shapes that are useful to architects, including WRectangle, Channel, Angle, Tee, and Wide Flange, as shown in Figure 12.2. Clicking any of these shape buttons lets you create the shape by dragging in one of the viewports. After a shape is created, several new rollouts appear.

Note

Within the Create menu, the various shapes are listed as Shapes and Extended Shapes, but within the Create panel of the Command Panel, the sub-categories are listed as Splines and Extended Splines. They both have the same elements, but are only called differently.

FIGURE 12.1

The shape primitives in all their 2D glory: Line, Circle, Arc, NGon, Text, Egg, Rectangle, Ellipse, Donut, Star, Helix, and Section

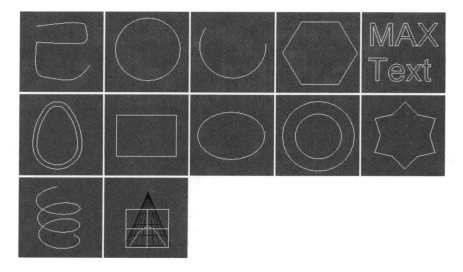

FIGURE 12.2

The extended shape primitives: WRectangle, Channel, Angle, Tee, and Wide Flange

Above the Shape buttons are two check boxes: AutoGrid and Start New Shape. AutoGrid creates a temporary grid, which you can use to align the shape with the surface of the nearest object under the cursor at the time of creation. This feature is helpful for starting a new spline on the surface of an object.

For more details on AutoGrid, see Chapter 7, "Transforming Objects, Pivoting, Aligning, and Snapping."

The Start New Shape option creates a new object with every new shape drawn in a viewport. Leaving this option unchecked lets you create compound shapes, which consist of several shapes used to create one object. Because compound shapes consist of several shapes, the shapes are automatically converted to be an Editable Spline object, and you cannot edit them using the Parameters rollout. For example, if you want to create a target from several concentric circles, keep the Start New Shape option unselected to make all the circles part of the same object.

Just as with the Geometric primitives, every shape that is created is given a name and a color. You can change either of these in the Name and Color rollout.

Most of the shape primitives have several common rollouts: Rendering, Interpolation, Creation Method, Keyboard Entry, and Parameters, as shown in Figure 12.3. I cover these rollouts initially and then present the individual shape primitives.

FIGURE 12.3

These rollouts are common for most of the shape primitives.

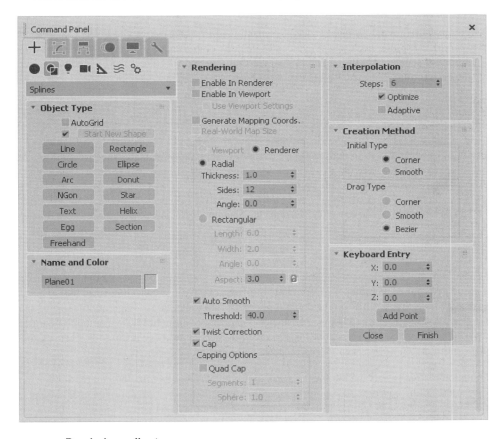

Rendering rollout

The Rendering rollout includes options for making a spline a renderable object. Making a spline a renderable object converts the spline into a 3D object that is visible when you render the scene. For renderable objects, you can choose to make the spline Radial or Rectangular. For the Radial option, you can specify a Thickness, the number of Sides, and the Angle values; for the Rectangular option, you can specify Length, Width, Angle, and Aspect values.

The Radial Thickness is the diameter of the renderable spline. The number of Sides sets the number of sides that make up the cross section of the renderable spline. The lowest value possible is 3, which creates a triangle cross section. The Length and Width values set the size along the Y-axis and the X-axis, respectively, of the rectangular sides. The Angle value determines where the corners of the cross-section sides start, so you can set a three-sided spline to have a corner or an edge pointing upward. The Aspect value sets the ratio of the Length per Width. If the Lock icon to the right of the Aspect value is enabled, the aspect ratio is locked, and changing one value affects the other.

Note
By default, a renderable spline has a 12-sided circle as its cross section.

You can choose different rendering values for the viewport and for the renderer using the Viewport and Renderer options above the Radial option. Each of these settings can be enabled or disabled using the Enable in Renderer and Enable in Viewport options at the top of the Rendering rollout. Renderable splines appear as normal splines in the viewport unless the Enable in Viewport option is selected. The Use Viewport Settings option gives the option of setting the spline render properties differently in the viewport and the renderer.

The Auto Smooth option and Threshold value offer a way to smooth edges on the renderable spline. If the angle between two adjacent polygons is less than the Threshold value, the edge between them is smoothed. If it is greater than the Threshold value, the hard edge is preserved.

The Twist Correction option removes any twist caused by misaligned first and last vertices. The Cap option adds a cap to either end of the spline. For the Cap, you can select the option to make all polygons in the cap quads, define the number of segments in the cap and make the Cap rounded with the Sphere value. A setting of 0 makes a flat cap and a setting of 1 makes a hemispherical cap.

New Feature in 2020
The Spline Cap and Twist Correction features are new to 3ds Max 2020.

The Generate Mapping Coordinates option automatically generates mapping coordinates that are used to mark where a material map is placed, and the Real-World Map Size option allows real-world scaling to be used when mapping a texture onto the renderable spline.

To learn more about mapping coordinates and real-world scaling, see Chapter 18, "Adding Material Details with Maps."

Interpolation rollout

In the Interpolation rollout, you can define the number of interpolation steps or segments that make up the shape. The Steps value determines the number of segments to include between adjacent vertices. For example, a circle shape with a Steps value of 0 has only 4 segments and looks like a diamond. Increasing the Steps value to 1 makes a circle out of 8 segments. For shapes composed of straight lines (like the Rectangle and simple NGons), the Steps value is set to 0, but for a shape with many sides (like a Circle or Ellipse), the Steps value can have a big effect. Larger step values result in smoother curves.

The Adaptive option automatically sets the number of steps to produce a smooth curve by adding more interpolation points to the spline based on the spline's curvature. When the Adaptive option is enabled, the Steps and Optimize options become disabled. The Optimize option attempts to reduce the number of steps to produce a simpler spline by eliminating all the extra segments associated with the shape.

Note
The Section and Helix shape primitives have no Interpolation rollout.

Figure 12.4 shows the number 5 drawn with the Line primitive in the Front viewport. The line has been made renderable so you can see the cross sections. The images from left to right show the line with Steps values of 0, 1, and 3. The fourth image has the Optimize option enabled. Notice that it uses only one segment for the straight edges. The fifth image has the Adaptive option enabled.

FIGURE 12.4

Using the Interpolation rollout, you can control the number of segments that make up a line.

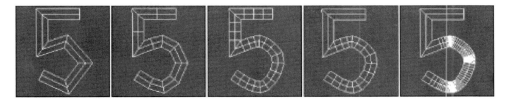

Creation Method and Keyboard Entry rollouts

Most shape primitives also include Creation Method and Keyboard Entry rollouts. (Text, Section, and Star are the exceptions.) The Creation Method rollout offers options for specifying different ways to create the spline by dragging in a viewport, such as from edge to edge or from the center out. Table 12.1 lists the various creation method options for each of the shapes and each of the extended shapes.

TABLE 12.1 Shape Primitive Creation Methods

Primitive Object	Primitive Object Name	Number of Viewport Clicks to Create	Default Creation Method	Other Creation Method
	Line	2 to Infinite	Corner Initial, Bézier Drag	Smooth, Initial, Corner, or Smooth Drag
	Circle	1	Center	Edge
	Arc	2	End-End-Middle	Center-End-End
	NGon	1	Center	Edge
	Text	1	none	none
-	Egg	2	none	none
-	Freehand	1	none	none
	Section	1	none	none
	Rectangle	1	Edge	Center
	Ellipse	1	Edge	Center
	Donut	2	Center	Edge
	Star	2	none	none
	Helix	3	Center	Edge
-	WRectangle	2	Edge	Center
-	Channel	2	Edge	Center
-	Angle	2	Edge	Center
-	Tee	2	Edge	Center
-	Wide Flange	2	Edge	Center

Some shape primitives, such as Star, Text, and Section, don't have any creation methods because 3ds Max offers only a single way to create these shapes.

The Keyboard Entry rollout offers a way to enter exact position and dimension values. After you enter the values, click the Create button to create the spline or shape in the active viewport. The settings are different for each shape.

The Parameters rollout includes such basic settings for the primitive as Radius, Length, and Width. You can alter these settings immediately after an object is created. However, after you deselect an object, the Parameters rollout moves to the Modify panel, and you must do any alterations to the shape there.

Line

The Line primitive includes several creation method settings, enabling you to create hard, sharp corners or smooth corners. You can set the Initial Type option to either Corner or Smooth to create a sharp or smooth corner for the first point created.

After clicking where the initial point is located, you can add points by clicking in the viewport. Dragging while creating a new point can make a point a Corner, Smooth, or Bézier based on the Drag Type option selected in the Creation Method rollout. The curvature created by the Smooth option is determined by the distance between adjacent vertices, whereas you can control the curvature created by the Bézier option by dragging with the mouse a desired distance after the point is created. Bézier corners have control handles associated with them, enabling you to change their curvature.

Tip

Holding down the Shift key while clicking creates points that are constrained vertically or horizontally. This makes it easy to create straight lines that are at right angles to each other. Holding down the Ctrl key snaps new points at an angle from the last segment, as determined by the Angle Snap setting.

After creating all the points, you exit Line mode by clicking the right mouse button. If the last point is on top of the first point, a dialog box asks whether you want to close the spline. Click Yes to create a closed spline or No to continue adding points. Even after creating a closed spline, you can add more points to the current selection to create a compound shape if the Start New Shape option isn't selected. If the first and last points don't correspond, an open spline is created.

Figure 12.5 shows several splines created using the various creation method settings. The left spline was created with all the options set to Corner, and the second spline with all the options set to Smooth. The third spline uses the Corner Initial type and shows where dragging has smoothed many of the points. The last spline was created using the Bézier option.

FIGURE 12.5

The Line shape can create various combinations of shapes with smooth and sharp corners.

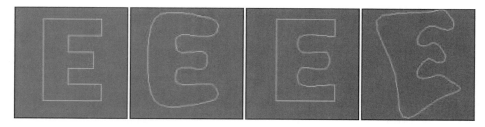

In the Keyboard Entry rollout, you can add points by entering their X, Y, and Z coordinates and clicking the Add Point button. You can close the spline at any time by clicking the Close button or keep it open by clicking the Finish button.

Rectangle

The Rectangle shape produces simple rectangles. In the Parameters rollout, you can specify the Length and Width and also a Corner Radius. Holding down the Ctrl key while dragging creates a perfect square shape.

Circle

The Circle button creates—you guessed it—circles. The only adjustable parameter in the Parameters rollout is the Radius. All other rollouts are the same, as explained earlier. Circles created with the Circle button have only four vertices.

Ellipse

Ellipses are simple variations of the Circle shape. You define them by Length and Width values. Holding down the Ctrl key while dragging creates a perfect circle (or you can use the Circle shape).

Arc

The Arc primitive has two creation methods. Use the End-End-Middle method to create an arc shape by clicking and dragging to specify the two end points and then dragging to complete the shape. Use the Center-End-End method to create an arc shape by clicking and dragging from the center to one of the end points and then dragging the arc length to the second end point.

Other parameters include the Radius and the From and To settings, where you can enter the value in degrees for the start and end of the arc. The Pie Slice option connects the end points of the arc to its center to create a pie-sliced shape, as shown in Figure 12.6. The Reverse option lets you reverse the arc's direction.

FIGURE 12.6

Enabling the Pie Slice option connects the arc ends with the center of the arc.

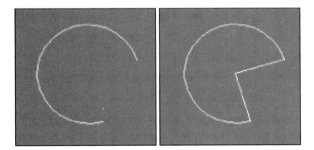

Donut

As another variation of the Circle shape, the Donut shape consists of two concentric circles; you create it by dragging once to specify the outer circle and again to specify the inner circle, or vice versa. The parameters for this object are simply two radii.

NGon

The NGon shape lets you create regular polygons by specifying the number of Sides and the Corner Radius. You also can specify whether the NGon is Inscribed or Circumscribed, as shown in Figure 12.7. Inscribed polygons are positioned within a circle that touches all the outer polygon's vertices. Circumscribed polygons are positioned outside a circle that touches the midpoint of each polygon edge. The Circular option changes the polygon to a circle that inscribes the polygon.

FIGURE 12.7

An inscribed pentagon and a circumscribed pentagon

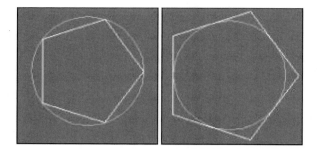

Star

The Star shape also includes two radii values; the larger Radius value defines the distance of the outer points of the Star shape from its center, and the smaller Radius value is the distance from the center of the star to the inner points. The Points setting indicates the number of points. This value can range from 3 to 100. The Distortion value causes the inner points to rotate relative to the outer points and can be used to create some interesting new star types. Range of distortion (-180 to 180) The Fillet Radius 1 and Fillet Radius 2 values adjust the Fillet for the inner and outer points. Figure 12.8 shows a sampling of what is possible with the Star shapes.

FIGURE 12.8

The Star primitive can be changed to create some amazing shapes.

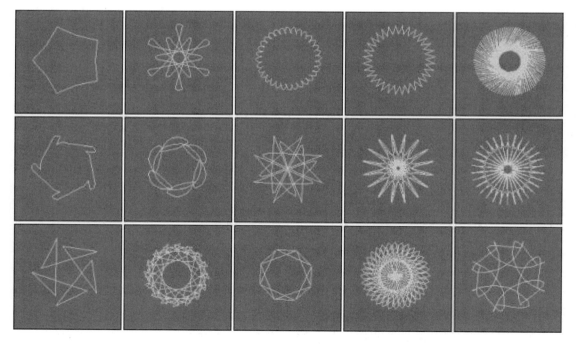

Text

You can use the Text primitive to add outlined text to the scene. In the Parameters rollout, you can specify a Font by choosing one from the drop-down list at the top of the Parameters rollout. Under the Font drop-down list are six icons, shown in Table 12.2. The left two icons are for the Italic and Underline styles. Selecting either of these styles applies the style to all the text. The right four icons are for aligning the text to be left, centered, right, or justified.

TABLE 12.2 Text Font Attributes

Icon	Description
I	Italic
U	Underline
☰	Align Left
☰	Centered
☰	Align Right
☰	Justify

Note

The list of available fonts includes only the Windows TrueType fonts and Type 1 PostScript fonts installed on your system and any extra fonts located in the font path listed in the Configure System Paths dialog box. You need to restart 3ds Max before the fonts in the font path are recognized.

The size of the text is determined by the Size value. The Kerning (the space between adjacent characters) and Leading (the space between lines of text) values can actually be negative. Setting the Kerning value to a large negative number actually displays the text backward. Figure 12.9 shows an example of some text and an example of Kerning values in the 3ds Max interface.

FIGURE 12.9

The Text shape lets you control the space between letters, known as kerning.

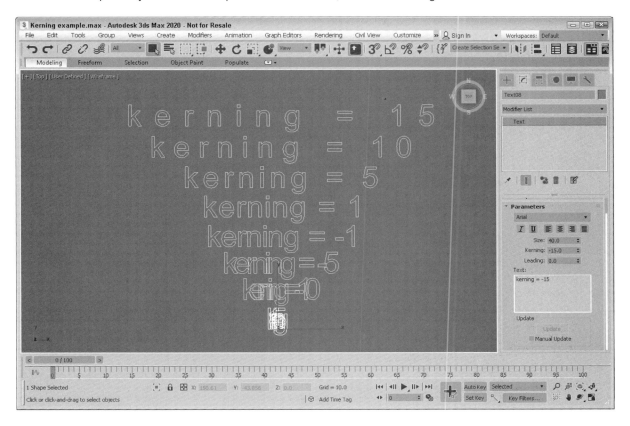

You can type the text to be created in the text area. You can cut, copy, and paste text into this text area from an external application if you right-click the text area. After setting the parameters and typing the text, the text appears as soon as you click in one of the viewports. The text is updated automatically when any of the parameters (including the text) are changed. To turn off automatic updating, select the Manual Update toggle. You can then update with the Update button.

If you open the Character Map application, you can see a complete list of special characters. The Character Map application, shown in Figure 12.10, can be opened in Windows by selecting Start→All Programs→Accessories→System Tools→Character Map. To enter special characters into the text area in 3ds Max, choose the special character by clicking it in the Character Map dialog box and then clicking the Select button. Click the Copy button to copy the character to the Windows clipboard, and in 3ds Max, use the Ctrl+V paste command to add it to the text area.

Tip

If the text object is selected, and you click the drop-down list to access the available fonts, you can use the up and down arrow keys to scroll through the font choices; the selected font is automatically displayed in the viewport.

FIGURE 12.10

The Character Map application shows all the special characters that are available.

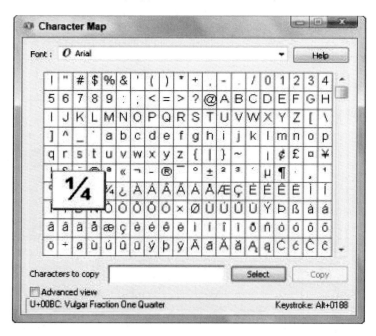

Helix

A Helix is like a spring coil shape, and it is the one shape of all the Shape primitives that exists in 3D. Helix parameters include two radii for specifying the inner and outer radius. These two values can be equal to create a coil or unequal to create a spiral. Parameters also exist for the Height and number of Turns. The Bias parameter causes the Helix turns to be gathered all together at the top or bottom of the shape. The CW and CCW options let you specify whether the Helix turns clockwise or counterclockwise.

Figure 12.11 shows a sampling of Helix shapes. The first Helix has equal radii values, the second one has a smaller second radius, the third Helix spirals to a second radius value of 0, and the last two Helix objects have Bias values of 0.8 and –0.8.

FIGURE 12.11

The Helix shape can be straight or spiral shaped.

Egg

The Egg shape creates two concentric egg-shaped outlines positioned one within the other. The distance between the two shapes is defined by the Thickness value; you can eliminate the inner shape by deselecting the Outline option. The size of the egg is set by the Length and Width values, but these two values are locked so that the egg shape is maintained. The Angle value lets the shape rotate about its center, where the pivot is located.

Section

Section stands for cross section. The Section shape is a cross section of the edges of any 3D object through which the Section's cutting plane passes. The process consists of dragging in the viewport to create a cross-sectioning plane. You can then move, rotate, or scale the cross-sectioning plane to obtain the desired cross section. In the Section Parameters rollout is a Create Shape button. Clicking this button opens a dialog box where you can name the new shape. You can use one Section object to create multiple shapes.

Note

You can make sections only from intersecting a 3D object. If the cross-sectioning plane doesn't intersect the 3D object, it won't create a shape. You cannot use the Section primitive on shapes, even if it is a renderable spline.

The Section Parameters rollout includes settings for updating the Section shape. You can update it when the Section plane moves, when the Section is selected, or Manually (using the Update Section button). You also can set the Section Extents to Infinite, Section Boundary, or Off. The Infinite setting creates the cross-section spline as though the cross-sectioning plane were of infinite size, whereas the Section Boundary limits the plane's extents to the boundaries of the visible plane. The color swatch determines the color of the intersecting shape.

To give you an idea of what the Section shape can produce, Figure 12.12 shows the shapes resulting from sectioning two Cone objects, including a circle, an ellipse, a parabola, and a hyperbola. The shapes have been moved to the sides to be more visible.

FIGURE 12.12

You can use the Section shape primitive to create the conic sections (circle, ellipse, parabola, and hyperbola) from a set of 3D cones.

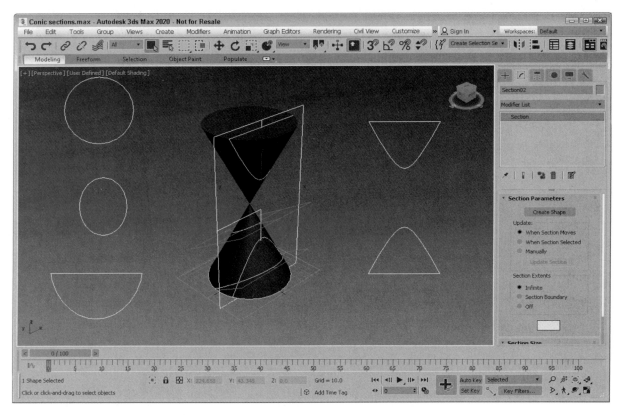

Freehand

The Freehand shape lets you draw unconstrained in the viewport. Using the Granularity and Threshold values, you can set how many knots are included in the drawn spline. The Granularity is the number of position samples taken before a knot is created and the Threshold defines the distance the cursor moves before a knot is created. You can also display the knots (or vertices) for the spline with the Show Knots option.

If the Constrain option is enabled, then you select a scene object with the Pick Object button. Once picked, the drawn spline stays on the surface of the picked object and ends if the cursor leaves the object.

Tutorial: Drawing a company logo

One of the early uses for 3D graphics was to animate corporate logos, and although 3ds Max can still do this without any problems, it now has capabilities far beyond those available in the early days. The Shape tools can even be used to design the logo. In this example, you'll design and create a simple logo using the Shape tools for the fictitious company named Expeditions South.

To use the Shape tools to design and create a company logo, follow these steps:

1. Create a four-pointed star by selecting the Create→Shapes→Star menu and dragging in the Top view to create a shape. Change the parameters for this star as follows: Radius1 = **60**, Radius2 = **20**, and Points = **4**.

2. Select and move the star shape to the left side of the viewport with the Select and Move tool.

318

3. Now click the Text button in the Command Panel, and change the font to **Impact** and the Size to **50**. In the Text area, type **Expeditions South**, and include a line return and several spaces between the two words so they are offset. Click in the Top viewport to place the text.

4. Use the Select and Move button (W) to reposition the text next to the Star shape.

5. Click the Line button, and create several short highlighting lines around the bottom point of the star.

The finished logo is now ready to extrude and animate. Figure 12.13 shows the result.

FIGURE 12.13

A company logo created entirely in 3ds Max using shapes

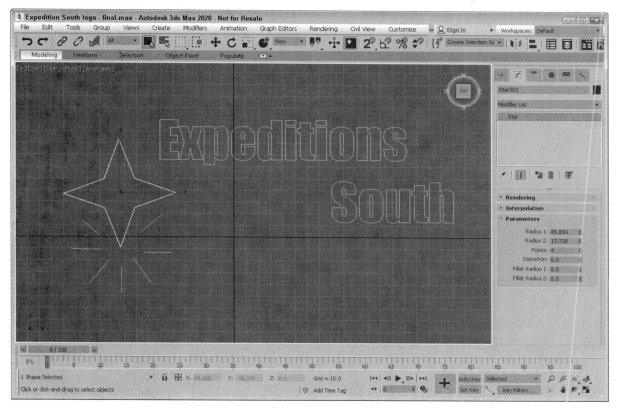

Tutorial: Viewing the interior of a tooth

As an example of the Section primitive, you'll explore a section of a Tooth model. The model includes the outer enamel and the interior nerve.

To create a spline from the cross section of a tooth, follow these steps:

1. Open the Tooth section.max file from the Chap 12 directory in the downloaded content set.

   This file includes a physical model of a tooth.

2. Select Create→Shapes→Section, and drag a plane in the Front viewport that is large enough to cover the tooth.

   This plane is your cross-sectioning plane.

3. Select the Select and Rotate button on the main toolbar (or press the E key), and rotate the cross-sectioning plane to cross the Tooth at the desired angle.

4. In the Section Parameters rollout of the Modify panel, click the Create Shape button, give the new shape the name **Tooth Section**, and then click the OK button.

5. From the Select From Scene dialog box (opened with the H key), select the new section by name, separate it from the model, and reposition it to be visible.

Figure 12.14 shows the resulting model and section in a maximized viewport.

FIGURE 12.14

You can use the Section shape to view the interior area of the tooth.

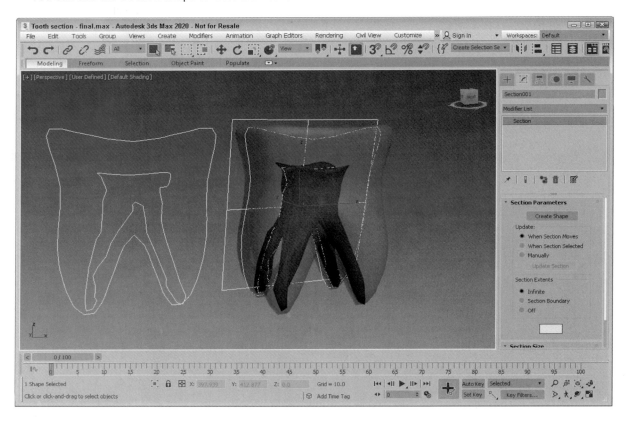

Editing Splines

After you create a shape primitive, you can edit it by modifying its parameters, but the parameters for shapes are fairly limited. For example, the only parameter for the Circle shape is Radius. All shapes can be converted to Editable Splines, or they can have the Edit Spline modifier applied to them. Doing either enables a host of editing features. Before you can use these editing features, you must convert the shape primitive to an Editable Spline (except for the Line shape). You can do so by right-clicking the spline shape in the viewport and choosing Convert to→Convert to Editable Spline from the pop-up quad menu or by right-clicking the base object in the Modifier Stack and selecting Convert to Editable Spline in the pop-up menu.

Editable Splines versus the Edit Spline modifier

After you convert the spline to an Editable Spline, you can edit individual subobjects within the spline, including Vertices, Segments, and Splines. The difference between applying the Edit Spline modifier and converting the shape to an Editable Spline is subtle. Applying the Edit Spline modifier maintains the shape parameters and enables the editing features found in the Geometry rollout. However, an Editable Spline loses the ability to change the base parameters associated with the spline shape.

Note

When you create an object that contains two or more splines (such as when you create splines with the Start New Shape option disabled), all the splines in the object are automatically converted into Editable Splines.

Another difference is that the shape primitive base name is listed along with the Edit Spline modifier in the Modifier Stack. Selecting the shape primitive name makes the Rendering, Interpolation, and Parameters rollouts visible, and the Selection, Soft Selection, and Geometry rollouts are made visible when you select the Edit Spline modifier in the Modifier Stack. For Editable Splines, only a single base object name is visible in the Modifier Stack, and all rollouts are accessible under it.

Note

Another key difference is that subobjects for the Edit Spline modifier cannot be animated.

Making splines renderable

Splines normally do not show up in a rendered image, but using the Renderable option in the Rendering rollout and assigning a thickness to the splines makes them appear in the rendered image. Figure 12.15 shows a rendered image of the Expeditions South logo after all shapes have been made renderable and assigned a Thickness of 3.0.

FIGURE 12.15

Using renderable splines with a Thickness of 3.0, the logo can be rendered.

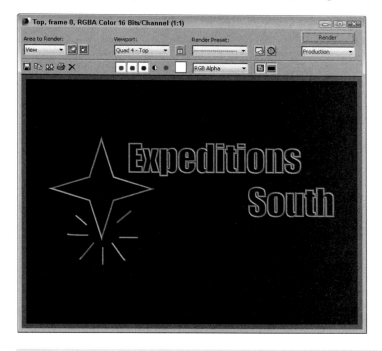

Note

The settings in the Rendering and Interpolation rollouts are the same as those used for newly created shapes, which are covered earlier in this chapter.

Selecting spline subobjects

When editing splines, you must choose the subobject level to work on. For example, when editing splines, you can work with Vertex (1), Segment (2), or Spline (3) subobjects. A spline object can have multiple splines as part of the single object. The Spline subobject mode lets you access the individual splines within the spline object.

Before you can edit spline subobjects, you must select them. To select the subobject type, click the small drop-down arrow icon to the left of the Editable Spline object (or the Edit Spline modifier) in the Modifier Stack. This lists all the subobjects available for this object. Click the subobject in the Modifier Stack to select it. Alternatively, you can click the icons under the Selection rollout, shown in Figure 12.16. You also can select the different subobject modes using the 1, 2, and 3 keyboard shortcuts. When you select a subobject, the selection in the Modifier Stack and the associated icon in the Selection rollout are highlighted.

Note

The subobject button is highlighted when selected to remind you that you are in Subobject Edit mode. Remember, you must exit this mode before you can select another object.

FIGURE 12.16

The Selection rollout provides icons for entering the various subobject modes.

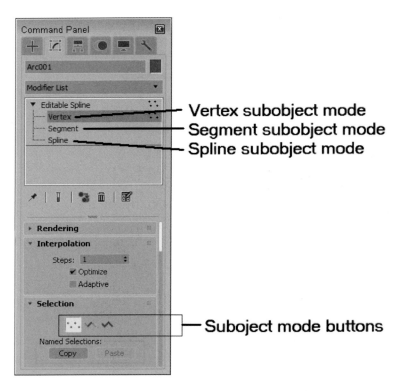

You can select many subobjects at once by dragging an outline over them in the viewports. You also can select and deselect subobjects by holding down the Ctrl key while clicking them. Holding down the Alt key removes any selected vertices from the selection set.

After selecting several vertices, you can create a named selection set by typing a name in the Named Selection Sets drop-down list in the main toolbar. You can then copy and paste these selection sets onto other shapes using the buttons in the Selection rollout.

The Lock Handles option allows you to move the handles of all selected vertices together when enabled, but each handle moves by itself when disabled. With the Lock Handles and the All options selected, all selected handles move together. The Alike option causes all handles on one side to move together.

The Area Selection option selects all the vertices within a defined radius of where you click. This is helpful for dense mesh objects with lots of vertices that are close together. The Segment End option, when enabled, allows you to select a vertex by clicking the segment. The closest vertex to the segment that you clicked is selected. This feature is useful when you are trying to select a vertex that lies near other vertices. The Select By button opens a dialog box with Segment and Spline buttons on it. These buttons allow you to select all the vertices on either a spline or segment that you choose.

The Selection rollout also has the Show Vertex Numbers option to display all the vertex numbers of a spline or to show the numbers of only the selected vertices. This can be convenient for understanding how a spline is put together and to help you find noncritical vertices. The Selected Only option displays the Vertex Numbers only for the selected subobjects when enabled.

Note

The vertex order is critical in determining the direction in which cross sections are swept when using the Loft and Sweep commands. You can identify the first vertex in a spline because it is yellow when one of the subobject modes is enabled.

Figure 12.17 shows a simple star shape that was converted to an Editable Spline. The left image shows the spline in Vertex Subobject mode. All the vertices are marked with small squares, and the starting point is marked with a yellow square. The middle image has the Show Vertex Numbers option enabled. For the right image, the vertex numbers are shown after the Reverse button was used (in Spline Subobject mode).

FIGURE 12.17

Several spline shapes displayed with vertex numbering turned on

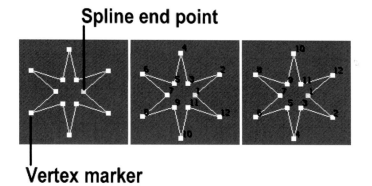

At the bottom of the Selection rollout, the Selection Information is displayed. This information tells you the number of the spline (or segment) and vertex selected, or the number of selected items and whether a spline is closed.

Note

The Soft Selection rollout allows you to alter adjacent nonselected subobjects (to a lesser extent) when selected subobjects are moved, creating a smooth transition. See Chapter 11, "Accessing Subobjects and Modifiers and Using the Modifier Stack," for the details on this rollout.

Controlling spline geometry

Much of the power of editing splines is contained within the Geometry rollout, shown in Figure 12.18, including the ability to add new splines, attach objects to the spline, weld vertices, use Boolean operations, Trim and Extend splines, and many more. Some Geometry buttons may be disabled, depending on the subobject type that you've selected. Many of the features in the Geometry rollout can be used in all subobject modes. Some of these features do not even require that you be in a Subobject mode. These features are covered first.

Tip

FIGURE 12.18

For Editable Splines, the Geometry rollout holds most of the features.

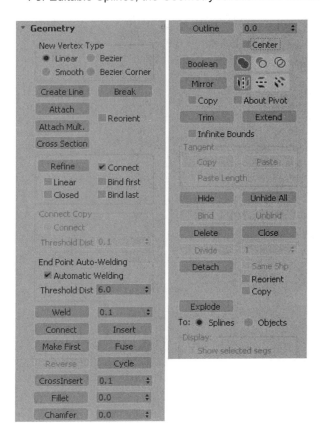

Create line

While editing splines, you can add new lines to a spline by clicking the Create Line button and then clicking in one of the viewports. You can add several lines at the same time, and all these new splines are part of the same object. Right-click in the viewport to exit this mode. Any new lines are their own spline, but you can weld them to the existing splines.

Break

Selecting a vertex and clicking the Break button in Vertex Subobject mode breaks the segment at the selected vertex by creating two separate end points. You also can use the Break button in Segment Subobject mode by clicking anywhere along the segment to add two separated vertices on the segment, thereby breaking the segment into two.

Attach and Attach Multiple

The Attach button lets you attach any existing splines to the currently selected spline. The cursor changes when you're over the top of a spline that can be attached. Clicking an unselected object makes it part of the current object. The Reorient option aligns the coordinate system of the spline being attached with the selected spline's coordinate system.

For example, using the Boolean button requires that all the splines be part of the same object. You can use the Attach button to attach several splines into the same object.

The Attach Mult. button enables several splines to be attached at once. When you click the Attach Mult. button, the Attach Multiple dialog box (which looks much like the Select by Name dialog box) opens. Use this dialog box to select the objects you want to attach to the current selection. Click the Attach button in the dialog box when you're finished. You can use both the Attach and Attach Mult. buttons in all three subobject modes.

Note

If the spline object that is being attached has a material applied to it, a dialog box appears that gives you options for handling the materials. These options include Match Material IDs to Material, Match Material to Material IDs, and Do Not Modify Material IDs or Material. Applying materials is covered in Chapter 17, "Creating and Applying Standard Materials with the Slate Material Editor."

Cross Section

The Cross Section button works just like the Cross Section modifier by creating splines that run from one cross-section shape to another. For example, imagine creating a baseball bat by positioning circular cross sections for each diameter change and connecting each cross section from one end to the other. All the cross sections need to be part of the same Editable Spline object, and then using the Cross Section button, you can click from one cross section to another. The cursor changes when the mouse is over a shape that can be used. When you're finished selecting cross-section shapes, you can right-click to exit Cross Section mode.

The type of vertex used to create the new splines that run between the different cross sections is the type specified in the New Vertex Type section at the top of the Geometry rollout.

Caution

Although the splines that connect the cross sections are positioned alongside the cross-section shape, they are not connected.

After the splines are created, you can use the Surface modifier to turn the splines into a 3D surface.

Auto Welding end points

To work with surfaces, you typically need a closed spline. When you enable the Automatic Welding option in the End Point Auto-Welding section and specify a Threshold, all end points within the threshold value are welded together, thus making a closed spline. This provides a quick way to close all splines in the selected object.

Insert

The Insert button adds vertices to a selected spline. Click the Insert button and then click the spline to place the new vertex. After placing the new vertex, you can reposition the new vertex and its attached segments and then click to set it in place. A single click adds a Corner type vertex, and a click-and-drag adds a Bézier type vertex.

After positioning the new vertex, you can add another vertex next to the first vertex by dragging the mouse and clicking. To add vertices to a different segment, right-click to release the currently selected segment but stay in Insert mode. To exit Insert mode, right-click in the viewport again, or click the Insert button to deselect it.

Tutorial: Working with cross sections to create a doorknob

You can work with cross sections in several ways. You can use the Cross Section feature for Editable Splines, the Cross Section modifier, or the Loft compound object. All these methods have advantages, but the first is probably the easiest and most forgiving method.

To create a simple doorknob using the Editable Spline Cross Section button, follow these steps:

1. Select the Tools→Grid and Snaps→Grid and Snap Settings menu, and select Grid Points in the Snaps panel of the Grid and Snap Settings dialog box. Close the Grid and Snap Settings dialog box; then click the Snap toggle button on the main toolbar (or press the S key) to enable grid snapping.

2. Select the Create→Shapes→Circle menu command, and drag from the center grid point in the Top viewport to create a small circle. Repeat this step to create two more circles—one the same size and one much larger.

3. Select the Create→Shapes→Rectangle menu command, and hold down the Ctrl key while dragging in the Top viewport to create a square that is smaller than the first circle. Repeat this step to create another square the same size. Aligning the squares is easier if you select the Center option in the Creation Method rollout. Press the S key to disable snapping.

4. Click the Select and Move (W) button on the main toolbar, and drag the shapes in the Left viewport upward in this order: square, square, small circle, large circle, small circle so that the square is on the bottom and the small circle is on the top. Separate the squares by a distance equal to the width of a door, and spread the circles out to be the width of a doorknob.

5. Select the bottom-most square shape, and then right-click and select Convert To→Convert to Editable Spline in the pop-up quad menu.

6. In the Geometry rollout, click the Attach button and then select the other shapes one at a time to add them to the selected Editable Spline object. Then right-click in the viewport to exit the Attach tool.

7. Orbit the Perspective viewport until all shapes are visible and easily selectable. Then select each and rotate each of the cross sections in the Top viewport so their first vertices (the yellow one when selected) are lined up horizontally and vertically so each is on top of the others. This helps prevent any twisting that may occur when the cross sections try to align the first vertices.

8. Select the Linear option in the New Vertex Type section in the Geometry rollout and then click the Cross Section button. Click the lowest square shape in the Perspective viewport, followed by the higher square shape and then the lower small circle. This creates a spline that runs linearly between these lowest three cross-section shapes. Right-click in the Perspective viewport to exit Cross Section mode.

9. Select the Bezier option in the New Vertex Type section and then click the Cross Section button again. Click the lowest circle shape in the Perspective viewport, followed by the larger circle shape and then the higher small circle. This creates a spline that runs smoothly between the last three cross-section shapes. Right-click in the Perspective viewport to exit Cross Section mode.

Tip

After a spline outline is constructed, you can use the Surface modifier to add a surface to the object.

Figure 12.19 shows the splines running between the different cross sections. A key benefit to the Editable Spline approach is that you don't need to order the cross-section shapes exactly. You just need to click them in the order that you want.

FIGURE 12.19

The Cross Section feature of Editable Splines can create splines that run between several cross-section shapes.

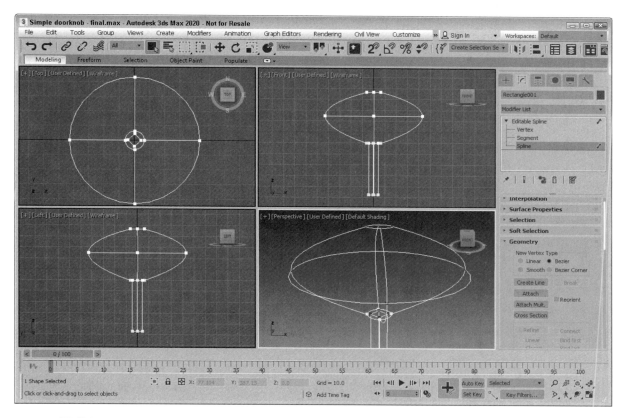

Editing vertices

To edit vertices, click the Vertex subobject in the Modifier Stack or select the vertex icon from the Selection rollout (keyboard shortcut 1). After the Vertex subobject type is selected, you can use the transform buttons on the main toolbar to move, rotate, and scale the selected vertex or vertices. Moving a vertex around causes the associated spline segments to follow.

With a vertex selected, you can change its type from Corner, Smooth, Bézier, or Bézier Corner by right-clicking and selecting the type from the pop-up quad menu.

Caution

The New Vertex Type section in the top of the Geometry rollout sets only the vertex type for new vertices created when you Shift-copy segments and splines or new vertices created with the Cross Section button. These options cannot be used to change the vertex type for existing vertices.

Selecting the Bézier or Bézier Corner type vertex reveals green handles on either side of the vertex. Dragging these handles away from the vertex alters the curvature of the segment. Bézier type vertices have both handles in the same line, but Corner Bézier type vertices do not.

Note

Holding down the Shift key while clicking and dragging a handle causes the handle to move independently of the other handle, turning it into a Bézier Corner type vertex instead of a plain Bézier. You can use it to create sharp corner points.

Figure 12.20 shows how the Bézier and Bézier Corner handles work. The first image shows all vertices of a circle selected, where you can see the handles protruding from both sides of each vertex. The second image shows what happens to the circle when one of the handles is moved. The handles for Bézier vertices move together, so moving one upward causes the other to move downward. The third image shows a Bézier Corner vertex, where the handles can move independently to create sharp points. The fourth image shows two Bézier Corner vertices moved, with the Lock Handles and Alike options enabled. This causes the handles to the left of the vertices to move together. The final image has the Lock Handles and All options selected, causing the handles of all selected vertices to move together.

FIGURE 12.20

Moving the vertex handles alters the spline around the vertex.

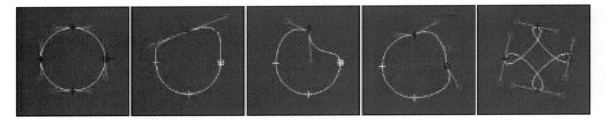

The pop-up quad menu also includes a command to Reset Tangents. This option makes the tangents revert to their original orientation before the handles were moved.

Refine

The Refine button lets you add vertices to a spline without changing the curvature, giving you more control over the details of the spline. With the Refine button selected, just click a spline where you want the new vertex, and one is added.

The Connect option adds a new segment that connects each two successive points added with the Refine tool. These segments don't actually appear until the Refine button is disabled. This provides a method for copying part of an existing spline. When the Connect option is enabled, the Linear, Closed, Bind First, and Bind Last options become enabled. The Linear option creates Corner type vertices, resulting in linear segments. The Closed option closes the spline by connecting the first and last vertices. The Bind First and Bind Last options bind the first and last vertices to the center of the selected segment. Refine is available only for Vertex and Segment Subobject modes.

Weld and Fuse

When two end point vertices are selected and are within the specified Weld Threshold, they can be welded into one vertex and moved to a position that is the average of the welded points using the Weld button. Several vertices can be welded simultaneously. Another way to weld vertices is to move one vertex on top of another. If they are within the threshold distance, a dialog box asks whether you want them to be welded. Click the Yes button to weld them.

Caution
The Weld button can be used only to weld spline end points.

The Fuse button is similar to the Weld command, except that it doesn't delete any vertices. It just positions the two vertices on top of one another at a position that is the average of the selected vertices.

In Figure 12.21, the left image shows a star shape with all its lower vertices selected. The middle image is the same star shape after the selected vertices have been welded together, and the right image shows the star shape with the selected vertices fused. The Selection rollout shows five selected vertices for the fused version.

FIGURE 12.21

Using the Fuse and Weld buttons, several vertices in the star shape have been combined.

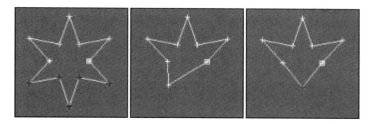

You can use the Fuse button to move the selected vertices to a single location. This is accomplished by selecting all the vertices to relocate and clicking the Fuse button. The average point between all the selected vertices becomes the new location. You can combine these vertices into one after they've been fused by clicking the Break button to make the fused points into end points and then clicking the Weld button.

Connect

The Connect button lets you connect end vertices to one another to create a new line. This works only on end vertices, not on connected points within a spline. To connect the ends, click the Connect button, drag the cursor from one end point to another (the cursor changes to a plus sign when it is over a valid end point), and release. The first image in Figure 12.22 shows an incomplete star drawn with the Line primitive, the middle image shows a line being drawn between the end points (notice the cursor), and the third image is the resulting star.

FIGURE 12.22

You can use the Connect button to connect end points of shapes.

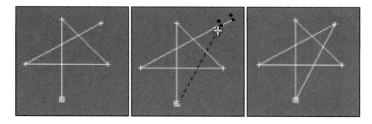

Make First

The Show Vertex Numbers option in the Selection rollout displays the number of each vertex. The first vertex is identified by the yellow square when selected. The Make First button lets you change which vertex you want to be the first vertex in the spline. To do this, select a single vertex, and click the Make First button. If more than one vertex is selected, 3ds Max ignores the command. If the selected spline is an open spline, an end point must be selected before you can use the Make First command.

Note

The vertex number is important because it determines the starting location for path animations and where Loft objects start.

Cycle

If a single vertex is selected, the Cycle button causes the next vertex in the Vertex Number order to be selected. The Cycle button can be used on open and closed splines and can be repeated around the spline. The exact vertex number is shown at the bottom of the Selection rollout. This is very useful for locating individual vertices in groups that are close together, such as groups that have been fused.

CrossInsert

If two splines that are part of the same object overlap, you can use the CrossInsert button to create a vertex on each spline at the location where they intersect. The distance between the two splines must be closer than the Threshold value for this to work. Note that this button does not join the two splines; it only creates a vertex on each spline. Use the Break and Weld buttons to join the splines. Figure 12.23 shows how you can use the CrossInsert button to add vertices at the intersection points of two elliptical splines. Notice that each ellipse now has eight vertices.

FIGURE 12.23

The CrossInsert button can add vertices to any overlapping splines of the same object.

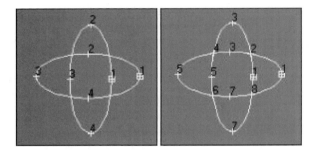

Fillet

The Fillet button is used to round the corners of a spline where two edges meet. To use the Fillet command, click the Fillet button and then drag on a vertex in the viewport. The more you drag, the larger the Fillet. You also can enter a Fillet value in the Fillet spinner for the vertices that are selected. The Fillet has a maximum value based on the geometry of the spline. Figure 12.24 shows the Fillet command applied to an 8-pointed star with values of 10, 15, and 20. Notice that each selected vertex has split into two.

Caution

Be careful not to apply the Fillet command multiple times to the selected vertices. If the new vertices cross over each other, the normals will be misaligned, which will cause problems when you use modifiers.

Note

You can fillet several vertices at once by selecting them, clicking the Fillet button, and dragging the Fillet distance.

FIGURE 12.24

The Fillet tool can round the corners of a shape.

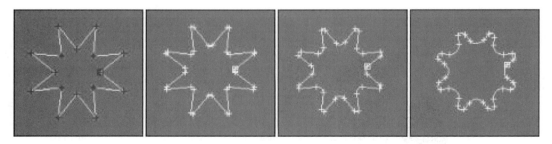

Chamfer

The Chamfer button works much like the Fillet button, except that the corners are replaced with straight-line segments instead of smooth curves. This keeps the resulting shape simpler and maintains hard corners. To use the Chamfer command, click the Chamfer button, and drag on a vertex to create the Chamfer. You also can enter a Chamfer value in the rollout. Figure 12.25 shows chamfers applied to the same 8-pointed shape with the same values of 10, 15, and 20.

FIGURE 12.25

Chamfers alter the look of spline corners.

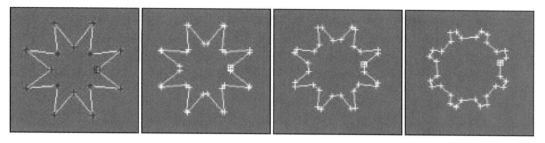

Tangent Copy and Tangent Paste

If you spend considerable time positioning the handles for the Bézier or Bézier Corner vertices just right, it can be tricky to repeat these precise positions again for other handles. Using the Tangent Copy and Tangent Paste buttons, you can copy the handle positions between different handles. To do so, simply select a handle that you want to copy, click the Copy button, select the vertex to which you want to copy the handle, and click the Paste button. The Paste Length button copies the handle length along with its orientation, if enabled.

Hide/Unhide All

The Hide and Unhide All buttons hide and unhide spline subobjects. They can be used in any Subobject mode. To hide a subobject, select the subobject and click the Hide button. To unhide the hidden subobjects, click the Unhide All button.

Bind/Unbind

The Bind button attaches an end vertex to a segment. The bound vertex then cannot be moved independently, but only as part of the bound segment. The Unbind button removes the binding on the vertex and lets it move independently again. To bind a vertex, click the Bind button and then drag from the vertex to the segment to bind to. To exit Bind mode, right-click in the viewport or click the Bind button again.

For Figure 12.26, a circle shape is created and converted to an Editable Spline object. The right vertex is selected and then separated from the circle with the Break button. Then, by clicking the Bind button and dragging the vertex to the opposite line segment, the vertex is bound to the segment. Any movement of the spline keeps this vertex bound to the segment.

FIGURE 12.26

The Bind button attaches one end of the circle shape to a segment.

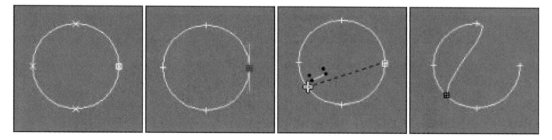

Delete

The Delete button deletes the selected subobject. You can use it to delete vertices, segments, or splines. This button is available in all subobject modes. Pressing the Delete key when the subobject is selected has the same effect.

Show Selected Segments

The Show Selected Segs option causes any selected segments to continue to be highlighted in Vertex Subobject mode as well as Segment Subobject mode. This feature helps you keep track of the segments that you are working on when moving vertices.

Tutorial: Making a ninja star

If you're involved with fighting games, either creating or playing them, chances are good that when you look at the Star primitive, you think, "Wow, this is perfect for creating a ninja star weapon." If not, just pretend.

To create a ninja star using splines, follow these steps:

1. Select the Tools→Grid and Snaps→Grid and Snap Settings menu, and select Grid Points in the Snaps panel of the Grid and Snap Settings dialog box. Close the Grid and Snap Settings dialog box and then click the Snaps Toggle button (or press the S key) on the main toolbar to enable grid snapping.

2. Select the Create→Shapes→Circle menu command, and drag from the center grid point in the Top viewport to create a circle.

3. Select the Create→Shapes→Star menu command, and drag again from the center of the Top viewport to center align the star with the circle. Make the star shape about three times the size of the circle, and set the number of Points to **10**.

4. With the star shape selected, right-click in the Top viewport, and select Convert To→Convert to Editable Spline. In the Modify panel, click the Attach button, and click the circle shape. Then click the Vertex icon in the Selection rollout (or press 1) to enter Vertex Subobject mode.

5. Click the Create Line button in the Geometry rollout; then click the circle's top vertex and bottom vertex to create a vertical line that divides the circle. Right-click to end the line, and right-click again to exit Create Line mode.

6. Select the top vertex of the line that you just created. (Be careful not to select the circle's top vertex; you can use the Cycle button to find the correct vertex.) Right-click the vertex, and select the Bézier vertex type from the quad menu. Then drag its lower handle until it is on top of the circle's left vertex. Repeat this step for the bottom vertex, and drag its handle to the circle's right vertex to create a yin-yang symbol in the center of the ninja star.

7. While holding down the Ctrl key, click all the inner vertices of the star shape. Click the Chamfer button, change the value until the chamfer looks like that in Figure 12.27, and click the Chamfer key again to deselect it.

Figure 12.27 shows the resulting ninja star.

FIGURE 12.27

The completed ninja star, ready for action (or extruding)

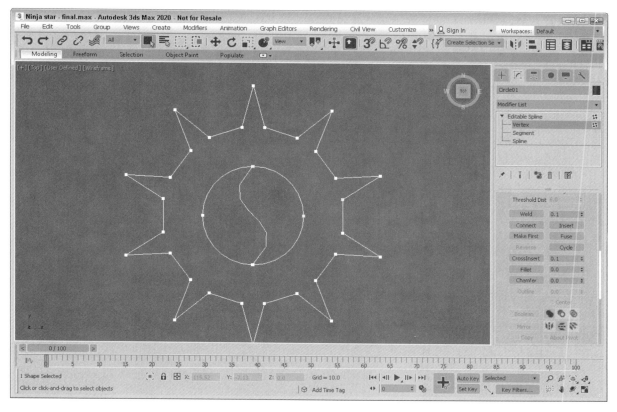

Editing segments

To edit a segment, click the Segment subobject in the Modifier Stack, or select the Segment icon from the Selection rollout to enter Segment Subobject mode. Clicking either again exits this mode. *Segments* are the lines or edges that run between two vertices. Many of the editing options work the same way as when you're editing Vertex subobjects. You can select multiple segments by holding down the Ctrl key while clicking the segments, or you can hold down the Alt key to remove selected segments from the selection set. You also can copy segments when they're being transformed by holding down the Shift key. The cloned segments break away from the original spline but are still part of the Editable Spline object.

You can change segments from straight lines to curves by right-clicking the segment and selecting Line or Curve from the pop-up quad menu. Line segments created with the Corner type vertex option cannot be changed to Curves, but lines created with Smooth and Bézier type vertex options can be switched back and forth.

Connect Copy

When you create a copy of a segment by moving a segment with the Shift key held down, you can enable the Connect Copy option to make segments that join the copied segment with its original. For example, if you have a single straight horizontal line segment, dragging it upward with the Copy Connect option enabled creates a copy that is joined to the original, resulting in a rectangle. Be aware that the vertices that connect to the original segment are not welded to the original segment.

Divide

When you select a segment, the Divide button becomes active. This button adds the number of vertices specified to the selected segment or segments. These new vertices are positioned based on the curvature of the segment, with more vertices being added to the areas of greater curvature. Figure 12.28 shows the hexagon shape after all four segments were selected, a value of 1 was entered into the spinner, and the Divide button was clicked.

FIGURE 12.28

The Divide button adds segments to the spline.

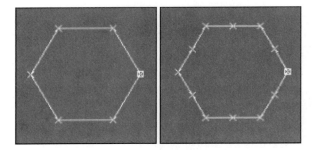

Detach

The Detach button separates the selected subobjects from the rest of the object (opposite of the Attach button). When you click this button, the Detach dialog box opens, enabling you to name the new detached subobject. When segments are detached, you can select the Same Shape option to keep them part of the original object. The Reorient option realigns the new detached subobject to match the position and orientation of the current active grid. The Copy option creates a new copy of the detached subobject.

You can use Detach on either selected Spline or Segment subobjects.

Tutorial: Using Connect Copy to create a simple flower

Connect Copy is one of the features that you'll use and wonder how you ever got along without it. For this tutorial, you create a simple flower from a circle shape using the Connect Copy feature.

To create a simple flower using the Connect Copy feature, follow these steps:

1. Select Create→Shapes→Circle, and drag in the Top viewport to create a simple circle shape.
2. Right-click the circle, and select Convert to→Convert to Editable Spline to convert the shape.
3. In the Modifier Stack, select the Segment Subobject mode (keyboard shortcut, 2), and enable the Connect option in the Connect Copy section.
4. Select one of the circle segments, and with the Shift key held down, drag it outward away from the circle with the Move tool. Then repeat this step for each segment.

Figure 12.29 shows the results. With the Connect Copy option, you don't need to worry about the connecting lines.

FIGURE 12.29

The Connect Copy feature joins newly copied segments to the original.

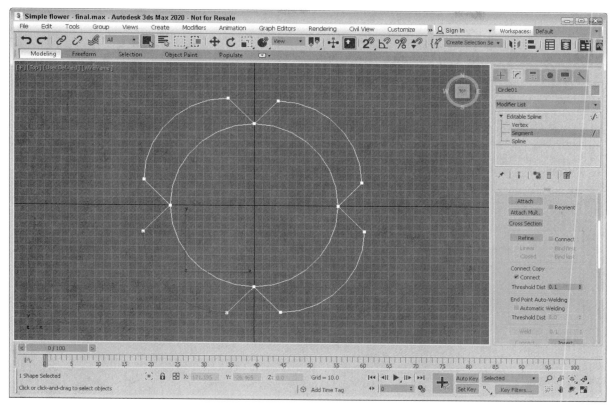

Surface Properties

For segment and spline subobjects, you can access a Surface Properties rollout that lets you assign a Material ID to the subobject. These Material IDs are used with the Multi/Sub-Object Material available in the Material Editor. For example, suppose you've created a road from a bunch of splines that are part of the same object. You can assign one Material ID for the lines at the edge of the road that will be the curb and a different Material ID for the yellow lines running down the middle of the road. Separate materials then can be applied to each of the parts using the matching Material IDs.

You can find information on Material IDs in Chapter 17, "Creating and Applying Standard Materials with the Slate Material Editor."

Using the Select ID button and drop-down list, you can locate and select all subobjects that have a certain Material ID. Simply select the Material ID that you are looking for, and click the Select ID button. All segments (or splines) with that Material ID are selected. Beneath the Select ID button is another drop-down list that lets you select segments by material name. The Clear Selection option clears all selections when the Select ID button is clicked. If disabled, all new selections are added to the current selection set.

Editing Spline subobjects

To edit a spline, click the Spline subobject in the Modifier Stack, or select the Spline icon from the Selection rollout. Transforming a spline object containing only one spline works the same way in Subobject mode as it does in a normal transformation. Working in Spline Subobject mode lets you move splines relative to one another. Right-clicking a spline in Subobject mode opens a pop-up quad menu that lets you convert it between Curve and Line types. The Curve type option changes all vertices to Bézier type, and the Line type option makes all vertices Corner type. Spline Subobject mode includes many of the buttons previously discussed, as well as some new ones in the Geometry rollout.

Reverse

The Reverse button is available only for Spline subobjects. It reverses the order of the vertex numbers. For example, a circle's vertices that are numbered clockwise from 1 to 4 are numbered counterclockwise after you click the Reverse button. The vertex order is important for splines that are used for animation paths or loft compound objects.

Outline

The Outline button creates a spline that is identical to the one selected and offset by an amount specified by dragging or specified in the Offset value. The Center option creates an outline on either side of the selected spline, centered on the original spline. When the Center option is not selected, an outline is created by offsetting a duplicate of the spline on only one side of the original spline. To exit Outline mode, click the Outline button again, or right-click in the viewport. Figure 12.30 shows an arc (first image) that has had the Outline feature applied (second image). The third image shows the arc without the Center option enabled, and the fourth image shows the arc with the Center option enabled.

FIGURE 12.30

The Outline button creates a duplicate copy of the original spline and offsets it.

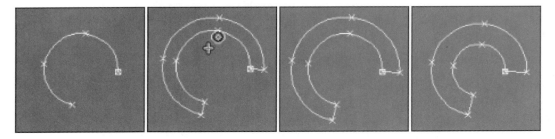

Boolean

The Boolean operations are still available, but the Shape Boolean command, covered later in this chapter, is more robust and has additional features.

Mirror

You can use the Mirror button to mirror a spline object horizontally, vertically, or along both axes. To use this feature, select a spline object to mirror and then locate the Mirror button. To the right of the Mirror button are three smaller buttons, shown in Table 12.3, each of which indicates a direction—Mirror Horizontally, Mirror Vertically, and Mirror Both. Select a direction and then click the Mirror button. If the Copy option is selected, a new spline is created and mirrored. The About Pivot option causes the mirroring to be completed about the pivot-point axes.

TABLE 12.3 Mirror Button Options

Button	Description
	Mirror Horizontally
	Mirror Vertically
	Mirror Both

Figure 12.31 shows a little critter that has been mirrored horizontally, vertically, and both. The right image was horizontally mirrored with the About Pivot option disabled. Notice that the eye spline was mirrored about its own pivot.

FIGURE 12.31

Mirroring a shape is as simple as selecting a direction and clicking the Mirror button.

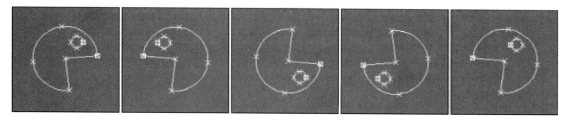

Trim and Extend

The Trim button cuts off any extending portion between two overlapping splines. The splines must be part of the same object. To use the Trim feature, select the spline that you want to keep, click the Trim button, and then click the segment to trim. The spline you click is trimmed back to the nearest intersecting point of the selected object. This button works only in Spline Subobject mode. The trimming command is dependent on the viewport that is active. When the Perspective or a Camera view is active, this command uses the Top viewport to trim.

Figure 12.32 shows a circle intersected by two ellipse shapes. The Trim button was used to cut the center sections of the ellipse shapes away.

FIGURE 12.32

You can use the Trim button to cut away the excess of a spline.

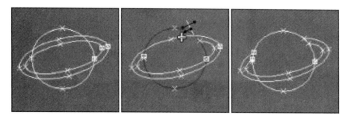

The Extend button works in the reverse manner compared to the Trim button. The Extend button lengthens the end of a spline until it encounters an intersection. (There must be a spline segment to intersect.) To use the Extend command, click the Extend button and then click the segment to extend. The spline you click is extended. To exit Extend mode, right-click in the viewport or click the Extend button again.

The Infinite Bounds option works for both the Trim and Extend buttons. When enabled, it treats all open splines as though they were infinite for the purpose of locating an intersecting point. The Extend command, like Trim, is dependent on the active viewport.

Close

The Close button completes an open spline and creates a closed spline by attaching a segment between the first and last vertices. You can check which vertex is first by enabling the Show Vertex Numbers in the Selection rollout. This is similar to the Connect feature (accessible in Vertex Subobject mode), but the Connect feature can connect the end point of one spline to the end point of another as long as they are part of the same Editable Spline object. The Close feature works only in Spline Subobject mode and connects only the end points of each given spline.

Explode

The Explode button performs the Detach command on all subobject splines at once. It separates each segment into a separate spline. You can select to explode all spline objects to separate Splines or Objects. If you select to explode to Objects, a dialog box appears, asking you for a name. Each spline uses the name you enter with a two-digit number appended to distinguish among the different splines.

Tutorial: Spinning a spider's web

Now that you're familiar with the many aspects of editing splines, this tutorial will try to mimic one of the best spline producers in the world—the spider. The spider is an expert at connecting lines together to create an intricate pattern. (Luckily, unlike the spider, which depends on its web for food, you won't go hungry if this example fails.)

To create a spider web from splines, follow these steps:

1. Select Create→Shapes→Circle, and drag in the Front viewport to create a large circle for the perimeter of the web. (Pretend that the spider is building this web inside a tire swing.) Right-click the circle, and select Convert To→Convert to Editable Spline to convert the circle shape.

2. Select the Spline subobject in the Modifier Stack (or press the 3 key) to enter Spline Subobject mode.

3. Click the Create Line button in the Geometry rollout, and click in the center of the circle and again outside the circle to create a line. Then right-click to end the line. Repeat this step until 12 or so radial lines extend from the center of the circle outward.

4. Select and right-click the 2D Snaps Toggle in the main toolbar. In the Grid and Snap Settings dialog box, enable the Vertex and Edge/Segment options, and close the dialog box. While you're still in Create Line mode, click the circle's center, and create lines in a spiral pattern by clicking each radial line that you intersect. Right-click to end the line when you finally reach the edge of the circle. Then right-click again to exit Create Line mode.

5. Select the circle shape, and click the Trim button. Then click each line segment on the portion that extends beyond the circle. This trims the radial lines to the edge of the circle. Click the Trim button again when you are finished to exit Trim mode.

6. Change to Vertex Subobject mode by clicking Vertex in the Modifier Stack (or by pressing 1). Turn off the Snaps Toggle, select all the vertices in the center of the circle, and click the Fuse button.

Figure 12.33 shows the finished spider web. (I have a new respect for spiders.)

FIGURE 12.33

A spider web made from Editable Splines

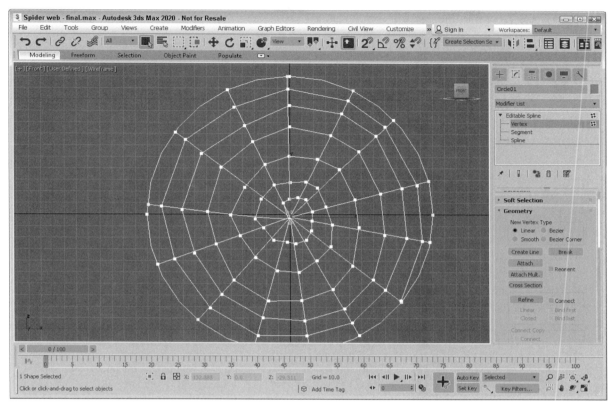

Using Shape Booleans

Although the Boolean button is still available for Editable Splines, the Shape Boolean feature located in the Compound Shapes category of the Create panel is the preferred way to work with boolean shapes. You can also find this command in the Create, Shapes, Shape Boolean menu. Be aware that the Shape Boolean menu is only available if a shape object is selected.

Boolean operations work with two or more splines that overlap. Several different operations can happen, but the main ones are: combine the splines to create a single spline (Union), you can subtract the overlapping area from one of the splines (Subtract), or you can throw away everything except the overlapping area (Intersection). There are also options to Merge, Attach and Insert shapes.

You also can use Booleans to combine or subtract 3D volumes, which are covered in Chapter 15, "Working with Compound Objects." The interface for shape booleans and 3D booleans is the same.

The Boolean button works on overlapping closed splines and has several different options—Union, Intersect, Subtract, and Symmetrical Difference (which is the opposite of the Intersect result), Merge (which combines the shapes and places a point at each intersecton), Attach, and Insert. Any overlapping splines can be added to the set using the Add Operands button and then picking the spline. The selected spline is added to the Operands list and you can then choose one of the Boolean buttons.

To use the Boolean feature, select one of the splines, and select one of the Shape Boolean operations from the Create panel or the Create menu. Then click the Add Operands button, and select the second spline and choose the Boolean button from the Operand Parameters panel. Depending on which Boolean operation you chose, the two areas are combined, the second spline acts to cut away the overlapping area on the first, or only the overlapping area remains. To exit Boolean mode, right-click in the viewport.

Note

Boolean operations can be performed on both open and closed splines that exist within a 2D plane.

Figure 12.34 shows the results of applying the Spline Boolean operators on a circle and star shape. The first image consists of the circle and star shapes without any Boolean operations applied. The second image shows the result of the Union feature; the third (circle selected first) and fourth (star selected first) use the Subtraction feature; and the fifth image uses the Intersection feature.

FIGURE 12.34

Using the Boolean operations on two overlapping shapes

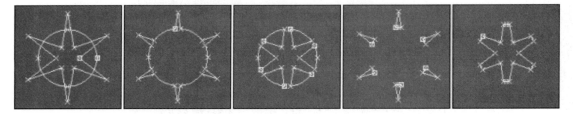

Tutorial: Creating a Not Permitted Sign

It is common for signs to describe something as not permitted by placing it in a circle with a diagonal bar going through it. Using Shape Booleans, this task is easy to create.

To create a not permitted sign using splines, follow these steps:

1. Select Create→Shapes→Circle, and drag in the Front viewport to create a large circle, then create and position a long rectangle shape with the Create→Shapes→Rectangle command so it crosses through the circle at an angle.

2. Select the circle shape and choose the Create→Shapes→Shape Boolean menu command. Click on the Add Operands button in the Boolean Parameters rollout and click on the rectangle spine, then choose the Subtract button in the Operand Parameters rollout. This removes the rectangle from the circle shape.

3. Select the Create→Shapes→Circle menu command and drag in the Front viewport from the center of the existing circle and drag to create another circle that is slightly larger than the first.

4. Select the outer circle shape and choose the Create→Shapes→Shape Boolean menu command. Click on the Add Operands button and click on the inner circle, then choose the Merge button. This adds the two shapes together to create the desired shape.

5. With the shape selected, choose the Modifiers→Mesh Editing→Extrude menu command to add some depth to the shape.

Figure 12.35 shows the extruded not permitted sign. Now just think of what you don't want and place it in the circle.

FIGURE 12.35

A not permitted sign created using Shape Booleans

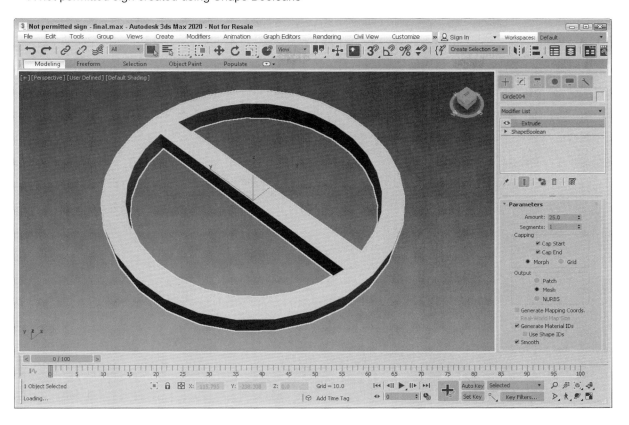

Summary

As this chapter has shown, there is much more to splines than just points, lines, and control handles. Splines in 3ds Max are one of the fundamental building blocks and the pathway to advanced modeling skills.

This chapter covered the following spline topics:

* Understanding the various shape primitives
* Editing splines
* Working with the various spline subobjects
* Using Shape Booleans

The next chapter continues your voyage down the modeling pathway with perhaps the most common modeling types—meshes and polys.

Chapter 13

Modeling with Polygons

IN THIS CHAPTER

Creating Editable Poly objects

Working with the poly subobject modes

Editing poly geometry

Changing surface properties like NURMS

Meshes (or, more specifically, polygon meshes) are perhaps the most popular and the default model type for most 3D programs. You create them by placing polygonal faces next to one another so the edges are joined. The polygons can then be smoothed from face to face during the rendering process. Using meshes, you can create almost any 3D object, including simple primitives such as a cube or a realistic dinosaur.

Meshes have lots of advantages. They are common, intuitive to work with, and supported by a large number of 3D software packages. In this chapter, you learn how to create and edit mesh and poly objects.

Understanding Poly Objects

Before continuing, you need to understand exactly what a Poly object is, how it differs from a regular mesh object, and why it is the featured modeling type in the Autodesk® 3ds Max® 2020 software. To understand these issues, you'll need a quick history lesson. Initially, 3ds Max supported only mesh objects, and all mesh objects had to be broken down into triangular faces. Subdividing the mesh into triangular faces ensured that all faces in the mesh object were coplanar, which prevented any hiccups with the rendering engine.

Over time, the rendering engines have been modified and upgraded to handle polygons that weren't subdivided (or whose subdivision was invisible to the user), and doing such actually makes the model more efficient by eliminating all the extra edges required to triangulate the mesh. Also, users can work with polygon objects (specifically four-sided quads) more easily than individual triangular faces. To take advantage of these new features, the Editable Poly object was added to 3ds Max.

As development has continued, many new features have been added to the Editable Poly object, while the Editable Mesh remained mainly for backward compatibility. But the Editable Mesh object type still exists, and there may be times when you'll want to use each type, as shown in Figure 13.1. Editable Mesh objects split all polygons into triangular faces, but the Editable Poly object maintains four-sided (or more) polygon faces. Another key difference is found in the subobjects. Editable Meshes can work with Vertex, Edge, Face, Polygon, and Element subobjects, and Editable Poly objects can work with Vertex, Edge, Border, Polygon, and Element subobjects.

FIGURE 13.1

Editable Mesh objects have triangular faces; the Editable Poly object uses faces with four or more vertices.

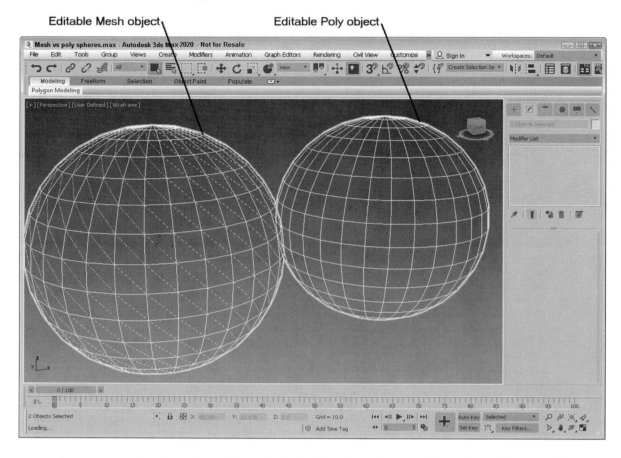

Some game and render engines still require that all faces be coplanar, and for such conditions you'll want to continue to use the Editable Mesh object. Another case where the Editable Mesh object is helpful is in performing certain face-oriented operations. In addition, normal meshes have a smaller memory footprint, which enables them to render more quickly, especially if you have many of them. Regardless, 3ds Max lets you convert seamlessly between these two modeling types.

Although many of the same features are available for both object types, the advanced features available for the Editable Poly object make it the preferred object type to use for mesh modeling. This chapter focuses on working with Editable Poly objects. Although the specific features of the Editable Mesh object aren't covered, most of these same commands apply equally to the Editable Mesh object. However, the Graphite Modeling tools (covered in Chapter 14, "Using the Graphite Modeling Tools and Painting with Objects") can be used only on Editable Poly objects.

Creating Editable Poly Objects

The Create panel has no method for making mesh objects; mesh objects must be converted from another object type or produced as the result of a modifier. Object types that you can convert include shapes, primitives, Booleans, patches, and NURBS. Many models that are imported appear as mesh objects, but they can easily be converted to Editable Poly objects.

Note
You can even convert spline shapes to Editable Poly objects, whether they are open or closed. Closed splines are filled with a polygon, whereas open splines are only a single edge and can be hard to see.

Before you can use many of the mesh editing functions discussed in this chapter, you need to convert the object to an Editable Poly object, collapse an object with modifiers applied, or apply the Edit Poly modifier.

Converting objects

To convert an object into an Editable Poly object, right-click the object, and choose Convert To→Convert to Editable Poly from the pop-up quad menu. You also can convert an object by right-clicking the object within the Modifier Stack and selecting one of the convert options from the pop-up menu or select it from the Modifiers→Conversion menu.

Collapsing to a mesh object

When an object is collapsed, it loses its parametric nature and the parameters associated with any applied modifiers. Only objects that have had modifiers applied to them can be collapsed. Objects are made into an Editable Poly object when you use the Collapse To option, available from the right-click pop-up menu in the Modifier Stack, or when you use the Collapse utility, found in the Utilities panel.

Most objects collapse to Editable Poly objects, but some objects, such as the compound objects, give you an option of which object type to collapse to.

Applying the Edit Poly modifier

Another way to enable the mesh editing features is to apply the Edit Poly modifier to an object. You apply this modifier by selecting the object and choosing Modifiers→Mesh Editing→Edit Poly or selecting Edit Poly from the Modifier drop-down list in the Modify panel.

The Edit Poly modifier is different from the Editable Poly object in that, as an applied modifier, it maintains the parametric nature of the original object. For example, you cannot change the Radius value of a sphere object that has been converted to an Editable Poly, but you could if the Edit Poly modifier were applied.

Editing Poly Objects

After an object has been converted to an Editable Poly, you can alter its shape by applying modifiers, or you can work with the mesh subobjects. You can find the editing features for these objects in the Modify panel, but a better place to look for the Editable Poly features is in the Graphite Modeling tools.

This chapter presents many of the editing features found in the Modify panel. Many of these same features also are available in the Graphite Modeling Tool's Ribbon, which is covered in Chapter 14, "Using the Graphite Modeling Tools and Painting with Objects." The mesh-related modifiers are covered in Chapter 16, "Deforming Surfaces and Using the Mesh Modifiers."

Editable Poly Subobject modes

Before you can edit poly subobjects, you must select them. To select a Subobject mode, select Editable Poly in the Modifier Stack, click the small drop-down arrow icon to its left to display a hierarchy of subobjects, and then click the subobject type with which you want to work. Another way to select a subobject type is to click the appropriate subobject button in the Selection rollout. The subobject button in the Selection rollout and the subobject listed in the Modifier Stack both get highlighted when selected. You also can type a number from 1 to 5 to enter Subobject mode with 1 for Vertex, 2 for Edge, 3 for Border, 4 for Polygon, and 5 for Element.

The Vertex Subobject mode lets you select and work with all vertices in the object. Edge Subobject mode makes all edges that run between two vertices available for selection. The Border Subobject mode lets you select all edges that run around an opening in the object, such as a hole. The Polygon Subobject mode lets you work with individual polygon faces, and the Element Subobject mode picks individual objects if the object includes several different elements.

To exit subobject edit mode, click the highlighted subobject button again. Remember, you must exit this mode before you can select another object.

Note

Selected subobject edges appear in the viewports in red to distinguish them from edges of the selected object, which appear white when displayed as wireframes.

After you're in a Subobject mode, you can click a subobject (or drag over an area to select multiple subobjects) to select it and edit the subobject using the transformation buttons on the main toolbar. You can transform subobjects just like other objects.

For more information on transforming objects, see Chapter 7, "Transforming Objects, Pivoting, Aligning, and Snapping."

When working with Editable Poly objects in Subobject mode, you can use Press and Release keyboard shortcuts. These shortcuts are identified in bold in the Editable Poly group of the Keyboard panel of the Customize User Interface dialog box. When using these keyboard shortcuts, you can access a different editing mode without having to exit Subobject mode. For example, if you press and hold Alt+C while in Polygon Subobject mode, you can make a cut with the Cut tool, and when you release the keyboard keys you'll return to Polygon Subobject mode.

You can learn more about the Customize User Interface dialog box in Chapter 49, "Customizing the Interface," and Appendix C, "Keyboard Shortcuts," has a complete list of the Press and Release keyboard shortcuts.

You can select multiple subobjects at the same time by dragging an outline over them. You also can select multiple subobjects by holding down the Ctrl key while clicking them. Holding down the Alt key removes any selected vertices from the current selection set.

With one of the transform buttons selected, hold down the Shift key while clicking and dragging a subobject to clone it. During cloning, the Clone Part of Mesh dialog box appears, enabling you to Clone to Object or Clone to Element. Using the Clone to Object option makes the selection an entirely new object, and you are able to give the new object a name. If the Clone to Element option is selected, the clone remains part of the existing object but is a new element within that object.

If you hold down the Ctrl key while choosing a different Subobject mode, the current selection is maintained for the new subobject type. For example, if you select all the polygons in the top half of a model using the Polygon Subobject mode and click the Vertex Subobject mode while holding down the Ctrl key, all vertices in the top half of the model are selected. This works only for the applicable subobjects. If the selection of polygons doesn't have any borders, holding down the Ctrl key while clicking the Border Subobject mode selects nothing.

You also can hold down the Shift key to select only those subobjects that lie on the borders of the current selection. For example, selecting all the polygons in the top half of a model using the Polygon Subobject mode and clicking the Vertex Subobject mode with the Shift key held down selects only those vertices that surround the selection, not the interior vertices.

Subobject selection

The Selection rollout, shown in Figure 13.2, includes options for selecting subobjects. The By Vertex option is available in all but the Vertex Subobject mode. It requires that you click a vertex in order to select an edge, border, polygon, or element. It selects all edges and borders that are connected to a vertex when the vertex is selected. The Ignore Backfacing option selects only those subobjects with normals pointing toward the current viewport. For example, if you are trying to select some faces on a sphere, only the faces on the side closest to you are selected. If this option is off, faces on both sides of the sphere are selected. This option is helpful if many subobjects are on top of one another in the viewport.

Tip
The select commands in the Edit menu also work with subobjects. For example, in Vertex Subobject mode, you can select Edit→Select All (or press Ctrl+A) to select all vertices.

FIGURE 13.2

The Selection rollout includes options for determining which subobjects are selected.

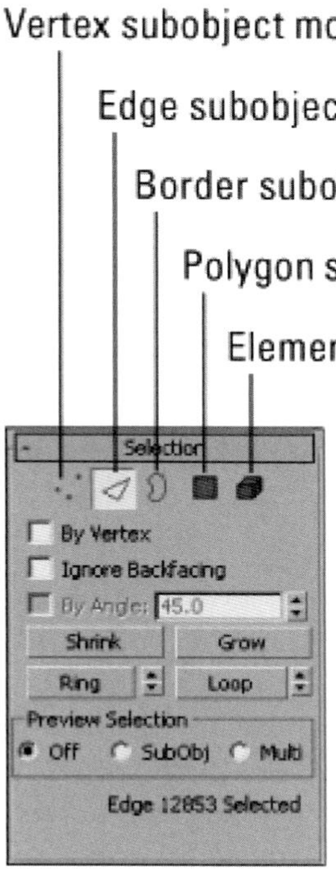

Vertex subobject mode

    Edge subobject mode

        Border subobject mode

            Polygon subobject mode

                Element subobject mode

The By Angle option selects adjacent polygons that are within the specified threshold. The threshold value is defined as the angle between the normals of adjacent polygons. For example, if you have a terrain mesh with a smooth, flat lake area in its middle, you can select the entire lake area if you set the Planar Threshold to 0 and click the lake.

The Selection rollout also includes four buttons: Shrink, Grow, Ring, and Loop. Use the Grow button to increase the current selection around the perimeter of the current selection, as shown in Figure 13.3. Click the Shrink button to do the opposite.

FIGURE 13.3

Using the Grow button, you can increase the subobject selection.

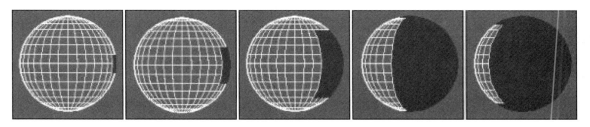

The Ring and Loop buttons are available only in Edge and Border Subobject modes. Use Ring and Loop to select all adjacent subobjects lined up horizontally and vertically around the entire object. Ring selection looks for parallel edges, and Loop selection looks for all edges around an object that are lined up end to end as the initial selection. For example, if you select a single edge of a sphere, the Ring button selects an entire row of edges going around the sphere that are lined up parallel to one another, and the Loop button selects the entire line of edges lined up around the sphere.

Next to the Ring and Loop buttons is a set of up/down arrows. These arrows are used to shift the current ring and/or loop selection to the immediate adjacent ring or loop. Holding down the Ctrl key adds the adjacent ring or loop to the current selection, and holding down the Alt key removes the adjacent selection. Figure 13.4 shows how the Ring and Loop buttons work. The first sphere shows a selection made using the Ring button; the second sphere has increased this selection by holding down the Ctrl key while clicking the up arrow next to the Loop button. The third sphere shows a selection made using the Loop button; the fourth sphere has increased this selection by holding down the Ctrl key while clicking the up arrow next to the Ring button.

FIGURE 13.4

The Ring and Loop buttons can select an entire row and/or column of edges.

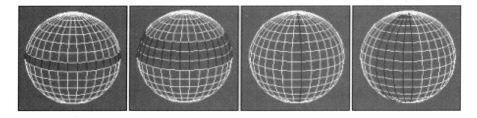

Note

For Editable Poly objects, the Hide Selected, Unhide All, Copy, and Paste buttons are located at the bottom of the Edit Geometry rollout.

The Soft Selection rollout allows you to alter adjacent nonselected subobjects when selected subobjects are moved, creating a smooth transition. For the details on this rollout, see Chapter 11, "Accessing Subobjects and Modifiers and Using the Modifier Stack."

Tutorial: Modeling a clown head

Now that you know how to select subobjects, you can use the transform tools to move them. In this example, you'll quickly deform a sphere to create a clown face by selecting, moving, and working with some vertices.

To create a clown head by moving vertices, follow these steps:

1. Select Create→Standard Primitives→Sphere, and drag in the Front viewport to create a sphere object. Then right-click the sphere, and select Convert To→Editable Poly in the pop-up quad menu.

2. Open the Modify panel. Now make a long, pointy nose by pulling a vertex outward from the sphere object. Click the small drop-down arrow icon to the left of the Editable Poly object in the Modifier Stack, and select Vertex in the hierarchy (or press the 1 key). This activates the Vertex Subobject mode. Enable the Ignore Backfacing option in the Selection rollout, and select the single vertex in the center of the Front viewport. Make sure that the Select and Move button (W) is selected, and in the Left viewport, drag the vertex along the X-axis until it projects outward from the sphere.

3. Next, create the mouth by selecting and indenting a row of vertices in the Front viewport below the protruding nose. Holding down the Ctrl key makes selecting multiple vertices easy. Below the nose, select several vertices in a circular arc that make a smile. Then move the selected vertices along the negative X-axis in the Left viewport.

4. For the eyes, select Create→Standard Primitives→Sphere, and enable the AutoGrid option. Then drag in the Front viewport to create two eyes above the nose.

This clown head is just a simple example of what is possible by editing subobjects. Figure 13.5 shows the clown head in a shaded view.

FIGURE 13.5

A clown head created from an editable poly by selecting and moving vertices

Editing geometry

Much of the power of editing meshes is contained within the Edit Geometry rollout, shown in Figure 13.6. Features contained here include, among many others, the ability to create new subobjects, attach subobjects to the mesh, weld vertices, chamfer vertices, slice, explode, and align. Some Edit Geometry buttons are disabled, depending on the Subobject mode that you select. The features detailed in this section are enabled for the Editable Poly object before you enter a Subobject mode.

FIGURE 13.6

The Edit Geometry rollout includes many general-purpose editing features.

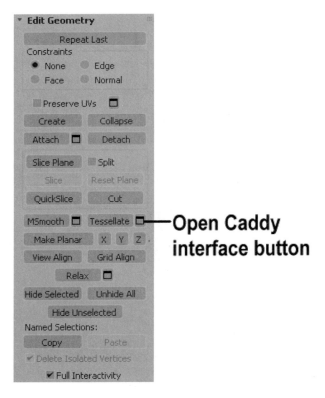

Open Caddy interface button

Many of the buttons for the Editable Poly include a small icon to the right of the button that opens a settings caddy. These caddy interfaces appear around the selected subobject and allow you to change the settings and immediately see the results in the viewports. The OK button (a check-mark icon) applies the settings and closes the caddy interface, and the Apply button (a plus-sign icon) applies the settings and leaves the dialog box open. These caddy interfaces are included next to the buttons such as MSmooth, Tessellate, and Relax.

Editable Poly objects include all their common buttons in the Edit Geometry rollout and all subobject-specific buttons in a separate rollout named after the Subobject mode, such as Edit Vertices or Edit Edges.

Repeat Last

The first button in the Edit Geometry rollout is the Repeat Last button. This button repeats the last subobject command. This button does not work on all features, but it's very convenient for certain actions. For example, if you select and extrude a polygon to make it stick out from the surface, you can get other polygons to be extruded by the same amount by selecting them and clicking the Repeat Last button.

Tip

The tooltip for this button displays the last repeatable command.

Enabling constraints

The Constraints options limit the movement of subobjects to a specified subobject. The available constraints are None, Edge, Face, and Normal. For example, if you select and move a vertex with the Edge constraint enabled, the movement is constrained to the adjacent edges. This lets you move, rotate, and scale vertices, edges, and polygons while making sure they stay with the surface of the current object.

Tutorial: Creating a quick flying saucer

Primitive shapes can be quickly changed in many ways to create other simple shapes, and the Constraints settings help keep the moved subobjects in check.

To create a simple flying saucer, follow these steps:

1. Select Create→Standard Primitives→Sphere, and drag in the Top viewport to create a Sphere object.
2. Right-click the sphere object, and select Convert To→Convert to Editable Poly in the pop-up quad menu.
3. Open the Modify panel, choose the Polygon Subobject mode and then drag over to select the two middle rows of polygons in the Left viewport.
4. In the Constraints section of the Edit Geometry panel, select the Normal option, and drag the selected polygons outward with the Move tool in the Top viewport roughly twice the size of the sphere.
5. In the Constraints section, select the Edge option, and drag the Y-axis with the Select and Scale tool downward in the Left viewport to pull the polygons close together.
6. Exit Subobject mode, and scale the entire object by dragging on the Y-axis in the Left viewport with the Select and Scale tool.

The resulting flying saucer is shown in Figure 13.7, ready to be populated with aliens.

FIGURE 13.7

The Constraints options keep the subobjects within the specified bounds.

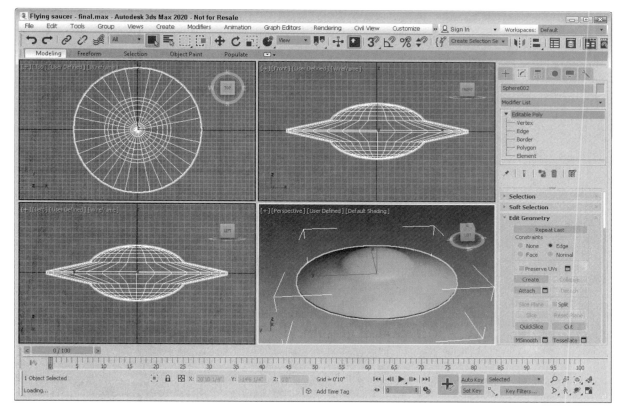

Preserve UVs

UV coordinates define how a texture map is applied to an object's surface. These UV coordinates are tied closely to the surface subobject positions, so moving a subobject after a texture is applied moves the texture also. This could cause discontinuities to the texture map. The Preserve UVs option lets you make subobject changes without altering the UV coordinates for an existing texture.

The Settings button, located to the right of the Preserve UVs option, opens the Preserve Map Channels dialog box, which lets you select a Vertex Color and Texture Channel to preserve. Figure 13.8 shows two block objects with a brick texture map applied. The inner vertices on the left block were scaled outward without the Preserve UVs option selected; the right block had this option enabled.

FIGURE 13.8

The Preserve UVs option lets you make subobject changes after texture maps have been applied.

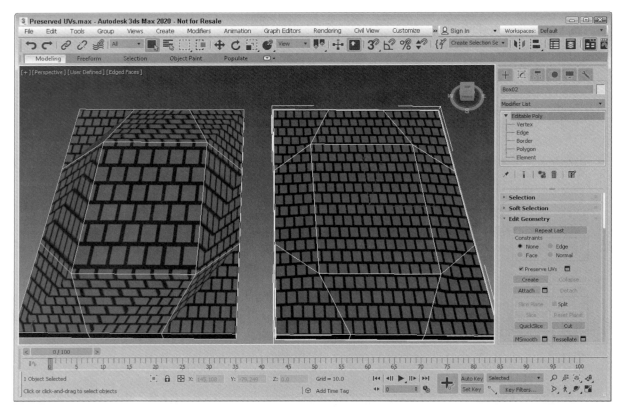

Create

The Create button lets you create new subobjects, including vertices, edges, and polygons. This is done in Vertex mode by simply clicking to place vertices. New edges are created by connecting vertices to existing vertices. When the cursor is over a vertex in Edge or Border Subobject mode, it changes to a crosshair, and you can click to create a polygon edge by clicking two vertices one after the other.

Tip

New vertices and polygons are placed by default on the construction grid, but you can specify to create new vertices and polygons on the object surface by enabling the Face or Edge snap options.

You also can use the Create button to create new polygons. To create a new polygon, click the Create button, which highlights all vertices in the selected mesh. Next, click a vertex to start the polygon; after you click two more vertices, a new face is created. You also can create a new vertex not based on any existing vertices by simply clicking to place vertices. Right click to exit polygon create mode.

As you create new polygons, the normal is determined by the direction in which you create the polygon using the right-hand rule. If you bend the fingers of your right hand in the direction (clockwise or counterclockwise) in which the vertices are clicked, your thumb points in the direction of the normal. If the normal is pointing away from you, the back side of the polygon will be visible, and the lighting could be off.

Polygons aren't limited to three vertices. You can click as many times as you want to add additional vertices to the polygon. Click the first vertex again, or double-click to complete the polygon.

Note
If you click the Create button with no Subobject mode selected, the Polygon Subobject mode is automatically selected.

Collapse

The Collapse button is used to collapse all the selected subobjects to a single subobject located at the averaged center of the selection. This button is similar to the Weld button, except that the selected vertices don't need to be within a Threshold value to be combined. This button works in all Subobject modes.

Attach and Detach

The Attach button is available with all Subobject modes, even when you are not in a Subobject mode. Use the Attach button to add objects to the current Editable Poly object. You can add primitives, splines, patch objects, and other mesh objects. Any object attached to a mesh object is automatically converted into an Editable Poly and inherits the object color of the object to which it is attached. Any objects added to a poly object can be selected individually using the Element Subobject mode.

Caution
If you attach an object that is smoothed using NURMS, the NURMS are lost when the object is attached.

To use this feature, select the main object, and click the Attach button. Move the mouse over the object to be attached; the cursor changes over acceptable objects. Click the object to select it. Click the Attach button again or right-click in the viewport to exit Attach mode.

Note
If the object that you click to attach already has a material applied that is different from the current Editable Poly object, a dialog box appears, giving you options to Match Material IDs to Material, Match Material to Material IDs, or Do Not Modify Material IDs or Material. Materials and Material IDs are discussed in more detail in Chapter 17, "Creating and Applying Standard Materials with the Slate Material Editor."

Clicking the Attach List button opens the Attach List dialog box (which looks just like the Select From Scene [H] dialog box), where you can select the objects to attach. The list contains only objects that you can attach.

Attaching objects is different from grouping objects because all attached objects act as a single object with the same object color, name, and transforms. You can access individual attached objects using the Element Subobject mode.

Use the Detach button to separate the selected subobjects from the rest of the object. To use this button, select the subobject, and click the Detach button. The Detach dialog box opens, enabling you to name the new detached subobject. You also have the options to Detach to Element or Detach as Clone.

Slicing and cutting options

The Slice Plane button lets you split the poly object along a plane. When you click the Slice Plane button, a yellow slice plane gizmo appears on the selected object. You can move, rotate, and scale this gizmo using the transform buttons. After you properly position the plane and set all options, click the Slice button to finish slicing the mesh. All intersected faces split in two, and new vertices and edges are added to the mesh where the Slice Plane intersects the original mesh.

The Slice Plane mode stays active until you deselect the Slice Plane button or until you right-click in the viewport; this feature enables you to make several slices in one session. The Slice Plane button is enabled for all Subobject modes. A Reset Plane button is located next to the Slice Plane button. Use this button to reset the slice plane to its original location. You use the Split option to double the number of vertices and edges along the Slice Plane, so each side can be separated from the other. When used in Element mode, the Split option breaks the sliced object into two separate elements.

Caution
Although the Slice Plane feature can be used in all Subobject modes, cuts are made only to the object in Vertex, Edge, and Border modes.

The QuickSlice button lets you click anywhere on an Editable Poly object where you want a slicing line to be located. You can then move the mouse, and the QuickSlice line rotates about the point you clicked. When you click the mouse again, a new vertex is added at every place where the QuickSlice line intersects an object edge. This is a very convenient tool for slicing objects because the slice line follows the surface of the object, so you can see exactly where the slice will take place.

For the QuickSlice and Cut tools, you can enable the Full Interactivity option (located near the bottom of the Edit Geometry rollout). With this option enabled, the slice lines are shown as you move the mouse about the surface. With Full Interactivity disabled, the resulting lines are shown only when the mouse is clicked.

For Editable Poly objects, the Cut button is interactive. If you click a polygon corner, the cut edge snaps to the corner, and a new edge extends from the corner to a nearby corner. The cursor changes when you are over corners and edges and within the polygon interior. As you move the mouse around, the edge moves until you click where the edge should end. If you click in the middle of an edge or face, new edges appear. The Cut tool lets you add successive connected edges by clicking new endpoints. If you right-click once, the current cut is released, allowing you to start a new cut. Click in a new position to create a new cut, or right-click a second time to exit the Cut tool. The Cut tool is great for cutting holes into an existing object.

Tutorial: Detaching a plane model into separate objects

Sometimes when importing models, you'll find that the entire model is a single mesh. This makes it hard to apply materials and to animate, but using the QuickSlice and Detach operations, you can quickly separate objects into the different objects.

To separate a plane model into separate objects, follow these steps:

1. Open the Prop plane object.max file from the Chap 13 directory in the downloaded content set.

2. Before you can separate the model into parts, you'll want to convert it to an Editable Poly object. Select the model, right-click it, and select the Convert To→Convert to Editable Poly in the pop-up quad menu.

3. Open the Modify panel, and select the Element subobject mode. Often separate objects are grouped as separate elements, which makes them easy to separate. Click on both halves of the propeller and choose the Detach button. A dialog box appears where you can name this detached object as **Propeller**.

4. Repeat this step for many of the other elements including the wings, tail and canopy.

5. To separate out the Prop shaft pieces, drag over the front of the plane in the Front viewport and then hold down the Alt key and select the fuselage to remove it from the selection set, then use the Detach button to detach the objects as **Prop Shaft**.

6. Once all the elements are separated, you can then switch to the Polygon subobject mode and select the front cowling of the fuselage. If there isn't a clean break that you can select, then use the QuickSlice tool and click on the location where you want the break to be and a row of edges will be placed where you want them. Detach the front of the fuselage as **Cowling**.

7. Once separated into separate objects, you can assign each object a different object color to keep them straight.

Figure 13.9 shows the separated plane.

FIGURE 13.9

Using the Detach and QuickSlice features, you can slice and separate mode parts.

MSmooth

Both the MSmooth and Tessellate buttons have caddy interfaces, as shown in Figure 13.10. The MSmooth setting for Smoothness rounds all the sharp edges of an object.

FIGURE 13.10

The Caddy interfaces for the MSmooth and Tessellate buttons let you interactively set the Smoothness and Tension values.

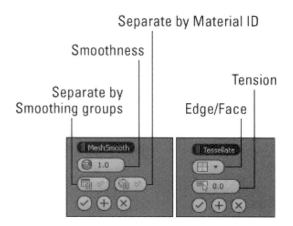

The MSmooth button can be used to smooth the selected subobjects in the same way as the MeshSmooth or TurboSmooth modifiers. This button can be used several times, but each time, the density of the polygons is increased. The Smoothness value determines which vertices are used to smooth the object. The higher the value, the more vertices are included and the smoother the result. You also can select that the smoothing is separated by Smoothing Groups or by Materials.

Figure 13.11 shows a simple diamond-shaped hedra that has been MeshSmoothed using the MSmooth button and then tessellated three consecutive times.

FIGURE 13.11

Using MSmooth reduces the sharp edges, and tessellating adds more editable faces.

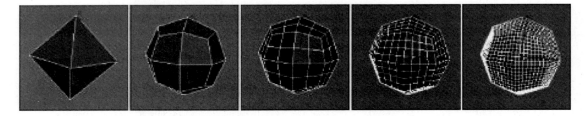

Tessellate

Tessellation is used to increase the density of the object's faces or edges. When modeling, you may want more details in a select area. This is where the tessellation command comes in. Tessellation can be applied to individual selected subobjects or to the entire object.

You can use the Tessellate button to increase the resolution of a mesh by splitting a face or polygon into several faces or polygons. You have two options to do this: Edge and Face.

The Edge method splits each edge at its midpoint. For example, a rectangular face would be split into four smaller rectangles, one at each corner. The Tension spinner to the right of the Tessellate button specifies a value that is used to make the tessellated face concave or convex.

The Face option creates a new vertex in the center of the face and also creates new edges, which extend from the center vertex to each original vertex. For a square polygon, this option would create four triangular faces. Figure 13.12 shows the faces of a cube that has been tessellated once using the Edge option and then again using the Face-Center option.

FIGURE 13.12

A cube tessellated twice, using each option once

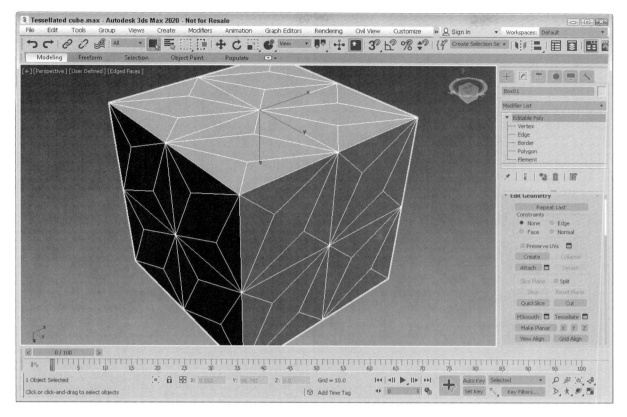

Make Planar

A single vertex or two vertices don't define a plane, but three or more vertices do. If three or more vertices are selected, you can use the Make Planar button to make these vertices coplanar (which means that all vertices are on the same plane). Doing so positions the selected vertices so they lie in the same plane. This is helpful if you want to build a new polygon face or if you need to flatten an area that is deformed. Polygonal faces need to be coplanar for most rendering and game engines. This button works in all Subobject modes. The X, Y, and Z buttons let you collapse the current object or subobject selection to a single plane lying on the specified axis.

View and Grid Align

The View and Grid Align buttons move and orient all selected vertices to the current active viewport or to the current construction grid. These buttons also can be used in all Subobject modes. This causes all the selected face normals to point directly at the grid or view.

Relax

The Relax button works just like the Relax modifier by moving vertices so they are as far as possible from their adjacent vertices according to the Amount value listed in the Relax caddy interface. The caddy interface also includes an Iterations value, which determines the number of times the operation is performed. You also can select to hold all Boundary and Outer points from being moved. This is a great feature for removing any pinching or areas where the vertices are too close together to work with.

Hide, Copy, and Paste

The Hide button hides the selected subobjects. You can make hidden objects visible again with the Unhide All button.

After selecting several subobjects, you can create a named selection set by typing a name in the Named Selection Sets drop-down list in the main toolbar. You can then copy and paste these selection sets onto other shapes.

At the bottom of the Selection rollout is the Selection Information, which is a text line that automatically displays the number and subobject type of selected items.

Editing Vertex subobjects

When working with the Editable Poly objects, after you select a Vertex Subobject mode (keyboard shortcut, 1) and select vertices, you can transform them using the transform buttons on the main toolbar. All vertex-specific commands are found within the Edit Vertices rollout, shown in Figure 13.13.

FIGURE 13.13

When the Vertex Subobject mode is selected, these vertex commands become available.

Remove

The Remove button lets you delete the selected vertices or other subobjects. You also can delete the selected subobjects with the Delete key, but there is a big difference between the two. The Delete key removes the selected subobjects and all the polygons that are attached to them, thereby creating a hole in the object. The Remove key eliminates the subobject while keeping the surrounding polygons. The Remove button automatically adjusts the surrounding subobjects to maintain the mesh integrity.

Figure 13.14 shows a sphere object with several vertex subobjects selected. The middle image is an Editable Mesh that used the Delete feature, and the right image is an Editable Poly that used the Remove feature.

FIGURE 13.14

Deleting vertices also deletes the adjoining faces and edges, but Remove maintains the mesh.

The Remove button also is available in Edge Subobject mode. If you hold down the Ctrl key when clicking the Remove button when an edge is selected, vertices at either end of the deleted edge are also removed.

Break

You use the Break button to create a separate vertex for adjoining faces that are connected by a single vertex.

In a normal mesh, faces are all connected by vertices: Moving one vertex changes the position of all adjoining faces. The Break button enables you to move the vertex associated with each face independent of the others. The button is available only in Vertex Subobject mode.

Figure 13.15 shows a hedra object. The Break button was used to separate the center vertex into separate vertices for each face. After the Break button is used, the face vertices can be manipulated independently, as the figure shows.

FIGURE 13.15

You can use the Break button to give each face its own vertex.

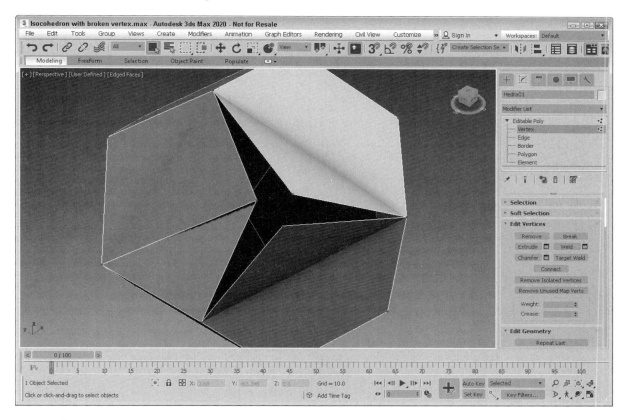

Extrude

The Extrude button copies and moves the selected subobject perpendicular to the surface a given distance and connects the new copy with the original one. This works in several Subobject modes. For example, extruding a single vertex raises a spike off the surface, extruding four edges forms a fenced area like a box with no lid, and extruding a four-sided polygon raises a box from the surface. To use this feature, select a subobject, click the Extrude button, and then drag in a viewport. The cursor changes when it's over the selected subobject. Release the button when you've reached the desired distance. To exit Extrude mode, click the Extrude button again or right-click in the viewport.

Alternatively, you can click the small button next to the Extrude button to open the caddy settings. These settings include extrusion Height and Width values. The Height defines how far from the surface the vertex, edge, border, or polygon is raised; the Width value defines how far the base of the extrusion spreads out at the base. The Height value also can be negative to push the subobject toward the center of the object, creating a dimple.

Several options are available if a polygon selection is made. The Group option extrudes all selected polygons in the direction of the average of all the normals for the group (the normal runs perpendicular to the face); the Normal Local option moves each individual polygon along its local normal. You can extrude By Polygons, which extrudes each individual polygon along its normal as a separate extrusion.

Figure 13.16 shows a GeoSphere object with all edges selected and extruded. The Group Normal option averages all the normals and extrudes the edges in the averaged direction. The Local Normal option extrudes each edge along its own normal.

FIGURE 13.16

Subobjects can be extruded along an averaged normal or locally.

Weld and Chamfer

Vertices and edges that are close to or on top of one another can be combined with the Weld command. The Weld and Chamfer buttons both have caddy settings that you can access. The Weld caddy includes settings for the weld Threshold value and displays the number of vertices before and after the welding process, which is very useful for checking whether a weld was successful.

Tip

If the Weld feature isn't working because its Threshold value is too low, try using the Collapse button.

The Target Weld button lets you click a single vertex and move the cursor over an adjacent vertex. A rubber-band line stretches from the first selected vertex to the target weld vertex, and the cursor changes to indicate that the vertex under the cursor may be selected. Clicking the target vertex welds the two vertices together.

Note

When two vertices are welded, the new vertex is positioned at a location that is halfway between both vertices, but when Target Weld is used, the first vertex is moved to the location of the second.

The Chamfer button—which is enabled in Vertex, Edge, and Border Subobject modes—lets you cut the edge off a corner and replace it with a face. Using the caddy settings, you can interactively specify a Chamfer Amount and the number of segments. You can also choose a Standard Chamfer or a Quad Chamfer. The latter uses quad with four edge faces. The caddy settings also include an Open Chamfer option, which cuts a hole in the polygon face instead of replacing it with a new polygon. Figure 13.17 shows two plane objects that have been chamfered with the Open option enabled. The left plane had all its interior vertices selected, and the right plane had a selection of interior edges selected.

New Feature in 2020

The Chamfer feature for Editable Poly objects has been overhauled and improved in 3ds Max 2020. New features include weighted chamfer, the ability to inset faces and end point bias.

FIGURE 13.17

Enabling the Open Chamfer option in the caddy interface removes a polygon instead of replacing it.

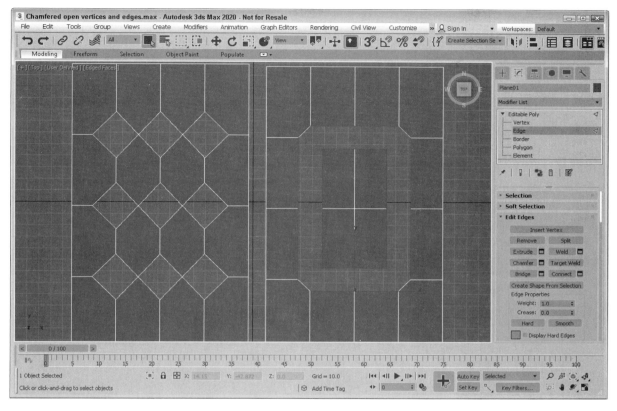

Connect

The Connect button can be used to add new edges to subobjects. In Vertex Subobject mode, the Connect button connects selected vertices on the opposite side of a face. If the selected vertices don't share a face, nothing happens. This provides a quick way to tessellate a selected area.

In Edge and Border Subobject mode, the Connect button adds new edges that connect the two selected edges within a shared face. It also includes a small icon that makes the caddy settings available, which includes the Segments setting. This value is the number of edge segments to add between the selected edges or borders. It also includes Pinch and Slide values. The Pinch value moves the segments closer together or farther away from each other; the Slide value moves the segments along the original edge.

Remove Isolated and Unused Map Vertices

The Remove Isolated Vertices button deletes all isolated vertices. Vertices become isolated by some operations and add unneeded data to your file. You can search for and delete them quickly with this button. Good examples of isolated vertices are those created using the Create button but never attached to an edge.

The Remove Unused Map Vertices button removes any leftover mapping vertices from the object.

Weight and Crease

The Weight settings control the amount of pull that a vertex has when NURMS subdivision or a MeshSmooth modifier is used. The higher the Weight value, the more resistant a vertex is to smoothing. For edge and border subobjects, the Weight value is followed by a Crease value that determines how visible the edge is when the mesh is smoothed. A value of 1.0 ensures that the crease is visible.

Editing Edge subobjects

Edges are the lines that run between two vertices. Edges can be *closed*, which means that each side of the edge is connected to a face, or *open*, which means that only one face connects to the edge. When a hole exists in a mesh, all edges that are adjacent to the hole are open edges. Mesh edges, such as those in the interior of a shape that has been converted to a mesh, also can be *invisible.* These invisible edges appear as dotted lines that run across the face of the polygon. These invisible edges are found only in Editable Mesh objects; Editable Poly objects don't have these invisible edges because each polygon face can have any number of edges.

You can select multiple edges by holding down either the Ctrl key while clicking the edges or the Alt key to remove selected edges from the selection set. You also can copy edges using the Shift key while transforming the edge. The cloned edge maintains connections to its vertices by creating new edges.

Many of the Edge subobject options work in the same way as the Vertex subobject options. Figure 13.18 shows the Edit Edges rollout.

FIGURE 13.18

All the Edge-specific commands are available when the Edge Subobject mode is enabled.

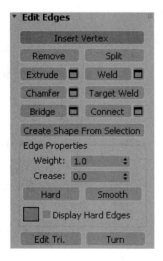

Split and Insert Vertex

The Split button adds a new vertex at the middle of the edge and splits the edge into two equal sections. This button is handy when you need to increase the resolution of a section quickly.

The Insert Vertex button lets you add a new vertex anywhere along an edge. The cursor changes to crosshairs when it is over an edge. Click to create a vertex. In Edge, Border, Polygon, or Element Subobject mode, this button also makes vertices visible.

Bridge Edges

The Bridge button for edges allows you to create a new set of polygons that connect the selected edges. If two edges are selected when the Bridge button is pressed, they are automatically connected with a new polygon. If no edges are selected, you can click the edges to bridge after clicking the Bridge button. The selected edges on either side of the bridge can be different in number.

You also can access the caddy settings for the Bridge feature. This caddy offers the options to Use Specific Edges or to Use Edge Selection. The Use Specific Edges option has two buttons for each edge. If you click one of these buttons, you can select an edge in the viewport. The Use Edge Selection option lets you drag a marquee in the viewport to select the edges. The Bridge Edges caddy also includes options for setting the number of Segments; the Taper, Bias, and Smooth values; and a Bridge Adjacent value, which increases the triangulation for angles above the given threshold. A Reverse Triangulation option also is available. For Edge selection, there is only the Smooth value, and for a Border selected, two Twist values rotate the bridge segments clockwise or counterclockwise relative to the connected edge and border selections.

Figure 13.19 shows a simple example of some edges that have been bridged. The letters before bridging are on the top and after bridging on the bottom.

FIGURE 13.19

Selecting two opposite edges and clicking the Bridge button in Edge Subobject mode creates new connecting polygons.

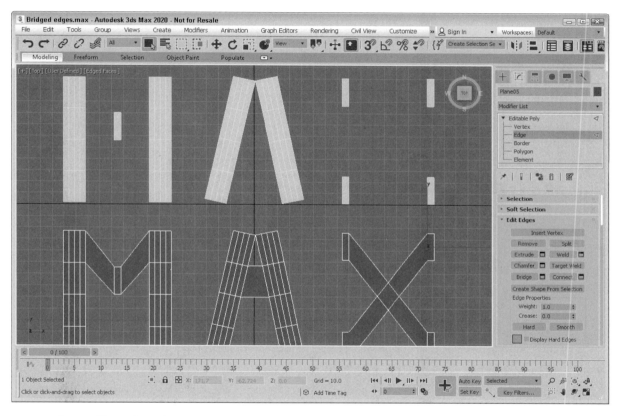

Create Shape from Selection

The Create Shape from Selection button creates a new spline shape from selected edges. The Create Shape dialog box appears, as shown in Figure 13.20, enabling you to give the new shape a name. You also can select options for Smooth or Linear shape types.

FIGURE 13.20

The Create Shape dialog box lets you name shapes created from selected edge subobjects.

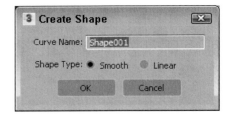

Edit Edge Properties

Each edge can have its own Weight and Crease settings. These settings define how likely the edge is to be smoothed or pronounced. Lower edge weights make the edge smoother, but higher edge weights will repel any smoothing. For example, if the MSmooth is used on an object, then the edges with higher weights will not get smoothed as much as those with lower weights. To make the edge stand out even more, you can set the Crease value higher. This makes the edge stand out with a noticeable crease.

Another way to set edge properties is by controlling the smoothing groups on either side of the edge. The Hard button causes no smoothing over the edge and the Smooth button renders the polygons on either side of the edge as smooth. The Display Hard Edges highlights all hard edges on the object using the designated color to the left of this option. These lines are only seen when the Edged Faces option is enabled in the viewport.

Edit Triangulation

For the Editable Poly object, the Edge, Border, Polygon, and Element subobjects include the Edit Triangulation button, which is labeled Edit Tri. The Edit Triangulation button shows the triangulation edges for all polygon faces and lets you change the internal edges of the polygon by dragging from one vertex to another. When this button is clicked, all hidden edges appear. To edit the hidden edges, just click a vertex and then click again where you want the hidden edge to go. If you're dealing with multiple four-sided polygons, the Turn button is quicker.

Turn

The Turn button rotates the hidden edges that break up the polygon into triangles (all polygonal faces include these hidden edges). For example, if a quadrilateral (four-sided) face has a hidden edge that runs between vertices 1 and 3, the Turn button changes this hidden edge to run between vertices 2 and 4. This affects how the surface is smoothed when the polygon is not coplanar.

This button is available for all Subobject modes except Vertex. When enabled, all subobjects that you click are turned until the button is disabled again. Figure 13.21 shows the top face of a Box object with a hidden edge across it diagonally. The Turn button was used to turn this hidden edge.

Tip

Surfaces can deform only along places where there are edges, so as you create your models, be aware of where you place edges and how the edges flow into one another. When building characters, it is important to have the edges follow the muscle flow to deform properly.

FIGURE 13.21

The Turn feature is used to change the direction of edges.

Editing Border subobjects

Editable Poly objects do not need the Face subobject that is found in the Editable Mesh objects because they support polygon faces. Instead, they have a Border subobject. Border subobjects are polygons with no faces that are actually holes within the geometry. The Border rollout is shown in Figure 13.22.

FIGURE 13.22

Many of the Border subobject commands are the same as those for Edges.

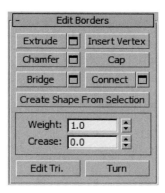

Cap

The Cap button causes the existing border selection to be filled in with a single coplanar polygon. After this feature is used, the Border subobject is no longer identified as a Border subobject. Be aware that capping a large hole may create a polygon face with multiple edges. Once capped, you may want to use the Connect tool to subdivide the new face into many smaller polygons.

Bridge

The Bridge feature joins two selected Border subobjects with a tube of polygons that connect the two borders. The two selected borders must be part of the same object and need not have an equal number of segments.

The Bridge caddy, shown in Figure 13.23, lets you specify twist values for each edge; the number of segments; and the Taper, Bias, and Smooth values.

FIGURE 13.23

The Bridge caddy interface for edges lets you specify options such as the number of segments, the Taper, and whether the bridge twists.

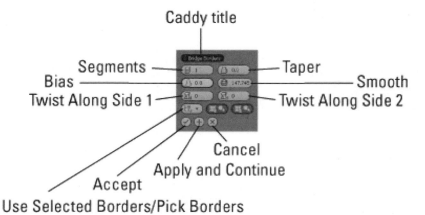

Tutorial: Bridging a forearm

The Bridge tool is great for working with two border selections, allowing you to create a smooth set of polygons that flow between them. For this example, you'll create a forearm by bridging a hand model with a simple cylinder. The polygons of the hand and the cylinder that are to be joined have already been removed.

To create a forearm object by bridging a cylinder with a hand model, follow these steps:

1. Open the Forearm bridge.max file from the Chap 13 directory in the downloaded content set.

2. With the body parts selected, open the Modify panel, and select the Border Subobject mode. Then press and hold the Ctrl key, and click the borders for the hand and cylinder elements.

3. With both facing Border subobjects selected, click the settings icon next to the Bridge button in the Edit Borders rollout.

4. In the Bridge caddy, select the Use Border Selection option, and set the Segments value to **6**. Then click the green check mark to accept the settings.

Figure 13.24 shows the resulting forearm object.

FIGURE 13.24

The Bridge feature can be used to quickly connect body parts such as this forearm.

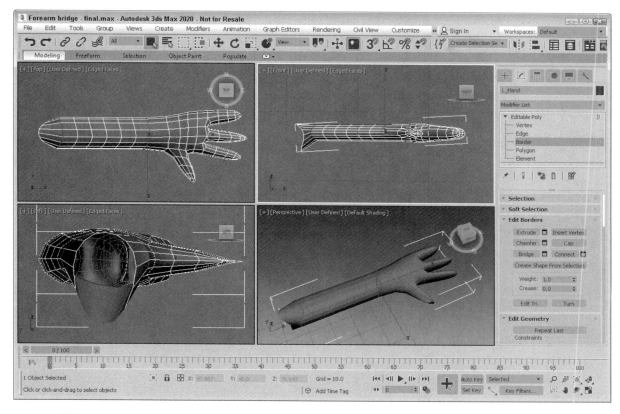

Editing Polygon and Element subobjects

Like the other Subobject modes, Editable Polys can be edited at the polygon and element subobject level. The buttons for these modes are found in the Edit Polygons and Edit Elements rollouts. Figure 13.25 shows the Edit Polygons rollout.

FIGURE 13.25

The Polygon subobjects commands are found in the Edit Polygons rollout.

Insert Vertex

In Polygon and Element Subobject mode, the Insert Vertex button adds a vertex on the surface of the object where you clicked and automatically connects it with edges of the adjacent faces.

Outline and Inset

The Outline button (in Polygon Subobject mode) offsets the selected polygon a specified amount. This increases (or decreases) the size of the selected polygon. The Inset button creates another polygon set within the selected polygon and connects their edges. For both these buttons, caddy settings are available that include the Outline or Inset Amount values. If multiple polygons are selected, you can inset the selection as a Group or by individual polygons.

Bevel

The Bevel button extrudes the Polygon subobject selection and then lets you size the extruded polygon face. To use this feature, select a polygon, click the Bevel button, drag up or down in a viewport to the Extrusion height, and release the button. Drag again to specify the polygon's outline value. The Outline amount determines the relative size of the extruded face. If multiple polygons are selected, you can apply this feature to the group, to the group using the Local Normal, or individually to each polygon with the By Polygon option.

Figure 13.26 displays a poly dodecahedron. Each face has been locally extruded with a value of 20 and then locally beveled with a value of –10.

FIGURE 13.26

The top faces of this dodecahedron have been individually extruded and beveled.

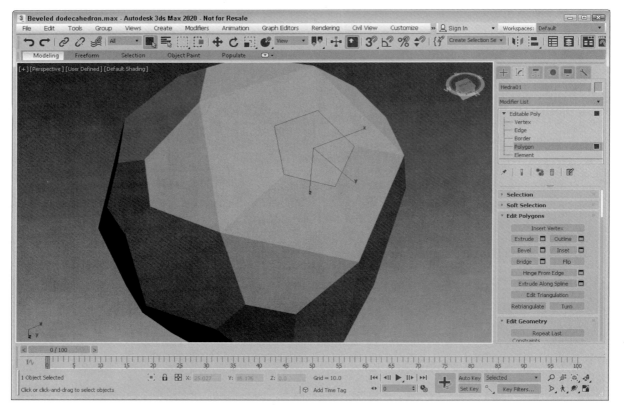

Flip

The Flip button flips the normal vectors for the selected subobjects. The Flip button is available only in Polygon and Element Subobject modes.

Retriangulate

The Retriangulate button automatically computes all the internal edges for you for the selected subobjects.

Hinge From Edge

The Hinge From Edge button rotates a selected polygon as though one of its edges were a hinge. The angle of the hinge depends on the distance that you drag with the mouse, or you can use the available caddy settings. In the caddy settings, shown in Figure 13.27, you can specify an Angle value and the number of segments to use for the hinged section.

FIGURE 13.27

The Hinge From Edge caddy interface lets you select a hinge.

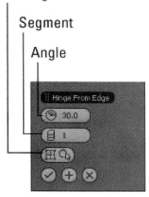

Pick Hinge

Segment

Angle

By default, one of the polygon's edges will be used as the hinge about which the section rotates, but in the caddy settings, you can click the Pick Hinge button and select an edge (which doesn't need to be attached to the polygon). Figure 13.28 shows a sphere primitive with four polygon faces that have been hinged around an edge at the sphere's center.

FIGURE 13.28

Several polygon faces in the sphere have been extruded along a hinge.

Extrude Along Spline

The Extrude Along Spline button can be used to extrude a selected polygon along the spline path. The caddy settings, shown in Figure 13.29, include a Pick Spline button that you can use to select the spline to use. You also can specify the number of segments, the Taper Amount and Curve, and a Twist value. You also have an option to Align the extrusion to the face normal or to specify the amount to rotate about the normal.

FIGURE 13.29

The Extrude Along Spline settings dialog box

Taper Curve

Segments Taper amount

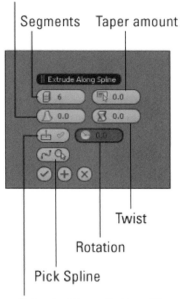

Twist

Rotation

Pick Spline

Extrude Along Spline Align

Tutorial: Building an octopus

The thing about an octopus that makes it unique is the fact that it has eight tentacles. Creating these tentacles can be easily accomplished with the Extrude Along Spline feature.

To create an octopus using the Extrude Along Spline feature, follow these steps:

1. Open the Octopus.max file from the Chap 13 directory in the downloaded content set.

   This file includes the base of an octopus created from a squashed sphere primitive that has been converted to an Editable Poly. Eight splines surround the object.

2. Select the octopus object to automatically open the Modify panel. In the Selection rollout, click the Polygon subobject button (keyboard shortcut, 4) and enable the Ignore Backfacing option in the Selection rollout.

3. Click the Shading viewport label, and select the Edged Faces option from the pop-up menu (or press the F4 key).

   This makes the polygons easier to see.

4. Click a single face object at the base of the sphere object that is close to one end of the splines, and click the Extrude Along Spline caddy settings button to open the Extrude Along Spline caddy interface.

5. Click the Pick Spline button, and select the spline to the side of the face. Set the Segments to **6** and the Taper Amount to **–1.0.** Make sure that the Extrude Along Spline Align option isn't selected, and click the green check mark to accept the settings.

6. Repeat Steps 4 and 5 for each spline surrounding the octopus.

7. In the Subdivisions Surface rollout, enable the Use NURMS Subdivision option, and set the Display Iterations value to **2** to smooth the entire octopus.

Figure 13.30 shows the resulting octopus.

FIGURE 13.30

The tentacles of this octopus were created easily with the Extrude Along Spline feature.

Surface properties

Below the Edit Geometry rollout are several rollouts of options that enable you to set additional properties, such as vertex colors, material IDs, Smoothing Groups, and Subdivision Surface.

Vertex Surface properties

The Vertex Properties rollout in Vertex Subobject mode lets you define the Color, Illumination, and Alpha value of object vertices. The color swatches enable you to select Color and Illumination colors for the selected vertices. The Alpha value sets the amount of transparency for the vertices. After you assign colors, you can recall vertices with the same color by selecting a color (or illumination color) in the Select Vertices By section and clicking the Select button. The RGB (red, green, and blue) values match all colors within the Range defined by these values.

You can find more information on vertex colors in Chapter 22, "Painting in the Viewport Canvas and Rendering Surface Maps."

Polygon and Element Surface properties

For Polygon and Element subobjects, the rollouts shown in Figure 13.31 include Material IDs, Smoothing Groups, and Vertex Colors options. The Material IDs settings are used by the Multi/Sub-Object material type to apply different materials to faces or polygons within an object. By selecting a polygon subobject, you can use these option settings to apply a unique material to the selected polygon. The Select ID button selects all subobjects that have the designated Material ID, or you can select subobjects using a material name in the drop-down list under the Select ID button.

You can find more information on the Multi/Sub-Object material type in Chapter 20, "Creating Compound Materials and Using Material Modifiers."

FIGURE 13.31

Polygon and Elements subobjects includes settings for Material IDs, Smoothing Groups, and Vertex Colors.

You use the Smoothing Groups option to assign a polygon or multiple polygons to a unique smoothing group. To do this, select a polygon, and click a Smoothing Groups number. The Select By SG button, like the Select By ID button, opens a dialog box where you can enter a Smoothing Groups number, and all subobjects with that number are selected. The Clear All button clears all Smoothing Groups number assignments, and the Auto Smooth button automatically assigns Smoothing Groups numbers based on the angle between faces as set by the value to the right of the Auto Smooth button.

The Vertex Colors rollout also includes options for setting vertex Color, Illumination, and Alpha values for the selected subobject.

Subdivision Surface

Editable Poly objects include an extra rollout called Subdivision Surface that automatically smoothes the object when enabled. The Subdivision Surface rollout, shown in Figure 13.32, applies a smoothing algorithm known as NURMS, which stands for Non-Uniform Rational MeshSmooth. It produces similar results to the MSmooth button but offers control over how aggressively the smoothing is applied; the settings can be different for the viewports and the renderer.

FIGURE 13.32

The Subdivision Surface rollout includes controls for NURMS subdivision.

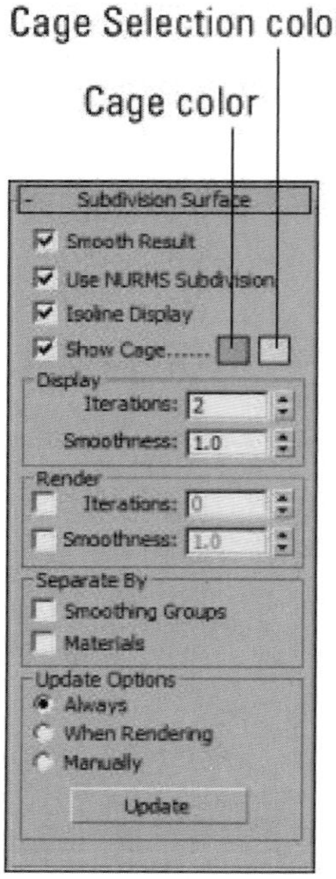

Cage Selection color

Cage color

To enable NURMS subdivision, you need to enable the Use NURMS Subdivision option. The Smooth Result option places all polygons into the same smoothing group and applies the MeshSmooth to the entire object. Applying NURMS with a high Iterations value results in a very dense mesh, but the Isoline Display option displays a simplified number of edges, making the object easier to work with. The process of smoothing adds many edges to the object, and the Isoline Display option displays only the isolines. The Show Cage option makes the surrounding cage visible or invisible. The two color swatches to the right of the Show Cage option let you set the color of the cage and the selection.

The Iterations value determines how aggressive the smoothing is. The higher the Iterations value, the more time it takes to compute and the more complex the resulting object. The Smoothness value determines how sharp a corner must be before you add extra faces to smooth it. A value of 0 does not smooth any corners, and a maximum value of 1.0 smoothes all polygons.

Caution

Each smoothing iteration quadruples the number of faces. If you raise the number of Iterations too high, the system can become unstable quickly.

The two check boxes in the Render section can be used to set the values differently for the Display and Render sections. If disabled, both the viewports and the renderer use the same settings. The smoothing algorithm can be set to ignore smoothing across Smoothing Groups and Materials.

If the Show Cage option is enabled (at the bottom of the Edit Geometry rollout), an orange cage surrounds the NURMS object and shows the position of the polygon faces that exist if NURMS is disabled. This cage makes selecting the polygon faces easier.

Tutorial: Modeling a tooth

If you've ever had a root canal, you know how much pain dental work can cause. Luckily, modeling a tooth isn't painful at all, as you'll see in this example.

To model a tooth using NURMS, follow these steps:

1. Select Create→Standard Primitives→Box, and drag in the Top viewport to create a Box object. Set its dimensions to **140** x **180** x **110** with Segments of **1** x **1** x **1**. Then right-click, and select Convert To→Convert to Editable Poly from the pop-up quad menu.

2. Click the Polygon icon in the Selection rollout to enable Polygon Subobject mode. Select the Top viewport, and press B to change it to the Bottom viewport. Then click the box's bottom polygon in the Bottom viewport.

3. Click the Select and Scale button (R), and scale the bottom polygon down by 10 percent.

4. Drag over the entire object to select all polygons, and click the Tessellate button in the Edit Geometry rollout once to divide the polygon into more polygons. Then click the Window/Crossing button in the main toolbar to enable the Window selection method, and drag over the bottom of the Box object in the Left viewport to select just the bottom polygons. Click the Tessellate button again.

5. Select the Vertex Subobject mode in the Selection rollout, press and hold the Ctrl key, and select the vertices at the center of each quadrant. Then move these vertices downward in the Left viewport a distance about equal to the height of the Box.

6. Select the Bottom viewport, and press T to change it back to the Top viewport. Select the single vertex in the center of the polygon with the Ignore Backfacing option enabled in the Selection rollout, and drag it slightly downward in the Left viewport.

7. Disable the Ignore Backfacing option in the Selection rollout, and select the entire second row of vertices in the Left viewport. With the Select and Scale tool, scale these vertices toward the center in the polygon in the Top viewport.

8. In the Subdivision Surface rollout, enable the Use NURMS Subdivision option, and set the Iterations value to **2**.

Figure 13.33 shows the completed tooth.

FIGURE 13.33

The organic look of this tooth is accomplished with NURMS.

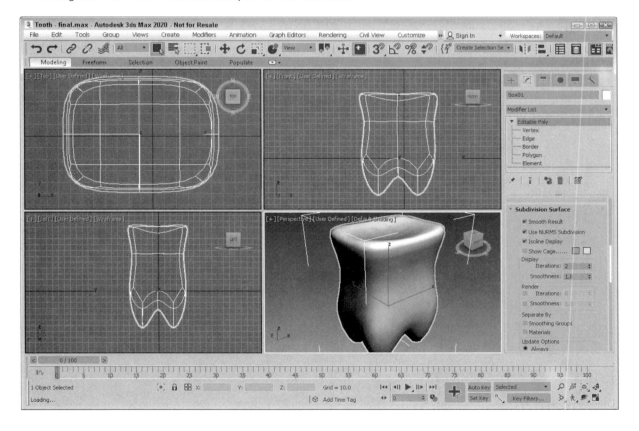

Summary

Meshes are probably the most common 3D modeling types. You can create them by converting objects to Editable Meshes or Editable Poly objects or by collapsing the Stack. Editable Poly objects in 3ds Max have a host of features for editing meshes, as you learned in this chapter. More specifically, this chapter covered the following topics:

* Creating Editable Poly objects by converting other objects or applying the Edit Poly modifier
* The features for editing Editable Poly objects
* How to select and use the various mesh subobject modes
* Editing mesh objects using the various features found in the Edit Geometry rollout
* Changing surface properties using features like NURMS

This chapter provided an introduction to mesh and polygon objects and showed how to edit them, but the next chapter steps it up with coverage of the Graphite Modeling tools, which work only with Editable Poly objects.

Chapter 14

Using the Graphite Modeling Tools and Painting with Objects

IN THIS CHAPTER

Working with the Graphite Modeling tools

Using the Freeform tools

Selecting specific subobjects

Painting with objects

The previous chapter covered everything you need to know about modeling with polygons, but the problem with the polygon workflow is the ping-pong effect of moving back and forth between the current model and the Command Panel. Although you can float the Command Panel or even use the quad menus to access most of these commands, the Autodesk® 3ds Max® 2020 software presents an entirely new workflow based on the new Ribbon interface.

The Ribbon sits conveniently above the viewports, but you can pull off and float any of the individual panels as needed. The Ribbon panels are dynamic, so only those tools that work with the current selection are presented. This places the tools you need right in front of you when you need them.

The Ribbon is populated with all the features for working with Editable Poly objects that are found in the Command Panel, but it also includes many features that cannot be found in the Command Panel. It includes a large number of brand-new tools for selecting and working with polygon objects. These tools collectively are called the Graphite Modeling tools.

The Ribbon also is home to several additional panels of tools that allow you to paint deformations into your models, make unique selections, and select and paint with objects using brushes. The best part of these new tools is that they all eliminate the ping-pong effect. Our necks thank you, Autodesk.

Working with the Graphite Modeling Tools

 The Ribbon interface, shown in Figure 14.1, can be turned on and off using a button on the main toolbar or using the Customize→Show UI→Show Ribbon menu. When enabled, the toolbar button is highlighted and the Ribbon appears in the same state it was in the last time it was opened. By double-clicking the Ribbon title bar, you can switch among these displays—just the top tabs; just the tabs and panel titles; just the tabs, panel titles, and panel buttons; or the entire panel. This is great if you want to keep the Ribbon around but hide most of the buttons.

You can learn more details on working with the Ribbon interface in Chapter 1, "Exploring the Interface."

When expanded, the Graphic Modeling Tools interface shows several panels of tools. Each of these panels can be separated from the Ribbon and floated independently. You also can rearrange the panels by dragging them to a new position on the Ribbon. If the icons displayed within a panel don't fit, a small downward-pointing arrow offers access to the additional tools.

All the panels and tools that make up the Ribbon are adaptable and change depending on the subobject mode you select. Only the relevant tools are visible, which saves space and makes it easier to locate the tool you want to use.

If you don't like the layout of the tools in the Ribbon, you can customize your own panels of tools, which is covered in Chapter 49, "Customizing the Interface."

FIGURE 14.1

The Ribbon holds several panels of modeling tools.

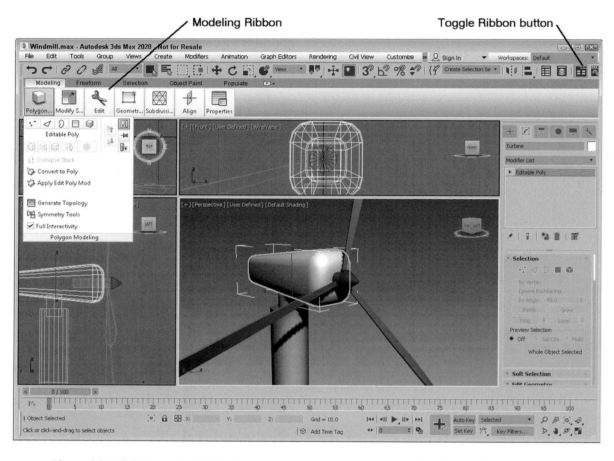

If you right-click the Ribbon's title bar, you can access a pop-up menu of options to hide and show different tabs and panels. There is also a Ribbon Configuration menu that includes options for customizing the Ribbon, loading and saving custom configurations, and switching to a vertically oriented Ribbon.

Using the Polygon Modeling panel

The Graphite Modeling tools are exclusive to Editable Poly objects or objects with the Edit Poly modifier applied. If you select any other type of object with the Ribbon open, all the buttons are disabled. However, two options are available in the Polygon Modeling panel of the Graphite Modeling Tools tab. These options are to Convert to Poly and Apply Edit Poly Mod. These commands convert the selected object to an Editable Poly object or apply an Edit Poly modifier and automatically open the Modify panel.

After an Editable Poly object or an object with the Edit Poly modifier applied is selected, the subobject modes can be selected from the top of the Graphite Modeling Tools panel. The Polygon Modeling panel, shown in Figure 14.2, includes options that work with the Modifier Stack, including collapsing the stack, pinning the stack, and moving up and down between modifiers. The Show End Result button lets you see the object with all stack modifiers applied. Another button in this panel lets you toggle the Command Panel off.

FIGURE 14.2

The Ribbon's Polygon Modeling panel includes options for determining which subobjects are selected.

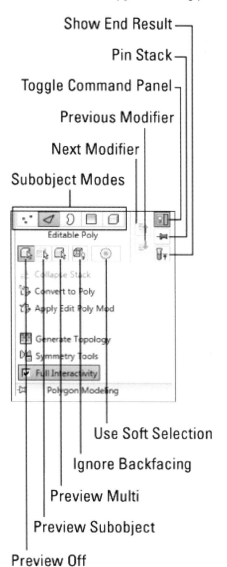

The Polygon Modeling panel also includes buttons for turning on and off subobject selection preview and for ignoring backfacing. It has three Preview buttons, but only one can be enabled at a time. Preview Off disables preview mode. Preview Subobject causes the different subobjects to be highlighted as you move the mouse over them. The Preview Multi button highlights all subobjects regardless of the current subobject mode. For example, if Preview Multi is enabled and the Polygon Subobject mode is enabled, moving the mouse over an edge highlights that edge, and if you select it, Edge Subobject mode is enabled. This provides a nice way to select different subobjects without having to change between the different subobject modes.

The Use Soft Selection button enables you to select a feathered group of subobjects. When enabled, the Soft panel appears, as shown in Figure 14.3.

Using the Soft panel, you can enable Edit mode, which lets you interactively change the Falloff, Pinch, and Bubble settings. When the Edit button is enabled, the cursor looks like two circles. This is Falloff mode, and dragging the mouse changes the amount of falloff for the soft selection. If you click in the viewport, the cursor changes to a peak indicating Pinch edit mode. Click again to access Bubble edit mode, which looks like an upside-down letter *U*. Continue to click to cycle through these edit modes. These values also can be set in the Soft panel using the various spinner controls.

Using the Soft Selection features is covered in Chapter 11, "Accessing Subobjects and Modifiers and Using the Modifier Stack."

The Paint button in the Soft panel opens the PaintSS panel, also shown in Figure 14.3, with buttons for blurring; reverting; and opening the Painter Options dialog box, which holds the brush settings. You also can set the Value, Size, and Strength of the soft selection brush.

The Use Edge Distance option lets you select adjacent edges to the current selection rather than using a falloff amount.

FIGURE 14.3

The Soft and PaintSS panels let you control how soft selections are made.

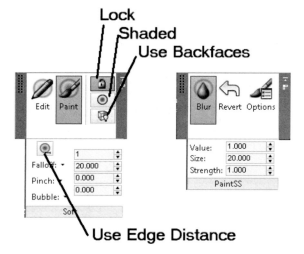

Generate Topology

Located at the bottom of the Ribbon's Polygon Modeling panel is an option to Generate Topology. This option opens the Topology pop-up panel, showing several patterns. Selecting a pattern applies the selected pattern to the selected object or to the object's subobject selection if you hold down the Shift key.

The Size, Iterations, and Smooth values let you configure the selected pattern. The ScrapVerts button removes any vertices with two edges going to it. The Plane button creates a simple plane object where the S value sets the resolution. Figure 14.4 shows four planes with different topologies applied.

FIGURE 14.4

Changing a plane's topology gives you a unique set of faces to work with.

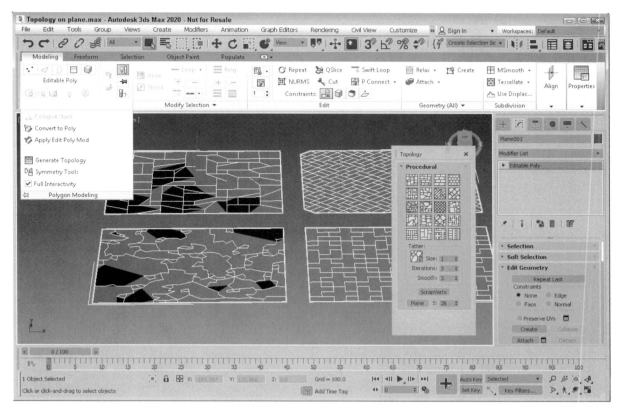

Using the Symmetry tools

The Polygon Modeling panel also holds a link for accessing the Symmetry tools. This command opens the Symmetry Tools dialog box, shown in Figure 14.5. Using this dialog box, you can select an object with the top button and automatically copy the changes on either side of any axis to the opposite side with the + To − and − To + buttons. You also can use the Flip Symmetry button to switch the moved subobjects to the opposite side of the model.

The Copy Selected button lets you copy the position of an entire object or of just the selected vertices and paste them (with the Paste button) onto another object with the same number of vertices.

FIGURE 14.5

The Symmetry Tools dialog box lets you mirror subobject movements across an axis.

Tutorial: Building a Skateboard wheel

Starting with a simple sphere, you can quickly create a symmetrical skateboard wheel using the Symmetry tools.

To create a skateboard wheel, follow these steps:

1. Use the Create→Standard Primitives→Sphere menu, and drag in the Front viewport to create a sphere object.
2. Open the Graphite Modeling Tools by clicking its button in the main toolbar. Then select the Convert to Poly option from the Polygon Modeling panel in the Ribbon.
3. Select the Symmetry Tools option in the Polygon Modeling panel, click the Pick Main Model button in the Symmetry Tools dialog box, and pick the sphere object.
4. Select the Vertex subobject mode in the Polygon Modeling panel, and drag over the top four rows of vertices in the Top viewport. Then drag with the Select and Move tool downward in the Top viewport.
5. Back in the Symmetry Tools dialog box, enable the Z Axis, and click the − To + button to symmetrically copy the moved vertices, as shown in Figure 14.6.

FIGURE 14.6

The skateboard wheel is symmetrical and ready to roll.

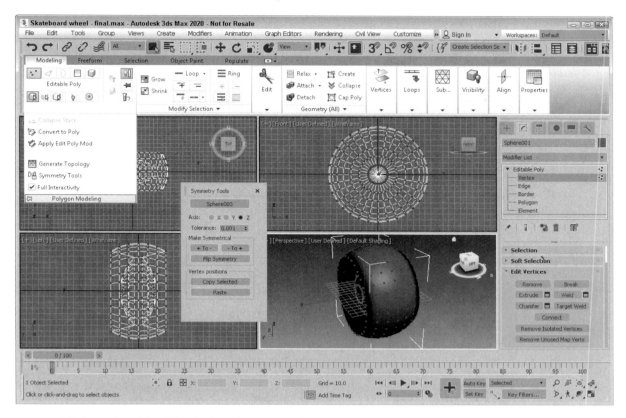

Using the Modify Selection panel

The Ribbon also holds tools for making Loop and Ring selections. These tools are found in the Modify Selection panel, shown in Figure 14.7. When a loop or ring is selected, there are also buttons for growing and shrinking the adjacent rows or columns. There are also tools called Loop Mode and Ring Mode that cause the entire edge loop or edge ring to be automatically selected when a single edge is picked when enabled. A text label appears in the viewport when Loop Mode is enabled. Tools called Dot Loop and Dot Ring let an edge loop and edge ring with gaps be selected. The number of loops or rings to skip is set using the Dot Gap value, where a setting of 3 selects one loop and then skips three before selecting another. As a flyout tool under the Dot Loop is Dot Loop Cylinder, which selects the top and bottom edges of a cylinder using the gap settings.

Note

The Modify Selection panel is only visible when a subobject mode is enabled.

FIGURE 14.7

The Ribbon's Modify Selection panel includes tools for working with loops and rings.

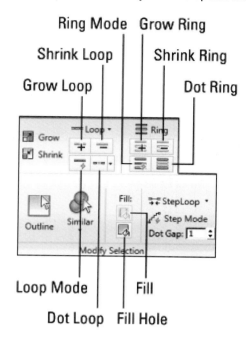

The Modify Selection panel also includes these tools: Outline, which selects all subobjects surrounding the current selection; Similar, which selects all subobjects that are similar to the current selection, including Edge Count, Edge Length, Face Count, Face Areas, Topology, and Normal Direction; and Fill, which selects all subobjects that are within the selected subobjects or within the current outlined selection. The Fill option lets you pick two vertices that are diagonally across from each other, and then all interior vertices between the two vertices are selected to make a square area.

The StepLoop option lets you select two subobjects within the same loop, and then all subobjects between the two are selected. When Step Mode is enabled, you can pick two subobjects in the same loop and all subobjects between the two are selected. This continues for all additional selections within the same loop until Step Mode is disabled again.

Tip

Using the Shift key, you can select loops of subobjects even more quickly. If you select a single vertex, edge, or polygon and then hold down the Shift key and click an adjacent subobject, the entire loop or ring of subobjects is automatically selected.

Editing geometry

When no subobject modes are selected, several Ribbon panels are available, as shown in Figure 14.8. These panels contain many of the same features that are found in the Edit Geometry rollout in the Command Panel. Features contained here include, among many others, the ability to create new subobjects, attach subobjects to the mesh, weld vertices, chamfer vertices, slice, explode, and align. These panels are available regardless of the subobject mode that is selected, but some of these tools are disabled, depending on the subobject mode that you select.

Details on using many of the Ribbon tools are discussed in Chapter 13, "Modeling with Polygons."

FIGURE 14.8

The Ribbon includes many general-purpose editing features that are always available.

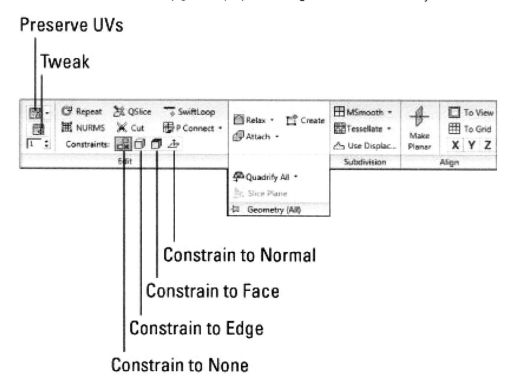

Many of the tools, such as Relax, MSmooth, Tessellate, PConnect, and Quadrify All, include options for accessing caddy settings by clicking the small arrow to the right of the button. You also can open the caddy settings by holding down the Shift key and clicking the Tool button. These options open a caddy of settings that are located in the viewport next to the selected subobject. These caddies allow you to change the settings and immediately see the results in the viewports. The OK (check-mark icon) button applies the settings and closes the dialog box, and the Apply (plus-sign icon) button applies the settings and leaves the dialog box open. Similar caddies are available for other subobject-specific buttons, such as Extrude, Bevel, Outline, and Inset.

Preserve UVs

The Preserve UVs Settings option opens a dialog box that lets you select a Vertex Color and Texture Channel to preserve. Beneath the Preserve UVs button in the Edit panel is a Tweak button. This button lets you move the UVs for the given object by using a paintbrush. The value beneath this button is the map channel that you can adjust.

Cutting holes with Paint Connect (PConnect)

Within the Edit panel are the QuickSlice (QSlice) and Cut tools, which work the same as their Command Panel counterparts, but the Edit panel also includes several new methods for cutting and slicing objects.

Another way to slice an object is with the SwiftLoop button. This button lets you click any open space between edges to add a loop between the adjacent edges. Once placed, you can drag to slide the loop to its actual location.

Another handy Ribbon tool for cutting holes is PConnect, which stands for Paint Connect. This tool lets you paint the location of a hole on the surface. As you paint, every edge that you cross is marked, and new edges are created between adjacent marks. By holding down the Shift key, you can make the connections happen in exactly the middle of each edge. Holding down the Ctrl key lets you paint new edges between adjacent vertices. Clicking a vertex with the Alt key held down removes the vertex, and Ctrl+Alt removes an edge.

385

Ctrl+Shift is used to remove an entire edge loop, and Shift+Alt is used to paint two parallel lines between edges.

Tip

Help videos are available for some tools, such as PConnect. These videos are located in the tooltips for the tool and appear when you hold the mouse over the tool.

NURMS

The Edit panel also includes a NURMS button. NURMS stands for Non-Uniform Rational MeshSmooth. It produces similar results to the MSmooth button but offers control over how aggressively the smoothing is applied; the settings can be different for the viewports and the renderer. When NURMS is enabled, the Use NURMS panel appears, as shown in Figure 14.9. This panel holds the settings for the NURMS tool and works the same as the options found in the Command Panel.

FIGURE 14.9

The Ribbon's Use NURMS panel includes controls for NURMS subdivision.

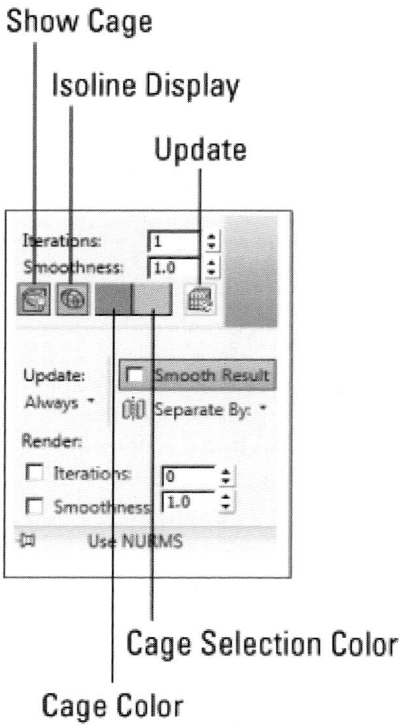

Show Cage

Isoline Display

Update

Cage Selection Color

Cage Color

Tutorial: Smoothing an Ice Cube

NURMS can give a general smoothing to an object, which is just what we need to build an ice cube.

To create an ice cube, follow these steps:

1. Use the Create→Standard Primitives→Box menu command, and drag in the viewport to create a rectangular box object.
2. Open the Graphite Modeling Tools by clicking its button in the main toolbar. Then select the Convert to Poly option from the Polygon Modeling panel in the Ribbon.
3. Select the Vertex subobject mode, and drag over the lower four vertices in the Perspective viewport. Then drag inward with the Select and Scale tool to make the base of the box smaller than the top.

4. Select the Edge subobject mode, drag over all the edges of the box object, and then Shift+click the Chamfer button in the Edges panel to open the Chamfer caddy. Enter the value of **1** for the Edge Chamfer Amount and the Connect Edge Segments, and click the OK button to apply the settings.

5. In the Edit panel, click the NURMS button, and set the Iterations value in the Use NURMS panel to 1. This smoothes all the selected edges, as shown in Figure 14.10.

FIGURE 14.10

A simple ice cube made smooth with the NURMS feature

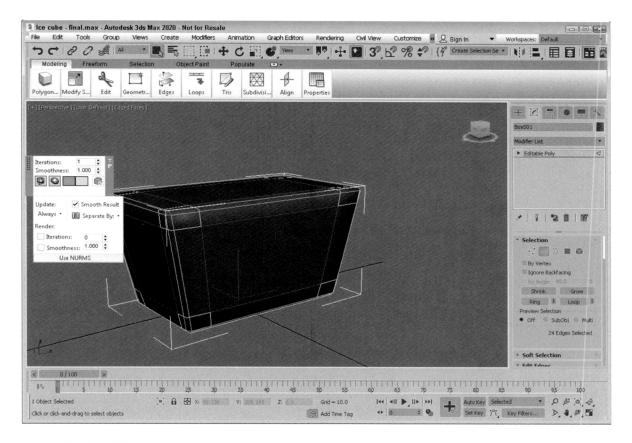

Quadrify All

The Ribbon's Geometry (All) panel also includes a Quadrify All tool for converting triangles to quads. You have options to Quadrify All, Quadrify Selection, Select Edges from All, and Select Edges from Selection. This is an awesome tool if you like to work with edge loops and edge rings.

Figure 14.11 shows a bookshelf model that has been built using triangular faces on the left. Using the Quadrify All command, the face on the right is aligned to much neater rows and columns of four-sided polys. This allows the edge loop features to be used.

FIGURE 14.11

The Quadrify All command greatly simplifies this model, making it easier to work with.

Subdivision panel

MeshSmooth (MSmooth) and Tessellate within the Subdivision panel are buttons for smoothing and tessellating the object. Both the MSmooth and Tessellate tools include caddies, as shown in Figure 14.12. The MSmooth setting for Smoothness rounds all the sharp edges of an object. Tessellation can be done using Edges or Faces, and the Tension setting controls how tight the adjacent faces are.

FIGURE 14.12

The caddies for the MSmooth and Tessellate buttons let you interactively set the Smoothness and Tension values.

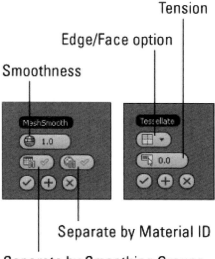

Use Displacement

The Use Displacement tool opens the Displacement panel, shown in Figure 14.13, when enabled. Using this panel, you can specify the subdivision method that is used and the settings for the displacement.

You can learn more about using displacement maps in Chapter 20, "Creating Compound Materials and Using Material Modifiers."

FIGURE 14.13

The Displacement panel includes all the subdivision settings.

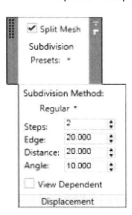

Editing Vertex subobjects

When working with the Editable Poly objects, after you select a Vertex subobject mode (keyboard shortcut, 1) and select vertices, you can transform them using the transform buttons on the main toolbar. All vertex-specific commands are found within the Vertices panel, shown in Figure 14.14, but some new tools also appear in the Geometry (All) panel, including Collapse, Detach, and Cap Poly.

FIGURE 14.14

When the Vertex subobject mode is selected, these vertex commands become available.

Editing Edge and Border subobjects

All the edge editing tools are located in the Edges and Borders panels, shown in Figure 14.15. Many of the Edge subobject options work in the same way as the Vertex subobject options.

FIGURE 14.15

All the Edge-specific commands are available when the Edge subobject mode is enabled. Many of the Border subobject commands are the same as those for Edges.

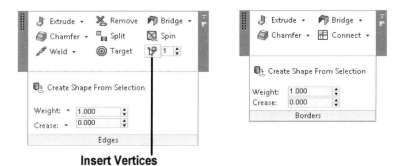

Insert Vertices

Editing Polygon and Element subobjects

Like the other subobject modes, Editable Polys can be edited at the Polygon and Element subobject level. The buttons for these modes are found in the Polygons and Elements panels. Figure 14.16 shows the Ribbon's Polygons and Elements panels.

FIGURE 14.16

The Polygon and Element subobjects commands are found in the Polygons and Elements panels.

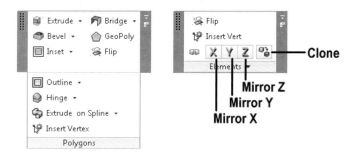

GeoPoly

The GeoPoly button moves the vertices of the selected polygon to make a regular polygon whose vertices are all equally spaced. The shape of the resulting polygon depends on the number of vertices included in the polygon face.

Hinge

The Hinge button rotates a selected polygon as though one of its edges were a hinge. The angle of the hinge depends on the distance that you drag with the mouse, or you can use the available caddy. By default, one of the polygon's edges will be used as the hinge about which the section rotates, but in the caddy, you can click the Pick Hinge button and select an edge (which doesn't need to be attached to the polygon).

Tutorial: Adding a handle to a mug

Creating a cup is fairly easy using the Lathe modifier, but adding the handle is another story. It could be created as half a torus and Boolean connected to the cup, but this example shows how to use the poly modeling features to hinge the handle on.

To add a handle to a mug, follow these steps:

1. Open the Mug.max file from the Chap 14 directory in the downloaded content set.

   This file includes a simple cup created using the Lathe modifier, and it has been converted to an Editable Poly object. Open the Graphite Modeling Tools by clicking its button in the main toolbar.

2. Select the mug object. Then select the QSlice button from the Edit panel, click in the Front viewport about one-third up from the bottom of the cup, orient the line to be horizontal, and click again to slice the mug horizontally.

3. Click the Cut button in the Edit panel, and cut edges into each corner of one of the rectangular polygon faces near the bottom of the cup in the Front viewport to create an eight-sided polygon for the mug handle. Then cut another horizontal edge in the middle of the cup directly above the cut polygon to be the hinge edge for the handle. Right-click in the viewport to exit the Cut tool.

4. Select the Polygon subobject mode in the Polygon Modeling panel, and click the interior of the cut polygon just created. Shift-click the Hinge button in the Polygons panel to open the caddy settings. In the Hinge From Edge caddy, click the Pick Hinge button, click the midline edge that was cut above the cut polygon, set the Segments to **10** and the Angle to **185**, and then click the green Accept Caddy button.

Figure 14.17 shows the resulting handle on the mug. You can add a NURMS command to smooth the mug handle.

FIGURE 14.17

The handle for this mug was created with the Hinge feature.

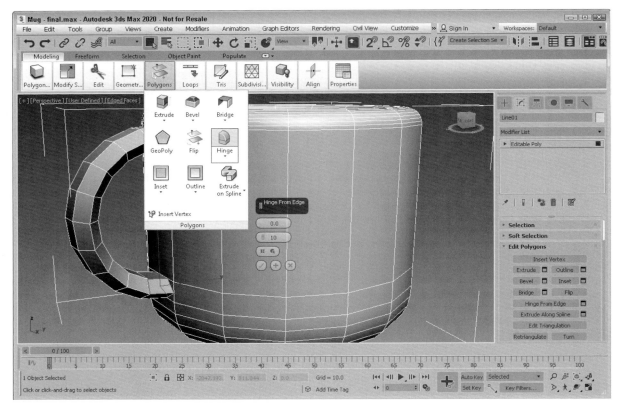

Mirroring elements

Within the Elements panel are commands for mirroring the current selection about the X, Y, or Z axes. These commands move the element to the opposite side, but clicking the Clone button creates a clone of the selected element.

Surface properties

In the Properties panel are several settings for additional properties, such as vertex colors, material IDs, and Smoothing Groups. For Polygon and Element subobjects, the Properties panel, shown in Figure 14.18, includes Material IDs and Smoothing Groups options.

FIGURE 14.18

The Properties panel includes settings for Material IDs, Smoothing Groups, and Vertex Colors.

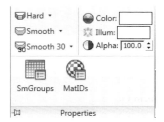

Using the Freeform Tools

The second tab in the Graphite Modeling Tools includes an assortment of Freeform tools for sculpting and modeling surfaces as if working with clay. These tools are divided into three panels—PolyDraw, Paint Deform, and Defaults. The PolyDraw panel is shown in Figure 14.19.

FIGURE 14.19

Using the PolyDraw tools

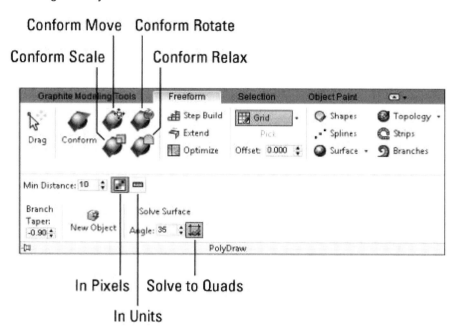

The various PolyDraw tools let you create and extend the surface subobjects using tools that work like a common drawing program.

Drag

The Drag tool lets you move selected subobjects around by simply dragging them with the mouse. Clicking a subobject automatically selects it and lets you drag it to a new position. Holding down the Shift key or the Ctrl key lets you drag edges or polygons regardless of the subobject mode that is selected. Pressing Shift+Ctrl lets you drag entire edge loops, and pressing the Alt key lets you move the subobject in a direction that is perpendicular to the view axis, which is great for dragging subobjects off to the side of an object.

Conform

The Conform brushes let you push all the vertices of the selected object toward a selected underlying object. When the Freeform panel is first accessed, you can select the Draw On: Surface option to the right of the Step Build button. When Surface is selected, the Pick button becomes active. Using this button, you can select the object to which you want the selected object to be conformed.

With the object to conform to selected, you can select the Conform button and change the options for this brush using the Conform Options panel, shown in Figure 14.20. Within the Conform Options panel, you can set the Strength, Falloff, and Conform values. The Conform value sets the rate at which the vertices move toward the conform object. Higher values make the vertices move immediately, and smaller values, such as 0.1, cause the movement to be gradual. The Strength Percent value sets how much of the full strength is applied.

FIGURE 14.20

The Conform panel holds the options for using the Conform brushes.

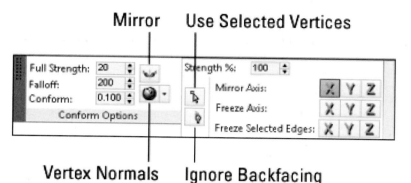

The Mirror option causes the vertices' movement to be applied equally on either side of the mirror axis. The Vertex Normals/View drop-down list lets you control how the vertices move. The View option causes the vertices to move toward the target into the current view, and the Vertex Normals option moves the vertices along their normals toward the conform object. You also have options to move only the selected vertices, to Ignore Backfacing vertices, to select the mirror and freeze axes, and to freeze specific selected edges so no movement happens.

Within the PolyDraw panel are four brushes for defining how the vertices are transformed. The Conform Move brush moves vertices under its brush range. Vertices also can be rotated and scaled using the Conform Rotate and Conform Scale brushes, and the Conform Relax brush smoothes the moved vertices by moving them gradually toward their original position.

Step Build

The Step Build tool lets you click to place new vertices on the surface of the object. The location of these new vertices depends on the current Grid, on the current object's Surface, or within the Selection using the On selections located to the right of the Step Build button. If you select the On: Surface option, you can use the Pick button to select the object whose surface you want to create vertices on. After freestanding vertices are created, you can hold down the Shift key and drag over these new vertices to create a polygon. The Ctrl key lets you remove polygons, and the Alt key lets you remove vertices. With these controls, you can quickly remove and rebuild polygons to create new shapes.

Extend

The Extend tool works on border subobjects and lets you add new polygons to fill the hole by dragging the border vertices. If you press and hold the Shift key while dragging an edge, you can pull the edge away to create a new polygon. Dragging with the Shift+Ctrl keys lets you drag two adjacent edges out. Pressing the Ctrl key and clicking deletes the polygon, and dragging with the Alt key held down moves the polygon perpendicular to the view axis.

Optimize

The Optimize tool quickly collapses subobjects. Click an edge with this tool to remove it and to combine its two end points. The Shift key is used to target weld two vertices into a single one, and the Alt key removes vertices. The Shift+Ctrl key combo can remove an entire edge loop at once.

Draw On and Pick

The Draw On drop-down list gives you three options for specifying the object that is drawn on: Grid, Surface, and Selection. If Surface is selected, you can use the Pick button to choose the surface object. You also can set an Offset, which is the distance above the surface on which the drawn objects appear.

Tip

It is best to keep a non-zero offset value when drawing on the surface of an object, especially if the drawn object overlaps an edge. This keeps the surfaces from interpenetrating.

Tutorial: Matching a road to a rolling terrain

If you've created a rough rolling landscape using a Plane object and the Noise modifier, matching a road running across the surface can be tricky if you have to select and move each individual vertex. Instead, you can use the Conform brushes to move the vertices so they match the underlying hills.

To conform a road to an underlying terrain, follow these steps:

1. Open the Conforming road.max file from the Chap 14 directory in the downloaded content set.

2. With the road selected, click the Graphite Modeling Tools button on the main toolbar to make the Ribbon appear, and select the Freeform tab.

3. Select the Draw On: Surface option from the drop-down list to the right of the Step Build button, click the Pick button, and click the hilly object. Then select the straight road object, and click the Conform button.

4. In the Conform Options panel, set the Full Strength to **10** and the Falloff to **70**, and change the viewport to the Top view.

5. With the Conform brush, and click and slowly rotate the brush in small circles over the road object in the Top viewport to move its vertices to conform to the hilly landscape, as shown in Figure 14.21.

FIGURE 14.21

The Conform Move brush is used to make the road match the underlying terrain.

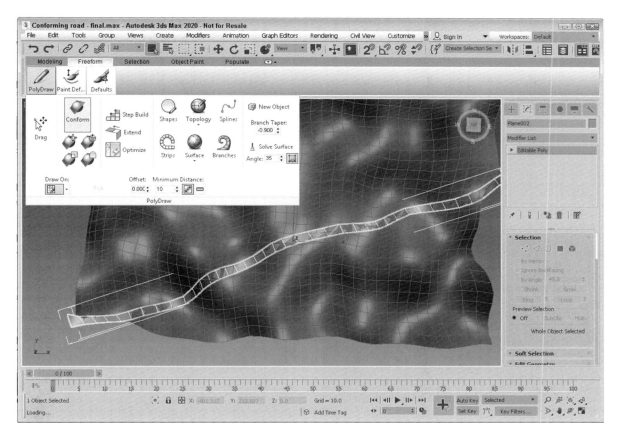

Shapes and Solve Surface

The Shapes tool lets you draw polygonal shapes directly on the surface of an object. Figure 14.22 shows three polygons drawn on the surface of a torus. You also can delete drawn polygons with the Ctrl key. The completed polygons likely will have multiple vertices, but you can reduce the polygons to tris and quads using the Solve Surface button.

FIGURE 14.22

Using the Shapes tool, you can draw polygons that conform to the surface of the underlying object.

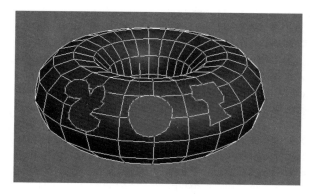

Splines

The Splines tool lets you draw spline objects that follow the surface of the underlying object. The Ctrl key can be used to delete drawn splines.

Surface, Topology, Strips, and Branches

The Surface tool covers the object with a mesh of quads by painting over the object. You can delete any polygon by clicking it with Ctrl key held down. The Topology tool lets you draw a series of parallel lines followed by a set of perpendicular lines to form quads. The Auto Weld option automatically welds vertices together to form a mesh. The Ctrl key extends a line from the nearest end point.

The Strips tool draws a consecutive row of quads that flow across the surface of the object. The Shift key extends the strip from the nearest edge. The Branches tool extends a tapered branch from a single polygon. For the branches, you can set a Taper amount; the Minimum Distance value sets the distance between the segments. This is useful for creating tentacles, as shown in Figure 14.23.

FIGURE 14.23

Using the Branches tool, you can drag out extending arms from polygons.

Tutorial: Carving a pumpkin

When carving a pumpkin for Halloween, you should follow the round curvature of the pumpkin or the face will look skewed. In 3ds Max, this is easy to accomplish using the Freeform tools and selecting the pumpkin object as the Draw-On surface.

To draw the face details on the surface of a pumpkin, follow these steps:

1. Open the Pumpkin face.max file from the Chap 14 directory in the downloaded content set. This simple pumpkin model has been converted to an Editable Poly object.

2. Open the Graphite Modeling tools, click the Freeform tab, and select the Surface option as the Draw-On shape by selecting it from the drop-down list to the right of the Step Build tool. Then click the Pick button located under the selected Draw-On object, and click the pumpkin object.

3. Click the Shapes button, and draw the eyes, nose, and mouth for the jack o' lantern. Right-click to exit Shapes mode.

Figure 14.24 shows the resulting face that follows the curvature of the pumpkin.

FIGURE 14.24

The PolyDraw tools let you draw on the surface of an object.

Using the Paint Deform tools

The Paint Deform tools let you sculpt the surface of an object by pushing, pulling, and modeling the object in organic ways like it was clay. Whenever any of the tools on the Paint Deform panel, shown in Figure 14.25, is selected, its settings for the brush's Size and Strength (and sometimes Offset, depending on the tool) appear in the Paint Options panel to the side of the Paint Deform panel.

FIGURE 14.25

The Paint Deform tools are used to sculpt an object's surface with gradual changes.

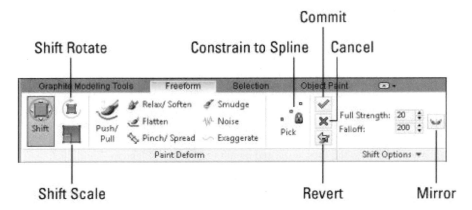

Shift/Shift Rotate/Shift Scale

The Shift tool lets you drag all subobjects within the brush radius, and all subobjects within the falloff radius are moved to a lesser extent. You can control the brush size and falloff in the Options panel, or you can press and hold the Ctrl key and drag to alter the brush's radius. Shift+dragging changes the brush's falloff indicated by the inner white circle.

The Shift Rotate and Shift Scale brushes are used to rotate and/or scale all the vertices within the brush's range.

Push/Pull

The Push/Pull tool also has Size and Strength values, but it is different in that the brush follows the surface of the object. Dragging over an area pulls the vertices within the brush's radius outward, and holding down the Ctrl key pushes the vertices inward. The Shift key relaxes the area under the brush.

Relax/Soften

The Relax/Soften brush removes any extreme changes in the surface such as the hard edges of a cube. The Alt key lets you relax the surface without changing its volume, and the Ctrl key causes the surface to revert to its previous state.

Flatten and Pinch/Spread

The Flatten brush pulls any bends out of the mesh causing the object to be working into a flat plane. The Ctrl key causes the object to revert to its previous state, and the Shift key accesses the Relax/Soften brush. The Pinch/Spread tool causes vertices to be pulled in closer to one another, and holding down the Alt key has the opposite effect and pushes them away.

Smudge, Noise, and Exaggerate

The Smudge tool pushes the surrounding vertices away from the center of the brush. Dragging over the same vertices multiple times moves them each time. Holding down the Alt key causes the vertices to be moved only along the surface, not along the normal. The Noise tool randomly moves the vertices about to create a random noise pattern. The Exaggerate tool pushes vertices farther in the current direction to emphasize the details.

Constrain to Spline

If you want more control over the precise surface area that is changed using the Paint Deform tools, you can select the Constrain to Spline option and use the Pick button to select a spline near the surface area to change.

Revert

At any point, you can use the Revert tool to gradually change the object back to its last saved point. You can set a save point by clicking the Commit button, which looks like a green check mark. The Cancel button (a red X) removes the recent changes from the object.

Using the Selection Tools

The next tab offers several additional Selection tools. These tools make it possible to locate specific subobjects by looking for certain criteria such as concavity, normals, and symmetry. Figure 14.26 shows the panels for this tab. These tools are only available if a subobject mode is selected.

FIGURE 14.26

The Selection tab includes panels for selecting specific subobject selections.

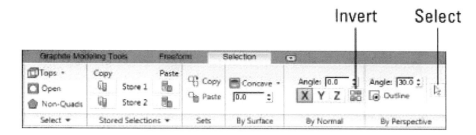

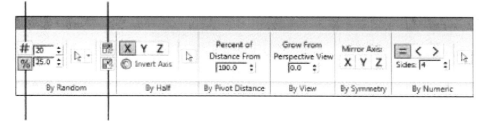

Selecting Tops, Open, and Non-Quads

For the Polygon and Element subobject modes, the first three tools in the Selection tab are the Tops, Open, and Non-Quads tools. The Tops tool selects all vertices resulting from the extruded sections. This quickly lets you grab all extended sections and extend or reduce them as needed. The Open tool selects all open borders, and the Non-Quads tool finds all polygons that are not quadrilaterals, including all tris and all polygons with more than four corners. This is a valuable tool when working with edge loops.

For Edge and Border subobject modes, there is a Hard option that locates all edges that aren't smoothed between adjacent faces. In all subobject modes, there is also a Patterns options that selects subobjects using a predefined pattern such as Checker.

Note

The Non-Quads option is only available when the Polygon or Element subobject mode is selected.

Copying and pasting selections

The Stored Selections panel lets you copy a selection of subobjects into two available stores. These copied selections can be restored at any time by clicking the Paste button. Additional buttons let you combine the two selection stores, subtracting one from another and getting only the intersecting selection between the two. The Clear button removes the selection from the store.

The Copy and Paste Sets buttons, in the Sets panel, let you copy a selection set from the main toolbar and paste it as needed.

Selecting by criteria

The remaining selection criteria let you locate subobjects using a variety of different methods.

By Surface, Normal, and Perspective

The By Surface panel lets you specify a degree of concavity, and the tool locates all the concave areas in the current object. Negative values also can be used to find convex regions. Figure 14.27 shows the selected concave regions.

FIGURE 14.27

The By Surface tool can be used to find the concave regions of an object.

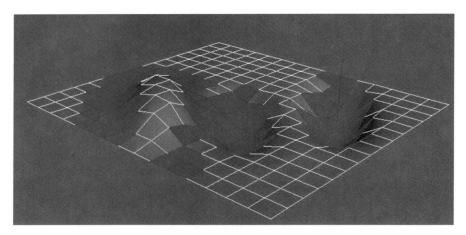

The By Normal panel lets you choose an axis and a value, and all subobjects within the Angle value for the selected axis are selected. This is a great way to quickly determine which polygons are facing away from the current view. The Invert button can find all normals pointing toward the negative axis side.

The By Perspective panel selects those polygons that are within the Angle value to the view axis. If the Outline button is enabled, only the outer borders of polygons are selected. Click the Select button to see all the selected polygons meeting the criteria.

By Random, Half, and Pivot Distance

The By Random panel lets you randomly select polygons within the current object. You can set to randomly select a given number or a percentage of the total. The Select button makes the random selection, or you can randomly select within the current selection. Additional buttons grow or shrink the selection. Figure 14.28 shows a random selection.

FIGURE 14.28

The By Random tool can be used to make a random selection of polygons.

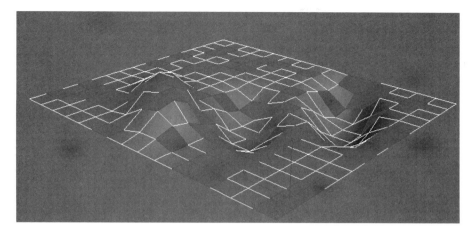

The By Half panel lets you quickly choose half of the available polygons as determined by axis. The Invert Axis button lets you choose the negative side of the axis. The By Pivot Distance chooses those polygons that are farthest away from the current pivot point, creating a circular selection area. Reducing the distance value creeps the selection closer to the pivot's location.

By View, Symmetry, and Numeric

The By View panel selects those polygons closest to the current view camera. Increasing this value extends the selection farther into the scene, as shown in Figure 14.29.

FIGURE 14.29

The By View tool is used to select those polygons closest to the current view.

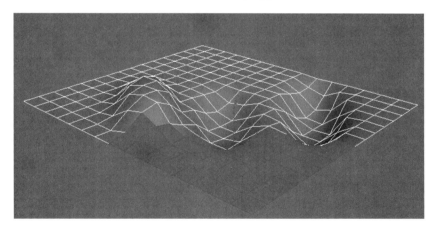

Regardless of the current selection, you can make it symmetrical about any of the three axes using the By Symmetry panel. The By Numeric panel, which is only available in Vertex and Polygon modes, lets you select all vertices that have a given number of edges or polygons that have a given number of sides. The Equal, Less Than, and Greater Than buttons are used to mathematically determine those subobjects.

By Color

The By Color panel, available in Vertex mode, lets you locate any vertices that have a given color or Illuminate vertex color setting according to the specified RGB values.

Using the Object Paint Tools

Next to the Selection tab is the Object Paint tab. This tab includes tools that let you select and paint with a specific object. This is great for spreading objects around a scene. The tab also includes several options for randomizing the size, orientation, and placement of the painted objects.

The final tab in the Modeling Ribbon is the Populate tab, which is covered in Chapter 41, "Creating Crowds and Using Populate."

Selecting an object to paint with

Within the Paint Objects panel, shown in Figure 14.30, are two paint modes for painting and filling. These buttons are toggles, and they turn the paint mode on and off. While each mode is active, you can switch between objects and subobjects as needed.

FIGURE 14.30

The Paint Objects panel lets you paint or fill with objects.

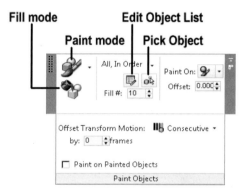

Before you can begin painting with an object, you need to select an object using the Pick Object button. Simply click the Pick Object button, and select an object in the scene. The selected object is highlighted and added to the list of paint objects. If you select the Pick Object button again, you can add another object to the list of paint objects. Clicking the Edit Object List button opens the list of current paint objects, as shown in Figure 14.31. Using this list, you can change the order of the items, pick new items, add items using the Select Objects dialog box, add the selected scene object, or remove the selected item from the list.

Note
Edits to the Paint Objects list can be made only while the Paint and Fill modes are disabled.

FIGURE 14.31

The Paint Objects list lets you manage the objects that you're painting with.

Painting with objects

After a paint object is selected, you can click the Paint button to enable Paint mode and then drag in the viewport, as shown in Figure 14.32. Each stroke drawn with the brush lays down a new curve. The Undo command can be used to remove the last stroke, but all strokes are not added to the scene until the Commit button in the Brush Settings panel is clicked. The Cancel button removes all strokes drawn since the last commit.

Caution

Painting with objects dramatically increases the overall polygon count of the scene, especially if you are painting with a complex object. Try to keep the paint object as simple as possible to avoid unwieldy scenes.

FIGURE 14.32

After an object is selected, simply drag in the viewport to paint with the selected object.

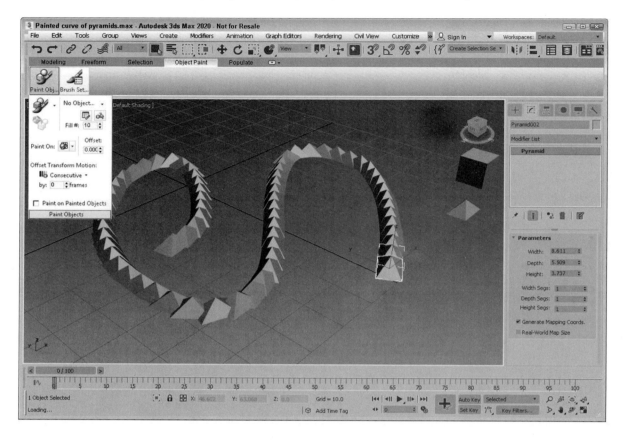

Before a painted set of objects is committed to the scene, you can use the settings in the Brush Settings panel, shown in Figure 14.33, to change the alignment, spacing, rotation, and scale of the objects. You have several options for randomly scattering the objects. The default alignment for the painted objects is to match the picked object, or you can align the object to the X-, Y-, or Z-axis, or flip it about the specified axis with the Flip Axis button.

FIGURE 14.33

The Brush Settings panel lets you change the position, rotation, and scale of the painted objects.

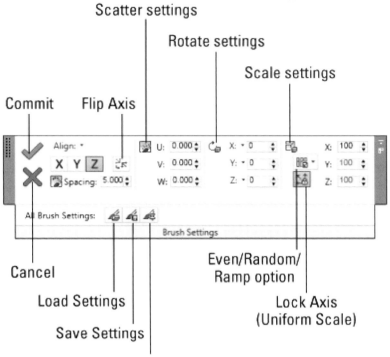

The Spacing value lets you change how far away each object is from its neighbor. The Scatter settings let you move the objects in the U, V, or W direction. The Rotation setting uniformly rotates all the objects together, or you can allow random rotations by clicking the small arrow to the right of each axis and enabling the Random option.

For the Scale settings, you can enable the Lock Axis (Uniform Scaling) option to scale all objects evenly, or you can enable the Random option to randomly scale the objects within a set range of values. The Ramp option scales the objects gradually from the start of the stroke to the end of the stroke to a given size. Figure 14.34 shows several lines of pyramids with different settings. The top line is the default. The second line has an altered spacing value, the third line is oriented about the Y-axis, and the final line uses a ramp scaling.

FIGURE 14.34

Painted objects can be altered by spacing, orientation, and scaling.

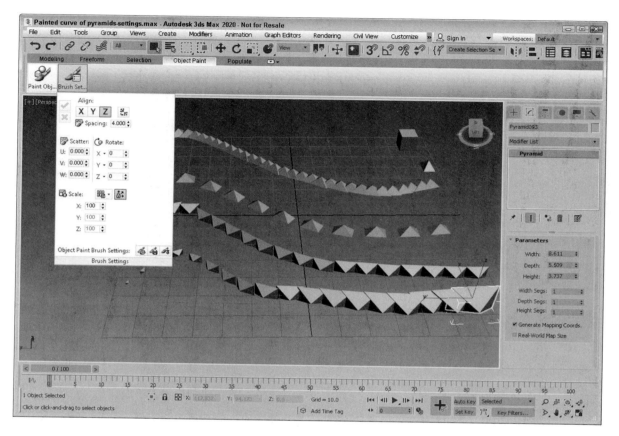

Painting with multiple objects

If you have multiple objects in the Paint Objects list, you can choose which objects to paint with, using the option in the Paint Objects panel. The first option is to paint with just the most recently picked object. The second option is to paint with all objects in order, and the last option is to randomly paint with all objects. Figure 14.35 shows each of these options.

FIGURE 14.35

When painting with multiple objects, you can choose to paint the objects in order or randomly.

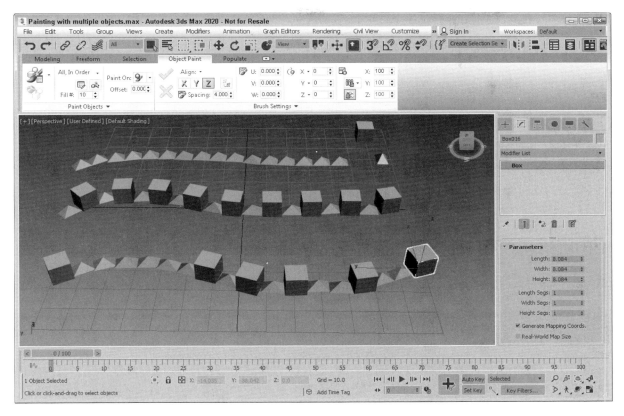

Painting on objects

The Object Paint feature lets you paint on the default construction grid, on the selected object, or on the entire scene. These options are available in the Paint Objects panel. Figure 14.36 shows painting some cones on a simple sphere. The cones are aligned by default to the surface normals of the sphere.

When the Scene option is selected, the painted objects are placed on the default grid unless a scene object is encountered, and then it is placed on top of the scene object. The Offset value can be used to move the painted objects onto or off the surface of the underlying object.

FIGURE 14.36

Objects also can be painted on the surface of another object in the scene.

Using the Paint Fill mode

The Paint Fill mode allows you to place the paint object at regular intervals along a selected edge. Before the Paint Fill mode is enabled, you need to have a paint object selected and an edge or an edge loop selected. Once selected, the Paint Fill button simply places the paint object along the edge. The Fill Number value determines the number of objects that are placed along the selected edge loop. Figure 14.37 shows a sphere filled with several cone objects using this mode.

FIGURE 14.37

The Paint Fill mode places objects along a selected edge loop.

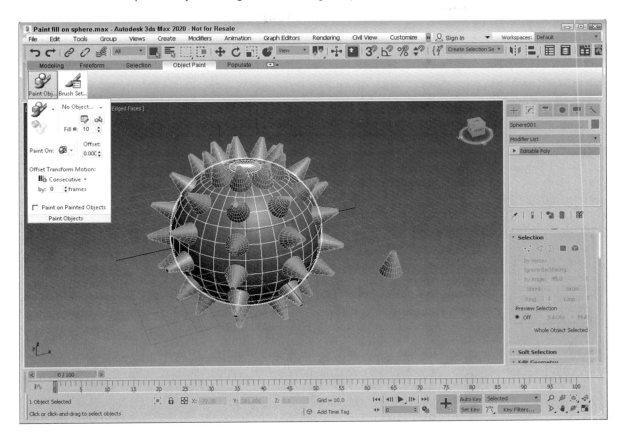

Painting with animated objects

If the object that you are painting with has some animation associated with it, you can specify an offset for the motion of the object. The options for how the animated object plays are in the Paint Objects panel and include Consecutive and Random options. The Consecutive options plays the animation on each painted object in order offset by the By Frames value. For example, if you have a box that spins, and you set the By Frames value to 2, the first box in the painted line starts spinning at frame 0, the second starts at frame 2, and so on. The Random option randomly starts the animation for each object.

Tutorial: Painting a scar

Although medical companies are searching for an easy way to remove scars, we're going to use the Object Paint feature to add one to our character.

To add a character scar using the Object Paint feature, follow these steps:

1. Open the Little Dragon with scar.max file from the Chap 14 directory in the downloaded content set.
2. Click the Graphite Modeling Tools button on the main toolbar to make the Ribbon appear, and select the Object Paint tab.
3. Click the Pick Object button in the Paint Objects panel and then click the scar object. Select the Paint On option, and select the Little Dragon's face object.
4. Select the Paint with Objects in List option under the Paint mode button, then click the Paint button, and drag across Little Dragon's face to place the scar.

5. In the Brush Settings panel, enable the Lock Axis (Uniform Scale) button, set the Scale X value to **40**, and drag the Spacing value until the scar is equally spaced out at a value around 11.5. Figure 14.38 shows the applied scar.

6. Click the Commit button in the Brush Settings panel to apply the scar to the face.

FIGURE 14.38

Applying scars on the surface of a character is easy with the Object Paint feature.

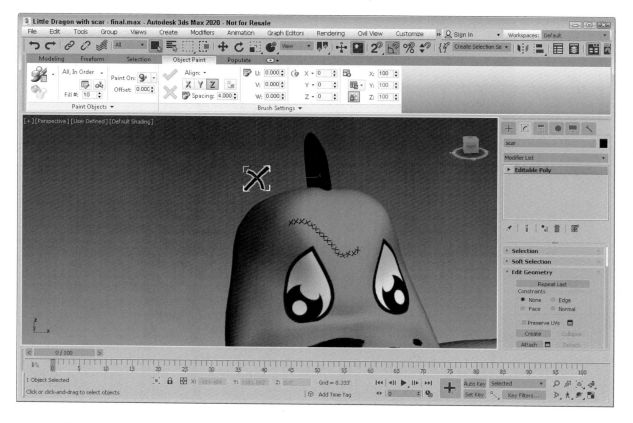

Summary

When modeling with polygons, the Graphite Modeling tools become your best friends. With all the tools at your fingertips, you can model faster and with greater ease. In addition to the base modeling features, the Freeform modeling tools, the Selection tools, and the Object Paint features make modeling a delight. More specifically, this chapter covered the following topics:

* Accessing and using the Graphite Modeling Tools

* Modeling in the various subobject modes

* Using the Freeform tools to sculpt surfaces

* Making specific selections with the Selection tools

* Painting with objects and filling edges with the Paint Objects panel

The next chapter covers a miscellaneous set of modeling objects collectively called the Compound objects.

Chapter 15
Working with Compound Objects

IN THIS CHAPTER

Understanding compound objects

Morphing objects

Creating Terrain objects

Working with BlobMesh objects

Using ProBoolean and ProCutter objects

So far, we've covered a variety of modeling types, including shapes, meshes, and polys. The Compound Objects subcategory includes several additional modeling types that don't seem to fit anywhere else. As you will see in this chapter, these modeling types provide several new and unique ways to model objects, such as working with Boolean objects, morphing objects, and cutting an object into several pieces.

The compound objects would be the miscellaneous set of modeling types that are very good at one specific task. Most compound objects are identified in the Modifier Stack by the compound object name, but if the object needs more editing, you can convert it to an Editable Poly and use the various polygon tools to finish the work.

Understanding Compound Object Types

The Compound Objects subcategory includes several unique object types. You can access these object types with the Create→Compound menu or by clicking the Geometry category button in the Create panel and selecting Compound Objects in the subcategory drop-down list. All the object types included in the Compound Objects subcategory are displayed as buttons at the top of the Create panel. They include the following:

Note
Many of the features found initially as compound objects have been added to the Editable Poly toolkit. For example, the Scatter compound object for positioning objects on the surface of another object is available in the Object Paint feature. Object Paint actually has more features and is easier to use than the Scatter compound object. The Conform, Connect and ShapeMerge compound objects features have also been added to Editable Poly object. These compound objects still remain for legacy support.

* **Morph:** Consists of two or more objects with the same number of vertices. The vertices are interpolated from one object to the other over several frames.

* **Scatter:** Randomly scatters a source object about the scene. You also can select a Distribution object that defines the volume or surface where the objects scatter.

* **Conform:** Wraps the vertices of one object onto another. You can use this option to simulate a morph between objects with different numbers of vertices.

* **Connect:** Connects two objects with open faces by joining the holes with additional faces.

* **BlobMesh:** Creates a metaball object that flows from one object to the next like water.

* **ShapeMerge:** Lets you embed a spline into a mesh object or subtract the area of a spline from a mesh object.

* **Boolean:** Created by performing Boolean operations on two or more overlapping objects. The operations include Union, Subtraction, Intersection, and Cut.

* **Terrain:** Creates terrains from the elevation contour lines like those found on topographical maps.

* **Loft:** Sweeps a cross-section shape along a spline path.

* **Mesher:** Creates an object that converts particle systems into mesh objects as the frames progress. This makes assigning modifiers to particle systems possible.

* **ProBoolean:** Replaces the original Boolean compound object with the ability to perform Boolean operations on multiple objects at a time.

* **ProCutter:** Cuts a single stock object into multiple objects using several cutter objects.

Note

When two or more objects are combined into a single compound object, they use a single object material. The Multi/Sub-Object Material type can be used to apply different materials to the various parts.

Not all of the Compound objects are covered in this chapter.

Morphing Objects

Morph objects are used to create a Morph animation by interpolating the vertices in one object to the vertex positions of a second object. The original object is called the *Base object*, and the second object is called the *Target object*. The Base and Target objects must have the same number of vertices. One Base object can be morphed into several targets.

Caution

To ensure that the Base and Target objects have the same number of vertices, create a copy of one object and modify it to be a target. Be sure to avoid such modifiers as Tessellate and Optimize, which change the number of vertices.

To morph a Base object into a Target, select the Base object, and select Create→Compound→Morph. Then click the Pick Target button in the Pick Targets rollout, shown in Figure 15.1, and select a Target object in the viewport. The cursor changes to a set of crosshairs when it is over an acceptable object. Unavailable objects (that have a different number of vertices) cannot be selected. Pick Target options include Copy, Instance, Reference, and Move. (The Move option deletes the original object that is selected.) The Target object appears under the Current Targets rollout in the Morph Targets list.

FIGURE 15.1

A Morph rollout lets you pick targets and create morph keys.

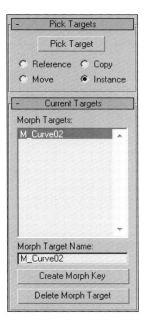

Each Morph object can have several Target objects. You can use the Pick Target button to select several targets, and the order in which these targets appear in the list is the order in which they are morphed. To delete a Target object, select it from the list, and click the Delete Morph Target button. Beneath the list is a Name field where you can change the name of the selected Target object.

Creating Morph keys

With a Target object name selected in the Morph Targets list, you can drag the Time Slider to a frame and set a Morph key by clicking the Create Morph Key button, found at the bottom of the rollout. This option sets the number of frames used to interpolate among the different morph states.

Note
If the Morph object changes dramatically, set the Morph Keys to include enough frames to interpolate smoothly.

If a frame other than 0 is selected when a Target object is picked, a Morph Key is automatically created.

Morph objects versus the Morpher modifier

The Autodesk® 3ds Max® 2020 software includes two different ways to morph an object. You can create a Morph object or apply the Morpher modifier to an existing object. The Morph object is different from the Morph modifier, but the results are the same; however, some subtle differences exist between these two.

A Morph object can include multiple Morph targets, but it can be created only once. Each target can have several Morph keys, which makes it easy to control. For example, you could set an object to morph to a different shape and return to its original form with only two Morph keys.

The Morpher modifier, on the other hand, can be applied multiple times and works well with other modifiers, but the control for each modifier is buried in the Modifier Stack. The Parameters rollout options available for the Morpher modifier are much more extensive than for the Morph object, and they include channels and support for a Morph material.

You can find more information on the Morph modifier in Chapter 33, "Using Animation Layers and Animation Modifiers."

For the best of both worlds, apply the Morph modifier to a Morph object.

Tutorial: Morphing an octopus

Although this example is fairly simple, it demonstrates a powerful technique that can be very helpful as you begin to animate characters. One of the key uses of morphing is to copy a character and move it about to create a new pose. You can then morph between the different poses to create smooth actions, gestures, or face motions.

To morph an octopus, follow these steps:

1. Open the Octopus morph.max file from the Chap 15 directory in the downloaded content set.

2. Select the octopus object, and hold down the Shift key while dragging to the right in the Top viewport. In the Clone Options dialog box that opens, select Copy, and set the Number of Copies to **2**. Name one copy **pose01** and the other **pose02**.

3. Select the object named "pose01," and open the Modify panel. Zoom in around the various arms, and enable Vertex subobject mode. Turn on the Use Soft Selection option in the Soft Selection rollout and enable the Edge Distance option with a value that covers the whole arm. You may need to adjust the Falloff value also. Then select the vertices at the end of the arm, and drag it to a new position. Repeat this action for the other arms. Click the Vertex subobject button again to exit subobject mode.

4. Select the "pose02" object and repeat Step 3 for this object to create another pose.

5. Select the original head object, and choose Create→Compound→Morph to make this object into a morph object. In the Pick Targets rollout, select the Copy option, and click the Pick Target button. Then click the "pose01" object, or press the H key and select it from the Pick Object dialog box. Then click the "pose02" object. Both targets are now added to the list. Click the Pick Target button again to disable pick mode.

6. In the Morph Targets list, select the "pose01" object, and click the Create Morph Key button. Then drag the Time Slider (below the viewports) to frame 50, select the "pose02" object in the Morph Targets list, and press the Create Morph Key button again.

7. Click the Play Animation button (in the Time Controls section at the bottom of the 3ds Max window) to see the morph. The octopus object morphs when you move the Time Slider between frames 0 and 50. Figure 15.2 shows different stages of the morph object.

FIGURE 15.2

An octopus being morphed into different poses

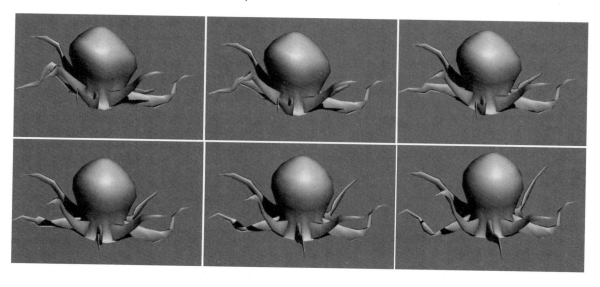

Creating a Terrain Object

The Terrain object is a great object that enables you to create terrains from splines representing elevation contours. These contour splines can be created in 3ds Max or imported using a format like AutoCAD's DWG. If the splines are created in 3ds Max, make sure that each contour spline is a separate object. The splines all must be closed splines. If all the splines have an equal number of vertices, the resulting terrain object is much cleaner. You can ensure this by copying and scaling the base spline.

To create a terrain, create splines at varying elevations, select all the splines, and click the Terrain button. The Terrain button is available only if closed splines are selected. You can use the Pick Operand button in the Pick Operand rollout to select additional splines to add to the Terrain object. All splines in the object become operands and are displayed in the Operands list.

The Form group includes three options that determine how the terrain is formed: Graded Surface, Graded Solid, and Layered Solid. The Graded Surface option displays a surface grid over the contour splines; the Graded Solid adds a bottom to the object; and the Layered Solid displays each contour as a flat, terraced area. The Stitch Border option causes polygons to be created to close open splines by creating a single edge that closes the spline. The Retriangulate option optimizes how the polygons are divided to better represent the contours.

The Display group includes options to display the Terrain mesh, the Contour lines, or Both. You also can specify how you want to update the terrain.

The Simplification rollout lets you alter the resolution of the terrain by selecting how many vertical and horizontal points and lines to use. Options include using all points (no simplification), half of the points, a quarter of the points, twice the points, or four times the points.

Coloring elevations

The Color by Elevation rollout, shown in Figure 15.3, displays the Maximum and Minimum Elevations for the current Terrain object. Between these is a Reference Elevation value, which is the location where the landmass meets the water. Entering a Reference Elevation and clicking the Create Defaults button automatically creates several separate color zones. You can add, modify, or delete zones using the Add, Modify, or Delete Zone buttons.

FIGURE 15.3

The Color by Elevation rollout lets you change the color for different elevations.

You can access each color zone from a list. To change a zone's color, select it, and click the color swatch. You can set colors to Blend to the Color Above or to be Solid to Top of Zone.

Tutorial: Creating an island with the Terrain compound object

In this tutorial, you create a simple island. The Color by Elevation rollout makes distinguishing the water from the land easy.

To create an island using the Terrain object, follow these steps:

1. Select Create→Shapes→Ellipse, and drag in the Top view to create several ellipses of various sizes representing the contours of the island.

    The first ellipse you create should be the largest, and the ellipses should get progressively smaller.

2. In the Left view, select and move the ellipses up and down so the largest one is on the bottom and the smallest one is on top. You can create two smaller hills by including two ellipses at the same level.

Note

If you create the ellipses in the proper order from largest to smallest, you can use the Select All command. If not, select the splines in the order that they'll be connected from top to bottom before clicking the Terrain button.

3. Use the Edit→Select All (Ctrl+A) menu command to select all the ellipses, and select Create→Compound→Terrain.

    The ellipses automatically join together. Joining all the ellipses forms the island.

4. In the Color by Elevation rollout, select a Reference Elevation of **5**, and click the Create Defaults button.

    This automatically creates color zones for the island. The elevation values for each zone are displayed in a list within the Color by Elevation rollout. Selecting an elevation value in the list displays its color in the color swatch.

5. Select each elevation value individually, and set all Zones to Blend to the Color Above option for all zones, except for the Zone with the lightest blue.

This creates a distinct break between the sea and the land of the island.

Figure 15.4 shows the final terrain.

FIGURE 15.4

A Terrain island created with the Terrain compound object

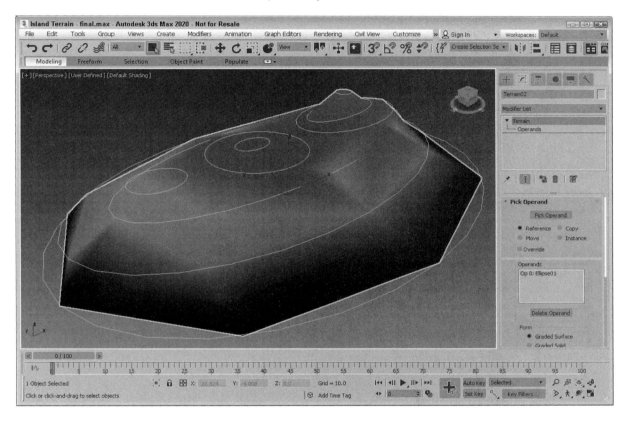

Working with BlobMesh Objects

BlobMesh objects are simple spheres. If you have only one of them, they aren't interesting at all, but if you get them together, they run into each other much like the metal mercury. This makes them an ideal choice for modeling flowing liquids and soft organic shapes.

BlobMesh objects are used as sets of objects rather than as individual objects. If you click the BlobMesh button in the Compound Objects subcategory and then create a BlobMesh in the viewports, it appears as a sphere with the radius set using the Size parameter. The real benefit comes from clicking the Pick or Add buttons below the Blob Objects list and selecting an object in the scene.

Note

The Pick, Add, and Remove buttons become enabled only in the Modify panel.

The object that is picked is added to the Blob Objects list, and each vertex of the object gets a BlobMesh added to it. If the BlobMesh objects are large enough to overlap, the entire object is covered with these objects, and they run together to form a flowing mass of blobs.

Setting BlobMesh parameters

The Size value sets the radius of the BlobMesh object. Larger sizes result in more overlapping of surrounding objects. For particle systems, the Size is discounted, and the size of the particles determines the size of the BlobMesh objects. The Tension value sets how loose or tight the surface of the BlobMesh object is. Small tension values result in looser objects that more readily flow together.

The Evaluation Coarseness value sets how dense the BlobMesh objects will be. By enabling the Relative Coarseness option, the density of the objects changes as the size of the objects changes. The Evaluation Coarseness values can be different for the viewport and the render engine.

When a BlobMesh object is selected and applied to the picked object, each vertex has an object attached to it, but if you apply a selection modifier, such as the Mesh Select modifier, to the picked object, only the selected subobjects get a BlobMesh object. You also can use the Use Soft Selection option to select those subobjects adjacent to the selected subobjects. The Minimum Size value is the smallest BlobMesh object that is used when Soft Selection is enabled.

The Large Data Optimization option is a quicker, more efficient way of rendering a huge set of BlobMesh objects. The benefit from this method comes when more than 2,000 BlobMesh objects need to be rendered. If the viewport updates are slow because of the number of BlobMesh objects, you can select to turn them Off in Viewport.

When BlobMesh objects are applied to a particle system, they can be used as part of a Particle Flow workflow. The Particle Flow Parameters rollout includes a list of events to apply to the BlobMesh objects. Particle Flow is covered in detail in Chapter 42, "Creating Particles and Particle Flow."

Tutorial: Creating icy geometry with BlobMesh

The BlobMesh object can be combined with a geometry object to create the effect of an object that has been frozen in ice. Using the BlobMesh's Pick feature, you can select a geometry object, and a BlobMesh is placed at each vertex of the object. I suggest using an object with a fairly limited number of vertices.

To create the effect of an object covered in ice, follow these steps:

1. Open the Icy birdbath.max file from the Chap 15 directory in the downloaded content set.

2. With the birdbath selected, choose Create→Compound→BlobMesh, and create a simple BlobMesh by simply clicking in the Top viewport. Set the Size value to **20.0** and the Tension value to **0.01**. Then right-click to exit BlobMesh mode, open the Modify panel, click the Pick button in the Parameters rollout, and select the birdbath object.

3. Select Rendering→Material Editor→Compact Material Editor, and select the first sample slot. Change the Diffuse color to a light blue, and set the Opacity to 20. Then increase the Specular Level to **90** and the Glossiness to **40**, and apply the material to the BlobMesh001 object by dragging the sample slot material and dropping it on the BlobMesh001 object.

4. Render the Perspective viewport to see the final result. The birdbath is embedded in ice.

Figure 15.5 shows the resulting birdbath, all ready to be defrosted.

FIGURE 15.5

BlobMesh objects can be used to cover objects in ice.

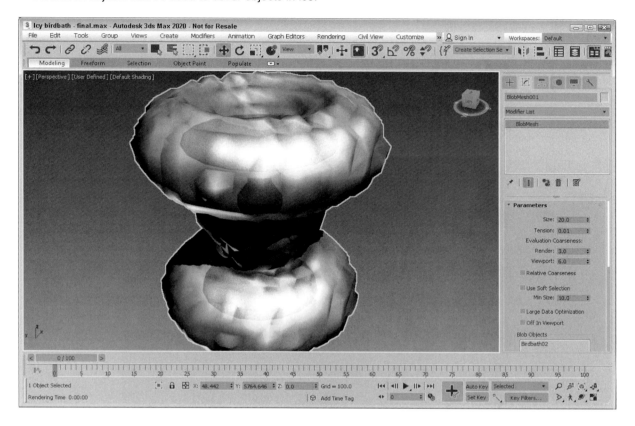

Working with ProBoolean and ProCutter Objects

The original Boolean compound object worked well enough for combining, subtracting, and intersecting objects, but it had some limitations that have been overcome with the ProBoolean and ProCutter compound objects. The original Boolean could combine only two operands, but the ProBoolean object can perform multiple Boolean operations simultaneously. ProBoolean also can subdivide the result into quad faces. The results of the ProBoolean and ProCutter objects are much cleaner and more accurate than the original Boolean object.

The original Boolean compound object still is available for backward compatibility, but if you perform a new Boolean operation, you really should use the ProBoolean object.

Using ProBoolean

When two objects overlap, you can perform different Boolean operations on them to create a unique object. The ProBoolean operations include Union, Intersection, Subtraction, Merge, Attach, and Insert. Two additional options are available: Imprint and Cookie.

Note

The ProBoolean compound object provides a different workflow from the standard Boolean compound object, but both are acceptable. The Boolean has a cleaner workflow and provides better results since it is based on newer algorithms.

The Union operation combines two objects into one. The Intersection operation retains only the overlapping sections of two objects. The Subtraction operation subtracts the overlapping portions of one object from another. The Merge operation combines objects without removing the interior faces and adds new edges where the objects overlap. Figure 15.6 shows the original object and the first four possible Boolean operators.

Note

Unlike many CAD packages that deal with solid objects, the Booleans in 3ds Max are applied to surfaces, so if the surfaces of the two objects don't overlap, all Boolean operations (except for Union) will have no effect.

FIGURE 15.6

Object before any operations and with Boolean operations, Union, Intersection, Subtraction, and Merge with the Imprint option enabled

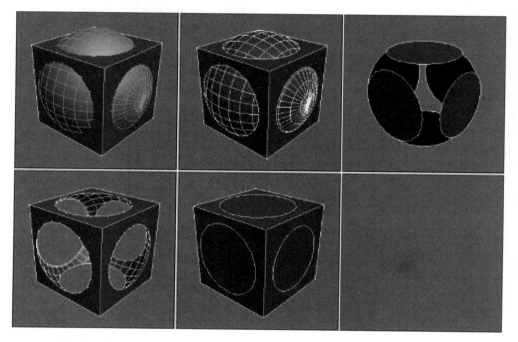

The Attach operation combines the objects like Union but keeps them as separate elements of the same compound object. For example, if you were to look inside a compound object created with Union, you would not see the interior polygons of the combined object, but with Attach, the interior polygons would still be there.

The Insert operation subtracts the second object from the first and then combines the two objects into one. If the subtracted volume makes a dent or hole into the first object, that hole remains after the two are combined, but if the second surface has access to the first object through a hole, the first surface covers the subtracted volume. In Figure 15.7, two tube objects have been overlapped with a box object. The left tube is capped at the bottom, forming a closed volume, but the right tube is open at the bottom. When made into a ProBoolean object with the Insert operations, the closed volume is subtracted, but the open volume is not.

FIGURE 15.7

The Insert operation maintains closed volumes only when subtracted.

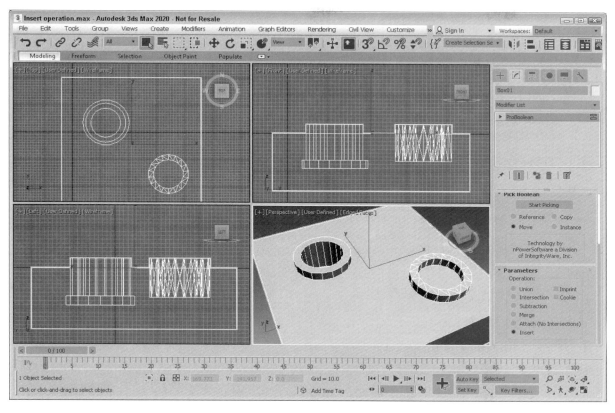

The Cookie option causes the operation to cut the original object without adding any of the faces from the picked object to the original object. The Imprint option causes the outline of the operation to appear on the original object.

All Boolean operations are added in the order in which they are applied to a list in the Parameters rollout. You can select any of the operations in the list at any time and change the operation. For example, if you select the Subtraction operation from the list and then change the operation type to Union and click the Change Operation button, the Subtraction changes to a Union. With an operation selected in the list, the Extract Selected button restores the original object. When using this button, you can choose to Remove, Copy, or Instance the operation.

The order in which the operations are applied affects the result. You can reorder the operations in the list by selecting an operation, choosing its position in the list, and clicking the Reorder Ops button.

You also can apply Boolean operations to shapes using the Boolean operators available for Editable Meshes in the Geometry rollout. Chapter 12, "Drawing and Editing 2D Splines and Shapes," covers these 2D Boolean operators.

The materials that get applied to a ProBoolean result can be set to use the Operand Material or to retain the Original Material. If you use the Apply Operand Material with the subtraction operation, the surface that touches the picked object retains the removed object's material, and the rest of the object has the original object's material. If the Retain Original Material option is selected, the entire result gets the original object's material.

In the Advanced Options rollout are options for updating the scene and for reducing the complexity of the object. The Decimation value is the percentage of edges to remove from the result. If you plan to smooth the object or convert it to an Editable Poly object, you want to enable the Make Quadrilaterals option, which causes the polygon reduction to avoid triangles in favor of quads. You also can set the Quad Size, and you can select how planar edges are handled.

Note

If you plan to deform the mesh as part of a skinned object or a cloth simulation, the resulting mesh must be clean, or the deformation will have problems.

Tutorial: Creating a keyhole

What was it that Alice saw when she looked through the keyhole? The ProBoolean feature is the perfect tool for cutting a keyhole through a doorknob plate.

To use the ProBoolean object to create a keyhole, follow these steps:

1. Open the Doorknob.max file from the Chap 15 directory in the downloaded content set.
2. Select the panel object, and choose the Create→Compound→ProBoolean menu command. In the Parameters rollout, select the Subtraction option.
3. In the Pick Boolean rollout, click the Start Picking button, and select the Box and Cylinder objects positioned where the keyhole should be. Then right-click to exit pick mode. A keyhole has been cut into the panel object.

Figure 15.8 shows the finished keyhole.

FIGURE 15.8

A keyhole built using the ProBoolean object

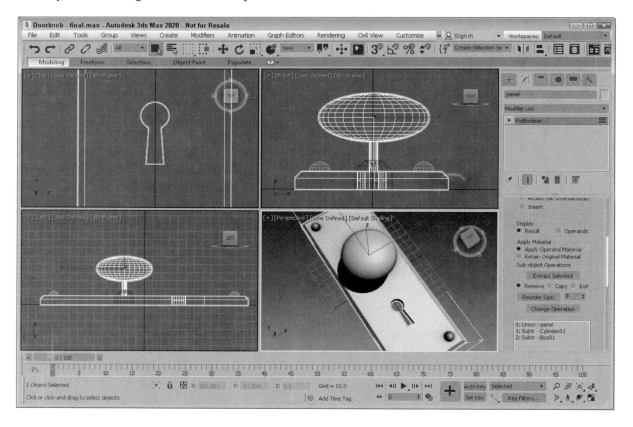

Tips for working with Booleans

Working with Boolean objects can be difficult. If you try to perform a Boolean operation on an ill-suited object, the results could end up being erratic. As you prepare objects for Boolean operations, keep the following points in mind:

* Avoid meshes with long, skinny polygon faces. All faces should have roughly equal lengths and widths. The ratio of edge length to width should be less than 4 to 1.

* Avoid curved lines where possible. Curved lines have the potential of folding back on themselves, which causes problems. If you need to use a curve, try not to intersect it with another curve; keep the curvature to a minimum.

* Unlink any objects not involved in the Boolean operation. Linked objects, even if they don't intersect, can cause problems.

* If you're having difficulty getting a Boolean operation to work, try applying the XForm modifier (found in the Modifiers List) to combine all the transformations into one. Then collapse the Stack, and convert the objects to Editable Mesh objects. This technique removes any modifier dependencies.

* Make sure that your objects are completely closed surfaces with no holes, overlapping faces, or unwelded vertices. You can check these criteria by applying the STL-Check modifier or by looking at all sides of the objects in a viewport with Smooth Shading enabled.

* Make sure that all surface normals are consistent: Inconsistent normals cause unexpected results. You can use the Normal modifier to unify and flip all normals on an object. The Show Normals option in the viewport can also help.

* Collapsing the Stack after all Boolean operations have been performed eliminates dependencies on the previous object types.

Tutorial: Creating a Lincoln Log set

In the household where I grew up, we had sets of Legos and a lesser-known construction set known as Lincoln Logs that let you to create buildings using notched logs that fit together. Using a Subtraction operation, you can create your own virtual set of Lincoln Logs.

To use Boolean objects to create a log cabin, follow these steps:

1. Open the Lincoln logs booleans.max file from the Chap 15 directory in the downloaded content set.

   This file contains some simple primitives.

2. Select the Cylinder object, and choose the Create→Compound→ProBoolean menu command. In the Parameters rollout, select the Subtraction option. Then click the Start Picking button and click on all four box objects in the viewport.

   When finished, you should have a cylinder with four notches.

3. Select and right click on the cylinder and choose the Convert To→Convert to Editable Poly menu from the quad menu.

4. Clone the single log by choosing Edit→Clone and then selecting the Copy option. Move the cloned log along the negative Y-axis a distance of 160. The easiest way to do so is to right-click the Select and Move button to open the Move Transform Type-In dialog box. In the Absolute World field, enter **−160** as the Y-axis value.

   This step positions two logs next to one another to form the bottom layer of the house.

5. Select both logs, and open the Array dialog box by choosing Tools→Array. In the Incremental Move row, enter **10** for the Z-axis. In the Incremental Rotate row, enter **90** for the Z-axis. In the Array Dimensions section, enter a Count value of **16**, and click the OK button.

   This step stacks several layers of logs.

6. Open the Modify panel, click the Attach button, and then select every log to combine them all into a single object. Click the Attach button again to exit attach mode when you're finished.

7. Select the Create→Standard Primitives→Box menu command, and create a Box object with the following dimensions: Length **40**, Width **40**, and Height **80**. Then position the Box in the Top viewport where the front door should be.

8. Return to the Compound Objects subcategory, select the logs, and click the ProBoolean button. Then select the Subtraction option again, click the Start Picking button and select the Box where the door should be.

9. To add a roof, select the Extended Primitives subcategory and click the Prism button. Drag in the Left view to create a prism object that covers the logs.

Figure 15.9 shows our ProBoolean log cabin, ready for the virtual pioneers.

FIGURE 15.9

A log cabin built using ProBoolean objects.

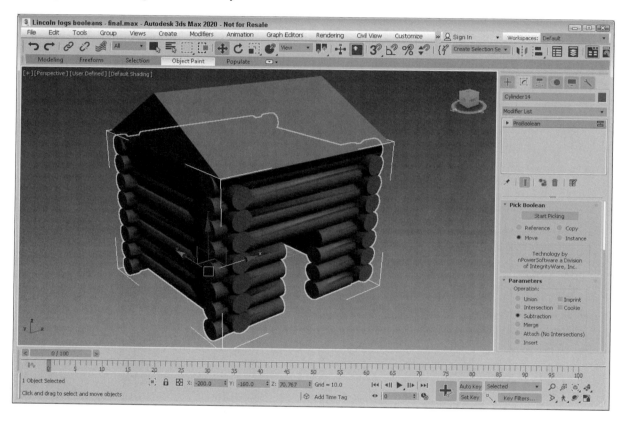

Using ProCutter

The original Boolean compound object included a Cut option. This feature has been replaced with ProCutter, which offers many more features than the original option. ProCutter allows you to cut a single object (known as the Stock object) with multiple cutter objects.

You can pick both the Stock and Cutter objects using the buttons found in the Cutter Picking Parameters rollout. You have four options with each selection: Reference, Move, Copy, and Instance. The Auto Extract Mesh option automatically replaces the selected stock object with the extracted result. The Explode by Elements option works only when the Auto Extract Mesh option is enabled. It separates each cut element into an object.

Within the Cutter Parameters rollout, you can select from three Cutting Options. The Stock Outside Cutter option keeps the portion of the stock object that is on the outside of the cutter. The Stock Inside Cutter option is the opposite, keeping the stock portion inside the cutters. The Cutters Outside Stock option maintains those portions of the cutters that are outside the stock.

All selected cutters and stock objects are added to a list in the Cutter Parameters. Using the Extract Selected button, you can restore any cutter or stock object that has been operated on as a Copy or Instance. Materials and Decimation also work the same as for ProBoolean objects.

Tip
The ProCutter features are helpful for dividing an object that will be animated exploding into pieces.

Tutorial: Creating a jigsaw puzzle

The ProCutter is useful for creating destructive scenes such as shattering glass and breaking down buildings, but it also can be used for constructive cutting, such as creating a jigsaw puzzle.

To use the ProCutter compound object to divide an object into a jigsaw puzzle, follow these steps:

1. Open the ProCutter puzzle.max file from the Chap 15 directory in the downloaded content set. This file includes a simple box mapped with a scenic image and several extruded lines that mark the jigsaw puzzle's edges.
2. Select one of the extruded lines that mark where the cuts should be (known as a cutter), and choose the Create→Compound→ProCutter menu command.
3. Click the Pick Cutter Objects button, and select all the remaining cutter objects. In the Cutter Parameters rollout, enable the Stock Outside Cutter, along with the Stock Inside Cutter options. Select the Retain Original Material option also.
4. In the Cutter Picking Parameters rollout, enable the Auto Extract Mesh and Explode By Elements options; then select the Pick Stock Objects button, and choose the Box object in the viewport. Each separate piece is given a different object color and clears its material.
5. To reapply the materials, open the Compact Material Editor, and drag the image material in the first sample slot onto each puzzle piece.

Note
If you don't want to see the cutter lines after the puzzle is cut, you can select and hide them.

Figure 15.10 shows the final puzzle with one piece moved away from the others.

FIGURE 15.10

A puzzle cut using the ProCutter compound object

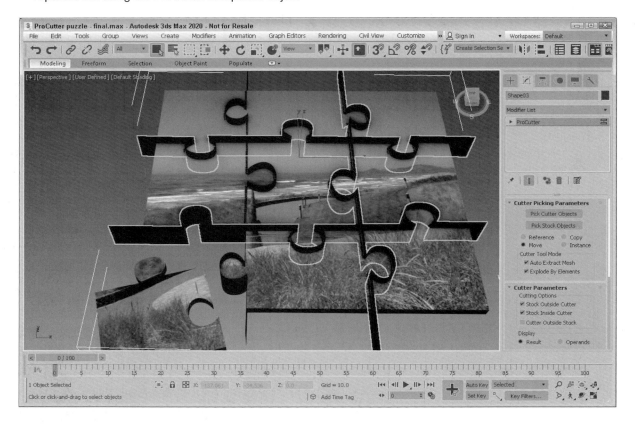

Summary

Compound objects add several additional modeling types to your bulging modeling toolkit. From morph objects to complex deformed lofts, you can use these special-purpose types to model many different objects. This chapter covered these topics:

* The various compound object types

* Morphing objects with the same number of vertices

* Creating a Terrain object using splines

* Using the BlobMesh object to simulate water

* Modeling with the ProBoolean and ProCutter objects

Once the basic shape of a model is looking good, you can focus on some small modeling changes to add details to the model. You can also deform the entire model using one of the many modifiers.

Chapter 16

Deforming Surfaces and Using the Mesh Modifiers

IN THIS CHAPTER

Using the Selection Modifiers

Maintaining primitive objects with the Edit Mesh and Edit Poly modifiers

Using spline modifiers

Changing mesh geometry with modifiers

Deforming objects with the Parametric Deformer and FFD modifiers

Editing mesh normals

Working with Subdivision Surfaces

Using the Paint Deformation brush

A majority of the available modifiers are used to help modeling objects. This chapter covers most of the available modeling modifiers. These modifiers are divided into different categories based on the type of base object that they work with and the type of features they enable. These categories include Edit/Convert, Selection, Spline, Geometry (Convert to Mesh), Geometry (Parametric), Free Form Deformers, Conform, NURBS, Patch, Animation, Cloth, Hair and Fur, Simulation, Cameras, Subdivision Surfaces, and UVs, Maps, and Materials.

When an Editable Poly object is selected, three specific deformation brushes may be selected in the Paint Deformation rollout. Using these brushes, you can deform the surface of an object by dragging over the surface with the selected brush.

In addition to the editing features available for Editable Mesh and Editable Poly objects and the Paint Deformation brushes, you also can modify mesh geometries using modifiers. The Modifiers menu includes a submenu of modifiers that are specific to mesh (and poly) objects. These modifiers are found in the Geometry (Convert to Mesh) submenu and can be used to enhance the features available for these objects.

Using Selection Modifiers

The first modifiers available in the Modifiers menu are the Selection modifiers. You can use these modifiers to select subobjects for the various object types. You can then apply other modifiers to these subobject selections. Any modifiers that appear above a Selection modifier in the Modifier Stack are applied to the subobject selection.

Selection modifiers are available for every modeling type, including Mesh Select, Poly Select, Patch Select, Spline Select, Volume Select, FFD Select, and Select by Channel. There is also a Surface Select in the NURBS Editing category. You can apply the Mesh Select, Poly Select, Patch Select, and Volume Select modifiers to any 3D object, but you can apply the Spline Select modifier only to spline and shape objects, the FFD Select modifier only to the FFD Space Warps objects, and the NURBS Surface Select modifier (found in the NURBS Editing submenu) only to NURBS objects. Any modifiers that appear above one of these Selection modifiers in the Modifier Stack are applied only to the selected subobjects.

Each of the Selection modifiers is covered for the various modeling types in its respective chapter.

When a Selection modifier is applied to an object and a subobject is selected, the transform buttons on the main toolbar become inactive. If you want to transform the subobject selection, you can do so with the XForm modifier.

Object-Space versus World-Space modifiers

If you view the modifiers listed in the Modifier List, they are divided into two categories: Object-Space and World-Space modifiers (except for the selected set of modifiers that appear at the very top for quick access). Object-Space modifiers are more numerous than World-Space modifiers. For most World-Space modifiers, there is also an Object-Space version. World-Space modifiers are all identified with the abbreviation *WSM*, which appears next to the modifier's name.

Object-Space modifiers are modifiers that are applied to individual objects and that use the object's Local Coordinate System, so as the object is moved, the modifier goes with it.

World-Space modifiers are based on World-Space coordinates instead of on an object's Local Coordinate System, so after a World-Space modifier is applied, it stays put, no matter where the object with which it is associated moves.

Another key difference is that World-Space modifiers appear above all Object-Space modifiers in the Modifier Stack, so they affect the object only after all the other modifiers are applied.

All Space Warps are also applied using World-Space coordinates, so they also have the WSM letters next to their name. You can get more information on Space Warps in Chapter 43, "Using Space Warps."

Volume Select modifier

Among the Selection modifiers, the Volume Select modifier is unique. It selects subobjects based on the area defined by the modifier's gizmo. The Volume Select modifier selects all subobjects within the volume from a single object or from multiple objects. One of the benefits of using this modifier is that the subobjects within the volume can change as the object is moved during an animation sequence. It also can work with several objects at once.

In the Parameters rollout for the Volume Select modifier, you can specify whether subobjects selected within a given volume should be Object, Vertex, or Face subobjects. Any new selection can Replace, be Added to, or be Subtracted from the current selection. You can use the Invert option to select the subobjects outside of the current volume. You also can choose either a Window or Crossing Selection Type.

The actual shape of the gizmo can be a Box, Sphere, Cylinder, or Mesh Object. To use a Mesh Object, click the button beneath the Mesh Object option and then click the object to use in a viewport. In addition to selecting by a gizmo-defined volume, you also can select subobjects based on certain surface characteristics, such as Material IDs, Smoothing Groups, or a Texture Map including Mapping Channel or Vertex Color. This makes it possible to quickly select all vertices that have a Vertex Color assigned to them.

The Alignment options can Fit or Center the gizmo on the current subobject selection. The Reset button moves the gizmo to its original position and orientation, which typically is the bounding box of the object. The Auto Fit option automatically changes the size and orientation of the gizmo as the object it encompasses changes.

Note

The Volume Select modifier also includes a Soft Selection rollout. Soft Selection lets you select adjacent subobjects to a lesser extent. The result is a smoother selection over a broader surface area.

You can see another example of how a Selection modifier can be used to select and apply a modifier to a subobject selection in Chapter 11, "Accessing Subobjects and Modifiers and Using the Modifier Stack."

Using Primitive Maintenance Modifiers

Included among the Mesh Editing modifiers are two unique modifiers that can be applied to primitive objects, allowing them to maintain their parametric nature.

Note
Using the Edit Mesh or Edit Poly modifiers increases the file size and memory required to work with the object. You can reduce the overhead by collapsing the modifier stack.

Edit Mesh modifier

All mesh objects are by default Editable Mesh objects. This modifier enables objects to be modified using the Editable Mesh features while maintaining their basic creation parameters.

When an object is converted to an Editable Mesh, its parametric nature is eliminated. However, if you use the Edit Mesh modifier, you can still retain the same object type and its parametric nature while having access to all the Editable Mesh features. For example, if you create a sphere and apply the Edit Mesh modifier and then extrude several faces, you can still change the radius of the sphere by selecting the Sphere object in the Modifier Stack and changing the Radius value in the Parameters rollout.

If the Sphere base object is selected after a modifier has been applied, the Topology Dependence warning dialog box appears. This dialog box doesn't prevent you from making any changes to the base object parameters, but it reminds you that changes you make to the base object's parameters may disrupt the changes to the surface accomplished by the modifiers. The warning dialog box also includes a Hold button to load the current scene in a buffer in case you don't like the results.

Caution
One drawback of the Edit Mesh modifier is that its subobjects cannot be animated.

Edit Poly modifier

The Edit Poly modifier lets you work with primitive objects using the operators found in the Editable Poly rollouts. A huge benefit of this modifier is that you can remove it at any time if the changes don't work out.

The Edit Poly modifier includes two separate modes: Model and Animate. You can select these modes in the Edit Poly Mode rollout, shown in Figure 16.1.

FIGURE 16.1

The Edit Poly Mode rollout lets you switch between Model and Animate modes.

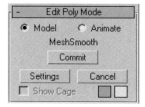

Model mode lets you access the same features available for Editable Poly objects. Animate mode lets you animate subobject changes made with the features used to edit the object. To animate these subobject changes, you use the Auto Key or Set Key button to set the keys.

The Commit button lets you freeze the changes and set the keyframe for the current change. The current change is listed directly above the Commit button. The Settings button lets you access the dialog box used to make the changes. The Cancel button cancels the last change, and the Show Cage option displays an orange cage around the object; you can change the color of the cage using the color swatch. The cage is useful when using the MeshSmooth modifier to see the original shape of the object before being smoothed.

The differences between the features available for the Edit Poly modifier and the Editable Poly object are subtle. In the Selection rollout is a Get Stack Selection button. Clicking this button passes the subobject selection up from the stack. Also, the Edit Poly modifier doesn't include the Subdivision Surfaces rollout, but you can use the MeshSmooth modifier to get this functionality.

Using Spline Modifiers

In the Modifiers menu is a whole submenu of modifiers that apply strictly to splines. You can find these modifiers in the Modifiers→Patch/Spline Editing menu.

Spline-specific modifiers

Of the modifiers that work only on splines, several of these duplicate functionality that is available for Editable Splines, such as the Fillet/Chamfer modifier. Applying these features as modifiers gives you better control over the results because you can remove them using the Modifier Stack at any time.

Edit Spline modifier

The Modifiers→Patch/Spline Editing→Edit Spline modifier (mentioned at the start of the chapter) makes spline objects so they can be edited. It has all the same features as the Editable Spline object. It shows up in the Modifier Stack above the base object. The key benefit of the Edit Spline modifier is that it enables you to edit spline subobjects while maintaining the parametric nature of the primitive object.

Spline Select modifier

This modifier enables you to select spline subobjects, including Vertex, Segment, and Spline. You can copy and paste named selection sets. The selection can then be passed up the Stack to the next modifier. The Spline Select modifier provides a way to apply a modifier to a subobject selection.

The Modifiers→Selection Modifiers→Spline Select modifier lets you select objects from any of the subobject modes available in the Editable Spline object. It also includes buttons for selecting subobjects based on the other subobject modes. For example, if you select Vertex subobject mode, two buttons available in the Select Vertex rollout are Get Segment Selection and Get Spline Selection. Clicking either of these buttons gets all the vertices that are part of the other subobject mode.

You also can Copy and Paste selection sets using the Copy and Paste buttons.

Delete Spline modifier

You can use the Delete Spline modifier to delete spline subobjects. This is helpful if you want to remove a spline from an object so it doesn't render without destroying the curvature of the other splines in the object. To do this, simply select the splines to remove from the object in Spline subobject mode and then apply the Delete Spline modifier. The selection will be passed up the stack.

Normalize, Optimize and Relax Spline modifiers

The Normalize Spline modifier adds or removes points to the spline as needed. These points are spaced regularly based on the Segment Length value. Figure 16.2 shows a simple flower shape with the Spline Select modifier applied so you can see the vertices. The Normalize Spline modifier was then applied with Segment Length values of 1, 5, 10, and 15. Notice that the shape is changing with fewer vertices.

FIGURE 16.2

The Normalize Spline modifier relaxes the shape by removing vertices.

The Optimize Spline modifier reduces the total number of vertices in a spline. This can be done by reducing the number of knots in a spline by a certain percentage or to a maximum number of knots. Once the optimization parameters are set, clicking the Make Spline button creates an Editable Spline object that matches the parameters. The Statistics for the optimized spline are shown at the bottom of the rollout. They show the number of shapes, splines and knots.

The Spline Relax modifier pulls all outlying knots towards the center of the spline shape. This is useful if you want to smooth out extraneous knots or take the randomness out of a drawn circle. In the Relax rollout, you can set the Amount of smoothness to apply up to a value of 1 and the number of Iterations to apply.

Fillet/Chamfer modifier

You can use the Fillet/Chamfer modifier to Fillet or Chamfer the corners of shapes. Fillet creates smooth corners, and Chamfer adds another segment where two edges meet. Parameters include the Fillet Radius and the Chamfer Distance. Both include an Apply button. The results of this modifier are the same as if you were to use the Fillet or Chamfer features of an Editable Spline.

Renderable Spline modifier

The Renderable Spline modifier lets you make any selected spline renderable. The Parameters rollout includes the same controls that are available for Editable Splines including Thickness, Sides, and Angle values.

Sweep modifier

The Sweep modifier works just like the loft compound object, letting you follow a spline path with a defined cross section, except that the Sweep modifier is a modifier, making it easier to apply and remove from splines and shapes. Another benefit of the Sweep modifier is that it has several Built-In Sections available that you can choose, or you can pick your own. The built-in sections include many that are useful for architectural structures including Angle, Bar, Channel, Cylinder, Half Round, Pipe, Quarter Round, Tee, Tube, Wide Flange, Egg, and Ellipse.

Using the Merge From File button, you can choose a shape from another file. You also can set the number of interpolation steps. The Sweep Parameters rollout includes options for mirroring, offsetting, smoothing, aligning, and banking the generated sweep. The Union Intersecting option causes self-intersecting portions of the path to be combined using a union Boolean command. You also can select to have mapping coordinates generated on the sweep object.

Tutorial: Plumbing with pipes

If you want to create a shape that renders in the scene, you can use the Renderable Spline option or you can apply the Sweep modifier. In this example, you apply the Sweep modifier to a line that defines the path of a bathroom sink drain.

To create a pipe that follows a spline, follow these steps:

1. Open the Bathroom sink.max file from the Chap 16 folder in the downloaded content set.

    This file includes a simple bathroom sink and a line that defines its drain path.

2. With the spline selected, choose the Modifiers→Patch/Spline Editing→Sweep menu command to apply the Sweep modifier.

3. In the Section Type rollout, choose the Cylinder option from the Built-In Section drop-down list. Then set the Radius value to **10** in the Parameters rollout.

Figure 16.3 shows the resulting sink, complete with a drain created using a cylinder cross section.

FIGURE 16.3

The resulting drain pipe was created using the Sweep modifier.

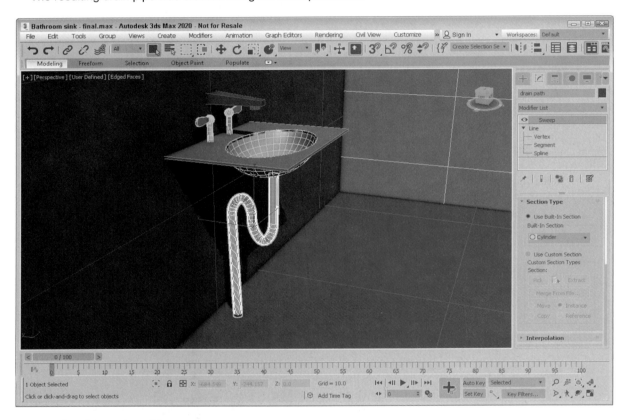

Trim/Extend modifier

The Trim/Extend modifier lets you trim the extending end of a spline or extend a spline until it meets another spline at a vertex. The Pick Locations button turns on Pick mode, where the cursor changes when it is over a valid point. Operations include Auto, Trim Only, and Extend Only with an option to compute Infinite Boundaries. You also can set the Intersection Projection to View, Construction Plane, or None.

Spline Mirror modifier

The Spline Mirror modifier lets you mirror a selected spline about any specific axis. The Mirror subobject can be selected in the Modifier Stack and moved as needed to define the center and axis about which the mirror takes place. There are also options to Slice Along Mirror, Flip and Weld Seam.

Using the Shape Check utility

The Shape Check utility is helpful in verifying that a shape doesn't intersect itself. Shapes that have this problem cannot be extruded, lofted, or lathed without problems. To use this utility, open the Utilities panel (the icon for the Utilities panel looks like a Wrench) and click the More button. Select Shape Check from the Utilities dialog box list and click OK.

Note

The Shape Check utility is found in the Utilities panel and not in the Modifiers menu.

The Shape Check rollout includes only two buttons: Pick Object and Close. Click the Pick Object button, and click the shape you want to check. Any intersection points are displayed as red squares, as shown in Figure 16.4, and the response field displays "Shape Self-Intersects." If the shape doesn't have any intersections, the response field reports "Shape OK."

Note

You can use the Shape Check utility on normal splines and on NURBS splines.

FIGURE 16.4

The Shape Check utility can identify spline intersections.

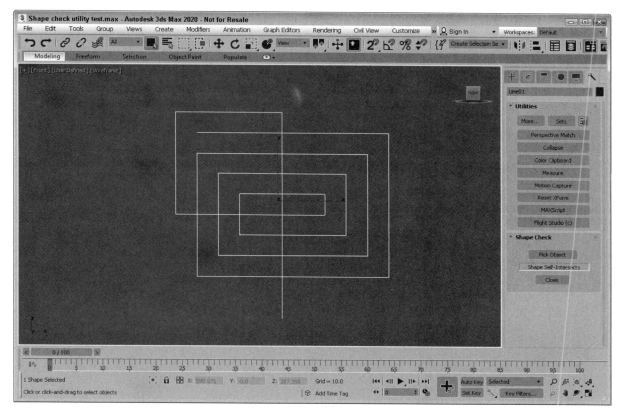

Moving splines to 3D

Although splines can be rendered, the real benefit of splines in the Autodesk® 3ds Max® 2020 software is to use them to create 3D objects and for animation paths. You can use splines in several ways as you model 3D objects, including Loft objects and modifiers. One way to use splines to make 3D objects is with modifiers.

Note

As you build splines that will be used to create mesh objects, remember that the number of vertices in a spline determines the number of segments in the final mesh. For example, if you have a spline with 10 points that is extruded 5 times, you'll end up with more than 50 polygons, but the same spline with 80 points would have more than 400 polygons.

Using splines to create an animation path is covered in Chapter 32, "Understanding Animation and Keyframes." General information on working with modifiers is covered in Chapter 11, "Accessing Subobjects and Modifiers and Using the Modifier Stack."

Extruding splines

Because splines are drawn in a 2D plane, they already include two of the three dimensions. By adding a Height value to the shape, you can create a simple 3D object. The process of adding Height to a shape is called *extruding*.

To extrude a shape, you need to apply the Extrude modifier. To do so, select a spline object and choose Modifiers→Mesh Editing→Extrude, or select the Extrude modifier from the Modifier Stack drop-down list. In the Parameters rollout, you can specify an Amount, which is the height value of the extrusion; the number of Segments; and the Capping options (caps fill in the surface at each end of the Extruded shape). You also can specify the final Output to be a Patch, Mesh, or NURBS object. Figure 16.5 shows capital *E*s that model the various vertex types extruded to a depth of 10.0.

FIGURE 16.5

Extruding simple shapes adds depth to the spline.

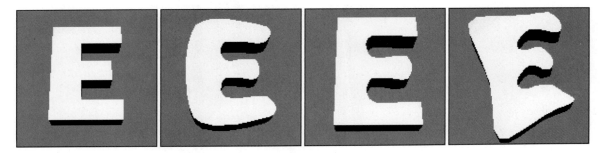

Tutorial: Routing a custom shelf

In Woodshop 101, you use a router to add a designer edge to doorframes, window frames, and shelving of all sorts. In Woodshop 3D, the Boolean tools work nicely as you customize a bookshelf.

To create a custom bookshelf using spline Boolean operations, follow these steps:

1. Open the Bookshelf.max file from the Chap 16 folder in the downloaded content set.

   This file includes a triangle shape drawn with the Line primitive that is overlapped by three circles. All these shapes have been converted and combined into a single Editable Spline object.

2. Select the shape, open the Modify panel, and select the Spline subobject mode (or press the 3 key) and select the triangle shape.

3. Select the Subtraction Boolean operation (the middle icon) in the Geometry rollout, and click the Boolean button. Then select each of the circles and right-click to exit Boolean mode. Then click Spline in the Modifier Stack again to exit subobject mode.

4. Back in the Modify panel, select the Extrude modifier from the Modifier drop-down list and enter an Amount of **1000**. Select Zoom Extents All to resize your viewports, switch to the Perspective viewport, and view your bookshelf.

Figure 16.6 shows the finished bookshelf in the Perspective viewport ready to hang on the wall.

FIGURE 16.6

The finished bookshelf created with spline Boolean operations and the Extrude modifier

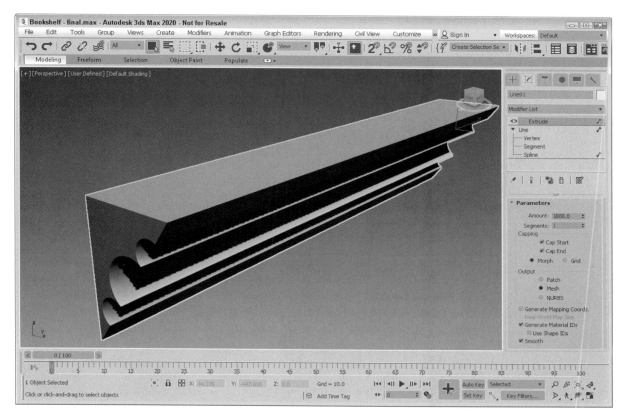

Lathing splines

Another useful modifier for 2D splines is the Lathe. This modifier rotates the spline about an axis to create an object with a circular cross section (such as a baseball bat). In the Parameters rollout, you can specify the Degrees to rotate (a value of 360 makes a full rotation) and Cappings, which add ends to the resulting mesh. Additional options include Weld Core, which causes all vertices at the center of the lathe to be welded together, and Flip Normals, which realigns all the normals.

The Direction option determines the axis about which the rotation takes place. The rotation takes place about the object's pivot point.

Caution

If your shape is created in the Top view, lathing about the screen Z-axis produces a thin disc without any depth.

Tutorial: Lathing a crucible

As an example of the Lathe modifier, you create a simple crucible, although you could produce any object that has a circular cross section. A *crucible* is a thick porcelain cup used to melt chemicals. I chose this as an example because it is simple (and saying "crucible" sounds much more scientific than "cup").

To create a crucible using the Lathe modifier, follow these steps:

1. Open the Crucible.max file from the Chap 16 folder in the downloaded content set.

   This file includes a rough profile cross-section line of the crucible that has been converted to an Editable Spline.

2. Select the line, and select the Modifiers→Patch/Spline Editing→Lathe menu command. Set the Degrees value in the Parameters rollout to **360**. Because you'll lathe a full revolution, you don't need to check the Cap options. In the Direction section, select the Y button (the Y-axis), and you're finished.

Figure 16.7 shows the finished product. You can easily make this into a coffee mug by adding a handle. To make a handle, simply loft an ellipse along a curved path.

FIGURE 16.7

Lathing a simple profile can create a circular object.

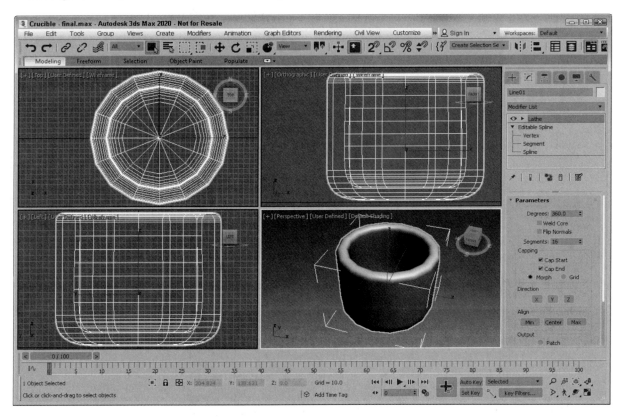

Bevel and Bevel Profile modifiers

Another common set of modifiers that can be used with splines and shapes are the Bevel and Bevel Profile modifiers.

Note

Neither the Bevel nor Bevel Profile modifiers are found in the Modifiers menu. To apply them, use the Modifier List found in the Modifier Stack. They are among the Object-Space modifiers.

Using the Bevel modifier, you can extrude and outline (scale) the shape in one operation. With the Bevel modifier, you can set the Height and Outline values for up to three different bevel levels. The Capping options let you select to cap either end of the beveled shape. The Cap Type can be either Morph or Grid. The Morph type is for objects that will be morphed. You can specify that the Surface use Linear or Curved Sides with a given number of segments. You also can select to Smooth Across Levels automatically. The Keep Lines from Crossing option avoids problems that may result from crossing lines.

The Bevel Profile modifier lets you select a spline to use for the bevel profile using the Classic mode. The Improved mode offers several distinct bevel presets, each with its own settings. The Improved mode also offers access to a Bevel Profile Editor, shown in Figure 16.8.

FIGURE 16.8

The Bevel Profile Editor lets you create custom bevels.

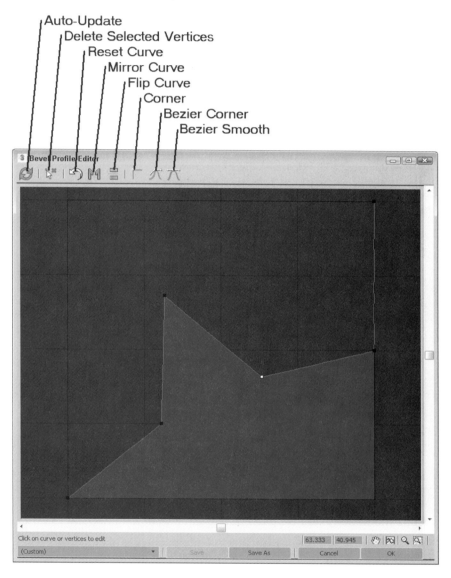

The Bevel Profile Editor shows the bevel applied to the current object. If you click on the line you can add a vertex and then drag the vertices to change the bevel. For each selected vertex, you can change it between a corner, Bezier Corner and Bezier Smooth tangent type. You can also mirror and flip the profile using the buttons at the top. In the lower left corner of the editor is a drop-down list of preset bevel curves. Click the Okay button to close the editor window and apply the bevel profile.

Tutorial: Modeling unique rings

You can create a simple ring using a Tube or Torus primitive object (or with an extruded donut shape), but if you want the ring to have a unique profile, the Bevel and Bevel Profile modifiers are what you need.

To create a couple of unique rings with the Bevel and Bevel Profile modifiers, follow these steps:

1. Select Create→Shapes→Donut, and drag in the Top viewport to create two donut objects that are positioned side by side. Set the Radius 1 value to **80** and the Radius 2 value to **75** for both rings.

2. Select the ring on the left in the Top viewport, open the Modify panel, and select the Bevel modifier from the Modifier List drop-down list in the Modifier Stack. In the Bevel Values rollout, set the Start Outline to **0**, the Height values for Levels 1, 2, and 3 to **20**, the Outline for Level 1 to **15**, and the Outline value for Level 3 to **–15**. Then enable the Smooth Across Levels option.

3. Select Create→Shapes→Line, and draw a profile curve in the Front viewport from bottom to top that is about the same height as the first ring. This curve doesn't have to be a closed spline.

4. Select the donut shape on the right, open the Modify panel, and choose the Bevel Profile modifier from the Modifier List in the Modifier Stack. In the Parameters rollout choose the Classic option, then click the Pick Profile button in the Classic rollout and select the profile curve.

Figure 16.9 shows the finished rings.

FIGURE 16.9

Bevels applied to a shape can give a unique profile edge.

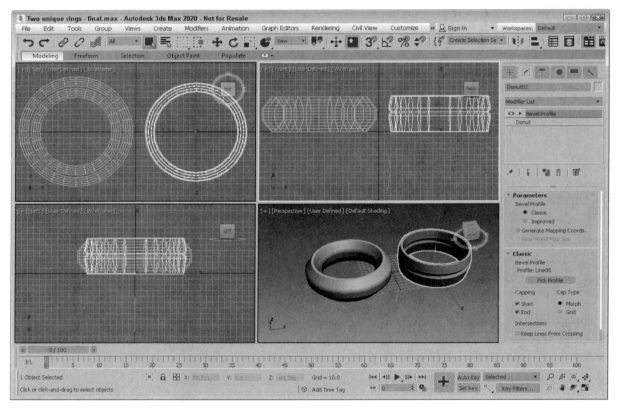

CrossSection modifier

The CrossSection modifier is one of two modifiers that collectively are referred to as the *surface tools*. The surface tools provide a way to cover a series of connected cross sections with a surface. The CrossSection modifier connects the vertices of several cross-sectional splines together with additional splines in preparation for the Surface modifier. These cross-sectional splines can have different numbers of vertices. Parameters include different spline types such as Linear, Smooth, Bézier, and Bézier Corner.

The second half of the surface tools is the Surface modifier.

Using Edit Geometry Modifiers

Most of the Mesh Editing modifiers (found in the Modifiers→Mesh Editing menu) are used to change the geometry of objects. Some of these modifiers, such as Extrude and Tessellate, perform the same operation as features available for the Editable Mesh or Editable Poly objects. Applying them as modifiers separates the operation from the base geometry.

You can find a more general explanation of modifiers in Chapter 11, "Accessing Subobjects and Modifiers and Using the Modifier Stack."

Cap Holes modifier

The Cap Holes modifier patches any holes found in a geometry object. Sometimes when objects are imported, they are missing faces. This modifier can detect and eliminate these holes by creating a face along open edges.

For example, if a spline is extruded and you don't specify Caps, the extruded spline has holes at its end. The Cap Holes modifier detects these holes and creates a Cap. Cap Holes parameters include Smooth New Faces, Smooth with Old Faces, and Triangulate Cap. Smooth with Old Faces applies the same smoothing group as that used on the bordering faces.

Chamfer modifier

The Chamfer modifier smoothes any hard edges by adding in additional edges that run parallel to the selected edge. The modifier gives you several Mitering options including Quad, Uniform and Tri and you can set an End Bias value. The End Bias moves the end point towards the next edge. The Amount value sets how deep the chamfer goes into the adjacent polygons and the Segments is the number of new parallel edges that are added in. Notice how the chamfer added to this duck model, in Figure 16.10, has smoothed the hard edge and made the whole object smoother (and safer for babies).

New Feature in 2020

Several new features have been added to the Chamfer modifier in 3ds Max 2020.

FIGURE 16.10

The Chamfer modifier smoothes hard edges.

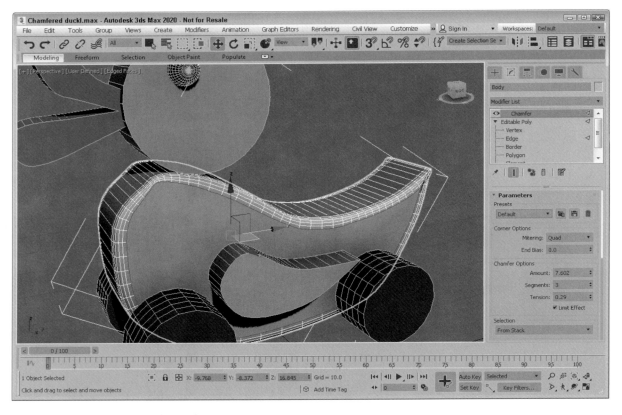

Delete Mesh modifier

You can use the Delete Mesh modifier to delete mesh subobjects. Subobjects that you can delete include Vertices, Edges, Faces, and Objects. The nice part about the Delete Mesh modifier is that it remains in the Modifier Stack and can be removed to reinstate the deleted subobjects.

The Delete Mesh modifier deletes the current selection as defined by the Mesh Select (or Poly Select) modifier. It can be used to delete a selection of Vertices, Edges, Faces, Polygons, or even the entire mesh if no subobject selection exists. The Delete Mesh modifier has no parameters.

Note

Even if the entire mesh is deleted using the Delete Mesh modifier, the object still remains. To completely delete an object, use the Delete key.

Extrude modifier

The Extrude modifier can be applied only to spline or shape objects, but the resulting extrusion can be a Patch, Mesh, or NURBS object. This modifier copies the spline, moves it a given distance, and connects the two splines to form a 3D shape. Parameters for this modifier include an Amount value, which is the distance to extrude, and the number of segments to use to define the height. The Capping options let you select a Start Cap and/or an End Cap using either a Morph or Grid option. The Morph option divides the caps into long, thin polygons suitable for morph targets, and the Grid option divides the caps into a tight grid of polygons suitable for deformation operations. The Cap fills the spline area and can be made as a Patch, Mesh, or NURBS object. Only closed splines that are extruded can be capped. You also can have mapping coordinates and Material IDs generated automatically. The Smooth option smoothes the extrusion.

Chapter 12, "Drawing and Editing 2D Splines and Shapes," includes a good example of the Extrude modifier.

Face Extrude modifier

The Face Extrude modifier extrudes the selected faces in the same direction as their normals. Face Extrude parameters include Amount and Scale values and an option to Extrude From Center. Figure 16.11 shows a mesh object with several extruded faces. The Mesh Select modifier was used to select the faces, and the extrude Amount was set to **30**.

FIGURE 16.11

Extruded faces are moved in the direction of the face normal.

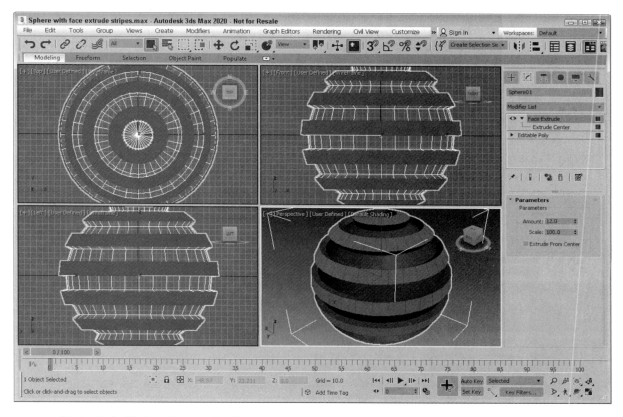

Tutorial: Extruding a bullet

As a simple example that uses a couple of Mesh modifiers, you'll create a single bullet using a hemisphere object. You can create this simple object in other ways, but this offers some good practice.

To create a bullet using the Face Extrude modifier, follow these steps:

1. Select Create→Standard Primitives→Sphere, and drag in the Top viewport to create a sphere object. Set the Radius value to **60** and the Hemisphere value to **0.5** to create half a sphere.

2. Right-click on the sphere object, and select Convert To→Editable Poly in the pop-up quad menu to convert the hemisphere to an Editable Poly object.

3. Select the Top viewport, click on the viewport label, and select Bottom from the pop-up menu (or press the B key) to switch to the Bottom view.

4. In the Selection rollout, click the Vertex button to enter Vertex subobject mode and enable the Ignore Backfacing option. Then select the single vertex in the center of the hemisphere, and press the Delete key.

5. In the Selection rollout, click the Border button to enter Border subobject mode, and then click the bottom edge of the hemisphere in the Front viewport to select the border of the hole that was created by deleting the center vertex. Then click the Cap button in the Edit Borders rollout.

6. Select the Polygon button in the Selection rollout to enter Polygon subobject mode and select the bottom polygon subobject in the Perspective viewport after rotating the object around. Then select Modifiers→Mesh Editing→Face Extrude to apply the Face Extrude modifier to the selected polygon face. Set the Amount value to **200**.

7. Select Create→Standard Primitives→Cylinder, and drag in the Bottom viewport to create a thin Cylinder object that is just wider than the extruded hemisphere. Then move the new Cylinder object until it is positioned at the end of the bullet object.

Figure 16.12 shows the completed simple bullet.

FIGURE 16.12

A simple bullet can be created by extruding one face of a hemisphere.

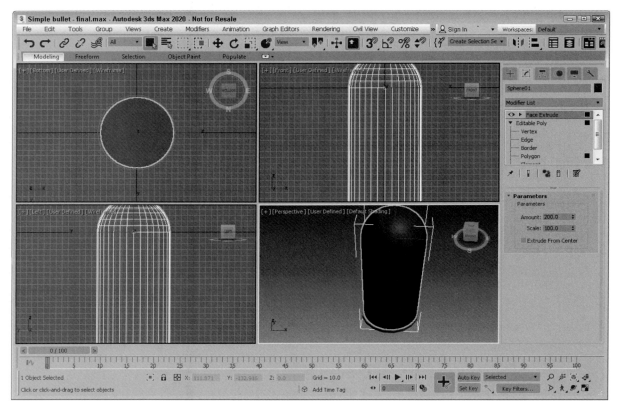

ProOptimizer modifier

Although the latest game consoles are getting much better at handling lots of polygons, sometimes you'll want to reduce a high-resolution model. For example, if you have a high-res statue in your scene, you might want the same statue to be lower-res when used as part of the background. 3ds Max offers an excellent modifier that enables you to reduce the total number of polygons in a model while maintaining its shape.

When first applied, the ProOptimizer modifier doesn't do anything. To use it, you need to select the settings first and then click the Calculate button. The settings found in the Optimization Options rollouts let you define which vertices can be removed. For example, you can set the optimization to Crunch Borders, Protect Borders, or Exclude Borders. A border is an edge connected to a single face. The Crunch Borders option makes the borders fair game for being optimized. This can yield the greatest amount of reduction but also can change the surface of the model. The Protect Borders option minimizes the amount of reduction at the borders, and the Exclude Borders option removes any border faces from being considered for reduction. The last choice limits the amount of reduction that is possible, but rigidly maintains the surface.

Other settings let you specify the Material Boundaries, Textures, and/or UV Boundaries off-limits. Additional settings protect any applied Vertex Colors and Normals. The Merge tools cause all vertices and/or faces within a given Threshold to be merged before optimizing the mesh. This helps to eliminate any extra vertices or co-planar faces that could cause problems. Finally, the Sub-Object Selection setting lets you preserve a given selection of vertices and makes it possible to optimize only a portion of the model.

The Symmetry options cause the modifier to equally reduce polygons on either side of a designated axis to maintain visual symmetry. There is also a Tolerance value for looking for symmetrical edges. Within the Advanced Options rollout, the Favor Compact Faces resists eliminating a face if it causes sharp pointed faces as a result. You also can select to Prevent Flipped Normals and the Lock Vertex Position option prevents any of the remaining vertices from being moved from their original locations, thus maintaining the model's shape.

After all the settings have been configured, click the Calculate button to run the optimization pass. The Statistics panel (located under the Calculate button) shows the number of points and faces in the model. By dialing down the Vertex % value, you can interactively reduce the total number of faces.

Figure 16.13 shows an alligator model that has been optimized using the ProOptimizer modifier. Notice the dramatic reduction in the number of faces from the left to the right. In the Modify panel, you can see that the number of faces has been reduced from 21,776 to 3,070 faces. (I guess that would make the gator on the right "lean and mean.")

Note

The Optimize and MultiRes modifiers are older versions of the ProOptimizer modifier and are maintained for compatibility with older files.

If you have an entire folder full of models that you want to optimize, you can use the Batch ProOptimizer utility. This utility is available from the Utilities panel and opens a dialog box where you can specify the source files to optimize, all the optimization settings, and the out filenames and locations. Once configured, you can simply click the OK button to optimize several model files in a batch process.

FIGURE 16.13

You can use the ProOptimizer modifier to reduce the complexity of the alligator model.

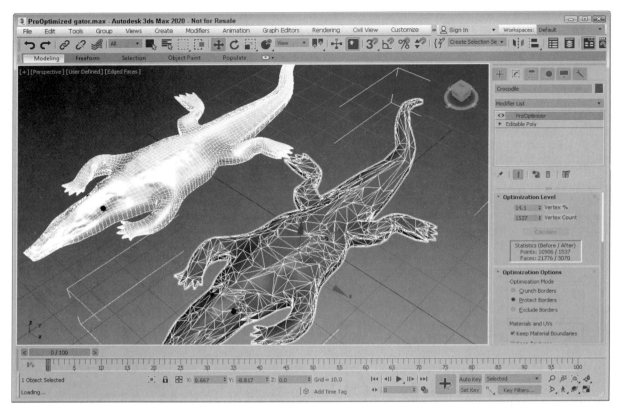

Tutorial: Creating a low-res mug

You can use the ProOptimizer modifier to dynamically dial down the resolution to exactly what you want. In this example, you use the ProOptimizer modifier on a high-res hand model of a mug. The mug weighs in at 5,668 polygons, which is a little heavy for any game engine.

To create an optimized mug, follow these steps:

1. Open the ProOptimizer mug.max file from the Chap 16 folder in the downloaded content set.

   This file contains a simple mug model.

2. With the mug selected, choose ProOptimizer from the Modifier List to apply the modifier to the mug model.

3. In the Optimization Options rollout, enable the Protect Borders option, the Merge Vertices option with the Threshold to **0.05**, and the Merge Faces option with the Threshold set to 0.5. Then click the Calculate button.

4. Create a copy of the mug by holding down the Shift key and dragging the mug to the right. In the Clone Options dialog box that appears, select Copy, name the clone **Mug – Lo**, and click OK.

5. With the cloned mug selected, set the Vertex % to **10**.

   Notice that the number of faces has dropped from 5,668 to 558.

Figure 16.14 shows the results of the ProOptimizer modifier. If you look closely, you can see that the mug on the right isn't as smooth, but it still looks pretty good and the game engine won't complain.

Caution

Reducing mesh density on the fly should not be done with animated objects such as characters because the mesh can become chaotic and dirty, resulting in poor deformations. However, the ProOptimizer modifier does work very well with static background objects.

FIGURE 16.14

You can use the ProOptimizer modifier to dynamically dial back the complexity of a mesh.

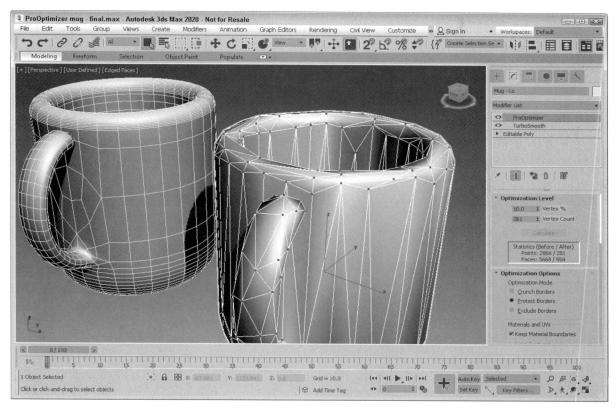

Using the Level of Detail utility

As a scene is animated, some objects are close to the camera, and others are far from it. Rendering a complex object that is far from the camera doesn't make much sense. Using the Level of Detail (LOD) utility, you can have 3ds Max render a simpler version of a model when it is farther from the camera and a more complex version when it is close to the camera.

To open the utility, click the More button in the Utilities panel and select the Level of Detail utility. A single rollout is loaded into the Utilities panel, as shown in Figure 16.15. To use this utility, you need to create several versions of an object and group them together. The Create New Set button lets you pick an object group from the viewports. The objects within the group are individually listed in the rollout pane.

If you select a listed object, you can specify the Threshold Units in pixels or as a percentage of the target image. For each listed item, you can specify minimum and maximum thresholds. The Image Output Size values are used to specify the size of the output image, and the different models used are based on the size of the object in the final image. The Display in Viewports check box causes the appropriate LOD model to appear in the viewport.

FIGURE 16.15

The Level of Detail utility (split into two parts) can specify how objects are viewed, based on given thresholds.

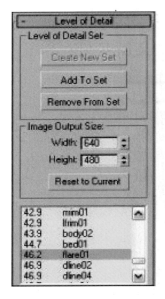

Quadify Mesh modifier

If you look at the individual shape of polygons in most models, you find either a triangle with three sides or a rectangle with four sides. Rectangle polygons are called quads, and they generally are nice to work with because they line up neatly into equal rows and columns. Triangles, on the other hand, flip back and forth even when lined up, making them more difficult to work with.

You can convert quads into triangles in many ways. This can be done in 3ds Max using a Triangulate command. This command simply divides the quads in half, but moving from triangles to quads isn't as easy and takes some clever calculating. But 3ds Max has figured it out and made it a feature with the Quadify Mesh modifier, shown in Figure 16.16.

The single parameter for this modifier is the Quad Size %. Larger values result in bigger and fewer quads, and smaller values increase the number of quads.

Tip

The Quadify Mesh modifier works well with shapes that have been extruded.

FIGURE 16.16

The Quadify Mesh modifier converts the model into regular-shaped quad faces.

Smooth modifier

You can use the Smooth modifier to auto-smooth an object. This automates the creation of different smoothing groups based on the angular threshold. Smooth parameters include options for Auto Smooth and Prevent Indirect Smoothing along with a Threshold value. The Parameters rollout also includes a set of 32 Smoothing Groups buttons labeled 1 through 32. These same Smoothing Groups are available as options for the Polygon and Element subobjects.

Symmetry modifier

The Symmetry modifier allows you to mirror a mesh object across a single axis. You also can select to Slice Along Mirror and weld along the seam with a defined Threshold. The gizmo for this modifier is a plane, which matches the selected axis and the arrow vector that extends from the plane.

Tutorial: Creating symmetrical antlers

Using symmetry, you can create one half of a model and then use the mirror tool to create the other half. If you need to see the changes as you make them, you can use the Symmetry modifier. In this example, I've taken the antlers off an elk model so you can practice putting them back on.

To create a set of symmetrical antlers, follow these steps:

1. Open the Elk with short antlers.max file from the Chap 16 folder in the downloaded content set.

    This file contains an elk model with its antlers removed.

2. With the elk object selected, select the Modifiers→Mesh Editing→Symmetry menu command to apply the Symmetry modifier. Its default setting has the X-axis selected, which places a plane running

down the center of the elk model, but you need to enable the Y-axis option and the Flip option so the symmetry extends outward from the head.

3. In the Modifier Stack, select the Editable Poly object and enable the Polygon subobject mode. Then rotate the view until you're looking at the elk from behind its head. Select a polygon on the left side of the elk where the antler should be located.

4. In the Edit Polygons rollout, click the settings dialog box button for the Bevel tool. Set the Height value to **2.0**, the Outline Amount to **-0.2**, and click OK.

5. With the polygon still selected, move the beveled polygon outward away from the elk's head.

6. Disable the polygon subobject mode and click on the Symmetry modifier in the Modifier Stack to see the symmetrical antler.

Tip

If you toggle the Show End Result button in the Modifier Stack, you can see the results of the Symmetry modifier in real time.

Figure 16.17 shows the results of the Symmetry modifier.

FIGURE 16.17

When you use the Symmetry modifier, you have to model certain objects only once.

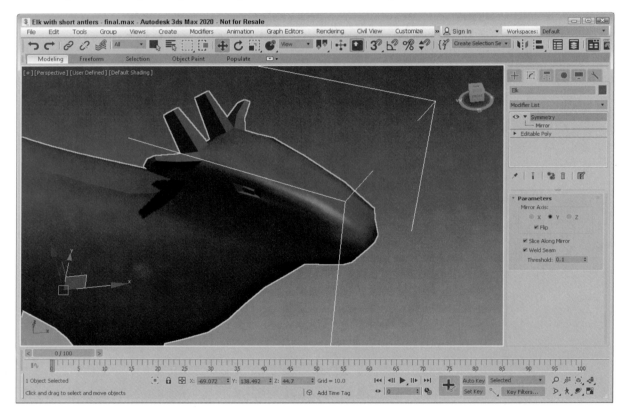

Tessellate modifier

You use the Tessellate modifier to subdivide the selected faces for higher-resolution models. You can apply tessellation to either Triangle or Polygonal faces. The Edge option creates new faces by dividing the face from the face center to the middle of the edges. The Face-Center option divides each face from the face center to the corners of the face. The Tension setting determines whether the faces are convex or concave. The Iterations setting is the number of times the modifier is applied.

Caution

Applying the Tessellate modifier to an object with a high Iterations value produces objects with many times the original number of faces.

Vertex Weld modifier

The Vertex Weld modifier is a simple modifier that welds all vertices within a certain Threshold value. This is a convenient modifier for cleaning up mesh objects.

The Vertex Paint modifier is covered in Chapter 22, "Painting in the Viewport Canvas and Rendering Surface Maps."

Using Parametric Deformer Modifiers

Perhaps the most representative group of modifiers are the Parametric Deformers. These modifiers affect the geometry of objects by pulling, pushing, and stretching them. They all can be applied to any of the modeling types, including primitive objects.

Note

In the upcoming figures, you might start to get sick of seeing the hammer model used over and over, but using the same model enables you to more easily compare the effects of the various modifiers, and it's more interesting to look at than a simple box.

Affect Region modifier

The Affect Region modifier can cause a local surface region to bubble up or be indented. Affect Region parameters include Falloff, Pinch, and Bubble values. The Falloff value sets the size of the affected area. The Pinch value makes the region tall and thin, and the Bubble value rounds the affected region. You also can select the Ignore Back Facing option. Figure 16.18 shows the Affect Region modifier applied to a Quad Patch with a Falloff value of 80 on the left and with a Bubble value of 1.0 on the right. The height and direction of the region are determined by the position of the modifier gizmo, which is a line connected by two points.

Note

The Affect Region modifier accomplishes the same effect as the Soft Selection feature, but Affect Region applies the effect as a modifier, making it easier to discard.

FIGURE 16.18

The Affect Region modifier can raise or lower the surface region of an object.

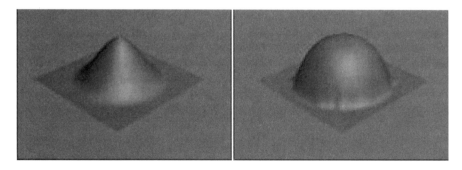

Bend modifier

The Bend modifier can bend an object along any axis. Bend parameters include the Bend Angle and Direction, Bend Axis, and Limits. The Bend Angle defines the bend in the vertical direction, and the Direction value defines the bend in the horizontal direction.

Limit settings are the boundaries beyond which the modifier has no effect. You can set Upper and Lower Limits relative to the object's center, which is placed at the object's pivot point. Limits are useful if you want the modifier applied to only one half of the object. The Upper and Lower Limits are visible as a simple plane on the modifier gizmo. For example, if you want to bend a tall cylinder object and have the top half remain straight, you can simply set an Upper Limit for the cylinder at the location where you want it to stay linear.

Note

Several modifiers have the option to impose limits on the modifier, including Upper and Lower Limit values.

The hammer in Figure 16.19 shows several bending options. The left hammer shows a Bend value of 75 degrees around the Z-axis, the middle hammer also has a Direction value of 60, and the right hammer has an Upper Limit of 8.

FIGURE 16.19

The Bend modifier can bend objects about any axis.

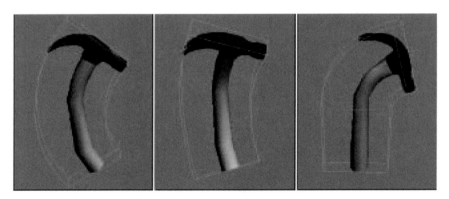

Displace modifier

The Displace modifier offers two unique sets of features. It can alter an object's geometry by displacing elements using a gizmo, or it can change the object's surface using a grayscale bitmap image. The Displace gizmo can have one of four different shapes: Planar, Cylindrical, Spherical, or Shrink Wrap. This gizmo can be placed exterior to an object or inside an object to push it from the inside.

Caution

In order for the Displace modifier to work, the surface requires a dense mesh, which can lead to very high polygon counts.

The Displace modifier parameters include Strength and Decay values. You also can specify the dimensions of the gizmo. A cylindrical-shaped gizmo can be capped or uncapped. The alignment parameters let you align the gizmo to the X-axis, Y-axis, or Z-axis, or you can align it to the current view. The rest of the parameters deal with displacing the surface using a bitmap image. Figure 16.20 shows a Quad Patch with the Plane-shaped gizmo applied with a Strength value of 25. To the right is a Quad Patch with the Sphere-shaped gizmo.

FIGURE 16.20

You can use the Displace modifier's gizmo as a modeling tool to change the surface of an object.

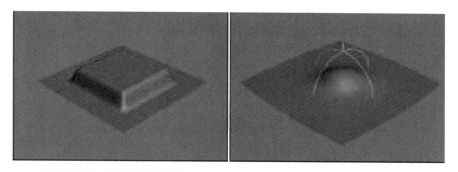

The Displace modifier also can alter an object's geometry using a grayscale bitmap image. This is similar to a Displacement map, which is covered in Chapter 18, "Adding Material Details with Maps." In many ways, the Displace gizmo works like the Conform compound object. You can learn about the Conform compound object in Chapter 15, "Working with Compound Objects."

Lattice modifier

The Lattice modifier changes an object into a lattice by creating struts where all the edges are located or by replacing each joint with an object. The Lattice modifier considers all edges as struts and all vertices as joints.

The parameters for this modifier include several options to determine how to apply the effect. These options include Apply to Entire Object, to Joints Only from Vertices, to Struts Only from Edges, or Both (Struts and Joints). If the Apply to Entire Object option isn't selected, the modifier is applied to the current subobject.

For struts, you can specify Radius, Segments, Sides, and Material ID values. You also can specify to Ignore Hidden Edges, to create End Caps, and to Smooth the Struts.

For joints, you can select Tetra, Octa, or Icosa types with Radius, Segments, and Material ID values. There are also controls for Mapping Coordinates.

Note

Although the joints settings enable you to select only one of three different types, you can use the Scatter compound object to place any type of object instead of the three defaults. To do this, apply the Lattice modifier and then select the Distribute Using All Vertices option in the Scatter Objects rollout.

Figure 16.21 shows the effect of the Lattice modifier. The left hammer has only joints applied, the middle hammer has only struts applied, and the right hammer has both applied.

FIGURE 16.21

The Lattice modifier divides an object into struts, joints, or both.

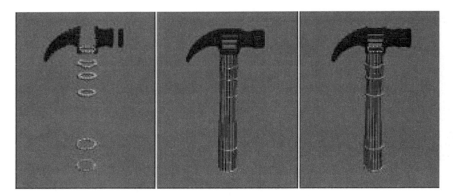

Mirror modifier

You can use the Mirror modifier to create a mirrored copy of an object or subobject. The Parameters rollout lets you pick a mirror axis or plane and an Offset value. The Copy option creates a copy of the mirrored object and retains the original selection.

Note

The Mirror modifier works the same as the Mirror command found in the Tools menu, but the modifier is handy if you want to be able to quickly discard the mirroring changes.

Noise modifier

The Noise modifier randomly varies the position of object vertices in the direction of the selected axes. Noise parameters include Seed and Scale values, a Fractal option with Roughness and Iterations settings, Strength about each axis, and Animation settings.

The Seed value sets the randomness of the noise. If two identical objects have the same settings and the same Seed value, they look exactly the same even though a random noise has been applied to them. If you alter the Seed value for one of them, they will look dramatically different.

The Scale value determines the size of the position changes, so larger Scale values result in a smoother, less rough shape. The Fractal option enables fractal iterations, which result in more jagged surfaces. If Fractal is enabled, Roughness and Iterations become active. The Roughness value sets the amount of variation, and the Iterations value defines the number of times to complete the fractal computations. More iterations yield a wilder or chaotic surface but require more computation time.

If the Animate Noise option is selected, the vertices positions will modulate for the duration of frames. The Frequency value determines how quickly the object's noise changes, and the Phase setting determines where the noise wave starts and ends.

Figure 16.22 shows the Noise modifier applied to several sphere objects. These spheres make the Noise modifier easier to see than on the hammer object. The left sphere has Seed, Scale, and Strength values along all three axes set to 1.0, the middle sphere has increased the Strength values to 2.0, and the right sphere has the Fractal option enabled with a Roughness value of 1.0 and an Iterations value of 6.0.

FIGURE 16.22

The Noise modifier can apply a smooth or wild look to your objects.

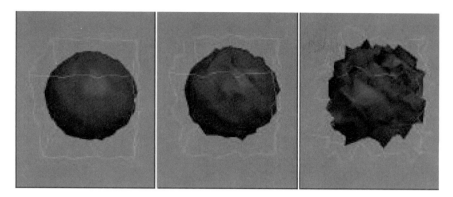

Push modifier

The Push modifier pushes an object's vertices inward or outward as if they were being filled with air. The Push modifier also has one parameter: the Push value. This value is the distance to move with respect to the object's center.

The positive Push value pushes the vertices outward away from the center, and a negative Push value pulls the vertices in toward the center. The Push modifier can increase the size of characters or make an object thinner by pulling its vertices in. Figure 16.23 shows the hammer pushed with 0.05, 0.1, and 0.15 values.

FIGURE 16.23

The Push modifier can increase the volume of an object.

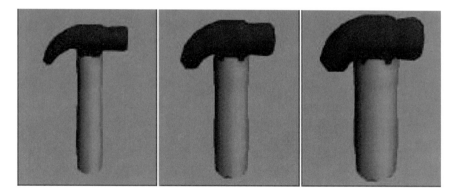

Preserve modifier

The Preserve modifier works to maintain Edge Lengths, Face Angles, and Volume as an object is deformed and edited. Before an object is modified, make an additional copy. Then edit one of the copies. To apply the Preserve modifier, click the Pick Original button; then click the unmodified object, and finally click the modified object. The object is modified to preserve the Edge Lengths, Face Angles, and Volume as defined in the Weight values. This helps prevent the topology of the modified object from becoming too irregular.

Note

In order for the Preserve modifier to keep objects and modifiers in check, it must be placed above the objects and modifiers it is watching.

The Iterations option determines the number of times the process is applied. You also can specify to apply to the Whole Mesh, to Selected Vertices Only, or to an Inverted Selection.

Relax modifier

The Relax modifier tends to smooth the overall geometry by separating vertices that lie closer than an average distance. Parameters include a Relax Value, which is the percentage of the distance that the vertices move. Values can range between 1.0 and −1.0. A value of 0 has no effect on the object. Negative values have the opposite effect, causing an object to become tighter and more distorted.

Tip

As you model, it is common for meshes to have sections that are too tight, which are dense locations in the mesh where a large concentration of vertices are close together. The Relax modifier can be used to cause the areas that are too tight to be relaxed.

The Iterations value determines how many times this calculation is computed. The Keep Boundary Points Fixed option removes any points that are next to an open hole. Save Outer Corners maintains the vertex position of corners of an object.

Ripple modifier

The Ripple modifier creates ripples across the surface of an object. This modifier is best used on a single object; if several objects need a ripple effect, use the Ripple Space Warp. The ripple is applied via a gizmo that you can control. Parameters for this modifier include two Amplitude values and values for the Wave Length, Phase, and Decay of the ripple.

The two amplitude values cause an increase in the height of the ripples opposite one another. Figure 16.24 shows the Ripple modifier applied to a simple Quad Patch with values of 10 for Amplitude 1 and a Wave Length value of 50. The right Quad Patch also has an Amplitude 2 value of 20.

FIGURE 16.24

The Ripple modifier can make small waves appear over the surface of an object.

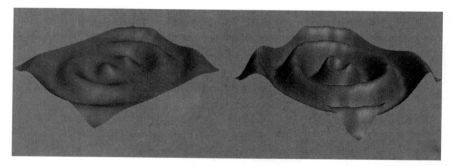

Shell modifier

When a mesh subobject is deleted, it leaves a hole in the surface that allows the inside of the object to be seen. This inside section doesn't have normals pointing the right direction, so the object appears blank unless the Force 2-Sided option in the Render Settings dialog box is selected. The Shell modifier makes an object into a shell with a surface on the inside and outside of the object.

For the Shell modifier, you can specify Inner and Outer Amount values. This is the distance from the original position that the inner or outer surfaces are moved. These values together determine how thick the shell is. The Bevel Edges and Bevel Spline options let you bevel the edges of the shell. By clicking on the Bevel Spline button, you can select a spline to define the bevel shape.

For each Material ID, you can use the Material ID for the inner section or the outer section. The Auto Smooth Edge lets you smooth the edge for all edges that are within the Angle threshold. The edges also can be mapped using the Edge Mapping options. The options include Copy, None, Strip, and Interpolate. The Copy option uses the same mapping as the original face, None assigns new mapping coordinates, Strip maps the edges as one complete strip, and Interpolate interpolates the mapping between the inner and outer mapping.

The last options make selecting the edges, the inner faces, or the outer faces easy. The Straighten Corners option moves the vertices so the edges are straight.

Tutorial: Making a character from a sphere

Creating a little game character from a sphere is a good example of how the Shell modifier can be used.

To use the Shell modifier to create a character, follow these steps:

1. Open the Gobbleman shell.max file from the Chap 16 folder in the downloaded content set.

   This file includes a simple sphere object that has had several faces deleted.

2. With the sphere object selected, select Modifiers→Parametric Deformers→Shell to apply the Shell modifier. Set the Outer Amount to **5.0**.

   This makes the hollow sphere into a thin shell. Notice that the lighting inside the sphere is now correct.

Figure 16.25 shows the resulting shell.

FIGURE 16.25

The Shell modifier can add an inside to hollow objects.

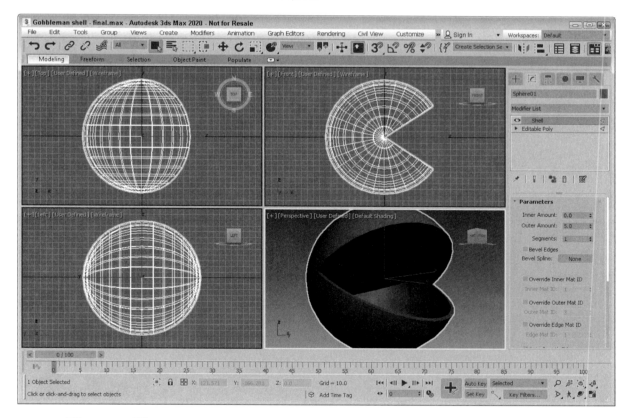

Slice modifier

You can use the Slice modifier to divide an object into two separate objects. Applying the Slice modifier creates a Slice gizmo. This gizmo looks like a simple plane and can be transformed and positioned to define the slice location. To transform the gizmo, you need to select it from the Stack hierarchy.

Note
You can use the Slice modifier to make objects slowly disappear a layer at a time.

The Slice parameters include four slice type options. Refine Mesh simply adds new vertices and edges where the gizmo intersects the object. The Split Mesh option creates two separate objects. The Remove Top and Remove Bottom options delete all faces and vertices above or below the gizmo intersection plane.

Using Triangular or Polygonal faces, you also can specify whether the faces are divided. Figure 16.26 shows the top and bottom halves of a hammer object. The right hammer is sliced at an angle.

FIGURE 16.26

The Slice modifier can cut objects into two separate pieces.

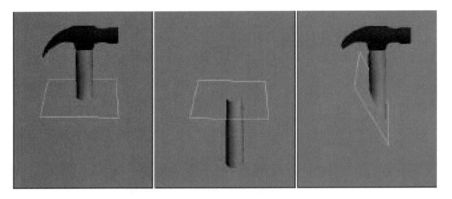

Note
Editable meshes also have a Slice tool that can produce similar results. The difference is that the Slice modifier can work on any type of object, not only on meshes.

Skew modifier

The Skew modifier changes the tilt of an object by moving its top portion while keeping the bottom half fixed. Skew parameters include Amount and Direction values, a Skew Axis, and Limits. Figure 16.27 shows the hammer on the left with a Skew value of 2.0, in the middle with a Skew value of 5, and on the right with an Upper Limit of 8.

FIGURE 16.27

You can use the Skew modifier to tilt objects.

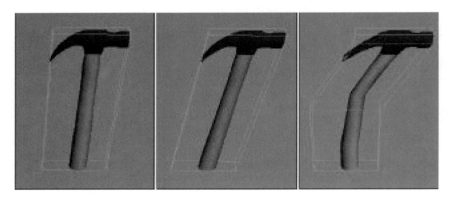

Stretch modifier

The Stretch modifier moves one axis in one direction while moving the other axes in the opposite direction, like pushing in on opposite sides of a balloon. Stretch parameters include Stretch and Amplify values, a Stretch Axis, and Limits.

The Stretch value equates the distance the object is pulled, and the Amplify value is a multiplier for the Stretch value. Positive values multiply the effect, and negative values reduce the stretch effect.

Figure 16.28 shows a Stretch value of 0.2 about the Z-axis applied to the hammer; the middle hammer also has an Amplify value of 2.0; and the right hammer has an Upper Limit value of 8.

FIGURE 16.28

The Stretch modifier pulls along one axis while pushing the other two.

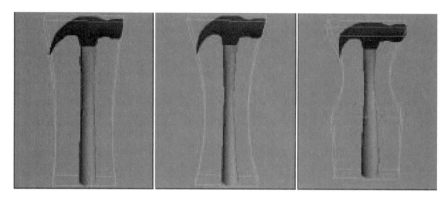

Spherify modifier

The Spherify modifier distorts an object into a spherical shape. The single Spherify parameter is the percent of the effect to apply. Figure 16.29 shows the hammer with Spherify values of 10, 20, and 30 percent.

The Spherify modifier is different from the Push modifier. Although they both are applied to the entire object, the Push modifier forces all vertices continuously outward, and the Spherify modifier uses a sphere shape as a limiting boundary. The visible difference is that the Spherify modifier creates a bulging effect.

FIGURE 16.29

The Spherify modifier pushes all vertices outward like a sphere.

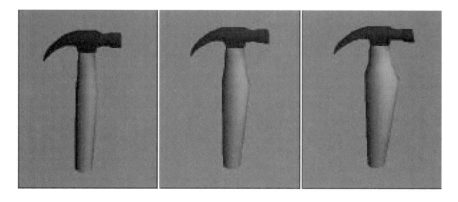

Tutorial: Making a fat crocodile

A good way to use the Spherify modifier is to add bulges to an object. For example, in this tutorial, you make a plump crocodile even fatter by applying the Spherify modifier.

To fatten up a crocodile character with the Spherify modifier, follow these steps:

1. Open the Fat crocodile.max file from the Chap 16 folder in the downloaded content set.

2. With the crocodile selected, select the Modifiers→Parametric Deformers→Spherify menu command to apply the Spherify modifier to the crocodile.

   The bulge appears around the object's pivot point.

3. In the Parameters rollout, set the Percent value to **15**.

Figure 16.30 shows the plump crocodile.

Note

The drawback of the Spherify modifier is that you have no control over its placement because there isn't a gizmo that you can position. One way around this problem is to use the Volume Select modifier to select a specific volume that is passed up the stack to the Spherify modifier.

FIGURE 16.30

The Spherify modifier can fatten up a crocodile.

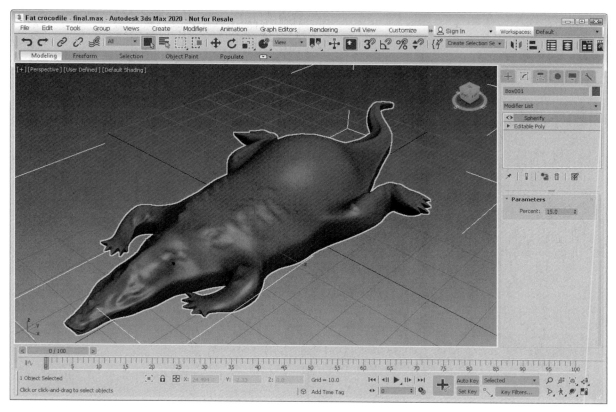

Squeeze modifier

The Squeeze modifier takes the points close to one axis and moves them away from the center of the object while it moves other points toward the center to create a bulging effect. Squeeze parameters include Amount and Curve values for Axial Bulge and Radial Squeeze, and Limits and Effect Balance settings.

The Effect Balance settings include a Bias value, which changes the object between the maximum Axial Bulge or the maximum Radial Squeeze. The Volume setting increases or decreases the volume of the object within the modifier's gizmo.

Axial Bulge is enabled with an Amount value of 0.2 and a Curve value of 2.0 in the left hammer in Figure 16.31; the middle hammer has also added Radial Squeeze values of 0.4 and 2.0; and the right hammer has an Upper Limit value of 8.

FIGURE 16.31

The Squeeze modifier can bulge or squeeze along two different axes.

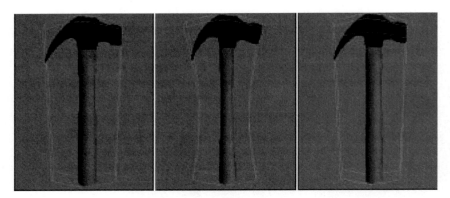

Twist modifier

The Twist modifier deforms an object by rotating one end of an axis in one direction and the other end in the opposite direction. Twist parameters include Angle and Bias values, a Twist Axis, and Limits.

The Angle value is the amount of twist in degrees that is applied to the object. The Bias value causes the twists to bunch up near the Pivot Point (for negative values) or away from the Pivot Point (for positive values).

The left hammer in Figure 16.32 shows a twist angle of 120 about the Z-axis, the middle hammer shows a Bias value of 20, and the right hammer has an Upper Limit value of 8.

FIGURE 16.32

The Twist modifiers can twist an object about an axis.

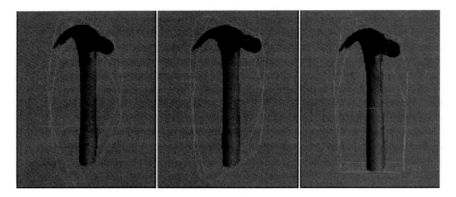

Taper modifier

The Taper modifier scales one end of an object. The tapered end is the end opposite the Pivot Point. Taper parameters include the Amount and Curve, Primary and Effect Axes, and Limits. The Amount value defines the amount of taper applied to the affected end. The Curve value bends the taper inward (for negative values) or outward (for positive values). You can see the curve clearly if you look at the modifier's gizmo. For example, you can create a simple vase or a bongo drum with the Taper modifier and a positive Curve value.

The Primary Axis defines the axis about which the taper is applied. The Effect axis can be a single axis or a plane, and the options change depending on your Primary Axis. This defines the axis or plane along which the object's end is scaled. For example, if the Z-axis is selected as the Primary Axis, then selecting the XY

Effect plane scales the object equally along both the X-axis and the Y-axis. Selecting the Y Effect axis scales the end only along the Y-axis. You also can select a Symmetry option to taper both ends equally. Taper limits work just like the Bend modifier.

The left hammer in Figure 16.33 shows a taper of 1.0 about the Z-axis; the middle hammer has a Curve value of –2; and the right hammer has the Symmetry option selected.

FIGURE 16.33

The Taper modifier can proportionally scale one end of an object.

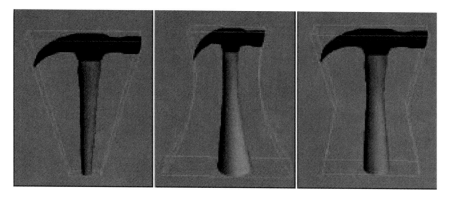

Tutorial: Creating a yo-yo

The Taper modifier can be used to create a variety of simple objects quickly, such as a yo-yo.

To create a yo-yo using the Taper modifier, follow these steps:

1. Select Create→Standard Primitives→Sphere, and drag in the Front viewport to create a sphere object.
2. With the sphere object selected, choose Modifiers→Parametric Deformers→Taper to apply the Taper modifier. Set the Taper Amount to **4.0** about the Primary Z-Axis and **XY** as the Effect plane, and enable the Symmetry option.

Figure 16.34 shows the resulting yo-yo; just add a string.

FIGURE 16.34

The Taper modifier can be used to create a simple yo-yo.

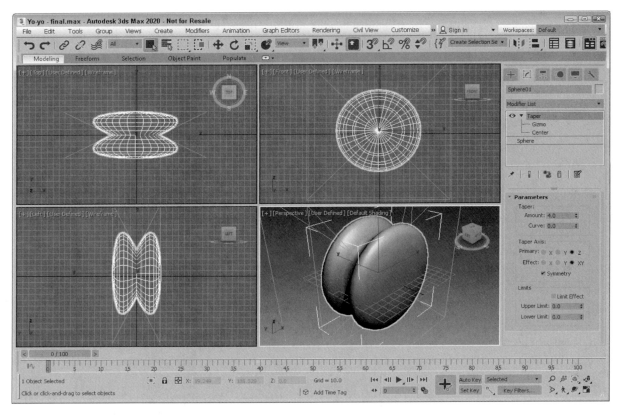

Substitute modifier

The Substitute modifier lets you place an object in the scene and substitute it with a higher-resolution object during render time. The substitute object may come from the scene or from an XRef file. To remove the substitute object, simply remove the Substitute modifier from the stack.

XForm modifier

The XForm modifier enables you to apply transforms such as Move, Rotate, and Scale to objects and/or subobjects. This modifier is applied by means of a gizmo that can be transformed using the transform buttons on the main toolbar. The XForm modifier has no parameters.

The XForm modifier solves a tricky problem that occurs during modeling. The problem happens when you scale, link, and animate objects in the scene, only to notice that the objects distort as they move. The distortion is caused because the transforms are the last action in the stack to be performed. So, when the object was first scaled and modifiers were applied, if you used the XForm modifier to do the scale transformation in the stack before the modifiers were applied, the object's children won't inherit the scale transformation.

Note

XForm is short for the word *transform*.

Wave modifier

The Wave modifier produces a wavelike effect across the surface of the object. All the parameters of the Wave Parameter are identical to the Ripple modifier parameters. The difference is that the waves produced by the Wave modifier are parallel, and they propagate in a straight line. Figure 16.35 shows the Wave modifier applied to a simple Quad Patch with values of 5 for Amplitude 1 and a Wave Length value of 50. The right Quad Patch also has an Amplitude 2 value of 20.

FIGURE 16.35

The Wave modifier produces parallel waves across the surface of an object.

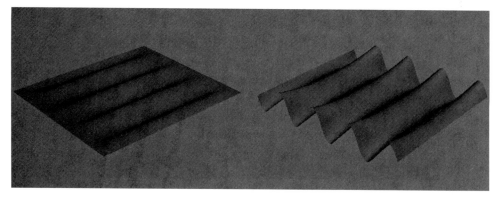

Tutorial: Waving a flag

The Wave modifier can add a gentle wave to an object such as a flag. If you animate the Phase value, you can show a flag unfurling in the breeze.

For a more realistic-looking flag, you can apply a Cloth modifier. See Chapter 47, "Working with Hair and Cloth," for more information on the Cloth modifier.

To animate a flag waving with the Wave modifier, follow these steps:

1. Open the Waving US flag.max file from the Chap 16 folder in the downloaded content set.

   This file includes a simple flag and flagpole made from primitive objects.

2. With the flag selected, select the Modifiers→Parametric Deformers→Wave menu command to apply the Wave modifier to the flag.

3. Notice how the waves run from the top of the flag to the bottom. You can change this by rotating the gizmo. Click the drop-down arrow icon to the left of the Wave modifier in the Modifier Stack and select the Gizmo subobject. With the Select and Rotate tool (E), rotate the gizmo 90 degrees and then scale the gizmo with the Select and Scale tool (R) so it covers the flag object. Click the Gizmo subobject again to deselect gizmo subobject mode.

4. Set Amplitude 1 to **25**, Amplitude 2 to **0**, and the Wave Length to **50**. Then click the Auto Key button (N), drag the Time Slider to frame 100, and set the Phase value to **4**. Click the Auto Key button (N) again to exit key mode.

Figure 16.36 shows the waving flag.

FIGURE 16.36

The Wave modifier can gently wave a flag.

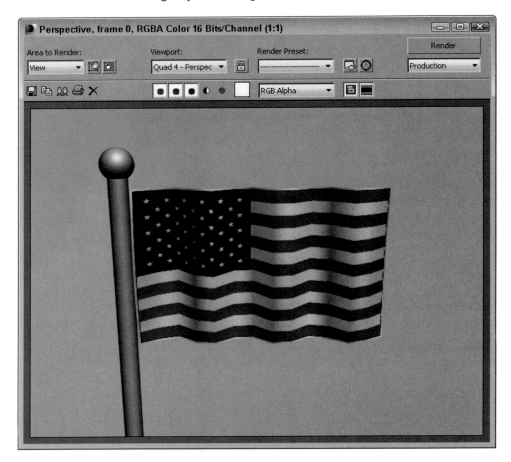

Using Free Form Deformer Modifiers

The Free Form Deformers category of modifiers causes a lattice to appear around an object. This lattice is bound to the object, and you can alter the object's surface by moving the lattice control points. Modifiers include FFD (Free Form Deformation) and FFD (Box/Cyl).

FFD (Free Form Deformation) modifier

The Free Form Deformation modifiers create a lattice of control points around the object. The object's surface can deform the object when you move the control points. The object is deformed only if it is within the volume of the FFD lattice. The three different resolutions of FFDs are 2 x 2 x 2, 3 x 3 x 3, and 4 x 4 x 4.

You also can select to display the lattice or the source volume, or both. If the Lattice option is disabled, only the control points are visible. The Source Volume option shows the original lattice before any vertices were moved.

The two deform options are Only In Volume and All Vertices. The Only In Volume option limits the vertices that can be moved to the interior vertices only. If the All Vertices option is selected for the FFD Box and FFD Cylinder modifiers, the Falloff value determines the point at which vertices are no longer affected by the FFD. Falloff values can range between 0 and 1. The Tension and Continuity values control how tight the lines of the lattice are when moved. Falloff, Tension and Continuity values are only available for the FFD Box and FFD Cylinder modifiers.

The three buttons at the bottom of the FFD Parameters rollout help in the selection of control points. If the All X button is selected, when a single control point is selected, all the adjacent control points along the X-axis are also selected. This feature makes selecting an entire line of control points easier. The All Y and All Z buttons work in a similar manner in the other dimensions.

Use the Reset button to return the volume to its original shape if you make a mistake. The Conform to Shape button sets the offset of the Control Points with Inside Points, Outside Points, and Offset options.

To move the control points, select the Control Points subobject. This enables you to alter the control points individually.

FFD (Box/Cyl) modifiers

The FFD (Box) and FFD (Cyl) modifiers can create a box-shaped or cylinder-shaped lattice of control points for deforming objects. The Set Number of Points button enables you to specify the number of points to be included in the FFD lattice. Figure 16.37 shows how you can use the FFD modifier to distort the hammer by selecting the Control Point's subobjects. The left hammer is distorted using a 2 x 2 x 2 FFD, the middle hammer has a 4 x 4 x 4 FFD, and the right hammer is surrounded with an FFD (Cyl) modifier.

FIGURE 16.37

The FFD modifier changes the shape of an object by moving the lattice of Control Points that surround it.

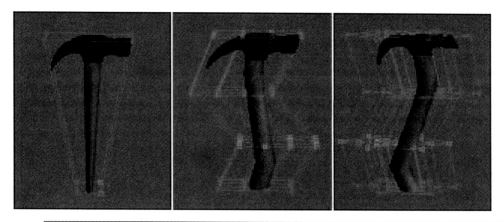

The FFD (Box) and FFD (Cyl) lattices are also available as Space Warps. To learn more about Space Warps, see Chapter 43, "Using Space Warps."

Tutorial: Modeling a tire striking a curb

The FFD modifiers are great for changing the shape of a soft-body object being struck by a solid object. Soft-body objects deform around the rigid object when they make contact. In this tutorial, you deform a tire hitting a curb.

To deform a tire striking a curb using an FFD modifier, follow these steps:

1. Open the Tire hitting a curb.max file from the Chap 16 folder in the downloaded content set.
   This file includes a simple tube object and a curb.
2. With the tire selected, choose the Modifiers→Free Form Deformers→FFD Cylinder menu option.
   A cylinder gizmo appears around the tire.
3. Click the drop-down arrow icon next to the FFD name in the Modifier Stack, and select the Control Points subobject from the Modifier Stack. Then select all the center control points in the Left viewport, and scale the control points outward with the Select and Scale tool (R) to add some roundness to the tire.

4. Then select all the control points in the lower-left corner of the Front viewport, and move these points diagonally up and to the right until the tire's edge lines up with the curb.

Figure 16.38 shows the tire as it strikes the hard curb.

FIGURE 16.38

This tire is being deformed via an FFD modifier.

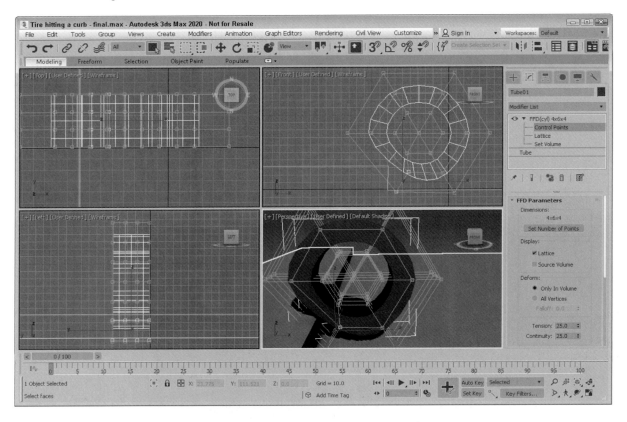

Miscellaneous Modifiers

Several of the Mesh Editing modifiers are unique, special-purpose modifiers. The Edit Normals modifier, for example, lets you change the direction of face normals, which doesn't really change the geometry, but it can have a big impact on how the object is smoothed and shaded.

Edit Normals

The Edit Normals modifier enables you to select and move normals. Normals appear as blue lines that extend from the vertex of each face. The Edit Normal modifier includes a Normal subobject that you can select and change its direction. To move (or rotate) a normal, you can use the transform tools on the main toolbar. If your viewport has shading enabled, you can see the effect of moving the normals.

Tip

The Edit Normals modifier can be used to create the illusion of surface geometry. For example, by selecting and flipping specific localized normals, you can create what look like dents in a metal plate.

You can select normals by Normal, Edge, Vertex, or Face. Selecting normals by Face, for example, selects all normals attached to a face when you click on a face. Moving a normal then moves all selected normals together.

You also have options to Ignore Backfacing and to Show Handles. Handles appear as small squares at the top of the normal vector. The Display Length value defines the length of the normals displayed in the viewports. Figure 16.39 shows all the normals extending from a simple Sphere object.

FIGURE 16.39

The Edit Normals parameters let you work with normals.

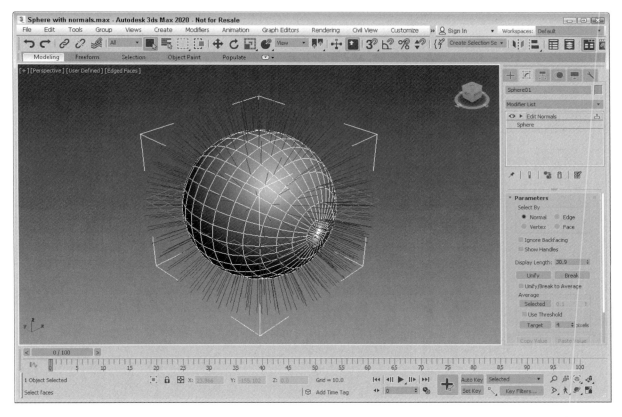

The modifier also includes buttons to Unify and to Break selected normals. Unify causes all selected normals to be combined to a single normal, and Break splits unified normals into their separate components again. The direction of the unified normal is an average of the surface points when unified or an average of the normals if the Unify/Break to Average option is enabled.

The Selected button in the Average section averages all selected normals that are within the specified value if the Use Threshold value is enabled. If it is not enabled, all selected normals are averaged. The Average Target button lets you interactively select a normal and then click on another normal to average. The Target normal must be within the specified Target value.

The direction of selected normals can be copied and pasted between normals. The Specify button marks a normal as a Specified normal. These normals appear cyan in the viewport and ignore any smoothing group information associated with the vertex. Explicit normals are green in the viewport, which denotes a normal that has deviated from its regular position. The Make Explicit button can be used to make normals explicit, thereby removing them from the normal computation task. The Reset button returns a normal to its regular type and position. At the bottom of the Parameters rollout is an information line that displays which normal is selected or how many normals are selected.

Normal modifier

The Normal modifier is the precursor to the Edit Normals modifier. It enables object normals to be flipped or unified. When some objects are imported, their normals can become erratic, producing holes in the geometry. By unifying and flipping the normals, you can restore an object's consistency. This modifier includes only two options: Unify Normals and Flip Normals.

STL Check modifier

The STL Check modifier checks a model in preparation for exporting it to the StereoLithography (STL) format. STL files require a closed surface: Geometry with holes or gaps can cause problems. Any problems are reported in the Status area of the Parameters rollout.

This modifier can check for several common errors, including Open Edge, Double Face, Spike, or Multiple Edge. Spikes are island faces with only one connected edge. You can select any or all of these options. If found, you can have the modifier select the problem Edges or Faces, or neither, or you can change the Material ID of the problem area.

Subdivision Surface Modifiers

The Modifiers menu also includes a submenu of modifiers for subdividing surfaces. These include the Crease, Crease Set, MeshSmooth, OpenSubDiv, TurboSmooth and HSDS modifiers. You can use these modifiers to smooth and subdivide the surface of an object. Subdividing a surface increases the resolution of the object, allowing for more detailed modeling.

Crease and Crease Set modifier

Whenever an object is smoothed by means of one of the Subdivision Surface modifiers, the entire surface is smoothed. The Crease Set modifier lets you select and define which edges and vertices remain unsmoothed by passing them up the stack. The Crease Set includes sub-objects for selecting and adding to a set those sub-object to remain unsmoothed. There are also options to grow and shrink the selection and to select entire edge loops and rings.

The Crease modifier lets you set the amount of crease that is applied from 0 to 1. You can also access the Crease Explorer that includes a list of all the available crease sets.

HSDS modifier

You use the HSDS (Hierarchical SubDivision Surfaces) modifier to increase the resolution and smoothing of a localized area. It works like the Tessellate modifier, except that it can work with small subobject sections instead of the entire object surface. The HSDS modifier lets you work with Vertex, Edge, Polygon, and Element subobjects. After a subobject area is selected, you can click the Subdivide button to subdivide the area. Each time you press the Subdivide button, the selected subobjects are subdivided again, and each subdivision level appears in the list above the Subdivide button.

Using the subdivision list, you can move back and forth between the various subdivision hierarchy levels. When edges are selected, you can specify a Crease value to maintain sharp edges. In the Advanced Options rollout, you can select to Smooth Result, Hide, or Delete Polygon. The Adaptive Subdivision button opens the Adaptive Subdivision dialog box, in which you can specify the detail parameters. This modifier also includes a Soft Selection rollout.

MeshSmooth modifier

The MeshSmooth modifier smoothes the entire surface of an object by applying a chamfer function to both vertices and edges at the same time. This modifier has the greatest effect on sharp corners and edges. With this modifier, you can create a NURMS object. NURMS stands for Non-Uniform Rational MeshSmooth. NURMS can weight each control point. The Parameters rollout includes three MeshSmooth types: Classic,

468

NURMS, and Quad Output. You can set it to operate on triangular or polygonal faces. Smoothing parameters include Strength and Relax values.

Settings for the number of Subdivision Iterations to run and controls for weighting selected control points are also available. Update Options can be set to Always, When Rendering, and Manually using the Update button. You also can select and work with either Vertex or Edge subobjects. These subobjects give you local control over the MeshSmooth object. Included within the Local Control rollout is a Crease value, which is available in Edge subobject mode. Selecting an Edge subobject and applying a 1.0 value causes a hard edge to be retained while the rest of the object is smoothed. The MeshSmooth modifier also makes the Soft Selection rollout available. The Reset rollout is included to quickly reset any crease and weight values.

OpenSubDiv modifier

The OpenSubDiv modifier converts the object into an Open Sub-Division type object. This object type is unique in that it can dynamically increase the number of polygons in an area in order to make the object smooth. It has an Iterations value that is used to set the general density of the polygons. You can also set the Operating Mode to GPU Display, which offloads the display of the object to the system's GPU processor. The Boundary conditions mark the areas where the edge creases are maintained.

Caution
Be careful when adjusting the Iterations value because even a simply object with a high Iteration value will quickly become unyielding and will crash your system.

If you enable the Adaptive button, then the number of display of the polygons will decrease as the camera moves further away from the object. This insures quick display of the object when not viewed up close.

Note
The Adaptive button will not be enabled until the GPU Display mode is selected and the changes won't be visible unless the Iterations are set to 2 or higher.

TurboSmooth modifier

The TurboSmooth modifier works just like the MeshSmooth modifier, except that it is much faster and doesn't require as much memory.

Tutorial: Smoothing a birdbath

One effective way to model is to block out the details of a model using the Editable Poly features and then smooth the resulting model using the TurboSmooth modifier. This gives the model a polished look and increases the resolution.

To create a smoothed birdbath object, follow these steps:

1. Open the Birdbath.max file from the Chap 16 folder in the downloaded content set.

   This file includes a simple birdbath created by selecting and scaling rows of cylinder vertices. The water is simply an inverted cone.

2. Select and clone the existing birdbath as a copy by pressing the Shift key and moving the birdbath.

3. Select the cloned birdbath, and apply the TurboSmooth modifier with the Modifiers→Subdivision Surfaces→TurboSmooth menu.

4. In the TurboSmooth rollout, set the Iterations value to **2**.

   Notice that the entire birdbath on the right is smooth and the resolution is greatly increased, as shown in Figure 16.40.

FIGURE 16.40

The TurboSmooth modifier can make a model flow better.

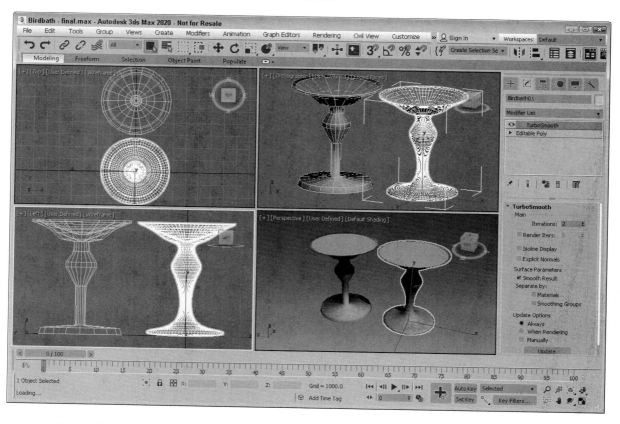

The Basics of Deformation Painting

The first thing to remember about the Paint Deformation feature is that it is available only for Editable Poly objects (or objects with the Edit Poly modifier applied). When an Editable Poly object is selected, the Paint Deformation rollout appears at the very bottom of the Modify panel in the Command Panel.

Painting deformations

At the top of the Paint Deformation rollout are three buttons used to select the type of deformation brush to use. These three brushes are the Push/Pull brush, the Relax brush, and the Revert brush.

When one of these brushes is selected, the mouse cursor changes to a circular brush, shown in Figure 16.41, which follows the surface of the object as you move the mouse over the object. A single line points outward from the center of the circle in the direction of the surface normal. Dragging the mouse affects the surface in a certain manner, depending on the brush that is selected.

FIGURE 16.41

The Paint Deformation brush looks like a circle that follows the surface.

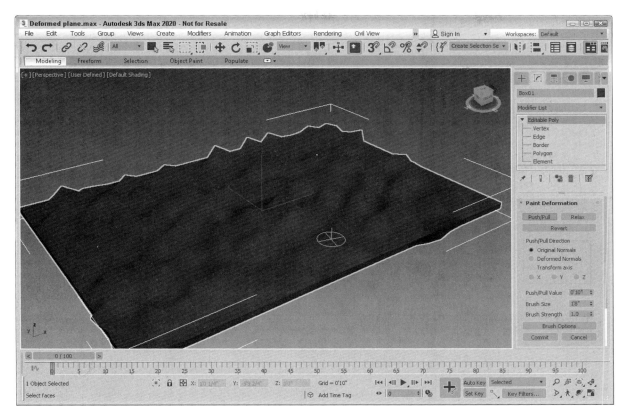

Dragging the Paint Deformation brush over the object surface deforms the surface by moving the vertices within the brush's area. The direction that the vertices are moved follows the surface normals by default, or you can have the deformation follow a deformed normal or along a specified transform axis. For example, if you select to deform vertices along the X-axis, all vertices underneath the brush are moved along the X-axis as the brush is dragged over the surface.

Note

Deformations created using the Paint Deformation brushes cannot be animated. For objects with the Edit Poly modifier applied, the Paint Deformation brushes are disabled when in Animate mode.

The Push/Pull value determines the distance that the vertices are moved, thereby setting the amount of the deformation. The Brush Size sets the size (or radius) of the brush and determines the area that is deformed. The Brush Strength value sets the rate at which the vertices are moved. For example, if the Push/Pull Value is set to 100 mm, a Brush Strength value of 1.0 causes the vertices directly under the brush center to move 100 mm; a Brush Strength value of 0.4 causes the same vertices to move only 40 mm.

Tip

Holding down Shift+Alt while dragging in the viewports lets you interactively change the Brush Strength value.

Accessing brush presets

If you right-click the main toolbar away from any of the buttons, you can access the Brush Presets toolbar, shown in Figure 16.42, from the pop-up menu.

FIGURE 16.42

The Brush Presets toolbar lets you quickly select from a selection of predefined brushes.

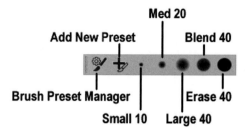

The first toolbar button opens the Brush Preset Manager, shown in Figure 16.43. From this interface, you can choose to create preset brushes for each of the different features that use brushes—Vertex Paint, Paint Deformation, Paint Soft Selection, , and Paint Skin Weights. The Add button works the same as the Add New Preset toolbar button. It opens a dialog box where you can name the new preset. The new preset is then added to the list of presets.

Note

The brushes in the Brush Presets toolbar become active only when one of the features that use brushes is selected.

FIGURE 16.43

The Brush Preset Manager lets you create new preset brushes.

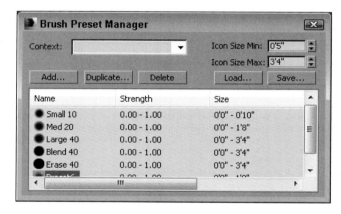

When a brush preset is selected on the Brush Presets toolbar, you can change its attributes using the Painter Options dialog box. The Paint Deformation rollout also includes a button for accessing the Painter Options dialog box. Any changes to the brush attributes are automatically updated in the Brush Preset Manager. The Load and Save buttons in the Brush Preset Manager dialog box let you save and load brush preset sets. The brush options are covered later in this chapter.

Using the Deformation Brushes

The Push/Pull brush may be used to pull vertices away from the object surface or to indent the surface by moving the surface toward the object's center. The difference is determined by the Push/Pull value. Positive values pull vertices, and negative values push vertices.

Tip

Holding down the Alt key while dragging reverses the direction of the Push/Pull brush, causing a pull brush to push and vice versa.

Controlling the deformation direction

By default, dragging over vertices with the Push/Pull brush causes the affected vertices to be moved inward or outward along their normals. If you drag over the same vertices several times, they are still deformed using the original face normals.

The Deformed Normals option causes the vertices to be moved in the direction of the normal as the normals are deformed. Using the Original Normals option causes the deformed area to rise from the surface like a hill with a gradual increasing height. The Deformed Normals option causes the deformed area to bubble out from the surface.

The Transform Axis option causes the vertices to be moved in the direction of the selected transform axis. This option is useful if you want to skew or shift the deformed area.

Limiting the deformation

If a subobject selection exists, the vertices that are moved are limited to the subobject area that is selected. You can use this to your advantage if you want to make sure that only a certain area is deformed.

Committing any changes

After you make some deformation changes, the Commit and Cancel buttons become active. Pressing the Commit button makes the changes permanent, which means that you can no longer return the vertices to their original location with the Revert brush. The Cancel button rejects all the recent deformation changes.

Using the Relax and Revert brushes

The Relax brush provides a much more subtle change. It moves vertices that are too close together farther apart, causing a general smoothing of any sharp points. It works the same way as the Relax feature for the Editable Poly object and the Relax modifier.

The Revert brush is used to return to their original position any vertices that have moved. For example, if you pushed and pulled several vertices, the Revert brush can undo all of these changes for the area under the brush cursor.

Tip

Holding down the Ctrl key while dragging with the Push/Pull brush lets you temporarily access the Revert brush.

Tutorial: Adding veins to a forearm

The Paint Deformation feature is very useful in adding surface details to organic objects such as the veins of a forearm.

To add veins to a forearm object, follow these steps:

1. Open the Forearm with veins.max file from the Chap 16 folder in the downloaded content set.

   The polygons that make up the forearm object have been selected and tessellated to increase its resolution.

2. With the forearm object selected, open the Modify panel and in the Paint Deformation rollout, click the Relax button. Set the Brush Size to **1.0**, and drag over the entire forearm.

   This smoothes out some of the vertical lines that run along the forearm.

3. Click the Push/Pull button, and set the Push/Pull Value to **0.15**, the Brush Size to **0.08**, and the Brush Strength to **0.5**. Then draw in some veins extending from the elbow toward the hand.

4. Lower the Brush Strength value to **0.25**, and extend the vein farther down the arm. Then drop the Brush Strength to **0.1**, and finish the veins.

5. With the Push/Pull brush still selected, hold down the Alt key and drag near the wrist to indent the surface around the area where the hand tendons are located.

Figure 16.44 shows the resulting forearm.

FIGURE 16.44

The Paint Deformation brushes are helpful in painting on raised and indented surface features.

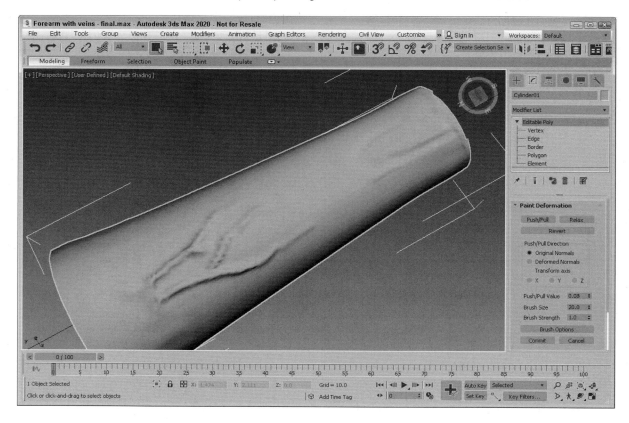

Setting Painter Options

At the bottom of the Paint Deformation rollout is a button labeled Brush Options. Clicking this button opens the Painter Options dialog box, shown in Figure 16.45. Using this dialog box, you can set several customized brush options, including the sensitivity of the brush.

The Painter Options dialog box is also used by the brushes to paint vertex colors for the Vertex Paint modifier and to paint skin weights as part of the Skin modifier. The Vertex Paint modifier is covered in Chapter 22, "Painting in the Viewport Canvas and Rendering Surface Maps," and the Skin modifier is covered in Chapter 39, "Skinning Characters."

FIGURE 16.45

The Painter Options dialog box includes a graph for defining the minimum and maximum brush strengths and sizes.

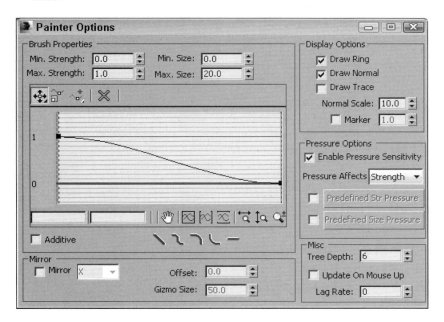

The Min/Max Strength and Min/Max Size values determine the minimum and maximum weight values and paint gizmo sizes. You can define the brush falloff using the curve. This keeps the weights from making an abrupt change (muscles tend to look funny when this happens). Under the curve are several buttons for quickly defining the shape of the falloff curve, including Linear, Smooth, Slow, Fast, and Flat.

The Display Options section includes options that determine the look of the painting gizmo. The Draw Ring, Draw Normal, and Draw Trace options make a ring; the surface normal or an arrow showing the trace direction appears. The Normal can be scaled, and the Marker option displays a small circular marker at the end of the normal.

The Pressure Options let you paint using a graphics tablet with the pressure applied to affect the Strength, Size, or a combination. You can Enable Pressure Sensitivity for the brush gizmo. The options include None, Strength, Size, and Both. Using the graph, you can predefine Strength and Size pressure curves and then select to use them.

The Mirror option paints symmetrically on the opposite side of the gizmo across the specified axis. You also can set an Offset and the Gizmo Size. This is handy for muscles that you want to deform symmetrically.

In the Miscellaneous section, the Tree Depth, Update on Mouse Up, and Lag Rate options control how often the scene and the painted strokes are updated.

Summary

With the modifiers contained in the Modify panel, you can alter objects in a vast number of ways. Modifiers can work with every aspect of an object, including geometric deformations, materials, and general object maintenance. In this chapter, you looked at the various modifiers used to help modeling.

The Paint Deformation feature is a welcome addition, allowing you to add surface details using an intuitive and easy-to-use interface. 3ds Max includes several unique modifiers that apply specifically to mesh objects.

This chapter covered these topics:

* Using the Edit Mesh and Edit Poly modifiers to edit a primitive object using the Editable Mesh and Editable Poly features while maintaining its parametric nature
* Applying modifiers to splines
* How several mesh modifiers are used to edit surface geometry
* Using the Parametric Deformer and FFD modifiers
* How to edit normals and what the effects are
* Working with the Subdivision Surface modifiers to smooth mesh objects
* Using the various Paint Deformation brushes
* Setting brush options with the Painter Options dialog box

If you're tired of the dull gray default objects, then you'll be happy to know that the next chapter shows how to create and add materials to objects.

Part IV
Applying Materials and Textures

IN THIS PART

Chapter 17

Creating and Applying Standard Materials with the Slate Material Editor

Materials are used to dress, color, and paint objects. Just as materials in real life can be described as scaly, soft, smooth, opaque, or blue, materials applied to 3D objects can mimic properties such as color, texture, transparency, shininess, and so on. In this chapter, you learn the basics of working with materials and all the features of the Slate Material Editor.

After you're familiar with the Material Editor, this chapter gives you a chance to create some simple original materials and apply them to objects in the scene. The simplest material is based on the Standard material type, which is the default material type.

Materials come in libraries, and just like the school library, finding exactly what you want can be tough. That's where the various material interfaces come in. Using the Material/Map Browser and the Material Explorer interface, you can browse through the available materials and quickly filter and find just what you need.

Understanding Material Properties

Before jumping into the Material Editor, let's take a close look at the type of material properties that you will deal with. Understanding these properties will help you as you begin to create new materials.

Until now, the only material property that has been applied to an object has been the default object color, randomly assigned by the Autodesk® 3ds Max® 2020 software. The Material Editor can add a whole new level of realism using materials that simulate many different types of physical properties.

Colors

Color is probably the simplest material property and the easiest to identify. However, unlike the object color defined in the Create and Modify panels, there isn't a single color swatch that controls an object's color.

Consider a basket of shiny red apples. When you shine a bright blue spotlight on them, all the apples turn purple because the blue highlights from the light mix with the red of the apple's surface. So, even if the apples are assigned a red material, the final color in the image might be very different because the light makes the color change.

Within the Material Editor are several different color swatches that control different aspects of the object's color. The following list describes the types of color swatches that are available for various materials:

* **Ambient:** Defines an overall background lighting that affects all objects in the scene, including the color of the object when it is in the shadows. This color is locked to the Diffuse color by default so that they are changed together.

* **Diffuse:** The surface color of the object in normal, full, white light. The normal color of an object is typically defined by its Diffuse color.

* **Specular:** The color of the highlights where the light is focused on the surface of a shiny material.

* **Self-Illumination:** The color that the object glows from within. This color takes over any shadows on the object.

* **Filter:** The transmitted color caused by light shining through a transparent object.

* **Reflect:** The color reflected by a raytrace material to other objects in the scene.

* **Luminosity:** Causes an object to glow with the defined color. It is similar to Self-Illumination color but can be independent of the Diffuse color.

If you ask someone the color of an object, he or she would respond by identifying the Diffuse color, but all these properties play an important part in bringing a sense of realism to the material. To get a sense of the contribution of each color, try applying very different, bright materials to each of these color swatches and notice the results.

Opacity and transparency

Opaque objects are objects that you cannot see through, such as rocks and trees. Transparent objects, on the other hand, are objects that you can see through, such as glass and clear plastic. The materials in 3ds Max include several controls for adjusting these properties, including Opacity and several Transparency controls.

Opacity is the amount that an object refuses to allow light to pass through it. It is the opposite of transparency and is typically measured as a percentage. An object with 0 percent opacity is completely transparent, and an object with 100 percent opacity doesn't let any light through.

Transparency is the amount of light that is allowed to pass through an object. Because this is the opposite of opacity, transparency can be defined by the opacity value. Several options enable you to control transparency, including Falloff, Amount, and Type. These options are discussed later in this chapter.

Reflection and refraction

A *reflection* is what you see when you look in the mirror. Shiny objects reflect their surroundings. By defining a material's reflection values, you can control how much it reflects its surroundings. A mirror, for example, reflects everything, but a rock won't reflect at all.

Reflection Dimming controls how much of the original reflection is lost as the surroundings are reflected within the scene.

Refraction is the bending of light as it moves through a transparent material. Think of how the background image is distorted when you look through a fishbowl full of water. The amount of refraction that a material produces is expressed as a value called the Index of Refraction. The *Index of Refraction* is the amount that light bends as it goes through a transparent object. For example, a diamond bends light more than a glass of water, so it has a higher Index of Refraction value. The default Index of Refraction value is 1.0 for objects that don't bend light at all. Water has a value of 1.3, glass a value of around 1.5, and solid crystal a value of around 2.0.

Shininess and specular highlights

Shiny objects, such as polished metal or clean windows, include highlights where the lights reflect off their surfaces. These highlights are called *specular highlights* and are determined by the Specular settings. These settings include Specular Level, Glossiness, and Soften values.

The Specular Level is a setting for the intensity of the highlight. The Glossiness determines the size of the highlight: Higher Glossiness values result in a smaller highlight. The Soften value thins the highlight by lowering its intensity and increasing its size.

A rough material has the opposite properties of a shiny material and almost no highlights. The Roughness property sets how quickly the Diffuse color blends with the Ambient color. Cloth and fabric materials have a high Roughness value; plastic and metal Roughness values are low.

Note

Specularity is one of the most important properties that we sense to determine what kind of material the object is made from. For example, metallic objects have a specular color that is the same as their diffuse color. If the colors are different, the objects look like plastic instead of metal.

Other properties

3ds Max uses several miscellaneous properties to help define standard materials, including Diffuse Level and Metalness.

The Diffuse Level property controls the brightness of the Diffuse color. Decreasing this value darkens the material without affecting the specular highlights. The Metalness property controls the metallic look of the material. Some properties are available only for certain material types.

Note

Before proceeding, you need to understand the difference between a material and a map. A *material* is an effect that permeates the 3D object, but most *maps* are 2D images (although procedural 3D maps also exist) that can be wrapped on top of the object. Materials can contain maps, and maps can be made up of several materials. In the Material Editor, materials appear as blue nodes, and maps appear as green nodes in the View pane. Usually, you can tell whether you're working with a material or a map by looking at the default name. Maps show up in the name drop-down list as Map and a number (Map #1), and materials are named a number and Default (7- Default).

Working with the Slate Material Editor

The Material Editor is the interface with which you define, create, and apply materials. You can access the Material Editor by choosing Rendering→Material Editor→Slate Material Editor, clicking the Material Editor button on the main toolbar, or using the M keyboard shortcut.

Note

The M keyboard shortcut opens the Material Editor to the last mode that was used, either Slate or Compact.

The Material Editor comes in two flavors: regular and extra strength. The Material Editor from earlier versions of 3ds Max is still there, but now it is called the Compact Material Editor, and the new Material Editor interface is called the Slate interface. You can choose either from the Rendering→Material Editor menu or switch between them using the Modes menu in the Material Editor.

Note

Although the Slate Material Editor and the Compact Material Editor share most controls, the Slate Material Editor has more features and is the focus of our discussion. The Compact Material Editor is maintained for backward compatibility and is easier to use for existing users.

Using the Slate Material Editor controls

The Slate Material Editor, shown in Figure 17.1, consists of four panels: the Material/Map Browser panel, the Node View panel, the Navigator panel, and the Parameter Editor panel. Of these panels, only the Material View panel is open at all times. The others can be closed and reopened using the Tools menu. If you drag the panel title away from the interface, the panel floats independently. If you drag a floating panel over the interface, several arrow icons appear. Dropping a panel on one of these arrows positions the floating panel to the side of the panel in the direction of the arrow. This interface gives you the power to set up the Slate Material Editor just as you want.

Tip

You also can use a keyboard shortcut to show or hide the various panels: O for the Material/Map Browser, P for the Parameter Editor, and N for the Navigator panel. But note that these shortcuts work only if the Keyboard Shortcut Override Toggle is enabled on the main toolbar.

FIGURE 17.1

The Slate Material Editor has four unique panels.

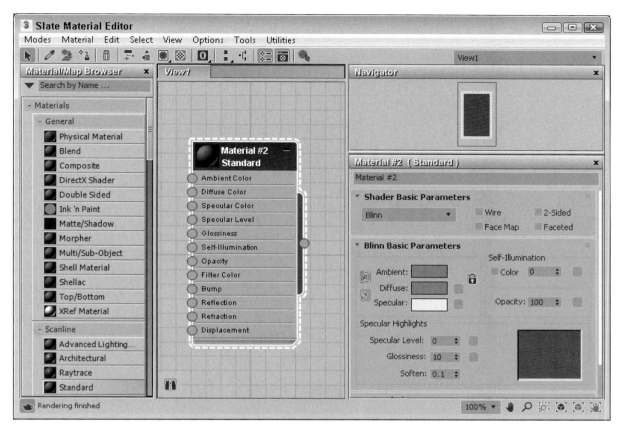

At the top of the default Slate Material Editor window is a menu of options. The menu commands found in these menus offer most of the same functionality as the toolbar buttons, but the menus are often easier to find than the buttons with which you are unfamiliar.

Below the menus are several toolbar buttons. These buttons are defined in Table 17.1.

TABLE 17.1 Slate Material Editor Buttons

Toolbar Button	Name	Description
	Select Tool	Enables a tool for selecting, moving, and working with material trees and nodes.
	Pick Material From Object	Enables you to select a material from an object in the scene and load the material into the Node View panel.
	Put Material to Scene	If a new material is created with the same name as an applied material, this command replaces the applied material with the new one.
	Assign Material to Selection	Applies the selected object with the selected material.
	Delete Selected	Removes any modified properties and resets the material properties to their defaults. The selected node is also deleted from the Node View panel.
	Move Children	Locks the position of the children nodes so they

		move with the material block. If disabled, the children nodes remain in place as the material block moves.
	Hide Unused Nodeslots	Condenses the material block so that only the nodes that are being used are visible.
	Show Shaded Material in Viewport, Show Realistic Material in Viewport	Displays 2D material maps and hardware maps on objects in the viewports.
	Show Background in Preview	Displays a checkered background image (or a custom background) behind the material, which is helpful when displaying a transparent material.
	Material ID Channel	Sets the Material ID for the selected material.
	Layout All - Vertical, Layout All - Horizontal	Aligns and places all material blocks in a vertical column or a horizontal row.
	Layout Children	Moves and orients all children nodes to be next to their respective material blocks.
	Material/Map Browser	Opens the Material/Map Browser panel, which displays all the available materials and maps.
	Parameter Editor	Toggles the Parameter Editor panel on and off.
	Select by Material	Selects all objects using the current material and opens the Select Objects dialog box with those objects selected.

Loading the Material Node View panel

When the Slate Material Editor is first opened, the Node View panel is blank. You can add material nodes to the node view by double-clicking them in the Material/Map Browser or by dragging them from the Material/Map Browser onto the Node View panel. This loads the selected node into the Node View.

If your scene has some objects with materials already applied, you can use the Material→Pick from Object menu or select the Pick Material from Object (the eyedropper) tool on the toolbar and click an object in the viewport. The applied material for that object is loaded in the Node View. If the selected object doesn't have an applied material, nothing is loaded.

You also can get all the applied materials in the current scene using the Material→Get All Scene Materials menu. This loads all applied materials. If all the scene materials make it tough to find what you are looking for, you can use the Edit→Clear View menu to clear the Node View panel. This doesn't remove any assigned materials; it only clears the Node View panel.

Navigating the Material Node View panel

All current materials for the open scene are displayed as material node blocks in the Material Node View panel of the Slate Material Editor. At the top of each material node block are the material name and type. You can change the material name using the Name field at the top of the Parameter Editor or by right-clicking the material name and selecting Rename from the pop-up menu. Beneath the material name are all the parameters that are available for this material. Each of these parameters has a corresponding parameter in the Parameter Editor panel. These parameters are displayed when you double-click a node in the Node View panel.

If you drag the material title (where the name is located), you can move the material node block around within the Node View panel. You also can reduce the size of the material block by clicking the Hide Unused Nodeslots toolbar button. If multiple material blocks are present, you can use the Layout All buttons to align them in a column or a row. Figure 17.2 shows some material node blocks.

FIGURE 17.2

The Node View panel can hold multiple material blocks.

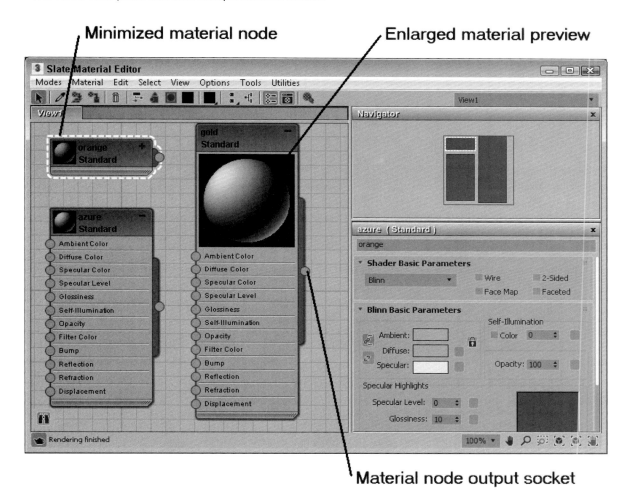

The Navigator panel shows all the material blocks and provides a way to quickly drag to view other sets of nodes. The red outline corresponds to the viewable area in the Node View panel. Navigating the Node View panel is accomplished using the navigation tools at the lower-right corner of the Slate Material Editor. These tools include a Zoom value list and Pan, Zoom, Zoom Region, Zoom Extents, Zoom Extents Selected, and Pan to Selected tools.

Tip

You also can pan the view by dragging with the middle mouse button and zoom by scrubbing the mouse scroll wheel, just like you can in the viewports.

In addition to the navigation tools in the lower-right corner of the Material Editor, the View menu includes several options for navigating the Node View panel, including options to Show/Hide the Grid (G), show scrollbars, and lay out all nodes (L).

If you right-click the Node View tab, you can access a menu to rename or delete the current view. You also can create a new view. This new view appears as another tab at the top of the Material Editor. The new view is navigated independently of the other views and can hold a completely different set of materials. With several views created, you can drag the tabs to reorder the panels as desired.

Selecting and applying materials

A material node block can be selected by simply clicking its title. When selected, the title bar is outlined in white in both the Material Node View and Navigator panels. Selecting a material node doesn't make its parameters appear in the Parameters panel, but if you double-click a material node, its parameters are loaded in the Parameters panel. This node is identified by a dashed white line in the Node View and Navigator panels.

Selected materials are applied to the object selected in the viewport using the Assign Material to Selection button in the Material Editor toolbar, using the Material→Assign Material to Selection (A) menu or by using the right-click pop-up menu. You also can apply a material to a scene object by dragging the material node's output socket and dropping the material on a viewport object whether it is selected or not.

Tip

Although they aren't listed in the menu options, you can use the Undo (Ctrl+Z) and Redo (Ctrl+Y) commands to undo and redo actions done in the Slate Material Editor.

Holding down the Ctrl key while clicking material node blocks or dragging an outline over multiple nodes lets you select multiple nodes at once. When multiple nodes are selected, you can move them all together within the Node View panel. Pressing the Delete key deletes the selected material node.

The Node View panel is a temporary placeholder for materials and maps. An actual scene can have hundreds of materials. By loading a material into a material node, you can change its parameters, apply it to other objects, or save it to a library for use in other scenes. When a file is saved, all materials in the Material Editor are saved with the file.

Changing the material preview

Next to the material name is a preview of the material. If you double-click the preview, the preview is enlarged to show more detail. You can change the Sample Type object displayed in the material block to be a sphere, cylinder, or box using the Preview Object Type menu in the right-click pop-up menu.

The right-click pop-up menu also includes options to show the background in the preview, to show a backlight, and to change the preview tiling for applied maps. The Open Preview Window option opens the material preview in a separate window, as shown in Figure 17.3. Within this floating window, you can resize the preview to be larger, revealing more details. The Show End Result button shows the material with all materials and maps applied.

Tip

The Material Preview panel also can be docked to the Material Editor.

FIGURE 17.3

Material previews can be opened in a floating window.

Show End Result

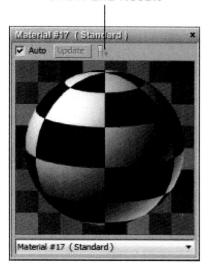

When you assign a material to an object in the scene, the material becomes "hot." A *hot material* is automatically updated in the scene when the material parameters change. Hot materials have corner brackets displayed around their material preview. These brackets turn white when the object with that material is selected in the viewport. You can "cool" a material by making a copy of its material node. To copy a material node, simply drag it with the Shift key held down. This detaches the material node from the object in the scene to which it is applied, so that any changes to the material aren't applied to the object.

Whenever a material is applied to an object in the scene, the material is added to a special library of materials that get saved with the scene. Materials do not need to be seen in the Node View panel to be in the scene library. You can see all the materials included in the scene library in the Material/Map Browser by selecting the Scene Materials rollout.

Selecting objects by material

If you want to select all the objects in your scene with a specific material applied (like the shiny, gold material), select the material in the Node View panel and click the Select by Material button in the toolbar, or use the Utilities→Select Objects by Material menu. This command opens the Select Objects dialog box with all the objects that have the selected material applied. Clicking the Select button selects these objects in the viewport.

Setting Slate Material Editor preferences

You open the Slate Material Editor Options dialog box, shown in Figure 17.4, by selecting the Options→Preferences menu. The top option lets you choose how the nodes are oriented. You also can select to hide the Additional Parameters, use anti-aliased fonts, and set the number of default materials in a multi-subobject material and the grid spacing. The Bitmap Editor Path lets you set where to look for maps. By default, this is set to the maps directory where 3ds Max is installed, but you can change it to your current project folder.

FIGURE 17.4

The Slate Material Editor Options dialog box offers many options for controlling the Slate Material Editor window.

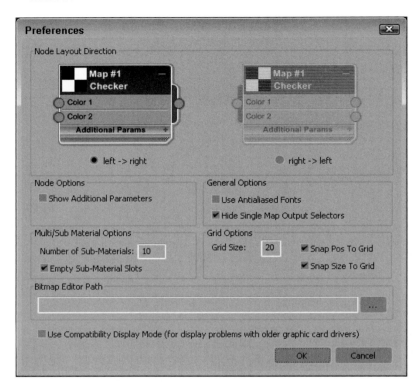

Removing materials and maps

If you accidentally apply an unwanted material to an object, you can replace the material with another material by applying a new material onto the object. If you want to view the object color within the viewport, open the Display panel in the Command Panel, and in the Display Color rollout, select the Object color option for Wireframe and Shaded. The Material Color options display the material color in the viewports.

If you apply a material or map to an object that doesn't look just right, and tweaking it won't help, you can always return to square one by removing the material or any mappings that have been applied to the object. The tool to remove materials and maps is the UVW Remove utility. You can access this utility by clicking the More button in the Utilities panel in the Command Panel and selecting UVW Remove from the list of utilities.

This utility includes a single rollout that lists the number of objects selected. It also includes two buttons. The UVW button removes any mapping coordinates from the selected objects, and the Materials button removes any materials from the selected objects. This button restores the original object color to the selected objects. Alternatively, you can select the Set Gray option, which makes the selected object gray when the materials are removed.

Using utilities

Within the Utilities menu are several additional commands. The Render Map command lets you render out the selected map node. Once rendered, you can save the results to a file. The Render Map dialog box also lets you render out animated maps.

The Clean MultiMaterial utility removes any unused maps from the material tree, and the Instance Duplicate Map identifies and uses instances of duplicate maps throughout the scene. Both these features can be used to reduce the file size of the scene file.

More information on using the Clean MultiMaterial utility is covered in Chapter 20, "Creating Compound Materials and Using Material Modifiers." The Instance Duplicate Map utility is presented in Chapter 18, "Adding Material Details with Maps."

Using the Fix Ambient utility

Standard material types always have their Ambient and Diffuse colors locked together. If you have older files with unlocked Diffuse and Ambient colors, the Fix Ambient utility can be used to locate and fix all materials in the scene with this condition. To access this utility, open the Utilities panel, click the More button, and select the Fix Ambient utility. Clicking the Find All button opens a dialog box that lists all materials in the scene with unlocked Diffuse and Ambient color values.

Tutorial: Coloring Easter eggs

Everyone loves spring, with its bright colors and newness of life. One of the highlights of the season is the tradition of coloring Easter eggs. In this tutorial, you use virtual eggs—no messy dyes and no egg salad sandwiches for the next two weeks.

To create your virtual Easter eggs and apply different colors to them, follow these steps:

1. Open the Easter eggs.max file from the Chap 17 directory in the downloaded content set.

    This file contains several egg-shaped objects.

2. Open the Slate Material Editor by choosing Rendering→Material Editor→Slate Material Editor (or press the M key).

3. Double-click the Standard material in the Material/Map Browser (it is located in the Scanline group). Double-click the material node in the Node View, and click the Diffuse color swatch in the Parameter Editor panel. From the Color Selector that appears, drag the cursor around the color palette until you find the color you want and then click OK.

Tip

If you are having trouble locating a material or map, just type its name in the Search by Name text field at the top of the Material/Map Browser.

4. In any viewport, select an egg and then click the Assign Material to Selection button in the Material Editor, or simply drag from the material node's output socket to the viewport object.

5. Repeat Steps 3 and 4 for all the eggs.

Figure 17.5 shows the assortment of eggs that we just created.

FIGURE 17.5

These eggs have been assigned materials with different Diffuse colors.

Near the Material Editor menu command in the Rendering menu are commands for accessing the Material/Map Browser and the Material Explorer.

Using the Standard Material

Standard materials are the default 3ds Max material type. They provide a single uniform color determined by the Ambient, Diffuse, Specular, and Filter color swatches. Standard materials can use any one of several different shaders. *Shaders* are algorithms used to compute how the material should look, given its parameters.

Standard materials have parameters for controlling highlights, opacity, and self-illumination. They also include many other parameters sprinkled throughout many different rollouts. With all the various rollouts, even a Standard material has an infinite number of possibilities.

Using Shading Types

3ds Max includes several different shader types. These shaders are all available in a drop-down list in the Shader Basic Parameters rollout at the top of the Parameter Editor panel in the Material Editor. Each shader type displays different options in its respective Basic Parameters rollout. Figure 17.1 shows the basic parameters for the Blinn shader. Other available shaders include Anisotropic, Metal, Multi-Layer, Oren-Nayar-Blinn, Phong, Strauss, and Translucent Shader.

The Material/Map Browser holds all the various material types. Other material types are covered in Chapter 20, "Creating Compound Materials and Using Material Modifiers."

The Shader Basic Parameters rollout also includes several options for shading the material, including Wire, 2-Sided, Face Map, and Faceted, as shown in Figure 17.6. Wire mode causes the model to appear as a wireframe model. The 2-Sided option makes the material appear on both sides of the face and is typically used in conjunction with the Wire option or with transparent materials. The Face Map mode applies maps to each single face on the object. Faceted ignores the smoothing between faces.

Note

Using the Wire option or the 2-Sided option is different from the wireframe display option in the viewports. The Wire and 2-Sided options define how the object looks when rendered.

The Material/Map Browser holds all the various material types. The other material types are covered in Chapter 20, "Creating Compound Materials and Using Material Modifiers," and Chapter 19, "Using Specialized Material Types."

FIGURE 17.6

Basic parameter options include (from left to right) Wire, 2-Sided, Face Map, and Faceted.

Blinn shader

This shader is the default. It renders simple circular highlights and smoothes adjacent faces.

The Blinn shader includes color swatches for setting Ambient, Diffuse, Specular, and Self-Illumination colors. To change the color, click the color swatch, and select a new color in the Color Selector dialog box.

Note

You can drag colors among the various color swatches in the Parameters panel. When you do so, the Copy or Swap Colors dialog box appears, which enables you to copy or swap the colors.

You can use the Lock buttons to the left of the color swatches to lock the colors together so that both colors are identical and a change to one automatically changes the other. You can lock Ambient to Diffuse and Diffuse to Specular.

The small, square buttons to the right of the Diffuse, Specular, Self-Illumination, Opacity, Specular Level, and Glossiness controls are shortcut buttons for adding a map in place of the respective parameter. Clicking these buttons opens the Material/Map Browser, where you can select the map type.

When a map is loaded and active, it appears in the Maps rollout, and an uppercase letter *M* appears on its button. When a map is loaded but inactive, a lowercase *m* appears. After you apply a map, these buttons open to make the map the active level and display its parameters in the rollouts. Figure 17.7 shows these map buttons.

For more on maps and the various map types, see Chapter 18, "Adding Material Details with Maps."

FIGURE 17.7

The Blinn Basic Parameters rollout lets you select and control properties for the Blinn shader.

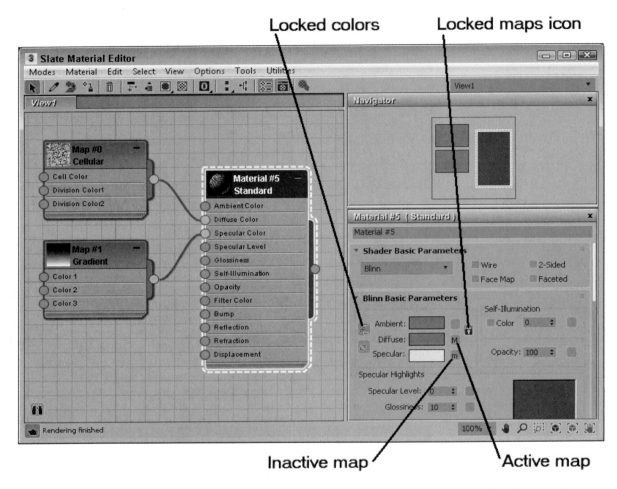

You also can lock the Ambient and Diffuse maps together with the lock icon to the right of the map buttons. The Ambient and Diffuse colors are locked together by default.

Self-Illumination can use a color if the Color option is enabled. If this option is disabled, a spinner appears that enables you to adjust the amount of default color used for illumination. Materials with a Self-Illumination value of 100 or a bright color like white lose all shadows and appear to glow from within. This happens because the self-illumination color replaces the ambient color, but a material with self-illumination can still have specular highlights. To remove the effect of Self-Illumination, set the spinner to 0 or the color to black. Figure 17.8 shows a sphere with Self-Illumination values (from left to right) of 0, 25, 50, 75, and 100.

FIGURE 17.8

Increasing the Self-Illumination value reduces the shadows in an object.

The Opacity spinner sets the level of transparency of an object. A value of 100 makes a material completely opaque, while a value of 0 makes the material completely transparent. Use the Show Background in Preview button on the Material Editor toolbar to enable a patterned background image to make it easier to view the effects of the Opacity setting. Figure 17.9 shows materials with Opacity values of 10, 25, 50, 75, and 90.

FIGURE 17.9

The Opacity value sets how transparent a material is.

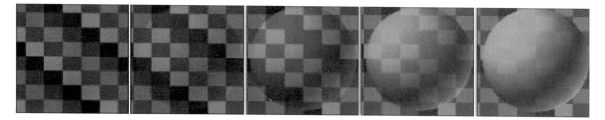

Specular highlights are the bright points on the surface where the light is reflected at a maximum value. The Specular Level value determines how bright the highlight is. Its values can range from 0, where there is no highlight, to 100, where the highlight is at a maximum. The graph to the right of the values displays the intensity per distance for a cross section of the highlight. The Specular Level defines the height of the curve or the value at the center of the highlight where it is the brightest. This value can be overloaded to accept numbers greater than 100. Overloaded values create a larger, wider highlight.

The Glossiness value determines the size of the highlight. A value of 100 produces a pinpoint highlight, and a value of 0 increases the highlight to the edges of the graph. The Soften value doesn't affect the graph, but it spreads the highlight across the area defined by the Glossiness value. It can range from 0 (wider) to 1 (thinner). Figure 17.10 shows a sampling of materials with specular highlights. The left image has a Specular Level of 20 and a Glossiness of 10; the second image has the Specular Level increased to 80; the third image has the Specular Level overloaded with a value of 150; and the last two images have the Glossiness value increased to 50 and 80, respectively.

FIGURE 17.10

You can control specular highlights by altering brightness and size.

Phong shader

The Phong shader creates smooth surfaces like Blinn without the quality highlights, but it renders more quickly than the Blinn shader does. The parameters for the Phong shader are identical to those for the Blinn shader. The differences between Blinn and Phong are very subtle, but Blinn can produce highlights for lights at low angles to the surface, and its highlights are generally softer.

Tip

The Blinn shader is typically used to simulate softer materials like rubber, but the Phong shader is better for hard materials like plastic.

Anisotropic shader

The Anisotropic shader is characterized by noncircular highlights. The Anisotropy value is the difference between the two axes that make up the highlight. A value of 0 is circular, but higher values increase the difference between the axes, and the highlights are more elliptical.

Most of the parameters for this shader are the same as those for the Blinn shader, but several parameters of the Anisotropic type are unique. The Diffuse Level value determines how bright the Diffuse color appears. This is similar to Self-Illumination, but it doesn't affect the specular highlights or the shadows. Values can range from 0 to 400.

Compared with the Blinn shader, the Specular Highlight graph looks very different. That is because it displays two highlight components that intersect at the middle. The Specular Level value still controls the height of the curve, and the Glossiness still controls the width, but the Anisotropy value changes the width of one axis relative to the other, creating elliptical highlights. The Orientation value rotates the highlight. Figure 17.11 compares the Specular Highlight graphs for the Blinn and Anisotropic shaders.

Tip

Because the Anisotropy shader can produce elliptical highlights, it is often used on surfaces with strong grooves and strands, like fabrics and stainless steel objects.

FIGURE 17.11

The Specular Highlight graph for the Blinn and Anisotropic shaders

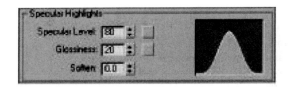

Figure 17.12 shows several materials with the Anisotropic shader applied. The first three images have Anisotropic values of 30, 60, and 90, and the last two images have Orientation values of 30 and 60.

FIGURE 17.12

Materials with the Anisotropic shader applied have elliptical highlights.

Multi-Layer shader

The Multi-Layer shader includes two Anisotropic highlights. Each of these highlights can have a different color. All parameters for this shader are the same as the Anisotropic shader, described previously, except that there are two Specular Layers and one additional parameter: Roughness. The Roughness parameter defines how well the Diffuse color blends into the Ambient color. When Roughness is set to a value of 0, an object appears the same as with the Blinn shader, but with higher values, up to 100, the material grows darker.

Figure 17.13 shows several materials with a Multi-Layer shader applied. The first two images have two specular highlights, each with an Orientation value of 60 and Anisotropy values of 60 and 90. The third image has an increased Specular Level of 110 and a decrease in the Glossiness to 10. The fourth image has a change in the Orientation value for one of the highlights to 20, and the final image has a drop in the Anisotropy value to 10.

Tip

The Multi-Layer shader is useful to give a material a sense of surface depth. For example, it can give the illusion of a layer of shellac on wood or a layer of wax on tile.

FIGURE 17.13

Materials with a Multi-Layer shader applied can have two crossing highlights.

Oren-Nayar-Blinn shader

The Oren-Nayar-Blinn shader is useful for creating materials for matte surfaces such as cloth and fabric. The parameters are identical to the Blinn shader, with the addition of the Diffuse Level and Roughness values.

Metal shader

The Metal shader simulates the luster of metallic surfaces. The Highlight curve has a shape that is different from that of the other shaders. It is rounder at the top and doesn't include a Soften value. It also can accept a much higher Specular Level value (up to 999) than the other shaders. Also, you cannot specify a Specular color. All other parameters are similar to those of the Blinn shader. Figure 17.14 shows several materials with the Metal shader applied. These materials differ in Specular Level values, which are (from left to right) 50, 100, 200, 400, and 800.

Note

For the Metal shader, the specular color is always the same as the material's diffuse color.

FIGURE 17.14

A material with a Metal shader applied generates its own highlights.

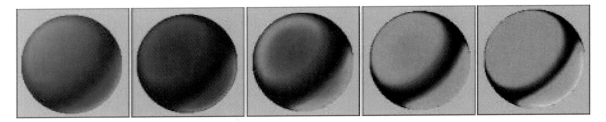

Strauss shader

The Strauss shader provides another alternative for creating metal materials. This shader has only four parameters: Color, Glossiness, Metalness, and Opacity. Glossiness controls the entire highlight shape. The Metalness value makes the material appear more metal-like by affecting the primary and secondary highlights. Both of these values can range between 0 and 100.

Tip

The Strauss shader is often better at making metal than the Metal shader because of its smoothness value and the ability to mix colors with the Metalness property.

Translucent shader

The Translucent shader allows light to easily pass through an object. It is intended to be used on thin, flat plane objects, such as a bed sheet used for displaying shadow puppets. Most of the settings for this shader are the same as the others, except that it includes a Translucent color. This color is the color that the light becomes as it passes through an object with this material applied. This shader also includes a Filter color and an option for disabling the specular highlights on the backside of the object.

Tutorial: Making curtains translucent

The Translucent shader can be used to create an interesting effect. Not only does light shine through an object with this shader applied, but shadows also are visible.

To make window curtains translucent, follow these steps:

1. Open the Translucent curtains.max file from the Chap 17 directory in the downloaded content set.

   This file contains a simple scene of a tree positioned outside a window.

2. Open the Material Editor by choosing Rendering→Material Editor→Slate Material Editor by clicking the Material Editor button on the main toolbar or by pressing the M key.

3. In the Material/Map Browser panel of the Material Editor, double-click the Standard material; double-click the material node; and in the Name field, name the material **Curtains**. Select the Translucent Shader from the Shader Basic Parameters rollout. Click the Diffuse color swatch, and select a light blue color. Click the OK button to exit the Color Selector.

4. Click the Translucent Color swatch, change its color to a light gray, and set the Opacity to **75**.

5. Drag the Curtains material's output node socket onto the curtain object in the Left viewport or select the curtains in the viewport, and use the Assign Material to Selection button in the Material Editor toolbar.

Figure 17.15 shows the resulting rendered image. Notice that the tree's shadow is cast on the curtains.

FIGURE 17.15

These translucent window curtains show shadows.

Accessing Other Parameters

In addition to the basic shader parameters, several other rollouts of options can add to the look of a material.

Extended Parameters rollout

The Material Editor includes several settings, in addition to the basic parameters, that are common for most shaders. The Extended Parameters rollout, shown in Figure 17.16, includes Advanced Transparency, Reflection Dimming, and Wire controls. All shaders include these parameters.

FIGURE 17.16

The Extended Parameters rollout includes Advanced Transparency, Reflection Dimming, and Wire settings.

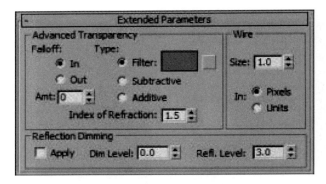

You can use the Advanced Transparency controls to set the Falloff to be In, Out, or a specified Amount. The In option increases the transparency as you get farther inside the object, and the Out option does the opposite. The Amount value sets the transparency for the inner or outer edge. Figure 17.17 shows two materials that use the Transparency Falloff options on a gray background and on a patterned background. The two materials on the left use the In option, and the two on the right use the Out option. Both are set at Amount values of 100.

Tip

If you look closely at a glass sphere, you'll notice that the glass is thicker when you look through the edge of the sphere than through the sphere's center. This can be created using the In option in the Advanced Transparency section.

FIGURE 17.17

Materials with the In and Out Falloff options applied

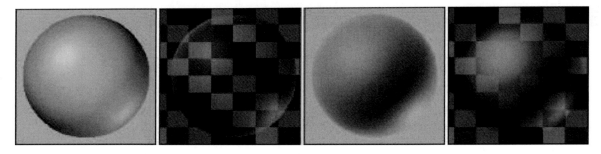

The three transparency types are Filter, Subtractive, and Additive. The Filter type multiplies the Filter color with any color surface that appears behind the transparent object. With this option, you can select a Filter color to use. The Subtractive and Additive types subtract from or add to the color behind the transparent object.

The Index of Refraction is a measure of the amount of distortion caused by light passing through a transparent object. Different physical materials have different Index of Refraction values. The amount of distortion also depends on the thickness of the transparent object. The Index of Refraction for water is 1.3 and for glass is 1.5. The default of 1.0 has no effect.

The Wire section lets you specify a wire size or thickness. Use this setting if the Wire option or the 2-Sided option is enabled in the Shaders Basic Parameters rollout. The size can be measured in either Pixels or Units. Figure 17.18 shows materials with different Wire values from 1 to 5 pixels.

FIGURE 17.18

Three materials with Wire values of (from left to right) 1, 2, 3, 4, and 5 pixels

Reflection Dimming controls how intense a reflection is. You enable it by using the Apply option. The Dim Level setting controls the intensity of the reflection within a shadow, and the Refl Level sets the intensity for all reflections not in the shadow.

SuperSampling rollout

Pixels are small square dots that collectively make up the entire screen. At the edges of objects where the material color changes from the object to the background, these square pixels can cause jagged edges to appear. These edges are called *artifacts* and can ruin an image. *Anti-aliasing* is the process through which these artifacts are removed by softening the transition between colors.

3ds Max includes anti-aliasing filters as part of the rendering process. SuperSampling is an additional anti-aliasing pass that can improve image quality that is applied at the material level. You have several SuperSampling methods from which to choose. The SuperSampling method can be defined in the Material Editor, or you can choose the settings in the Scanline Renderer rollout of the Render Setup dialog box by enabling the Use Global Settings option.

Note

Anti-aliasing happens before raytracing when rendering, so even if the anti-aliasing material option is enabled, the reflections and/or refractions will still be aliased unless anti-aliasing is turned on in the Render Settings.

For more about the various anti-aliasing filters, see Chapter 27, "Rendering a Scene and Enabling Quicksilver."

SuperSampling is calculated only if the Anti-Aliasing option in the Render Setup dialog box is enabled. The Raytrace material type has its own SuperSampling pass that is required in order to get clean reflections.

Note

Using SuperSampling can greatly increase the time it takes to render an image.

In a SuperSampling pass, the colors at different points around the center of a pixel are sampled. These samples are then used to compute the final color of each pixel. The SuperSampling settings can be set globally in the Render Setup dialog box or for each material individually by disabling the Use Global Settings option. These four SuperSampling methods are available:

* **Adaptive Halton:** Takes semi-random samples along both the pixel's X-axis and Y-axis. It takes from 4 to 40 samples.

* **Adaptive Uniform:** Takes samples at regular intervals around the pixel's center. It takes from 4 to 26 samples.

* **Hammersley:** Takes samples at regular intervals along the X-axis, but takes random samples along the Y-axis. It takes from 4 to 40 samples.

* **Max 2.5 Star:** Takes four samples along each axis.

The first three methods enable you to select a Quality setting. This setting specifies the number of samples to be taken. The more samples taken, the higher the resolution, but the longer it takes to render. The two

Adaptive methods (Adaptive Halton and Adaptive Uniform) offer an Adaptive option with a Threshold spinner. This option takes more samples if the change in color is within the Threshold value. The SuperSample Texture option includes maps in the SuperSampling process along with materials.

Tip
To get good reflections and refractions, enable SuperSampling for final renders.

Maps rollout

A *map* is a bitmap image that is wrapped about an object. The Maps rollout includes a list of the maps that you can apply to an object. Using this rollout, you can enable or disable maps, specify the intensity of the map in the Amount field, and load maps. Clicking the Map buttons opens the Material/Map Browser, where you can select the map type.

Find out more about maps in Chapter 18, "Adding Material Details with Maps."

Tutorial: Coloring a dolphin

As a quick example of applying materials, you'll take a dolphin model and position it over a watery plane. You then apply custom materials to both objects.

To add materials to a dolphin, follow these steps:

1. Open the Dolphin.max file from the Chap 17 directory in the downloaded content set.

   This file contains a simple plane object and a dolphin mesh.

2. Open the Material Editor by choosing Rendering→Material Editor→Slate Material Editor, clicking the Material Editor button on the main toolbar, or pressing the M key.

3. In the Material/Map Browser panel, double-click the Standard material; double-click the Standard material node; and in the Name field in the Parameter Editor panel, rename the material **Dolphin Skin**. Click the Diffuse color swatch, and select a light gray color. Then click the Specular color swatch, and select a light yellow color. Click the OK button to exit the Color Selector. In the Specular Highlights section, increase the Specular Level to **45**.

4. Double-click the Standard material in the Material/Map Browser again, double-click this new node to access its parameters, and name it **Ocean Surface**. Click the Diffuse color swatch, and select a light blue color. Set the Specular Level and Opacity values to **80**. In the Maps rollout, click the No Map button to the right of the Bump option. In the Material/Map Browser that opens, double-click the Noise selection in the Maps rollout (in the General group). Double-click the Noise map node to access the Noise parameters, enable the Fractal option, and set the Size value to **15** in the Noise Parameters rollout.

5. Select the dolphin body in the viewport and, with the Dolphin Skin material selected in the Material Editor, click the Assign Material to Selection button in the Material Editor toolbar. Then do the same for the ocean surface.

Note
This model also includes separate objects for the eyes, mouth, and tongue. These objects could have different materials applied to them, but they are so small in this image that you won't worry about them.

6. Choose Rendering→Environment (keyboard shortcut 8), click the Background Color swatch, and change it to a light sky blue.

Figure 17.19 shows the resulting rendered image.

FIGURE 17.19

A dolphin over the water with applied materials

Using the Material/Map Browser

Now that you know how to apply materials to objects, the easiest way to get materials is from the Material/Map Browser. The Autodesk® 3ds Max® 2020 software ships with several libraries of materials that you can access using the Material/Map Browser. There are several places that you can open the Material/Map Browser. The first is the Rendering→Material/Map Browser menu command. If the Slate Material Editor is open, you can open the Material/Map Browser with the Tools→Material/Map Browser menu, but it is probably already visible because it is open and positioned to the left of the Slate Material Editor. Within the Compact Material Editor, click to the Get Materials button and the Material/Map Browser panel appears, as shown in Figure 17.20.

The Material/Map Browser also opens anytime you need to access a bitmap, material, or map including the map buttons in the Material Editor, the Environment and Effects dialog box, and the Background dialog box.

Note
The Material/Map Browser can be docked to the left edge of the Slate Material Editor or pulled away from the interface as a floating dialog box. Double-click the title bar to re-dock it back to the interface.

The Material/Map Browser is the place where all your materials are stored. They are stored in sets called libraries. These libraries are saved along with the scene file or they can be saved as a separate file if you want to load and access them within a different scene. Within the Material/Map Browser, each library is contained with a separate rollout. These rollouts, called groups, hold any loaded material libraries, the available default Material and Map types, Scene Materials, and Sample Slots, which hold the temporary material slots used by the Compact Material Editor.

FIGURE 17.20

The Material/Map Browser lets you select new materials from a library of materials.

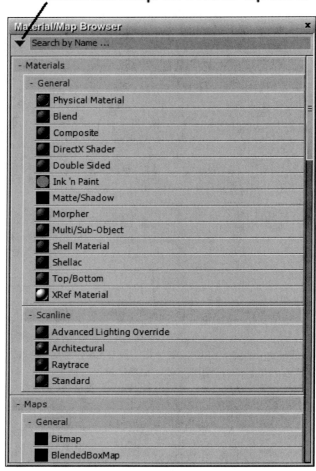

The text field directly above the material sample slot in the Material/Map Browser is a Search by Name field. By typing a name in this field, you can search and select materials.

Tip

When browsing materials in the Material/Map Browser, you can use the keyboard to move up and down the material list. The left and right keyboard keys are used to open and close rollouts and the Enter key adds the selected material to the View pane.

Working with libraries

Any time you adjust a material or a map parameter, a new material is created and the material's sample slot is updated. Although newly created materials are saved along with the scene file, you can make them available for reuse by including them in a library.

As more and more libraries get added to the Material/Map Browser, it can be difficult to locate the specific library that you want to use. To help with this problem, you can right-click the library title and choose the Edit Group Color option. This opens a Color Selector where you can pick a color for the library rollout. There are also options for changing how the materials in the library are displayed. The options include Small, Medium and Large Icons, Icons and Text, or just Text.

The right-click pop-up menu also offers an option to create a new library. New Libraries can be saved and loaded also using the right-click pop-up menu. You can populate a new material library rollout by simply dragging materials and maps and dropping them on the new material library's rollout.

Note

3ds Max ships with several different material libraries and several architecture material sets. You can find all these libraries in the matlibs directory. Some libraries are only available when the Arnold renderer is enabled.

Save a layout with the current materials using the Material/Map Browser Options→Additional Options→Save Layout As. These files are saved as files with the .mpl extension.

Tutorial: Loading a custom material library

To practice loading a material library, I've created a custom library of materials using various textures created with Kai's Power Tools.

To load a custom material library into the Slate Material Editor, follow these steps:

1. Choose Rendering→Material Editor→Slate Material Editor (or press the M key) to open the Material Editor. Then make sure the Material/Map Browser is active or select the Tools→Material/Map Browser menu to open the Material/Map Browser.

2. Click the Material/Map Browser Options button to the left of the Search field at the top of the Material/Map Browser and select Open Material Library from the pop-up menu.

3. Select and open the KPT samples.mat file from the Chap 17 directory in the downloaded content set.

   The library loads into the Material/Map Browser.

4. In the Search field (above the sample slot), type **Bug** to locate and select the bug eyes material.

Figure 17.21 shows the Material/Map Browser with the custom material library open.

FIGURE 17.21

The Material/Map Browser also lets you work with saved custom material libraries.

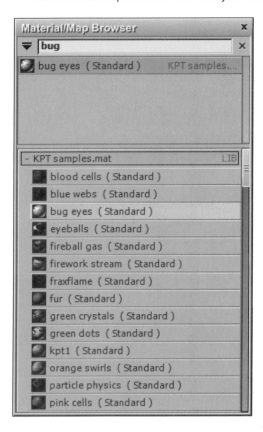

Using the Material Explorer

The Scene Explorer makes working with scene objects much easier. So easy in fact that the 3ds Max team has looked for other places where a similar interface can be used, and the first stop is with materials. The result is the Material Explorer.

The Material Explorer, shown in Figure 17.22, lets you quickly view all the scene materials along with their hierarchies and all their properties in a single interface. It also lets you sort the materials by their various properties and even make changes to multiple materials at once. You can access the Material Explorer with the Rendering→Material Explorer command.

FIGURE 17.22

The Material Explorer shows the layered material as a hierarchy.

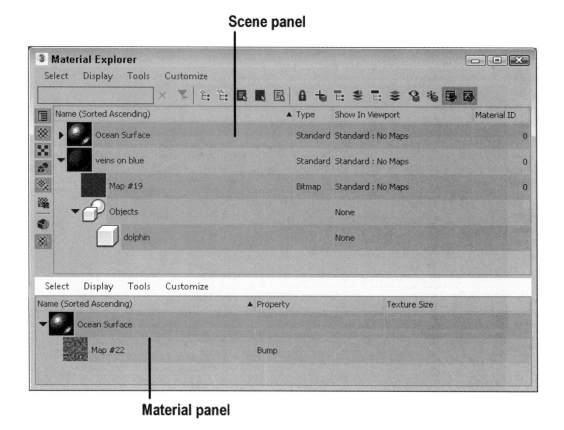

Scene panel

Material panel

The Material Explorer works exactly like the Scene Explorer and has configurable columns. It is divided into two panels. The top panel shows all the materials in the current scene; the bottom panel shows the hierarchy of the selected material or submaterial.

In addition to the menus, several toolbar buttons run horizontally under the menus and several display buttons run vertically down the left side of the interface. These buttons are described in Tables 17.2 and 17.3.

To learn more about the Scene Explorer interface, see Chapter 6, "Selecting Objects and Setting Object Properties."

TABLE 17.2 Material Explorer Display Buttons—Vertical

Toolbar Button	Name	Description
	Display Thumbnails	Displays thumbnails of the various material components as displayed in the material preview
	Display Materials	Includes materials within the hierarchy
	Display Maps	Includes maps within the material hierarchy
	Display Objects	Includes the applied object in the material hierarchy
	Display Sub-Materials/Maps	Includes any applied submaterials/ maps within the material hierarchy

	Display Unused Map Channels	Includes all unused map channels as part of the hierarchy
	Sort by Object	Sorts the materials based on the objects they are applied to and lists the materials underneath
	Sort by Material	Sorts the materials and lists the objects underneath

TABLE 17.3 Material Explorer Toolbar Buttons—Horizontal

Toolbar Button	Name	Description
	Find text field	Allows searching for specific named materials and maps
	Remove Filter	Clears the search field
	Toggle Display Toolbar	Turns the display toolbar on and off
	Select All Materials	Selects all materials in the scene
	Select All Maps	Selects all maps in the scene
	Select All	Selects all entries in the list
	Select None	Deselects all entries in the list
	Select Invert	Inverts the current selection
	Lock Cell Editing	Locks all cells so they cannot be edited
	Create New Layer	Creates a new material layer
	Pick Parent	Select the parent for the selected child
	Add to Active Layer	Add the selected node to the active layer
	Select Children	Select all the children for this parent
	Make Selected Layer Active	Make the current layer active
	Hide All Layers	All layers are hidden
	Freeze All Layers	All layers are frozen
	Sync to Material Explorer	Causes the material selected in the top panel to be visible in the lower panel
	Sync to Material Level	Displays the entire hierarchy for the selected material in the lower panel when enabled

The Material Explorer lets you apply materials directly to scene objects by simply dragging and dropping the material thumbnail onto the object. You also can drag and drop maps onto other materials and channels. The Material Explorer also works with the Material Browser; for example, you can drag the material type and drop it on the Type column in the Material Explorer to change a material type.

Summary

Materials can add much to the realism of your models. Learning to use the Material Editor enables you to work with materials. This chapter covered the following topics:

Material libraries, the Material/Map Browser, and the Material Explorer let you work with materials and, more importantly, find them. This chapter covered the following topics:

* Understanding various material properties

* Working with the Slate Material Editor buttons and material nodes

* Using various material types

* Using the various material parameters

* Understanding the basics of using Standard materials

* Learning to use the various shaders

* Exploring the other material rollouts

* Applying materials to a model

* Using the Material/Map Browser and material libraries

* Using the Material Explorer to quickly see all materials in a scene

This chapter should have been enough to whet your appetite for materials, yet it really covered only one part of the equation. The other critical piece for materials is maps, and you'll dive into those in the next chapter.

Chapter 18

Adding Material Details with Maps

IN THIS CHAPTER

Understanding mapping

Connecting maps to material nodes

Applying maps to material properties using the Maps rollout

Using the Bitmap Path Editor

Creating textures with Photoshop

In addition to using materials, another way to enhance an object is to use a map—but not a roadmap. In the Autodesk® 3ds Max® 2020 software, maps are bitmaps that can be applied to the surface of an object. Some maps wrap an image onto objects, but others, such as displacement and bump maps, modify the surface based on the map's intensity. For example, you can use a diffuse map to add a label to a soup can or a bump map to add some texture to the surface of an orange.

Several external tools can be very helpful when you create texture maps. These tools include an image-editing package such as Photoshop, a digital camera, and a scanner. With these tools, you can create and capture bitmap images that can be applied as materials to the surface of the object. Maps are great, but with these maps, you'll still need to stop and ask for directions.

Understanding Maps

To understand a material map, think of this example. Cut the label off of a soup can, scan it into the computer, and save the image as a bitmap. You can then create a cylinder with roughly the same dimensions as the can, load the scanned label image as a material map, and apply it to the cylinder object to simulate the original soup can. This is exactly how maps work in 3ds Max.

Different map types

Different types of maps exist. Some maps wrap images about objects, while others define areas to be modified by comparing the intensity of the pixels in the map. An example of this is a *bump map*. A standard bump map would be a grayscale image; when mapped onto an object, those areas of the object that are beneath the lighter-colored sections of the bump map would be raised to show a bump when rendered, and those areas on the object beneath darker sections would show no bump when rendered. The highest raised bumps match the areas where the map is white, the lowest areas are black, and everything in between is a gradient of gray. This enables you to easily create surface textures, such as the rivets on the side of a machine, without having to model them.

Still other uses for maps include *environment* maps for backgrounds and *projection* maps that are used with lights.

For information on environment maps, see Chapter 27, "Rendering a Scene and Enabling Quicksilver." Chapter 45, "Adding Volume Light and Lens Effects," covers projector maps.

Maps that are used to create materials are all applied using the Material Editor. The Material/Map Browser provides access to all the available maps. These maps have many common features.

Enabling the global viewport rendering setting

To see applied maps in the viewports, select the Materials→Shaded Materials with Maps menu command from the Shading viewport menu.For more accurate maps that show highlights, you can enable the Materials→Realistic Materials with Maps option from the Shading viewport menu. This is especially helpful when the scene objects use the Arch & Design materials.

Using Real-World maps

When maps are applied to scene objects, they are applied based on the object's UV coordinates, which control the scale, orientation, and position of the applied map. But each bitmap can be sized along each axis to stretch the map over the surface. Another way to stretch a texture map is to resize the geometric object that the map is applied to. This is the default behavior of maps, but another option is available.

When a geometric object is created, you can enable the Real-World Map Size parameter in the Modify panel, which is generally near the Generate Mapping Coords option. This option is also available when the UVW Mapping modifier is applied to an object. When enabled, this option lets you specify the size of the applied texture using Width and Height values. When this option is enabled, it causes the texture maps to maintain their sizes as geometry objects are resized. You can set the dimensions of applied texture maps in the Coordinates rollout in the Material Editor.

Tip

You can select to have Real-World mapping enabled for all new objects by default by enabling the Use Real-World Texture Coordinates option in the General panel of the Preference Settings dialog box.

Working with Maps

Maps are used along with materials and are not applied to objects by themselves. You can access material maps from the Material/Map Browser. To open the Material/Map Browser if it isn't visible in the Slate Material Editor, select the Tools→Material/Map Browser menu or press the O key in the Slate Material Editor. You also can open the Material/Map Browser by clicking any of the map buttons found throughout the Parameter Editor panel, including those found in the Maps rollout. Figure 18.1 shows the Material/Map browser with some of its available standard maps.

FIGURE 18.1

Use the Material/Map Browser to list all the maps available for assigning to materials.

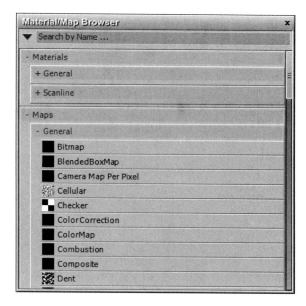

In the Material/Map Browser all available maps are displayed by default in the Maps/General rollout, but if you have the Quicksilver Hardware, ART or the Arnold renderer enabled, rollouts for Autodesk Material Library maps and Arnold maps are also displayed. If you right-click within the Material/Map Browser and the scroll bar, and choose the Show Incompatible option from the pop-up menu, all map rollouts, including Arnold maps, are displayed even if the Arnold renderer isn't enabled.

To load a map node into the Node View panel of the Material Editor, simply double-click it or drag the material from the Material/Map Browser to the Node View panel. All map nodes are easily identified by their green title bars in the Node View and Navigator panels.

Connecting maps to materials

A map by itself in the Node View panel cannot be applied to objects in the scene. To add a map to a material, it must be connected to one of the material properties. This is done by dragging on the map node's output socket and dropping the connecting line on the input socket for the material property where the map is being applied. For example, Figure 18.2 shows a connection between the Checker map node and the Standard material node's Diffuse Color. Once a connection is made, the material preview is updated to show that the applied map and the sockets at either end of the connection are highlighted green.

Tip

If you drag from a map node's output socket and drop anywhere on the blank Node View, a pop-up menu lets you choose to select and create a material, map, controller, or sample slot material node.

FIGURE 18.2

Map nodes need to be connected to material properties in order to show up in the material.

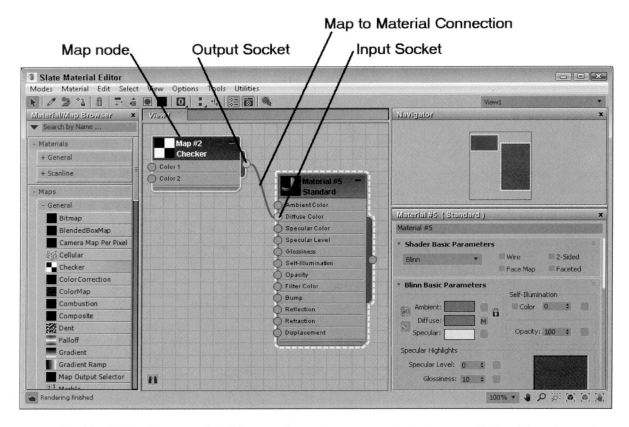

Double-clicking the map node's title opens the map's parameters in the Parameter Editor. Maps also can be applied by clicking a map button in the Parameter Editor and selecting a map type from the Material/Map Browser that opens.

A single map node can be connected to several different material parameters on the same or on different nodes. Map nodes also can be connected to other map nodes. For example, Figure 18.3 shows a Marble and a Noise map connected to a Checker map node, which is connected to a standard material node.

FIGURE 18.3

Map nodes also can be connected to other map nodes.

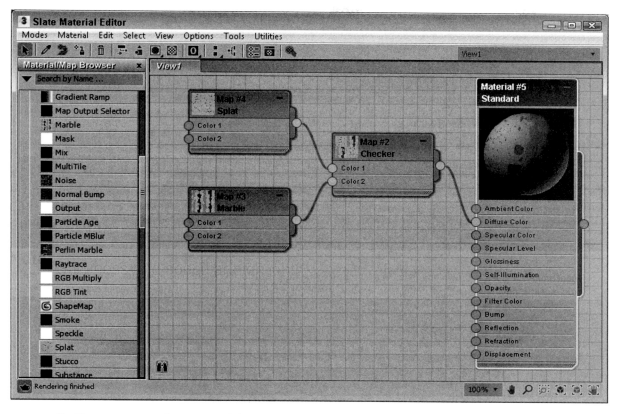

To disconnect a map from a material node, simply select the connection line and press the Delete key. This removes the connection but leaves both the map and material nodes in place.

Using the Maps rollout

You can quickly see all the maps that are connected to the current material in the Maps rollout, shown in Figure 18.4.

FIGURE 18.4

The Maps rollout can turn maps on or off.

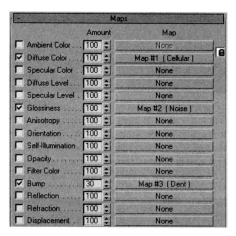

When a material node is selected, the Maps rollout is available in the Parameters panel. The Maps rollout is where you can see all the different maps available for the given material. To add a map, click the Map button; this opens the Material/Map Browser, where you can select the map to use. The selected map appears as a node and is automatically connected to the material. The Amount spinner sets the intensity of the map, and an option to enable or disable the map is available. For example, if a bump map is applied to an object, you can reduce the height of the bumps with the Amount value.

The available maps in the Maps rollout depend on the type of material and the Shader that you are using. Raytrace materials have many more available maps than the Standard material. Some of the common mapping types found in the Maps rollout are discussed in Table 18.1.

TABLE 18.1 Material Properties for Maps

Material Property	Description
Ambient Color	Replaces the ambient color component of the base material. You can use this feature to make an object's shadow appear as a map. Diffuse Color mapping (discussed next) also affects the Ambient color. A lock button in the Maps rollout enables you to lock these two mappings together.
Diffuse Color	Replaces the diffuse color component of the base material. This is the main color used for the object. When you select a map such as Wood, the object appears to be created out of wood. As mentioned previously, Diffuse Color mapping also can affect the Ambient color if the lock button is selected.
Diffuse Level	Changes the diffuse color level from 0, where the map is black, to a maximum, where the map is white. This mapping is available only with the Anisotropic, Oren-Nayar-Blinn, and Multi-Level Shaders.
Diffuse Roughness	Sets the roughness value of the material from 0, where the map is black, to a maximum, where the map is white. This mapping is available only with the Oren-Nayar-Blinn and Multi-Layer Shaders.
Specular Color	Replaces the specular color component of the base material. This option enables you to include a different color or image in place of the specular color. It is different from the Specular Level and Glossiness mappings, which also affect the specular highlights.
Specular Level	Controls the intensity of the specular highlights from 0, where the map is black, to 1, where the map is white. For the best effect, apply this mapping along with the Glossiness mapping.
Glossiness	Defines where the specular highlights will appear. You can use this option to make an object appear older by diminishing certain areas. Black areas on the map show the non-glossy areas, and white areas are where the glossiness is at a maximum.
Self-Illumination	Makes certain areas of an object glow, and because they glow, they won't receive any lighting effects, such as highlights or shadows. Black areas represent areas that have no self-illumination, and white areas receive full self-illumination.
Opacity	Determines which areas are visible and which are transparent. Black areas for this map are transparent, and white areas are opaque. This mapping works in conjunction with the Opacity value in the Basic Parameters rollout. Transparent areas, even if perfectly transparent, still receive specular highlights.
Filter Color	Colors transparent areas for creating materials such as colored glass. White light that is cast through an object using filter color mapping is colored with the filter color.
Anisotropy	Controls the shape of an anisotropy highlight. This mapping is available only with the Anisotropic and Multi-Layer Shaders.
Orientation	Controls an anisotropic highlight's position. Anisotropic highlights are elliptical, and this mapping can position them at a different angle. Orientation mapping is available only with the Anisotropic and Multi-Layer Shaders.
Metalness	Controls how metallic an area looks. It specifies metalness values from 0, where the map is black, to a maximum, where the map is white. This mapping is available only

	with the Strauss Shader.
Bump	Uses the intensity of the bitmap to raise or indent the surface of an object. The white areas of the map are raised, and darker areas are lowered. Although bump mapping appears to alter the geometry, it actually doesn't affect the surface geometry.
Reflection	Reflects images off the surface as a mirror does. The three types of Reflection mapping are Basic, Automatic, and Flat Mirror. Basic reflection mapping simulates the reflection of an object's surroundings. Automatic reflection mapping projects the map outward from the center of the object. Flat-Mirror reflection mapping reflects a mirror image off a series of coplanar faces. Reflection mapping doesn't need mapping coordinates because the coordinates are based on world coordinates and not on object coordinates. Therefore, the map appears different if the object is moved, which is how reflections work in the real world.
Refraction	Bends light and displays images through a transparent object, in the same way a room appears through a glass of water. The amount of this effect is controlled by a value called the Index of Refraction. This value is set in the parent material's Extended Parameters rollout.
Displacement	Changes the geometry of an object. The white areas of the map are pushed outward, and the dark areas are pushed in. The amount of the surface displaced is based on a percentage of the diagonal that makes up the bounding box of the object. Displacement mapping isn't visible in the viewports unless the Displace NURBS (for NURBS [Non-Uniform Rational B-Splines] objects) or the Displace Mesh (for Editable Meshes) modifiers have been applied.

Tutorial: Aging objects for realism

I don't know whether your toolbox is well worn like mine; it must be the hostile environment that it is always in (or all the things I keep dropping in and on it). Rendering a toolbox with nice specular highlights just doesn't feel right. This tutorial shows a few ways to age an object so that it looks older and worn.

To add maps to make an object look old, follow these steps:

1. Open the Toolbox.max file from the Chap 18 directory in the downloaded content set.

   This file contains a simple toolbox mesh created using extruded splines.

2. Press the M key to open the Slate Material Editor. Locate and double-click the Standard material in the Material/Map Browser and then double-click the material node to open its parameters. Select the Metal shader from the drop-down list in the Shader Basic Parameters rollout. In the Metal Basic Parameters rollout, set the Diffuse color to a nice, shiny red and increase the Specular Level to **97** and the Glossiness value to **59**. Name the material **Toolbox**.

3. In the Maps rollout of the Material/Map Browser, double-click the option for the Splat map, and connect this node to the Glossiness parameter. Double-click the Splat map node. In the Splat Parameters rollout, set the Size value to **100**, change Color #1 to a rust color, and change Color #2 to white.

4. Double-click the Dent option in the Material/Map Browser, and connect its node to the Bump parameter. Double-click the Dent map node, and in the Dent Parameters rollout, set the Size value to **200**, Color #1 to black, and Color #2 to white.

5. Create another standard material node, and name it **Hinge**. Select the Metal shader from the Shader Basic Parameters rollout for this material also, and increase the Specular Level in the Metal Basic Parameters rollout to **26** and the Glossiness value to **71**. Also change the Diffuse color to a light gray. Click the map button next to the Glossiness value, and double-click the Noise map in the Material/Map Browser. This automatically creates and connects the Noise map node. Double-click the Noise map node, and in the Noise Parameters rollout, set the Noise map to Fractal with a Size of **10**.

6. Drag the output socket for the "Toolbox" material to the toolbox object and the "Hinge" material to the hinge and the handle objects.

Note

Bump and glossiness mappings are not visible until the scene is rendered. To see the material's results, choose Rendering→Render.

Figure 18.5 shows the well-used toolbox.

FIGURE 18.5

This toolbox shows its age with Glossiness and Bump mappings.

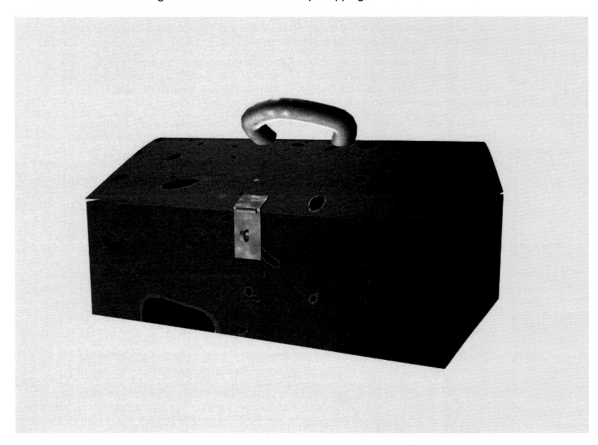

Understanding map types

Within the Material/Map Browser is a wide variety of map types. Tables 18.2 and 18.3 lists the common standard and scanline maps and briefly explains their features. Each of these maps has multiple parameters that you can change to alter the look of the map. For example, the Checker map includes color swatches for changing the colors or the checkerboard and a Soften value to blur the lines between.

TABLE 18.2 Standard Maps

Material Property	Description
Advanced Wood	An advanced procedural map that creates realistic wood textures including a variety of presets.
Bitmap	A bitmap image file. A large variety of formats are supported.
Blended Box Map	Allows maps to be applied from all 6 directions like a box surrounding the object.
Camera Map Per Pixel	Project a map from the location of a camera.

Cellular	A 3D image composed of patterns of small objects referred to as *cells*.
Checker	A 2D checkerboard image with two colors that alternate every row and column.
Color Correction	Used to color correct and change the colors of a map.
ColorMap	Applies a solid color as a map.
Composite	Combines a specified number of maps into a single map using the alpha channel.
Dent	A 3D map that works as a bump map to create indentations across the surface of an object.
Falloff	A 3D map that creates a grayscale image based on the direction of the surface normals.
Gradient	A 2D gradient image using three colors where the color gradually shifts to each color.
Gradient Ramp	A 2D gradient image using multiple selected colors.
Marble/Perlin Marble	A marbled material with random colored veins; a second alternative with a different look.
Mask	Used to select one map to use as a mask for another one to display holes.
Mix	Used to combine two maps or colors.
MultiTile	Allows multiple textures to be applied to an object at once.
Noise	A 3D map that randomly alters the surface of an object using two colors.
Normal Bump	Lets you alter the appearance of the details on the surface using a Normal map.
OSL Map	A shell for OSL maps that can be loaded externally and used.
Output	Provides a way to add the functions of the Output rollout to maps that don't include an Output rollout.
Particle Age/ Particle MBlur	Works with particle systems to change the particles' color as they age or blur them as they increase in velocity.
Raytrace	Used as an alternative to the raytrace material.
RGB Multiply	Multiplies the RGB values for two separate maps and combines them to create a single map.
RGB Tint	Includes color swatches for tinting the red, green, and blue channel values of a map.
ShapeMap	Adds shapes and splines drawn in 3ds Max to any texture, shapes can be filled and lines colored using map parameters.
Smoke	A map that creates random fractal-based patterns such as those you would see in smoke.
Speckle	Produces small, randomly positioned specks.
Splat	Used to create the look of covering an object with splattered paint.
Stucco	Generates random patches of gradients that create the look of a stucco surface.
Substance	Procedural map generation that creates textures with small file sizes.
Swirl	A 2D image that swirls two colors into a whirlpool effect.
TextMap	Picks a text object to display as a map where you can change the fill and line colors.
Tiles	A 2D image that creates patterns of bricks and tiles.
Vector Map	Applies loaded vector map as a texture. Supported file types include Adobe AI, SVG, AutoCAD Pattern files and PDF.
Vertex Color	Makes the vertex colors assigned to an Editable Poly object visible when the object is rendered.
Waves	Creates wavy, watery-looking maps.

TABLE 18.3 Scanline Maps

Material Property	Description
Flat Mirror	Reflects the surroundings using a coplanar group of faces.
Reflect/Refract	Provides another way to create reflections and refractions on objects.
Thin Wall Refraction	Simulates the refraction caused by a piece of glass, such as a magnifying glass.

Using OSL Maps

In addition to the main list of Standard maps in the Material/Map Browser, you can find a category listed as OSL maps. OSL stands for Open Shading Language, which is a language that makes freely available shaders that can be edited as needed. The default OSL maps will work like any other map and can be viewed and rendered without any trouble. Additional OSL maps can be downloaded and used within 3ds Max.

New Feature in 2020

Several additional OSL maps have been added to 3ds Max 2020 including Falloff, Simple Gradient, Threads, Waveform and Weave..

Accessing Map parameters

Many maps have several rollouts in common. These include Coordinates, Noise, and Time. In addition to these rollouts, each individual map type has its own parameters rollout.

The Coordinates rollout

Every map that is applied to an object needs to have mapping coordinates that define how the map lines up with the object. For example, with the soup-can-label example mentioned earlier, you probably would want to align the top edge of the label with the top edge of the can, but you could position the top edge of the map at the middle of the can. Mapping coordinates define where the map's upper-right corner is located on the object.

All map coordinates are based on a UVW coordinate system that equates to the familiar XYZ coordinate system, except that it is named uniquely so it's not confused with transformation coordinates. For UVW coordinates, U represents the horizontal coordinate, V is the vertical coordinate, and W is along the surface normal. To keep them straight, remember that the UVW coordinate system applies to surfaces and the XYZ coordinate system applies to spatial objects. These coordinates are required for every object to which a map is applied. In most cases, you can generate these coordinates automatically when you create an object by selecting the Generate Mapping Coordinates option in the object's Parameter rollout.

Note

Editable meshes don't have any default mapping coordinates, but you can generate mapping coordinates using the UVW Map modifier.

In the Coordinates rollout for 2D Maps, shown in Figure 18.6, you can specify whether the map will be a texture map or an environment map. The Texture option applies the map to the surface of an object as a texture. This texture moves with the object as the object moves. The Environ option creates an environment map, which often shows up on the object as reflections. Environment maps are locked to the world and not to an object. This causes the texture to change as the object is moved. Moving an object with an environment map applied to it scrolls the map across the surface of the object.

FIGURE 18.6

The Coordinates rollout lets you offset and tile a map.

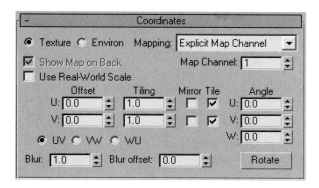

Different mapping types are available for both the Texture and Environ options. Mapping types for the Texture option include Explicit Map Channel, Vertex Color Channel, Planar from Object XYZ, and Planar from World XYZ. The Explicit Map Channel option is the default. It applies the map using the designated Map Channel. The Vertex Color Channel uses specified vertex colors as its channel. The two planar mapping types place the map in a plane based on the Local or World coordinate systems.

The Environ option includes Spherical Environment, Cylindrical Environment, Shrink-Wrap Environment, and Screen mapping types. The Spherical Environment mapping type is applied as though the entire scene were contained within a giant sphere. The same applies for the Cylindrical Environment mapping type, except that the shape is a cylinder. The Shrink-Wrap Environment plasters the map directly on the scene as though it were covering it like a blanket. All four corners of the bitmap are pulled together to the back of the wrapped object. The Screen mapping type just projects the map flatly on the background.

The Show Map on Back option causes planar maps to project through the object and be rendered on the object's back.

The U and V coordinates define the X and Y positions for the map. For each coordinate, you can specify an Offset value, which is the distance from the origin. The Tiling value is the number of times to repeat the image and is used only if the Tile option is selected. If the Use Real-World Scale option is selected, the Offset fields change to Height and Width, and the Tiling fields change to Size. The Mirror option inverts the map. The UV, VW, and WU options apply the map onto different planes.

Tiling is the process of placing a copy of the applied map next to the current one and so on until the entire surface is covered with the map placed edge to edge. You often will want to use tiled images that are seamless or that repeat from edge to edge.

Tip

Tiling can be enabled within the material itself or in the UVW Map modifier.

Figure 18.7 shows an image tile that is seamless. Notice how the horizontal and vertical seams line up. This figure shows three tiles positioned side by side, but because the opposite edges line up, the seams between the tiles aren't evident.

FIGURE 18.7

Seamless image tiles are a useful way to cover an entire surface with a small map.

The Material Editor includes an option in the Preview window that you can use to check the Tiling and Mirror settings. The Preview UV Tiling option is available in the right-click pop-up menu if you click on the preview sphere in the material node. You can switch to 2x2, 3x3, or 4x4.

You also can rotate the map about each of the U, V, and W axes by entering values in the respective fields, or by clicking the Rotate button, which opens the Rotate Mapping Coordinates dialog box, shown in Figure 18.8. Using this dialog box, you can drag the mouse to rotate the mapping coordinates. Dragging within the circle rotates about all three coordinates, and dragging outside the circle rotates the mapping coordinates about their center point.

FIGURE 18.8

The Rotate Mapping Coordinates dialog box appears when you click the Rotate button in the Coordinates rollout.

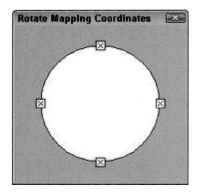

The Blur and Blur Offset values affect the blurriness of the image. The Blur value blurs the image based on its distance from the view, whereas the Blur Offset value blurs the image regardless of its distance.

Tip

You can use the Blur setting to help make tile seams less noticeable.

The Noise rollout

You can use the Noise rollout to randomly alter the map settings in a predefined manner. Noise can be thought of as static you see on the television added to a bitmap. This feature is helpful for making textures more grainy, which is useful for certain materials.

The Amount value is the strength of the noise function applied; the value ranges from 0 for no noise to 100 for maximum noise. You can disable this noise function at any time, using the On option.

The Levels value defines the number of times the noise function is applied. The Size value determines the extent of the noise function based on the geometry. You also can Animate the noise. The Phase value controls how quickly the noise changes over time.

The Time rollout

Maps, such as bitmaps, that can load animations also include a Time rollout for controlling animation files. In this rollout, you can choose a Start Frame and the Playback Rate. The default Playback Rate is 1.0; higher values run the animation faster, and lower values run it slower. You also can set the animation to Loop, Ping-Pong, or Hold on the last frame.The Output rollout

The Output rollout includes settings for controlling the final look of the map. The Invert option creates a negative version of the bitmap. The Clamp option prevents any colors from exceeding a value of 1.0 and prevents maps from becoming self-illuminating if the brightness is increased.

The Alpha From RGB Intensity option generates an alpha channel based on the intensity of the map. Black areas become transparent and white areas opaque.

Note

For materials that don't include an Output rollout, you can apply an Output map, which accepts a submaterial.

The Output Amount value controls how much of the map should be mixed when it is part of a composite material. You use the RGB Offset value to increase or decrease the map's tonal values. Use the RGB Level value to increase or decrease the saturation level of the map. The Bump Amount value is used only if the map is being used as a bump map; it determines the height of the bumps.

The Enable Color Map option enables the Color Map graph at the bottom of the Output rollout. This graph displays the tonal range of the map. Adjusting this graph affects the highlights, midtones, and shadows of the map. Figure 18.9 shows a Color Map graph.

FIGURE 18.9

The Color Map graph enables you to adjust the highlights, midtones, and shadows of a map.

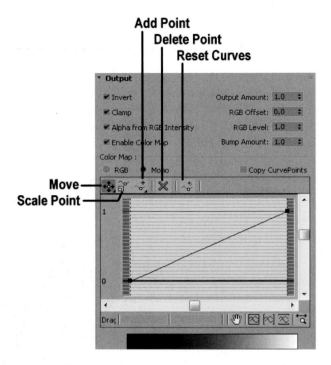

The left end of the graph equates to the shadows, and the right end is for the highlights. The RGB and Mono options let you display the graphs as independent red, green, and blue curves or as a single monocolor curve. The Copy CurvePoints option copies any existing points from Mono mode over to RGB mode, and vice versa. The buttons across the top of the graph are used to manage the graph points.

The buttons above the graph include Move (with flyout buttons for Move Horizontally and Move Vertically), Scale Point, Add Point (with a flyout button for adding a point with handles), Delete Point, and Reset Curves. Along the bottom of the graph are buttons for managing the graph view. The two fields at the bottom left contain the horizontal and vertical values for the current selected point. The other buttons are to Pan and Zoom the graph.

Using OSL Maps

Beneath the General category of maps in the Material/Map Browser is a category of maps called OSL. OSL stands for Open Shading Language. This is an open-source shading language that is easy to modify and allows you to create your own custom shaders. Within the OSL category are several preset shaders that work just like any other map, but each OSL map has a magnify button that opens the text for the shading language in a browser. These text files can be edited to change the behavior of the map.

New Feature in 2020

Several new OSL maps have been added to 3ds Max 2020. Also, OSL maps can now be viewed in the viewports.

Within the General category of maps is also an OSL Map item. This map lets you load and use any .osl file downloaded from the Internet. Just use this map like normal and click the OSL button to load an OSL file. Loaded files can then be edited as needed.

Creating Textures with External Tools

Several external tools can be valuable when you create material textures. These tools can include an image-editing program like Photoshop, a digital camera or camcorder, and a scanner. With these tools, you can create or capture images that can be applied as maps to a material using the map channels.

After the image is created or captured, you can apply it to a material by clicking a map shortcut button or by selecting a map in the Maps rollout. This opens the Material/Map Browser, where you can select the Bitmap map type and load the image file from the File dialog box that appears.

Creating material textures using Photoshop

When you begin creating texture images, Photoshop becomes your best friend. Using Photoshop's filters enables you to quickly create a huge variety of textures that add life and realism to your textures.

Table 18.3 is a recipe book of several common textures that you can create in Photoshop. The table provides only a quick sampling of some simple textures. Many other features and effects are possible with Photoshop.

TABLE 18.3 Photoshop Texture Recipes

Texture	Technique	Create in Photoshop	Apply in 3ds Max as
	Faded color	Decrease the image saturation value (Image→Adjustments→Hue/Saturation) by 20% to 30%.	Diffuse map
	Surface scratches	Apply the Chalk & Charcoal filter (Filter→Sketch→Chalk & Charcoal) with a Stroke Pressure of 2 to a blank white image and then apply the Film Grain (Filter→Artistic→Film Grain) filter with maximum Grain and Intensity.	Bump map
	Stains on fabric	Use the Dodge and Burn tools to add stains to a fabric bitmap.	Diffuse map
	Surface relief texture	Apply Dark Strokes filter (Filter→Brush Strokes→Dark Strokes) to a texture bitmap, and save the image as a separate bump image.	Diffuse map (original texture), Bump map (Dark Strokes version)
	Planar hair	Apply the Fibers filter (Filter→Render→Fibers).	Diffuse, Bump, and Specular maps
	Clouds or fog background	Apply the Clouds filter (Filter→Render→Clouds).	Diffuse map
	Nebula or plasma cloud	Apply the Difference Clouds filter (Filter→Render→Difference Clouds). Then switch black and white color positions, and apply the Difference Clouds filter again.	Diffuse map

	Rock wall	Apply the Clouds filter (Filter→Render→Clouds) and then apply the Bas Relief (Filter→Sketch→Bas Relief) filter.	Diffuse and Bump maps
	Burlap sack	Apply the Add Noise filter (Filter→Noise→Add Noise), followed by the Texturizer filter (Filter→Texture→Texturizer) with the Burlap setting.	Diffuse and Bump maps
	Tile floor	Apply the Add Noise filter (Filter→Noise→Add Noise), followed by the Stained Glass filter (Filter→Texture→Stained Glass).	Diffuse map
	Brushed metal	Apply the Add Noise filter (Filter→Noise→Add Noise), followed by the Angled Strokes filter (Filter→Brush Strokes→Angled Strokes).	Diffuse and Bump maps
	Frosted glass	Apply the Clouds filter (Filter→Render→Clouds); then apply the Glass (Filter→Distort→Glass) filter, and select the Frosted option.	Diffuse map
	Pumice stone	Apply the Add Noise filter (Filter→Noise→Add Noise), followed by the Chalk & Charcoal filter (Filter→Sketch→Chalk & Charcoal).	Diffuse and Bump maps
	Planet islands	Apply the Difference Clouds filter (Filter→Render→Difference Clouds) and then apply the Note Paper (Filter→Sketch→Note Paper) filter.	Diffuse and Shininess maps
	Netting	Apply the Mosaic Tiles filter (Filter→Texture→Mosaic Tiles), followed by the Stamp (Filter→Sketch→Stamp) filter.	Diffuse and Opacity maps
	Leopard skin	Apply the Grain filter (Filter→Texture→Grain) with the Clumped option, followed by the Poster Edges (Filter→Artistic→Poster Edges) filter applied twice.	Diffuse and Opacity maps

Capturing digital images

Digital cameras and camcorders are inexpensive enough that they really are necessary items when creating material textures. Although Photoshop can be used to create many unique and interesting textures, a digital image of riverbed stones is much more realistic than anything that can be created with Photoshop. The world is full of interesting textures that can be used when creating images.

Avoiding specular highlights

Nothing can ruin a good texture taken with a digital camera faster than the camera's flash. Taking a picture of a highly reflective surface like the surface of a table can reflect back to the camera, thereby ruining the texture.

You can counter this in several ways. One technique is to block the flash and make sure that you have enough ambient light to capture the texture. Taking pictures outside can help with this because you don't need the flash. Another technique is to take the image at an angle, but this might skew the texture. A third technique is to take the image and then crop away the unwanted highlights.

Tip

The best time to take outdoor photos that are to be used for materials is on an overcast day. This eliminates the direct shadows from the sun, which are very difficult to remove. It also makes the light gradient across the surface much cleaner for tiling the photo.

Adjusting brightness

Digital images that are taken with a digital camera are typically prelit, meaning that they already have a light source lighting them. When these prelit images are added to a 3ds Max scene that includes lights, the image gets a double dose of light that typically washes out the images.

You can remedy this problem by adjusting the brightness of the image prior to loading it into 3ds Max. For images taken in normal indoor light, you'll want to decrease the brightness value by 10 percent to 20 percent. For outdoor scenes in full sunlight, you may want to decrease the brightness even more.

You can find the Brightness/Contrast control in Photoshop in the Image→Adjustments→Brightness/Contrast menu.

Scanning images

In addition to taking digital images with a digital camera, you can scan images from other sources. For example, the maple-leaf tutorial was scanned from a real leaf found in my yard.

When scanning images, use the scanner's descreen option to remove any dithering from the printed image. If you place the image on a piece of matte black construction paper, the internal glare from the scanning bulb gets a more uniform light distribution.

Caution

Most magazine and book images are copyrighted and cannot be scanned and used without permission.

Tutorial: Creating a fishing net

Some modeling tasks can be solved more easily with a material than with geometry changes. A fishing net is a good example. Using geometry to create the holes in the net would be tricky, but a simple Opacity map makes this complex modeling task easy.

To create a fishing net, follow these steps:

1. Before working in 3ds Max, create the needed texture in Photoshop. In Photoshop, select File→New, enter the dimensions of **512** pixels x **512** pixels in the New dialog box, and click OK to create a new image file.
2. Select the Filter→Texture→Mosaic Tiles menu command to apply the Mosaic Tiles filter. Set the Tile Size to **30** and the Grout Width to **3**, and click OK. Then select the Filter→Sketch→Stamp menu command to apply the Stamp filter with a Light/Dark Balance value of **49** and a Smoothness value of **50**.
3. Choose File→Save As, and save the file as **Netting.tif**.

   A copy of this file is available in the Chap 18 directory in the downloaded content set.
4. Open the Fish net.max file from the Chap 18 directory in the downloaded content set.

   This file includes a fishing-net model created by stretching half a sphere with the Shell modifier applied.

5. Select the Rendering→Material Editor→Slate Material Editor menu command (or press the M key) to open the Material Editor. Double-click the Standard option in the Material/Map Browser to create a material node. Name the material **net**.

6. Locate and double-click the Bitmap option in the Material/Map Browser to add a node to the Node View. In the Select Image Bitmap File dialog box that opens, locate and select the Netting.tif image file. Drag a connection wire from the output socket of the Bitmap node and drop it on the input node of the Opacity parameter of the material node. Then drag the output socket for the material node and drop it on the net object in the viewports.

7. If you were to render the viewport, the net would look rather funny because the black lines are transparent instead of the white spaces. To fix this, double click the Bitmap node, open the Output rollout, and enable the Invert option. This inverts the texture image.

Note

Although you can enable the maps to display in the viewport, the transparency of the map is not displayed until you render the scene.

Figure 18.10 shows the rendered net.

FIGURE 18.10

A fishing net, completed easily with the net texture applied as an Opacity map

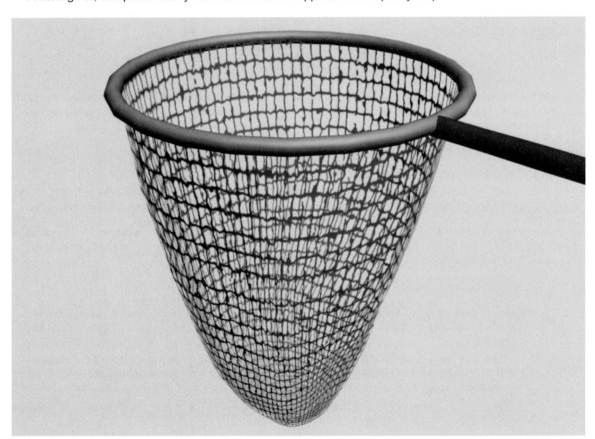

Summary

Now that you know about maps, you have a one-two punch in your materials corner. Learning to use these maps will make a big difference in the realism of your materials.

In this chapter, you learned about the following:

* Connecting map nodes in the Slate Material Editor to material nodes
* The various mapping possibilities provided in the Maps rollout
* How to create materials using external tools such as Photoshop and a digital camera

So far you've learned about the Standard material and using maps, but there are several additional specialized materials listed in the Material/Map Browser. The next chapter looks at these specialized materials and shows how they can be used.

Chapter 19
Using Specialized Material Types

IN THIS CHAPTER

Using Ink 'n Paint materials

Using the architectural, Autodesk Material Library and Physical materials

Using the DirectX Shader material

Creating Shaders in the ShaderFX Editor

Applying Substance textures

Randomizing Substance textures

You probably noticed back in the chapter on compound materials that a number of intriguing material types listed in the Material/Map Browser were not discussed. These material types are important, but they can't be classified as compound materials.

These specialized materials are sort of like your waffle iron. They're meant to be pulled out of your arsenal when you have a specific need. You wouldn't cook eggs on a waffle iron, would you?

The specialized materials included here are the Ink 'n Paint material, the Architectural materials, and the DirectX Shader material. These materials let you render the scene objects as a cartoon, use building materials, and view objects as they'll be seen in a game. For the DirectX Shader material, you can edit and create new shaders using the ShaderFX Editor.

Even more special material types are listed after you enable the Arnold renderer. These materials are so specialized that they require a change in the rendering engine. These Arnold materials have a vast array of properties that you can use to get a specific, realistic look.

With these specialized materials, you can create some really advanced, realistic results. Unfortunately, they don't make waffles.

Substance materials are made possible using a Substance map type found in the Material Editor. Although using these Substance materials tends to produce materials that look the same as other material types, the procedural aspect of these materials makes them unique. Most materials with this level of detail require a bitmap texture, and the larger the bitmap, the better the detail. These bitmaps, even when saved using a compressed format, can easily run several megabytes in size. Procedural textures, however, are created using a code, and the result is file sizes that are usually measured in kilobytes instead of megabytes.

So the benefit to using procedural textures is that you can get the detail of an amazing material without the cost in file size. Another huge advantage is that you can easily make changes to the textures by inputting different values into the code. This gives you lots of variety without having to duplicate, reload, and save in memory several different bitmaps. For certain conditions, such as 3d for the Web or for games, these advantages are huge.

Using the Ink 'n Paint Material

Although it may seem silly, many different production houses use the Autodesk® 3ds Max® 2020 software to create 2D line-drawn cartoons. This is accomplished using the Ink 'n Paint material. Traditionally, cartoons have been drawn by hand using a paper and pen. Then animation houses found that, using computers, you can fill in a cartoon feature easily, but using a 3D program like 3ds Max with its ability to animate using keyframes simplifies the animation task even further. The difference is in how the objects are rendered; the Ink 'n Paint material controls this.

With the Ink 'n Paint material selected, several rollouts appear, including the Basic Material Extensions. This rollout includes options for making the material 2-Sided, enabling a Face Map, and making the material Faceted. Other options cause the background to be foggy when not painting and make the alpha channel opaque. Maps are available for Bump and Displacement.

Controlling paint and ink

The Paint Controls rollout includes settings for how the paint (or colors inside the ink outline) is applied. You can specify colors for the Lighted, Shaded, and Highlight colors. The Lighted color is used for sections of the material that face the scene lights, the Shaded color is used for sections that are in the shadows, and the Highlight color is for the specular highlights. For each color, you also can select a map with an amount value. The Paint Levels value sets the number of colors that are used to color the material. The Glossiness value determines the size of the highlight. Figure 19.1 shows materials with Paint Level values of 2-6.

FIGURE 19.1

Use the Paint Level value to set the number of colors used in the material.

The Ink Controls rollout includes an Ink option that you can use to turn off the outlining ink completely. You also can set the Ink Quality to values between 1 and 3. The higher-quality values trace the edges better, but require more time to complete. The width of the ink strokes can be set to Variable Width or a Clamped width. Clamping the variable width keeps the ink lines from getting too thin. For each option, you can select a Minimum ink width, and if Variable Width is enabled, you can choose a Maximum ink width. For the Variable Width option, the stroke changes so that the minimum setting is used in lighted areas and the maximum width is used in shaded areas. This helps to accentuate the lighting. You also can apply the ink width as a map. Figure 19.2 shows the Ink 'n Paint material applied to a cube. The first image shows a standard material; the second image has the Ink option disabled. The last three images have Width values of 1, 10, and 30.

Tip

Placing a noise map on the Ink Width property is a great way to give the ink outline a hand-drawn look.

FIGURE 19.2

Changing the Variable Width value can have a big impact on the resulting image.

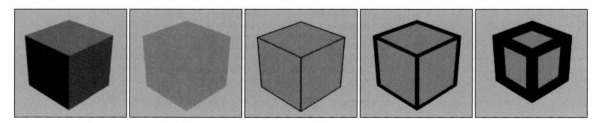

The rest of the options in the Ink Controls rollout are enabled to control where the ink is applied to the object. Options include Outline, Overlap, Underlap, Smoothing Group, and Material ID. For each of these options (except for Smoothing Group), you can alter a Bias value that can adjust intersecting edges. Each of these options can be applied as a map.

Tutorial: Cartooning a chicken

As an example of the Ink 'n Paint material, you'll render a cartoonish chicken model as a cartoon that is fit for the Sunday papers.

To apply the Ink 'n Paint material to a chicken model, follow these steps:

1. Open the Cartoon chicken.max file from the Chap 19 directory in the downloaded content set.

   This file includes a simple chicken model.

2. Open the Material Editor by choosing the Rendering→Material Editor→Slate Material Editor menu command (or by pressing the M key).

3. Double click on the Ink 'n Paint material to add a material node to the Material Editor work area, then double click on this node to make its parameters visible. Click on the Lighted Paint color swatch in the Parameters panel to open the Color Selector dialog box. Select the Sample Screen Color eyedropper tool and click on one of the unique colors in the viewport, such as the yellow in the beak, then drag from the material nodes output socket to each object in the scene that uses that color to apply the new Ink 'n Paint material.

4. Repeat step 3 for each color used in the chicken model.

5. Select the Rendering→Render menu command to render the scene.

Figure 19.3 shows the resulting cartoon.

FIGURE 19.3

Cartooning made easy with the Ink 'n Paint material

Using Architectural, Autodesk and Physical Materials

If you look in the Material/Map Browser, you'll notice an Architectural material. This material is specifically designed for creating realistic materials that can be applied to buildings and interiors. The Architectural material uses predefined templates to create almost any type of material that you'd find in a building, including ceramic, fabric, metal, glass, stone, wood, and water. The real benefit of using the Architectural materials is how they interact with real-world lights. Using Photometric lights and Radiosity produces realistic results. Figure 19.4 shows the available settings for this material type.

Note

If you've selected the Quicksilver renderer, the Autodesk Material Library become available. These materials are a more powerful and extensive set of architectural materials.

FIGURE 19.4

Architectural materials settings include many common material properties.

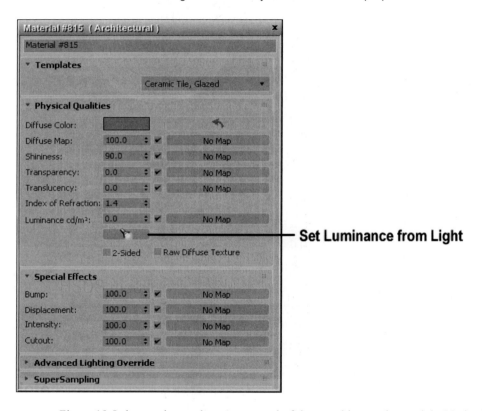

Figure 19.5 shows a house that uses several of these architectural materials. Notice how the shingles on the roof are consistent.

FIGURE 19.5

Architectural materials make adding textures to a building easy.

If you switch the rendering engine in the Render Setup dialog box to the Quicksilver Hardware Renderer, then the Material/Map Browser reveals a large number of specific real-world materials within the Autodesk Material Library section. These materials are even a step-up from the Architectural materials. These materials include the associated maps required to make the material look realistic, but the real benefit of these materials is the variety, which includes presets for Ceramic, Concrete, Fabric, Flooring, Glass, Metal, Paint, Roofing, Stone, Stucco, Wood and more. Figure 19.6 shows the Material Editor with Golden Sand selected.

FIGURE 19.6

Golden Sand is just one of the many materials available in the Autodesk Material Library.

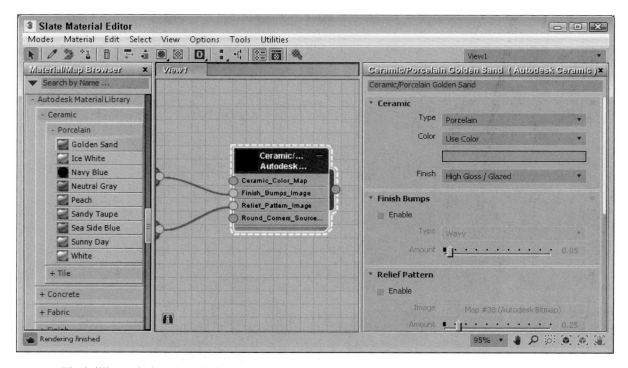

The brilliant minds at Autodesk keep improving the material selection and the latest improvement to appear is the Physical Material type. This material includes a huge list of available presets from Glossy Varnished Wood to Candle Wax and Polished Gold. Figure 19.7 shows a simple scene with several of these materials. This material is only available in the Material/Map Browser when the ART renderer or the Arnold renderer are selected in the Render Setup dialog box.

If you want more information on the ART and Arnold renderers, check out Chapter 30, "Rendering with ART and Arnold."

FIGURE 19.7

Physical Material along with the Arnold renderer produces realistic results.

Using the DirectX Material

A *shader* is an algorithm that defines the surface color for all pixels in the scene. Surface Shaders come to this conclusion by factoring in such details as texture files, reflection, refraction, and atmospheric effects. Atmospheric Shaders change the light properties as they move through a volume. Shaders can be programmed and applied to objects and scenes, and renderers like Arnold render the scene based on these shaders.

The Material Editor includes support for DirectX 9, 10 and 11 shaders using the DirectX Shader material. These shaders are saved as .fx files and are available only if the Direct3D display driver is enabled. FX files are text files created using the Higher-Level Shader Language (HLSL). The DirectX Shader also can load in NVIDIA CgFX files and mental image's MetaSL shaders, saved with the .XMSL extension.

Note

The DirectX Shader material is available only if you are using the DirectX display driver and the ART or Arnold renderer is enabled in the Render Setup dialog box.

In the Maps/fx directory where 3ds Max is installed are several example DirectX shaders. You can load these example shaders using the button in the DirectX Shader rollout. The available parameters are different for each shader.

FX shaders are commonly used in games, and applying an FX shader to an object displays the shader in the viewport. This display looks the same as if the object were rendered in the game engine. However, if you render an object with an FX shader applied, it will appear as default gray unless you specify a material to appear using the Software Render Style rollout.

DirectX Shader

The DirectX Manager rollout appears in the Material Editor when a standard material is selected and the Direct3D display driver is enabled. It lets you display the current material in the viewport as a DirectX shader when the Show Hardware Map in Viewport option is enabled. The current material also can be saved as an .fx material file. Many game engines render using DirectX, so this option lets you view your materials in the viewport as they will appear within the game.

Caution

The DirectX Manager rollout isn't available by default. It appears only when the Direct3D display driver is selected.

At the bottom of the DirectX Manager rollout is a drop-down list for selecting to use the available DirectX shaders. The two available DirectX shaders are LightMap and Metal Bump. These shaders are generic, so they can be used on many different types of objects. The Light Map shader includes a parameter for loading a custom light map, and the Metal Bump shader includes parameters for specifying two texture maps: specular; and normal, bump, and reflection maps.

Accessing Arnold materials and maps

With the Arnold renderer selected, The Material/Map Browser makes several Arnold specific materials and maps available, as shown in Figure 19.8. These materials take advantage of the advanced rendering features included in the Arnold rendering engine.

FIGURE 19.8

The Arnold renderer makes several specialized materials and maps available.

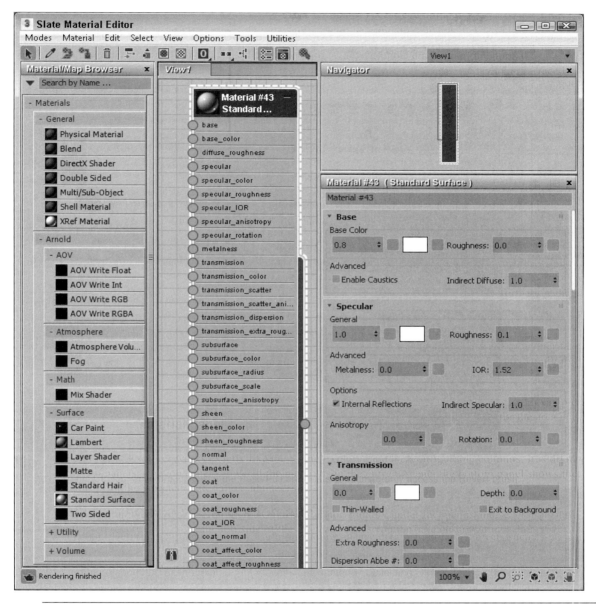

If you want more information on the ART and Arnold renderers, check out Chapter 30, "Rendering with ART and Arnold."

Combining bump and displacement maps

When Arnold is enabled, you also can take advantage of the Utility Bump and Displace Combiner materials. These materials let you combine up to three different bump or displacement maps into a single material. Each map has its own multiplier that is used to set its intensity.

Accessing the ShaderFX Editor

All geometry objects can be dressed up with materials and textures to give them a much more realistic look, but some objects have the benefit of being rendered in real-time using a shader. Shaders are different from materials in that they are actually pieces of programmed code that the real-time rendering engine uses to change its look based on environmental triggers.

One of the most popular places where shaders are used is in the game industry. Since game assets are rendered using the game engine as the player progresses, shaders are used to change the look of the object based on different criteria. For example, a shader could include glowing lights that are turned on when the environment is dark or damage applied to a car's material as it gets bumped into by the other cars.

Shaders are created using a programming language that defines what happens to the visual look of the shader under different various conditions. Although several shader languages exist, 3ds Max supports the Higher-Level Shading Language (HLSL). This language is quite complex, so several tools have been created that allow you to generate shaders without knowing the code language. These visual editors let you define the shader by connecting different nodes together. Shaders created using one of these visual editors can then be exported to a rendering engine that can read and display the shader code.

Autodesk® 3ds Max® 2020 has its own version of a visual shader creator called ShaderFX. Using this editor, you can create custom shaders and even view them in the viewports.

Don't look for the ShaderFX Editor in any of the menus. You can only access it after applying the DirectX Shader material to an object. The DirectX Shader material is found in the default Material/Map Browser in the Slate Material Editor. Double click on this material and it will load into the Material Editor window. At the top of the Parameters panel is a drop-down list with ShaderFX as an option and a button to Open ShaderFX. To use the shader created by ShaderFX, you need to select the ShaderFX option from the drop-down list and then click the Open ShaderFX to access the visual editor. Figure 19.9 shows the ShaderFX Editor.

FIGURE 19.9

The ShaderFX Editor is a simple node-based editor.

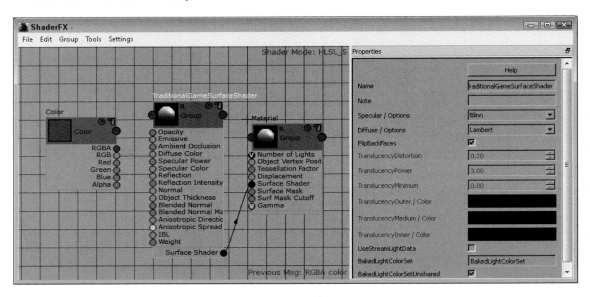

The ShaderFX Editor is made up on two panels. The left panel is the Work Area. This is where all the shader nodes are places and connected with each other. The right panel is the Properties panel. This panel displays the settings for the selected node. The editor also includes a menu. Using the File→Export Graph option you can export the current shader for use in other rendering engines.

Working with shader nodes

By default, two shader nodes appear when you launch the ShaderFX Editor. These two nodes are the Traditional Game Surface Shader and the Material nodes. These two nodes provide an interface to 3ds Max for displaying and working with the defined shader.

Each shader node consists of a title followed by a list of parameters, as shown in Figure 19.10. At the top of each node are icons for displaying a preview, showing/hiding the parameters and outputting the results.

FIGURE 19.10

Shader nodes have input and output channels.

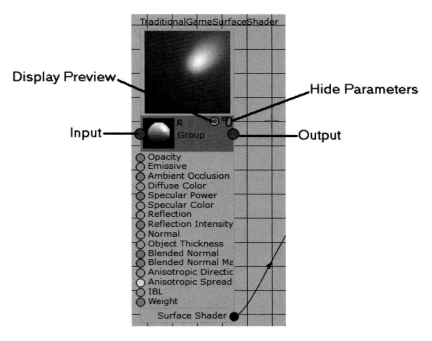

Within the Work Area, you can select and drag the nodes about by dragging on their title area.

Tip

You can pan the work area by dragging with the middle mouse button and zoom in and out by scrubbing the mouse scroll button.

New nodes are added to the work area by right clicking in the work area and selecting the node type from the pop-up menu. The available hardware node types include Flow Control, Inputs Application, Inputs Common, Lighting, Math, Matrices, Patterns, Textures, Values, and Various.

Each node parameter is color-coded based on the type of data that it is expecting. For example, single float values are green and RGB colors are light blue. You cannot connect a single float value into an RGB color input that is expecting three values, one for each color or an error will occur. Connecting inputs and outputs with the same color will help prevent this from happening.

Nodes can be deleted by simply selecting them and pressing the Delete key.

Tutorial: Building a Shader Tree

When several nodes are connected together, the whole mess is called a Shader tree. In this example, we will create a simple shader tree and assign it to an object in the scene.

To create and assign a Shader tree to an object, follow these steps:

1. Open the Command Panel and click the Sphere button, then drag in the viewport to create a simple sphere object.

2. Open the Slate Material Editor (or press the M key), and double-click the DirectX Shader material in the Material/Map Browser. Double-click the new material node to access its parameters; then select the ShaderFX option from the drop-down list and click on the Open ShaderFX button. The ShaderFX Editor opens with two nodes.

3. Right click on the work area of the ShaderFX Editor and choose the HW Shader Nodes→Textures→Texture Map option. This adds another node to the work area.

4. Click on the new Texture Map node to access its parameters and click on the My Texture/Path button. In the file dialog box that opens, locate and load the Laser stripe.jpg file from the Chap 21 folder.

5. Right click on the work area of the ShaderFX Editor again and choose the HW Shader Nodes→Textures→UV Panner option. This adds another node to the work area.

6. Drag from the yellow output at the bottom right of the UV Panner node to the upper left corner of the Texture Map node. Then drag from the light blue Color output on the Texture Map node to the Diffuse Color input on the Traditional Game Surface Shader node. Figure 19.11 shows the resulting shader tree. Then close the ShaderFX Editor.

Tip

Since the UV Panner is an animated node, you can see the results in the preview pane of the Texture Map node if you enable the Settings→Play Animated Shaders.

FIGURE 19.11

The resulting shader tree consists of several interconnected nodes.

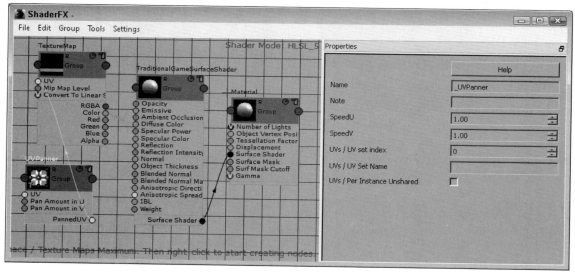

7. With the DirectX Shader node selected in the Material Editor, select the sphere object and click the Assign Material to Selection button at the top of the Material Editor. This assigns the defined shader to the sphere object.

8. Click the Play Animation button to see the resulting animated material.

Figure 19.12 shows the resulting shader.

FIGURE 19.12

This animated shader sends a stripe up along the surface of the sphere.

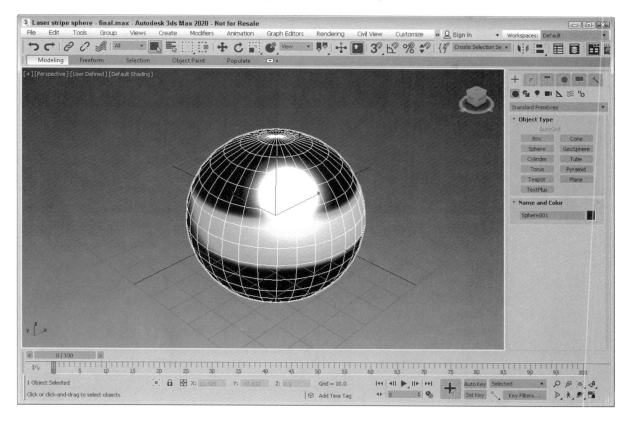

Selecting and Applying Substance Textures

The Substance textures feature isn't a whole new way of creating textures with its own interface and rendering engine. Substance textures work with the existing materials in the Material Editor, and unless you were looking for them, you might miss them. Substance textures are available in the Material/Map Browser under the Maps category and can be used with the normal set of default materials.

The Substance map node includes output nodes for attributes such as Diffuse, Specular, Normal, Bump, Displacement, Height, Opacity, and Glossiness. The material tree, shown in Figure 19.13, uses a Substance map to control the Diffuse and Normal channels of a Standard material.

FIGURE 19.13

Substance materials are made possible using the Substance map node.

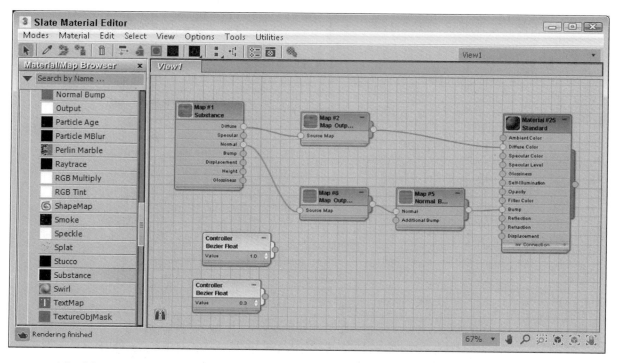

After it's correctly linked, the resulting material is applied to an object in the viewport as usual by selecting the object and clicking the Assign Material to Selection button in the Material Editor or by simply dragging the output for the Standard material node to the object. The applied material is then rendered using the default rendering methods. Figure 19.14 shows the resulting Old Painted Planks Substance material applied to a cube.

FIGURE 19.14

Substance materials have lots of detail despite their small file size.

Loading Substance textures

The available Substance textures are limited to the included set that ships with the Autodesk® 3ds Max® 2020 software, but within the Substance Package Browser rollout is a button to Get Substance From Marketplace. This button opens a web browser to the Allegorithmic website where you can browse and purchase additional texture sets. Allegorithmic is the company that created the Substance textures.

3ds Max includes more than 80 preinstalled Substance textures. To choose from the installed texture sets, simply click the Load Substance button in the Substance Package Browser rollout of the Material Editor. This opens to the Maps\Substance folder where 3ds Max is installed. The available textures are divided into two folders: Noises and Textures.

Linking Substance maps

Because Substance textures are available only as map nodes in the Material Editor, you need to link them to a base material in order to apply them to objects. To do this, simply create a Substance node by selecting and double-clicking it in the Material/Map Browser. A base material node, such as Standard, also must be added to the Slate Material Editor view pane.

After a Substance node is in the Material Editor's view window, select it and use the Load Substance button in the Substance Package Browser rollout to load a specific texture type. Then drag from the Substance node's Diffuse channel to the Diffuse Color channel on the Standard node. A Map Output node is automatically added to the material tree between the Substance and Standard nodes.

The other channels—Normal, Bump, and Displacement on the Substance node—also can be mapped to the Standard node. If you link the Normal map to the Standard node, you need to go through a Normal Bump map before linking to the Standard node's Bump channel.

Any Substance textures that have a Displacement channel can link directly to the Displacement channel on the Standard material node. The Height channel in Substance textures also can be linked to the Standard node's Displacement channel. The Height channel combines the Bump and Displacement information into a single channel. If a Substance texture is linked to both the Bump and Displacement channels, you can use the Relief Balance setting in the Parameters rollout to change the weighting between these two channels.

Some Substance textures such as Fencing have an Opacity channel. Connecting this channel to the Opacity channel on the Standard node causes the area between the fence links to show what is behind them. The Autumn Leaves texture has an Opacity channel also. Figure 19.15 shows an autumn leaves texture on top of a grass texture.

FIGURE 19.15

Some textures use the Opacity channel to allow textures beneath to show through.

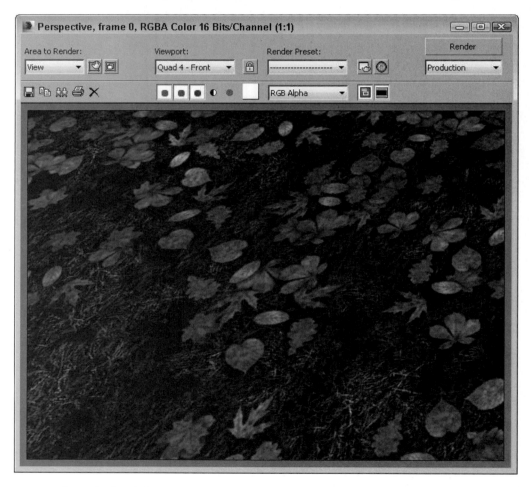

Several other Substance textures have an Emissive channel. This can be connected to the Self-Illumination channel to create a glow.

After the Substance map node is linked in, you can apply it to a scene object and render to see its results.

Randomizing Substance Textures

Although the Substance node has several different rollouts available, including Texture Size, Coordinates, and Noise, each specific texture has its own set of parameters with several options for randomizing the texture. For example, Crumpled Paper has a Crumpled Paper Parameters rollout with settings for Crumple, Dirt, Paper Color, and so on.

Almost every Substance texture also has a Random Seed setting that controls the randomness of the texture. By changing this value, you can create a uniquely different texture, even though all the other settings are identical. Also, two textures with the same Random Seed value have the same results. Figure 19.16 shows the Cracked Plaster texture with identical settings except for the Random Seed value.

FIGURE 19.16

Substance textures can be made to appear different by changing the Random Seed value.

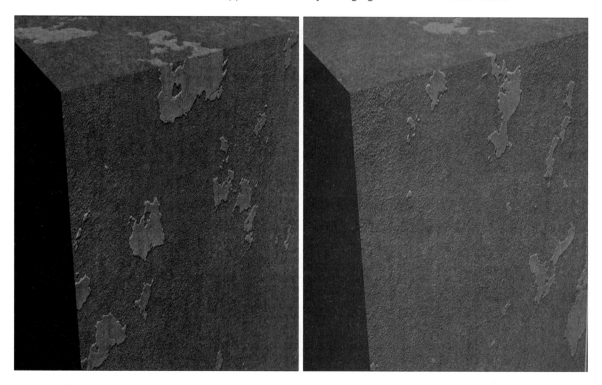

Tutorial: Applying Substance Textures to a Scene

The Substance textures are especially good at creating worn, old-looking textures. It is easy to add the randomness of rust and corrosion to surfaces, so we use several Substance textures to create an abandoned city scene.

To create a scene using Substance textures, follow these steps:

1. Open the Wall and crates.max file from the Chap 19 directory in the downloaded content set.

   This file contains several primitive shapes positioned to create a simple scene.

2. Select the plane object in front that represents the ground, and open the Slate Material Editor by choosing Rendering→Material Editor→Slate Material Editor (or by pressing the M key).

3. Double-click the Standard material in the Material/Map Browser, and then locate and double-click the Substance and Normal Bump maps in the Maps folder. Double-click the Substance node in the Node View pane, and click the Load Substance button. Locate and select and load the Dry Ground 02 map from the file dialog box.

4. Connect the Diffuse Color channels from the Substance node to the Standard node, and then connect the Normal channel of the Substance node to the input of the Normal Bump map and the output of the Normal Bump map node to the Bump mode of the Standard node. Between nodes, a Map Output node is automatically created. Double-click the Standard material node in the Node View, and click the Assign Material to Selection button in the Material Editor to apply the selected material.

5. Select the background wall object in the viewport, and repeat Steps 3 and 4, applying the Brick Wall 03 Substance texture.

6. Select the top crate object, and repeat Steps 3 and 4 to apply the Aircraft Metal Substance texture. Then repeat for the other crates.

Figure 19.17 shows the resulting scene.

FIGURE 19.17

Substance textures can add lots of detail to a scene.

Summary

This chapter covered an assortment of specialized materials that enable specific types of effects. These specialized materials include Ink 'n Paint, architectural, and DirectX Shader. For the DirectX Shader material, the ShaderFX Editor provides a visual way to connect shader nodes together to make shaders.

This chapter covered the basics of using the Substance map node to create highly detailed textures without the file size overhead. Textures created with Substance can be exported to a game engine using a plug-in tool that is available from Allegorithmic.

In this chapter, you've learned about the following:

* Creating a cartoon rendering with the Ink 'n Paint materials

* Using Architectural, Autodesk and Physical Materials

* Viewing game assets with the DirectX Shader

* Accessing the ShaderFX Editor

* Creating and applying a Shader tree

* Selecting and applying Substance materials to objects

* Randomizing Substance textures using the Parameter values

There are several materials that are made by combing several materials together. These are called Compound Materials and they are covered in the next chapter.

Chapter 20
Creating Compound Materials and Using Material Modifiers

IN THIS CHAPTER

Creating and using compound materials

Using material IDs to apply multiple materials

Using Matte/Shadow materials

Working with material modifiers

Displacing surface with a bitmap

Now that you've learned to create materials using the Standard material type, you get a chance to see the variety of material types that you can create in the Autodesk® 3ds Max® 2020 software. You can select all the various 3ds Max materials from the Material/Map Browser. Open this browser automatically by selecting Rendering→Material/Map Browser.

Although many of these materials are called compound materials, they are really just collections of materials that work together as one. Just like a mesh object can include multiple elements, materials also can be made up of several materials. Using material IDs, you can apply multiple materials to the subobject selections of a single mesh object. In this way, you can use a compound material that holds all different materials that a car model needs, including glass, rubber, car body, chrome, and so on, and then just apply this one material to the car. The material IDs are used to mark which subobjects get which material.

The chapter concludes with a quick look at the various modifiers that are applied to materials.

Using Compound Materials

Compound materials combine several different materials into one. You select a compound object type by double-clicking the material type from the Material/Map Browser. Most of the entries in the Material/Map Browser are compound objects.

Compound materials usually include several different levels. For example, a Top/Bottom material includes a separate material for the top and the bottom. Each of these submaterials can then include another Top/Bottom material, and so on. The links between these different submaterials are clearly visible in the View Node panel as connected nodes.

Each compound material includes a customized rollout in the Parameter Editor for specifying the submaterials associated with the compound material.

Note
Some of the material types work closely with specific objects and other 3ds Max features such as Advanced Lighting Override, DirectX Shader, Shell and XRef Material. These materials are covered in their respective chapters.

Physical Material

The first available compound material is the newest and probably the most complex. The Physical Material is based on real-world materials and includes several distinct layers. This material gets the greatest effect when rendered in a physical-based renderer such as ART or Arnold. Although you can use the Scanline renderer to render an object with a Physical Material applied, it will not have the details and look that is possible in the other renderers.

Cross Reference

Enabling and using the ART and Arnold renderers is covered in Chapter 30, "Rendering with ART and Arnold."

The Physical Material node had a large number of settings that you can configure, but the easiest way to use this material is with the Presets rollout, shown in Figure 20.1. Some of the available presets include Glossy Paint, Rough Concrete, Glazed Ceramic, Rubber, Frosted Glass, Sandblasted Silver, Red Sports Car Paint and Candle Wax.

FIGURE 20.1

The Physical Material node includes a Presets rollout where you can choose from a huge list of available real-world materials.

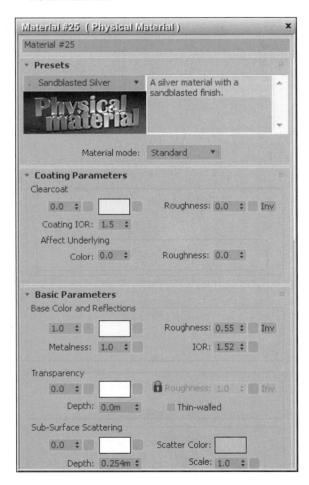

The Physical Material node also has parameters for a clear coat that appears above the base color. You can also set the Sub-Surface Scattering parameters to make the material appear semi-transparent to light. This material also supports a wide range of maps.

Blend

The Blend material blends two separate materials on a surface. The Blend Basic Parameters rollout, shown in Figure 20.2, includes separate nodes for each of the two submaterials. The check boxes to the right of these buttons enable or disable each submaterial. The Interactive option enables you to select one of the submaterials to be viewed in the viewports.

The Mask button (which appears below the two submaterial buttons) lets you load a map to specify how the submaterials are mixed. Gray areas on the map are well blended, white areas show Material 1, and black areas show Material 2. As an alternative to a mask, the Mix Amount determines how much of each submaterial to display. A value of 0 displays only Material 1, and a value of 100 displays only Material 2. This value can be animated, allowing an object to gradually change between materials.

FIGURE 20.2

The Blend material can include a mask to define the areas that are blended.

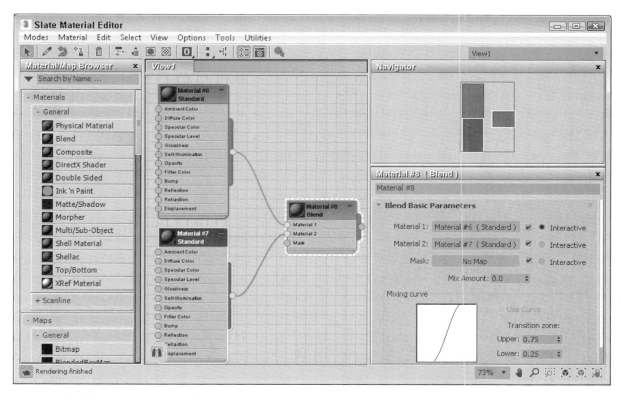

The Mixing curve defines the transition between edges of the two materials. The Upper and Lower spinners help you control the curve.

Composite

The Composite material mixes up to ten different materials by adding, subtracting, or mixing the opacity. The Composite Basic Parameters rollout, shown in Figure 20.3, includes buttons for the base material and nine additional materials that can be composited on top of the base material. The materials are applied from top to bottom.

FIGURE 20.3

Composite materials are applied from top to bottom, with the last layer placed on top of the rest.

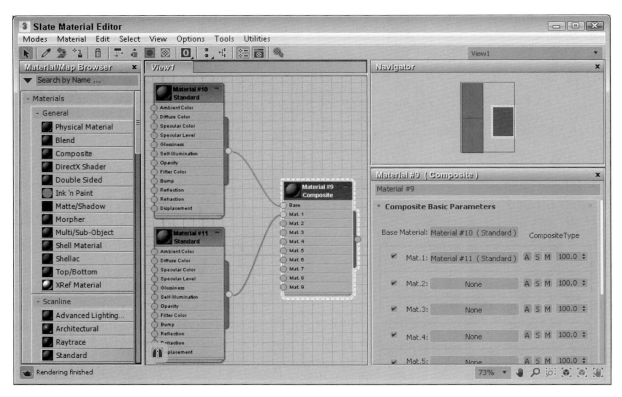

You enable or disable each material using the check box to its left. The buttons labeled with the letters *A, S,* and *M* specify the opacity type: Additive, Subtractive, or Mix. The Additive option brightens the material by adding the background colors to the current material. The Subtractive option has the opposite effect and subtracts the background colors from the current material. The Mix option blends the materials based on their Amount values.

To the right of the A, S, and M buttons is the Mix amount. This value can range from 0 to 200. At 0, none of the materials below it will be visible. At 100, full compositing occurs. Values greater than 100 cause transparent regions to become more opaque.

Double Sided

The Double Sided material specifies different materials for the front and back of object faces. You also have an option to make the material translucent. This material is for objects that have holes in their surface. Typically, objects with surface holes do not appear correctly because only the surfaces with normals pointing outward are visible. Applying the Double Sided material shows the interior and exterior of such an object.

The Double Sided Basic Parameters rollout includes two buttons, one for the Facing material and one for the Back material. The Translucency value sets how much of one material shows through the other.

Multi/Sub-Object

You can use the Multi/Sub-Object material to assign several different materials to a single object via the material IDs. You can use the Mesh or Poly Select modifier to select each subobject area to receive the different materials.

At the top of the Multi/Sub-Object Basic Parameters rollout, shown in Figure 20.4, is a Set Number button that lets you select the number of subobject materials to include. This number is displayed in a text field to the left of the button. Each submaterial is displayed as a separate area on the sample object in the sample slots. Using the Add and Delete buttons, you can selectively add submaterials to or delete submaterials from the list.

Tip

You can set the number of materials that are included by default in the Multi/Sub-Object material using the Options→Preferences dialog box in the Slate Material Editor. Nodes for each material are also included by default when a Multi/Sub-Object material node is created, but you can ensure that no extra material nodes are included by enabling the Empty Sub-Material Slots option in the Preferences dialog box.

FIGURE 20.4

The Multi/Sub-Object material defines materials according to material IDs.

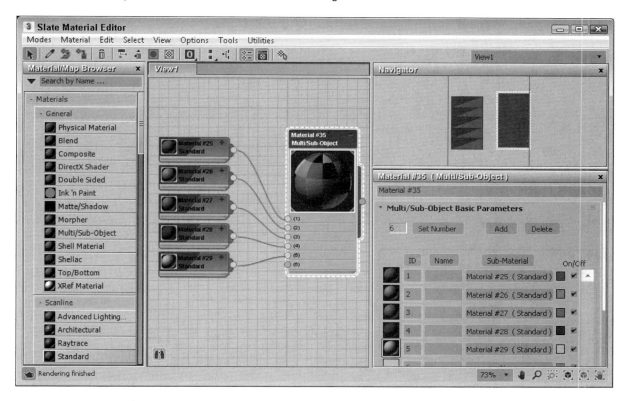

Each submaterial includes a sample preview of the submaterial and an index number listed to the left, a Name field where you can type the name of the submaterial, a button for selecting the material, a color swatch for creating solid color materials, and a check box for enabling or disabling the submaterial. You can sort the submaterials by clicking the ID, Name, or Sub-Material buttons at the top of each column.

After you apply a Multi/Sub-Object material to an object, convert the object to an Editable Mesh or Poly, or use the Mesh or Poly Select modifier to make a subobject selection and match the Material IDs in the Surface Properties rollout to the material for the subobject selection. In the Material section for this subobject selection, choose a material ID to associate with a submaterial ID, or select the material by name from the drop-down list.

Tutorial: Creating a patchwork quilt

When I think of patches, I think of a 3D 3ds Max object type, but for many people *patches* instead brings to mind small scraps of cloth used to make a quilt. Because they have the same name, this example uses 3ds

Max patches to create a quilt. You can then use the Multi/Sub-Object material to color the various patches appropriately.

To create a quilt using patches, follow these steps:

1. Open the Patch quilt.max file from the Chap 20 directory in the downloaded content set.

   This file contains a quilt made of patch objects that have been combined into one object.

2. Open the Material Editor by choosing Rendering→Material Editor→Slate Material Editor (or press M), and double-click the Multi/Sub-Object material in the Material/Map Browser panel.

   The Multi/Sub-Object material node loads into the Node View. Double-click the new node to make the Multi/Sub-Object Basic Parameters rollout appear in the Parameter Editor panel. Click the Set Number button, and enter the value of **10**.

3. Double-click the Standard material in the Material/Map Browser to create a separate node for each of the submaterials included in the Multi/Sub-Object material. Then drag from the output socket of each Standard material node to the input socket in the Multi/Sub-Object node for each of the submaterials.

4. In the Multi/Sub-Object Basic Parameters rollout, click the color swatches to the right of the Material button to open the Color Selector. Select different colors for each of the first ten material ID slots.

5. Drag the Multi/Sub-Object material node's output socket, and drop it onto the patch object in the viewports. Close the Material Editor. Initially, the entire quilt is only one color because all the Material IDs haven't been set.

6. In the Modify panel, select the Patch subobject, and scroll to the bottom of the Modify panel to the Surface Properties rollout.

7. Assign each patch a separate material ID by clicking a patch and changing the ID number in the rollout field.

Figure 20.5 shows the finished quilt. Because it's a patch, you can drape it over objects easily.

FIGURE 20.5

A quilt composed of patches and colored using the Multi/Sub-Object material

Morpher

The Morpher material type works with the Morpher modifier to change materials as an object morphs. For example, you can associate a blushing effect with light red applied to the cheeks of a facial expression to show embarrassment. You can use this material only on an object that has the Morpher modifier in its Stack. The Morpher modifier includes a button called Assign New Material in the Global Parameters rollout for loading the Material Editor with the Morpher material type.

Discover more about the Morpher modifier in Chapter 33, "Using Animation Layers and Animation Modifiers."

For the Morpher material, the Choose Morph Object button in the Morpher Basic Parameters rollout lets you pick a morpher object in the viewports and then open a dialog box used to bind the Morpher material to an object with the Morpher modifier applied. The Refresh button updates all the channels. The base material is the material used before any channel effects are used.

The Morpher material includes 100 channels that correlate to the channels included in the Morpher modifier. Each channel can be turned on and off. At the bottom of the parameters rollout are three Mixing Calculation options that can be used to determine how often the blending is calculated. The Always setting can consume lots of memory and can slow down the system. Other options are When Rendering and Never Calculate.

Shellac

The Shellac material is added on top of the Base material. The Shellac Basic Parameters rollout includes only two buttons for each material, along with a Color Blend value. The Blend value has no upper limit.

Top/Bottom

The Top/Bottom material assigns different materials to the top and bottom of an object. The Top and Bottom areas are determined by the direction in which the face normals point. These normals can be according to the World or Local Coordinate System. You also can blend the two materials.

The Top/Bottom Basic Parameters rollout includes two buttons for loading the Top and Bottom materials. You can use the Swap button to switch the two materials. Using World coordinates enables you to rotate the object without changing the material positions. Local coordinates tie the material to the object.

The Blend value can range from 0 to 100, with 0 being a hard edge and 100 being a smooth transition. The Position value sets the location where the two materials meet. A value of 0 represents the bottom of the object and displays only the top material. A value of 100 represents the top of the object, and only the Bottom material is displayed.

Tutorial: Surfing the waves

There's nothing like hitting the surf early in the morning, unless you consider hitting the virtual surf early in the morning. As an example of a compound material, you apply the Top/Bottom material to a surfboard.

To apply a Top/Bottom compound material to a surfboard, follow these steps:

1. Open the Surfboard.max file from the Chap 20 directory in the downloaded content set.

    This file contains a surfboard model and an infinite plane to represent the ocean.

2. Apply the Ocean Surface material, which is already created in the Material Editor, to the plane object by dragging the material node's output socket from the Material Editor to the Plane object.

3. In the Material/Map Browser, select and double-click the Top/Bottom material to create a new material node.

4. Double-click the new material node, and name it **Surfboard**. Then click the Top Material node, name the material **Surfboard Top**, and change the Diffuse color to White. Double-click the Surfboard node again and then click the Bottom Material node. Give this material the name **Surfboard Bottom**, and change the Diffuse color to Black.

5. Drag this material node's output socket to the surfboard object.

Figure 20.6 shows the resulting image.

FIGURE 20.6

A rendered image of a surfboard with the Top/Bottom compound material applied

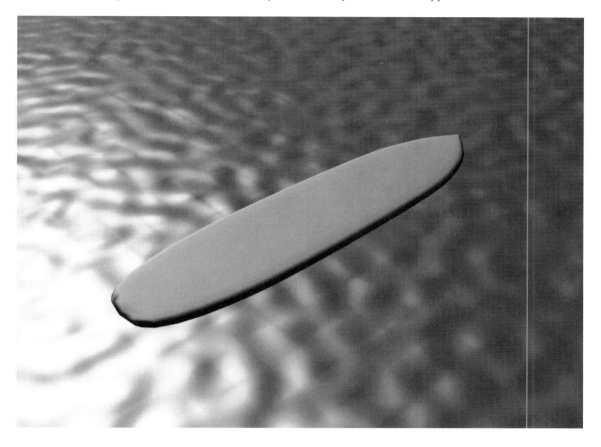

Applying Multiple Materials

Most complex models are divided into multiple parts, each distinguished by the material type that is applied to it. For example, a car model would be separated into windows, tires, and the body, so that each part can have a unique material applied to it.

Using material IDs

Sometimes you may want to apply multiple materials to a single part. Selecting subobject areas and using material IDs can help you accomplish this task.

Many of the standard primitives have material IDs automatically assigned: Spheres get a single material ID, boxes get six (one for each side), and cylinders get three (one for the cylinder and one for each end cap). In addition to the standard primitives, you can assign material IDs to Editable Mesh objects. You also can assign these material IDs to any object or subobject using the Material modifier. These material IDs correspond to the various materials specified in the Multi/Sub-Object material.

Note

Don't confuse these material IDs with the material effect IDs, which are selected using the Material Effect flyout buttons under the sample slots. Material IDs are used only with the Multi/Sub-Object material type, whereas the effect IDs are used with the Render Effects and Video Post dialog boxes for adding effects such as glows to a material.

Tutorial: Mapping die faces

As an example of mapping multiple materials to a single object, consider a die. Splitting the cube object that makes up the die into several different parts wouldn't make sense, so you'll use the Multi/Sub-Object material instead.

To create a die model, follow these steps:

1. Open the Pair of dice.max file from the Chap 20 directory in the downloaded content set.

   This file contains two simple cube primitives that represent a pair of dice. I also used Adobe Photoshop and created six images with the dots of a die on them. All these images are the same size.

2. Open the Material Editor, and double-click the Multi/Sub-Object material from the Material/Map Browser. Then double-click the material node, and name the material **Die Faces**.

3. In the Multi/Sub-Object Basic Parameters rollout, click the Set Number button, and enter a value of **6**.

4. Name the first material **face 1**, and click the material button to the right that is currently labeled None to open the Material/Map Browser. Select the Standard material type, and click OK. Then click the material button again to view the material parameter rollouts for the first material. Click the map button to the right of the Diffuse color swatch to open the Material/Map Browser again, and double-click the Bitmap map type. In the Select Bitmap Image File dialog box, choose the dieface1.tif image from the Chap 20 directory in the downloaded content set, and click Open.

5. Back in the Material Editor, return to the Multi/Sub-Object Basic Parameters rollout by double-clicking the material node, and repeat Step 4 for each of the die faces, using the bitmap corresponding to each face.

6. When the Multi/Sub-Object material is defined, select the cube object, and click the Assign Material to Selection button.

Note

Because the cube object used in this example is a box primitive, you didn't need to assign the material IDs to different subobject selections. The box primitive automatically assigned a different material ID to each face of the cube. When material IDs do need to be assigned, you can specify them in the Surface Properties rollout for editable meshes.

Figure 20.7 shows a rendered image of two dice being rolled.

Tip

If you enable the Materials→Shaded Materials with Maps menu command in the Shading Viewport label, the subobject materials are visible.

FIGURE 20.7

These dice have different bitmaps applied to each face.

Using the Clean MultiMaterial utility

All compound materials have submaterials that are used to add layers of detail to the material, but if these submaterials aren't used, they can take up memory and disk space. For example, if you have a Multi/Sub-Object material with ten materials, and the scene only uses three of the materials, the other seven materials aren't needed and can be eliminated.

You can locate and eliminate unused submaterials in the scene using the Clean MultiMaterial utility. This utility can be accessed from the Utilities panel in the Command Panel by clicking the More button or from the Utilities menu in the Material Editor.

Clicking the Find All button finds all submaterials that aren't used and presents them in a list where you can select the ones to clean.

Using the Matte/Shadow Material

Another commonly used material is the Matte/Shadow material. You can apply Matte/Shadow materials to objects to make portions of the model invisible. This lets any objects behind the object or in the background show through. This material also is helpful for compositing 3D objects into a photographic background. Objects with Matte/Shadow materials applied also can cast and receive shadows. The effect of these materials is visible only when the object is rendered.

Note

The Matte/Shadow material type is unavailable if the Arnold renderer is enabled.

Matte/Shadow Basic Parameters rollout

You can apply a Matte/Shadow material by double-clicking Matte/Shadow from the Material/Map Browser. Matte/shadow materials include only a single rollout: the Matte/Shadow Basic Parameters.

The Opaque Alpha option causes the matte material to appear in an alpha channel. This essentially is a switch for turning Matte objects on and off.

You can apply atmospheric effects such as fog and volume light to Matte materials. The At Background Depth option applies the fog to the background image. The At Object Depth option applies the fog as though the object were rendered.

Find out about Atmospheric effects in Chapter 44, "Using Atmospheric and Render Effects."

The Receive Shadows option enables shadows to be cast on a Matte object. You also can specify the Shadow Brightness and color. Increasing Shadow Brightness values makes the shadow more transparent. The Affect Alpha option makes the shadows part of the alpha channel.

Matte objects also can have reflections. The Amount spinner controls how much reflection is used, and the Map button opens the Material/Map Browser.

Tutorial: Adding 3D objects to a scene

A common use of the Matte/Shadow map is to add 3D objects to a background image. For example, if you add a Plane object that is aligned with the ground plane in the background image with a Matte/Shadow material applied to it, the Plane object can capture the shadows of the 3D objects, but the Matte/Shadow material allows the background to be seen through the Plane object. The result is that the 3D object appears added to the background image scene.

To use a Matte/Shadow material to add an object to a background image, follow these steps:

1. Open the Xylophone on shadow matte.max file from the Chap 20 directory in the downloaded content set.

    This file contains a background image of a cow statue taken in front of the Boston Convention Center. The scene also includes a xylophone positioned on the Plane object. The Plane object has a Scale multiplier of 4 to make a complete ground plane when the scene is rendered.

2. Open the Slate Material Editor by pressing the M keyboard shortcut. Double-click the Matte/Shadow option in the Material/Map Browser. Name the material **Shadow plane**, and apply it to the Plane object in the Perspective viewport.

3. Select the Create→Lights→Omni menu, and click in the Top viewport to create a light. In the General Parameters rollout, enable the Shadows option. Then, in the Shadow Parameters rollout, click the Color swatch. In the Color Selector that opens, select the Sample Screen Color eyedropper tool, and click the cow's shadow color to select it. Then adjust the Density setting to 0.7 to make the shadow slightly transparent.

4. Click the Perspective viewport's rendering label in the upper-left corner of the viewport, and select the Lighting and Shadows→Shadows option.

    This makes the shadows appear in the viewport on the Plane object.

5. Move the Omni light in the Left and Top views until the shadows are projected along the same path as the cow's shadows.

Tip

One way to help align the shadows is to make the shadows of parallel geometry objects, like the xylophone's vertical leg and the cow's leg, parallel.

6. To see the final result, you need to render the image. To do this, select Rendering→Render Setup (or press F10) to open the Render Setup dialog box. Click the Render button at the bottom of the dialog box, or press the F9 key.

   The image is rendered in the Rendered Frame Window.

Figure 20.8 shows the resulting rendered image.

FIGURE 20.8

A rendered xylophone fits into the scene because its shadows are cast on an object with a Matte/Shadow material applied.

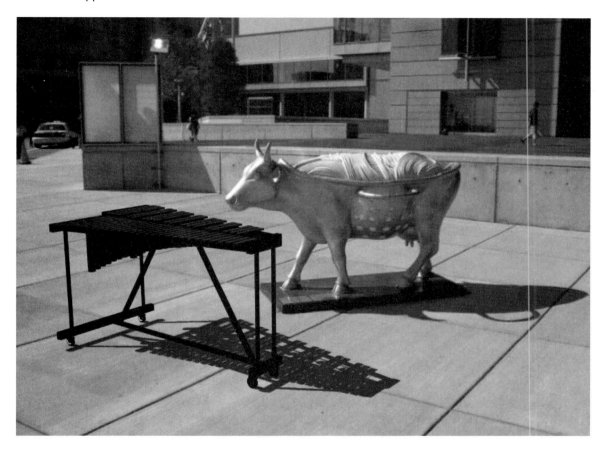

Material Modifiers

Of the many available modifiers, most modifiers change the geometry of an object, but several work specifically with materials and maps, including the Material, MaterialByElement, Disp Approx, and Displace Mesh (WSM) modifiers in the Surface category. In this section, you get a chance to use several material-specific modifiers.

Material modifier

The Material modifier lets you change the material ID of an object. The only parameter for this modifier is the Material ID. When you select a subobject and apply this modifier, the material ID is applied only to the subobject selection. This modifier is used in conjunction with the Multi/Sub-Object Material type to create a single object with multiple materials.

MaterialByElement modifier

The MaterialByElement modifier enables you to change material IDs randomly. You can apply this modifier to an object with several elements. The object needs to have the Multi/Sub-Object material applied to it.

The parameters for this modifier can be set to assign material IDs randomly with the Random Distribution option or according to a desired Frequency. The ID Count is the minimum number of material IDs to use. You can specify the percentage of each ID to use in the fields under the List Frequency option. The Seed option alters the randomness of the materials.

Tutorial: Creating random marquee lights with the MaterialByElement modifier

The MaterialByElement modifier enables you to change material IDs randomly. In this tutorial, you reproduce the effect of lights randomly turning a marquee on and off by using the Multi/Sub-Object material together with the MaterialByElement modifier.

To create a randomly lighted marquee, follow these steps:

1. Open the Marquee Lights.max file from the Chap 20 directory in the downloaded content set.

   This file includes some text displayed on a rectangular object surrounded by spheres that represent lights.

2. Open the Slate Material Editor, and double-click the Multi/Sub-Object material from the Material/Map Browser. Give the material the name **Random Lights**.

3. Double-click the Multi/Sub-Object material node, and in the Multi/Sub-Object Basic Parameters rollout, click the Set Number button, and change its value to **2**. Then click the Material 1 button and select the Standard material type; in the Material name field, give the material the name **Light On**. Select the material button in the Multi/Sub-Object Basic Parameters rollout, and set the Diffuse color to yellow; then enable the Color option next to the Self-Illumination value, and set the Self-Illumination color to yellow. Double-click the main material node again.

4. Name the second material **Light Off**, click the material button to right of the name field and select the Standard material type. Click the material button again, and select a gray Diffuse color. Then double-click the Multi/Sub-Object material node.

5. Select all the spheres, and click the Assign Material to Selection button to assign the material to the spheres.

6. With all the spheres selected, open the Modify panel, and select the MaterialByElement modifier from the Modifier List drop-down list. In the Parameters rollout, select the Random Distribution option, and set the ID Count to **2**.

Figure 20.9 shows the marquee with its random lights. (I've always wanted to see my name in lights!)

Tip

If you want to have the lights randomly flash, simply animate the changing Seed value.

FIGURE 20.9

This marquee is randomly lighted, thanks to the MaterialByElement modifier.

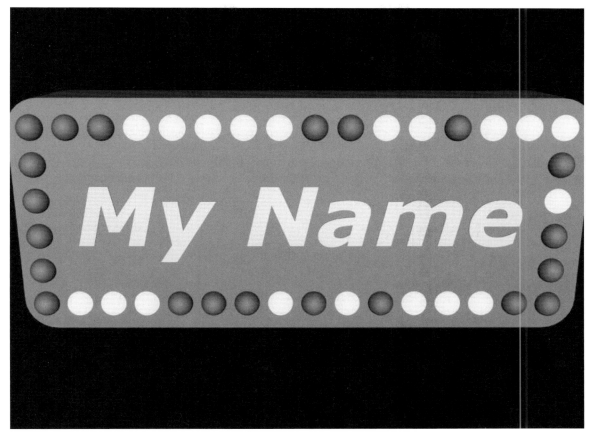

Comparing Displacement maps and Displace modifiers

You can change the geometry of an object in several ways using a bitmap. One way is to use the Displace modifier (found in the Modifiers→Parametric Deformers menu). The Displace modifier lets you specify a bitmap and a map to use to alter the object's geometry. Black areas on the bitmap are left unmoved, gray areas are indented, and white areas are indented a greater distance. Several controls are available for specifying how the image is mapped to the object and how it tiles, and buttons are available for setting its alignment, including Fit, Center, Bitmap Fit, Normal Align, View Align, Region Fit, Reset, and Acquire.

Note

3ds Max also supports Vector Displacement maps, which are found in the Maps rollout of the Material/Map Browser. Vector Displacement maps require the Arnold renderer, and they allow displacement in any direction, not just along surface normals like other displacement methods. Autodesk's Mudbox provides a good way to create this type of map saved using the EXR file format.

Another way to displace geometry with a bitmap is to use a displacement map. Displacement maps can be applied directly to Editable Poly and Mesh, NURBS, and Patch objects. If you want to apply a displacement map to another object type, such as a primitive, you first need to apply the Modifiers→Surface→Disp Approx modifier, which is short for Displacement Approximation. This modifier includes three default presets for Low, Medium, and High that make it easy to use.

More details on working with maps are covered in Chapter 18, "Adding Material Details with Maps."

One drawback to using displacement maps is that you cannot see their result in the viewport, but if you apply the Modifiers→Surface→Displace Mesh (WSM) modifier, the displacement map becomes visible in the viewports. If you change any of the displacement map settings, you can update the results by clicking the Update Mesh button in the Displacement Approx rollout.

Note

The Displace modifier requires a dense mesh in order to see the results of the displacement map, but the Disp Approx modifier creates the required density at render time.

Tutorial: Displacing geometry with a bitmap

When faced with how to displace an object using a bitmap, 3ds Max once again comes through with several ways to accomplish the task. You can use the Displace modifier or you can use a Displacement map in the Material Editor. The method you choose depends on the pipeline. You can choose to keep the displacement in the Modifier Stack or on the material level. This simple tutorial compares using both of these methods.

To compare the Displace modifier with a displacement map, follow these steps:

1. Create two square plane objects side by side in the Top viewport, using the Create→Standard Primitives→Plane menu command. Then set the Length and Width Segments to **150** for the left plane object and to **20** for the right plane object.

Tip

When displacing geometry using a bitmap, make sure the object faces that will be displaced have sufficient resolution to represent the displacement.

2. Select the first plane object, and apply the Displace modifier with the Modifiers→Parametric Deformers→Displace menu command. In the Parameters rollout, set the Strength value to **2** and click the Bitmap button. In the Select Displacement Image dialog box, select the Tulip logo.tif file from the Chap 20 directory in the downloaded content set, and click Open.

3. Select the second plane object, and open the Slate Material Editor by selecting Rendering→Material Editor→Slate Material Editor. In the Material Editor, double-click the Standard material in the Material/Map Browser to add a Standard material node to the Node View panel; then double-click the new node to access its parameters. In the Parameter Editor panel, open the Maps rollout, set the Displacement Map Amount value to **10**, and click the Displacement map button. Double-click the Bitmap option in the Material/Map Browser, and load the same Tulip logo.tif file from the Chap 20 directory in the downloaded content set. Then apply the material to the second plane object by pressing the Assign Material to Selection button, and close the Material Editor.

4. With the second plane still selected, choose the Modifiers→Surface→Displace Mesh (WSM) menu command. In the Displacement Approx. rollout, enable the Custom Settings option, and click the High subdivision preset.

Figure 20.10 shows the resulting displacement on both plane objects.

FIGURE 20.10

Objects can be displaced using the Displace modifier or a displacement map.

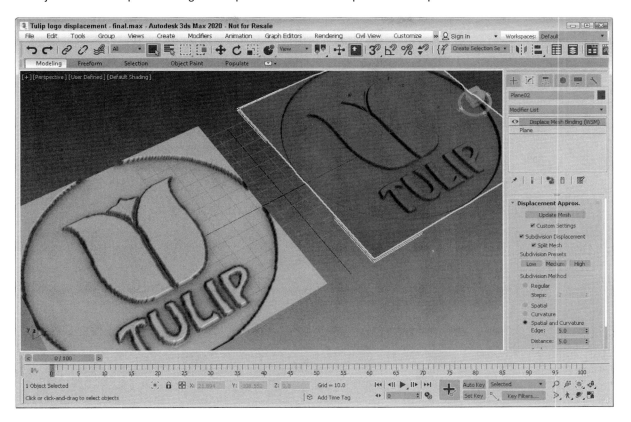

Summary

This chapter introduced several compound materials that you can create in 3ds Max. The chapter presented various material types, including compound and Multi/Sub-Object materials and the Matte/Shadow material. The chapter also showed off a number of key material modifiers, including the Displace Mesh modifier.

The following topics were covered in this chapter:

* Various compound material types

* Applying multiple materials to an object with material IDs

* Using the Matte/Shadow material to hide objects

* Exploring several material modifiers, including the Material and MaterialByElement

* Comparing the different displacement methods

After a map is placed on the surface of an object, you can use the Unwrap UV interface to manipulate how the texture is placed on the surface. This lets you control exactly which portions of a texture get mapped to the individual subobjects. This is the topic of the next chapter.

Chapter 21

Unwrapping UVs and Mapping Textures

IN THIS CHAPTER

Working with mapping modifiers

Applying decals with the UVW Map modifier

Using the Unwrap UVW modifier and the Edit UVWs window

Throughout the modeling chapter, as you created objects, the Generate Mapping Coordinates option appeared for almost all objects. Now you find out what mapping coordinates are and how to use them.

Mapping coordinates define how a texture map is aligned to an object. These coordinates are expressed using U, V, and W dimensions. U is a horizontal direction, V is a vertical direction, and W is depth.

When you enable the Generate Mapping Coordinates option for new objects, the Autodesk® 3ds Max® 2020 software takes its best guess at where these coordinates should be located. For example, a Box primitive applies a texture map to each face. This works well in some cases, but you won't have to wait long until you'll want to change the coordinates.

Control over the mapping coordinates is accomplished using many different modifiers, including the granddaddy of them all—UVW Unwrap.

Mapping Modifiers

Among the many modifiers found in the Modifiers menu are several that are specific to material maps. These modifiers are mainly found in the UV Coordinates submenu and are used to define the coordinates for positioning material maps. These modifiers include the UVW Map, UVW Mapping Add, UVW Mapping Clear, UVW XForm, MapScaler (WSM), Projection, Unwrap UVW, Camera Map (WSM), and Camera Map.

The Projection modifier is used to create normal maps and is covered in Chapter 23, "Creating Baked Textures and Normal Maps."

UVW Map modifier

The UVW Map modifier lets you specify the mapping coordinates for an object. Primitives, Loft Objects, and NURBS can generate their own mapping coordinates, but you need to use this modifier to apply mapping coordinates to mesh and poly objects and patches.

Note

Objects that create their own mapping coordinates apply them to Map Channel 1. If you apply the UVW Map modifier to Map Channel 1 of an object that already has mapping coordinates, the applied coordinates overwrite the existing ones.

You can apply the UVW Map modifier to different map channels. Applying this modifier places a map gizmo on the object. You can move, scale, or rotate this gizmo. To transform a UVW Map gizmo, you must select it from the subobject list. Gizmos that are scaled smaller than the object can be tiled.

Many different types of mappings exist, and the parameter rollout for this modifier lets you select which one to use. The Length, Width, and Height values are the dimensions for the UVW Map gizmo. You also can set tiling values in all directions.

Note

It is better to adjust the tiling within the UVW Map modifier than in the Material Editor because changes in the Material Editor affect all objects that have that material applied, but changes to a modifier affect only the object.

The Alignment section offers eight buttons for controlling the alignment of the gizmo. The Fit button fits the gizmo to the edges of the object. The Center button aligns the gizmo center with the object's center. The Bitmap Fit button opens a File dialog box where you can align the gizmo to the resolution of the selected bitmaps. The Normal Align button lets you drag on the surface of the object, and when you release the mouse button, the gizmo origin is aligned with the normal of the polygon that you are over when you release the mouse. The View Align button aligns the gizmo to match the current viewport. The Region Fit button lets you drag a region in the viewport and match the gizmo to this region. The Reset button moves the gizmo to its original location. The Acquire button aligns the gizmo with the same coordinates as another object.

Most objects can automatically generate mapping coordinates—with the exception of meshes. For meshes, you need to use the UVW Map modifier. The UVW Map modifier includes seven different mapping options. Each mapping option wraps the map in a different way. The options include Planar, Cylindrical, Spherical, Shrink Wrap, Box, Face, and XYZ to UVW.

Figure 21.1 displays a brick map applied to a mug using cylindrical mapping.

FIGURE 21.1

The UVW Map modifier lets you specify various mapping coordinates for material maps.

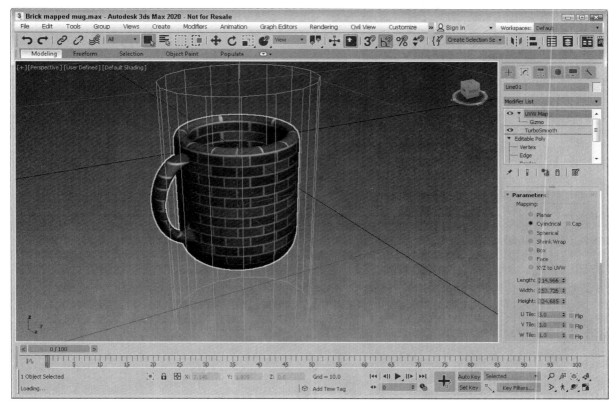

Tutorial: Using the UVW Map modifier to apply decals

After mapping coordinates have been applied either automatically or with the UVW Map modifier, you can use the UVW XForm modifier to move, rotate, and scale the mapping coordinates.

In this tutorial, you use the UVW Map modifier to apply a decal to a rocket model. To use the UVW Map modifier, follow these steps:

1. Open the Nasa decal on rocket.max file from the Chap 21 directory in the downloaded content set.

   This file includes a model of a rocket with the appropriate materials applied. This chapter's folder in the downloadable content also includes a 300 x 600 image of the word *NASA* in black capital letters on a white background. The background color of this image was set to be transparent, and the image was saved as a .gif.

Note
The GIF format, typically used for Web pages, can easily make areas of the image transparent. These transparent areas become the alpha channel when loaded into 3ds Max.

2. Open the Slate Material Editor (or press the M key), and double-click the Standard material in the Material/Map Browser. Double-click the new material node to access its parameters; then name the material **NASA Logo**, click the Diffuse color swatch, and select a white color. Double-click the Bitmap map button in the Material/Map Browser. Locate the nasa.gif image from the Chap 21 directory in the downloaded content set, click Open, and connect the Bitmap node to the Diffuse Colorchannel of the Standard material. The bitmap image loads. Double-click the new node to view the Bitmap parameters. In the Coordinates rollout, enter a value of **-90** in the W Angle field. The letters rotate vertically. Then, in the Bitmap Parameters rollout, select the Image Alpha option.

3. Select the lower white section of the rocket in the viewport, and open the Modify panel. At the top of the Modify panel, click the Modifier List, and select the UVW Map modifier. Select the Cylindrical Mapping option, but don't select the Cap option in the Parameters rollout.

4. With the cylinder section selected, select the logo material in the Material Editor, and click the Assign Material to Selection button or drag from the material's output socket to the rocket cylinder. To see the applied logo, make sure the Materials→Shaded Materials with Maps or the Realistic Materials with Maps menus in the Shading viewport label are enabled.

Tip

When a bitmap is applied to an object using the UVW Map modifier, you can change the length, width, and tiling of the bitmap by using the UVW Map manipulator. If you enable the Select and Manipulate button on the main toolbar, the manipulator appears as green lines. When you move the mouse over the top of these green lines, they turn red, and you can drag them to alter the map dimensions. Use the small green circles at the edges of the map to change the tiling values. As you use the manipulator, the map is updated in real time within the viewports if you have enabled materials to display in the viewport.

Figure 21.2 shows the resulting rendered image.

FIGURE 21.2

You can use the UVW Map modifier to apply decals to objects.

UVW Mapping Add and Clear modifiers

The UVW Mapping Add and the UVW Mapping Clear modifiers are added to the Modifier Stack when you add or clear a channel using the Channel Info utility.

UVW XForm modifier

The UVW XForm modifier enables you to adjust mapping coordinates. It can be applied to mapping coordinates that are automatically created or to mapping coordinates created with the UVW Map modifier.

The parameter rollout includes values for the UVW Tile and UVW Offsets. You also can select the Map Channel to use.

Map Scaler modifier

The Map Scaler modifier is available as both an Object-Space modifier and a World-Space modifier. The World-Space version of this modifier maintains the size of all maps applied to an object if the object itself is resized. The Object-Space version ties the map to the object, so the map scales along with the object. The Wrap Texture option wraps the texture around the object by placing it end to end until the whole object is covered.

Camera Map modifier

The Camera Map modifier creates planar mapping coordinates based on the camera's position. It comes in two flavors—one applied using Object-Space and another applied using World-Space.

The single parameter for this modifier is Pick Camera. To use this modifier, click the Pick Camera button, and select a camera. The mapping coordinates are applied to the selected object.

Using the Unwrap UVW Modifier

The Unwrap UVW modifier lets you control how a map is applied to a subobject selection. It also can be used to unwrap the existing mapping coordinates of an object. You then can edit these coordinates as needed. This is accomplished by applying a texture map to an object. By selecting each face of the object, you then use the Edit UVWs interface to move and orient the polygon in UVW space to determine how the polygon is set over the applied texture map. For example, imagine a bitmap texture that shows all the different sides of a die. In the UVW Editor, you could select and manipulate each face's UVs to align to the die sides included in the bitmap.

You also can use the Unwrap UVW modifier to apply multiple planar maps to an object. You accomplish this task by creating planar maps for various sides of an object and then editing the mapping coordinates in the Edit UVWs interface.

Another common way the Edit UVW interface is used is to make game assets. Most game engines like to have all the textures for a model combined into a single bitmap. This efficiently uses memory and eases the loaded time for the assets. To texture a model, a game artist creates a single, high-resolution bitmap that holds the textures for the entire model. Matching bitmaps also are created for specular, bump, and normal channels. The Edit UVWs interface is then used to move and position all the polygons to match the bitmap textures. After the coordinates are positioned correctly, they can be used for the other channels also.

Selecting UVW subobjects

The Unwrap UVW modifier lets you control precisely how a map is applied to an object. The Unwrap UVW modifier has Vertex, Edge, and Polygon subobject modes. In subobject mode, you can select a subobject, and the same selection is displayed in the Edit UVWs interface, and vice versa. This synchronization between the Edit UVWs window and the viewports helps ensure that you're working on the same subobjects all the time.

The Selection rollout, shown in Figure 21.3, includes a button with a plus sign and one with a minus sign. These buttons grow or shrink the current selected subobject selection. When Edge subobjects are selected, the Ring and Loop buttons and the Point-to-Point Edge Selection button become active. The Point-to-Point Edge Selection button is a great way to create seams. When you select two points, all the edges connecting those two points are selected.

In Polygon subobject mode, the Select by Planar Angle and Select by Smoothing Group buttons are active. The Select by Element button selects all the subobjects in the current element, which is an easier way to quickly select all subobjects of a defined element. You also can select to Ignore Backfacing. The

Symmetrical Selection button lets you automatically select all subobjects mirrored about the X, Y, or Z axis. For models that are symmetrical, this is a huge time-saver.

FIGURE 21.3

The Selection panel for the UVW Unwrap modifier lets you work with Vertex, Edge, and Face subobjects.

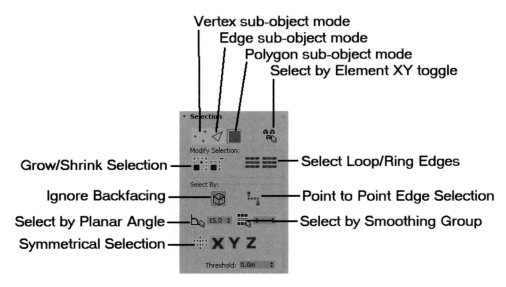

Vertex sub-object mode
Edge sub-object mode
Polygon sub-object mode
Select by Element XY toggle

Grow/Shrink Selection ——— ——— Select Loop/Ring Edges

Ignore Backfacing ——— ——— Point to Point Edge Selection

Select by Planar Angle ——— ——— Select by Smoothing Group

Symmetrical Selection ——— X Y Z

The Select menu commands let you convert selections between vertices, edges, and faces. Additional options in the Select menu let you select all inverted and overlapped faces, allowing you to find potential problem areas.

Accessing the Edit UVWs interface

The Edit UVs rollout includes a button named Open UV Editor. This button opens the Edit UVWs interface. Within the UVW Editor, you have complete control over the various UVs, and the subobject selected in the viewport also is selected within the UVW Editor. The UVW Editor also can select and display the applied texture bitmap; the mapped bitmap also is shown in the viewport.

Tweaking vertices in the viewport

The Edit UVs rollout also includes a Tweak in View button. When this button is enabled, you can drag a single vertex in the viewports to move the texture mapping. This doesn't cause the vertex's actual position in the scene to move, only the mapping. The effects of this mode are apparent only in the viewport when the texture is visible. Texture maps can be made visible in the viewports using the Materials→Shaded Materials with Maps or Realistic Materials with Maps in the Shading viewport label.

Using the Quick Planar Map

One of the easiest ways to isolate mapping surfaces is with the Quick Planar Map button, found in the Edit UVs rollout under the Tweak in View button. If you select a set of faces either in the viewports or in the Edit UVWs window and click this button, a planar map based on the X, Y, Z or an Averaged Normals is separated and the selected area is marked with a map seam. A button to Display Quick Planar Map shows the orientation of the mapping plan when enabled. Quick Planar Maps are one of the easiest ways to select, orient, and group mapping areas together.

Saving and loading mapping coordinates

You also can load and save the edited mapping coordinates using the Save and Load buttons in the Channel rollout, and the Reset UVWs button resets all the mapped coordinates. Saved mapping coordinate files have the .uvw extension and can be loaded for use on another object in another scene. For example, suppose you have a game level with multiple crates. If you create a texture that has all the different sides and correctly map the texture for one cube object, all other crates in the scene could be correctly mapped by simply saving the mapping coordinates for the completed one and loading them into the others.

Each map can hold up to 99 mapping channels, and the Map Channel value lets you tell them apart. If a single object has multiple instances of the Unwrap UVW modifier applied, each instance can have a different Map Channel value. Video game objects use these map channels to add wear and tear to a character as the game progresses. Also, you can make the map channel a vertex color channel by enabling the Vertex Color Channel option.

The Configure rollout lets you set whether the seams of the planar maps are displayed, as well as their width. The options include Map Seams, Peel Seams, and either Thick or Thin. Map seams mark the different map clusters and show up as green lines. Peel seams mark the edges that get split when a mesh is peeled and show up as blue lines. By making the seams visible, you can easily tell where the textures don't match the object's creases.

Using the Edit UVWs Interface

Although several features for working with UVs are available in the various panels in the main interface, the main work is accomplished in the Edit UVWs interface. All changes made in the Edit UVWs window, shown in Figure 21.4, are automatically reflected in the viewports.

FIGURE 21.4

The Edit UVWs interface lets you control how different planar maps line up with the model.

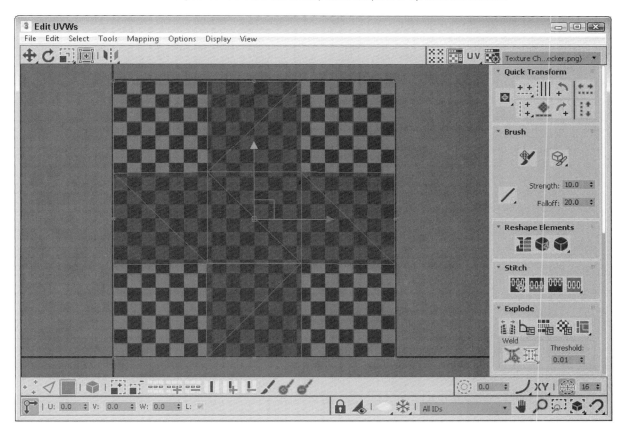

The Edit UVWs dialog box has multiple icons surrounding the main window. Along the top edge are buttons for selecting and transforming the selected subobjects. Table 21.1 shows and describes the buttons along the top edge.

TABLE 21.1 Top Edit UVW Interface Buttons

Buttons	Name	Description
	Move, Move Horizontal, Move Vertical	Moves the selected vertices when dragged.
	Rotate	Rotates the selected vertices when dragged.
	Scale, Scale Horizontal, Scale Vertical	Scales the selected vertices when dragged.
	Freeform Mode	Displays a gizmo that you can use to transform the subobject selection.
	Mirror Horizontal, Mirror Vertical, Flip Horizontal, Flip Vertical	Mirrors or flips the selected vertices about the center of the selection.
	Show Multi-Tile	Toggles the display of the multi-tile maps.
	Show Map	Toggles the display of the map in the dialog box.
	Coordinates	Displays the vertices for the UV, UW, and WU axes.
	Show Options	Opens the Options dialog box.
CheckerPatt...	Pick Texture drop-down list	Displays a drop-down list of all the maps applied to this object. You can display new maps by using the Pick Texture option.

The gizmo is simply a highlighted rectangle that surrounds the current selection. Move the selection by clicking in the gizmo and dragging; Shift+dragging constrains the selection to move horizontally or vertically. The plus sign in the center marks the rotation and scale center point. Scale the selection by dragging one of its handles. Ctrl+dragging a handle maintains the aspect ratio of the selection. Click+dragging the middle handles rotates the selection. Ctrl+dragging snaps to 5-degree positions, and Alt+dragging snaps to 1-degree positions.

Within the Pick Texture drop-down list is a Checker Pattern option. This option applies a checker map to the mesh without your having to assign a material. The checker pattern makes easy work of looking for stretching on the model. The drop-down list also includes any textures that are applied to the object. The selected texture appears in the background of the window.

Selecting subobjects within the dialog box

Under the main window in the Edit UVWs dialog box is a toolbar of icons for selecting different subobjects. Table 21.2 shows and describes these buttons.

TABLE 21.2 Selection Buttons

Buttons	Name	Description
	Vertex Subobject mode	Allows Vertex UV selection
	Edge Subobject mode	Allows Edge UV selection
	Polygon Subobject mode	Allows Polygon UV selection
	Select by Element Toggle	Selects all subobjects within the selected element
	Grow Selection	Adds adjacent subobjects to the current selection
	Shrink Selection	Subtracts adjacent subobjects from the current selection
	Loop	Selects a loop of edges placed end to end
	Grow Loop	Adds to the selected edge loop
	Shrink Loop	Subtracts from the selected edge loop
	Ring	Selects a ring of edges parallel to each other
	Grow Ring	Adds to the selected edge ring
	Shrink Ring	Subtracts from the selected edge ring
	Paint Selection	Allows selection by painting over the subobjects
	Enlarge Brush Size	Increases the radius of the selection brush
	Shrink Brush Size	Decreases the radius of the selection brush

Within the Selection toolbar are the Selection modes with buttons for selecting Vertex, Edge, and Face subobjects. Selected subobjects in the Edit UVWs interface are highlighted red. Using the + (plus) and – (minus) buttons, you can expand or contract the current selection. The paintbrush icon button is used to Paint Select subobjects, and the Expand and Shrink Brush buttons to its immediate right allow you to increase and decrease the brush size. The Select Element option selects all subobjects in the given cluster. This happens only when the Select Face subobject mode is enabled.

The Loop button automatically selects all edges that form a loop with the current selected edges. Edge loops are edges that run end to end. You also can use the Ring button to select adjacent edges that are parallel to each other. The Grow and Shrink buttons next to the Loop and Ring buttons are used to select the adjacent edges on either side of the current selection in the loop or ring direction.

Tip
The Loop and Ring buttons also can be used on vertices and faces if two or more are selected.

To the right of the selection buttons located under the main view in the Edit UVWs dialog box are several buttons for enabling and configuring soft selections. Table 21.3 shows and describes these buttons.

TABLE 21.3 Soft Selection Buttons

Buttons	Name	Description
	Enable Soft Selection	Toggles Soft Selection on and off
0.0	Soft Selection Falloff value	Sets the amount of falloff for soft selection
	Linear Falloff, Smooth Falloff, Slow Out Falloff, Fast Out Falloff	Changes the falloff type
XY UV	Soft Selection Falloff Space	Switches falloff coordinates between XY and UV space
	Limit Soft Selection by Edges	Limits falloff to a specified number of edges
16	Edge Limit value	Sets the number of edges to include in the falloff

Using these controls, you can enable Soft Selection with a specified Falloff value. The UV and XY options let you switch between texture coordinates and object coordinates for the falloff. The Edge Distance lets you specify the Soft Selection falloff in terms of the number of edges from the selection instead of a falloff value. You also can choose the falloff profile as Smooth, Linear, Slow Out, or Fast Out.

Navigating the main view

Along the bottom edge of the Edit UVWs dialog box are several more buttons for viewing the coordinate values, changing the display options, and navigating the main view. Table 21.4 shows and describes these buttons.

TABLE 21.4 Lower Toolbar Buttons

Buttons	Name	Description
	Absolute/Relative Toggle	Lets you enter U, V, and W values as absolute values or relative offsets.
U: 0.0 V: 0.0 W: 0.0	U, V, W values	Displays the coordinates of the selected vertex. You can use these values to move a vertex.
	Lock Selected Subobjects	Locks the selected components and prevents additional ones from being selected.
	Display Only Selected Faces	Displays vertices for only the selected faces.
	Hide/Unhide Selected Subobjects	Allows selected subobjects to be hidden and unhidden.
	Freeze/Unfreeze Selected Subobjects	Allows selected subobjects to be frozen and unfrozen.

	All IDs drop-down list	Filters selected material IDs.
	Pan View	Lets you drag to pan the view.
	Zoom View	Lets you drag to zoom in on the view.
	Zoom To Region	Lets you drag to zoom in a specific region.
	Zoom Extents View, Zoom to Selection, Zoom to Subobject Selection	Lets you zoom to include all the mapping coordinates, just the current selection, or any elements with a selected subobject.
	Snaps Toggle, Snaps Settings	Snaps to the closest pixel corner or to the closest grid intersection.

The buttons in the lower-right corner of the Edit UVWs dialog box work just like the Viewport Navigation buttons described in earlier chapters, including buttons to snap to grid and snap to pixel. You also can navigate about the Edit UVWs window using the scroll wheel to zoom in and out of the window and dragging the scroll wheel to pan the view.

Using the Quick Transform buttons

One of the first tasks when unwrapping an object is to select and separate off different areas of polygons that have unique UVs. Each unique set of faces in the Edit UVWs dialog box is called a cluster. For example, if you have a car with a matching texture, the same texture can be mapped for each of the wheel hubs by placing each of the clusters for the wheel hub on top of each other over the bitmap section that shows the hub details. The same can be done for the side panels on the car, but one of the sides must be flipped horizontally. This is where the Quick Transform functions come in. They allow you to select and manipulate the different clusters.

The Quick Transform rollout, shown in Figure 21.5, in the Edit UVWs dialog box includes buttons for aligning, rotating, and spacing subobjects.

FIGURE 21.5

The Quick Transform rollout includes buttons for aligning, rotating, and spacing subobjects.

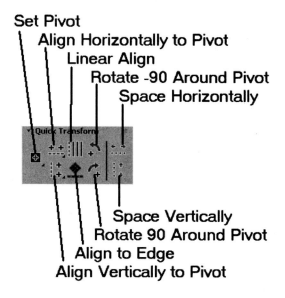

Set Pivot
Align Horizontally to Pivot
Linear Align
Rotate -90 Around Pivot
Space Horizontally

Space Vertically
Rotate 90 Around Pivot
Align to Edge
Align Vertically to Pivot

The Set Pivot button includes options as flyouts to set the pivot at the center or at any of the four corners of the selection. You also can move the pivot to a precise location by dragging it when the Freeform Mode button on the top toolbar is selected. The pivot is marked by an orange set of crosshairs, and it is the point about which the selection is rotated and scaled.

The Align feature includes buttons to align the selected vertices or edges to the current pivot or to an average of the selected subobjects. The Align to Pivot option is the default, and the average align is available as a flyout. If you press and hold the Shift key, the entire edge loop is aligned. When used on an edge ring, all the individual edges are oriented to be perfectly parallel. The Linear Align button aligns the vertices or edges to a straight line that stretches between the two endpoints. The Align to Edge button is available only in Edge subobject mode. It rotates the entire cluster until the selected edge is either vertically or horizontally aligned.

Caution

The align features are designed to be used only on vertices and edges, but you can select and use them on faces. However, this only collapses the polygons to a single line because it acts to align all vertices in the selection.

The Rotate +90 and Rotate –90 buttons rotate the selected subobjects 90 degrees in the Edit UVW interface. The rotation is about the set pivot. The Space Horizontal and Space Vertical buttons are used to equally space all the selected vertices or edges. The Shift key applies this to the entire edge loop.

Using the Brush tool

The Brush tool in the UV Editor is used to move subobjects and to relax subobjects. It only affects those vertices within the defined brush area. The size of the brush area is determined by the Strength and Falloff settings. The falloff, just like with the Soft Selection, can be either Linear, Smooth, Slow Out or Fast Out.

The UV Paint Movement tool will move the subobjects that are painted over, but the Relax tool will relax the vertices depending on their Polygon Angles, Edge Angles, or Centers.

Straightening and Relaxing UV clusters

Within the Reshape Elements rollout are three buttons for reshaping the selected cluster of UVs. The Straighten Selection button is available only in Polygon subobject mode. It realigns the selected polygons into a rectangular grid with all polygons oriented vertically and horizontally.

If your mapping coordinates are too tight, and you're having a tough time moving them, you can use the Relax feature to space the vertices equally. The second button is Relax Until Flat. It causes all vertices within the cluster to move in order to remove any tension from the cluster. It also tries to make all faces in the cluster roughly the same size. Relaxing a cluster removes stretching that occurs across the surface of the area.

If the Relax Until Flat option pushes the cluster too far, the third button allows you to apply a custom set of relax settings. You can access the Relax Tool dialog box using the flyout button under the Relax:Custom button or the Tools→Relax menu command. This tool works like the Relax modifier, pushing close vertices away and pulling far vertices closer together. Selecting this menu option opens the Relax Tool dialog box, shown in Figure 21.6, which offers three relax methods: Relax by Face Angles, Relax by Edge Angles, and Relax by Centers. The Iterations value is the number of times to apply the relax algorithm. The Amount value is how aggressive the movements of the vertices are, and the Stretch value controls how much vertices are allowed to move. You also have options to Keep Boundary Points Fixed and Save Outer Corners.

FIGURE 21.6

The Relax Tool dialog box includes custom settings for the Relax feature.

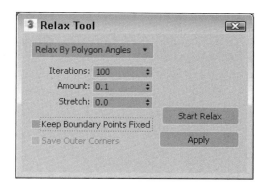

Stitching and welding

If you select some faces within the Edit UVWs dialog box and drag them away from the other faces, you notice that some edges connecting the selected faces with the non-selected faces remain. You can use the buttons in the Stitch and Explode rollouts, shown in Figure 21.7, to break these remaining edges and to stitch broken clusters back together again.

FIGURE 21.7

The Stitch and Explode rollouts include buttons for breaking, welding, and stitching vertices.

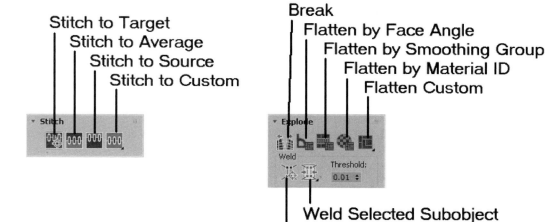

The Stitch button is used in Vertex mode to match the selected vertices along a border to their corresponding vertices. Clicking one of the Stitch buttons moves both faces to a new location, depending on the button you use. The Stitch to Target moves the selected subobjects to the location of their matching subobject. The Stitch to Average moves both to a location midway between each, and the Stitch to Source moves the matching subobjects to the location of the selected subobjects.

If you enable the Display→Show Shared Sub-objects menu command, the shared edges of the selected vertices are shown as blue lines. This indicates where the subobject face would be moved to when stitched.

The Stitch Custom button moves the subobjects based on the Stitch Settings dialog box's configuration. You can access the Stitch Settings dialog box, shown in Figure 21.8, using the flyout button under the Stitch Custom button or using the Tools→Stitch Selected menu command. The Stitch Tool dialog box includes options to align and scale the moved cluster, and the Bias value determines how close the selection moves to or away from the target subobject. A value of 0 moves the target to the source, and a value of 1 moves the source to the target.

FIGURE 21.8

The Stitch Tool dialog box includes a Bias value to determine how far the selected subobjects move.

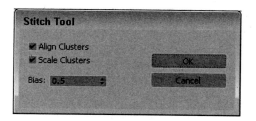

The Break button in the Explode rollout breaks the selected subobjects from their surrounding faces so they can be moved without stretching out the edges of the adjacent faces. This works best if the Break tool is used before moving the selected faces. When the Break tool is used on vertices or edges, two vertices or edges are created. This allows the independent vertices or edges to be moved away from each other. When Break is used on faces, a new cluster is created.

The opposite of Break is Weld, and the Explode rollout includes several ways to weld subobjects together. The Target Weld button enables a mode that lets you drag vertices or edges and drop them on their matching separated subobject. The cursor changes to a bold set of crosshairs when a matching subobject is under the cursor.

The Weld Selected Subobject button welds selected subobjects only if they are located within the Threshold value.

If an edge is selected, its shared edge is highlighted in blue. Click the Weld All Selected Seams button once to automatically select the blue shared edges. If both shared edges are selected, clicking the Weld All Selected Seams button again welds the two edges by moving both to an average location. The Weld Any Match with Selected button welds the selected subobject without requiring that both shared edges are selected, and it can be used on all vertices, edges, and faces.

The Edit menu also includes Copy, Paste, and Paste Weld commands. The Copy and Paste commands let you copy a mapping and paste it to another set of faces. The Paste Weld command welds vertices as it pastes the mapping.

Separating into clusters using flattening methods

The Explode rollout also includes several methods for automatically separating the object into clusters. This process is called flattening. The Flatten by Face Angle breaks the faces up using an angle threshold of 60 degrees. Any adjacent faces that have normals greater than this threshold are split along the edge they share. This breaks up a cube object into six separate faces.

You also can split up the UVs based on Smoothing Group values and Material IDs. Both of these are helpful if you've already applied smoothing groups or materials to the object. The Flatten Custom breaks the UVs into clusters based on the Flatten Mapping dialog box, which is accessed using the flyout button under the Flatten: Custom button or with the Mapping→Flatten Mapping menu.

The Flatten Mapping dialog box, shown in Figure 21.9, lets you break the mesh into clusters based on the angle between adjacent faces. This option is good for objects that have sharp angles, like robots or machines. The Spacing value sets the distance between adjacent clusters.

FIGURE 21.9

The Flatten Mapping dialog box includes a Face Angle Threshold value for determining how clusters are separated.

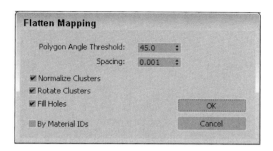

Figure 21.10 shows the UVs for a backhoe bucket that was separated into clusters using the Flatten Mapping method. All the clusters have been automatically aligned within the square texture area, and some smaller pieces have been placed within the holes created by the larger pieces. This makes the most of the available texture space.

FIGURE 21.10

The Flatten Mapping method was used to break this backhoe bucket object into several clusters.

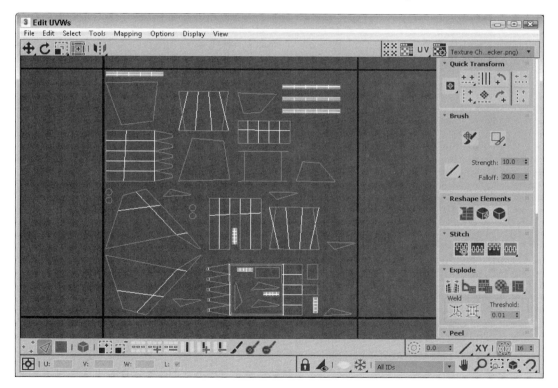

In addition to the flattening options in the Explode rollout, the Mapping menu includes two additional auto-mapping options: Normal and Unfold Mapping.

The Normal Mapping option lets you select to map a mesh using only specific views, including Top/Bottom, Back/Front, Left/Right, Box, Box No Top, and Diamond. These views are based on the direction of the normals from the faces of the mesh. It is helpful for thin models like butterfly wings or coins.

The Unfold Mapping option is unique because it starts at one face and slowly unwraps all the adjacent faces into a single segment if possible. Figure 21.11 shows a simple cylinder that has been unwrapped using this method. The advantage of this mapping is that it results in a map with no distortions. It includes two options: Walk to Closest Face and Walk to Farthest Face. You almost always want to use the Walk to Closest Face option.

Note

Within a single mesh, multiple different mapping methods can be used. For example, a car's wheel uses cylindrical mapping, but its hood might use planar mapping. Even a subobject selection can use different mapping methods.

FIGURE 21.11

The Unfold Mapping option splits the model and unfolds it by adjacent faces into a single segment.

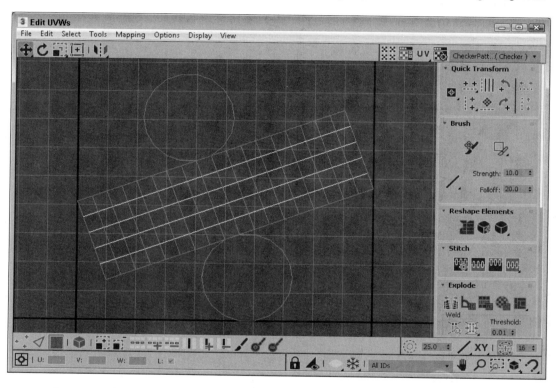

Arranging and grouping clusters

Within the Edit UVWs dialog box is a square area that represents the texture bitmap. The goal is to use as much of this texture space as possible to ensure that the maximum amount of detail from the texture is used on the model. The various buttons in the Arrange Elements rollout help with this task.

After all the various clusters are separated from each other, the Pack Custom button packs all the clusters to fit within the texture space using the settings in the Pack dialog box, shown in Figure 21.12. This dialog box is accessed using the flyout under the Pack Custom button or the Tools→Pack UVs menu command.

FIGURE 21.12

The Pack dialog box includes settings for crunching all clusters within the given texture space.

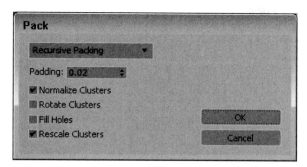

The Pack UVs menu command lets you combine UVs into a smaller space. Packed UVs are easy to move and work with because they use a smaller-resolution bitmap. Three packing algorithms are available in the Pack dialog box. The Linear Packing method is fast but not very efficient, the Recursive Packing option is more efficient, although it takes longer, and the Non-Convex is an improved version of the Recursive Packing that is pretty quick. Within the Pack dialog box, the Padding value sets the amount of space between segments, and the Normalize Clusters option fits all clusters into the given space. The Rotate Clusters option allows segments to be rotated to fit better, and the Fill Holes option places smaller segments within open larger segments.

The Arrange Elements rollout also includes a Rescale Elements button that scales all clusters relative to one another. The Pack Together button fits all clusters within the texture space without normalizing. The Pack Normalize fits all clusters and allows scaling while fitting. You can select to enable or disable rescaling and rotating, and the padding value is the space between the clusters.

Within the Element Properties rollout are buttons to create, destroy, and select groups. For each group, you can set a Rescale Priority that determines which groups are rescaled during packing. To create a group of clusters, select two or more clusters using the Face subobject mode, and click the Group Selected button. The Selected Groups label identifies each group with a number when selected.

Accessing the Unwrap Options

The Options→Preferences menu command (Ctrl+O) opens the Unwrap Options dialog box, shown in Figure 21.13, and lets you set the Line and Selection Colors as well as the preferences for the Edit UVWs dialog box. You can load and tile background images at a specified map resolution or use the Use Custom Bitmap Size option. It also has options to constantly update, display units as pixels, show selected vertices in the viewport, and snap to the center pixel.

FIGURE 21.13

In the Unwrap Options dialog box, you can set the preferences for the Edit UVWs dialog box.

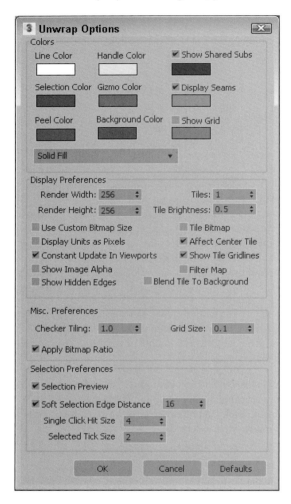

The Display Preferences section lets you specify the exact size of the loaded bitmap. This only affects how the bitmap is displayed in the interface and doesn't change the actual bitmap file dimensions. The Tile Bitmap option places the bitmap end to end for the specified number of tiles. The Constant Update option causes the viewport to update along with the texture map. You can also show Tile Gridlines and the texture alpha channel. The Show Hidden Edges option lets you make the hidden edges visible or invisible.

Tutorial: Controlling the mapping of a covered wagon

The covered wagon model is strong enough to carry the pioneers across the plains, but you can add a motivating slogan to the wagon using the Unwrap UVW modifier. In this tutorial, you add and edit the mapping coordinates for the covered wagon using the Unwrap UVW modifier.

To control how planar maps are applied to the side of a covered wagon, follow these steps:

1. Open the Covered wagon.max file from the Chap 21 directory in the downloaded content set.

   This file includes a covered wagon model. The Chap 21 directory also includes a 256 x 256 image, created in Photoshop, of the paint that you want to apply to its side. The file is saved as Oregon or bust.tif. (Note that the spelling in the image is rough.) This material has already been applied to the wagon covering, but the orientation isn't correct.

2. With the covered section selected, choose Modifiers→UV Coordinates→Unwrap UVW. In the Edit UVs rollout, click the Open UV Editor button.

    The Edit UVWs interface opens. Select the Polygon subobject mode and drag over all the polygons to select them all.

3. In the Edit UVWs interface, choose Mapping→Normal Mapping. In the NormalMapping dialog box, select the Left/Right Mapping option from the drop-down list, and click OK.

    The left and right views of the wagon's top are displayed in the Edit UVWs interface.

4. From the drop-down list at the top right corner of the interface, select the Oregon or bust.tif image as the background image.

    The texture appears in the window.

5. Drag the mouse over all the UVs in the upper half of the Edit UVWs interface. This represents one side of the wagon's cover, now move the selected UVs and position them on top of the unselected UVs located at the bottom half of the Edit UVWs interface.

    By matching these two UV sections together, you can apply the same texture to both sides of the covering.

6. Select all the UV faces, and with the Move tool, drag them to the center of the Edit UVWs window. Click and hold over the Scale tool, and select the Vertical Scale tool. Drag in the window to vertically scale the vertices until they fit over the texture. Then horizontally scale the vertices slightly until the background texture is positioned within the wagon's covering. You'll be able to see the alignment on the model in the viewports as you make changes. When you're finished, click the X button in the upper-right corner to close the Edit UVWs window.

    Figure 21.14 shows the UVW mapping for the wagon cover.

FIGURE 21.14

The Edit UVWs interface lets you transform the mapping coordinates by moving vertices.

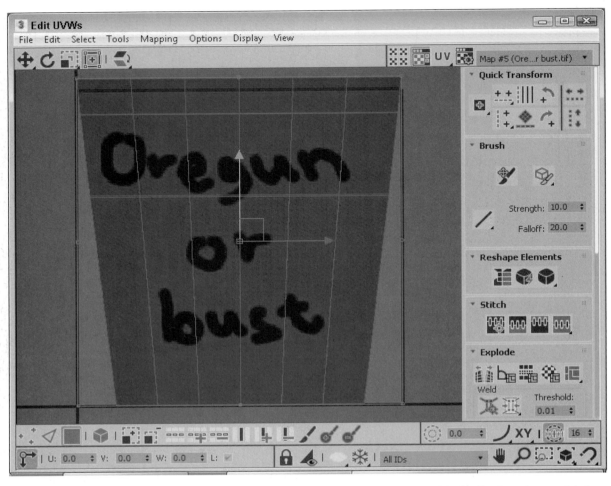

7. Select the Materials→Realistic Materials with Maps menu command in the Shading viewport label to see the map on the covered wagon.

Figure 21.15 shows the results of the new mapping coordinates.

FIGURE 21.15

The position of the covered wagon's texture map has been set using the Unwrap UVW modifier.

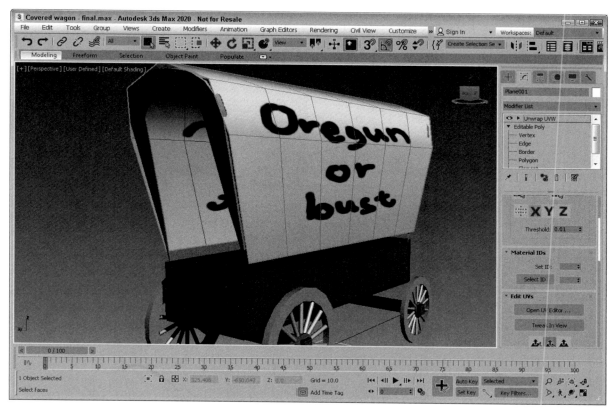

Rendering UV templates

After all the UV coordinates are mapped onto a model, you can paint the desired textures in an external paint program like Photoshop and load the texture back into the Edit UVWs window, where they can be aligned to the correct UVs. Using the Tools→Render UVW Template menu command, you can create a template that can be saved and loaded into Photoshop, showing you exactly where the UV boundaries are.

The Tools→Render UVW Template menu command opens the Render UVs dialog box, shown in Figure 21.16. This dialog box lets you set the template's dimensions, set the template's Fill and Edges colors, and show overlaps and seams. The Fill mode can be set to None, Solid, Normal, and Shaded, providing more information about the object.

FIGURE 21.16

The Render UVs dialog box lets you render and save a template for painting textures.

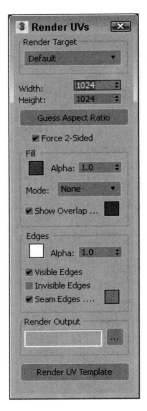

Clicking the Render UV Template button renders the template into the Render Frame Buffer window, where it can be saved to the needed image format. Then, within Photoshop, you can make the template layer a background layer that you can turn on and off to show the lines that you need to stay within when creating a texture.

Mapping multiple objects

If your model is divided into several different pieces, you're in luck because 3ds Max allows you to apply the Unwrap UVW modifier to several pieces at once. When loaded into the Edit UVWs window, the wireframe for each piece is displayed using its object color, which makes identifying the various separate pieces easy.

Tutorial: Creating a mapping for a fighter plane

As a final example of unwrapping, you add a Navy logo to a prop plane. The logo was created and saved as a PNG file, which allowed the background to be saved as transparent. This allows the shiny metallic material to show through the applied logo.

For this tutorial, you add the logo to one of the wings. The tricky part of this tutorial, which you didn't have for the earlier rocket tutorial, is that the wing's ailerons are separated from the wing. This makes it possible to animate the ailerons.

Note

I realize that all you military-aircraft enthusiasts out there know that the logo actually belongs on the fuselage and not on the wing, but the fuselage isn't divided into separate parts like the wing, so I'm taking creative license for the sake of the tutorial.

To add a texture map to several pieces of an airplane, follow these steps:

1. Open the Prop plane with aileron.max file from the Chap 21 directory in the downloaded content set.

2. In the Perspective viewport, select the right wing and its aileron. Two objects should be selected.

3. Select the Modifiers→UV Coordinates→Unwrap UVW menu to apply the modifier to all the selected pieces. Then click the Open UV Editor button in the Edit UVs rollout to open the Edit UVWs window. This applies the Unwrap UVW modifier to both objects.

4. From the drop-down list at the top right of the Edit UVWs interface, select the Pick Texture option. The Material/Map Browser opens. Double-click the Bitmap option, and select the Navy logo.png image from the Chap 21 directory in the downloaded content set. Zoom out until the entire Navy texture is visible.

   The texture appears in the window.

5. In the viewport, select the right wing object by itself and reopen the Edit UVWs interface. Select Polygon subobject mode in the Edit UVWs interface at the lower left corner, and click in the viewport on the center of the right wing. Make sure to click on the model in the viewport and not on the UVs in the Edit UVWs dialog box. Click the Grow button (the one with the plus sign) at the bottom of the Edit UVWs window to grow the selection. Keep clicking the Grow button until the entire top section of the right wing is selected, but the aileron is not selected since it is a separate object. Select the Tools→Break command in the Edit UVWs interface. Then zoom out in the Edit UVWs window, and move the separated wing to the top of the window so it doesn't overlap any other sections. Click on the Polygon subobject mode button again to exit subobject selection mode.

   The top half of the selected wing UVs are separated from the rest of the wing object.

Tip

Another way to select the faces is to use the Planar Angle value, which selects all the polygons on one side of the wing.

6. In the viewport, select the right aileron object and reopen the Edit UVWs interface. Select the center of the right aileron in the viewport, and click the Grow button until the entire part is selected. Then move the aileron down to the bottom of the Edit UVWs window. Click on the Polygon subobject mode button again to exit subobject selection mode.

7. In the viewport, select both the right wing and right aileron objects and reopen the Edit UVWs interface. Drag over the UVs for the aileron at the bottom of the Edit UVWs interface and move it upward until it fits in the opening where the wing's UVs are at the top of the interface. It should fit just like it does in the actual model. The combined set of UVs is for the top of the right wing where the logo needs to go.

Tip

If you need to seamlessly fit two separate parts together, you can use the Welding feature in Vertex subobject mode to weld the vertices on each part.

8. Select all the remaining UVs with the Polygon subobject mode, and move them to the bottom of the window, where they don't overlap the logo. Select the top wing and aileron UVs positioned together, and click the Rotate 90 Around Pivot button to rotate the wing to align with the logo. Scale and position the UVs over the logo. Figure 21.17 shows the resulting alignment with the aileron object selected so you can see the orientation. Then close the Edit UVWs window.

FIGURE 21.17

The UVs are positioned to match the loaded bitmap.

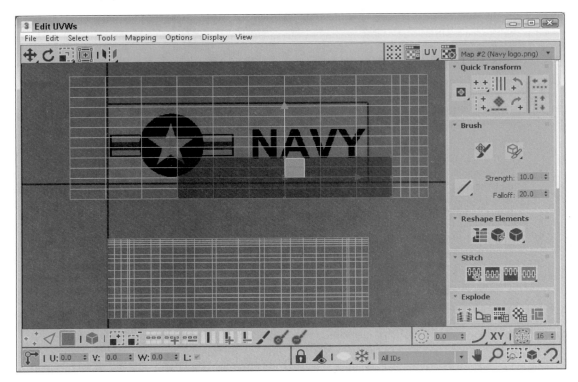

9. Press the M key to open the Material Editor. Double click on the Standard material in the Material/Map browser and double click on the material node to view its parameters. Click the mapping button next to the Diffuse Color in the Parameters panel, and double-click the Bitmap type in the Material/Map Browser. Select the T-28 Trojan logo.png file from the Chap 21 directory in the downloaded content set. Apply this material to the right wing and aileron. Select the Bitmap node and disable the Tile options in the Coordinates rollout of the Parameters panel.

10. Select the Materials→Realistic Materials with Maps menu command in the Shading viewport label to see the map on the plane.

Figure 21.18 shows the results of the plane mapping. After the map is applied to the plane and visible in the viewports, you can open the Edit UVWs window again and tweak the mapping coordinates.

FIGURE 21.18

The logo map is positioned on the wing and spreads over to the ailerons also, even though they are separate parts.

The Unwrap UVW modifier includes several additional advanced mapping methods, including Pelt, Peel, and Spline mapping.

Summary

The Edit UVWs window gives you control over the mapping coordinates for your models so the beautiful texture maps you've created can be placed where they need to be.

In this chapter, you learned about the following:

* Understanding the basics of mapping coordinates
* Using mapping modifiers
* Applying labels with the UVW Map modifier
* Controlling mapping coordinates with the Unwrap UVW modifier
* Rendering a UV template

Material textures aren't the only place that you can paint. Using the Viewport Canvas, you can paint directly on objects in the viewport and it is covered next.

Chapter 22

Painting in the Viewport Canvas and Rendering Surface Maps

IN THIS CHAPTER

Using the Viewport Canvas

Using the Vertex Paint modifier

Rendering surface maps

3ds Max is a 3D tool, and creating scenes with 3ds Max is quite a bit different from the traditional painting programs. Sometimes when you're working on a scene, especially when applying textures, you'll ache to return to those simple, older painting programs of yesteryear. Happily, 3ds Max includes a mode that lets you simply throw paint around just like those old paint programs.

This paint mode is called the Viewport Canvas, and it turns the entire active viewport into a 2D surface; even better, when you are finished painting, your masterpiece is automatically transferred to the current object as a texture map.

If the ability to paint directly in the viewport doesn't interest you, you'll be happy to know that you can use the Autodesk® 3ds Max® 2020 software to render out a surface map that you can load into Photoshop or your favorite image-editing package and use as a template for your textures.

Using the Viewport Canvas

The Viewport Canvas lets you easily apply a painted texture to the selected object. It also has a feature that lets you choose the type of brush you paint with. The Canvas also includes a standard paint brush that is configurable and a Clone brush for copying anything viewed in the active viewport, along with several other brushes.

To activate the Viewport Canvas, simply select the Tools→Viewport Canvas menu command and the Viewport Canvas panel appears, as shown in Figure 22.1.

Tip

The Viewport Canvas panel can be docked to the left or right side of the interface by right-clicking the palette's title bar and selecting the dock location from the pop-up menu.

FIGURE 22.1

The Viewport Canvas panel turns the viewport into a 2D painting canvas.

Current Color ——

Open Color Palette ——

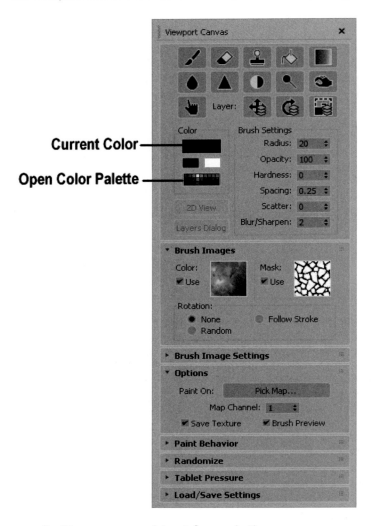

Setting up an object for painting

The Canvas requires some setup before you can use it. The Viewport Canvas can be used on any object that has mapping coordinates. Mapping coordinates are applied to a primitive object by enabling the Generate Mapping Coordinates option in the Command Panel, or by adding the UVW Mapping or UVW Unwrap modifier to an object.

With an object selected, select one of the brushes in the Canvas palette; if the object has a material with a bitmap applied, you can begin painting right away. If the object doesn't have a material or a texture applied, the Assign Material dialog box appears, as shown in Figure 22.2.

FIGURE 22.2

The Assign Material dialog box lets you choose the channel to paint on.

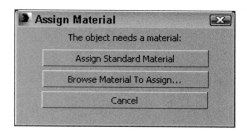

The Assign Standard Material button pops up a list of available channels where the texture may be applied using the Standard material. The Browse Material to Assign button opens the Material/Map Browser, where you can choose the type of material to use. The list of available channels includes Ambient Color, Diffuse Color, Specular Color, Specular Level, Glossiness, Self-Illumination, Opacity, Filter Color, Bump, Reflection, Refraction, and Displacement. After choosing a texture channel, the Create Texture dialog box appears, as shown in Figure 22.3, where you can set the size of the texture map and specify where the texture is saved, or you can select an existing texture. If a new texture is created, you need to specify a path and name for the new texture file. The texture is automatically mapped to the selected channel for the object's material. The Color setting is used for the texture's initial background color.

FIGURE 22.3

The Create Texture dialog box automates the process of applying a material with a texture.

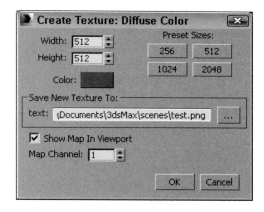

Using the Canvas brushes

After the texture is set up, you can begin painting by clicking the Paint brush icon and dragging in the viewport over the selected object. Using the settings under the brush icons, you can set the Color, Radius, Opacity, Hardness, Spacing, Scatter, and Blur/Sharpen values. The available settings change depending on the brush selected.

When a brush is selected, it is highlighted in the Viewport Canvas palette and remains active until another brush is selected or until you right-click in the viewport. Table 22.1 lists the available brushes and layer tools.

TABLE 22.1 Viewport Canvas Brush Icons

Palette Icon	Name	Description
	Paint	Applies paint using the selected color or image.

	Erase	Removes paint applied from any of the brushes on the selected layer but not from the background layer.
	Clone	Copies paint from another part of the current texture. Click with the Alt key held down to set the copy point. A green dot marks the copy location. Within the Paint Behavior rollout, you can select the Clone Source to be the Current Layer, All Layers, the Viewport, or the Screen.
	Fill	Fills the entire texture with paint. If you're painting with an image, you can set the image to Tile or to 3D Wrap using the options in the Paint Behavior rollout.
	Gradient	Lets you drag to extend a gradient across the surface of the object. Full color (or image) is painted where you first click and full transparency is painted where you stop dragging. The orientation of the gradient is aligned to the direction that you drag.
	Blur	Blurs the area under the brush.
	Sharper	Sharpens the area under the brush.
	Contrast	Increases the contrast of the area under the brush.
	Dodge	Lightens the area under the brush.
	Burn	Darkens the area under the brush.
	Smudge	Distorts and smudges the area under the brush.
	Move Layer	Allows the current layer to be moved.
	Rotate Layer	Allows the current layer to be rotated about the texture's center.
	Scale Layer	Allows the current layer to be scaled about the texture's center.

Caution

The Background layer cannot be moved, rotated, or scaled.

If the Paint brush is selected, you can immediately switch to the Erase brush by holding down the Shift key. Holding down the Ctrl key lets you click to select a different color from the current texture. Pressing the Spacebar causes a straight line to be drawn from the last painted location to the current cursor location. The Spacebar shortcut also works with the Erase brush.

With any of the standard brushes, you can hold down the Ctrl+Shift keys to drag and change the brush radius. The Alt+Shift keys change the brush Opacity, and the Ctrl+Alt keys change the brush's Hardness value.

Clicking the Color swatch opens a Color Selector dialog box where you can choose a new color. You also can load a custom color palette using the Open Color Palette button or quickly switch between black and white colors, which is helpful for painting value maps.

At any time while painting, you are free in the viewport to rotate the model around to paint in a different location.

If you don't like the results after the painted texture has been applied to the object, click the Undo button on the main interface to remove the last set of changes. Clicking the Redo button reapplies the recent changes.

Painting with images

In the Brush Images rollout are two swatches for selecting an image and a mask to paint with. Clicking the swatches opens a palette of presets, as shown in Figure 22.4. Clicking the Browse Custom Maps Directory button opens Windows Explorer to the Viewport Canvas→Custom Brushes folder where 3ds Max is installed. Within this folder are all the custom images and masks that are included in the palette.

FIGURE 22.4

Custom images and masks can be added to the available presets in the Viewport Canvas.

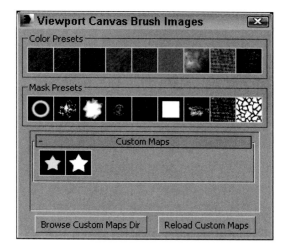

Placing a new image or an image with an alpha channel (for masks) in this folder makes them appear in the Viewport Canvas Brush Images dialog box and allows you to select and paint with them. The Reload Custom Maps button reloads any images placed in the folder so they can be seen.

To use an image or a mask, simply enable the Use option next to each. You also have several Rotation options including None, Random, and Follow Stroke. Figure 22.5 shows examples of each painting with stars. The left eye shows the None option, and all stars are oriented exactly the same. The right eye shows the Random option with each star oriented differently; the lower line shows the Follow Stroke option with the stars aligned in the direction of the brush stroke.

FIGURE 22.5

The image rotation options determine how the stars are oriented.

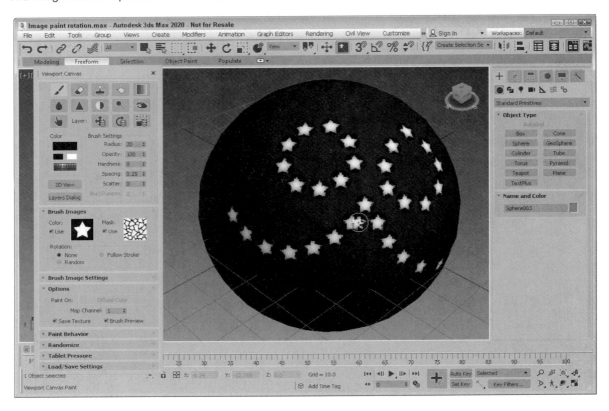

Within the Brush Image Settings rollout are some options for setting how the image is projected onto the object. The options include Hit Normal and From Screen. The Hit Normal option applies the image as if projected down onto the normal of the object. The From Screen option applies the texture as if projected from the screen position onto the object. Other options make the image fit within the brush size and offer tiling options of None, Tile, and Across Screen.

Using paint layers

The Viewport Canvas palette lets you paint on layers just like you can in Photoshop. To access the layers, click the Layers Dialog button in the Viewport Canvas dialog box. The Layers palette, shown in Figure 22.6, shows the current layers in a stack. Each layer can be named, and the layers are placed on the object with the top layers appearing on top of the lower layers. You also can select a blend method and an Opacity value. The buttons at the bottom of the Layers palette let you create new layers, duplicate the current layer, or delete a layer. The small light bulb icon to the left of the layer name lets you turn a layer on or off.

FIGURE 22.6

The Layers palette lets you work with layers for the current texture.

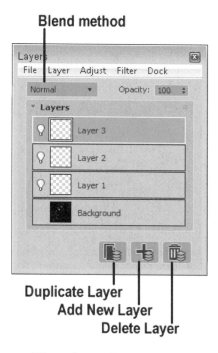

When a layer other than the Background layer is selected, you can use the Layer tool icons located at the top of the Viewport Canvas palette. These Layer tools let you move, rotate, and/or scale the current layer.

Caution
Be aware that you cannot erase any paint applied to the Background layer using the Erase tool.

The Layers dialog box includes some menus that have many of the same commands found in Photoshop. For example, you can use the Layer menu to add layer masks, merge a layer down, flip a layer either horizontally or vertically, or flatten all visible layers.

The Adjust menu includes options for changing the Brightness, Contrast, Hue, Saturation, Levels, or Color Balance. You also have an Auto Levels option. The Filter menu includes filters for blurring, sharpening, finding edges, median, threshold, high pass, and distort.

Within the File menu are options for pasting from the clipboard, loading a bitmap into the current layer, saving the current bitmap (as a flattened file), or saving the image with all layers to a PSD file that can be reopened in Photoshop.

Tip
If you want to maintain the various layers after you're finished painting, be sure to save your texture using an image format that supports layers, such as PSD.

To exit Canvas paint mode, select the Tools→Viewport Canvas menu again. If you've added any layers to your texture and you exit the Canvas, the Save Texture Layers dialog box, shown in Figure 22.7, opens. Using this dialog box, you can choose to continue painting, save the file as a PSD file, which maintains the various layers, flatten all layers and save the texture, save and replace, save, flatten and then save again, or simply discard. If you plan to revisit the layers of your texture, save it as a PSD file, but if you're happy with the results, you can flatten and save the texture.

FIGURE 22.7

The Save Texture Layers dialog box lets you maintain the various texture layers.

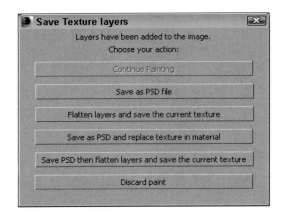

Painting in 2D

Painting directly on a 3D object has its advantages, but when painting over a random surface, the results can be irregular. The Viewport Canvas dialog box includes a 2D painting mode that displays the current texture in a rectangular window and lets you paint directly on the texture. This is helpful for textures that need to be projected onto an object such as those with text.

To access the 2D painting mode, simply click the 2D View button under the colors and the current texture is opened in a window, as shown in Figure 22.8. The buttons at the top of this window let you see the UV wireframes, fit the texture in the view, or view the image at its actual size. Any changes made to the texture in 2D painting mode are immediately reflected on the selected object in the viewports.

Clicking the 2D View button again toggles the 2D painting mode off, and the 2D Paint window closes.

FIGURE 22.8

2D painting mode lets you paint directly on the texture without any perspective distortion.

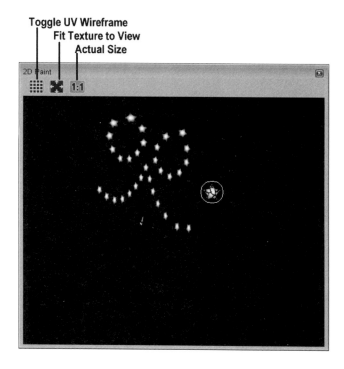

Using the paint options

When the Viewport Canvas palette is open, several additional rollouts hold the options that affect the various brushes and texture. Within the Options rollout, you can select which map type and channel to paint on. There is also an option to Save Texture. If this option is disabled, any paint applied to the texture isn't saved. This lets you try a different look without saving it. You also can do this by painting on a new layer that could be deleted. The Brush Preview option shows the outline and size of the brush as you paint.

Within the Paint Behavior rollout are options to have the paint affect any areas of the object within the brush's spherical radius or to apply the paint through the entire object with the Depth option. The Mirroring options let you mirror all paint strokes across the X, Y, or Z axes. The Clone Source option defines where the Clone brush pulls from including Current Layer, All Layers, or Screen. When using a Fill brush, you can set it to Tile end to end or to wrap around the object. There are also 3 settings for Stroke Smoothing. The higher the value the smoother the result.

The Randomize rollout, shown in Figure 22.9, lets you set the minimum and maximum values for several different settings including Brush Radius, Opacity, Spacing, Scatter, and Color.

FIGURE 22.9

Using the Randomize rollout, you can add variety to the texture.

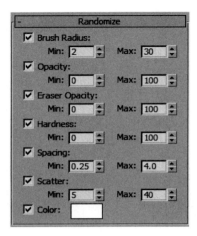

Using the Tablet Pressure rollout, you can select which attributes are affected by an increase in the tablet pressure. The options include Brush Radius, Opacity, Hardness, and Scatter.

The Load/Save Settings rollout includes buttons for saving and loading the Viewpoint Canvas settings. Settings are saved using a simple text file. There is also a button for saving the current settings as the default.

Tutorial: Face painting

To show off the Viewport Canvas, we do some face painting on a woman's head model.

To paint on a character's face using the Viewport Canvas, follow these steps:

1. Open the Face painting.max file from the Chap 22 directory in the downloaded content set.

   This file includes a woman's head mesh.

2. Rotate the head model so the cheek is clearly visible, apply the UVW Map modifier to make sure the object has mapping coordinates with the Planar map, and click the View Align button.

3. Open the Viewport Canvas panel with the Tools→Viewport Canvas menu command.

4. Select the model, and click the Paint Brush icon in the Viewport Canvas dialog box, then select the Material #2→Diffuse Color option in the pop-up menu. In the Create Texture dialog box that pops up next, select the 512 x 512 preset and click the file button next to the Save New Texture field to open the file dialog box. Save the file as Face Painting texture.tif, and click the OK button.

5. Set the brush Radius to **10**, the blending mode to **Normal**, and the Opacity to **100,** and choose a bright red color. Then click the Brush icon, and draw a heart on the woman's cheek. Then click the Paint Brush icon again to apply the texture to the model.

Figure 22.10 shows the resulting applied texture.

FIGURE 22.10

The Viewport Canvas painting mode can be used to apply paint directly to a model.

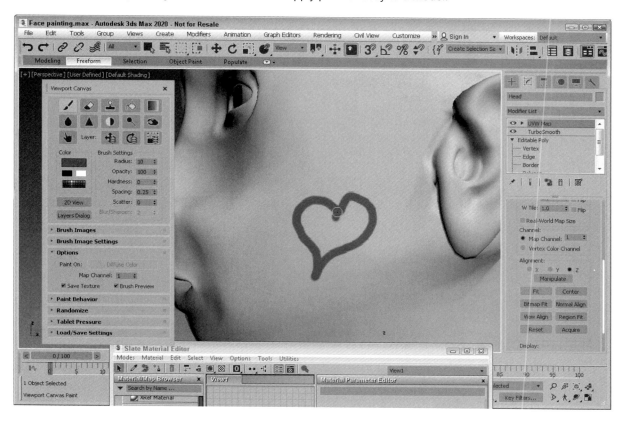

Using Vertex Colors

Another similar feature to the Viewport Canvas is the Vertex Paint modifier. This modifier lets you apply colors to a model without the weight of a texture. When creating models for games, the size of the texture map can be prohibitive. I mean, what model that weighs in at 16KB or less wants to carry around a 2MB texture map? The solution that much of the gaming world relies on is to apply a single color to a vertex. Having each vertex remember its color (or even several colors) requires very little additional information for the mesh and can create some good shading. Colors are then interpolated across the face of the polygon between two different colors on adjacent vertices.

The results aren't as clean and detailed as texture maps, but for their size, vertex colors are worth the price.

Assigning vertex colors

Vertex colors can be assigned in the Surface Properties rollout for Editable Mesh and Editable Patch objects, and in the Vertex and Polygon Properties rollouts for Editable Poly objects. They also can be assigned in Face, Polygon, and Element subobject modes using a little rollout section called Edit Vertex Colors. Within this section are two color swatches for selecting Color and Illumination values. The Alpha value sets the alpha transparency value for the vertex.

Painting vertices with the Vertex Paint modifier

Another, more interactive way to color vertices is with the Vertex Paint modifier. This modifier lets you paint on an object by specifying a color for each vertex. If adjacent vertices have different colors assigned, a gradient is created across the face. The benefit of this coloring option is that it is very efficient and requires almost no memory.

The Vertex Paint modifier lets you specify a color and paint directly on the surface of an object by painting the vertices. The color is applied with a paintbrush-shaped cursor. The modifier can be applied multiple times to an object, giving you the ability to blend several layers of vertex paints together. You can find this modifier in the Modifiers→Mesh Editing submenu.

Note

After the Vertex Paint modifier has been applied to an object, the Paintbox automatically reappears whenever the object is reselected.

Applying this modifier opens a Vertex Paint dialog box, shown in Figure 22.11. At the top of the Paintbox are four icons that can be used to show the visible results of the painting in the viewports. The options include Vertex Color Display-Unshaded, Vertex Color Display-Shaded, Disable Vertex Color Display, and Toggle Texture Display On/Off.

FIGURE 22.11

The Vertex Paint dialog box for the Vertex Paint modifier includes a wealth of features.

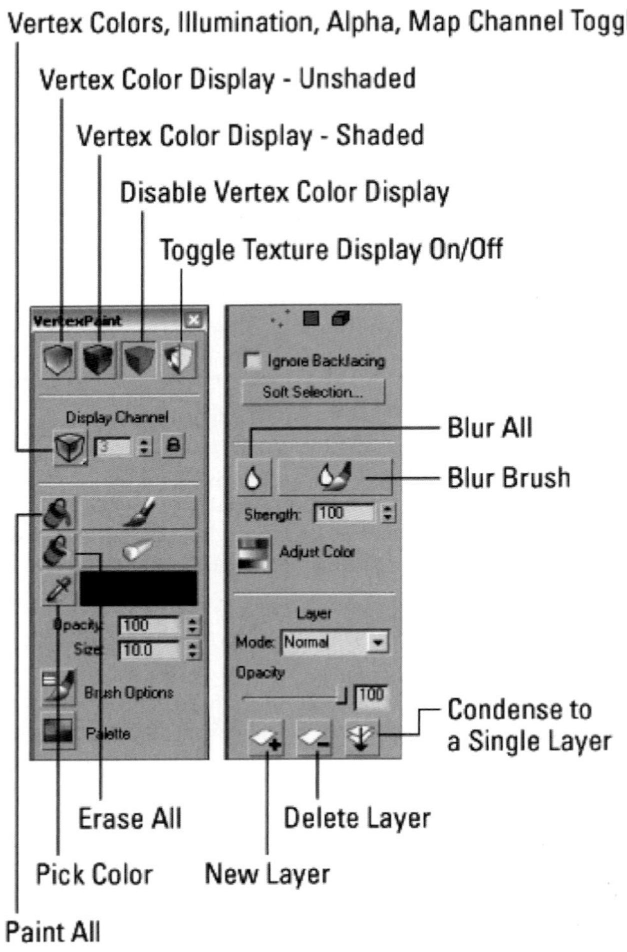

Vertex Colors, Illumination, Alpha, Map Channel Toggle

Vertex Color Display - Unshaded

Vertex Color Display - Shaded

Disable Vertex Color Display

Toggle Texture Display On/Off

Blur All

Blur Brush

Condense to a Single Layer

Erase All

Delete Layer

Pick Color

New Layer

Paint All

The Vertex Color icon flyout lets you work on the Vertex Color, Illumination, Alpha, or any one of the 99 available map channels. The lock icon locks the display to the selected channel, or you could be looking at a different channel from the one you are painting.

The large Paint and Erase buttons let you add or remove vertex colors using the color specified in the color swatch. You also can select colors from objects in the viewports using the eyedropper tool and then set the Opacity.

The Size value determines the size of the brush used to paint. 3ds Max supports pressure-sensitive devices such as a graphics tablet, and you can set the brush options using the Painter Options dialog box, opened with the Brush Options button. When painting on the surface of an object, a blue normal line appears. This line guides you as you paint so you know you're on the correct surface.

The Painter Options dialog box also is used by the Skin modifier and the Paint Deformation tool. It is described in detail in Chapter 13, "Modeling with Polygons."

Beneath the Brush Options button is a Palette button that opens the Color Palette dialog box, shown in Figure 22.12. The Color Palette holds custom colors and lets you copy and paste colors between the different swatches. Collections of colors can be saved by right-clicking the Color Palette and selecting the Save As command. Color palettes are saved as Color Clipboard files with the .ccb extension.

FIGURE 22.12

The Color Palette can display colors as a list or as swatches.

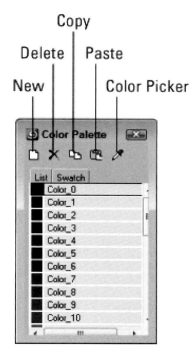

The Paintbox also includes three subobject selection icons. These icons can be used to select certain Vertices, Faces, or Elements to be painted. This limits the painting to the selected subobjects only. You also can select to Ignore Backfacing and use Soft Selection.

The Blur brush button lets you blur colors across polygons using a brush that works just like the Paint and Erase brushes.

The Adjust Color dialog box lets you change all the painted colors applied to an object using HSV or RGB color sliders. The Preview option makes the color adjustment visible in the viewports if selected.

Colors can be mixed between layers using the various blending modes. Clicking the New Layer button adds a new instance of the Vertex Paint modifier to the Modifier Stack; the Delete Layer button does the opposite. Click the Condense to a Single Layer button to merge all the consecutive Vertex Paint modifiers to a single instance using the selected blending mode.

Beneath the layer options are buttons to Display, Cull and Color. The Display button displays a number value that corresponds to the color. This provides a quick way to see which vertices have the same color value. The Cull button hides all the backfacing vertice values and the Color lets you change the color of the number values.

Tutorial: Marking an octopus

As an example of using the Vertex Paint modifier, we'll color some spots on an octopus.

To color spots on an octopus using the Vertex Paint modifier, follow these steps:

1. Open the Octopus - vertex colors.max file from the Chap 22 directory in the downloaded content set. This file includes an octopus model.

2. Select the octopus model, and choose Modifiers→Mesh Editing→Vertex Paint to apply the Vertex Paint modifier.

3. In the Paintbox that opens, choose the Vertex Color Display-Shaded button at the top of the Vertex Paint dialog box, select the red color, and click the Paint button. Then drag the mouse over the surface of the Perspective viewport.

Figure 22.13 shows the resulting color.

FIGURE 22.13

The Vertex Paint modifier can apply color to an object by assigning a color to its vertices.

The Assign Vertex Color utility

The Assign Vertex Color utility works a little differently. It converts any existing material colors to vertex colors. To use this utility, select the utility from the Utilities list that opens when you click the More button in the Utility panel, select an object, choose a Channel, choose a Light Model (Lighting + Diffuse, Lighting Only, or Diffuse Only), and click the Assign to Selected button.

Rendering Surface Maps

3ds Max can be used to create some useful maps based on the geometry of the object. For example, a Cavity map is generated by looking at the concavity of the model. This appears as a grayscale map where the convex portions are white and the concave portions are black. Such a map can then be reused as a diffuse dirt submaterial map for showing those areas that are tight and indented. These tightly concave areas are the likely areas for non-specular highlights and for dirt to appear on old and weathered models.

To access the Render Surface Map panel, shown in Figure 22.14, select the Rendering→Render Surface Map. This feature works only on Editable Poly objects that have mapping coordinates. If any of these conditions are missing, a warning dialog box appears when you click one of the mapping buttons.

Tip

If you change the mapping coordinates for the model, but the surface maps are still using the old mapping coordinates, then try collapsing the modifier stack down to an Editable Poly object and the new mapping coordinates will be used.

FIGURE 22.14

The Render Surface Map panel can create several different types of maps.

The Width and Height values set the resolution of the rendered bitmap. The Size button has several presets. You also can set the Map Channel and the Seam Bleed values. When you click the map type, the map is generated and displayed in the Rendered Frame Window.

3ds Max can use the Render Surface Map feature to render several types of surface maps, including the following:

* **Cavity Map:** The Cavity Map creates a grayscale map that highlights convex areas in white and concave areas in black. The Contrast value defines the difference between black and white values.

* **Density Map:** The Density Map option creates a map that shows the areas where the vertices are closest together as white and farther apart as black.

* **Dust Map:** The Dust Map option creates a grayscale map that identifies the areas that face upward as white and the underneath areas as black, as if the dust were to settle from above and land on all the white areas.

* **SubSurface Map:** The SubSurface Map is used to identify those areas of the mesh that are thickest as black and the thinnest areas of the mesh as white. This map type is used to show how likely light would pass through a given area. The Blur value is used to blur those areas between black and white.

* **Occlusion Map:** The Occlusion is used to those parts of an object that are occluded from light in the scene by other parts.

* **Selection to Bitmap:** This map is used to identify a specific subobject selection. Each selected vertex is displayed as a white dot.

* **Texture Wrap:** The Texture Wrap feature lets you load in a texture that is wrapped about the object in a way to eliminate all seams. This texture could be a simple skin or hide texture. The Tile value is the number of times the texture is repeated end to end to cover the surface.

* **Bitmap Select:** This feature allows you to select specific subobject selections based on the color of the applied bitmap. For example, if you load a bitmap with white lines running through it, all polygon faces that touch those lines are selected.

Figure 22.15 shows the Cavity map for the crocodile model.

FIGURE 22.15

The Cavity map shows areas that are convex and concave.

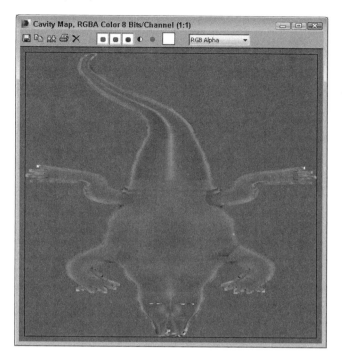

Using Surface Maps

Once you've rendered out a surface map, you can pull the map into Photoshop and edit it to create the desired texture map you want or you can apply it directly to the object it was pulled from. These maps easily can be applied in the Material Editor to the Diffuse channel, but they also can be useful as Specular, Glossiness and Specular Color maps.

Figure 22.16 shows a simple render of an octopus using default lighting and next to it is the same model with the cavity map rendered with a contrast of 6 applied to the model's Diffuse channel. Notice how the increase in Contrast makes the character look old.

FIGURE 22.16

An octopus rendered normally in default lighting and with a cavity map applied.

Applying the various surface maps as a texture can simulate different lighting effects without needing the lights. This can be beneficial for models used in certain environments such as real-time games by not requiring the lighting effects to be rendered as the game is being played.

Tutorial: Generating and Applying Surface Maps

Under default lighting a detailed object like a human heart can lose many of its details. Generating and applying a surface map can add the details back in without changing the lighting and enable you to see through areas of the model.

To generate and apply a cavity map to an object, follow these steps:

1. Open the Head surface map.max file from the Chap 22 directory in the downloaded content set.

    This file contains a head model with the Unwrap UVW modifier applied.

2. Before you can render a surface map, the object has to be an Editable Poly object and it must have mapping coordinates.

3. Right-click the Editable Poly in the Modifier Stack and select the Collapse All option in the pop-up menu. This collapses the modifiers so you can render surface maps.

4. Select the Rendering→Render Surface Map menu option. Set the Size to 512, the Map Channel to 1, and the Contrast value to 3. Then click the Occlusion Map button. The resulting Cavity Map is displayed in the Rendered Frame Window. Click the Save Image button and save the image as Head-occlusion map.tif. In the TIF Image Control dialog box, select the 8-bit Color option and click OK.

5. Before closing the Render Surface Map dialog box, click the SubSurface Map button and save the rendered image as Head-subsurf map.tif.

6. Press the M key to open the Slate Material Editor. Locate and double-click the Standard material in the Material/Map Browser, and then select the Bitmap option from the Maps rollout of the Material/Map Browser. Select Head-occlusion map.tif from the file dialog box that opens. Then connect this node to the Diffuse Color parameter in the Standard material node. Double-click the Bitmap option in the Material/Map Browser again and load the Head-subsurf map.tif file. Connect this node to the Opacity parameter of the Standard material node. Then apply this material to the head object.

Figure 22.17 shows the head model after changing the background color to something lighter. Notice how the head model's details looks unique and other-worldly like an alien.

FIGURE 22.17

Adding surface maps as an object's materials makes the details pop.

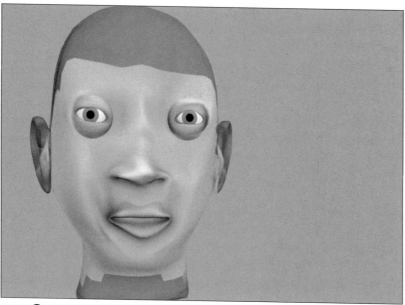

Summary

This chapter covered a couple of key features for applying materials. The Viewport Canvas lets you paint directly in the viewport and have the results transferred to the selected object. Surface maps can be rendered to provide a map that gives information about the surface of the object.

Surface maps can be rendered to provide a template for the object's texture map or can be applied directly as a map.

In this chapter, you learned about the following:

* Painting on objects with the Viewport Canvas
* Using Vertex Colors to paint models
* Rendering a variety of surface maps
* Using surface maps as a material for the object

In the next chapter, you learn how to bake textures onto objects and work with normal maps.

Chapter 23

Creating Baked Textures and Normal Maps

IN THIS CHAPTER

Using channels

Using the Render to Texture interface

Creating normal maps

3D games pose an interesting dilemma—creating interactive scenes that are displayed in real time with the highest quality graphics. To achieve this, game developers use a number of tricks designed to speed up the rendering time. One of these tricks is pre-rendering textures that include all the lighting information and applying these pre-rendered textures as texture maps. This allows advanced lighting solutions such as global illumination to be included within a game without requiring extra time to render such a complex solution. The process of applying pre-rendered textures as maps is called *baking* a texture.

Rendering textures is a significant part of the rendering process, and baking a texture doesn't remove this step; it simply completes the step beforehand, so that the game engine doesn't need to do the texture calculations.

Another common efficiency trick is to use normal maps. Normal maps calculate the lighting results used to light small details that stick out from the surface of an object. These details are then re-created using a normal map that is applied back onto a simplified version of the object. The normal map allows these details to be simulated without the extra polygons used to create them. By allowing simple base objects to have details such as bolts and rivets without the extra polygons, the objects can be redrawn quickly without losing their visual quality.

This chapter covers some of the features found in the Autodesk® 3ds Max® 2020 software that enable the amazing graphics that are found in the latest real-time games.

Using Channels

When 3D models are used in games, the color and material information for the models is stored in channels. One channel can hold the diffuse colors and another can hold the bump map details. A model can also have several channels for the some information like the diffuse colors and can switch between them as needed. The game engine then knows that if it wants to change the color of a group of vertices because of an explosion that has happened, it just looks in the preset channel, finds the vertices it needs, changes the color, and then goes on with the game.

Working with channels is a very efficient way to interface with the gaming engine, but a sloppy game developer can introduce a model to the game engine with all sorts of unneeded or exaggerated channels. If this happens, the game engine can ignore the extra channels and get the wrong information, which can cause your hero to march off into battle without a weapon. Worse, it can crash the system.

To prevent problems and to streamline the number of channels that are included with game models, 3ds Max includes a Map Channel Info editor that you can use to manipulate the various channel data. This editor, shown in Figure 23.1, can be opened using the Tools→Channel Info menu command.

FIGURE 23.1

The Map Channel Info dialog box lets you edit channel data.

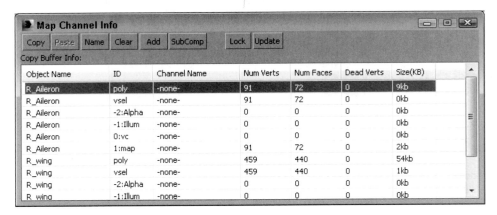

Using the Map Channel Info dialog box

The Map Channel Info dialog box shows lots of information, including the Object Name; its ID; its Channel Name; the number of Vertices, Faces, and Dead Vertices (unattached vertices); and its File Size. With this information, you can quickly determine which channels are taking up the most space and eliminate them.

All objects include some default channels for mesh, which holds the geometry; vsel, which holds the selected vertices; -2:Alpha, which holds the alpha channel information; -1:Illum, which holds illumination values; and channel 0:vc, which holds vertex color information. Objects also include at least one default map channel (even if it is empty). These channels cannot be deleted.

Vertex colors are covered along with painting on objects in Chapter 20, "Creating Compound Materials and Using Material Modifiers."

The interface lets you Copy and Paste selected channels. You can give each channel a name with the Name button. Beneath the Copy button, text appears that lists the information currently copied in the Copy Buffer. Selected channels can be copied only between channels that have the same number of vertices.

The Clear button clears out the selected channels, but you cannot clear a map channel if there is another map channel above it. The Add button adds a new map channel to the object. Objects can hold as many as 99 map channels. The Clear and Add buttons also apply UVW Mapping Clear or UVW Mapping Add modifiers to the Modifier Stack. The Paste command also adds a modifier. These modifiers are convenient because they can easily be removed or reordered in the Stack. If changes have been made in the Modifier Stack, the Update button reflects these changes in the Map Channel Info dialog box.

The SubComp button shows the channel components if they exist. For example, map channels can be broken into X, Y, and Z components, and other channels such as Alpha have R, G, and B components. The Lock button holds the current channels even if another object is selected.

Select by Channel modifier

After new channels have been created, you can recall them at any time using the Select by Channel modifier. This modifier is found in the Modifiers→Selection Modifiers→Select by Channel menu command. Using this modifier, you can choose to Replace, Add, or Subtract a given channel from the selection. The available channels for the selection are listed by their channel name in a drop-down list.

Rendering to a Texture

When working with a game engine, game designers are always looking for ways to increase the speed and detail of objects in the game. One common way to speed game calculations is to pre-render the textures used in a game and then to save these textures as texture maps. The texture map takes more memory to save but can greatly speed the rendering time required by the game engine. This process of pre-rendering a texture is called *texture baking*.

Caution

If you bake a texture into an object and then render it with the rest of the scene, the object gets a double dose of light.

Texture baking can be accomplished in 3ds Max using the Rendering→Render to Texture menu command (or by pressing the 0 key). This opens the Render to Texture dialog box. In several ways, the Render to Texture dialog box, shown in Figure 23.2, resembles the Render Setup dialog box, including a Render button at the bottom edge of the interface.

FIGURE 23.2

The General Settings rollout of the Render to Texture dialog box includes settings for all objects.

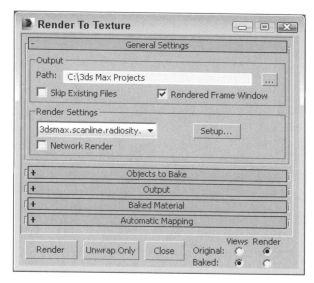

To create a baked texture, select a Texture Element from the Output rollout and click the Render button. Clicking the Render button creates the baked texture for the selected object and saves it in the directory specified in the General Settings rollout. It also applies an Automatic Flatten UVs modifier to the Modifier Stack and applies a Shell material to the object. The Shell material contains the object's original material along with the new baked material. You can select which material is displayed in the viewport and which is rendered using the options to the bottom right of the Render to Texture dialog box.

Note

You also can select Use an Existing Channel for mapping coordinates. This is a good choice because the Auto Unwrap option doesn't have the control over the UV coordinates that you may want, and sometimes it does a poor job of unwrapping the mesh.

The interface also includes an Unwrap Only button. This button can be used to flatten the UVW Coordinates for the selected objects and to automatically create a map channel.

General Settings

The General Settings rollout includes an output path where the baked texture is saved. The file is saved by default using the Targa file format. The Skip Existing Files option renders only those elements that don't already exist in the designated directory. The Rendered Frame Window option displays the resulting map in the Rendered Frame Window along with saving the image as a file. For the render pass, you can select which rendering settings to use, including the Arnold rendering engine. The Setup button opens the Render Setup panel, where you can change the render settings. There is also a Network Render option for rendering over a network.

Network rendering is covered in Chapter 29, "Batch and Network Rendering."

Selecting objects to bake

In the Objects to Bake rollout, shown in Figure 23.3, a list displays exactly which objects, subobjects, and channels will be included in the rendered texture. The Padding value defines the overlap in pixels of the texture. The Objects to Bake rollout also includes a Presets list that lets you save and reload defined settings.

FIGURE 23.3

The Objects to Bake rollout of the Render to Texture dialog box lets you specify which objects are baked into the texture map.

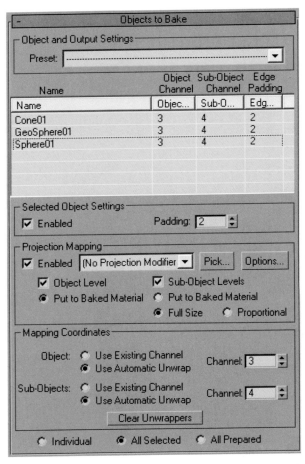

The Projection Mapping section lets you enable the creation of a normal map using a Projection modifier. These settings are covered in detail in the "Creating Normal Maps" section that appears later in this chapter.

The Mapping Coordinates section lets you choose to use the mapping coordinates of the Object or the Subobject selection contained within a specified channel, or you can select to use the Use Automatic Unwrap feature, which automatically flattens the mapping coordinates. If the Use Automatic Unwrap option is selected, you can set the mapping options in the Automatic Mapping rollout. By default, unwrap mapping uses channel 3, but you can change this channel if you wish. If a different mapping uses channel 3 and you don't change this, the new mapping replaces the old one. The Clear Unwrappers button removes any existing Unwrap UVW modifiers from the object's stack.

You can select to bake an Individual object, All Selected objects, or All Prepared objects, which are all objects with at least one texture element.

Output settings

The Output rollout, shown in Figure 23.4, lists the texture elements that are included in the texture map. The Enable option can be used to disable the selected texture element, or elements can be deleted with the Delete button.

FIGURE 23.4

The Output rollout of the Render to Texture dialog box lets you choose which texture elements are baked.

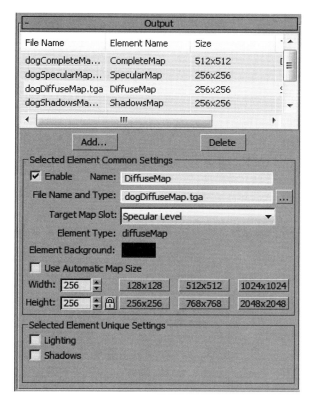

Clicking the Add button lets you select the type of texture elements that you can render. You'll want to use different maps depending on the purpose of the map, and you may want to render several at a time. The available types are CompleteMap, SpecularMap, DiffuseMap, ShadowsMap, LightingMap, NormalsMap, BlendMap, AlphaMap, and HeightMap. There are also several maps for the Arnold renderer. You also can change the map size or use the Automatic Map Size option, which bases the map size on the object size. Some map elements present a list of components to include in the map. These components appear below the size settings.

More on rendering with Arnold is covered in Chapter 30, "Rendering with ART and Arnold."

Baked Material and Automatic Mapping settings

The Baked Material and Automatic Mapping rollouts, shown in Figure 23.5, provide a way to keep the existing object material using the Shell material. The Clear Shell Materials button removes the Shell materials for the baked objects and restores their original materials.

FIGURE 23.5

The final two rollouts of the Render to Texture dialog box include settings for handling the baked material and how the texture is mapped.

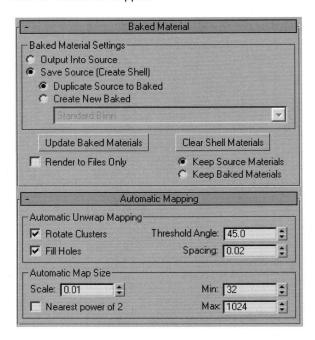

In the Automatic Mapping rollout, you can set how the mapping is applied. If the Use Automatic Unwrap option in the Objects to Bake rollout is enabled, the object to be baked has the Automatic Flatten UVs modifier applied. For this type, you can set the Threshold Angle (which is the difference between the normals of adjacent faces; if the angular value is greater than the Threshold Angle value, a hard edge is created between the faces), the Spacing (which is the amount of space between different map pieces), and whether map pieces can be rotated and used to fill in holes of larger map pieces.

The size of the texture map depends on the size of the object, but you can set a Scale value for greater resolution and set Min and Max values to keep the maps within reason. By default, maps are saved to the /images directory, but you can select a different directory if you prefer. The Nearest Power of 2 option causes the map to be optimized for use in memory to a square pixel size that is a power of 2, such as 8 x 8, 16 x 16, 32 x 32, or 64 x 64.

Note

Most game engines require square texture maps because they are efficiently loaded into memory. Main characters can use texture maps that are 1024 x 1024 or even 2048 x 2048, but background characters and props usually only have textures maps that are 256 x 256 or 512 x 512, so clustering is important to get as much into the textures as you can.

Tutorial: Baking the textures for a covered wagon model

To practice baking textures, you'll bake a complete map of just the covered wagon shell.

To bake a covered wagon texture, follow these steps:

1. Open the Covered wagon.max file from the Chap 23 directory in the downloaded content set.
2. Select Rendering→Render to Texture (or press the 0 key) to open the Render to Texture dialog box.
3. Select the covered object part of the wagon. In the Render to Texture dialog box, set the Threshold Angle to **75** in the Automatic Mapping rollout, and make sure that the Rendered Frame Window option in the General Settings rollout is enabled. In the Output rollout, click the Add button and double-click the CompleteMap option. Set the Map Size to **512x512**, select the Diffuse Color option as the Target Map Slot, and click the Render button.

Figure 23.6 shows the resulting texture map. If you look in the Modify panel, you'll see that the Automatic Flatten UVs modifier has been applied to the object. If you look at the material applied to the object, you'll see that it consists of a Shell material.

FIGURE 23.6

A texture map created with the Render to Textures panel

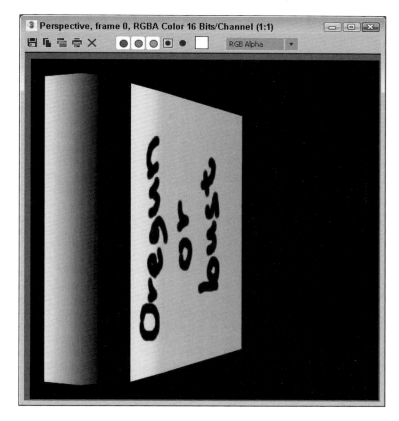

Creating Normal Maps

Normal maps are becoming more common in games because they offer a way to increase the bump details of a model by mapping high-detail bump information onto a low-resolution model. Normal maps are created using the Render to Texture dialog box and applied to an object using the Normal Bump map type found in the Material Editor.

The Normal Bump map type is typically applied as a bump map in the Maps rollout and includes a separate button, shown in Figure 23.7, to apply an additional bump map.

Caution

Normal maps can be displayed in the viewports only if the DirectX display driver is selected.

FIGURE 23.7

Although normal maps are created using the Render to Texture dialog box, they are applied using the Material Editor.

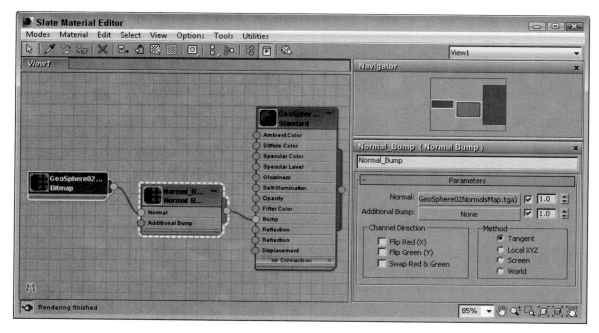

Using the Projection modifier

The Projection modifier is used to create a normal map. It works by being applied to a low-resolution object, and then you pick a high-resolution object that is similar to the low-resolution one. The Projection modifier surrounds the object with a cage that can be manipulated to include all the object details. The Projection modifier is applied using the Modifiers→UV Coordinates→Projection menu command.

Within the Modifier Stack, the Projection modifier includes three subobject modes: Cage, Face, and Element. The Geometry Selection rollout includes a list of objects, a Pick button, and a Pick List for selecting the high-resolution object to be used.

The Cage rollout includes settings for displaying and pushing the cage out from the surface of the object. A Tolerance setting is used for wrapping the cage about the surface. The Selection Check rollout informs you if the Material IDs or Geometry faces are overlapping.

Setting Projection Mapping options

With a Projection modifier applied to a selected object, the Projection Mapping option can be enabled in the Objects to Bake rollout of the Render to Texture dialog box. The object can actually include several Projection modifiers, so a drop-down list lets you select the one to use, or you can use the Pick button to select a target object in the viewports.

The Options button in the Render to Texture dialog box opens the Projection Options dialog box, shown in Figure 23.8. Using this dialog box, you can set the projection method, determine how to resolve how vertices get projected, and define the Map Space.

FIGURE 23.8

The Projection Options dialog box lets you specify how the projection values are determined.

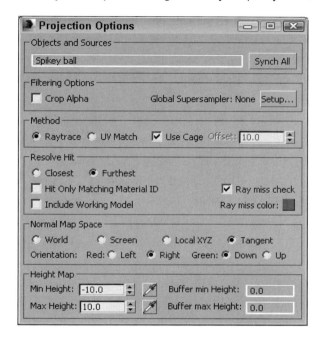

At the top of the Projection Options dialog box is the Source object. The Synch All button causes each object to use its active source for the projection. The two projection methods are Raytrace, which traces each normal line from its source to its target, and UV Match, which works by matching the UV coordinates between the source and the target objects.

For transparent objects, two projection rays may hit the same point. The Resolve Hit options let you set which one is selected, either the Closest or the Furthest. Most projections use the Tangent Map Space, but you can select to use the World, Screen, or Local Map Spaces also.

Note

Before the Projection modifier is applied, you need to have the high-res object and the low-res object positioned at the same place.

Tutorial: Creating a normal map for a Spikey sphere

For this example, I've created two sphere objects, extruded the vertices on one of them, and called it Spikey ball. The other is a plain GeoSphere. The Spikey ball sphere weighs in at 1280 polygons, while the normal GeoSphere is only 320 polygons. Although the Spikey ball includes many more polygons, many of these details can be reclaimed using a normal map.

To create a normal map for the Spikey ball model, follow these steps:

1. Open the Spikey ball.max file from the Chap 23 directory in the downloaded content set.

2. Select and move the Spikey ball object over the top of the low-res sphere object in the Top viewport.

3. Select the normal GeoSphere object, and select the Rendering→Render to Texture menu command (or press the 0 key) to open the Render to Texture dialog box. In the General Settings rollout, select the 3dsmax Scanline renderer, no advanced lighting option as the Render Settings preset. In the Select Preset Categories dialog box that appears, click the Load button.

4. In the Objects to Bake rollout, click the Pick button, select the Spikey ball object in the Add Targets dialog box that appears, and click the Add button. Then enable Projection Mapping.

5. In the Output rollout, click the Add button and select the Normals map. From the Target Map Slot drop-down list, select the Bump option. Click the 512 x 512 button to set the map size, and enable the Output into Normal Bump option.

6. Click the Render button at the bottom of the Render to Texture dialog box.

7. To see the normal map when rendered, open the Material Editor, use the Pick Material from Object eyedropper tool, and click the Geosphere object. Then set the low-res sphere's render material to be the baked material.

8. Drag the normal Geosphere away from the Spikey ball object, and render the Perspective viewport.

Figure 23.9 shows the resulting normal map rendered on the GeoSphere.

FIGURE 23.9

The normal map for the Spikey ball can be applied as a bump map to reclaim the high-res details.

Tutorial: Creating a normal map for an optimized gator

For this example, we reuse the optimized gator created earlier when mesh modifiers were covered. The ProOptimizer modifier was used to reduce this high-res gator model from 34,000 faces to fewer than 1,000. This kind of reduction represents a huge step down for the quality of the model, but it enables it to be lightning fast when used in a game. The normal map makes it possible to use the lower poly model while recovering much of the display quality.

To create a normal map for the optimized gator model, follow these steps:

1. Open the ProOptimized gator.max file from the Chap 23 directory in the downloaded content set.

2. Select and move the low-res gator model over the top of the high-res gator model in the Top viewport.

3. With the low-res gator model selected, choose the Rendering→Render to Texture menu command (or press the 0 key) to open the Render to Texture dialog box. In the General Settings rollout, select the 3dsmax Scanline renderer, no advanced lighting option as the Render Settings preset. In the Select Preset Categories dialog box that appears, click the Load button.

4. In the Objects to Bake rollout, click the Pick button, select the crocodile001 object (the high-res gator model) in the Add Targets dialog box that appears, and click the Add button. Then enable Projection Mapping.

5. In the Output rollout, click on the Add button and select the Normals map. From the Target Map Slot drop-down list, select the Bump option. Click the 512 x 512 button to set the map size, and enable the Output into Normal Bump option.

6. Click the Render button at the bottom of the Render to Texture dialog box.

7. Drag the low-res gator model away from the high-res model.

8. To see the normal map when rendered, open the Slate Material Editor and use the Material→Get All Scene Materials menu command. This loads all the material nodes including the normal map. There are separate materials for the skin and eyes. Locate and select the green normal map material node. The normal map is connected to this material's Bump map. Locate the Bump map in the Maps rollout of the Parameter Editor panel and increase the Bump Amount to 100, and then drag the green material's output socket and drop it on the low-res gator's skin.

9. Render the two gators side by side.

Figure 23.10 shows the resulting normal map rendered on the right gator.

FIGURE 23.10

The normal map for the gator can be applied as a bump map to reclaim the high-res details.

Summary

If you're working with games, you'll want to use these features to help keep your models small and fleet. This chapter covers the following topics:

* Discovering what channels the models have

* Learning how to bake textures

* Creating normal maps using the Projection modifier

The next chapter takes on the topic of cameras so you can finally control exactly what you are seeing.

Part V

Working with Cameras and Lights

IN THIS PART

Chapter 24
Configuring and Aiming Cameras

Do you remember as a kid when you first got your own camera? After taking the usual pictures of your dog and the neighbor's fence, you quickly learned how much fun you could have with camera placement, such as a picture of a flagpole from the top of the flagpole or your mom's timeless expression when she found you inside the dryer. Cameras in the Autodesk® 3ds Max® 2020 software also can offer all kinds of amusing views of your scene.

The benefit of cameras is that you can position them anywhere within a scene to offer a custom view. Camera views let you see the scene from a different position such as from the top, front, or left. You can open camera views in a viewport, and you also can use them to render images or animated sequences. Cameras in 3ds Max also can be animated (without damaging the camera, even if your mischievous older brother turns on the dryer).

In the Camera Parameters rollout is a section for enabling multi-pass camera effects. These effects include Motion Blur and Depth of Field. Essentially, these effects are accomplished by taking several rendered images of a scene and combining them with some processing. It is common when creating a 3d scene to start with all the objects and then to add a background image after the scene is finished. This works fine when the scene is in the sky or in space where there isn't a ground plane, but it can cause problems if the background image has some definite landmarks.

Using the Perspective Match feature, you can align the scene to the background image so they match without requiring a lot of rework. You can also use Matte/Shadow materials to make the scene objects cast shadows onto the background.

Learning to Work with Cameras

If you're a photography hobbyist or like to take your video camera out and shoot your own footage, many of the terms in this section will be familiar to you. The cameras used in 3ds Max to get custom views of a scene behave in many respects just like real-world cameras.

3ds Max and real-world cameras both work with different lens settings, which are measured and defined in millimeters. You can select from a variety of preset stock lenses, including 35mm, 80mm, and even 200mm. 3ds Max cameras also offer complete control over the camera's focal length, field of view, and perspective for wide-angle or telephoto shots. The big difference is that you never have to worry about setting flashes, replacing batteries, or loading film.

Light coming into a camera is bent through the camera lens and focused on the film, where the image is captured. The distance between the film and the lens is known as the *focal length*. This distance is measured in millimeters, and you can change it by switching to a different lens. On a camera that shoots 35mm film, a lens with a focal length of 50mm produces a view similar to what your eyes would see. A lens with a focal length less than 50mm is known as a wide-angle lens because it displays a wider view of the scene. A lens longer than 50mm is called a telephoto lens because it has the ability to give a closer view of objects for more detail, as a telescope does.

Field of view is directly related to focal length and is a measurement of how much of the scene is visible. It is measured in degrees. The shorter the focal length, the wider the field of view.

When you look at a scene, objects appear larger if they are up close than they would be lying at a farther distance. This effect is referred to as *perspective* and helps you interpret distances. As mentioned, a 50mm lens gives a perspective similar to what your eyes give. Images taken with a wide field of view look distorted because the effect of perspective is increased.

Creating a camera object

To create a camera object, you can use the Create→Cameras menu, or you can open the familiar Create panel and click the Cameras category button. The three types of cameras that you can create are a Free camera, Target camera, and Physical camera.

Camera objects are visible as icons in the viewports, but they aren't rendered. The camera icon looks like a box with a smaller box in front of it, which represents the lens or front end of the camera. All of the camera types include a rectangular cone that shows where the camera is pointing.

Free camera

The Free camera object offers a view of the area that is directly in front of the camera and is the better choice if the camera will be animated. When a Free camera is initially created, it points at the negative Z-axis of the active viewport.

Target camera

A Target camera always points at a controllable target point some distance in front of the camera. Target cameras are easy to aim and are useful for situations where the camera won't move. To create this type of camera, click a viewport to position the camera and drag to the location of its target. The target can be named along with the camera. When a target is created, 3ds Max automatically names the target by attaching ".target" to the end of the camera name. You can change this default name by typing a different name in the Name field. Both the target and the camera can be selected and transformed independently of each other.

Physical camera

The physical camera object is similar to the Target camera, but it is based on real-world values including Film Width, Focal Length, Zoom, Aperture, Shutter Speed, Exposure Gain and White Balance. This enables you to simulate the results from an actual camera or to recreate the look of an existing camera set-up. There is also a panel of Depth of Field effects with Circular and Bladed apertures; a panel for Perspective Control with Lens Shift and Tilt Correction settings, and panel for Lens Distortion effects including an option to load a gel (texture) to place over the camera.

Note

Previous versions of 3ds Max had a mr Photographic Exposure Control setting that included real-world photography settings like Shutter Speed, Aperture and Film Speed. This control is still there, but if selected, it is applied equally to all scene cameras. The Physical Camera object lets you change the exposure controls for each camera independently.

Creating a camera view

You can change any viewport to show a camera's viewpoint. To do so, click the viewport's Point-of-View label, and select Cameras and the camera's name from the pop-up menu. Any movements done to the camera are reflected immediately in the viewport.

Another way to select a camera for a viewport is to press the C key. This keyboard shortcut makes the active viewport into a camera view. If several cameras exist in a scene, the Select Camera dialog box appears, from which you can select a camera to use. You also can select a camera and press the C key to make that camera's view appear in the active viewport. Figure 24.1 shows two Target cameras pointing at a spaceship. The two viewports on the right are the views from these cameras.

FIGURE 24.1

A spaceship as seen by two different cameras

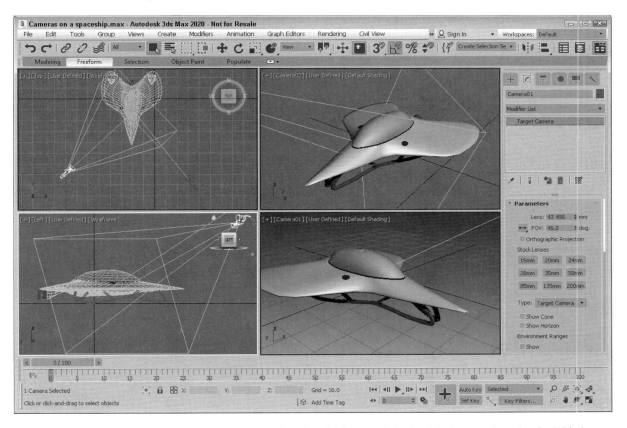

You can turn off the camera object icons using the Display panel. In the Display panel, under the Hide by Category rollout, select the Cameras option. When selected, the camera icons are not visible in the viewports.

Note

Cameras are usually positioned at some distance from the rest of the scene. Their distant position can make scene objects appear very small when the Zoom Extents button is used. If the visibility of the camera icons is turned off, Zoom Extents does not include them in the zoom. You also can enable the Ignore Extents option in the camera's Object Properties dialog box.

Tutorial: Setting up an opponent's view

There is no limit to the number of cameras that you can place in a scene. Placing two cameras in a scene showing a game of checkers lets you see the game from the perspective of either player.

To create a new aligned view from the opponent's perspective, follow these steps:

1. Open the Checkers game.max file from the Chap 24 directory in the downloaded content set.

2. Select Create→Cameras→Target Camera, and drag in the Top viewport to create the camera. Then give the new camera the name **Opponents Camera**.

3. Position the new target camera behind the opponent's pieces roughly symmetrical to the other camera by dragging the camera upward in the Front viewport.

4. With the new camera selected, select and drag the target point and position it near the other camera's target point somewhere below the center of the board.

To see the new camera view, click the Point-of-View label in the Perspective viewport and choose Cameras→Opponents Camera (or select the camera and the Perspective viewport, and press the C key). Figure 24.2 shows the view from this camera.

FIGURE 24.2

Positioning an additional camera behind the opponent's player's pieces offers a view from their perspective.

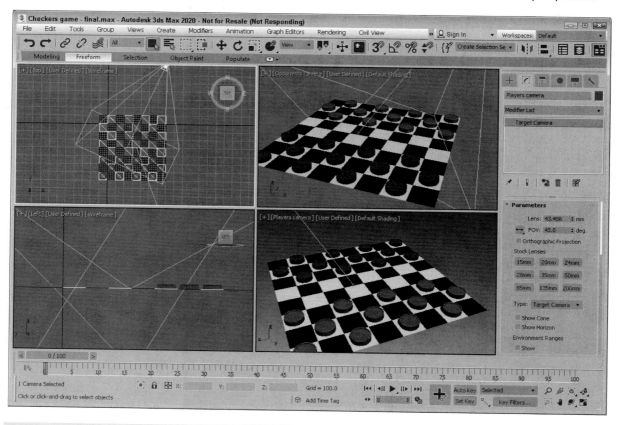

Tip

Because viewports can be resized, the view you see in the viewport isn't necessarily the view that will be rendered. Using the Safe Frames feature found in the Safe Frames panel of the Viewport Configuration dialog box, you can see a border around exactly what will be rendered.

Controlling a camera

I was once on a ride at Disneyland when a person behind me decided to blatantly disregard the signs not to take photographs. As he leaned over to snap another picture, I heard a fumbling noise, a faint, "Oh no," and then the distinct sound of his camera falling into the depths of the ride. (That was actually more enjoyable than the ride. It served him right.) As this example shows, controlling a camera can be difficult. This chapter offers many tips and tricks for dealing with the cameras in 3ds Max, and you won't have to worry about dropping them.

You control the camera view in a viewport by means of the Camera Navigation controls located in the lower-right corner of the screen. These controls replace the viewport controls when a camera view is selected and are different from the normal Viewport Navigation controls. The Camera Navigation controls are identified and defined in Table 24.1.

Note
Many of these controls are identical to the controls for lights.

You can constrain the movements when panning a camera view to a single axis when dragging with the middle mouse button by holding down the Shift key. Dragging with the Ctrl key causes the movements to increase rapidly and with the Alt key causes the view to move slowly. For example, holding down the Ctrl key while dragging the Perspective tool magnifies the amount of perspective applied to the viewport.

You can undo changes in the normal viewports using the Undo View Change (Shift+Z) command in the POV viewport label, but you undo camera object changes with the regular Undo command in the main toolbar because it involves the movement of an object.

TABLE 24.1 Camera Navigation Control Buttons

Control Button	Name	Description
	Dolly Camera, Dolly Target, Dolly Camera + Target	Moves the camera, its target, or both the camera and its target closer to or farther away from the scene in the direction it is pointing.
	Perspective	Increases or decreases the viewport's perspective by dollying the camera and altering its field of view.
	Roll Camera	Spins the camera about its local Z-axis.
	Zoom Extents All, Zoom Extents All Selected	Zooms in on all objects or the selected objects by reducing the field of view until the objects fill the viewport.
	Field of View	Changes the width of the view, similar to changing the camera lens or zooming without moving the camera.
	Truck Camera, 2D Pan Zoom Mode, Walk Through	The Truck Camera button moves the camera perpendicular to the line of sight, the 2D Pan Zoom slides the camera to the side and zooms in on the scene and the Walk Through button enables a mode in which you can control the camera using the arrow keys and the mouse.
	Orbit Camera, Pan Camera	The Orbit Camera button rotates the camera around the target, and the Pan Camera button rotates the target around the camera.
	Maximize Viewport Toggle	Makes the current viewport fill the viewport area. Clicking this button a second time returns the display to several viewports.

Note
If a Free camera is selected, the Dolly Target and Dolly Camera + Target buttons are not available.

Aiming a camera

In addition to the Camera Navigation buttons, you can use the Transformation buttons on the main toolbar to reposition the camera object. To move a camera, select the camera object and click the Select and Move button (W). Then drag the Move gizmo to move the camera.

Using the Select and Rotate (E) button changes the direction in which a camera points, but only Free cameras rotate in all directions. When applied to a Target camera, the rotate transformation spins only the camera about the axis pointing to the target. You aim Target cameras by moving their targets.

Caution

Don't try to rotate a Target camera so that it is pointing directly up or down, or the camera will flip.

Select the target for a Target camera by selecting its camera object, right-clicking to open the pop-up menu, and selecting Select Camera Target.

Tutorial: Watching a rocket

Because cameras can be transformed like any other geometry, they also can be set to watch the movements of any other geometry. In this tutorial, you aim a camera at a distant rocket and watch it as it flies past us and on into the sky.

To aim a camera at a rocket as it hurtles into the sky, follow these steps:

1. Open the Following a rocket.max file from the Chap 24 directory in the downloaded content set.

   This file includes a rocket mesh.

2. Select Create→Cameras→Target Camera, and drag in the Front viewport from the middle to the bottom of the viewport to create a camera. Then select the camera's target and move it in the Left viewport to be positioned right on the rocket. Set the Field of View value to **2.0** degrees. The corresponding Lens value is around 1031mm.

3. With the camera target selected, click the Select and Link button in the main toolbar, and drag from the target to the rocket object.

4. To view the scene from the camera's viewpoint, click the Point-of-View viewport label for the Perspective viewport and choose Cameras→Camera01 from the pop-up menu (or press the C button). Then click the Play Animation button to see how well the camera follows the target.

Figure 24.3 shows some frames from this animation.

FIGURE 24.3

Positioning the camera's target on the rocket enables the camera to follow the rocket's ascent.

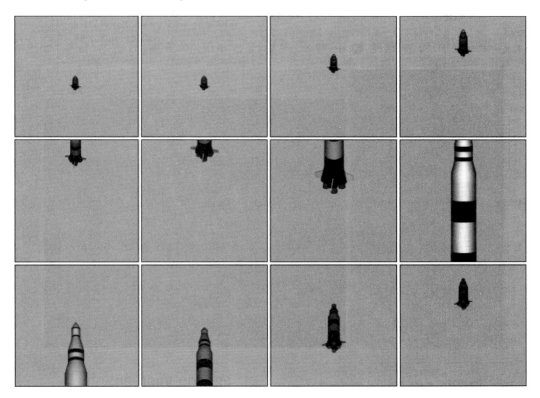

Aligning cameras

 Another way to aim a camera is with the Tools→Align→Align Camera menu command or by clicking the Align Camera button on the main toolbar (under the Align flyout). After selecting this command, click an object face and hold down the mouse button; the normal to the object face that is currently under the cursor icon is displayed as a blue arrow. When you've located the point at which you want the camera to point, release the mouse button. The camera is repositioned to point directly at the selected point on the selected face along the normal. The Align Camera command requires that a camera be selected before the command is used.

The Align Camera command does the same thing for cameras that the Place Highlight command does for lights. A discussion of the Place Highlight command appears in Chapter 25, "Using Lights and Basic Lighting Techniques."

Cameras can be positioned automatically to match any Perspective view that a viewport can display, including light views. The Create→Cameras→Create Physical Camera From View (Ctrl+C) menu command creates a new Free camera if one doesn't already exist, matches the view of the current active Perspective viewport, and makes the active viewport a camera view. This provides you with the ability to position the view using the Viewport Navigation Controls, and it automatically makes a camera that shows that view. If a camera already exists in the scene and is selected, this command uses the selected camera for the view.

Caution
The Create Camera View command doesn't work on Orthogonal views like Top, Front, and Left.

Tutorial: Seeing the dragon's good side

Using the Align Camera tool, you can place a camera so that it points directly at an item or the face of an object, such as the dragon's good side (if a dragon has a good side). To align a camera with an object point, follow these steps:

1. Open the Little Dragon.max file from the Chap 24 directory in the downloaded content set.

2. Select Create→Cameras→Free Camera, and click in the Top viewport to create a new Free Camera in the scene.

3. With the camera selected, choose Tools→Align→Align Camera or click the Align Camera flyout button on the main toolbar.

   The cursor changes to a small camera icon.

4. Right click to select the Perspective viewport and click the cursor on the dinosaur's face just under its eye in the Perspective viewport.

   This point is where the camera will point.

5. To see the new camera view, click the Point-of-View viewport label and choose Cameras→Camera01 (or press C).

   Although the camera is pointing at the selected point, you may need to change the field of view or dolly the camera to correct the zoom ratios.

Figure 24.4 shows your dinosaur from the newly aligned camera.

FIGURE 24.4

This new camera view of the dinosaur shows his best side.

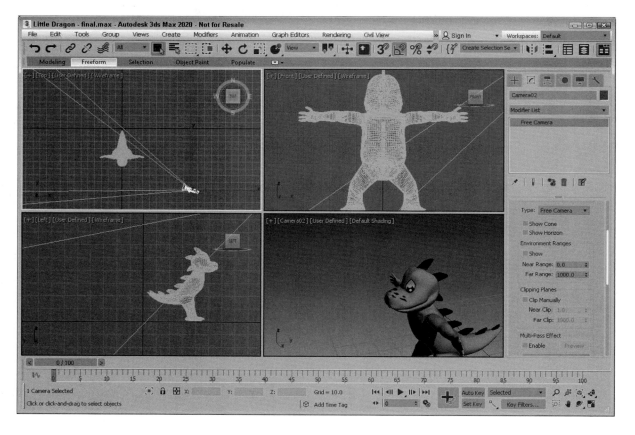

The Align Camera command points a camera at an object only for the current frame. It does not follow an object if it moves during an animation. To have a camera follow an object, you need to use the Look At Constraint, which is covered in Chapter 35, "Animating with Constraints and Simple Controllers."

Setting Camera Parameters

When a camera is first created, you can modify the camera parameters directly in the Create panel as long as the new camera is selected. After the camera object has been deselected, you can make modifications in the Modify panel's Parameters rollout for the camera.

Lens settings and field of view

The first parameter in the Parameters rollout sets the Lens value or, more simply, the camera's focal length in millimeters.

The second parameter, FOV (Field of View), sets the width of the area that the camera displays. The value is specified in degrees and can be set to represent a Horizontal, Vertical, or Diagonal distance using the flyout button to its left, as shown in Table 24.2.

TABLE 24.2 Field of View Buttons

Button	Description
↔	Horizontal distance
↕	Vertical distance
↗	Diagonal distance

The Orthographic Projection option displays the camera view in a manner similar to any of the orthographic viewports such as Top, Left, or Front. This eliminates any perspective distortion of objects farther back in the scene and displays true dimensions for all edges in the scene. This type of view is used heavily in architecture.

Professional photographers and film crews use standard stock lenses in the course of their work. These lenses can be simulated in 3ds Max by clicking one of the Stock Lens buttons. Preset stock lenses include 15, 20, 24, 28, 35, 50, 85, 135, and 200mm lengths. The Lens and FOV fields are automatically updated on stock lens selection.

Tip

On cameras that use 35mm film, the typical default lens is 50mm.

Camera type and display options

The Type option enables you to change a Free Camera to a Target Camera and then change back at any time.

The Show Cone option enables you to display the camera's cone, showing the boundaries of the camera view when the camera isn't selected. (The camera cone is always visible when a camera is selected.) The Show Horizon option sets a horizon line within the camera view, which is a dark gray line where the horizon is located.

Environment ranges and clipping planes

You use the Near and Far Range values to specify the volume within which atmospheric effects like fog and volume lights are to be contained. The Show option causes these limits to be displayed as rectangles within the camera's cone.

You use clipping planes to designate the closest and farthest object that the camera can see. In 3ds Max, they are displayed as red rectangles with crossing diagonals in the camera cone. If the Clip Manually option is disabled, the clipping planes are set automatically with the Near Clip Plane set to 3 units. Figure 24.5 shows a camera with Clipping Planes specified. The front Clipping Plane intersects the spaceship and chops off its front end. The far Clipping Plane is far behind the car.

Tip

Clipping planes can be used to create a cutaway view of your model.

FIGURE 24.5

A camera cone displaying Clipping Planes

Camera Correction modifier

To understand the Camera Correction modifier, you first need to understand what two-point perspective is. Default cameras in 3ds Max use three-point perspective, which causes all lines to converge to a vanishing point off in the distance, but two-point perspective causes all vertical lines to remain vertical.

The visual effect of this modifier is that extra-tall objects appear to bend toward the camera when corrected. For example, if you have a camera pointed at a skyscraper, correcting the camera with the Camera Correction modifier makes the top of the building appear closer rather than having it recede away.

The Camera Correction modifier has an Amount value that lets you specify how much correction to apply and a Direction value that orients the angle of vertical lines in the scene. There is also a Guess button, which automatically sets the correction values for you based on the Z-axis vertical.

Caution

The Camera Correction modifier doesn't appear in the Modifier List in the Modifier Stack, but you can select it from the Modifiers→Cameras menu or from the quad menu when a camera is selected.

Creating multi-pass camera effects

All cameras have the option to enable them to become multi-pass cameras. A multi-pass camera creates and blends several passes of the view from the camera's perspective to create the desired effect. You can find these settings in the Parameters rollout when a camera object is selected. Multi-pass cameras are created by checking the Enable button and selecting the effect from the drop-down list. The current available effects include Depth of Field, and Motion Blur. For each, an associated rollout of parameters opens.

Caution

Preview of the multi-pass camera effects in the viewport does not work when the Nitrous display drivers are enabled.

The Multi-Pass Effect section of the Parameters rollout also includes a Preview button. This button makes the effect visible in the viewports for the current frame. This feature can save you a significant amount of time that normally would be spent test-rendering the scene. The Preview button is worth its weight in render speed. Using this button, you can preview the effect without having to render the entire sequence.

Caution

The Preview button does not work unless the Camera view is the active viewport.

The Render Effects Per Pass option causes any applied Render Effect to be applied at each pass. If disabled, any applied Render Effect is applied after the passes are completed.

You also can apply these multi-pass effects as Render Effects. See Chapter 44, "Using Atmospheric and Render Effects."

Using the Depth of Field effect

The Depth of Field Parameters rollout, shown in Figure 24.6, appears when the Depth of Field option is selected in the Multi-Pass Effect section of the Parameters rollout. It includes settings for controlling the Depth of Field multi-pass effect.

FIGURE 24.6

Use the Depth of Field Parameters rollout to set the number of passes.

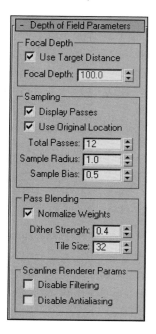

You can select to use the Target Distance (which is the distance to the camera's target), or you can specify a separate Focal Depth distance. This location is the point where the camera is in focus. All scene objects closer and farther from this location are blurred to an extent, depending on their distance from the focal point.

Note

Even Free cameras have a Target Distance. This distance is displayed at the bottom of the Parameters rollout.

Within the Depth of Field Parameters rollout, you also have the option to display each separate pass in the Rendered Frame Window with the Display Passes option and to use the camera's original location for the first rendering pass by enabling the Use Original Location option.

The Total Passes is the number of times the scene is rendered to produce the effect, and the Sample Radius is the potential distance that the scene can move during the passes. By moving the scene about the radius value and re-rendering a pass, the object becomes blurred more away from the focal distance.

Note

The Depth of Field effect is applied only to rendered scene objects. It is not applied to any background images.

The Sample Bias value moves the blurring closer to the focal point (for higher values) or away from the focal point (for lower values). If you want to highlight the focal point and radically blur the other objects in the scene, set the Sample Bias to 1.0. A Sample Bias setting of 0 results in a more even blurring.

The Normalize Weights option allows you to control how the various passes are blended. When enabled, you can avoid streaking along the object edges. The Dither Strength value controls the amount of dither taking place. Higher Dither Strength values make the image grainier. The Tile Size value also controls dither by specifying the dither pattern size.

With lots of passes specified, the render time can be fairly steep. To lower the overall rendering time, you can disable the Anti-alias and filtering computations. These speed up the rendering time at the cost of image quality.

Tutorial: Applying a Depth of Field effect to a row of windmills

In the dry plains of Southwest America, the wind blows fiercely. Rows of windmills are lined up in an effort to harness this energy. For this example, you use the Depth of Field effect to display the windmills.

To apply a Depth of Field effect to a row of windmills, follow these steps:

1. Open the Depth of field windmills.max file from the Chap 24 directory in the downloaded content set.

   This file includes a windmill object duplicated multiple times and positioned in a row.

2. Select Create→Cameras→Target Camera, and drag in the Top viewport from the lower-left corner to the center of the windmills. In the Left viewpoint, select the camera and move it upward, and then select the Camera Target and also move it upward to the upper third of the windmill's height, so the entire row of windmills can be seen. If the windmills don't fill the camera view, adjust the Field of View (FOV) setting.

Tip

You can select both the camera and its target by clicking on the line that connects them.

3. Select the Perspective viewport, click on the Point-of-View viewport label, and select Cameras→Camera01 (or just press the C key) to make this viewport the Camera view.

4. With the Camera selected, open the Modify panel, enable the Multi-Pass Effect option, and then select Depth of Field in the drop-down list.

5. In the Depth of Field Parameters rollout, enable the Use Target Distance option and set the Total Passes to **15**, the Sample Radius to **3.0**, and the Sample Bias to **1.0**.

6. Select the Camera viewport, and select the Rendering→Render command.

   This shows the Depth of Field effect in the viewport.

Figure 24.7 shows the resulting Depth of Field effect in the viewport for the row of windmills.

FIGURE 24.7

Multi-pass camera effects can be viewed in the viewport using the Preview button.

Using the Motion Blur effect

Motion Blur is an effect that shows motion by blurring objects that are moving. If a stationary object is surrounded by several moving objects, the Motion Blur effect blurs the moving objects and the stationary object remains in clear view, regardless of its position in the scene. The faster an object moves, the more blurry it becomes.

This blurring is accomplished in several ways, but with a multi-pass camera, the camera renders subsequent frames of an animation and then blurs the images together.

The Motion Blur Parameters rollout, shown in Figure 24.8, appears when the Motion Blur option is selected in the Multi-Pass Effect section of the Parameters rollout. Many of its parameters work the same as the Depth of Field effect.

FIGURE 24.8

For the Motion Blur effect, you can set the number of frames to include.

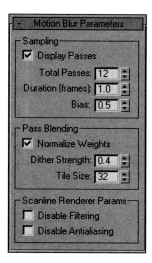

The Display Passes option displays the different frames as they are being rendered, and Total Passes is the number of frames that are included in the averaging. You also can select the Duration, which is the number of frames to include in the effect. The Bias option weights the averaging toward the current frame. Higher Bias values weight the average more toward the latter frames, and lower values lean toward the earlier frames.

The remaining options all work the same as for the Depth of Field effect.

Tutorial: Using a Motion Blur multi-pass camera effect

The Motion Blur effect works only on objects that are moving. Applying this effect to a stationary 2D shape does not produce any noticeable results. For this tutorial, you apply this effect to a speeding spaceship model.

To apply a Motion Blur multi-pass effect to the camera looking at a car mesh, follow these steps:

1. Open the Motion Blur spaceship.max file from the Chap 24 directory in the downloaded content set. This file includes a spaceship and a camera. The spaceship is animated.

2. Click the Select by Name button on the main toolbar to open the Select From Scene dialog box (or press the H key). Double-click the Camera01 object to select it.

3. With the camera object selected, open the Modify panel. In the Multi-Pass Effect section of the Parameters rollout, click the Enable check box and select the Motion Blur effect from the drop-down list.

4. In the Motion Blur Parameters rollout, set the Total Passes to **10**, the Duration to **1.0**, and the Bias to **0.9**.

5. Drag the Time Slider to frame 57. This is the location where the spaceship just passes the camera.

6. With the Camera viewport active, select the Rendering→Render command.

Figure 24.9 shows the results of the Motion Blur effect. This effect has been exaggerated to show its result.

FIGURE 24.9

Using the Motion Blur multi-pass effect for a camera, you can blur objects moving in the scene.

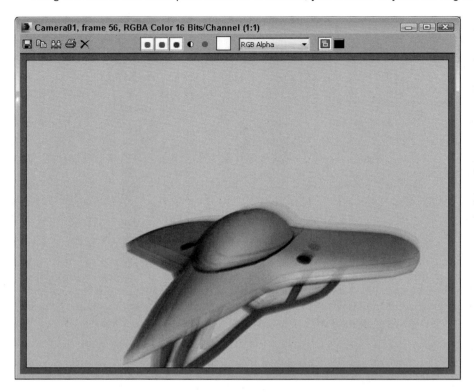

Using the Camera Sequencer

If a single scene includes multiple cameras then you control which camera views the scene for which frames using the Camera Sequencer. This tool is accessible via the Rendering menu and it opens underneath the viewports, as shown in Figure 24.10.

Caution

The Camera Sequencer shares space with the State Sets feature. You can resize the Camera Sequencer by dragging the dividing line, but be aware if your window is too small, that the Camera Sequencer feature will not be visible.

FIGURE 24.10

The Camera Sequencer appears beneath the viewports and lets you control which camera view is used for which frames.

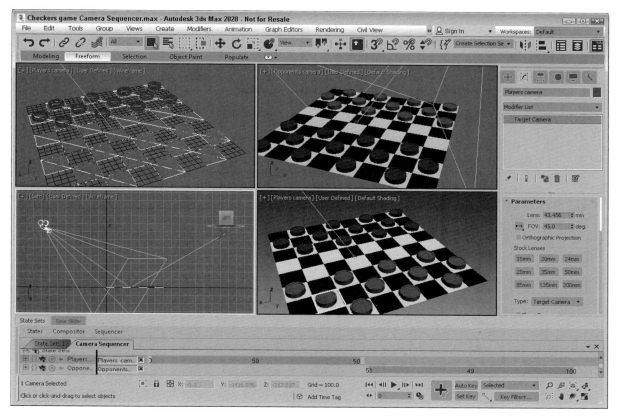

Within the Camera Sequencer interface is a list of each camera in the scene. The frame ranges for each camera are automatically divided between the available cameras. You can alter these ranges by dragging on either end or by dragging the center of the range bar to move the entire range.

If you right click on the range bar, then a pop-up menu offers you options to Rename, Clone, Set Range, Set Color, Select Camera, and Lock Camera Animation. Once the ranges are set, you can render the entire animation and the designated cameras are used.

Loading a Background Image and Camera

Before you can match the background image to the scene, you'll need to load the background image and place a camera in the scene. The Perspective Match feature actually positions and orients the scene camera to match the background. If a good match is made, the virtual camera ends up positioned relative to the background image in a location that is similar to the actual camera used to take the background.

Choosing the right background image will make a huge difference in your ability to complete a successful match. You'll want to look for a background that has a definite vanishing point. It also helps to have an image with lots of straight lines that you can gauge the perspective such as buildings. Figure 24.11 shows two example backgrounds. The one on the left would be tough to match because it doesn't have any straight lines that you can match. The right image, however, has several vertical lines in the building that you can line up to and the street curbs recede into the distance making it easy to make a good match.

FIGURE 24.11

Choosing the right background image can make all the difference in creating a good perspective match.

Although you can load the background image into the Perspective viewport, the better choice is to add a camera to the scene and to use a camera viewport. The camera should be a Free Camera so that it is free to rotate in order to make the perspective match.

Once a Free Camera is added to the scene and the camera viewport is selected, you can use the Background panel in the Viewport Configuration dialog box to add the background to the viewport. You can access the Background panel of the Viewport Configuration dialog box using the Viewport Background→Configure Viewport Background menu in the Shading viewport label for the Camera viewport or you can use the Views→Viewport Background→Configure Viewport Background (Alt+B) menu.

Within the Background panel, select the Use Files options and make sure the Match Rendering Output option is selected, then click the Files button and load the background image. Click Ok to close the Viewport Configuration dialog box and the background image will appear in the viewport. The Use Files option works great and gives you some control over the background image, but if you want the background image rendered then you'll need to apply it as an Environment Map.

To set the background image as an Environment Map, open the Environment and Effects dialog box using the Rendering→Environment menu (or press the 8 key). In the Common Parameters rollout, enable the Use Map option and click the Environment Map button. The Material/Map Browser opens where you can select any of the available maps or choose the Bitmap option and locate and load the background image. This loads the image okay, but the scaling of the background image will likely be off.

To correct the scaling, open the Slate Material Editor and drag the Environment Map button to the Material Editor. This displays the Environment Map as a node. Double click on that node and in the Coordinates rollout, select the Environ option and change the Mapping option to Screen. This scales the background image and fits it in the screen. It also syncs up the image to the screen so that if the screen changes, the background image changes with it.

Perspective Matching the Background Image

Once the background image is loaded, you can select the Perspective Match feature from the Tools→Perspective Match menu. The controls for the Perspective Match utility appear in the Utilities panel of the Command Panel. If you enable the Show Vanishing Lines button in the Perspective Match Controls rollout, then two red, green and blue lines matching the scene axes appear.

By selecting the lines and dragging them, you can move them about the scene and dragging either endpoint changes its orientation. Move and orient each of the gizmo lines until they match the background image. The red lines correspond to the X-axis, the green the Y-axis and the blue, the Z-axis. Typically, you'll want the blue lines to be vertical in the scene. Figure 24.12 shows a background image with the gizmo lines placed to align with the background's features.

FIGURE 24.12

Once the gizmo lines are correctly positioned and oriented, the camera will be in position to match the background.

After the gizmo lines are correctly positioned, the camera is oriented correctly, but its position relative to the scene might be off. You can use the Camera Adjustment controls in the Perspective Match Controls rollout to adjust the camera's position. The grid plane is the best way to judge when the scene matches. Another way to control the camera's position is by specifying an object in the scene to act as the camera's pivot point. To select such an object, click the Pick Anchor Point button and select an object in the scene.

Tip

If the default Home Grid isn't visible, then you can turn it on and off using the Tools→Grids and Snaps→Show Home Grid or press the G key in the active viewport.

After the camera is adjusted, you can add objects to the scene and their perspective will match the background image, as shown in Figure 24.13.

FIGURE 24.13

Objects added to a perspective matched background will have the same perspective.

The final thing to check before rendering the scene is to make sure that the aspect ratio in the Render Setup dialog box matches the aspect ratio for the background image. If you set the Width and Height values equal to the pixel values for the background image, then the aspect ratio will be the same.

Tutorial: Perspective Matching a Background

The real power of the Perspective Match feature is for those times when you'll want to add CG elements to a background photo. For this example, we'll place the xylophone on the porch of a house at sunset.

To perspective match the background image to this scene, follow these steps:

1. Open the Xylophone on porch.max file from the Chap 24 directory in the downloaded content set.

2. Select the Create→Cameras→Free Camera menu and click on the Top viewport to add a free camera to the scene. Select the Perspective viewport and switch it to the camera view by pressing C with the camera selected.

3. Open the Environment and Effects dialog box with the Rendering→Environment menu or by pressing the 8 key. Enable the Use Map option and click on the Environment Map button. Double click on the Bitmap option in the Material/Map Browser and locate and load the Sunset at Cabin.jpg file.

4. Select the Rendering→Material Editor→Slate Material Editor menu (or press the M key) to load the Material Editor. Drag the Environment Map from the Environment and Effects dialog box to the Slate Material Editor. Select the Instance option in the Instance (Copy) dialog box that appears. Double click on the new Map node that appears in the Material Editor. In the Coordinates rollout, select the Environ option and set the Mapping option to Screen. Then, close the Material Editor and the Environment and Effects dialog box.

5. Maximize the Camera viewport and choose the Tools→Perspective Match menu. Enable the Show Vanishing Lines option in the Perspective Match Controls rollout, then position the red, green and blue lines to match the X, Y and Z-axis in the viewport with the blue lines running vertically.

6. Select the Tools→Grids and Snaps→Show Home Grid menu if the home grid isn't visible in the scene. Use the Camera Adjustment settings to match the home grid to the porch plane.

7. Select the Rendering→Render Setup menu and set the Width and Height to 800 x 600, which matches the aspect ratio of the original background image.

Figure 24.14 shows the finished image after being matched.

FIGURE 24.14

Using the Perspective Match feature enabled this xylophone to be placed naturally on the porch of this cabin.

Summary

Cameras can offer a unique look at your scene. You can position and move them anywhere. In this chapter, you discovered how cameras work and how to control and aim them at objects. With multi-pass camera effects, you can add Depth of Field and Motion Blur effects. You also were introduced to the Perspective Match feature, which enables scene objects to be matched with the background image. The feature is easy to use by simply moving the guides into place.

In this chapter, you've accomplished the following:

* Learned the basics of cameras

* Created a camera object and view

* Discovered how to control a camera

* Aimed a camera at objects

* Changed camera parameters

* Learned to correct camera perspective with the Camera Correction modifier

* Used a multi-pass camera to create a Depth of Field effect

* Used a multi-pass camera to create a Motion Blur effect

* Loading the background image as an Environment Map

* Using the Perspective Match feature

Now that cameras and perspective matching have been covered, we can move onto lights and the basics of lighting a scene, so Let there be Light.

Chapter 25

Using Lights and Basic Lighting Techniques

IN THIS CHAPTER

Learning lighting basics

Understanding the various light types

Creating and positioning light objects

Viewing a scene from a light

Altering light parameters

Using projector maps

Positioning the Sun and Setting the Lighting Environment

Positioning the compass

Choosing a location

Lights play an important part in the visual process. Have you ever looked at a blank page and been told it was a picture of a polar bear in a blizzard or looked at a completely black image and been told it was a rendering of a black spider crawling down a chimney covered in soot? The point of these two examples is that with too much or too little light, you really can't see anything.

Light in the 3D world figures into every rendering calculation, and 3D artists often struggle with the same problem of too much or too little light.

With the inclusion of Photometric lights, the lighting solution for outdoor scenes got much closer to real-world lights. The Sun Positioner and Physical Sky systems are real-world equivalents for external lighting providing realistic lighting based on the sun. Using these systems, you can simulate the lighting conditions based on time of year, time of day, and geographical location. You also can animate the hours of the day changing to simulate a lighting study for a location.

This chapter covers creating and controlling lights in the Autodesk® 3ds Max® 2020 software.

Understanding the Basics of Lighting

Lighting plays a critical part of any 3ds Max scene. Understanding the basics of lighting can make a big difference in the overall feeling and mood of your rendered scenes. Most 3ds Max scenes typically use one of two types of lighting: natural light or artificial light. *Natural light* is used for outside scenes and uses the sun and moon for its light source. *Artificial light* is usually reserved for indoor scenes where light bulbs provide the light. However, when working with lights, you'll sometimes use natural light indoors, such as sunlight streaming through a window, or artificial light outdoors, such as a streetlight. So, it is important to know how to work with both types.

Natural and artificial light

Natural light is best created using lights that have parallel light rays coming from a single direction: You can create this type of light using a Direct light. The intensity of natural light is also dependent on the time, date, and location of the sun: you can control this intensity precisely using the software's Sunlight or Daylight systems.

The weather also can make a difference in the light color. In clear weather, the color of sunlight is pale yellow; in clouds, sunlight has a blue tint; and in dark, stormy weather, sunlight is dark gray. The colors of light at sunrise and sunset are more orange and red. Moonlight is typically white.

Artificial light is typically produced with multiple lights of lower intensity. The Omni light is usually a good choice for indoor lighting because it casts light rays in all directions from a single source. Standard white fluorescent lights usually have a light green or light blue tint.

A standard lighting method

When lighting a scene, not relying on a single light is best. A good lighting method includes one key light and several secondary lights.

A spotlight is good to use for the main key light. It should be positioned in front of and slightly above the subject, and it should usually be set to cast shadows, because it will be the main shadow-casting light in the scene.

The secondary lights fill in the lighting gaps and holes. You can position these at floor level on either side of the subject, with the intensity set at considerably less than the key light and set to cast no shadows. You can place one additional light behind the scene to backlight the subjects. This light should be very dim and also cast no shadows. From the user's perspective, all the objects in the scene will be illuminated, but the casual user will identify only the main spotlight as the light source because it casts shadows.

Figure 25.1 shows the position of the lights on an elk model that are included in the standard lighting model using a key light, two secondary lights, and a backlight. This model works for most standard scenes, but if you want to highlight a specific object, additional lights are needed.

FIGURE 25.1

A standard lighting model includes a key light, two secondary lights, and a backlight.

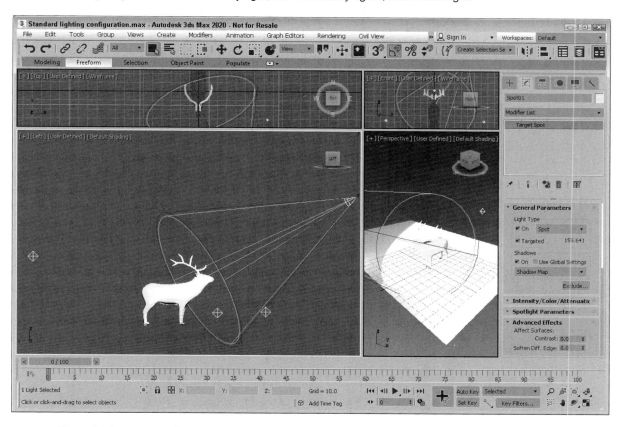

Figure 25.2 shows an elk model that was rendered using different levels of the standard lighting model. The upper-left image uses the default lighting with no lights. The upper-right image uses only the key light. This makes a shadow visible, but the details around the head are hard to define. The lower-left image includes the secondary lights, making the head details more easily visible and adding some highlights to the antlers. The bottom-right image includes the backlight, which highlights the back end of the model and casts a halo around the edges if viewed from the front.

FIGURE 25.2

An elk model rendered using default lighting, a single key light, two secondary lights, and a backlight

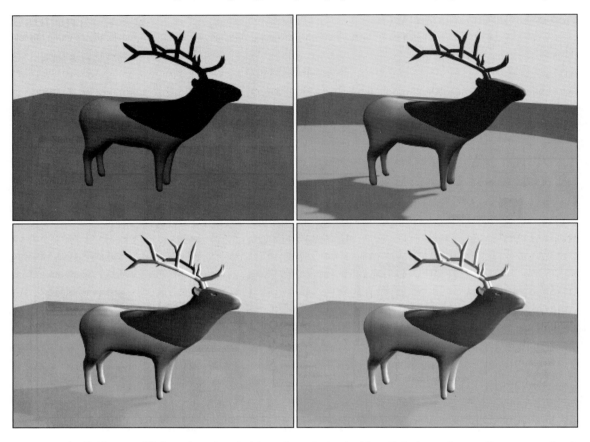

The final type of light to keep in mind is *ambient light*. Ambient light is not from a direct source but is created by light that is deflected off walls and objects. It provides overall lighting to the entire scene and keeps shadows from becoming completely black. Global Lighting (including ambient light) is set in the Environment panel.

Shadows

Shadows are the areas behind an object where the light is obscured. 3ds Max supports several types of shadows, including Area Shadows, Shadow Maps, and Ray Traced Shadows.

Area Shadows create shadows based on an area that casts a light. It doesn't require lots of memory and results in a soft shadow that is created from multiple light rays that blur the shadows. Shadow maps are actual bitmaps that the renderer produces and combines with the finished scene to produce an image. These maps can have different resolutions, but higher resolutions require more memory. Shadow maps typically create fairly realistic, softer shadows, but they don't support transparency.

3ds Max calculates raytraced shadows by following the path of every light ray striking a scene. This process takes a significant amount of processing cycles but can produce very accurate, hard-edged shadows. Raytracing enables you to create shadows for objects that shadow maps can't, such as transparent glass. The Shadows drop-down list also includes an option called Advanced Raytraced Shadows, which uses memory more efficiently than the standard Raytraced Shadows. Another option is the Arnold Shadow Map.

You can learn more about raytracing and Arnold in Chapter 30, "Rendering with ART and Arnold."

Figure 25.3 shows several images rendered with the different shadow types. The image in the upper left includes no shadows. The upper-right image uses Area Shadows. The lower-left image uses a Shadow Map, and the lower-right image uses Advanced Ray Traced Shadows. The last two images took considerably longer to create.

FIGURE 25.3

Images rendered with different shadow types, including no shadow (upper left), Area Shadows (upper right), a Shadow Map (lower left), and Advanced Ray Traced Shadows (lower right)

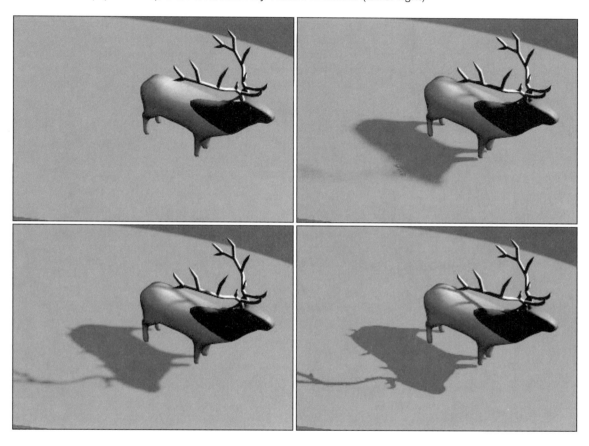

Getting to Know the Light Types

3ds Max includes several different types of lights. The main difference in these types is how the light rays are cast into the scene. Light can come from the default lights that are present when no other user-created lights have been added to the scene. Light also can come from ambient light, which is light that bounces off other objects. 3ds Max includes several standard light objects that can be added where needed to a scene, including Omni, Direct, Spot, and Skylights, each having its own characteristics. 3ds Max also includes a category of Photometric lights, which are based on real-world lights and some lights that work with Arnold. Understanding these sources of light will help you know where to look to control the lighting.

Default lighting

So you get 3ds Max installed, and you eagerly start the application, throw some objects in a scene, and render it . . . and you'll be disappointed in the output, because you forgot to put lights in the scene. Right? Wrong! 3ds Max is smart enough to place default lighting in the scene that does not have any light sources.

The default lighting disappears as soon as a light is created in a scene (even if the light is turned off). When all the lights in a scene are deleted, default lighting magically reappears. So you can always be sure that your objects are rendered using some sort of lighting. Default lighting actually consists of two lights: The first light, the key light, is positioned above and to the left, and the bottom light, the fill light, is positioned below and to the right.

The Per-View Presets panel of the Viewport Setting and Preference dialog box has an option to enable default lighting for any viewport or set the default lighting to use only one light, the key light. You can open this dialog box by choosing Views→Viewport Per-View Settings or by clicking the Presets viewport label and selecting Per-View Presets from the pop-up menu.

If you want to access the default lights in your scene, you can use the Create→Lights→Standard Lights→Add Default Lights to Scene menu command to convert the default lights into actual light objects that you can control and reposition. This command opens a simple dialog box where you can select which lights to add to the scene and set the Distance Scaling value. This feature lets you start with the default lights and modify them as needed.

Caution

The Create→Lights→Standard Lights→Add Default Lights to Scene menu command is enabled only if the Default Lights and 2 Lights options are selected in the Viewport Setting and Preference dialog box.

Ambient light

Ambient light is general lighting that uniformly illuminates the entire scene. It is caused by light that bounces off other objects. Using the Environment tab of the Environment and Effects dialog box, you can set the ambient light color. You also can set the default ambient light color in the Rendering panel of the Preference Settings dialog box. This is the darkest color that can appear in the scene, generally in the shadows.

In addition to these global ambient settings, each material can have an ambient color selected in the Material Editor.

Caution

Don't rely on ambient light to fill in unlit sections of your scene. If you use a heavy dose of ambient light instead of placing secondary lights, your scene objects appear flat, and you won't get the needed contrast to make your objects stand out.

Standard lights

Within the Create panel, the available lights are split into three subcategories: Standard, Photometric, and Arnold. Each subcategory has its own unique set of properties. The Standard light types include Omni, Spot (Target and Free), Direct (Target and Free), and Skylight.

Omni light

The Omni light is like a light bulb: it casts light rays in all directions. The two default lights are Omni lights.

Spotlight

Spotlights are directional: they can be pointed and sized. The two spotlights available in 3ds Max are a Target Spot and a Free Spot. A Target Spot light consists of a light object and a target marker at which the spotlight points. A Free Spot light has no target, which enables it to be rotated in any direction using the Select and Rotate transform button. Spotlights always are displayed in the viewport as a cone with the light positioned at the cone apex.

Both Target Spot and Target Direct lights are very similar in functionality to the Target Camera object, which you learn about in Chapter 24, "Configuring and Aiming Cameras."

Direct light

Direct lights cast parallel light rays in a single direction, like the sun. Just like spotlights, direct lights come in two types: a Target Direct light and a Free Direct light. The position of the Target Direct light always points toward the target, which you can move within the scene using the Select and Move button. A Free Direct light can be rotated to determine where it points. Direct lights are always displayed in the viewport as cylinders when selected.

Skylight

The Skylight is like a controllable ambient light. You can move it about the scene just like the other lights, and you can select to use the Scene Environment settings or select a Sky Color.

Photometric lights

The standard 3ds Max lights rely on parameters like Multiplier, Decay, and Attenuation, but the last time I was in the hardware store looking for a light bulb with a 2.5 Multiplier value, I was disappointed. Lights in the real world have their own set of measurements that define the type of light that is produced. Photometric lights are lights that are based on real-world light measurement values such as Intensity in Lumens and temperatures in degrees Kelvin.

If you select the Lights menu or the Lights category in the Create panel, you'll notice another subcategory called Photometric. Photometric lights are based on photometric values, which are the values of light energy. The Photometric light types include a Target light, Free light and the Sun Positioner.

To make choosing the right light easier, 3ds Max includes a Templates rollout for photometric lights that lets you set the configuration for a number of different common real-world lights, including 40, 60, 75, and 100W light bulbs, a number of Halogen spotlights, recessed lights, fluorescent lights, and even street and stadium lights.

Note

Whenever a photometric light is created, a warning dialog box appears, informing you that it is recommended that the Logarithmic Exposure Control be enabled. It also offers you an option to enable this setting. You can learn more about this feature in Chapter 44, "Using Atmospheric and Render Effects."

The Sun Positioner light works with the Sun & Sky Environment system to add outdoor, natural light to the scene. This system is covered later in the chapter.

Arnold light

The Armold subcategory includes a single Arnold light. If the Arnold renderer is enabled in the Render Setup dialog box, then you can select and use the Arnold Light object. Arnold Light objects have several settings that are unique from the other light types.

The Arnold Light has a Shape property that lets you change the shape of the light object without having to create a new light object. The options include Point, Distant, Spot, Quad, Disc, Cylinder, Skydome, Photometric and Mesh. If the Mesh option is selected, you can Pick a scene object to act as a light.

For more details on the Arnold renderer, see Chapter 30, "Rendering with ART and Arnold."

Arnold Lights also have Color, Intensity, Exposure and Normalize Energy settings. The Normalize Energy option lets you change the size of the light object without changing the amount of emitted light. This provides another way to soften the shadows. There are also settings to enable shadows and Atmospheric Shadows, along with Shadow Color and Density.

Controls for settings the amount of realism in the light verses the time it takes to render include the number of Samples, Volume Samples, and Max. Bounces.

Creating and Positioning Light Objects

3ds Max, in its default setup, can create many different types of light. Each has different properties and features. To create a light, just select Create→Lights and choose the light type or click the Lights category button in the Create panel. Then click the button for the type of light you want to create and drag in a viewport to create it. Most light types are created with a single click, but you create Target lights by clicking at the light's position and dragging to the position of the target.

Transforming lights

Lights can be transformed just like other geometric objects. To transform a light, click one of the transformation buttons and then select and drag the light.

Target lights can have the light and the target transformed independently, or you can select both the light and target by clicking the line that connects them. Target lights can be rotated around the X and Y axes only if the light and target are selected together. A target light can spin about its local Z-axis even if the target isn't selected. Scaling a target light increases its cone or cylinder. Scaling a Target Direct light with only the light selected increases the diameter of the light's beam, but if the light and target are selected, the diameter and distance are scaled.

An easy way to select or deselect the target is to right-click the light and choose Select Target from the pop-up menu. All transformations work on free lights.

Viewing lights and shadows in the viewport

Lighting effects and shadows can be displayed in the viewports if you are using the Nitrous or the Direct3D display driver. You can check to see if your video card supports interactive lights and shadows using the Help→3ds Max Resources and Tools→Diagnose Video Hardware menu command. This command runs a script and returns the results to the MAXScript Listener window.

If your graphics card supports viewport shadows, you can enable them using the Lighting and Shadows submenu under the viewport Shading viewport label or in the Per-View Presets panel of the Viewport Setting and Preference dialog box. If you're using the Nitrous display driver, the same menu also includes an option for enabling Ambient Occlusion.

The Viewport Lighting and Shadows quad menu and the Views menu also include options to lock and unlock the selected light, and to display the effects of the selected light. Figure 25.4 shows a snowman with shadows enabled in the viewport.

FIGURE 25.4

By enabling viewport shadows, you can view shadows in real time.

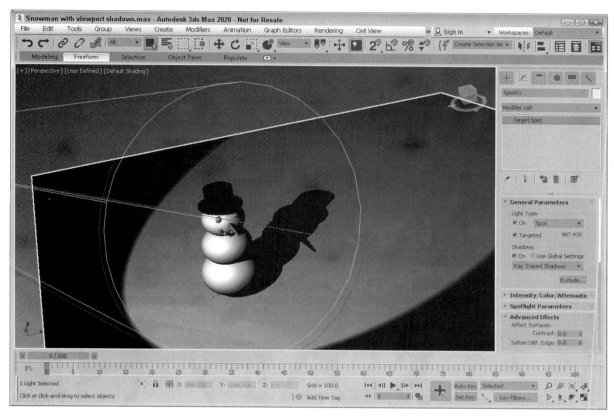

Listing lights

The Tools→Light Lister menu command opens the Light Lister dialog box, shown in Figure 25.5, where you can see at a quick glance all the details for all the lights in the scene. This dialog box also lets you change the light settings. It includes two rollouts: Configuration, which lets you select to see All Lights, the Selected Lights, or the General Settings that apply to all lights; and Lights, which holds details on each individual light.

FIGURE 25.5

The Light Lister dialog box includes a comprehensive list of light settings in one place.

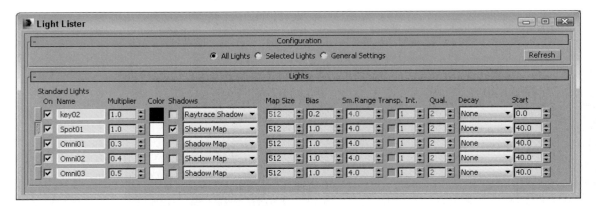

If the General Settings option is selected, a separate rollout opens with all the typical settings, including Multiplier, Color, Shadows, Map Size, and so on. You can apply these changes to the Selected Lights or to All Lights. The Light Lister provides an easy way to change the parameters of many lights at once.

If either the All Lights option or the Selected Lights option is selected in the Configuration rollout, the parameters are listed in the Lights rollout. Using this rollout, you can change the settings for any of the listed lights that affect all lights. The Refresh button updates the Light Lister dialog box if a new light has been added to the scene or if any parameters have been altered in the Modify panel.

Note

If several lights are instanced, only one of the instanced lights appears in the Light Lister dialog box, but each of the instanced lights can be selected from a drop-down list.

Placing highlights

 The Place Highlight tool enables you to control the position and orientation of a light in order to achieve a highlight in a precise location. To use this tool, you must select a light object in the scene and then choose Tools→Align→Place Highlight, or click the Place Highlight flyout button (located under the Align button) on the toolbar. The cursor changes to the Place Highlight icon. Click a point on the object in the scene where you want the highlight to be positioned, and the selected light repositions itself to create a specular highlight at the exact location where you clicked. The light's position is determined by the Angle of Incidence between the highlight point and the light.

Tip

If you click and drag on the object surface, a small blue vector points from the surface of the object. The light is positioned inline with this vector when the mouse button is released. This is helpful when trying to precisely place a light.

Tutorial: Lighting the snowman's face

You can use the Place Highlight tool to position a light for our snowman. To place a highlight, follow these steps:

1. Open the Snowman.max file from the Chap 25 directory in the downloaded content set.

   This file contains a simple snowman created using primitive objects.

2. Select the Create→Lights→Standard Lights→Target Spotlight menu command, and position the spot light below and to the left of the Snowman model in the Top viewport.

3. To place the highlight so it shows the Snowman's face, select the spot light and then choose Tools→Align→Place Highlight. Then click and drag on the Snowman's face in the Perspective viewport where the highlight should be located, just above his right eye.

Figure 25.6 shows the results.

FIGURE 25.6

The snowman, after the lights have been automatically repositioned using the Place Highlight command

Viewing a Scene from a Light

You can configure viewports to display the view from any light, with the exception of an Omni light. To do so, click the viewport Point-of-View label and then select Lights and the light name at the top of the pop-up menu.

Note

The keyboard shortcut for making the active viewport a Light view is the $ (the dollar sign that appears above the 4) key. If more than one light exists, and none is selected, the Select Light dialog box appears and lets you select which light to use. This can be used only on spot and direct lights.

Light viewport controls

When a viewport is changed to show a light view, the Viewport Navigation buttons in the lower-right corner of the screen change into Light Navigation controls. Table 25.1 describes these controls.

Note

Many of these controls are identical for viewports displaying lights or cameras.

TABLE 25.1 Light Navigation Control Buttons

Toolbar Button	Name	Description
	Dolly Light, Dolly Target, Dolly Spotlight + Target	Moves the light, its target, or both the light and its target closer to or farther away from the scene in the direction it is pointing.
	Light Hotspot	Adjusts the angle of the light's hotspot, which is displayed as a blue cone.
	Roll Light	Spins the light about its local Z-axis.
	Zoom Extents All, Zoom Extents All Selected	Zooms in on all objects or the selected objects until they fill the viewport.
	Light Falloff	Changes the angle of the light's falloff cone.
	Truck Light, 2D Pan Zoom Mode	Moves the light perpendicular to the line of sight.
	Orbit Light, Pan Light	Orbit rotates the light around the target. Pan Light rotates the target around the light.
	Maximize Viewport Toggle	Makes the current viewport fill the screen. Clicking this button a second time returns the display to several viewports.

If you hold down the Ctrl key while using the Light Hotspot or Falloff buttons, 3ds Max changes the size at a much faster rate. Holding down the Alt key causes the size to change at a much slower rate. The Hotspot cone cannot grow any larger than the Falloff cone, but if you hold down the Shift key, trying to make the size of the hotspot larger than the falloff causes both to increase, and vice versa.

You can constrain any light movements to a single axis by holding down the Shift key. The Ctrl key causes the movements to increase rapidly.

For Free lights, an invisible target is determined by the distance computed from the other light properties. You can use the Shift key to constrain rotations to be vertical or horizontal.

Manipulating Hotspot/Beam and Falloff/Field cones

When the Select and Manipulate mode is enabled on the main toolbar, clicking on the ends of the Hotspot/Beam and Falloff/Field cones make them appear green for a selected spotlight. When you click and drag on these lines, the lines turn red, allowing you to drag the lines and make the Hotspot/Beam and/or Falloff/Field angle values greater. These manipulators provide visual feedback as you resize the spotlight cone.

Tutorial: Lighting a lamp

To practice using lights, you'll try to get a lamp model to work as it should.

To add a light to a lamp model, follow these steps:

1. Open the Lamp.max file from the Chap 25 directory in the downloaded content set.

   This file includes a lamp mesh surrounded by some plane objects used to create the infinite walls at render time and floor. It looks like a standard living room lamp that you could buy in any department store.

2. Select the Create→Lights→Stanard Lights→Omni menu command, and click in the Perspective viewport.

3. Use the Select and Move transform button (W) to position the light object inside the lamp's light bulb.

The resulting image is shown in Figure 25.7. Notice that the light intensity is greater at places closer to the light.

FIGURE 25.7

The rendered lighted-lamp image

Altering Light Parameters

Lights affect every object in a scene and can really make or break a rendered image, so it shouldn't be surprising that each light comes with many controls and parameters. Several different rollouts work with lights.

If you're looking for a light switch to turn lights on and off, look no further than the Modify panel. When a light is selected, several different rollouts appear. The options contained in these rollouts enable you to turn the lights on and off, select a light color and intensity, and determine how a light affects object surfaces.

General parameters

The Light Type drop-down list in the General Parameters rollout lets you change the type of light instantly, so you can switch from Omni light to Spotlight or Directional Light with little effort. You also can switch between targeted and untargeted lights. To the right of the Targeted option is the distance in scene units between the light and the target. This feature provides an easy way to look at the results of using a different type of light. When you change the type of light, you lose the settings for the previous light.

The General Parameters rollout also includes some settings for shadows. Shadows can be turned on or off easily. In this rollout, you can defer to the global settings by selecting the Use Global Settings option. This option helps to maintain consistent settings across several lights. It applies the same settings to all lights, so changing the value for one light changes that same value for all lights that have this option selected.

You also can select whether the shadows are created using Area Shadows, a Shadow Map, regular or advanced Ray Traced shadows. A new rollout appears, depending on the selection you make.

The Exclude button opens the Exclude/Include dialog box, where you can select objects to be included in or excluded from illumination and/or shadows. The pane on the left includes a list of all the current objects in the scene. To exclude objects from being lit, select the Exclude option, select the objects to be excluded from the pane on the left, and click the double-arrow icon pointing to the right to move the objects to the pane on the right.

Figure 25.8 shows the Exclude/Include dialog box. This dialog box also recognizes any Selection Sets you've previously defined. You select them from the Selection Sets drop-down list.

FIGURE 25.8

The Exclude/Include dialog box lets you set which objects are excluded from or included in being illuminated.

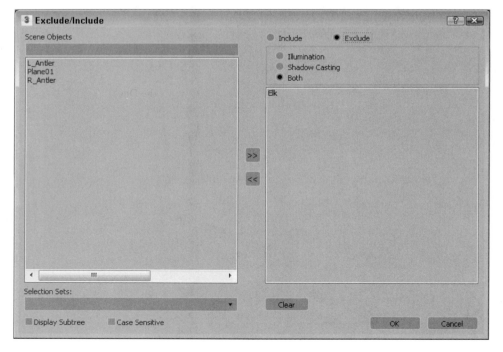

As an example of the Exclude/Include feature, Figure 25.9 shows the elk model with the antlers (left) and its body (right) excluded from shadows and illumination.

FIGURE 25.9

Using the Exclude/Include dialog box, you can exclude objects from casting shadows.

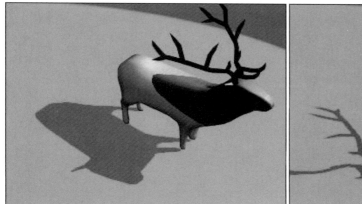 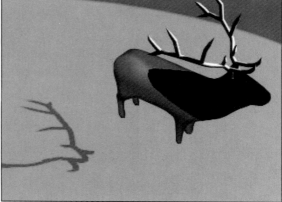

The Intensity/Color/Attenuation rollout

In the Intensity/Color/Attenuation rollout, the Multiplier value controls the light intensity. A light with a Multiplier set to 2 is twice as bright as a light with its Multiplier set to 1. Higher Multiplier values make a light appear white regardless of the light color. The Multiplier value also can be negative. A negative value can be used to pull light from a scene but it should be used with caution.

Tip
Adding and positioning another light typically is better than increasing the multiplier as brighter lights tend to wash out the surface details.

To the right of the Multiplier value is a color swatch. Clicking the color swatch opens a Color Selector where you can choose a new light color.

Attenuation is a property that determines how light fades over distance. An example of this is a candle set in a room. The farther you get from the candle, the less the light shines.

You use three basic parameters to simulate realistic attenuation. Near Attenuation sets the distance at which the light begins to fade, and Far Attenuation sets the distance at which the light falls to 0. Both these properties are ranges that include Start and End values. The third parameter sets the Decay value, which simulates attenuation using a mathematical formula to compute the drop in light intensity over distance.

Selecting the Use option enables the Near and Far Attenuation values; both have Start and End values that set the range for these attenuation types. The Show option makes the attenuation distances and decay values visible in the viewports. The three types of decay from which you can choose are None, Inverse, and Inverse Square. The Inverse type decays linearly with the distance away from the light. The Inverse Square type decays exponentially with distance.

Note
The Inverse Square type approximates real lights the best, but it is often too dim for computer graphic images. You can compensate for this by increasing the Multiplier value.

Spotlight and directional light parameters

The Spotlight Parameters rollout includes values to set the angular distance of both the Hot Spot/Beam and Falloff/Field cones. The Show Cone option makes the Hotspot and Falloff cones visible in the viewport when the light is not selected. The Overshoot option makes the light shine in all directions like an Omni light, but light projection effects and shadows occur only within the Falloff cone. You also can set the light shape to be circular or rectangular. For a rectangular-shaped spotlight, you can control the aspect ratio. You can use the Bitmap Fit button to make the aspect ratio match a particular projection bitmap.

The Directional Light Parameters rollout, which appears for Directional light types, is identical to the Spotlight Parameters rollout and also includes settings for the Hot Spot/Beam and Falloff/Field values.

Advanced Effects

Options in the Affect Surfaces section of the Advanced Effects rollout control how light interacts with an object's surface. The Contrast value alters the contrast between the diffuse and the ambient surface areas. The Soften Diffuse Edge value blurs the edges between the diffuse and ambient areas of a surface. The Diffuse and Specular options let you disable these properties. When the Ambient Only option is turned on, the light affects only the ambient properties of the surface.

Find more detail on the Diffuse, Specular, and Ambient properties in Chapter 17, "Creating and Applying Standard Materials with the Slate Material Editor."

You can use any light as a projector; you find this option in the Advanced Effects rollouts. Selecting the Map option enables you to use the light as a projector. You can select a map to project by clicking the button to the right of the map option. You also can drag a material map directly from the Material/Map Browser onto the Projector Map button. Projector maps can be simple images, animated images, or black-and-white masks to cast shadows.

Tutorial: Creating a stained-glass window

When a light that uses raytraced shadows shines through an object with transparent materials, the Filter color of the material is projected onto objects behind. In this tutorial, you create a stained-glass window and shine a light through it using raytraced shadows.

To create a stained-glass window, follow these steps:

1. Open the Stained glass window.max file from the Chap 25 directory in the downloaded content set.

   This file includes a stained-glass window for a fish market. (Don't ask me why a fish market has a stained-glass window.)

2. Select the Create→Lights→Standard Lights→Target Spotlight menu command, and drag in the Left view from a position to the right and above the window to the window.

   This creates a target spotlight that shines through the stained-glass window onto the floor behind it.

3. In the General Parameters rollout, make sure that the On option is enabled in the Shadows section and select Ray Traced Shadows from the drop-down list.

Figure 25.10 shows the stained-glass window with the colored shadow cast on the scene floor.

FIGURE 25.10

A stained-glass window effect created with raytraced shadows

Shadow parameters

All light types have a Shadow Parameters rollout that you can use to select a shadow color by clicking the color swatch. The default color is black. The Dens setting stands for "Density" and controls how dark the shadow appears. Lower values produce lighter shadows, and higher values produce dark shadows. This value also can be negative.

The Map option, like the Projection Map, can be used to project a map along with the shadow color. The Light Affects Shadow Color option alters the Shadow Color by blending it with the light color if selected.

In the Atmosphere Shadows section, the On button lets you determine whether atmospheric effects, such as fog, can cast shadows. You also can control the Opacity and the degree to which atmospheric colors blend with the Shadow Color.

When you select a light and open the Modify panel, one additional rollout is available: the Atmospheres & Effects rollout. This rollout is a shortcut to the Add Atmosphere or Effect dialog box, where you can specify atmospheric effects such as fog and volume lights.

Note
The only effects that can be used with lights are Volume Light and Lens Effects.

Chapter 44, "Using Atmospheric and Render Effects," covers atmospheric effects.

If the Area Shadows option is selected in the General Parameters rollout, the Area Shadows rollout appears, which includes several settings for controlling this shadow type. In the drop-down list at the top of the rollout, you can select from several Basic Options, including Simple, Rectangle Light, Disc Light, Box Light, and Sphere Light. You can select dimensions depending on which option is selected. You also can set the Integrity, Quality, Spread, Bias, and Jitter amounts.

For the Shadow Map option, the Shadow Map Params rollout includes values for the Bias, Size, and Sample Range. The Sample Range value softens the shadow edges. You also can select to use an Absolute Map Bias and 2 Sided Shadows.

If the Ray Traced Shadows option is selected in the Shadow Parameters rollout, the Ray Traced Shadow Parameters rollout appears below it. This simple rollout includes only two values: Bias and 3ds Max Quadtree Depth. The Bias settings cause the shadow to move toward or away from the object that casts the shadow. The 3ds Max Quadtree Depth determines the accuracy of the shadows by controlling how long the ray paths are followed. There is also an option to enable 2 Sided Shadows, which enables both sides of a face to cast shadows, including backfacing objects.

For the Advanced Ray Traced Shadows options, the rollout includes many more options, including Simple, 1-Pass, or 2-Pass Antialias. This rollout also includes the same quality values found in the Area Shadows rollout.

Note
Depending on the number of objects in your scene, shadows can take a long time to render. Enabling Ray Traced shadows for a complex scene can greatly increase the render time.

Optimizing lights

If you select either the Area Shadows type or the Advanced Ray Traced shadow type, a separate Optimizations rollout appears. This rollout includes settings that help speed up the shadow rendering process. Using this rollout, you can enable Transparent Shadows. You also can specify a color that is used at the Antialiasing Threshold. You also can turn off anti-aliasing for materials that have SuperSampling or Reflection/Refraction enabled. Or you can have the shadow renderer skip coplanar faces with a given threshold.

Photometric light parameters

Several of the light rollouts for photometric lights are the same as those for the standard lights, but several key parameters are unique for photometric lights, such as the ability to choose a light distribution model and a shape type.

Distribution options

The Distribution options are listed in a drop-down list in the General rollout. Both Free and Target photometric lights can be set to one of four distribution types. Each of these types appears as a different icon in the viewports:

* **Uniform Spherical:** This distribution type emanates light equally in all directions from a central point, like the standard Omni light.

* **Uniform Diffuse:** This distribution type spreads light equally in all directions for only one hemisphere, such as when a light is positioned against a wall.

* **Spotlight:** This distribution type spreads the light in a cone shape, like a flashlight or a car's headlight.

* **Photometric Web:** This distribution type can be any arbitrary 3D representation and is defined in a separate file that can be obtained from the light manufacturer and loaded into the light object. Once loaded, the distribution graph is visible in the Distribution (Photometric Web) rollout.

The Uniform Spherical option distributes light equally in all directions. The Uniform Diffuse option has its greatest distribution at right angles to the surface it is emitted from and gradually decreases in intensity at increasing angles from the normal. For both options, the light gradually becomes weaker as the distance from the light increases.

The Spotlight option concentrates the light energy into a cone that emits from the light. This cone of light energy is directional and can be controlled with the Hotspot and Falloff values.

The Photometric Web option is a custom option that lets you open a separate file describing the light's emission pattern. These files have the .ies, .cibse, or .ltli extensions. Light manufacturers have this data for the various real-world lights they sell. You load these files using the Choose Photometric File button found in the Distribution (Photometric Web) rollout. You also can specify the X-, Y-, and Z-axis rotation values.

Color options

The Color section of the Intensity/Color/Attenuation rollout, shown in Figure 25.11, includes two ways to specify a light's color. The first is a drop-down list of options. The options found in the list include standard real-world light types such as Fluorescent (Cool White), Mercury, Quartz Metal Halide, and Halogen.

FIGURE 25.11

The Intensity/Color/Attenuation rollout for photometric lights uses real-world intensity values.

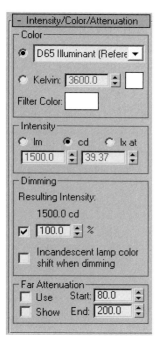

In addition to a list of available light types, you can specify a color based on temperature expressed in degrees Kelvin. Temperature-based colors run from a cool 1,000 degrees, which is a mauve-pink color, through light yellow and white (at 6,000 degrees Kelvin) to a hot light blue at 20,000 degrees Kelvin. Typical indoor lighting is fairly low on the Kelvin scale at around 3,300 degrees K. Direct sunlight is around 5,500 degrees K. Thunderbolts, arc welders, and electric bolts run much hotter, from 10,000 to 20,000 degrees Kelvin.

You also can set a Filter Color using the color swatch found in this section. The Filter Color simulates the color caused by colored cellophane or gels placed in front of the light.

Intensity and Attenuation options

The Intensity options can be specified in Lumens, Candelas, or Lux at a given distance. Light manufacturers have this information available. You also can specify a Multiplier value, which determines how effective the light is. There are also settings for specifying the intensity due to a dimming effect, and the Incandescent lamp color shift when dimming option causes the light from an incandescent light to turn more yellow as it is dimmed. This effect is common as you get farther from a light bulb.

All real-world lights have attenuation, and Far attenuation values also can be set for photometric lights. This helps to speed up rendering times for scenes with lots of lights by limiting the extent of the cast light rays.

Light shapes

In addition to the distribution type, you also can select the light shape, which has an impact on how shadows are cast in the scene using the settings in the Shape/Area Shadows rollout. Selecting a different-shaped light causes the light to be spread over a wider area, so in most cases the Point light results in the brightest intensity with sharper shadows, and lights covering a larger area are less intense and have softer shadows. The available photometric light shapes include the following:

* **Point:** This shape emits light from a single point like a light bulb.
* **Line:** This shape emits light from a straight line like a fluorescent tube.
* **Rectangle:** This shape emits light from an area like a bank of fluorescent lights.
* **Disc:** This shape emits light from a circular area like the light out of the top of a shaded lamp.
* **Sphere:** This shape emits light from a spherical shape like a Chinese lantern.
* **Cylinder:** This shape emits light from a cylindrical shape like some kinds of track lighting.

For each shape you can set the shape's dimensions in the Shape/Area Shadows rollout. The rollout also lets you switch between the different shapes. If you need to see the actual light shape, you can enable the Light Shape Visible in Rendering option in the Shape/Area Shadows rollout.

Positioning the Sun and Setting the Lighting Environment

The Sun Positioner and Physical Sky system, accessed through the Lights category of the Create panel, create a light that simulates the sun for a specific geographic location, date, time, and compass direction.

Note
The Daylight system can be created using the Create→Lights menu, but the Sunlight system cannot be created using a menu.

To create this system, open the Create panel and click the Lights category button. Then click the Sun Positioner button, and drag the mouse in a viewport. A Compass helper object appears on the ground plan grid. Click again to create a Direct light representing the sun and drag to set its height above the ground plane. Figure 25.12 shows the Compass helper created as part of the Sunlight system.

FIGURE 25.12

The Compass helper provides an orientation for positioning the sun in a Sun Positioner system.

Note

The Compass helper object's orientation aligns with the ViewCube's directions.

Using the Compass helper

The Compass helper is useful when working with a Sun Positioning system. It can be used to define the map directions of North, East, South, and West. The Sunlight system uses these directions to orient the system light. This helper is not renderable and is created automatically when you define a Sun Positioning object. The Compass helper object is found in the Create→Helpers menu.

After you create a Sun Positioning system, you can alter the point that the sun is pointing at by transforming the Compass helper. Doing so causes the direct light object to move appropriately. The light's position in the sky is controlled by the Time, Date, and Location parameters, but if you want to move the light independent of these parameters, you can select the Manual option and move the light using the transform tools.

Note

You can change the settings for the light that is the sun by selecting the light from the Select from Name dialog box and opening the Modify panel. The sunlight object uses raytraced shadows by default.

Once created, the light parameters for the Sun Positioning system, most of the settings including light position, light intensity and shadows, are located in the Modify panel, but you also can find several settings including the current Latitude, Longitude, Azimuth and Altitude values, the Time, Date and Location settings when the light object is selected.

Understanding Azimuth and Altitude

Azimuth and Altitude are two values that help define the location of the sun in the sky. Both are measured in degrees. *Azimuth* refers to the compass direction and can range from 0 to 360, with 0 degrees being North, 90 degrees being East, 180 degrees being South, and 270 degrees being West. *Altitude* is the angle in degrees between the sun and the horizon. This value ranges typically between 0 and 90, with 0 degrees being either sunrise or sunset and 90 degrees when the sun is directly overhead.

Specifying date and time

The Time section of the Control Parameters rollout lets you define a time and date. The Time Zone value is the number of offset hours for your current time zone. You also can set the time to be converted for Daylight Saving Time.

Specifying location

Clicking the Get Location button in the Control Parameters rollout opens the Geographic Location dialog box, shown in Figure 25.13, which displays a map or a list of cities. Selecting a location using this dialog box automatically updates the Latitude and Longitude values. In addition to the Get Location button, you can enter Latitude and Longitude values directly in the Control Parameters rollout.

FIGURE 25.13

The Geographic Location dialog box lets you specify where you want to use the Sun Positioning system. You have many different cities to choose from.

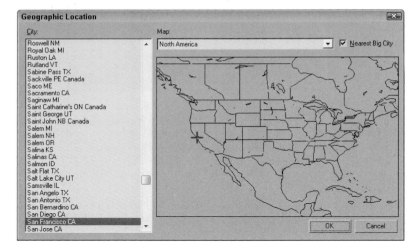

Loading Weather Data

The Sun Positioning system also includes an option to load in a custom weather data file by clicking on the small button next to the Weather Data File option. This opens the Configure Weather Data dialog box, shown in Figure 25.14. Weather Data Files have an .EPW extension and can be obtained from several different sources on the Web for different worldwide locations. These data files include information on the weather conditions including cloud cover for a given time period. They can be configured for a specific date and time or animated to match the animated sunlight movement.

FIGURE 25.14

The Configure Weather Data dialog box lets you load in weather data for a given date and time.

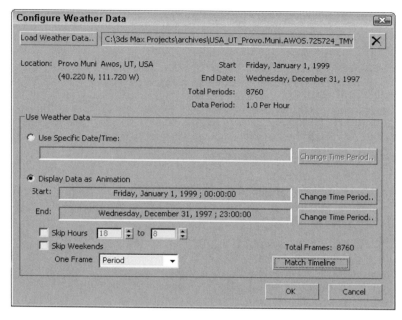

Configuring Sunlight and Skylight

When a Sun Positioning system is added, you still need to set up the sun and sky environment. This is easily done by simply clicking on the Install Sun & Sky Environment button. This automatically adds the Physical Sun & Sky Environment material to the Environment Map panel in the Environment and Effects panel. You can verify this by opening the Environment panel using the Rendering→Environment menu command.

Caution

The light effects added by the Physical Sun & Sky Environment material can only be seen when a physically-based renderer like ART or Arnold is used. Rendering the scene using the Scanline renderer renders it as if no lights are present.

If you open the Physical Sun & Sky Environment in the Material Editor, then you'll have access to the sun and sky settings, as shown in Figure 25.15.

FIGURE 25.15

The settings for the Physical Sun & Sky Environment material are available in the Material Editor.

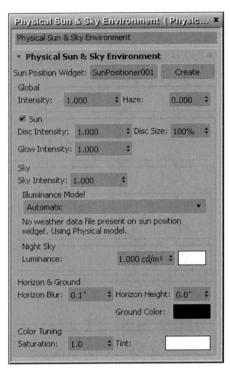

Using these settings you can control the Intensity and Haze around the sun, but the default settings are pretty close. You can also change the Illuminance Model. The options are Automatic, Physical (Preetham) and Measured (Perez All-Weather). Each of these sky model options has controls for setting how overcast the sky is.

Tutorial: Animating a day in 20 seconds

You can animate the Sun Positioner and Physical Sky system to show an entire day from sunrise to sundown in a short number of frames. In this tutorial, you focus on an old tree positioned somewhere in Phoenix, Arizona, on Christmas. The tree certainly won't move, but watch its shadows.

To use the Sun Positioner and Physical Sky system to animate shadows, follow these steps:

1. Open the Sun Positioner.max file from the Chap 25 directory in the downloaded content set.

2. Add a Sun Positioning System by selecting the Lights category in the Create panel and clicking the Sun Positioner button. Then drag in the Top viewport to create the Compass helper, and click again to create the light. In the Sun Position rollout, enter **12/25** and the current year for the Date and an early morning hour for the Time.

3. Click the Get Location button, locate Phoenix, AZ in the City list, and click OK. Rotate the compass helper in the Top view so that north is pointing toward the top of the viewport.

4. If the Sun & Sky Environment hasn't been installed, click on the Install button. Change the Renderer to the ART Renderer in the Render Setup dialog box.

5. Click the Auto Key button (or press the N key), and move the Time slider to frame 100.

6. In the Control Parameters rollout, change the Time value to an evening hour. Then click the Auto Key button (N) again to disable animation mode.

Note

You can tell when the sun comes up and goes down by looking at the Altitude value for each hour. A negative Altitude value indicates that the sun is below the horizon.

Figure 25.16 shows a snapshot of this quick day. The upper-left image shows the animation at frame 20, the upper-right image shows it at frame 40, the lower-left image shows it at frame 60, and the final image shows it at frame 80.

FIGURE 25.16

Several frames of an animation showing a tree scene from sunrise to sunset

Summary

I hope you have found this chapter enlightening. (Sorry about the bad pun, but I need to work them in where I can.) 3ds Max has many different lights, each with plenty of controls. Learning to master these controls can take you a long way toward increasing the realism of the scene. If your scene includes outdoor lighting, then the Sun Positioner and Physical Sky system is the way to go. They provide realistic outdoor lighting, and you can specify the exact conditions required. In this chapter, you've accomplished the following:

* Learned the basics of lighting

* Discovered the software's standard and photometric light types

* Created and positioned light objects

* Learned to change the viewport view to a light

* Used raytraced shadows to create a stained-glass window

* Used the Sun Positioner and Physical Sky systems
* Added a compass helper
* Specified a location

In the next chapter, we look at the features that make indoor lighting realistic using Advanced Lighting, Light Tracing and Radiosity.

Chapter 26

Working with Advanced Lighting, Light Tracing, and Radiosity

IN THIS CHAPTER

Using advanced lighting

Understanding light tracing

Setting local advanced lighting parameters

Understanding radiosity

Setting global advanced lighting parameters

Using the Advanced Lighting materials

If you were to walk into a dark room and reach for the light switch, you would be confused if you found a separate switch that controlled the advanced lighting. But in the Autodesk® 3ds Max® 2020 software, the advanced lighting controls are worth the trouble. They enable you to take your lighting solution to the next level.

The advanced lighting controls in 3ds Max enable you to light scenes using two separate global illumination techniques known as light tracing and radiosity. Both solutions deal with the effect of light bouncing off objects and being reflected to the environment.

Light tracing is typically used for outdoor scenes where the light consists of a single powerful light source at a far distance from the scene. Light tracing includes support for color bleeding between surfaces. Another aspect of light tracing is that the shadows are softer.

Radiosity computes lighting solutions that are much more realistic than using standard lights. As you learn to use radiosity, you quickly discover that it is a complex system that takes lots of tweaking to get just right.

Selecting Advanced Lighting

You control the advanced lighting settings for the scene in the Advanced Lighting panel, which is part of the Render Setup dialog box. You can open this dialog box by selecting Rendering→Render Setup (or by pressing the 9 key). The Advanced Lighting panel includes a rollout with a single drop-down list where you can select the lighting plug-in to use. The options are None, Light Tracer, and Radiosity.

Note

The Light Tracer and Radiosity options are only available if the Scanline renderer is enabled. These features are not available if any of the other renderers are enabled.

Light Tracer and Radiosity are two different techniques for applying advanced lighting to a scene. Although they are fundamentally different, they both simulate a critical piece of the lighting puzzle that adds dramatically to the realism of the lights in the scene—light bouncing. When light strikes a surface in real life, a portion of the light bounces off the surface and illuminates other surfaces. Traditionally, 3ds Max hasn't worried about this, which required that users add more lights to the scene to account for this additional lighting. Both the Light Tracer and the Radiosity solutions include light bouncing in their calculations.

How light tracing works

The Light Tracer is a Global Illumination (GI) system that is similar to raytracing, but it focuses more on calculating how light bounces off surfaces in the scene. The results are fairly realistic without being computationally expensive, and its solutions are rendered much quicker than a radiosity solution.

The Light Tracer is similar in many ways to raytracing. Chapter 30, "Rendering with ART and Arnold," presents more information on raytracing.

The Light Tracer works by dividing the scene into sample points. These sample points are more heavily concentrated along the edges of objects in the scene. An imaginary light ray is then shot at each sample point, and the light intensity at the location of contact is recorded; then it is computed where the light ray would bounce to, and a reduced intensity value is recorded. One of the settings is how many times the light rays will bounce within the scene, and this value increases the amount of time required to compute the solution. When all the rays and light bounces have been computed, the total light intensity value for each sample point is totaled and averaged.

Caution

Transparent objects split each ray in two. One ray bounces, and the second ray is projected through the transparent object. Transparent objects in the scene quickly double the amount of time required to compute a solution.

The end result of a light tracing solution is that objects that are typically hidden in the shadows become much easier to see. Figure 26.1 shows a house model that was rendered using the standard lighting solution with raytraced shadows and then again using the Light Tracer opened side by side in the RAM Player. Notice that many of the details hidden in the shadows of one figure are visible in the other.

FIGURE 26.1

A house scene rendered using standard lighting (left) and light tracing (right)

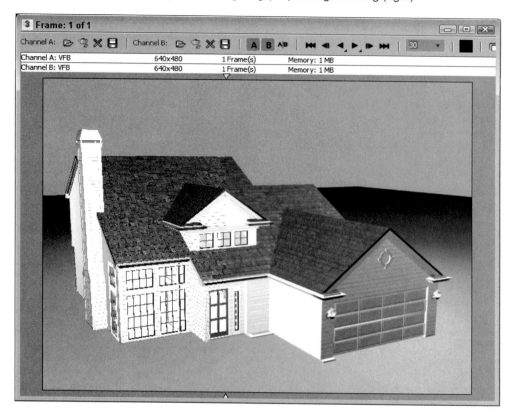

Enabling light tracing

To enable light tracing in a scene, select Rendering→Render Setup to open the Render Setup dialog box, then select the Advanced Lighting panel, as shown in Figure 26.2.

FIGURE 26.2

The Light Tracer Parameters rollout sets values for GI lighting.

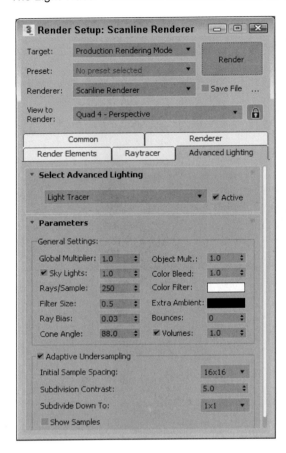

The Global Multiplier value increases the overall effect of the Light Tracer, much like increasing the multiplier of a light. The net result is to brighten the scene. You also can increase the multiplier of skylights with the Sky Lights values. The Object Multiplier sets the amount of light energy that bounces off the objects.

Color bleeding

Another characteristic of global illumination is color bleeding. As a light ray strikes the surface of an object and bounces, it carries the color of the object that is struck with it to the next object. The result of this is that colors from one object bleed onto adjacent objects. You can control this effect using the Color Bleed setting. You can greatly exaggerate the amount of color bleeding by increasing the Object Multiplier along with the Color Bleed value. You also can select colors to use for a color filter and for extra ambient light.

Note

The color bleeding effect doesn't happen unless the Bounces value is set to 2 or greater.

When using color bleeding, you also want to enable the Exposure Control to the scene. Exposure Control is found in the Environment panel (keyboard shortcut, 8), which can be opened with the Rendering→Environment... menu command.

Tip

When changing the Exposure Control settings, you can get a quick preview of the scene by clicking the Render Preview button in the Exposure Control rollout of the Environments and Effects dialog box.

The Exposure Control features are discussed in Chapter 44, "Using Atmospheric and Render Effects."

Figure 26.3 shows an example of color bleeding with several colored cylinders projecting from a gray Box object. The Object Multiplier value was set to 4.0, and the Color Bleed was set a maximum value of 25.0 with a Bounces value of 3. Using the Exposure Control settings, you can isolate the color bleed.

FIGURE 26.3

Color bleeding spreads color about the scene. Exposure Control can highlight it with Automatic (left) and Logarithmic (right).

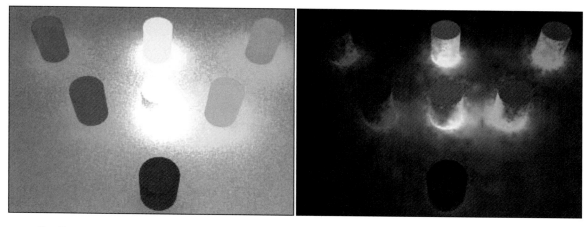

Quality versus speed

The big trade-off of global illumination is between quality and render time. The more rays per sample that you specify, the better the quality and the longer the render time. This is controlled with the Rays/Sample setting. The Rays/Sample setting and the number of Bounces dramatically increase the rendering time. The Ray Bias setting biases rays toward object edges versus flat areas.

Tip

If you want to see a preview of your scene using light tracing, set the Rays/Sample value to around 10 percent of its normal value and render the scene. The resulting image is grainy, but it shows a rough approximation of the scene lighting without having to change the Bounce value.

If you don't include enough rays in the scene, noise patterns appear within the scene. The Filter Size can help control the amount of noise that appears in the scene.

The number of Bounces value specifies the number of times the ray bounces before being dropped from the solution. A setting of 0 is the same as disabling the Light Tracer, and the maximum value of 10 requires a long time to compute. The Cone Angle defines the cone region within which the rays are projected. The Volumes option is a multiplier for the Volume Light and Volume Fog atmosphere effects.

Adaptive undersampling

With the Adaptive Undersampling option enabled, the Light Tracer focuses on the areas of most contrast, which usually occur along the edges of objects. When this option is enabled, you can specify the spacing of the samples and how finely the samples get subdivided. The Initial Sample Spacing options range from 1 x 1 to a very dense 32 x 32. The Subdivision Contrast affects the density for contrast edges between objects and shadows. This value is a minimum amount of contrast that is allowed. If the amount of contrast is greater than this value, the area is further subdivided into more samples. These high-contrast areas use the Subdivide Down To setting. The Show Samples option displays each sample as a red dot on the rendered image.

Using Local Advanced Lighting Settings

You can set advanced lighting settings locally for specific objects using the Object Properties dialog box, as shown in Figure 26.4.

At the top of the Advanced Lighting panel in the Object Properties dialog box is the number of selected objects and lights. The Object Properties dialog box is opened using Edit→Object Properties. This dialog box lets you specify whether this object should be excluded from the advanced lighting calculations. The properties can be set By Object or By Layer. If included, you can select whether the object casts shadows, whether it receives illumination, and how it handles radiosity. The Number Regathering Rays Multiplier option sets the number of rays cast by the selected object. For large, smooth surfaces, reducing artifacts by increasing this value can be helpful. The remaining settings in this panel deal with radiosity.

FIGURE 26.4

Use the Advanced Lighting panel in the Object Properties dialog box to disable advanced lighting.

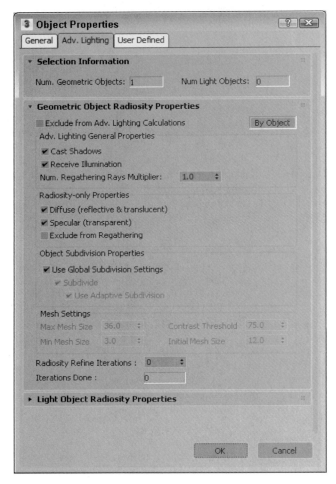

Tutorial: Viewing color bleeding

One of the easiest effects of the Light Tracer to see is color bleeding. Although this is often undesirable, it is a telltale sign of global illumination.

To compare the differences between a regular rendering and the Light Tracer, follow these steps:

1. Open the Hotplate.max file from the Chap 26 directory in the downloaded content set.

This file includes a simple model of a hotplate.

2. Open the Advanced Lighting panel in the Render Setup dialog box by pressing the 9 key. Select and enable the Light Tracer. In the Parameters rollout, set the Object Multiplier to **10**, the Color Bleed to **25**, and the Bounces to **1**.

3. In the Front viewport, select the plug, cord, and floor objects, and then select Edit→Object Properties to open the Object Properties dialog box for these objects. In the Object Properties dialog box, open the Advanced Lighting panel and enable the Exclude from Advanced Lighting Calculations option. Then click OK.

4. In the Render Setup dialog box, click the Render button.

    This renders the scene in the Rendered Frame Window.

Caution

Remember that selecting an advanced lighting option greatly increases the render time.

Figure 26.5 shows the scene rendered shows the rendered hotplate in the RAM Player. The left half is without Advance Lighting and the right half included Advanced Lighting. Notice how the bottom of the hotplate is affected by the floor tint and how the color bleeds onto the floor.

FIGURE 26.5

Color bleeding happens only when global illumination is enabled.

Understanding Radiosity

Imagine a scene that includes an umbrella with a light source directly overhead. If you rendered the scene, the object caught in the umbrella's shadow would be too dark for you to see clearly. To fix this situation, you would need to add some extra lights under the umbrella and set them to not cast shadows. Although this workaround provides the solution you want, it is interesting to note that this isn't the case in real life.

The difference between the workaround and real life has to do with the effect of light energy being reflected (or bounced) off the lit objects. It is this phenomenon that allows me to look down the hall and see whether my children's light is still on past bedtime. Even though I can't see the light directly, I know it is on because of the light that reflects off the other walls.

Radiosity is a lighting algorithm that is based on how heat or energy transfers across surfaces. Every time a bit of light energy, called a photon, strikes a surface, the light energy is reduced, but the light energy is bounced onto the surrounding faces. The greater the number of bounces that are computed, the more realistic the lighting solution, but the longer it takes to compute. So, using radiosity, the objects under the umbrella are visible even if they are in the shadows. Because of the way the light is computed, radiosity solutions are not capable of generating direct specular highlights.

Radiosity is mostly used to light indoor scenes because that is where the effect of light bouncing is most evident. Radiosity, along with light tracing, is another method for computing global illumination. Figure 26.6 shows the mech robot exhibit in a museum with and without radiosity. Notice how dark the shadows are in the normal lighting image.

FIGURE 26.6

This scene is lighted using normal lighting (left) and radiosity lighting (right).

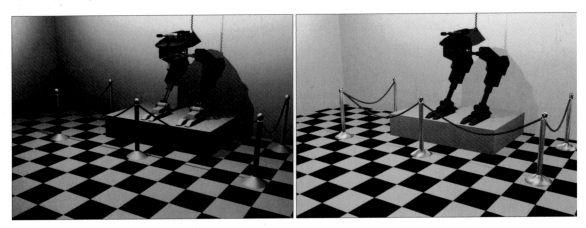

Lighting for radiosity

The Advanced Lighting panel in 3ds Max includes Light Tracer and Radiosity options. You can choose either option from the Advanced Lighting panel in the Render Setup dialog box. Pressing the 9 key opens the Render Setup dialog box with the Advanced Lighting panel open.

Radiosity lighting is not displayed until you have 3ds Max compute it by clicking the Start button in the Radiosity Processing Parameters rollout. After a radiosity solution is calculated, the results are saved as light maps. These maps are easy to apply to a scene and can be viewed within the viewports. However, when the geometry or lights of the scene change, you need to recalculate the lighting solution.

The Radiosity Processing Parameters rollout, shown in Figure 26.7, lets you set the quality of the radiosity solution. You also can specify the number of iterations to use for the scene and for the selected objects. These are different steps in the radiosity computation. The Initial Quality defines the accuracy of the rays that are bounced around the scene. This stage sets the brightness for the scene. The Refine Iterations improves the general quality of the lighting solution for each iteration. You can refine iterations only for the selected object. This lets you target the iterations instead of computing them for the entire scene.

FIGURE 26.7

The Radiosity Processing Parameters rollout includes buttons for computing a solution.

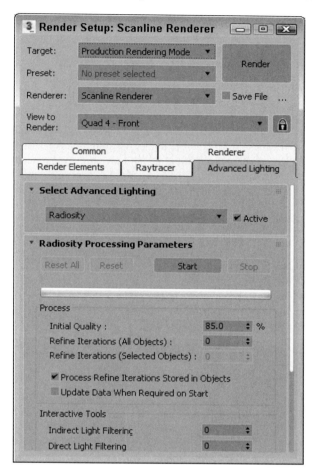

The Interactive Tools section lets you specify a Filtering value. A greater Filtering value eliminates noise between adjacent surfaces by averaging the lighting coming from all surrounding surfaces. The Setup button offers access to the Exposure Control rollout in the Environment panel. You also can turn off radiosity in the viewports.

Chapter 27, "Rendering a Scene and Enabling Quicksilver," includes coverage of the Exposure Control rollout.

Subdividing a mesh for radiosity

As you begin to play with radiosity, you'll quickly find that to get accurate results, you need to have good, clean models. If any models have long, thin faces, the results are unpredictable.

The Radiosity Meshing Parameters rollout includes an option to enable meshing and a Meshing Size value. This setting is the same as the Size value parameter for the Subdivide modifier, except that it is applied globally.

Tip

If you're creating an indoor room using the Box object, be aware that Box objects have only one external face surface with normals, so the interior of a Box object will not have the correct lighting. You can easily fix this by applying the Shell modifier to the Box object. This adds an interior set of faces to the Box object.

Using the Subdivide modifier

The Modifiers menu includes a submenu for Radiosity modifiers. This submenu includes only the Subdivide modifier and a World-Space version of the Subdivide modifier. This modifier accomplishes a simple task—creating a mesh that has regular, equally shaped triangular faces that work well when computing a radiosity solution.

Tip

Although this modifier was created to help with radiosity solutions, it also helps with other commands that require regular mesh faces, such as the Boolean and Terrain compound objects.

The Parameters rollout includes a Size value that determines the density of the mesh. The lower the value, the denser the mesh and the better the resulting radiosity solution, but the longer the solution takes. This same Subdivision Size setting also can be found (and set globally) in the Radiosity Meshing Parameters rollout of the Advanced Lighting panel. It is also found in the Advanced Lighting panel of the Object Properties dialog box. Figure 26.8 shows a simple cube with the Subdivide modifier applied and the Size value set to (from left to right) 50, 30, 25, 20, and 12.

Tip

If you drag the Size value, you'll probably want to set the Update option to Manual or you'll find yourself waiting while 3ds Max computes some seriously dense mesh, or you can just disable the Display Subdivision option.

FIGURE 26.8

The Subdivide modifier changes all mesh faces into regularly shaped triangular faces.

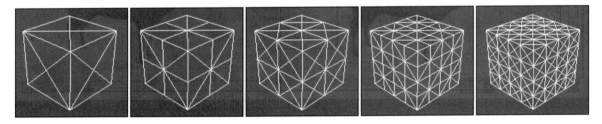

Tutorial: Preparing a mesh for radiosity

When it comes to meshes that have long, thin, and irregular faces, you don't have to look any further than Boolean compound objects. These objects typically are divided along strange angles, producing ugly meshes. The good news is that these meshes are easy to subdivide.

To subdivide an irregular mesh in preparation for a radiosity solution, follow these steps:

1. Open the Boolean object.max file from the Chap 26 directory in the downloaded content set.

    This file includes two copies of a Box object with an arch shape Boolean subtracted from it.

2. Select the top object, and choose the Modifiers→Radiosity→Subdivide menu command.

    This applies the Open Subdivide modifier to the object.

3. In the Parameters rollout, select the Manual update option, set the Size value to **5.0**, and click Update Now.

    If the Display Subdivision option is enabled, the changes are visible in the viewport.

4. Open the Advanced Lighting panel in the Render Setup dialog box by pressing the 9 key. Select Radiosity from the drop-down menu.

Figure 26.9 shows the two objects with and without the Subdivide modifier applied. The top object is ready for a radiosity solution.

FIGURE 26.9

Subdividing an irregular mesh prepares it for radiosity lighting.

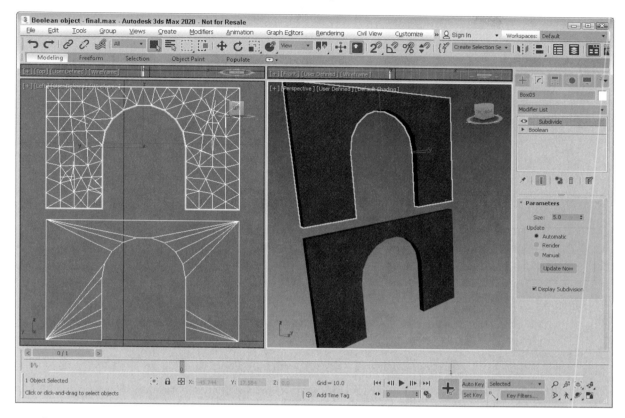

Painting with light

The Light Painting rollout (found in the Advanced Lighting panel of the Render Setup dialog box), shown in Figure 26.10, includes buttons for Adding Illumination, Subtracting Illumination, and Picking an Illumination value from the scene. Using these tools, you can paint lighting on the objects in the scene once a lighting solution is computed. The Clear button removes all the changes you've made using the Light Painting tool.

FIGURE 26.10

Because lighting is saved as a light map, you can add or subtract light from the scene using a brush tool.

Rendering parameters and statistics

Settings in the Rendering Parameters rollout, shown in Figure 26.11, are used during the rendering process. The Re-Use and Render Direct Illumination options give you the chance to reuse the existing radiosity solution when rendering or to recalculate it as part of the rendering process. This can save some time during rendering.

FIGURE 26.11

The Rendering Parameters and Statistics rollouts offer rendering options and statistics for radiosity solutions.

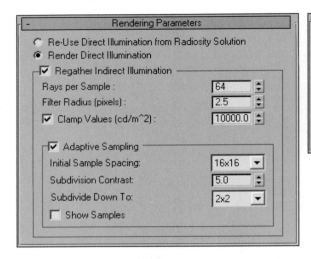
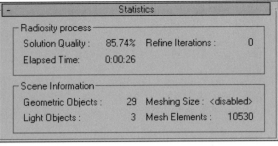

The Regather Indirect Illumination option enables a Light-Tracer-like step along with the radiosity solution and produces an image that has the best of both solutions. The regathering options are the same as those defined for the light tracer.

The Statistics rollout includes information about the radiosity process. Using this information, you can judge whether the settings are too high or too low.

Tutorial: Lighting a house interior with radiosity

Radiosity works best in indoor scenes or scenes that are mostly interior. The only light source for this scene is a Daylight system and a single Omni light in the front hallway.

To light a house interior with radiosity, follow these steps:

1. Open the House interior.max file from the Chap 26 directory in the downloaded content set.
2. Open the Advanced Lighting panel in the Render Setup dialog box by pressing the 9 key. Select Radiosity from the drop-down list in the Select Advanced Lighting rollout.
3. In the Radiosity Processing Parameters rollout, set the Refine Iterations (All Objects) value to **2** and click the Start button to have 3ds Max compute the radiosity solution. It will take some time to process the radiosity solution.
4. After the radiosity solution is done, enable the Render Direct Illumination, the Regather Indirect Illumination, and the Adaptive Sampling options in the Rendering Parameters rollout. Then click the Render button to begin the rendering process.

Figure 26.12 shows the finished rendered house interior. Notice that all surfaces are well lit even though the scene has a limited number of lights.

By way of comparison, this same house interior is rendered with Arnold in Chapter 30, "Rendering with ART and Arnold."

FIGURE 26.12

The radiosity solution for this scene adds to the lighting levels for the entire room.

Working with Advanced Lighting Materials

Applying an advanced lighting solution can have a direct impact on the materials in the scene. The Material Editor includes a material that is useful for working with advanced lighting: Advanced Lighting Override. You can find this material in the Scanline materials subcategory of the Material/Map Browser.

Advanced Lighting Override

The Advanced Lighting Override material type includes material parameters that override the global Advanced Lighting solution. These parameters let you set the amount of Reflectance, Color Bleed, Transmittance, Luminance, and Bump Map Scale that the material uses. This offers a way to make a specific material have its own defined lighting parameters.

Note

The Advanced Lighting Override material does not need to be applied to all objects that are to receive advanced lighting. It is used only to override the existing scene settings for certain materials.

As an example of how this material can be used, consider the hot plate example. Rather than excluding objects from the Advanced Lighting calculations, you can instead use the Light Tracer panel to set the global settings for the scene and apply the Advanced Lighting Override material to the hot plate coils with a higher Color Bleed value. This causes the coils to bleed, but not the rest of the scene.

Another common use for this material is to use the Luminance Scale value to cause self-illuminating materials to add light energy to the global illumination calculations. In other words, increasing this value makes self-illuminating materials act as lights to the scene.

The Advanced Lighting Override material parameters, shown in Figure 26.13, let you set the amount of Reflectance, Color Bleed, Transmittance, Luminance, and Indirect Light Bump Scale that the material uses.

FIGURE 26.13

The Advanced Lighting Override Material rollout defines how light interacts with the material.

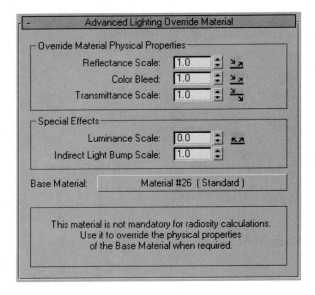

Summary

Advanced lighting offers a new way to shine lights on your scenes. With these features, many new lighting options are available. Advanced lighting enables two global illumination methods: light tracing and radiosity. In this chapter, you accomplished the following:

* Enabled advanced lighting
* Discovered light tracing
* Set local advanced lighting settings
* Discovered radiosity
* Used the Subdivide modifier
* Set local and global advanced lighting settings
* Used advanced lighting materials

In the next chapter, you learn to render a scene so that you can finally hang some new art on Mom's fridge.

Part VI

Rendering a Scene

IN THIS PART

Chapter 27

Rendering a Scene and Enabling Quicksilver

IN THIS CHAPTER

Setting render parameters and preferences

Using the Rendered Frame Window

Working with the ActiveShade window and the RAM Player

Rendering non-photorealistic scenes

After hours of long, hard work, the next step—rendering—is where the "rubber hits the road" and you get to see what you've worked on so hard. After modeling, applying materials, and positioning lights and cameras, you're finally ready to render the final output. Rendering deals with outputting the objects that make up a scene at various levels of detail.

The Autodesk® 3ds Max® 2020 software includes a Scanline Renderer that is optimized to speed up this process, and several settings exist that you can use to make this process even faster. Understanding the Render Setup dialog box and its functions can save you many headaches and computer cycles. However, other rendering options are available, including Quicksilver and Arnold.

The need for all these different rendering engines comes about because of a trade-off between speed and quality. For example, the renderer used to display objects in the viewports is optimized for speed, but the renderer used to output final images leans toward quality. Each renderer includes many settings that you can use to speed the rendering process or improve the quality of the results.

Working with Render Parameters

Commands and settings for rendering an image are contained within the Render Setup dialog box. This dialog box includes several tabbed panels.

After you're comfortable with the scene file and you're ready to render a file, you need to open the Render Setup dialog box, shown in Figure 27.1, by means of the Rendering→Render Setup menu command (F10) or by clicking the Render Setup button on the main toolbar. This dialog box has several panels: Common, Renderer, Render Elements, Raytracer, and Advanced Lighting. The Common panel includes commands that are common for all renderers, but the Renderer panel includes specific settings for the selected renderer.

The Common and Renderer panels for the Scanline Renderer are covered in this chapter. The Raytracer and Renderer panel for the Arnold renderer are covered in Chapter 30, "Rendering with ART and Arnold," the Render Elements panel is covered in Chapter 31, "Compositing with Render Elements and the Video Post Interface," and the Advanced Lighting panel is covered in Chapter 26, "Working with Advanced Lighting, Light Tracing, and Radiosity."

FIGURE 27.1

You use the Render Setup dialog box to render the final output.

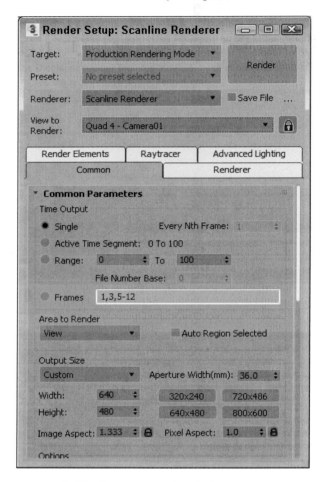

Initiating a render job

At the top of the Render Setup dialog box are several controls that are visible for all panels; these controls let you initiate a render job. The Target render modes, available in the drop-down list to the left of the Render button, are Production, Iterative, and Active Shade. There is also an A360 Cloud Rendering Mode and a Submit to Network Rendering option, which sends the job to be rendered over the network, is available as well. Each of these modes can use a different renderer with different render settings as defined using the Assign Renderer rollout.

Network rendering is covered in Chapter 29, "Batch and Network Rendering."

Note

If any objects in the rendered scene are missing mapping coordinates, a dialog box appears as you try to render the scene with options to Continue or Cancel. A similar dialog box appears for any missing external files or any missing XRefs with options to continue, cancel, or browse from the missing file.

Iterative rendering mode is different from production in that it doesn't save the render to a file, use network rendering, or render multiple frames. Using this mode, you can leave the settings in the Render Setup dialog boxes unchanged while still getting a test render out quickly. This makes it a good mode to use for quickly getting test renders.

The Preset drop-down list to the left of the Render button lets you save and load a saved preset of renderer settings. When saving or loading a preset, the Select Preset Categories dialog box, shown in Figure 27.2, opens (after you select a preset file in a file dialog box). In this dialog box, you can select which panels of settings to include in the preset. The panels listed depend on the selected renderer. All presets are saved with the .rps file extension.

FIGURE 27.2

The Select Preset Categories dialog box lets you choose which settings to include in the preset.

The View drop-down list includes all the available viewports and camera views. When the Render Setup dialog box opens, the currently active viewport appears in the View drop-down list. The one selected is the one that gets rendered when you click the Render button (or when you press Shift+Q). The Render button starts the rendering process. You can click the Render button without changing any settings, and the default parameters are used.

Tip

The little lock icon next to the viewport indicates that the selected viewport is always rendered when the Render button is clicked, regardless of the active viewport.

When you click the Render button, the Rendering dialog box appears. This dialog box, shown in Figure 27.3, displays all the settings for the current render job and tracks its progress. The Rendering dialog box also includes Pause and Cancel buttons for halting the rendering process. If the rendering is stopped, the Rendering dialog box disappears, but the Rendered Frame Window stays open.

Caution

If you close the Rendered Frame Window, the render job still continues. To cancel the rendering, click the Pause or Cancel button, or press the Esc key on your keyboard.

FIGURE 27.3

The Rendering dialog box displays the current render settings and progress of the render job.

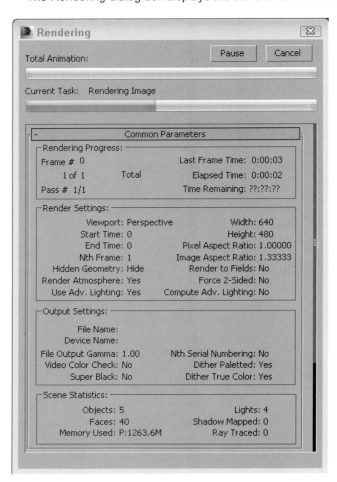

Tip

After you've set up the render settings for an image, you can re-render an image without opening the Render Setup dialog box by clicking the Render Production button on the main toolbar, by selecting the Rendering→Render menu command, by using the Shift+Q keyboard shortcut, or by selecting a render option from the Render Shortcuts toolbar. The F9 shortcut renders the last viewport again.

Common parameters

The Common Parameters rollout in the Render Setup dialog box includes the same controls regardless of the renderer being used.

Specifying range and size

The Time Output section defines which animation frames to include in the output. The Single option renders the current frame specified by the Time Slider. The Active Time Segment option renders the range of frames currently shown in the Time Slider. The Range option lets you set a unique range of frames to render by entering the beginning and ending frame numbers. These values can exceed the range of the Time Slider. The last option is Frames, where you can enter individual frames and ranges using commas and hyphens. For example, entering "1, 6, 8-12" renders frames 1, 6, and 8 through 12. The Every Nth Frame value is active for the Active Time Segment and Range options. It renders every nth frame in the active segment. For example, entering 3 would cause every third frame to be rendered. This option is useful for sped-up animations. The File Number Base is the number to add to or subtract from the current frame number for the reference numbers attached to the end of each image file. For example, a File Number Base value of 10 for a Range value of 1-10 would label the files as image0011, image0012, and so on.

Tip
Don't render long animation sequences using the .avi, .mpeg, or .mov formats. If the rendering has trouble, the entire file will be corrupt. Instead, choose to render the frames as individual images. These individual images can then be reassembled into a video format using the software's RAM Player, the Video Post interface, or an external package like Adobe Premiere.

Within the Viewport Configuration dialog box, you can set up a Blowup Region or Sub Region that only updates that particular region. This is convenient for updating only a small area of a busy scene and not having to wait for the entire scene to update in the viewport. The same can be done when rendering using the Area to Render settings. The options include View, Selected, Region, Crop and Blowup.

The Output Size section defines the resolution of the rendered images or animation. The drop-down list includes a list of standard film and video resolutions, including various 35mm and 70mm options, Anamorphic, Panavision, IMAX, VistaVision, NTSC (National Television Standards Committee), PAL (Phase Alternate Line), and HDTV standards. A Custom option allows you to select your own resolution.

Tip
Setting up the aspect ratio of the final rendering at the start of the project is helpful. Once an aspect ratio is established, you can use the Safe Frames panel in the Viewport Configuration dialog box to display the borders of the render region in the viewport.

Aperture Width is a property of cameras that defines the relationship between the lens and the field of view. Changing the Output Size by using the drop-down list alters the Aperture Value without changing the view by modifying the Lens value in the scene.

For each resolution, you can change the Width and Height values. Each resolution also has several preset buttons for setting these values.

Tip
You can set the resolutions of any of the preset buttons by right-clicking the button that you want to change. The Configure Preset dialog box opens, where you can set the button's Width, Height, and Pixel Aspect values.

The Image Aspect is the ratio of the image width to its height. You also can set the Pixel Aspect ratio to correct rendering on different devices. Both of these values have lock icons to their left that lock the aspect ratio for the set resolution. Locking the aspect ratio automatically changes the Width dimension whenever the Height value is changed, and vice versa. The Aperture Width, Image Aspect, and Pixel Aspect values can be set only when Custom is selected in the Output Size drop-down list.

Render options

Within the Options section are several toggle options that are generally disabled to speed up the rendering. The Options section includes the following options:

*   **Atmospherics:** Renders any atmospheric effects that are set up in the Environment dialog box.
*   **Effects:** Enables any Render Effects that have been set up.
*   **Displacement:** Enables any surface displacement caused by an applied displacement map.
*   **Video Color Check:** Displays any colors that cannot be displayed in the HSV (hue, saturation, and value) color space used by television in black.
*   **Render to Fields:** Enables animations to be rendered as fields. Fields are used by video formats. Video animations include one field with every odd scan line and one field with every even scan line. These fields are composited when displayed.
*   **Render Hidden Geometry:** Renders all objects in the scene, including hidden objects. Using this option, you can hide objects for quick viewport updates and include them in the final rendering.

* **Area Lights/Shadows as Points:** Rendering area lights and shadows can be time-consuming, but point lights render much more quickly. By enabling this option, you can speed the rendering process.

* **Force 2-Sided:** Renders both sides of every face. This option essentially doubles the render time and should be used only if singular faces or the inside of an object are visible.

* **Super Black:** Enables Super Black, which is used for video compositing. Rendered images with black backgrounds have trouble in some video formats. The Super Black option prevents these problems.

The Advanced Lighting section offers options to use Advanced Lighting or Compute Advanced Lighting when Required. Advanced lighting can take a long time to compute, so these two options give you the ability to turn advanced lighting on or off.

Advanced lighting is covered in more detail in Chapter 26, "Working with Advanced Lighting, Light Tracing, and Radiosity."

Bitmap Proxies

The Bitmap Performance and Memory Options section includes a Setup button to enable a feature that can downscale all maps for the current scene. Clicking the Setup button opens the Global Settings and Defaults for Bitmap Proxies dialog box, shown in Figure 27.4.

FIGURE 27.4

The Bitmap Proxies dialog box lets you replace all texture maps with proxy images.

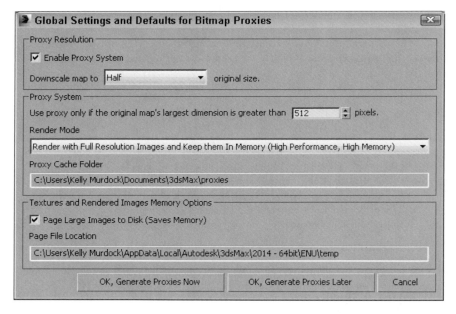

The Downscale map option lets you select to downscale all maps to Half, Third, Quarter, or Eighth, or to their current size. This lets you create your scene with high-quality maps and quickly reduce their sizes as needed without having to open and scale each individual map. The Proxy System lets you select to use a proxy image if the current map is larger than a specified size in pixels.

This dialog box also lets you set the Render Mode to be optimized for performance or memory. The options include Render with Proxies, Render with Full Resolution Images and Keep them [image maps] In Memory, and Render with Full Resolution Images and Free them from Memory. If you enable the Page Large Images to Disk, large textures and rendered images are split into pages, which frees up memory, but it can make the update of the scene slower when a new page has to be recalled. The page file location can be specified.

Choosing a Render Output option

The Render Output section enables you to output the image or animations to a file, a device, or the Rendered Frame Window. To save the output to a file, enable the Save File option, click the Files button, and select a location in the Render Output File dialog box. Supported formats include AVI, BMP, DDS, Postscript (EPS), JPEG, Kodak Cineon (CIN), Open EXR, Radiance Image File (HDRI), QuickTime (MOV), PNG, RLA, RPF, SGI's Format (RGB), Targa (TGA), and TIF. The Use Device option and the Devices button can output to a device such as a video recorder. The Rendered Frame Window option displays the render progress in a separate window, which is discussed later in this chapter. The Skip Existing Images option doesn't replace any images with the same filename, a feature that you can use to continue a rendering job that has been canceled.

Tip
Each of these output formats has its advantages. For example, Targa files are good for compositing because they have an alpha channel. TIF and EPS files are good for files to be printed. JPEG and PNG files are used for web images. DDS images are used in many game engines.

You also have an option to Put Image File List in Output Path, which creates a list of image files in the same location as the rendered file. You also have the choice of choosing the software's IFL standard or the Autodesk ME Image Sequence File (IMSQ). The Create Now button creates an image list instantly. This list is helpful to make sure all the image files have been rendered and are accounted for.

E-mail notifications

The process of rendering an animation (or even a single frame) can be brief or it can take several days, depending on the complexity of the scene. For complex scenes that will take a while to render, you can configure 3ds Max to send you an e-mail message when every so many frames are completed, when your entire rendering is complete, or if it fails. These options are in the Email Notifications rollout, shown in Figure 27.5.

In addition to the options, you can enter whom the e-mail is from, whom it is to, and an SMTP Server.

FIGURE 27.5

The Email Notifications rollout includes options for sending an e-mail message to report on rendering status.

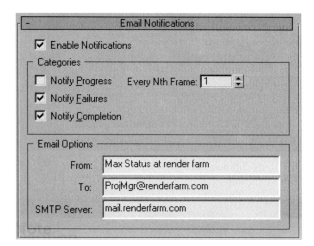

Adding pre-render and post-render scripts

The Scripts rollout includes File buttons for adding pre-render and post-render scripts. The scripts must be .ms scripts and are executed before and after the rendering of each file. These scripts can be used to compile information about the render or to do some post-processing work. Above each script file button is an Execute Now button that can be used to check the script before rendering.

Chapter 52, "Automating with MAXScript," covers the details of scripting in 3ds Max.

Assigning renderers

3ds Max performs rendering operations in several different places: The Render Setup dialog box renders to the Render Frame Window or to files; the material previews in the Material Editor also are rendered; and the ActiveShade windows show another level of rendering.

The plug-in nature of 3ds Max enables you to select the renderer to use to output images. To change the default renderer, look in the Assign Renderers rollout in the Common panel of the Render Setup dialog box (F10). The assigned renderer is also selectable from the top of the Render Setup dialog box. You can select different renderers for the Production, Material Editor, and ActiveShade modes. For each, you can select from the Scanline Renderer, ART Renderer, Arnold, Quicksilver Hardware Renderer, and the VUE File Renderer. This list also could contain additional renderers if a new render plug-in has been installed.

Note
The VUE File Renderer is used to create a VUE file, which is an editable text file that holds all the details of the scene.

The lock next to the Material Editor option indicates that the same renderer is used for both Production and Material Editor.

Scanline Renderer

The Scanline Renderer rollout, found in the Renderer panel and shown in Figure 27.6, is the default renderer rollout that appears in the Render Setup dialog box. If a different renderer is loaded, a different rollout for that renderer is displayed in the Renderer panel.

FIGURE 27.6

The Scanline Renderer rollout includes settings unique to this renderer.

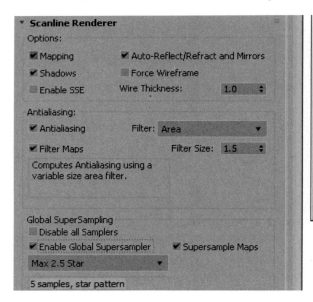
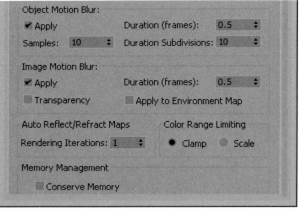

You can use the Options section at the top of the Scanline Renderer rollout to quickly disable various render options for quicker results. These options include Mapping, Shadows, Auto-Reflect/Refract and Mirrors, and Force Wireframe. For the Force Wireframe option, you can define a Wire Thickness value in pixels. The Enabled SSE option uses Streaming SIMD (Single Instruction, Multiple Data) Extensions to speed up the rendering process by processing more data per instruction.

Tip

Intel Pentium III and later processors include the SSE instructions and can benefit from enabling this option.

Anti-alias filters

Another way to speed up rendering is to disable the Anti-aliasing and Filter Maps features. Anti-aliasing smoothes jagged edges that appear where colors change. The Filter Maps option allows you to disable the computationally expensive process of filtering material and environment maps. The Filter drop-down list lets you select image filters that are applied at the pixel level during rendering. Below the drop-down list is a description of the current filter. The Filter Size value applies only to the Soften filter. Available filters include the following:

* **Area:** Does an anti-aliasing sweep using the designated area specified by the Filter Size value.
* **Blackman:** Sharpens the image within a 25-pixel area; provides no edge enhancement.
* **Blend:** Somewhere between a sharp and a coarse Soften filter; includes Filter Size and Blend values.
* **Catmull-Rom:** Sharpens with a 25-pixel filter and includes edge enhancement.
* **Cook Variable:** Can produce sharp results for small Filter Size values and blurred images for larger values.
* **Cubic:** Based on cubic-spline curves; produces a blurring effect.
* **Mitchell-Netravali:** Includes Blur and Ringing parameters.
* **Plate Match/MAX R2:** Matches mapped objects against background plates as used in 3ds Max R2.
* **Quadratic:** Based on a quadratic spline; produces blurring within a 9-pixel area.
* **Sharp Quadratic:** Produces sharp effects from a 9-pixel area.
* **Soften:** Causes mild blurring and includes a Filter Size value.
* **Video:** Blurs the image using a 25-pixel filter optimized for NTSC and PAL video.

SuperSampling

Global SuperSampling is an additional anti-aliasing process that you can apply to materials. This process can improve image quality, but it can take a long time to render; you can disable it using the Disable all Samplers option. SuperSampling can be enabled in the Material Editor for specific materials, but the SuperSampling rollout in the Material Editor also includes an option to Use Global Settings. The Global Settings are defined here in the Scanline Renderer rollout.

3ds Max includes anti-aliasing filters as part of the rendering process. You have several SuperSampling methods from which to choose.

SuperSampling is disabled if the Antialiasing option is disabled. Global SuperSampling can be enabled using the Enable Global SuperSampler option.

In a SuperSampling pass, the colors at different points around the center of a pixel are sampled. These samples are then used to compute the final color of each pixel. 3ds Max has four available SuperSampling methods: Max 2.5 Star, Hammersley, Adaptive Halton, and Adaptive Uniform.

You can find more information on each of these sampling methods in Chapter 17, "Creating and Applying Standard Materials with the Slate Material Editor."

Motion Blur

The Scanline Renderer rollout also offers two different types of motion blur: Object Motion Blur and Image Motion Blur. You can enable either of these using the Apply options.

Object Motion Blur is set in the Object Properties dialog box for each object. The renderer completes this blur by rendering the object over several frames. The movement of the camera doesn't affect this type of blur. The Duration value determines how long the object is blurred between frames. The Samples value specifies how many Duration units are sampled. The Duration Subdivision value is the number of copies rendered within each Duration segment. All these values can have a maximum setting of 16. The smoothest blurs occur when the Duration and Samples values are equal.

Image Motion Blur is also set in the Object Properties dialog box for each object. This type of blur is affected by the movement of the camera and is applied after the image has been rendered. You achieve this blur by smearing the image in proportion to the movement of the various objects. The Duration value determines the time length of the blur between frames. The Apply to Environment Map option lets you apply the blurring effect to the background as well as the objects. The Transparency option blurs transparent objects without affecting their transparent regions. Using this option adds time to the rendering process.

You can add two additional blur effects to a scene: the Blur Render Effect, found in the Rendering Effects dialog box (covered in Chapter 44, "Using Atmospheric and Render Effects") and the Scene Motion Blur effect, available through the Video Post dialog box (covered in Chapter 31, "Compositing with Render Elements and the Video Post Interface").

Other options

The Auto Reflect/Refract Maps section lets you specify a Rendering Iterations value for reflection maps within the scene. The higher the value, the more objects are included in the reflection computations and the longer the rendering time.

Color Range Limiting offers two methods for correcting over-brightness caused by applying filters. The Clamp method lowers any value above a relative ceiling of 1 to 1 and raises any values below 0 to 0. The Scale method scales all colors between the maximum and minimum values.

The Conserve Memory option optimizes the rendering process to use the least amount of memory possible. If you plan on using 3ds Max (or some other program) while it is rendering, you should enable this option.

Quicksilver Hardware Renderer

The Quicksilver Hardware Renderer takes advantage of the advanced graphics processing capabilities found in modern video cards. The advantage of this rendering option is speed. The Quicksilver Hardware Renderer can render scenes much faster than Arnold and at a better quality than the Scanline Renderer.

Caution
Some 3ds Max features don't work with Quicksilver including Exclude/Include for lights, Visibility in the Object Properties dialog box, Vertex Colors, multiple layers of transparency, and several map types including cellular, flat mirror, particle age, particle mblur, thin wall refraction, and non-regular noise.

The Quicksilver Hardware Renderer is available only if your video card supports Shader Model 3.0. If you select Help→3ds Max Resources and Tools→Diagnose Video Hardware, 3ds Max runs a utility that checks and reports the capabilities of your current graphics card. Look for the GPU Shader Model Support to be SM3.0 or later.

You can enable the Quicksilver Renderer by clicking the button to the right of the listed Renderer at the top of the Render Setup dialog box. Then select Quicksilver from the list that appears. When the Quicksilver Hardware Renderer is enabled, the Renderer panel displays the options available for this renderer, as shown in Figure 27.7. This renderer supports many of the same options found in the ART Renderer, making it easy to use, but it also includes options for enabling details such as the rendering level, lighting, and shadows. If you use any hardware shaders in the scene, you can specify their directory in the Hardware Shaders Cache Folder located at the bottom of the panel.

You can learn more about the Arnold rendering options in Chapter 30, "Rendering with ART and Arnold."

FIGURE 27.7

Many of the options for the Quicksilver Hardware Renderer are similar to the ART renderer.

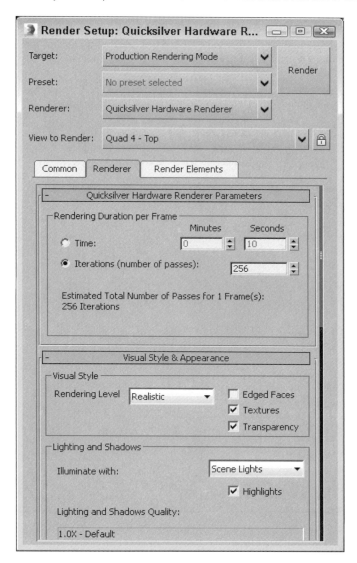

The Quicksilver Renderer is easy to use. You simply tell it how much time to take or how many iterations to make. Quicksilver then continually improves the rendered image until the time or iterations are completed. If the image quality isn't good enough, increase the time and/or iterations and render again.

Rendering stylized scenes

For many years, the goal of 3D graphics has been to make scenes as realistic as possible, but other types of art emphasize style over realism. These stylistic approaches give us the cubes of Picasso, the points of Seurat, and the surreal landscapes of Dali. Although no software title has a button to magically turn your scene into a classic piece of art, the new Nitrous display drivers found in 3ds Max allow you to display your scene as if it were drawn using acrylic, ink, or pastels.

The same stylized display options that are available in the viewports also are available as render options using the Quicksilver rendering engine. After Quicksilver is enabled, you can select one of the stylized render options from the Rendering Level drop-down list in the Visual Style & Appearance rollout.

When a stylized non-photorealistic rendering option is selected, such as the Colored Pencil style, clicking the Render button renders the scene in the Rendered Frame Window, as shown in Figure 27.8.

FIGURE 27.8

Non-photorealistic rendering methods can be specified in the Render Setup dialog box using the Quicksilver renderer.

Using the Rendered Frame Window

The Rendered Frame Window is a temporary window that holds any rendered images. Often when developing a scene, you want to test-render an image to view certain materials or transparency not visible in the viewports. The Rendered Frame Window, shown in Figure 27.9, enables you to view these test renderings without saving any data to the network or hard drive.

FIGURE 27.9

The Rendered Frame Window displays rendered images without saving them to a file.

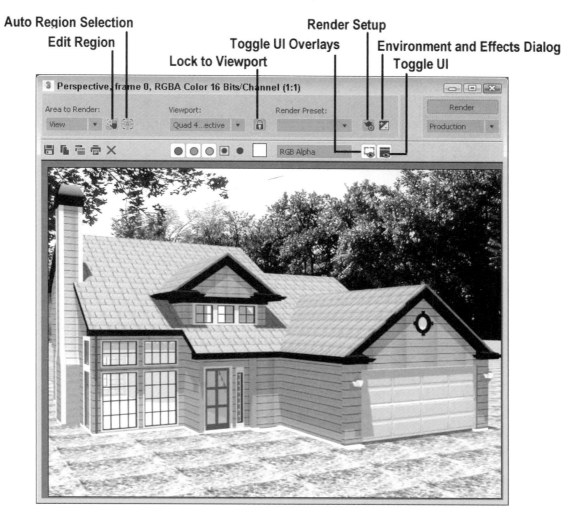

This window opens when you select the Rendered Frame Window option and click the Render button in the Render Setup dialog box. You also can view images from a local hard drive or a network drive in the simple version of the Rendered Frame Window using the File→View Image File menu command.

To zoom in on the rendered image, hold down the Ctrl key and click the window. Right-click while holding down the Ctrl key to zoom out. The Shift key enables you to drag and pan the image. You also can use the mouse wheel (if you have a scrolling mouse) to zoom and pan within the Rendered Frame Window, just like you can in the viewports.

Tip
You can even zoom and pan the image while it is rendering.

Using the Render Types

From the top of the Rendered Frame Window, the Area to Render drop-down list enables you to render subsections of the scene. The default setting is View. After you pick a selection from the list, click the Render button to begin the rendering. The available Render Types are described in Table 27.1.

TABLE 27.1 Render Type Options

Render Type	Description
View	Renders the entire view as shown in the active viewport.
Selected	Renders only the selected objects in the active viewport.
Region	Places a frame of dotted lines with handles in the active viewport. This frame lets you define a region to render. You can resize the frame by dragging the handles. When you have defined the region, click OK in the lower-right corner of the active viewport.
Crop	Similar to Region in that it uses a frame to define a region, but the Crop setting doesn't include the areas outside the defined frame.
Blowup	Takes the defined region and increases its size to fill the render window. The frame for Blowup is constrained to the aspect ratio of the final resolution.

At the top of the Rendered Frame Window are several controls and icon buttons, as described in Table 27.2.

TABLE 27.2 Rendered Frame Window Buttons

Buttons	Name	Description
	Edit Region	Activated when the Region option is selected. It lets you drag the center of the area to move it around or drag the edge and corner handles of the region to resize the render area. This option lets you precisely define the area that is rendered.
	Auto Region Selected	Automatically sets the render region to the current selection in the viewports.
	Lock To Viewport	Causes all renders to be done using the specified viewport, regardless of which viewport is active. Using this option you can lock the renders to the Perspective viewport even if you're using a different viewport.
		Opens the Render Setup dialog box.
	Environment and Effects Dialog	Opens this dialog box, which includes the Exposure Control settings.
	Save Image	Enables you to save the current rendered image.
	Copy Image	Copies the image in the Rendered Frame Window to the Windows clipboard where you can paste it into another application like Photoshop.
	Clone Rendered Frame Window	Creates another frame buffer dialog box. Any new rendering is rendered to this new dialog box, which is useful for comparing two images.
	Print Image	Sends the rendered image to the default printer.
	Clear	Erases the image from the window.

704

	R, G, B, Alpha, Monochrome channels	The next five buttons enable the red, green, blue, alpha, and monochrome channels. The alpha channel holds any transparency information for the image. The alpha channel is a grayscale map, with black showing the transparent areas and white showing the opaque areas. Next to the Display Alpha Channel button is the Monochrome button, which displays the image as a grayscale image.
RGB Alpha	Channel Display drop-down list	Lets you select the channel to display. The color swatch at the right shows the color of the currently selected pixel. You can select new pixels by right-clicking and holding on the image. This temporarily displays a small dialog box with the image dimensions and the RGB value of the pixel directly under the cursor. The color in the color swatch can then be dragged and dropped in other dialog boxes such as the Material Editor.
	Toggle UI Overlays	Causes the frame that marks the region area to be visible when rendered.
	Toggle UI	Hides the top selection of controls in the Rendered Frame Window.

Previewing with ActiveShade

The ActiveShade window gives a quick semi-rendered look at the current scene. You can open an ActiveShade display within a viewport by clicking the viewport Shading label in the upper-left corner of each viewport and choosing ActiveShade from the pop-up menu. Using the ActiveShade button on the main toolbar (which is a flyout under the Render Production button), you can open a floating ActiveShade window, which is similar to the Rendered Frame Window without all the buttons.

The benefit of ActiveShade is that it shows a rendered view of the scene that is automatically updated every time a change is made. This is helpful when you are tweaking lights or making subtle changes.

Note

The ActiveShade window used to be quite valuable, but now that 3ds Max can render lights and shadows in the viewport, the ActiveShade window isn't as helpful.

Only one ActiveShade viewport can be open at a time.

Tip

You can drag materials from the Material Editor and drop them directly on objects in the ActiveShade window.

Summary

This chapter covered the basics of producing output using the Render Setup dialog box. Although rendering a scene can take a long time to complete, 3ds Max includes many settings that can speed up the process and helpful tools such as the Rendered Frame Window and ActiveShade.

In this chapter, you accomplished the following:

* Discovering how to control the various render parameters
* Switching to a different renderer
* Rendering non-photorealistic scenes using the Quicksilver Renderer
* Using the Rendered Frame Window and ActiveShade

The next chapter covers the Save State interface and shows how you can use states to render multiple takes on a scene.

Chapter 28

Managing Render States

IN THIS CHAPTER

Working with State Sets

Accessing compositing tools in the State Sets interface

The Autodesk® 3ds Max® 2020 software has an interface that is kinda like a time machine. Using the State Sets interface, you can save all the settings for the current scene that can be quickly recalled at any time. This lets you stop and save the state of a scene at critical decision points so you can recall them and try a different path. The State Sets interface keeps track of view changes, settings in the interface and also rendering settings.

Another useful way to use State Sets is to create states for several different styles for the current scene. For example, you could have one state that shows the scene from one angle and another with different lighting and a third with a stylized render. Each of these states can then be recalled and presented to a client without having to make all the changes individually. See just like a time machine.

Using State Sets

Another more convenient way to work with render passes is with the State Sets panel, shown in Figure 28.1. This panel is opened using the Rendering→State Sets menu command and includes an interface where you can define a custom state. This state can have some objects hidden, have some lights and shadows disabled, alter the number of frames to display or the rendering method, or specify a specific render element to include.

FIGURE 28.1

The State Sets panel lets you define several individual states, each with unique settings.

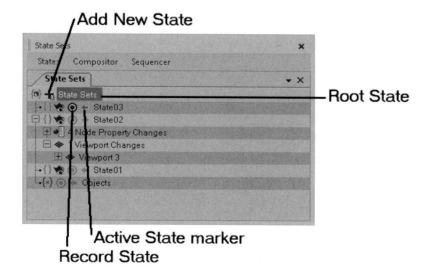

After several states are defined, you can select to render individual selected states or render them all. For example, suppose you have a single file that includes all the animations for a specific game character one after another, such as a walk cycle in frames 1 through 10 and a run cycle from 11 through 15, and so on. Using the State Sets panel, you can define a separate state for each individual range of frames in the file and then render them all at once using the panel's States→Render All States menu command.

At the top of the State Sets interface is the root state that includes all states; to its right is the Add New State button. Clicking this button creates a new state that appears under the root state. Initially, new states are simply named State with a number, but you can right click the state name and choose Rename from the pop-up menu to change it.

To the left of the state name are two toggle buttons used to enable/disable the state and to enable/disable the rendering of that state. To the right of the state name are buttons to begin/end recording and a green arrow marking the current active state. Only one state can be active at a time.

Recording states

Clicking the Begin Recording State button to the right of the state name begins the recording of any changes made to the scene. The record button is highlighted when recording is active. Any changes are listed in a hierarchy under the state name where you can expand the list to see the specific commands. Some changes make the new value, such as color swatches and frame number, visible in the State Sets interface where you can easily change them. Clicking the record button again ends the recording of changes. If you want to make some additional changes after the recording has stopped, simply click the Record button again and the new changes are added to the state.

Using templates and nesting states

If a certain state has most of the changes you need, you can make it into a template with the States→Create Template menu command. These template states are then available for selection using the States→Add State Template menu command. State templates are added to the bottom of the current hierarchy.

States also can be nested within one another using the States→Add Sub-State menu command. The Delete State menu is used to delete the current selected state. Another option is scripted states, which open a dialog box where you can add a script to be performed when the state is applied and another when the state is disabled.

New states are added to the top of the hierarchy, but you can reorder the states by dragging them above or below their current position. You also can drop a state within another state to nest them.

Rendering states

To render a current state, simply select that state, right-click, and choose the Render Selected State option from the pop-up menu. Several states can be selected at once using the Ctrl and Shift keys. You also can render all states starting at the top of the hierarchy and continuing downward with the States→Render All States menu command. This renders only those states with the Enable Render toggle turned on.

With the States→Render Outputs menu command, a new tabbed panel appears with path and naming information for the render output. The Browse button is used to change the location where the renders are saved, and using the Set Path button, you can add an output command to all states. The state name typically is included as part of the filename. This is a huge timesaver from saving the file path for each render pass in the Render Setup dialog box.

Tutorial: Presenting several stylized rendered options

Suppose you have a client who wants to have a scene rendered using one of the non-photorealistic rendering options available in 3ds Max, but the client doesn't know which look is right and has asked you to create a rendered copy of the scene using each of the available styles. The client also anticipates that the scene will change, so it would be nice to have a setup that can quickly render all these variants. The answer to this all-too-common problem is using state sets.

To add setup and render images using each of the available render styles, follow these steps:

1. Open the Elk-state sets.max file from the Chap 28 directory in the downloaded content set.

   This file contains a simple scene of an elk on a hill.

2. Open the Render Setup dialog box with the Rendering→Render Setup menu command (or by pressing F10), and choose the Quicksilver Hardware Renderer option for the Production render in the Assign Renderer rollout. This sets the root state to use the Quicksilver Renderer.

3. Select the Rendering→State Sets menu to open the State Sets interface. Select State01, rename it **Graphiterender**, and then click the Begin Recording State button.

4. Select the Renderer panel in the Render Setup dialog box, and choose the Graphite option from the Rendering Level drop-down list in the Visual Style & Appearance rollout. Then click the Record State button in the State Sets interface again.

5. Click the Add New State button in the State Sets interface, and rename the state Ink render; then click the Begin Recording State button for this state, change the Rendering Level in the Render Setup dialog box to Ink, and click again on the Recording State button to stop the recording.

6. Repeat Step 5 for each of the remaining render styles in the Rendering Level drop-down list, including Color Ink, Acrylic, Tech, Color Pencil, and Pastel.

7. In the State Sets interface, select the States→Render Outputs menu, set the output path to where you want to save the renders files, and press the Set Path button. This adds the output file changes to each state.

8. Select the States→Render All States menu command to create and save all the stylized rendered images.

Figure 28.2 shows the State Sets interface with all the different states and one of the resulting rendered images.

FIGURE 28.2

The State Sets interface can automate multiple renders.

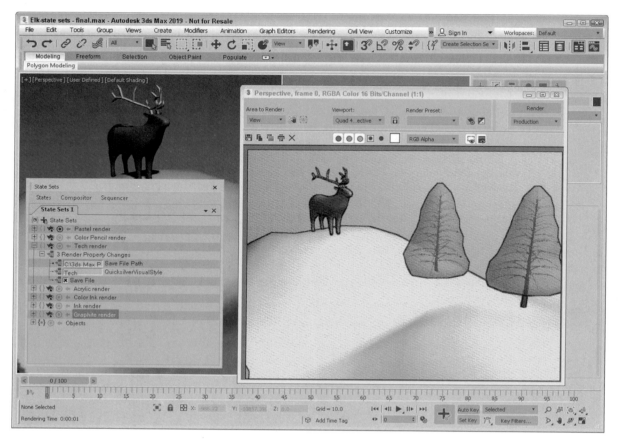

Accessing Compositor View

After a state is set up and rendered, you can use the Compositor→Compositor View menu command to view a node-based hierarchy of the states, as shown in Figure 28.3. All rendered and saved images appear as small thumbnails. If your states include any render elements, you can view those using the Compositor→Autobuild Compositor View (RE) menu command.

FIGURE 28.3

The Compositor View includes some basic image-compositing features.

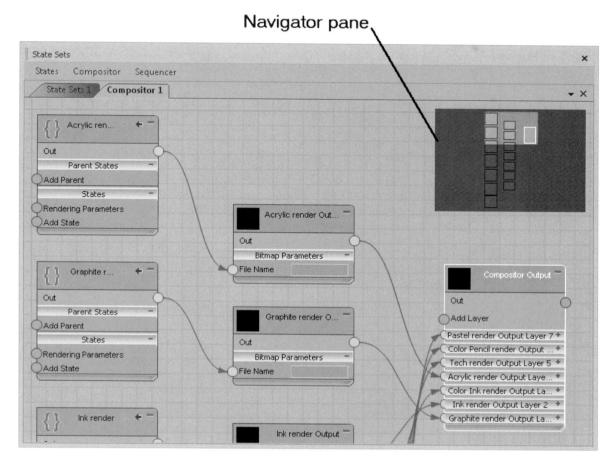

Navigation about the Compositor window works just like in the Slate Material Editor. Dragging with the scroll button pans the view, the scroll wheel zooms in and out, and a Navigator pane in the upper-right corner shows and lets you move about the entire hierarchy. Additional menu commands in the Compositor menu of the State Sets interface are Select All, Auto-Layout Nodes, Zoom Extents, and Create Color Correct Node.

Connection lines are drawn between input and output sockets for each node. You can disconnect any connection line by right clicking on it and selecting Disconnect Edge. Connections are made by dragging from an output socket to an input socket on another node.

If you need to refresh the Compositor View, you can use the Compositor→Refresh menu command, or you can close the Compositor View pane by clicking the X in the upper-right corner and reopening it. Any nodes that are not connected are deleted when the Compositor View pane is reopened.

Changing node parameters

Each node that holds a render displays a small thumbnail. Clicking the small plus or minus sign in the upper-right corner of each node lets you toggle the node's parameters on and off. For bitmap nodes, the single parameter is the image's filename.

For the composite node, you can click the plus or minus sign to the right of each layer name to open its parameters. For each layer, you can select to turn its layer and its mask on and off, and set its blending mode and opacity. Each of these parameters also has an input socket that you can use.

Adding new nodes

From within the Composite View, you can right click and choose Add State from the pop-up menu to add a new state node. You can also choose Add Sub-State and select a specific state node to add to the view.

You also can add a Color Correct node to the hierarchy with the Compositor→Create Color Correction Node menu command. This node can then be connected between a bitmap and the composite nodes. Its parameters include Hue Shift, Saturation, Contrast, and Brightness. The Hue Shift parameter provides a quick, easy way to change the color of a rendered object.

Exporting to Photoshop

The composited results can be exported to Photoshop as a PSD file using the Compositor→Create PSD menu command. This opens a simple dialog box where you can browse to set the save folder and filename. It also includes a Render button to initialize the export. The resulting image can then be opened in Photoshop. Each render state appears in Photoshop as a separate layer, as shown in Figure 28.4.

FIGURE 28.4

Compositor results can be exported to Photoshop, where each render state is a layer.

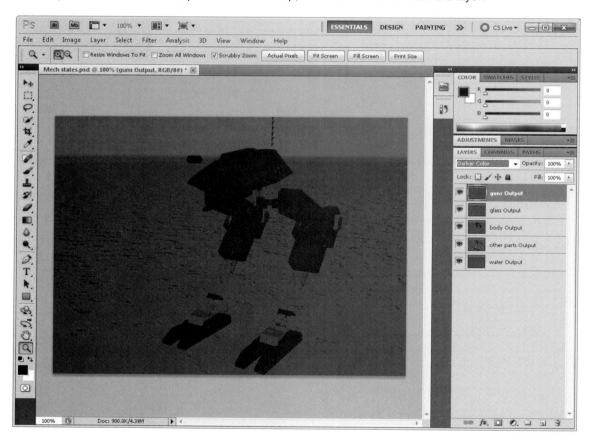

Linking data to After Effects

Selecting the Compositor→Compositor Link menu command opens a new Compositor Link tabbed panel. Clicking the Create Link button opens a file dialog box where you can specify a folder and filename for a file that is shared between 3ds Max and After Effects. This file has an .sof extension.

Note

Before a live link between 3ds Max and After Effects can be made, you need to copy the Adobe.AfterFX.dll file and the Autodesk.Plugins.Adobe.AfterFX.SceneIO.dll file from the 3ds Max/ExternalPlugins folder to the After Effects/Support Files folder. You also need to copy the Autodesk.Plugins.Adobe.AfterFX.SceneIO.Loader.aex file from the 3ds Max/ExternalPlugins folder to the After Effects/Support Files/Plug-ins/AEGP folder. Create an AEGP folder if there isn't one already.

Within After Effects, you can access the Open Compositor Link (Autodesk) menu command to open a file dialog box where you can locate the link .sof file created in 3ds Max. This loads the render states into After Effects, where you can edit them as needed. If changes are made in 3ds Max, you can use the Update to Link button in the Compositor Link pane to pass the changes on to After Effects.

Lights, cameras, and objects can be recorded in the Objects state. Any objects identified within this state are transferred to After Effects when a link is created. You can add objects to the Objects state by selecting them and clicking the Record State button or by clicking the Record State button and unhiding those objects that you want to pass on to After Effects. You also can control the type of objects that are passed on using the options listed in the Compositor Link pane. Plane objects are passed as Solids, and all other geometry objects are passed as Null objects.

Summary

This chapter covered the State Sets interface and showed how you can use it to composite scenes in 3ds Max. It also showed how State Sets can be used to save your scene settings at different alternatives. In this chapter, you learned how to do the following:

* How to use State Sets
* How to composite with the State Sets interface

The next chapter shows how to use batch and network rendering options to render while you sleep.

Chapter 29
Batch and Network Rendering

The Autodesk® 3ds Max® 2020 software can help you create some incredible images and animations, but that power comes at a significant price—time. Modeling scenes and animation sequences takes enough time on its own, but after you're finished, you still have to wait for the rendering to take place, which for a final rendering at the highest detail settings can literally take days. Because the time rendering takes is directly proportional to the amount of processing power you have access to, 3ds Max lets you use network rendering to add more hardware to the equation and speed up those painfully slow jobs.

This chapter shows you how to set up 3ds Max to distribute the rendering workload across an entire network of computers, helping you finish big rendering jobs in record time.

Batch Rendering Scenes

If you work all day modeling, texturing, and animating sequences, only to find that most of your day is shot waiting for a sequence to be rendered, happily you have several solutions. You can get a second system and use it for rendering while you work on the first system, you can use the network rendering feature to render over the network, or you have a third possibility: you can use the Batch Render tool.

Unless you are working around the clock (which is common for many game productions), you can set up a batch rendering queue before you leave for the evening, using the Batch Render tool. This queue runs through the night, giving you a set of takes to review in the morning.

Using the Batch Render tool

The Batch Render window, shown in Figure 29.1, is accessed from the Rendering menu. Render tasks can be added to the list by clicking the Add button. Render tasks can be disabled by selecting the check box to the left of the task name.

FIGURE 29.1

The Batch Render window lets you define render tasks to be run as a batch process.

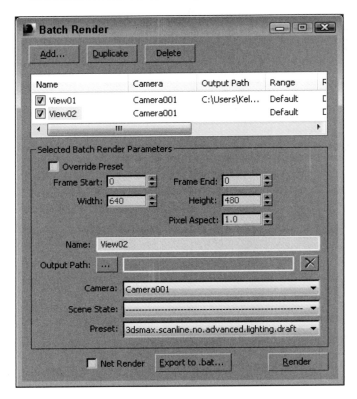

For each task, you can set the task's Name, Output Path, Camera, and render Preset. The Scene State drop-down list lets you select any scene states that are defined using the Rendering→State Sets command.

Each new task added to the Batch Render window uses the default render parameters for its frame range, dimensions, and pixel aspect, but if you enable the Override Preset option, you can customize each of these parameters for the selected task.

The Batch Render queue can be rendered over the network by enabling the Net Render option at the bottom of the Batch Render window.

Caution

Within the Batch Render dialog box is an option to select a Scene State. Scene States have been replaced by the State Sets feature, but this dialog box has not been updated to use State Sets.

Creating a stand-alone executable

After a batch render queue is established, you can click the Export to .bat button to open the Batch Render Export to Batch File dialog box, where you can save the batch file as a .bat file. This saved file can be executed from the command line or by using an agent.

Using the A360 Cloud Rendering System

Autodesk has a cloud rendering service that you can take advantage of and 3ds Max include a built-in link to this service. You can access and submit a scene to be rendered over the A360 system by selecting it from the Target drop-down list in the Render Setup dialog box, shown in Figure 29.2.

FIGURE 29.2

The A360 Cloud Rendering Mode option lets you submit your scene to be rendered externally

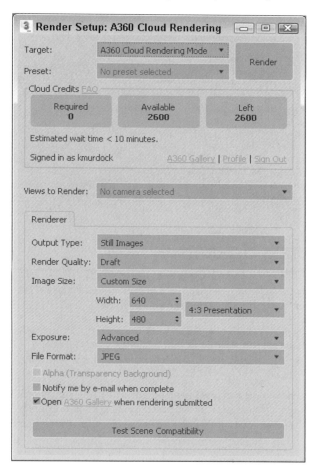

Before you can use this service, which runs on credits, you'll need to login first. The render settings also includes a button to Test Scene Compatibility. This button checks to make sure that all the settings are correct before sending the scene off to be rendered in the cloud and can save you some headaches.

Understanding Network Rendering

When you use network rendering to render your animation, 3ds Max divides the work among several machines connected via a network, with each machine rendering some of the frames. The increase in speed depends on how many machines you can devote to rendering frames: add just one computer, and you double the rate at which you can render. Add seven or eight machines, and instead of missing that important deadline by a week, you can get done early and take an extra day off.

Machines connected to handle network rendering are often referred to collectively as a *rendering farm*. The basic process during a network rendering goes like this: one machine manages the entire process and distributes the work among all the computers in the farm. Each machine signals the managing computer when it is ready to work on another frame. The manager then sends or "farms out" a new frame, which gets worked on by a computer in the rendering farm, and the finished frame gets saved in whatever format you've chosen.

The software in 3ds Max that makes network rendering possible is called Backburner. You may have noticed that it was installed when 3ds Max was installed. 3ds Max has several features to make the network

717

rendering process easier. If one of the computers in your rendering farm crashes or loses its connection with the manager, the manager reclaims the frame that was assigned to the down computer and farms it out to a different machine. You can monitor the status of any rendering job you have running, and you can even have 3ds Max e-mail you when a job is complete.

Note
One additional caveat to using network rendering is that you have no guarantee that the frames of your animation will be rendered in order. Each participating computer renders frames as quickly as possible and saves them as separate files, so you cannot use network rendering to create .avi or .mov files. Instead, you have to render the scene with each frame saved as a separate bitmap file, and then use the RAM Player, Video Post, or a third-party program (such as Adobe Premiere) to combine them into an animation file format such as .avi.

Note
A licensed version of 3ds Max is required to do network rendering, but the good news is that only one machine in your farm needs to have an *authorized* copy of 3ds Max installed. No authorization whatsoever is needed on machines used for network rendering only. Simply install 3ds Max, and each network rendering machine gets its authorization from the computer that launched the render job.

Setting Up a Network Rendering System

Before getting into the details of setting up 3ds Max for network rendering, it's important to understand the different parts of the network rendering system. This list shows the major players involved:

* **Manager:** The *manager* is a program (manager.exe) that acts as the network manager. It's the network manager's job to coordinate the efforts of all the other computers in your rendering farm. Only one machine on your network needs to be running the manager, and that same machine also can be used to render.

* **Server:** A rendering server is any computer on your network used to render frames of your animation. When you run the server program (server.exe), it contacts the network manager and informs it that this particular computer is available to render. The server starts up 3ds Max when the manager sends a frame to be rendered.

* **3ds Max:** At least one computer in your rendering farm must have an authorized copy of 3ds Max running, although it does not need to be the same computer that is running the manager. It is from this machine that you initiate a rendering job.

* **Monitor:** The Monitor (monitor.exe) is a special program that lets you monitor your rendering farm. You can use it to check the current state of jobs that are running or that have been queued. You also can use it to schedule network rendering times. The Monitor is completely independent from the actual rendering process, so you can use it on one of the machines in your rendering farm, or you can use it to remotely check the status of things by connecting over the network.

We address the task of setting up the network rendering system in three stages. The first thing you need is a functioning network, so we first go through the steps of how to get it working. Next, we look at setting up the Max software on each computer, and finally I describe how to tell Max where to find scene data it needs and where to put the finished scenes.

Setting up the network

To communicate with the different machines on the network, Max uses *TCP/IP* (Transmission Control Protocol/Internet Protocol), a very common network protocol. It's so common, in fact, that if your computers are already set up with some sort of network, you might already have TCP/IP installed and configured properly. If so, you are saved from several hours of work. Each machine on a network is identified by an IP address, which is a series of four numbers, each between 0 and 255, separated by periods, such as:

```
192.1.17.5
```

Each computer on a network has a unique IP address as well as a unique, human-readable name. To use 3ds Max to do network rendering, you need to find the IP address and name of each computer used on the network.

The precise details of what constitutes a "correct" IP address are too long and boring to go into here, but for a nonpublic network, all the addresses start with 192 or 10. The second and third numbers can be anything between 0 and 255 inclusive, and the last number can be between 1 and 254 inclusive (0 and 255 have special meanings). As an example, I've chosen to set up my home network as follows:

```
Machine Name    IP Address
Dungar          10.0.0.1
Ramrod          10.0.0.2
Slazenger       10.0.0.3
Romulus        10.0.0.4
```

Notice that each address is unique and that because it's a private network, the first number in each address is 10.

Tutorial: Locating TCP/IP and gathering IP addresses

To get the computers to talk to one another, you need to make sure that each computer has the TCP/IP protocol installed, and you need to gather the IP addresses for all the machines on the network.

To see whether you already have TCP/IP installed and to find out the IP address and name of each computer, follow these steps:

1. Right-click the Computer icon in Windows Explorer, and select Properties from the pop-up menu. The System information window opens. About halfway down the window is the Full computer name. Write the computer name in a list of all the computers that you are using to render via a network. Each computer on your network needs a unique name, so if the field is blank, enter a unique name and add the computer to your list.

2. Open the Control Panel by going to the Windows taskbar and choosing Start→Control Panel.

3. Click the Network and Internet icon, and then click the Network and Sharing Center link to open the network connections window. This window displays a list of network connections.

4. Select the network connection identified as Local Area Connection, and select the View Status link. The Local Area Connection Status dialog box displays the status of the current connection. If you click the Details button, you see a dialog box that lists the current IP address being used if your network is already installed. You also can click the Properties button to open a dialog box, which lists the installed protocols. At the top of the Local Area Connection Properties dialog box, you should see your network adapter (such as the brand and model of your Ethernet card). If it is not listed, you need to get it set up before proceeding. Under the network adapter is a list of installed network protocols, similar to the list in Figure 29.3. Look through the list until you find the Internet (TCP/IP) Protocol. If you find it, double-click it to bring up the TCP/IP Properties dialog box. If you don't see the TCP/IP protocol anywhere, you have to add it yourself to do network rendering.

Note

Two different versions (version 4 or version 6) of TCP/IP may be installed on your computer. 3ds Max's network rendering can work with either version.

FIGURE 29.3

A list of the network protocols installed on this computer

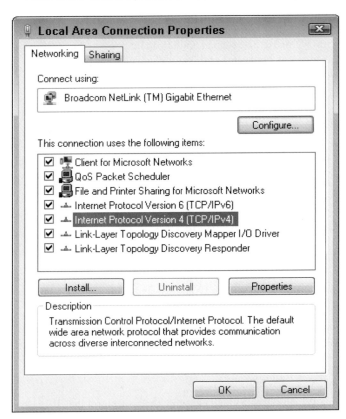

5. The Internet (TCP/IP) Protocol Properties dialog box has an important piece of information, called IP Address, shown in Figure 29.4. Do one of the following:

* If Obtain an IP address automatically is selected, you don't have to worry about the exact IP address of this computer because each time the computer connects to the network, it gets an IP address from a server, and the address it gets may be different every time.

* If Specify an IP address is selected, add the IP address to your written list of computers on the network. In the IP Address panel shown in the figure, the IP address has been statically assigned, which means that this computer will always have the same IP address.

FIGURE 29.4

You can find the IP Address in the Internet Protocol (TCP/IP) Properties dialog box.

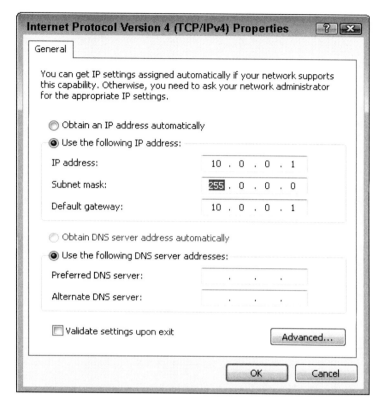

6. Repeat this procedure for each computer that will participate in your rendering farm. If you're lucky and all your machines have TCP/IP installed and are ready to go, you can skip the next section and move on to setting up Max on your network. If not, follow the steps for configuring TCP/IP in the next section.

Tutorial: Installing and configuring TCP/IP

If your computer doesn't already have TCP/IP installed, you have to install it to render via a network in Max.

Caution

Be careful about changing the TCP/IP settings of computers that are on a public network or are part of a large corporate network. Incorrect settings can not only keep your computers from communicating properly but also can cause problems in many other computers on the network.

If the computers you plan on using for your rendering farm are part of a public or corporate network, seek the assistance of the network administrator. If you're in charge of the computers yourself, dig out your Windows installation DVD-ROM, because you're going to need it.

To install and configure TCP/IP, follow these steps:

1. Go to the Networking panel on the Local Area Connection Properties dialog box. To get there, choose Start→Control Panel. Double-click the Network and Internet icon, and select the Network and Sharing Center links to open the network connections window. Click the View Status link for the network connection identified as Local Area Connection, and select Properties from the pop-up menu.

2. Click the Install button to open the Select Network Component Type dialog box. Select Protocol, and click the Add button. From the list of protocols, select Internet (TCP/IP) Protocol and click OK. At this point, Windows asks whether you want to use DHCP. DHCP is a server that automatically assigns an IP address to your computer.

3. If you know for sure that there is a DHCP server running, choose Yes. If you're setting up the network yourself or you have no idea what DHCP is, choose No.

   If DHCP is already set up and working, it can save you lots of time, but if not and if you're setting up the network yourself, I recommend steering away from it. You can change the settings to use DHCP at a later time if you're feeling ambitious. Now Windows starts looking for the files that it needs to install, and it pops up a dialog box asking for their location.

4. The default path it lists is probably the right one, so you can just click OK and continue. If Windows guessed wrong, or if you have the DVD-ROM in a different drive, correct the path and then click OK. After Windows finishes copying the files, TCP/IP is installed but not configured.

5. To configure the protocol, select it and click the Properties button.

   The Internet (TCP/IP) Protocol Properties dialog box (refer to Figure 29.4) opens. In the General panel is a section where you choose whether you want to have an IP address assigned automatically using DHCP.

6. If you're not using DHCP, choose Specify an IP address and enter an IP address for this machine. Refer to the beginning of this section if you need help choosing a valid IP address. (If you're a little confused about what numbers to use, you should be safe using the same numbers that I did.)

Caution

It's extremely important that you choose an IP address that is unique on the network. (Duplicate IP addresses are a great way to guarantee a nonfunctional rendering farm.)

7. Below the IP address section, enter the following as a subnet mask:

   `255.255.255.0`

   The subnet mask is used in conjunction with the IP address to identify different networks within the entire domain of every network in the world. If you do have to change this, be sure to change it in the 3dsnet.ini file that each rendering server creates (see "Configuring the Network Manager and Servers" later in this chapter).

8. Now select the Obtain DNS server address automatically option unless you know the IP address of your network's DNS server.

   This setting allows you to configure the TCP/IP protocol to check your local DNS when looking up addresses.

9. Click OK in the TCP/IP Properties dialog box to close it, and then click Close on the Local Area Connection Properties dialog box to close it.

   Windows needs to be shut down and restarted for the changes to take effect. This computer now has the proper network setup for your rendering farm.

Remember to repeat these instructions for every computer you want to use in your rendering farm. Each computer must be properly connected to the network and have TCP/IP installed and configured. This may seem like lots of work, but fortunately it's a one-time investment.

Tutorial: Setting up 3ds Max on the networked computers

If you've made it this far, you'll be happy to know that the worst is behind you. We've covered the most difficult parts of setting up a network rendering system; by comparison, everything else is relatively simple.

At this point, you should have a complete list of all the computers used in your network rendering system. Each computer should have a unique name and a unique IP address, and all should have TCP/IP installed and configured. You are now ready to move on to the actual Max installation for your rendering servers.

You have to set up 3ds Max on each computer in your rendering farm. Fortunately, this is as simple as a normal installation. To set up 3ds Max on each computer, follow these steps:

1. Run the setup.exe program on the Max installation CD-ROM or from the downloaded setup file.

Note

You don't need to have a DVD-ROM drive in every computer in your rendering farm. After your computers are networked, you can map a drive from your current machine to a computer that has the 3ds Max DVD-ROM in its DVD-ROM drive. In Windows Explorer, choose Tools→Map Network Drive and enter the path to the computer and drive with the CD-ROM.

2. Move past the first few introduction screens until you get to the Setup Type screen. Choose the minimum number of components so 3ds Max installs only the minimum number of files it needs to be able to render. You also need to choose a destination directory where you want 3ds Max to be installed. If possible, accept the displayed default destination and click Next.

Tip

Installing 3ds Max in the same directory on every computer can save you some maintenance headaches later on. Managing bitmap and plug-in directories is much easier if each machine has the same directory layout.

3. Continue with the rest of the installation as you would do for a normal installation of 3ds Max.

After the installation files get copied over, you'll probably have to reboot your computer for the changes to take effect.

Configuring shared directories

The last step in building your rendering farm is to tell 3ds Max where it can find the information it needs to render a scene. 3ds Max must be able to find textures and other information, and it must know where to put each frame that it renders.

Tutorial: Sharing directories

Instead of copying needed files to every machine in your rendering farm, you can share your directories across the network, which means that other computers on the network can use the files in that directory.

To make a directory shared, follow these steps:

1. Open Windows Explorer by selecting Start on the Windows taskbar and then choosing Computer.
2. Find the directory you want to share, right-click it, and choose Share With→Advanced Sharing.

   The File Sharing dialog box opens for that directory.
3. In the File Sharing dialog box, select the machines you want to share the file with. Under the Permission Level, you can choose each machine to be a Reader, Contributor, Co-owner, or Remove. For network rendering, choose the Co-owner option and click the Share button.

   After sharing is enabled, a dialog box appears informing you that sharing has been set up, as shown in Figure 29.5.

FIGURE 29.5

Sharing a directory so other computers on the network can use it

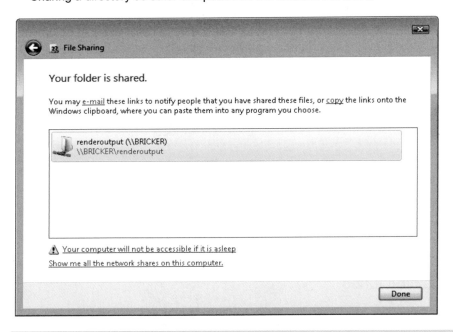

Tip
Other computers will refer to the shared directory by its shared name instead of its actual name, so to keep things simple, accept the default of using the actual name for the shared name.

Caution
Enabling the Allow network users to change my files option gives all network users full control access, letting everyone on the entire network read and write or erase the files in your shared directory. For now, leaving it this way until you're sure everything is configured properly is best. Later, however, restricting access to only those accounts that should have access would be a good idea.

4. Click Done in the File Sharing dialog box to close it, and you're back at Windows Explorer. If you press F5, Windows refreshes the display and your directory now has a little icon to show that the folder is shared. Figure 29.6 shows the "renderoutput" directory denoted as a shared directory.

Other computers can access your shared directory by specifying the full name of the directory's location. In the example, the "renderoutput" directory is on a computer named "Bricker," so the full path to that directory is as follows:

```
\\bricker\renderoutput
```

FIGURE 29.6

Other computers can now access the shared "renderoutput" directory.

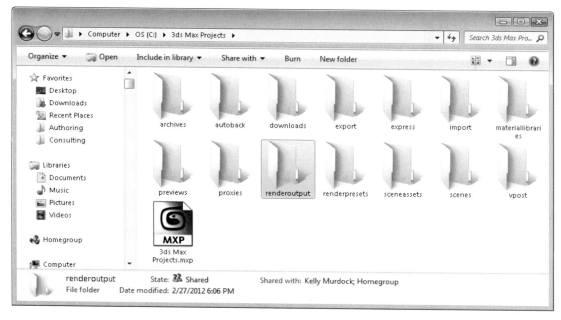

Tutorial: Choosing shared directories

Now you need to decide where to put the shared files and the output files. These directories are the place from which all machines in your rendering farm read files and images and to which they all write finished frames of your animation. The following procedure takes you through the steps for doing this using the directories I've set up on my own rendering farm.

To set up your shared directories, follow these steps:

1. First, decide which drives to use and then share them.

   On my network, Bricker has plenty of disk space, so I used the maps, images, scenes, and renderoutput directories that were already there from when I installed Max.

Tip

Use the maps, images, scenes, and renderoutput directories for all the scenes that you render via a network. In each directory, you can create other directories to organize files for your different scenes, but putting all needed files in the same place facilitates maintenance.

2. On each computer in your rendering farm, map a drive to your shared maps, images, scenes, and renderoutput directories as described in the preceding section. If possible, choose the same drive letters on all machines. I used the letter *Z* for the renderoutput directory on each computer.

Congratulations! You've made it through the installation and setup of your network rendering system. And no matter how long it took you, it was time well spent. The ability to render via a network will easily save you more time than you invested in setting up your network.

Starting the Network Rendering System

Now that the network is setup and you have software and shared files available, you are ready to start rendering over the network.

Tutorial: Initializing the network rendering system

The very first time you start your rendering farm, you need to help 3ds Max do a little initialization.

To initialize the network rendering system, follow these steps:

1. Start the network manager on one machine in your rendering farm. This program, Manager.exe, is in the Backburner directory (which is found by default where in the Autodesk folder). You can start the manager by selecting it and pressing Enter in Windows Explorer. After it starts up, you first see the Backburner Manager General Properties dialog box. This dialog box appears only the first time you run the Manager.exe program or if you choose Edit→General Settings. I cover its settings later in the chapter. After setting these properties, click OK, and the Manager window, shown in Figure 29.7, runs.

FIGURE 29.7

Starting the network manager

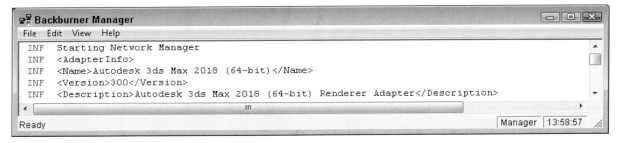

2. Now start a network server on each computer that you plan to use for rendering. To do this, find and start the Server.exe program just like you did with Manager.exe. When you start this program for the first time, the Backburner Server General Properties dialog box appears. This dialog box is covered later in the chapter. Click OK, and the Network Server window appears, as shown in Figure 29.8.

FIGURE 29.8

Starting a network server. Notice that the server is already looking for the manager.

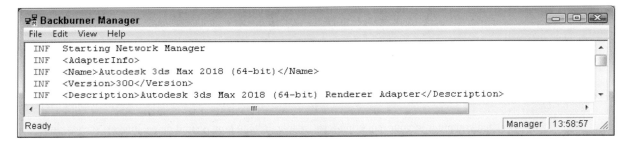

When the server finds the manager, it displays a message that the registration is accepted. The Network Manager window also shows a similar message.

If the server had trouble connecting to the manager, you need to follow these two additional steps:

1. If automatic detection of the manager fails, the server keeps trying until it times out. If it times out, or if you just get tired of waiting, choose Edit→General Settings to open the Backburner Server General Properties dialog box, shown in Figure 29.9. In this dialog box, uncheck the Automatic Search box and type in the name or IP address of the computer that is running the network manager. In this case, the server tried but couldn't quite find the manager, so it had to be told that the manager was running on the computer whose IP address is 150.150.150.150.

FIGURE 29.9

Manually choosing the manager's IP address

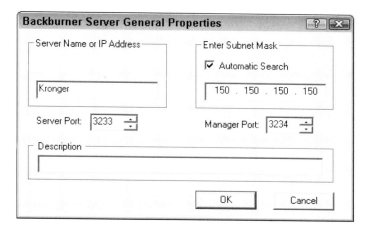

2. Click OK to close the Backburner Server General Properties dialog box, and then click Close to shut down the server (doing so forces the server to save the changes you've made). Restart the server the same way you did before, and now the server and manager are able to find each other.

Note

The network manager does not need to have a computer all to itself, so you also can run a network server on the same computer and use it to participate in the rendering.

Tutorial: Completing your first network rendering job

Your rendering farm is up and running and just dying to render something, so let's put those machines to work.

To start a network rendering job, follow these steps:

1. Start 3ds Max, and create a simple animation scene.

   This should be as simple as possible because all you're doing here is verifying that the rendering farm is functional.

2. In 3ds Max, choose Rendering→Render Setup (F10) to bring up the Render Setup dialog box. In the Time Output section of this dialog box, be sure that Range is selected so that you really do render multiple frames instead of the default single frame.

3. In the Render Output section of the Render Setup dialog box, click Files to open the Render Output File dialog box. In the Save In section, choose the output drive that can be accessed over the network and directory that you created earlier.

4. In the File name section of the Render Output File dialog box, type the name of the first frame. 3ds Max automatically numbers each frame for you. Choose a bitmap format from the Save as type list (remember, an animation format will not work).

5. Click Save to close the Render Output File dialog box. (Some file formats might ask you for additional information for your files; if so, just click OK to accept the default options.) Back in the Render Scene dialog box, 3ds Max displays the full path to the output directory.

6. In the Render Output section of the Render Setup dialog box, select the Submit to Network Rendering from the drop-down list to the right of the Render button, as shown in Figure 29.10.

FIGURE 29.10

The Net Render option must be enabled to start a network rendering job.

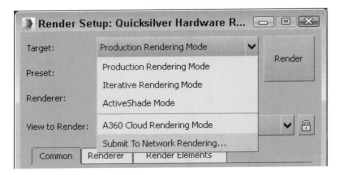

A Network Job Assignment dialog box opens, like the one shown in Figure 29.11.

FIGURE 29.11

Use the Network Job Assignment dialog box to locate the manager to handle the rendering job.

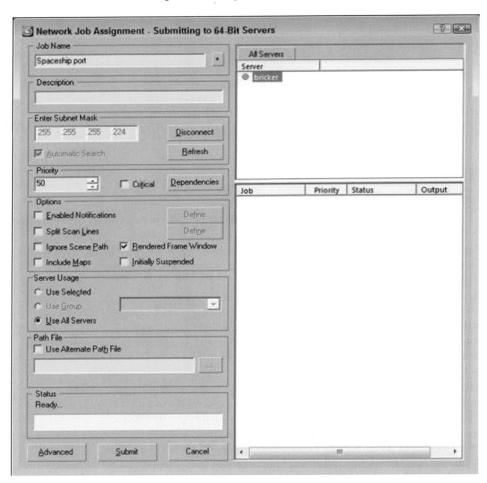

7. In the Enter Subnet section of the Network Job Assignment dialog box, click Connect if the Automatic Search box is checked. If it isn't checked, or if your servers had trouble finding the manager in the "Initializing the network rendering system" tutorial earlier in this chapter, type the IP address of the machine running the manager and then click Connect.

8. 3ds Max then searches for any available rendering servers, connects with it, and adds its name to the list of available servers. Click the server name once, and click Submit.

Tip

If you try to submit the same job again (after either a failed or a successful attempt at rendering), 3ds Max complains because that job already exists in the job queue. You can remove the job using the Monitor, or you can click the + button on the Network Job Assignment dialog box, and 3ds Max adds a number to the job name to make it unique.

After you've submitted your job, notices appear on the manager and the servers (like the ones shown in Figures 29.12 and 29.13) that the job has been received. Soon 3ds Max starts up on each server, and you see a Rendering dialog box showing the progress of the rendering task. As you can see, this displays useful information such as what frame is being rendered and how long the job is taking. When the entire animation has been rendered, you can go to your output directory to get the bitmap files that 3ds Max generated. The render servers and the render manager keep running, ready for the next job request to come in.

FIGURE 29.12

The network manager detects the new job.

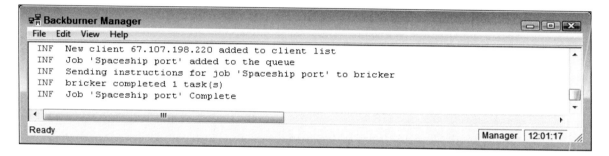

FIGURE 29.13

One of the network servers receives the command to start a new job.

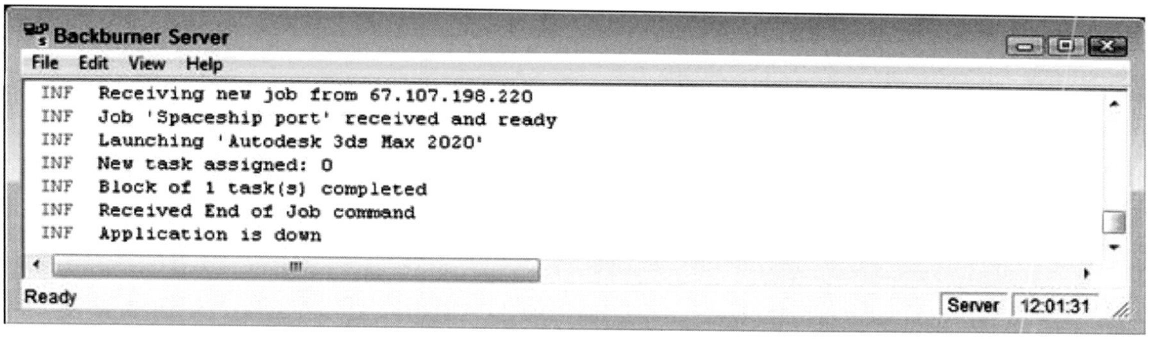

Job assignment options

The Network Job Assignment dialog box (refer to Figure 29.11) has an important section that you didn't use for your first simple render job; it's called Options.

The Options section has the following settings:

* **Enabled Notifications:** This option lets you tell 3ds Max when to notify you that certain events have occurred. If you check the Enable Notifications option, the Define button becomes active. The Define button opens a Notifications dialog box, shown in Figure 29.14.

FIGURE 29.14

The Notifications dialog box lets you specify which type of notifications to receive.

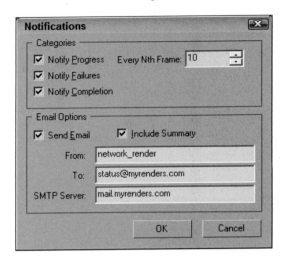

* **Split Scan Lines:** This option breaks a rendered image into strips that can be rendered separately. The Define button lets you specify the Strip Height, Number of Strips, and any Overlap.

* **Ignore Scene Path:** Use this option to force the servers to retrieve the scene file via TCP/IP. If disabled, the manager copies the scene file to the server.

* **Rendered Frame Window:** Use this option if you want to be able to see the image on the server as it gets rendered.

* **Include Maps:** Checking this box makes 3ds Max compress everything that it needs to render the scene (including the maps) into a single file and send it to each server. This option is useful if you're setting up a rendering farm over the Internet, although it takes more time and network bandwidth to send all that extra information.

* **Initially Suspended:** This option pauses the rendering before it starts so that you can manually start it when the network is ready.

* **Use Selected/Use Group/Use All Servers:** The Server Usage options makes the selected server, a group of servers, or all servers listed in the Server panel fair game for rendering.

* **Use Alternate Path File:** This option lets you specify an alternate path for map and other files, which is entered in the text field below the check box.

Configuring the Network Manager and Servers

You can configure both the manager and servers using their respective General Properties dialog boxes. You open these dialog boxes by choosing Edit→General Settings.

The network manager settings

The rendering manager has some options that let you modify how it behaves. You specify these options in the Network Manager General Properties dialog box, shown in Figure 29.15. To open this dialog box, select Edit→General Settings in the Manager window.

FIGURE 29.15

The Backburner Manager General Properties dialog box

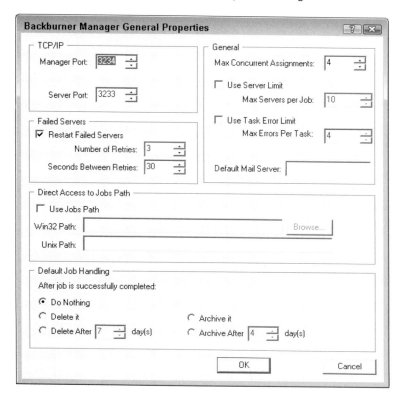

This dialog box includes the following sections:

* **TCP/IP:** Here you can change the communications ports used by the manager and the servers. In general, leaving these alone is a good idea. If some other program is using one of these ports, however, 3ds Max won't be able to render via the network, so you need to change them. If you change the Server Port number, be sure to change it to the same number on all your rendering servers. If you change the Manager Port number, you also need to change two files on your hard drive to match: queueman.ini (in your 3dsmax directory) and client.ini (in your 3dsmax\network directory). Both have lines for the Manager Port, and you can edit these files with any text editor or word processor.

* **General:** The 3ds Max Concurrent Assignments field is used to specify how many jobs the rendering manager sends out at a time. If you make this number too high, the manager might send out jobs faster than the servers can handle them. The default value here is fine for most cases.

Note

The network manager can automatically attempt to restart servers that failed, giving your rendering farm much more stability.

* **Failed Servers:** Usually, 3ds Max doesn't send more frames to a server that previously failed. If you check the Restarts Failed Servers box, 3ds Max tries to give the server another chance. The Number of Retries field tells 3ds Max how many times it should try to restart a server before giving up on the particular server for good, and the Seconds Between Retries field tells 3ds Max the number of minutes it should wait before trying to give the failing server another job.

Note

The rendering manager writes the configuration settings to a file on the disk that gets read when the manager loads. If you make changes to any of the settings, shutting down the manager and starting it up again to guarantee that the changes take effect is best.

* **Direct Access to Jobs Path:** This setting lets you specify a different path that exists anywhere on the network, regardless of where the manager is running. Separate fields are available for Win31 paths and Unix paths. If the Use Jobs Path option is selected, the specified path is used.

* **Default Job Handling:** This section lets you tell the Manager what to do after the rendering job is finished. The options include Do Nothing, Delete or Archive, and Delete or Archive after a specified number of days. This option gives you a way to clean out your queue. Note that this only deletes or archives the submitted job not the actual file.

The network servers settings

As you may have guessed, the Properties button on the Network Server window serves a similar purpose as the one on the Network Manager window: It enables you to specify the behavior of the network server. Clicking this button displays the Backburner Server General Properties dialog box.

This dialog box has the following section:

* **TCP/IP:** The port numbers serve the same function as they do for the rendering manager, described in the previous section. If you change them in the manager properties, change them here. If you change them here, change them in the manager properties.

Note

The Manager Name or IP Address setting lets you override automatic detection of the rendering manager and specify its exact location on the network. Generally, letting 3ds Max attempt to find the manager itself is best; if it fails, override the automatic detection by clearing the Automatic check box. If you happen to be running multiple managers on the same network, the servers connect to the first one they find. In this case, you have to manually choose the correct server.

Keep in mind that the server properties aren't shared among your servers, so if you want something to change on all your servers, you have to make that change on each machine.

Note

As with the rendering manager settings, if you change anything in the Network Servers Properties dialog box, be sure to shut down the server and restart it.

Logging Errors

Both the Network Manager and Network Server windows have a Logging button that you can click to access the Logging Properties dialog box, where you can configure how log information gets handled. This dialog box, shown in Figure 29.16, looks the same for managers and servers. You can access this dialog box with the Edit→Log Settings command.

FIGURE 29.16

The logging options for managers and servers let you tell 3ds Max where to report what.

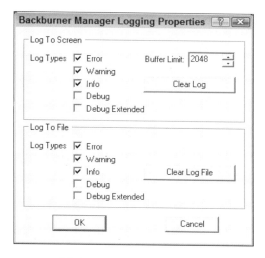

3ds Max generates the following types of messages:

* **Error:** Anything that goes wrong and is serious enough to halt the rendering of a frame.

* **Warning:** A problem that 3ds Max can still work around. If a server fails, for example, a warning is generated, but 3ds Max continues the rendering job by using other servers.

* **Info:** A general information message, such as notification that a job has arrived or that a frame is complete.

* **Debug:** A lower-level message that provides information to help debug problems with the rendering farm.

* **Debug Extended:** The same as the Debug option with more details.

3ds Max displays the type of message and the message itself in two locations: in the list window and in a log file (in your 3dsmax\network directory). The Logging Properties dialog box lets you choose whether each type of message gets reported to the screen, the log file, both places, or neither place. You also can use the Clear buttons to get rid of old messages.

Using the Monitor

The Monitor is a powerful utility that helps you manage your rendering farm and all the jobs in it. If you use network rendering frequently, the Monitor quickly becomes your best friend. You start it the same way that you start a rendering server or manager: go to the 3dsmax directory, find QueueManager.exe, and double-click it. Every computer that has 3ds Max installed on it also has a copy of the Monitor, so you can use it from any machine on your network. The main screen is shown in Figure 29.17.

FIGURE 29.17

The Monitor makes managing a rendering farm quick and easy.

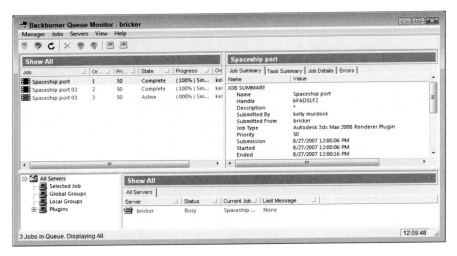

When the Monitor starts up, it automatically searches for the rendering manager and connects to it. (If you have more than one manager running, you have to choose which one to connect to.)

The main screen is divided into three panes. The top-left pane shows the job queue and their Priority and Status, and the top-right pane shows information about whatever you have selected in the left pane. You can use the tabs at the top of this pane to select the information that you want to view. The information tabs include Job Summary, Frames Summary, Advanced (which shows the rendering parameters), Render Elements, and Log.

The bottom pane lists all the available servers. Next to each server in the left pane is an icon that reflects its current status. Green icons mean that the job or server is active and hard at work. Yellow means the server is idle. Red means that something has gone wrong and gray means that a job has been inactivated or that a server is assigned to a job but is absent. When a job is complete, it can be deleted from the queue.

Jobs

If you choose a job in the top-left pane, the top-right pane displays information about the selected job. The panels in the top-right pane are as follows:

* **Job Summary:** Lists some of the rendering options you chose before you submitted the job. Among other things, the example in the figure shows that the job was rendered to 640 x 480 pixels.

* **Task Summary:** Lists the details of rendering each frame in the animation, including the time required to render and the server used.

* **Job Details:** Lists advanced settings from the Render Scene dialog box and gives limited information about the scene itself.

* **Errors:** Displays important messages from the job log. Whereas the log file on each server lists events for a particular server, this pane lets you see all the messages relating to a particular job.

When you point at a job in the top-left pane and right-click, a small pop-up menu appears. On this menu, you can delete a job from the queue or you can choose to activate or deactivate it. If you deactivate a job, all the servers working on that job save their work in progress to disk and then move on to the next job in the queue. This feature is very useful when you have a lower-priority job that you run when no other jobs are waiting; when something more important comes along, you deactivate the job so that you can later activate it when the servers are free again.

One last useful feature for jobs is that you can reorder them by dragging a job above or below other jobs. Jobs higher on the screen are rendered before lower ones, which enables you to "bump up" the priority of a particular job without having to deactivate other ones.

Note
A file with a Critical priority is rendered immediately.

Servers

If you right-click a server and select Properties from the pop-up menu, the Server Properties dialog box, shown in Figure 29.18, opens. This dialog box contains information about the selected server.

FIGURE 29.18

The Server Properties dialog box displays information about the server.

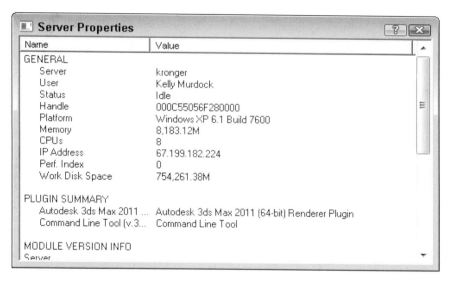

Many other features are available in the right-click pop-up menu. Using this pop-up menu, you can assign the server to a selected job, remove the server from its selected job, display specific server information, create a server group, or view the Week Schedule, as shown in Figure 29.19. Using the Week Schedule dialog box, you can set the active rendering period for a server.

FIGURE 29.19

The Week Schedule dialog box can set the time during the week when a server is available for rendering.

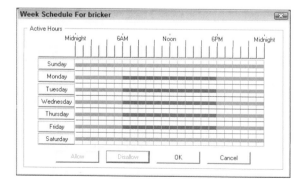

The Week Schedule dialog box lets you decide when a particular machine is available for rendering. (For example, you can have your co-worker's computer automatically become available for rendering after he or she goes home for the night.)

Click and drag with your mouse over different hours to select a group of times. Alternatively, you can click a day of the week to select the entire day or click a time to select that time for every day. In the example shown in Figure 29.19, the server is scheduled to render in the evenings and on the weekends.

After you've selected a group of times, click Allow to make the server available for rendering during that time or click Disallow to prevent rendering. When you're finished, click OK to close the Server Properties dialog box and return to the Monitor dialog box.

If you have several jobs going at once but suddenly need to get one finished quickly, you can take servers off one job and put them on another. To remove a server, right-click its name in the left pane of the Monitor dialog box and choose Delete Server (Ctrl+Enter) from the pop-up menu. The icon next to the server turns black, indicating that it has been unassigned. To assign this server to another job, right-click the server name in the list of servers for the job you want to assign it to, and choose Assign to Selected Jobs.

Summary

If your goal is to spend more time modeling and less time waiting for rendering jobs to complete, the network rendering services provided by 3ds Max can help you take a step in the right direction. After the initial complexities of setting up a rendering farm are out of the way, network rendering can be a great asset in helping you reach important deadlines, and it lets you enjoy your finished work sooner. Even if you can afford to add only one or two computers to your current setup, you'll see a tremendous increase in productivity—an increase that you can't truly appreciate until you've completed a job in a fraction of the time it used to take!

In this chapter, you accomplished the following:

* Used the Batch Render window to create a batch render queue
* Set up a network suitable for network rendering with Max
* Set up a 3ds Max rendering farm
* Used the rendering manager and servers to carry out rendering jobs
* Used the Monitor to control job priority
* Made 3ds Max notify you when problems occur or when jobs finish

Now that you have the basics of rendering down, the next chapter steps the rendering quality up with an advanced rendering plug-in called Arnold.

Chapter 30
Rendering with ART and Arnold

IN THIS CHAPTER

Enabling the ART or Arnold renderer

Working with the ART Renderer

Working with Arnold lights

Lighting with the Arnold skydome

Using the specialized Arnold materials

Configuring the number of Arnold Samples

The Autodesk® 3ds Max® 2020 software includes a plug-in architecture that lets you replace or extend any part of the software including the rendering engine. Autodesk RayTracer (ART) and Arnold are specialized plug-in rendering engines that offer many advanced features. These engines take the rendering in 3ds Max to a new level, enabling you to render your scenes with amazing accuracy. It also includes a host of advanced rendering features including caustics, subsurface scattering, and image-based lighting (IBL) that are physically realistic.

The Arnold rendering engine is a physically-based ray tracing system. Ray tracing works by following the light photons as they move from the light sources and bounce off the various scene surfaces until they finally end up at the camera's source. Each surface is defined by a shader that described how the light photons interact with the surface. Some photons are reflected, some are refracted, some are absorbed and others are scattered about the scene. Each time a light photon interacts with a surface, a portion of its total energy is lost. The ray tracing engine simply follows each light beam until its energy is depleted. Keeping track of the energy level of each photon is how the realistic effects are accomplished because that is how light in the real world works also.

Although the Arnold renderer is awesome, it can take lots of tweaking to get results to look just right. This is where the ART engine comes in. It lets you start a render, and it automatically adjusts the settings to give you great results based on the amount of time you give it. In other words, ART is Arnold for dummies.

Enabling ART and Arnold

If you're accustomed to using the Scanline Renderer and you're wondering if the ART or Arnold rendering engines are worth using, the answer is yes. Actually, you should try a couple of test renderings first and play with the different settings, but in working with Arnold I've been amazed at its results. One of the chief benefits of Arnold is its speed. The whole engine was designed to be interactive. It can render a fully raytraced scene in a fraction of the time without sacrificing quality.

Note
Arnold is an external rendering engine that plugs into 3ds Max. It was developed by a company named Solid Angle, so its development was separate from 3ds Max.

ART and Arnold can use a lot of different materials, but the best results occur when specific physical-based materials are used and some materials can cause problems. For example, if ART finds a material it doesn't recognize, it will just ignore it, which results in a blank render. Arnold, however, will try to approximate the material, which sometimes is good and often is bad. To really get the best use of these rendering engines, you should outfit the scene objects with the correct materials. The ART engine works best with the Arch & Design and those materials in the Autodesk Material Library and Arnold works well with the Physical Material and the Arnold specific materials and maps located in the Arnold sections of the Material/Maps Browser.

Some of the available Arnold materials are presented in Chapter 19, "Using Specialized Material Types."

Lights also can be an issue for ART and Arnold. Since these renderers are both physical-based, the Photometric lights will work with both systems. For better results and more features, an Arnold light is also available. The Arnold light includes support for Area Lights, Depth of Field, and Motion Blur. It also includes some specialized light features with functionality, like atmospheric shadows, that are unavailable in the Scanline Renderer.

More information on using Arnold lights is presented in Chapter 25, "Using Lights and Basic Lighting Techniques."

To choose ART or Arnold as the renderer for your scene, simply select either from the list of available renderers in the Renderer setting at the top of the Render Setup dialog box. The available settings for the selected renderer will be shown as panels in the Render Setup dialog box and other dialogs, like the Material/Map Browser, will also be updated with new materials when the renderer is selected.

Once selected as your Production renderer, you don't need to modify any other settings for the renderer to work. The Arnold settings in the Material Editor, Lights category, and Modifier list enable additional features that Arnold can take advantage of, but they aren't required to render the scene.

Note
You may have noticed the Raytracer rollout in the Render Setup dialog box when the Scanline Renderer is enabled. There is also a Raytrace material in the Standard set of the materials. These settings and materials are used to enable and configure raytracing in the Scanline Renderer, but the Arnold results are much better than these raytrace options, so the focus is to learn Arnold instead of the older raytrace options.

Working with Autodesk RayTracer (ART)

Having an advanced raytracer doesn't do you any good if you it is too difficult to use. This is the idea behind the Autodesk Raytracer (ART) rendering engine. When selected in the Render Setup dialog box, the ART renderer enables physically-based results.

Within the ART Renderer panel, the Render Quality is determined by a simple slide toggle, as shown in Figure 30.1. There are also options to render a given amount of time or a specified number of Iterations. The Rendering Method options include Fast Path Tracing and Advanced Path Tracing. The Fast Path Tracing option is good enough for most renders and is relatively quick using Indirect Illumination. The Advanced Path Tracing option results in much better results with minimal noise, but can take a long time to render.

FIGURE 30.1

The ART Renderer panel includes simple settings for the Render Quality.

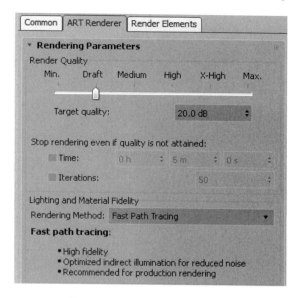

Within the Filtering rollout, you can specify the Filter Strength again with a slider. Fully Filtered will take a lot longer, but it will effectively remove all the noise from the scene. You can also enter a Filter Diameter value for Anti-Aliasing.

The trick behind using this renderer is that you need to use the available physically-based lights and materials. For lights, the photometric lights work great and for materials, try using the Physical Materials, Arch & Design materials or materials from the Autodesk Material Library. Even at the lower quality settings, the results are quite good, as shown in Figure 30.2.

Tip

If ART keeps throwing errors about the Environment being unsupported, then add a Sun Positioner to the scene. This takes care of the environment settings.

FIGURE 30.2

The ART Renderer produces good quality raytraced images.

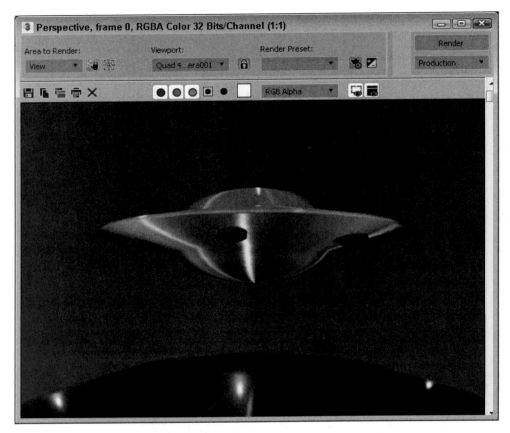

Tutorial: Starting ART

Once you get the hang of using ART, you'll find it is great to start the renderer with the unlimited setting before running off to lunch and check out the results when you get back.

To use the ART renderer to render a scene, follow these steps:

1. Open the Mech on water.max file from the Chap 30 directory in the downloaded content set.

   This file includes a mech model positioned over some water. The scene already has materials applied from the Autodesk Material Library.

2. Open the Render Setup dialog box (F10), and switch the Renderer to the ART renderer.

3. In the ART Renderer panel of the Render Setup dialog box, drag the Render Quality slider to High and click the Render button.

4. Let the scene render for a while and when it looks good enough, click the Cancel button in the Rendering dialog box. The rendered results are displayed in the Rendered Frame Window where you can save the image if you desire.

Figure 30.3 shows the scene rendered with ART.

FIGURE 30.3

Using the Unlimited option lets the ART renderer keep running as long as it can.

Working with Arnold

The big difference between ART and Arnold is that Arnold gives you access to all the gritty configuration settings. By tweaking these settings, you can enable specific effects, speed up the render time, and get the exact results you want, but the cost is the time it takes to figure out these settings.

Tip

One way to tell the difference between the Scanline and Arnold rendering is that the Scanline Renderer processes the image in horizontal lines that progress from the top of the image to the bottom, and the Arnold renderer processes square sections of the image (called buckets), usually working on the easiest areas first and the most complex areas last.

Arnold is a complete solution, so it is natural that Arnold implementation in 3ds Max also includes Arnold-specific lights, Arnold materials and maps and even an Arnold Properties Modifier for turning Arnold options on and off at the object level. Using the lights and materials that are made to work with Arnold gives you access to better results and advanced features like area lights and multi-layer materials.

Using Arnold Lights and Shadows

If you look in the Lights panel of the Command Panel, you'll see an Arnold subcategory with a single light simply called an Arnold light. This light has a Shape setting, as shown in Figure 30.4, that lets you change how the light source looks and how it casts light into the scene. Area lights, like the Quad or Disc options, spread light from a defined area much like a light card used in filming. Other light shapes like Cylinder and Mesh project light from a 3D shape. The benefit of 3d shaped lights is that the broader light source creates soft shadows. Specular highlights are also affected by the light shape. Other common light types, like Point, Distant and Spot light shapes are also available. There is also a Photometric light that mimics real-world lights and a Skydome light that acts like the sun and is useful for lighting outdoor scenes.

More on lighting basics is covered in Chapter 25, "Using Lights and Basic Lighting Techniques."

Caution
Although the Arnold light is available in the Command Panel, it is not available from the Create→Lights menu.

If an Arnold area light is added to the scene, you can set the area shape and dimensions in the Shape rollout. The dimensions will change depending on the shape selected.

FIGURE 30.4

The various rollouts for Arnold light lets you define the light settings for individual lights.

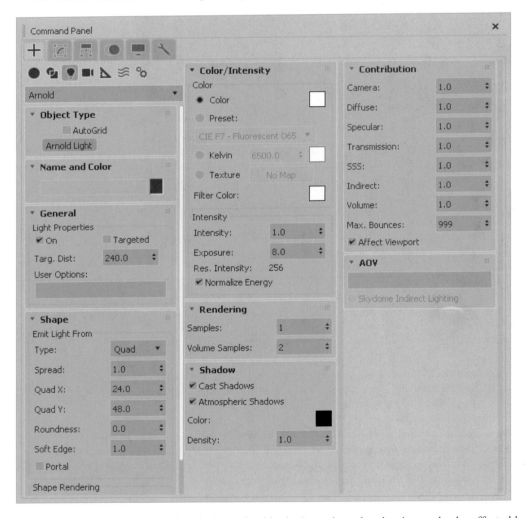

The brightness of the light is determined by its Intensity value, but it can also be affected by its Color value. For Photometric lights, you can set the Kelvin temperature or the Type of real-world light or you can add a texture in front of the light using a map.

The required Intensity value to light the scene will be greatly affected by the shape of the light. For example, a single area light, like Quad or Disc, requires an Intensity value of around 5 to provide enough brightness to light the scene. The same applies for Point, Spot, Cylinder and Mesh lights. Under the same conditions, Distant and Skydome lights require an Intensity value of 0.01 and 0.001 respectively to get similar results. Photometric lights are based on real-world values and need appropriate values for correct lighting results.

The Exposure value is another way to control the brightness of the light. It works much like an f-stop value for cameras by doubling the brightness for each stop increase. The Normalize Energy option lets you control the light intensity by changing the light object's size.

The Samples gives you control over the number of light rays that are traced through the scene. More Samples means more rays which means less noise in the final image, but also means more render time.

Enabling Arnold Shadows

In the Shadow rollout is an option to enable Cast Shadow and Atmospheric Shadows. You can also set the Shadow Color and Density. Atmospheric shadows will obscure a light object that comes behind it. This is a good effect if you are using any light shape besides Point. It is also not available for Distant and Skydome lights.

Using Arnold Portal

Underneath the Shape setting is an option for making the Arnold light a Portal. A Portal light source is used to define the exact area of a window that allows light into an interior scene and provides a way to focus light streaming into an interior space from an external source using a designated portal.

When creating a Portal light, simply position the light to the area where the window is located, and all light rays entering the interior space are focused on those areas defined by the Portal. When using a Portal light source, make sure that the direction of the Portal points into the interior section you wish to light. The gizmo for the Sky Portal is a simple rectangular box with an arrow pointing in the direction of the light rays. If the light rays are pointing outward, you will need to flip the light to change its direction.

Figure 30.5 shows a house interior lighted using Arnold and two Portal lights.

FIGURE 30.5

Sky Portals can focus all rays coming from an external light source.

Using Arnold Skydome and Physical Sky

If you want to quickly create an outdoor scene and render it using Arnold, there are two options--the Sun Positioner and the Arnold Skydome. This Arnold Skydome light works like the Sun Positioner system, but it doesn't have the daylight settings for controlling its position. For the Skydome, you can select the Format to be Lat-Long, Mirrored Ball or Angular. These are used to provide Image-Based Lighting (IBL) to the scene.

For the Sun Positioner system (found in the Lights panel of the Command Panel under the Photometric subcategory) click to add the system to the scene, and in the Modify panel, click the Install Sun & Sky Environment button. This will automatically set the environment map in the Environment panel and configure the Exposure Control settings.

Caution

The Sun Positioner system is different from the Daylight System located in the Create→Lights and in the Create→Systems menus. Both systems let you define the sun's placement for locations and times of the day, but the Sun Positioner is a newer system with more features.

If you look in the Environment panel of the Environment and Effects dialog box, you can select the Physical Sun & Sky Environment material applied as an environment map. This material defines how the sky and ground planes look. The default settings look pretty good.

The Sun Positioner system includes controls for positioning the sun in the sky based on a physical location, day of the year, and time of the day. By changing these controls, you can manipulate the sun's position in the sky relative to the scene objects, as shown in Figure 30.6.

More details on setting up the lighting environment are covered in Chapter 25, "Using Lights and Basic Lighting Techniques."

FIGURE 30.6

The Daylight can be endowed with Physical Sun & Sky Environment.

Selecting the sun light object and moving it about the scene or changing the Time and Date parameters automatically updates the background and the scene. This provides a good way to precisely position the sun where you want it, as shown in Figure 30.7.

FIGURE 30.7

The Physical Sun & Sky Environment system can be viewed and interactively updated in the viewport.

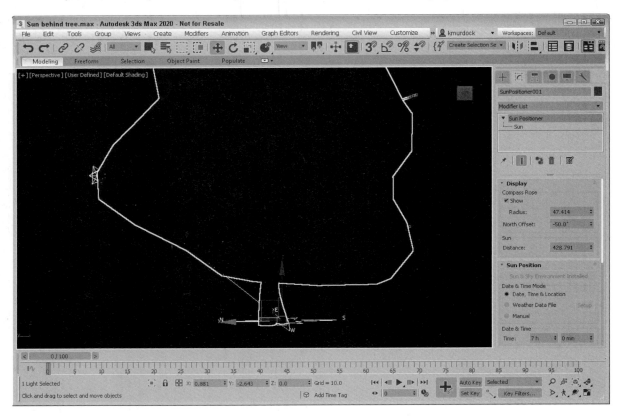

For the Arnold Skydome light, you can alter the Environment Lighting and Reflections (IBL) using the Environment, Background & Atmosphere rollout in the Arnold Renderer panel of the Render Setup dialog box. From here you can access the Environment and Effects dialog box using the Open Environment Settings button. You can also select a custom Background map and a Scene Atmosphere map, as shown in Figure 30.8.

FIGURE 30.8

The Environment, Background & Atmosphere rollout lets you make changes for the Arnold renderer.

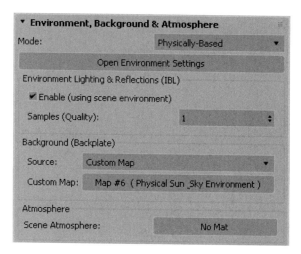

Using Arnold materials

If the Arnold renderer is enabled, you can open the Material/Map Browser in the Material Editor or separately using the Rendering menu and be greeted with many additional material and map types and the Autodesk Material Library, which includes hundreds of preset materials organized by category, as shown in Figure 30.9.

Note

If you enable the Show Incompatible option in the Material/Map Browser pop-up menu, you can view the list of Arnold materials even if the Arnold renderer isn't enabled.

FIGURE 30.9

The Material/Map Browser includes many additional Arnold materials and maps when enabled.

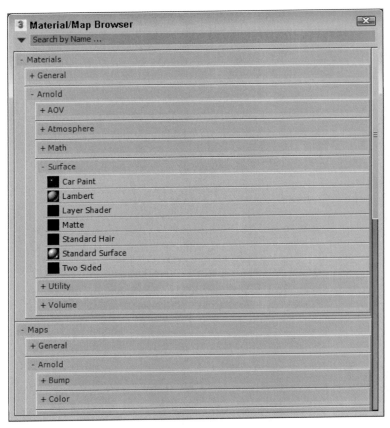

The Arnold materials are divided into several categories including AOV, Atmosphere, Math, Surface, Utility, and Volume. There are also several categories of Arnold Maps including Bump, Color, Conversion, Cryptomatte, Environment, Math, Shading State, Surface, Texture, User Data, Utility, and Volume. Each of these categories have different properties that you can connect within the Material Editor.

Within the Arnold→Surface material category is the Standard Surface material. This is the common basic material with a basic set of material properties including Coating, Color, Caustics, Specular Reflections, Transparency, Sub-Surface Scattering, Emission, Opacity and Maps, as shown in Figure 30.10.

FIGURE 30.10

The Arnold Standard Surface material holds a basic set of material properties.

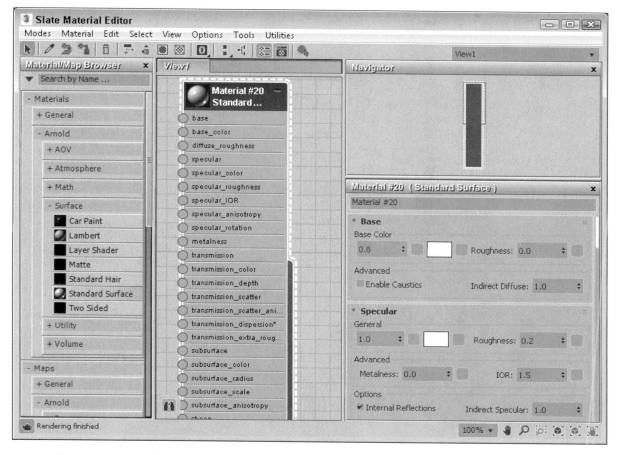

Tutorial: Adding materials to a cartoon chicken

For this tutorial, we'll revisit the cartoon chicken, but give him a much-needed makeover. I'm thinking something metallic.

To add Arnold materials to render a model, follow these steps:

1. Open the Metal cartoon chicken.max file from the Chap 30 directory in the downloaded content set.

   This file already has some materials applied, but we'll switch them to use the Arnold materials.

2. Open the Render Setup dialog box (F10), and switch the Renderer to the Arnold renderer.

3. Select Rendering→Material Editor or press the M key to open the Slate Material Editor. Within the Material/Map Browser, locate and double click on the Standard Surface material in the Materials→Arnold→Surface category.

4. Double click on the Standard Surface material node and change the Base Color property to a dark red, then apply this material to the chicken's comb and throat. Create another Standard Surface material and set its Base Color to white color, then set the Metalness value in the Specular Reflections section to 1.0 and apply it to the white sections of the chicken.

5. Create another Standard material with a gold color and apply it to the beak and feet of the chicken. For the last Standard material, set the transparency value (which is called Transmission in Arnold) to 0.85 and apply the material to the glass lens over the eyes.

6. Select the Rendering→Render menu command to render the metal chicken to the Rendered Frame Window.

Figure 30.11 shows the scene rendered with Arnold.

FIGURE 30.11

Metal materials show off the advantages of the Arnold renderer.

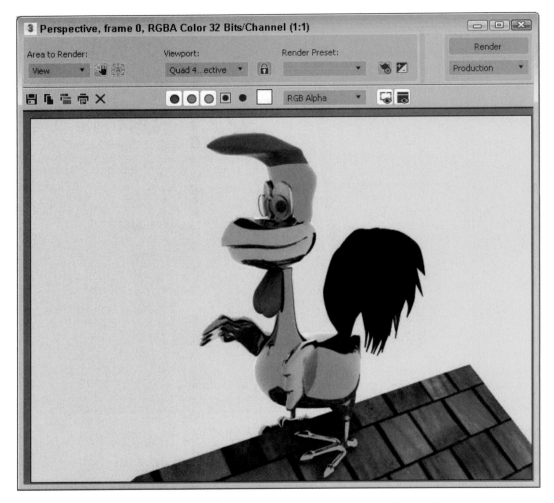

Using the Physical Material and the materials in the Autodesk Material Library

There are several sets of advanced physical-based materials available for the different renderer settings. The most recent category is the Physical Material, found in the General category of the Material/Map Browser. This material is available for the ART, Scanline, and Arnold renderers.

Caution
The Physical Material is not available for the Quicksilver renderer.

This material consists of a single material node that lets you choose the specific material from top of the Properties panel. Within the available presets are categories for Finishes, Non-Metallic, Transparent, and Metals. Each preset has its own set of parameters that you can tweak, as shown in Figure 30.12.

FIGURE 30.12

The Physical Material includes many presets.

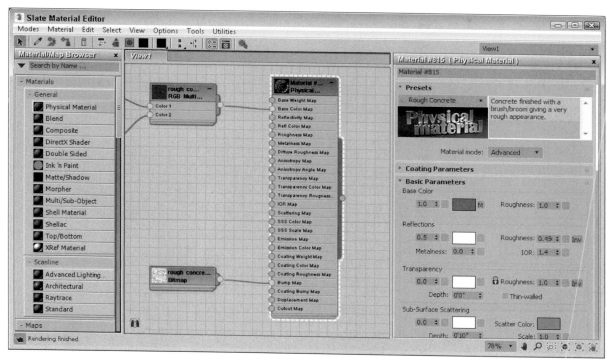

The Autodesk Material Library materials are a set of advanced materials designed for architects to add to buildings and surfaces found in architectural renderings. These materials are based on physical properties. For example, the Satin Varnished Wood template includes slightly blurred reflections and a high glossiness to give it a realistic look. This set of materials is available for the Quicksilver and ART renderers, but is not available for Arnold.

The materials contained in the Autodesk Material Library are categorized into groups and subgroups. Each material has several parameters allowing you to change the color, glossiness, reflectivity, and so on. The available parameters change based on the selected material.

Many of the Autodesk Material Library materials are already defined and available in its own rollout. Some of the available presets include Pearl Finish, Satin Varnished Wood, Glazed Ceramic, Glossy Plastic, Masonry, Leather, Frosted Glass, Translucent Plastic Film, Brushed Metal, and Patterned Copper, but many more exist.

For each template, several rollouts of parameters are available for defining the Diffuse, Reflection, Refraction, Translucency, Anisotropy, Fresnel Reflections, Bumps, and Self-Illumination properties along with a full selection of maps. Figure 30.13 shows a sampling of the available materials.

FIGURE 30.13

The materials in the Autodesk Material Library collection include a broad set of physical properties.

Using the Car Paint material

When cars are painted in the factory, the paint is composed of two layers that give the car its unique look. The undercoat is called the flake layer, and it shines through the top layer. Within the Physical Material is a preset called the Red Sports Car Paint. This material uses the Coating values to simulate the flake and top layers. Figure 30.14 shows a mech model painted with this material.

FIGURE 30.14

Cars rendered with the Red Sports Car Paint shader use multiple layers, just like real cars.

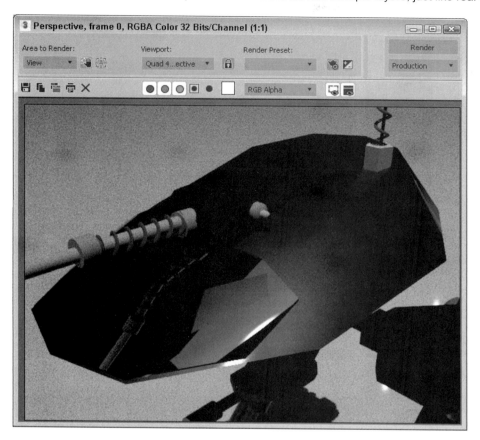

Using the Subsurface Scattering materials

Typically found within the various Arnold materials are settings for Subsurface Scattering (found under the Subsurface rollout). These properties are used to shade human skin and other membrane-like objects. Skin has an interesting property that allows it to become slightly translucent when it is placed in front of a strong light source. The ears in particular are a good example of this because they allow light to penetrate and highlight their features. Most organic objects including leaves, rubber, milk, and skin can benefit from these materials.

Using the Arnold Properties Modifier

When the Arnold renderer is selected, it makes the Arnold Properties Modifier available. This modifier can be applied to specific scene objects and lets you set which render attributes to enable. The available options include Visibility and whether the selected object will render reflections, transmissions, shadows, displacement, subdivision, motion blur, and sub-surface scattering.

By default, all these settings are controlled using the Arnold Renderer panel in the Render Setup dialog box, but the Arnold Properties modifier lets you enable or disable any of these options for a specific object. Sometimes an animation will include scene elements that you don't want to render such as the title of a scene shown as text that floats in and then out of the scene. These elements can be removed from the render calculations saving render time and preserving a certain look.

Tutorial: Using Arnold to create a disco ball

When using the Arnold renderer, you can see the results of the light beams as they are reflected around the room to help you determine the correct settings you need, but these beams themselves can be used to make a good disco ball effect.

To create a disco ball effect using Arnold, follow these steps:

1. Open the Disco ball.max file from the Chap 30 directory in the downloaded content set.

   This file includes a simple room where we'll add lights and a faceted sphere in the center.

2. Open the Render Setup dialog box (F10), and select Arnold in the Renderer drop-down list.

3. Select Create→Standard Primitives→Sphere, and click in the Top viewport to create a sphere, and then position it toward the top of the room. Set the Segments value to 64 and disable the Smooth option so you can see each separate face of the disco ball.

4. Press the M key to open the Slate Material Editor, double-click on the Materials→Arnold→Surface→Standard Surface material in the Material/Map Browser. Select the Standard Surface node and in the Properties panel, set the Metalness value in the Specular Reflections rollout to 1, then enable the Caustics option in the Basic Parameters rollout and apply the material to the sphere object.

5. Click on the Lights category in the Create Command Panel and choose the Arnold light, then click four times around the sphere in the top view to create four lights. Then move these lights in the Front view until they are positioned around the sphere object. Set each light to Point shape type with an Intensity value of 50, a Samples value of 2, and disable Shadows.

6. Select the Rendering→Render menu command to render the scene in the Rendered Frame Window.

Figure 30.15 shows the resulting disco scene with thousands of lights visible on the walls.

FIGURE 30.15

This disco ball simply reflects the light beams around the room.

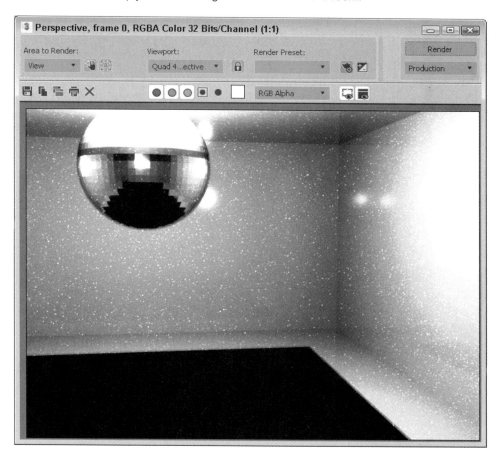

Arnold Render Settings

In addition to the light, material and object property settings, you can set many of the global properties for the Arnold Renderer in the Render Setup dialog box. Within the Arnold Renderer panel, shown in Figure 30.16, you can set the number of Samples and the Ray Depth for Diffuse, Specular, Transmission, SSS and Volume Indirect. You can also set the Ray Limit Total, which increases the total number of rays that are cast into the scene.

Caution

Be very careful in how you tweak the number of Samples or Rays because even slight changes can increase the render time dramatically and high values can easily crash your system. You can quick cancel a render request using the Escape key.

FIGURE 30.16

The Arnold Renderer panel lets you change the number of Samples for the scene.

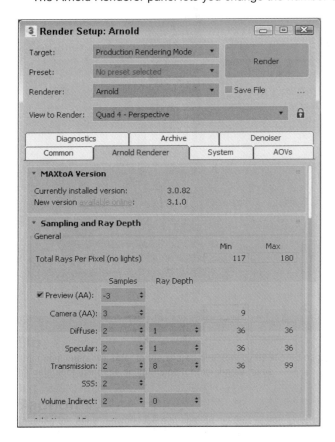

Summary

If you're looking for a rendering option that perfectly calculates reflections, refractions, and transparencies, the Arnold renderer is what you need. ART is another option for rendering when you aren't sure how to configure Arnold.

In this chapter, you accomplished the following:

* Learned to enable the ART and Arnold renderers

* Used the ART renderer to progressively render complex scenes

* Created Arnold lights and shadows

* Added the Arnold skydome

* Accessed Arnold materials

* Configuring the number of Samples

After you have some great looking renders, you'll need to know how to composite them into a scene. The next chapter covers some simple compositing techniques including 3ds Max's own Video Post.

Chapter 31

Compositing with Render Elements and the Video Post Interface

IN THIS CHAPTER

Learning about post-processing

Compositing with Photoshop, Premiere, and After Effects

Using render elements

Using the Video Post interface

Working with sequences

Understanding the various filter types

Adding and editing events

Specifying event ranges

After you've completed your scene and rendered it, you're finished, right? Well, not exactly. You still have post-production to complete: that's where you work with the final rendered images to add some additional effects. This phase of production typically takes place in another package, such as Photoshop or After Effects, and understanding how to interact with these packages can be a lifesaver when your client wants some last-minute changes (and they always do).

You can set the Autodesk® 3ds Max 2020® software to render any part in the rendering pipeline individually. These settings are called *render elements*. By rendering out just the Specular layer or just the shadow, you have more control over these elements in your compositor.

If you don't have access to a compositing package, or even if you do, 3ds Max includes a simple interface that can be used to add some post-production effects. This interface is the Video Post interface.

You can use the Video Post window to composite the final rendered image with several other images and filters. These filters let you add lens effects like glows and flares, and other effects like blur and fade, to the final output. The Video Post window provides a post-processing environment within the 3ds Max interface.

Note
Many of the post-processing effects, such as glows and blurs, also are available as render effects, but the Video Post window is capable of much more. Render effects are covered in Chapter 44, "Using Atmospheric and Render Effects."

Using External Compositing Packages

Before delving into the Video Post interface, let's take a quick look at some of the available compositing packages. Several of these packages have direct links into 3ds Max that can be used to give you a jump on the post-production process.

Compositing enables motion graphics, editing, and visual effects, which doesn't sound too different from what 3ds Max does, except for that funny word—*compositing*. If you think of the final rendered image produced using 3ds Max as just an image that needs to be combined with other elements such as text, logos, other images, or even a DVD menu, you're starting to see what post-production teams know. Compositing is the process of combining several different elements into a finished product. Positioning these elements can even be done in 3D by placing images behind or in front of other images or in time by working with animations.

Compositing with Photoshop

Perhaps the most common tool for compositing images is Photoshop. Photoshop can bring multiple images together in a single file and position them relative to one another. Working with layers makes applying simple filters and effects to the various element pieces easy.

Figure 31.1 shows Photoshop with several separate pieces, each on a different layer.

FIGURE 31.1

Photoshop is an important compositing tool for static images.

To composite images in Photoshop, you need to load all the separate images into Photoshop and then select the portions of the images that you want to combine. When saving image files in 3ds Max, be sure to include an alpha channel. You can see the alpha channel in the Rendered Frame window if you click the Display Alpha Channel button, as shown in Figure 31.2.

FIGURE 31.2

The Rendered Frame window can display an image's alpha channel.

In Photoshop, you can see an image's alpha channel if you select the Channels panel in the Layers palette. Selecting the alpha channel and using the Magic Wand tool makes selecting the rendered object easy. After it's selected, you can copy and paste the rendered image onto your background image as a new layer.

Note

Not all image file formats support an alpha channel. When rendering images to be composited, be sure to use an alpha channel format such as RLA, RPF, PNG, or TGA.

After all your images have been positioned on the background image, you can apply a filter, such as a Gaussian Blur, to smooth the edges between the composite images.

Video editing with Premiere Pro

Photoshop works with still images, but if you work with animations, Adobe has Premiere Pro to help with your video editing needs. The editing that Premiere Pro makes possible includes patching several animation clips together, adding sound, color-correcting the frames, and adding transitions between animation clips.

Within Premiere Pro, various animation clips can be imported (or dragged directly from Windows Explorer) into the Project panel. From here, the clips can be dropped onto the Timeline in the desired order. The Monitor panel shows the current animation or individual animation clips.

Sound clips can be dropped in the Timeline in the Audio track. The Title menu also can be used to add text to the animation. Another common activity in Premiere Pro is to add transition effects between clips. This is done by clicking the Effects tab in the Project panel, selecting a transition effect, dragging the effect to the Timeline, and dropping it between two animation clips.

When the entire sequence is completed, you can render it using the Sequence→Render Work Area menu command. The completed animation file then can be saved using the File→Export menu command.

Figure 31.3 shows the Premiere Pro interface with the animation clips loaded and positioned in the Timeline panel.

FIGURE 31.3

Premiere Pro can be used to combine several animation sequences.

Video compositing with After Effects

If you need to add a little more to your animations than just transitions, you should look into Adobe's After Effects. After Effects lets you composite 2D and 3D clips into a single image or animation. You can paint directly on the animation frames, add lights and cameras, and create visual effects such as Distort, Shatter, and Warp.

After Effects includes a library of resources much like those found in Premiere. These resources can be positioned on a Composition pane. Effects applied to the loaded animation clip are listed in the Effects panel along with all the effects settings.

After Effects includes many of the same tools used in Photoshop and Illustrator. These tools let you paint and select portions of the animation clip as if it were a still image, but the results can be added or removed over time.

Tutorial: Adding animation effects using After Effects

Some effects are much easier to add using a package like After Effects than to create in 3ds Max. A good example is adding a blurry look and the waves coming from a heat source to a melting snowman animation.

To add video effects using After Effects, follow these steps:

1. Open After Effects, and drag the Melting snowman.avi file from this chapter's folder in the downloadable content to the Project panel.

2. Select Composition→New Composition, select the NTSC DV option from the Preset list, and click OK.

3. Drag the Melting snowman.avi file from the Project panel, and drop it on the Composition pane.

4. With the animation selected in the Composition pane, select Effect→Distort→Wave Warp. The Wave Warp effect appears in a panel. Set the Wave Height to **4**, the Wave Width to **30**, the Direction to **Vertical**, and the Wave Speed to **1**. This adds a heat wave effect to the entire animation.

5. Select Effect→Blur & Sharpen→Gaussian Blur, and set the Blurriness value to **3.0**.

6. In the Timeline panel, drag the Work Area End icon so that it coincides with the end of the animation.

7. Select Composition→Make Movie. In the Render Queue dialog box that opens, click the Render button to render the animation with its effects.

Figure 31.4 shows the After Effects interface with the animation clip loaded.

FIGURE 31.4

After Effects can add special effects to an animation sequence.

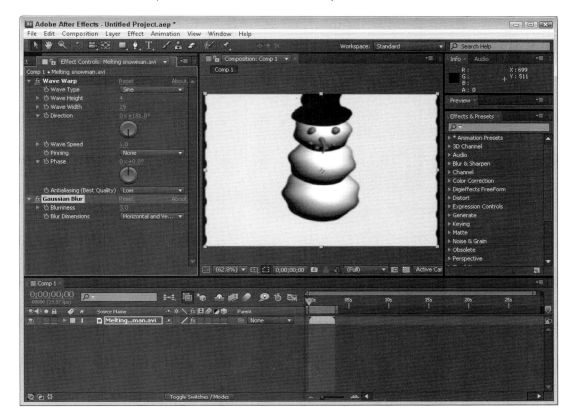

Using Render Elements

If your production group includes a strong post-processing team that does compositing, there may be times when you just want to render certain elements of the scene, such as the alpha information or a specific atmospheric effect. Applying individual elements to a composite image gives you better control over the elements. For example, you can reposition or lighten a shadow without having to re-render the entire scene or animation sequence.

Using the Render Elements panel of the Render Setup dialog box, shown in Figure 31.5, you can render a single effect and save it as an image. This figure has all the available Render Elements added to it.

FIGURE 31.5

You can use the Render Elements rollout to render specific effects.

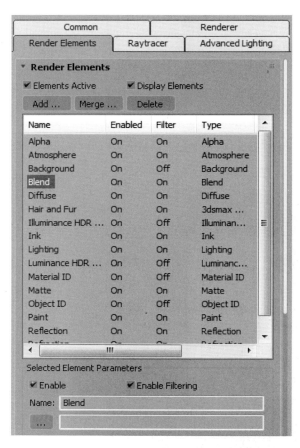

You can select and render several render elements at the same time. The available render elements include Alpha, Atmosphere, Background, Blend, Diffuse, Hair and Fur, Illuminance HDR Data, Ink, Lighting, Luminance HDR Data, Material ID, Matte, Object ID, Paint, Reflection, Refraction, Self-Illumination, Shadow, Specular, Velocity, and Z Depth. You also can use the Shift and Ctrl keys to select multiple elements from the list at once.

If the Arnold rendering engine is enabled, many more Render Elements are also available, including all the parameters that are included as part of the Arch&Design materials, along with an Arnold Shader Element and the Arnold Labeled Element.

The Add button opens the Render Elements dialog box, where you can select the elements to include. The Merge button lets you merge the elements from another 3ds Max scene, and the Delete button lets you delete elements from the list. To be included in the rendered image, the Elements Active option must be checked. The Display Elements option causes the results to be rendered separately and displayed in the Rendered Frame Window.

The Enable check box can turn off individual elements; Enable Filtering enables the anti-aliasing filtering as specified in the Scanline Renderer rollout in the Renderer panel. A separate Rendered Frame Window is opened for each render element that is enabled when the rendering starts.

Each element in the list can be given a unique list name, and clicking the Browse button opens a file dialog box where you can give the rendered element a filename. 3ds Max automatically appends an underscore and the name of the element on the end of the filename. For example, if you name the file **myScene** and select to render the Alpha element, the filename for this element is **myScene_alpha**.

When you select the Blend and Z Depth render elements, an additional rollout of parameters appears. You can use the Blend render element to combine several separate elements together. The Blend Element Parameters rollout includes check boxes for each render element type. The Z Depth render element includes parameters for setting Min and Max depth values.

Figure 31.6 shows the resulting image in the Rendered Frame Window for the Alpha render element.

FIGURE 31.6

The Alpha render element shown in the Rendered Frame Window

The Render Elements rollout also can output files that Autodesk's Composite product can use. These files have the .cws extension. Composite is a compositing product that can work with individual elements to increase the highlights, change color hues, darken and blur shadows, and do many other things without having to re-render the scene.

Completing Post-Production with the Video Post Interface

Post-production is the work that comes after the scene is rendered. It is the time when you add effects, such as glows and highlights, as well as add transitional effects to an animation. For example, if you want to include a logo in the lower-right corner of your animation, you can create and render the logo and composite several rendered images into one during post-production.

Video Post interface is another post-processing interface within 3ds Max that you can use to combine the current scene with different images, effects, and image-processing filters. Compositing is the process of combining several different images into a single image. Each element of the composite is included as a separate event. These events are lined up in a queue and processed in the order in which they appear in the queue. The queue also can include looping events.

The Video Post interface, like the Render Setup dialog box, provides another way to produce final output. You can think of the Video Post process as an artistic assembly line. As the image moves down the line, each item in the queue adds an image, drops a rendered image on the stack, or applies a filter effect. This process continues until the final output event is reached.

The Video Post interface, shown in Figure 31.7, includes a toolbar, a pane of events and ranges, and a status bar. You can open it by choosing Rendering→Video Post.

FIGURE 31.7

The Video Post interface lets you composite images with your final rendering.

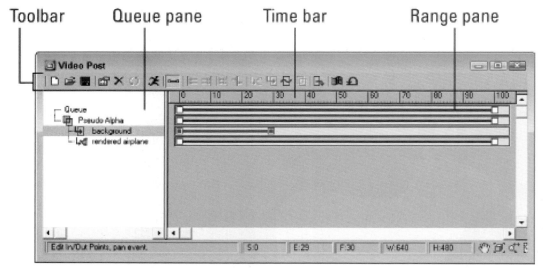

Within the Video Post interface, each event is displayed as a track in the Queue pane to the left. To the right is the Range pane, where the range for each track is displayed as lines with square boxes at each end. You can edit these ranges by dragging the squares on either end. The time bar, above the Range pane, displays the frames for the current sequence, and the status bar at the bottom of the interface includes information and view buttons.

The Video Post toolbar

At the top of the Video Post interface is a toolbar with several buttons for managing the Video Post features. Table 31.1 shows and explains these buttons.

TABLE 31.1 Video Post Toolbar Buttons

Toolbar Button	Name	Description
	New Sequence	Creates a new sequence
	Open Sequence	Opens an existing sequence
	Save Sequence	Saves the current sequence
	Edit Current Event	Opens the Edit Current Event dialog box where you can edit events
	Delete Current Event	Removes the current event from the sequence
	Swap Events	Changes the position in the queue of two selected events
	Execute Sequence	Runs the current sequence
	Edit Range Bar	Enables you to edit the event ranges
	Align Selected Left	Aligns the left ranges of the selected events
	Align Selected Right	Aligns the right ranges of the selected events
	Make Selected Same Size	Makes the ranges for the selected events the same size
	Abut Selected	Places event ranges end-to-end
	Add Scene Event	Adds a rendered scene to the queue
	Add Image Input Event	Adds an image to the queue
	Add Image Filter Event	Adds an image filter to the queue
	Add Image Layer Event	Adds a compositing plug-in to the queue when two events are selected
	Add Image Output Event	Sends the final composited image to a file or device
	Add External Event	Adds an external image-processing event to the queue
	Add Loop Event	Causes other events to loop

The Video Post Queue and Range panes

Below the toolbar are the Video Post Queue and Range panes. The Queue pane is on the left; it lists all the events to be included in the post-processing sequence in the order in which they are processed. You can rearrange the order of the events by dragging an event in the queue to its new location.

You can select multiple events by holding down the Ctrl key and clicking the event names, or you can select one event, hold down the Shift key, and click another event to select all events between the two.

Each event has a corresponding range that appears in the Range pane to the right. Each range is shown as a line with a square on each end. The left square marks the first frame of the event, and the right square marks the last frame of the event. You can expand or contract these ranges by dragging the square on either end of the range line.

If you click the line between two squares, you can drag the entire range. If you drag a range beyond the given number of frames, then additional frames are added.

The time bar is at the top of the Range pane. This bar shows the number of total frames included in the animation. You also can slide the time bar up or down to move it closer to a specific track by dragging it.

The Video Post status bar

The status bar includes a prompt line, several value fields, and some navigation buttons. The fields to the right of the prompt line include Start, End, and Total Frames of the selected track, and the Width and Height of the image. The navigation buttons include (in order from left to right) Pan, Zoom Extents, Zoom Time, and Zoom Region.

Working with Sequences

All the events that are added to the Queue pane make up a *sequence*. You can save these sequences and open them at a later time. The Execute Sequence button (Ctrl+R), found on the toolbar, starts the compositing process.

Note

The keyboard shortcuts for the Video Post interface work only if the Keyboard Shortcut Override Toggle on the main toolbar is enabled.

To save a sequence, click the Save Sequence button on the toolbar to open the Save Sequence dialog box, where you can save the queue sequence. Sequences are saved along with the 3ds Max file when the scene is saved, but they also can be saved independently of the scene. By default, these files are saved with the .vpx extension in the vpost directory.

Note

Saving a sequence as a VPX file maintains the elements of the queue, but it resets all parameter settings. Saving the file as a 3ds Max file maintains the queue order along with the parameter settings.

You can open saved sequences using the Open Sequence button on the toolbar. When a saved sequence is opened, all the current events are deleted. Clicking the New Sequence button also deletes any current events.

The Execute Sequence toolbar button opens the Execute Video Post interface, shown in Figure 31.8. The controls in this dialog box work exactly the way those in the Render Scene dialog box work.

Note

The time and resolution settings in the Execute Video Post dialog box are unique from those in the Render Setup dialog box.

FIGURE 31.8

The Execute Video Post interface includes the controls for producing the queue output.

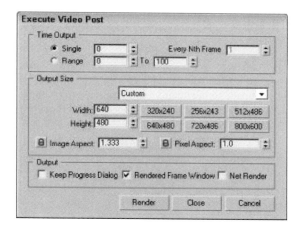

The Time Output section enables you to specify which frames to render, and the Output Size section lets you specify the size of the output. The Custom selection lets you enter Width and Height values, or you can use one of the presets in the drop-down list or one of the preset resolution buttons. This dialog box also includes controls for entering the Image Aspect and Pixel Aspect ratios.

The Output options let you select to keep the Progress dialog box open, to render to the Rendered Frame Window, and/or to use network rendering. When you're ready to render the queue, click the Render button.

Adding and Editing Events

The seven event types that you can add to the queue are Image Input, Scene, Image Filter, Image Layer, Loop, External, and Image Output. If no events are selected, then adding an event positions the event at the bottom of the list. If an event is selected, the added event becomes a subevent under the selected event.

Every event dialog box, such as the Add Image Input Event dialog box shown in Figure 31.9, includes a Label field where you can name the event. This name shows up in the queue window and is used to identify the event.

FIGURE 31.9

The Add Image Input Event dialog box lets you load an image to add to the queue.

Each event dialog box includes a Video Post Parameters section. This section contains VP Start Time and VP End Time values for defining precisely the length of the Video Post range. It also includes an Enabled option for enabling or disabling an event. Disabled events are grayed out in the queue.

To edit an event, you simply need to double-click its name in the Queue pane (or select it and press Ctrl+E) to open an Edit Event dialog box.

Adding an image input event

The Add Image Input Event dialog box lets you add a simple image to the queue. For example, you can add a background image using this dialog box rather than the Environment dialog box. To open the Add Image Input Event dialog box, click the Add Image Input Event button (Ctrl+I) on the toolbar.

Tip

If you don't name the image event, the filename appears in the Queue pane as the name for the event.

The Files button in this dialog box opens the Select Image File for Video Post Input dialog box, where you can locate an image file to load from the hard disk or network. The Devices button lets you access an external device such as a video recorder. The Options button becomes enabled when you load an image. The Cache option causes the image to be loaded into memory, which can speed up the Video Post process by not requiring the image to be loaded for every frame.

The Image Driver section of the Add Image Input Event dialog box lets you specify the settings for the image driver, such as the compression settings for an AVI file. Clicking the Setup button opens a dialog box of options available for the selected format, but note that the Setup button is not active for all formats.

Click the Options button open the Image Input Options dialog box, shown in Figure 31.10. This dialog box lets you set the alignment, size, and frames where the image appears. The Alignment section of the Image Input Options dialog box includes nine presets for aligning the image. Preset options include top-left corner, top centered, top-right corner, left centered, centered, right centered, bottom-left corner, bottom centered, and bottom-right corner. You also can use the Coordinates option to specify in pixels the image's upper-left corner.

FIGURE 31.10

The Image Input Options dialog box lets you align and set the size of the image.

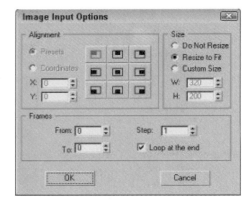

In the Size section of this dialog box, you can control the size of the image, using the Do Not Resize, Resize to Fit, or Custom Size options. Note that you lose the Alignment options if the Resize to Fit option is selected. The Custom Size option lets you enter Width and Height values.

The Frames section applies only to animation files. The From and To values define which frames of the animation to play. The Step value lets you play every nth frame as specified. The Loop at the End value causes the animation to loop back to the beginning when finished.

Adding scene events

A scene event is the rendered scene that you've built in 3ds Max. When you click the Add Scene Event button on the toolbar, the Add Scene Event dialog box shown in Figure 31.11 opens. This dialog box lets you specify the scene ranges and define the render options.

FIGURE 31.11

The Add Scene Event dialog box lets you specify which viewport to use to render your scene.

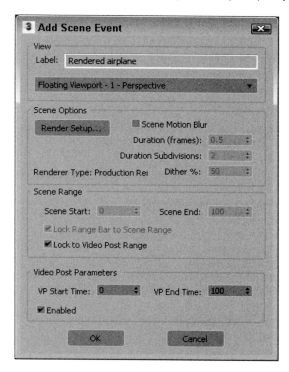

Below the Label field where you can name the event is a drop-down list that lets you select which viewport to use to render your scene. The active viewport is selected by default. The Render Setup button opens the Render Setup panel, where the Render button has been replaced with OK and Cancel buttons at the bottom of the dialog box because the rendering is initiated with the Execute Sequence button.

Caution

When selecting a viewport to render, make sure not to select a floating viewport if it is not open.

For more information about the Render Setup panel, see Chapter 27, "Rendering a Scene and Enabling Quicksilver."

The Scene Options section of the Add Scene Event dialog box also includes an option for enabling Scene Motion Blur. This motion blur type is different from the object motion blur that is set in the Object Properties dialog box. Scene motion blur is applied to the entire image and is useful for blurring objects that are moving fast. The Duration (frames) value sets how long the blur effect is computed per frame. The Duration Subdivisions value is how many computations are done for each duration. The Dither % value sets the amount of dithering to use for blurred sections.

In the Scene Range section, the Scene Start and Scene End values let you define the range for the rendered scene. The Lock Range Bar to Scene Range option maintains the range length as defined in the Time Slider, though you can still reposition the start of the rendered scene. The Lock to Video Post Range option sets the range equal to the Video Post range.

Adding image filter events

The Add Image Filter Event button (Ctrl+F) on the toolbar opens the Add Image Filter Event dialog box, shown in Figure 31.12, where you can select from many filter types. The available filters are included in a drop-down list under the Label field.

Below the filter drop-down list are two buttons: About and Setup. The About button gives some details about the creator of the filter. The Setup button opens a separate dialog box that controls the filter. The dialog box that appears depends on the type of filter that you selected in the drop-down list.

FIGURE 31.12

The Add Image Filter Event dialog box lets you select from many filter types.

Several, but not all, filters require a mask such as the Image Alpha filter. To open a bitmap image to use as the mask, click the Files button in the Mask section and select the file in the Select Mask Image dialog box that opens. A drop-down list lets you select the channel to use. Possible channels include Red, Green, Blue, Alpha, Luminance, Z Buffer, Material Effects, and Object. The mask can be Enabled or Inverted. The Options button opens the Image Input Options dialog box for aligning and sizing the mask.

Contrast filter

You use the Contrast filter to adjust the brightness and contrast. Selecting this filter and clicking the Setup button opens the Image Contrast Control dialog box. This simple dialog box includes values for Contrast and Brightness. Both values can be set from 0 to 1. The Absolute option computes the center gray value based on the highest color value. The Derived option uses an average value of the components of all three colors (red, green, and blue).

Fade filter

You can use the Fade filter to fade out the image over time. You can select it from the drop-down list. Clicking the Setup button opens the Fade Image Control dialog box where you select to fade either In or Out. The fade takes place over the length of the range set in the Range pane.

Image Alpha filter

The Image Alpha filter sets the alpha channel as specified by the mask. This filter doesn't have a setup dialog box.

Lens Effects filters

There are four Lens Effects filters available including Flare, Focus, Glow and Highlight. Each of these filters opens the Lens Effects dialog box with the Setup button.

Several Lens Effects filters are also included in the drop-down list. These filters use an advanced dialog box with many options, which is covered in Chapter 45, "Adding Volume Lights and Lens Effects."

Negative filter

The Negative filter inverts all the colors, as in the negative of a photograph. The Negative Filter dialog box includes a simple Blend value.

Pseudo Alpha filter

The Pseudo Alpha filter sets the alpha channel based on the pixel located in the upper-left corner of the image. This filter can make an unrendered background transparent. When this filter is selected, the Setup button is disabled because it doesn't have a setup dialog box.

Simple Mix Compositor filter

The Simple Mix Compositor filter combines two images together at different ratios. The Mix-In Ratio % is set in the Setup dialog box and defines how much of each picture is used.

Simple Wipe filter

The Simple Wipe filter removes the image by replacing it with a black background. The length of the wipe is determined by the event's time range. The Simple Wipe Control dialog box, shown in Figure 31.13, lets you wipe from the left to the right or from the right to the left. You also can set the mode to Push, which displays the image, or to Pop, which erases it.

FIGURE 31.13

The Simple Wipe Control dialog box lets you select which direction to wipe the image.

Starfield filter

The Starfield filter creates a starfield image. By using a camera, you can motion blur the stars. The Stars Control dialog box, shown in Figure 31.14, includes a Source Camera drop-down list that you can use to select a camera.

FIGURE 31.14

The Stars Control dialog box lets you load a custom database of stars.

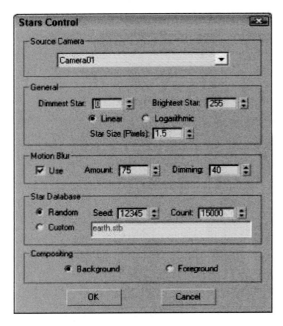

The General section sets the brightness and size of the stars. You can specify brightness values for the Dimmest Star and the Brightest Star. The Linear and Logarithmic options use two different algorithms to compute the brightness values of the stars as a function of distance. The Star Size value sets the size of the stars in pixels. Size values can range from 0.001 to 100.

The Motion Blur settings let you enable motion blurring, set the blur Amount, and specify a Dimming value.

The Star Database section includes settings for defining how the stars are to appear. The Random option displays stars based on the Count value, and the random Seed determines the randomness of the star's positions. The Custom option reads a star database specified in the Database field.

Note
3ds Max includes a starfield database named earth.stb that includes the stars as seen from Earth.

You also can specify whether the stars are composited in the background or foreground.

Tutorial: Creating space backdrops

Space backgrounds are popular backdrops, and 3ds Max includes a special Video Post filter for creating starfield backgrounds. You would typically want to use the Video Post interface to render the starfield along with any animation that you've created, but in this tutorial, you render a starfield for a single planet that you've created and outfitted with a planet material.

To create a starfield background, follow these steps:

1. Open the Planet with starfield background.max file from the Chap 31 directory in the downloaded content set.

    This file includes a simple space scene with a camera because the Starfield filter requires a camera.

2. Choose Rendering→Video Post to open the Video Post interface. A Scene Event must be added to the queue in order for the render job to be executed. Click the Add Scene Event button, type **planet scene** in the Label field, Select Camera01 as the Source Camera, and click OK.

    This adds the event to the Queue pane.

3. Click the Add Image Filter Event button (or press Ctrl+F) to open the Add Image Filter Event dialog box, and in the Label field type the name **starfield bg**. Select Starfield from the drop-down list, and click the Setup button to open the Stars Control dialog box, set the Star Size to **3.0** and the Count to **150,000**, and click OK.

4. Click the Execute Sequence button (or press Ctrl+R), select the Single output time option and an Output Size, and click the Render button.

Figure 31.15 shows the resulting space scene.

FIGURE 31.15

A space scene with a background, compliments of the Video Post interface

Adding image layer events

In addition to the standard filters that can be applied to a single image, several more filters, called *layer events*, can be applied to two or more images or rendered scenes. The Add Layer Event button (Ctrl+L) is available on the toolbar only when two image events are selected in the Queue pane. The first image (which is the selected image highest in the queue) becomes the source image, and the second image is the compositor. Both image events become subevents under the layer event.

Note

If the layer event is deleted, the two subevent images remain.

The dialog box for the Add Image Layer Event is the same as the Add Image Filter Event dialog box shown earlier, except that the drop-down list includes filters that work with two images.

Simple Wipe compositor

The Simple Wipe compositor is similar to the Simple Wipe filter, except that it slides the image in or out instead of erasing it. Its setup dialog box looks just like that of the Simple Wipe dialog box.

Other layer filters

The remaining layer filters include simple methods for compositing images and some simple transitions. None of these other filters has a Setup dialog box.

You can use the Alpha compositor to composite two images, using the alpha channel of the foreground image. The Cross Fade Transition compositor fades one image out as it fades another image in. You can use the Pseudo Alpha compositor to combine two images if one doesn't have an alpha channel. This compositor uses the upper-left pixel to designate the transparent color for the image. The Simple Additive Compositor combines two images based on the intensity of the second image. The Simple Mix Compositor combines the two images based on a defined ratio.

Adding external events

The Add External Event button on the toolbar lets you use an external image-processing program to edit the image. This button is available only when an image event is selected, and the image event becomes a sub-event under the external event. The Add External Event dialog box, shown in Figure 31.16, includes a Browse button for locating the external program. It also includes a Command Line Options field for entering text commands for the external program. Many external programs use the clipboard to do their processing, so the Write image to clipboard and Read image from clipboard options make this possible.

FIGURE 31.16

The Add External Event dialog box lets you access an external program to edit images.

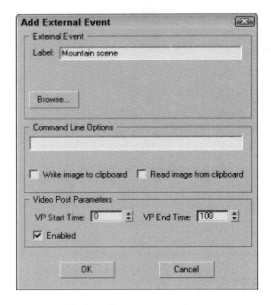

Using loop events

The Add Loop Event button is enabled on the Video Post toolbar when any single event is selected. This button enables an event to be repeated a specified number of times or throughout the Video Post range. The Add Loop Event dialog box, shown in Figure 31.17, includes a value field for the Number of Times to repeat the event, along with Loop and Ping Pong options. The Loop option repeats from beginning to end until the Number of Times value is reached. The Ping Pong option alternates playing the event forward and in reverse. You can name Loop events using the Label field.

FIGURE 31.17

The Add Loop Event dialog box lets you play an event numerous times.

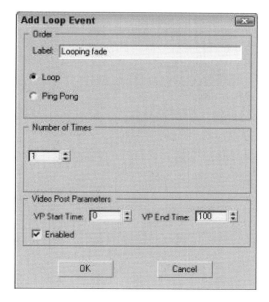

Adding an image output event

If you've added all the events you need and configured them correctly, and you click the Execute Sequence button and nothing happens, chances are good that you've forgotten to add an Image Output event. This event creates the file that all the events use to output to and should appear last in the queue.

The Add Image Output Event dialog box (Ctrl+O) looks similar to the Add Image Input Event dialog box shown earlier. The output can be saved to a file or to a device using any of the standard file types.

Note

If you don't give the output event a name, the filename automatically becomes the event name.

Working with Ranges

The Range pane in the Video Post interface is found to the right of the Queue pane. It displays the ranges for each event. These turn red when selected. The beginning and end points of the range are marked with squares. You can move these points by dragging the squares. This moves the beginning and end points for all selected events.

Note

Before you can move the ranges or drag the end points of a range, you need to select the Edit Range Bar button from the toolbar. The button is highlighted blue when active.

When two or more events are selected, several additional buttons on the toolbar become enabled, including Swap Events, Align Selected Left, Align Selected Right, Make Selected Same Size, and Abut Selected. (These buttons were shown earlier in Table 31.1.)

The Swap Events button is enabled only if two events are selected. When clicked, it changes the position of the two events. Because the order of the events is important, this can alter the final output.

The Align Selected Left and Align Selected Right buttons move the beginning or end points of every selected track until they line up with the first or last points of the top selected event.

The Make Selected Same Size button resizes any bottom events to be the same size as the top selected event. The Abut Selected button moves each selected event under the top event until its first point lines up with the last point of the selected event above it.

Figure 31.18 shows four image events that have been placed end-to-end using the Abut Selected button. Notice that the queue range spans the entire distance.

FIGURE 31.18

You can use the Abut Selected button to position several events end-to-end.

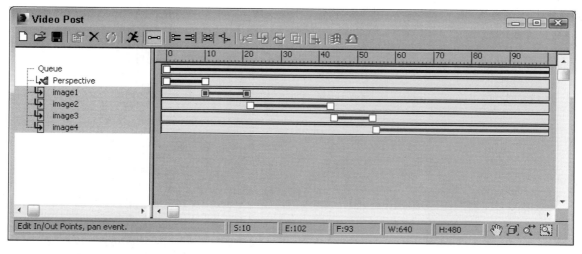

Adding backgrounds and filters using Video Post

As an example of the Video Post interface in action, you'll composite a background image of a waterfall with a rendered scene of an airplane model. Then you'll add some filter effects.

To composite an image with the Video Post interface, follow these steps:

1. Open the Airplane over waterfall.max file from the Chap 31 directory in the downloaded content set.

   This file includes an airplane model. The directory also includes an image called waterfall.tif that is used later.

2. Open the Video Post interface by choosing Rendering→Video Post.

3. Add a background image to the queue by clicking the Add Image Input Event button (or by pressing Ctrl+I). Click the Files button. Locate the waterfall.tif image from the Chap 31 directory in the downloaded content set, and click Open. Then click OK again to exit the Add Image Input Event dialog box.

4. Next, add the rendered image by clicking the Add Scene Event button and selecting the Perspective viewport. Name the event **rendered airplane**. Click the Render Setup button to open the Render Setup dialog box, and select the Renderer panel. Disable the Antialiasing option in the Scanline Renderer rollout in the Renderer panel, and click OK. Click OK again to exit the Add Scene Event dialog box.

5. Select both the background (waterfall.tif) and the rendered airplane (Perspective) events, and click the Add Image Layer Event button (or press Ctrl+L). Select the Pseudo Alpha option, and click OK.

   This composites the background image and the rendered image together by removing all the green background from the rendered scene.

6. To run the processing, click the Execute Sequence button on the toolbar (or press Ctrl+R) to open the Execute Video Post interface, select the Single Time Output range option, click the 640 x 480 size button, and click Render.

Figure 31.19 shows the final composited image.

FIGURE 31.19

The airplane in this image is rendered, and the background is composited.

Summary

Post-production is an important, often overlooked, part of the production pipeline. Using compositing packages, as simple as Photoshop and Premiere Pro or as advanced as After Effects, enables you to make necessary edits after rendering.

The render elements in 3ds Max enable you to pick apart the rendering details of your scene. Rendering using render elements allows you to have more control over individual scene elements in the compositing tool.

Using the Video Post interface, you can composite several images, filters, and effects together. All these compositing elements are listed as events in a queue. Using this interface provides, along with the Render Setup dialog box, more ways to create output. In this chapter, you learned about the following:

* The post-production process
* How Photoshop can be used to composite images
* How Premiere and After Effects can be used to composite animations
* How to use render elements
* The Video Post interface
* How to work with sequences
* The various filter types
* How to add and edit events and manipulate their ranges

This concludes the Rendering part of the book. The next part, "Animating Objects and Scenes," finally gets to the animation portion of the book. We'll begin with a chapter on animation basics and understanding keyframes.

Part VII
Animating Objects and Scenes

IN THIS PART

Chapter 32

Understanding Animation and Keyframes

The Autodesk® 3ds Max® 2020 software can be used to create some really amazing images, but I bet more of you go to the movies than go to see images in a museum. The difference is in seeing moving images versus static images.

In this chapter, I start discussing what is probably one of the main reasons you decided to learn 3ds Max in the first place—animation. 3ds Max includes many different features to create animations. This chapter covers the easiest and most basic of these features—keyframe animation.

Along the way, you'll examine all the various controls that are used to create, edit, and control animation keys, including the Time Controls, the Track Bar, and the Motion panel. Keyframes can be used to animate object transformations, but they also can be used to animate other aspects of the scene, such as materials. If you get finished with this chapter in time, you may have time to watch a movie.

Using the Time Controls

Before jumping into animation, you need to understand the controls that make it possible. These controls collectively are called the Time Controls and can be found on the lower interface bar to the left of the key controls and the Viewport Navigation Controls. The Time Controls also include the Time Slider found directly under the viewports.

The Time Slider provides an easy way to move through the frames of an animation. To do this, just drag the Time Slider button in either direction. The Time Slider button is labeled with the current frame number and the total number of frames. The arrow buttons on either side of this button work the same as the Previous and Next Frame (Key) buttons.

The Time Control buttons include buttons to jump to the Start or End of the animation, or to step forward or back by a single frame. You also can jump to an exact frame by entering the frame number in the frame number field. The Time Controls are presented in Table 32.1.

TABLE 32.1 Time Controls

Toolbar Button	Name	Description
	Go to Start	Sets the time to frame 0.
	Previous Frame/Key	Decreases the time by one frame or selects the previous key.
	Play Animation, Play Selected	Cycles through the frames; this button becomes a Stop Animation button when an animation is playing.
	Next Frame/Key	Advances the time by one frame or selects the next key.
	Go to End	Sets the time to the final frame.
	Key Mode Toggle	Toggles between key and frame modes; with Key Mode on, the icon turns light blue and the Previous Frame and Next Frame buttons change to Previous Key and Next Key.
0	Current Frame field	Indicates the current frame; a frame number can be typed in this field for more exact control than the Time Slider.
	Time Configuration	Opens the Time Configuration dialog box where settings like Frame Rate, Time Display, and Animation Length can be set.

The default scene starts with 100 frames, but this is seldom what you actually need. You can change the number of frames at any time by clicking the Time Configuration button, which is to the right of the frame number field. Clicking this button opens the Time Configuration dialog box, shown in Figure 32.1. You also can access this dialog box by right-clicking any of the Time Control buttons.

FIGURE 32.1

The Time Configuration dialog box lets you set the number of frames to include in a scene.

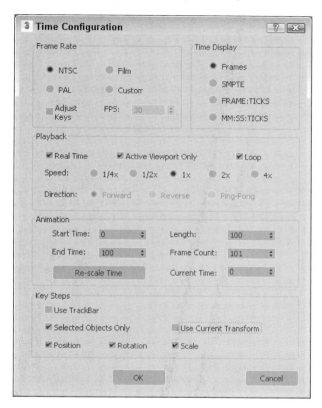

Setting frame rate

Within this dialog box, you can set several options, including the Frame Rate. *Frame rate* provides the connection between the number of frames and time. It is measured in frames per second. The options include standard frame rates such as NTSC (National Television Standards Committee, around 30 frames per second), Film (around 24 frames per second), and PAL (Phase Alternate Line, used by European countries, around 25 frames per second), or you can select Custom and enter your own frame rate. The Adjust Keys options scales keys to whole frames.

The Time Display section lets you set how time is displayed on the Time Slider. The options include Frames, SMPTE (Society of Motion Picture Technical Engineers), Frame:Ticks, or MM:SS:Ticks (Minutes and Seconds). *SMPTE* is a standard time measurement used in video and television. A *Tick* is 1/4800 of a second.

Setting speed and direction

The Playback section sets options for how the animation sequence is played back. The Real Time option skips frames to maintain the specified frame rate. The Active Viewport Only option causes the animation to play only in a single viewport, which speeds up the animation. The Loop option repeats the animation over and over. The Direction options are available only if the Real Time option is disabled. If the Loop option is set, you can specify the Direction as Forward, Reverse, or Ping-Pong (which repeats, playing forward and then reverse). The Speed setting can be 1/4, 1/2, 1, 2, or 4 times normal.

The Time Configuration dialog box also lets you specify the Start Time, End Time, Length, Frame Count, and Current Time values. These values are all interrelated, so setting the Length and the Start Time, for example, automatically changes the End Time. These values can be changed at any time without destroying any keys. For example, if you have an animation of 500 frames and you set the Start and End Time to 30 and

50, the Time Slider controls only those 21 frames. Keys before or after this time are still available and can be accessed by resetting the Start and End Time values to 0 and 500.

The Re-scale Time button fits all the keys into the active time segment by stretching or shrinking the number of frames between keys. You can use this feature to resize the animation to the number of frames defined by Start and End Time values.

The Key Steps group lets you set which key objects are navigated using key mode. If you select Use Track Bar, key mode moves through only the keys on the Track Bar. If you select the Selected Objects Only option, key mode jumps only to the keys for the currently selected object. You also can filter to move between Position, Rotation, and Scale keys. The Use Current Transform option locates only those keys that are the same as the current selected transform button.

Using Time Tags

To the right of the Prompt Line is a field marked Add Time Tag. Clicking this field pops up a menu with options to Add or Edit a Time Tag. Time Tags can be set for each frame in the scene. Once set, the Time Tags are visible in the Time Tag field whenever that time is selected. These are useful for identifying key frames.

Working with Keys

It isn't just a coincidence that the largest button in the entire 3ds Max interface has a key on it. Creating and working with keys is how animations are accomplished. Keys define a particular state of an object at a particular time. Animations are created as the object moves or changes between two different key states. Complex animations can be generated with only a handful of keys.

You can create keys in numerous ways, but the easiest is with the Key Controls found on the lower interface bar. These controls are located to the right of the Time Controls. Table 32.2 displays and explains all these controls. Closely related to the Key Controls is the Track Bar, which is located under the Time Slider.

TABLE 32.2 Key Controls

Toolbar Button	Name	Description
	Set Keys (K)	Creates animation keys in Set Keys mode.
Auto Key	Toggle Auto Key Mode (N)	Sets keys automatically for the selected object when enabled.
Set Key	Toggle Set Key Mode	Sets keys as specified by the key filters for the selected object when enabled.
Selected	Selection Set drop-down list	Specifies a selection set to use for the given keys.
	Default In/Out Tangents for New Keys	Assigns the default tangents that are used on all new keys.
Key Filters...	Open Filters Dialog box	Contains pop-up options for the filtering keys.

3ds Max includes two animation modes: Auto Key (N) and Set Key (K). You can select either of these modes by clicking the respective buttons at the bottom of the interface. When active, the button turns bright red, and the track bar and the border around the active viewport also turns red to remind you that you are in animate mode. Red also appears around a spinner for any animated parameters.

Auto Key mode

With the Auto Key button enabled, every transformation or parameter change creates a key that defines where and how an object should look at that specific frame.

To create a key, drag the Time Slider to a frame where you want to create a key and then move the selected object or change the parameter, and a key is automatically created. When the first key is created, 3ds Max automatically goes back and creates a key for frame 0 that holds the object's original position or parameter. Upon setting the key, 3ds Max then interpolates all the positions and changes between the keys. The keys are displayed in the Track Bar.

Each frame can hold several different keys, but only one for each type of transform and each parameter. For example, if you move, rotate, scale, and change the Radius parameter for a sphere object with the Auto Key mode enabled, separate keys are created for position, rotation, scaling, and a parameter change.

Caution

Be sure to turn Auto Key mode off when you are finished creating keys; if you leave it on, it automatically creates keys that you didn't mean to set.

Set Key mode

The Set Key Mode button offers more control over key creation and sets keys only when you click the Set Keys button (K). It also creates keys only for the key types enabled in the Set Key Filters dialog box. You can open the Set Key Filters dialog box, shown in Figure 32.2, by clicking the Key Filters button. Available key types include All, Position, Rotation, Scale, IK Parameters, Object Parameters, Custom Attributes, Modifiers, Materials, and Other (which allows keys to be set for manipulator values).

FIGURE 32.2

Use the Set Key Filters dialog box to specify the types of keys to create.

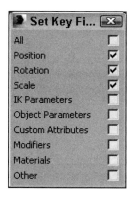

Tutorial: Rotating a windmill's blades

The best way to learn is to practice, and there's no better time to practice than now. For this quick example, you animate a set of blades on a windmill.

To animate a set of windmill blades rotating, follow these steps:

1. Open the Rotating windmill blades.max file from the Chap 32 directory in the downloaded content set.

2. Click the Auto Key button (or press the N key) at the bottom of the 3ds Max window, and drag the Time Slider to frame 100.

3. Select the "Shaft" object at the top of the windmill in the Left viewport. Then click the Select and Rotate button on the main toolbar (or press E key), and rotate the "shaft" object about its Y-axis for a full revolution (360 degrees). The blades and nosecone are attached to the center shaft and will rotate with the shaft object.

   Since the windmill blades are rotated a full revolution, the blades will be in the same location at frame 100 as they were at the start. This makes the animation loop smoothly as it repeats over and over.

4. Click the Play Animation button in the Time Controls to see the animation.

Figure 32.3 shows frame 50 of this simple animation.

FIGURE 32.3

Frame 50 of this simple windmill animation

Another way to create keys is to select the object to be animated and right-click the Time Slider button. This opens the Create Key dialog box, shown in Figure 32.4, where you can set Position, Rotation, and Scale keys for the currently selected object. You can use this method only to create transform keys.

FIGURE 32.4

The Create Key dialog box enables you to create a Position, Rotation, or Scale keys quickly.

If a key already exists, you can clone it by dragging the selected key on the Track Bar with the Shift key held down. Dragging the Track Bar with the Ctrl and Alt keys held down changes the active time segment.

Copying parameter animation keys

If a parameter is changed while the Auto Key mode is enabled, keys are set for that parameter. You can tell when a parameter has a key set because the arrows to the right of its spinner are outlined in red when the Time Slider is on the frame where the key is set. If you change the parameter value when the spinner is highlighted red, the key value is changed (and the Auto Key mode doesn't need to be enabled).

If you highlight and right-click the parameter value, a pop-up menu of options appears. Using this pop-up menu, you can Cut, Copy, Paste, and Delete the parameter value. You also can select Copy Animation, which copies all the keys associated with this parameter and lets you paste them to another parameter. Pasting the animation keys can be done as a Copy, an Instance, or a Wire. A Copy is independent; an Instance ties the animation keys to the original copy so that they both are changed when either changes; and a Wire lets one parameter control some other parameter.

Caution

To copy a parameter value, be sure to select and right-click the value. If you right-click the parameter's spinner, the value is set to 0.

The right-click pop-up menu also includes commands to let you Edit a wired parameter, show the parameter in the Track View, or show the parameter in the Parameter Wiring dialog box.

Parameter wiring and the Parameter Wiring dialog box are discussed in more detail in Chapter 34, "Wiring Parameters."

Deleting all object animation keys

Individual keys can be selected and deleted using the Track Bar or the right-click pop-up menu, but if an object has many keys, this can be time consuming. To delete all animation keys for the selected object quickly, choose the Animation→Delete Selected Animation menu command.

Using the Track Bar

The 3ds Max interface includes a simple way to work with keys: with the Track Bar, which is situated directly under the Time Slider. The Track Bar displays a rectangular marker for every key for the selected object. These markers are color-coded, depending on the type of key. Position keys are red, rotation keys are green, scale keys are blue, and parameter keys are dark gray.

Caution

The Track Bar/Time Slider can be pulled away from the interface and made into a floating toolbar by dragging on the double line at the left end of the Track Bar. You can also resize the toolbar to make it easier to see the keys.

The current frame also appears in the Track Bar as a light blue, transparent rectangle, as shown in Figure 32.5. The icon at the left end of the Track Bar is the Open Mini Curve Editor button, which opens a mini Track View.

For more on the Track View interface, see Chapter 36, "Editing Animation Curves in the Track View."

FIGURE 32.5

The Track Bar displays all keys for the selected object.

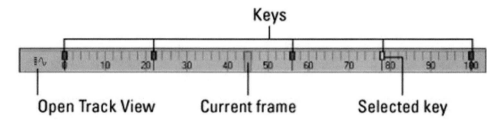

Using the Track Bar, you can move, copy, and delete keys. The Track Bar shows key markers only for the currently selected object or objects, and each key marker can represent several different keys. When the mouse is moved over the top of these markers, the cursor changes to a set of crosshairs, and you can select a marker by clicking it (selected key markers turn white). Using the Ctrl key, you can select multiple key markers at the same time. You also can select multiple key markers by clicking an area of the Track Bar that contains no keys and then dragging an outline over all the keys you want to select. If you move the cursor over the top of a selected key, the cursor is displayed as a set of arrows enabling you to drag the selected key to the left or right. Holding down the Shift key while dragging a key creates a copy of the key. Pressing the Delete key deletes the selected key.

Tip

Because each marker can represent several keys, you can view all the keys associated with the marker in a pop-up menu by right-clicking the marker.

Note

The marker pop-up menu also offers options for deleting selected keys or filtering the keys. In addition, there is a Go to Time command, which automatically moves the Time Slider to the key's location when selected.

To delete a key marker with all of its keys, right-click to open the pop-up menu and choose Delete Key→All, or select the key marker and press the Delete key.

Viewing and Editing Key Values

At the top of the marker's right-click pop-up menu is a list of current keys for the selected object (or if there are too many keys for a marker, they are placed under the Key Properties menu). When you select one of these keys, a key information dialog box opens. This dialog box displays different controls, depending on the type of key selected. Figure 32.6 shows the dialog box for the Position key. There are slight variations in this dialog box, depending on the key type.

FIGURE 32.6

Key dialog boxes enable you to change the key parameters.

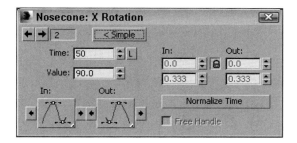

Note
You also can access key-specific dialog boxes in the Motion panel for a selected object by clicking the Parameters button.

Within each of these key dialog boxes is a Time value that shows the current frame. Next to the Time value are two arrows that enable you to move easily to the other keys in the scene. The dialog box also includes several text fields, where you can change the key parameters.

Most of the key dialog boxes also include flyout buttons for selecting Key Tangents. Key Tangents determine how the animation moves into and out of the key. For example, if the In Key Tangent is set to Slow, and the Out Key Tangent is set to Fast, the object approaches the key position in a slow manner but accelerates as it leaves the key position. The arrow buttons on either side of the Key Tangent buttons can copy the current Key Tangent selection to the previous or next key.

The available types of Tangents are detailed in Table 32.3.

TABLE 32.3 Key Tangents

Toolbar Button	Name	Description
	Smooth	Produces straight, smooth motion; this is the default type.
	Linear	Moves at a constant rate between keys.
	Step	Causes discontinuous motion between keys; it occurs only between matching In-Out pairs.
	Slow	Decelerates as you approach the key.
	Fast	Accelerates as you approach the key.
	Custom	Lets you control the Tangent handles in function curves mode.

	Custom – Locked Handles	Lets you control the Tangent handles in function curves mode with the handles locked.

Using the Motion Panel

You have yet another way to create keys: by using the Motion panel. The Motion panel in the Command Panel includes settings and controls for animating objects. At the top of the Motion panel are two buttons: Parameters and Motion Paths.

Setting parameters

The Parameters button on the Motion panel lets you assign controllers and create and delete keys. *Controllers* are custom key-creating algorithms that can be defined through the Parameters rollout, shown in Figure 32.7. You assign these controllers by selecting the position, rotation, or scaling track and clicking the Assign Controller button to open a list of applicable controllers that you can select.

For more information on controllers, see Chapter 35, "Animating with Constraints and Simple Controllers."

FIGURE 32.7

The Parameters section of the Motion panel lets you assign controllers and create keys.

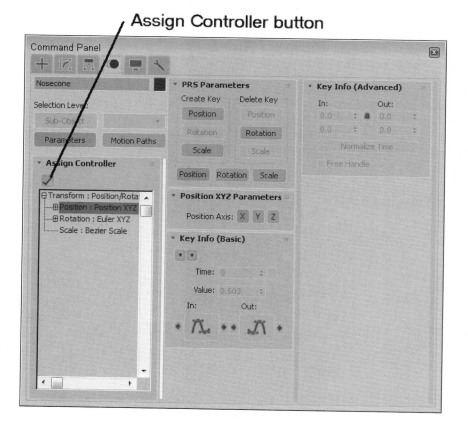

When a keyable object is selected, below the Assign Controller rollout is the PRS Parameters rollout where you can create and delete Position, Rotation, and Scale keys. You can use this rollout to create Position,

Rotation, and Scale keys whether or not the Auto Key or Set Key buttons are enabled. Additional rollouts may be available, depending upon the selected controller.

Below the PRS Parameters rollout are two Key Info rollouts: Basic and Advanced. These rollouts include the same key-specific information that you can access using the right-click pop-up menu found in the Track Bar.

Using Motion Paths

A *Motion Path* is the actual path that the animation follows. When you click the Motion Paths button in the Motion panel, the animation path is shown as a spline with each key displayed as a square and each frame shown as a white dot. You can then edit the trajectory and its nodes by clicking the Sub-Object button at the top of the Motion panel, shown in Figure 32.8. The only subobject available is Keys. With the Sub-Object button enabled, you can use the transform buttons to move and reposition the key nodes. You also can add and delete keys with the Add Key and Delete Key buttons.

FIGURE 32.8

The Motion Paths rollout in the Motion panel enables you to see the animation path as a spline.

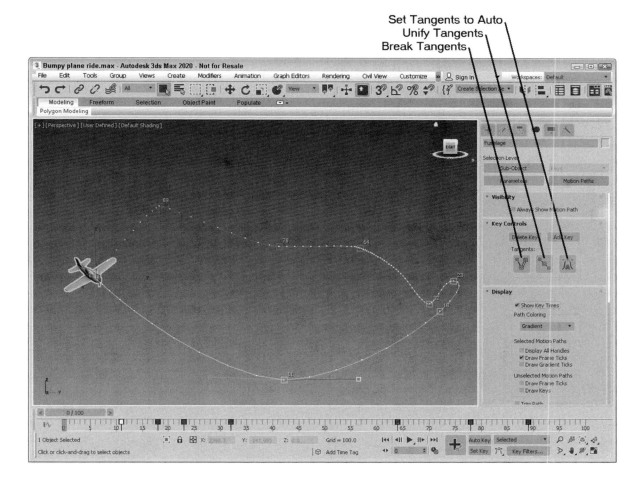

To edit the curvature at each key point, you can use the Tangent buttons to break, unify or auto set the tangents. Once a tangent is broken, you can drag each side of the tangent handle independently to create a sharp point in the motion path. Unified tangent handles are locked to the same line and auto tangents automatically adjusts the tangent handles to make the curve smooth.

The Display rollout has a Show Key Times option that displays frame numbers along the trajectory path where every animation key is located. Enabling this option displays the frame numbers next to any key along a motion path. You can make the motion path visible for any object by enabling the Motion Path option in the Object Properties dialog box. You can also change the coloring of the motion path. The available options are Gradient, Object Color, Uniform and Velocity. The Velocity option changes the path's color to correspond with the speed of the object. You can also trim the path by a given offset value.

Within the Conversion Tools rollout, you can convert the motion path to a normal editable spline with the Convert To button. You also can convert an existing spline into a motion path with the Convert From button.

To use the Convert From button, select an object, click the Convert From button, and then click a spline path in the scene. This creates a new trajectory path for the selected object. The first key of this path is placed at the spline's first vertex, and the final key is placed as the spline's final vertex position. Additional keys are spaced out along the spline based on the spline's curvature as determined by the Samples value listed in the Sample Range group. All these new keys are roughly spaced between the Start and End times, but smaller Bézier handles result in more closely packed keys.

Click the Collapse button at the bottom of the Conversion Tools rollout to reduce all transform keys into a single editable path. You can select which transformations to collapse, including Position, Rotation, and Scale, using the options under the Collapse button. For example, an object with several Controllers assigned can be collapsed, thereby reducing the complexity of all the keys.

Note

If you collapse all keys, you cannot alter their parameters via the controller rollouts.

Tutorial: Making an airplane follow a looping path

Airplanes that perform aerobatic stunts often follow paths that are smooth. You can see this clearly when watching a sky writer. In this example, I've created a simple looping path using the Line spline primitive, and you'll use this path to make a plane complete an aerobatic loop.

To make an airplane follow a looping path, follow these steps:

1. Open the Looping airplane.max file from the Chap 32 directory in the downloaded content set.

   This file includes a simple looping spline path and an airplane.

2. With the airplane selected, open the Motion panel and click the Motion Paths button. Then click the Convert From button in the Conversion Tools rollout, and select the path in the Front viewport.

3. If you drag the Time Slider, you'll notice that the plane moves along the path, but it doesn't rotate with the path. To fix this, click the Key Mode Toggle button in the Time Controls to easily move from key to key. Click the Key Filters button, select only Rotation, and then click the Set Key button (or press the spacebar key) to enter Set Key mode.

4. Before moving the Time Slider, click the Set Keys button to create a rotation key at frame 0. Then click the Select and Rotate button, click the Next Key button, rotate the plane in the Front viewport to match the path, and click the large Set Keys button (or press the K key) to create a rotation key. Click the Next Key button to move to the next key, and repeat this step until rotation keys have been set for the entire path.

5. Drag the Time Slider, and watch the airplane circle about the loop.

3ds Max provides an easier way to make the plane follow the path using the Path constraint. To learn more about constraints, see Chapter 35, "Animating with Constraints and Simple Controllers."

Figure 32.9 shows the plane's motion path.

FIGURE 32.9

When you use a spline path, the position keys are automatically set for this plane.

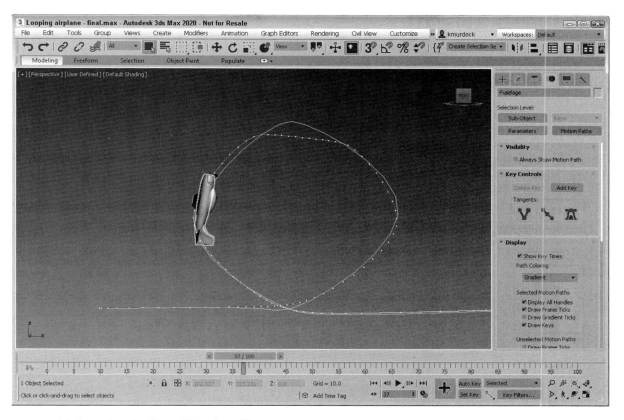

Using the Follow/Bank utility

When an object travels along a path that defines its trajectory, it maintains its same orientation without rotating. Imagine a roller coaster car; it rotates and banks as it moves around the track. This rotation and banking motion can be added to an object following a path using the Follow/Bank utility. You can access this utility by opening the Utilities panel and clicking the More button. Double-click the Follow/Bank utility to load it into the Utilities panel.

Caution

The Follow/Bank utility aligns the local X-axis of the object with the local Z-axis of the spline when the utility is applied, so you need to correctly orient the object's pivot point before applying the utility. If you don't, the object will be aligned at right angles to the path.

The Follow/Bank utility lets you enable a Bank option and set its Amount and Smoothness. Another option allows the object to turn upside down (not recommended for a traditional roller coaster car). Click the Apply Follow button to add the keys to cause the object to follow and bank. The Samples section determines how many keys are created.

Using Ghosting

As you're trying to animate objects, using the ghosting feature can be very helpful. This feature displays a copy of the selected object being animated before and/or after its current position. To enable ghosting, choose Views→Show Ghosting. The Show Ghosting command displays the position of the selected object in the previous several frames, the next several frames, or both. This command uses the options set in the

Preference Settings dialog box. Access this dialog box by choosing Customize→Preferences. In the Viewports panel of this dialog box is a Ghosting section.

You use this Ghosting section to set how many ghosted objects are to appear; whether the ghosted objects appear before, after, or both before and after the current frame; and whether frame numbers should be shown. You also can specify every Nth frame to be displayed. You also have an option to display the ghost object in wireframe (it is displayed as shaded if this option is not enabled) and an option to Show Frame Numbers. Objects before the current frame are colored yellow, and objects after are colored light blue.

Figure 32.10 shows a duck toy object that is animated to travel in a bumpy circle with ghosting enabled. The Preference settings are set to show five ghosting frames at every other frame before and after the current frame.

FIGURE 32.10

Enabling ghosting lets you know where an object is and where it's going.

Animating Objects

Many different objects in 3ds Max can be animated, including geometric objects, cameras, lights, and materials. In this section, you'll look at several types of objects and parameters that can be animated.

Animating cameras

You can animate cameras using the standard transform buttons found on the main toolbar. When animating a camera that actually moves in the scene, using a Free camera is best. A Target camera can be pointed by moving its target, but you risk it being flipped over if the target is ever directly above the camera. If you want to use a Target camera, attach both the camera and its target to a Dummy object using the Select and Link button and move the Dummy object.

Two useful constraints when animating cameras are the Path constraint and the Look At constraint. You can find both of these in the Animation→Constraints menu. The Path constraint can make a camera follow a spline path, and the Look At constraint can direct the focus of a camera to follow an object as the camera or the object moves through the scene.

For more on constraints, including these two, see Chapter 35, "Animating with Constraints and Simple Controllers."

Tutorial: Animating darts hitting a dartboard

As a simple example of animating objects using the Auto Key button, you'll animate several darts hitting a dartboard.

To animate darts hitting a dartboard, follow these steps:

1. Open the Dart and dartboard.max file from the Chap 32 directory in the downloaded content set.

2. Click the Auto Key button (or press the N key) to enable animation mode. Drag the Time Slider to frame 25, and click the Select and Move button on the main toolbar (or press the W key).

3. Select the first dart in the Left viewport, and drag it to the left until its tip just touches the dartboard.

   This step creates a key in the Track Bar for frames 0 and 25.

4. Click the Select and Rotate button on the main toolbar, set the reference coordinate system to Local, and constrain the rotation to the Y-axis. Then drag the selected dart in the Front viewport to rotate it about its local Y-axis.

   This step also sets a key in the Track Bar.

5. Select the second dart, and click the Select and Move button again. Right-click the Time Slider to make the Create Key dialog box appear. Make sure that the check boxes for Position and Rotation are selected, and click OK.

   This step creates a key that keeps the second dart from moving before it's ready.

6. With the second dart still selected, drag the Time Slider to frame 50 and move the dart to the dartboard as described in Step 3. Then repeat Step 4 to set the rotation key for the second dart.

7. Repeat Steps 3, 4, and 5 for the last two darts at frames 75 and 100.

8. Click the Auto Key button (or press the N key) again to disable animation mode, maximize the Perspective viewport, and click the Play Animation button to see the animation. Figure 32.11 shows the darts as they're flying toward the dartboard.

FIGURE 32.11

One frame of the dart animation

Animating lights

The process for animating lights includes many of the same techniques as those for animating cameras. For moving lights, use a Free Spot light or attach a Target Spot light to a Dummy object. You also can use the Look At and Path controllers with lights.

If you need to animate the Sun at different times in the day, use the Daylight or Sunlight systems, which are discussed in Chapter 25, "Using Lights and Basic Lighting Techniques."

To flash lights on and off, enable and disable the On parameter at different frames and assign a Step Tangent. To dim lights, just alter the Multiplier value over several frames.

Animating materials

Materials can be animated if their properties are altered while the Auto Key button is active. 3ds Max interpolates between the values as the animation progresses. The material must stay the same for the entire animation: You cannot change materials at different keys; you can only alter the existing material parameters.

If you want to change materials as the animation progresses, you can use a material that combines multiple materials, such as the Blend material. This material includes a Mix Amount value that can change at different keyframes. The next tutorial shows how to use the Blend material in this manner.

Several maps include a Phase value, including all maps that have a Noise rollout. This value provides the means to animate the map. For example, using a Noise map and changing the Phase value over many keys animates the noise effect.

Note

Another common way to animate materials is with the Controller nodes that are applied in the Slate Material Editor. Using these Controller nodes, you can alter material parameters for different frames and even access the Curve Editor for these controllers.

Tutorial: Dimming lights

Animation keys are typically used to animate objects, but this example shows how other parameters, such as material properties can also be animated. For example, if you want to change material properties in a scene gradually over a set number of frames, such as dimming a light, you can easily accomplish this task with the Blend material.

To create a light that dims with time, follow these steps:

1. Open the Dimming light.max file from the Chap 32 directory in the downloaded content set.

   This file contains a simple lamp object with a sphere to represent a light bulb.

2. Open the Slate Material Editor by pressing the M key, and double-click the Blend material in the Material/Map Browser to add a Blend node with two other materials connected to it. Double-click the new Blend node, and give the material the name **Dimming Light**.

3. Double click the first attached material and name the material **Light On**. Set the Diffuse color in the Blinn Basic Parameters rollout to yellow and the Self-Illumination to yellow after enabling the Color option.

4. Double click the second attached material node for the Blend node, name the second material **Light Off**, and select a gray Diffuse color.

5. Select the light bulb object, and click the Assign Material to Selection button to assign the material to the bulb object.

6. With the Time Slider at frame 0, click the Auto Key button (or press the N key). Drag the Time Slider to frame 100, and change the Mix Amount to **100**. Click the Auto Key button again to deactivate it. The material changes gradually from the "Light On" material to the "Light Off" material. When you drag the Time Slider, you won't see the material change, but if you render the scene, you can see the dimming light.

Figure 32.12 shows a simple lamp object with a dimming sphere in its center. The actual dimming effect isn't visible in the viewport—only when the image is rendered.

FIGURE 32.12

This lamp object dims as the animation proceeds.

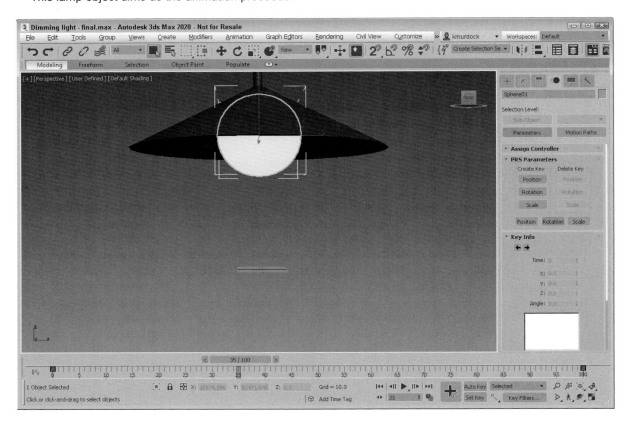

Working with Previews

More than likely, your final output will be rendered using the highest-quality settings with all effects enabled, and you can count on this taking a fair amount of time. After waiting several days for a sequence to render is a terrible time to find out that your animation keys are off. Even viewing animation sequences in the viewports with the Play Animation button cannot catch all problems.

New Feature in 2020

The Preview Animation feature has been overhauled and improved.

One way to catch potential problems is to create a sample preview animation. Previews are test animation sequences that render quickly to give you an idea of the final output. The Tools→Preview - Grab Viewport menu includes several commands for creating, renaming, and viewing previews. The rendering options available for previews are the same as the shading options available in the viewports, but seeing the animation gives you a sense of the timing.

Creating previews

You create previews by choosing Tools→Preview - Grab Viewport→Create Preview Animation to open the Make Preview dialog box, shown in Figure 32.13. You also can access this command from the viewport General label in the upper-left corner of the viewport.

New Feature in 2020

The Make Preview dialog box is improved in 3ds Max 2020.

FIGURE 32.13

The Make Preview dialog box lets you specify the range, size, and output of a preview file.

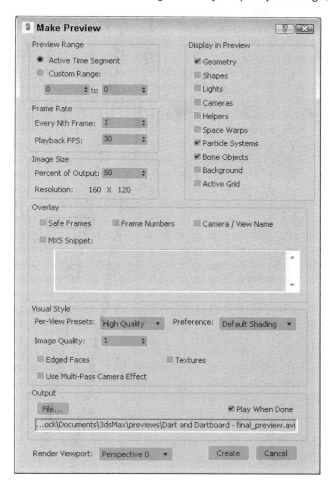

In the Make Preview dialog box, you can specify what frames to include using the Active Time Segment or Custom Range options. You also can choose Every Nth Frame or select a specific frame rate in the Playback FPS field. The image size is determined by the Percent of Output value, which is a percentage of the final output size. The resolution is also displayed.

The Display Filter section offers a variety of options to include in the preview. These options include Geometry, Shapes, Lights, Cameras, Helpers, Space Warps, Particle Systems, Bone Objects, Background and Active Grid. Because the preview output is rendered like the viewports, certain selected objects such as Lights and Cameras actually display their icons as part of the file. The Overlay options prints the safe frame borders, frame numbers, the camera/view name, and a MaxScript Snippet in the upper-left corner of each frame. This is a nice feature that helps in identifying the results.

The Visual Style section includes the same shading options used to display objects in the viewports, including High Quality, Standard, Performance, DX Mode and User Defined. It also includes the various Shading Stylized render options in the Preference drop-down list. There are also options to toggle on or off Edged Faces, and Textures.

If the viewport uses one of the multi-pass camera effects such as depth-of-field or motion blur, you can enable these effects for the preview also.

The File button opens a file dialog box where you can name and save the animation preview. There are also a number of output options that you can select. Output options include all the various output file types including AVI, BMP, CIN, JPG, PNG, MOV, TGA, DDS, and TIF.

At the bottom of the dialog box is a Render Viewport drop-down list, where you can select which viewport to use to create your preview file. The Create button starts the rendering process. When a preview is being rendered, the viewports are replaced with a single image of the current render frame, and the Status bar is replaced by a Progress bar and a Cancel button.

Tip

You can use the Esc key on your keyboard to cancel a rendering job.

If you cancel the rendering, the Make Preview alert box offers the options Stop and Play; Stop and Don't Play; and Don't Stop.

Viewing previews

When a preview file is finished rendering, the default Media Player for your system loads and displays the preview file if the Play When Done option is enabled.

At any time, you can replay the preview file using the Tools→Preview - Grab Viewport→Play Preview Animation menu command. This command loads the latest preview file and displays it in the Media Player.

Renaming previews

To save a preview file by renaming it, choose Tools→Preview - Grab Viewport→Save Preview Animation As. This command opens the Save Preview As dialog box, where you can give the preview file a name. There is also an option to open the current folder where the previews are saved.

Using the RAM Player

Just as you can use the Rendered Frame Window to view and compare rendered images, the RAM Player enables you to view rendered animations in memory. With animations loaded in memory, you can selectively change the frame rates. Figure 32.14 shows the RAM Player interface, which you open by choosing Rendering→Compare Media in RAM Player. You see two images of the rendered house placed on top of each other with half of each showing. One was rendered using its default materials; the other was rendered using Color Pencil.

FIGURE 32.14

The RAM Player interface lets you load two different images or animations for comparison.

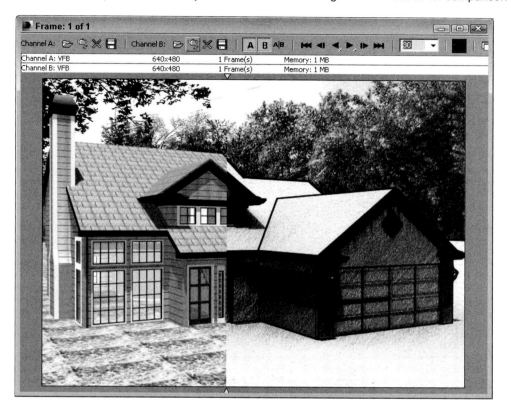

The buttons at the top of the RAM Player interface window, shown in Table 32.4, enable you to load an image to two different channels named A and B. The two Open Channel buttons open a file dialog box where you can select the file to load. Notice that the image on the right side of the RAM Player is a different frame from the left side.

TABLE 32.4 RAM Player Interface Buttons

Button	Description
	Open Channel
	Open Last Rendered Image
	Close Channel
	Save Channel
A|B	Horizontal/Vertical Split Screen
	Double Buffer

The Open Last Rendered Image button in the RAM Player interface window provides quick access to the last rendered image using the RAM Player Configuration dialog box. The Close Channel button clears the channel. The Save Channel button opens a file dialog box for saving the current file.

Caution
All files that load into the RAM Player are converted to 24-bit images.

The Channel A and Channel B (toggle) buttons enable either channel or both. The Horizontal/Vertical Split Screen button switches the dividing line between the two channels to a horizontal or vertical line. When the images are aligned one on top of the other, two small triangles mark where one channel leaves off and the other begins. You can drag these triangles to alter the space for each channel.

The frame controls let you move between the frames. You can move to the first, previous, next, or last frame and play the animation forward or in reverse. The drop-down list to the right of the frame controls displays the current frame rate setting.

You can capture the color of any pixel in the image by holding down the Ctrl key while right-clicking the image. This puts the selected color in the color swatch. The RGB value for this pixel is displayed in the blue title bar.

The Double Buffer button synchronizes the frames of the two channels.

Tip
You can use the arrow keys and Page Up and Page Down keys to move through the frames of the animation. The A and B keys are used to enable the two channels.

Tutorial: Using the RAM Player to Combine Rendered Images into a Video File

Not only can the RAM Player be used to load and compare images, but it also can handle animation files. It also can load in multiple frames of an animation that were saved as individual image files and save them back out as an animated file.

To combine multiple rendered image files into a video file using the RAM Player, follow these steps:

1. Select the Rendering→Compare Media in RAM Player menu command to open the RAM Player. Then click the Open Channel A button, and locate and open the Exploding Planet - frame 10.tif file from the Chap 32 directory in the downloaded content set.

   This file is the first rendered frame of a ten-frame animation.

2. The Image File List Control dialog box opens. Using this dialog box, you can specify the Start and End Frames to load. You also can select to load every nth frame. Select 0 as the Start Frame and 9 as the End Frame, and click OK.

3. The RAM Player Configuration dialog box opens next, letting you set the Resolution and Memory Usage for the loaded files. Click OK to accept the default values.

   All the image files that share the same base name as the selected file are loaded into the RAM Player sequentially. Press the Play button to see the loaded animation.

4. Click the Save Channel A button to access the file dialog box, where you can save the loaded animation using the AVI, MPEG, or MOV video format.

Figure 32.15 shows one frame of the loaded animation files.

FIGURE 32.15

This rendered image is just one of a series of rendered animation frames that can be viewed in the RAM Player.

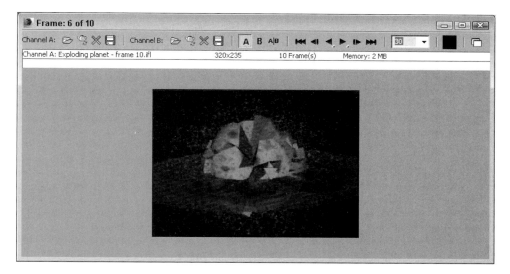

Summary

This chapter covered the basics of animating objects in 3ds Max, including working with time and keys. You also learned about the two key creation modes and editing keys. Several animation helps are available, such as motion paths and ghosting. This chapter also discussed how to animate materials and how to create preview animations. In this chapter, you learned how to do the following:

*   Control time and work with keys

*   Use the two key creation modes

*   Work with the Track Bar and the Motion panel

*   View and edit key values

*   Use motion paths and ghosting

*   Animate cameras, lights, and materials

*   Create preview animations

*   Use the RAM Player

In the next chapter, you learn about more animation features including animation layers and the various animation modifiers.

Chapter 33

Using Animation Layers and Animation Modifiers

IN THIS CHAPTER

Using animation layers

Saving and loading animation files

Using the Point Cache modifier

Using the Morpher modifier

Adding secondary animation with the Flex modifier

Animating geometry deformations

Using other animation modifiers

Just as layers can be used to organize a scene by placing objects on different layers, you also can separate the various animation motions into different layers. This gives you great control over how motions are organized and blended together.

If you've worked to animate some object in the Autodesk® 3ds Max® 2020 software and are pleased with the result, you can save the animation clip and reuse it on similar objects. Several animation clips can be mixed together to create an entirely new animation sequence.

Another way to work with animations is with the animation modifiers. Modifiers can be used to deform and otherwise alter the geometry of objects, but they also can be used to affect other aspects of an object, including animated changes. One such important animation modifier is the Point Cache modifier. This modifier lets the movement of each vertex in the scene be saved to a cached file for immediate recall and for animating multiple objects simultaneously.

The Modifiers menu also includes an Animation submenu that contains many such modifiers. These modifiers are unique in that each of them changes with time. They can be useful as an alternate to controllers, but their resulting effects are very specific. Included with this submenu are modifiers such as Morpher, which allows an object to move through several different preset shapes.

The last part of this chapter covers some miscellaneous animation modifiers that haven't been covered yet. These modifiers deal with the motion of objects, including Flex, Melt, and PathDeform modifiers.

Using the Animation Layers Toolbar

Behind the scenes, animation layers add several new controller tracks to objects that are visible in the Motion panel and in the Track View interface, but the front end is accessible through a simple toolbar. The Animation Layers toolbar, shown in Figure 33.1, is similar in many ways to the Layers toolbar.

The Layers toolbar is covered in Chapter 6, "Selecting Objects and Setting Object Properties."

FIGURE 33.1

The Animation Layers toolbar includes icons for defining and merging layers.

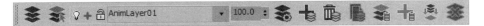

You can open the Animation Layers toolbar by right-clicking the main toolbar away from the buttons and selecting Animation Layers from the pop-up menu. Each of the toolbar buttons is labeled and explained in Table 33.1.

TABLE 33.1 Animation Layers Toolbar Controls

Toolbar Button	Name	Description
	Enable Animation Layers	Turns the animation layers system on
	Select Active Layer Objects	Selects the objects in the viewport that are on the active animation layer
	Layer Selection drop-down list	Presents a selection list of all the available animation layers
	Animation Layer Weight	Displays the weight value for the current animation layer
	Animation Layer Properties	Opens the Animation Layer Properties dialog box
	Add Animation Layer	Adds another animation layer
	Delete Animation Layer	Deletes the current animation layer
	Copy Animation Layer	Copies the current animation layer
	Paste Active Animation Layer	Pastes the keys from the current animation layer to the selected object
	Paste New Layer	Pastes the copied animation layer keys to a new layer
	Collapse Animation Layer	Combines and deletes the current animation layer with the layer above it
	Disable Animation Layer	Turns the current animation layer off

When a new animation layer is created using the Add Animation Layer button, a dialog box appears where you can name the new layer and select to duplicate the active controller type or use the default controller type. After you click the OK button, a new entry is added to the Animation Layer Selection List. The default name of the animation layer as AnimLayer with a number. The original layer is named Base Layer. To the left of the animation layer name is a small light bulb icon that indicates whether the animation layer is enabled or disabled.

Caution
You cannot rename animation layers after they are created.

Each layer can have a weight assigned to it. These weight values control how much influence the current animation layer has. The weight value also can be animated. For example, if a car is animated moving forward 100 meters over 50 frames, weighting the animation layer to 30 causes the car to move forward only 30 meters over the 50 frames.

Working with Animation Layers

Animation layers are good for organizing motions into sets that can be easily turned on and off, but you also can use them to blend between motions to create an entirely new set of motions.

Note
Animations can be divided into *primary* motions, which are the major motions, and *secondary* motions, which are derivative motions that depend on the animation of other parts. Using animation layers, you can separate primary motions, like foot placement, from secondary motions, like a swinging arm, and adjust them independently.

Enabling animation layers

The first button on the Animation Layers toolbar is the Enable Animation Layers button. Clicking this button opens a dialog box, shown in Figure 33.2, where you can filter the type of keys to include in the animation layer.

FIGURE 33.2

The Enable Animation Layers dialog box lets you limit which type of keys are included.

The base animation layer can be disabled using the Disable Animation Layer button. This button is available only when you collapse all its layers. This disables the existing animation layer.

Note
If Animation Layers are enabled for an object whose animation is loaded into the Motion Mixer, a dialog box automatically appears, asking if you want to create a new map file for the animation.

Setting animation layers properties

The button to the right of the Weight value opens the Layer Properties dialog box, shown in Figure 33.3. This dialog box lets you specify the type of controller to which the layers are collapsed. The options include

Bézier (for Position and Scale tracks) or Euler (for Rotation tracks), Linear or TCB, and Default. You also can specify a range to collapse.

FIGURE 33.3

The Layer Properties dialog box lets you set the controller type to collapse to.

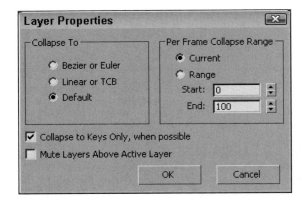

Collapsing animation layers

By collapsing layers, you combine the animation keys on each layer into a single set of keys that includes all the various motions. Be careful when collapsing; the results can be unexpected. Collapsing animation layers is accomplished with the Collapse Animation Layers button.

Animation layers provide a simple yet effective way to blend animations, but if you need more blending options, check out the Graph Editors→Motion Mixer.

Tutorial: Using animation layers for a plane takeoff

Have you ever been to a small airport and watched the commuter planes take off? Sometimes they leave the ground and then return to the ground and then finally take off. It's like they need to get a good bounce to overcome gravity. This is a good example of when animation layers come in handy.

To animate a jet's takeoff using animation layers, follow these steps:

1. Open the Plane take-off.max file from the Chap 33 directory in the downloaded content set.

2. Open the Animation Layers toolbar by right-clicking the main toolbar away from the buttons and choosing Animation Layers from the pop-up menu.

3. Select the airplane object, click the Enable Animation Layers button, enable the Position track, and then click OK to close the pop-up dialog box. This adds a Base Layer to the Selection list of the Animation Layers toolbar.

4. Click the Auto Key button, drag the Time Slider to frame 100, and move the jet to the far end of the runway with the Select and Move tool. Then disable the Auto Key button. Set the Weight value for this layer to **0**.

5. In the Animation Layers toolbar, click the Add Animation Layer button to add a new layer. In the Create New Animation Layer dialog box that appears, select the Duplicate the Active Controller Type option and click OK. A new layer labeled AnimLayer01 is added to the Selection list.

6. Click the Auto Key button again, drag the Time Slider to frame 100, and move the jet upward away from the runway. Then disable the Auto Key button again.

7. Select the Base Layer from the Selection list in the Animation Layers toolbar and set its Weight value to **100**. Then drag the Time Slider, and notice that the jet moves up at an angle over the 100 frames. This motion is caused by blending the two animation layers together.

8. Click the Auto Key button again, drag the Time Slider to frame 0, and set the Weight value for AnimLayer01 to **0**; drag the Time Slider to frame 30, and set the Weight to **60**; drag the Time Slider

to frame 50, and set the Weight back to **0**; and finally drag the Time Slider to frame 80, and set the Weight to **80**. Then disable the Auto Key button again.

Dragging the Time Slider shows the jet bounce down the runway before taking off, as shown in Figure 33.4. This gives a single weight value for controlling the vertical height of the airplane as it takes off.

FIGURE 33.4

The Animation Layers feature provides a single parameter for controlling the plane's height.

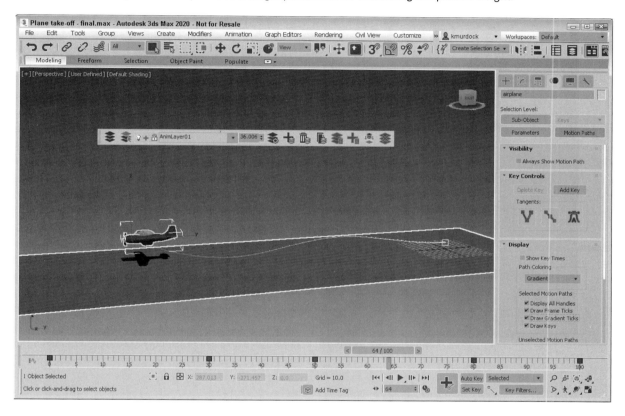

Saving and Loading Animation Files

Before an animation can be reused on another model, it must be saved. Saving a sequence to the local hard disk makes it accessible for other 3ds Max scenes.

Note
The Animation→Load Animation and Animation→Save Animation menu commands are active only when an object is selected.

Saving general animations

The animation of objects can be saved using the XML Animation File (XAF) format. To save the animation for the selected object, open the Save XML Animation File dialog box, shown in Figure 33.5, using the Animation→Save Animation.

FIGURE 33.5

The Save XML Animation File dialog box is used to save animations of the selected object.

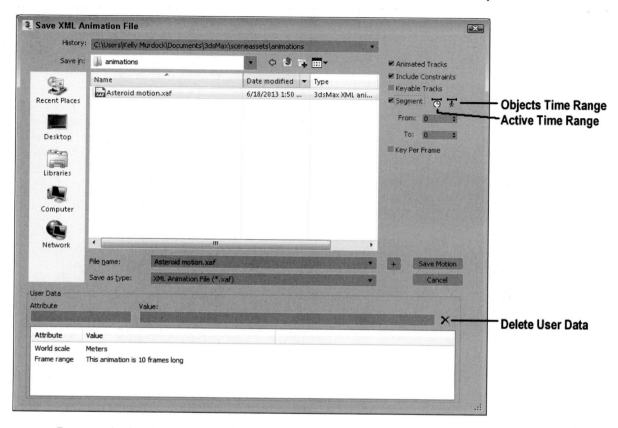

For general animations, you can select to include animated tracks, constraints, only keyable tracks, and a specific segment or range. The User Data fields let you enter notes or specific data used by plug-ins about the animation sequence.

The Load XML Animation File dialog box, shown in Figure 33.6, is opened using the Animation→Load Animation menu command. It looks like a normal file dialog box, but it has some additional features.

The Relative and Absolute options determine whether the animation is loaded relative to the object's current location or whether it is loaded into the frames where it was saved. The Replace and Insert options let the new keys overwrite the existing ones or move them out to insert the loaded keys. You can even select the frame where the new keys are loaded. The Load Motion button lets you load an existing mapping file named the same as the animation file if one exists or lets you create a new one.

FIGURE 33.6

The Load XML Animation File dialog box lets you load animation files from one scene and apply them to another.

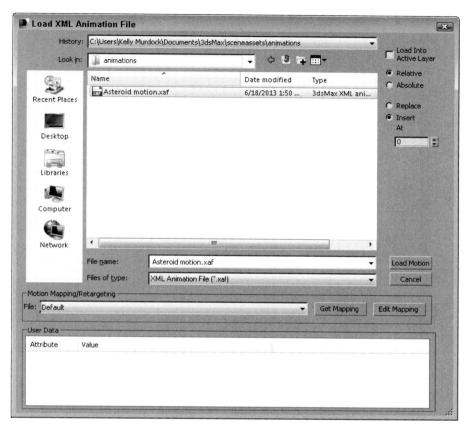

Mapping animated objects

Mapping files defines a relationship between objects in the saved animation file and objects in the current 3ds Max file. These relationships allow the animation keys to be transferred from one scene object to another. For example, if you save a dancing robot animation sequence and then want to load it onto a dinosaur model, the mapping file tells the animation that the robot's left arm needs to map to the dinosaur's left arm in the target scene because they might be called by different names in both files. If the objects have the same names in both files, the mapping file isn't needed.

Mapping files are listed in the drop-down list for easy selection, or you can use the Get Mapping button to select a different mapping file to load. Mapping files are saved with the .xmm file extension. The Edit Mapping button is active when a mapping associated with the file in the Load Animation dialog box exists. It opens the Map Animation dialog box, where you can define the mapping between objects in the two scenes.

Using the Map Animation dialog box

The Map Animation dialog box, shown in Figure 33.7, includes several rollouts. The Motion Mapping Parameters rollout includes options for allowing 3ds Max to make its best guess at mapping objects. If the scenes are fairly similar, this option may be just the ticket. The Exact Names, Closest Names, and Hierarchy buttons allow 3ds Max to attempt the mapping on its own. This works especially well on bipeds that use the default naming conventions. You also can select to have 3ds Max look at the various controllers that are used when trying to match up objects.

FIGURE 33.7

The Map Animation dialog box lets you map objects to receive animation.

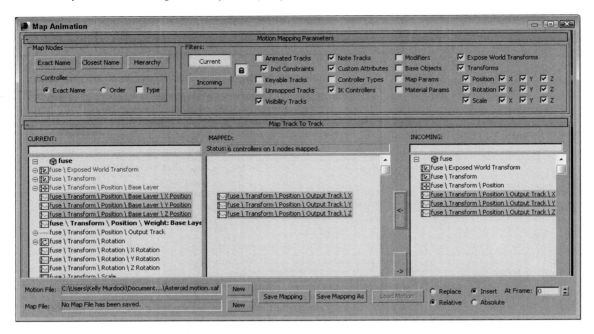

The Filters section lets you filter out the tracks that you don't want to see. The Lock button applies the selected filters to both the Current and Incoming lists.

The Map Track to Track rollout consists of three lists. The left list contains all the tracks for the current scene objects, the middle list contains all the mapped tracks, and the right list contains all the tracks from the incoming animation file. Select tracks and click the button with the left-pointing arrow to add tracks to the Mapped list; click the other arrow button to remove them.

At the bottom of the Map Animation dialog box are buttons for saving the current mapping file.

Retargeting animations

The Retargeting rollout, shown in Figure 33.8, lets you specify how the scale changes between certain mapped objects. Scale values can be entered for the mapped nodes as Absolute or Multiply Derived Scale values for each axis. Derived scale values can be obtained from a specific origin object. After the settings are right, the Set button applies the scaling to the selected mapping.

FIGURE 33.8

Use the Retargeting rollout to specify how the scale changes between mapped objects.

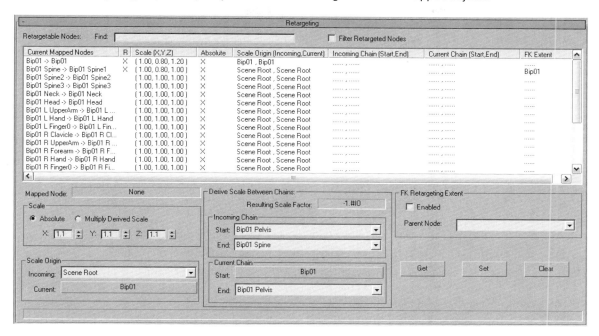

Using the Animation Modifiers

Animation is more than just moving an object from here to there. All objects move not only with major transformations but with lots of secondary motions also. When a human character walks, the motions of his arms and legs are major, but the secondary motions of his swinging hips and bobbing shoulders make the walk realistic. Many of the animation modifiers enable these key secondary motions.

All the animation modifiers presented in this chapter are located in the Modifiers→Animation submenu.

Also included among the Animation modifiers are several Skin modifiers, which are used to make an object move by attaching it to an underlying skeleton. The Skin modifiers are covered in Chapter 39, "Skinning Characters."

Baking Animation Keys with the Point Cache Modifier

When you add keys to an object to control its animation using modifiers, the modifiers remain with the object and can be revisited and altered as needed. However, if you have multiple objects that follow the same set of keys, such as for a crowd scene, including a set of modifiers for each object can increase the overhead many times over. A simple solution is to bake all the keys into the object, allowing all the keys to be pulled from an external file. This frees the resources required to animate multiple objects and makes the animated keys portable. The Point Cache modifier makes this possible.

You also can use the Point Cache modifier when playback in the viewport is too slow because 3ds Max needs to compute the vertex positions of a huge number of vertices. Reading their position from a separate file increases the playback speed. You also can use the file on a cloned object to control its motion at a different speed.

The Point Cache modifier records the movement of every vertex of an object to a file. Point Cache files have the .xml extension, but they also can be saved using the older .pc2 extension. To create a Point Cache file, apply the Point Cache modifier, click the New button in the Parameters rollout, and name a new file on the hard drive. Then set the range of the animation to capture and click the Record button. Once recorded, the total number of points along with the sample rate and range are displayed for the active cache.

Caution

Point Cache files can be loaded and used only on objects with the same number of vertices as the original used to record the file.

If you select the Disable Modifiers Below button, all modifiers below the Point Cache in the Modifier Stack are disabled. You can enable the Relative Offset option and set the Strength value to cause the cached animation to be exaggerated or even reversed. In the Playback Type section, you can control the range of the animation.

Tutorial: Trees in a hurricane

As an example of using the Point Cache modifier, you'll use a tree that is bending under violent forces such as a hurricane and duplicate it many times.

To create a forest of trees in a hurricane, follow these steps:

1. Open the Bending tree.max file from the Chap 33 directory in the downloaded content set.

    This file includes an animated tree swaying back and forth using the Bend modifier.

2. Select the tree and choose the Modifiers→Cache Tools→Point Cache menu to apply the Point Cache modifier to the tree.

3. In the Parameters rollout, click the New button, create a file named "Bending tree.xml," and click the Save button to create the cached animation file. Then click the Record button to save all the animation data to the file.

4. Select the tree and delete its Bend modifier. Then use the Tools→Array dialog box to create several rows of trees. Be sure to create the trees as copies and not instances.

5. Select a random tree and change the Playback Type to **Custom Start** in the Parameters rollout and change the Start Frame to **-2** to cause some random motion. Then repeat this step for several other trees.

6. Press the Play Animation button to see the results.

Figure 33.9 shows several of the trees being moved about by the storm.

FIGURE 33.9

Using the Point Cache modifier, you can animate a whole forest of trees.

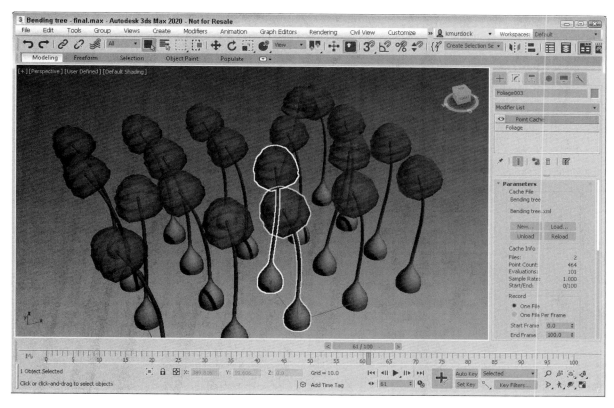

Morpher modifier

The Morpher modifier lets you change a shape from one form into another. You can apply this modifier only to objects with the same number of vertices.

Tip

In many ways, the Morpher modifier is similar to the Morph compound object, which is covered in Chapter 15, "Working with Compound Objects."

The Morpher modifier can be very useful for creating facial expressions and character lip-synching. 3ds Max makes 100 separate channels available for morph targets, and channels can be mixed. You can use the Morpher modifier in conjunction with the Morph material. For example, you could use the Morpher material to blush a character for an embarrassed expression.

Tip

When it comes to making facial expressions, a mirror and your own face can be the biggest help. Coworkers may look at you funny, but your facial expressions will benefit from the exercise.

The first task before using this modifier is to create all the different morph targets. Because the morph targets need to contain the same number of vertices as the base object, make a copy of the base object for each morph target that you are going to create. As you create these targets, be careful not to add or delete any vertices from the object.

Note

Because morph targets deal with each vertex independently, you cannot mirror morph targets, so if you want a morph target for raising the left eyebrow and a morph target for raising the right eyebrow, you need to create each morph target by hand.

After all your morph targets are created, select a channel in the Channel Parameters rollout, shown in Figure 33.10, and use the Pick Object from Scene button to select the morph target for that channel. Another option for picking is to use Capture Current State. After a morph target has been added to a channel, you can view it in the Channel List rollout.

FIGURE 33.10

The Morpher modifier's rollouts

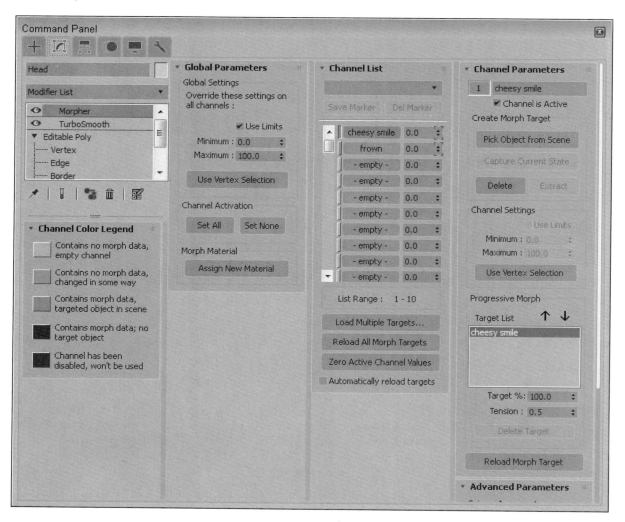

As you animate, you can specify the amount of each morph target to include in the frame using the value to the right of the channel name in the Channel List rollout. The slim color bar to the left of the channel name designates the status of the channel. You can find information on what each color represents in the Channel Color Legend rollout.

The Channel Parameters rollout also includes a Progressive Morph section. This feature lets you define an intermediate step for how the morph is to progress, with the final step being the morph target. Using these intermediate steps, you can control how the object morphs.

Tutorial: Morphing facial expressions

The Morpher modifier is very helpful when you're trying to morph facial expressions, such as those to make a character talk. With the various facial expressions added to the different channels, you can quickly morph between them. In this example, you use the Morpher modifier to change the facial expressions of the general character.

Tip

When creating facial expressions, be sure to enable the Use Soft Selection feature, which makes modifying the face meshes much easier.

To change facial expressions using the Morpher modifier, follow these steps:

1. Open the Morphing facial expressions.max file from the Chap 33 directory in the downloaded content set.

   This file includes a head model. The model has been copied twice, and the morph targets have already been created by modifying the subobjects around the mouth.

2. Select the face on the left, and select the Modifiers→Animation→Morpher menu command to apply the Morpher modifier.

3. In the Channel List rollout, select channel 1, click the Pick Object from Scene button, and select the middle face object. Then select the second empty channel from the Channel List rollout; in the Channel Parameters rollout, again click the Pick Object from Scene button and select the face on the right.

   If you look in the Channel List rollout, you'll see "cheesy smile" in Channel 1 and "frown" in Channel 2.

4. Click the Auto Key button (or press the N key), drag the Time Slider to frame 50, and then increase the "cheesy smile" channel in the Channel List rollout to **50**. Drag the Time Slider to frame 100, and increase the "frown" channel to **100** and the "cheesy smile" channel to **0**. Then return the Time Slider to **50**, and set the "frown" channel to **0**.

5. Click the Play Animation button in the Time Controls to see the resulting animation.

Figure 33.11 shows the three facial expressions. The Morpher modifier is applied to the left face.

Tip

Be sure to keep the morph target objects around. You can hide them in the scene or select them and save them to a separate file with the File→Save As→Save Selected menu command.

FIGURE 33.11

Using the Morpher modifier, you can morph one facial expression into another.

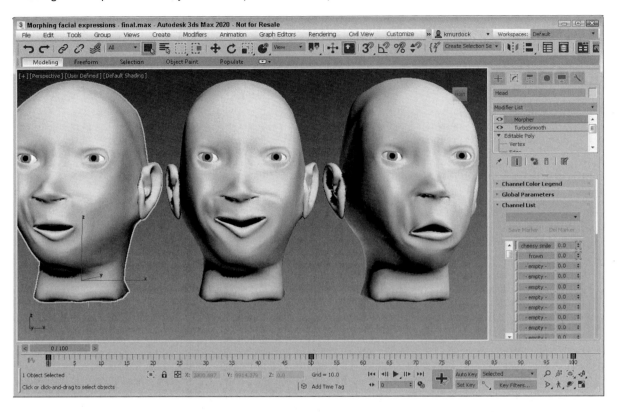

Using More Animation Modifiers

The Modifiers→Animation menu includes several more modifiers that weren't covered in the previous chapter. Using these modifiers, you can add secondary animation to an object that is already moving, melt an object, and make an object bend to follow a patch or surface.

Using the Flex modifier

The Flex modifier can add soft body dynamic characteristics to an object. The characteristic of a *soft body* is one that moves freely under a force. Examples of soft body objects are clothes, hair, and balloons. The opposite of soft body dynamics is *rigid body* dynamics. Think of a statue in the park. When the wind blows, it doesn't move. The statue is an example of a rigid body. On the other hand, the flag flying over the library moves all over when the wind blows. The flag is an example of a soft body.

Soft body objects also can be defined and simulated using MassFX, which is covered in Chapter 46, "Simulating Physics-Based Motion with MassFX."

Another way to think of soft bodies is to think of things that can flex. Objects such as a clothesline flex under very little stress, but other objects like a CD flex only a little when you apply a significant force. The settings of the Flex modifier make it possible to represent all kinds of soft body objects.

Figure 33.12 shows many of the rollouts available for the Flex modifier.

FIGURE 33.12

The Flex modifier rollout lets you control the flex settings.

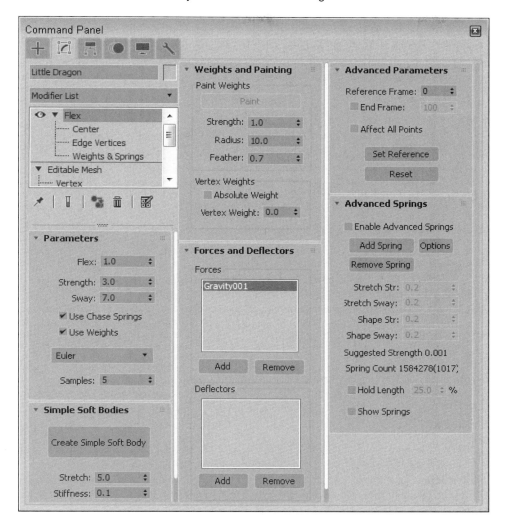

Flex subobjects

In the Modifier Stack, the Flex modifier has three subobjects that you can access: Center, Edge Vertices, and Weights and Springs. The Center subobject is a simple box gizmo that marks the center of the flex effect. Portions of the object that are farther from the center move a greater distance. The Edge Vertices subobject can be selected to control the direction and falloff of the flex effect.

Setting flex strength

The Parameters rollout includes a Flex value, which controls the amount of bending the object does; a Strength value, which controls the rigidity of the object; and a Sway value, which controls how long the flexing object moves back and forth before coming to a stop. An antenna on a car is an example of an object that has fairly high Flex and Sway values and a low Strength value.

Chase Springs cause an object to return to its original position when the force is removed. A twig on a tree is an example of an object with Chase Springs. The Use Chase Springs option lets you disable these springs. A piece of cloth is an example of when you would want Chase Springs disabled.

Selecting the Weights and Springs subobject lets you apply weights to certain selected springs. You can disable these weights using the Use Weights option. If you disable these weights, the entire object acts together.

The Flex modifier offers three solution methods to compute the motions of objects. These are presented in a drop-down list. The Euler solution is the simplest method, but it typically requires five samples to complete an accurate solution. The Midpoint and Runge-Kutta4 solutions are more accurate and require fewer samples, but they require more computational time. Setting the Samples value higher produces a more accurate solution.

Creating simple soft bodies

In the Simple Soft Bodies rollout, use the Create Simple Soft Body button to automatically set the springs for the selected object to act like a soft body. You also can set the amount of Stretch and Stiffness the object has. For cloth, you want to use a high Stretch value and a low Stiffness value, but a racquetball would have both high Stretch and Stiffness values.

Tip
You can manually set the spring settings for the object using the Advanced Springs rollout.

Painting weights

When you select the Weights and Springs subobject mode, the spring vertices are displayed on the object. The vertices are colored to reflect their weight. By default, the vertices that are farthest from the object's pivot point have the lowest weight value. Vertices with the greatest weight value (closest to 1) are colored red, and spring vertices with the lowest weight value (closest to -1) are blue. Vertices in between these two values are orange and yellow. The higher-weighted vertices move the greatest distance, and the lower-weighted vertices move the least.

Selecting the Weights and Springs subobject also enables the Paint button in the Weights and Painting rollout. Clicking this button puts you in Paint mode, where you can change the weight of the spring vertices by dragging a paint gizmo over the top of the object in the viewports. As you paint the spring vertices, they change color to reflect their new weights.

The Strength value sets the amount of weight applied to the vertices. This value can be negative. The Radius and Feather settings change the size and softness of the Paint brush. Figure 33.13 shows a dinosaur model with the Flex modifier applied. The dinosaur has had some weights painted so its tail, arms, and neck move under the influence of the Flex modifier. The blue vertices mark those vertices with low weight values that don't move as much.

FIGURE 33.13

Use the Paint button to change the weight of the spring vertices.

The weights applied using the Paint button are relative to the existing vertex weight. If you select the Absolute Weight option, the Vertex Weight value is applied to the selected vertices.

Adding Forces and Deflectors

To see the effect of the Flex modifier, you need to add some motion to the scene. The flex object flexes only when it is moving. One of the easiest ways to add motion to the scene is with Space Warps.

The Forces and Deflectors rollout includes two lists: one for Forces and one for Deflectors. Below each are Add and Remove buttons. Using these buttons, you can add and remove Space Warps from the list. The Forces list can use any of the Space Warps in the Forces subcategory (except for Path Follow). The Deflector list can include any of the Space Warps in the Defectors subcategory.

Tip

When you add Space Warps to the Forces and Deflectors list for the Flex modifier, they do not need to be bound to the object.

Manually creating springs

The final two rollouts for the Flex modifier are Advanced Parameters and Advanced Springs. The Advanced Parameters rollout includes settings for controlling the Start and End frames where the Flex modifier has an effect. The Affect All Points option ignores any subobject selections and applies the modifier to the entire object. The Set Reference button updates all viewports if any changes were made, and the Reset button resets all the vertices' weights to their default values.

You can use the Advanced Springs rollout to manually add and configure springs to the object. Clicking the Options button opens a dialog box where you can select the type of spring to add to the selected vertices, including Edge and Shape springs. Edge springs are applied to vertices at the edges of an object, and Shape springs are applied between vertices. For these advanced springs, you can set the Stretch Strength, Stretch Sway, Shape Strength, and Shape Sway.

Melt modifier

The Melt modifier simulates an object melting by sagging and spreading edges over time. Melt parameters include Amount and Spread values, Solidity (which can be Ice, Glass, Jelly, or Plastic), and a Melt Axis.

Figure 33.14 shows the Melt modifier applied to the snowman model (it was inevitable).

FIGURE 33.14

The Melt modifier slowly deforms objects to a flat plane.

PatchDeform and SurfDeform modifiers

Among the animation modifiers are several that are similar in function but that work on different types of objects. The PatchDeform modifier uses patches, and the SurfDeform modifier deforms an object according to a NURBS surface.

In the Parameters rollout for each of these modifiers is a Pick Patch (or Surface) button that lets you select an object to use in the deformation process. After the object is selected, you can enter the Percent and Stretch values for the U and V directions, along with a Rotation value.

Note

The PatchDeform modifier is also available as a World Space Modifier (WSM). WSMs are similar to the normal Object Space Modifiers (OSM), except that they use World Space coordinates instead of Object Space coordinates. The most noticeable differences are that WSMs don't use gizmos and that the OSM moves the patch to the object, while the WSM causes the object to move to the patch.

Tutorial: Deforming a plane going over a hill

Have you seen those commercials that use rubber cars to follow the curvature of the road as they drive? In this tutorial, you use the PatchDeform modifier to bend a plane over a hill made from a patch.

To deform a plane according to a patch surface, follow these steps:

1. Open the Plane bending over a hill.max file from the Chap 33 folder in the downloaded content set.

   This file contains a simple hill made from patch objects and a plane model.

2. Select the plane model, and choose Modifiers→Animation→Patch Deform. Then click the Pick Patch button in the Parameters rollout, and select the hill object.

   This applies the World Space Patch Deform modifier to the airplane object and moves the airplane to align with the hill.

3. Set the V Percent value to **50** in the Parameters rollout. Select the Gizmo subobject under the PatchDeform modifier in the Modifier Stack and move it so it aligns with the hill object.

4. Click the Auto Key button (or press the N key) to enable key mode, and drag the Time Slider to frame 100. Then set the V Percent value to **-50**, and click the Auto Key button again to disable key mode.

4. Click the Play button (or press the / key) to see the plane deform over the hill.

Figure 33.15 shows the results of this tutorial.

FIGURE 33.15

The plane model hugs the hill, thanks to the PatchDeform modifier.

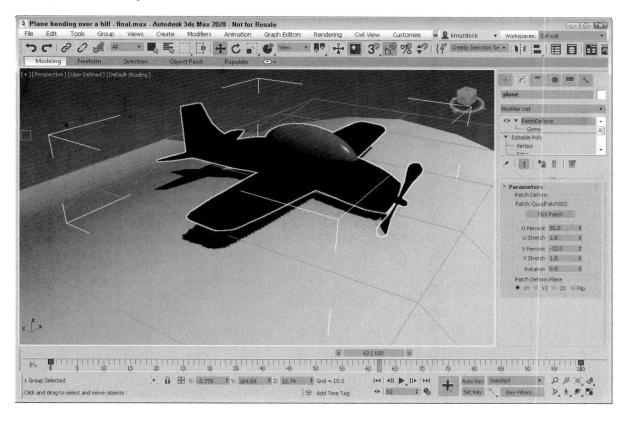

PathDeform modifier

The PathDeform modifier uses a spline path to deform an object. The Pick Path button lets you select a spline to use in the deformation process. You can select either an open or closed spline. The Parameters rollout also includes spinners for controlling the Percent, Stretch, Rotation, and Twist of the object. The Percent value is the distance the object moves along the path.

Note

If you use the PathDeform modifier, you can benefit from using the Follow/Bank utility, which gives you control over how the object follows and banks along the path.

Figure 33.16 shows some text wrapped around a spline path.

FIGURE 33.16

The text in this example has been deformed around a spline path using the PathDeform modifier.

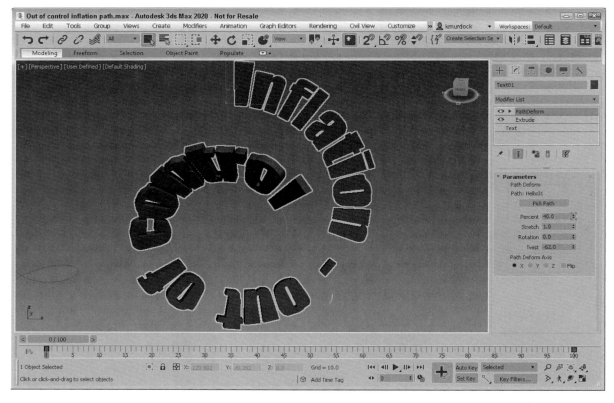

LinkedXForm modifier

The Linked XForm modifier passes all transformations of one object onto another, but not vice versa. The object that controls the transformation is designated as the Control Object and is selected via the Pick Control Object button (which is the only control in the Parameters rollout for this modifier). After the Control object is selected, the Control Object controls the selected object's transforms, but the object that is being controlled can move independently of the control object without affecting the control object.

SplineIK Control modifier

The SplineIK Control modifier can be applied only to spline objects. In the Spline IK Control Parameters rollout, you can click the Create Helpers button, which adds a dummy object to every vertex on the spline.

These dummy objects make it much easier to control the spline without having to enter vertex subobject mode. You also can specify how the dummy objects are linked and how they are displayed.

Summary

The Animation Layers feature provides a simple way to organize animated motions into an easy-to-manage method. The Animation Layers toolbar includes all the tools you need to manage these unique layers. The Animation→Load Animation option, along with its mapping and retargeting features, lets you reuse saved animations by applying them to other scenes. This chapter also introduced several of the available animation modifiers and showed you how to use them. In this chapter, you learned about the following:

* Using the Animation Layers toolbar
* Creating new layers and using weights
* Collapsing animation layers
* Loading saved XML animation files
* Remapping animation tracks between objects
* Retargeting to adjust for a change in scale
* Using the Point Cache modifier
* Using the Morpher modifier to deform a face
* Using the Flex modifier and the other deformation modifiers
* Using the various miscellaneous animation modifiers

The next chapter takes a close look at wiring parameters so that one parameter can control another, and creating custom parameters.

Chapter 34
Wiring Parameters

IN THIS CHAPTER

This chapter looks at a unique way to drive animations based on object parameters. Parameters of one object can be wired to parameters of another object so that when one parameter changes, the wired parameter changes with it. For example, you can wire the On/Off parameter of a light to the movement of a switch. All parameters that can be animated can be wired.

As long as you are working with parameters, the Autodesk® 3ds Max® 2020 software includes several helpful tools for viewing and working with the available parameters, including the Parameter Collector. If the Parameter Collector doesn't gather the exact parameters that you need, you can create your own custom parameters also.

Wiring Parameters

When parameters are wired together, the value of one parameter controls the value of the parameter to which it is wired. This is a powerful animation technique that lets a change in one part of the scene control another aspect of the scene. Another way to use wired parameters is to create custom animation controls such as a slider that dims a light source that animators can use as needed.

Using the Parameter Wiring dialog box

You can access the Parameter Wiring dialog box in several places. The Animation→Wire Parameters (Ctrl+5) menu makes a pop-up menu of parameters appear. Selecting a parameter from the menu changes the cursor to a dotted line (like the one used when linking objects). Click the object that you want to wire to, and another pop-up menu lets you choose the parameter to wire to. The Parameter Wiring dialog box appears with the parameter for each object selected from a hierarchy tree.

You also can wire parameters by selecting an object and right-clicking to access the quad menu and selecting Wire Parameters. The Wire Parameters option is disabled if multiple objects are selected.

The Parameter Wiring dialog box (Alt+5), shown in Figure 34.1, displays two tree lists containing all the available parameters. This tree list looks very similar to the Track View and lets you connect parameters in either direction or to each other. If you used the Wire Parameters feature to open the Parameter Wiring dialog box, the parameter for each object is already selected and highlighted in blue.

FIGURE 34.1

The Parameter Wiring dialog box lets you connect parameters so they affect one another.

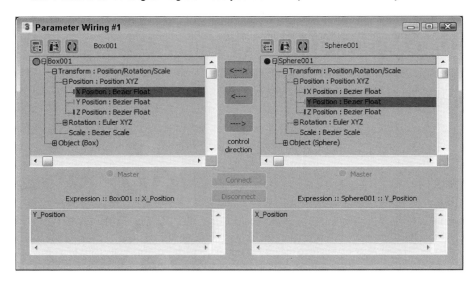

The three arrow buttons between the two tree lists let you specify the connection direction. These buttons connect the parameter in one pane to the selected parameter in the opposite pane. The direction determines whether the parameter in the left pane controls the parameter in the right pane, or vice versa. You also can select the top bidirectional button to make the parameters mutually affect each other. Below each tree list is a text area where you can enter an expression. An *expression* is a mathematical statement that follows a specific syntax for defining how one parameter controls the other. These expressions can be any valid expression that is accepted in the Animation Controller dialog box or in MAXScript.

After the parameters are selected, click the direction button and enter an expression if needed. Then click the Connect button to complete the wiring. Based on the connection direction, the Master radio button indicates which object controls the other. You also can use this dialog box to disconnect existing wired parameters. You can use the icon buttons at the top of the dialog box, shown in Table 34.1, to Show All Tracks, to find the next wired parameter, and to refresh the tree view.

TABLE 34.1 Parameter Wiring Dialog Box Icons

Button	Description
📋	Show All Tracks
🔍	Next Wired Parameter
( )	Refresh Tree View

After the wiring is completed, the Parameter Wiring dialog box remains open. You can try out the wiring by moving the master object. If the results aren't what you wanted, you can edit the expression and click the Update button (the Connect button changes to an Update button).

Note

When you select objects to be wired, the order in which the objects are selected doesn't matter because in the Parameter Wiring dialog box you can select the direction of control.

After you've made a connection in the Parameter Wiring dialog box, the parameter for the controlling object turns green, and the parameter of the object being controlled turns red. If you make a bidirectional connection, both tracks turn green.

Manipulator helpers

To create general-use controls that can be wired to control various properties, 3ds Max includes three manipulator helpers. These helpers are Cone Angle, Plane Angle, and Slider. They are available as a subcategory named Manipulators under the Helpers category of the Create panel or in the Create→Helpers→Manipulators menu. These manipulators, like Dummy objects, are not rendered along with the scene.

For the Cone Angle helper, you can set the Angle, Distance, and Aspect settings. The default cone base is a circle, but you can make it a square. The Plane Angle helper includes settings for Angle, Distance, and Size. This manipulator is helpful for controlling light and camera parameters.

The Plane Angle helper places a straight line that you can manipulate through an angle. It is useful for rotating objects about a point and makes the rotation angle parameter available.

You can name the Slider helper in the Label field. This name appears in the viewports above the Slider object. You also can set a default value along with maximum and minimum values. To position the object, you can set the X Position, Y Position, and Width settings. You also can set a snap value for the slider. The Hide option lets you hide the Slider if it isn't needed any more.

Once created, you can use these manipulator helpers when the Select and Manipulate button on the main toolbar is enabled (this button must be disabled before the manipulator helpers can be created). The advantage of these helpers is in wiring parameters to be controlled using the helpers.

Tutorial: Controlling a dragon's bite

One way to use manipulator helpers and wired parameters is to control within limits certain parameters that can be animated. This gives your animation team controls they can use to quickly build animation sequences. In this example, you use a slider to control a dragon's jaw movement.

To create a slider to control a dragon's bite, follow these steps:

1. Open the Biting Little Dragon.max file from the Chap 34 directory in the downloaded content set.

   This file includes the Little Dragon model. For this model, the head and upper teeth have been joined into a single object, and the pivot point for this object has been moved to where the jaw hinges.

2. Select Create→Helpers→Manipulators→Slider, and click and drag in the Perspective view above the dragon to create a slider object. Name the slider **Dragon Bite** in the Label field in the Command Panel, and set the Maximum value to **60**.

3. With the Slider selected, choose Animation→Wire Parameter→Parameter Wire Dialog (or press the Alt+5).

   The Parameter Wiring dialog box appears.

4. In the Parameter Wiring dialog box, select the Slider's value in the left selection list. It is located at Objects→sliderManipulator001→Object (sliderManipulator)→value. In the right pane, select the Y Rotation of the dragon's head, located at Objects→head→Transform:Position/Rotation/Scale→Rotation:Euler XYZ→Y Rotation. Then, click the direction arrow that points from the Slider to the head so that the slider is set to control the head. Then click the Connect button. This connects the slider value to the rotation of the crocodile's jaw. Then enable the Select and Manipulate button on the main toolbar, and drag the Slider.

   The crocodile's jaw spins around erratically. This happens because the rotation values are in radians and you need them in degrees.

5. Select the Y Rotation track in the right selection pane of the Parameter Wiring dialog box. In the expression text area under the head object, change where it says **value** to the expression

degToRad(value) and click the Update button. Figure 34.2 shows the Parameter Wiring dialog box. Then drag the Slider again.

The degToRad() expression is a special function that converts the value from radians to degrees. Now the values are in degrees, and the range is correct.

FIGURE 34.2

The Parameter Wiring dialog box lets you enter expressions that get calculated as you move the manipulator.

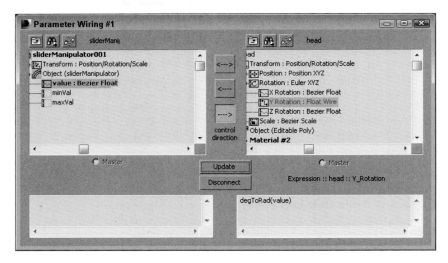

6. Click the Select and Manipulate button on the main toolbar if it is not already enabled, and drag the slider to the right.

    The dragon's mouth opens in proportion to the slider's value.

Figure 34.3 shows the dragon biting using the slider control.

FIGURE 34.3

A slider control is wired to open the dragon's mouth.

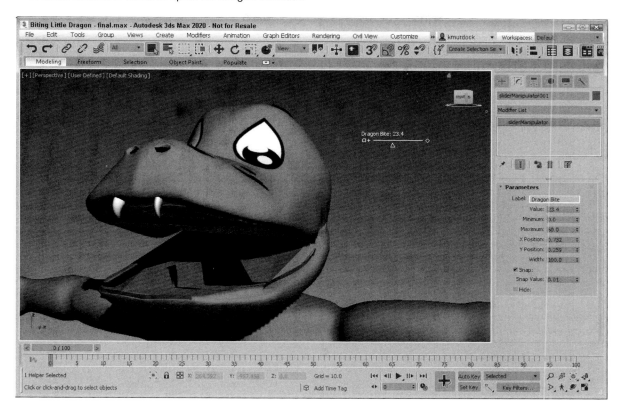

Collecting Parameters

To help in organizing the various parameters that you use to animate a scene, you can use the Parameter Collector to gather all custom and animated parameters used in the scene. The Parameter Editor can be used to create custom attributes and parameters, but the custom attributes are attached to the specific object or element that was selected when the attribute was created. This can make finding the custom attributes difficult, but 3ds Max includes another tool that you can use to collect all these custom attributes into a single location—the Parameter Collector.

Open the Parameter Collector dialog box, shown in Figure 34.4, with the Animation→Parameter Collector (Alt+2) menu command. This dialog box lets you gather a set of parameters into a custom rollout that can be opened and accessed from anywhere. This provides a convenient way to compile and look at only the parameters that you need to animate a certain task. Under the menus are several toolbar buttons, explained in Table 34.2.

FIGURE 34.4

The Parameter Collector dialog box is used to gather several different parameters into a custom rollout.

Select parameter marker   Select rollout marker   Properties

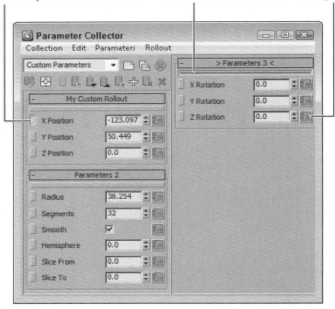

TABLE 34.2 Parameter Collector Toolbar Buttons

Toolbar Button	Name	Description
	Collection Name	Enters a name for the current collection or selects an existing collection.
	New Collection	Creates a new collection of parameters.
	Duplicate Collection	Creates a copy of the existing collection.
	Delete Collection	Deletes the current collection.
	Multiple Edits	Toggle button that allows multiple parameters to be changed at once when enabled.
	Absolute/Relative	Toggle button that maintains the current value in Absolute mode and resets the value to 0 when the mouse is released in Relative mode.
	Key Selected	Creates a key for the selected parameters when the Auto Key mode is enabled.
	Reset Selected	Sets the selected parameter values to 0.
	Move Parameters Down	Moves the selected parameters downward in the rollout order.
	Move Parameters Up	Moves the selected parameters upward in the rollout order.
	Add to Selected Rollout	Opens a Track View Pick dialog box where you can select the parameter to add to the selected rollout.
	Add to New Rollout	Opens a Track View Pick dialog box where you can select the parameter to add to a new rollout.
	Delete Selected	Deletes the selected parameters.

	Delete All	Deletes all parameters in the current collection.
	Properties	Opens the Key Info dialog box for the selected parameter if a key is set. This button is found to the right of a parameter.

Within the Parameter Collection dialog box, you can create and name new rollouts, add parameters to these rollouts using a Track View Pick dialog box, and save multiple rollouts into collections that can be recalled.

Collections are named by typing a new name in the drop-down list located in the upper-left corner of the interface. This drop-down list holds all available collections. The Collection menu (or the toolbar buttons) may be used to create, rename, duplicate, or delete a collection.

A Parameter Collection can include multiple rollouts. The current active rollout is marked with a blue bar directly under the rollout title and with brackets that surround the rollout name. Using the Rollout menu (or the toolbar buttons) you can create a new rollout, or rename, reorder, or delete existing rollouts.

Parameters are added to the rollouts using the Parameters→Add to Selected and the Parameters→Add to New Rollout menu commands. Both of these commands open a Track View Pick dialog box where you can select the specific parameter to add to the current rollout. To the left of each parameter is a small box that you can use to select the parameter. You also can select parameters using the Edit menu commands.

If multiple parameters are selected, you can change the values for all selected parameters at the same time by enabling the Edit→Multiple Edits option. With the Multiple Edits option enabled, changing any parameter value also changes all other selected parameters.

Caution
Multiple parameters' values can be changed together only if they are of the same type.

At the bottom of the Edit menu is the Edit Notes menu command. Using this menu command, you can change the parameter's name, link it to a URL, and type some notes about how this parameter works.

The Parameter Collector dialog box also can be used to create animation keys. To create a key for the selected parameters, you'll need to enable the Auto Key button and then select the Parameters→Key Selected or Parameters→Key All menu commands. If a key exists for the selected parameter, the Properties button to the right of the parameter becomes active and displays the Key Info dialog box when clicked.

Using the Collection→Show Selected Keys in Track Bar menu command, you can see the keys for the selected parameters regardless of whether the objects are selected in the viewports.

Adding Custom Parameters

Another useful way to expand the number of parameters is to create custom parameters. These custom parameters can define some aspect of the scene that makes sense to you. For example, if you create a model of a bicycle, you can define a custom parameter for the pedal rotation. You can add your own custom parameters using the Parameter Editor dialog box, shown in Figure 34.5. You can open this modeless panel by choosing Animation→Parameter Editor (or by pressing the Alt+1 keys).

FIGURE 34.5

You can use the Parameter Editor dialog box to create custom parameters.

Pick Explicit Track

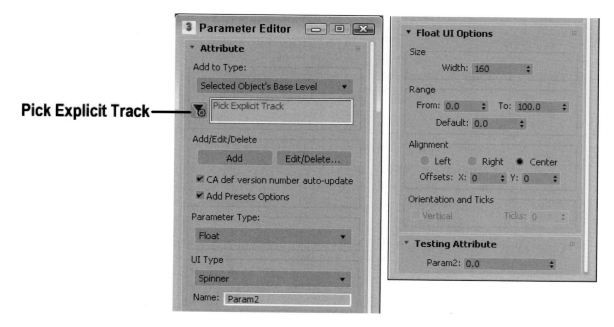

The Add to Type drop-down list at the top of the Attribute rollout in the Parameter Editor dialog box lets you select where the custom attribute shows up. Custom attributes can be created for an object, for the selected modifier, for the object's material, or for any track found in the Track View. The Pick Explicit Track button opens a dialog box where you can select a specific track.

The Add button creates the custom attribute and adds it to a rollout named Custom Attributes for the specified element. If the specified element is selected, you can click the Edit/Delete button to open the Edit Attributes/Parameters dialog box, shown in Figure 34.6. All custom attributes associated with the selected element are displayed.

Note

Custom attributes show up in a rollout named Custom Attributes positioned beneath all the other rollouts, but if you add the Attribute Holder modifier to the object before creating the new attribute, the Custom Attributes rollout appears under the Attribute Holder modifier.

FIGURE 34.6

The Edit Attributes/Parameters dialog box lets you edit or delete custom attributes.

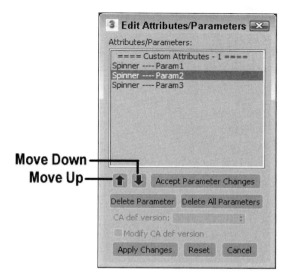

Move Down
Move Up

The Edit Attributes/Parameters dialog box lets you select and reorder the custom attributes within their rollout. Selecting a custom attribute also loads its settings into the Parameter Editor where they can be changed.

The Parameter Type drop-down list lets you choose the parameter format. Possibilities include Angle, Array, Boolean (true or false), Color, Float (a decimal point number), fRGBA, Integer, Material, Node, Percent, String, TextureMap, and WorldUnits. The UI Type drop-down list defines how the parameter is displayed in the rollout. How the parameter looks depends on the type of parameter. Float and integer values can be spinners or sliders, Boolean values can be check boxes or Check buttons, array values are drop-down lists, combo Box or List Box, nodes are pick buttons (allowing you to select an object in the viewports), color and RGB values are color pickers, and texture maps are map buttons. You also can name the parameter.

The Options rollout changes depending on which parameter type was selected. These rollouts contain settings for the interface's Width, value ranges, default values, Alignment (left, right, or center), and list items.

The Testing Attribute rollout shows what the interface element will look like and lets you change the attribute to see how the custom parameter works.

The value of custom attributes becomes apparent when you start wiring parameters.

Summary

This chapter covered the feature used to wire parameters together in the Parameter Wire dialog box, which opens up a whole new way to control objects. This chapter also covered how you can create new parameters with the Parameter Collector dialog box. All of these tools give you lots of control over the scene using parameters and expressions. In this chapter, you accomplished the following:

* Wired parameters

* Gathered and edited several parameters at once with the Parameter Collector

* Created custom parameters with the Parameter Editor

The next chapter shows how to automate the creation of animation keys with constraints and the basic controllers.

Chapter 35

Animating with Constraints and Simple Controllers

IN THIS CHAPTER

Using constraints

Attaching an object to the surface of an object

Making an object travel along a path with the Path constraint

Controlling the weighted position and orientation of objects

Shifting between two controlling objects using the Link constraint

Following objects with the LookAt constraint

Understanding the controller types

Assigning controllers using the Motion panel and the Track View

Setting default controllers

When you first begin animating and working with keys, having the Autodesk® 3ds Max® 2020 software figure out all the frames between the start and end keys seems amazing, especially if you've ever animated in 2D by drawing every frame. But soon you realize that animating with keys can be time-consuming for complex realistic motions, and again, 3ds Max comes to the rescue. You can use animation constraints and controllers to automate the creation of keys for certain types of motions.

Constraints and controllers store and manage the key values for all animations in 3ds Max. When you animate an object using the Auto Key button, the default controller is automatically assigned. You can change the assigned controller or alter its parameters using the Motion panel or the Track View.

This chapter explains how to work with constraints and some simple controllers. For example, you can use the Noise controller to add random motion to a flag blowing in the wind or use the Surface constraint to keep a bumper car moving over the surface.

Restricting Movement with Constraints

The trick of animating an object is to make it go where you want it to go. Animating objects deals not only with controlling the motion of the object but also with controlling its lack of motion. Constraints are a type of animation controller that you can use to restrict the motion of an object.

Using these constraints, you can force objects to stay attached to another object or follow a path. For example, the Attachment constraint can be used to make a robot's feet stay connected to a ground plane as it moves. The purpose of these constraints is to make animating your objects easier.

Using constraints

You can apply constraints to selected objects using the Animation→Constraints menu. The constraints contained within this menu include Attachment, Surface, Path, Position, Link, LookAt, and Orientation.

All constraints have the same controller icon displayed in the Motion panel or the Track View.

 After you select one of the constraints from the Animation→Constraints menu, a dotted link line extends from the current selected object to the mouse cursor. You can select a target object in any of the viewports to apply the constraint. The cursor changes to a plus sign when it is over a target object that can be selected. Selecting a constraint from the Constraints menu also opens the Motion panel, where the settings of the constraint can be modified.

You also can apply constraints using the Assign Controller button found in the Motion panel and in the Track View window.

Find out more about the Track View window in Chapter 36, "Editing Animation Curves in the Track View."

Working with the constraints

Each constraint is slightly different, but learning how to use these constraints will help you control the animated objects within a scene. You can apply several constraints to a single object. All constraints that are applied to an object are displayed in a list found in the Motion panel. From this list, you can select which constraint to make active and which to delete. You also can cut and paste constraints between objects.

Attachment constraint

The Attachment constraint determines an object's position by attaching it to the face of another object. This constraint lets you attach an object to the surface of another object. For example, you could animate the launch of a rocket ship with booster rockets that are attached with the Attachment constraint. The booster rockets would move along with the ship until the time when they are jettisoned.

The pivot point of the object that the constraint is applied to is attached to the target object. At the top of the Attachment Parameters rollout is a Pick Object button for selecting the target object to attach to. You can use this button to change the target object or to select the target object if the Animation→Constraints menu wasn't used. There is also an option to align the object to the surface. The Update section enables you to manually or automatically update the attachment values.

Note

The Attachment constraint shows up in the Position track of the Assign Controller rollout as the Position List controller. To minimize the effect of other controllers, set their Weight values in the Position List rollout to 0.

The Key Info section of the Attachment Parameters rollout displays the key number and lets you move between the various keys. The Time value is the current key value. In the Face field, you can specify the exact number of the face to attach to. To set this face, click the Set Position button and drag over the target object. The A and B values represent Barycentric coordinates for defining how the object lies on the face. You can change these coordinate values by entering values or by dragging the red cross-hairs in the box below the A and B values. The easiest way to position an object is to use the Set Position button to place the object and then to tweak its position with the A and B values. The Set Position button stays active until you click it again.

The TCB section sets the Tension, Continuity, and Bias values for the constraint. You also can set the Ease To and Ease From values.

Tutorial: Attaching eyes to a melting snowman

When part of a model is deformed, such as applying the Melt modifier to a snowman's body, smaller parts like the eyes either get left behind or get the full weight of the modifier applied to them. If the Melt modifier weren't applied to these items, they would stay floating in the air while the rest of the snowman melted about them. This problem can be fixed with the Attachment constraint, which causes the eyes to remain attached to the snowball as it melts.

The tutorial where the Melt modifier is applied to the snowman is included in Chapter 33, "Using Animation Layers and Animation Modifiers."

To constrain the solid objects to a melting snowman, follow these steps:

1. Open the Melting snowman.max file from the Chap 35 directory in the downloaded content set.

   This file includes the melting snowman file from another chapter with the Melt modifier applied to all objects.

2. Select the left eye object in the scene. In the Modifier Stack, select the Melt modifier and click the Remove Modifier from the Stack button to throw that modifier away.

3. With the left eye still selected, select Animation→Constraints→Attachment Constraint. A connecting line appears in the active viewport. Click the top snowball to select it as the attachment object. This moves the eye object to the top of the snowball where the snowball's first face is located.

4. In the Attachment Parameters rollout, change the Face value until the eye is positioned where it should be. This should be around face 315. Then change the A and B values (or drag in the Position graph) to position the eye where it looks good.

5. Repeat Step 5 for the right eye and for any other objects in the scene that you want to attach.

6. Click the Play button (/) and notice that the snow melts, but the eye objects stay the same size.

Figure 35.1 shows the resulting melted snowman. Notice how the carrot nose still melts because it is not attached to the head.

FIGURE 35.1

The Attachment constraint sticks one object to the surface of another.

Surface constraint

The Surface constraint moves an object so that it is on the surface of another object. The object with Surface constraint applied to it is positioned so that its pivot point is on the surface of the target object. You can use this constraint only on certain objects, including spheres, cones, cylinders, toruses, quad patches, loft objects, and NURBS objects.

In the Surface Controller Parameters rollout is the name of the target object that was selected after the menu command. The Pick Surface button enables you to select a different surface to attach to. You also can select specific U and V Position values. Alignment options include No Alignment, Align to U, Align to V, and a Flip toggle.

Note

Don't be confused because the rollout is named Surface Controller Parameters instead of Surface Constraint Parameters. The developers at Autodesk must have missed this one.

Tutorial: Rolling a tire down a hill with the Surface constraint

Moving a vehicle across a landscape can be a difficult procedure if you need to place every rotation and position key, but with the Surface constraint, it becomes easy. In this tutorial, you use the Surface constraint to roll a tire down a hill.

To roll a tire down a hill with the Surface constraint, follow these steps:

1. Open the Tire rolling on a hill.max file from the Chap 35 directory in the downloaded content set.

This file includes a patch grid hill and a wheel object made from primitives. The center hub and spokes are all linked to the outer tire, so they all move together.

2. Create a dummy object from the Helpers category, and position the dummy object's pivot point at the bottom of the tire and the top of the hill. Then, link the tire object to the dummy object as a child. This causes the tire to move along with the dummy object.

3. Select the dummy object, choose Animation→Constraints→Surface Constraint, and select the hill object.

4. In the Surface Controller Parameters rollout, select the Align to V and Flip options to position the dummy and tire at the top of the hill. Set the V Position value to **50** to move the tire to the center of the hill.

5. Click and select the Linear option for the Default In/Out Tangents button to the right of the Set Key button, then click the Auto Key button (or press the N key), drag the Time Slider to frame 100, and change the U Position to **100**. Then, select the tire object and drag counterclockwise in the Left viewport for a full revolution (360 degrees).

Changing the U Position value causes the tire to move along the surface of the hill. Changing the rotation of the tire makes it rotate as it moves down the hill.

6. Click the Auto Key button again to deactivate it, and click the Play Animation button to see the tire move and rotate down the hill.

Figure 35.2 shows the tire as it moves down the hill.

FIGURE 35.2

The Surface constraint can animate one object moving across the surface of another.

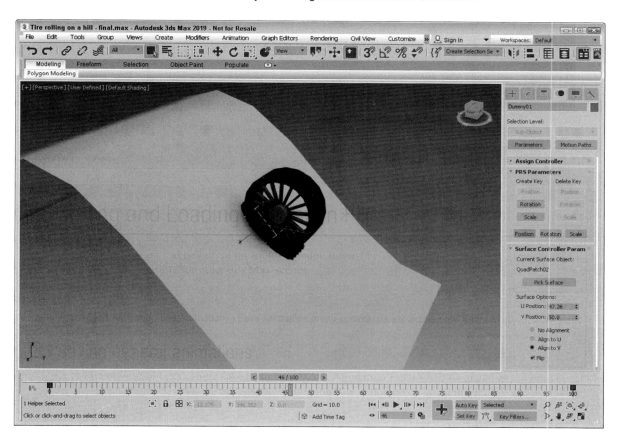

Path constraint

The Path constraint lets you select a spline path for the object to follow. The object is locked to the path and follows it even if the spline is changed. This is one of the most useful constraints because you can control the exact motion of an object using a spline. With the software's spline features, you can control very precisely the motions of objects that are constrained with the Path constraint. A good example of this constraint is an animated train following a track. Using a spline to create the train tracks, you can easily animate the train using the Path constraint.

When you choose the Animation→Constraints→Path Constraint menu command, you can select a single path for the object to follow. This path is added to a list of paths in the Path Parameters rollout.

The Path Parameters rollout also includes Add and Delete Path buttons for adding and deleting paths to and from the list. If two paths are added to the list, the object follows the position centered between these two paths. By adjusting the Weight value for each path, you can make the object favor a specific path.

The Path Options include a % Along Path value for defining the object's position along the path. This value ranges from 0 at one end to 100 at the other end. The Follow option causes the object to be aligned with the path as it moves, and the Bank option causes the object to rotate to simulate a banking motion.

The Bank Amount value sets the depth of the bank, and the Smoothness value determines how smooth the bank is. The Allow Upside Down option lets the object spin completely about the axis, and the Constant Velocity option keeps the speed regular. The Loop option returns the object to its original position for the last frame of the animation, setting up a looping animation sequence. The Relative option lets the object maintain its current position and does not move the object to the start of the path. From its original position, it follows the path from its relative position. At the bottom of the Path Parameters rollout, you can select the axis to use.

Tutorial: Creating a spaceship flight path

Another way to use splines is to create animation paths. As an example, you use a Line spline to create an animation path. You can use splines for animation paths in two ways. One way is to create a spline and have an object follow it using either the Path constraint or the Path Follow Space Warp. The first vertex of the spline marks the first frame of the animation. The other way is to animate an object and then edit the Trajectory path.

In this tutorial, you use a simple path and attach it to a spaceship model. To attach an object to a spline path, follow these steps:

1. Open the Spaceship and asteroids.max file from the Chap 35 directory in the downloaded content set.

    This file contains the spaceship model and several asteroid objects. The spaceship hull is the main object you need to select. All other objects, such as the canopy, are linked to this hull object.

2. Select Create→Shapes→Line, and click and drag in the Top viewport to create an animation path that moves the spaceship through the asteroids. Right-click when the path is complete. Then select the Modify panel, click the Vertex button in the Selection rollout to enable Vertex subobject mode, and edit several vertices in the Front viewport. Then right-click to exit vertex subobject mode.

3. With the spaceship selected, choose Animation→Constraints→Path Constraint. Then click the animation path to select it as the path to follow. Select the Follow option in the Path Parameters rollout of the Motion panel, and choose the X-Axis option.

4. Click the Play Animation button in the Time Controls to see the spaceship follow the path.

Figure 35.3 shows the spaceship as it moves between the asteroids.

FIGURE 35.3

The spaceship object has been attached to a spline path that it follows.

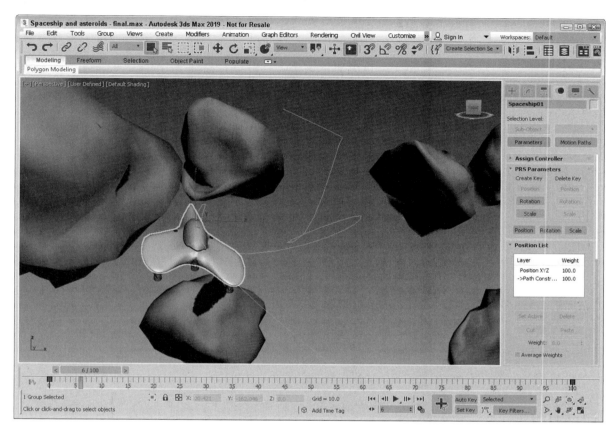

Position constraint

You can use the Position constraint to tie the position of an object to the weighted position of several target objects. For example, you could animate a formation of fighter jets by animating one of the jets and using Position constraints on all adjacent jets.

The Position constraint menu option lets you select a single target object, enabling you to place the pivot points of the two objects on top of one another. To add another target object, click the Add Position Target button in the Position Constraint rollout in the Motion panel. This button enables you to select another target object in the viewports; the target name appears within the target list in the rollout.

If you select a target name in the target list, you can assign a weight to the target. The constrained object is positioned close to the object with the higher weighted value. The Weight value provides a way to center objects between several other objects. The Keep Initial Offset option lets the object stay in its current location, but centers it relative to this position.

Figure 35.4 shows a rock positioned between four tree objects using the Position constraint. Notice how the weight of the downhill tree object is weighted higher than the other targets and the sled rock is close to it.

FIGURE 35.4

You can use the Position constraint to control the position of an object in relation to its targets.

Link constraint

The Link constraint can transfer hierarchical links between objects. This constraint can cause a child's link to be switched during an animation. Any time you animate a complex model with a dummy object, the Link constraint makes it possible to switch control from one dummy object to another during the animation sequence. This keeps the motions of the dummy objects simple.

The Link Params rollout includes Add Link and Delete Link buttons, a list of linked objects, and the Start Time field. To switch the link of an object, enter for the Start Time the frame where you want the link to switch, or drag the Time Slider and click the Add Link button. Then select the new parent object. The Delete key becomes active when you select a link in the list.

Note

If you create a link using the Link constraint, the object is not recognized as a child in any hierarchies.

All links are kept in a list in the Link Params rollout. You can add links to this list with the Add Link button, create a link to the world with the Link to World button, or delete links with the Delete Link button. The Start Time field specifies when the selected object takes control of the link. The object listed in the list is the parent object, so the Start Time setting determines when each parent object takes control.

The Key Mode section lets you choose a No Key option. This option does not write any keyframes for the object. If you want to set keys, you can choose the Key Nodes options and set keys for the object itself (Child option) or for the entire hierarchy (Parent option). The Key Entire Hierarchy sets keys for the object and its parents (Child option) or for the object and its targets and their hierarchies (Parent option).

This constraint also includes the PRS Parameters and Key Info rollouts.

Caution

You cannot use Link constraints with Inverse Kinematics systems.

Tutorial: Skating a figure eight

For an animated object to switch its link from one parent to another halfway through an animation, you need to use the Link constraint. Rotating an object about a static point is easy enough: simply link the object to a dummy object, and rotate the dummy object. The figure-eight motion is more complex, but you can do it with the Link constraint.

To move an object in a figure eight, follow these steps:

1. Open the Figure skater skating a figure eight.max file from the Chap 35 directory in the downloaded content set.

   This file includes a figure skater model imported from Poser and two dummy objects. The figure skater is linked to the first dummy object (the one initially closest to the skater).

2. Click the Auto Key button (or press the N key), drag the Time Slider to frame 100, and rotate the first dummy object two full revolutions in the Top viewport.

3. Select the second dummy object, and rotate it two full revolutions in the opposite direction. Click the Auto Key button again to deactivate it.

Tip

If you enable the Angle Snap Toggle button on the main toolbar, it is easier to rotate objects exactly two revolutions.

4. With the figure skater selected, choose Animation→Constraints→Link Constraint. Then click the first dummy object (the top one in the Top viewport).

   The Link constraint is assigned to the figure skater.

5. Drag the Time Slider to frame 25, then in the Link Params rollout, click the Add Link button and pick the second dummy object (the bottom one in the Top viewport). This switches control of the skater to the second dummy object. Then drag the Time Slider to frame 75, and with the Add Link button still active, click again on the first dummy object. In the Motion panel when done, it should say Frame 0, Dummy01, Frame 25, Dummy02 and Frame 75, Dummy01.

6. Click the Play Animation button (or press the / key) to see the animation play.

Tip

Another way to accomplish this same motion is to create a spline of a figure eight and use the Path constraint.

Figure 35.5 shows the skater as she makes her path around the two dummy objects.

FIGURE 35.5

With the Link constraint, the figure skater can move in a figure eight by rotating about two dummy objects.

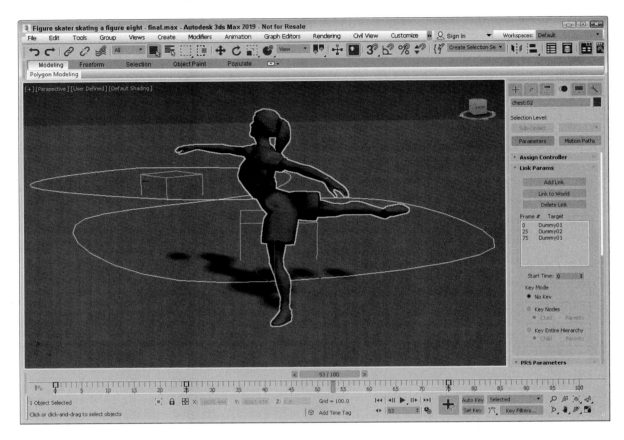

LookAt constraint

The LookAt constraint won't move an object, but it rotates the object so it is always orientated toward the target object. For example, you could use the LookAt constraint to animate a character's head that is watching a flying bumblebee. It is also very useful to apply to camera objects that follow a specific object throughout the animation.

After you select a target object, a single line extends from the object and points at the target object. This line, called the Viewline, is visible only within the viewports.

The LookAt Constraint rollout, like many of the other constraints, includes a list of targets. With the Add and Delete LookAt Target buttons, you can add and remove targets from the list. If several targets are on the list, the object is centered on a location between them. Using the Weight value, you can cause the various targets to have more of an influence over the orientation of the object. The Keep Initial Offset option prevents the object from reorienting itself when the constraint is applied. Any movement is relative to its original position.

You can set the Viewline Length, which is the distance that the Viewline extends from the object. The Viewline Length Absolute option draws the Viewline from the object to its target, ignoring the length value.

The Set Orientation button lets you change the offset orientation of the object using the Select and Rotation button on the main toolbar. If you get lost, the Reset Orientation button returns the orientation to its original position. You can select which local axis points at the target object.

The Upnode is an object that defines the up direction. If the LookAt axis ever lines up with the Upnode axis, the object flips upside-down. To prevent this, you can select which local axis is used as the LookAt axis and which axis points at the Upnode. The World is the default Upnode object, but you can select any object as the Upnode object by deselecting the World object and clicking the button to its right.

To control the Upnode, you can select the LookAt option or the Axis Alignment option, which enables the Align to Upnode Axis option. Using this option, you can specify which axis points toward the Upnode.

Caution

The object using the LookAt constraint flips when the target point is positioned directly above or below the object's pivot point.

When you assign the LookAt constraint, the Create Key button for rotation changes to Roll. This is because the camera is locked to point at the assigned object and cannot rotate; rather, it can only roll about the axis.

You can use the LookAt constraint to let cameras follow objects as they move around a scene. It is the default transform controller for Target camera objects.

Orientation constraint

You can use the Orientation constraint to lock the rotation of an object to another object. You can move and scale the objects independently, but the constrained object rotates along with the target object. A good example of an animation that uses this type of constraint is a satellite that orbits the Earth. You can offset the satellite and still constrain it to the Earth's surface. Then, as the Earth moves, the satellite follows.

In the Orientation Constraint rollout, you can select several orientation targets and weight them in the same manner as with the Position constraint. The target with the greatest weight value has the most influence over the object's orientation. You also can constrain an object to the World object. The Keep Initial Offset option maintains the object's original orientation and rotates it relative to this original orientation. The Transform Rule setting determines whether the object rotates using the Local or World Coordinate Systems.

Using the Walkthrough Assistant

One alternative to using the Path and LookAt constraints is to use the Walkthrough Assistant. This tool is accessed from the Animation menu. It opens up a utility panel with several rollouts, as shown in Figure 35.6. Using this panel, you can create a new camera, select a path, and set the viewport to use the created camera. You can then use the View Controls rollout to cause the view to tilt to the left or right as you move through the path. This automates the process of getting a camera to follow a path.

FIGURE 35.6

The Walkthrough Assistant automates several constraints into a single interface.

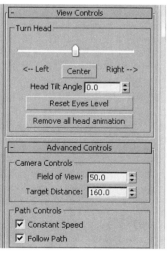

The Walkthrough Assistant also includes a Render Preview that you use to see the results. If you drag the Time Slider to a different frame and click the Render Preview pane, the preview is updated. At specific frames, you can drag the Turn Head slider to change where the camera is looking. You can even tilt the camera up and down as well as side to side.

In the Advanced Controls rollout (which appears only after a camera has been created) are options for changing the Field of View and the Target Distance, which is useful if you're using a Depth of Field effect. You also can set the camera to move at a constant speed and an option to cause the camera to follow the path.

Understanding Controller Types

Controllers are used to set the keys for animation sequences. Every object and parameter that is animated has a controller assigned, and almost every controller has parameters that you can alter to change its functionality. Some controllers present these parameters as rollouts in the Motion panel, and others use a Properties dialog box.

3ds Max has five basic controller types that work with only a single parameter or track and one specialized controller type that manages several tracks at once (the Transform controllers). The type depends on the type of values the controller works with. The types include the following:

* **Transform controllers:** A special controller type that applies to all transforms (position, rotation, and scale) at the same time, such as the Position, Rotation, Scale (PRS) controllers

* **Position controllers:** Control the position coordinates for objects, consisting of X, Y, and Z values

* **Rotation controllers:** Control the rotation values for objects along all three axes

* **Scale controllers:** Control the scale values for objects as percentages for each axis

* **Float controllers:** Used for all parameters with a single numeric value, such as Wind Strength and Sphere Radius

* **Point3 controllers:** Consist of color components for red, green, and blue, such as Diffuse and Background colors

Note
Understanding the different controller types is important. When you copy and paste controller parameters between different tracks, both tracks must have the same controller type.

Float controllers work with parameters that use float values, such as a sphere's Radius or a plane object's Scale Multiplier value. Float values are numbers with a decimal value, such as 2.3 or 10.99. A Float controller is assigned to any parameter that is animated. After it is assigned, you can access the function curves and keys for this controller in the Track View and in the Track Bar. Because Float and Point3 controllers are assigned to parameters and not to objects, they don't appear in the Animation menu.

Assigning Controllers

Any object or parameter that is animated is automatically assigned a controller. The controller that is assigned is the default controller. The Animation panel in the Preference Settings dialog box lists the default controllers and lets you change them. You also can change this automatic default controller using the Track View window or the transformation tracks located in the Motion panel.

Automatically assigned controllers

The default controllers are automatically assigned for an object's transformation tracks when the object is created. For example, if you create a simple sphere and then open the Motion panel (which has the icon that looks like a wheel), you can find the transformation tracks in the Assign Controller rollout. The default Position controller is Position XYZ, the default Rotation controller is Euler XYZ, and the default Scale controller is the Bézier Scale controller.

The default controller depends on the type of object. For example, the Barycentric Morph controller is automatically assigned when you create a Morph compound object, and the Master Point controller is automatically assigned to any vertices or control points subobjects that are animated.

Note

Because controllers are automatically assigned to animation tracks, they cannot be removed; they can only be changed to a different controller. There is no function to delete controllers.

Assigning controllers with the Animation menu

The easiest way to assign a controller to an object is with the Animation menu. Located under the Animation menu are four controller submenus consisting of Transform Controllers, Position Controllers, Rotation Controllers, Scale Controllers, and MCG Controllers.

Note

Although constraints are contained within a separate menu, they control the animating of keys just like controllers.

When a controller is assigned to an object using the Animation menu, the existing controller is not removed, but the new controller is added as part of a list along with the other controllers. You can see all these controllers in the Motion panel.

For example, Figure 35.7 shows the Motion panel for a sphere object that has the default Position XYZ controller assigned to the Position track. If you choose Noise from the Assign Controller dialog box, then the Position List controller is added to the Position track, of which Position XYZ and Noise are two available controllers. This lets you animate multiple motions such as the shimmy of a car with a bad carburetor as it moves down the road.

FIGURE 35.7

The Motion panel displays all transform controllers applied to an object.

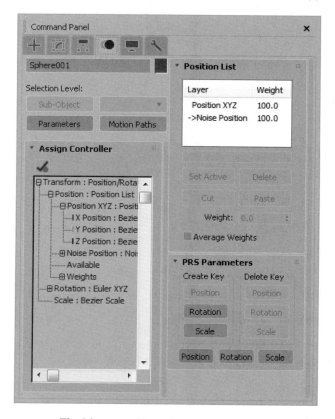

The List controller makes it possible to add several controllers to a single track. It also allows you to set Weights for each of its controllers. Using the Position List rollout, you can set the active controller and delete controllers from the list. You also can Cut and Paste controllers to other tracks.

Assigning controllers in the Motion panel

The top of the Motion panel includes two buttons: Parameters and Motion Paths. Clicking the Parameters button makes the Assign Controller rollout available.

To change a transformation track's controller, select the track and click the Assign Controller button positioned directly above the list. An Assign Controller dialog box opens that is specific to the track you selected.

For more about the Motion Paths button, see Chapter 32, "Understanding Animation and Keyframes."

For example, Figure 35.8 shows the Assign Position Controller dialog box for selecting a controller for the Position track. The arrow mark (>) shows the current selected controller. At the bottom of the dialog box, the default controller type is listed. Select a new controller from the list, and click OK. This new controller now is listed in the track, and the controller's rollouts appear beneath the Assign Controller rollout.

FIGURE 35.8

The Assign Position Controller dialog box lets you select a controller to assign.

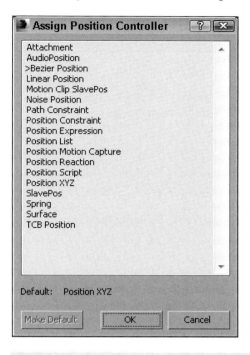

Note

Transformation controllers can be applied in the Motion panel, but the Track View can be used to apply controllers to all parameters including transforms.

Assigning controllers in the Track View

You also can use the Track View to assign controllers. To do this, locate and select the track to apply a controller to, and then click the Assign Controller button on the Controllers toolbar, choose the Edit→Controller→Assign (keyboard shortcut, C if the Keyboard Shortcut Override Toggle on the main toolbar is enabled) menu command, or right-click the track and select Assign Controller from the pop-up menu. An Assign Controller dialog box opens in which you can select the controller to use.

Chapter 36, "Editing Animation Curves in the Track View," covers the details of the Track View.

You also can use the Controller toolbar to copy and paste controllers between tracks, but you can paste controllers only to similar types of tracks. When you paste controllers, the Paste dialog box lets you choose to paste the controller as a copy or as an instance. Changing an instanced controller's parameters changes the parameters for all instances. The Paste dialog box also includes an option to replace all instances. This option replaces all instances of the controller, whether or not they are selected.

Setting default controllers

When you assign controllers using the Track View, the Assign Controller dialog box includes the option Make Default. With this option, the selected controller becomes the default for the selected track.

You also can set the global default controller for each type of track by choosing Customize→Preferences, selecting the Animation panel, and then clicking the Set Defaults button. The Set Controller Defaults dialog box opens, in which you can set the default parameter settings, such as the In and Out curves for the controller. To set the default controller, select a controller from the list and click the Set Defaults button to open a controller–specific dialog box where you can adjust the controller parameters. The Animation panel also includes a button to revert to the original settings.

Note

Changing a default controller does not change any currently assigned controllers.

Examining Some Simple Controllers

Now that you've learned how to assign controllers, let's look at some simple controllers.

Earlier in the chapter, I mentioned several specific controller types. These types define the type of data that the controller works with. This section covers the various controllers according to the types of tracks with which they work.

Note

Looking at the function curves for a controller provides a good idea of how you can control it, so many of the figures that follow show the various function curves for the different controllers.

Each of these controllers has a unique icon to represent it in the Track View. This makes them easy to identify.

Bézier controller

The Bézier controller is the default controller for many parameters. It enables you to interpolate between values using an adjustable Bézier spline. By dragging its tangent vertex handles, you can control the spline's curvature. Tangent handles produce a smooth transition when they lie on the same line, or you can create an angle between them for a sharp point. Figure 35.9 shows the Bézier controller assigned to a Position track.

FIGURE 35.9

The Bézier controller produces smooth animation curves.

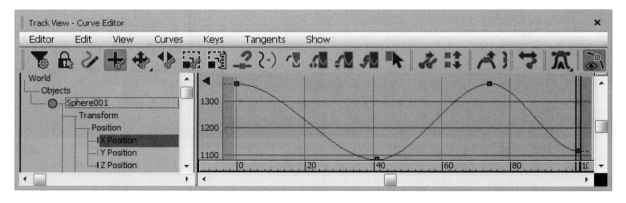

The Bézier controller parameters are displayed in the Motion panel under two rollouts: Key Info (Basic) and Key Info (Advanced).

At the top of the Key Info (Basic) rollout are two arrows and a field that shows the key number. The arrows let you move between the Previous and Next keys. Each vertex shown in the function curve represents a key. The Time field displays the frame number where the key is located. The Time Lock button next to the Time field can be set to prevent the key from being dragged in Track View. The value fields show the values for the selected track; the number of fields changes depending on the type of track that is selected.

At the bottom of the Key Info (Basic) rollout are two flyout buttons for specifying the In and Out curves for the key. The arrows to the sides of these buttons move between the various In/Out curve types. The curve types include Smooth, Linear, Step, Slow, Fast, Custom, and Tangent Copy.

The In and Out values in the Key Info (Advanced) rollout are enabled only when the Custom curve type is selected. These fields let you define the rate applied to each axis of the curve. The Lock button changes the two values by equal and opposite amounts. The Normalize Time button averages the positions of all keys. The Constant Velocity option interpolates the key between its neighboring keys to provide smoother motion.

Linear controller

The Linear controller interpolates between two values to create a straight line by changing its value at a constant rate over time.

The Linear controller doesn't include any parameters and can be applied to time or values. Figure 35.10 shows the curves from the previous example after the Linear controller is assigned—all curves have been replaced with straight lines.

FIGURE 35.10

The Linear controller uses straight lines.

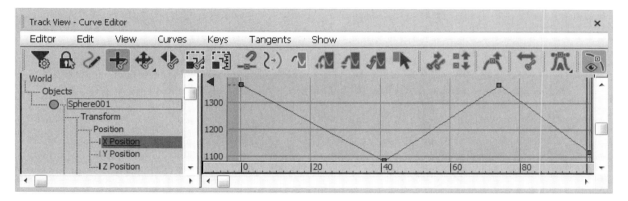

Noise controller

The Noise controller applies random variations in a track's values. In the Noise Controller dialog box, shown in Figure 35.11, the Seed value determines the randomness of the noise and the Frequency value determines how jagged the noise is. You also can set the Strength along each axis: The > (greater than) 0 option for each axis makes the noise values remain positive.

FIGURE 35.11

The Noise controller properties let you set the noise strength for each axis.

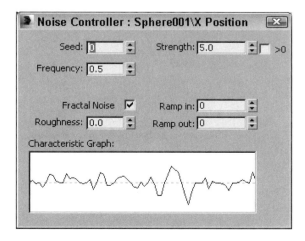

You also have an option to enable Fractal Noise with a Roughness setting.

The Ramp in and Ramp out values determine the length of time before or until the noise can reach full value. The Characteristic Graph gives a visual look at the noise over the range. Figure 35.12 shows the Noise controller assigned to the Position track. If you need to change any Noise properties, right-click the Noise track and select Properties from the pop-up menu.

FIGURE 35.12

The Noise controller lets you randomly alter track values.

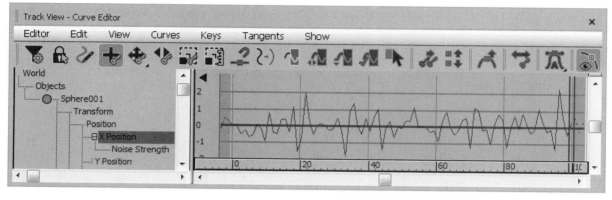

Spring controller

The Spring controller is similar in many ways to the Flex modifier in that it adds secondary motion associated with the wiggle of a spring after a force has been applied and then removed. When the Spring controller is applied, a panel with two rollouts appears. These rollouts, shown in Figure 35.13, let you control the physical properties of the spring and the forces that influence it.

FIGURE 35.13

The Spring controller rollouts can add additional springs and forces.

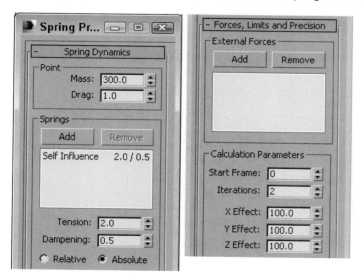

In the Spring Dynamics rollout, you can change the Mass and Drag values. Higher mass values result in greater secondary motion as the object is moved, and the Drag value controls how quickly the bouncing motion stops. You can add multiple springs, each with its own Tension and Damping values to be applied Relative or Absolute.

The Forces, Limits, and Precision rollout lets you add forces that affect the spring motion. The Add button lets you identify these forces, which are typically Space Warps, and you can limit the effect to specific axes.

Tutorial: Wagging a tail with the Spring controller

One of the best uses of the Spring controller is to gain the secondary motion associated with an existing motion. For example, if a character moves, an appendage such as a tail can easily follow if you apply a Spring controller to it.

To wag a row of spheres using the Spring controller, follow these steps:

1. Open the Dog wagging tail.max file from the Chap 35 directory in the downloaded content set.

   This file contains a dog made of primitives and a linked row of spheres with the top sphere animated rotating back and forth.

2. Select the smallest sphere, select the Position XYZ track in the Motion panel, and choose the Spring option from the Assign Controller dialog box. This moves the sphere to its parent. Choose the Select and Move button (or press the W key), and return the sphere to its original position.

3. Repeat Step 2 for the remaining spheres, moving from smallest to largest.

4. Click the Play Animation button (or press the / key) to see the resulting motion.

Figure 35.14 shows a frame of the final motion. Notice that the spheres aren't lined up exactly. The smallest sphere is moving the greatest distance because all the springs are adding their effect.

FIGURE 35.14

The Spring controller adds secondary motion to the existing motion of the largest sphere.

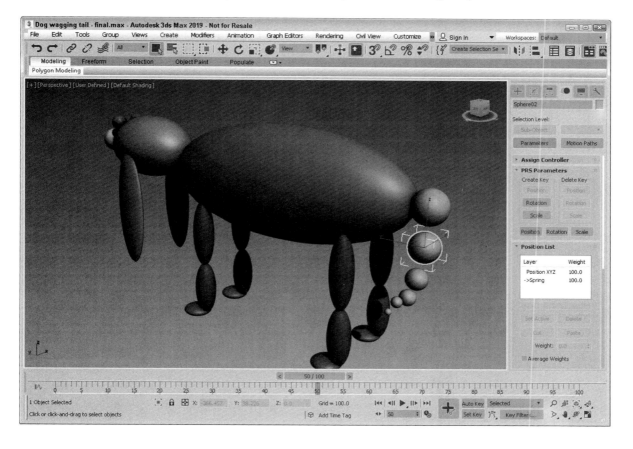

Position XYZ controller

The Position XYZ controller splits position transforms into three separate tracks, one for each axis. Each axis has a Bézier controller applied to it, but each component track can be assigned a different controller. The Position XYZ Parameters rollout lets you switch between the component axes.

Rotation Euler XYZ controller

The Rotation tracks can also be split into separate X, Y, and Z tracks and each can use a variety of controllers, many of them common to the Position track. The default Bézier controller is also used to produce smooth animations for rotations too.

Scale XYZ controller

3ds Max has one controller that you can use only in Scale tracks. The Scale XYZ controller breaks scale transforms into three separate tracks, one for each axis. This feature enables you to precisely control the scaling of an object along separate axes. It is a better alternative to using Select and Non-Uniform Scale from the main toolbar because it is independent of the object geometry.

The Scale XYZ Parameters rollout lets you select which axis to work with. This controller works the same way as the other position and rotation XYZ controllers.

Summary

Using the Animation→Constraints menu, you can apply constraints to objects. This menu also lets you select a target object. You can use the various constraints to limit the motion of objects, which is helpful as you begin to animate. If you're an animator, you should thank your lucky stars for controllers. Controllers offer power flexibility for animating objects—and just think of all those keys that you don't have to set by hand.

In this chapter, you accomplished the following:

* Constrained an object to the surface of another object using the Attachment and Surface constraints
* Forced an object to travel along a path with the Path constraint
* Controlled the position and orientation of objects with weighted Position and Orientation constraints
* Shifted between two different controlling objects using the Link constraint
* Followed objects with the LookAt constraint
* Learned about the various controller types
* Discovered how to assign controllers using the Motion panel and the Track View
* Saw a few examples of using controllers

In the next chapter, you learn to use the Track View to display and manage all the animation details of the current scene.

Chapter 36
Editing Animation Curves in the Track View

IN THIS CHAPTER

Learning the Track View interface

Understanding the Track View Curve Editor and Dope Sheet layouts

Working with keys and time ranges

Adjusting curves

Filtering tracks

Assigning controllers

Optimizing animation keys

Using out-of-range types

Adding notes to a track

Synching animation to a sound track

As you move objects around in a viewport, you often find yourself eyeballing the precise location of an object in the scene. If you've ever found yourself wishing that you could precisely see all the values behind the scene, you need to find the Track View. The Track View can be viewed using three different layouts: Curve Editor, Dope Sheet, and Track Bar. Each of these interfaces offers a unique view into the details of the scene.

These Track View layouts can display all the details of the current scene, including all the parameters and keys. This view lets you manage and control all these parameters and keys without having to look in several different places.

The Track View in Autodesk® 3ds Max® 2020 also includes additional features that enable you to edit key ranges, add and synchronize sound to your scene, and work with animation controllers using curves. And, yes, you can precisely change values, giving your eyeballs a rest.

Learning the Track View Interface

Although the Track View can be viewed using different layouts, the basic interface elements are the same. They all have menus, toolbars, a Controller pane, a Key pane, and a Time Ruler. Figure 36.1 shows these interface elements. You can hide any of these interface elements using the Show UI Elements option in the pop-up menu that appears when you right-click the title bar. Also, you can quickly hide the Controller pane by clicking the triangle icon in the upper-left corner of the Key pane.

FIGURE 36.1

The Track View interface offers a complete hierarchical look at your scene.

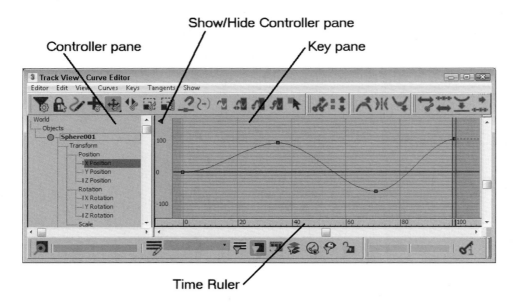

Show/Hide Controller pane

Controller pane

Key pane

Time Ruler

The Track View layouts

The Track View includes three different layouts: a Curve Editor, a Dope Sheet, and the Track Bar. The Curve Editor layout displays all parameter and motion changes as graphs that change over time. You manipulate these curves just like normal splines by selecting and dragging the keys marked as small squares. You also can use the Dope Sheet layout to coordinate key ranges between the different parameter tracks. And the Track Bar layout offers a way to quickly view the Track View within the viewports.

You can open the Curve Editor and Dope Sheet layouts using the Graph Editors menu. You can open the Curve Editor window by choosing Graph Editors→Track View–Curve Editor or by clicking the Curve Editor button on the main toolbar. You open the Dope Sheet interface in a similar manner by choosing Graph Editors→Track View–Dope Sheet.

You also use the Graph Editors menu to access the Schematic View interface. For more on the Schematic View interface, see Chapter 10, "Organizing Scenes with Layers, Containers, XRefs and the Schematic View."

After the Track View opens, you can give it a unique name using the Track View–Curve Editor field found on the Name: Track View toolbar. If the Name toolbar isn't visible, right-click the title bar and select Show Toolbars→Name: Track View from the pop-up menu. These named views are then listed in the Graph Editors→Saved Track Views menu. Any saved Track Views that are named are saved along with the scene file, and multiple track views can be named and included with the scene.

To open the Track Bar layout in the viewports, expand the Track Bar using the Open Mini Curve Editor button at the left end of the Timeline. Close the Track Bar layout by clicking the Close button on this same toolbar. Figure 36.2 shows this Track View.

FIGURE 36.2

The Track Bar offers quick access to the Track View.

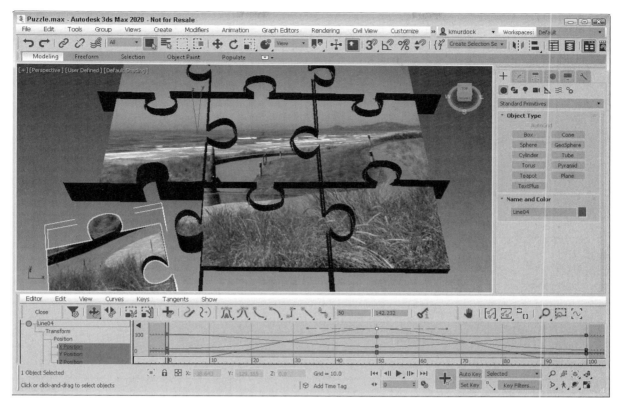

After you open a Track View, you can switch between the Curve Editor and the Dope Sheet using the Editor menu or by right-clicking the menu bar or toolbar (away from the buttons) and selecting a new layout from the Load Layout menu. You also can save customized layouts using the Save Layout or Save Layout As menu commands.

Track View menus and toolbars

In many cases, the menus and the toolbars provide access to the same functionality. One difference is the Editor menu, which lets you switch the current interface between the Curve Editor and the Dope Sheet layouts. The Curve Editor menus include Editor, Edit, View, Curves, Keys, Tangents, and Show. The Dope Sheet menu loses the Tangents menu and adds a Time menu in its place. The Track Bar menus are the same as the Curve Editor menus and can display both the Curve Editor and the Dope Sheet.

The Track View consists of several toolbars. You can open these toolbars by right-clicking the toolbar (away from the buttons) and selecting the Show Toolbars submenu. The available toolbars depend on the layout (Curve Editor or Dope Sheet). All these toolbars can be docked, floated, and hidden. You also can add and delete new toolbars using the right-click pop-up menu.

Note

Depending on the size of the Track View window, you may need to drag the toolbar to the left to see the buttons at the right end of the toolbar.

Key Controls toolbar

The Key Controls toolbar is the one of the first toolbars to learn for the Curve Editor. Table 36.1 describes these buttons.

TABLE 36.1 Key Controls Toolbar Buttons

Toolbar Button	Name	Description
	Move Keys, Move Keys Horizontal, Move Keys Vertical	Enables you to move the selected keys or limit their movement to horizontal or vertical
	Draw Curves	Creates a curve by dragging the mouse; Curve Editor layout only
	Add/Remove Keys	Lets you add new or remove keys to a track. Hold the Shift key to remove keys.
	Region Keys tool	Lets you drag to select move and/or scale multiple keys at once
	Slide Keys tool	Lets you move the selected keys left or right
	Retime tool	Lets you change the timing for selected tracks
	Retime All tool	Lets you change the timing for all tracks

Navigation toolbar

The Navigation toolbar for the Curve Editor lets you pan and zoom to focus on a specific area of the Track View window. Table 36.2 describes these buttons. This same toolbar is available in the Dope Sheet.

TABLE 36.2 Navigation Toolbar Buttons

Status Bar Button	Name	Description
	Pan	Pans the view
	Frame Horizontal Extents, Frame Horizontal Extents Keys, Frame Horizontal Extents Selected Keys	Displays the entire horizontal track or keys
	Frame Value Extents, Frame Value Extents Range, Frame Value Extents Selected Keys	Displays the entire vertical track or the range of values
	Frame Horizontal and Value Extents	Scales the entire track in both the time range and value extents
	Zoom, Zoom Time, Zoom Values	Zooms in and out of the view
	Zoom Region	Zooms within a region selected by dragging the mouse
	Isolate Curve	Temporarily hides all curves except for those with selected keys

Key Tangents toolbar

The Key Tangents toolbar is another of the default Curve Editor toolbars. It is used to set the In and Out curve types. Table 36.3 describes these buttons.

TABLE 36.3 Key Tangents Toolbar Buttons

Toolbar Button	Name	Description
	Set Tangents to Auto, Set In Tangents to Auto, Set Out Tangents to Auto	Sets curve to approach and leave the key in an automatic manner
	Set Tangents to Spline, Set In Tangents to Spline, Set Out Tangents to Spline	Sets curve to approach and leave the key in a custom manner defined by the handle positions
	Set Tangents to Fast, Set In Tangents to Fast, Set Out Tangents to Fast	Sets curve to approach and leave the key in an ascending manner
	Set Tangents to Slow, Set In Tangents to Slow, Set Out Tangents to Slow	Sets curve to approach and leave the key in a descending manner
	Set Tangents to Stepped, Set In Tangents to Stepped, Set Out Tangents to Stepped	Sets curve to approach and leave the key in a stepping manner
	Set Tangents to Linear, Set In Tangents to Linear, Set Out Tangents to Linear	Sets curve to approach and leave the key in a linear manner
	Set Tangents to Smooth, Set In Tangents to Smooth, Set Out Tangents to Smooth	Sets curve to approach and leave the key in a smooth manner

Tangent Actions toolbar

The Tangent Actions toolbar is another of the default toolbars in the Curve Editor. It is used to quickly show, break, unify and lock tangent handles. Table 36.4 describes these buttons.

TABLE 36.4 Tangent Actions Toolbar Buttons

Toolbar Button	Name	Description
	Show Tangents Toggle	Turns the tangent lines on and off
	Break Tangents	Allows the tangent handles on either side of the key to move independently
	Unify Tangents	Moves the tangent handles so they form a straight line
	Lock Tangents Toggle	Locks the tangents of all points. Can be toggled on and off

Key Entry toolbar

The Key Entry toolbar in the Curve Editor displays the current frame and value for the selected key. If multiple keys are selected, only the frame value is displayed. You also can enter values in these fields to change the current frame and/or value for the selected key or keys. This is a huge timesaver if you want to change multiple keys to the same value.

The same frame and value fields are available in the Dope Sheet in the Key Stats toolbar.

Keys toolbar

The Keys toolbar is the first of the default toolbars in the Dope Sheet, but it also exists in the Curve Editor. Table 36.5 describes these buttons. Notice that some of the buttons are the same as those found on the other toolbars.

TABLE 36.5 Keys Toolbar Buttons

Toolbar Button	Name	Description
	Edit Keys	Enables edit keys mode, Dope Sheet layout only
	Edit Ranges	Enables edit ranges mode, Dope Sheet layout only
	Filters	Opens the Filters dialog box, where you can specify which tracks will appear
	Move Keys, Move Keys Horizontal, Move Keys Vertical	Enables you to move the selected keys or limit their movement to horizontal or vertical
	Slide Keys	Enables you to slide the selected keys
	Scale Keys	Enables you to scale the selected keys
	Scale Values	Enables you to scale the selected values, Curve Editor layout only
	Add/Remove Keys	Enables you to add new keys to a track. Hold Shift key to remove keys
	Draw Curves	Creates a curve by dragging the mouse; Curve Editor layout only
	Simplify Curve	Reduces the complexity of the curve; Curve Editor layout only

Time toolbar

The Time toolbar is another of the default toolbars for the Dope Sheet. It is used to work with time ranges. Table 36.6 describes these buttons.

TABLE 36.6 Time Toolbar Buttons

Toolbar Button	Name	Description
	Select Time	Enables you to select a block of time by clicking and dragging
	Delete Time	Deletes the selected block of time

	Reverse Time	Reverses the order of the selected time block
	Scale Time	Scales the current time block
	Insert Time	Inserts an additional amount of time
	Cut Time	Deletes the selected block of time and places it on the clipboard for pasting
	Copy Time	Makes a copy of the selected block of time and places it on the clipboard for pasting
	Paste Time	Inserts the current clipboard time selection

Display toolbar

The Display toolbar is another of the default toolbars for the Dope Sheet. Table 36.7 describes these buttons.

TABLE 36.7 Display Toolbar Buttons

Toolbar Button	Name	Description
	Lock Selection	Prevents any changes to the current selection
	Snap Frames	Causes moved tracks to snap to the nearest frame
	Show Keyable Icons	Displays a key icon next to all tracks that can be animated
	Modify Subtree	Causes changes to a parent to affect all tracks beneath the parent in the hierarchy; Dope Sheet layout only
	Modify Child Keys	Causes changes to child keys when parent keys are changed; Dope Sheet layout only

Track Selection toolbar

At the bottom edge of the Dope Sheet window are three toolbars that appear by default. These toolbars are the Track Selection, Key Stats, and Navigation. Using these toolbars, you can locate specific tracks, see information on the various keys, and navigate the interface.

In the Track Selection toolbar is the Zoom Selected Object button and the Select by Name field, in which you can type a name to locate any tracks with that name.

Note

In the Select by Name field, you also can use wildcard characters such as * (asterisk) and ? (question mark) to find several tracks.

The Key Stats toolbar includes Key Time and Value Display fields that display the current time and value. You can enter values in these fields to change the value for the current time. You also can enter an expression in these fields in which the variable n equals the key time or value. For example, to specify a key value that is 20 frames from the current frame, enter $n+20$ (where you supply the current value in place of n). You also can include any function valid for the Expression controller, such as sin( ) or log( ). Click the Show Selected Key Stats button to display the key value in the Key pane.

Table 36.8 describes the buttons found in the Track Selection and Key Stats toolbars.

TABLE 36.8 Track Selection and Key Stats Toolbar Buttons

Toolbar Button	Name	Description
	Zoom Selected Object	Places current selection at the top of the hierarchy
	Edit Track Set	Opens a dialog box where selected sets of tracks can be edited
	Filter Selected Tracks Toggle	Toggles to show only the selected tracks in the Controller pane
	Filter Selected Objects Toggle	Toggles to show the tracks for the selected objects in the Controller pane
	Filter Animated Tracks Toggle	Toggles to show only the animated tracks in the Controller pane
	Filter Active Layer Toggle	Toggles to show only the active layer tracks in the Controller pane
	Filter Keyable Tracks Toggle	Toggles to show only the tracks that have keys in the Controller pane
	Filter Visible Objects Toggle	Toggles to show only the visible object tracks in the Controller pane
	Filter Unlocked Attributes Toggle	Toggles to show only the tracks with unlocked attributes in the Controller pane
	Show Selected Key Statistics	Displays the frame number and values next to each key

Other toolbars

The other toolbars provide access to features that also are available through the menus. These toolbars are hidden by default, but they can be made visible using the Show Toolbars menu command in the right-click pop-up menu. Many of these additional toolbars offer the same icons as mentioned above.

Controller and Key panes

Below the menus (and below the topped docked toolbars) are two panes. The left pane, called the Controller pane, presents a hierarchical list of all the tracks. The right pane is called the Key pane, and it displays the time range, keys, or curves, depending on the layout. You can pan the Controller pane by clicking and dragging on a blank section of the pane: the cursor changes to a hand to indicate when you can pan the pane.

Tip

Using the triangle icon in the upper-left corner of the Key pane, you can quickly hide the Controller pane.

Each track can include several subtracks. To display these subtracks, click the plus sign (+) to the left of the track name. To collapse a track, click the minus sign (–). You also can use the Show menu to Auto Expand a selected hierarchy. Under the Show→Auto Expand menu are options for Selected Objects Only, Transforms, XYZ Components, Limits, Keyable, Animated, Base Objects, Modifiers, Materials, and Children. For example, if the Auto Expand→Transforms option is enabled, the Transform tracks for all objects is automatically expanded in the Track View. Using these settings can enable you to quickly find the track you're looking for.

The Show→Auto Select→Animated toggle automatically selects all tracks that are animated. You also can auto select Position, Rotation, and/or Scale tracks. The Show→Track View - Auto Scroll menu command can be set for Selected and/or Objects tracks. This command automatically moves either the Selected tracks or the Objects track to the top of the Controller pane.

Note

You also can select, expand, and collapse tracks using the right-click pop-up quad menu.

The Controller pane includes many different types of tracks. Every scene includes many global tracks such as World, Sound, Global Tracks, Video Post, Anim Layer Control Manager, SME, Retimer Manager, Environment, Render Effects, Render Elements, Renderer, Global Shadow Parameters, Scene Materials, Medit Materials (for materials in the Material Editor), and Objects, as shown in Figure 36.3. If the global tracks aren't visible, you can enable them in the Filters dialog box.

Caution

If you open the Track View interface and you only see the World track, then it could be that you have filtered all the other tracks away. Click on the Filters button or choose the Views→Filters menu and select the All button in the Show column.

FIGURE 36.3

Several tracks are available by default.

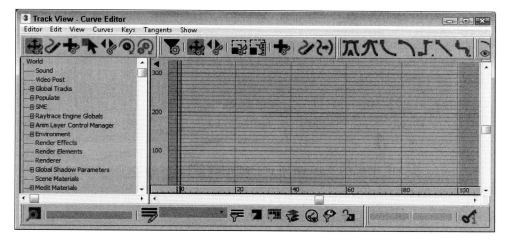

The Shift, Ctrl, and Alt keys make selecting and deselecting multiple tracks possible. To select a contiguous range of tracks, select a single track and then select another track while holding down the Shift key. This selects the two tracks and all tracks in between. Hold down the Ctrl key while selecting tracks to select multiple tracks that are not contiguous. The Alt key selects all tracks at the same level. If multiple animated tracks are selected, all the animation curves for the selected tracks are displayed in the Key pane.

Below the Key pane is the Time Ruler, which displays the current frame range as specified in the Time Configuration dialog box. The current frame is marked with a light blue time bar. This time bar is linked to the Time Slider, and moving one updates the other automatically.

At the top right of the Key pane (above the vertical scroll bar) is a split tab that you can use to split the Controller and Key pane into two separate views, as shown in Figure 36.4. Using this feature, you can look at two different sections of the tree at the same time. This makes it easy to copy and paste keys between different tracks.

Tip

You can drag the Time Ruler vertically in the right pane.

FIGURE 36.4

Drag the tab above the vertical scroll bar to split the Track View into two views.

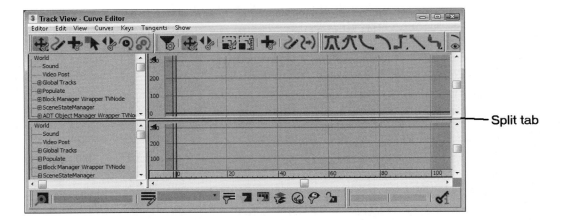

Working with Keys

Keys define the main animation points in an animation. 3ds Max interpolates all the positions and values between the key points to generate the animation. Using the Track View, you can edit these animation keys with precision.

Chapter 32, "Understanding Animation and Keyframes," covers key creation in more detail.

In the Curve Editor, keys are shown as small squares positioned along the curve, and the color of the curves denotes the type of track it represents. Position curves are red, Rotation curves are green, and Scale curves are blue. Parameter curves are gray. In the Dope Sheet, keys are shown as colored boxes that extend across the applicable tracks, as shown in Figure 36.5. Parent tracks (such as an object's name) are colored gray. Selecting a parent key selects all its children keys. Any selected keys appear white. A track title that includes a key is highlighted yellow in the Controller pane.

Caution

If the Key pane is not wide enough, a key is shown as a thick, black line.

FIGURE 36.5

In the Dope Sheet, Position keys are red, Rotation keys are green, Scale keys are blue, and Parameter keys are gray.

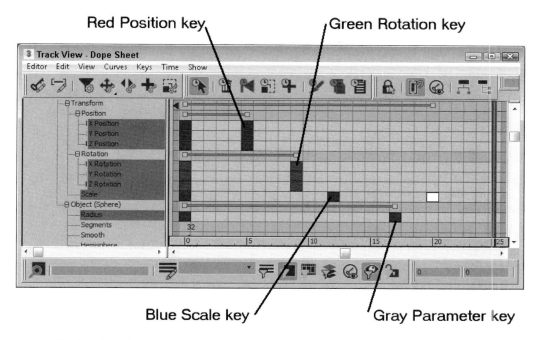

Selecting keys

Before you can move and edit keys, you need to be able to select them. Just like selecting keys on the Track Bar, you select keys by clicking them. Selected keys turn white. To select multiple keys, hold down the Ctrl key while clicking several keys, or drag an outline over several keys to select them. Click away from the keys to deselect all the selected keys.

 With a key or multiple keys selected, you can lock the selection with the Lock Selection button on the Tools toolbar. The spacebar is the keyboard shortcut for this button. With the selection locked, you cannot select any new keys.

Tip

If you want to access a specific parameter in the Track View, you can right-click the parameter in the Command Panel and select the Show in Track View command from the pop-up menu, and the Track View loads with the parameter visible.

 There is also a Key Selection toolbar that has two buttons. The first, Select Next Key, selects the next key on the current curve or you can hold down the Shift key and click this button to choose the previous key.

 The Grow Key Selection button increases the total key selection by selecting keys on either side of the current key or you can hold down the Shift key to shrink the current selection.

Double click on a curve to quickly select all curve keys at once.

Using soft selection

The Keys menu also includes a Use Soft Select option. This feature is similar to the soft selection found in the Modify panel when working on a subobject selection, except that it works with keys causing adjacent keys to move along with the selected keys, but not as much. The Keys→Soft Select Settings menu command opens a simple toolbar where you can enable soft selection and set the Range and Falloff values. The Range value sets how many frames the soft selection covers.

When enabled, all keys within a specified range are also selected and moved to a lesser degree than the selected key. When enabled, the curve is displayed with a gradient for the Curve Editor layout and as a gradient across the key markers in the Dope Sheet layout. This shows the range and falloff for the curve.

Adding and deleting keys

You can add a key by clicking the Add Keys button (or pressing the A key) and clicking the location where the new key should appear. Each new key is set with the interpolated value between the existing keys. This can be done whether the Auto Key button at the bottom of the 3ds Max interface is on or off.

To delete keys, select the keys and press the Delete key on the keyboard. If a key is deleted, the curve changes to account for the missing key. You can also remove keys by holding down the Shift key and clicking on an existing key with the Add Keys tool.

Moving, sliding, and scaling keys

The Move Keys button (keyboard shortcut M) lets you select and move a key to a new location. You can clone keys by pressing and holding the Shift key while moving a key. Using the flyout buttons, you can select to restrict the movement horizontally or vertically. You also can move the selected key to the Time Slider's location (the current frame) with the Keys→Align to Cursor menu command.

The Slide Keys button in the Dope Sheet lets you select a key and move all adjacent keys in unison to the left or right. If the selected key is moved to the right, all keys from that key to the end of the animation slide to the right. If the key is moved to the left, all keys to the beginning of the animation slide to the left.

The Scale Keys button, also in the Dope Sheet, lets you move a group of keys closer together or farther apart. The scale center is the current frame. You can use the Shift key to clone keys while dragging.

If the Edit→Snap Frames menu command or button (keyboard shortcut S) is enabled (found in the Display toolbar in the Dope Sheet and in the Keys toolbar for the Curve Editor), the selected key snaps to the nearest key as it is moved. This makes aligning keys to the same frame easy.

If you want to move keys only a little bit, then the Nudge tool might be just what you need. When nudging a key, it will not pass any adjacent keys which keeps all the keys in order. You can find the Nudge tool on the Keys Only or the New Keys Tools toolbars. The Nudge tool moves the selected keys to the right and holding the Shift key moves them to the left.

Flattening selected keys

Sometimes keys can get out of bounds causing unwanted noise in an animation and sometimes it is easier to just reset a bunch of keys so you can edit them manually. For this case, the Flatten tool, found on the Keys Only or the New Key Tools toolbar, is just the ticket. The Flatten tool resets all the selected keys to the value of the first key selected. You can also reset all the selected keys to their average with the Flatten to Average Values tool.

Using the Region tool

Within the Curve Editor, the Region tool is used to move, slide, and scale a set of selected keys. To use this tool, simply drag in the Key pane over the keys you want to select. A box is displayed around all the selected keys with handles at each edge, as shown in Figure 36.6. The selected keys can be moved to a different location by clicking within the region and dragging to a new location. You can drag the selected keys up or down to change their values or left and right to change their frame.

FIGURE 36.6

The Region Keys tool places handles around each side of the selected keys for sliding and scaling keys.

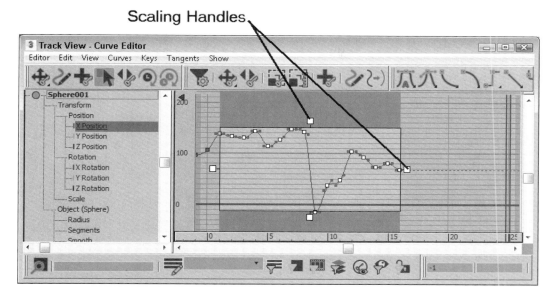

Dragging on either of the side handles scales the range of the selected keys, and dragging on the top or bottom handles scales their values. The side of the selected handle is the side of the region that moves during a scale operation, but if you press and hold the Shift key, the region is scaled equally from both sides. If you scale the region over any non-selected keys, those non-selected keys are simply deleted. The Region Keys tool remains active until another tool, such as the Move tool is selected. If you click and drag outside of the selected region, a new region is selected.

Using the Retime tool

As you work with keys and curves, you may want to move and scale only part of a section of a curve without affecting the sections beyond. Suppose you have a top that moves into position as the spinning starts and slowly stops spinning as it falls over. If the keys for the start and the end of the animation are fine, but you need to move the middle keys, you could use the Retime tool to do this easily.

When the Retime tool is selected, you can double-click any curve to place a retime marker, which appears as a vertical bar. If you place two markers, as shown in Figure 36.7, all the keys within those two markers scale as you drag either marker left or right, and all keys outside of these markers slide without any change to their scale.

FIGURE 36.7

The Retime tool lets you scale keys within a specific area.

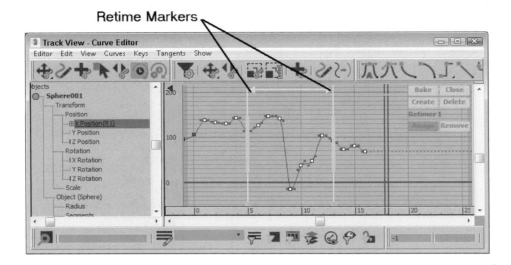

In the upper-right corner of the Track View are several yellow buttons that let you control the retime markers. The Bake button commits the current changes and places the keys on the Track Bar. The Close button exits the Retime tool. The Create and Delete buttons add and removes markers, and the Assign and Remove buttons add and remove curves from the current Key pane.

Editing keys

To edit the key parameters for any controller, right-click the key in the Curve Editor or click the Properties button or use the Edit→Properties menu in the Dope Sheet; this opens the Key Info dialog box for most controllers. You also can access this dialog box by right-clicking a track and selecting Properties from the pop-up menu. These commands also can be used when multiple keys on the same or on different tracks are selected.

Using the Randomize Keys utility

The Randomize Keys utility lets you generate random time or key positions with an offset value. To access this utility, select the track of keys that you want to randomize and choose Edit→Track View Utilities to open the Track View Utilities dialog box. From this dialog box, select Randomize Keys from the list of utilities and click OK; the Randomize Keys utility dialog box opens, as shown in Figure 36.8.

FIGURE 36.8

Use the Randomize Keys utility to create random key positions and values.

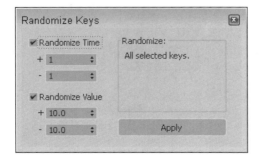

In this dialog box, you can specify positive and negative shift values for both Time and Value. Click the Apply button to apply the randomization process.

Using the Euler Filter utility

Euler rotations are easy to understand and use. They provide rotations about each of the three axes or can be thought of as yaw, pitch, and roll, but they have an inherit flaw—they are susceptible to Gimbal flipping and Gimbal lock. Gimbal flipping can occur when the rotation is directed straight up or straight down. This causes the object to instantly flip 180 degrees to continue its rotation. Gimbal lock can occur when two Euler rotation angles are aligned, causing the object to lose a degree of freedom.

To counter these problems, 3ds Max has the ability to use Quaternions instead of Euler angles. Quaternions are vector-based instead of angle-based, so they aren't susceptible to the Gimbal flipping and lock problems, but many animators find Quaternions difficult to understand and use, so they stick to Euler rotations and are watchful for potential problems.

The Track View has a utility that can help if you're dealing with Euler rotations. The Euler Filter utility analyzes the current frame range and corrects any Gimbal flipping that it detects. Selecting this utility from the Track View Utilities dialog box opens a simple dialog box where you can set the range to analyze. You also have an option to Insert Keys if Needed.

Displaying keyable icons

If you're not careful, you could animate a track that you didn't mean to (especially with the Auto Key mode enabled). By marking a track as non-keyable or keyable, you can control which tracks can be animated. To do this, choose the View→Keyable Icons menu command. This places a small key icon to the left of each track that can be animated. You can then click the icon to change it to a non-keyable track. Figure 36.9 shows the Curve Editor with this feature enabled.

FIGURE 36.9

The Keyable Icons feature displays an icon next to all tracks that can be keyed.

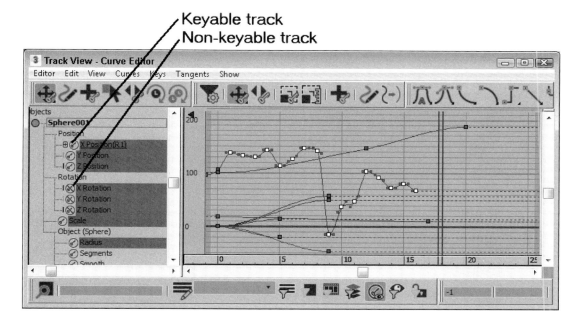

Editing Time

In some cases, directly working with keys isn't what you want to do. For example, if you need to change the animation length from six seconds to five seconds, you want to work in the Dope Sheet's Edit Ranges mode. To switch to this mode in the Dope Sheet, click the Edit Ranges button on the Keys toolbar. In this mode, the key ranges are displayed as black lines with square markers on either end, as shown in Figure 36.10.

FIGURE 36.10

Click the Edit Ranges button to display the key ranges in the Key pane.

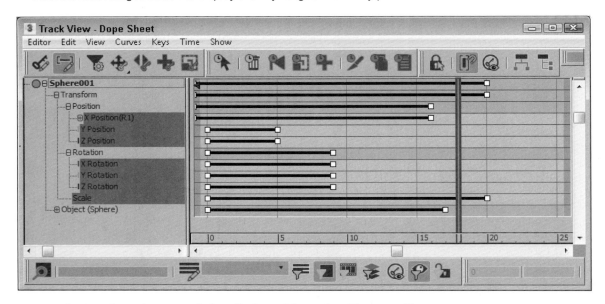

Selecting time and the Select Keys by Time utility

Before you can scale, cut, copy, or paste time, you need to select a track and then select a time block. This can be done back in the Edit Keys mode. To select a section of time, click the Select Time button and drag the mouse over the time block.

The Select Keys by Time utility lets you select all the keys within a given time block by entering the frame or time values. To use this utility, click the Track View Utilities button in the Tools toolbar or select the Edit→Track View Utilities menu command to open the Track View Utilities dialog box, and choose the Select Keys by Time utility from the list. Then in the Select Keys by Time dialog box, enter the Start Time and End Time values to complete the selection.

Deleting, cutting, copying, and pasting time

After you select a block of time, you can delete it by clicking the Delete Time button. Another way to delete a block of time is to use the Cut Time button, which removes the selected time block but places a copy of it on the clipboard for pasting. The Copy Time button also adds the time block to the clipboard for pasting, but it leaves the selected time in the track.

After you copy a time block to the clipboard, you can paste it to a different location within the Track View. The track where you paste it must be of the same type as the one from which you copied it. This is a great way to copy transform keys from one object to another.

All keys within the time block are also pasted, and you can select whether they are pasted relatively or absolutely. *Absolute* pasting adds keys with the exact values as the ones on the clipboard. *Relative* pasting adds the key value to the current initial value at the place where the key is pasted.

You can enable the Exclude Left End Point and Exclude Right End Point buttons on the Extras toolbar when pasting multiple sections next to each other. By excluding either end point, the time block loops seamlessly.

Reversing, inserting, and scaling time

The Reverse Time button flips the keys within the selected time block.

The Insert Time button lets you insert a section of time anywhere within the current track. To insert time, click and drag to specify the amount of time to insert; all keys beyond the current insertion point slide to accommodate the inserted time.

The Scale Time button scales the selected time block. This feature causes all keys to be pushed closer together or farther apart. The scaling takes place around the current frame.

Setting ranges

The Position Ranges button on the Ranges toolbar enables you to move ranges without moving keys. In this mode, you can move and scale a range bar independently of its keys, ignoring any keys that are out of range. For example, this button, when enabled, lets you remove the first several frames of an animation without moving the keys. The Recouple Ranges button (also on the Ranges toolbar) can be used to line up the keys with the range again. The left end of the range aligns with the first key, and the right end aligns with the last key.

Editing Curves

When an object is moving through the scene, estimating the exact point where its position changes direction can sometimes be difficult. Animation curves provide this information by presenting a controller's value as a function of time. The slope of the curve shows the value's rate of change. Steep curves show quick movements. Shallow lines are slow-moving values. Each key is a vertex in the curve. Curves are visible only in the Curve Editor and the Track Bar layout.

Curves mode lets you edit and work with these curves for complete control over the animation parameters.

Inserting new keys and moving keys

Curves with only two keys have slow in and out tangents, making the animation start slow, speed up, and then slow down. You can add more curvature to the line with the addition of another key. To add another key, click the Add Keys button, and then click the curve where you want to place the key.

Tip
Keep the total number of keys to a minimum. More keys make editing more difficult.

If the curve contains multiple curves, such as a curve for the X, Y, and Z Position or RGB color values, a point is added to each curve. The Move Keys button enables you to move individual keys by dragging them. It also includes flyouts for constraining the key movement to a horizontal or vertical direction.

To scale keys, use the Region Keys tool or the Retime tool.

Tutorial: Animating a monorail

As an example of working with curves, you'll animate a monorail that moves around its track, changing speeds and stopping for passengers.

To animate the monorail using curves, follow these steps:

1. Open the Monorail.max file from the Chap 36 directory in the downloaded content set.

    This file contains a simple monorail setup made from primitives.

2. Click the Play button, and watch the train move around the track.

   As a default, the Path Constraint's Percent track has a Linear Float controller that causes the train to move at a constant speed. To refine the animation, you need to change it.

3. Open the Track View–Curve Editor by first selecting the train object in the scene and then right-clicking and choosing Curve Editor from the pop-up quad menu. The Track View-Curve Editor window opens and shows the Percent track along with a straight linear curve. Select and right-click the Percent track, and select Assign Controller from the pop-up menu to open the Assign Float Controller dialog box. Select the Bézier Float controller, and click OK. Click on the Set Tangents to Auto button in the Key Tangents toolbar. Notice how the linear curve changes.

4. Click the Play button. The train starts slowly (represented by the flattish part of the curve), accelerates (the steeper part of the curve), and slows down again (another flattish part).

Tip

When "reading" curves, remember that a steep curve produces fast animation, a shallow curve produces slow animation, a horizontal curve produces no movement or value change, and a straight curve produces a constant animation.

5. You need the train to stop for passengers at the station, so click the Add Keys button and add a key to the curve somewhere around frame 140 when the train is at the dock.

6. Choose the Move Keys Horizontal button from the Move Keys flyout and select the newly created key. Hold down the Shift key, and drag left to copy the key to frame 120.

   The curve is flat, so the train stops at the station.

Figure 36.11 shows the final curve after you've completed the editing.

FIGURE 36.11

The finished Percent curve for the train's position along the path

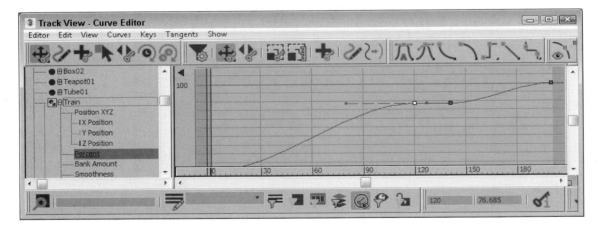

Drawing curves

If you know what the curve you want is supposed to look like, you can actually draw it in the Key pane with the Draw Curves button enabled. This mode adds a key for every change in the curve. You may want to use the Reduce Keys optimization after drawing a curve.

Tip

If you make a mistake, you can just draw over the top of the existing curve to make corrections.

Figure 36.12 shows a curve that was created with the Draw Curves feature.

FIGURE 36.12

Drawing curves results in numerous keys.

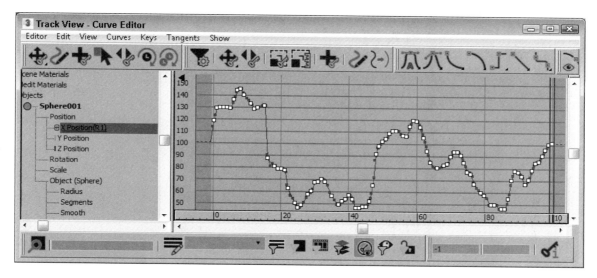

Reducing keys

The Curves→Simplify Curve menu command allows you to optimize the number of keys used in an animation. Certain IK (inverse kinematics) methods and the MassFX system calculate keys for every frame in the scene, which can increase your file size greatly. By optimizing with the Simplify Curve command, you can reduce the file size and complexity of your animations.

Using the Simplify Curve command opens the Simplify Curve dialog box. The threshold value determines how close to the actual position the solution must be to eliminate the key. Figure 36.13 shows the same curve created with the Draw Curves feature after it has been optimized with a Threshold value of 3.0 using the Simplify Curve feature.

FIGURE 36.13

The Simplify Curve feature optimizes the curve by reducing keys.

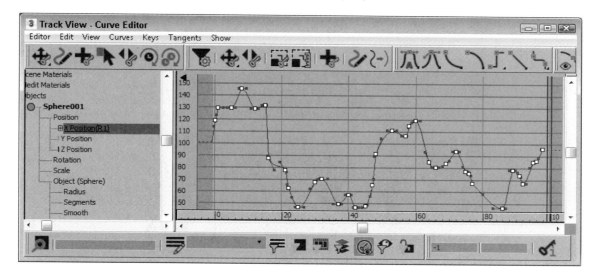

Using Buffer Curves

The Buffer Curves feature lets you save the selected curve in a buffer. The saved curve can then be recalled or viewed to see how it compares with the changes. Buffer curves are shown as dotted black lines, as shown in Figure 36.14.

FIGURE 36.14

Buffer curves can be saved for recall at a later time.

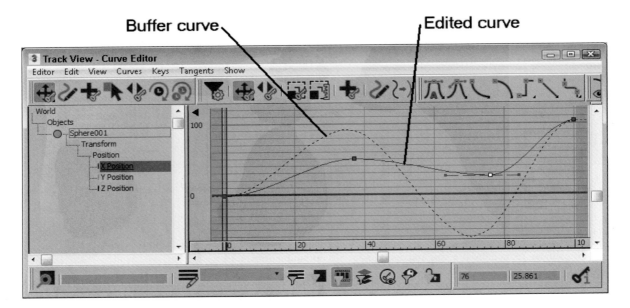

You can enable buffer curves with the Curves→Buffer Curves menu or using the Buffer Curves toolbar. The Use Buffer Curves button turns the curves on and off. You can also keep them on and hide the current buffer curves with the Show/Hide Buffer Curves button. The Swap Curves with Buffer button makes the buffer curves the current edited curve and vice versa. The Snapshot button remembers the current curve in the buffer and the Revert to Buffer Curve button resets the current curve to the buffer curve.

Buffer curves are great for keeping track of any changes that might happen while editing, but it also is good for comparing an existing edit to an existing curve.

Working with tangents

Animation curves for the Bézier controller have tangents associated with every key. To view and edit these tangents, select the View→All Tangents menu command. These tangents are lines that extend from the key point with a handle on each end. By moving these handles, you can alter the curvature of the curve around the key.

You can select the type of tangent from the Key Tangents toolbar. These can be different for the In and Out portion of the curve. You also can select them using the Key dialog box opened by right-clicking a key. The default tangent type for all new keys is set using the button to the left of the Key Filters button at the bottom of the 3ds Max interface (not the Track View interface). Using this button, you can quickly select from any of the available tangent types.

You open the Key dialog box, shown in Figure 36.15, by selecting a key and right-clicking it. It lets you specify two different types of tangent points: Continuous and Discontinuous. *Continuous* tangents are points with two handles on the same line. The curvature for continuous tangents is always smooth. *Discontinuous* tangents have any angle between the two handle lines. These tangents form a sharp point.

By default, all tangents are continuous and move together, but you can break them apart to be discontinuous by clicking the Break Tangents button in the Tangent Actions toolbar. This lets each handle be moved independently. Broken tangents can be locked together again with the Unify Tangents button, so they move together. This unifies the tangent handles even if they don't form a straight line.

Tip

Holding down the Shift key while dragging a handle lets you drag the handle independently of the other handle without having to use the Break Tangents button.

FIGURE 36.15

The Key dialog box lets you change the key's Time, Value, and In and Out tangent curves.

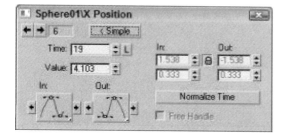

The Lock Tangents button in the Key Info dialog box lets you change the handles of several keys at the same time. If this button is disabled, adjusting a tangent handle affects only the key of that handle.

Tutorial: Animating a flowing river

The default auto-tangent types create a curve that has ease-in and ease-out built into the curve. This causes the animation to start slowly, speed up, and then slow to a stop. While this may be a good starting point for many animations, it won't work for those that should have a constant speed. This example shows how to create a river with a material animated to a constant speed.

To create a flowing river, follow these steps:

1. Open the River.max file from the Chap 36 directory in the downloaded content set.

   This file contains a river surface made from a loft. The V Offset for the River Water material's diffuse channel has been animated to simulate flowing water (yes, this river has a checkered past . . .).

2. Click the Play button.

   The river flow starts out slow, speeds up, and then slows to a stop.

Note

The river flows using a checker texture map. Make sure to select a viewport with textures enabled to see the flowing effect.

3. Open the Track View–Curve Editor, and locate and select the V Offset track for the river's material. (You can find this track under the Scene Materials→River Water→Maps→Diffuse Color: Map #2 (Checker)→Coordinates→V Offset track.) Click the Frame Horizontal Extents Selected Keys and the Frame Value Extents Selected Keys buttons in the Navigation toolbar to zoom in on the V Offest track's keys. These commands are also found in the View→Frame menu.

4. You have two easy options for creating an animation with a constant speed. The first changes the entire controller type; the second changes the individual key's tangent types.

   Option 1: Right-click over the V Offset track and choose Assign Controller from the pop-up menu to open the Assign Float Controller dialog box. Select Linear Float, and click OK.

   or

Option 2: Select both keys by clicking one key, holding down the Ctrl button, and clicking the other, or by dragging an outline around both keys. Click the Set Tangents to Linear button in the Key Tangents toolbar.

Whichever method you use, the line between the two keys is now straight.

5. Click the Play button.

The river now flows at a constant speed.

6. To increase the speed of the flow, select Move Keys Vertical from the Move Keys button flyout, and select and move the end key higher in the graph.

The river flows faster.

Figure 36.16 shows the river as it flows along.

FIGURE 36.16

The Checkered River flows evenly.

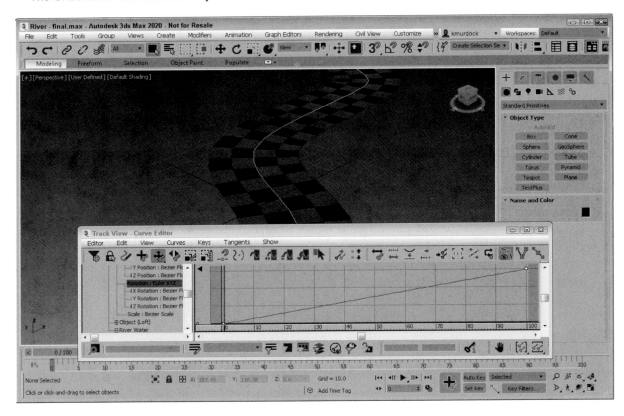

Applying out-of-range, ease, and multiplier curves

Out-of-range curves define what the curve should do when it is beyond the range of specified keys. For example, you could tell the curve to loop or repeat its previous range of keys. To apply these curves, select a track and select the Out-of-Range Types menu command from the Edit→Controller menu. This opens a dialog box, shown in Figure 36.17, where you can select from the available curve types.

FIGURE 36.17

The Param Curve Out-of-Range Types dialog box lets you select the type of out-of-range curve to use.

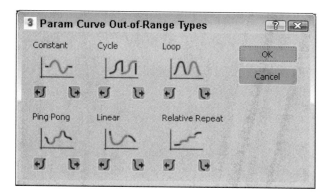

Note

You also can apply an out-of-range curve to a select range of frames using the Create Out-of-Range Keys utility. This utility is available via the Edit→Track View Utilities menu.

By clicking the buttons below the types, you can specify a curve for the beginning and end. This Out-of-Range dialog box includes six options:

* **Constant:** Holds the value constant for all out-of-range frames
* **Cycle:** Repeats the track values as soon as the range ends
* **Loop:** Repeats the range values, like the Cycle option, except that the beginning and end points are interpolated to provide a smooth transition
* **Ping Pong:** Repeats the range values in reverse order after the range end is reached
* **Linear:** Projects the range values in a linear manner when out of range
* **Relative Repeat:** Repeats the range values offset by the distance between the start and end values

You can apply ease curves (choose Curves→Apply Ease Curve, or press Ctrl+E) to smooth the timing of a curve. You can apply multiplier curves (Curves→Apply Multiplier Curve, Ctrl+M) to alter the scaling of a curve. You can use ease and multiplier curves to automatically smooth or scale an animation's motion. Each of these buttons adds a new track and curve to the selected controller track.

Ease and Multiplier curves add another layer of control on top of the existing animation and allow you to edit the existing animation curves without changing the original animation keys. For example, if you have a standard walk cycle, you can use an ease curve to add a limp to the walk cycle or you can reuse the walk cycle for a taller character by adding a multiplier curve.

Note

Not all controllers can have an ease or multiplier curve applied.

You can delete these tracks and curves using the Curves→Remove Ease Curve/Multiplier menu command. You also can enable or disable these curves with the Curves→On/Off Ease Curve/Multiplier menu command.

After you apply an ease or multiplier curve, you can assign the type of curve to use with the Ease Curve Out-of-Range Types button. This button opens the Ease Curve Out-of-Range Types dialog box, which includes the same curve types as the Out-of-Range curves, except for the addition of an Identity curve type.

Note

In the Ease Curve Out-of-Range Types dialog box is an Identity option that isn't present in the Parameter Curve Out-of-Range Types dialog box. The Identity option begins or ends the curve with a linear slope that produces a gradual, constant rate increase.

When editing ranges, you can make the range of a selected track smaller than the range of the whole animation. These tracks then go out of range at some point during the animation. The Ease/Multiplier Curve Out-of-Range Types buttons are used to tell the track how to handle its out-of-range time.

Tutorial: Animating a wind-up teapot

As an example of working with multiplier curves, you'll create a wind-up teapot that vibrates its way across a surface.

To animate the vibrations in the Track View, follow these steps:

1. Open the Wind-up teapot.max file from the Chap 36 directory in the downloaded content set.

   This file contains a teapot with legs.

2. Click the Play button.

   The teapot's key winds up to about frame 39 and then runs down again as the teapot moves around a bit. To add the random movement and rotation to make the vibrations, you use Noise controllers and Multiplier curves to limit the noise.

3. Open the Track View–Curve Editor, and navigate down to the Wind-up Key's X Rotation track, located at Objects, Teapot Group, Key, Transform, Rotation, X Rotation. Take a moment to observe the shape of the curve, shown in Figure 36.18.

   The key is "wound up" in short spurts and then runs down, slowing until it stops. The vibration, then, should start midway and then taper off as the key runs down.

FIGURE 36.18

The rotation of the Wind-up Key object

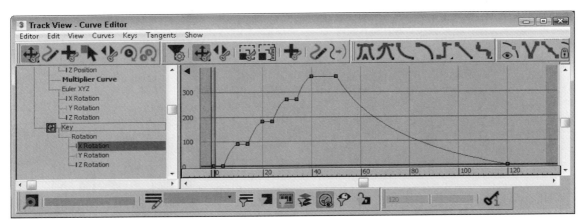

When adding the Noise controller, you should assign a List controller first to retain the ability to transform the object independently of the Noise.

Note

Assigning controllers through the Animation menu automatically creates a List controller first.

4. Select the Teapot Group's Position track, and press the C key on the keyboard or double click on the Position track to access the Assign Controller dialog box. Choose Position List and click OK. Under the Position track are now the X, Y, and Z Position tracks and an Available track. Select the Available track, access the Assign Controller dialog box again, and choose Noise Position and click OK. The Noise Controller dialog box opens. Close it, and click Play.

Note

The teapot vibrates the entire animation. You add a multiplier curve to correct the situation.

5. Select the Noise Position track, and choose Curves→Apply - Multiplier Curve. Select the Multiplier Curve track. Assign the first key a value of **0**. Right-click the first key, and change its Out tangent to Stepped so it holds its value until the next key. Click the Add Keys button, and add a key at frame 50 with a value of **1**. Move the last key to frame 120, and set a value of **0** with Smooth tangents for both In and Out. The Multiplier curve should now look like the curve in Figure 36.19. Select the Noise Position track.

    The noise curve now conforms to the multiplier track with loudest noise where the multiplier curve is at 1.

FIGURE 36.19

The Multiplier curve keeps the Noise track in check.

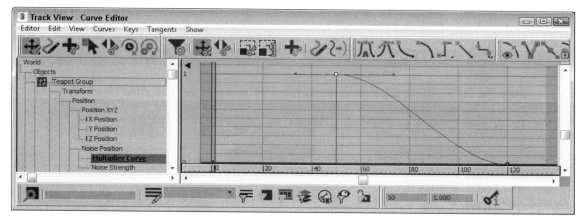

6. With the Noise Position track still selected, right-click and choose Properties. In the Noise Controller dialog box, set the X and Y Strength to **30**, set the Z strength to **2.0**, and check the >0 check box for the Z-Strength to keep the teapot from going through the floor. Close the dialog box, and click Play.

    The animation is much better. Next, you add some noise to the Rotation track.

7. Select the Teapot Group's Rotation track, right-click, and choose Assign Controller or press the C key to bring up the Assign Controller dialog box. Select Rotation List, and click OK. Select the Available track, access the Assign Controller dialog box again, and choose Noise Rotation and click Ok.

    Click Play. Again, the noise is out of control.

8. This time, select the Noise Strength track and add a multiplier curve.

    You already have a perfectly good multiplier curve, so you can instance it into the new track.

9. Select the Position Multiplier Curvetrack, right-click, and choose Copy. Now select the rotation Noise Strength Multiplier Curve track, right-click, and choose Paste. Choose Instance, and close the dialog box.

10. Click the Play button, and watch the Teapot wind up and then vibrate itself along until it winds down.

Figure 36.20 shows the teapot as it dances about, compliments of a controlled noise controller.

FIGURE 36.20

The wind-up teapot moves about the scene.

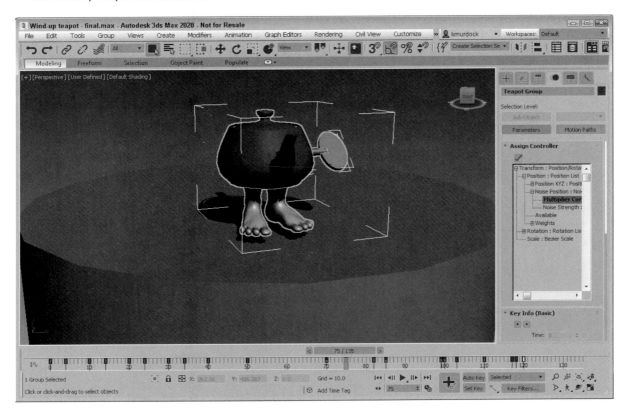

Filtering Tracks and Creating Track Sets

With all the information included in the Track View, finding the exact tracks you need can be difficult. The Filters button on the Controllers toolbar (or in the View menu) can help. Clicking this button (or pressing the Q keyboard shortcut) opens the Filters dialog box, shown in Figure 36.21.

FIGURE 36.21

The Filters dialog box lets you focus on the specific tracks.

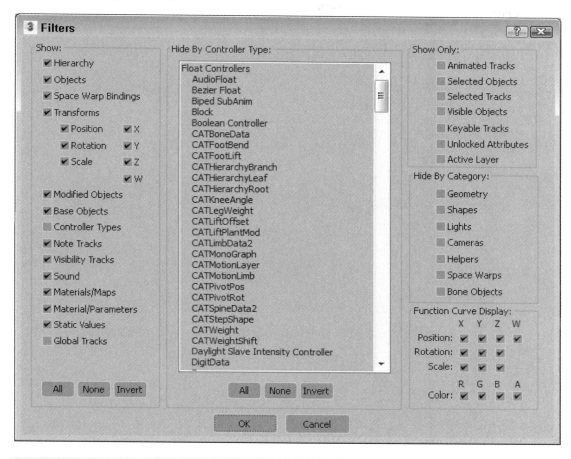

Tip

Right-clicking the Filters button reveals a quick list of filter items.

Using the Filters dialog box

Using this dialog box, you can limit the number of tracks that are displayed in the Track View. The Show section contains many display options. The Hide by Controller Type pane lists all the available controllers. Any controller types selected from this list do not show up in the Track View. You also can elect to not display objects by making selections from the check boxes in the Hide By Category section.

The Show Only group includes options for displaying only the Animated Tracks, Selected Objects, Selected Tracks, Visible Objects, Keyable Tracks, Unlocked Attributes, Active Layer, or any combination of these. For example, if you wanted to see the animation track for a selected object, select the Animated Tracks option and click OK; then open the Filters dialog box again, choose Selected Objects, and click OK.

You also can specify whether the curve display includes the Position, Rotation, and Scale components for each axis or the RGB color components.

Creating a track set

A selection of tracks can be saved into a track set by clicking the Edit Track Set button located on the Track Selection toolbar. This button opens the Track Sets Editor, shown in Figure 36.22. Clicking the Create a New Track Set button in the Track Sets Editor creates a new track set containing all the currently selected tracks and lists it in the editor window. Selected tracks can be added, removed, and selected using the other editor buttons.

FIGURE 36.22

The Track Sets Editor dialog box lets you name track selections for easy recall.

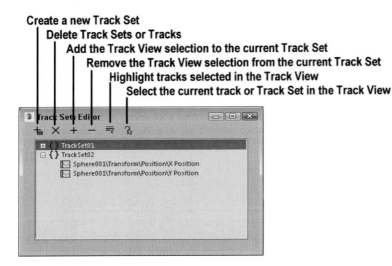

After a track set is created, its tracks can be instantly selected by choosing the track set's name from the drop-down list located next to the Edit Track Set button at the bottom of the Track View window.

Working with Controllers

Controllers offer an alternative to positioning keys manually. Each controller can automatically control a key's position or a parameter's value. The Edit→Controller menu includes several commands for working with controllers. The right-click Copy and Paste commands for controllers let you move existing controllers between different tracks, and the Assign Controller command lets you add a new controller to a track.

Chapter 35, "Animating with Constraints and Simple Controllers," covers all the various controllers used to automate animated sequences.

Although the commands are named Copy and Paste, they can be used to copy different tracks. Tracks can be copied and pasted only if they are of the same type. You can copy only one track at a time, but that single controller can be pasted to multiple tracks. A pasted track can be a copy or an instance, and you have the option to replace all instances. For example, if you have several objects that move together, using the Replace All Instances option when modifying the track for one object modifies the tracks for all objects that share the same motion.

All instanced copies of a track change when any instance of that track is modified. To break the linking between instances, you can use the Edit→Controller→Make Unique menu command.

Selecting the Controller→Assign Controller menu command opens the Assign Controller dialog box, where you can select the controller to apply. If the controller types are similar, the keys are maintained, but a completely different controller replaces any existing keys in the track.

Using visibility tracks

When an object track is selected, you can add a visibility track using the Edit→Visibility Track→Add menu command or with the Visibility value in the Object Properties dialog box. This track enables you to make the object visible or invisible. The selected track is automatically assigned the Bézier controller, but you can change it to an On/Off controller if you want that type of control. You can use curves mode to edit the visibility track.

Adding note tracks

You can add note tracks to any track and use them to attach information about the track in the Dope Sheet. The Edit→Note Track→Add menu command is used to add a note track, which is marked with a yellow triangle and cannot be animated.

After you've added a note track in the Controller pane, use the Add Keys button to position a note key in the Key pane by clicking in the Note track. This adds a small note icon. Right-clicking the note icon opens the Notes dialog box, where you can enter the notes, as shown in Figure 36.23. Each note track can include several note keys.

The Notes dialog box includes arrow controls that you can use to move between the various notes. The field to the right of the arrows displays the current note key number. The Time value displays the frame where a selected note is located, and the Lock Key option locks the note to the frame so it can't be moved or scaled.

You can use the Edit→Note Track→Remove menu command to delete a selected note track.

FIGURE 36.23

The Notes dialog box lets you enter notes and position them next to keys.

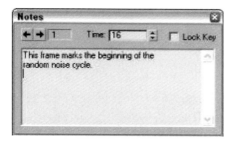

Using the ProSound Plug-in

3ds Max includes an audio plug-in called ProSound for adding multiple audio tracks to a scene. You need to initialize the plug-in before you start, using the Animation panel in the Preference Settings dialog box. You can access this dialog box with the Customize→Preferences menu.

To select the ProSound plug-in, click the Assign button and select the ProSound option. Once enabled, you can access ProSound by right-clicking the Sound track in the Track View and choosing the Properties menu. This opens the ProSound dialog box, as shown in Figure 36.24.

FIGURE 36.24

The ProSound dialog box lets you configure sounds to play during the animation.

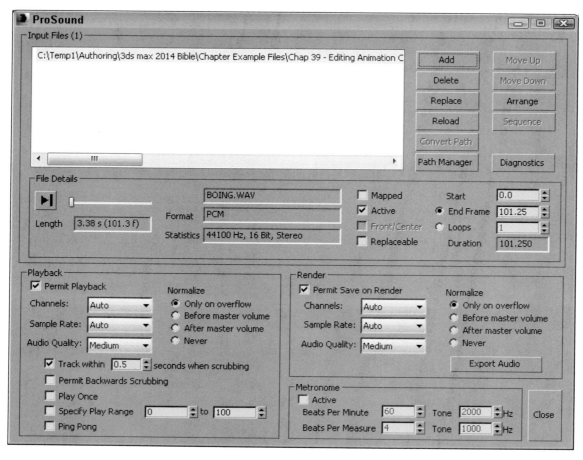

Using the ProSound dialog box, you can load and arrange multiple sound files. ProSound supports loading in WAV and AVI sound files. For the selected sound file, you can view its Length, Format, and Statistics. You also can set the sound's Start and End frame. Multiple other options are available for controlling how the audio file plays and how it interacts with other sounds.

The ProSound dialog box also includes some Metronome settings for keeping a defined beat. For a metronome, you can specify the beats per minute and the beats per measure. The first option sets how often the beats occur, and the second option determines how often a different tone is played. This dialog box also contains an Active option for turning the metronome on and off.

Tutorial: Adding sound to an animation

As an example of adding sound to an animation, you'll work with a hyper pogo stick and synchronize its animation to a sound clip.

To synchronize an animation to a sound clip, follow these steps:

1. Open the Hyper pogo stick.max file from the Chap 36 directory in the downloaded content set.

2. Open the Preference Settings dialog box with the Customize→Preferences menu command, and select the Animation panel. Click the Assign button in the Sound Plug-In section, and double-click the ProSound option and click OK.

3. In the Track View–Dope Sheet window, select and right-click the Sound track and select Properties from the pop-up menu to open the ProSound dialog box. In this dialog box, click the Add button.

Then locate the Boing.wav file from the Chap 36 directory in the downloaded content set, and click Open. Make sure the Permit Playback option is selected.

4. Expand the Boing.wav track to see the Waveform, as shown in Figure 36.25.

FIGURE 36.25

Sounds loaded into the sound track appear as waveforms.

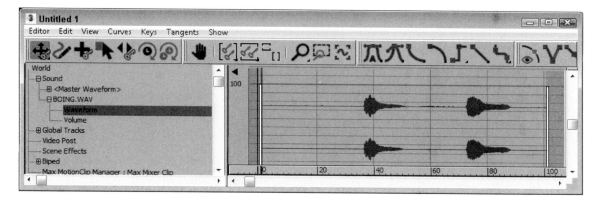

Note
The ProSound dialog box includes a play button that lets you play the sound before loading it.

5. Enter a Start frame value of **2** in the ProSound dialog box for the audio file to align with the pogo stick's upward motion.

6. Click the Play Animation button, and the sound file plays with the animation.

Figure 36.26 shows the sound track under the Track Bar for this example.

FIGURE 36.26

To help synchronize sound, the audio track can be made visible under the Track Bar.

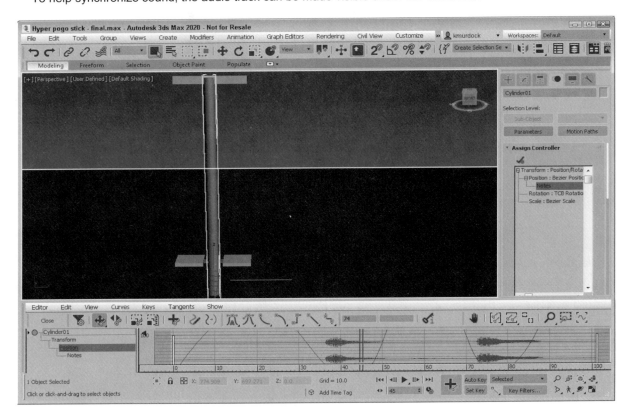

Summary

Using the Track View, you have access to all the keys, parameters, and tracks in a scene in one convenient location. Different features are available in the different layouts. Using the Track View interface, you can fine tune your animations for each specific track. In this chapter, you accomplished the following:

* Learned the Track View interface elements

* Learned about the different Track View layouts, including the Curve Editor, Dope Sheet, and Track Bar

* Discovered how to work with keys, times, and ranges

* Controlled and adjusted curves

* Selected specific tracks using the Filters dialog box

* Assigned controllers

* Explored the different out-of-range types

* Added notes to a track

* Added multi-track audio with ProSound

Now that you have figured out the Track View interface, the next chapter uses this interface to dive into the complex controllers.

Chapter 37

Exploring the Complex Controllers

IN THIS CHAPTER

Examining controllers

Using the Numerical Expression Evaluator

Understanding the Expression controller interface

If you look under the hood of animated objects, you'll find a group of controllers that are making it all work. Understanding how the various controllers work (and more importantly, how to configure them) is an important key in becoming an effective animator. This chapter dives deeper into the more complex controllers and takes a closer look at the Expression controller that makes scripted animations possible.

Expressions are looks that you make in the mirror when you're trying to wake up, but in the Autodesk® 3ds Max® 2020 software, they are a series of equations that define how an object acts. 3ds Max expressions can be as simple as adding two numbers together or as complex as several lines of MAXScript. But expressions enable you to create customized animated reactions.

Although 3ds Max expressions can be used with any 3ds Max spinner, they are mainly used within MAXScript scripts or in the Expression controller. The Expression controller is a specialized controller that lets you control the object's behavior using scripted expressions.

Examining Complex Controllers

Earlier chapters showed the basics on controllers, but this reference fills in the ones that we initially skipped, so let's look at some more complex controllers. 3ds Max includes a vast assortment of controllers, and you can add more controllers as plug-ins.

Transform controllers

Multi-track transform controllers work with the Position, Rotation, and Scale tracks all at the same time. You access them by selecting the Transform track in the Motion panel and then clicking the Assign Controller button.

Note
Some controllers are assigned through the Animation menu and others through the Assign Controller dialog box. Look to both if you cannot find a specific controller.

Note
Each of the available constraints is listed again in the appropriate controller submenu.

Position/Rotation/Scale Transform controller

The Position/Rotation/Scale Transform controller is the default controller for all transforms. This controller includes a Bézier controller for the Position and Scale tracks and a Euler XYZ controller for the Rotation track.

The PRS Parameters rollout, shown in Figure 37.1, lets you create and delete keys for Position, Rotation, and Scale transforms. The Position, Rotation, and Scale buttons control the fields that appear in the Key Info rollouts positioned below the PRS Parameters rollout.

FIGURE 37.1

The PRS Parameters rollout is the default transform controller.

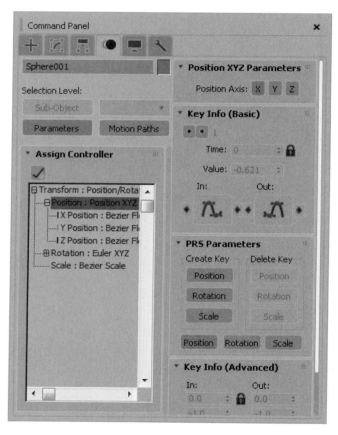

Script controller

The Script controller (named the Transform Script controller in the Motion panel) is similar to the Expression controller, except that it can work with the MAXScript lines of code for controlling the scene. Right-clicking a track with the Script controller assigned and selecting Properties opens the Script Controller dialog box. Script controllers are available for all transform tracks, including Transform, Position, Rotation, and Scale. The flexibility of the Script controller is quite robust. The Script controller is covered in more detail at the end of this chapter, as is the Expression controller.

For more information on MAXScript, see Chapter 52, "Automating with MAXScript."

XRef controller

If you have a defined motion used by an object in another file that you want to access, you can use the XRef controller. This controller can be assigned only to the Transform track. When this controller is assigned, a file dialog box opens where you can select the XRef file; then in the XRef Merge Object dialog box, you can select a specific object that has the controller and motion you want to use.

In the Parameters rollout for the XRef controller is a button to open the XRef Objects dialog box with the XRef Record highlighted. The Parameters rollout also lists the XRef file, object, and status.

XRefs are covered in detail in Chapter 10, "Organizing Scenes with Layers, Containers, XRefs and the Schematic View."

Position track controllers

Position track controller types include some of the common default controllers and can be assigned to the Position track. They typically work with three unique values representing the X-, Y-, and Z-axes. These controllers can be assigned by selecting a Position track and clicking the Assign Controller button. Many of the controllers found in this menu also are found in the Rotation and Scale Controllers menu.

Audio controller

The Audio controller can control an object's transform, color, or parameter value in response to the amplitude of a sound file. The Audio Controller dialog box, shown in Figure 37.2, includes Choose Sound and Remove Sound buttons for loading or removing sound files. You can access the Audio Controller dialog box by right-clicking on the track where the Audio Controller is assigned and selecting the Properties menu command from the pop-up menu, or simply double-click on the track.

FIGURE 37.2

The Audio Controller dialog box lets you change values based on the amplitude of a sound file.

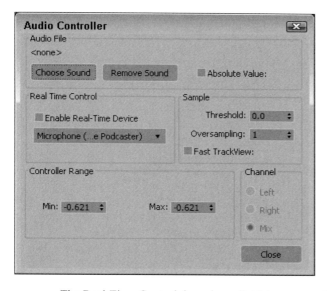

The Real Time Control drop-down list lets you specify a device to control the system. To control the sound input, you can specify a Sample Threshold and Oversampling rate. You also can set Min and Max range values for the controller. The Channel options let you specify which channel to use: Left, Right, or Mix.

Motion Clip Slave controller

The Motion Clip Slave controller lets a linked motion clip that is loaded and defined in the Motion Mixer control the object's transform.

Motion Capture controller

The Motion Capture controller allows you to control an object's transforms using an external device such as a mouse, keyboard, joystick, or MIDI device. This controller works with the Motion Capture utility to capture motion data.

After you assign the Motion Capture controller to a track, right-click the track and select Properties from the pop-up menu to open the Motion Capture dialog box, shown in Figure 37.3. This dialog box lets you select the devices to use to control the motion of the track values. Options include None, Keyboard, Mouse, Joystick, and MIDI devices.

FIGURE 37.3

The Motion Capture controller lets you control track values using external devices.

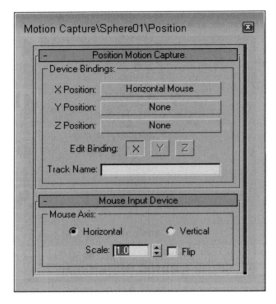

For the Keyboard control, the Keyboard Input Device rollout appears, as shown in Figure 37.4. The Assign button lets you select a keyboard key to track. The other settings control the Envelope Graph, which defines how quickly key presses are tracked.

FIGURE 37.4

The Keyboard Input Device rollout lets you select which key press is captured.

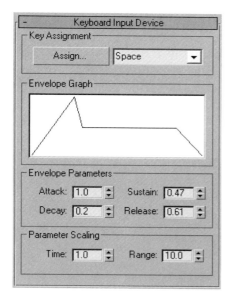

The Motion Capture dialog box defines only which device controls which values. The actual capturing of data is accomplished using the Motion Capture utility. Selecting the Motion Capture utility in the Utilities panel displays the Motion Capture rollout. This rollout includes buttons to Start, Stop, and Test the data-capturing process.

Before you can use the Start, Stop, and Test buttons, you need to select the tracks to capture from the Tracks list. The Record Range section lets you set the Preroll, In, and Out values, which are the frame numbers to include. You also can set the number of Samples Per Frame. The Reduce Keys option removes any unnecessary keys, if enabled.

Tutorial: Drawing with a pencil with the Motion Capture controller

Some motions, such as drawing with a pencil, are natural motions for our hands, but they become very difficult when you're trying to animate using keyframes. This tutorial uses the Motion Capture controller and utility to animate the natural motion of drawing with a pencil.

To animate a pencil drawing on paper, follow these steps:

1. Open the Drawing with a pencil.max file from the Chap 37 directory in the downloaded content set.

   This file has a pencil object positioned on a piece of paper.

2. Select the pencil object, open the Motion panel, and select the Position track for the pencil object. Then click the Assign Controller button, and double-click the Position Motion Capture selection.

   The Motion Capture dialog box opens.

3. Click the X Position button, and double-click the Mouse Input Device selection. Then click the Y Position button, and double-click the Mouse Input Device selection again. In the Mouse Input Device rollout, select the Vertical option. This sets the X Position to the Horizontal Mouse movement and the Y Position to the Vertical Mouse movement. Close the Motion Capture dialog box.

4. Open the Utilities panel, and click the Motion Capture button. In the Motion Capture rollout, select the Position track, and get the mouse ready to move. Then click the Start button in the Record Controls section, and move the mouse as if you were drawing with the mouse. The pencil object moves in the viewport along with your mouse movements.

   The Motion Capture utility creates a key for each frame. It quits capturing the motion when it reaches frame 100.

5. Click the Play Animation button (or press the / key) to see the results.

Figure 37.5 shows the scene after the Motion Capture controller has computed all the frames.

FIGURE 37.5

The Motion Capture controller and utility let you animate with a mouse, keyboard, joystick, or MIDI device.

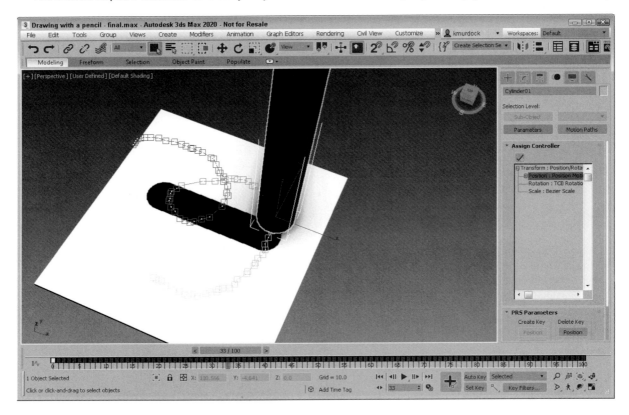

TCB controller

The TCB controller produces curved animation paths similar to the Bézier controller, but it uses the values for Tension, Continuity, and Bias to define their curvature. The benefit of this controller is that it enables objects to be rotated without having the problem of Gimbal lock, which can happen when the Euler XYZ controller is used. Gimbal lock can occur when two of the rotation axes become aligned, causing the object to lose one of its degrees of freedom.

The parameters for this controller are displayed in a single Key Info rollout. Like the Bézier controller rollouts, the TCB controller rollout includes arrows and Key, Time, and Value fields. It also includes a graph of the TCB values; the red plus sign represents the current key's position, while the rest of the graph shows the regular increments of time as black plus signs. Changing the Tension, Continuity, and Bias values in the fields below the graph changes its shape. Right-clicking the track and selecting Properties from the pop-up menu opens the TCB graph dialog box, shown in Figure 37.6.

FIGURE 37.6

This dialog box shows, and lets you control, a curve defined by the Tension, Continuity, and Bias values.

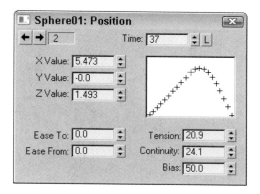

The Tension value controls the amount of curvature: High Tension values produce a straight line leading into and away from the key, and low Tension values produce a round curve. The Continuity value controls how continuous, or smooth, the curve is around the key: the default value of 25 produces the smoothest curves, whereas high and low Continuity values produce sharp peaks from the top or bottom. The Bias value controls how the curve comes into and leaves the key point, with high Bias values causing a bump to the right of the key and low Bias values causing a bump to the left.

The Ease To and Ease From values control how quickly the key is approached or left.

Note

Enabling the trajectory path by clicking the Motion Paths button in the Motion panel lets you see the changes to the path as they are made in the Key Info rollout.

Figure 37.7 shows a TCB curve assigned to the Position track of an object.

FIGURE 37.7

The TCB controller offers a different way to work with curves.

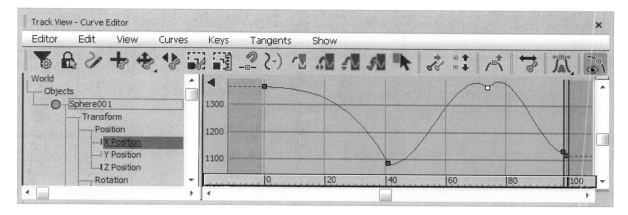

Reaction controller

The Reaction controller changes its values as a reaction to another controller. This controller is different from the Attachment controller in that the motions don't need to be in the same direction. For example, you can have one object rise as another object moves to the side.

Don't confuse the Reaction controller with the reactor plug-in, which computes motion based on physical dynamics. The reactor plug-in is covered in Chapter 46, "Simulating Physics-Based Motion with MassFX."

After the Reaction controller is assigned to a track, you can define the reactions using the Reaction Manager dialog box, shown in Figure 37.8. Selecting and right-clicking the track with this controller assigned and selecting Properties from the pop-up menu opens this dialog box. You also can open the Reaction Manager dialog box using the Animation→Reaction Manager menu.

FIGURE 37.8

The Reaction Manager dialog box lets you set the parameters of a reaction.

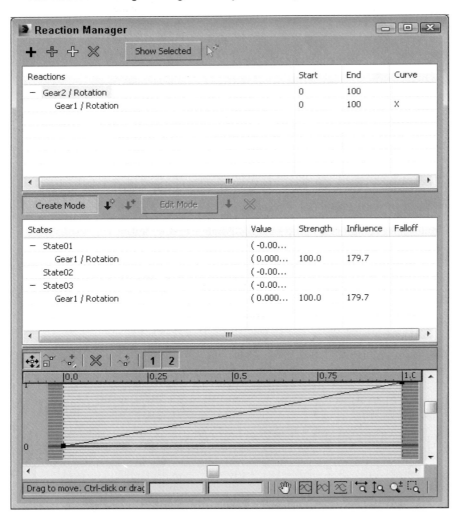

The Reaction Manager is made up of two lists and a graph of function curves. The top list holds all the object values that are involved in reactions. These are listed in a hierarchy with the master object listed above the slave object. A single master object can control several slave parameters.

The buttons above the Reactions list let you add new masters, slaves, and selected objects to the list. The cursor changes after you click any of these buttons, allowing you to click an object in the viewport and select a value from a pop-up menu.

For the slave objects selected in the Reactions list, you can set states using the buttons above the States list. To set a state, click the Create Mode button, drag the Time Slider to the appropriate frame, and change the slave object's value. Then click the Create State button to create the target object state. Several unique states can be defined for each slave object.

State values can be changed by accessing the Edit Mode button or by editing the curves displayed at the bottom of the Reaction Manager dialog box.

Rotation and Scale track controllers

The Rotation and Scale track controller types include some of the common default controllers and can be assigned to the Rotation and Scale tracks. They typically work with three unique values representing the X-, Y-, and Z-axes. These controllers can be assigned using the Assign Controller button. Many of the controllers found in this menu also are found in the Position Controllers menu. Only the controllers unique to the Rotation and Scale tracks are covered here.

Euler XYZ Rotation controller

The Euler XYZ Rotation controller lets you control the rotation angle along the X-, Y-, and Z-axes based on a single float value for each frame.

The main difference is that Euler rotation gives you access to the function curves. Using these curves, you can smoothly define the rotation motion of the object.

Note

Euler XYZ Rotation values are in radians instead of degrees. If using these as part of an expression, be sure to use radians and not degrees. Radians are much smaller values than degrees. A full revolution is 360 degrees or 2 times Pi radians, so 1 degree equals about 0.0174 radians.

The Euler Parameters rollout lets you choose the Axis Order, which is the order in which the axes are calculated. You also can choose which axis to work with.

Caution

The Euler XYZ controller is susceptible to Gimbal lock, which occurs when two of the three axes align to each other, causing the object to lose a degree of freedom. This can be minimized by making the axis that rotates the least the middle axis. You also can use the Euler Filter utility in the Track View to avoid Gimbal lock, or you can use the Quaternion controller instead.

Quaternion (TCB) Rotation controller

The Quaternion (TCB) Rotation controller lets you control the rotation angle based on Tension, Continuity, and Bias values to define their curvature. Unique to this controller is the Rotation Windup option, located beneath the TCB values. This option allows angular values greater than 180 degrees, which is useful if the object is rotating for multiple revolutions.

Smooth Rotation controller

The Smooth Rotation controller automatically produces a smooth rotation. This controller doesn't add any new keys; it simply changes the timing of the existing keys to produce a smooth rotation. It does not have any parameters.

Parameter controllers

Other controllers are used to affect the animated changes of parameters whether they are float, Point3, or other parameter types. Many of these controllers combine several controllers into one, such as the List and Block controllers. Others include separate interfaces, such as the Waveform controller for defining the controller's functions.

Most of these special-purpose controllers can be assigned only by using the Track View window. The Motion panel contains only the tracks for transformations.

Boolean controller

The Boolean controller, like the On/Off controller, can hold one of two states: 0 for off and 1 for on. But, unlike the On/Off controller, the Boolean controller changes only when a different state is encountered. The easiest place to change this controller's state is in the Dope Sheet interface.

Limit controller

The Limit controller sets limits for the motion or parameters of the selected controller. It is applied on top of the existing controller and opens the Limit Controller dialog box, shown in Figure 37.9, when applied.

FIGURE 37.9

The Limit Controller dialog box lets you set upper and lower limits for the current controller value.

The upper limit is the maximum value to which the controller can be set, and the lower limit is the minimum value that the controller uses. Controller values may exceed the upper and lower limit values, but the object's motion stops at the limit values when the Limit controller is enabled. The Smoothing value provides a range that gradually alters the value as it approaches the limit value.

After a Limit controller is applied to a parameter track, you can quickly change its upper and lower limit values by right-clicking the object in the Track View and accessing the Limit Controller options in the quad menu.

Tip

You can disable all limits at once using the Animation→Toggle Limits menu command.

List controller

You can use the List controller to apply several controllers at once. This feature enables you to produce smaller, subtler deviations, such as adding some noise to a normal Path controller.

When the List controller is applied, the default track appears as a subtrack along with another subtrack labeled Available. By selecting the Available subtrack and clicking the Assign Controller button, you can assign additional controllers to the current track.

All subtrack controllers are included in the List rollout of the Motion panel. You also can access this list by right-clicking the track and selecting Properties from the pop-up menu. The order of the list is important because it defines which controllers are computed first.

The Set Active button lets you specify which controller you can interactively control in the viewport; the active controller is marked with an arrow, which is displayed to the left of the name. You also can cut and paste controllers from and to the list. Because you can use the same controller type multiple times, you can distinguish each one by entering a name in the Name field.

On/Off controller

The On/Off controller works on tracks that hold a binary value, such as the Visibility track; you can use it to turn the track on and off or to enable and disable options. In the Track View, each On section is displayed in blue, with keys alternating between on and off. No parameters exist for this controller. Figure 37.10 shows a Visibility track that has been added to a sphere object. This track was added using the Edit→Visibility

Track→Add menu command. You can add keys with the Add Keys button. Each new key toggles the track on and off.

Note
You also can add a Visibility track by changing the Visibility value in the Object Properties dialog box.

FIGURE 37.10

The On/Off controller lets you make objects appear and disappear.

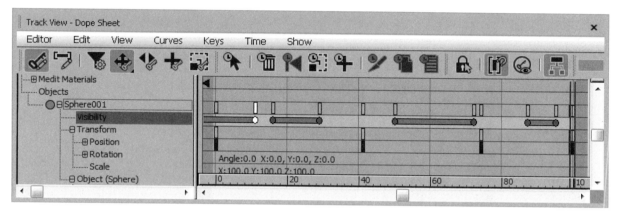

Tutorial: Animating a hazard light

As an example of working with the Track View, you'll animate a flashing hazard light in this tutorial.

To animate a flashing hazard light using curves, follow these steps:

1. Open the Hazard.max file from the Chap 37 directory in the downloaded content set.

   This file contains a hazard barrier with a light.

2. Select the Omni01 light object, open the Graph Editors→Track View–Curve Editor, and locate the Omni light's Multiplier track. (You can find this track under the Objects→Omni01→Object (Omni Light)→Multiplier menu command.) Zoom in on the Multiplier track by selecting the View→Frame→Frame Value Extents menu. Then, click the Add Keys button (or press the A key), and create a key on the dotted line (its current multiplier value) at frame 0 and at frame 15.

3. Select the new key created at frame 0, right click on the key and set its value to 0. Then, select and right click on the key at frame 15 and set its value of **1.2**. Select the View→Frame→Frame Value Extents menu to better see the shape of the curve. Click the Play button.

   These two key values cause the light to come on slowly over 15 frames, but we want the light to immediately turn on and off, so we need to change the tangent types to Stepped.

4. Select both keys, and click the Set Tangents to Stepped button in the top toolbar of the Track View window. Click the Play button.

   The light now turns on at frame 15.

5. Select the key at frame 0, and choose the Move Keys button. Hold down the Shift key and drag right to copy the key at frame 30.

   The Multiplier curve now has a value of 0 at frame 0 and jumps to 1.2 at frame 15, then drops back to 0 at frame 30. This causes the light to turn on at frame 15 and to turn off at frame 30. You can continue to make the light blink on and off by copying keys in this manner, but using the Parameter Curve Out of Range feature to complete the animation is easier.

6. Select the Edit→Controller→Out-of-Range Types menu command. The Param Curve Out-of-Range dialog box appears. Choose Cycle, and click OK.

   Figure 37.11 shows the Stepped tangents.

FIGURE 37.11

The curve with Stepped in and out tangents and a Cycle Parameter Curve Out-of-Range type.

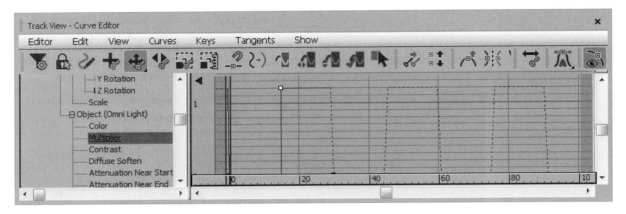

7. Click Play.

   The light flashes off and on.

Figure 37.12 shows the hazard light as it repeatedly blinks on and off.

FIGURE 37.12

The hazard light flashing on.

Waveform controller

The Waveform controller can produce regular periodic waveforms, such as a sinusoidal wave. Several different waveform types can make up a complete waveform. The Waveform Controller dialog box, shown in Figure 37.13, includes a list of all the combined waveforms. To add a waveform to this list, click the Add button.

When you select a waveform in the list, you can give it a name and edit its shape using the buttons and values. Preset waveform shapes include Sine, Square, Triangle, Sawtooth, and Half Sine. You also can invert and flip these shapes.

FIGURE 37.13

The Waveform Controller dialog box lets you produce sinusoidal motions.

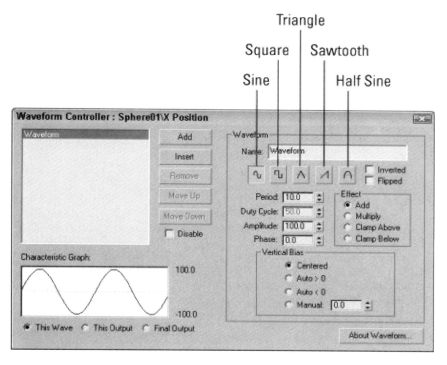

The Period value defines the number of frames required to complete one full pattern. The Amplitude value sets the height of the wave, and the Phase value determines its location at the start of the cycle. The Duty Cycle value is used only for the square wave to define how long it stays enabled.

You can use the Vertical Bias options to set the values range for the waveform. Options include Centered, which sets the center of the waveform at 0; Auto > 0, which causes all values to be positive; Auto < 0, which causes all values to be negative; and Manual, which lets you set a value for the center of the waveform.

The Effect options determine how different waveforms in the list are combined. They can be added, multiplied, clamped above, or clamped below. The Add option simply adds the waveform values together, and the Multiply option multiplies the separate values. The Clamp Above and Clamp Below options force the values of one curve to its maximum or minimum while not exceeding the values of the other curve. The Characteristic Graph shows the selected waveform, the output, or the final resulting curve. Figure 37.14 shows the Characteristic Graph for each Effect option when a sine wave and a square wave are combined.

FIGURE 37.14

Combining sine and square waves with the Add, Multiply, Clamp Above, and Clamp Below Effect options.

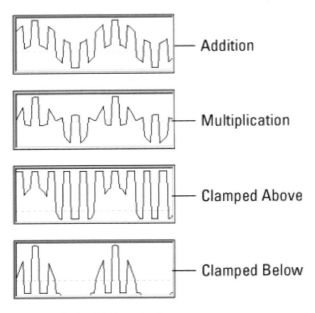

— Addition

— Multiplication

— Clamped Above

— Clamped Below

Color RGB controller

You can use the Color RGB controller to animate colors. Color values are different from regular float values in that they include three values that represent the amounts of red, green, and blue (referred to as RGB values) present in the color. This data value type is known as Point3.

The Color RGB controller splits a track with color information into its component RGB tracks. You can use this controller to apply a different controller to each color component and also to animate any color swatch in 3ds Max.

Cubic Morph controller

You can assign the Cubic Morph controller to a morph compound object. You can find the track for this object under the Object's track. A subtrack of the morph object is the Morph track, which holds the morph keys.

The Cubic Morph controller uses Tension, Continuity, and Bias values to control how targets blend with one another. You can access these TCB values in the Key Info dialog box by right-clicking any morph key or by right-clicking the Morph track and selecting Properties from the pop-up menu.

Note
You also can access the TCB values by right-clicking the keys in the Track Bar.

Barycentric Morph controller

The Barycentric Morph controller is automatically applied when a morph compound object is created. Keys are created for this controller based on the morph targets set in the Modify panel under the Current Targets rollout for the morph compound object. You can edit these keys using the Barycentric controller Key Info dialog box, which you can open by right-clicking a morph key in the Track View or in the Track Bar.

The main difference between the Cubic Morph controller and the Barycentric Morph controller is that the latter can have weights applied to the various morph keys.

The Barycentric Morph controller Key Info dialog box includes a list of morph targets. If a target is selected, its Percentage value sets the influence of the target. The Time value is the frame where this key is located. The TCB values and displayed curve control the Tension, Continuity, and Bias parameters for this controller. The Constrain to 100% option causes all weights to equal 100 percent; changing one value changes the other values proportionally if this option is selected.

Block controller

The Block controller combines several tracks into one block so you can handle them all together. This controller is located in the Global Tracks track. If a track is added to a Block controller, a Slave controller is placed in the track's original location.

To add a Block controller, select the Available track under the Block Control track under the Global Tracks track, and click the Assign Controller button. From the Assign Constant Controller dialog box that opens, select Master Block and click OK. Right-click the Master Block track and select Properties to open the Master Block Parameters dialog box, shown in Figure 37.15.

FIGURE 37.15

The Master Block Parameters dialog box lists all the tracks applied to a Block controller.

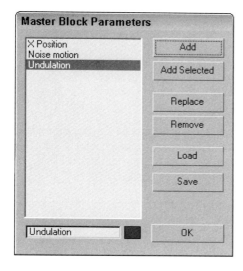

In the Master Block Parameters dialog box, you can add a track to the Block controller with the Add button. All tracks added are displayed in the list on the left. You can give each track a name by using the Name field. You also can use the Add Selected button to add any selected tracks. The Replace button lets you select a new controller to replace the currently selected track. The Load and Save buttons enable you to load or save blocks as separate files.

The Add button opens the Track View Pick dialog box, shown in Figure 37.16. This dialog box displays all valid tracks in a darker color to make them easier to see, while graying out invalid tracks.

FIGURE 37.16

The Track View Pick dialog box lets you select the tracks you want to include in the Block controller.

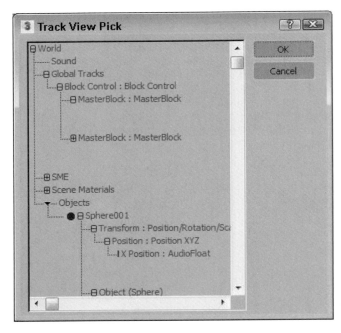

Select the tracks that you want to include, and click OK. The Block Parameters dialog box opens, shown in Figure 37.17, in which you can name the block, specify Start and End frames, and choose a color. Click OK when you're finished with this dialog box.

FIGURE 37.17

The Block Parameters dialog box lets you name a block.

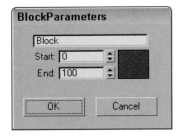

Back in the Master Block Parameters dialog box, click the Load button to open a file dialog box where you can load a saved block of animation parameters. The saved block files have the .blk extension. After the parameters have loaded, the Attach Controls dialog box opens, as shown in Figure 37.18. This dialog box includes two panes. The Incoming Controls pane on the left lists all motions in the saved block. By clicking the Add button, you can add tracks from the current scene, to which you can copy the saved block motions.

FIGURE 37.18

The Attach Controls dialog box lets you attach saved tracks to the Block controller.

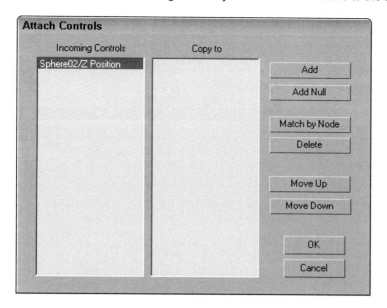

Because the saved motions in the Incoming Controls pane will match up with the Copy to entries in the right pane, the Add Null button adds a space in place of a specific track if you don't want a motion to be copied. The Match by Node button matches tracks by means of the Track View Pick dialog box.

IK controller

The IK controller works on a bones system for controlling the bone objects of an IK (inverse kinematics) system. The IK controller includes many different rollouts for defining joint constraints and other parameters.

Find out more about the IK controller in Chapter 38, "Understanding Rigging, Kinematics, and Working with Bones."

Master Point controller

The Master Point controller controls the transforms of any point or vertex subobject selections. The Master Point controller is a controller that you can select and add, but instead it gets added as a track to an object whose subobjects are transformed. Subtracks under this track are listed for each subobject. The keys in the Master track are colored green.

Right-clicking a green master key opens the Master Track Key Info dialog box, shown in Figure 37.19. This dialog box shows the Key number with arrows for selecting the previous or next key, a Time field that displays the current frame number, and a list of all the vertices. Selecting a vertex from the list displays its parameters at the bottom of the dialog box.

FIGURE 37.19

The Master Track Key Info dialog box lets you change the key values for each vertex.

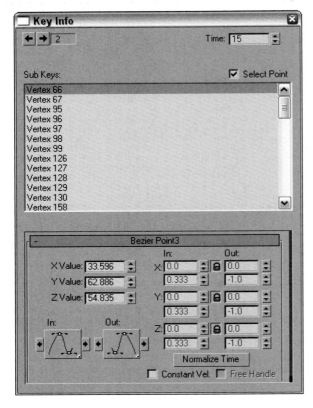

Tutorial: Animating a checkers move

As an example of working with curves, you'll animate a checkers move in this tutorial. It is often easiest to block in the animation using keyframing and then to refine the animation in the Track View.

To animate a white checker making its moves, follow these steps:

1. Open the Checkers.max file from the Chap 37 directory in the downloaded content set.

   This file contains a simple checkerboard with one white piece and three red pieces.

   The white checker is on the wrong-colored square to start, so first you move it into place and then to each successive position.

2. Turn on Auto Key. Move the time slider to frame 25, and move the white checker to the square with the "1" text object (visible in the Top viewport). Move the time slider to 50, and move the white piece to the "2" position. At frame 75, move it to the "3" position, and at 100, move it to the "4" position. Turn off Auto Key.

3. Right-click over the white piece, and choose Curve Editor to open the Track View–Curve Editor. The white piece's X, Y, and Z Position tracks should be highlighted, and you should be able to see the curves in the graph editor. If not, find them by choosing Objects→White Piece→Transform→Position→X, Y and Z Position. Click the Frame Horizontal Extents Selected Keys and the Frame Value Extents Selected Keys buttons in the Navigation toolbar to zoom in on the V Offest track's keys. These commands are also found in the View→Frame menu.

   Keeping in mind that RGB (red, green, blue) = XYZ, you can see the white checker's movement across the board. From 0 to 25, it moves only in the X direction. From 25 to 100, it moves diagonally across the board as indicated by the slope of both the X and Y curves. Note that when the object goes back the other way in the X direction at frame 75, the curve goes in the opposite direction, as shown in Figure 37.20. If the curves are not smooth, then press the Set Tangents to Auto button in the Key Tangents toolbar.

FIGURE 37.20

The blocked-in animation curves for the white piece

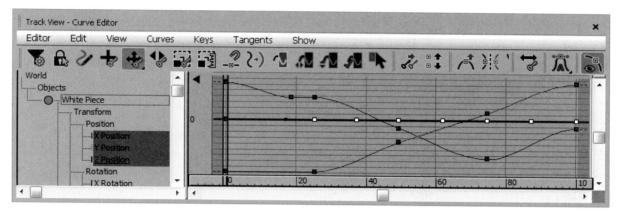

4. Click the Play button.

   The white piece slides sloppily around the board. Next, you create keys to make it hop over the red pieces.

5. Click the Add Keys button, or select Add Keys from the right-click menu. Click to insert keys on the Z-position track between the second, third, fourth, and fifth keys.

6. Select Move Keys Vertical from the Move Keys flyout, and select the three new keys. Move them up about 50 units, as shown in Figure 37.21.

FIGURE 37.21

The new keys are moved up.

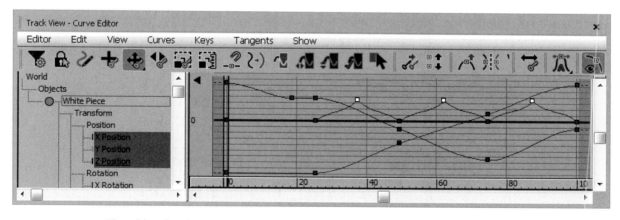

The white piece hops over the red pieces, but the motion is not correct. The In and Out tangents should be fast so the piece does not spend much time on the board.

7. With the new keys still selected, hold down the Shift key, and move the handles on the keys to make them discontinuous tangent types, as shown in Figure 37.22. You can make the tangent handles visible by enabling the Show Tangents button in the Curves toolbar.

907

FIGURE 37.22

The In and Out tangents corrected for the new keys

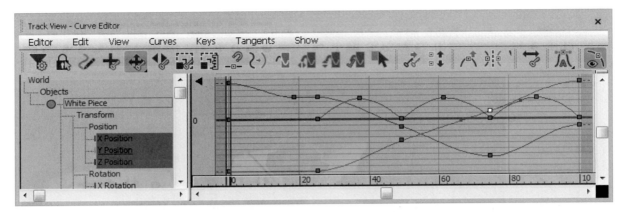

The hopping looks better, but the sliding motion would look better with a slight pause before hopping over the first red piece.

8. Select Move Keys Horizontal from the Move Keys flyout, and choose the second X-position track key. Hold down the Shift key, and move the key a few frames to the left to make a copy of the original key. Click the play button.

The red pieces should disappear as the white piece hops over them.

9. Scroll down the Controller pane on the left, and select the three red pieces. Under Tracks on the menu bar, choose Edit→Visibility Track→Add.

A Visibility track has been added to each of the red pieces, directly below the root name.

With visibility, a value of 0 is invisible, and a value of 1 is visible. You could change the tangent types to Stepped to turn the red pieces invisible from one frame to the next, but changing the entire track to an On/Off controller helps to visualize what is happening.

10. Select the Visibility track for Red Piece 01. Right-click, and choose Assign Controller. Choose On/Off.

Nothing seems to have happened. The Off/On controller is best used in Dope Sheet mode.

11. Choose Dope Sheet from the Editor menu. The On/Off controller track is represented with a blue bar. Blue indicates "on" or visible. Click the Add Keys button, and click to add a key at frame 50. The blue bar stops at the key at frame 50. Click Play.

The first red piece disappears at frame 50. You can copy and paste tracks to save yourself a bit of work.

12. Select Red Piece 01's Visibility track in the Controller pane. Right-click, and choose Copy. Select the Visibility tracks for Red Piece 02 and Red Piece 03 (hold down the Ctrl key to add to the selection). Right-click, and choose Paste.

Paste as a Copy because the other pieces should disappear at different times.

13. Move Red Piece 02's key to 75 and Red Piece 03's key to 100. Click Play.

Things are looking pretty good, but the animation would look better if the whole thing were faster.

14. Click the Edit Ranges button and the Modify Subtree button. A World track bar appears at the top of the Key Pane. Click and drag the rightmost end of the range bar to frame 75 to scale all the tracks at one time. Click Play.

The animation is quite respectable as the white piece slides into the correct square and then hops over and captures the three red pieces.

Figure 37.23 shows the checkerboard.

FIGURE 37.23

The checker pieces on the checkerboard

Working with Expressions in Spinners

Although much of this chapter focuses on the advanced controllers, the next section shows a special controller that lets you enter scripts or expressions. The term *expression* refers to a mathematical expression or simple formula that computes a value based on other values. But the Expression Controller Interface isn't the only place where you can play with expressions. Expressions also can be entered into spinner controls using the Numerical Expression Evaluator, shown in Figure 37.24. This simple dialog box is accessed by selecting a spinner and pressing Ctrl+N.

Another place that commonly uses expressions is the Parameter Wiring dialog box, which is covered in Chapter 34, "Wiring Parameters."

FIGURE 37.24

The Numerical Expression Evaluator dialog box lets you enter expressions for a spinner.

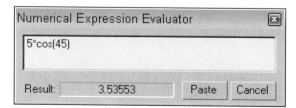

To use this evaluator, just type the expression in the field; the result is displayed in the result field. The result field is updated as you type the expression. If you make a mistake, the Result is blanked out. The Paste button places the result value in the spinner, and the Cancel button closes the dialog box without a change.

Understanding the Expression Controller Interface

These expressions can be simple, as with moving a bicycle based on the rotation of the pedals, or they can be complex, as with computing the sinusoidal translation of a boat on the sea as a function of the waves beneath it.

You can use almost any value as a variable in an expression, from object coordinates and modifier parameters to light and material settings. The results of the expression are computed for every frame and used to affect various parameters in the scene. You can include the number of frames and time variables in the expression to cause the animation results to repeat for the entire sequence.

Note
Although it is not as powerful as the Script controller, the Expression controller is much faster than the Script controller because it doesn't require any compile time.

Of all the controllers that are available, the Expression controller has limitless possibilities that could fill an entire book. This section covers the basics of building expressions and includes several examples.

The Expression controller is just one of the many controllers that are available for automating animations. This controller enables you to define how the object is transformed by means of a mathematical formula or expression, which you can apply to any of the object's tracks. It shows up in the controller list, based on the type of track to which it is assigned, as a Position Expression, Rotation Expression, Scale Expression, Float Expression, or Point3 Expression controller.

Before you can use the Expression controller on a track, you must assign it to a track. You can assign controllers to the Position, Rotation, and Scale tracks using the Motion panel or the Track View, or you can apply a controller by right-clicking on a track in the Track View and selecting the Assign Controller option. After you assign a controller, the Expression Controller dialog box immediately opens, or you can access this dialog box at any time by right-clicking the track and selecting Properties from the pop-up menu. For example, select an object in your scene, open the Motion panel, and select the Position track. Then click the Assign Controller button at the top of the Assign Controller rollout, and select Position Expression from the list of Controllers. This causes the Expression Controller dialog box to appear.

You can use this dialog box to define variables and write expressions. The dialog box, shown in Figure 37.25, includes four separate panes used to display a list of Scalar and Vector variables, build an expression, and enter a description of the expression.

FIGURE 37.25

You can use the Expression controller to build expressions and define their results.

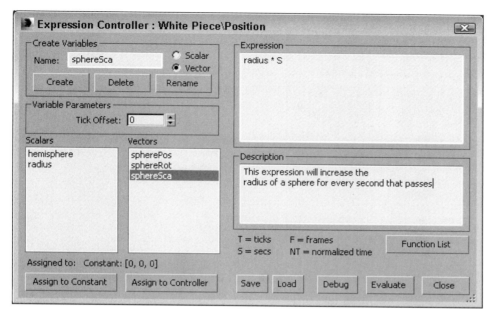

Defining variables

Variables are placeholders for different values. For example, creating a variable for a sphere's radius called "r" would simplify an expression for doubling its size from "take the sphere's radius and multiply it by two," to simply "r times 2."

To add variables to the list panes in the Expression Controller dialog box, type a name in the Name field, select the Scalar or Vector option type, and click the Create button; the new variable appears in the Scalars or Vectors list. To delete a variable, select it from the list and click the Delete button. The Tick Offset value is the time added to the current time and can be used to delay variables.

You can assign any new variable either to a constant or to a controller. Assigning a variable to a constant does the same thing as typing the constant's value in the expression. Constant variables are simply for convenience in writing expressions. The Assign to Controller button opens the Track View Pick dialog box, shown in Figure 37.26, where you can select the specific controller track for the variable, such as the position of an object.

FIGURE 37.26

The Track View Pick dialog box displays all the tracks for the scene. Tracks that you can select are displayed in black.

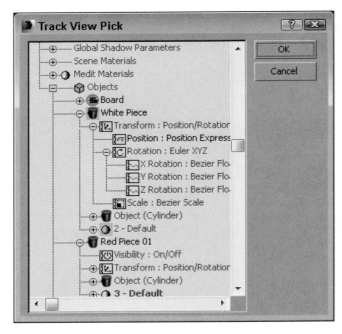

Assigning a variable to a controller enables you to animate the selected object based on other objects in the scene. To do this, create a variable and assign it to an animated track of another object. For example, if you create a Vector variable named boxPos and assign it to the Position track for a box object, then within the expression you can use this variable to base the motion of the assigned object on the box's position.

Building expressions

You can type expressions directly into the Expression pane of the Expression Controller dialog box. To use a named variable from one of the variable lists (Scalars or Vectors), type its name in the Expression pane. Predefined variables (presented later in the chapter) such as F and NT do not need to be defined in the variable panes. The Function List button opens a list of functions, shown in Figure 37.27, where you can view the functions that can be included in the expression. This list is for display only; you still need to type the function in the Expression pane.

Tip

One way to learn the syntax for different expressions is to enable the MacroRecorder in the MAXScript Listener window and look in the upper pane while performing a task in the viewports.

FIGURE 37.27

The Function List dialog box lets you view all the available functions that you can use in an expression.

abs(x)	absolute value of x
acos(x)	arccosine of x
asin(x)	arcsine of x
atan(x)	arctangent of x
ceil(x)	smallest integer greater than or equal to x
comp(v,i)	i'th component of v
cos(x)	trigonometric cosine of x
cosh(x)	hyperbolic cosine of x
degToRad(x)	x converted from degrees to radians
e	the constant e (2.71828...)
exp(x)	e raised to the power x
floor(x)	largest integer less than or equal to x
if(c,t,f)	conditional: value is t if c is true, else value is f
length(v)	length of v
ln(x)	natural logarithm (base e) of x
log(x)	common logarithm (base 10) of x
max(x,y)	maximum of x and y
min(x,y)	minimum of x and y
mod(x,y)	remainder of x divided by y
noise(x,y,z)	3D noise function
pi	the constant pi (3.1415...)
pow(x,y)	x raised to the power y
radToDeg(x)	x converted from radians to degrees
sin(x)	trigonometric sine of x
sinh(x)	hyperbolic sine of x
sqrt(x)	square root of x
tan(x)	trigonometric tangent of x
tanh(x)	hyperbolic tangent of x
TPS	number of ticks per second
unit(v)	unit vector in the direction of v
vif(c,v1,v2)	"vector if" - value is v1 if c is true, else v2

Close

Note

The Expression pane ignores any white space, so you can use line returns and spaces to make the expression easier to see and read.

Debugging and evaluating expressions

After typing an expression in the Expression pane, you can check the values of all variables at any frame by clicking the Debug button. This opens the Expression Debug window, shown in Figure 37.28. This window displays the values for all variables, as well as the return value. The values are automatically updated as you move the Time Slider.

FIGURE 37.28

The Expression Debug window offers a way to test the expression before applying it.

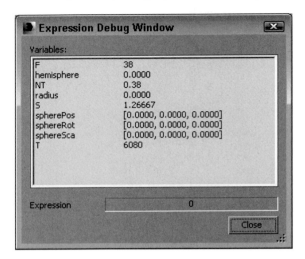

The Evaluate button in the Expression Controller dialog box commits the results of the expression to the current frame segment. If the expression contains an error, an alert dialog box warns you of the error. Replacing the controller with a different one can erase the animation resulting from an Expression controller.

Managing expressions

You can use the Save and Load button to save and recall expressions. Saved expressions are saved as files with an .xpr extension. Expression files do not save variable definitions.

Caution

If you load a saved expression into the Expression Controller dialog box, you need to reassign all variables before you can use the loaded expression.

Tutorial: Creating following eyes

As a quick example, you'll start with a simple expression. The Expression controller is very useful for setting the eye pupil objects to move along with a ball's motion. This same functionality can be accomplished using a manipulator and wiring the parameter, but you show it with the Float Expression controller.

To make eye pupil objects follow a moving ball object, follow these steps:

1. Open the Following eyes.max file from the Chap 37 directory in the downloaded content set.

   This file includes a face taken from a Greek model, along with a ball that is animated to move back and forth.

2. Select the right eye pupil object named "pupil_l" (the right pupil has been linked to move with the left pupil), and open the Motion panel. Select the Position track, and click the Assign Controller button. Select the Position Expression controller, and click OK.

   The Expression Controller dialog box opens.

3. Create a new vector variable named **ballPos** by typing its name in the Name field, selecting the Vector option, and clicking the Create button.

4. With the ballPos variable selected, click the Assign to Controller button. In the Track View Pick dialog box, locate and select Objects and click the "+" to the left. Then choose Sphere01, and click the drop-down icon to its left. Finally, choose the Transform: PRS, Position: Position XYZ, as shown in Figure 37.29.

FIGURE 37.29

Select the Position track for the Sphere01 object in the Track View Pick dialog box.

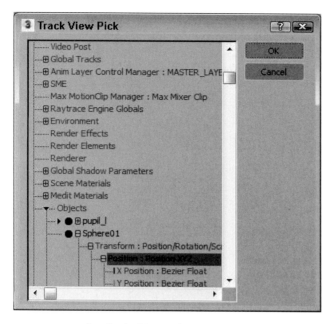

5. In the Expression pane, erase the existing expression and type the following:

```
[ -3.1 + ballPos.x/20, -2.9, 41.0 ]
```

This moves the pupils to follow the sphere along a horizontal path. Then click the Debug button. The Expression Debug window appears, in which you can see the variable values change as items in the scene change. With the expression complete, you can drag the Time Slider back and forth and watch the pupil follow the ball from side to side. If you're happy with the motion, click the Evaluate button and then the Close button to exit the interface.

This is a simple example, but it demonstrates what is possible. Figure 37.30 shows the resulting face.

FIGURE 37.30

The Expression controller was used to animate the eyes following the ball in this example.

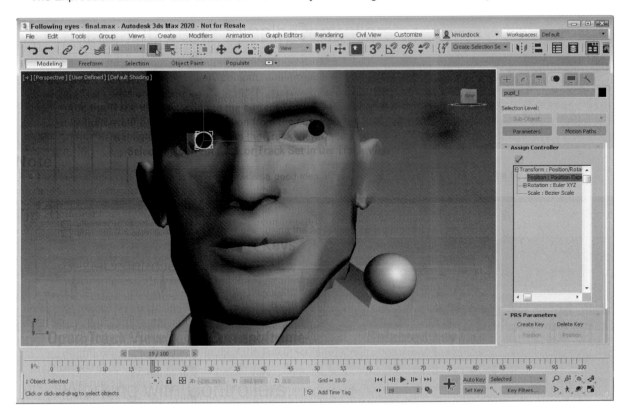

Using Expression Controllers

You can use expressions to control the transforms of objects. You can access these transforms from the Track View or from the Motion panel. You also can use expressions to control object parameters such as a box's length or material properties such as the amount of illumination applied to a material. You can access all these parameters from the Track View.

Animating transforms with the Expression controller

After you assign a controller to a transform track, the Expression pane in the Expression Controller dialog box includes the current values of the selected object. Position transforms display the X, Y, and Z coordinates of the object; Rotation transforms display the rotation value in radians; and Scale transforms display values describing the relative scaling values for each axis.

Note
Radians are another way to measure angles. A full revolution equals 360 degrees, which equates to 2 x pi radians. The Expression dialog box includes the `degToRad` and `radToDeg` functions to convert back and forth between these two measurement systems.

Animating parameters with the Float Expression controller

To assign the Float Expression controller, select an object with a parameter or Modifier applied and open the Track View. Find the track for the parameter that you want to change, and click the Assign Controller button. Select the Float Expression controller from the list, and click OK.

Note

The actual controller type depends on the parameter selected. Many parameters use float expressions, but some use Transform controllers.

After you assign the Float Expression controller, the Expression Controller dialog box opens, or you can open it by right-clicking the track and selecting Properties from the pop-up menu to load the dialog box. Within this dialog box, the Expression pane includes the current value of the selected parameter.

Tutorial: Inflating a balloon

The Push Modifier mimics filling a balloon with air by pushing all its vertices outward. In this tutorial, you'll use a balloon model to see how you can use the Float Expression controller to control the parameters of a modifier.

To inflate a balloon using the Float Expression controller, follow these steps:

1. Open the Balloon and pump.max file from the Chap 37 directory in the downloaded content set.

    This file includes a pump created from primitives and the balloon model with the Push modifier applied.

2. Open the Track View by choosing Graph Editors→Track View - Curve Editor. Select the balloon and navigate the balloon object's tracks until you find the Push Value track (found under Objects→b3→Modified Object→Push→Push Value). Select the Push Value track, and right click and select the Assign Controller option from the quad menu. From the list of controllers, select Float Expression and click OK.

    The Expression Controller dialog box opens.

3. In the Expression pane, you should see a single scalar value of 0. Modify the expression to read like this:

    ```
    2 * NT
    ```

    Click the Debug button to see the value results. With the Expression Debug window open, drag the Time Slider and notice that the balloon inflates.

Note

If you use a parameter such as Radius as part of an Expression, then the parameter is unavailable in the Modify panel if you try to change it by hand.

Figure 37.31 shows the balloon as it is being inflated.

FIGURE 37.31

A balloon being inflated using an Expression controller to control the Push modifier.

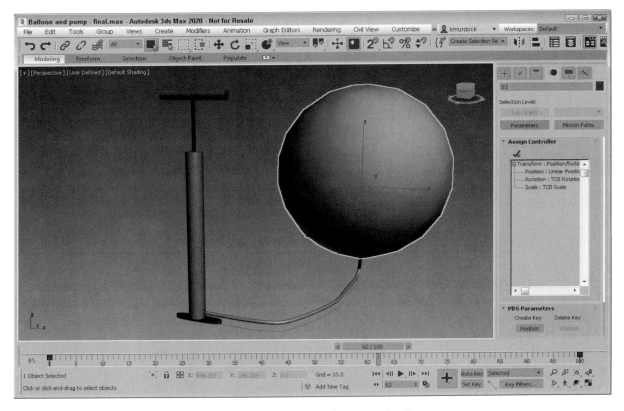

Animating materials with the Expression controller

You can locate the material's parameter in the Track View and assign the Expression controller to it to control material parameters. Some of these parameters are scalar values, but any material parameter set with a color swatch has a Point3 return type.

When using material parameters and color values, be sure not to combine them in expressions with vector values.

Summary

This chapter covered the available animation controllers and showed you how to use several of them. This chapter also covered the basics of using the Expression controller. Using mathematical formulas to control the animation of an object's transformation and parameters offers lots of power. You also can use the values of one object to control another object.

In this chapter, you learned about the following:

* Examining the various controllers in several different categories
* Building expressions in the Expression Controller dialog box
* Understanding expressions and what they can do
* What the available operators, variables, and functions do
* Controlling object transformations and parameters

The next chapter dives into working with characters, rigging, and bones.

Part VIII
Working with Characters

IN THIS PART

Chapter 38

Understanding Rigging, Kinematics, and Working with Bones

IN THIS CHAPTER

Creating a rigging workflow

Building a bones system

Setting bone parameters and IK Solvers

Making linked objects into a bones system

Understanding Forward and Inverse Kinematics

Learning to work with the available IK solvers

What does a graveyard have in common with animated characters? The answer is bones. Bones are used as an underlying structure attached to a character skin that is to be animated. By using a bones structure, you can produce complex character motions by simply animating the bones and not having to move all the vertices associated with a high-resolution character mesh.

Although the Autodesk® 3ds Max® 2020 software includes a prebuilt skeleton with its Biped and CAT systems, at times you may want to build a custom bones system because not all characters stand on two feet. Have you ever seen a sci-fi movie in which the alien was less than humanlike? If your character skeleton can't be created easily by modifying a biped or CAT skeleton, you must use the traditional manual methods of rigging by building a skeleton from scratch using bone objects.

This chapter focuses on the process of manually rigging a character that, depending on the complexity of your character, could end up being even easier than working with bipeds. It also gives you a clear idea of what is involved in rigging a character.

Understanding Rigging

A rigged character consists of two parts—an underlying skeleton made of individual linked bones and a high resolution skin mesh that defines how the character looks. The skeleton is the part of the rig that is animated and holds all the animation keys, and the skin is bound to the skeleton and follows the bones as they move. When the animation is complete, the skeleton is hidden and the skin alone is rendered.

The rig's skeleton consists of bones that are organized in a linked hierarchy. A linked hierarchy attaches, or links, one object to another and makes it possible to transform the attached object by moving the one to which it is linked. The arm is a classic example of a linked hierarchy: when the shoulder rotates, so do the elbow, wrist, and fingers. Establishing linked hierarchies can make moving, positioning, and animating many connected objects easy.

A bones system is a unique case of a linked hierarchy that has a specific structure. You can create a structure of bones from an existing hierarchy, or you can create a bones system and attach objects to it. A key advantage of a bones system is that you can use IK (Inverse Kinematics) Solvers to manipulate and animate the structure. These IK Solvers enable the parents to rotate when the children are moved. In this way, the IK Solver maintains the chain integrity.

After the bone structure is created, it needs to be edited to fit the skin mesh that it will control. You also need to define the limits of each bone and joint. This helps prevent the skeleton from moving in unrealistic ways. Applying IK systems is another way to control the motion of the bones and joints. This process of creating a skeleton structure and defining its limits is called *rigging*. Rigging also involves building specialized animation controls.

After you've edited a system of bones, you can cover the bones with objects that have the Skin modifier applied. This modifier lets the covering object move and bend with the bones structure underneath. The process of attaching a model to a bones system and setting the various skin weights is called *skinning*.

The Skin modifier is covered, along with other aspects of skinning a character, in Chapter 39, "Skinning Characters."

After a character is rigged and skinned, the character is ready to be animated.

A typical rigging workflow

The goal of creating a rigged character is that it's easy to animate. In a typical studio environment, this involves several key players. First on the scene is the modeler. The modeler builds the character as a high-resolution mesh complete with materials. This model, which is typically a single object, is known as the skin for the character, and it usually is posed using a standard T-pose, as shown in Figure 38.1, with feet flat at shoulder width, eyes and head looking forward, and arms stretched out straight with palms forward.

FIGURE 38.1

The Little Dragon character displayed in a standard T-pose

The second player in this process is the rigger. This person takes the skin model and builds a skeleton made of bones that is used to control the attached skin. For a standard human character, this skeleton includes a separate bone for every straight section of the character, similar to the way a real set of bones would work. At the places where the model needs to bend, a joint is placed, such as at the knees and elbows. Each bone also can be constrained to allow only proper motion. For example, the elbow joint is limited to bend only within a 180 degree range, which keeps it from bending unrealistically backward.

Another task for the rigger is to build animation controls that let the animator quickly create all the animation keys for specific motions. These controls could include morph targets used to create facial expressions or other special controls for manipulating the hands and feet.

The rigger also attaches the skin to the skeleton and defines the skin weights. Skin weights associate each skin vertex with a specific bone or set of bones. These skin weights determine how the skin moves and bends with the animated bones. For example, the vertices in the center of the thigh are weighted to move with the thigh bone, but vertices near the knee must be weighted to move with the thigh and shin bones, so the weighting is split between those two bones.

After the character is fully rigged and skinned, it is turned over to an animator who uses the skeleton and the animation controls to complete the animations for the character. As the animator is working with the character rig, he or she may request additional controls and adjustments from the rigger, so all team members are still involved until the final animations are completed.

Building a Bones System

To create a skeleton, you need to build a hierarchy of objects that are linked together. This can be done using primitives and the Select and Link tool, but an easier way to build this hierarchy is to use a *bones system*. A bones system consists of many bone objects that are linked together. These bone objects are normally not rendered, but you can set them to be renderable, like splines. You also can assign an IK Solver to the bones system for controlling their motion.

To create a bones system, select Create→Systems→Bone IK Chain, click in a viewport at the location for the start of the bone, and click again to place the bone's end. Each successive click adds another bone linked to the bone chain. Each bone has a thick end that represents the bone's head and a tapered small end, which is the tail. When you're finished adding bones, right-click to exit bone creation mode. In this manner, you can create a long chain of bone objects all linked together. Each bone is named simply Bone followed by a three-digit number.

When you right-click to exit bone creation mode, an End bone is added to the bone chain. This End bone doesn't have any length and is used to mark the end of the chain. It also is used by the IK solver and is a necessary part of the bone chain.

These bones are actually linked joints. Moving one bone pulls its neighbors in the chain along with it. Bones also can be rotated, scaled, and stretched. Scaling a bones system affects the distance between the bones. The pivot point for rotating and scaling bones is the bone's head.

Caution
Bones should never be scaled after keys are set without the XForm modifier applied, or all animation keys will behave erratically.

To branch the hierarchy of bones, simply click the bone where you want the branch to start while still in bone creation mode (the Bones button in the Command Panel is still highlighted). A new branching bone is created automatically with its head located where the tail of the clicked bone is. Then position and click in the viewport to place the new bone's tail. Continue to click to add new bones to the branch, and right-click to end the new branch. Right-click a second time to turn bone creation mode off.

Assigning an IK Solver

When you first create a bone chain in the IK Chain Assignment rollout of the Create panel, you can select from four IK Solvers: History Dependent, IKHISolver, IKLimb, and SplineIK Solver. You can assign each of these solver types to children and to the root bone using the available options. You need to select both the Assign to Children and the Assign to Root options to assign the IK Solver to all bones in the system. If the Assign to Children option is deselected, the Assign to Root option is disabled. More on these IK solvers is presented later in this chapter.

Setting bone parameters

The Bone Parameters rollout includes parameters for setting the size of each individual bone, including its Width and Height. You also can set the percentage of Taper applied to the bone. By changing these parameters, you can make the bone appear long and flat, which is helpful for manipulating Plane objects.

Tip

Because bones are simple geometry objects, you can apply an Edit Poly modifier to it and edit the bone shape to be whatever you'd like. However, custom bone geometry doesn't always work with the Bone Tools.

Fins can be displayed on the front, back, and/or sides of each bone by enabling the Fin options. For each fin, you can specify its size and start and end taper values. Including fins on your bones makes correctly positioning and rotating the bone objects easier. It is often helpful to have the Front (or Back) Fin enabled, so you can tell when the bone is upside down. Figure 38.2 shows a simple bones system containing two bones with fins enabled.

FIGURE 38.2

This bone includes fins that make understanding its orientation easier.

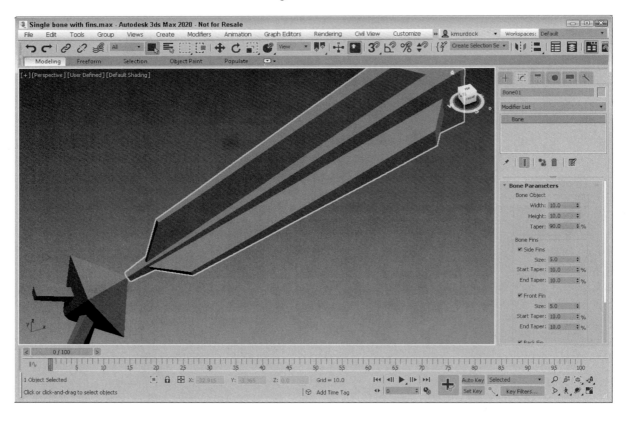

At the bottom of the Bone Parameters rollout is an option to Generate Mapping Coordinates. Bones are renderable objects, so this option lets you apply texture maps to them.

Tip

Although bones can be rendered, they are most often hidden for the final render, but applying a bright color or texture to the bones helps them to be seen easily, and it's the sharp reminder to hide all bones when doing final test renders.

Defining joint constraints

The next step is to define the joint constraints, which you specify in the Sliding Joints and Rotational Joints rollouts. These rollouts are found in the Hierarchy panel of the Command Panel if you click the IK button when a bone is selected.

By default, each joint has six degrees of freedom, meaning that the two objects that make up the joint can each move or rotate along the X-, Y-, or Z-axis. The axis settings for all other sliding and rotational joints are identical. Defining joint constraints enables you to constrain these motions to prevent unnatural motions, such as an elbow bending backward. To constrain an axis, select the bone object that includes the pivot point for the joint, locate in the appropriate rollout the section for the axis that you want to restrict, and deselect the Active option. If an axis's Active option is deselected, the axis is constrained. You also can limit the motion of joints by selecting the Limited option.

When the Limited option is selected, the object can move only within the bounds set by the From and To values. The Ease option causes the motion of the object to slow as it approaches either limit. The Spring Back option lets you set a rest position for the object; the object returns to this position when pulled away. The Spring Tension sets the amount of force that the object uses to resist being moved from its rest position. The Damping value sets the friction in the joint, which is the value with which the object resists any motion.

Note

As you enter values in the From and To fields, the object moves to that value to show visually the location specified. You also can press and hold the left mouse button on the spinners for the From and To values to cause the object to move temporarily to its limits. These settings are based on the current Reference Coordinate system.

Naming bones

A critical, but often forgotten, step in rigging is naming the bones. When bones are created, they are automatically given a name, which is simply "Bone" and a number. As the skeleton gets more complex, these numbered names make it hard to find anything. Naming each bone using the Name field in the Command panel is essential. You also can use the Rename Objects dialog box in the Tools menu to quickly rename multiple bones.

Tutorial: Creating a bones system for an alligator

To practice creating a bones system, you'll take a trip to the Deep South to gator country. The main movement for this gator is going to be in its tail, so you need the most bones there. The front legs are smaller and can be controlled with only two simple bones. You also won't worry about fingers.

To create a bones system for an alligator, follow these steps:

1. Open the Alligator bones.max file from the Chap 38 directory in the downloaded content set.

   This file includes an alligator model.

2. Select Create→Systems→Bone IK Chain, and in the IK Chain Assignment rollout, select IK Limb from the IK Solver drop-down list. Then set the Width and Height values to **10**, enable the Side Fins, and set the Size to **5**.

3. In the Top viewport, click once at the top of the tail, again at a spot between the gator's two back legs, again at the mid-abdomen between the two front legs, again at the base of the neck, and finally at the end of the nose. Then right-click to end the bones chain.

4. While still in Bones mode, click below the first bone without clicking on the bone and create an additional five bones that run down the length of tail. Then right-click to end the chain, and right-click again to exit bone creation mode. Select the first bone in the tail chain, and link it to the first joint in the upper bone chain with the Select and Link tool.

Tip

If you can't see the bones to make the link, hide the body so the bones are clearly visible.

5. Click the Bones button in the Create panel, and select the bone just below the back set of legs in the Top viewport (the cursor changes to a cross-hair when it is over a bone). This creates a new branch from the end of the clicked bone. Then click to place new bones at the gator's knee, ankle, and tip of the foot. Right-click to end the chain. Repeat to create bones for the opposite leg, then repeat this step for each of the arms.

6. Click the Select Objects button on the main toolbar to exit Bones mode, and select and name each bone object so it can be easily identified later.

Tip

The Schematic View window is a good interface for quickly labeling bones.

Figure 38.3 shows the completed bones system for the alligator.

FIGURE 38.3

This bones system for an alligator was easy to create.

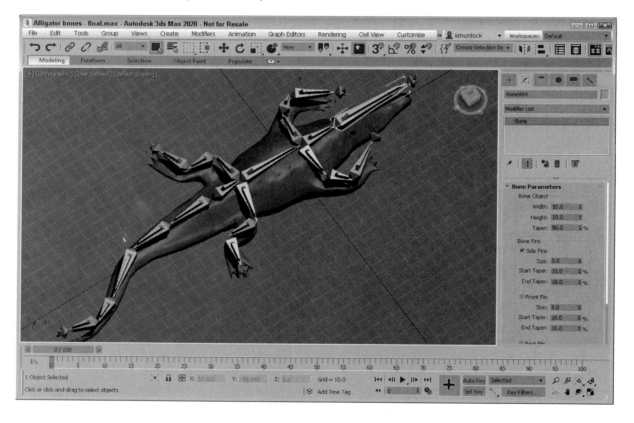

Using the Bone Tools

After you've created a bones system, you can use the Bone Tools to edit and work with the bones system. You access these tools from a panel that is opened using the Animation→Bone Tools menu command. Figure 38.4 shows this panel of tools, which includes three separate rollouts: Bone Editing Tools, Fin Adjustment Tools, and Object Properties.

FIGURE 38.4

The Bone Tools dialog box includes several buttons for working with bones systems.

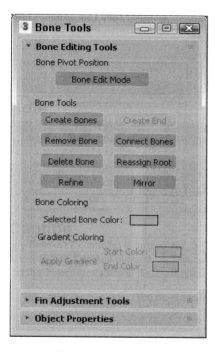

Reordering bones

You can use the transform buttons on the main toolbar to move, rotate, and scale a bone along with all its children, but if you want to transform the parent without affecting any of the children, you need to open the Bone Tools panel using the Animation→Bone Tools menu command. Bone Edit Mode lets you move and realign a bone without affecting its children.

Clicking the Remove Bone button removes the selected bone and reconnects the bone chain by stretching the child bone. If you hold down the Shift key while removing a bone, the parent is stretched. Clicking the Delete Bone button deletes the selected bone and adds an End bone to the last child.

Caution

Using the Delete key to delete a bone does not add an End bone, and the bone chain does not work correctly with an IK Solver.

If a bone exists that isn't connected to another bone, you can add an End bone to the bone using the Create End button. The bone chain must end with an End bone in order to be used by an IK Solver.

The Connect Bones button lets you connect the selected bone with another bone. After clicking this button, you can drag a line from the selected bone to another bone to connect the two bones.

Use the Reassign Root button to reverse the chain and move the End bone from the parent to the last child.

Refining and mirroring bones

As you start to work with a bones system that you've created, you may discover that the one long bone for the backbone of your monster is too long to allow the monster to move like you want. If this happens, you can refine individual bones using the Refine button. This button appears at the bottom of the Bone Tools section of the Bone Editing Tools rollout.

Clicking the Refine button enables you to select bones in the viewport. Every bone that you select is divided into two bones at the location where you click. Click on the Refine button again to exit Refine mode.

The Mirror button lets you create a mirror copy of the selected bones. This button makes the Bone Mirror dialog box appear, where you can select the Mirror Axis and the Bone Axis to Flip. You also can specify an Offset value. In the previous example, you easily could have created the arms and legs manually for one side of the gator and then used the Mirror button to create its opposite.

Coloring bones

Bones, like any other objects, are assigned a default object color, and materials can be applied from the Material Editor. For each separate bone, its object color can be changed in the Modify panel or in the Bone Tools dialog box.

You also can apply a gradient to a bone chain using the Bone Tools dialog box. This option is available only if two or more bones are selected. The Start Color is applied to the chain's head, and the End Color is applied to the last selected child. The colors are applied or updated when the Apply Gradient button is clicked. Figure 38.5 shows a long, spiral bone chain with a white-to-black gradient applied.

Tip

Gradient coloring along long straight bone chains is helpful in visually showing where the end of the chain is located. It is great to use for tails, whips, and long braided strands.

FIGURE 38.5

A white-to-black gradient was applied to this spiral bone chain.

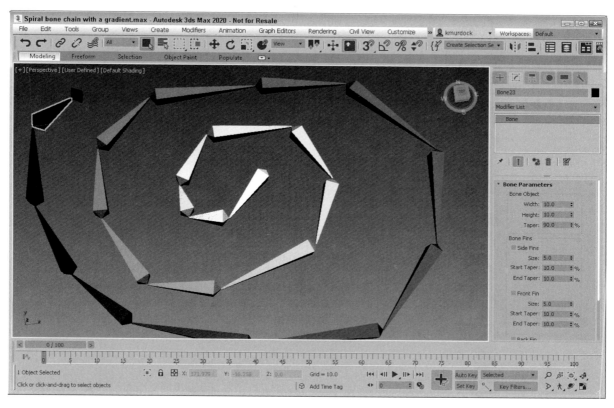

Adjusting fins

The Fin Adjustment Tools rollout includes the same parameters as those found in the Bone Parameters rollout. You can specify the dimensions and taper of a bone and its fins. But you also can specify that the parameters are applied using Absolute or Relative values. Relative values are based on the parameters of the bone that is above the current bone in the chain.

This rollout also includes Copy and Paste buttons that you can use to copy the bone parameters from one bone to another.

Making objects into bones

You can make any object act like a bone. To make an object into a bone, you need to open the Object Properties rollout in the Bone Tools dialog box. The Object Properties rollout includes a setting for Bone On. If enabled, the object acts like a bone. When the Bone On/Off option is enabled, the remaining Bone controls become available. The Auto-Align option causes the pivot points of adjacent bones to be aligned automatically. The Freeze Length option causes a bone to keep its length as the bones system is moved. If the Freeze Length is disabled, you can specify a Stretch type. None prevents any stretching from occurring, and Scale changes the size along one axis, but Squash causes the bone to get wider as its length is decreased and thinner as it is elongated. You also can select to stretch an axis and choose whether to Flip the axis.

You can use the Realign button to realign a bone to its original orientation. Click the Reset Stretch or Reset Scale button to reset the stretch or scale value to its original value.

Forward Kinematics versus Inverse Kinematics

The biggest advantage of using a bones system is that it moves according to kinematic principles. *Kinematics* is a branch of mechanics that deals with the motions of a system of objects, so *Inverse Kinematics* is its evil twin brother that deals with the non-motion of a system of objects, right? Well, not exactly.

In 3ds Max, a system of objects is a bunch of objects that are linked together. After a system is built and the parameters of the links are defined, the motions of all the pieces below the parent object can be determined as the parent moves, using kinematics formulas. This is called Forward Kinematics.

Chapter 9, "Grouping, Linking, and Parenting Objects," covers linking objects and creating kinematics chains.

Forward Kinematics is really just a complex word for a simple concept. When a bones system is set up using linked objects, moving any bone in the system automatically affects all its children. For example, when the shoulder bone is rotated, the arm, forearm, and hand bones all move together along with the shoulder. This works great for some motions like walking because you can get all the motions of the legs moving and the arms swinging by simply rotating the top arm and leg bones.

Forward Kinematics works great for some types of motion, but imagine animating a character reaching for a light switch. To animate this motion, you'd need to rotate the hip, shoulder, elbow, and wrist to get the hand even close, and then you'd have to work with the fingers until the motion looks good, but it is hard to correctly position the finger even close to the light switch unless multiple bones are rotated. This is a motion where Inverse Kinematics is a better choice.

Inverse Kinematics (IK) is different from Forward Kinematics in that it switches the control of the bones from parent controlling child to child controlling parent. Inverse Kinematics determines all the motions of objects in a system when the last object in the hierarchy chain is moved. The position of the last object, such as a finger or a foot, is typically the one you're concerned with. With IK, you can use these solutions to animate the system of objects by moving the last object in the system. So, using IK, you can drag the hand to the exact position you want, and all other parts in the system follow.

Forward Kinematics is an automatic result of a linked system of bones, so nothing needs to be added to the system to use FK. An IK system, on the other hand, needs to be set up. This is done by simply selecting the start and end bones and then enabling IK for the bone chain that runs between these two. Once established, you use the End Effector object to control the IK system.

Creating an Inverse Kinematics System

Before establishing an IK system, you need to decide where you need one. For human characters, it is helpful to have an IK system for each of the different limbs. IK also is used on long bending extremities like tails and antennae. The spine is another good place to include an IK chain.

3ds Max includes several different IK options, and the one to use depends on the type of motion you need. These IK solvers are set using the drop-down list found in the IK Chain Assignment rollout of the Command Panel when creating a bone chain. To apply the selected IK Solver, make sure the Assign to Children option is enabled.

If you forget to enable the Assign to Children option or if you want to switch the IK solver, you also can apply an IK solver with the selected bone as the root using the Animation→IK Solvers options. After selecting an option, click the end bone for the IK and an IK chain is created from the initial selection to the picked bone, except for the IK Limb solver, which always is applied to two bones. These are the four available IK solvers:

* **History Independent (HI):** This is the most versatile of the solvers. It can work with a long bone chain, and it uses a Swivel Angle setting to control how the bone chain twists as it moves.

* **History Dependent (HD):** This solver places a point object at each joint in the chain. It isn't as easy to use as the HI solver, but it does allow for sliding joints.

* **IK Limb:** This solver works only on two bones and is good for simple chains such as the upper and lower arm.

* **Spline IK:** This solver lets you select a spline that moves parallel to a bone chain, and it places dummy object handles at regular intervals along the spline.

Note
If you look at the top of the Hierarchy panel in the Command Panel, you'll notice an IK button. The Interactive IK and Apply IK buttons are older IK methods that remain for compatibility with older files, but the IK solvers are newer and better.

You can tell when an IK solver is applied to a bone chain because a blue point object appears at the pivot point for the end bone. This point is the goal, which is given the name IK Chain and a number. At this same pivot is a green point object called the End Effector. If you select and move the goal object, the solver works to make the End Effector match the position and orientation of the goal, and the other bones realign to accommodate the motion. When the goal is selected, a white line is displayed between the start and end bones.

When an IK chain is selected, its settings are found in the Motion panel. Using these settings, you can enable or disable the IK chain and change the chain thresholds and iterations. It also lists the Start and End Joints and lets you pick new start and end bones if you wish.

History Independent (HI) IK solver

The History Independent (HI) IK solver is the best option to use when the bone chain you want to control has lots of links. This solver looks at each keyframe independently when making its solution, which makes it speedy. You can animate linked chains with this IK solver applied by positioning the goal object; the solver then inserts a keyframe at the pivot point of the last object in the chain to match the goal object.

The first rollout in the Motion panel after the Assign Controller rollout is the IK Solver rollout. Using this rollout, you can select to switch between the IK HI solver and the IK Limb solver. The Enabled button lets you disable the solver. By disabling the solver, you can use Forward Kinematics to move the objects. To return to the IK solution, simply click the Enabled button again. The IK for FK Pose option enables IK control even if the IK solver is disabled. This lets you manipulate the hierarchy of objects using Forward Kinematics while still working with the IK solution. If both the IK for FK Pose and the Enabled buttons are disabled, the goal can move without affecting the hierarchy of objects.

If the goal ever gets moved away from the end link, clicking the IK/FK Snap button automatically moves the goal to match the end links position. Auto Snap automatically keeps the goal and the end link together. The Set as Preferred Angle button remembers the angles for the IK system. These angles can be recalled at any time using the Assume Preferred Angle button.

If you want to change the start and end bone objects, you can click the Pick Start Joint or Pick End Joint button and choose the new bone.

Tip
The best way to select an object using the Pick Start Joint and Pick End Joint buttons is to open the Select by Name dialog box by pressing the H key. Using this dialog box, you can select an exact object.

Caution
If you select a child as the start joint and an object above the child as the end joint, moving the goal has no effect on the IK chain.

The IK Solver Properties rollout includes the Swivel Angle value. The swivel angle defines the plane that includes the joint objects and the line that connects the starting and ending joints. This plane is key because it defines the direction in which the joint moves when bent. If the IK chain is selected, you also can display and manipulate the swivel angle by enabling the Select and Manipulate button on the main toolbar. By moving the swivel angle control, you can set the direction in which the chain bends as the End Effector is moved. For example, a knee could be set to point outward for a bull-legged cowboy character. Figure 38.6 shows a multi-bone chain with the HI IK solver applied. The chain has been duplicated and shown at different swivel angles.

FIGURE 38.6

Adjusting the swivel angle changes the plane within which the chain moves.

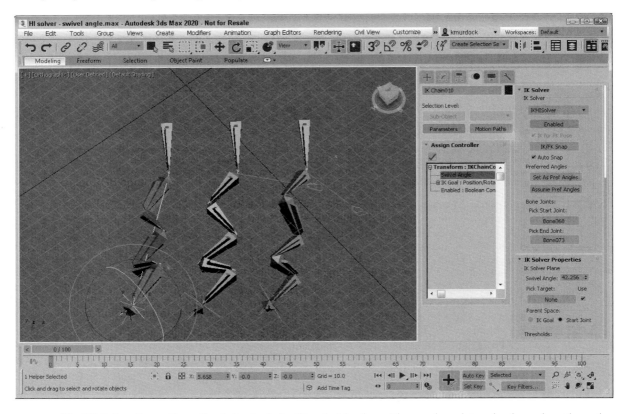

The IK Solver Properties rollout also includes Threshold values. These values determine how close the end joint and the goal must be before the solution is pronounced valid. You can set thresholds for Position and Rotation. The Iterations value sets the number of times the solution is tried.

Tip

Setting the Iterations value to a higher number produces smoother (less jerky) results, but it increases the time required to find a solution.

The IK Display Options rollouts can enable, disable, and set the size of the gizmos used when working with IK solvers. Using this rollout, you can Enable the End Effector, the Goal, the Swivel Angle Manipulator, and the IK solver (which is the line connecting the start and end joints).

History Dependent (HD) IK solver

The History Dependent (HD) IK solver takes into account the previous keyframes as it makes a solution. This solver makes having very smooth motion possible, but the cost of time to compute the solution is increased significantly.

This IK solver places an End Effector at the pivot point of the last bone, but it also places a point object at every bone in the chain. Moving the End Effector changes the position of all the bones in the chain, and you can move the other point objects at each joint to create a sliding effect. This IK solver shows up as a controller in the Motion panel when the IK chain is selected. The settings are contained in a rollout named IK Controller Parameters, which is visible in the Motion panel if you select one of the End Effector gizmos. The End Effector gizmo is the object that you move to control the IK chain. It is displayed as a set of crossing axes.

Any parameter changes affect all bones in the current structure. In the Thresholds section, the Position and Rotation values set how close the End Effector must be to its destination before the solution is complete. In the Solution section, the Iterations value determines the maximum number of times the solution is attempted. The Start Time and End Time values set the frame range for the IK solution.

The Show Initial State option displays the initial state of the linkage and enables you to move it by dragging the End Effector object. The Lock Initial State option prevents any linkage other than the End Effector from moving.

The Update section enables you to set how the IK solution is updated with Precise, Fast, and Manual options. The Precise option solves for every frame, Fast solves for only the current frame, and Manual solves only when the Update button is clicked. The Display Joints options determine whether joints are Always displayed or only When Selected.

When you first create a bones system, an End Effector is set to the last joint automatically. In the End Effectors section, at the bottom of the IK Controller Parameters rollout, you can set any joint to be a Positional or Rotational End Effector. To make a bone an End Effector, select the bone and click the Create button. If the bone is already an End Effector, the Delete button is active. You also can link the bone to another parent object outside of the linkage with the Link button. The linked object then inherits the transformations of this new parent.

Click the Delete Joint button in the Remove IK section to delete a joint. If a bone is set to be an End Effector, the Position or Rotation button displays the Key Info parameters for the selected bone.

Tutorial: Animating a spyglass with the HD IK solver

A telescoping spyglass is a good example of a kinematics system that you can use to show off the HD solver. The modeling of this example is easy because it consists of a bunch of cylinders that gradually get smaller.

To animate a spyglass with the HD IK solver, follow these steps:

1. Open the Spyglass.max file from the Chap 38 directory in the downloaded content set.

   This file includes a simple spyglass made from primitive objects. The pieces of the spyglass are linked from the smallest section to the largest section. At the end of the spyglass is a dummy object linked to the last tube object.

2. To define the joint properties, select the largest tube object, open the Hierarchy panel, and click the IK button. In the Object Parameters rollout, select the Terminator, Bind Position, and Bind Orientation options to keep this joint from moving.

3. With the largest tube section selected, make the Z Axis option active in the Sliding Joints rollout, and disable all the axes in the Rotational Joints rollout. Click the Copy button for both Sliding Joints and Rotational Joints in the Object Parameters rollout.

4. Select each remaining tube object individually, and click the Paste buttons for both the Sliding Joints and Rotational Joints.

   This enables the local Z-axis sliding motion for all tube objects.

5. Select the largest tube section again, and choose Animation→IK Solvers→HD Solver. Then drag the dotted line to the dummy object at the end of the spyglass.

6. Select the second largest tube object, and back in the Hierarchy panel for the Sliding Joint Z Axis, select the Limited option with values from **0.0** to **−80**. Click the Copy button for the Sliding Joints in

935

the Object Parameters rollout. Select tubes 3 through 6 individually, and click the Paste button for the Sliding Joints to apply these same limits to the other tube objects.

7. Click the Auto Key button (or press N), drag the Time Slider to frame 100 (or press End), select the Select and Move button on the main toolbar (or press W), and drag the dummy object away from the largest tube object.

Figure 38.7 shows the end tube segment collapsing within the spyglass.

FIGURE 38.7

The HD IK solver is used to control the spyglass.

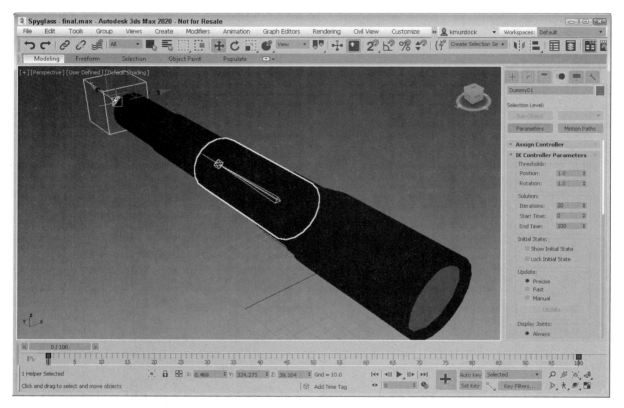

IK Limb solver

The IK Limb solver was specifically created to work with limbs. It is used on chains of two bones such as an upper leg and lower leg. Only two of the bones in the chain actually move. The goal for these joints is located at the pivot point for the second bone. This solver is ideal for game character rigging.

This solver works by considering the first joint as a spherical joint that can rotate along three different axes, such as a hip or shoulder joint. The second joint can bend only in one direction, such as an elbow or knee joint.

The rollouts and controls for the IK Limb solver, including the swivel angle, are exactly the same as those used for the HI solver covered earlier in this chapter.

Tutorial: Animating a spider's leg with the IK Limb solver

As an example of the IK Limb solver, you should probably animate a limb, so I created a simple spider skeleton with not two limbs, but eight. I created this skeleton fairly quickly using four bones for the abdomen; then I created one limb and cloned it three times. Then I used the Bone Tools to connect the leg bones to the abdomen bones, and finally I selected and mirrored the bones on all four legs to get the opposite legs. The hardest part was naming all the bones.

To animate a spider skeleton's leg using the IK Limb solver, follow these steps:

1. Open the Spider skeleton.max file from the Chap 38 directory in the downloaded content set.

2. Click the Select by Name button on the main toolbar (or press the H key) to open the Select From Scene dialog box. Double-click the RUpperlegBone01 object to select the upper leg bone object.

3. With the upper leg bone selected, choose Animation→IK Solvers→IK Limb Solver. A dotted line appears in the viewport. Press the H key again to open the Pick Object dialog box, and double-click the RFootBone01 object to select it.

    This bone corresponds to the foot bone, which is the end of the limb hierarchy.

4. With the IK Chain01 object selected, click the Auto Key button (or press the N key) and drag the Time Slider to frame 100 (or press End). With the Select and Move button (or by pressing the W key), move the IK chain in the viewport.

    The leg chain bends as you move the End Effector.

Figure 38.8 shows the spider's leg being moved via the IK Limb solver. The IK Limb solver provides a simple and quick way to add an Effector to the end of a limb, giving you good control for animating the spider's walk cycles.

FIGURE 38.8

You can use the IK Limb solver to control limbs such as legs and arms.

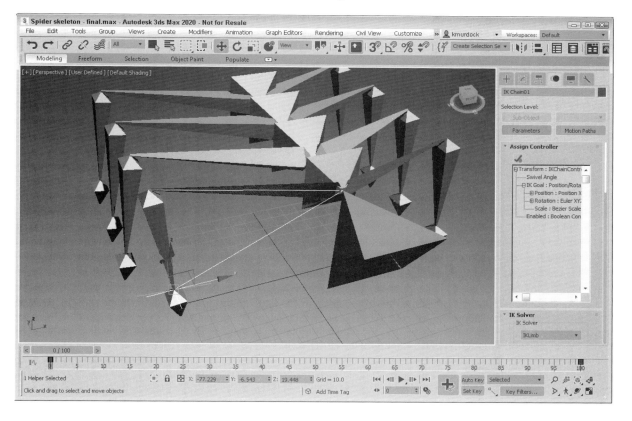

Spline IK solver

The IK Limb solver works well for arms and legs that have a joint in the middle, but it doesn't work well for tails. Tails are unique because they require multiple bones to deform correctly. The Spline IK solver works well for tails, but it also works well for rigging tentacles, chains, and rope.

To use the Spline IK solver, you need to create a chain of bones and a spline path. By selecting the first and last bone and then selecting the spline, the bone chain moves to the spline. Each control point on the spline has a dummy object associated with it. By moving these dummy objects, you can control the position of the bones. At either end of the spline are manipulators that you can use to twist and rotate the bones.

The easiest way to use this IK solver is to select SplineIKSolver from the drop-down list in the IK Chain Assignment rollout while you're creating the bone structure. After the bone structure is complete, the Spline IK Solver dialog box, shown in Figure 38.9, appears. If the Auto Create Spline option is enabled, a spline to match the bone chain is created. With this dialog box, you also can select a name for the IK chain, specify the curve type, and set the number of spline knots. The curve type options include Bézier, NURBS Point, and NURBS CV. You also can select to Create Helpers and to display several options.

FIGURE 38.9

The Spline IK Solver dialog box automatically creates the spline for you.

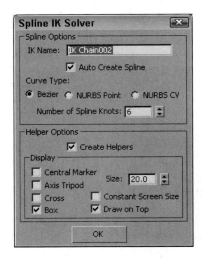

Another way to use this IK solver is with an existing bone structure. To do this, you need a spline curve in the scene that matches how you want the bone chain to look. Select the first bone where you want the solver to be applied, and choose Animation→IK Solvers→Spline IK Solver. In the viewports, a dragging line appears; move the line to the last bone that you want to include, and click to pick it; then drag and click a second time to the spline that you want to use.

The bone structure then assumes the shape of the spline curve. A helper object is positioned at the location of each curve vertex. These helper objects let you refine the shape of the curve.

Tutorial: Building an IK Spline alligator

The IK Spline solver is perfect for creating long, winding objects like snakes or an alligator's tail. For this example, you take an existing bone structure and, using the Spline IK solver, make it match a spline.

To create a bone structure for an alligator that follows a spline using the IK Spline solver, follow these steps:

1. Open the Alligator spline IK.max file from the Chap 38 directory in the downloaded content set.

   This file includes an alligator model, a simple bone chain and a spline.

2. With the first bone in the tail chain selected, choose Animation→IK Solvers→Spline IK Solver.

   A dotted line appears in the viewport extending from the first bone.

3. Drag and click the cursor on the last bone in the bone tail chain.

4. Another dotted line appears; drag and click the spline, and the bone structure moves to match the spline's curve.

Figure 38.10 shows the bone structure for the gator's tail. You can now control the gator's tail by moving the dummy objects along the spline.

FIGURE 38.10

The IK Spline solver is perfect for creating objects such as snakes and animal tails.

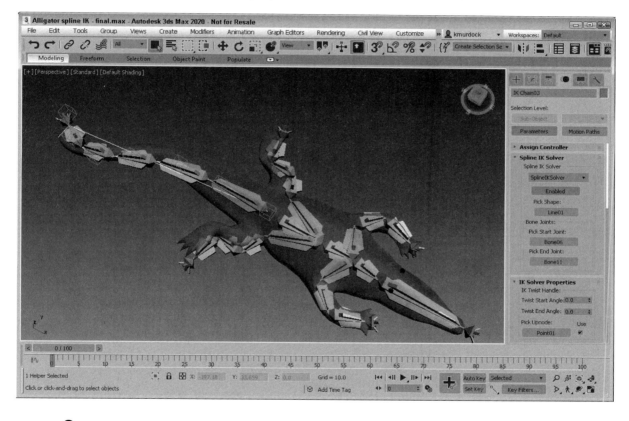

Summary

Understanding the benefits of a bones system helps if you ever need to customize a rig or create a new rig from scratch. Inverse Kinematics (IK) provides a unique way to control and animate hierarchical structures by transforming the child node. In this chapter, you learned how to create and work with bones systems and the Bone Tools. This chapter covered the following topics:

* Creating bones systems

* Setting bone parameters and the IK Solver

* Using the Bone Tools

* Making objects into bones systems
* The basic concepts behind IK
* Creating and animating an IK system

Now that you've learned the process for rigging a character and using IK, we look at using the skin modifier to make the character mesh move along with the skeleton.

Chapter 39

Skinning Characters

IN THIS CHAPTER

Planning your character

Using the Skin modifier

Painting skin weights

Using the Weight Tool

Working with the Skin Wrap and Skin Morph modifiers

In the taxidermy world, skinning an animal usually involves removing its skin, but in the Autodesk® 3ds Max® 2020 software, skinning a character involves adding a skin mesh to a group of controlling bones. Skinning a character also involves defining how the skin deforms as the bones are moved using skin weights.

A character skin created in 3ds Max can be any type of object and is attached to a biped or a bones skeleton using the Skin modifier. The Skin modifier isn't alone. 3ds Max includes other modifiers like the Skin Wrap and Skin Morph modifiers that make your skin portable. This chapter covers how the various Skin modifiers are attached to a skeleton and used to aid in animating your character.

Understanding Your Character

What are the main aspects of your character? Is it strong and upright, or does it hunch over and move with slow, twisted jerks? Before you begin modeling a character, you need to understand the character. It is helpful to sketch the character before you begin. This step gives you a design that you can return to as needed.

Tip
The sketched design also can be loaded and planar mapped to a plane object to provide a guide to modeling.

You have an infinite number of reference characters available to you (just walk down a city street if you can't think of anything new). If you don't know where to start, try starting with a human figure. The nice thing about modeling a human is that an example is close by (try looking in the mirror).

We all know the basic structure of humans: two arms, two legs, one head, and no tail. If your character is human, starting with a human character and changing elements as needed is the easiest way to go. As you begin to model human figures, being familiar with anatomy is helpful. Understanding the structure of muscles and skeletal systems helps explain the funny bumps you see in your elbow and why muscles bulge in certain ways.

Tip
If you don't have the ideal physique, a copy of *Gray's Anatomy* (the book, not the TV show) can help. With its detailed pictures of the underlying muscular and skeletal systems, you'll have all the details you need without having to pull back your own skin.

The curse and blessing of symmetry

The other benefit of the human body is that it is symmetrical. You can use this to your benefit as you build your characters, but be aware that unless you're creating a band of killer robots, it is often the imperfections in the characters that give them, well, character. Positioning an eye a little off normal might give your character that menacing look you need.

Dealing with details

When you start to model a human figure, you quickly realize that the body includes lots of detail, but before you start naming an object "toenail lint on left foot," look for details you won't need. For example, modeling toes is pointless if your character will be wearing shoes and won't be taking them off. (In fact, I think shoes were invented so that modelers wouldn't have to model toes.)

At the same time, details in the right places add to your character. Look for the right details to help give your character life—a pirate with an earring, a clown with a big, red nose, a tiger with claws, a robot with rivets, and so on.

Animated Skin Modifiers

Of all the animation modifiers, several specifically are used to deal with skin. The Skin modifier is a key modifier for enabling character animation. The Skin Morph modifier lets you deform a skin object and create a morph target. It is designed to help fix problem areas, such as shoulders and hips that have trouble with the standard Skin modifier. The Skin Wrap modifier offers a way to animate a low-res proxy and then apply the same animation to a high-res wrapped object.

Understanding the skinning process

Unless you like animating using only a skeleton or a biped by itself, a bones system will have a skin attached to it. Any mesh can be made into a skin using the Skin modifier. The Skin modifier is used to bind a skin mesh to a bones system and to define the associations between the skin vertices and the bones. The first step is to bind the skin mesh to the skeleton object. With a skin attached to a bones system, you can move the bones system and the skin follows, but just how well it follows the skeleton's motion depends on a process called skinning.

Creating a bones system is covered in more detail in Chapter 38, "Understanding Rigging, Kinematics, and Working with Bones."

Skinning is where you tell which parts of the skin mesh to move with which bones. Obviously, you'd want all the skin vertices in the hand to move with the hand bone, but the skin vertices around the waist and shoulders are trickier.

Each skin area that surrounds a bone gets encompassed by a capsule-shaped envelope. All vertices within this envelope move along with the bone. When two of these envelopes overlap, their surfaces blend together like skin around a bone joint. Most of the skinning process involves getting the skin vertices into the right envelope. The Skin modifier includes several tools to help make this easier, including the Skin Weight table and a painting weights feature.

Binding to a skeleton

After you have both a skeleton and a skin mesh, you need to bind the skin to the skeleton before you can move the skin using the underlying bones. The binding process is fairly easy: simply select the skin mesh, and apply the Skin modifier to it using the Modifiers→Animation→Skin menu command.

Tip

Before binding a skeleton to a skin mesh, take some time to match the size and dimensions of the bones close to the skin mesh. When the skin mesh is bound to the skeleton, the envelopes are created automatically. If the bones match the skin, the new envelopes are pretty close to what they need to be.

In the Parameters rollout, click the Add button above the bones list. This opens a Select Bones dialog box where you can select the bones to use to animate this skin. The selected bones appear in the bones list. The text field directly under the bones list lets you locate specific bones in the list by typing the name. Only one bone at a time may be selected from the list. The Remove button removes the selected bone from the bone list.

Tutorial: Attaching skin to a CAT rig

For human figures, using a biped or Character Animation Toolkit (CAT) skeleton saves lots of time. For this example, because he's close to a human in form, you'll bind a CAT skeleton to the Marvin Moose model.

We are getting a little ahead of ourselves here. The Character Animation Toolkit (CAT) is covered in Chapter 40, "Animating Characters with CAT."

To bind the skin of a model to a CAT skeleton, follow these steps:

1. Open the Marvin Moose CAT rig.max file from the Chap 39 directory in the downloaded content set. This file includes the CAT rig built in the CAT chapter for the Marvin Moose character.

2. With the CAT rig aligned to the skin mesh, select Unfreeze All from the right-click quad menu. Select the moose model, disable the See-Through option in the Object Properties dialog box, and choose Modifiers→Animation→Skin to apply the Skin modifier.

3. In the Parameters rollout, click the Add button. The Select Bones dialog box opens. Click the Select All button to select all the bones, and click Select. Don't include the Character001, or the Leg Platform objects. All the selected bones show up in the Bones list for the Skin modifier.

4. In the Parameters rollout, select one of the bones in the list and click the Edit Envelopes button. Zoom in on the highlighted bone, select the cross-section handles for this bone, and pull them outward until all the vertices surrounding the bone and included within the envelope. For some bones like the left shin bone, set the Radius values to **75** for both ends. Figure 39.1 shows the moose skin with all the CAT rig bones added.

FIGURE 39.1

Bone references are added to the skin modifier.

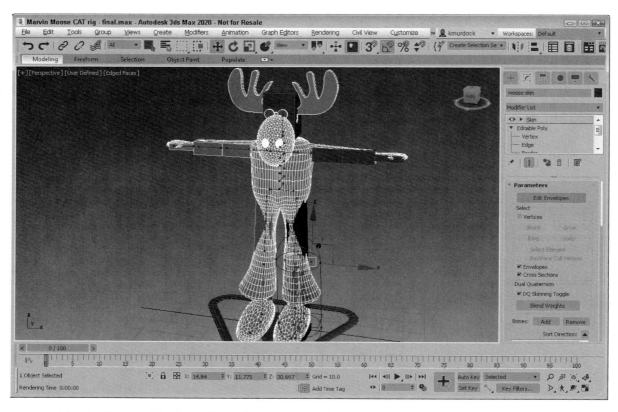

After the Skin modifier is applied and bound to the skin mesh, every bone includes an area of influence called an envelope that defines the skin vertices that it controls. If any of the skin mesh vertices are outside of the bone's envelope or are included in an envelope for the wrong bone, the vertices are left behind when the bone is moved. This causes an odd stretching of the skin that is easy to identify.

To check the envelopes, select and rotate several of the skeleton's key bones. If any envelope problems exist, they are easy to spot, as shown in Figure 39.2. The incorrect stretching of the vertices for the shoe simply means that the envelopes need to be adjusted.

FIGURE 39.2

If the envelopes are off for any of the skin vertices, the skin stretches incorrectly.

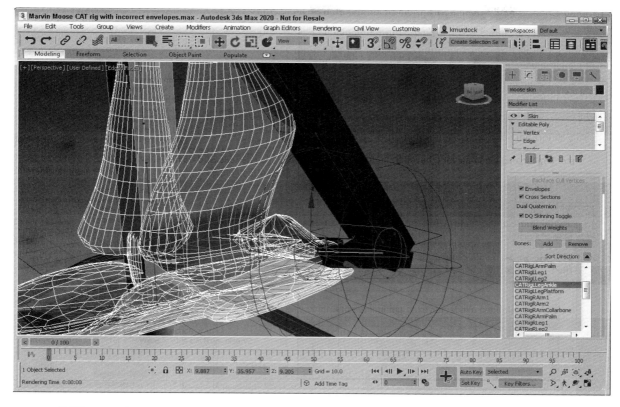

Editing envelopes

When the Skin modifier is selected, the Parameters rollout includes an Edit Envelopes button that places you in a special mode that lets you edit the envelope for the selected bone in the bone list. This mode is also enabled by selecting the Envelope subobject under the Skin modifier at the top of the Modifier Stack.

When the Edit Envelopes mode is enabled, the entire skin mesh is colored with a gradient of colors to visually show the influence of the envelope, similar to a heat map. Areas of red are completely inside the envelope's influence, areas of green are somewhat affected by the bone, blue areas are minimal, and areas of gray are completely outside the envelope's influence.

Figure 39.3 shows a simple loft object surrounding three bone objects with a Skin modifier applied. The Add Bone button was used to include the three bones within the Skin modifier list. The skin has been set to See Through in the Object Properties dialog box, so the bones are visible. The first bone was selected in the bone list, and the Edit Envelope button was clicked, revealing the envelope for the first bone.

FIGURE 39.3

Envelopes define which Skin vertices move with the underlying bone.

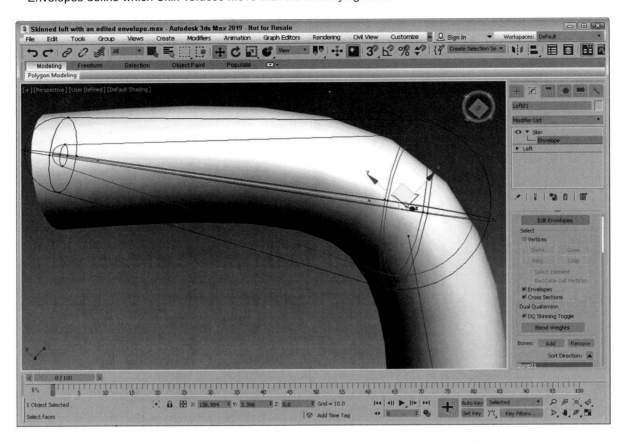

An envelope consists of two capsule-shaped volumes within each other called the Inner and Outer Envelopes. Any vertices within the inner envelope are controlled exclusively by that bone. Any vertices positioned between the inner and outer envelopes are controlled by a falloff where the influence is shared between bones.

At either end of these envelopes are four small handles that can be dragged to change the cross-section radius. The cross-section area changes to pink when selected. The radius of the selected cross section is displayed in the Radius field within the Envelope Properties section of the Parameters rollout. The Squash value determines the amount of squash applied to the object for bones that can stretch. You can change an envelope's size by changing its Radius value or by dragging the cross-section handles. Within the Select section of the Parameters rollout, you can choose to select and edit Vertices, Envelopes, and/or Cross Sections. If you choose the Vertices option, selected object vertices are shown as small squares, and the Shrink, Grow, Ring, and Loop buttons become active. These buttons allow you to select a desired set of vertices easily. If the Select Element option is enabled, all vertices in the element are selected, and the Backface Cull Vertices option prevents vertices on the backside of the object from being selected.

If a cross section is selected, you can add a different cross-section shape using the Add button. This button lets you select a cross-section shape within the viewports. The Remove Cross Section button removes an added cross section from the envelope.

Note

The orientation of the envelope spline is set by the longest dimension of the bone. This works well for arm and leg bones, but the pelvis or clavicle may end up with the wrong orientation.

The Envelope Properties section of the Parameters rollout (just below the Radius and Squash values) includes five icon buttons, shown in Table 39.1. The first toggles between Absolute and Relative. All vertices that fall within the outer envelope are fully weighted when the Absolute toggle is set, but only those within both envelopes are fully weighted when the Relative toggle is selected.

TABLE 39.1 Envelope Properties

Button	Name	Description
A **R**	Absolute/Relative	Toggles between Absolute and Relative
(icon)	Envelope Visibility	Makes envelopes remain visible when another bone is selected
(icons)	Falloff Linear, Falloff Sinual, Falloff Fast Out, Falloff Slow Out	Sets Falloff curve shape
(icon)	Copy Envelope	Copies envelope settings to a temporary buffer
(icons)	Paste Envelope, Paste to All Bones, Paste to Multiple Bones	Pastes envelope settings to the selected bone, to all bones, or to multiple bones chosen from a dialog box

The second icon button enables envelopes to be visible even when not selected. This helps you see how adjacent bones overlap. The third icon button sets the Falloff curve for the envelopes. The options within this flyout are Linear, Sinual, Fast Out, and Slow Out. The last two icon buttons can be used to Copy and Paste envelope settings to other bones. The flyout options for the Paste button include Paste (to a single bone), Paste to All Bones, and Paste to Multiple Bones (which opens a selection dialog box).

Working with weights

For a selection of vertices, you can set its influence value (called its Weight value) between 0 for no influence and 1.0 for maximum influence. This provides a way to blend the motion of vertices between two or more bones. For example, the vertices on the top of a character's shoulder could have a weight value of 1.0 for the shoulder bone, a weight value of 0.5 for the upper arm bone, and a weight value of 0 for all other bones. This lets the shoulder skin area move completely when the shoulder bone moves and only halfway when the upper arm moves.

Note
The shading in the viewport changes as vertices are weighted between 0 and 1. Weight values around 0.125 are colored blue, weight values around 0.25 are colored green, weight values around 0.5 are colored yellow, and weight values around 0.75 are colored orange.

The Absolute Effect field lets you specify a weight value for the selected vertices. The Rigid option makes the selected vertices move only with a single bone. The Rigid Handles causes the handles of the selected vertices for a patch object to move only with a single bone. This is important if the character is wearing a hard item such as armor plates. By enabling this option, you can be sure that the armor plate doesn't deform. The Normalize option requires that all the weights assigned to the selected vertices add up to 1.0.

The other buttons found in the Weight Properties section of the Parameters rollout are defined in Table 39.2. Include Vertices and Exclude Vertices buttons let you remove the selected vertices from those being affected by the selected bone. The Select Exclude Verts button selects all excluded vertices.

TABLE 39.2 Envelope Properties

Button	Name	Description
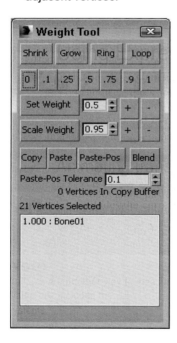	Exclude Selected Vertices	Excludes the selected vertices from the influence of the current bone.
	Include Selected Vertices	Includes all selected vertices in the bone's influence.
	Select Excluded Vertices	Selects all excluded vertices.
	Bake Selected Verts	Bakes the vertex weights into the model so they aren't changed with the envelope. Baked vertices can be changed using the Weight Table or the Absolute Effect value.
	Weight Tool	Opens the Weight Tool interface.

Using the Weight Tool

The Weight Tool button in the Parameters rollout opens the Weight Tool dialog box, shown in Figure 39.4. The Shrink, Grow, Ring, and Loop buttons work the same as those in the Select section, and they let you quickly select precise groups of vertices. The value buttons on the second row allow you to change weight values with a click of the button or by adding or subtracting from the current value.

FIGURE 39.4

The Weight Tool dialog box includes buttons for quickly altering weight values and for blending the weights of adjacent vertices.

The Copy, Paste, and Paste-Position buttons let you copy weights between vertices quickly. The Paste Position button pastes the given weight to the surrounding vertices based on the Paste Position Tolerance value. The Blend button quickly blends all the surrounding vertices from the current weight value to 0, creating a smooth blend weight. The Weight Tool dialog box also lists the number of vertices in the copy buffer and currently selected. The list at the bottom of the dialog box lists the weight and bone for the selected vertices.

Using the Weight Table

The Weight Table button opens the Weight Table interface, shown in Figure 39.5. This table displays all the vertices for the skinned object by ID in a column on the left side of the interface. All bones are listed in a row along the top. For each vertex and bone, you can set a weight.

FIGURE 39.5

The Weight Table lets you specify weight values for each vertex and for each bone.

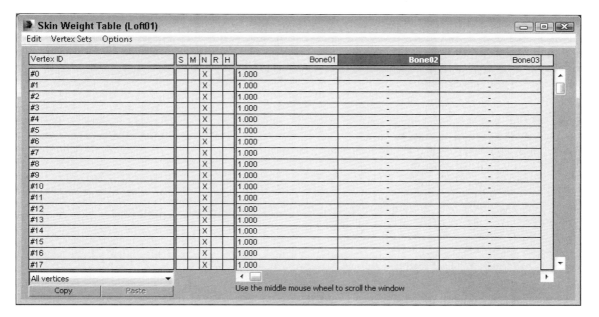

The Edit menu includes commands to Copy and Paste weights. It also includes commands to Select All, Select Invert, and Select None. A selection of vertices can be combined into a Vertex Set and named. The Vertex Sets menu lets you create and delete these sets. There is also an option to Remove Zero Weights to make the table easier to view.

The Options menu lets you flip the interface so that bones are displayed in the first column and the vertex IDs are along the top row. The Update On Mouse Up option limits the updates until the mouse is released. Several options for showing and hiding interface elements are included. The Show Affected Bones option lists only the bones that are affected. The Show Attributes option displays a column of attributes labeled S, M, N, R, and H. The Show Exclusions option makes a check box available in each cell. When checked, the vertex is excluded. The Show Global option makes a drop-down list available that enables you to set an attribute for all vertices. The Show Set Sets UI makes available two buttons for creating and deleting vertex sets.

The S attribute is marked if a vertex is selected, the M attribute marks a vertex weight that has been modified, the N attribute marks a normalized weight, the R attribute marks rigid vertices, and an H attribute marks a vertex with rigid handles.

To set a weight, just locate the vertex for the bone, click in the cell, and type the new value. If you click a cell and drag to the left or right, the weight value changes. Weight values can be dragged between cells. Right-clicking a cell sets its value to 0, and right-clicking with the Ctrl key held down sets its value to 1.0.

After the vertex weights are set, you can click the Bake Selected Vertices to lock down the weight values. Changes to envelopes do not affect baked vertices.

Painting weights

Using the Paint Weights button, you can paint with a brush over the surface of the skin object. The Paint Strength value (in the Painter Options dialog box) sets the value of each brush stroke. This value can be positive (up to 1.0) for vertices that will move with the bone or negative (to -1.0) for vertices that will not move with the bone.

To the right of the Paint Weights button is the Painter Options button (it has three dots on it), which opens the Painter Options dialog box. Using this dialog box, you can set the brush strength and size.

The Painter Options dialog box also is used by the Vertex Paint modifier and the Paint Deformation tool. It is described in detail in Chapter 16, "Deforming Surfaces and Using the Mesh Modifiers."

Tutorial: Applying skin weights

In this example, you change the skin weights of the Marvin Moose character in an attempt to fix the problem you saw with his ankle and shoe. Basically, you want to add all the skin vertices that need to move with the selected bone to have an influence of 1 and remove all those that don't move with the selected bone or set their weight to 0.

To apply the skin weights to a character, follow these steps:

1. Open the Marvin Moose CAT rig with skin weights.max file from the Chap 39 directory in the downloaded content set.

2. With the mesh skin selected, open the Modify panel and choose the CATRigLLegAnkle bone in the bone list. Then click the Edit Envelopes button at the top of the Parameters rollout. This displays the envelopes around the foot object. Zoom in on the foot object, and make sure that shading is enabled in the viewport, so you can see the weight shading.

3. Make sure the Cross Sections option in the Select section of the Parameters rollout is selected. Then select the cross-section handles on the outer envelope and pull them in toward the foot. Rotate the view until the side of the shoe is visible to make sure that the toe and heel of the shoe are still covered.

4. Click the Vertices option in the Parameters rollout, and disable the Backface Cull Vertices option. Drag over all the vertices that are contained in the lower part of the shoe, and enable the Rigid option. Then select the vertices above the shoe in the shin. You can use the Loop button to select all vertices about the upper part of the shoe and press and hold the Alt key to remove any vertices that are part of the lower shoe.

5. In the Weight Properties section of the Parameters rollout, click the Weight Tool button. In the Weight Tool dialog box, click the .25 button. This changes the weight of the selected vertices. Then click the Blend button to smooth the transition areas.

6. Click the Edit Envelopes button to exit Edit Envelopes mode. Then select and rotate the upper leg to see whether the problem has been fixed. As I rotated my model, I noticed that some vertices on the back of the shoe were left behind, which means that they aren't being influenced by the foot bone.

Caution

Be sure to undo the upper leg rotation before making changes to the envelopes, or the envelopes will move along with the rotation.

7. Select the skin object, enter Edit Envelopes mode again, and click the Paint Weights button. The cursor changes to a round brush. Click the Painter Options button (which is next to the Paint Weights button with three dots). This opens the Painter Options dialog box. Set the Max Strength value to **1.0**, and close the dialog box. Then paint in the viewport over the vertices on the back of the shoe heel, including those areas that aren't shaded red. Figure 39.6 shows the correct shoe with its envelopes.

FIGURE 39.6

Vertices' weights can be fixed with the Weight Tool and by Painting Weights.

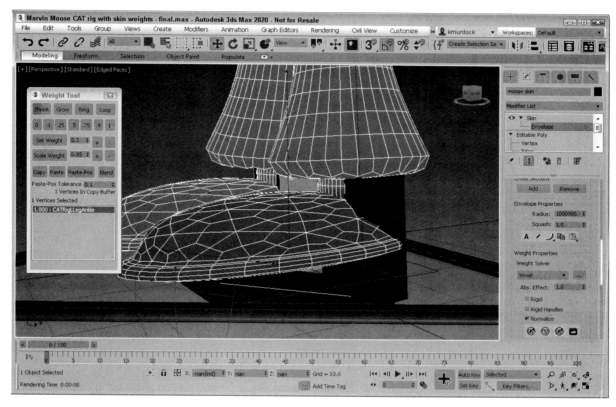

8. Exit Edit Envelopes mode again, and check the changes by rotating the upper leg bone.

Mirror settings

Most characters have a natural symmetry, and you can use this symmetry to mirror envelopes and vertex weights between different sides of a model. You can use this feature by clicking the Mirror Mode button in the Mirror Parameters rollout. This button is active only when the Edit Envelopes button is active.

In Mirror Mode, you see an orange plane gizmo that marks the symmetrical line for the model. You can move and orient this plane using the Mirror Offset, Mirror Plane, and Mirror Threshold controls. Once oriented, 3ds Max computes the matching vertices based on the volumes from the mirror plane. All vertices on one half of the character appear blue, and all the matching vertices appear green. All vertices that cannot be matched appear red.

If you drag over bones or vertices in the viewports, you can select them. Clicking the Mirror Paste button pastes the envelopes and vertex weights of the selected vertices to their matches. Or you can select the Paste Green to Blue Bones button or one of its neighbors to copy all the green bones or vertices to their matches, or vice versa.

The Display Projection drop-down list projects the position of the selected vertices onto the Mirror Plane so you can compare their locations relative to each other. With lots of vertices in your skin, you want to enable the Manual Update, or the viewport refreshes become slow.

Caution

Mirror Mode will not work for CAT rigs because CAT has its own coordinates system for left and right.

Tutorial: Mirroring skin weights

Now that you have spent the time correcting the skin weights for the left shoe, you use the Mirror Weights feature to apply the same weights to the opposite foot.

To apply the skin weights to a character, follow these steps:

1. Open the Spider skeleton.max file from the Chap 39 directory in the downloaded content set.

2. Select the skinned model and with Edit Envelopes mode turned on, select the RFootBone01 bone from the list in the Parameters rollout, and click the Mirror Mode button in the Mirror Parameters rollout with the X Mirror Plane. An orange gizmo is added to the viewport that divides the model, as shown in Figure 39.7.

3. Drag over all the vertices that make up the right front spider foot, and then click the Paste Green to Blue Verts button. All the vertex weights on the left side of the model are copied to the right side. The pasted vertices turn yellow once pasted.

FIGURE 39.7

In Mirror mode, matched bones and vertices appear green and blue.

Display and Advanced settings

The Display rollout controls which features are visible within the viewports. Options include Show Colored Vertices, Show Colored Faces, Color All Weights, Show All Envelopes, Show All Vertices, Show All Gizmos, Show No Envelopes, Show Hidden Vertices, Draw On Top Cross Sections, and Draw On Top Envelopes.

With these display options, you can turn on and off the weight color shading on vertices and faces. You also can select to show all envelopes, vertices, and gizmos. The Color All Weights option is unique. It assigns every bone a different color and shows how the weights blend together between bones.

The Advanced Parameters rollout includes an option to Back Transform Vertices. This option avoids applying transform keys to the skin because the bones control the motion. The Rigid Vertices and Rigid Patch Handles options set the vertices so that they are controlled by only one bone. This rollout also includes buttons to Reset Selected Vertices, Reset Selected Bones, and Reset All Bones. This is handy if you really mess up a skinning job, because it gives you a chance to start over. It also includes buttons for saving and loading envelopes. The envelopes are saved as files with the .env extension.

The Animatable Envelopes option lets you create keys for envelopes. Weight All Vertices is a useful option that automatically applies a weight to the nearest bone of all vertices that have no weight. The Remove Zero Weights button removes the vertex weight of any vertex that has a value lower than the Zero Limit. This can be used to remove lots of unnecessary data from your model if you need to put it on a diet.

Using deformers

Above the Advanced Parameters rollout is the Gizmos rollout. You use this rollout to apply deformers to selected skin object vertices. Three deformers are available in the Gizmos rollout: a Joint Angle Deformer, a Bulge Angle Deformer, and a Morph Angle Deformer.

Each of these deformers is unique. They include the following features:

* **Joint Angle Deformer:** Deforms the vertices around the joint between two bones where the skin can bunch up and cause problems. This deformer moves vertices on both the parent and child bones.

* **Bulge Angle Deformer:** Moves vertices away from the bone to simulate a bulging muscle. This deformer works only on the parent bone.

* **Morph Angle Deformer:** Can be used on vertices for both the parent and child bones to move the vertices to a morph position.

All deformers added to a skin object are listed in the Gizmo rollout. You can add deformers to and remove deformers from this list using the Add and Remove Gizmo buttons. You also can Copy and Paste the deformers to other sets of vertices. Before a deformer gizmo can be applied, you need to select vertices within the skin object. To select vertices, enable the Vertices check box of the Parameters rollout and drag over the vertices in the viewport to select the vertices.

The parameters for the deformer selected in the Gizmos rollout's list appear when the deformer is selected in the Deformer Parameters rollout. This rollout lists the Parent and Child bones for the selected vertices and the Angle between them. This rollout changes depending on the type of deformer selected.

For the Joint and Bulge Angle Deformers, a new rollout labeled Gizmo Parameters or Deformer Parameters appears. The Gizmo Parameters rollout includes buttons to edit the control Lattice and to edit the deformer Key Curves. The Edit Lattice button lets you move the lattice control points in the viewports. The Edit Angle Keys Curves opens a Joint Graph window that displays the transformation curves for the deformation.

Using the Skin Wrap modifiers

If you've created a high-resolution model that you want to animate as a skin object, but the mesh is too complex to move around, the Skin Wrap modifier might be just what you need. The Skin Wrap modifier may be applied to a high-resolution mesh, and with the Parameters rollout you can select a low-resolution control object. Any movements made by the low-resolution control object automatically are applied to high-resolution mesh.

Tip

Skin Wrap is also very useful for animating clothes on a character.

The Skin Wrap modifier has two available Deformation Engines: Face Deformation and Vertex Deformation. Each vertex contained with the control object acts as a control vertex. The Vertex Deformation option moves the vertices closest to each control vertex when the control vertex is moved, and the Face Deformation option moves the faces that are closest.

For each deformation mode, you can set a Falloff value, which moves vertices farther from the moved control vertex to a lesser extent to ensure a smoother surface. For the Vertex Deformation mode, you also can set a Distance Influence value and Face Limit values to increase the extent of influence for a control vertex.

The Reset button can be used to reset the control object to the high-resolution mesh object. This is useful if you need to realign the control object to the Skin Wrap object. When you're finished animating the control object, the Convert to Skin button may be used to transfer the animation keys to the high-resolution objects.

The Advanced Parameters rollout includes a button for mirroring the selected vertices to the opposite side of the control object.

The Modifiers menu also includes a Skin Wrap Patch modifier, which works the same as the Skin Wrap modifier, but it allows the control object to be a patch.

Tutorial: Making a simple squirt bottle walk

Creating a bones structure and applying a Skin modifier works well for characters with structure, but to animate the motion of an amorphous object such as a squirt bottle, the Skin Wrap modifier works much better than bones.

To animate a squirt bottle object walking using the Skin Wrap modifier, follow these steps:

1. Open the Walking squirt bottle.max file from the Chap 39 directory in the downloaded content set.

   This file includes a squirt bottle model with all its parts attached together. The file also includes a simple box object that is roughly the same shape as the squirt bottle. The box object has been animated walking forward.

2. With the squirt bottle object selected, choose the Modifiers→Animation→Skin Wrap menu command to apply the Skin Wrap modifier.

3. In the Parameters rollout, click the Add button and select the Box object. The Box object is added to the Skin Wrap list in the Parameters rollout. Click again on the Add button to disable it.

4. Select the Box object and hide it in the scene.

5. Click the Play Animation button, and the hi-res bottle follows the same animation as the box object.

Figure 39.8 shows the squirt bottle as it moves through the scene.

FIGURE 39.8

Not all animated objects need a bone structure.

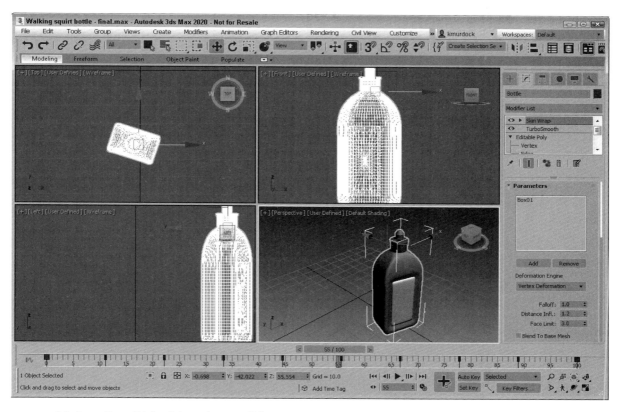

Using the Skin Morph modifier

Using the deformation options found in the Skin modifier, you can deform any part of the skin, but this feature relies on using gizmos found in an already complex envelope-editing mode. Skin Morph offers another way to deform a skin object using the underlying bones. The Skin Morph modifier is applied on top of a Skin modifier and lets you pick which bones to use in the deformation. For example, for bulging muscles, the forearm bone is rotated and should be added to the list in the Parameters rollout.

After selecting a bone from the Parameters rollout list, select the frame where the deformation is at a maximum. Then click the Create Morph button in the Local Properties rollout to create the morph target. Morph targets can be given names to make them easy to select later. The Edit button in the Local Properties rollout then lets you move the vertices for the deformation.

Tutorial: Bulging arm muscles

Perhaps the most common bulging deformation for characters is making the bicep muscle bulge as the forearm is raised. This effect can be simplified using the Skin Morph modifier.

To bulge an arm muscle using the Skin Morph modifier, follow these steps:

1. Open the Bulging bicep.max file from the Chap 39 directory in the downloaded content set.

   This file includes a rough arm model with a Skin modifier applied attached to a four-bone chain.

2. With the arm skin selected, choose the Modifiers→Animation→Skin Morph menu command to apply the Skin Morph modifier on top of the Skin modifier.

3. In the Parameters rollout, click the Add Bone button and select the forearm bone object from the Select Bones dialog box that appears.

The bone object is added to the Skin Morph list in the Parameters rollout.

4. Select and rotate the forearm bone to the location where the skin deformation is at its maximum. Then select the skin and choose the forearm bone object from the list in the Parameters rollout, and click the Create Morph button in the Local Properties rollout. In the Morph Name field, name the morph **Bulging bicep**.

Tip

The forearm bone can be hard to see under the skin, but you can always select the bone using the Select From Scene dialog box, which is opened by pressing H.

5. In the Local Properties rollout, click the Edit button. This enables the Points subobject mode. Then enable the Edge Limit option in the Selection rollout, so the vertices on the forearm aren't accidentally selected. Drag over the points on the front of the bicep muscle in the Left viewport, and enable the Use Soft Selection option in the Selection rollout. Set the Radius value to **0.5**, and scale the points to the right (along the X axis) in the Left viewport to form a bulge. Click the Edit button again to exit Edit mode.

6. Select and rotate the forearm bone to see the bulging bicep muscle.

Figure 39.9 shows the arm muscle as it bulges along with the rotating forearm.

FIGURE 39.9

Using the Skin Morph modifier, you can set a muscle to bulge as the forearm is rotated.

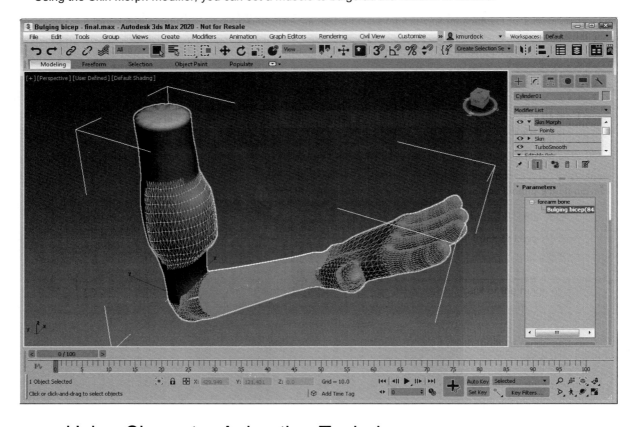

Using Character Animation Techniques

When it comes to character animation, several techniques can really help. Keeping these points in mind as you animate characters can make a difference:

* **Use the Character Animation Toolkit (CAT):** Manual rigging is useful for those cases where CAT won't work, but the tools and features found in CAT make it silly to look elsewhere for human characters.

* **Use dynamics:** Dynamic packages like MassFX can provide incredibly realistic motion based on physical properties. Learning to use this powerful tool for even the most basic animation sequences is worth the effort.

* **Learn by example:** If you're working on a cartoon character, by all means, watch cartoons. Traditional cartoons understand and invented the language of cartoon motion, including squash and stretch, exaggerated motion, or scaling eyes large to indicate surprise. If your character motion is more realistic, find the motion, videotape it, and watch it over and over to catch the subtle secondary motion.

* **Use background animations:** The viewport background can load animation clips, which can make positioning characters to match real motion easy. This is a poor man's motion capture system.

* **Include secondary motion:** The primary motion of a character is often the main focus, but you can enhance the animation by looking for secondary motion. For example, when a person walks, you see his legs take the steps and his arms moving opposite the legs' motion, but secondary motion includes his hair swishing back and forth and shoelaces flopping about.

* **Use the Flex modifier:** The Flex modifier gives soft bodies, such as tails, hair, ears, and clothing, the realistic secondary motion needed to make them believable.

* **Use the Morpher modifier:** The Morpher modifier can be used to morph a character between two poses or to morph its face between the different phonemes as the character talks.

* **Use IK:** Having a character move by positioning its foot or hand is often much easier than pushing it into position.

* **Use the Spring Controller:** Another good way to get secondary motion is to use the Spring Controller. This controller works well with limbs.

* **Add randomness with the Noise Controller:** Often, perfect animation sequences don't look realistic, and using the Noise Controller can help to make a sequence look more realistic, whether the Noise Controller is applied to a walking sequence or to the subtle movement of the eyes.

* **Use manipulators:** Manipulators can be created and wired to give you control over the animation values of a single motion, such as opening and closing the character's eyes.

Summary

Characters are becoming more and more important in the 3ds Max world and can be saved as separate files just like 3ds Max scene files. Combining a detailed skin mesh with a skeleton of bones lets you take advantage of the character animation features. This chapter covered the following topics:

* Designing your character before building
* Working with the Skin modifier
* Reusing animations with Skin Wrap
* Bulging muscles with Skin Morph

The next chapter delves into the Character Animation Toolkit (CAT) and shows how it can be used to quickly create skeletons and animate characters.

Chapter 40

Animating Characters with CAT

IN THIS CHAPTER

Learning the basics of creating characters

Creating and editing a CAT preset rig

Creating a custom CAT rig

Animating using CAT

Earlier versions of the Autodesk® 3ds Max® 2020 software have always had a great way to create and animate characters, but until recently, it was available only as a separate plug-in known as Character Studio. Happily, Character Studio has been integrated into 3ds Max to the point that it isn't distinguishable as a separate package. Character Studio was a good first step, and it still exists in 3ds Max, but it has lots of shortcomings that make it difficult to work with.

Another plug-in package known as Character Animation Toolkit, or CAT for short, has been embraced by many 3ds Max animators, and now CAT is embedded within 3ds Max. CAT offers a simple interface that gets great results whether you're building your own custom rig or animating an existing preset rig.

Although 3ds Max includes other features for rigging characters, if you plan on animating a character, CAT is definitely the way to go. It's an incredible time-saver.

Character Creation Workflow

A typical workflow for creating characters in 3ds Max involves first creating a skin mesh object. After the skin mesh is complete, you can create a skeletal rig to drive its animation. The skeleton consists of a set of bones that provide an underlying structure to the character. Animating these bones provides an easy way to give life to the character.

With a skeletal rig created, position the rig within the skin mesh and match the bone links to the relative size and position inside the skin mesh. The bones do not need to be completely within the skin mesh, but the closer they are to the skin mesh, the more accurate the movements of the character are.

After the rig is sized and matched to the skin mesh, use the Skin modifier to attach the skin mesh to the rig. This automatically sets all the skin weights that govern which skin parts move with which bones. You also can use the Skin modifier settings to deform the skin at certain bone angles, such as bulging a muscle when the arm is flexed.

The next step is to animate the rig using its animation tools, which can include walk, run, and jump cycles using keys. Along the way, you can save, load, and reuse animation sequences, including motion capture files. Animated sequences can be combined and mixed together to form a smooth-flowing animation using the Motion Mixer.

Although this workflow for characters is quite straightforward, there are many ways to speed it up. Creating a skeletal rig can be time consuming and many days can be shaved off a project if you can use an existing skeletal rig that only needs to be modified. This is one place where CAT comes in. The other huge improvement comes on the animation side. CAT includes several tools for making walk cycles easy.

Creating a CAT Rig

Creating a hierarchical skeleton used to control the animation of the mesh skin draped over it is quite easy using CAT. The skeleton can be set to be invisible in the final render and exists only to make the process of animating easier. Although 3ds Max includes a robust set of tools that can be used to create a skeleton of bones, CAT features automate this entire process and even includes a number of prebuilt skeletons.

For some characters, modifying a prebuilt skeleton is more work than building a custom skeleton. For these occasions, you can manually create a skeleton structure. Building a rig system by hand is covered in Chapter 38, "Understanding Rigging, Kinematics, and Working with Bones."

Using prebuilt CAT rigs

To add a prebuilt CAT rig to the scene, simply open the Create panel and select the Helpers category. When you choose the CAT Objects subcategory, you have three options: CAT Parent, CAT Muscle, and Muscle Strand. Click the CAT Parent button and drag in the viewport to place the CAT Parent object. This parent is simply an icon used to control the global position of the rig and is not rendered.

Tip

It is best to place the CAT Parent object at the origin of the scene.

While the CAT Parent object is selected, you can choose a preset rig from the list in the CAT Rig Load Save rollout in the Modify panel or you can select a preset before creating a CAT parent to make the specified preset. The size of the rig is determined by how far you drag in the viewport, or you can set the size using the CATUnits Ratio value in the Command Panel. These custom rigs include a variety of human-shaped and animal-shaped rigs, such as the Alien rig shown in Figure 40.1.

FIGURE 40.1

CAT includes several default preset rigs such as this alien character.

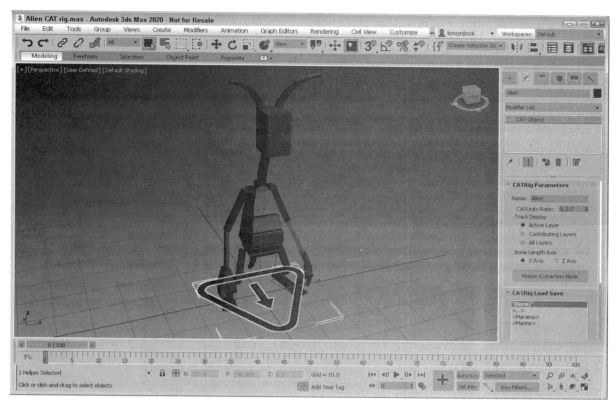

One key advantage of CAT rigs over the rigs available in Character Studio and Biped is that they aren't limited in their structure. Although the available CAT rig presets includes a Base Human and even a Bip01 rig that are used for animating human characters, some of the other presets are Dragon, Horse, Lizard, Spider, and Centipede. These different rigs have multiple legs, arms, and wings and all are easily controlled.

Modifying prebuilt CAT rigs

With a prebuilt rig added to the scene, you can use the Transform tools to select and move the bones to match the skin mesh. The arms are automatically set up as a Forward Kinematics (FK) chain, so rotating the upper arm bone automatically rotates the rest of the arm bones with it. The legs are set up with an Inverse Kinematics (IK) chain, so you can position the legs by dragging the feet, and the rest of the leg bones follow. By default, all CAT prebuilt rigs have stretchy bones, so if you select and move a bone, the selected and attached bones stretch to maintain the joint connection.

Tip

If you double-click a bone, the bone and all its children are selected. For example, double-clicking a collar bone selects the entire arm, making it easy to move the whole arm into place.

A CAT rig keeps track of the different types of body parts and presents the appropriate set of parameters depending on which bone is selected. For example, if you select any part of the arm, the Limb Setup rollouts appear in the Modify panel. This rollout has specific parameters for the arms and legs, as shown in Figure 40.2. CAT also recognizes spines, tails, palms, digits (fingers and toes), hubs (head and pelvis), and generic bones.

FIGURE 40.2

The Limb Setup rollout appears when any arm bone is selected.

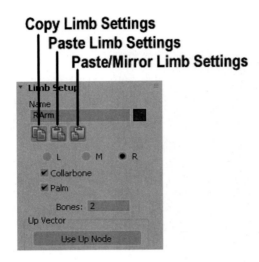

Using these settings, you can specify whether the selected arm is the left, middle, or right arm. You also can choose whether this arm has a collarbone or a palm. The Bones value determines the number of bones that make up the arm. The default is 2, but you can change this to be any whole number from 1 to 20. Because arms can be above or below the head, the Up Vector lets you determine which bone points up.

Beneath the Name are three icons used to copy the settings between two bones. The Copy and Paste Settings buttons let you transfer the settings and orientation of one bone to another. The Paste/Mirror Settings mirror the position of one bone to the opposite side. This lets you set up one arm just right and then quickly copy the settings to the opposite arm.

For spines, you can set the number of bones in the chain, the length and size of the spine bones, and the spine curvature using a simple graph. Tails have these same parameters, plus a height and taper value. For palms and ankles, you can specify the length, width, height, and number of digits.

Individual bones that make up the arms and legs also have a Segments value in the Bone Setup rollout that you can set. Within the Limb Setup is a value for setting the number of bones in the limb, but you can use the Segments value to set the number of segments for each individual bone. This allows you to change a long bone, like the thigh bone, into a series of segments that can rotate independent of each other. This allows for twisting bones, as shown for the forearm in Figure 40.3. You also can set the weight curve for bones with many segments.

FIGURE 40.3

Increasing the number of bone segments allows the bone to twist like this forearm.

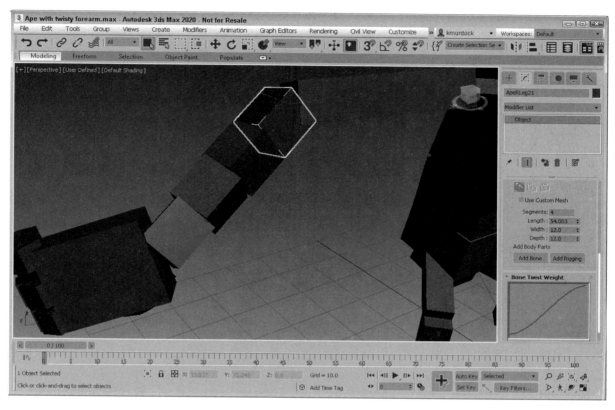

If you want to save the preset rig after you've made some changes, simply select the CAT Parent object and click the Save Preset Rig button in the CAT Rig Load Save rollout. Rigs are saved by default to the CATRigs folder using the .rg3 extension. Once saved, the rigs appear in the list with the other CAT rig presets for easy recall.

Rigs also are easy to delete. Simply select the CAT Parent object and press the Delete key.

Using custom meshes

The bones that make up a CAT rig by default are simple box objects. This allows them to move quickly with a minimum amount of lag, but the bones don't need to be overly simple. The power of modern computers allows complex scenes to be animated without any lag. If you select a bone and enable the Use Custom Mesh option in the Setup rollout, you can access and modify the existing bone to be more representative of the mesh.

To edit an existing bone, simply apply a modifier to the bone object or an Edit Poly modifier and edit the bone as you wish. You also can attach another mesh to the Edit Poly modifier. After editing is done, simply collapse the changes to the base bone object, and then you can switch back and forth between normal box bones and the custom mesh using the Use Custom Mesh option. Figure 40.4 shows the preset for a gnou.

Tip

If you've gone to all the trouble of building a skin mesh, you can quickly use the same skin mesh as the skeleton by stripping down its details and using it as a custom mesh.

FIGURE 40.4

Simplified meshes can be used as custom rig bones like this gnou preset.

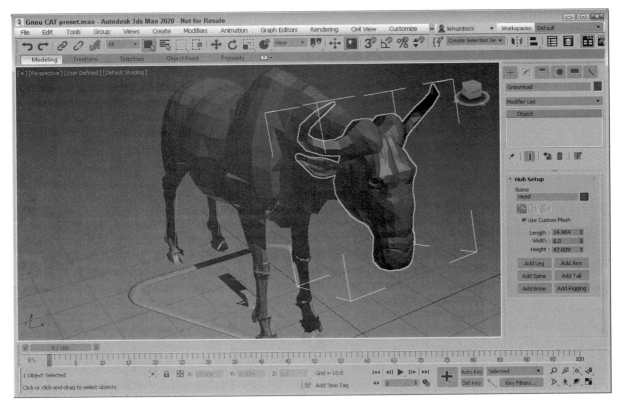

Tutorial: Editing the head bone

Starting with the alien preset CAT rig, this example edits the alien's head bone to show more character and to demonstrate how custom meshes can be used.

To edit the head bone on a CAT rig, follow these steps:

1. Open the Custom alien head bone.max file from the Chap 40 directory in the downloaded content set. This file includes the Alien CAT preset rig.

2. Select the head bone and open the Modify panel. Enable the Use Custom Mesh option in the Hub Setup rollout.

3. Apply the TurboSmooth modifier from the Modifier list and scale the head object up to match the other bones.

4. Open the Create panel and select the Sphere button. Enable the AutoGrid option at the top of the Object Type rollout and drag on the surface of the head to create two eyes and a nose.

5. Open the Modify panel again, select the head bone, and apply the Edit Poly modifier to the object. Then click the Attach button and pick the two eyes and nose objects to combine them to the head object.

6. Right-click in the Modifier Stack, select the Collapse All option from the pop-up menu, and click Yes in the warning dialog box that appears.

Figure 40.5 shows the custom alien head bone. You can switch back to the default box bone by disabling the Use Custom Mesh option.

FIGURE 40.5

Bones can be replaced with custom mesh objects.

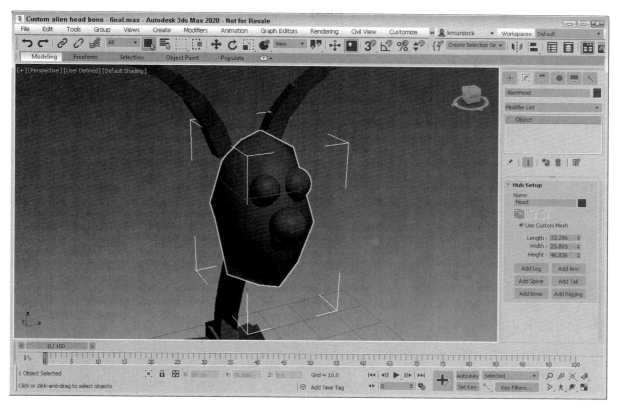

Building a custom CAT rig

When building a custom CAT rig, start by left clicking in the viewport with the CAT Parent object, but make sure the None option is selected in the CAT Rig Load Save rollout. Then, right-click in the viewport to exit CAT Parent creation mode. This creates the parent without any rig. Position the CAT Parent object so both feet of the skin mesh that you are building the skeleton for are contained within the parent icon's outline.

Tip

If you make the skin mesh object frozen and enable the See-Through option in the Object Properties dialog box, you can easily place the rig bones where they need to be.

The first step in creating a custom CAT rig is to create the pelvis using the Create Pelvis button beneath the list of presets. The pelvis object appears as a simple box object above the CAT Parent object. You can then use the Transform tools to move, scale, and rotate the pelvis into place to match the skin mesh.

With the pelvis object in place and selected, you then have options in the Hub Setup rollout, shown in Figure 40.6, to add legs, arms, and a spine or a tail if you want. The Add Leg button adds to the pelvis a leg with two bones that extends to the floor and an ankle. You can position the leg bones by dragging the foot into position. When one of the legs is in position, you can select the pelvis again and click the Add Leg button again to create the opposite leg. The opposite leg is created using the same settings and position as the first.

FIGURE 40.6

The Hub Setup rollout includes buttons for creating connected legs, arms, and spine.

With the pelvis selected again, click the Add Spine button. This adds a set of spine bones with another hub object on top. The top hub object is used to connect the arms and the neck. The neck is simply another set of spine bones with a hub object on top for the head.

Note

For creatures with multiple arms and legs, the difference between the limbs is that a leg extends from the hub to the ground, and the arm hangs loosely.

You can then select the pelvis, shoulder, or head hub objects and use the Add Tail button to add a tail, wings, or ponytail as needed. Figure 40.7 shows a custom CAT rig created with only a few clicks. The rig includes IK chains on both legs and FK chains on the arms. Fingers and toes could be added easily by selecting the palm or ankle objects and specifying the number of digits.

FIGURE 40.7

Custom CAT rigs are easily created using the CAT tools.

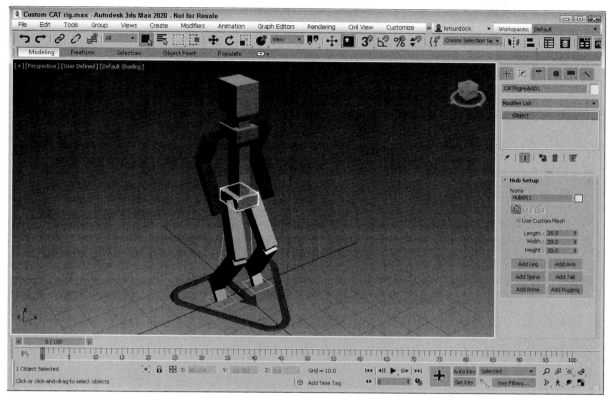

Naming CAT bones

When bones are added to a CAT rig, they are automatically named using the text entered into the Name field. The default name of the entire rig is taken from the name entered into the Name field when the CAT Parent is selected, and names for each body part are associated with the various object parts such as RLeg and LArm. Each bone within a chain is given a default number, so the upper arm object is labeled as 1 and the lower arm is 2.

If you change the name field, all bone names that are affected by the name change are automatically updated, so changing the Name field in the CAT Parent to Reuben automatically changes all bones to start with this name. This makes keeping track of all the various bones much easier and intuitive. It also helps when you start to animate the rig.

Tutorial: Building a custom CAT rig to match a skin mesh

For this example, we use the CAT tools to create a custom rig that matches a mesh skin. The chosen mesh skin is none other than Marvin Moose.

To create a custom CAT rig, follow these steps:

1. Open the Marvin Moose skin.max file from the Chap 40 directory in the downloaded content set. This file includes the Marvin Moose skin mesh positioned at the origin.

2. Select and right-click the moose skin mesh object, and select the Object Properties option from the quad menu. Enable the See-Through and Freeze options in the Object Properties dialog box and click OK.

3. Click the Helpers category, open the CAT Objects subcategory in the Create panel, click the CAT Parent button, select the None option in the CAT Rig Load Save rollout, and drag in the viewport to create the object. Make the CAT Parent just big enough to contain the moose's feet.

4. With the CAT Parent object selected, open the Modify panel and click the Create Pelvis button. Then select and resize the pelvis object and position it to match the skin mesh.

5. With the pelvis object selected, click the Add Leg button to create the left leg. Select and rotate the ankle so the foot is flat against the ground. Scale the foot to roughly match the skin mesh's foot.

6. Select the pelvis, and click Add Leg to create the opposite leg; then click the Add Spine button to create the spine and an object for the shoulders. Position and scale the shoulder hub object. Select the shoulder hub object, and click the Add Arm button. Position the arm bones to match the skin mesh.

7. Select the shoulder hub object, and click Add Arm to create the opposite arm; then click the Add Spine button to create another spine and an object for the head. Select one of the new spine bones, and change the name to **Neck** and the number of bones to **2**. Then scale the head bone to match the moose's head and horns, and name the head bone **Head**.

Figure 40.8 shows the completed custom CAT rig ready to be skinned and animated.

FIGURE 40.8

The moose's skin has been rigged using the CAT tools.

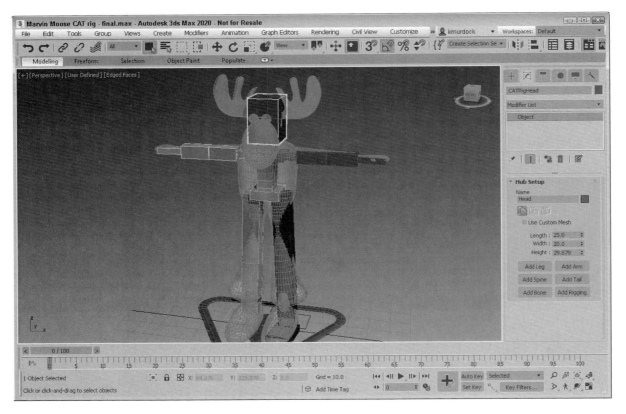

Animating a CAT Rig

The best rig in the world doesn't do you much good if you can't animate it well. Luckily, CAT's animation tools are excellent, just like its rigging tools. CAT uses the concept of animation layers to hold its animation keys. This allows you to blend between different motions.

Animation layers are discussed further in Chapter 33, "Using Animation Layers and Animation Modifiers."

All the animation controls for CAT rigs are found in the Motion panel whenever any of the rig's bones are selected. Within the Layer Manager rollout, shown in Figure 40.9, is a list of the available animation layers. To create a new animation layer, select from four different types, using the Add Layer drop-down list at the bottom right of the list. The four animation layer types include the following:

* **Absolute Layer:** Holds animation key data that defines full motions

* **Local Adjustment Layer:** Holds relative key data relative to the local coordinate system of the above layer

* **World Adjustment Layer:** Applies relative motion in world space that is independent of the previous layers

* **CAT Motion Layer:** Creates procedural-based looping motion such as walk cycles

FIGURE 40.9

The Layer Manager rollout holds the various animation layers.

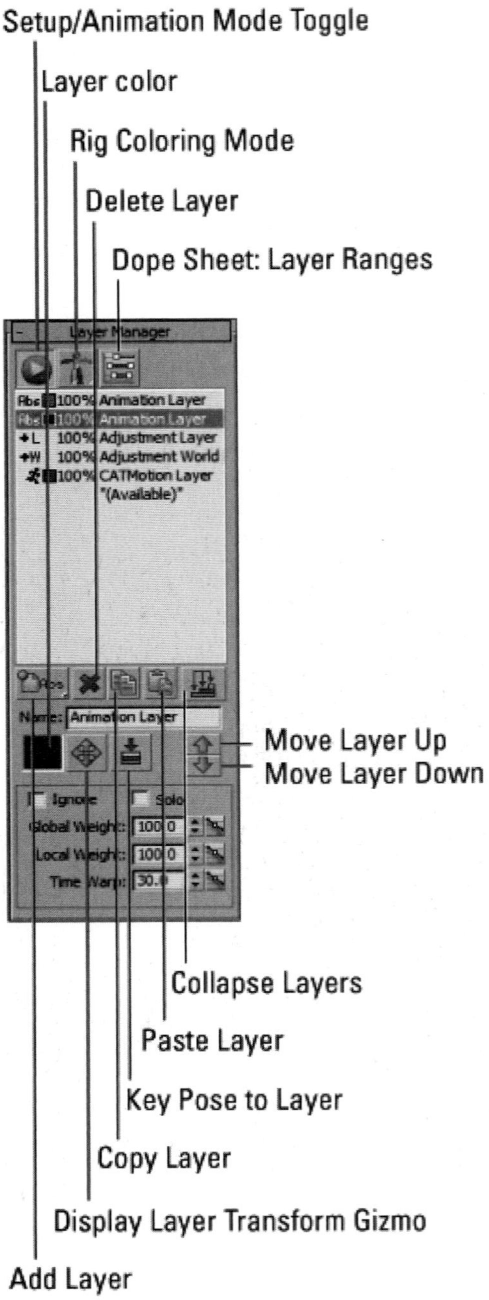

Setup/Animation Mode Toggle

Layer color

Rig Coloring Mode

Delete Layer

Dope Sheet: Layer Ranges

Move Layer Up
Move Layer Down

Collapse Layers

Paste Layer

Key Pose to Layer

Copy Layer

Display Layer Transform Gizmo

Add Layer

After an animation layer is added and selected, you need to press the Setup/Animation Mode toggle button to begin adding keys to the selected animation layer. You do this using the Auto or Set Key modes to create the keys like normal. Any time a new animation layer is added, it is automatically placed above the currently selected animation layer.

Tip

If you're animating some of the rig bones using Auto Key and the keys don't appear on the Track Bar, check to make sure you have clicked the Setup/Animation Mode toggle button to enable animation.

The selected animation layer can be removed from the list using the Delete Layer button. You also can copy and paste layers between different rigs using the Copy and Paste Layer buttons. The Move Layer Up and Move Layer Down buttons are used to reorder the selected layer. Each layer can be given a unique name using the Name field.

The Ignore option disables the selected animation layer, and the Solo option disables all animation layers except for the current selection.

Blending absolute animation layers

When several absolute animation layers are added, the layers are evaluated according to their order in the Layer Manager list from top to bottom. Each layer can have a Global Weight value that determines how much of the animation layer is blended.

For example, if the top absolute animation layer contains keys for a character raising an arm, and a second absolute animation layer is added that has keys of the character waving its hand, these two can be blended to create the combined motion of the character raising its hand and waving by setting the Global Weight for the second absolute animation layer to 50 percent, as shown in Figure 40.10. If the second layer is set to 0 percent, the character simply raises its arm; if the Global Weight of the second layer is set to 100 percent, the layer takes over the entire animation and the character waves its hand without raising its arm.

FIGURE 40.10

If the Layer Manager contains multiple animation layers, you can blend between them using the Global Weight values.

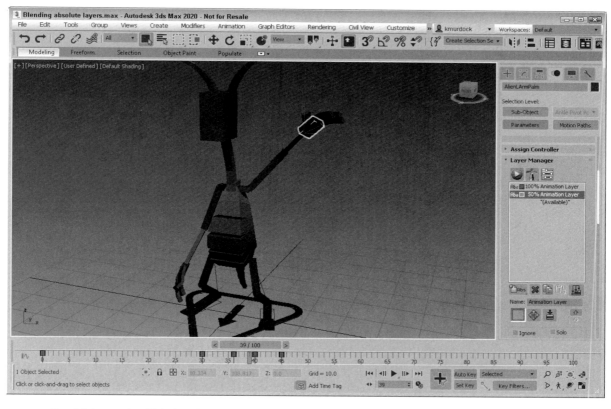

In addition to the Global Weight value, which affects all parts of the rig, you also can set a Local Weight value for specific bones. For example, if an absolute animation layer is set that moves the rig to a specific pose, you select the left collar bone and set its Local Weight to be a percentage of the final pose. The Local Weight setting lets you control individual bones and limbs differently from the global animation layer.

Another helpful tool as you begin to animate your rig is the Rig Coloring Mode option located to the right of the Setup/Animation Mode toggle button. If you switch the Rig Coloring Mode to display Animation Layer Colors (which is available as a flyout), the color of the rig matches the color of the animation layer; if two layers are blended together, the color of the rig also is mixed. If a specific bone is given a different Local Weight, that bone is colored the same as the animation layer that is controlling it.

Tip

If you plan to blend layers, setting each layer to use a primary color makes it easier to see where layers are blended when the Layer Colors option is enabled.

Clicking the Dope Sheet: Layer Ranges button opens the Dope Sheet with the ranges of the various animation layers displayed. This provides an easy way to modify the ranges for the different animation layers. You also can access the Curve Editor for each of the Global and Local Weights using the button to the right of the respective weight values.

You can use the Display Layer Transform Gizmo button for each layer. This creates a simple helper object that is linked to the character. It can be moved and rotated to control the entire rig. The gizmo is normally placed at floor level between the rig's feet, but if you hold down the Ctrl key while clicking this button, the gizmo is placed at the current bone; if you hold down the Alt key, the gizmo is placed at the world origin. This gizmo is available only for absolute animation layers.

Using adjustment animation layers

If you have your animation layers working just right with the motion you like, but your animation needs a little more exaggeration or a hand needs to reach just a little farther to grab a doorknob, you can return to the base absolute layer and make the change, or you can apply an adjustment layer.

There are two different adjustment layer types: Local and World. The difference is in how they are affected by the previous animation layer. Local adjustment layers add the adjustment layers changes onto the above layer's motion, so if a local adjustment layer has a hand reach forward a little more, the motion is added to the existing motion.

World adjustment layers work in world space and cause the hand to reach to a specific location in the world. This still blends with the previous layer's motion, but it also moves the selected object to a global position.

Creating a walk cycle with a CAT Motion layer

Keyframing absolute animation layers is okay, but it can be tedious. The CAT Motion layer is where the fun really begins. Adding a CAT Motion layer to the list automatically applies a walk cycle to the rig. This is done without having to create any keys; just press the Play button, and you see the default walk cycle.

When a CAT Motion layer is selected in the Layer Manager rollout, the CAT Motion Editor button appears to the right of the layer color swatch. This button opens the CAT Motion dialog box, shown in Figure 40.11. Using this dialog box, you can adjust the parameters of the walk cycle.

FIGURE 40.11

The CAT Motion dialog box lets you alter the walk cycle parameters.

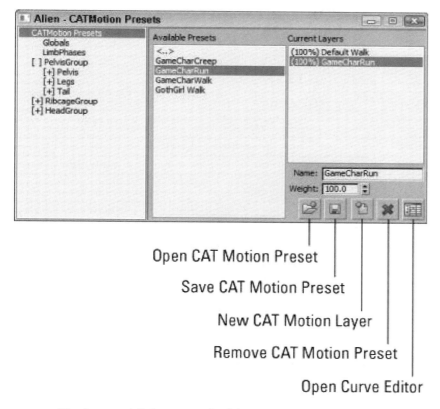

Open CAT Motion Preset

Save CAT Motion Preset

New CAT Motion Layer

Remove CAT Motion Preset

Open Curve Editor

The first panel (leftmost panel) of the CAT Motion dialog box presents a list of available presets. The buttons at the lower-right corner of the dialog box let you open saved presets. You also can name and save custom presets. The CAT Motion dialog box has its own set of animation layers that are listed in the rightmost pane. Double-clicking a preset in the middle panel opens a simple dialog box with options to load the preset into a new layer or into the existing layer.

The layers work just like those in the Layer Manager rollout with weights assigned to each layer. For example, if you add the default walk cycle and then a new run cycle set to 50 percent, the run cycle is blended with the walk cycle, creating a slower run cycle. You also can open the Curve Editor to change the shape of the weight curves.

Setting global parameters

Clicking the Globals option in the leftmost pane of the CAT Motion dialog box opens a panel of global CAT motion parameters, shown in Figure 40.12. At the top of the global parameters are settings for changing the Start and End frame of the walk cycle. Note that these settings are different from the Start and End time settings in the Time Configuration dialog box, and they affect only the CAT rig motion.

FIGURE 40.12

The Globals panel of the CAT Motion dialog box lets you change the walk tempo, speed, and direction.

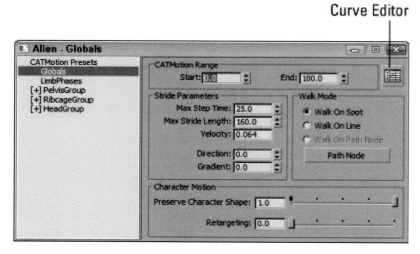

The Max Step Time defines how quickly each step is taken. Low values make the character walk crazy fast, and higher values make a slower, more casual walk. The Max Stride Length sets the distance of each step. These two values together determine the Velocity, but you can't alter the Velocity setting.

The character by default is pointing in the direction indicated on the CAT Parent object, but you can alter the direction that the character is walking by altering the Direction value. A setting of 0 makes the character walk in the direction he is pointing, a value of 90 makes the character shuffle to his right, a value of 180 makes the character walk backward, and a value of 270 makes the character shuffle to the left. The Gradient setting controls the angle that the character is pointing. Negative values make the character point forward as if walking down a hill, and large, positive values make the character walk as if going up a hill.

Walking along a path

The default is to have the character walk in place, but if you select the Walk On Line option, the character walks forward in a straight line. When the range of frames is reached, the character returns to its starting position and walks the line again. If the Direction value is changed, the character moves straight in the specified direction.

If you click the Path Node button, you can choose a scene object that the character will follow. For example, if you make the Path Node a dummy object, the character is positioned and walks on top of the dummy object. You can then animate the movement of the dummy object in the scene, and the character follows it.

To make the character walk along a path, you simply need to create a path using the Line tool. Then select the dummy object and use the Animation→Constraints→Path Constraint menu command to attach the dummy object to the path. If you enable the Follow option in the Motion panel after making the link, the character turns to stay on the path.

When having a character follow a path with tight corners, the character may become distorted as it attempts to stay on the path. The Preserve Character Shape setting in the CATRig - Globals dialog box lets you minimize the distortion. A setting of 0 allows the distortion caused by tight corners. The Retargeting slider blends the leg motion more or less as you change the slider.

Controlling footsteps and limbs

The Limb Phases panel of the CAT Motion dialog box, shown in Figure 40.13, controls the placement of footsteps and the swing of each leg and arm. The footprints that appear when a character is walking are tied to the rig, so altering the footsteps also alters the rig. If the footsteps don't appear when you select a walk cycle, you can use the Create button to make them appear. You also can delete them with the Delete button.

If you move or rotate any of the footsteps, you can use the Reset button to remove any changes for All footsteps or for just the selected footsteps.

FIGURE 40.13

The Limb Phases panel lets you set how the arms and legs swing relative to each other.

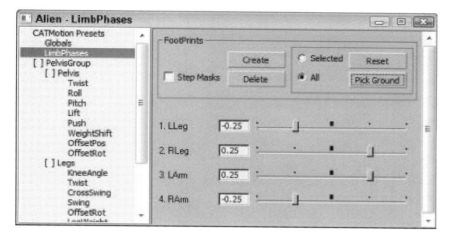

The sliders for each leg and arm at the bottom of the Limb Phases panel let you alter how the arms and legs swing relative to each other. By default, the opposite leg and arm swing together, but you can alter these sliders to give the walk cycle a different look.

Matching footsteps to the ground

After footsteps appear, you can select them and use the Pick Ground button. This lets you select a ground plane object. This ground object needs to be a single object, but once selected, the footsteps are moved vertically to align to the ground plane, causing the character to walk along the surface of the ground.

Controlling secondary motions

The remaining panels in the CAT Motion dialog box are used to control the motions of the other rig groups, such as the pelvis, head, and ribcage. If you open the Pelvis group, you see several parameters that you can access including Twist, Roll, Pitch, Lift, Push, Weight Shift, and positional and rotation offsets. Each of these parameters shows an animation curve that you can use to exaggerate or calm the selected motion, such as the Twist parameter shown in Figure 40.14.

FIGURE 40.14

Using the parameter curves in the CAT Motion dialog box, you can control motions such as the twisting of the pelvis.

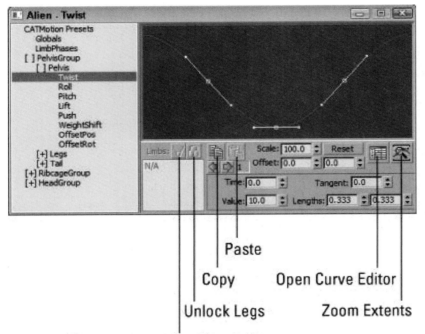

Paste

Copy   Open Curve Editor

Unlock Legs   Zoom Extents

Toggle Independent Leg Settings

Tutorial: Animating a character walking along a path

In this example, you take the default alien character rig and make it walk along a drawn path.

To animate a character walking along a path, follow these steps:

1. Open the Wandering alien.max file from the Chap 40 directory in the downloaded content set. This file includes the default CAT alien rig and a random path.

2. With the alien rig's parent object selected, open the Motion panel and add a CAT Motion animation layer to the Layer Manager list by clicking and holding the Add Layers button and selecting the bottom option from the drop-down list. Then click the Animation Mode button. If you click the Play button, you can see the alien walk in place.

3. Click the Helpers category in the Create panel, select the Standard subcategory, and create a dummy object near the alien. Right-click to exit creation mode.

4. Select any part of the rig again, and click the CAT Motion Editor button (it looks like a cat paw print) in the Motion panel. Select the Globals option in the left pane, and click the Path Node button.

5. Select the dummy object in the scene and select the Animation→Constraints→Path Constraint menu command, and click the path. With the dummy object selected, move and rotate the dummy object so the alien is facing the starting end of the path. Open the Motion panel, and enable the Follow option in the Path Parameters rollout.

6. Click the Play button; the character walks along the path, and footsteps mark each step the alien takes.

7. Select the File→Import→Merge menu command, and merge the QuadPatch02 object in the Hilly surface.max file into the current scene. Zoom out or click the Zoom Extents All button to see the whole surface. This curvy surface appears as a ground plane. Select the character rig and in the Limb Phases panel of the CAT Motion dialog box, click the Pick Ground button and select the ground plane. The footsteps rise to match the ground plane, as shown in Figure 40.15.

FIGURE 40.15

Constraining the path node to a path makes the character walk along the path.

Summary

This chapter serves as an introduction to CAT and covers all aspects of working with CAT rigs, including its presets, custom rigs, and animation. The following topics were covered:

* Learning the basic workflow for creating characters

* Creating and editing CAT rigs

* Creating a custom CAT rig

* Animating CAT rigs

* Creating walk cycles with the CAT Motion dialog box

Now that you have your characters moving around the scene nicely, the next chapter shows how you can use the Populate feature to place and control crowds of people in a scene.

Chapter 41

Creating Crowds and Using Populate

Nobody likes crowds (except for certain types of bugs), but with the Crowd and Populate features in Autodesk® 3ds Max® 2020, controlling crowds can be lots of fun. The fun comes when you realize that you would spend weeks animating by hand all the actions that are possible with these animation tools.

Characters included in a crowd simulation are called delegates, and these delegates can have assigned behaviors that tell them to follow a certain object or a certain path and to avoid designated objects. As you begin to simulate crowds, you'll delight in how much it is like taking the whole family shopping together, except the delegates actually do what you say.

The Populate feature lets you designate areas for animated characters to be placed. These areas can be Flow areas where the characters walk and move or Idle areas where they are standing and chatting. The tools also include all the people that you'll need, so it becomes incredibly easy to add crowds to your scene.

Creating Crowds

Crowd systems are composed of two helper objects called Crowd and Delegate, but the system also can use other scene objects that can be avoided or followed.

Using Crowd and Delegate helpers

Crowd and Delegate helper objects are created using the Create→Helpers→Crowd or Delegate menu commands. The Crowd object looks like a simple diamond-shaped object, and the Delegates are sideways pyramids. Objects can be linked to a behavior object that travels along with it.

Several of the tools you need to define the crowd system are accessed from the Setup rollout, shown in Figure 41.1.

FIGURE 41.1

Use the Setup rollout to define the crowd system.

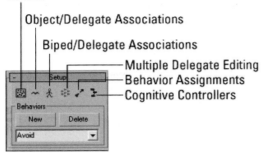

Scattering delegates

If only a single delegate is created, it is not much of a crowd. Multiple delegates can be created using the various cloning methods or using the Scatter button located in the Setup rollout in the Modify panel when the Crowd object is selected.

The various cloning methods are covered in Chapter 7, "Cloning Objects and Creating Object Arrays."

 Clicking the Scatter button causes the Scatter Objects dialog box, shown in Figure 41.2, to open. This dialog box includes several panels used to randomize the Position, Rotation, and Scale of the cloned delegates. In the Clone panel, you can select the Object to Clone and How Many clones to create. After the settings are correct, click the Generate Clones button to create the duplicates.

FIGURE 41.2

The Scatter Objects dialog box lets you quickly create crowds of objects.

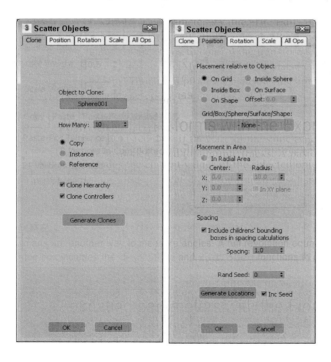

Setting delegate parameters

Each selected delegate has parameters that can be set in the Modify panel, including its dimensions, Speed, Acceleration, and Turning values. With the Crowd object selected, you can click the Multiple Delegate Editing button to open a dialog box, shown in Figure 41.3, which lets you change the parameters of multiple delegates simultaneously.

FIGURE 41.3

The Edit Multiple Delegates dialog box lets you quickly set the parameters of multiple delegates.

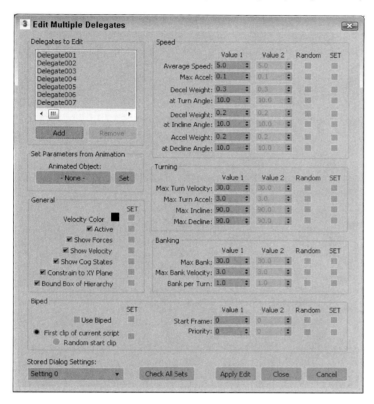

In the upper-left corner, you can select the delegate objects to change using the Add button. You can then specify two values for the parameters. If the Random option is specified, the parameter falls somewhere between the two values. You also can save groups of settings using the drop-down list in the lower-left corner of the interface.

Assigning behaviors

Within the Setup rollout of the Crowd object is a New button that lets you add new behaviors that can be used with the crowd system. All behaviors that are added to the crowd system appear in a drop-down list. By typing a new name in the list, you can name each behavior. The available behaviors include the following:

* **Avoid:** Prevents collisions between scene objects and other delegates

* **Orientation:** Controls the direction the delegates face

* **Path Follow:** Forces delegates to move only along a designated path

* **Repel:** Forces delegates to move away from a target object

* **Scripted:** Makes a delegate behave using a MAXScript

* **Seek:** Moves delegates toward a target object

* **Space Warp:** Uses a Space Warp object to control behavior

* **Speed Vary:** Changes the speed of delegates as they move about the scene
* **Surface Arrive:** Moves delegates toward a surface
* **Surface Follow:** Moves delegates across a surface
* **Wall Repel:** Uses a grid object to repel delegates
* **Wall Seek:** Uses a grid object to attract delegates
* **Wander:** Makes delegates move randomly

When a behavior is selected from the Setup rollout, a custom rollout of parameters for the selected behavior appears. Using these parameters, you can govern how the behavior acts and select which objects are targets.

After the behavior's parameters are set, you can assign specific delegates or teams of delegates to use certain behaviors in the Behavior Assignments and Teams dialog box, shown in Figure 41.4. This dialog box is opened using the Behavior Assignment button found in the Setup rollout.

FIGURE 41.4

The Behavior Assignments and Teams dialog box lets you organize teams of delegates and assign them to behaviors.

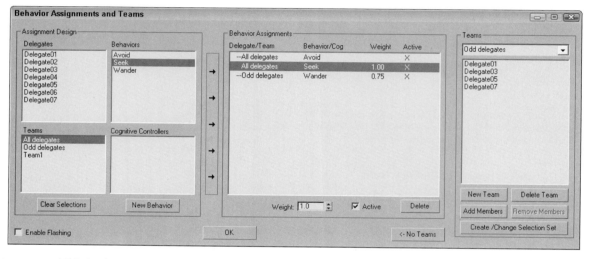

All behavior assignments are listed in the center pane, and each assignment can be given a weight. If the delegate has two conflicting behaviors to follow, it follows the one with the greatest weight value.

Solving the simulation

The final step in the process is to solve the simulation. This step creates keyframes for all the motion in the scene. To solve the simulation, click the Solve button in the Solve rollout, or click the Step Solve button to solve for a single frame at a time. By default, the solution saves keyframes for every frame, but you can increase the Positions and Rotations values to compute the simulation faster.

Tutorial: Rabbits in the forest

This example uses the delegate primitives linked to rabbit meshes to navigate through a forest. Moving several objects through an array of objects can be time-consuming when done by hand, but the Crowd system makes it easy.

To move a group of rabbit delegates through an array of trees, follow these steps:

1. Open the Bunny in the forest.max file from the Chap 41 directory in the downloaded content set. This file includes several pine trees and a single bunny.

2. Select the Create→Helpers→Crowd menu command, and drag in the viewport to create a Crowd object. Make the Crowd object big enough to be easily selected.

3. Select the Create→Helpers→Delegate menu command, and drag in the Left viewport to create a Delegate object positioned outside the trees. Rotate the bunny to set its orientation towards the goal, and then link the bunny to the delegate object. Then Shift+drag the delegate and bunny objects to create six total delegate bunnies as Copies positioned about the scene.

4. Select the crowd object, click the New button in the Setup rollout of the Modify panel, select the Avoid behavior, and name it **Avoid trees**. In the Avoid Behavior rollout, click the Multiple Selection button. In the Select dialog box that appears, choose all tree objects and click the Select button. Enable the Display Hard Radius option, and then decrease the Hard Radius value to 0.2 so the bunnies can travel through the trees.

5. Click the New button again. This time, select the Seek behavior and name it **Seek hole**. In the Seek Behavior rollout, click the None button and choose the red box object that represents the hole.

6. In the Setup rollout, click the Behavior Assignments button to open the Behavior Assignments and Teams dialog box. Click the New Team button, select all delegate objects, and click OK. Then select the Team0 team and the **Avoid trees** behavior, and click the center New Assignment button (the long, thin button with the right-pointing arrows). Select the Team0 team again with the **Seek hole** behavior, and click the New Assignment button again. Both assignments are listed in the center pane; click OK.

7. Set the End Solve value equal to the animation length of **300** frames. In the Solve rollout, click the Solve button.

   The crowd system solves the movement of all delegates as they move toward the goal.

8. Click the Play Animation button to see the resulting solution.

Figure 41.5 shows the position of the delegates after the simulation has ended. Notice the random position of the various delegates.

FIGURE 41.5

The Crowd simulation automatically figures out how to move the delegate bunnies to reach the goal while avoiding the trees.

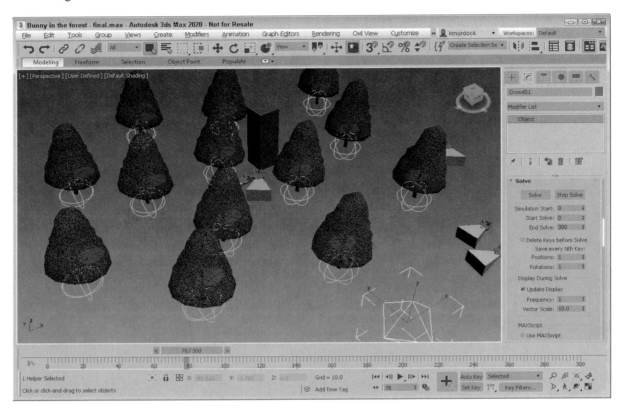

Creating a Crowd of Bipeds

Working with delegates is simple enough, but you probably don't have much need for an animation of several pyramids moving about the scene. The real advantage of the crowd system comes when you're working with objects and bipeds.

Associating delegates with objects

 The second button in the Setup rollout is called the Object/Delegate Associations button. It is used to open a dialog box of the same name. Using this dialog box, shown in Figure 41.6, you can specify which scene objects get associated with which delegates.

FIGURE 41.6

The Object/Delegate Associations dialog box lets you link objects to delegates.

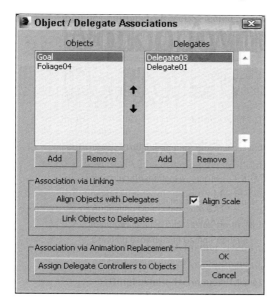

Click the Add buttons to add Objects and Delegates to their respective lists. The arrow buttons between the columns can be used to reorder items in the list. After all objects are matched with their correct delegates, click the Align Objects button or the Link Objects button to complete the linking.

Creating Bipeds

Bipeds are complete human skeletons made of bone objects that are linked and ready-made for animating characters. You can create a Biped object using the Create→Systems→Biped menu command. It is also available in the Systems subcategory in the Create panel. To create a Biped object, simply drag in the viewport. All the settings and parameters for the Biped object are located in the Motion panel.

Caution

Biped objects have been replaced by the CAT system and only remain as a legacy feature for system such as Crowd that use the Biped objects.

Associating delegates with objects

The third button in the Setup rollout is called the Biped/Delegate Association button. It is used just like the Object/Delegate Association button to open a dialog box to associate biped objects with delegates, as shown in Figure 41.7.

FIGURE 41.7

The Associate Bipeds with Delegates dialog box lets you associate bipeds to delegates.

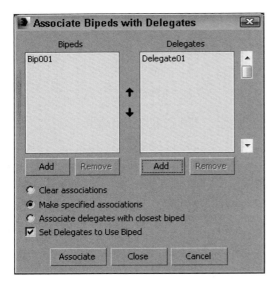

Using Populate

The Crowd feature gives you a lot of power for controlling how the characters move through the scene, but for those times when you just want a random bunch of people to populate your scene, you can use the amazing Populate feature. This feature lets you define areas where the characters walk and/or stand idly by.

Creating Flow and Idle Areas

You can find the Populate features in the Modeling Ribbon or you can access them directly using the Animation→Populate→Populate Tools menu. The Populate Tools are shown in Figure 41.8.

FIGURE 41.8

The Populate Tools ribbon lets you define areas where characters will appear.

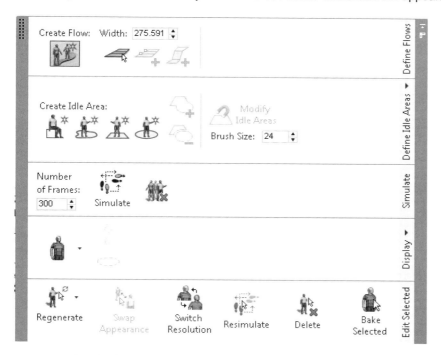

Using the first two tabs, you can add either Flow areas or Idle areas to the scene. Flow areas are those areas where the people walk and are indicated by the arrows lines. To create a Flow area, simply click in the viewports like you are creating a line. You can set the Width of the Flow using the Width setting.

Once a Flow area is added, you can use the Edit Flow button to tweak the Flow's path, or use the Add to Flow button to extend an existing Flow area. The Create Ramp button lets you select a Flow section and raise or lower it. Settings for the selected Flow area in addition to Width include the Lane Spacing and Direction. Each flow has a Portal where the characters emerge and the Portal settings include Density and Speed of people, whether they are Running or not and the mix of Gender.

Idle areas are created using the Free Area, Rectangle and/or Circle shapes. You can also create a seat idle area. These areas contain people that aren't moving, but are sitting, standing around chatting, looking around or talking on the cell phone. There are also modes for adding and subtracting away from these areas. The Modify Idle Areas button lets you edit the Idle areas using a Brush.

Note

Keep in mind that the Flow areas aren't rendered and you'll need to add some surfaces underneath the characters unless you like the idea of them floating in the air.

Adding People to the Flow areas

Once the Flow and Idle areas are defined, you can add people to the scene using the Simulate button. The resolution of the characters is determined by the setting in the Display panel. The options include Stick, Custom Skin, and Textured Skin. After the Simulation is done, clicking the Play button will show the people walking and interacting.

If you find that some of the people are too similar, you can select a person to change and click the Regenerate button. This will replace the selected person with a new character or you can select a character from the Crowd Styles Customization palette, which is available as a flyout under the Regenerate button.

Figure 41.9 shows the Crowd Styles Customization palette. You can also select swap character appearances if two or more are selected or simply delete the selected character.

FIGURE 41.9

The Crowd Styles Customization palette lets you change the look of crowd characters.

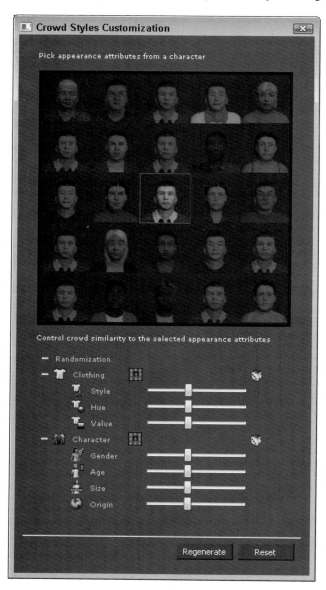

Tutorial: Adding a Crowd with Populate

For this tutorial, we'll take a trip to a busy city center where a clock tower sits majestically. In this example, you'll use the Populate feature to make the Clock Tower a busy, bustling place. This gives you a chance to work with the Populate system to animate crowds of characters.

The first thing to consider is setting up the scene. For this sequence, we need to load the Clock Tower model. Luckily for us, it is already modeled and ready to go. The people also we don't need to worry about modeling or even animating as the Populate system will take care of all of that for us. We only need to focus on defining the areas where the people will move. Populate lets you designate areas for walking people and areas where the people stand idle.

To add a crowd around a clock tower with Populate, follow these steps:

1. Open the Clock Tower scene.max file from the Chap 41 directory in the downloaded content set.
2. Minimize the viewports so the Top viewport is visible.
3. Open the Modeling Ribbon and click on the Populate tab or select the Animation@@>Populate@@>Populate Tools menu command. Click on the Create Flow button and drag four connected paths that surround the outer ring of the clock tower.
4. Select the Create Circular Idle Area button and create several idle areas around the clock tower that don't interfere with the flow areas.
5. Click the Simulate button. The people appear walking and moving about the scene.
6. Click the Play button to see the people moving about the scene.

Figure 41.10 shows the scene after the Populate crowds have been added.

FIGURE 41.10

A crowd of people walk and mingle around a clock tower thanks to Populate.

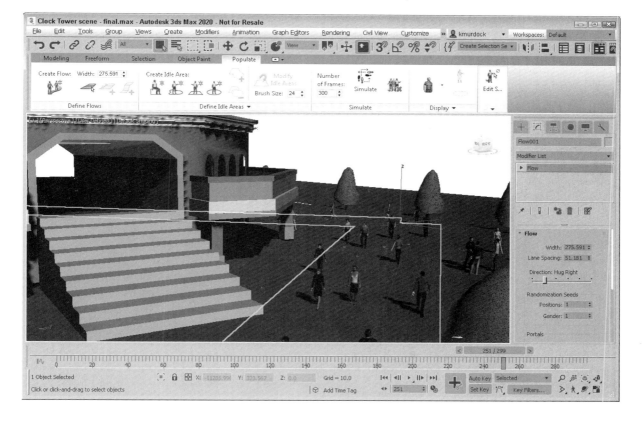

Summary

The Crowd features of Character Studio are useful for animating the motion of delegates following specified behaviors. The following topics were covered:

* Creating crowds with helper objects
* Assigning behaviors
* Using bipeds with crowds
* Adding people with Populate

The next chapter shows how to work with particles and the Particle Flow interface.

Part IX

Adding Special Effects

IN THIS PART

Chapter 42
Creating Particles and Particle Flow

Every object that you add to the scene slows down the Autodesk® 3ds Max® 2020 software to a small degree because 3ds Max needs to keep track of them all. If you add thousands of objects to a scene, not only does 3ds Max slow down noticeably, but also the objects become difficult to identify. For example, if you had to create thousands of simple snowflakes for a snowstorm scene, the system would become unwieldy, and the number wouldn't get very high before you ran out of memory.

Particle systems are specialized groups of objects that are managed as a single entity. By grouping all the particle objects into a single controllable system, you can easily make modifications to all the objects with a single parameter. This chapter discusses using these special systems to produce rain and snow effects, fireworks sparks, sparkling butterfly wings, and even fire-breathing dragons.

The particle features within the Autodesk® 3ds Max® 2020 software enable you to do some great effects. Particle systems like Super Spray are great for certain applications, but they suffer from an inflexibility. Each system has lots of parameters, but after the parameters are set, the particles follow the set parameters without changing. The Particle Flow system is an event-driven system that constantly tests its particles for certain criteria and alters its actions based on these defined criteria. This gives you the ability to program the particles to react in unique and different ways not possible with the other systems.

Understanding the Various Particle Systems

A *particle* is a small, simple object that is duplicated en masse, like snow, rain, or dust. Just as in real life, 3ds Max includes many different types of particles that can vary in size, shape, texture, color, and motion. These different particle types are included in various particle systems.

When a particle system is created, all you can see in the viewport is a single gizmo known as an *emitter icon*. An emitter icon is the object (typically a gizmo, but it can be a scene object) where the particles originate. Selecting a particle system gizmo makes the parameters for the particle system appear in the Modify panel.

3ds Max includes the following particle systems:

* **Particle Flow Source:** Particles that can be defined using the Particle Flow window and controlled using actions and events.

* **Spray:** Simulates drops of water. These drops can be Drops, Dots, or Ticks. The particles travel in a straight line from the emitter's surface after they are created.

* **Snow:** Similar to the Spray system, with the addition of some fields to make the particles Tumble as they fall. You also can render the particles as Six Pointed shapes that look like snowflakes.

* **Blizzard:** An advanced version of the Snow system that can use the same mesh object types as the Super Spray system. Binding the system to the Wind Space Warp can create storms.

* **PArray:** Can use a separate distribution object as the source for the particles. For this system, you can set the particle type to Fragment and bind it to the PBomb Space Warp to create explosions.

* **PCloud:** Confines all generated particles to a certain volume. A good use of this system is to reproduce bubbles in a glass or cars on the road.

* **Super Spray:** An advanced version of the Spray system that can use different mesh objects, closely packed particles called MetaParticles, or an instanced object as its particles. Super Spray is useful for rain and fountains. Binding it to the Path Follow Space Warp can create waterfalls.

The Particle Flow window is a complete interface for controlling the motions of particles. Using the interface, you can program particles to respond to different events.

Creating a Particle System

You can find all the various particle systems under the Create panel and also in the Create menu. To access these systems, click the Geometry category and select the Particle Systems subcategory from the drop-down list. All the particle systems then appear as buttons. Or you can select them from the Create→Particles menu.

With the Particle Systems subcategory selected, click the button for the type of particle system that you want to use, and then click in a viewport to create the particle system emitter icon. The emitter icon is a gizmo that looks like a simple primitive and that defines the location in the system where the particles all originate. Attached to the icon is a single line that indicates the direction in which the particles move when generated. This line points by default toward the construction grid's negative Z-axis when first created. Figure 42.1 shows the emitter icons for each particle system type, including, from left to right, Particle Flow Source; Super Spray; Spray, Snow, and Blizzard (which all have the same emitter icon); PArray; and PCloud.

FIGURE 42.1

The emitter icons for each particle system type

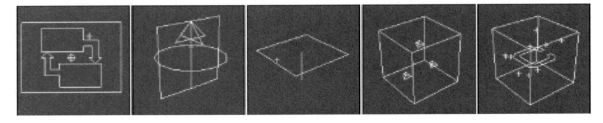

You can transform these icons using the standard transform buttons on the main toolbar. Rotating an emitter changes the direction in which the particles initially move.

After an icon is created, you can set the number, shape, and size of the particles and define their motion in the Parameter rollouts. To apply a material to the particles, simply apply the material to the system's icon. This material is applied to all particles included in the system.

Note
Be aware that the particles are displayed as simple objects such as ticks or dots in the viewports. To see the actual resulting particles, you need to render the scene file.

You can set the parameters for the 3ds Max particle systems in the Create panel when they are first created or in the Modify panel at any time. The simpler systems, Spray and Snow, have a single Parameters rollout.

Using the Spray and Snow Particle Systems

All I can say about the Spray and Snow particle systems is that when it rains, it pours. The Spray Parameters rollout includes values for the number of particles to be included in the system. These values can be different for the viewport and the renderer. By limiting the number of particles displayed in the viewport, you can make the viewport updates quicker. You also can specify the drop size, initial speed, and variation. The Variation value alters the spread of the particles' initial speed and direction. A Variation value of 0 makes the particles travel in a straight line away from the emitter.

Spray particles can be Drops, Dots, or Ticks, which affect how the particles look only in the viewport. Drops appear as streaks, Dots are simple points, and Ticks are small plus signs. You also can set how the particles are rendered—as Tetrahedron objects or as Facing objects (square faces that always face the viewer).

The Timing values determine when the particles appear and how long the particles stay around. The Start Frame is the first frame where particles begin to appear, and the Life value determines the number of frames in which the particles are visible. When a particle's lifetime is up, it disappears. The Birth Rate value lets you set how many new particles appear in each frame; you can use this setting or select the Constant option. The Constant option determines the number of particles created at each frame by dividing the total number of particles by the Life value.

The Emitter dimensions specify the width and height of the emitter gizmo. You also can hide the emitter with the Hide option.

Note
The Hide option hides the emitter only in the viewports. Emitters are never rendered.

The parameters for the Snow particle system are similar to the Spray particle system, except for a few unique settings. Snow can be set with a Tumble and Tumble Rate. The Tumble value can range from 0 to 1, with 1 causing a maximum amount of rotation. The Tumble Rate determines the speed of the rotation.

The Render options are also different for the Snow particle system. The three options are Six Point, Triangle, and Facing. The Six Point option renders the particle as a six-pointed star. Triangles and Facing objects are single faces.

Tutorial: Creating rain showers

One of the simplest uses for particle systems is to simulate rain or snow. In this tutorial, you use the Spray system to create rain and then learn how to use the Snow system to create snow.

To create a scene with rain using the Spray particle system, follow these steps:

1. Open the Simple rain.max file from the Chap 42 directory in the downloaded content set.

This file includes a tree model.

2. Select the Create→Particles→Spray menu command, and drag the icon in the Top viewport to cover the entire scene. Position the icon above the umbrella object, and make sure the vector is pointing down toward the tree.

3. Open the Modify panel, and in the Parameters rollout, set the Render Count to **100000** and the Drop Size to **1,** the Speed of **2** and the Variation to **0.1,** and select the **Drops** option; these settings make the particles appear as streaks. Select the Tetrahedron Render method, and set the Start and Life values to **0** and **100,** respectively.

Note
To cover the entire scene with an average downpour, set the number of particles to **1000** for a 100-frame animation.

4. Open the Material Editor (by pressing the M key), and create a material with a light blue Diffuse color and drag this material to the particle system icon.

Figure 42.2 shows the results of this tutorial.

FIGURE 42.2

Rain created with the Spray particle system

Tutorial: Creating a snowstorm

Creating a snowstorm is very similar to what you did in the preceding tutorial. To create a snowstorm, use the Snow particle system with the same number of particles and apply a white material to the particle system.

To create a scene with snow using the Snow particle system, follow these steps:

1. Open the Snowman in snowstorm.max file from the Chap 42 directory in the downloaded content set. This file includes a snowman created using primitive objects.

2. Select the Create→Particles→Snow menu command, and drag the icon in the Top viewport to cover the entire scene. Position the icon above the objects, and make sure that the vector is pointing down toward the scene objects.

3. Open the Modify panel, and in the Parameters rollout, set the Render Count to **1000** and the Flake Size to **6**, and use the Six Point Render option. Set the Start and Life values to **0** and **100**, respectively.

4. Open the Material Editor (by pressing the M key), and create a material with a white Diffuse color with some self-illumination and drag this material to the particle system gizmo.

Figure 42.3 shows the results of this tutorial.

FIGURE 42.3

A simple snowstorm created with the Snow particle system

Using the Super Spray Particle System

If you think of the Spray particle system as a light summer rain shower, the Super Spray particle system is like a fire hose. The Super Spray particle system is considerably more complex than its Spray and Snow counterparts. With this complexity comes a host of features that make this one of the most robust effects creation tools in 3ds Max.

Unlike the Spray and Snow particle systems, the Super Spray particle system includes several rollouts.

Super Spray Basic Parameters rollout

The Super Spray particle system emits all particles from the center of the emitter icon. The emitter icon is a simple circle set in a plane with an arrow that points in the direction in which the particles will travel. In the Basic Parameters rollout, shown in Figure 42.4, the Off Axis value sets how far away from the icon's arrow the stream of particles will travel. A value of 0 lines up the particle stream with the icon's arrow, and a value of 180 emits particles in the opposite direction. The Spread value can range from 0 to 180 degrees and fans the particles equally about the specified axis. The Off Plane value spins the particles about its center axis, and the Spread value sets the distance from this center axis that particles can be created. If all these values are left at 0, the particle system emits a single, straight stream of particles, and if all values are 180, particles go in all directions from the center of the emitter icon.

Note

To actually see the particles in the viewport, you need to drag the Time Slider or press the Play button.

FIGURE 42.4

The Basic Parameters rollout lets you specify where and how the particles appear in the viewports.

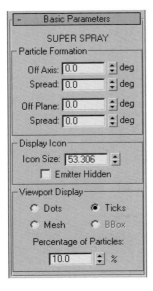

The emitter icon size can be set, or the emitter icon can be hidden in the viewport. You also can set the particles to be displayed in the viewport as Dots, Ticks, Meshes, or Bounding Boxes. The Percentage of Particles value is the number of the total particles that are visible in the viewport and should be kept low to ensure rapid viewport updates.

Particle Generation rollout

The Particle Generation rollout, shown in Figure 42.5, is where you set the number of particles to include in a system as either a Rate or Total value. The Use Rate value is the number of particles per frame that are generated. The Use Total value is the number of particles generated over the total number of frames. Set the Use Rate value if you want the animation to have a steady stream of particles throughout the animation; use the Use Total value if you want to set the total number of particles that will appear throughout the entire range of frames.

FIGURE 42.5

The Particle Generation rollout lets you control the particle motion.

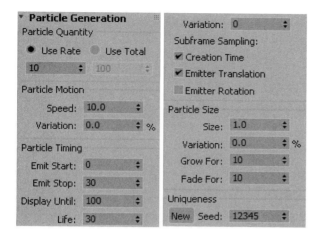

In the Particle Motion group, the Speed value determines the initial speed of particles. The Variation value alters this initial speed as a percentage of the Speed value. A high Variation value results in particles with all sorts of different speeds.

Note

Be sure to use the Variation values liberally to get more realistic particle behavior.

In the Particle Timing group, you can set when the emitting process starts and stops. Using the Display Until value, you also can cause the particles to continue displaying after the emitting has stopped. The Life value is how long particles stay around, which can vary based on another Variation setting.

When an emitter is animated (such as moving back and forth), the particles can clump together where the system changes direction. This clumping effect is called *puffing*. The Subframe Sampling options help reduce this effect. The three options are Creation Time (which controls emitting particles over time), Emitter Translation (which controls emitting particles as the emitter is moved), and Emitter Rotation (which controls emitting particles as the emitter is rotated). All three options can be enabled, but each one that is enabled adds the computation time required to the render.

Note

The Subframe Sampling options increase the rendering time and should be used only if necessary.

You can specify the particle size along with a Variation value. You also can cause the particles to grow and fade for a certain number of frames.

The Seed value helps determine the randomness of the particles. Clicking the New button automatically generates a new Seed value.

Note

If you clone a particle system and each system has the same Seed value, the two systems will be exactly the same. Two cloned particle systems with different Seed values will be unique.

Particle Type rollout

The Particle Type rollout, shown in Figure 42.6, lets you define the look of the particles. At the top of the rollout are three Particle Type options: Standard Particles, MetaParticles, and Instanced Geometry.

FIGURE 42.6

The Particle Type rollout (shown in four parts) lets you define how the particles look.

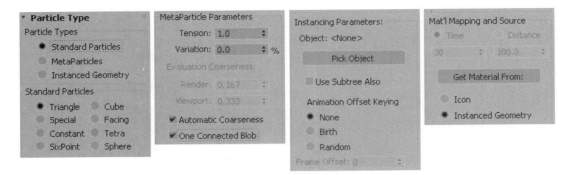

If you select Standard Particles as the particle type, you can select which geometric shape you want to use from the Standard Particles section. The options are Triangle, Special, Constant, SixPoint, Cube, Facing, Tetra, and Sphere.

The Special type consists of three intersecting planes, which are useful if you apply maps to them. The Facing type is also useful with maps; it creates a simple square face that always faces the viewer. The Constant type maintains the same pixel size, regardless of the distance from the camera or viewer. The Six Point option renders each particle as a 2D six-pointed star. All other types are common geometric objects.

Tutorial: Creating a fireworks fountain

For an example of the Super Spray particle system, you create a fireworks fountain. Fireworks are essentially just lots of particles with a short life span and a high amount of self-illumination. (Tell yourself that the next time you watch a fireworks display.)

To create a fireworks fountain using a particle system, follow these steps:

1. Open the Fireworks fountain.max file from the Chap 42 directory in the downloaded content set.

   This file includes a simple fountain base and the Gravity space warp to cause the particles to curve back toward the ground.

Tip

Some of the most amazing special effects are made possible by combining particle systems with Space Warps.

2. Select the Create→Particles→Super Spray menu command, drag in the Top viewport, and position the system at the top of the fireworks cylinder with the direction arrow pointing toward the sky.

3. Open the Modify panel, and set the Off Axis Spread to **44** and the Off Plane Spread to **90**. In the Particle Generation rollout, select the Use Total option, set the Total number of particles to **2000** with a Speed of **20** and a Variation of **100**. Set the Emit Start to **0** and the Emit Stop to **100**. Set the Display Until to **100** and the Life to **25** with a Variation of **20**. The Size of the particles should be **5**.

4. Open the Material Editor (by pressing the M key), double-click the Standard material in the Material/Map Browser, and then double-click the Standard node to access its parameters. Name the material **Spark** set its Diffuse color to yellow, and set its Self-Illumination color to yellow as well. Then drag the material from the Material Editor to the particle system's icon.

5. Select the Super Spray icon, right-click it to open the pop-up quad menu, and select the Object Properties menu option. In the Object Properties dialog box, select the Object Motion Blur option.

Caution

When viewing the animation, maximize a single viewport. If 3ds Max tries to update all four viewports at once with this many particle objects, the update is slow. The shortcut to maximize the viewport is Alt+W.

Figure 42.7 shows sparks emitting from the fireworks fountain.

FIGURE 42.7

The Super Spray particle system is used to create fireworks sparks.

Tutorial: Adding spray to a spray can

The Super Spray particle system is complex enough to warrant another example. What good is a spray can without any spray? In this tutorial, you create a spray can model and then use the Super Spray particle system to create the spray coming from it.

To create a stream of spray for a spray can, follow these steps:

1. Open the Spray can.max file from the Chap 42 directory in the downloaded content set.

   This file includes a simple spray can object created using a cylinder for the can base and the nozzle and a lathed spline for the top of the can.

2. Select the Particle Systems subcategory button from the drop-down list in the Create panel, and click the Super Spray button. Drag in the Top viewport to create the Super Spray icon, and position it at the mouth of the nozzle.

3. In the Basic Parameters rollout, set the Off Axis Spread to **20** and the Off Plane Spread to **90**. In the Particle Generation rollout, set the Use Rate to **1000**, the Speed to **20**, and the Life to **30**. Set the Size of the particles to **5**.

4. Open the Material Editor (by pressing M), and locate the Spray Mist material already created in the Sample Slots rollout in the Material/Map Browser. Then drag this material onto the Super Spray icon to apply this material to the Super Spray particle system.

Figure 42.8 shows the fine spray from an aerosol can.

FIGURE 42.8

Using a mostly transparent material, you can create a fine mist spray.

Using the MetaParticles option

The MetaParticles option in the Particle Type rollout makes the particle system release Metaball objects. *Metaballs* are viscous spheres that, like mercury, flow into each other when near. These particles take a little longer to render, but are effective for simulating water and liquids. The MetaParticles type is available for the Super Spray, Blizzard, PArray, and PCloud particle systems.

Selecting the MetaParticles option in the Particle Types section enables the MetaParticle Parameters group. In this group are options for controlling how the MetaParticles behave. The Tension value determines how easily objects blend together. MetaParticles with a high tension resist merging with other particles. You can vary the amount of tension with the Variation value. The Tension value can range between 0.1 and 10, and the Variation can range from 0 to 100 percent.

Because MetaParticles can take a long time to render, the Evaluation Coarseness settings enable you to set how computationally intensive the rendering process is. This can be set differently for the viewport and the renderer—the higher the value, the quicker the results. You also can set this to Automatic Coarseness, which automatically controls the coarseness settings based on the speed and ability of the renderer. The One Connected Blob option speeds the rendering process by ignoring all particles that aren't connected.

Tutorial: Spilling soda from a can

MetaParticles are a good option to use to create drops of liquid, like those from a soda can.

To create liquid flowing from a can, follow these steps:

1. Open the MetaParticles from a soda can.max file from the Chap 42 directory in the downloaded content set.

This file includes a soda can model positioned with the can on its side.

2. Select the Create→Particles→Super Spray menu command, and drag the icon in the Left viewport. Position the icon so that its origin is at the opening of the can and the directional vector is pointing outward.

3. With the Super Spray icon selected, open the Modify panel, and in the Basic Parameters rollout, set the Off Axis and Off Plane Spread values to **40**.

4. In the Particle Generation rollout, keep the default Rate and Speed values, but set the Speed Variation to **50** to alter the speed of the various particles. Set the Particle Size to **20**.

5. In the Particle Type rollout, select the MetaParticles option, set the Tension value to **1**, and make sure that the Automatic Coarseness option is selected.

6. Open the Material Editor (by pressing the M key), and drag the previously created Purple Soda material from the Sample Slots rollout in the Material/Map Browser to the particle system icon.

Figure 42.9 shows a rendered image of the MetaParticles spilling from a soda can at frame 25.

FIGURE 42.9

MetaParticles emitting from the opening of a soda can

Instanced Geometry

Using the Particle Type rollout, you can select any existing scene object to use as the particle with the Instanced Geometry option. To choose an object to use as a particle, click the Pick Object button and select an object from the viewport. If the Use Subtree Also option is selected, all child objects are also included.

Caution

Using complicated objects as particles can slow down a system and increase the rendering time.

The Animation Offset Keying options determine how an animated object that is selected as the particle is animated. The None option animates all objects the same, regardless of when they are born. The Birth option starts the animation for each object when it is created, and the Random option offsets the timing randomly based on the Frame Offset value. For example, if you have selected an animated bee that flaps its wings as the particle, and you select None as the Animation Offset Keying option, all the bees flap their wings in concert. Selecting the Birth option instead starts them flapping their wings when they are born, and selecting Random offsets each instance differently.

For materials, the Time and Distance values determine the number of frames or the distance traveled before a particle is completely mapped. You can apply materials to the icon that appears when the particle system is created. The Get Material From button lets you select the object from which to get the material. The options include the emitter icon and the Instanced Geometry.

Rotation and Collision rollout

In the Rotation and Collision rollout is an option to enable interparticle collisions. This option causes objects to bounce away from one another when their object boundaries overlap.

The Rotation and Collision rollout, shown in Figure 42.10, contains several controls to alter the rotation of individual particles. The Spin Time is the number of frames required to rotate a full revolution. The Phase value is the initial rotation of the particle. You can vary both of these values with Variation values.

Note
The Rotation and Collision rollout options also can increase the rendering time of a scene.

FIGURE 42.10

The Rotation and Collision rollout options can control how objects collide with one another.

You also can set the axis about which the particles rotate. Options include Random, Direction of Travel/MBlur, and User Defined. The Stretch value under the Direction of Travel option causes the object to elongate in the direction of travel. The User Defined option lets you specify the degrees of rotation about each axis.

Interparticle collisions are computationally intensive and can easily be enabled or disabled with the Enable option. You also can set how often the collisions are calculated. The Bounce value determines the speed of particles after collisions as a percentage of their collision speed. You can vary the amount of Bounce with the Variation value.

Tutorial: Basketball shooting practice

When an entire team is warming up before a basketball game, the space around the basketball hoop is quite chaotic—with basketballs flying in all directions. In this tutorial, you use a basketball object as a particle and spread it around a hoop. (Watch out for flying basketballs!)

To use a basketball object as a particle, follow these steps:

1. Open the Basketballs at a hoop.max file from the Chap 42 directory in the downloaded content set.

   This file includes basketball and basketball hoop models.

2. Select the Create→Particles→Super Spray menu command, and drag the icon in the viewport. Position the icon in the Front view so that its origin is above and slightly in front of the hoop and the directional vector is pointing down. (You need to rotate the emitter icon.)

3. Open the Modify panel, and in the Basic Parameters rollout, set the Off Axis Spread value to **90** and the Off Plane Spread value to **40**; this randomly spreads the basketballs around the hoop. In the Viewport Display group of the Basic Parameters rollout, select the Mesh option. Set the Percentage of Particles to **100** percent to see the position of each basketball object in the viewport.

Caution

Because the basketball is a fairly complex model, using the Mesh option severely slows down the viewport update. You can speed the viewport display using the BBox (Bounding Box) option, but you'll need to choose it after selecting the Instanced Geometry option.

4. In the Particle Generation rollout, select the Use Total option, and enter **30** for the value. (This number is reasonable and not uncommon during warm-ups.) Set the Speed value to **0.2** and the Life value to **100** because you don't want basketballs to disappear. Because of the low number of particles, you can disable the Subframe Sampling options. Set the Grow For and Fade For values to **0**.

5. In the Particle Type rollout, select the Instanced Geometry option and click the Pick Object button. Make sure that the Use Subtree Also option is selected to get the entire group, and then select the basketball group in the viewport. At the bottom of this rollout, select the Instanced Geometry option and click the Get Material From button to give all the particles the same material as the original object.

6. In the Rotation and Collision rollout, set the Spin Time to **100** to make the basketballs spin as they move about the scene. Set the Spin Axis Controls to Random. Also enable the Interparticle Collisions option, and set the Calculation Interval per Frame to **1** and the Bounce value to **100**.

   With the Interparticle Collisions option enabled, the basketballs are prevented from overlapping one another.

7. At the floor of the basketball hoop is a Deflector Space Warp. Move this deflector vertically upward about half the radius of the basketball to prevent the balls from sinking into the floor. Click the Bind to Space Warp button on the main toolbar, and drag from this floor deflector to the Super Spray icon.

   This makes the basketballs bounce off the floor.

Figure 42.11 shows a rendered image of the scene at frame 30 with several basketballs bouncing chaotically around a hoop.

FIGURE 42.11

Multiple basketball particles flying around a hoop

Object Motion Inheritance rollout

The settings on the Object Motion Inheritance rollout, shown in Figure 42.12, determine how the particles move when the emitter is moving. The Influence value defines how closely the particles follow the emitter's motion; a value of 100 has particles follow exactly, and a value of 0 means they don't follow at all.

FIGURE 42.12

The Object Motion Inheritance rollout sets how the particles inherit the motion of their emitter, and the Bubble Motion rollout defines how particles act like bubbles.

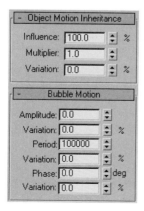

The Multiplier value can exaggerate or diminish the effect of the emitter's motion. Particles with a high multiplier can actually precede the emitter.

Bubble Motion rollout

The Bubble Motion rollout, also shown in Figure 42.12, simulates the wobbling motion of bubbles as they rise in a liquid. Three values define this motion, each with variation values. Amplitude is the distance that the particle moves from side to side. Period is the time that it takes to complete one side-to-side motion cycle. Phase defines where the particle starts along the amplitude curve.

Particle Spawn rollout

The Particle Spawn rollout, shown in Figure 42.13, sets options for spawning new particles when a particle dies or collides with another particle. If the setting is None, colliding particles bounce off one another, and dying particles simply disappear. The Die After Collision option causes a particle to disappear after it collides. The Persist value sets how long the particle stays around before disappearing. The Variation value causes the Persist value to vary by a defined percentage.

FIGURE 42.13

The Particle Spawn rollout (shown in two parts) can cause particles to spawn new particles.

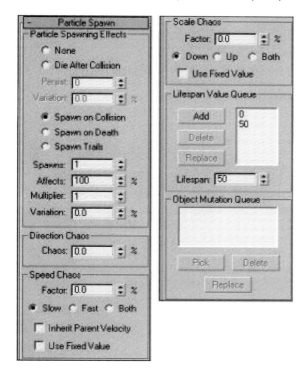

The Spawn on Collision, Spawn on Death, and Spawn Trails options all enable the spawn controls and define when particles spawn new particles. The Spawns value is the number of times a particle can spawn other particles. The Affects value is the percentage of particles that can spawn new particles; lowering this value creates some duds that do not spawn. The Multiplier value determines the number of new particles created.

Note

The Spawn Trails option causes every particle to spawn a new particle at every frame. This option can quickly create an enormous number of particles and should be used with caution.

The Chaos settings define the direction and speed of the spawned particles. A Direction Chaos value of 100 gives the spawned particles the freedom to travel in any direction, whereas a setting of 0 moves them in the same direction as their originator.

The Speed Chaos Factor is the difference in speed between the spawned particle and its originator. This factor can be faster or slower than the original. Selecting the Both option speeds up some particles and slows others randomly. You also can choose to have spawned particles use their parent's velocity or use the factor value as a fixed value.

The Scale Chaos Factor works similarly to the Speed Chaos Factor, except that it scales particles to be larger or smaller than their originator.

The Lifespan Value Queue lets you define different lifespan levels. Original particles have a lifespan equal to the first entry in the queue. The particles that are spawned from those spawned particles last as long as the second value, and so on. To add a value to the list, enter the value in the Lifespan spinner and click the Add button. The Delete button removes the selected value from the list, and the Replace button switches value positions.

If Instanced Geometry is the selected particle type, you can fill the Object Mutation Queue with additional objects to use at each spawn level. These objects appear after a particle is spawned. To pick a new object to add to the queue, use the Pick button. You can select several objects, and they are used in the order in which they are listed.

Load/Save Presets rollout

You can save and load each particle configuration using the Load/Save Presets rollout, shown in Figure 42.14. To save a configuration, type a name in the Preset Name field and click the Save button. All saved presets are displayed in the list. To use one of these preset configurations, select it and click the Load button.

FIGURE 42.14

The Load/Save Presets rollout enables you to save different parameter settings.

Note

A saved preset is valid only for the type of particle system used to save it. For example, you cannot save a Super Spray preset and load it for a Blizzard system.

3ds Max includes several default presets that can be used as you get started. These presets include Bubbles, Fireworks, Hose, Shockwave, Trail, Welding Sparks, and Default (which produces a straight line of particles).

Using the Blizzard Particle System

The Blizzard particle system uses the same rollouts as the Super Spray system, with some slightly different options. The Blizzard emitter icon is a plane with a line pointing in the direction of the particles (similar to the Spray and Snow particle systems). Particles are emitted across the entire plane surface.

The differences between the Blizzard and Super Spray parameters include dimensions for the Blizzard icon. In the Particle Generation rollout, you'll find values for Tumble and Tumble Rate. Another difference is the Emitter Fit Planar option under the Material Mapping group of the Particle Type rollout. This option sets particles to be mapped at birth, depending on where they appear on the emitter. The other big difference is that the Blizzard particle system has no Bubble Motion rollout because snowflakes don't make very good bubbles. Finally, you'll find a different set of presets in the Load/Save Presets rollout, including Blizzard, Rain, Mist, and Snowfall.

Using the PArray Particle System

The PArray particle system is a unique particle system. It emits particles from the surface of a selected object. These particles can be emitted from the object's surface, edges, or vertices. The particles are emitted from an object separate from the emitter icon.

The PArray particle system includes many of the same rollouts as the Super Spray particle system. The PArray particle system's emitter icon is a cube with three tetrahedron objects inside it. This system has some interesting parameter differences, starting with the Basic Parameters rollout, shown in Figure 42.15.

FIGURE 42.15

The Basic Parameters rollout for the PArray particle system lets you select the location where the particles form.

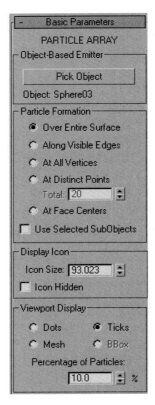

In the PArray system, you can select separate objects as emitters with the Pick Object button. You also can select the location on the object where the particles are formed. Options include Over Entire Surface, Along Visible Edges, At All Vertices, At Distinct Points, and At Face Centers. For the At Distinct Points option, you can select the number of points to use.

The Use Selected SubObject option forms particles in the locations selected with the Pick Object button, but only within the subobject selection passed up the Stack. This is useful if you want to emit particles only from a certain selection of a mesh, such as a dragon's mouth or the end of a fire hose. The other options in the PArray system's Basic Parameters rollout are the same as in the other systems.

The Particle Generation rollout includes a Divergence value. This value is the angular variation of the velocity of each particle from the emitter's normal.

Splitting an object into fragments

The Particle Type rollout for the PArray system contains a unique particle type: Object Fragments. This type breaks the selected object into several fragments. Object Fragment settings include a Thickness value. This value gives each fragment a depth. If the value is set to 0, the fragments are all single-sided polygons.

Also in the Particle Type rollout, the All Faces option separates each individual triangular face into a separate fragment. An alternative to this option is to use the Number of Chunks option, which enables you to divide the object into chunks and define how many chunks to use. A third option splits up an object based on the smoothing angle, which can be specified.

In the Material section of the Particle Type rollout, you can select material IDs to use for the fragment's Outside, Edge, and Backside.

The Load/Save Presets rollout includes a host of interesting presets, including the likes of Blast, Disintegrate, Geyser, and Comet.

Tutorial: Creating rising steam

In this tutorial, you create the effect of steam rising from a street vent. Using the PArray particle system, you can control the precise location of the steam.

To create the effect of steam rising from a vent, follow these steps:

1. Open the Street vent.max file from the Chap 42 directory in the downloaded content set.

    This file includes a street scene with a vent.

2. Select the Create→Particles→PArray menu command, and drag in the Front viewport to create the system.

3. In the Basic Parameters rollout, click the Pick Object button and select the Quadpatch object that is positioned directly beneath the vent object. Set the Particle Formation option to Over Entire Surface.

    Because the Plane object has only a single face, the particles travel in the direction of the Plane's normal.

4. In the Particle Generation rollout, set the Emit Stop value to **100** and the Life value to **60** with a Variation of **50**. Set the Particle Size value to **5.0** with a Variation of **30**.

5. In the Particle Type rollout, select the Standard Particles and the Constant options.

6. Open the Material Editor (by selecting the M key), double-click the Standard material in the Material/Map Browser, and then double-click the new node to access its parameters. Name the selected sample slot **steam**. Click the map button to the right of the Opacity value, and double-click the Mask map type from the Material/Map Browser. Double-click the new node and in the Mask Parameters rollout, click the map button and select the Noise map type. Then double-click the Mask node again, click the Mask button, and select the Gradient map type. Then, double-click the Gradient node, drag the black color swatch to the white color swatch, select Swap in the dialog box that appears, and enable the Radial option. Finally, drag the steam material to the PArray icon.

Figure 42.16 shows the steam vent at frame 60.

FIGURE 42.16

A Plane object positioned beneath the vent is an emitter for the particle system.

Using the PCloud Particle System

The PCloud particle system keeps all emitted particles within a selected volume. This volume can be a box, sphere, cylinder, or a selected object. The emitter icon is shaped as the selected volume. This particle system includes the same rollouts as the Super Spray system, with some subtle differences.

The options on the Basic Parameters rollout are unique to this system. This system can use a separate mesh object as an emitter. To select this emitter object, click the Pick Object button and select the object to use. Other options include Box, Sphere, and Cylinder Emitter. For these emitters, the Rad/Len, Width, and Height values are active for defining its dimensions.

In addition to these differences in the Basic Parameters rollout, several Particle Motion options in the Particle Generation rollout are different for the PCloud system as well. Particle Motion can be set to a random direction, a specified vector, or in the direction of a reference object's Z-axis.

The only two presets for this particle system in the Load/Save Presets rollout are Cloud/Smoke and Default.

Using Particle System Maps

Using material maps on particles is another way to add detail to a particle system without increasing its geometric complexity. You can apply all materials and maps available in the Material Editor to particle systems. To apply them, select the particle system icon and click the Assign Material to Selection button in the Material Editor.

For more details on using maps, see Chapter 18, "Adding Material Details with Maps."

Two map types are specifically designed to work with particle systems: Particle Age and Particle MBlur. You can find these maps in the Material/Map Browser. You can access the Material/Map Browser using the Rendering→Material/Map Browser menu command or from the Material Editor by clicking the Get Material button.

Using the Particle Age map

The Particle Age map parameters include three different colors that can be applied at different times, depending on the Life value of the particles. Each color includes a color swatch, a map button, an Enable check box, and an Age value for when this color should appear.

This map typically is applied as a Diffuse map because it affects the color.

Using the Particle MBlur map

The Particle MBlur map changes the opacity of the front and back of a particle, depending on the color values and sharpness specified in its parameters rollout. This results in an effect of blurred motion if applied as an Opacity map.

Note
MBlur does not work with the Constant, Facing, MetaParticles, or PArray object fragments.

Tutorial: Creating jet engine flames

The Particle Age and MBlur maps work well for adding opacity and colors that change over time, such as hot jets of flames, to a particle system.

To create jet engine flames, follow these steps:

1. Open the Jet airplane flames.max file from the Chap 42 directory in the downloaded content set.

   This file includes a spaceship model.

2. Select the Create→Particles→Super Spray menu command, and drag the icon in the viewport. Rotate and position the emitter icon so that its origin is right in one of the jet's exhaust ports and the directional vector is pointing outward, away from the jet.

3. Open the Modify panel, and in the Basic Parameters rollout, set the Off Axis Spread value to **20** and the Off Plane Spread value to **90**.

   These settings focus the flames shooting from the jet's exhaust.

4. In the Particle Generation rollout, select the Use Rate option, set the Rate to **1000**, set the Speed to **2**, set the Emit Stop to **100**, the Life value to **30**, and the Particle Size to **5.0**.

5. In the Particle Type rollout, select the Standard Particles option and select the Sphere type.

6. Open the Material Editor by pressing the M key, double-click the Standard material in the Material/Map Browser, and then double-click the new node to access its parameters. Name this material **Jet's Exhaust**, and click the map button to the right of the Diffuse color swatch.

7. From the Material/Map Browser that opens, select the Particle Age map. Double-click this node and in the Particle Age Parameters rollout, select dark red, dark yellow, and black as colors for the ages 0, 50, and 100.

   You should use darker colors because the scene is lighted.

8. Double-click the Standard node again to access the base material's parameters again, and then click the map button to the right of the Opacity setting. Select the Particle MBlur map.

9. Double-click the Particle MBlur map node and in the Particle MBlur Parameters rollout, make Color #1 white and Color #2 black with a Sharpness value of **0.1**. Then apply this material onto the particle system's icon.

10. With the Shift key held down, drag the Super Spray icon in the Front viewport to all three exhaust ports.

Figure 42.17 shows the jet at frame 30 with its fiery exhaust.

FIGURE 42.17

Realistic jet flames created using the Particle Age and MBlur maps

Controlling Particles with Particle Flow

The Particle Flow Source option in the Create→Particles menu is more than just a fancy particle system: it is an entire interface and paradigm that you can use to control particles throughout their life. This is accomplished using the Particle View window, where you can visually program the flow of particles.

The Particle View window

The Particle View window, shown in Figure 42.18, is opened by clicking the Particle View button in the Setup rollout when a Particle Flow Source icon is selected or by pressing the keyboard shortcut, 6. Pressing 6 opens the Particle View window even if its icon isn't selected.

FIGURE 42.18

The Particle View window lets you program the flow of particles using a visual editor.

Event display

Navigation pane

Parameters

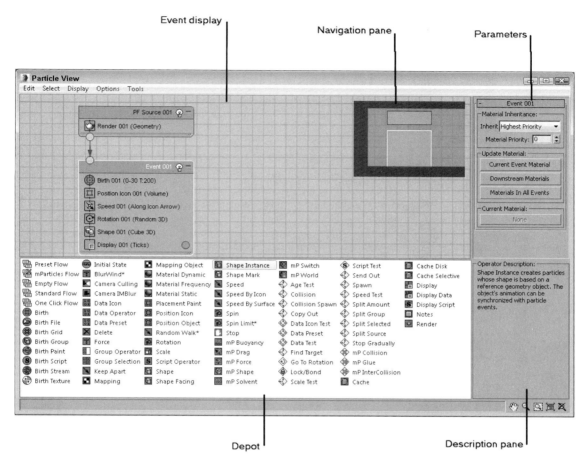

Depot

Description pane

The Particle View window is divided into four panes. The Event display contains all the event nodes. These nodes contain individual actions, and nodes can be wired to one another to define the flow. The Parameter pane in the upper-right corner displays the parameters in rollouts for the action selected in the Event display. The Depot pane is below the Event display pane and contains all the possible actions that can work with particles. In the lower right is a Description pane that offers a brief description of the action that is selected in the Depot pane. Except for the Event display pane, you can turn off the other panes using the Display menu.

In the lower-right corner of the window are several display tools. These tools can be used to navigate the Event display. The Display tools include Pan, Zoom, Zoom Region, Zoom Extents, and No Zoom. The No Zoom button eliminates all zooming and displays the nodes at their normal size.

Tip

With the Particle View window open, you can Pan the Event display by dragging the scroll wheel on the mouse.

The Standard Flow

When the Particle Flow (PF) Source icon is created and the Particle View window is first opened, two nodes appear in the Event display. These nodes are called a Standard Flow. These two nodes identify the Particle Flow Source and are wired to an event node containing a Birth action. The Birth action defines when particles Start and Stop and the Amount or Rate. You can change these values by selecting the Birth event in the Event display and changing its parameters in the rollout that appears in the Parameters pane.

1015

Several other default events appear in this default event node, including Position Icon, Speed, Rotation, Shape, and Display. Each of these events has parameters that you can alter that appear in the Parameters pane when the action is selected, and each event and action is identified with a number (such as 01) that appears next to its name. Each new event or action gets an incremented number. You also can rename any of these events if you right-click the event and select Rename from the pop-up menu.

A new Standard Flow can be created using the Edit→New→Particle System→Standard Flow menu command. An Empty Flow includes only the PF Source node. When a new Standard Flow (or Empty Flow) is created in the Particle Flow window, a PF Source icon is added to the viewports. And if the PF Source icon is deleted in the viewports, the associated event nodes are also deleted in the Particle Flow window.

Working with Actions

The Depot pane includes all the different actions that can affect particles. These actions can be categorized into Birth actions (identified with green icons), Operator actions (identified with blue icons), Test actions (identified with yellow icons), and Miscellaneous actions (which are also blue). Each of these categories also can be found in the Edit→New menu.

New events can be dragged from the Depot pane to the Event display. If you place them within an existing node, a blue line appears at the location where the action will appear when dropped. The particles are affected within a node in the order, from top to bottom, in which they appear. If a new action from the Depot is dragged over the top of an existing action in the Event display, a red line appears on top of the existing action. When you drop the new action, it replaces the existing action.

Actions also can be dropped away from an event node, making it a new event node. If an event is a new node, you can wire certain actions that are tested as true. For example, if you have a set of particles with a random speed assigned, you could use a Test event to determine which particles are moving faster than a certain speed and wire those particles to change size using the new event node.

Clicking the action's icon disables the action.

Combining Particle Flow with MassFX

Included within Particle Flow is a new set of particles known as mParticles. These particles differ from the normal set of particles in that they can take advantage of the MassFX simulation system to respond to real-world physics. mParticles are easily added to a simulation using the mParticles Flow node. These particles are already configured to work with MassFX.

To control the position of the mParticles, you can use the new Birth Grid, Birth Group and Birth Stream operators. Birth Grid places the mParticles on a grid that you can position. Care should be taken to not let the mParticle grids overlap one another. The Birth Group operator creates mParticles based on an existing mesh object and the Birth Stream operator lets you position the mParticles to start from a specific location.

Once mParticles are added to the simulation, you can use several different operators to apply forces to the areas that the mParticles move through. These forces included Drag, Buoyancy and Force. There are also operators for gluing mParticles together (Glue) and breaking them apart again (Solvent), as well as operators for detecting collisions and intercollisions.

There are also three modifiers that work with mParticles. The PFlow Collision Shape modifier lets standard mesh objects act as deflecting objects for mParticles. You can also hide and deform geometry faces as they interact with mParticles with the Particle Face Creator and Particle Skinner modifiers.

Tutorial: Creating an avalanche

One of the cautions that come with particles is that if you're not careful, you can quickly spawn enough particles to bring any system to its knees. Using the particle spawn feature is one of the worst offenders. For this tutorial, you use a Collision Spawn action to create an avalanche, but you must be sure to keep the number of spawned particles in check.

To use Particle Flow to make an avalanche effect, follow these steps:

1. Open the Avalanche.max file from the Chap 42 directory in the downloaded content set.

    This file includes a simple hillside covered with snow.

2. Select Create→Space Warps→Deflectors→SOmniFlect, and drag in the Top viewport to create a spherical deflector that covers the entire hill. Click the Select and Scale tool, and in the Y-axis, scale the SOmniFlect sphere down in the Left viewport until it is roughly the same thickness as the hill object. Then rotate the SOmniFlect until it is aligned parallel to the hill. This deflector is going to keep the snowball particles on top of the hill.

3. Select Create→Particles→Particle Flow Source, and drag in the Top viewport to create the emitter icon. Scale and rotate the emitter icon so it is positioned inside the SOmniFlect object on the uphill side pointing downhill. The emitter can be fairly small so it fits inside the deflector.

4. With the Particle Flow Source still selected, click the Particle View button in the Modify panel of the Command Panel to open the Particle View window (or press 6). In the Event 01 box, select the Shape01 action, and in the rollout that appears to the right, set the 3D Shape to **Sphere** and the Size to **5.0**. Select the Speed01 action, and set the Speed value to **100**. Click the colored dot to the right of the Display01 action, and select white as the new color.

5. Select the Collision Spawn action from the Depot pane, and drop it in Event01 beneath the Display01 action. Then select the Collision Spawn action, click the By List button in the Parameters pane, and select the SOmniFlect object. Then disable the Test True for Parent and Spawn Particles option, enable the Spawn On Each Collision option, and set the Offspring value to **10**.

Figure 42.19 shows an avalanche of snowballs as it rages down a snowy hillside.

FIGURE 42.19

Using the Collision Spawn and a well-placed deflector, you can create an avalanche effect.

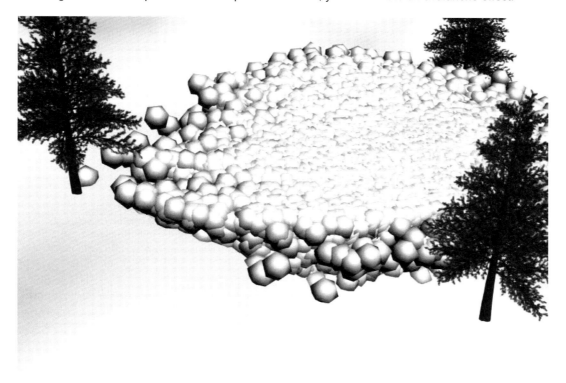

Using Particle Flow Helpers

In addition to the Standard Flow event, several other actions create icons in the viewports that are controlled using the action parameters. One of these helpers appears when the Find Target action is added to an event node. This helper is a simple sphere, but all particles in the scene are attracted to it. It also can be animated.

The Speed By Icon action creates an icon that forces particles to follow its trajectory path.

Wiring events

Each new event node that is created has an input that extends from the upper-left corner of the node, and each Test event that is added has an output that extends to the left of its icon, as shown in Figure 42.20. Test action outputs can be wired to event inputs by dragging from one to the other. The cursor changes when it is over each.

FIGURE 42.20

Event outputs can be wired to event inputs.

Event output

Event input

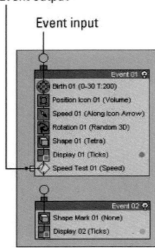

Once wired, all particles that are tested to be true are transferred to the new event node and are subject to the actions in the wired event node.

Tutorial: Moths chasing a light

Another cool feature that the Particle Flow interface makes available is the ability to have particles chase a target object. In this example, you use the Target event to make some annoying bugs follow a lantern's light.

To use Particle Flow to make several bugs chase a light, follow these steps:

1. Open the Moths chasing light.max file from the Chap 42 directory in the downloaded content set.

    This file includes a simple lantern created from primitives that is suspended from a chain and animated rocking back and forth. The file also includes a simple moth.

2. Select Create→Particles→Particle Flow Source, and drag in the Front viewport to create the emitter icon. Click the Particle View button to open the Particle View interface.

3. In the Event01 node, select the Birth01 action and set the Emit Stop to **100** with an Amount of **50**.

4. Drag the Position Object action from the Depot pane, and drop it on top of the Position Icon action to replace it. Select the new Position Object action, click the By List button under the Emitter Objects list in the Parameters pane, and select the Sphere01 object.

    This sphere surrounds the lantern and is the source of the moths. It has a material with an Opacity setting of 0 applied so that it is not visible in the scene.

5. Select the Rotation action in the Event01 node, and change the Orientation Matrix option to Speed Space Follow.

    This rotates the moths as they follow the swinging lantern.

6. Drag the Shape Instance action from the Depot pane, and drop it on top of the Shape action to replace it. Select the new Shape Instance action, click the Particle Geometry Object button in the Parameters pane, and select the moth object in the viewports.

7. Drag the Find Target action from the Depot pane, and drop it at the bottom of the Event01 node. This adds a new Find Target icon to the viewports. Select the Find Target icon in the viewports, and move it to the lantern flame's position. Select Group→Attach, and click the lantern object to add the Find Target icon to the lantern group. This makes the target move with the lantern. Back in the Particle Flow interface, select the Find Target event. In the Parameters pane, enable the Use Cruise Speed option, and then set the Speed to **1000** with a Variation of **50** and the Accel Limit to **5000** with an Ease In % of **50**. You also need to enable the Follow Target Animation option.

8. Drag the Material Dynamic action icon from the Depot pane to the Event Display pane, and drop it outside the Event01 node to create a new node called Event02. In the Parameters pane, enable the Assign Material option and click the material button. Select the Fire scene material from the Material/Map Browser. Then drag a wire from the Find Target action to Event02.

9. Drag the Age Test action from the Depot pane, and drop it below the Material Dynamic action. Then select Event Age from the drop-down list in the Parameters pane, and set the Test Value to **2**.

10. Finally, drag the Delete action from the Depot pane, and drop it away from the other events. Then wire the Age Test action to the new event node, and select the Selected Particles Only option in the Parameters pane.

Figure 42.21 shows several moths eagerly pursuing the swinging lantern.

FIGURE 42.21

All the moths in this scene are particles and are following a target linked to the lantern.

Debugging Test Actions

Any test action can be made to return a True or False value if you click the left (for True) or right (for False) side of the test action's icon in the Particle View interface. This lets you debug the particle flow. Tests set to be true show an icon with a green light, and tests set to be false show a red light icon.

Tutorial: Firing at a fleeing spaceship

Well, it is about time for a space scene, and we all know that lots of particles float around out in space—stars, asteroids, comets, and so forth. It's all great stuff to animate. For this scene, you use the Particle Flow feature to fire laser blasts on a fleeing spaceship.

To use Particle Flow to fire on a fleeing spaceship, follow these steps:

1. Open the Fleeing spaceship.max file from the Chap 42 directory in the downloaded content set.

   This file includes a spaceship model that has been animated as if it were fleeing.

2. Select Create→Particles→Particle Flow Source, and drag in the Front viewport to create the emitter. With the Select and Move (W) tool selected, move and rotate the emitter until it is aligned with the end of the laser gun. Then click the Select and Link button, and drag from the emitter to the gun object to bind the emitter to the gun.

   The emitter now moves with the animated laser gun.

3. With the Particle Flow Source icon selected, open the Modify panel and click the Particle View button in the Setup rollout (or press the 6 key) to open the Particle View interface.

4. In the Event01 node, select the Birth 03 event; in the Parameters panel, set the Emit Stop to **100** and the Amount to **50**. This produces a laser blast every two frames. Click the blue dot in the lower-right corner of the Event node, and select a red color from the Color Selector that appears.

5. Select Create→Standard Primitives→Cylinder, and drag in the Front viewport to create a Cylinder object. In the Hierarchy panel, select the Affect Pivot Only button and rotate the Cylinder's Pivot Point until its Y-axis points at the spaceship, then exit the Affect Pivot Only tool. Then in the Particle View window, drag the Shape Instance event from the depot and drop it on top of the Shape event in the Event 01 node. This replaces the Shape event with a Shape Instance event. In the Parameters rollout, click the Particle Geometry Object button and select the Cylinder object. Select the Rotation event, and delete it with the Delete key.

6. Select Create→Space Warps→Deflectors→SOmniFlect, and drag in the Top viewport to create a spherical deflector that encompasses the spaceship. Click the Select and Non-Uniform Scale button, and scale the X- and Y-axes until the deflector just fits around the spaceship. Then link the deflector Space Warp to the spaceship.

7. In the Particle View window, drag the Collision event from the depot to the bottom of the Event 01 node. In the Parameters rollout, below the Deflectors list, click the Add button and select the SOmniFlect01 object surrounding the spaceship.

8. Drag a Spawn event from the depot to the event display, and then connect the Collision event by dragging from its output to the input of the new Spawn event. Then click the color for the new event particles, and change it to orange. Select the Spawn 01 event, enable the Delete Parent option, and set the Offspring to **200** and the Variation % to **20**. Then set the Inherited % to **50** with a Variation % of **30**.

9. Drag the Delete event to the bottom of the Event 02 node, select the By Particle Age option, and set the Life Span to **20** and the Variation to **30**. Drag a Shape event to the Event 02 node, and set the 3D Shape to Sphere and the Size to **0.5**.

10. Finally, click the Play button to see the resulting animation.

Figure 42.22 shows the final Particle View flow, and Figure 42.23 shows a frame of the animation in the viewport. You can still do several things to improve this animation, such as adding a Glow Render Effect to the laser blasts and using the Particle Age material with some transparency to improve the explosion's look.

FIGURE 42.22

The Particle View window shows the flow of the particles in the animation.

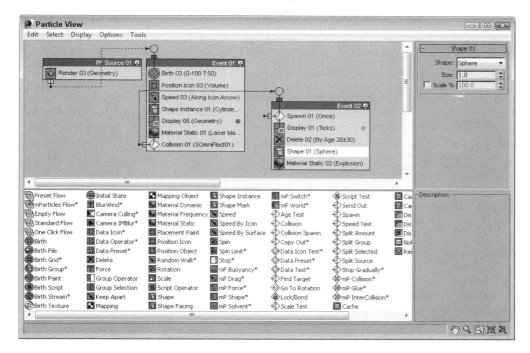

FIGURE 42.23

The spaceship is trying to outrun the laser blasts.

Tutorial: Creating a black hole using Particle Flow

Particle systems are one of the best sources for special effects, and with the Particle Flow interface, you can control them more easily. In this tutorial, we create an array of particle systems and have all their particles flow together to form a black hole.

To create a black hole using the Particle Flow interface, follow these steps:

1. Select Create→Particles→Particle Flow Source, and drag in the Left viewport to create the icon. The icon's direction arrow should point to the right in the Top viewport. With the icon selected, click the Affect Pivot Only button in the Hierarchy panel and move the icon to the origin location in the Top viewport. Then click the Affect Pivot Only button again to disable pivot mode. This centers the Particle Flow's icon pivot in the center of the scene.

2. Select Create→Standard Primitives→Sphere, and create a small sphere in the Top viewport at the grid origin in the center of the viewport at the same location as the Particle Flow icon's pivot.

3. Click the Auto Key button at the bottom of the interface, and drag the Time Slider to frame 100. Then select the sphere object in the Left viewport, and move it downward a little. Then select the Particle Flow Source icon in the Top viewport, and rotate it about 60 degrees. Then disable the Auto Key button to leave key mode. This makes the descending funnel for the black hole.

4. With the Particle Flow icon selected, click the Particle View button in the Modify panel (or press the 6 key) to open the Particle Flow window. Select the Birth event, and change the Emit Stop value to **100** and the Amount to **200**. Select the Shape event, and change the 3D Shape to Sphere and the Size value to **2.0**. In the Display event, change the Visible % to **10**.

5. Drag the Speed by Surface event from the Depot window, and drop it on top of the Speed event in the Event node. Select the Control Speed Continuously option from the drop-down list, enable the Speed

option, and set the Speed value to **100** and the Speed Variation to **20**. Then click the Add button, and select the small Sphere object in the Top viewport; you also can select the sphere from a dialog box using the By List button.

6. Select the Particle Flow icon; with the Shift key held down, rotate the icon about 52 degrees and enter **6** for the Number of Copies in the Clone Options dialog box that appears. This creates 6 Particle Flow icons that all feed particles into the center of the black hole.

This creates particle flow icons that surround and feed the black hole.

Figure 42.24 shows the resulting black hole with no materials or Render Effects applied after the Particle Amount is set to 500. For materials, I recommend using the Particle Age map along with a high Self-Illumination value and a Glow Render Effect.

This same structure can be modified to produce a tornado or hurricane.

FIGURE 42.24

One spiraling black hole accomplished with the Particle Flow interface

Summary

This chapter presented particle systems and showed how you can use them. The chapter also took a close look at each system, including Spray, Snow, Super Spray, Blizzard, PArray, and PCloud. This chapter covered these topics:

This chapter presented the Particle Flow interface and showed how it can be used to control the movement of particles. This chapter covered these topics:

* Learning about the various particle systems

* Creating a particle system for producing rain and snow

* Using the Super Spray particle system

* Working with MetaParticles

* Specifying an object to use as a particle and an object to use as an emitter

* Using the PArray and PCloud particle systems

* Using the Particle Age and Particle MBlur maps on particles

* Controlling and programming the flow of particles with the Particle Flow window

* Adding actions to the flow

* Responding to events

* Debugging the particle flow

The next chapter explains working with Space Warps to add forces to a scene. These forces can be used to control the motion of objects in the scene.

Chapter 43

Using Space Warps

IN THIS CHAPTER

Creating and binding Space Warps to objects

Understanding the various Space Warp types

Working with Space Warps and particle systems

Space Warps sound like a special effect from a science fiction movie, but actually they are nonrenderable objects that let you affect another object in many unique ways to create special effects.

You can think of Space Warps as the unseen forces that control the movement of objects in the scene such as gravity, wind, and waves. Several Space Warps, such as Push and Motor, deal with dynamic simulations and can define forces in real-world units. Some Space Warps can deform an object's surface; others provide the same functionality as certain modifiers.

Space Warps are particularly useful when combined with particle systems. This chapter includes some examples of Space Warps that have been combined with particle systems.

Creating and Binding Space Warps

Space Warps are a way to add forces to the scene that can act on scene objects. Space Warps are not renderable and must be bound to an object to have an effect. A single Space Warp can be bound to several objects, and a single object can be bound to several Space Warps.

In many ways, Space Warps are similar to modifiers, but modifiers typically apply to individual objects, whereas Space Warps can be applied to many objects at the same time and are applied using World Space Coordinates. This ability to work with multiple objects makes Space Warps the preferred way to alter particle systems and to add forces to dynamic hair and cloth systems.

Another nice feature of Space Warps is that they can be animated. Moving the Wave Space Warp gizmo over the surface of a bound plane object lets you control where the waves appear in the plane object. You also can animate its parameters to gradually increase the size of the waves.

Creating a Space Warp

Space Warps are found divided between several categories in the Objects menu including Forces, Space Warps and Deflectors. Selecting a Space Warp opens the Space Warps category (the icon is three wavy lines) in the Create panel. From the subcategory drop-down list, you can select from several different subcategories. Each subcategory has buttons to enable several different Space Warps, or you can select them using the Objects menu command. To create a Space Warp, click a button or select a menu option and then click and drag in a viewport.

When a Space Warp is created, a gizmo is placed in the scene. This gizmo can be transformed as other objects can: by using the standard transformation buttons. The size and position of the Space Warp gizmo often affect its results. After a Space Warp is created, it affects only the objects to which it is bound.

Binding a Space Warp to an object

A Space Warp's influence is felt only by its bound objects, so you can selectively apply gravity only to certain objects. For example, binding gravity to the ground plane wouldn't be helpful. The Bind to Space Warp button is on the main toolbar next to the Unlink Selection button. After clicking the Bind to Space Warp button, drag from the Space Warp to the object to which you want to link it, or vice versa.

All Space Warp bindings appear in the Modifier Stack. You can right-click on the binding in the Modifier Stack to copy and paste Space Warps between objects, or you can drag the binding from the Modifier Stack and drop it on other scene objects.

Some Space Warps can be bound only to certain types of objects. Each Space Warp has a Supports Objects of Type rollout that lists the supported objects. If you're having trouble binding a Space Warp to an object, check this rollout to see whether the object is supported.

Understanding Space Warp Types

Just as many different types of forces exist in nature, many different Space Warp types exist. These appear in several different subcategories, based on their function. The subcategories are Forces, Deflectors, Geometric/Deformable, Modifier-Based, and Particles & Dynamics.

Force Space Warps

The Forces subcategory of Space Warps is mainly used with particle systems and dynamic simulations. Space Warps in this subcategory include Motor, Push, Drag, Vortex, Path Follow, PBomb, Displace, Gravity, Wind, and Motion Field. Figure 43.1 shows the gizmos for these Space Warps.

FIGURE 43.1

The Force Space Warps: Motor, Push, Drag, Vortex, Path Follow, PBomb, Displace, Gravity, Wind, and Motion Field

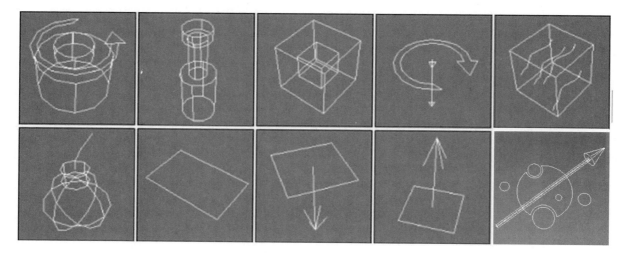

Motor

The Motor Space Warp applies a rotational torque to objects. This force accelerates objects radially instead of linearly. The Basic Torque value is a measurement of torque in Newton-meters, foot-pounds, or inch-pounds.

The On Time and Off Time options set the frames where the force is applied and disabled, respectively. Many of the Space Warps have these same values.

The Feedback On option causes the force to change as the object's speed changes. When this option is off, the force stays constant. You also can set Target Revolution units in revolutions per hour (RPH), revolutions per minute (RPM), or revolutions per second (RPS), which is the speed at which the force begins to change if the Feedback option is enabled. The Reversible option causes the force to change directions if the Target Speed is reached, and the Gain value is how quickly the force adjusts.

The motor force also can be adjusted with Periodic Variations, which cause the motor force to increase and then decrease in a regular pattern. You can define two different sets of Periodic Variation parameters: Period 1, Amplitude 1, Phase 1; and Period 2, Amplitude 2, Phase 2.

For particle systems, you can enable and set a Range value. The Motor Space Warp doesn't affect particles outside this distance. At the bottom of the Parameters rollout, you can set the size of the gizmo icon. You can find this same value for all Space Warps.

Figure 43.2 shows the Motor Space Warp twisting the particles being emitted from the Super Spray particle system in the direction of the icon's arrow.

FIGURE 43.2

You can use the Motor Space Warp to apply a twisting force to particles and dynamic objects.

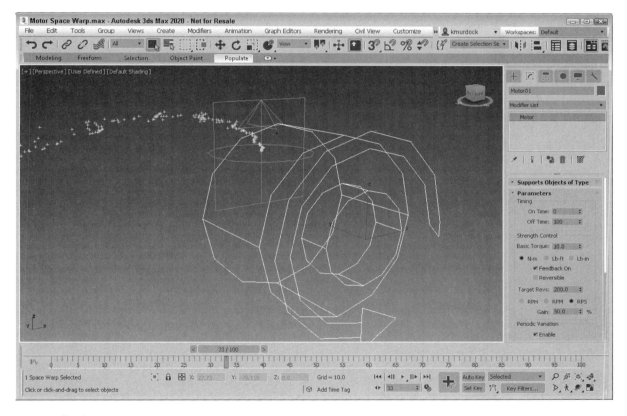

Push

The Push Space Warp accelerates objects in the direction of the Space Warp's icon from the large cylinder to the small cylinder. Many of the parameters for the Push Space Warp are similar to those for the Motor Space Warp. Using the Parameters rollout, you can specify the force Strength in units of Newtons or pounds.

The Feedback On option causes the force to change as the object's speed changes, except that it deals with Target Speed instead of Target Revolution like the Motor Space Warp does.

The push force also can be set to include Periodic Variations that are the same as with the Motor Space Warp. Figure 43.3 shows the Push Space Warp pushing the particles being emitted from the Super Spray particle system.

FIGURE 43.3

You can use the Push Space Warp to apply a controlled force to particles and dynamic objects.

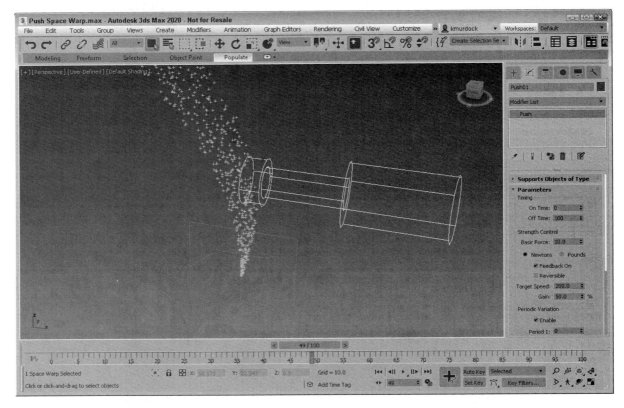

Vortex

You can use the Vortex Space Warp on particle systems to make particles spin around in a spiral like going down a whirlpool. You can use the Timing settings to set the beginning and ending frames where the effect takes place.

You also can specify Taper Length and Taper values, which determine the shape of the vortex. Lower Taper Length values wind the vortex tighter, and the Taper values can range between 1.0 and 4.0 and control the ratio between the spiral diameter at the top of the vortex versus the bottom of the vortex.

The Axial Drop value specifies how far each turn of the spiral is from the adjacent turn. The Damping value sets how quickly the Axial Drop value takes effect. The Orbital Speed is how fast the particles rotate away from the center. The Radial Pull value is the distance from the center of each spiral path that the particles can rotate. If the Unlimited Range option is disabled, Range and Falloff values are included for each setting. Both Orbital Speed and Radial Pull also have a Damping value. You also can specify whether the vortex spins clockwise or counterclockwise.

Figure 43.4 shows a Vortex Space Warp that is bound to a particle system.

FIGURE 43.4

You can use the Vortex Space Warp to force a particle system into a spiral like a whirlpool.

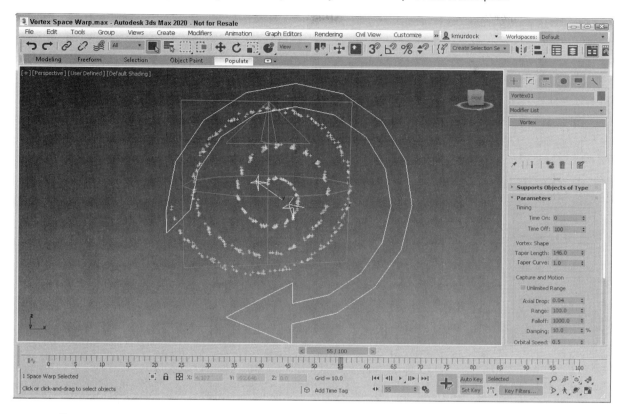

Drag

Drag is another common force that can be simulated with a Space Warp. The Drag Space Warp can be Linear, Spherical, or Cylindrical. This Space Warp causes particle velocity to be decreased, such as when simulating air resistance or fluid viscosity. Use the Time On and Time Off options to set the frame where the Space Warp is in effect.

For each of the Damping shape types—Linear, Spherical, and Cylindrical—you can set the drag, which can be along each axis for the Linear shape or in the Radial, Tangential, and Axial direction for the Spherical and Cylindrical shapes. If the Unlimited Range option is not selected, then the Range and Falloff values are available.

Figure 43.5 shows a Drag Space Warp surrounding a particle system. Notice how the particles are slowed and moved to the side as they pass through the Drag space warp.

FIGURE 43.5

You can use the Drag Space Warp to slow the velocity of particles.

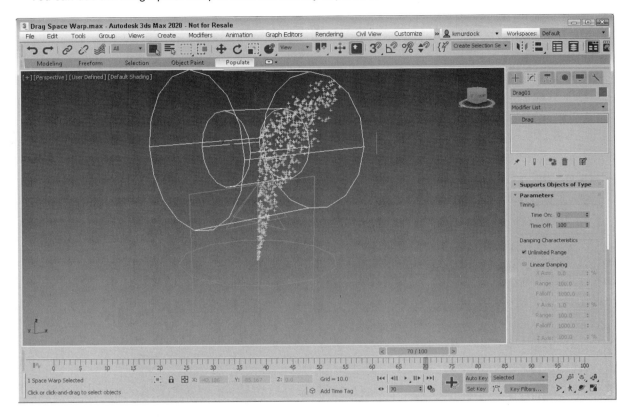

PBomb

The PBomb (particle bomb) Space Warp was designed specifically for the PArray particle system, but it can be used with any particle system. To blow up an object with the PBomb Space Warp, create an object, make it a PArray emitter, and then bind the PBomb Space Warp to the PArray.

You can find more information on the PArray particle system in Chapter 42, "Creating Particles and Particle Flow."

Basic parameters for this Space Warp include three blast symmetry types: Spherical, Cylindrical, and Planar. You also can set the Chaos value as a percentage, which defines how erratically the pieces move.

In the Explosion Parameters section, the Start Time is the frame where the explosion takes place, and the Duration defines how long the explosion forces are applied. The Strength value is the power of the explosion.

A Range value can be set to determine the extent of the explosion. It is measured from the center of the Space Warp icon. If the Unlimited Range option is selected, the Range value is disabled. The Linear and Exponential options change how the explosion forces die out. The Range Indicator option displays the effective blast range of the PBomb.

Figure 43.6 shows a box selected as an emitter for a PArray. The PBomb is bound to the PArray and not to the box object. The Speed value for the PArray has been set to 0, and the Particle Type is set to Fragments. Notice that the PBomb's icon determines the center of the blast.

FIGURE 43.6

You can use the PBomb Space Warp with the PArray particle system to create explosions.

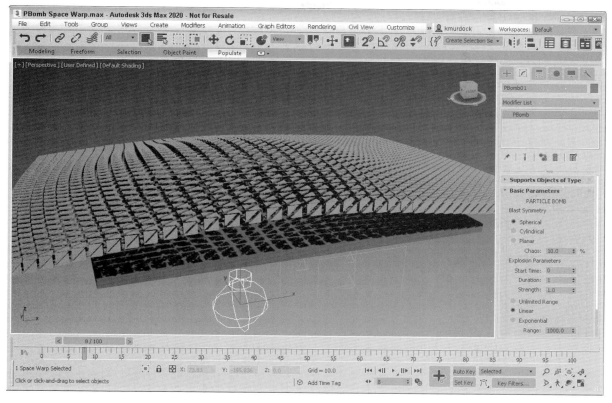

Path Follow

The Path Follow Space Warp causes particles to follow a path defined by a spline. The Basic Parameters rollout for this Space Warp includes a Pick Shape Object button for selecting the spline path to use. You also can specify a Range value or the Unlimited Range option. The Range distance is measured from the path to the particles.

The Path Follow Space Warp is similar to the Path Constraint, which is discussed in Chapter 35, "Animating with Constraints and Simple Controllers."

In the Motion Timing section, the Start Frame value is the frame where the particles start following the path, the Travel Time is the number of frames required to travel the entire path, and the Last Frame is where the particles no longer follow the path. There is also a Variation value to add some randomness to the movement of the particles.

The Basic Parameters rollout also includes a Particle Motion section with two options for controlling how the particles proceed down the path: Along Offset Splines and Along Parallel Splines. The first causes the particles to move along splines that are offset from the original, and the second moves all particles from their initial location along parallel path splines. The Constant Speed option makes all particles move at the same speed.

Also in the Particle Motion section is the Stream Taper value. This value is the amount by which the particles move away from the path over time. Options include Converge, Diverge, or Both. Converging streams move all particles closer to the path, and diverging streams do the opposite. The Stream Swirl value is the number of spiral turns that the particles take along the path. This swirling motion can be Clockwise, Counterclockwise, or Bidirectional. The Seed value determines the randomness of the stream settings.

Figure 43.7 shows a Path Follow Space Warp bound to a Super Spray particle system. A Helix shape has been selected as the path.

FIGURE 43.7

A Path Follow Space Warp bound to an emitter from the Super Spray particle system and following a Helix path

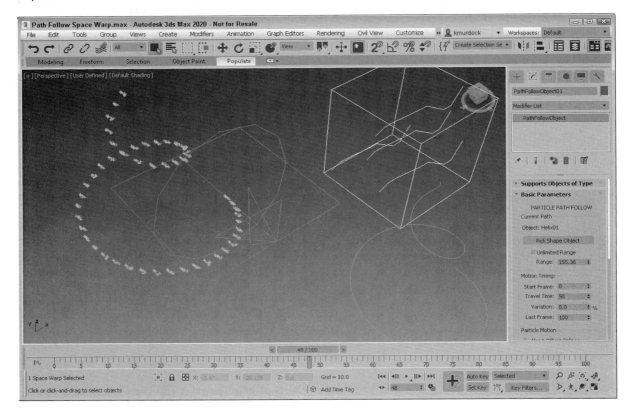

Gravity

The Gravity Space Warp adds the effect of gravity to a scene. This causes objects to accelerate in the direction specified by the Gravity Space Warp, like the Wind Space Warp. The Parameters rollout includes Strength and Decay values. Additional options make the gravity planar or spherical. You can turn on the Range Indicators to display a plane or sphere where the gravity is half its maximum value.

Wind

The Wind Space Warp causes objects to accelerate. The Parameters rollout includes Strength and Decay values. Additional options make the gravity planar or spherical. The Turbulence value randomly moves the objects in different directions, and the Frequency value controls how often these random turbulent changes occur. Larger Scale values cause turbulence to affect larger areas, but smaller values are wilder and more chaotic.

You can turn on the Range Indicators just like the Gravity Space Warp. Figure 43.8 shows the Wind Space Warp pushing the particles being emitted from a Super Spray particle system.

FIGURE 43.8

You can use the Wind Space Warp to blow particles and dynamic objects.

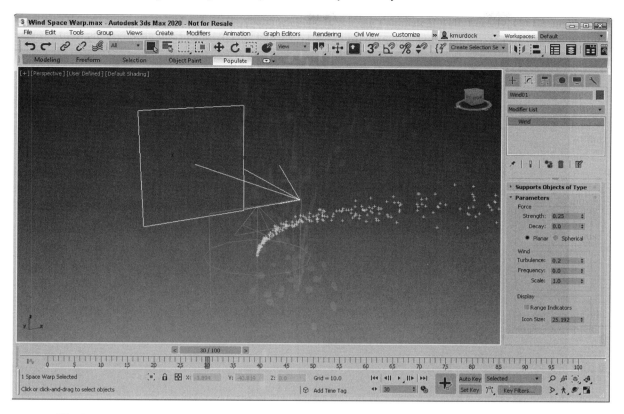

Displace

The Displace Space Warp is like a force field: it pushes objects and is useful when applied to a particle system. It also can work on any deformable object in addition to particle systems. The strength of the displacement can be defined with Strength and Decay values or with a grayscale bitmap.

The Strength value is the distance that the geometry is displaced and can be positive or negative. The Decay value causes the displacement to decrease as the distance increases. The Luminance Center is the grayscale point where no displacement occurs; any color darker than this center value is moved away, and any brighter areas move closer.

The Bitmap and Map buttons let you load images to use as a displacement map; the amount of displacement corresponds with the brightness of the image. The Bitmap option loads an image file, but the Map button can load any map type from the Material Editor. A Blur setting blurs the image. You can apply these maps with different mapping options, including Planar, Cylindrical, Spherical, and Shrink Wrap. You also can adjust the Length, Width, and Height dimensions; and the U, V, and W Tile values.

The Displace Space Warp is similar in function to the Displace modifier. The Displace modifier is discussed in Chapter 18, "Adding Material Details with Maps."

Figure 43.9 shows two Displace Space Warps with opposite Strength values.

FIGURE 43.9

The Displace Space Warp can raise or indent the surface of a patch object.

Motion Field

The Motion Field Space Warp is used in fluid simulations. It creates a force field where you can define both magnitude and direction. There is also a Turbulence component that is helpful.

The Motion Field Space Warp is discussed in Chapter 48, "Creating Fluid Simulations."

Deflector Space Warps

The Deflectors subcategory of Space Warps includes POmniFlect, SOmniFlect, SDeflector, UOmniFlect, UDeflector, and Deflector. You use them all with particle systems. This category includes several different types of deflectors starting with P, S, and U. The difference between these types is their shape. P-type (planar) deflectors are box-shaped, S-type (spherical) deflectors are spherical, and U-type (universal) deflectors include a Pick Object button that you can use to select any object as a deflector.

Tip

When you use a custom deflector object, the number of polygons makes a big difference. The deflector object uses the normal to calculate the bounce direction, so if the deflector object includes lots of polygon faces, the system slows way down. A solution to this, especially if you're using a simple plane deflector, because all its polygon normals are the same, anyway, is to use a simplified proxy object as the deflector and hide it so the particles look like they are hitting the complex object.

POmniFlect, SOmniFlect, and UOmniFlect

The POmniFlect Space Warp is a planar deflector that defines how particles reflect and bounce off other objects. The SOmniFlect Space Warp is just like the POmniFlect Space Warp, except that it is spherical in shape. The UOmniFlect Space Warp is another deflector, but this one can assume the shape of another object using the Pick Object button in the Parameters rollout. Its Parameters rollout includes a Timing section with Time On and Time Off values and a Reflection section.

The difference between this type of Space Warp and the other deflector Space Warps is the addition of refraction. Particles bound to this Space Warp can be refracted through an object. The values entered in the Refraction section of the Parameters rollout change the velocity and direction of a particle. The Refracts value is the percentage of particles that are refracted. The Pass Vel (velocity) is the amount that the particle speed changes when entering the object; a value of 100 maintains the same speed. The Distortion value affects the angle of refraction; a value of 0 maintains the same angle, and a value of 100 causes the particle to move along the surface of the struck object. The Diffusion value spreads the particles throughout the struck object. You can vary each of these values by using its respective Variation value.

Note

If the Refracts value is set to 100 percent, no particles are refracted.

You also can specify Friction and Inherit Velocity values. In the Spawn Effects Only section, the Spawns and Pass Velocity values control how many particle spawns are available and their velocity upon entering the struck object. Figure 43.10 shows each of these Space Warps bound to a Super Spray particle system. The Reflects value for each of the Space Warps is set to 50, and the remaining particles are refracted through the Space Warp's plane. Notice that the particles are also reflecting off the opposite side of the refracting object.

FIGURE 43.10

The POmniFlect, SOmniFlect, and UOmniFlect Space Warps reflecting and refracting particles emitted from the Super Spray particle system

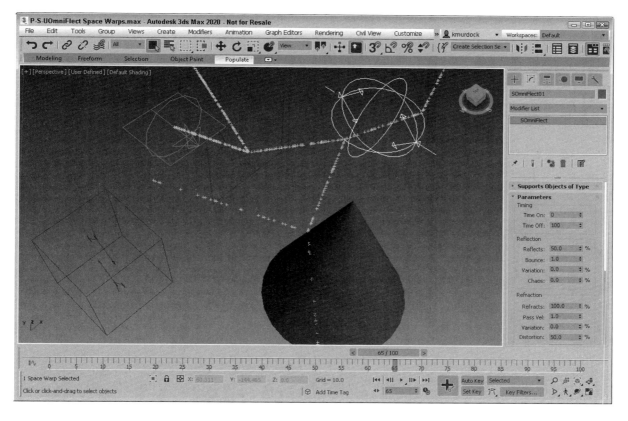

Deflector, SDeflector, and UDeflector

The Deflector and SDeflector Space Warps are simplified versions of the POmniFlect and SOmniFlect Space Warps. Their parameters include values for Bounce, Variation, Chaos, Friction, and Inherit Velocity. The UDeflector Space Warp is a simplified version of the UOmniFlect Space Warp. It has a Pick Object button for selecting the object to act as the deflector and all the same parameters as the SDeflector Space Warp.

Geometric/Deformable Space Warps

You use Geometric/Deformable Space Warps to deform the geometry of an object. Space Warps in this subcategory include FFD (Box), FFD (Cyl), Wave, Ripple, Displace, Conform, and Bomb. These Space Warps can be applied to any deformable object. Figure 43.11 shows the icons for each of these Space Warps.

FIGURE 43.11

The Geometric/Deformable Space Warps: FFD (Box), FFD (Cyl), Wave, Ripple, Displace, Conform, and Bomb

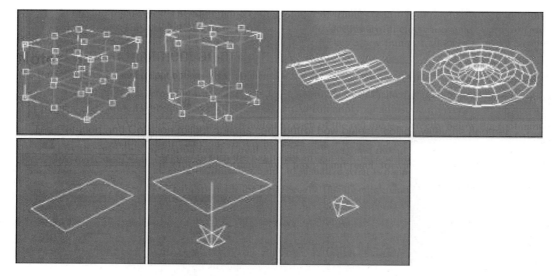

FFD (Box) and FFD (Cyl)

The FFD (Box) and FFD (Cyl) Space Warps show up as a lattice of control points in the shape of a box and a cylinder; you can select and move the control points that make up the Space Warp to deform an object that is bound to the Space Warp. The object is deformed only if the bound object is within the volume of the Space Warp.

These Space Warps have the same parameters as the modifiers with the same name found in the Modifiers→Free Form Deformers menu. The difference is that the Space Warps act in World coordinates and aren't tied to a specific object. This allows a single FFD Space Warp to affect multiple objects.

To learn about the FFD (Box) and FFD (Cyl) modifiers, see Chapter 16, "Deforming Surfaces and Using the Mesh Modifiers."

To move the control points, select the Space Warp object, open the Modify panel, and select the Control Points subobject, which lets you alter the control points individually.

FFD Select modifier

The FFD Select modifier is another unique selection modifier. It enables you to select a group of control point subobjects for the FFD (Box) or the FFD (Cyl) Space Warps and apply additional modifiers to the selection. When an FFD Space Warp is applied to an object, you can select the Control Points subobjects and apply modifiers to the selection. The FFD Select modifier lets you select a different set of control points for a different modifier.

Wave and Ripple

The Wave and Ripple Space Warps create linear and radial waves in the objects to which they are bound. Parameters in the rollout help define the shape of the wave. Amplitude 1 is the wave's height along the X-axis, and Amplitude 2 is the wave's height along its Y-axis. The Wave Length value defines the distance from one wave peak to the next wave peak. The Phase value determines how the wave starts at its origin. The Decay value sets how quickly the wave dies out. A Decay value of 0 maintains the same amplitude for the entire wave.

The Sides (Circles) and Segments values determine the number of segments for the X- and Y-axes. The Division value changes the icon's size without altering the wave effect. Figure 43.12 shows a Wave Space Warp applied to a simple Plane primitive. Notice that the Space Warp icon is smaller than the Plane object, yet it affects the entire object.

FIGURE 43.12

The Wave and Ripple Space Warps applied to a patch grid object

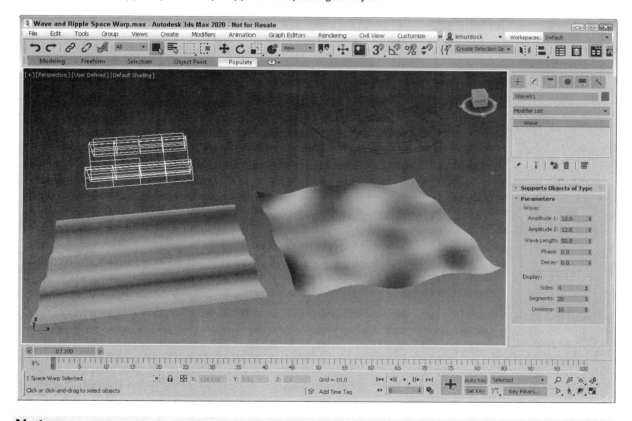

Note
Be sure to include enough segments in the bound object, or the effect won't be visible.

Tutorial: Creating pond ripples

For this tutorial, you position a patch object so it aligns with a background image and apply the Ripple Space Warp to it.

To add ripples to a pond, follow these steps:

1. Open the Pond ripple.max file from the Chap 43 directory in the downloaded content set.

   This file includes a background image of a bridge matched to a patch grid where the pond is located with a reflective material assigned to it.

Tip

If you're having trouble locating the patch grid, press F3 to switch to Wireframe mode.

2. Select the Create→Space Warps→Geometric/Deformable→Ripple menu command. Drag in the Perspective viewport to create a Space Warp object. In the Parameters rollout, set both Amplitudes to **4** and the Wave Length to **30**.

3. Click the Bind to Space Warp button, and drag from the patch object to the Space Warp.

Figure 43.13 shows the resulting image.

FIGURE 43.13

A ripple in a pond produced using the Ripple Space Warp

Conform

The Conform Space Warp pushes all object vertices until they hit another target object called the Wrap To Object, or until they've moved a preset distance. The Conform Parameters rollout includes a Pick Object button that lets you pick the Wrap To Object. The object vertices move no farther than this Wrap To Object.

You also can specify a Default Projection Distance and a Standoff Distance. The Default Projection Distance is the maximum distance that the vertices move if they don't intersect with the Wrap To Object. The Standoff Distance is the separation amount maintained between the Wrap To Object and the moved vertices. Another option, Use Selected Vertices, moves only a subobject selection.

The Conform Space Warp is similar in function to the Conform compound object that is covered in Chapter 15, "Working with Compound Objects."

Figure 43.14 shows some text being deformed with the Conform Space Warp. A warped quad patch has been selected as the Wrap To object.

FIGURE 43.14

The Conform Space Warp wraps the surface of one object around another object.

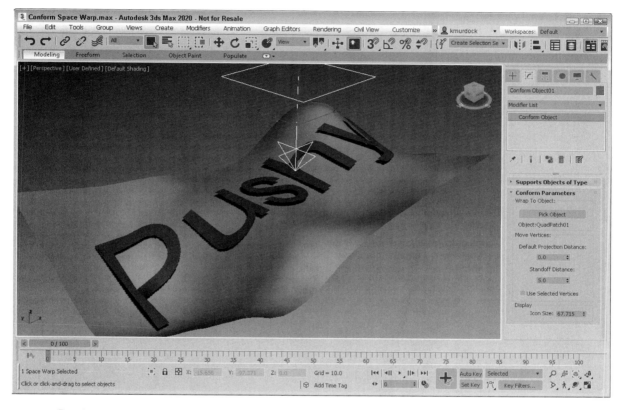

Bomb

The Bomb Space Warp causes an object to explode to its individual faces. The Strength value is the power of the bomb and determines how far objects travel when exploded. The Spin value is the rate at which the individual pieces rotate. The Falloff value defines the boundaries of faces affected by the bomb. Object faces beyond this distance remain unaffected. You must select Falloff On for the Falloff value to work.

The Min and Max Fragment Size values set the minimum and maximum number of faces caused by the explosion.

The Gravity value determines the strength of gravity and can be positive or negative. Gravity always points toward the world's Z-axis. The Chaos value can range between 0 and 10 to add variety to the explosion. The Detonation value is the number of the frame where the explosion should take place, and the Seed value alters the randomness of the event. Figure 43.15 shows a frame of an explosion produced by the Bomb Space Warp.

FIGURE 43.15

The Bomb Space Warp causes an object to explode.

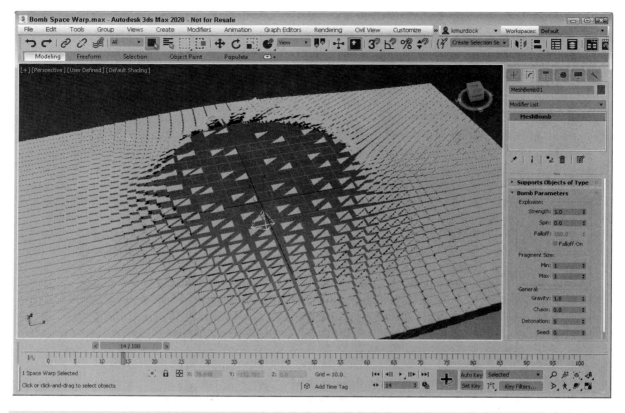

Note

The Bomb Space Warp's effect is seen over time. At frame 0, the object shows no effect.

Tutorial: Blowing a dandelion puff

You can use Space Warps with other types of objects besides particle systems. The Scatter compound object, for example, can quickly create many unique objects that can be controlled by a Space Warp. In this tutorial, you create a simple dandelion puff that can blow away in the wind.

To create and blow away a dandelion puff, follow these steps:

1. Open the Dandelion puff.max file from the Chap 43 directory in the downloaded content set.

    This file includes a sphere covered with a Scatter compound object representing the seeds of a dandelion.

2. Select the Create→Space Warps→Geometric/Deformable→Bomb menu command. Click in the Front viewport, and position the Bomb icon to the left and slightly below the dandelion object. In the Bomb Parameters rollout, set the Strength to **10**, the Spin to **0.5**, the Min and Max Fragment Size values to **24**, the Gravity to **0.2**, and the Chaos to **2.0**.

3. Click the Bind to Space Warp button on the main toolbar, and drag from the dandelion object to the Space Warp. Then press the Play button to see the animation.

Figure 43.16 shows one frame of the dandelion puff being blown away.

FIGURE 43.16

You can use Space Warps on Scatter objects as well as particle systems.

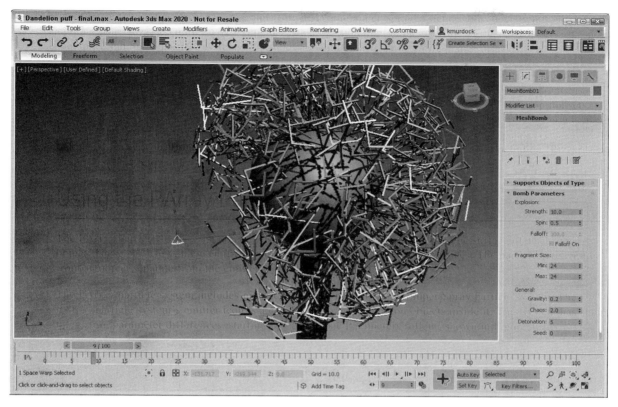

Modifier-Based Space Warps

Modifier-Based Space Warps produce the same effects as many of the standard modifiers, but because they are Space Warps, they can be applied to many objects simultaneously. Space Warps in this subcategory include Bend, Noise, Skew, Taper, Twist, and Stretch, as shown in Figure 43.17. All Modifier-Based Space Warp gizmos are simple box shapes. The parameters for all Modifier-Based Space Warps are identical to the modifiers (found in the Parametric Deformers category) of the same name. These Space Warps don't include a Supports Objects of Type rollout because they can be applied to all objects.

For details on the Bend, Noise, Skew, Taper, Twist, and Stretch modifiers and their parameters, see Chapter 16, "Deforming Surfaces and Using the Mesh Modifiers."

FIGURE 43.17

The Modifier-Based Space Warps: Bend, Noise, Skew, Taper, Twist, and Stretch

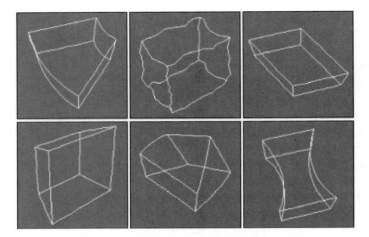

These Space Warps include a Gizmo Parameters rollout with values for the Length, Width, and Height of the gizmo. You also can specify the deformation decay. The Decay value causes the Space Warp's effect to diminish with distance from the bound object.

You can reposition the Modifier-Based Space Warp's gizmo as a separate object, but the normal modifiers require that you select the gizmo subobject to reposition it. Unlike modifiers, Space Warps don't have any subobjects.

Combining Particle Systems with Space Warps

To conclude this chapter, you'll look at some examples that use Space Warps along with particle systems. With all these Space Warps and their various parameters combined with particle systems, the possibilities are endless. These examples are only a small representation of what is possible.

Tutorial: Shattering glass

When glass shatters, it is very chaotic, sending pieces in every direction. For this tutorial, you shatter a glass mirror on a wall. The wall keeps the pieces from flying off, and most pieces fall straight to the floor.

To shatter glass, follow these steps:

1. Open the Shattering glass.max file from the Chap 43 directory in the downloaded content set.

   This file includes a simple mirror created from patch grid objects. The file also includes a simple sphere that is animated striking the mirror.

2. Select the Create→Particles→PArray menu command. Then drag in the Front viewport to create the PArray icon. In the Basic Parameters rollout, click the Pick Object button and select the first patch object representing the glass mirror. In the Viewport Display section, select the Mesh option. In the Particle Generation rollout, set the Speed and Divergence to 0. Also set the Emit Start to **30** and the Life value to **100**, so it matches the last frame. In the Particle Type rollout, select the Object Fragments option, and set the Thickness to **1.0**. Then in the Object Fragment Controls section, select the Number of Chunks option with a Minimum value of **30**. In the Rotation and Collision rollout, set the Spin Time to **100** and the Variation to **50**.

   These settings cause the patch to emit 30 object fragments with a slow, gradual rotation.

3. Select the Space Warps category button, and choose the Forces subcategory from the drop-down list. Click the PBomb button, and create a PBomb Space Warp in the Top viewport; then center it above the Mirror object. In the Modify panel, set the Blast Symmetry option to Spherical with a Chaos value

of **50** percent. Set the Start Time to **30** with a Strength value of **0.2**. Then click the Bind to Space Warp button, and drag from the PBomb Space Warp to the PArray icon.

4. Select the Create→SpaceWarps→Forces→Gravity menu command, and drag in the Top viewport to create a Gravity Space Warp. Make sure the Gravity Space Warp icon arrow is pointing down in the Front viewport. In the Modify panel, set the Strength value to **0.1**. Then bind this Space Warp to the PArray icon.

5. Select the Create→ SpaceWarps→Deflectors→POmniFlect menu command. Drag this Space Warp in the Top viewport, and make it wide enough to be completely under the mirror object. Rotate the POmniFlect Space Warp so that its arrows are pointing up at the mirror. Position it so that it lies in the same plane as the plane object that makes up the floor. In the Modify panel, set the Reflects value to **100** percent and the Bounce value to **0**. Bind this Space Warp to the PArray as well; this keeps the pieces from falling through the floor. Press Play to see the results.

Figure 43.18 shows the mirror immediately after being struck by a ball.

FIGURE 43.18

A shattering mirror

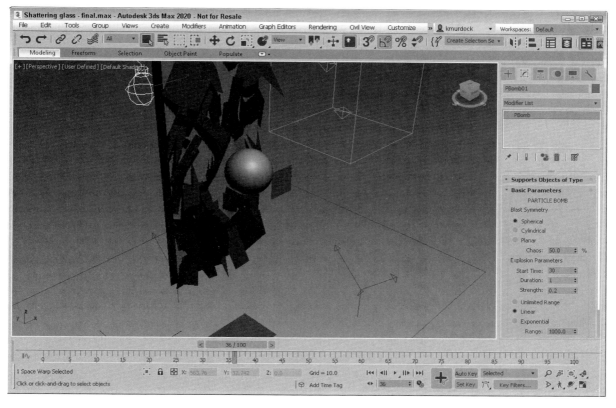

Tutorial: Making water flow down a trough

That should be enough destruction for a while. In this final example, you'll make some water particles flow down a trough. You accomplish this using the Path Follow Space Warp.

To make water flow down a trough, follow these steps:

1. Open the Water flowing down a trough.max file from the Chap 43 directory in the downloaded content set.

   This file includes a simple trough made from primitives and a spline path that the water will follow.

2. Select the Create→Particles→Super Spray menu command. Create a Super Spray object in the Left viewport and position its pointer where you want the particles to first appear. In the Viewport Display section, select the Ticks option. In the Particle Generation rollout, set the Speed to **10** and the Variation to **100**. Then set the Emit Start to **0** and the Display Until and Life values to **100** and the Size to **20**. In the Particle Type rollout, select MetaParticles and enable the Automatic Coarseness option.

3. Select the Space Warps category button, and choose Forces from the subcategory drop-down list. Click the Path Follow button, and create a Path Follow object; then click the Bind to Space Warp button on the main toolbar, and drag from the Path Follow icon to the Super Spray icon. Open the Modify panel, select the Path Follow icon, click the Pick Shape Object button, and select the path in the viewports. Set the Start Frame to **0** and the Travel Time to **100**.

Figure 43.19 shows the rendered result.

FIGURE 43.19

Water flowing down a trough using the Path Follow Space Warp

Summary

Space Warps are useful for adding forces and effects to objects in the scene. The Autodesk® 3ds Max® 2020 software has several different types of Space Warps, and most of them can be applied only to certain object types. In this chapter, you learned the following:

* How to create Space Warps
* How to bind Space Warps to objects
* How to use all the various Space Warps in several subcategories
* How to combine Space Warps with particle systems

You can have 3ds Max dynamically compute all the animation frames in a scene using the physics-based MassFX engine, which is covered in Chapter 46, "Simulating Physics-Based Motion with MassFX." In the next chapter, we delve into a special type of effects called atmospheric effects which includes fog, clouds and fire. We'll also look at the various Render Effects.

Chapter 44

Using Atmospheric and Render Effects

IN THIS CHAPTER

In the real world, an environment of some kind surrounds all objects. The environment does much to set the ambiance of the scene. For example, an animation set at night in the woods has a very different environment than one set at the horse races during the middle of the day. The Autodesk® 3ds Max® 2020 software includes dialog boxes for setting the color, background images, and lighting environment; these features can help define your scene.

This chapter also covers Exposure Controls and atmospheric effects, including the likes of clouds, fog, and fire. These effects can be seen only when the scene is rendered.

3ds Max also has a class of effects that you can interactively render to the Rendered Frame Window without using any post-production features, such as the Video Post dialog box. These effects are called *render effects*. Render effects can save you lots of time that you would normally spend rendering an image, touching it up, and repeating the process again and again.

The common thread among all these features is their location. All can be found within the Environment and Effects dialog box.

Adding an Environment Background

Whether it's a beautiful landscape or just clouds drifting by, the environment behind the scene can do much to make the scene more believable. In this section, you learn to define an environment using the Rendering→Environment (8) menu command.

Environment maps are used as background for the scene and also can be used as images reflected off shiny objects in the scene. Environment maps are displayed only in the final rendering and not in the viewports, but you can add a background to any viewport and even set the environment map to be displayed as the viewport backdrop.

Chapter 2, "Controlling and Configuring the Viewports," covers adding a background image to a viewport.

But there is more to an environment than just a background. It also involves altering the global lighting, controlling exposure, and introducing atmosphere effects.

Defining the rendered environment

You create environments in the Environment and Effects dialog box, shown in Figure 44.1, which you can open by choosing Rendering→Environment (or by pressing the 8 key). Several settings make up an environment, including a background color or image, global lighting, exposure control, and atmospheric effects.

FIGURE 44.1

The Environment and Effects dialog box lets you select a background color or image, define global lighting, control exposure, and work with atmospheric effects.

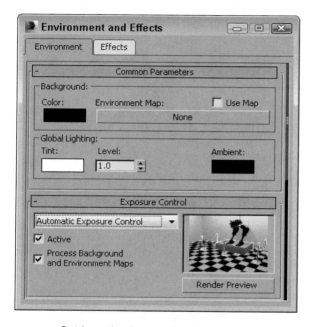

Setting a background color

The first color swatch in the Environment panel lets you specify a background color. This color appears by default if no environment map is specified or if the Use Map option is disabled (and is black by default). The background color can be animated, so you can set the background color to start black and slowly fade to white.

Using a background image

To select a background image to be used as an environment map, click the Environment Map button in the Environment panel to open the Material/Map Browser. If you want to load a bitmap image as the background image, double-click the Bitmap selection to open the Select Bitmap Image File dialog box. Locate the bitmap to use, and click Open. The bitmap name appears on the Environment Map button.

Tip

If the environment map that you want to use is already displayed in one of the Material Editor sample slots, you can drag it directly from the Material Editor and drop it on the Map button in the Environment panel. You also can drag and drop a filename from Windows Explorer onto the Map button.

To change any of the environment map parameters, you need to load the environment map into the Material Editor. You can do so by dragging the map button from the Environment panel into the Node View panel in the Material Editor. After releasing the material, the Instance (Copy) Map dialog box asks whether you want to create an Instance or a Copy. If you select Instance, any parameter changes that you make to the material automatically update the map in the Environment panel.

Once in the Material Editor, you can use the Environment Map to create a Spherical Environment map that is used to reflect realistically off objects in the scene.

For more information about the types of available mapping parameters, see Chapter 18, "Adding Material Details with Maps."

The background image doesn't need to be an image: you also can load animations. Supported formats include AVI, MPEG, MOV, and IFL files.

Figure 44.2 shows a scene with an image of the Golden Gate Bridge loaded as the environment map.

FIGURE 44.2

The results of a background image loaded into the Environment panel

Setting global lighting

The Tint color swatch in the Global Lighting section of the Environment panel specifies a color used to tint all lights. The Level value increases or decreases the overall lighting level for all lights in the scene. The Ambient color swatch sets the color for the ambient light in the scene, which is the darkest color that any shadows in the scene can be. You can animate all these settings.

Using Exposure Controls

The Exposure Control rollout of the Environment panel lets you control output levels and color rendering ranges. You can access the Environment panel from the Rendering→Environment menu command or by pressing the 8 key. Controlling the exposure of film is a common procedure when working with film and can result in a different look for your scene. Enabling the Exposure Controls can add dynamic range to your rendered images that is more comparable to what the eyes actually see. If you've worked with a Histogram in Photoshop, you'll understand the impact that the Exposure Controls can have.

The Active option lets you turn this feature on and off. The Process Background and Environment Maps option causes the exposure settings to affect the background and environment images. When this option is disabled, only the scene objects are affected by the exposure control settings. The Exposure Control rollout also includes a Render Preview button that displays the rendered scene in a tiny pane. The preview pane is small, but for most types of exposure control settings it is enough. When you click the Render Preview button, the scene is rendered. This preview is then automatically updated whenever a setting is changed.

Automatic, Linear, and Logarithmic Exposure Control

Selecting Automatic Exposure Control from the drop-down list automatically adjusts your rendered output to be closer to what your eyes can detect. Monitors are notoriously bad at reducing the dynamic range of the colors in your rendered image. This setting provides the needed adjustments to match the expanded dynamic range of your eyes.

When the Automatic Exposure Control option is selected, a new rollout appears in the Environment panel. This rollout includes settings for Brightness, Contrast, Exposure Value, and Physical Scale. You also can enable Color Correction, select a color, and select an option to Desaturate Low Levels. The Contrast and Brightness settings can range from 0 to 100. A Contrast value of 0 displays all scene objects with the same flat, gray color, and a Brightness value of 100 displays all scene objects with the same flat, white color. The Exposure Value can range from –5 to 5 and determines the amount of light allowed in the scene.

Another exposure control option is Linear Exposure Control. Although this option presents the same settings as the Automatic Exposure Control, the histogram values are a straight line across the light spectrum.

Tip

The tricky part is to know when to use which Exposure Control. For still images, the Automatic Exposure Control is your best bet, but for animations, you should use the Logarithmic Exposure Control. Automatic is also a good choice for any scenes that use many lighting effects. Using any of the exposure controls besides the Logarithmic Exposure Control when animating can lead to flickering. The Linear Exposure Control should be used for low dynamic range scenes such as nighttime or cloudy scenes.

The Logarithmic Exposure Control option replaces the Exposure Value setting with a Mid Tones setting. This setting controls the colors between the lowest and highest values. This exposure control option also includes options to Affect Indirect Only and Exterior Daylight. You should enable the Affect Indirect Only option if you use only standard lights in the scene, but if your scene includes an IES Sun light, enable the Exterior Daylight option to tone down the intensity of the light.

You should always use the Logarithmic Exposure Control setting when enabling the advanced lighting features because it works well with low-level light. You can learn more about the advanced lighting radiosity features in Chapter 26, "Working with Advanced Lighting, Light Tracing, and Radiosity."

Pseudo Color Exposure Control

As you work with advanced lighting solutions and with radiosity, determining whether interior spaces and objects have too much light or not enough light can be difficult, especially when comparing objects on opposite sides of the scene. This is where the Pseudo Color Exposure Control option comes in handy.

This exposure control option projects a band of colors (or grayscale) in place of the material and object colors that represent the illumination or luminance values for the scene. With these pseudo-colors, you can quickly determine where all the lighting is consistent and where it needs to be addressed.

In the Pseudo Color Exposure Control rollout, shown in Figure 44.3, you can select to apply the colors to show Illumination or Luminance. You also can select to use a Colored or Gray Scale style and to make the Scale Linear or Logarithmic. The Min and Max settings let you control the ranges of the colors, and a

Physical Scale setting is included. The color (or grayscale) band is shown across the bottom of the rollout with the values for each color underneath.

FIGURE 44.3

The Pseudo Color Exposure Control rollout can display illumination and luminance values as colors.

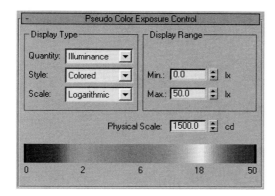

When this exposure control is used, the associated render element is automatically set in the Render Elements rollout of the Render Setup dialog box. If the scene is rendered, the appropriate (Illumination or Luminance) render element is also rendered.

See Chapter 31, "Compositing with Render Elements and the Video Post Interface," for more on render elements.

Physical Camera Exposure Control

If you're comfortable working with camera settings such as Shutter Speed, Aperture, and Film Speed, the Physical Camera Exposure Control puts these settings at your fingertips using real-world values. Although the Environment and Effects panel has a number of settings, shown in Figure 44.4, you can select to have the exposure control Use Physical Camera Controls.

FIGURE 44.4

The Physical Camera Exposure Control rollout works with real-world camera settings.

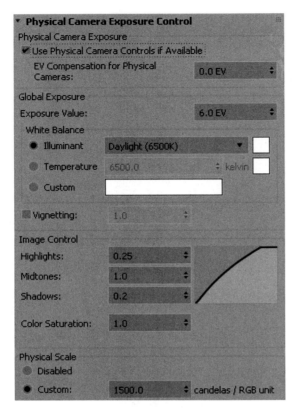

Tutorial: Using the Logarithmic Exposure Control

As you start to use the new photometric lights, you may find it difficult to get the settings just right. The results are oversaturation or undersaturation, but luckily the Logarithmic Exposure Control can quickly fix any problems that appear.

To adjust the effect of a photometric light using the Logarithmic Exposure Control, follow these steps:

1. Open the Array of chrome spheres.max file from the Chap 44 directory in the downloaded content set.

   This file contains lots and lots of chrome mapped spheres with advanced lighting enabled.

2. Choose Rendering→Render Setup (or press the F10 key) to open the Render Setup dialog box, and click the Render button.

   It takes a while to render, but notice the results, shown on the left in Figure 44.5.

3. Choose Rendering→Environment (or press the 8 key) to open the Environment and Effects dialog box. In the Exposure Control rollout, select Logarithmic Exposure Control from the drop-down list, and enable the Active and Process Background and Environment Maps options. Then click the Render Preview button.

4. In the Logarithmic Exposure Control Parameters rollout, set the Brightness value to **60**, set the Contrast value to **100**, and enable the Desaturate Low Levels option.

5. In the Render Scene dialog box, click the Render button again to see the updated rendering.

The image on the right in Figure 44.5 shows the rendered image with exposure control enabled.

FIGURE 44.5

This rendered image shows an image before and after exposure control was enabled.

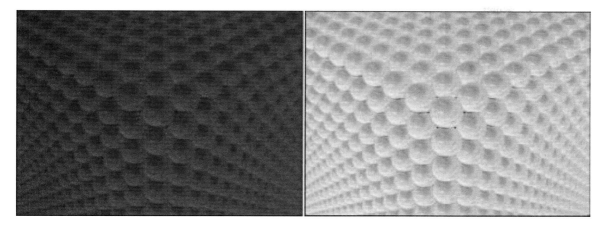

Creating Atmospheric Effects

The Environment and Effects dialog box (keyboard shortcut, 8) contains rollouts for adding atmospheric effects to your scene, but the first question is where. Atmospheric effects are placed within a container called an Atmospheric Apparatus gizmo, which tells the effect where it should be located. However, only the Fire and the Volume Fog effects need Atmospheric Apparatus gizmos. To create an Atmospheric Apparatus gizmo, select Create→Helpers→Atmospherics and choose the apparatus type.

The three different Atmospheric Apparatus gizmos are BoxGizmo, SphereGizmo, and CylGizmo. Each has a different shape similar to the primitives.

Working with the Atmospheric Apparatus

Selecting a gizmo and opening the Modify panel reveal two different rollouts: one for defining the basic parameters such as the gizmo dimensions, and another labeled Atmospheres & Effects, which you can use to Add or Delete an Environment Effect to the gizmo. Each gizmo parameters rollout also includes a Seed value and a New Seed button. The Seed value sets a random number used to compute the atmospheric effect, and the New Seed button automatically generates a random seed. Two gizmos with the same seed values have nearly identical results.

Adding effects to a scene

The Add button opens the Add Atmosphere dialog box, where you can select an atmospheric effect. The selected effect is then included in a list in the Atmospheres & Effects rollout. You can delete these atmospheres by selecting them from the list and clicking the Delete button. The Setup button is active if an effect is selected in the list. It opens the Environment and Effects dialog box. Adding Atmospheric Effects in the Modify panel is purely for convenience. They also can be added using the Environment and Effects dialog box.

In addition to the Modify panel, you can add atmospheric effects to the scene using the Atmosphere rollout in the Environment and Effects dialog box, shown in Figure 44.6. This rollout is pretty boring until you add an effect to it. You can add an effect by clicking on the Add button. This opens the Add Atmospheric Effect dialog box, which includes by default four atmospheric effects: Fire Effect, Fog, Volume Fog, and Volume Light. With plug-ins, you can increase the number of effects in this list. The selected effect is added to the Effects list in the Atmosphere rollout.

FIGURE 44.6

The Environment and Effects dialog box lets you select atmospheric effects.

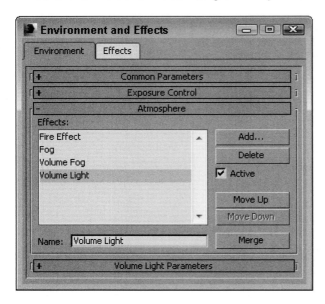

The effects are applied in the order in which they are listed, so the effects at the bottom of the list are layered on top of all other effects. To the right of the Effects pane are the Move Up and Move Down buttons, used to position the effects in the list. Below the Effects pane is a Name field where you can type a new name for any effect in this field. This enables you to use the same effect multiple times. The Merge button opens the Merge Atmospheric Effects dialog box, where you can select a separate 3ds Max file. You can then select and load any render effects contained in the other file.

Tip

The Merge button lets you create and save several different types of fire effects and then quickly merge them into a scene from an external file.

Using the Fire Effect

To add the Fire effect to the scene, select the Rendering→Environment (8) menu command and open the Environment panel; then click the Add button and select the Fire Effect selection. This opens the Fire Effect Parameters rollout, shown in Figure 44.7. At the top of the Fire Effect Parameters rollout is the Pick Gizmo button; clicking this button lets you select a gizmo in the scene. The selected gizmo appears in the drop-down list to the right. You can select multiple gizmos. To remove a gizmo from the list, select it and click the Remove Gizmo button.

FIGURE 44.7

The Fire Effect Parameters rollout lets you define the look of the effect.

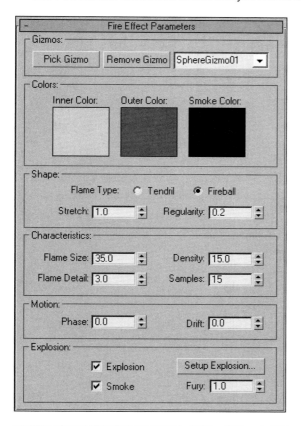

Note
The Fire effect renders only in non-orthographic views such as Perspective or a camera view.

The three color swatches define the color of the fire effect and include an Inner Color, an Outer Color, and a Smoke Color. The Smoke Color is used only when the Explosion option is set. The default red and yellow colors make fairly realistic fire.

The Shape section includes two Flame Type options: Tendril and Fireball. The Tendril shape produces veins of flames, and the Fireball shape is rounder and puffier. Figure 44.8 shows four fire effects. The left two have the Tendril shape, and the two on the right are set to Fireball. The difference is in the Density and Flame Detail settings.

FIGURE 44.8

The Fire atmospheric effect can be either Tendril or Fireball shaped.

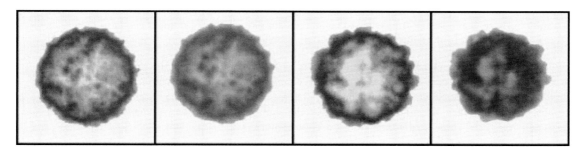

The Stretch value elongates the individual flames along the gizmo's Z-axis. Figure 44.9 shows the results of using the Stretch value. The Stretch values for these gizmos, from left to right, are 0.1, 1.0, 5.0, and 50.

FIGURE 44.9

The Stretch value can elongate flames.

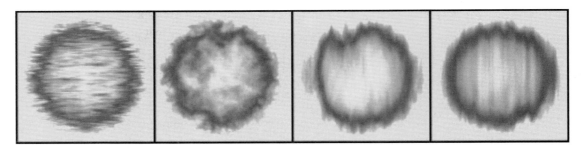

The Regularity value determines how much of the Atmospheric Apparatus is filled. The spherical gizmos in the previous figures were all set to 0.2, so the entire sphere shape wasn't filled. A setting of 1.0 adds a spherical look to the Fire effect because the entire gizmo is filled. For a more random shape, use a small Regularity value.

The Flame Size value affects the overall size of each individual flame (though this is dependent on the gizmo size as well). The Flame Detail value controls the edge sharpness of each flame and can range from 1 to 10. Lower values produce fuzzy, smooth flames, but higher values result in sharper, more distinct flames.

The Density value determines the thickness of each flame in its center; higher Density values result in flames that are brighter at the center, while lower values produce thinner, wispy flames. Figure 44.10 shows the difference caused by Density values of, from left to right, 10, 20, 50, and 100.

FIGURE 44.10

The Fire effect brightness is tied closely to the flame's Density value.

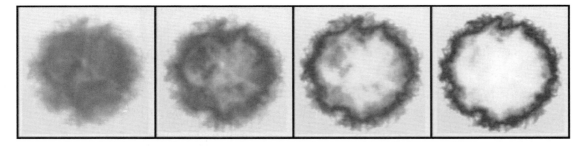

The Samples value sets the rate at which the effect is sampled. Higher sample values are required for more detail, but they increase the render time.

The Motion section includes options for setting the Phase and Drift of a fire effect. The Phase value determines how wildly the fire burns. For a wild, out-of-control fire, animate the Phase value to change rapidly. For a constant, steady fire, keep the value constant throughout the frames. The Drift value sets the height of the flames. High Drift values produce high, hot-burning flames.

When the Explosion check box is selected, the fire is set to explode. The Start and End Times for the explosion are set in the Setup Explosion Phase Curve dialog box that opens when the Setup Explosion button is clicked. If the Smoke option is checked, then the fire colors change to the smoke color for Phase values between 100 and 200. The Fury value varies the churning of the flames. Values greater than 1.0 cause faster churning, and values lower than 1.0 cause slower churning.

Tutorial: Creating the sun

You can use the Fire effect to create a realistic sun. The modeling part is easy—all it requires is a simple sphere—but the real effects come from the materials and the Fire effect.

To create a sun, follow these steps:

1. Open the Sun.max file from the Chap 44 directory in the downloaded content set.

    This file contains a simple sphere with a bright yellow material applied to it.

2. Select Create→Helpers→Atmospherics→Sphere Gizmo, and drag a sphere in the Front viewport that is larger than the "sun" sphere.

3. With the SphereGizmo still selected, open the Modify panel and click the Add button in the Atmospheres & Effects rollout or you can use the Add button located in the Atmosphere rollout in the Environment and Effects dialog box, which is opened using the Rendering→Environment menu command or by pressing the 8 key. Select Fire Effect from the Add Atmosphere dialog box, and click OK. Then select the Fire effect, and click the Setup button.

    The Environment and Effects dialog box opens.

4. Select the Fire Effect in the Atmosphere rollout, and in the Fire Effects Parameters rollout, leave the default colors as they are—Inner Color yellow, Outer Color red, and Smoke Color black. For the Flame Type, select Tendril with Stretch and Regularity values of **1**. Set the Flame Size to **30**, the Density to **15**, the Flame Detail to **10**, and the Samples to **15**.

Figure 44.11 shows the resulting sun after it's been rendered.

FIGURE 44.11

A sun image created with a simple sphere, a material with a Noise Bump map, and the Fire effect

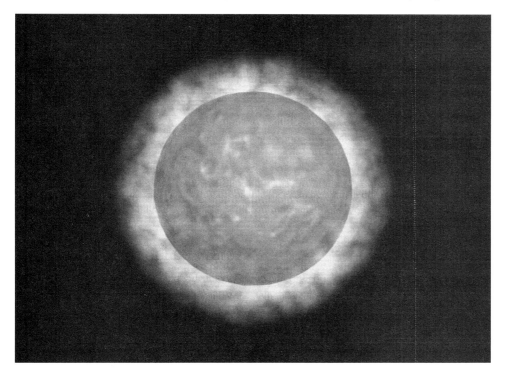

Tutorial: Creating clouds

Sky images are fairly easy to find, or you can just take your camera outside and capture your own. The trick comes when you are trying to weave an object in and out of clouds. Although you can do this with a

Shadow/Matte mask, it would be easier if the clouds were actual 3D objects. In this tutorial, you'll create some simple clouds using the Fire effect.

To create some clouds for a sky backdrop, follow these steps:

1. Open the Clouds.max file from the Chap 44 directory in the downloaded content set.

   This file includes several hemispherical-shaped Atmospheric Apparatus gizmos.

2. Choose Rendering→Environment (or press the 8 key) to open the Environment and Effects dialog box. Click the Background Color swatch, and select a light blue color. In the Atmosphere rollout, click the Add button, select Fire Effect from the Add Atmospheric Effect list, and click OK.

3. Name the effect **Clouds**, click the Pick Gizmo button, and select one of the Sphere Gizmo objects in the viewports. Repeat until all gizmos are selected, and then click on each of the color swatches. Change the Inner Color to a dark gray, the Outer Color to a light gray, and the Smoke Color to white. Set the Flame Type to Fireball with a Stretch of **1** and a Regularity of **0.2**. Set the Flame Size to **35**, the Flame Detail to **3**, the Density to **15**, and the Samples to **15**.

Tip

If you want to add some motion to the clouds, click the Auto Key button, drag the Time Slider to the last frame, and change the Phase value to 45 and the Drift value to 30. The clouds slowly drift through the sky. Disable the Auto Key button when you're finished.

Figure 44.12 shows the resulting rendered sky backdrop. By altering the Fire Effect parameters, you can create different types of clouds.

FIGURE 44.12

You can use the Fire atmospheric effect to create clouds.

Using the Fog Effect

Fog is an atmospheric effect that obscures objects or backgrounds by introducing a hazy layer; objects farther from view are less visible. The normal Fog effect is used without an Atmospheric Apparatus gizmo and appears between the camera's environment range values. The camera's Near and Far Range settings set these values.

In the Environment and Effects dialog box, the Fog Parameters rollout appears when the Fog effect is added to the Effects list. This rollout, shown in Figure 44.13, includes a color swatch for setting the fog color. It also includes an Environment Color Map button for loading a map. If a map is selected, the Use Map option turns it on or off. You also can select a map for the Environment Opacity, which affects the fog density.

The Fog Background option applies fog to the background image. The Type options include Standard and Layered fog. Selecting one of these fog background options enables its corresponding parameters.

FIGURE 44.13

The Fog Parameters rollout lets you use either Standard fog or Layered fog.

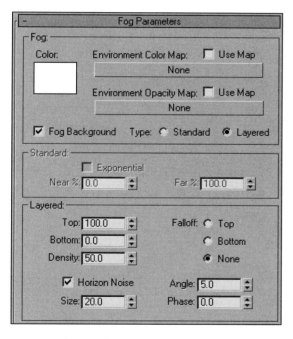

The Standard parameters include an Exponential option for increasing density as a function of distance. If this option is disabled, the density is linear with distance. The Near and Far values are used to set the range densities.

Layered fog simulates layers of fog that move from dense areas to light areas. The Top and Bottom values set the limits of the fog, and the Density value sets its thickness. The Falloff option lets you set where the fog density goes to 0. The Horizon Noise option adds noise to the layer of fog at the horizon as determined by the Size, Angle, and Phase values.

Figure 44.14 shows several different fog options. The upper-left image shows the scene with no fog, the upper-right image uses the Standard option, and the lower-left image uses the Layered option with a Density of 50. The lower-right image has the Horizon Noise option enabled.

FIGURE 44.14

A rendered image with several different Fog effect options applied

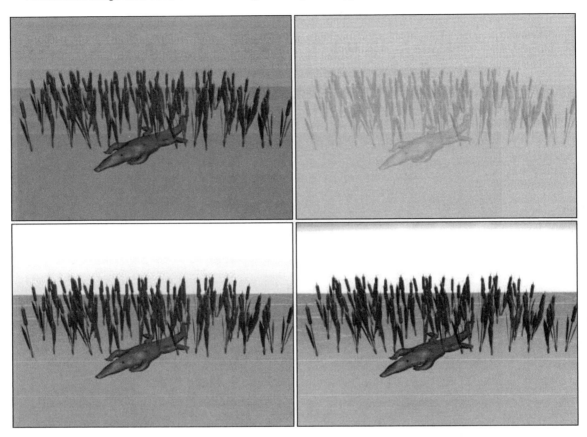

Using the Volume Fog effect

You can add the Volume Fog effect to a scene by clicking the Add button and selecting the Volume Fog selection. This effect is different from the Fog effect in that it gives you more control over the exact position of the fog. This position is set by an Atmospheric Apparatus gizmo. The Volume Fog Parameters rollout, shown in Figure 44.15, lets you select a gizmo to use with the Pick Gizmo button. The selected gizmo is included in the drop-down list to the right of the buttons. Multiple gizmos can be selected. The Remove Gizmo button removes the selected gizmo from the list.

Note
The Atmospheric Apparatus gizmo contains only a portion of the total Volume Fog effect. If the gizmo is moved or scaled it displays a different cropped portion of fog.

FIGURE 44.15

The Volume Fog Parameters rollout includes parameters for controlling the fog density and type.

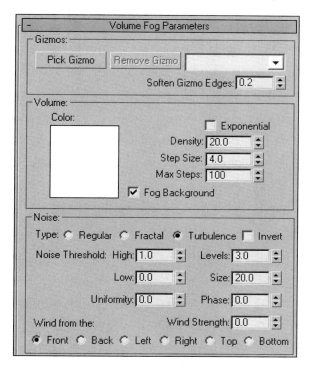

The Soften Gizmo Edges value feathers the fog effect at each edge. This value can range from 0 to 1.

Many of the settings for Volume Fog are the same as those for the Fog effect, but Volume Fog has several settings that are unique to it. These settings help set the patchy nature of Volume Fog. Step Size determines how small the patches of fog are. The Max Steps value limits the sampling of these small steps to keep the render time in check.

The Noise section settings also help determine the randomness of Volume Fog. Noise types include Regular, Fractal, and Turbulence. You also can select to Invert the noise. The Noise Threshold limits the effect of noise. Wind settings include direction and strength. The Phase value determines how the fog moves.

Tutorial: Creating a swamp scene

When I think of fog, I think of swamps. In this tutorial, you model a swamp scene. To use the Volume Fog effect to create the scene, follow these steps:

1. Open the Dragonfly in a foggy swamp.max file from the Chap 44 directory in the downloaded content set.

    This file includes several cattail plants and a dragonfly positioned on top of one of the cattails.

2. Select Create→Helpers→Atmospherics→Box Gizmo, and drag a box that covers the lower half of the cattails in the Left viewport.

3. Choose Rendering→Environment (or press the 8 key) to open the Environment and Effects dialog box. Click the Add button to open the Add Atmospheric Effect dialog box, and then select Volume Fog. Click OK. In the Volume Fog Parameters rollout, click the Pick Gizmo button and select the BoxGizmo in a viewport.

4. Set the Density to **0.5**, enable the Exponential option and select the Noise Type Turbulence. Then set the Uniformity to **1.0** and the Wind Strength to **10** from the Left.

Figure 44.16 shows the finished rendered image. Using Atmospheric Apparatus gizmos, you can position the fog in the exact place where you want it.

FIGURE 44.16

A rendered image that uses the Volume Fog effect

Using the Volume Light effect

The final Environment option is the Volume Light effect. This effect shares many of the same parameters as the other atmospheric effects. Although this is one of the atmospheric effects, it deals with lights and fits better in that section.

To learn about the Volume Light atmospheric effect, see Chapter 45, "Adding Volume Light and Lens Effects."

Adding Render Effects

In many cases, rendering a scene is only the start of the work to produce some final output. The post-production process is often used to add lots of different effects, as you'll see when I discuss the Video Post interface. But just because you can add it in post-production doesn't mean you have to add it in post-production. Render effects let you apply certain effects as part of the rendering process.

You can set up all render effects from the Rendering Effects panel, which you open by choosing Rendering→Effects. Figure 44.17 shows this dialog box. This dialog box also includes a panel with the Environment options.

FIGURE 44.17

The Effects panel lets you apply interactive post-production effects to an image.

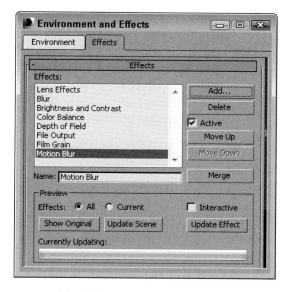

The Effects pane displays all the effects that are included in the current scene. To add a new effect, click the Add button to open the Add Effect dialog box, in which you can select from a default list of nine effects: Hair and Fur, Lens Effects, Blur, Brightness and Contrast, Color Balance, Depth of Field, File Output, Film Grain, and Motion Blur. You can delete an effect from the current list by selecting that effect and clicking the Delete button.

Below the Effects pane is a Name field. You can type a new name for any effect in this field; doing so enables you to use the same effect multiple times. The effects are applied in the order in which they are listed in the Effects pane. To the right of the Effects pane are the Move Up and Move Down buttons, which you use to reposition the effects in the list. The effects are added to the scene in the order that they are listed.

Caution

It is possible for one effect to cover another effect. Rearranging the order can help resolve this problem.

The Merge button opens a file dialog box, where you can select a separate 3ds Max file. If you select a 3ds Max file and click Open, the Merge Rendering Effects dialog box presents you with a list of render effects used in the opened 3ds Max file. You can then select and load any of these render effects into the current scene.

The Preview section holds the controls for interactively viewing the various effects. Previews are displayed in the Rendered Frame Window and can be set to view All the effects or only the Current one. The Show Original button displays the scene before any effects are applied, and the Update Scene button updates the rendered image if any changes have been made to the scene.

Note

If the Rendered Frame Window isn't open, any of these buttons opens it and renders the scene with the current settings in the Render Setup dialog box.

The Interactive option automatically updates the image whenever an effect parameter or scene object is changed. If this option is disabled, you can use the Update Effect button to manually update the image.

Caution

If the Interactive option is enabled, the image is re-rendered in the Rendered Frame Window every time a change is made to the scene. This can slow down the system dramatically.

The Currently Updating bar shows the progress of the rendering update.

The remainder of the Effects panel contains global parameters and rollouts for the selected render effect. These rollouts are covered in this chapter, along with their corresponding effects.

The Lens Effect render effect is covered in Chapter 45, "Adding Volume Lights and Lens Effects." The Hair and Fur render effect offers the ability to add hair and fur to models. The Hair and Fur features—including the render effect—are covered in Chapter 47, "Working with Hair and Cloth."

Using Render Effects

Within the Add Effect dialog box are several different render effects. Some of these effects are simple, and others are quite complex. If these selections aren't enough, 3ds Max enables you to add even more options to this list via plug-ins.

Blur render effect

The Blur render effect displays three different blurring methods in the Blur Type panel: Uniform, Directional, and Radial. You can find these options in the Blur Type tabbed panel in the Blur Parameters rollout, shown in Figure 44.18.

FIGURE 44.18

The Blur Parameters rollout lets you select a Uniform, Directional, or Radial blur type.

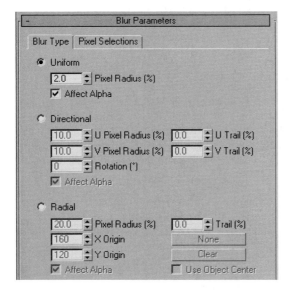

The Uniform blur method applies the blur evenly across the whole image. The Pixel Radius value defines the amount of the blur. The Directional blur method can be used to blur the image along a certain direction. The U Pixel Radius and U Trail values define the blur in the horizontal direction, and the V Pixel Radius and V Trail values blur in a vertical direction. The Rotation value rotates the axis of the blur.

The Radial blur method creates concentric blurred rings determined by the Radius and Trail values. When the Use Object Center option is selected, the None and Clear buttons become active. Clicking the None button lets you select an object about which you want to center the radial blur. The Clear button clears this selection.

Figure 44.19 shows a duck toy model. The actual rendered image shows the sharp edges of the polygons. The Blur effect can help this by softening all the hard edges. The left image is the original model, and the right image has a Directional blur applied.

FIGURE 44.19

The Blur effect can soften an otherwise hard model.

The Blur Parameters rollout also includes a Pixel Selections tabbed panel, shown in Figure 44.20 that contains parameters for specifying which parts of the image get blurred. Options include the Whole Image, Non-Background, Luminance, Map Mask, Object ID, and Material ID.

FIGURE 44.20

The Pixel Selections tabbed panel (shown in two parts) of the Blur Parameters rollout lets you select the parts of the image that get the Blur effect.

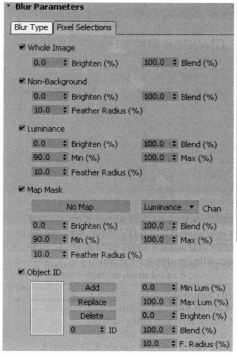
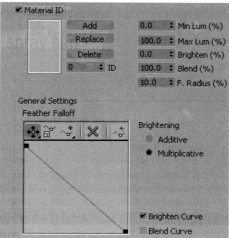

You can use the Feather Falloff curve at the bottom of the Blur Parameters rollout to define the Brighten and Blend curves. The buttons above this curve are for adding points, scaling the points, and moving them within the curve interface.

Brightness and Contrast render effect

The Brightness and Contrast render effect can alter these amounts in the image. The Brightness and Contrast Parameters rollout is a simple rollout with values for both the brightness and contrast that can range from 0 to 1. It also contains an Ignore Background option.

Color Balance render effect

The Color Balance effect enables you to tint the image using separate Cyan/Red, Magenta/Green, and Yellow/Blue channels. To change the color balance, drag the sliders in the Color Balance Parameters rollout. Other options include Preserve Luminosity and Ignore Background. The Preserve Luminosity option tints the image while maintaining the luminosity of the image, and the Ignore Background option tints the rendered objects but not the background image.

File Output render effect

The File Output render effect enables you to save the rendered file to a File or to a Device at any point during the render effect's post-processing. Figure 44.21 shows the File Output Parameters rollout.

FIGURE 44.21

The File Output Parameters rollout lets you save a rendered image before a render effect is applied.

Using the Channel drop-down list in the Parameters section, you can save out Whole Images, as well as grayscale Luminance, Depth, and Alpha images.

Film Grain render effect

The Film Grain effect gives an image a grained look, which hardens the overall look of the image. You also can use this effect to match rendered objects to the grain of the background image. This helps the objects blend into the scene better. Figure 44.22 shows the effect.

The Grain value can range from 0 to 10. The Ignore Background option applies the grain effect only to the objects in the scene and not to the background.

FIGURE 44.22

The Film Grain render effect applies a noise filter to the rendered image.

Motion Blur render effect

The Motion Blur effect applies a simple image motion blur to the rendered output. The Motion Blur Parameters rollout includes settings for working with Transparency and a value for the Duration of the blur. Objects that move rapidly within the scene are blurred.

Depth of Field render effect

The Depth of Field effect enhances the sense of depth by blurring objects close to or far from the camera. The Pick Cam button in the Depth of Field Parameters rollout, shown in Figure 44.23, lets you select a camera in the viewport to use for this effect. Multiple cameras can be selected, and all selected cameras are displayed in the drop-down list. A Remove button lets you remove cameras.

FIGURE 44.23

The Depth of Field Parameters rollout lets you select a camera or a Focal Point to apply the effect to.

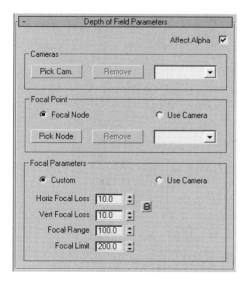

In the Focal Point section, the Pick Node button lets you select an object to use as the focal point. This object is where the camera focuses. Objects far from this object are blurred. These nodes are also listed in a drop-down list. You can remove objects from the list by selecting them and clicking the Remove button. The Use Camera option uses the camera's own settings to determine the focal point.

In the Focal Parameters section, if you select the Custom option, you can specify values for the Horizontal and Vertical Focal Loss, the Focal Range, and the Focal Limit. The Loss values indicate how much blur occurs. The Focal Range is where the image starts to blur, and the Focal Limit is where the image stops blurring.

Figure 44.24 shows a scene with some windmills. For this figure, I applied the Depth of Field effect using the Pick Node button and selecting the windmill blades. Then I set the Focal Range to 100, the Focal Limit to 200, and the Focal Loss values to 10 for both the Horizontal and Vertical.

The Depth of Field and Motion Blur effects also can be applied using a Multi-Pass camera, as discussed in Chapter 24, "Configuring and Aiming Cameras."

FIGURE 44.24

The Depth of Field effect focuses a camera on an object in the middle and blurs objects closer or farther away.

Summary

Creating the right environment can add lots of realism to any rendered scene. Using the Environment and Effects dialog box, you can work with atmospheric effects. Atmospheric effects include Fire, Fog, Volume Fog, and Volume Light.

Render effects are useful because they enable you to create effects and update them interactively. This gives a level of control that was previously unavailable. This chapter explained how to use render elements and render effects and described the various types.

This chapter covered these topics:

* Creating Atmospheric Apparatus gizmos for positioning atmospheric effects
* Working with the Fire atmospheric effects
* Creating fog and volume fog effects
* Applying render effects
* Working with the remaining render effects to control brightness and contrast, film grain, blurs, and more

The next chapter delves into effects made possible by the volume light feature.

Chapter 45

Adding Volume Light and Lens Effects

IN THIS CHAPTER

Using the Volume light effect

Using the Lens Effects to add glows, rays, and streaks

Working with the Lens Effects filters in the Video Post interface

Lights are an essential part of every scene, and although you can see the result, you don't often get to see the actual light beam in the scene. A Volume Light is different. It is an atmospheric effect that allows the actual light beam to be visible by having the light shine through some fog.

One of the most used render effects in Autodesk® 3ds Max® 2020 found in the Effects panel of the Environment and Effects dialog box is Lens Effects. With this effect, you can make objects glow and add effects like stars, streaks, and sparkles. And, just like a dance recital, we can all use more sparkles.

Lens Effects also can be added to a scene using the Video Post interface. The effects found here also include a tool for adding flares to the rendered scene.

Using Volume Lights

When light shines through fog, smoke, or dust, the beam of the light becomes visible. The effect is known as a *Volume Light*. To add a Volume Light to a scene in Autodesk® 3ds Max® 2020, choose Rendering→Environment (or press the 8 key) to open the Environment and Effects dialog box. Then click the Add button in the Atmosphere rollout to open the Add Atmospheric Effect dialog box, and select Volume Light. The parameters for the volume light are presented in the Volume Light Parameters rollout.

You also can access the Volume Light effect from the Atmospheres and Effects rollout in the Modify panel when a light is selected.

Chapter 44, "Using Atmospheric and Render Effects," covers the other atmospheric effects.

Volume light parameters

At the top of the Volume Light Parameters rollout, shown in Figure 45.1, is a Pick Light button, which enables you to select a light to apply the effect to. You can select several lights, which then appear in a drop-down list. You can remove lights from this list with the Remove Light button.

FIGURE 45.1

The Volume Light Parameters rollout in the Environment and Effects dialog box lets you choose which lights to include in the effect.

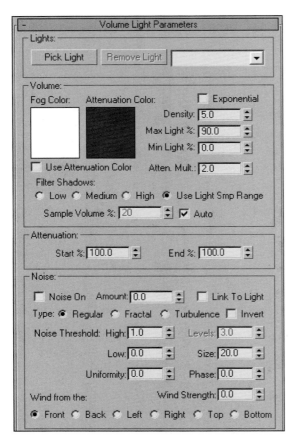

In the Volume section, the Fog Color swatch lets you select a color for the fog that is seen within the light. This color is combined with the color of the light. The Attenuation Color is the color the fog appears to have at a distance far from the light source. This color also combines with the Fog Color and is best set to a dark color.

The Density value determines the thickness of the fog. The Exponential option causes the density to increase exponentially with the distance. The Max and Min Light Percentage values determine the amount of glow that the volume light causes, and the Attenuation Multiplier controls the strength of the attenuation color.

You have four options for filtering shadows: Low, Medium, High, and Use Light Smp Range. The Low option renders shadows quickly but isn't very accurate. The High option takes a while but produces the best quality. The Use Light Smp Range option bases the filtering on the Sample Volume value and can be set to Auto. The Sample Volume can range from 1 to 10,000. The Low option has a Sample Volume value of 8; Medium, 25; and High, 50.

The Start and End Attenuation values are percentages of the Start and End range values for the light's attenuation. The attenuation property defines how the light dims over distance. These values have an impact only if attenuation is turned on for the light.

The Noise settings help to determine the randomness of Volume Light. Noise effects can be turned on and given an Amount. You also can Link the noise to the light instead of using world coordinates. Noise types include Regular, Fractal, and Turbulence. Another option inverts the noise pattern. The Noise Threshold limits the effect of noise. Wind settings affect how the light moves as determined by the wind's direction, Wind Strength, and Phase.

Figure 45.2 shows several volume light possibilities. The left image includes the Volume Light effect, the middle image enables shadows, and the right image includes some Turbulent Noise.

FIGURE 45.2

The Volume Light effect makes the light visible.

Tutorial: Showing UFO lights

One popular way to use volume lights is to display the headlights of cars or lights on a spaceship. For this tutorial, you're going to use a UFO spaceship model to light up the sky.

To display the lights on a UFO spaceship, follow these steps:

1. Open the UFO with lights.max file from the Chap 45 directory in the downloaded content set.

   This file includes a model of a UFO spaceship.

2. Select the Create→Lights→Standard Lights→Target Spotlight menu command, and drag in the Top viewport to create a spotlight object. Select and move the spotlight and the target to be positioned to look as if a light is shining out from one of the lights on the bottom of the UFO spaceship.

3. With the light selected, open the Modify panel, and in the Spotlight Parameters rollout, set the Hotspot/Beam value to **20** and the Falloff/Field to **25**, and in the Intensity/Color/Attenuation rollout, set the Decay setting to Inverse Square with a Start value of **3.0**. In the Atmospheres and Effects rollout, click the Add button, select Volume Light from the Add Atmosphere or Effect dialog box that appears, and click OK.

Note

When a light is added to the scene, the default lights are automatically turned off. To provide any additional lighting, add some Omni lights above the spaceship.

4. Select the Volume Light effect in the list within the Atmospheres and Effects rollout, and click the Setup button. The Environment and Effects dialog box opens, in which you can edit the Volume Light parameters for the newly created light. Set the Density value to **100** and the Fog Color to red.

5. Now create a copy of this spotlight for each of the other light spots. To do this, select the first spotlight object and move its pivot to the center of the spaceship, then use the Tools→Array dialog box to create 6 duplicates that circle the underside of the UFO spaceship.

Figure 45.3 shows the resulting UFO spaceship with its lights illuminated.

FIGURE 45.3

The spaceship now has menacing lights, thanks to spotlights and the Volume Light effect.

Tutorial: Creating laser beams

Laser beams are extremely useful lights. From your CD-ROM drive to your laser printer, lasers are found throughout a modern-day office. They also are great to use in fantasy and science fiction images. You can easily create laser beams using direct lights and the Volume Light effect. In this tutorial, you'll add some lasers to the spaceship model.

To add some laser beams to a scene, follow these steps:

1. Open the Spaceship laser.max file from the Chap 45 directory in the downloaded content set.

   This file includes a spaceship model.

2. Select the Create→Lights→Standard Lights→Directional menu command, and add a Free Direct light to the end of one of the laser guns in the Front viewport. Scale the light down until the cylinder is the size of the desired laser beam and rotate it so it points away from the laser gun.

3. With the light selected, open the Modify panel. In the Atmospheres and Effects rollout, click the Add button and double-click the Volume Light selection. Then select the Volume Light option in the list, and click the Setup button to open the Environment and Effects dialog box. Change the Fog Color to red and the Density value to 50, and make sure that the Use Attenuation Color is disabled. Then mirror the light to the laser on the opposite side of the ship.

4. With the direct light added to the scene, the default lights are deactivated, so you need to add some Omni lights above the spaceship to illuminate it. To do this, select the Create→Lights→Standard Lights→Omni menu command, and click above the spaceship in the Front view three times to create three lights. Set the Multiplier on the first light to **1.0**, and position it directly above the spaceship. Set

the other two lights to **0.5**, and position them on either side of the spaceship and lower than the first light.

Figure 45.4 shows the resulting laser beams shooting forth from the spaceship.

FIGURE 45.4

You can create laser beams using direct lights and the Volume Light effect.

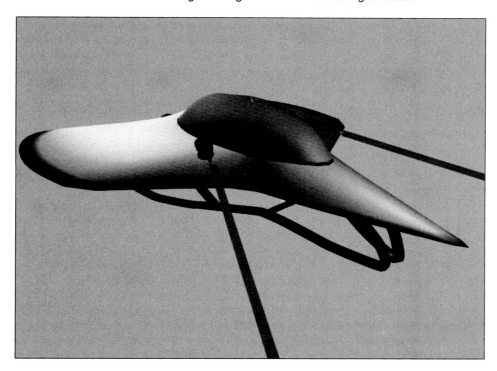

Creating Lens Effects

Of the available render effects, Lens Effects will be used perhaps more often than all the others combined. The Lens Effects option includes several different effects itself, ranging from glows and rings to streaks and stars.

Lens Effects simulate the types of lighting effects that are possible with actual camera lenses and filters. When the Lens Effects selection is added to the Effects list and selected, several different effects become available in the Lens Effects Parameters rollout, including Glow, Ring, Ray, Auto Secondary, Manual Secondary, Star, and Streak. When one of these effects is included in a scene, rollouts and parameters for that effect are added to the panel as well.

Several of these Lens Effects can be used simultaneously. To include an effect, open the Environment and Effects dialog box with the Rendering→Effects menu command, click the Add button, and select Lens Effects from the available effects. Then go to the Lens Effects Parameters rollout, select the desired effect from the list on the left, and click the arrow button pointing to the right. The pane on the right lists the included effects. Use the left-pointing arrow button to remove effects from the list.

Global Lens Effects Parameters

Under the Lens Effects Parameters rollout in the Rendering Effects panel is the Lens Effects Globals rollout. All effects available in Lens Effects use the two common tabbed panels in this rollout: Parameters and Scene. These two tabbed panels are shown side by side in Figure 45.5.

FIGURE 45.5

The Parameters tabbed panel of the Lens Effects Globals rollout lets you load and save parameter settings. The Scene tabbed panel lets you set the effect's Size and Intensity.

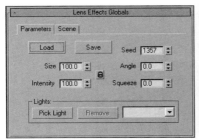
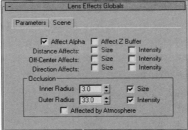

The Global Parameters tabbed panel

The Parameters panel of the Lens Effects Globals rollout includes Load and Save buttons for loading and saving parameter settings specified in the various rollouts. These settings are saved as LZV files.

The Size value determines the overall size of the effect as a percentage of the rendered image. Figure 45.6 shows the center of the Star Lens Effects with an Intensity value of 500 and Size values of 5, 10, 20, 50, and 100. The Size value increases the entire effect diameter and also the width of each radial line.

FIGURE 45.6

These Star Lens Effects vary in size.

The Intensity value controls the brightness and opacity of the effect. Large values are brighter and more opaque, and small values are dimmer and more transparent. The Size and Intensity values can be locked together. Intensity and Size values can range from 0 to 500. Figure 45.7 shows a glow effect with Intensity values of (from left to right) 50, 100, 200, 350, and 500.

FIGURE 45.7

Lens Effects also can vary in intensity, like these glows.

The Seed value provides the randomness of the effect. Changing the Seed value changes the effect's look. The Angle value spins the effect about the camera's axis. The Squeeze value lengthens the horizontal axis for positive values and lengthens the vertical axis for negative values. Squeeze values can range from –100 to 100. Figure 45.8 shows a Ring effect with Squeeze values of (from left to right) –30, –15, 0, 10, and 20.

FIGURE 45.8

These Ring effects vary in Stretch values.

All effects are applied to light sources, and the Pick Light button lets you select a light in the viewport to apply the effect to. Each selected light is displayed in a drop-down list. You can remove any of these lights with the Remove Light button.

The Global Scene tabbed panel

The second Lens Effects Globals tabbed panel common to all effects is the Scene panel. This rollout includes an Affect Alpha option that lets the effect work with the image's alpha channel. The alpha channel holds the transparency information for the rendered objects and for effects if this option is enabled. If you plan on using the effect in a composite image, enable this option.

Tip

Click the Display Alpha Channel button in the Rendered Frame Window to view the alpha channel.

The Affect Z Buffer option stores the effect information in the Z Buffer, which is used to determine the depth of objects from the camera's viewpoint.

The Distance Affects option alters the effect's Size and/or Intensity based on its distance from the camera. The Off-Center Affects option is similar, except that it affects the effect's size and intensity based on its Off-Center distance. The Direction Affects options can affect the size and intensity of an effect based on the direction in which a spotlight is pointing.

The Occlusion settings can be used to cause an effect to be hidden by an object that lies between the effect and the camera. The Inner Radius value defines the area that an object must block in order to hide the effect. The Outer Radius value defines where the effect begins to be occluded. You also can set the Size and Intensity options for the effect. The Affected by Atmosphere option allows effects to be occluded by atmospheric effects.

Glow

The Glow Element rollout, shown in Figure 45.9, appears when the Glow effect is selected from the list. It includes parameters for controlling the look of the Glow Lens Effect. This rollout has two tabbed panels: Parameters and Options.

FIGURE 45.9

The Glow Element rollout lets you set the parameters for the Glow effect.

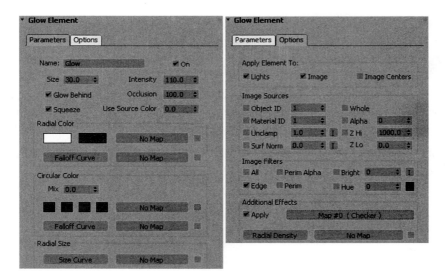

The Glow Element Parameters tabbed panel

In the Parameters tabbed panel of the Glow Element rollout, there is a Name field. Several glow effects can be added to a scene, and each one can have a different name. The On option turns each glow on and off.

The Parameters panel also includes Size and Intensity values. These work with the Global settings to determine the size of the glow and can be set to any positive value. The Occlusion and Use Source Color values are percentages. The Occlusion value determines how much of the occlusion set in the Scene panel of the Lens Effects Globals rollout is to be used. If the Use Source Color is at 100 percent, the glow color is determined by the light color; if it is set to any value below 100, the colors specified in the Source and Circular Color sections are combined with the light's color.

This panel also includes Glow Behind and Squeeze options. Glow Behind makes the glow effect visible behind objects. Squeeze enables any squeeze settings specified in the Parameters panel.

If the Use Source Color value is set to 0 percent, only the Radial Color swatches determine the glow colors. Radial colors proceed from the center of the glow circle to the outer edge. The first swatch is the inner color, and the second is the outer color. The Falloff Curve button opens the Radial Falloff function curve dialog box, shown in Figure 45.10, where you can use a curve to set how quickly or slowly the colors change.

FIGURE 45.10

The Radial Falloff dialog box lets you control how the inner radial color changes to the outer radial color.

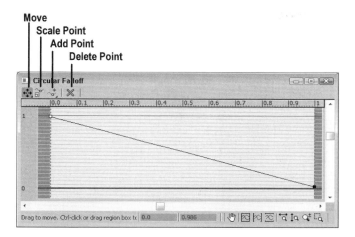

The Circular Color swatches specify the glow color around the glow circle starting from the top point and proceeding clockwise. The Mix value is the percentage to mix the Circular colors with the Radial Colors; a value of 0 displays only the Radial colors, and a value of 100 displays only the Circular colors. You also can access the Falloff Curve dialog box for the Circular Color Falloff curve.

You also can control the Radial Size using a curve by clicking the Size Curve button. Clicking the Size Curve button accesses the Radial Size dialog box. Figure 45.11 shows several glow effects where the radial size curve has been altered. The curves are, from left to right, roughly a descending linear curve, a v-shaped curve, a wide u-shaped curve, an m-shaped curve, and a sine curve.

FIGURE 45.11

These glow effects are distorted using the Radial Size function curves.

All these colors and function curves have map buttons (initially labeled No Map) that enable you to load maps. Useful maps to use include Falloff, Gradient and Gradient Ramp, Noise, and Swirl. You can enable a map by using the check box to its immediate right.

The Glow Element Options tabbed panel

The Options panel of the Glow Element rollout defines where to apply the glow effect. In the Apply Element To section, the first option is to apply a glow to the Lights. These lights are selected in the Lights section of the Lens Effects Globals rollout using the Pick Light button. The other two options—Image and Image Centers—apply glows using settings contained in the Options panel.

In the Image Sources section, you can apply glows to specific objects using the Object ID option and settings. Object IDs are set for objects in the Object Properties dialog box. If the corresponding Object ID is selected and enabled in the Options panel, the object is endowed with the Glow Lens Effect.

The Material ID option and settings work in a manner similar to Object IDs, except that they are assigned to materials in the Material Editor. You can use Material IDs to make a subobject selection glow.

The Unclamp option and settings enable colors to be brighter than pure white. Pure White is a value of 1. The Unclamp value is the lowest value that glows. The Surf Norm (Surface Normal) option and value let you set object areas to glow based on the angle between the surface normal and the camera. The "I" button to the right inverts the value.

Figure 45.12 shows an array of spheres with the Surf Norm glow enabled. Because the glows multiply, a value of only 2 was applied. Notice that the spheres in the center have a stronger glow.

FIGURE 45.12

The Surf Norm option causes objects to glow, based on the angle between their surface normals and the camera.

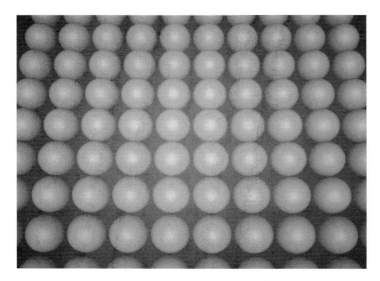

In the Image Sources section, options enable these glows to be applied to the Whole scene, the Alpha channel, or the Z buffer with specified Hi and Lo values.

The Image Filters section can further refine which objects to apply the glow effect to. Options include All, Edge, Perim (Perimeter) Alpha, Perim, Brightness, and Hue. The All option applies the effect to all pixels that are part of the source. The Edge, Perim Alpha, and Perim options apply the effect only to the edges, perimeter of the alpha channel, or perimeter of the source. The Brightness option includes a value and an "I" invert button. This applies the effect only to areas with a brightness greater than the specified value. The Hue option also includes a value and a color swatch for setting the hue, which receives the effect.

The Additional Effects section lets you apply a map to the Glow Lens Effect with an Apply option and a map button. You also can control the Radial Density function curve or add a map for the Radial Density.

Tutorial: Creating shocking electricity from a plug outlet

In addition to the light objects, lighting in a scene can be provided by self-illuminating an object and using a Glow render effect. Self-illuminating an object is accomplished by applying a material with a Self-Illumination value greater than 0 or a color other than black. You can create glows by using the Render Effects dialog box or the Video Post dialog box.

For more information on applying glows using the Video Post interface, see Chapter 31, "Compositing with Render Elements and the Video Post Interface."

Working with a faulty electrical outlet can be a shocking experience. In this tutorial, you create an electric arc that runs from an outlet to a plug. To create the effect of electricity, you can use a renderable spline with several vertices and apply the Noise modifier to make it dance around. You can set the light by using a self-illuminating material and a Glow render effect.

To create an electric arc that runs between an outlet and a plug, follow these steps:

1. Open the Electricity.max file from the Chap 45 directory in the downloaded content set.

   This file includes an outlet and an electric plug. A spline runs between the outline and the plug with a Noise modifier applied to it, which will be our electric arc.

2. Open the Material Editor, and double click on the Standard material in the Material/Map Browser, then double click on the new Standard node to reveal its parameters. Select a yellow Diffuse Color and an equally bright yellow for the Self-Illumination color. Set the Material Effects Channel to 1 by clicking the Material Effects ID button on the Material Editor toolbar and holding it down until a pop-up array of numbers appears, and then drag to the number 1 and release the mouse. Drag this new material to the electric arc and close the Material Editor.

3. Open the Effects panel of the Environment and Effects dialog box by choosing the Rendering→ Effects menu. Click the Add button, select the Lens Effects option, and click OK. Then select Lens Effects from the list, and double-click Glow in the Lens Effects Parameters rollout. Select Glow from the list; in the Glow Element rollout, set the Size to **1** and the Intensity value to **50**. Then open the Options panel, set the Material ID to **1**, and enable it.

Figure 45.13 shows the resulting electric arc.

FIGURE 45.13

You can create electricity using a simple spline, the Noise modifier, and the Glow render effect.

Tutorial: Creating neon

You also can use the Glow render effect to create neon signs. The letters for these signs can be simple renderable splines, as this tutorial shows.

To create a neon sign, follow these steps:

1. Open the Blues neon.max file from the Chap 45 directory in the downloaded content set.

   This file includes a simple sign that reads "Blues."

2. Open the Material Editor, double click on the Standard material in the Material/Map Browser, then double click on the new Standard node to reveal its parameters, and name it **Blue Neon**. Set its

Diffuse color to blue and its Self-Illumination color to dark blue. Set the Material Effects Channel to **1**, and apply the material to the sign.

3. Open the Effects panel of the Environment and Effects dialog box, and click the Add button. Double-click the Lens Effects option to add it to the Effects list. In the Lens Effects Parameters rollout, double-click the Glow option and select it in the list to enable its rollouts. In the Lens Effects Globals rollout, set the Size and Intensity values to **1**. In the Glow Element rollout, set the Size to **10** and the Intensity to **100**, and make sure that the Glow Behind option is selected. For the neon color, set the Use Source Color to **100**. Finally, open the Options panel, set the Material ID to **1**, and enable it.

Note

As an alternative to using the source color, you could set the Use Source Color value to 0 and set the Radial Color swatch to blue. This gives you more control over the glow color.

Figure 45.14 shows the rendered neon effect.

FIGURE 45.14

The glow of neon lights, easily created with render effects

Ring

The Ring Lens Effect is also circular and includes all the same controls and settings as the Glow Lens Effect. The only additional values are the Plane and Thickness values. The Plane value positions the Ring center relative to the center of the screen, and the Thickness value determines the width of the Ring's band.

Figure 45.15 shows several Ring effects with various Thickness values (from left to right): 1, 3, 6, 12, and 24.

FIGURE 45.15

Ring effects can vary in thickness.

Ray

The Ray Lens Effect emits bright, semitransparent rays in all directions from the source. It also uses the same settings as the Glow lens effect, except for the Num (Number) and Sharp values. The Num value is the number of rays, and the Sharp value can range from 0 to 10 and determines how blurry the rays are.

Figure 45.16 shows the Ray effect applied to a simple Omni light with increasing Num values: 6, 12, 50, 100, and 200. Notice that the rays aren't symmetrical and are randomly placed.

FIGURE 45.16

The Ray lens effect extends a given number of rays out from the effect center.

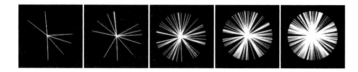

Star

The Star Lens Effect radiates semitransparent bands of light at regular intervals from the center of the effect. It uses the same controls as the Glow effect, with the addition of Width, Taper, Qty (Quantity), and Sharp values. The Width sets the width of each band. The Taper value determines how quickly the width angles to a point. The Qty value is the number of bands, and the Sharp value determines how blurry the bands are.

Figure 45.17 shows several Star lens effects with (from left to right) 3, 4, 5, 6, and 12 bands.

FIGURE 45.17

The Star lens effect lets you set the number of bands emitting from the center.

Streak

The Streak Lens Effect adds a horizontal band through the center of the selected object. It is similar to the Star effect, except it has only two bands that extend in opposite directions.

Figure 45.18 shows several Streak effects angled at 45 degrees with Width values of (from left to right) 2, 4, 10, 15, and 20.

FIGURE 45.18

The Streak effect enables you to create horizontal bands.

Auto Secondary

When a camera is moved past a bright light, several small circles appear lined up in a row proceeding from the center of the light. These secondary lens flares are caused by light refracting off the lens. You can simulate this effect by using the Auto Secondary Lens Effect.

Many of the settings in the Auto Secondary Element rollout are the same as in the Glow Element rollout described previously, but the Auto Secondary Element rollout has several unique values. Figure 45.19 shows this rollout.

FIGURE 45.19

The Auto Secondary Element rollout sets the parameters for this effect.

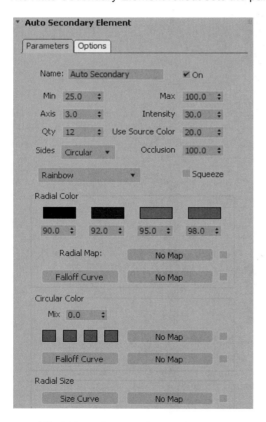

The Min and Max values define the minimum and maximum size of the flares. The Axis is the length of the axis along which the flares are positioned. Larger values spread the flares out more than smaller values. The actual angle of the flares depends on the angle between the camera and the effect object.

The Quantity value is the number of flares to include. The Sides drop-down list lets you select a Circular flare or flares with three to eight sides. Below the Sides drop-down list are several preset options in another drop-down list. These include options such as Brown Ring, Blue Circle, and Green Rainbow, among others.

You also can use four Radial Colors to define the flares. The color swatches from left to right define the colors from the inside out. The spinners below each color swatch indicate where the color should end.

Figure 45.20 shows the Auto Secondary Effect with the Rainbow preset and the Intensity increased to 50.

FIGURE 45.20

The Auto Secondary Effect displays several flares extending at an angle from the center of the effect.

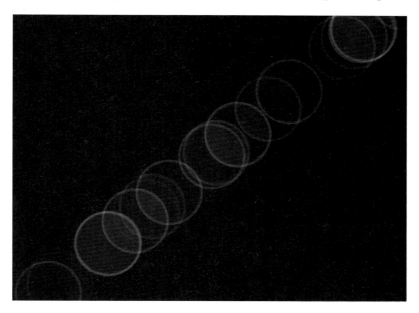

Manual Secondary

In addition to the Auto Secondary Lens Effect, you can add a Manual Secondary Lens Effect to add some more flares with a different size and look. This effect includes a Plane value that places the flare in front of (positive value) or behind (negative value) the flare source.

Figure 45.21 shows the same flares from the previous figure with an additional Manual Secondary Effect added.

FIGURE 45.21

The Manual Secondary Effect can add some randomness to a flare lineup.

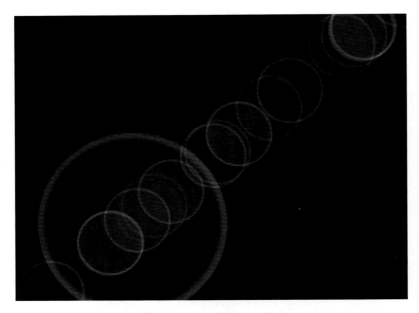

Tutorial: Making an apple sparkle

Objects can be highlighted with a lens effect to make them shine in the scene.

To make an object bright and shiny using Lens Effects, follow these steps:

1. Open the Shining apple.max file from the Chap 45 directory in the downloaded content set.

2. Open the Create panel, click the Lights category button, and select the Standard subcategory. Create several Omni lights, and position them around the scene to provide adequate lighting. Position a single light close to the apple's surface where you want the highlight to be located—make it near the surface, and set the Multiplier value to **0.5**.

3. Open the Effects panel of the Environment and Effects dialog box by choosing the Rendering→ Effects menu. Click the Add button, and select Lens Effects. Then, in the Lens Effects Parameters rollout, select the Glow effect in the left pane and click the button pointing to the right pane.

4. In the Parameters panel of the Lens Effects Globals rollout, click the Pick Light button and select the light close to the surface of the apple. Set the Size around **30** and the Intensity at **100**. Go to the Glow Element rollout, and in the Parameters panel, set the Use Source Color to **0**. Then, in the Radial Color section, click the second Radial Color swatch, and in the Color Selector dialog box, select a color like yellow and click the Close button.

5. Back in the Lens Effects Parameters rollout, select Star, and add it to the list of effects. It automatically uses the same light specified for the Glow effect. In the Star Element rollout, set the Quantity (Qty) value to **6**, the Size to **200**, and the Intensity to **100**.

Figure 45.22 shows the resulting apple with a nice shine.

FIGURE 45.22

This apple has had a sparkle added to it using the Glow and Star Lens Effects.

Working with Lens Effects Filters

Lens effects also can be added during post processing in the Video Post interface using the Lens Effects Filters. Several Lens Effect Filters are available in the Video Post interface, and you can access them by clicking the Add Image Filter Event button at the top of the interface.

The Add Image Filter Event dialog box's drop-down list has several Lens Effects filters. These filters include Lens Effects Flare, Focus, Glow, and Highlight. Each of these filters is displayed and discussed in the sections that follow, but several parameters are common to all of them. You can access the Lens Effects dialog boxes by clicking the Setup button in the Add Image Filter Event dialog box.

Many lens effects parameters in the various Lens Effects setup dialog boxes can be animated, such as Size, Hue, Angle, and Intensity. These are identified in the dialog boxes by green arrow buttons to the right of the parameter fields. These buttons work the same way as the animate buttons in the main interface work. To animate a parameter, just click the corresponding arrow button, move the Time Slider to a new frame, and change the parameter. Figure 45.23 shows how these buttons look in the Lens Effects Flare dialog box.

FIGURE 45.23

Green arrow buttons in the Lens Effects Flare dialog box identify the parameters that can be animated for this effect.

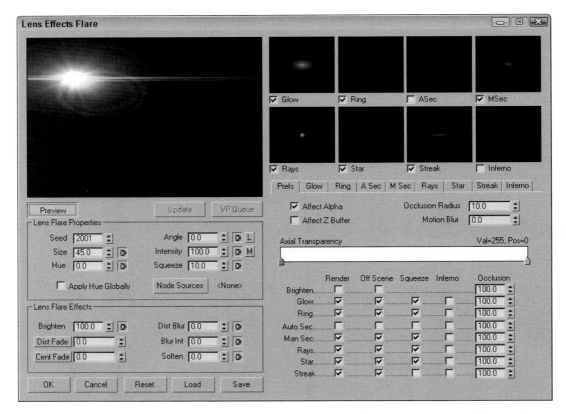

Each Lens Effects dialog box also includes a preview pane in the upper-left corner with three buttons underneath. Clicking the Preview button renders all enabled lens effects in the preview pane. The VP Queue button renders the current Video Post queue. Using the preview pane, you can get an idea of how the final output should look. With the Preview button enabled, any parameter changes in the dialog box are automatically updated in the preview pane. The Update button enables you to manually update the preview. You can right-click the Preview pane to change its resolution for faster updates at lower resolutions.

Tip

If the VP Queue button is enabled, a default Lens Effects image is displayed. Using this image, you can play around with the various settings while the Preview mode is enabled to gain an idea of what the various settings do.

You can save the settings in each Lens Effects dialog box as a separate file that can be recalled at any time. These saved files have an .lzf extension and can be saved and loaded with the Save and Load buttons at the bottom left of the dialog box.

Adding flares

The Lens Effects Flare dialog box includes controls for adding flares of various types to an image. This dialog box includes a main preview pane and several smaller preview panes for each individual effect. The check boxes below these smaller preview panes let you enable or disable these smaller panes.

Under the main preview pane are several global commands, and to their right is a series of tabbed panels that contain the settings for each individual effect type. The first panel is labeled Prefs and sets which effects are rendered (on and off scene), which are squeezed, which have the Inferno noise filter applied, and which have an Occlusion setting.

The settings for the individual flare types are included in the subsequent tabbed panels. They include Glow, Ring, A Sec, M Sec, Rays, Star, Streak, and Inferno. These tabbed panels include gradient color bars for defining the Radial Color, Radial Transparency, Circular Color, Circular Transparency, and Radial Size. Each of these tabbed panels has different settings, but Figure 45.24 shows the tabbed panels for the Glow and Ring effects.

FIGURE 45.24

The Glow and Ring tabbed panels are representative of all the different lens effect settings.

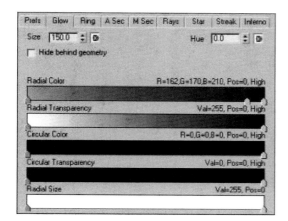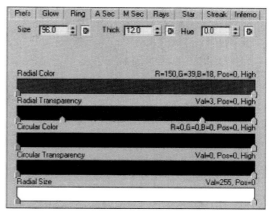

The gradient colors found in these tabbed panels are controlled by flags that appear under the gradient band. Double-clicking a flag opens a Color Selector dialog box where you can select a new color. Dragging these flags moves the gradient color. You can add a new flag to the band by clicking under the gradient away from the existing flags. The active flag is colored green. To delete flags, select them and press the Delete key. By right-clicking the gradient band, you can access a pop-up menu of options that let you access several options for the selected gradient color. You can even load and save gradients. Gradients are saved as files with the .dgr extension.

The rightmost tabbed panel, shown in Figure 45.25, is labeled Inferno and provides a way to add noise to any of the effects. The Prefs tabbed panel includes a check box for enabling the Inferno settings for each effect. Inferno noise can be set to three different states: Gaseous, Fiery, and Electric. If your effect is looking too perfect, you can add some randomness to it with the Inferno option.

FIGURE 45.25

The Inferno tabbed panel includes options for enabling noise for the various flare effects.

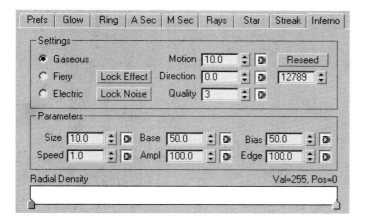

Adding focus

The Lens Effects Focus dialog box, shown in Figure 45.26, includes options for adding Scene Blur, Radial Blur, and Focal Node effects. If you click the Select button, the Select Focal Object dialog box opens and lets you choose an object to act as the focal point for the scene.

FIGURE 45.26

You can use the Lens Effects Focus dialog box to blur an image.

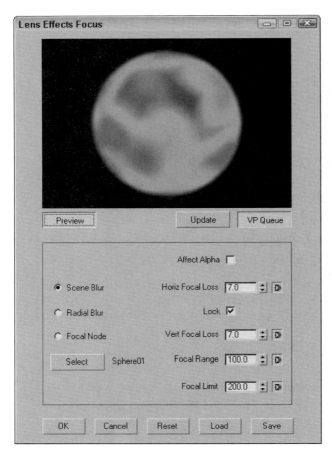

You also can set values for the Horizontal Focal Loss and Vertical Focal Loss or enable the Lock option to lock these two parameters together. The Focal Range and Focal Limit values determine the distance from the focal point where the blurring begins or reaches full strength. You also can set the blurring to affect the Alpha channel.

Adding glow

The Lens Effects Glow dialog box, shown in Figure 45.27, enables you to apply glows to the entire scene or to specific objects based on the Object ID or Effects ID. Other Source options include Unclamped, Surf Norm (Surface Normals), Mask, Alpha, Z High, and Z Lo. This dialog box also enables you to filter the glow using options such as All, Edge, Perimeter Alpha, Perimeter, Bright, and Hue.

Additional tabbed panels under the preview pane let you control the Preferences, Gradients, and Inferno settings. In the Preferences tabbed panel, you can set the color of the glow to be based on the Gradient tabbed panel-defined gradients, based on Pixel or a User-defined color. You also can set the Intensity in the Preference tabbed panel.

FIGURE 45.27

Use the Lens Effects Glow dialog box to make objects and scenes glow.

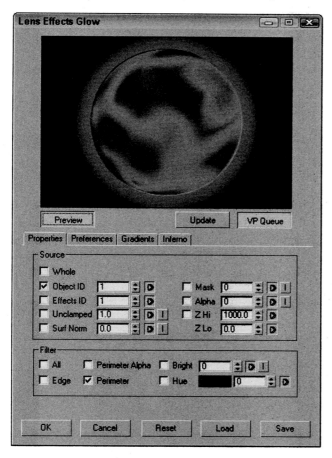

Adding highlights

The Lens Effects Highlight dialog box, shown in Figure 45.28, includes the same Properties, Preferences, and Gradient tabbed panels as the Glow dialog box, except that the effects it produces are highlights instead of glows. The Geometry tabbed panel includes options for setting the Size and Angle of the highlights and how they rotate away from the highlighted object.

FIGURE 45.28

Use the Lens Effects Highlight dialog box to add highlights to scene objects.

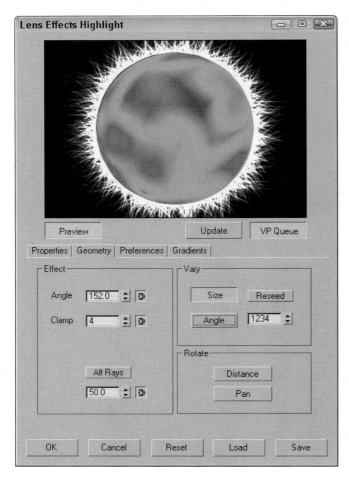

Tutorial: Making a halo shine

When it comes to glowing objects, I think of radioactive materials, celestial objects like comets and meteors, and heavenly objects like angels. In this tutorial, I'm leaning toward heaven in an attempt to create some glory. But because I couldn't locate an angel, you'll use a simple halo over a very good little dinosaur.

To add highlights to a halo using the Video Post interface, follow these steps:

1. Open the Glowing halo.max file from the Chap 45 directory in the downloaded content set.

   This file contains a cartoony dinosaur model and a halo. The halo object has been set to the G-Buffer Object Channel of 1 in its Object Properties dialog box.

2. Choose Rendering→Video Post to open the Video Post interface. Click the Add Scene Event button on the toolbar.

   The Add Scene Event dialog box appears.

3. Type a name for the event in the Label text field, choose the Quad 4 - Perspective viewport and click OK.

   The event is added to the Queue pane.

4. Select the halo object in the viewport and click the Add Image Filter Event button on the toolbar (or press Ctrl+F) to open the associated dialog box. Select Lens Effects Highlight from the drop-down list, and click the Setup button.

   The Lens Effects Highlight dialog box appears.

5. Click the VP Queue button followed by the Preview button to see the rendered scene. In the Properties tabbed panel, enable the Object ID option and set the Object ID to **1** to match the G-Buffer Object ID for the halo object. In the Filter section, enable the All option. In the Preferences tabbed panel, set the Size to **3.0**, Points to **4**, the Color option to **Pixel**, and Intensity to **100**. Then click OK.

6. Click the Execute Sequence button on the toolbar (or press Ctrl+R), and then click Render in the Execute Video Post dialog box.

Figure 45.29 shows the completed halo in all its shining glory.

FIGURE 45.29

Using the Lens Effects Highlight dialog box, you can add shining highlights to objects like this halo.

Summary

Using the Volume Light effect can give your scene the extra dazzle it needs.

One of the more useful render effects is the Lens Effects render effect. It can be used to add glows, stars, and flares to a scene or to specific scene objects. Flares, glows, and streaks also can be added in post production using the Lens Effects Filters available in the Video Post interface.

This chapter covered these topics:

* Used the Volume Light atmospheric effect
* Using the Lens Effects to create glows, rays, and stars
* Adding Lens Effects Filters in the Video Post interface

The next chapter delves into creating dynamic animations using the MassFX interface; I hope you're ready for some physics.

Part X

Using Dynamic Animation Systems

IN THIS PART

Chapter 46
Simulating Physics-Based Motion with MassFX

IN THIS CHAPTER

Exploring the MassFX tools

Setting object type and running simulations

Assigning object properties

Setting collision mesh type

Using constraints

Baking animation keys

Working with mCloth and ragdolls

When you speak of MassFX in the Autodesk® 3ds Max® 2020 software, you really are speaking of physics. Physics is one of the coolest arms of science because it deals with the science of matter and energy and includes laws that govern the motions and interactions between objects. For animators, this is great news because what you are trying to do is to animate the motions and interactions between objects.

So, should all animators study physics? The answer is absolutely. Understanding these laws through study and experience will sharpen your animating skills. But you also can take advantage of the work that other developers have done in understanding the laws of physics and turning them into a product that ships with 3ds Max. The other developers are the PhysX group at NVIDIA, and the product is MassFX.

The MassFX physics engine included in 3ds Max is the same engine commonly used in games to simulate game-world physics.

Using MassFX, you can simulate many physical properties like density, mass, and friction and automatically capture keyframes as the objects interact. It's like getting a physics degree for free.

The MassFX tools include everything you need to access the MassFX physics simulation engine. After physical properties are defined, you can define physical forces to act on these objects and simulate the resulting animation. Not only does MassFX make difficult physical motions realistic, but it also is fun to play with.

Understanding Dynamics

Dynamics is a branch of physics that deals with forces and the motions they cause, and regardless of your experience in school, physics is your friend—especially in the world of 3D. Dynamics in 3ds Max can automate the creation of animation keys by calculating the position, rotation, and collisions between objects based on physics equations.

Consider the motion of a simple yo-yo. Animating this motion with keys is fairly simple: Set rotation and position keys halfway through the animation and again at the end, and you're finished.

Now think of the forces controlling the yo-yo. Gravity causes the yo-yo to accelerate toward the ground, causing the string to unwind, which makes the yo-yo spin about its axis. When it reaches the end of the string, the rotation reverses and the yo-yo rises. Using Gravity and Motor Space Warps, you can simulate this motion, but setting the keys manually is probably easier for these few objects.

But before you write off dynamics, think of the motion of popcorn popping. With all the pieces involved, setting all the position and rotation keys for every piece would take a long time. For this system, using dynamics makes sense.

Dynamic tools let you specify objects to include in a simulation, the forces they interact with, and the objects to be involved in collisions. After the system is defined, the Dynamics utility automatically calculates the movement and collisions of these objects according to the forces involved, and then it sets the keys for you.

Note

MassFX isn't the first physics-based system available in 3ds Max. Early versions of 3ds Max had a Dynamics utility, and recently dynamics were possible using the reactor system in 3ds Max. These early systems have been replaced by MassFX, which is easier to use and much more robust than earlier versions.

Using MassFX

The MassFX plug-in was developed by a company named PhysX, which is now part of NVIDIA. MassFX is a complex piece of software with a huge assortment of features that enable you to define physical properties and forces and have the scene automatically generate the resulting animation keys as the objects interact while following the laws of physics.

The MassFX plug-in interface can be accessed from a simple toolbar that's opened using the right-click pop-up menu on any toolbar. The toolbar is named MassFX, shown in Figure 46.1, and it is surprisingly simple, consisting of only a few buttons. With the MassFX toolbar, you can open the MassFX Tools dialog box, define objects and constraints to include in the simulation, and control the simulation. The toolbar also includes several flyout tools that are described in Table 46.1. The MassFX commands are also available in the Animation→MassFX menu.

FIGURE 46.1

Use the MassFX toolbar to open the MassFX Tools dialog box and define and control simulations.

TABLE 46.1 MassFX Toolbar Buttons

Toolbar Button	Name	Description
	World Parameters, Simulation Tools, Multi-Object Editor, Display Options	Opens the MassFX Tools dialog box and displays the selected panel
	Set Selected as Dynamic Rigid Body, Set Selected as Kinematic Rigid Body, Set Selected as Static Rigid Body	Sets the object type for the selected object to dynamic, kinematic, or static
	Set Selected as mCloth Object, Remove mCloth from Selected	Adds or removes mCloth modifier to or from the selected object

	Create Rigid Constraint, Create Slide Constraint, Create Hinge Constraint, Create Twist Constraint, Create Universal Constraint, Create Ball & Socket Constraint	Adds a constraint to the scene that limits the motions of specific objects
	Create Dynamic Ragdoll, Create Kinematic Ragdoll, Remove Ragdoll	Adds all the ragdoll controls and constraints to the existing skeleton
	Reset Simulation	Resets the scene to a state as before the simulation started
	Start Simulation, Start Simulation without Animation	Begins the simulation
	Step Simulation	Moves the simulation forward one frame

The MassFX process

Before getting into the details of MassFX, I want to briefly explain the process involved in using this feature. MassFX works with geometry that is defined with certain physical properties. After these properties are defined, the MassFX engine can take over and determine how all the various objects interact with one another.

The first step is to assign each object to be included in the simulation an object type. These can be Dynamic, Kinematic, or Static. Dynamic objects are ones that are included in the simulation, Kinematic objects apply forces to the simulation and can be changed to a Dynamic object to interact with other objects, but Static objects don't move and provide the walls and floor for the objects to crash against.

In addition to the object type, you can specify values for the object's Density, Mass, Friction, and Bounciness using the MassFX Tools dialog box. More on these values is presented later in this chapter.

When an object is assigned a type, it automatically assumes a proxy shape that surrounds the object that is used to calculate any collisions. To keep the simulation calculations simple, this default proxy shape is usually a simple box or sphere. MassFX allows you to use a more complex collision mesh, but doing so compounds the simulation calculations and should be used sparingly. Within the Multi-Object Editor panel of the MassFX Tools dialog box are several options for defining the collision mesh, including a tool to generate a custom collision mesh based on the object's geometry.

Establishing the simulation properties

The MassFX Tools dialog box, shown in Figure 46.2, is opened using the first button in the MassFX toolbar or with the Animation→MassFX→Utilities→Show MassFX Tools menu command. The MassFX Tools dialog box includes four separate panels, and each of the flyout buttons for the first button provides access to the various panels.

The first panel, World Parameters, holds the settings for the World values. These World values are the global settings for the simulation. The global values defined in the World Parameters panel apply to all objects included in the simulation, but individual values can be changed for each individual object using the settings in the Multi-Object Editor panel.

FIGURE 46.2

The MassFX Tools dialog box includes World Parameters settings.

By default, the ground plane acts as a static object, and gravity is enabled. You can change the value of gravity and even make it a positive value so objects float upward using the settings in the World Parameters panel. You also can use the Pick Gravity button to select a Gravity Space Warp object in the scene to define the gravity value.

The number of Substeps determines the amount of calculations that the simulation goes through. Generally, the higher the Substeps value, the more accurate the simulation collisions and the longer it takes to complete. This is also impacted by the Solver Iterations. More Solver Iterations are needed for simulations involving a lot of constraints and collisions, but a value larger than 30 is overkill.

If you know that some objects, such as a projectile, will travel very fast, you can enable the Use High Velocity Collisions option. This option uses a different algorithm that prevents gross overlapping of surfaces for high velocity objects.

The Use Adaptive Force option causes the simulation to alter the force values within a range to prevent the problems caused by zero value forces when computing simulations. The Generate Shape Per Element option creates a separate collision mesh for each object element subobject when enabled. If disabled, the entire object gets a single collision mesh, which can result in less accurate collisions, but faster results.

Within the Advanced Settings rollout, you can set the Sleep, High Velocity Collisions, and Bounce settings to be Automatic or Manual. If the Manual option is selected, you can set the Sleep Energy or Minimum Speed threshold values. These settings are used to control the amount of jitter in the simulation. As objects get close to stopping, they can jitter around as small values are being computed. The Automatic options are generally good at controlling jitter, but if you find that some objects are still moving around too much, set the option to Manual and increase the threshold value.

You also can set the Contact Distance and Rest Depth values for collisions in the scene, which is the amount of overlap between collision meshes that is allowed during collisions and at rest. If the collision meshes are not accurate using bounding boxes, allowing some overlap will make the simulation look better.

Finally, the Engine rollout includes an option for enabling multithreading if your system includes multiple cores. The Hardware Acceleration option can be used for Nvidia cards to speed the simulation calculations.

Starting and stopping the simulation

After the world and object properties are set, you can click the Start Simulation button in the MassFX toolbar or from the Animation→MassFX→Simulation→Play Simulation menu and the results are immediately calculated and displayed within the viewport. The MassFX toolbar and the Animation→MassFX→Simulation→Reset Simulation menu also includes a button to reset the simulation and another button to step through the simulation one frame at a time.

As a flyout option under the Start Simulation button is an option to Start the Simulation without Animation. This runs the simulation without changing the current frame on the Track Bar.

These same simulation buttons are found in the Simulation Tools panel of the MassFX Tools dialog box, along with buttons to bake and unbake keys and Capture Transform. The Simulation Settings rollout includes several options that define what to do when the last frame of the animation is reached. The options are to simply continue the simulation, to stop the animation, or to loop the animation by resetting the simulation. Be aware that any motion beyond the last frame will not be captured if you bake the motion down into keyframes.

Tutorial: Filling a glass bowl

Imagine trying to animate a bunch of marbles falling into a glass bowl. If you were using keyframes, determining whether an object overlaps another would be difficult, but with this quick example you see the power of MassFX.

To animate marbles falling into a glass bowl, follow these steps:

1. Open the Glass bowl of marbles.max file from the Chap 46 directory in the downloaded content set.

   This file includes a glass bowl and several marbles positioned above its opening.

2. Right-click the main toolbar, and select MassFX Toolbar. Then select all the marbles located above the bowl, and click the Set Selected as Dynamic Rigid Body button.

3. Select the tabletop object, and click the Set Selected as Static Rigid Body button. This button is a flyout under the Set Selected as Dynamic Rigid Body button. Then click the Start Simulation button in the MassFX toolbar, and notice how the marbles all fall and spread out on the tabletop.

4. Select the bowl object, and set it as a static object by repeating Step 3. In the Modify panel, set the Shape Type in the Physical Shapes rollout to Original, and press the Start Simulation button again. This time, the marbles all fall into the bowl.

Figure 46.3 shows the bowl full of marbles positioned using MassFX.

FIGURE 46.3

MassFX can compute all the collisions between all these marbles.

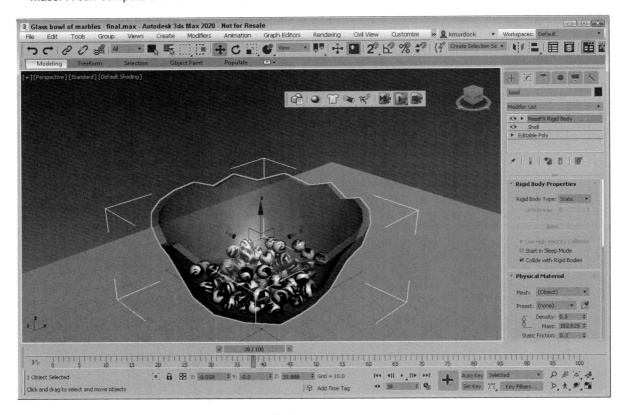

Setting Object Properties

One of the first steps in creating with a simulation is defining the object properties. For example, a simple sphere object in 3ds Max could represent a bowling ball, an orange, or a tennis ball. Each of these objects responds very differently when being animated to drop on the floor.

Setting the object type

All objects involved in a MassFX simulation must be assigned one of three object types. The available object types are:

* **Dynamic:** These objects move when they collide with other objects. Dynamic objects also are affected by gravity.

* **Kinematic:** These objects impact other dynamic objects in the scene, but they don't move when they collide with dynamic objects. They can be animated and often provide the force to the simulation. Kinematic objects are not affected by gravity.

* **Static:** These objects have an infinite mass and don't move when they collide with other objects. They frequently are used for ground and walls. They also are not affected by gravity.

These object types are available as buttons in the MassFX toolbar. To specify an object as one of these types, simply select the object and click the object type button in the toolbar. Any object with a defined object type is automatically added to the simulation. Any scene object without an object type is simply ignored. Kinematic objects can be animated using standard keyframe animation and then be changed to a dynamic object at a given frame using the Until Frame value located in the Rigid Body Properties rollout. This provides a way to add forces to the simulation such as a marble that is animated being shot at a group of marbles. Just before colliding, you can set the shot marble to become dynamic so it continues its motion and then interacts with the other marbles based on the physics.

When one of the object types is assigned to an object, it appears as a MassFX Rigid Body modifier in the Modifier Stack. This modifier has several subobjects, including Initial Velocity, Initial Spin, Center of Mass, and Mesh Transform. If you remove this modifier, the object is removed from the simulation and returned to its original state.

When an object's type is defined as a dynamic, kinematic, or static, the system automatically assigns some rough physical properties based on the object's size in the scene. Static objects, for example, are given a large mass, which essentially makes them immovable, and small dynamic objects are given very small mass values.

The Rigid Body Properties rollout also includes options for disabling collisions and computing for high-velocity objects.

When dynamic objects are placed in the scene, it is often difficult to get them positioned exactly on top of one another. Actually, it is better to have a small gap between objects so they don't overlap. Overlapping objects will repel one another when the simulation first starts causing motion before you want it. Within the Rigid Body Properties rollout is the Start in Sleep Mode option that causes all dynamic objects to freeze in their initial positions until hit by another object. Enabling this option helps prevent the initial movement of stacked objects when the simulation starts due to gravity.

Regardless of the object type and its default values, you can set the actual physical properties for the selected object using the settings in the Physical Material rollout in the Command Panel when the object is selected. You can access these same properties in the Multi-Object Editor panel of the MassFX Tools dialog box, shown in Figure 46.4. You also can select and set the properties for multiple selected objects at the same time.

FIGURE 46.4

The Multi-Object Editor panel of the MassFX Tools dialog box includes physical property values for the selected object.

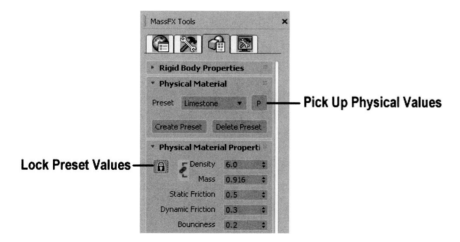

Density, mass, friction, and elasticity

Although many different physical properties are available, the MassFX system is mainly concerned with collisions between rigid body objects, which can be computed using a short list of physical properties including Density, Mass, Static Friction, Dynamic Friction, and Bounciness.

The Density property is the amount of mass per volume, so it's related to mass. The Mass property defines how heavy the object is. For example, a bowling ball has a higher mass value than a Ping-Pong ball, and a Ping-Pong ball is not as dense (or heavy) as a golf ball, even though they are roughly the same size.

Friction is the force of contact between two touching objects. It defines how resistant the object is to rolling or sliding along the floor. For example, when you slide an air-hockey puck over an air-hockey table, the friction is very low because of the jets of air, but moving a piece of sandpaper across a piece of wood has a very high value of friction that resists the movement.

Friction is actually defined by two properties. Static friction is the initial force required to start an object sliding across another, and dynamic friction is the amount of force required to keep the object sliding over the top of the other. Both of these properties are available in the MassFX system.

Note

A rigid body with a Mass value of 0 causes problems for all calculations, so the system automatically sets the Density or Mass value to 0.01 when you try to set it to 0.

Creating presets

If you've done some research and figured out the exact physical properties for a specific object, you can use the Create Preset button in the Physical Material rollout of the MassFX Tools dialog box to save and name the physical values for the selected object. All defined presets are populated in the drop-down list. This provides a great way to reuse values that you know are right. The eyedropper tool lets you quickly pick the properties from another object in the scene.

When a preset is used, all its properties are automatically locked, but you can unlock and change them using the Lock icon in the upper-left corner of the Physical Material Properties rollout of the MassFX Tools dialog box in the Multi-Object Editor panel.

Defining collision boundaries

Another common property that you can set pertains to how the object deals with collision detection. You can select the volume to use to determine when two objects collide with each other. If this sounds a bit funny because any collision volume that doesn't use the actual mesh would be inaccurate, you need to realize that a complex simulation with lots of collisions of complex objects could take a long time to compute. If MassFX has only to compute collisions based on the object's bounding box instead of the actual mesh object, the simulation runs much more quickly, and the inaccuracies aren't even noticeable.

Before deciding on the collision boundary to use, you need to determine whether an object is convex or concave. A *convex* object is one that you can penetrate with a ray and cross its mesh boundary only twice. *Concave* objects require more than two crossings with an imaginary ray. In other words, concave meshes have surface areas that bend inward like a doughnut and convex meshes don't, like a normal sphere. Convex meshes are much easier to use when calculating collisions than concave meshes.

Using convex meshes

A convex object can use several of the options found in the Physical Meshes rollout, including Sphere, Box, Capsule, and Convex. Use the shape that best fits the mesh you are working with.

The properties for the mesh types are listed in the Physical Shapes rollout located in the Command Panel. When an object type is applied to an object, the Convex mesh type is applied by default because it generally works for all objects. The Convex mesh type is created from a Geosphere object represented by 32 vertices. If you reduce the number of vertices—or switch to a Sphere, Box, or Capsule mesh type—you can speed up the simulation calculations.

When the Convex shape type is selected, an Inflation value shows up in the Physical Mesh Parameters rollout. Using the Inflation value, you can increase or decrease the size of the collision mesh. Additional settings are available in the Physical Mesh Parameters rollout for each type. For example, the Sphere mesh type has a Radius value and the Box mesh type has Length, Height, and Width values. If you make changes to the Convex mesh type that don't work, you can always regenerate the mesh using the Regenerate Selected button.

If the collision mesh shape is right, but its position is off, you can choose the Mesh Transform subobject mode in the Modifier Stack and change the mesh's position and orientation.

Using concave meshes

If you have a concave object that you want to use in the simulation, you should first try to use a convex collision mesh if possible, but if you can't (such as with the bowl in the preceding example), the best option is the Concave mesh type. For example, a doughnut-shaped object bouncing in a scene could easily use a flat, sphere-shaped convex collision mesh because it isn't likely that a smaller object will go through the doughnut's center. However, if you are dealing with a basketball rim, enabling a convex mesh won't work because objects will be going through the rim's center, so a concave collision mesh is required.

This Concave option lets you specify the maximum number of vertices per hull and several other settings and includes a Generate button to automatically create the collision mesh based on the object's geometry. Once generated, the number of hulls and vertices included in the collision mesh are displayed. If these numbers are still too high, you can change the Mesh Detail percentage, Min Hull Size, Max Vertices per Hull, or Improve Fitting option and try the Generate button again. Remember that having more vertices means better accuracy in the simulation, but also more time to complete.

The Original option works well for concave objects that are set to static and won't move during the simulation. It uses the actual mesh as the collision mesh.

The Custom option lets you create a collision mesh based on a selected piece of geometry in the scene. It includes buttons for picking the source object and extracting a mesh from the source object. The proxy mesh should be a low-resolution version of an object used here for collision detection.

Caution
Custom collision meshes are limited to objects with 256 vertices or fewer.

Setting initial motion

Every dynamic object has an initial motion setting that you can control to start the object in motion. This works whether gravity is enabled or disabled. The Initial Motion values are located in the Advanced rollout in the Command Panel and can be set to an Absolute or Relative value. If a kinematic object is in motion when it is converted to a dynamic object using the Until Frame setting in the Rigid Body Properties rollout of the MassFX Tools dialog box and it has an initial motion value, the Absolute setting uses only the initial motion value, and the Relative option adds the initial motion value to the existing keyframed motion that the object already has. The X, Y, and Z value denote a direction, and Speed sets how fast the object is moving. A setting for the initial spin of the object is available as well.

The Damping values are used to slow down the motion (Linear) and the rotation (Angular) of objects in motion with higher values causing the objects to rapidly decrease their speed.

Under the MassFX Rigid Body modifier that is applied to the selected object, you can select the Initial Velocity and Initial Spin subobject modes, and an arrow shows the current direction of the initial movement or spin. While either of these subobject modes is selected, you can use the Select and Rotate tool to change them in the viewport. This is easier than having to enter values in the rollout.

The modifier also includes an option for changing the Center of Mass location and for moving and manipulating the collision mesh's transform.

Displaying interactions

The Display Options panel of the MassFX Tools dialog box has a number of options that you turn on and off to see the simulation properties. The Display Physical Meshes option shows the collision mesh for all objects or only for the selected objects if the Selected Objects Only option is enabled.

If you turn on the Enable Visualizer option, a whole range of different properties are made visible as the simulation proceeds including arrows showing object speed, contact points between objects and collision meshes.

Tutorial: Knocking over milk cans

You can add motion to the simulation objects in two ways. One is to use the initial motion values in the Advanced rollout of the Multi-Object Editor panel of the MassFX Tools dialog box. The other is to set an object as a Kinematic object that is animated using standard keyframes and then switched to a Dynamic object at a given frame with the Until Frame option in the Rigid Body Properties rollout of the MassFX Tools dialog box.

To animate the milk can tipping carnival game, follow these steps:

1. Open the Tipping milk cans.max file from the Chap 46 directory in the downloaded content set.

   This file includes several milk cans positioned on a cylinder along with some balls.

2. Right-click the main toolbar, and select MassFX Toolbar. Then select all the milk cans located above the cylinder, and click the Set Selected as Dynamic Rigid Body button. Then click the Multi-Object Editor button on the MassFX toolbar, and set the Density value to **10** for all the milk cans.

3. Select the cylinder and the floor objects, and click the Set Selected as Static Rigid Body button. Select the Cylinder object and set the Mesh Type in the Physical Mesh rollout of the Multi-Object Editor panel to Convex.

4. Select the smallest ball object, and click the Set Selected as Dynamic Rigid Body button. Then open the Advanced rollout in the Multi-Object Editor panel and set the X Initial Velocity value to **-1** and the Speed value to **2490.** Then click the Start Simulation button in the MassFX toolbar, and notice how the first ball falls a little short. Click the Reset Simulation button in the MassFX toolbar.

5. Locate and click the Time Configuration button beneath the Play button at the bottom of the interface and in the Time Configuration dialog box, set the End Time to **300.**

6. Select the second ball object, and click the Set Selected as Kinematic Rigid Body button. Then enable the Until Frame option in the Rigid Body Properties rollout, and set it to **100.** Then set the Density to **3,** the X Initial Velocity value to **-1,** and the Speed value to **2490.** Then click the Start Simulation button in the MassFX toolbar, and notice how the second ball knocks them all down.

Figure 46.5 shows the large ball striking the milk cans.

FIGURE 46.5

Kinematic objects can be made to start at a later frame.

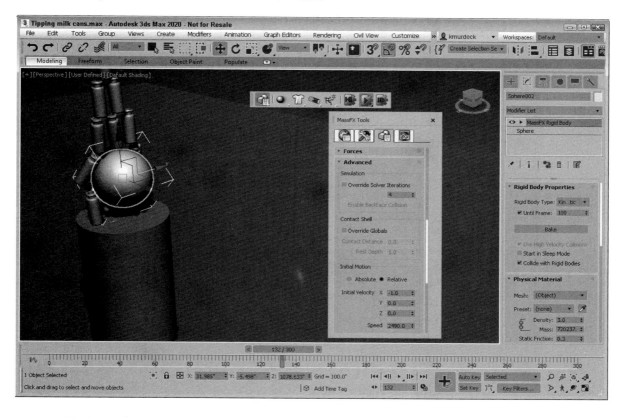

Using Constraints and Baking Keys

Constraints are ways to limit the amount of motion that an object can do. Using constraints can help control objects in the scene as they interact with other objects. Perhaps the simplest constraint isn't a constraint at all. If you set an object to be Static, it won't move.

Other constraints found in the MassFX toolbar include Rigid Constraint, Slide Constraint, Hinge Constraint, Twist Constraint, Universal Constraint, and Ball & Socket Constraint. You can also find these constraints in the Animation→MassFX→Constraints menu. To apply a constraint, you need two objects, and they both must be selected. The first object you select becomes the parent object and the second the child.

Note
You also can select a single object and apply a constraint and then pick the parent object later using the button in the Modify panel.

After the constraint is selected, a helper object titled UConstraint is added to the scene, and you can drag out the helper object to show the range of the applied constraint. When the helper constraint object is selected, the parameters for the constraint are displayed in the Command Panel.

Constraints can be made Breakable, and you can set the Max Force and/or Torque required to break the constraint. After a constraint is broken, it no longer has any effect on the simulation.

Within the Connection section of the General rollout, you can alter the objects that are used for the Parent and Child objects. Additional rollouts define the available translation, swing, twist, and spring properties for each type of constraint. For each, you can set the constraint to be Locked, Limited, or Free about each axis.

If the Limited option is enabled, you can set the limits using the values such as Limit Radius, Bounce, Spring, and Damping.

Although you can move the constraint helper to wherever it needs to be, it is often easier to define the constraint's location by positioning the object's pivot. Using the Child Attach Point subobject mode, you can change the constraint's pivot point. There is also a subobject mode for setting the Child Initial Twist.

Tutorial: Opening a door

Each of the constraints enables different types of motion and some situations require multiple constraints, such as the hinge on a door.

To use constraints to restrict the motion of a door, follow these steps:

1. Open the Door with a hinge.max file from the Chap 46 directory in the downloaded content set. This file includes a simple door between two walls.

2. Open the MassFX toolbar. Then select the door, ball and wall objects, and click the Set Selected as Dynamic Rigid Body button. Select the floor object, and set it as Static Rigid Body.

3. Select the far wall object and then the door object, and click the Create Hinge Constraint from the MassFX toolbar. Then drag in the viewport and click to set the constraint's size. Move the constraint gizmo so that its pivot is positioned at the point between the wall and the door, and rotate the constraint gizmo so it allows the door to swing away from the ball.

4. Select both wall objects, and set their Density value to **1000** in the Physical Material Properties rollout of the MassFX Tools dialog box. This prevents the walls from moving.

Caution
Static objects cannot be used as a parent or child for a constraint.

5. Select the ball object, and set its Density to **20**, the X Initial Velocity value to **-1**, and the Speed value to **4900.** This should be enough to give the door a good kick.

6. Click the Start Simulation button in the MassFX toolbar.

Figure 46.6 shows the door flying open.

FIGURE 46.6

Using the hinge constraint controls the motion of this door.

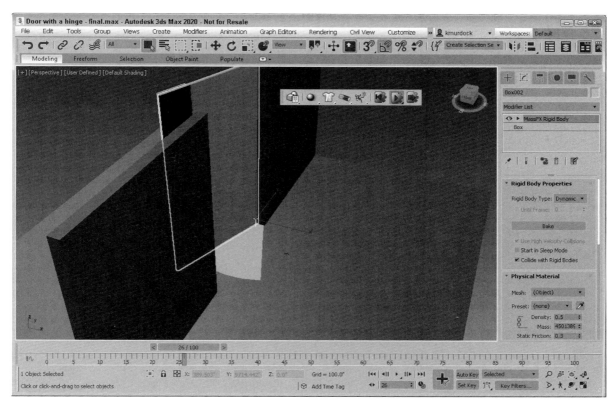

Capturing the Simulation Motion as Keys

After you're happy with the motion created by the simulation, you can capture the motion as keys using the Bake command. There is a Bake button in the Rigid Body Properties rollout of the Multi-Object Editor panel in the MassFX Tools dialog box. The Simulation Tools panel also has several different bake options, including options to Bake All, Bake Selected, Unbake All, and Unbake Selected.

After the motion is baked, the keys for each object in the simulation show up on the Track Bar, and the animation can be manipulated and edited using the animation tools and the Track View.

The Simulation Tools panel also includes another helpful button. The Capture Transforms button causes the current state of the selected objects to be reset as its new initial position. If the Reset Simulation button is then used, this new initial position is used instead of the object's original location. This is great because it give you a chance to let the objects settle at the start simulation and then reset them in this location before starting the simulation.

Within the Utilities rollout of the Simulation Tools panel, the Explore Scene button opens the Scene Explorer where it shows all the objects included in the simulation and their respective simulation properties including SimType, SimMode, SimEnabled, and SimBaked. From this scene, you can quickly enable and disable multiple objects or change their object type.

The Validate Scene button runs a quick check on the simulation, looking for potential problems that might be encountered when the scene is exported. It looks for issues such as non-uniform scaling and skewed objects. The Export Scene button opens a dialog box where you can save the exported simulation. Simulations are exported using the PhysX and APEX (PXPROJ) format. After you select a name, a dialog box of settings appears. Using this dialog box, you can select a folder name, select to include the FBX file, and export as XML or Binary.

Tutorial: Dropping a plate of donuts

All the great books have an element of tragedy, so consider a policeman carrying a dozen donuts on a plate when he stumbles and drops the plate. Donuts everywhere; how tragic! This animation sequence would be difficult or at least time-consuming if not for MassFX.

To use MassFX to animate a falling plate of donuts, follow these steps:

1. Open the Falling plate of donuts.max file from the Chap 46 directory in the downloaded content set.
   This file includes a simple plate of donuts created from primitives.

2. Open the MassFX toolbar, select all the donuts and the plate, and click the Set Selected as Dynamic Rigid Body button.

3. Because the floor object is aligned to the ground plane, just make sure that the Use Ground Collisions and the Directional Gravity options are enabled in the World Parameters panel of the MassFX Tools dialog box.

4. Click the Start Simulation button in the MassFX toolbar.

5. After the simulation is complete and looks fine, click the Bake All button in the Simulation Tools panel of the MassFX Tools dialog box, and the keys for the animation are added to the Track Bar.

6. When it completes, press the Play Animation button (or press the / key) to see the results.

Figure 46.7 shows the upturned plate of donuts. Oh, the tragedy, the horror, the creamy fillings.

FIGURE 46.7

Animating these falling donuts was easy with MassFX.

Working with mCloth and Ragdolls

The MassFX system also includes features for working with dynamic cloth and ragdoll simulations. These features are available on the MassFX toolbar as well. Dynamic cloth bends and flows as part of the simulation as it is affected by gravity, wind, and other colliding objects. The ragdoll features apply a dynamic set of collision meshes and constraints to a skeleton, biped, or CAT object, and is useful for simulating falls and crashes of characters.

Using mCloth

The Set Selected as mCloth Object button applies the mCloth modifier to the selected object. If the object has multiple faces, it acts like cloth, but if the selected object has only a few polygons, it's stiff and doesn't flow well during the simulation. It's best to use the mCloth modifier on dense object to get a good flow. Standard plane objects work very well.

More on Cloth is covered in Chapter 47, "Working with Hair and Cloth."

When an mCloth object is selected, you can access its parameters in the Modify panel. These settings include many of the same tools used on the other dynamic objects, including the ability to switch between Dynamic and Kinematic behaviors, baking keys, and capturing states. Space Warp forces also can be selected and applied to the simulation using the Forces rollout.

There are several differences between the properties for mCloth objects and standard rigid objects. mCloth objects include settings for Density, Stretchiness, Bendiness, Damping, and Friction. You also can set mCloth to check for Self Collisions (because cloth often folds back on itself) and Tearing.

In addition to the normal MassFX constraints, mCloth objects also have a unique way of constraining vertices using the Vertex Subobject mode. Within this mode, you can select and make groups. These groups can then be detached from other constraints, welded or made a node of another object, simply pinned so they don't move, or made to define a tear.

Tutorial: Hanging a flag on a flagpole

In this example, we connect a cloth flag to a flagpole and let some wind blow it about.

To use MassFX to animate a flag on a flagpole, follow these steps:

1. Open the Wind blowing flag.max file from the Chap 46 directory in the downloaded content set.

   This file includes a flagpole, a plane object for the flag, and a simple Wind Space Warp.

2. Open the MassFX toolbar, select the plane flag object, and click the Set Selected as mCloth Object button.

3. Select the Vertex Subobject mode in the Modifier Stack, and drag over the row of vertices closest to the flagpole in the Front viewport. Click the Make Group button in the Group rollout, and name the group **Flagpole** and click OK.

4. With the Flagpole group selected in the Modify panel, click the Pin button and exit Vertex Subobject mode.

5. Click the Add button in the Forces rollout, and pick the Wind Space Warp in the viewport.

6. Click the Start Simulation button in the MassFX toolbar.

Figure 46.8 shows the flag blowing in the breeze.

FIGURE 46.8

The mCloth feature in MassFX makes flying a flag easy.

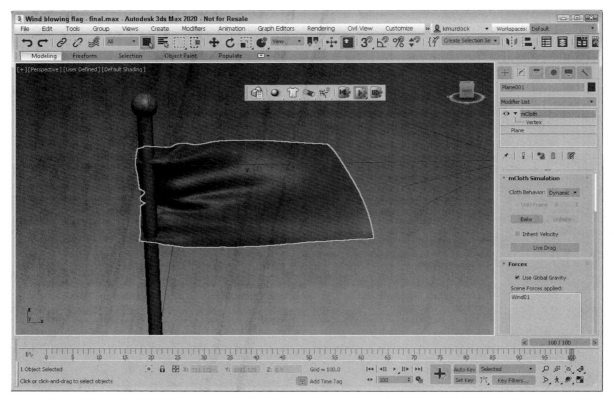

Creating a Ragdoll

To understand a ragdoll, imagine having a stuffed teddy bear falling off a bed in slow motion. The bear slowly hits the ground, and its limbs bounce around until it finally comes to a stop. This is how a ragdoll works in MassFX. It applies all the rigid body objects and constraints that let the character skeleton flop around realistically as part of a simulation. The benefit is that you don't have to create and configure all the constraints to get the character to fall realistically.

MassFX's ragdoll feature can be applied to any standard bone system, CAT, or Biped skeleton. Once applied, it can be used as a part of the simulation, including collisions with other objects. CAT and Biped objects work automatically, but if a bone system is used, you need to add all the bones to the system using the Pick or Add buttons in the Setup rollout.

Tutorial: Playing dodge ball

Playing with the rag doll object is just plain fun. For this example, we're going back to junior high to play some dodge ball.

To animate a ragdoll falling during a game of dodge ball, follow these steps:

1. Open the Dodgeball.max file from the Chap 46 directory in the downloaded content set.

    This file includes a default biped and three balls.

2. Open the MassFX toolbar, select the entire biped skeleton, and click the Create Dynamic Ragdoll button.

3. Select all three dodgeballs and make them Dynamic Rigid Bodies. Click the Multi-Object Editor to open the MassFX Tools dialog box, and set the Density to **5** in the Physical Material Properties rollout. Then set the Initial X Velocity to **-2** with a Speed of **2490** in the Advanced rollout.

4. Click the Start Simulation button in the MassFX toolbar.

Figure 46.9 shows the poor biped being pummeled by three balls.

FIGURE 46.9

The ragdoll feature lets you animate character falls and crashes.

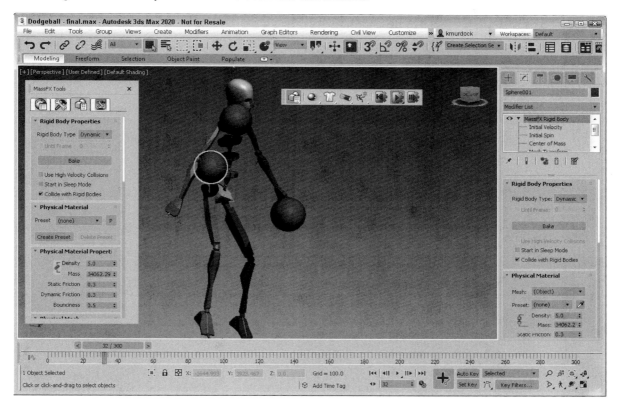

Summary

This chapter covered the basics of animating a dynamic simulation using MassFX. In this chapter, you accomplished the following:

* Understanding the principles of dynamics

* Accessing the MassFX tools

* Defining object types and properties

* Setting the collision mesh type

* Setting initial motion

* Working with constraints

* Baking animation keys

* Working with mCloth and ragdolls

The next chapter covers a very specific set of dynamic systems for creating and working with hair and cloth.

Chapter 47

Working with Hair and Cloth

IN THIS CHAPTER

Understanding hair

Growing hair and setting hair parameters

Styling hair

Enabling hair dynamics

Understanding cloth systems

Creating cloth panels using Garment Maker

Using the Cloth modifier

Simulating cloth dynamics

I want to start this chapter with a bold statement, "Bald Is Beautiful." I make this statement because I have a brother who, upon discovering his hair had started thinning, decided that bald was the way to go. But, after seeing far too many 3D models that were bald because of the unavailability of a decent hair plug-in, I also can declare that "Bald Is Boring."

Now that the Autodesk® 3ds Max® 2020 software has hair and fur, I expect to see the level of realism for a number of 3ds Max artists, including myself, take a quantum leap forward.

Adding hair to characters is a great leap forward, but another great resource is cloth. Creating cloth isn't that difficult. In fact, with a plane primitive, you can easily create a perfectly straight blanket, towel, or flag. However, animating cloth that drapes realistically over a character is very difficult and best left to dynamic engines like MassFX's mCloth.

Although you can still animate cloth collections using mCloth, 3ds Max also has a separate stand-alone cloth simulation system aptly named Cloth that you can use to create deformable cloth and to animate it as well.

The specialized hair and cloth systems can create believable, realistic hair, but the real benefit to making hair and cloth come alive is found in the dynamic abilities of both. Using these systems, you can simulate hair blown by the wind and cloth that drapes over and around underlying objects.

Understanding Hair

Although a section titled "Understanding Hair" sounds like it would be taken from a beauty salon guide, the way 3ds Max deals with hair is unique and needs some explanation. Hair, like particle systems, deals with thousands of small items that can bring even the most powerful computer screeching to a halt if not managed.

In 3ds Max, hair doesn't exist as geometry but is applied to scene objects as a separate modifier. This level of separation keeps the hair solution independent of the geometry and makes removing or turning off the hair solution as needed easy. It also keeps the viewport display from bogging down. The Hair and Fur modifier is a World Space Modifier (WSM), meaning that it is applied using the World Space coordinates instead of local ones.

The other half of the Hair and Fur solution is a render effect that allows the hair to be rendered. This render effect is applied and configured automatically when the Hair and Fur modifier is applied to an object. This causes the scene with hair to be rendered in two passes. The geometry is rendered first, followed by the hair.

Note

Hair can be rendered only when a Perspective or Camera view is selected. Hair cannot be rendered in any of the orthogonal views.

Another similarity to particle systems is that the hair follicles can be replaced by instanced geometry, so you can create a matchstick head character by replacing hairs with an instance of a matchstick.

Tip

Using instanced geometry with a fur system is a great way to create and position plants and ground cover.

The materials that are used on hair are defined in the Material Parameters rollout in the Modify panel instead of in the Material Editor. Many of the hair parameters have a square button to their right that lets you apply a map to the parameter.

When the Hair and Fur modifier is added to an object in the scene, a Hair and Fur render element becomes available that you can use to render out just the hair for compositing.

Working with Hair

Applying hair to an object is as easy as selecting an object and choosing the Hair and Fur WSM modifier, which is found in the Modifiers→Hair and Fur→Hair and Fur WSM. After hair is applied to an object, you can use the parameters in the Modify panel to change the hair's properties.

Growing hair

Hair can be grown on any geometry surface, including splines, by simply applying the Hair and Fur WSM to the selected object. When the Hair and Fur modifier is first applied to an object, it is applied to the entire surface of the selected object, and when it is applied to a set of splines, the hair appears between the first and last spline.

If you want to localize the hair growth to a specific area of the object, you can make a subobject selection using the controls in the Selection rollout. The available subobjects include Face, Polygon, Element, and Guides. After making a subobject selection, click the Update Selection to display the guide hairs only in the selected area in the viewports. Figure 47.1 shows hair grown on a girl's head.

Tip

After taking the time to make a subobject selection, create a selection set for the hair area for quick recall.

FIGURE 47.1

By making a subobject selection, you can control precisely where hair is grown.

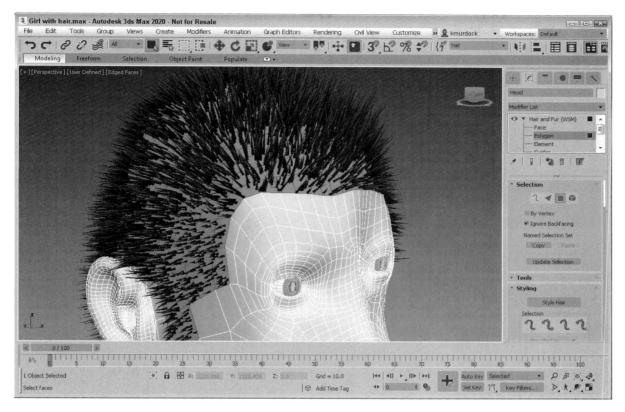

Applying hair to a single spline doesn't create any hair, but if multiple splines are included as part of the same Editable Spline object, the hair is interpolated between the various splines in the order they are attached following the spline's curvature. This provides a great way to add special features like a ringlet to an existing set of hair.

Setting hair properties

Several rollouts of properties can be used to change the look of the hair. The General Parameters rollout, shown in Figure 47.2, includes settings for the overall Hair Count, the number of Hair Segments between adjacent splines, the number of Hair Passes, Density, Scale, Cut Length, Random Scale, Root and Tip Thickness, and Displacement. The Hair Count value sets the total number of hairs for the given geometry. Higher values take longer to render but produce more realistic hair. The Hair Passes sets the number of render passes to use for determining hair transparency. The higher the Hair Passes value, the wispier the hair looks. The Rand Scale value provides a random amount of scaling for a percentage of hairs to look more natural. The Displacement value sets how far from the source object the hairs grow and can be a negative value.

Caution

Although the Hair Count value can accept huge numbers, adding a large number of hairs takes a long time to render and can really slow your system.

FIGURE 47.2

Hair properties can be altered using the General and Material Parameters rollouts.

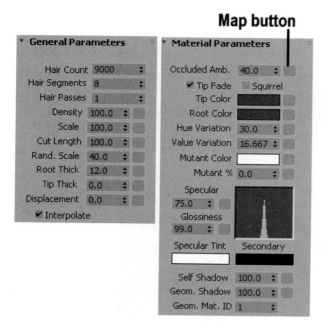

The buttons to the right of most of the parameters in the General Parameters rollout let you add a map to control the property with a grayscale bitmap. Figure 47.3 shows a simple plane object mapped with a bitmap used to control the hair density. The black areas of the bitmap have no hair growth, but the white areas have maximum growth. This provides a way to create patchy hair on a character or creature.

FIGURE 47.3

Many of the hair properties can be defined using maps.

The Material Parameters rollout, also shown in Figure 47.2, includes settings for the Occluded Ambient value, controlling the tip and root colors, Hue and Value Variations, the addition of mutant hairs, percentage to include and the mutant hair color, and specular and glossiness settings. The Occluded Ambient value sets the contrast for the hair lighting. Smaller values have a stronger contrast, and higher values appear more washed-out. The Hue and Value Variation values define a percentage that the hair color can deviate from the specified color to give the hair a more natural coloring. The Mutant percentage defines the percentage of hairs that are discolored and the Mutant Color swatch lets you set their color, such as white hairs.

Note

The Occluded Ambient property is applied as an effect, so it is actually a fake global illumination pass.

The Tip Fade option causes the hair to be more transparent toward the tip. The Squirrel option extends the Root Color farther up the hair strand. The Specular Tint and Secondary colors changes the color of specular highlights. The Tip Fade and Specular Tint options apply only to hairs rendered with Arnold. The Self Shadow value sets how much the individual hairs cast shadows on other hairs, and the Geometry Shadow settings determine how much shadow is contributed by other geometry objects. The Geometry Material ID is used to assign a material to geometry-rendered hair.

Many of the properties in the Material Parameters rollout also can be controlled using map buttons located to the right.

Caution

Map colors are multiplied with the base color value, so set the base color to white before using a map.

In addition to the General and Material Parameters, rollouts also are available for controlling the amount of Flyaway, Clumping, Frizz and Kink, and for the Multi-Strand nature of hair. The Flyaway settings define the percentage of hairs that stray from where they should be causing messy hair. The Clumping settings cause the hair to be collected together into clumps. Frizz causes the hair to curl at its tip or root. Kink parameters cause the hair to be zigzagged in shape; the Multi-Strand parameters cause hair to separate into groups like grass does. Figure 47.4 shows four areas with different hair properties applied, including normal straight hair with no frizz or kink, a section with Frizz, one with Kink, and Multi-Strand.

FIGURE 47.4

Changing hair properties can drastically alter the hair's look from normal to frizz, kink, and multi-strand.

Tutorial: Adding a spline fringe to a quilt

Quilts are designed to make you feel all warm and fuzzy, and what could be fuzzier than some soft fringe surrounding a quilt? Fringe makes grabbing the quilt easy.

To add hairy fringe to a quilt using splines, follow these steps:

1. Open the Patch quilt.max file from the Chap 47 directory in the downloaded content set.
2. Select the Create→Shapes→Line menu command, and draw a couple of simple splines that extend at right angles from every corner of the quilt in the Top view. These splines will be used to designate the start and end splines for the quilt fringe. Hair strands will be placed between these two simple splines. Start at one corner, and proceed in turn clockwise around each of the corners. Add a second line on top of the first line to complete the set of splines.
3. Select the first spline, and convert it to an Editable Spline object using the right-click quad menu. Then click the Attach button, and select the splines in the order they were created.
4. With all the splines selected, choose the Modifiers→Hair and Fur→Hair and Fur WSM menu command to apply the Hair and Fur modifier to the splines.
5. Open the Modify panel, and in the General Parameters rollout, set the Hair Count to **2500**, the Random Scale value to**10**, and the Tip and Root Thickness to **5.0**. Then open the Material Parameters rollout, and change the Tip and Root colors to white.

Figure 47.5 shows the resulting quilt with its fringe, all warm and fuzzy.

FIGURE 47.5

Hair or fringe can be added to a subobject selection or to the entire object.

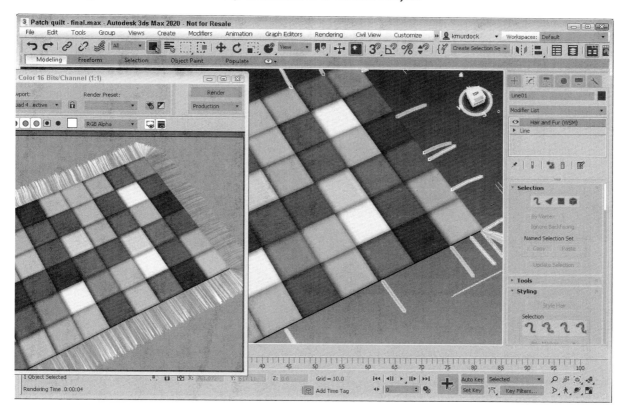

Styling Hair

The Display rollout (located way at the bottom of the rollouts) includes settings for controlling how many hairs are displayed in the viewports. To see all the hairs, you need to render the scene. You can select to Display Guides and actual hairs and set a color for each. The Override option causes the color specified in the color swatch to be used instead of the render color. You also can select a Percentage value of the total hairs to display in the viewport up to the number in the Max Hairs value. Enabling the As Geometry option causes the hairs to appear as geometry objects instead of lines.

Guide hairs control the position of all adjacent hairs. Guide hairs extend from each vertex in the attached object. Guide hairs are yellow by default, and viewport hairs are red, but you can change the color for each in the Display rollout.

Guide hairs provide a simple way to style, comb, and brush hair. By positioning the guide hairs, you can control what the rest of the hair looks like.

Note
No guide hairs are available when the Hair and Fur modifier is applied to a selection of splines because the splines act as guides.

Using the Style interface

In addition to the hair properties, you can change the look of hair by using the various hair-styling features. These features are found in the Styling rollout. The Style Hair button activates an interactive styling mode in the viewport where you can brush, comb, and manipulate the individual hairs. The Style Hair button is activated automatically when the Guides subobject mode is selected.

Within the Styling rollouts are several icon buttons. These buttons are described in Table 47.1. The brush size can be interactively set by holding down the Ctrl and Shift keys while dragging the mouse when in Brush mode, or you can change the brush size using the slider located under the Ignore Back Hairs option. The Distance Fade option causes the brushing effect to fade as it gets closer to the edge, resulting in a softer effect at the hair tips.

TABLE 47.1 Hair Styling Buttons

Toolbar Button	Name	Description
	Select Hair by Ends (Ctrl+1)	Selects the end vertex when you drag over hairs.
	Select Whole Guide (Ctrl+2)	Selects all vertices in the whole strand when you drag over hairs.
	Select Guide Vertices (Ctrl+3)	Selects specific vertices when you drag over hairs.
	Select Guide by Root (Ctrl+4)	Selects the entire hair by selecting only the root vertex when you drag over hairs.
Box Marker ▼	Marker Style drop-down list	Choose the style to mark the selected vertices. Options include Box, Plus, X, and Dot.
	Invert Selection (Shift+Ctrl+N)	Deselects the current selection, and selects all hairs not currently selected.
	Rotate Selection (Shift+Ctrl+R)	Rotates the current selection.
	Expand Selection (Shift+Ctrl+E)	Adds to the current selection set by increasing the selection area.
	Hide Selected (Shift+Ctrl+H)	Hides the selected vertices.
	Show Hidden (Shift+Ctrl+W)	Unhides all hidden vertices.
	Hair Brush (Ctrl+B)	Moves all selected vertices in the direction of the brush. Press Escape to exit this mode.
	Hair Cut (Ctrl+C)	Cuts the hair in length.
	Select (Ctrl+S)	Enters selection mode where you can select hairs by dragging over them.
✔ Distance Fade	Distance Fade (Shift+Ctrl+F)	Causes the brush effect to fade with distance. Available only in Hair Brush mode.
✔ Ignore Back Hairs	Ignore Back Hairs (Shift+Ctrl+B)	Causes only hairs facing the camera to be affected.
■·····│·····■	Brush Size slider (Ctrl+Shift+mouse drag)	Changes the size of the brush.

	Translate (Shift+Ctrl+1, Ctrl+T)	Moves the selected guides in the direction of the brush when you drag.
	Stand (Shift+Ctrl+2, Ctrl+N)	Stands the selected guides up straight.
	Puff Roots (Shift+Ctrl+3, Ctrl+P)	Causes small deviations to appear at the root of each selected guide.
	Clump (Shift+Ctrl+4, Ctrl+M)	Pulls the selected guides together to the center of the brush.
	Rotate (Shift+Ctrl+5, Ctrl+R)	Rotates and spins the selected guides about the center of the brush.
	Scale (Shift+Ctrl+6, Ctrl+E)	Scales the selected guides when you drag with the brush.
	Attenuate (Shift+Ctrl+A)	Scales the hairs based on the size of the polygon.
	Pop Selected (Shift+Ctrl+P)	Lengthens the selected hairs along the surface normal.
	Pop Zero Sized (Shift+Ctrl+Z)	Lengthens any zero length hairs along the surface normal.
	Recomb (Shift+Ctrl+M)	Combs the hair from the top downward.
	Reset Rest (Shift+Ctrl+T)	Relaxes the hair by averaging the position of hairs.
	Toggle Collisions (Shift+Ctrl+C)	Turns collisions of hairs on and off.
	Toggle Hair (Shift+Ctrl+I)	Toggles the display of hairs on and off.
	Lock (Shift+Ctrl+L)	Locks the selected vertices so they cannot be moved by other tools.
	Unlock (Shift+Ctrl+U)	Unlocks any locked vertices.
	Undo (Ctrl+Z)	Undoes the last command.
	Split Selected Hair Groups (Shift+Ctrl+-)	Separates the selected hairs into separate groups.
	Merge Selected Hair Groups (Shift+Ctrl+=)	Combines the selected hairs into groups.

When you're finished styling, click the Finish Styling button in the Styling rollout to exit Styling mode.

Tutorial: Creating a set of fuzzy dice

Fuzzy dice. What could be cooler?

To style the fur applied to a set of fuzzy dice, follow these steps:

1. Open the Fuzzy dice.max file from the Chap 47 directory in the downloaded content set.

2. Select one of the dice, and choose the Modifiers→Hair and Fur→Hair and Fur (WSM) menu command to apply hair to the selected die.

3. Open the General Parameters rollout, and set the Hair Count to **20000** and the Scale value to **50**. Then open the Material Parameters rollout, and change the Tip Color to white and the Root Color to Red.

Then open the Frizz Parameters rollout, and set the Frizz Root and Frizz Tip values to **0**. This makes the hair strands straight and red.

4. Open the Styling rollout, and click the Style Hair button. In the Utilities section, click the Pop Selected button to make all the hair stand out. Then select the Hair Brush icon, and drag downward in the viewport near the guides at each of the top corners. Click the Finish Styling button when completed.

5. Drag the Hair and Fur (WSM) modifier from the Modifier Stack, and drop it on the unselected die.

Figure 47.6 shows the resulting pair of dice.

FIGURE 47.6

Hair can be styled by changing the position and orientation of the guide hairs.

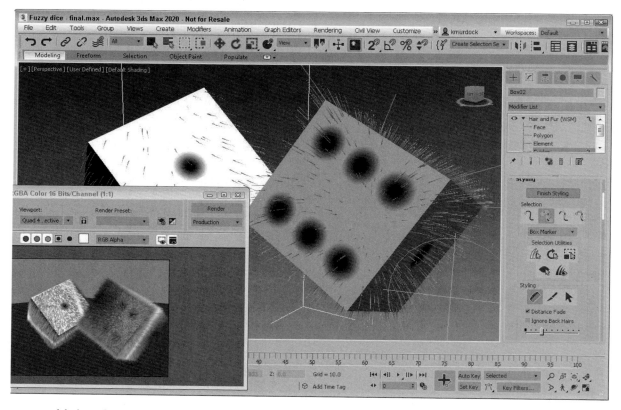

Using hair presets

If you have a specific set of parameters that create a unique hair look that you're happy with, you can save it using the Save presets button in the Tools rollout. Hair preset files are rendered on the spot and added to the Hair and Fur Presets dialog box, shown in Figure 47.7. To open this window of presets, click the Load button. To add a preset configuration to the current object, simply double-click it.

FIGURE 47.7

The Hair and Fur presets dialog box shows rendered thumbnails of the available presets.

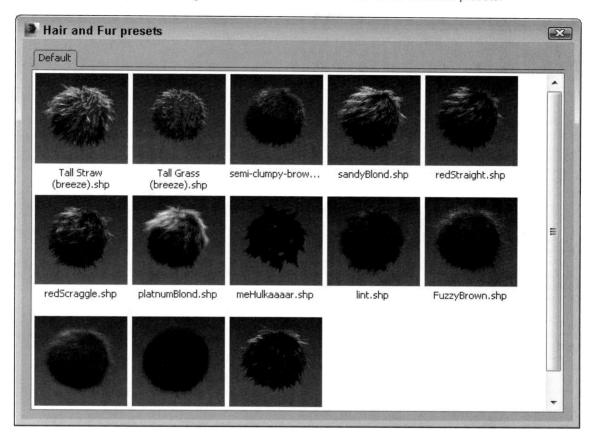

In addition to presets, hairdos—created by styling the hair—also can be copied and pasted onto other hair selections.

Tip

If you ever get into trouble styling hair, you can click the Regrow Hair button in the Tools rollout to reset all the styling to its original state.

Using hair instances

Although the default hair looks great, if you ever wanted to replace the hair splines with an instanced geometry, you can do so by using the Instance Node button in the Tools rollout. The X button to the right of the Instance Node button is used to remove the instance. Figure 47.8 shows a funny head created using a matchstick for a hair instance.

Tip

Be sure to adjust the Hair Count value before selecting an instanced object. Complex instanced objects should be used only with manageable numbers.

FIGURE 47.8

Mr. Matchstick head has all his hair replaced with matchsticks, an instance.

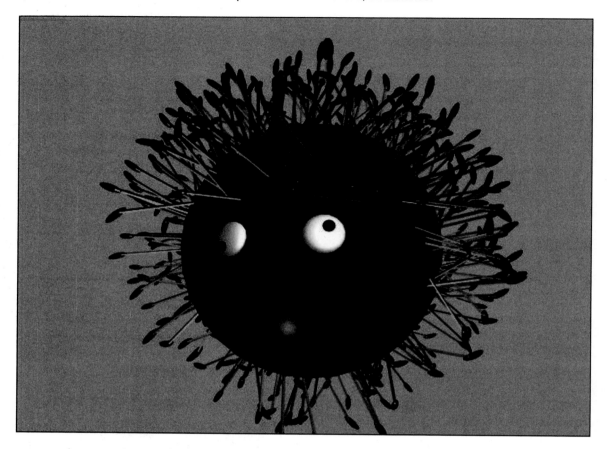

Rendering Hair

You can change the hair render settings by opening the Effects panel using the Rendering→Effects menu. The Render Settings button in the Tools rollout provides quick access to the Effects panel in the Environment and Effects dialog box. The Hair and Fur rollout includes settings for the Hairs and Lighting rendering methods.

Hairs can be rendered using three options. The Buffer method renders each hair individually and combines the effect. It uses minimal memory requirements. When the buffer method is selected, you can choose the Raytrace the Reflections/Refractions option for better quality. The Geometry method converts each hair into a geometry object that is rendered. The mr Prim method renders hair using the Arnold renderer. For this option, you can set the mr Voxel Resolution used to render hairs.

More on Arnold is covered in Chapter 30, "Rendering with ART and Arnold."

Motion Blur can be enabled to make hairs that are moving fast appear blurred. The Oversampling setting adds an anti-aliasing pass to the rendering. The options include Draft, Low, Medium, High, and Maximum. The Composite Method option depends on the rendering method that is selected. A list of occlusion objects and a setting for Shadow Density also are available.

By default, all lights (except for direct lights and daylight systems) are used to light hair, but if a spotlight is added to the scene, you can select the spotlight and click the Add Hair Properties button at the bottom of the

Hair and Fur rollout. This adds a Hair Light Attributes rollout to the Modify panel for the selected spotlight where you can set the Resolution and Fuzz of the shadow map for the hair object cast by the light.

Using Hair Dynamics

Being able to style hair is great, but have you ever left the barbershop and had the wind do its own styling job on your hair? The Dynamics rollout for Hair and Fur lets you define specific forces and let the hair fall where it may.

Tip
Be conservative with the forces that are applied to a hair system. Too many forces or too extreme forces can easily destroy any styling that you've created.

Making hair live

The Dynamics rollout is available only if the Hair and Fur (WSM) modifier has been applied to an object. At the top of the Dynamics rollout are three modes: None, Live, and Precomputed. If you select the Live mode, the hair around the growth object immediately becomes subject to gravity and other forces in real time, causing the hair to droop about the growth object. Moreover, if you move the object within the viewport, the hair flows about the object as if you were moving a real object with hair attached. Figure 47.9 shows a simple mouse with hair attached. The image on the left shows the hair particles with the None mode enabled, and the image on the right shows the hair after the Live mode is enabled. Notice how the hair particles fall around the mouse object.

FIGURE 47.9

The Live dynamic mode makes the hair react in real time to the scene forces.

If you press the Escape key while in Live mode, a dialog box appears, giving you the option to Freeze, Stop, or Continue. If you click the Freeze button, the hair stays in its current position.

The Precomputed mode lets you save the hair motions into a separate stat file that is set using the Stat Files button.

Setting properties

Only a few properties need to be defined to enable hair dynamics. In the Dynamics rollout, you find values for Gravity, Stiffness, Root Hold, and Dampen. These properties control how the hair behaves in response to the environment forces. The Gravity value can be negative if you want the hair to rise instead of fall. You can simulate space environments by setting the gravity to 0. The Stiffness value eliminates all dynamic movement if set to 1.0. If you want the hair to move only slightly as the object moves, a Stiffness value close to 1.0 should work. The Root Hold value is like stiffness but applies only to the root. The Dampen value causes motions to die out quickly.

All dynamic properties except Gravity can be controlled using a grayscale map, using the small button to the right of the value field.

Enabling collisions

The first type of dynamic force to address is to enable collisions between the hair and the other scene objects. To enable collisions between the growth object and its hair, simply enable the Use Growth Object option. This option is only enabled when Sphere or Polygon Collisions are enabled first. In addition to the growth object, other scene objects can be added to the list with which the hair will collide. To add other objects, click the Add button and pick the object to add in the viewport. Each collision object can use either a boundary Sphere to define its collision volume or a Polygon, which bases collisions on the actual surface geometry. The latter takes longer to compute, but it's more accurate.

Enabling forces

In addition to the ubiquitous gravity, you can enable collisions between the hairs and the growth object and any other scene objects. To add another scene object to the collision calculations, click the Add button and select the new collision object. The External Forces list lets you add Space Warps for additional forces such as Wind. You also can delete and replace any object added to the list.

Running a simulation

The Precomputed mode lets you save the hair dynamics to a separate stat file. If you want to capture the dynamic simulation, you first must specify a stat file by clicking the button to the right of the Stat Files section. With a stat file specified, click the Run button in the Simulation section to calculate the dynamic solution. The Start and End fields let you enter the range for the simulation. A separate stat file is generated for every frame of the animation.

If you enable the Precomputed mode option before you render, the stat file is read and used during the rendering process. You can delete all stat files quickly with the scarily named Delete all files button.

Tutorial: Simulating hair dynamics

Dynamic hair moves and flows around the other objects in the scene that are animated. As an example of this, you'll move a female character's head back and forth and simulate how the hair moves. I selected the Mohawk hairstyle because I'm hip and cool, a real rebel. Actually, the Mohawk is a simple style and gives you a chance to play with the Stiffness property.

To simulate the dynamics of a hair system, follow these steps:

1. Open the Girl with mohawk.max file from the Chap 47 directory in the downloaded content set.

   This file includes the head model. The hair modifier already has been added to this character and styled.

2. Click the Auto Key button, drag the Time Slider to frame 5, and move the character to the right in the Top viewport. Then drag the Time Slider to frame 10, and move the character back to the left. This simple motion should be enough to bend the hair over. Then disable the Auto Key mode.

3. With the head selected, open the Dynamics rollout and set the Stiffness value to **0.8**. This should keep the hair standing straight up. To check this, enable the Live mode and watch how the hair reacts.

4. Set the Simulation to run from 0 to 10 frames, and then open and specify a stat file location. The path of the stat file location is displayed. Then click the Run button to start the simulation. The precomputed values are saved to stat files.

5. Select the Precomputed mode option and drag through the animation frames to see how the hair reacts.

Figure 47.10 shows the hair bending to one side as the female head moves.

FIGURE 47.10

Using precomputed hair can save you a bundle of time when rendering.

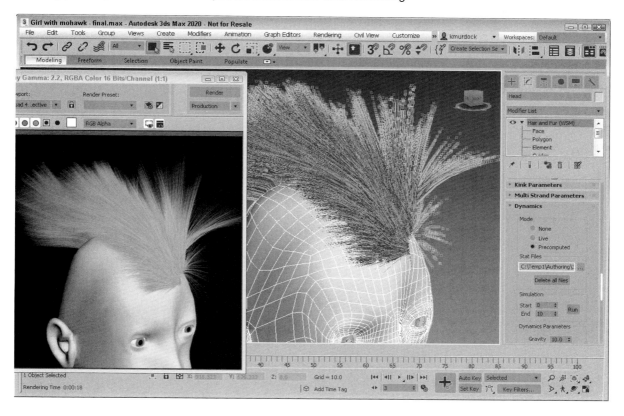

Understanding Cloth

If you drop a shirt over a chair, how does it land? It folds and bunches as it drapes over the solid object. Also notice how different types of cloth drape differently. Compare how silk reacts versus a terry cloth towel. Being able to accurately simulate how cloth reacts to scene objects and forces is yet another critical element that can really make a difference in your final scenes.

Cloth in 3ds Max is created using a modifier. This modifier can be applied to any object to make it a cloth object, or you can specifically create a set of clothes from flat plane objects cut to make up the front and back pieces of clothing using the Garment Maker modifier. During the process of making an object into a cloth object, the object is divided into multiple faces using one of several subdivision methods. This provides the mesh with enough resolution to be deformed accurately.

Cloth objects are endowed with characteristics that let them respond to external forces within the scene including gravity, wind, and collisions with other scene objects. After all the scene objects, forces, and cloth objects are added to the scene, you can start a simulation that defines how the cloth moves within the scene under the effect of the other scene objects and forces.

Creating Cloth

When you start to model a cloth object, keep in mind that the model must have enough resolution so that it can accurately fold over itself several times. If the resolution isn't defined enough, the bending of the cloth is not believable.

Note

Although dynamic cloth can be deformed under the effect of forces applied to it, cloth cannot apply a force back against the deforming object. This is a scenario that MassFX can handle because it deals with all objects in the scene. A trampoline, for example, is a good example of a deformable cloth surface that can apply a force back on the deforming object.

You can increase the resolution of a model in several ways. The Garment Maker modifier includes parameters that can increase the resolution of a mesh, or you can use the HSDS (for Hierarchical SubDivision Surfaces) modifier to increase an object's resolution.

Tip

When making cloth objects, use a single-sided Plane object instead of a Box object. Using a Box object needlessly doubles the number of polygons to be included in the simulation.

Using Garment Maker to define cloth

Clothes can be added to models using a method that is similar to the way real clothes are made. Each section of cloth, called a panel, is outlined using lines and splines in the Top viewport. Clothes patterns also can be imported and used. Include a break at each corner of the panel, or the modifier will round the corner. After you have all the various panels created, convert one of the panels to an Editable Spline object and attach all the panels into a single object. You can apply the Modifiers→Cloth→Garment Maker to the set of panels.

Tip

The easiest way to create cloth panels is to draw the entire cloth panel and then use the Break Vertex command in Vertex subobject mode to break each corner.

When the Garment Maker modifier is first applied, the panel outlines are made into mesh objects and subdivided using the Delaunay algorithm, as shown in Figure 47.11. This algorithm uses random triangulation and divides the triangles into roughly equal shapes. This helps keep the cloth from folding along common lines caused by regular patterns. You can alter the number of polygons in the subdivided mesh using the Density value. If the Auto Mesh option is disabled, you can update the density change using the Mesh It button. The Preserve option maintains the 3D shape of the object. If Preserve is disabled, the panel is made flat.

FIGURE 47.11

The Garment Maker modifier uses the Delaunay algorithm for subdividing cloth meshes.

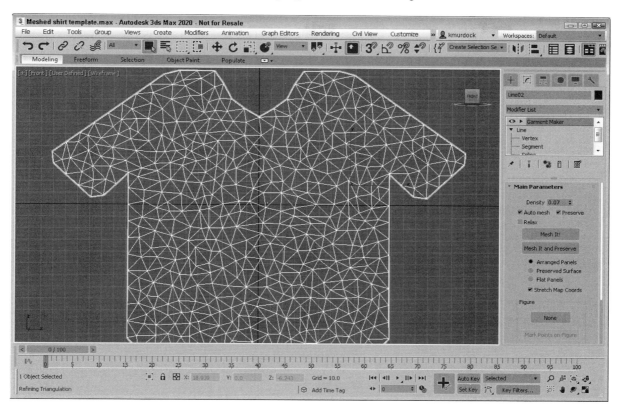

The next step is to position the panels so they surround the model that they will be draped over. The subject that is to receive the clothes can be selected by clicking the Figure button in the Main Parameters rollout. After an object is selected, clicking the Mark Points on Figure button makes a small figure outline appear in the upper-left corner of each viewport. This figure outline has markers corresponding to the chest, neck, pelvis, shoulders, and wrists. Dragging over the body of the character lets you mark corresponding locations on the figure.

After body parts are marked, you can enable the Panels subobject mode and select a cloth panel and use the Panel Position buttons in the Panel rollout to position the various cloth panels around the body. The Panel Position buttons include Front Center, Front Right, Back Center, Right Arm, and so on. You also can position the panels manually.

The Garment Maker modifier has three subobjects: Curves, Panels, and Seams. With the panels in place, you can stitch seams between the panels using the Curves subobject mode for flat drawn panels or with the Seams subobject mode for panels that are positioned about a figure. Each seam edge should be relatively the same length. The Seam Tolerance value sets how far apart the two edges can deviate. For each seam, you can set a Crease Angle and Strength. The Sewing Stiffness value determines the strength that the seams are pulled together.

After the seams are defined, you can apply the Cloth modifier to pull the panels together and simulate the cloth's motion.

Creating cloth from geometry objects

Any geometry object can be made into a cloth object using the Cloth modifier available by selecting the Modifiers→Cloth→Cloth menu command. Although the modifier is added to the object, it is set to Inactive by default. To activate the cloth, you need to open the Object Properties dialog box, shown in Figure 47.12.

FIGURE 47.12

The Object Properties dialog box includes all the parameters for the cloth and collision objects.

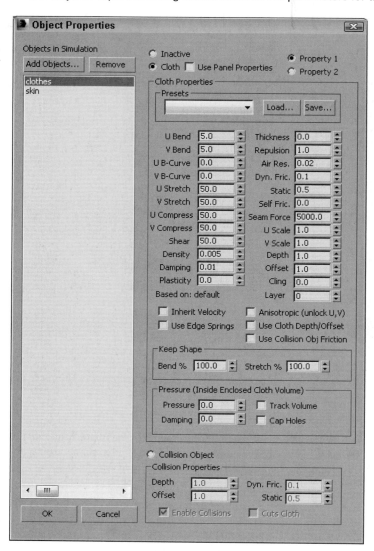

The list at the left holds all the objects that are involved in the cloth simulation. New objects can be added to the list using the Add Objects button. Each object can be set to Inactive, Cloth, or Collision Object; and properties can be set for each type. For cloth, the properties include Bend, Stretch, Shear, Density, Thickness, Friction, and Scale values. Defined cloth settings can be saved and recalled. Cloth property files are saved using the .sti file extension. 3ds Max also includes a sizable list of presets in a drop-down list, including Burlap, Cashmere, Cotton, Flannel, Rubber, Satin, Silk, Terrycloth, and Wool, among others.

In addition to objects added to the scene, a cloth simulation also can include forces. The Cloth Forces button in the Object rollout opens a dialog box where you can select which forces in the scene can be added to the cloth simulation. Gravity is added by default automatically to the simulation. You can set the Gravity value in the Simulation Parameters rollout along with several other parameters, including the Start and End Frames.

With all the objects and forces added, click the Simulate Local button to set the initial state of the cloth simulation. This drapes the cloth over the scene objects and pulls all defined seams from the Garment Maker modifier together. Sometimes the Simulate Local button moves the cloth panels together too fast, so you can use the Simulate Local (damped) button to add lots of damping to the scene to prevent problems from panels moving too fast.

Tutorial: Clothing a 3D model

If you've ever wanted to design and outfit a set of models with a custom-made line of clothing, here's your chance. Using the Garment Maker and the Cloth modifiers, you outfit the 3D model with a set of clothes. Watch for this new line next spring at your local retailers.

To create and apply a set of clothes to a 3D character, follow these steps:

1. Open the Bikini girl with clothes.max file from the Chap 47 directory in the downloaded content set.

2. Select the Create→Shapes→Line menu command, and draw in the Front viewport the front outline of a simple dress. Make sure the panels are closed splines. Then right-click in the viewport to exit Line mode. Move the dress pattern in front of the woman model in the Left viewport.

3. Select the Tools→Mirror menu command, and create mirrored copies of the dress pattern for the back side. Then convert one of the panels to an Editable Spline, and attach all the other panels together into a single object. In Vertex subobject mode, select all the corner vertices in all panels and click the Break button, then exit Vertex subobject mode.

4. With the clothes selected, choose the Modifiers→Cloth→Garment Maker modifier. This automatically subdivides all the cloth panels into multiple polygons. Make sure the dress is a closed spline or the Garment Maker modifier won't work.

5. In the Main Parameters rollout, click the Figure button and select the woman character. Then click the Mark Points on Figure button. A small figure appears in the upper-left corner of each viewport with the upper chest area marked in red. Drag the cursor over the character until the matching upper chest area is located, and click. Repeat for all the marked positions. When all points are marked, click the Mark Points on Figure button again to exit marking mode.

6. In the Modify panel, select the Panels subobject mode and choose the dress's front panel. Select the Top at Neck level, and click the Front Center button. This places the dress's front in front of the character. Repeat the placement for the other panels. If the placement isn't correct, manually move the panels until they are in front of and behind the character, as shown in Figure 47.13.

FIGURE 47.13

Using figure markers, you can approximate where the clothes are positioned on a character.

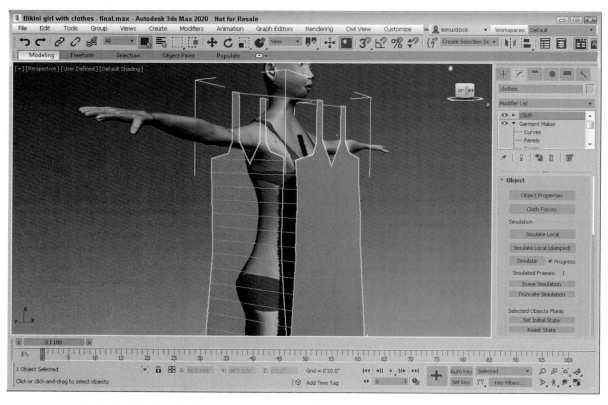

7. In the Modify panel, select the Seams subobject mode. In the Front viewport, select two edges that should be connected and click the Create Seam button. Repeat this for all seams including the sides of the dress and the top. Connecting lines are drawn between both sides of the seam. If the seam boundaries are crossed, click the Reverse Seam button.

8. With the seams defined, select Modifiers→Cloth→Cloth to apply the Cloth modifier. In the Object rollout, click the Object Properties button. In the Object Properties dialog box, select the clothes object and enable the Cloth option. Then click the Add Objects button, select the character mesh, and mark it as a Collision Object. Set the Offset value for the clothes to **0.3**, and click OK in the Object Properties dialog box.

9. Click the Simulate Local button. The cloth seams are pulled together, and the clothes are draped over the character.

Figure 47.14 shows the female character in a simple dress.

FIGURE 47.14

The Simulate Local button causes the clothes to be draped over the body.

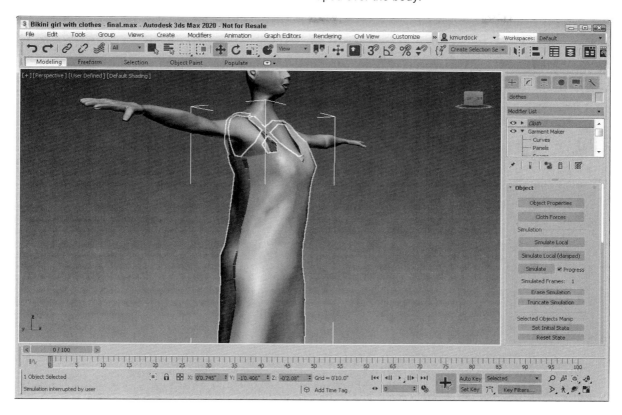

Simulating Cloth Dynamics

Hair isn't the only system that has the benefit of dynamic motion. Cloth also can benefit from dynamic simulations. The steps for setting up a dynamic cloth simulation are similar to those for hair. First, apply the cloth modifier and define the cloth properties and the environmental forces acting on the cloth. Then run the simulation.

Defining cloth properties and forces

To add objects (both cloth and collision objects) to the simulation, click the Object Properties button in the Object rollout. This opens the Object Properties dialog box. Clicking the Add Objects button lets you select scene objects to add to the simulation. Only objects added to the scene are included in the simulation. If an object isn't added, it's ignored. All objects added to the simulation are added to the list at the left. Selected objects in the list can be specified as Inactive, Cloth, or Collision Object. For Cloth and Collision Objects, you can set properties. You also can load and save cloth presets. Cloth presets are saved using the .sti extension.

Tip

If the cloth tends to pass through objects, you can increase the Offset value for the collision object.

After all the objects involved in the simulation are included and defined, you can set the simulation range in the Simulation Parameters rollout. The initial state for the object to be draped may be set using the Set Initial State button, which is located in the Object rollout. The Cloth Forces button, in the Object rollout, opens a simple dialog box where you can select to add additional forces to the simulation. Gravity is added by default, but you can change its value in the Simulation Parameters rollout.

Creating a cloth simulation

After completing the initial setup, clicking the Simulate button starts the simulation process. The objects are updated in the viewport as each frame is calculated. After every frame is calculated, you can see the entire dynamic simulation by dragging the Time Slider or clicking the Play Animation button. If you want to drape the cloth without running it over several frames, you can use the Simulate Local button. The Simulate Local (damped) button causes the simulation to run local with a large amount of damping, which is useful if the cloth tends to drape too fast. If you want to remove the current simulation because some properties have changed, click the Erase Simulation button, or you can remove all frames after the current simulation frame with the Truncate Simulation button.

Tip

If the simulation is taking too long to compute, you can cancel the simulation by pressing the Escape button.

If you need to change a cloth or force property, click Erase Simulation, make the change, and run the simulation again. Simulation motions can be saved as keys with the Create Keys button. Figure 47.15 shows a simple plane object that has been draped over a chair over the course of 100 frames.

FIGURE 47.15

After you've defined cloth and force properties, an executed simulation drapes the cloth over a chair.

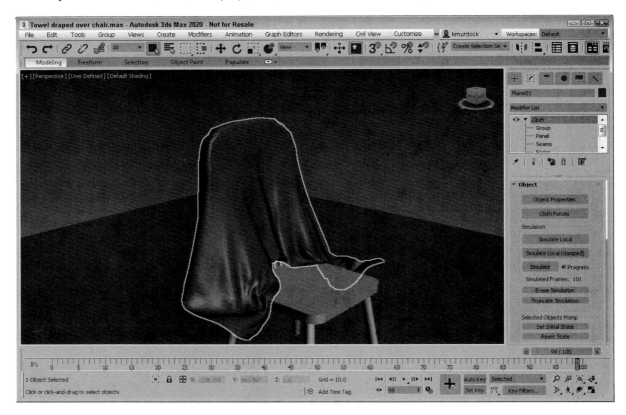

Viewing cloth tension

For cloth objects created using the Garment Maker modifier, you can view the tension in the cloth using the Tension option in the Simulation Parameters rollout. This option shows the areas of greatest tension in shaded colors. You also can select a seam to tear when enough force is applied to it.

Tutorial: Draping cloth over a jet

For a larger example of a cloth system, you'll drape a drop cloth over a jet.

To simulate the dynamics of a cloth object, follow these steps:

1. Open the Sheet over plane.max file from the Chap 47 directory in the downloaded content set.
   This file includes a plane model.

2. Choose the Create→Standard Primitives→Plane menu command, and drag in the Top viewport to create a plane object that covers the jet. Set the Length and Width values to **150** and the Length and Width Segments values to **100** to make the sufficient resolution for the cloth, and drag the plane object upward in the Front viewport, so it sits above the jet.

Note
You also can create the cloth from a rectangular spline that has the Garment Maker modifier applied to it. This approach uses the Delaunay tessellation, which is better for simulating cloth than the rectangular sections in the Plane object.

3. With the plane object selected, choose the Modifiers→Cloth→Cloth menu command to apply the Cloth modifier to the object.

4. Open the Modify panel, and click the Object Properties button in the Object rollout to open the Object Properties dialog box. Select the Plane01 object in the left list, and choose the Cloth option. Then select the Silk option from the Presets drop-down list, and set the Thickness to **0.5**.

5. With the Object Properties dialog box still open, click the Add Objects button, click the Select All button, and click the Add button. With all added objects in the left list selected, choose the Collision Object option and click OK to close the dialog box.

6. In the Simulation Parameters rollout, enable the End Frame option and set the end frame to **100**. Then click the Simulate button in the Object rollout. The plane object descends and covers the jet being draped as it falls.

Note
If you are draping the cloth over an object with some pointy edges, you may notice part of the object penetrating the cloth. To fix this, increase the density of the cloth or lighten the sharp edges.

Figure 47.16 shows the sheet draped over the plane.

FIGURE 47.16

Computing the dynamics of a cloth object is possible with a cloth system.

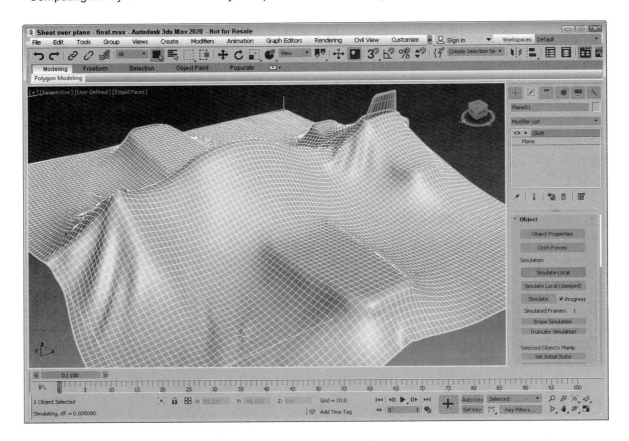

Summary

The Hair and Fur modifier allows you to add hair to scene objects. Using the parameters found in the various rollouts, you can define the type of hair that is created. The Cloth modifier enables a cloth dynamic system that can simulate the complex motion of cloth as it interacts with objects and forces in the scene.

Specifically, this chapter covered the following topics:

* Growing hair and setting hair parameters

* Styling hair and using the hair presets

* Enabling hair dynamics

* Understanding how cloth systems work

* Using the Garment Maker modifier

* Creating cloth objects from geometry objects using the Cloth modifier

* Simulating cloth dynamic motion by initiating a simulation process

In the next chapter, we'll look at the fluid simulation.

Chapter 48
Creating Fluid Simulations

IN THIS CHAPTER

It is easy to love the sound of a soft rain on the window pane or see a peaceful wave lap up onto a beach, but fluid flows can also include the sweet drip of syrup on a stack of hot pancakes or a splatter of fresh paint landing right on the carpet. All of these fluid flows are now possible in realistic detail within 3ds Max using their new fluid simulation system.

Understanding Fluids

If you consider the physics behind even the simplest fluid flow, you will quickly realize that these simulations take a lot of computing power. The fluid simulation approach used by 3ds Max uses caches to store the simulation data. This approach takes time to compute initially, but once done can be immediately recalled and viewed.

When setting up a fluid simulation, you have a lot of control over the initial properties of the fluid involved. Each physical property can be tweaked as needed to give it unique characteristics, but 3ds Max also includes many liquid presets that are defined to act as we would expect such as Water, Motor Oil and Ketchup. All the liquid properties are available and new liquids can be defined and saved for later recall.

Fluids are added to the scene using emitters that add new fluid droplets to the scene for each frame. This makes the fluid simulations very dynamic as new droplets are being added. Because new droplets are constantly being added, the total number of particles can quickly become huge, so the simulation lets you add Kill Planes to remove particles that stray too far away. This helps keep the simulation manageable.

At the other end of the simulation are the objects that the fluid will interact with. These collision objects also have settings that control the interaction including its shape, velocity and thickness. The default force in the simulation is gravity, but simulated fluids can also be impacted by Motion Field forces in the scene to push, drag and spin the fluid droplets around.

When setting up a simulation, the computations depend on the size of voxels. Voxels are the individual areas within which the fluid droplet's motion is computed. If you break the entire simulation into very small voxels, then the simulation will be extremely accurate, but it will take all day to compute. Larger voxels will be computed quicker, but will have less detail.

Tip

If your simulation calculations are taking too long, try reducing the voxel size.

When you click to start the solver, 3ds Max will open a warning dialog box if the Voxel Size value is too small.

Getting Started with Fluids

To complete a fluid simulation, the basic needs are a fluid emitter and an object for the fluid to interact with. Even a simple simulation like this will take some time to compute. It is also helpful to add a Kill Plane to the simulation. This Kill Plane will eliminate any particles beyond a certain point, which keeps the total number of droplets in check.

Starting with an Emitter

The first step in simulating fluids is to add some liquid to the scene. This is done using the Max Liquid emitter object found in the Create→Fluids→Liquid menu command. This places an emitter icon in the scene, but liquid won't be added in the scene until the simulation is run and the Time Slider is moved.

Once an emitter icon is added to the scene, there are a few parameters you can set including the shape of the icon. The options include Sphere, Box, Plane and Custom. There are also settings for defining the size of the emitter. Note that the emitter object is different from the emitter icon, which is only visible in the viewports. Figure 48.1 shows the emitter object as a sphere and the emitter icon. The emitter icon marks the starting location of the fluid stream and can be positioned and moved as needed.

FIGURE 48.1

The emitter object marks where the fluid stream starts from.

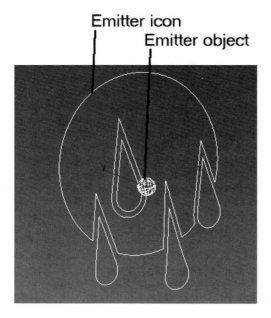

Within the Setup rollout for the emitter object is a button labeled Simulation View. Clicking this button opens the Simulation View dialog box, shown in Figure 48.1, that holds all the controls for working with the fluid simulation.

Tip

FIGURE 48.2

The Simulation View dialog holds all the settings for the fluid simulation.

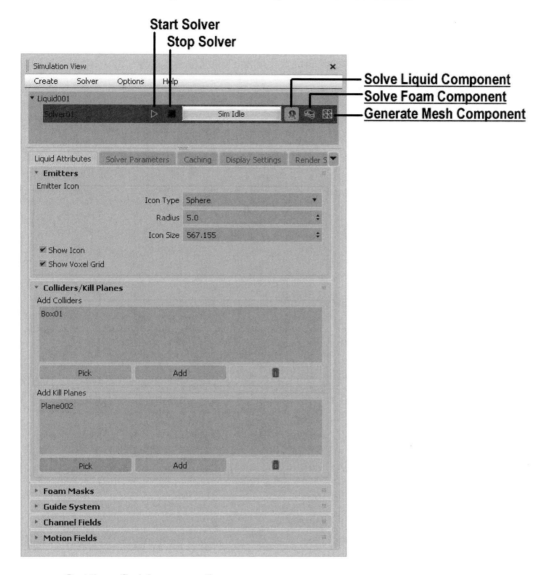

Setting fluid properties

The amount of liquid and its characteristics are controlled by the Liquid Properties, found in the Solver Parameters panel of the Simulation View dialog box. There are several settings for the emitter display found in the Liquid Attributes panel, but the real liquid parameters are in the next panel. From the top of the Liquid Parameters section, you can choose from one of many liquid presets including Blood, Carmel, Honey, Maple Syrup, Mercury, Putty and Water. Other parameters include Droplet size, Particle Distribution, Vorticity, Surface Tension, Viscosity, and Erosion.

If you change the liquid parameters to something unique, you can name the new liquid and save it as a preset using the Save button to the right of the Preset drop-down list. Saved presets can also be loaded and used.

Adding Collider objects

If the fluid doesn't interact with any other scene objects, then you'll simply have a fluid stream falling from the emitter straight down. This is sort of boring. To liven up the scene, you can add objects that collide with the fluid.

After an emitter object is added to the scene, you can scrub through the Time Slider to see the liquid droplets drop under the effect of gravity from the emitter icon to the bottom of the screen. If you place an object in the liquid stream and mark it as a Collider object, then the liquid droplets will collide and interact with the Collider object. If the Collider object has walls, then the object can act as a container that collects and holds the liquid.

Colliders objects are added to the simulation using the Liquid Attributes panel in the Simulation View dialog box. Simply click on the Pick button under the Colliders section and choose the scene object to make into a collider. The object name will be added to the list. The Add button opens a Scene Explorer window where you can choose the object by name. The selected Collider can be removed by selecting its name and clicking the trash icon next to the Add button.

Positioning Kill Planes

Directly underneath the Collider objects list is a list of Kill Plane objects. These planes mark the place where the liquid droplets disappear and are removed from the simulation. It is important to add these planes into the scene so that the number of liquid droplets doesn't become overly large and hard to manage.

Running the simulation

At the top of the Simulation View dialog box are controls for starting, pausing and stopping the simulation. Clicking the Start Solver button begins the simulation calculations. Each frame is calculated and saved in a cache for quick recall when the animation is played. The time required to calculate all frames of the simulation depends on the complexity of the scene and the number of frames, but you can pause or stop the simulation at any time using the Pause and Stop buttons.

Note

Even though the simulation calculations are running, the 3ds Max interface is still active and can be used. You can even render a frame while the simulation is being calculated.

Tutorial: Water over a sphere

For the first fluid simulation, we are going to start with a simple example. Simply place an emitter over a sphere object. The reason this example is so boring is that we don't want to overwhelm the system or set up something is going to take all night to calculate.

To simulate water running over a sphere, follow these steps:

1. Open the Water over sphere.max file from the Chap 48 directory in the downloaded content set. This scene has a simple sphere and plane objects.
2. Select the Create→Fluids→Max Liquid menu command to create a fluid emitter object. Select and move the emitter upwards to be positioned directly over the sphere.
3. Click on the Simulation View button to open the Simulation View dialog box. In the Liquid Attributes panel, click on the Pick button underneath the Colliders list and choose the Sphere object. Then, click on the Pick button underneath the Kill Plane list and choose the Plane object. Each object is added to the respective list.
4. Select the Solver Parameters panel and choose the Liquid Parameters category, then select the Water preset from the drop-down list at the top of the panel.
5. Click on the Start Solve button at the top of the Simulation View dialog box to begin the simulation calculations.
6. When the calculations are completed, click on the Play Animation button to see the results.

Figure 48.3 shows the water running down over the sphere.

FIGURE 48.3

This simple fluid simulation shows water running down over the top of a sphere object.

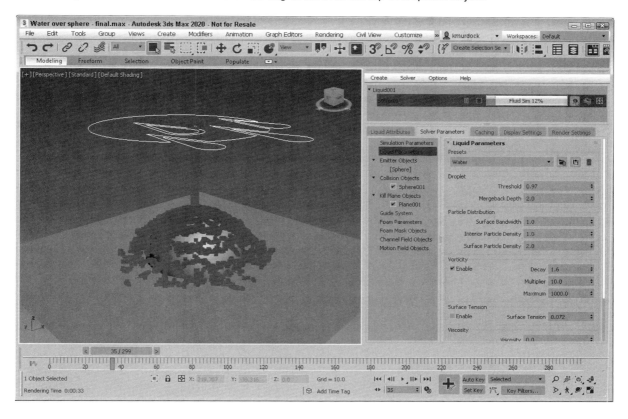

Setting Fluid Simulation Parameters

Within the Simulation View dialog box are several panels of settings used to set the global nature of the fluid simulation and how the fluid droplets are displayed in the viewports. For example, gravity is enabled by default, but you can change the gravity value, set it to 0 or even invert it as needed.

Changing the Solver Parameters

The Solver Parameters panel of the Simulation View dialog box includes all the global settings of the fluid simulation. Within this panel, you can set the number of frames to include in the simulation and also the System Scale. There is also a setting for the Master Voxel Size. This determines how accurate the simulation is, but small values will require more memory and computing power to complete the simulation.

The Gravity value is set by default to 9.8, which is the standard gravity value on Earth. Decreasing this value will cause fluid droplets to fall slower and making it a negative value will cause the droplets to flow upward. The Create Gravity Force button adds a helper object to the scene. Gravity will flow in the direction that this helper object is pointing.

The Transport and Time Steps values set how many iterations are used to compute where each droplet is moving and how its affected by the simulation settings. These settings are controlled by the Adaptivity value for each.

Setting the Fluid Display attributes

To control how the fluid droplets look in the viewport, you can use the attributes found in the Display Settings panel of the Simulation View dialog box, shown in Figure 48.4. The first setting here is the Display Type. The options include Bifrost Dynamic Mesh, Bifrost Cache Mesh, Point, Plane, Sphere, Custom and None.

FIGURE 48.4

Settings in the Display Settings panel of the Viewport View dialog box control how the fluid looks in the viewport.

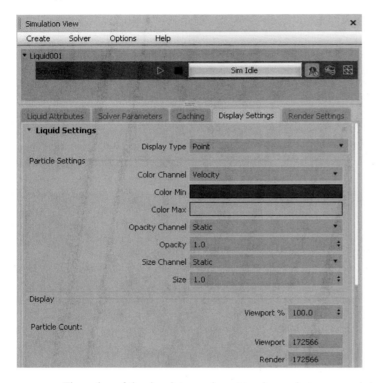

The color of the droplets can be set to change between a minimum and maximum value based on the Color Channel setting. The options are Static, Age, Air Distance, Churn, Curvature, Density, Droplet, Expansion Rate, Position, Static Friction Bandwidth, Static Friction Strength, UV, Velocity, and Vorticity. Channels can also be selected for Opacity and Size. For example, if the Color Channel is set to Velocity, then slower moving droplets will be the Color Min value and faster moving droplets will be the Color Max value.

You can also set the percentage of total droplets shown in the viewport and the total number shown in the Viewport and when Rendered.

Managing simulation caches

The Caching panel of the Simulation View dialog box has settings for the Cache Directory. There is also a button for clearing the cache and a setting for the Cache Limit.

Caution

Be sure to set a reasonable Cache Limit for your system. A Cache Limit that is too large will cause problems. Be aware that fluid simulations can very quickly fill up.

Adding Foam

For more realism, you can add foam particles into your fluid simulation. These particles show up in the simulation as foam on the surface, as bubbles when below the surface or as spray when they are above the surface. For foam particles, you can set how they look including their color and shape.

Enabling foam in the solver

Foam particles are not included in the simulation calculations by default, but you can enable them by clicking on the Solve Foam Component toggle button located to the right of the Start Solver buttons at the top of the Simulation View dialog box.

Setting foam properties

When foam is added to the simulation, you can set its properties in the Foam Parameters section of the Solver Parameters panel in the Simulation View dialog box. The Emission Rate sets how much foam is produced, but there are also settings for the Wind and Air Turbulence.

Within the Display Settings panel are parameters for the foam Display Type, Color, Opacity and Size. These parameters change how the foam particles appear in the viewport. You can define how the foam particles are rendered in the Render Settings panel.

Using foam masks

A foam mask object defines the volume where the foam will be generated. Any particles outside of the foam mask volume will not have any foam particles. This provides a way to limit where foam particles are created. These foam mask objects can be picked and added to a list using the controls found in the Liquid Attributes panel. Properties for the foam mask objects are located in the Solver Parameters panel.

Tutorial: Spilling cider down a set of stairs

For this example, we'll spill some cider down a set of stairs. This will show how the liquid moves over several collision objects and we'll enable foam particles to give it a unique look.

To simulate cider spilling down a set of stairs, follow these steps:

1. Open the Cider down stairs.max file from the Chap 48 directory in the downloaded content set. This scene has a simple set of stairs made from box objects and a plane object.

2. Select the Create→Fluids→Liquid menu command to create a fluid emitter object. Select and move the emitter upwards to be positioned over the stairs.

3. Click on the Simulation View button to open the Simulation View dialog box. In the Liquid Attributes panel, click on the Add button underneath the Colliders list and choose all the Box objects. Then, click on the Pick button underneath the Kill Plane list and choose the Plane object.

4. Select the Solver Parameters panel and choose the Simulation Parameters section, then set the Master Voxel Size to 2.0. This will speed up the simulation immensely.

5. Then, choose the Liquid Parameters category in the Solver Parameters panel and select the Beer preset.

6. Next, choose the Display Settings panel in the Simulation View dialog box and change the Color Min value to a dark yellow color and the Color Max to a lighter yellow color. In the Foam Settings section, keep the Foam color set to white.

7. At the top of the Simulation View dialog box, enable the Solve Liquid Component toggle and the Solve Foam Component toggle, then click on the Start Solve button to begin the simulation calculations.

8. When the calculations are completed, click on the Play Animation button to see the results.

Figure 48.5 shows the cider running down the stairs with foam flying all about.

FIGURE 48.5

Adding foam to a fluid simulation gives another level of realism.

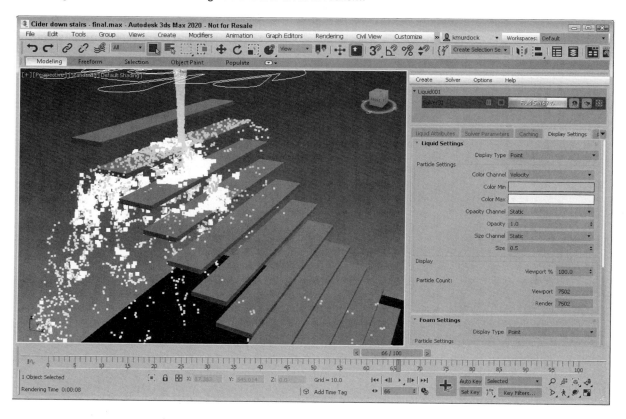

Controlling Fluids with Motion Fields

The gravity force is added to the liquid stream by default, but you can add more forces that affect the liquid using the Motion Fields Space Warp. To place a Motion Field object in the scene, select Create→Space Warps→Forces→Motion Field. You can also add a Motion Field using the Create menu in the Simulation View dialog box. The settings for this object are located in the Modify panel and include Magnitude, Direction, Turbulence and Velocity settings. Once a Motion Field object is positioned in the scene, you can add it to the simulation using the Pick list located at the bottom of the Liquid Attributes panel.

All the various Space Warp objects are covered in Chapter 43, "Using Space Warps."

The Display rollout for the Motion Fields object lets you see the force arrows for the given Motion Field. This provides a good way to visualize how the force will impact the fluid simulation. Figure 48.6 shows a directional force moving through a body of fluid.

FIGURE 48.6

Enabling the Display option for a Motion Field lets you see visually how the force will impact the fluid.

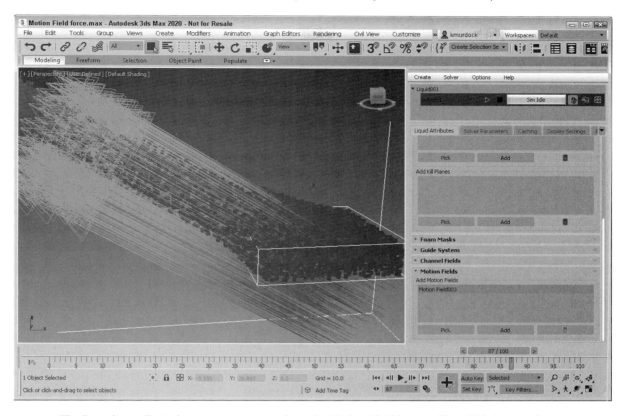

The Boundary rollouts lets you set an area where the Motion Field has an effect. This is a good way to control stray fluid droplets. You can also add a geometry object about which the force acts. Sometimes it is easy to mark the area where you want no effect. For that case, simply enable the Invert option.

Within the Direction rollout are several unique options that set the magnitude based on a picked object specified in the Geometry rollout. Concentric sets the force about the center of the Motion Field, Along Axis sets it in the direction of the object's axis, Around Axis sets the force to spin about the object's axis, and Away From Axis sets the force to accelerate away from the object's axis.

The Drag settings cause the fluid droplets to move toward the Motion Field icon. The Turbulence settings add chaos and randomness to the simulation.

Tutorial: Creating a whirlpool

You can create a static body of water with the Plane shaped emitter with the Gravity value set to 0. This provides a perfect fluid body to apply a force to. For this example, we'll create a vortex in the center of the body of water using a cylinder object and a Motion Field.

To create a whirlpool, follow these steps:

1. Select the Create→Fluids→Liquid menu command to create a fluid emitter object. Set the Emitter type to Plane with Length and Width values of 100.

2. Click on the Simulation View button to open the Simulation View dialog box. In the General Parameters section of the Solver Parameters panel, set the Gravity Magnitude to 0.

3. Create a long thin Cylinder object and place it in the center of the Plane-shaped Liquid object using the Create→Standard Primitives→Cylinder menu command.

4. Choose the Create→Space Warps→Forces→Motion Field menu command and place the Motion Field object in the center of the Plane-shaped Liquid object around the Cylinder object. In the Display rollout, enable the Velocity Grid option, then enable the Direction option in the Direction rollout, set the Magnitude to 0 and enable the Around Axis option. Then move the Motion Field object so it intersects the Liquid object.

5. In the Simulation View panel, select the Liquid Attributes panel, in the Motion Fields rollout, click the Pick button and choose the Motion Field object. Then, select the Solver Parameters panel and choose the Simulation Parameters section, then set the Master Voxel Size to 2.0.

6. Then, choose the Liquid Parameters category in the Solver Parameters panel and select the Water preset.

7. At the top of the Simulation View dialog box, enable the Solve Liquid Component toggle, then click on the Start Solve button to begin the simulation calculations.

8. When the calculations are completed, click on the Play Animation button to see the results.

Figure 48.7 shows the water vortex caused by the Motion Field.

FIGURE 48.7

A Motion Field Space Warp is used to create a whirlpool.

Using a Guide System

Using a fluid emitter places a 3d object in the scene and lets the fluid droplet emanate from that object, but you'll often want a single plane surface to act as the fluid, such as a lake or the ocean. For these occasions, the Guide System is what you need. To set up a Guide System, you create a single plane object to be the fluid surface and then several emitter objects that will interact with the surface. These emitter objects will likely also be Collider objects as well.

Caution

When creating a Guide Mesh, be sure to include enough segments to represent the details clearly.

With the Plane object and any Collider objects created, add the Liquid object to the scene and open the Simulation View dialog box. In the Liquid Attributes panel, pick the Plane object to add it to the Add Guide Emitter list and pick the Collider object to the Add Guide Mesh list. You'll also want to add the Collider object to the Collides list also. Once setup, any motion of the Collider object against the surface of the Place object will create a splash.

Tutorial: Ducks swimming at the sea

Ducks usually don't swim in the sea, especially toy ducks, but they provide a good chance to see the Guide System in action. For this scene, the toy duck is animated moving across the surface of the sea and a fluid simulation shows splashing where the duck passes through the water surface. The Wave modifier was also used to add a gentle wave to the sea object.

To simulate a duck swimming in the sea, follow these steps:

1. Open the Duck at sea.max file from the Chap 48 directory in the downloaded content set. This scene has a beach scene with a Plane object meeting the beach and a toy duck model animated moving through the Plane object.

2. Select the Create→Fluids→Liquid menu command to create a fluid emitter object in the Top viewport.

3. Click on the Simulation View button to open the Simulation View dialog box. In the Liquid Attributes panel, click on the Pick button underneath the Colliders list and choose all the Duck object. Beneath the Colliders list, pick the Sea object to add it to the Add Guide Emitter list and the Duck object again for the Add Guide Mesh list.

4. Select the Solver Parameters panel and choose the Simulation Parameters section, then set the Master Voxel Size to 2.0.

5. Then, choose the Liquid Parameters category in the Solver Parameters panel and select the Water preset.

6. At the top of the Simulation View dialog box, enable the Solve Liquid Component toggle and the Solve Foam Component toggle, then click on the Start Solve button to begin the simulation calculations.

7. When the calculations are completed, click on the Play Animation button to see the results.

Figure 48.8 shows the duck cutting through the sea causing quite a splash.

FIGURE 48.8

The Guide System lets you create an ocean scene.

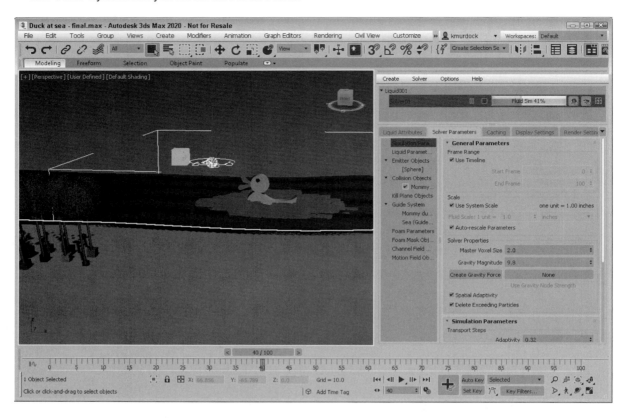

Rendering Fluids

The final panel in the Simulation View dialog box is the Render Settings panel. Using this panel, you can set how the Liquid and Foam will be rendered. For both liquid and foam, there is an option to render them as Armold Points or Arnold Surfaces. The Arnold renderer will need to be enabled in order to use these options.

More on Arnold is covered in Chapter 30, "Rendering with ART and Arnold."

Figure 48.9 shows the sea duck rendered.

FIGURE 48.9

Fluids can be rendered using any of the available renderers.

Summary

Adding fluid simulations to your scene can give it a realistic feel, but it can be computationally expensive. The best results will take time to tweak and computing time to represent accurately. The fluid simulation system offers several different ways to create fluid effects and is great for animated scenes.

Specifically, this chapter covered the following topics:

* Understanding fluids
* Creating a fluid emitter
* Setting fluid properties
* Working with Collider objects and Kill Planes
* Starting a simulation solver
* Adding foam and spray to the simulation
* Adding forces with the Motion Field object
* Using a Guide System
* Rendering fluids

Next up is a chapter showing how you can customize the interface, so if you don't like the way it is, you can change it.

Part XI

Extending 3ds Max

IN THIS PART

Chapter 49

Customizing the Interface

IN THIS CHAPTER

Using the Customize User Interface dialog box

Creating custom keyboard shortcuts, toolbars, quad menus, menus, and colors

Customizing the Ribbon

Customizing the Command Panel buttons

Loading and saving custom interfaces

When you get into a new car, one of the first things you do is to rearrange the seat and mirrors. You do this to make yourself comfortable. The same principle can apply to software packages: arranging or customizing an interface makes it more comfortable to work with.

Early versions of the 3ds Max software allowed only minimal changes to the interface, but later versions enable significant customization. The Autodesk® 3ds Max® 2020 interface can be customized to show only the icons and tools that you want to see.

Using the Customize User Interface Window

The Customize menu provides commands for customizing and setting up the interface for 3ds Max. The first menu item is the Customize User Interface menu command. This command opens the Customize User Interface dialog box. This dialog box includes six panels: Keyboard, Mouse, Toolbars, Quads, Menus, and Colors. You also can access this dialog box by right-clicking any toolbar away from the buttons and selecting Customize from the pop-up menu.

Customizing keyboard shortcuts

If used properly, keyboard shortcuts can increase your efficiency dramatically. The Keyboard panel of the Customize User Interface dialog box has been replaced with a separate Hotkey Editor interface, shown in Figure 49.1. In this interface, you can assign shortcuts to any command and define sets of shortcuts. You can assign keyboard shortcuts for any of the interfaces listed in the Group drop-down list. When an interface is selected from the Group drop-down list, all its commands are listed below, along with their current keyboard shortcuts. You can assign new hotkeys using the right panel and remove the hotkeys with the Remove button. You can save and load hotkey sets also.

Note
To access the defined keyboard shortcuts for the various interfaces, the Keyboard Shortcut Override Toggle button on the main toolbar must be enabled. If this button is disabled, only the keyboard shortcuts for the Main UI are active.

FIGURE 49.1

The Keyboard panel enables you to create keyboard shortcuts for any command.

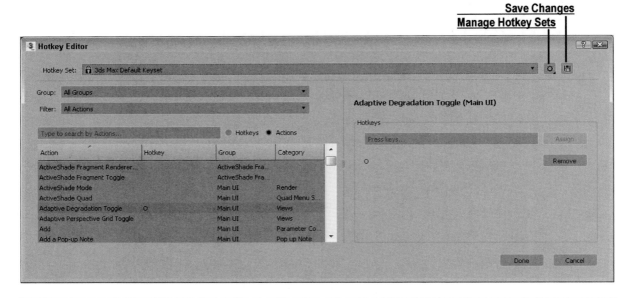

New Feature in 2020
The Hotkey Editor is new to 3ds Max 2020.

Groups are available for all the major features and commands in 3ds Max. You can use the Filter drop-down list to filter only select types of commands including Assigned, Customized and Unsaved Hotkeys. This helps you to quickly locate a specific set of commands. Entering a keyboard shortcut into the Search field lets you search for specific Hotkeys or Actions. You can Assign the hotkey to the selected command or Remove the hotkey from its current assignment.

Clicking on the Manage Hotkeys Sets buttons opens a pop-up menu where you can Save, Load or Reset the currently defined hotkey sets. Hotkey Sets are saved using the .hsx file extension. The Editable Poly commands that are marked in bold text can be activated by pressing and holding a keyboard shortcut. When you release the keyboard shortcut, the original mode is restored. For example, if you are working in Extrude mode, you can press and hold the Shift+Ctrl+B keyboard shortcut to access the Bevel command. When you release the keyboard shortcut, the Extrude mode is active again.

You can find a reference of the available default keyboard shortcuts in Appendix C, "Keyboard Shortcuts."

Defining the mouse actions

The Mouse panel, shown in Figure 49.2, in the Customize User Interface dialog box lets you define how to navigate the viewports using the mouse. If you update the interaction mode to Maya in the Interaction Mode panel of the Preference Settings dialog box, the Mouse panel settings are automatically updated. You can save the mouse interface settings to an external file using the Save button. These files have the .musx extension.

FIGURE 49.2

The Mouse panel lets you define the mouse controls for navigating the viewport.

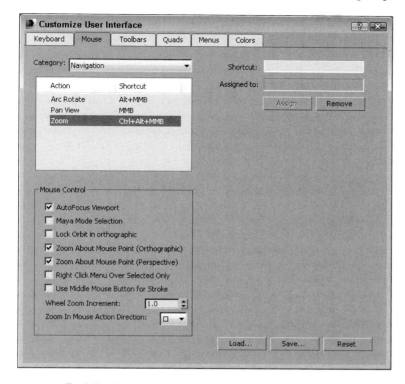

Enabling Maya compatibility

The AutoFocus Viewport option enables an option that makes the viewport under the mouse active when the spacebar is pressed. A similar feature is available in Maya and this option makes 3ds Max work like Maya.

The Maya Mode Selection option makes selecting objects in 3ds Max work just like Maya. Holding down the Shift key while clicking an object adds or subtracts the object from the selection set, and holding down the Ctrl key while clicking an object removes it from the selection set.

The Lock Orbit in Orthographic Views option prevents all orthographic views from being orbited, which is how these views work in Maya.

At the bottom of the Mouse panel is a Zoom In Mouse Action Direction drop-down list. Here you can choose which direction to drag to cause the zoom-in effect. This also lets you match the controls in Maya.

Panning, rotating, and zooming with the middle mouse button

If you're using a mouse that includes a middle button (this includes a mouse with a scrolling wheel), you can define how the middle button is used. The options are to use it to pan and zoom the viewports or to use the Strokes interface.

If the Use Middle Mouse Button for Stroke option is disabled, the middle mouse button pans the active viewport if the middle button is held down, zooms in and out by steps if you move the scrolling wheel, rotates the view if you hold down the Alt key while dragging, and zooms smoothly if you drag the middle mouse button with the Ctrl and Alt keys held down. You also can zoom in quickly using the scroll wheel with the Ctrl button held down, or more slowly with the Alt key held down. You can change the Wheel Zoom Increment value to speed up or slow down the zoom.

You also can set Zoom About Mouse Point options for the Orthographic and Perspective viewports to zoom about the mouse point. If this is disabled, you'll zoom about the center of the viewport. The Right Click

Menu Over Selected Only option causes the quad menus to appear only if you right-click on top of the selected object. This is a bad idea if you use the quad menus frequently.

Tip

I've found that using the middle mouse button along with the Alt key for rotating is the simplest and easiest way to navigate the viewport. Although Strokes is a clever idea, I always keep the Use Middle Mouse Button for Stroke option disabled.

Using strokes

The Use Middle Mouse Button for Stroke option lets you execute commands by dragging a predefined stroke in a viewport. With this option enabled, drag with the middle mouse button held down in one of the viewports. A simple dialog box identifies the stroke and executes the command associated with it. If no command is associated, a simple dialog box appears that lets you Continue (do nothing) or Define the stroke.

Another way to work with strokes is to enable the Strokes Utility. This is done by selecting the Utilities panel, clicking the More button, and selecting Strokes from the pop-up list of utilities. This utility makes a Draw Strokes button active. When the button is enabled, it turns blue. Then you can draw strokes with the left mouse button and access defined strokes with the middle mouse button.

If you select to define the stroke, the Define Stroke dialog box, shown in Figure 49.3, opens. You also can open this dialog box directly by holding down the Ctrl key while dragging a stroke with the middle mouse button. In the upper-left corner of this dialog box is a grid. Strokes are identified by the lines they cross on this grid as they are drawn. For example, an "HK" stroke would be a vertical line dragged from the top of the viewport straight down to the bottom.

FIGURE 49.3

The Define Stroke dialog box lets you define specific command strokes that are executed by drawing the stroke with the middle mouse button.

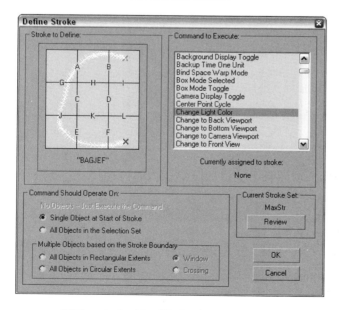

With a stroke identified, you can select a command in the upper-right pane. This is the command that executes when you drag the stroke with the middle mouse button in the viewport. For each command, you can set the options found below the stroke grid. These options define what the command is executed on.

All defined strokes are saved in a set, and you can review the current set of defined strokes with the Review button. Clicking this button opens the Review Strokes dialog box where all defined strokes and their commands are displayed, as shown in Figure 49.4.

FIGURE 49.4

The Review Strokes and Stroke Preferences dialog boxes list all defined strokes and their respective commands.

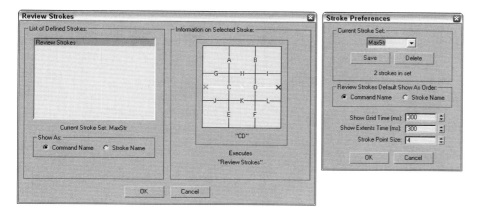

One of the commands available in the list of commands is Stroke Preferences. Using this command opens the Stroke Preferences dialog box, also shown in Figure 49.4, where you can save and delete different stroke sets, specify to list commands or strokes in the Review Strokes dialog box, set how long the stroke grid and extents appear, and set the Stroke Point Size.

Customizing toolbars

You can use the Customize User Interface dialog box's Toolbar panel to create custom toolbars. Figure 49.5 shows this panel.

FIGURE 49.5

The Toolbars panel in the Customize User Interface dialog box enables you to create new toolbars.

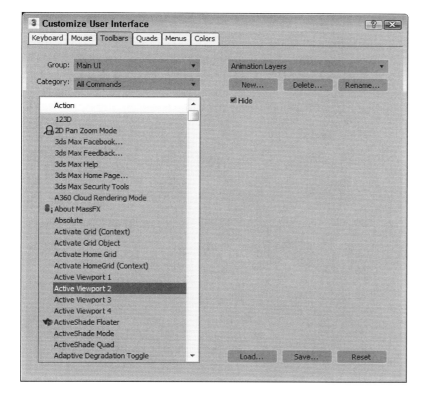

The Toolbars panel of the Customize User Interface dialog box includes the same Group and Category drop-down lists and command list as the Keyboard panel. Clicking the New button opens a simple dialog box where you can create and name the new toolbar. The Delete button lets you delete toolbars. You can delete only toolbars that you've created. The Rename button lets you rename the current toolbar. The Hide option makes the selected toolbar hidden.

Use the Load and Save buttons to load and save your newly created interface, including the new toolbar, to a custom interface file. Saved toolbars have the .cuix extension.

After you create a new toolbar, you can drag the commands in the Action list to either a new blank toolbar created with the New button or to an existing toolbar. By holding down the Alt key, you can drag a button from another toolbar and move it to your new toolbar. Holding down the Ctrl key and dragging a button retains a copy of the button on the first toolbar.

If you drag a command that has an icon associated with it, the icon appears on the new toolbar. If the command doesn't have an icon, the text for the command appears on the new toolbar.

Tutorial: Creating a custom toolbar

If you've been using 3ds Max for a while, you probably have several favorite commands that you use extensively. You can create a custom toolbar of all your favorite commands. To learn how to do this, you'll create a custom toolbar for the compound objects.

To create a custom toolbar for creating compound objects, follow these steps:

1. Open the Customize User Interface dialog box by choosing Customize→Customize User Interface.
2. Open the Toolbars panel, and click the New button. In the New Toolbar dialog box that appears, name the toolbar Compound Objects. After you click OK, a new blank toolbar appears.
3. Select the Main UI group and the Objects Compounds category from the drop-down lists on the left. Then drag each command in the Action list to the new blank toolbar.
4. Click the Save button to save the changes to the customized interface file. You can load the resulting toolbar from the Chap 49 folder. It is named Compound Objects toolbar.cui.

Note

Don't be alarmed if the toolbar icons show up gray. Gray icons are simply disabled. When the tool is enabled, they are shown in color.

Figure 49.6 shows the new toolbar. With the new toolbar created, you can float, dock, or edit this toolbar just like the other toolbars. Notice that some of the tools have icons and others have text names.

FIGURE 49.6

A new toolbar of compound objects created using the Customize User Interface dialog box

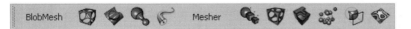

You can right-click any of the buttons on any of the existing toolbars, except for the main toolbar, to access a pop-up menu. This pop-up menu enables you to change the button's appearance, delete the button, edit the button's macro script, or open the Customize User Interface dialog box.

To learn more about editing macro scripts, see Chapter 52, "Automating with MAXScript."

Changing a button's appearance

Selecting the Edit Button Appearance command from the right-click pop-up menu opens the Edit Macro Button dialog box, shown in Figure 49.7. This dialog box enables you to quickly change the button's icon, tooltip, or text label. Each icon group shows both the standard icon and the grayed-out disabled version of the icon. Default buttons also can be changed. The Odd Only check box shows only the standard icons.

FIGURE 49.7

The Edit Macro Button dialog box provides a quick way to change an icon, tooltip, or text label.

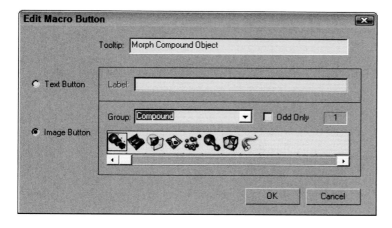

Note

If a text label doesn't fit within the toolbar button, you can increase the button width using the Fixed Width Text Buttons spinner in the General panel of the Preference Settings dialog box.

Tutorial: Adding custom icons

The 3ds Max interface uses two different sizes of icons. Large icons are 24 x 24 pixels, and small icons are 16 x 15 pixels. Large icons can be 24-bit color, and small ones must be only 16-bit. Multiple icons can be placed side by side in a single file. The easiest way to create some custom toolbars is to copy an existing set of icons into an image-editing program, make the modifications, and save them under a different name. You can find all the icons saved as BMPs and used by 3ds Max in the 3dsmax\UI\Icons directory.

To create a new group of icons, follow these steps:

1. Select a group of current icons to edit from the UI directory, and open them in Photoshop. I selected the Patches group, which includes all the files that start with the word *Patches*. This group includes only two icons. To edit icons used for both large and small icon settings and both active and inactive states, open the following four files: Patches_16a.bmp, Patches_16i.bmp, Patches_24a.bmp, and Patches_24i.bmp.

2. In each file, the icons are all included side by side in the same file, so the first two files are 32 x 15 and the second two are 48 x 24. Edit the files, being sure to keep each icon within its required dimensions. This includes space for two icons, an active and inactive state.

3. When you finish editing or creating the icons, save each file with the name of the icon group in front of the underscore character. My files were saved as Kels_16a.bmp, Kels_16i.bmp, Kels_24a.bmp, and Kels_24i.bmp, so they show up in Kels group in the Edit Macro Button dialog box. Copy these four edited files from the Chap 49 folder to the 3dsmax\UI\Icons directory.

4. After the files are saved, you need to restart 3ds Max. The icon group is then available within the Customize User Interface dialog box when assigned to a command.

Figure 49.8 shows the Edit Macro Button dialog box with my custom icon group named Kels open.

FIGURE 49.8

The Edit Macro Button dialog box with a custom icon group selected

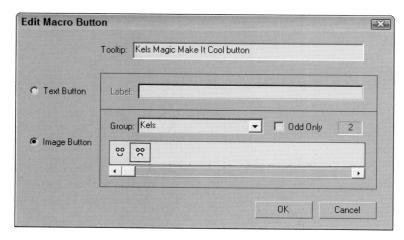

Customizing quad menus

The fourth panel in the Customize User Interface dialog box allows you to customize the quad menus. You can open quad menus by right-clicking the active viewport or in certain interfaces. Figure 49.9 shows this panel.

FIGURE 49.9

The Quads panel of the Customize User Interface dialog box lets you modify pop-up quad menus.

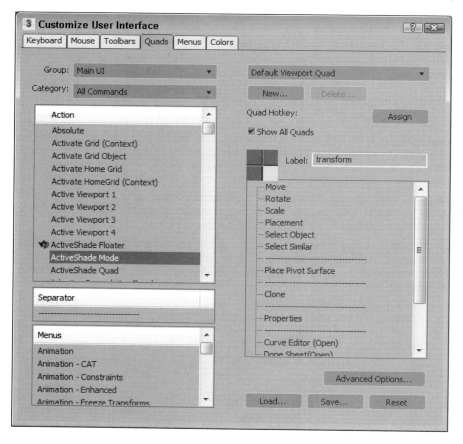

To the left of the panel are the Group and Category drop-down lists and a list of actions that are the same as those that appear in the Keyboard and Toolbars panels, but the Quads panel also includes a Separator and a list of Menu commands. Quad menus can include separators to divide the commands into different sections and menus that appear at the top of the standard interface.

The drop-down list at the top right of the Quads panel includes many different quad menu sets. These quad menus appear in different locations, such as within the ActiveShade window. Not only can you customize the default viewport quad menus, but you also can create your own named custom quad menus with the New button or you can rename an existing quad menu. The Quad Shortcut field lets you assign a keyboard shortcut to a custom quad menu.

Tip

Several quad menus have keyboard shortcuts applied to them. Right-clicking with the Shift key held down opens the Snap quad menu. Other shortcuts include Alt+right-click for the Animation quad menu, Ctrl+right-click for the Modeling quad menu, Shift+Alt+right-click for the MassFX quad menu, and Ctrl+Alt+right-click for the Lighting/Rendering quad menu.

If the Show All Quads option is disabled, it causes only a single quad menu to be shown at a time when unchecked. Although only one quad menu is shown at a time, the corner of each menu is shown, and you can switch between the different menus by moving the mouse over the corner of the menu. This is useful if you want to limit the size of the quad menu.

The four quadrants of the current quad menu are shown as four boxes. The currently selected quad menu is highlighted yellow, and its label and commands are shown in the adjacent fields. Click the gray boxes to select one of the different quad menus.

To add a command to the selected quad menu, drag an action, separator, or menu from the panes on the left to the quad menu commands pane on the right. You can reorder the commands in the quad menu commands pane by dragging the commands and dropping them in their new location. To delete a command, just select it and press the Delete key or select Delete Menu Item from the right-click pop-up menu.

If you right-click the commands in the right pane, a pop-up menu appears with options to delete or rename the command. Another command allows you to flatten a submenu, which displays all submenu commands on the top level with the other commands.

Custom quad menus can be loaded and saved as menu files (with the .mnux extension).

The Quads panel also includes an Advanced Options button. Clicking this button opens the Advanced Quad Menu Options dialog box, shown in Figure 49.10. Using this dialog box, you can set options such as the colors used in the quad menus.

FIGURE 49.10

The Advanced Quad Menu Options dialog box lets you change quad menu fonts and colors.

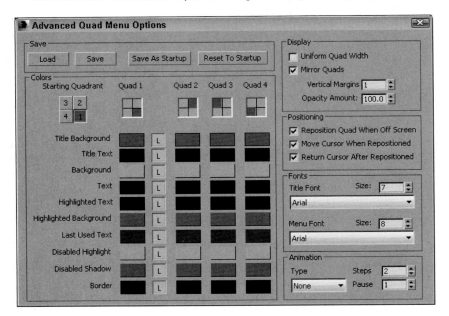

Changes to the Advanced Quad Menu Options dialog box affect all quad menus. You can load and save these settings to files (with the .qop extension). The Starting Quadrant determines which quadrant is first to appear when the quad menu is accessed. You can select to change the colors for each quad menu independent of the others. The column with the L locks the colors so they are consistent for all quad menus if enabled.

The remainder of the Advanced Quad Menu Options dialog box includes settings for controlling how the quad menus are displayed and positioned, as well as the fonts that are used.

The Animation section lets you define the animation style that is used when the quad menus appear. The animation types include None and Fade. The Fade style slowly makes the quad menus appear.

Tip

I personally don't like to wait for the quad menus to appear, so I keep the Animation setting set to None.

Customizing menus

The Menus panel of the Customize User Interface dialog box allows you to customize the menus used at the top of the 3ds Max window. Figure 49.11 shows this panel.

FIGURE 49.11

You can use the Menus panel of the Customize User Interface dialog box to modify menus.

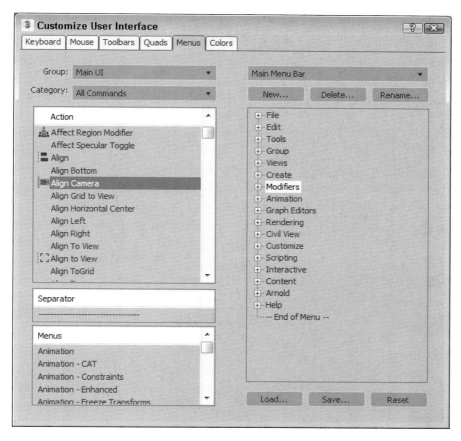

This panel includes the same Group and Category drop-down lists and the Action, Separator, and Menus panes found in the Quads panel. You can drag and drop these commands to the menu pane on the right. Menus can be saved as files (with the .mnux extension). In the menu pane on the right, you can delete menu items with the Delete key or by right-clicking and selecting Delete Item from the pop-up menu.

Tip

If you place an ampersand (&) character in front of a custom menu name letter, that letter is underlined and can be accessed using the Alt key; for example, Alt+E opens the Edit menu.

Tutorial: Adding a new menu

Adding a new menu is easy to do with the Customize User Interface dialog box. For this example, you tack another menu to the end of the Tools menu.

To add another menu item to the Tools menu, follow these steps:

1. Choose Customize→Customize User Interface to open the Customize User Interface dialog box.
2. Click the Menus tab to open the Menus panel.
3. In the top-left drop-down list, select Main UI from the Group drop-down list and Tools from the Category drop-down list.

   Expand the Tools menu in the right pane by clicking the plus sign to its left.
4. Locate the Cross Hair Cursor Toggle menu item in the Action list, drag it to the right, and drop it right after the Channel Info Editor menu item.

As you drag, a blue line indicates where the menu command will be located.

5. Click the Save button to save the menu as a file. You can find the customized menu from this example in the Chap 49 folder.

After you save the new menu file, you need to restart 3ds Max before you can see the changes. You can reset the default UI by choosing Customize→Revert to Startup UI Layout.

Customizing colors

Within 3ds Max, the colors often indicate the mode in which you're working. For example, red marks animation mode. Using the Colors panel of the Customize User Interface dialog box, you can set custom colors for all 3ds Max interface elements. This panel, shown in Figure 49.12, includes two panes. The upper pane displays the available items for the interface selected in the Elements drop-down list. Selecting an item in the list displays its color in the color swatch to the right. You also can set the color Intensity, invert the color, or make the Application Frame Light or Dark.

FIGURE 49.12

You can use the Colors panel of the Customize User Interface dialog box to set the colors used in the interface.

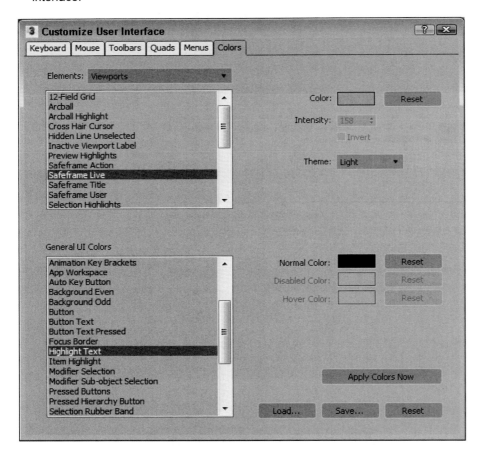

The lower pane displays a list of the custom colors that can be changed to affect the appearance of the interface. For example, Highlight Text isn't an element; it's an interface appearance. For these colors, you can set the Normal Color, Disabled Color, and Hover Color.

You can save custom color settings as files with the .clrx extension. You can use the Apply Colors Now button to immediately update the interface colors.

Customizing the Ribbon

The Ribbon interface also includes a dialog box for customizing its panels. To access this dialog box, right-click the Ribbon title bar and select the Ribbon Configuration→Customize Ribbon menu command. This opens the Customize Ribbon dialog box, shown in Figure 49.13.

FIGURE 49.13

The Customize Ribbon dialog box lets you repopulate the existing panels or even create your own panels.

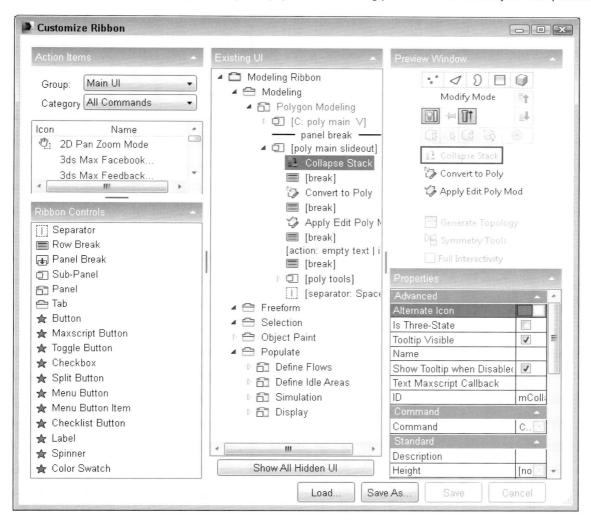

The Customize Ribbon dialog box is fairly easy to use. The top-left panels hold all the available commands grouped by category, similar to the Customize User Interface dialog box. The lower-left panel holds a set of controls that are unique to the Ribbon interface, including tabs, panels, and subpanels. The middle panel holds a hierarchical list of the current Ribbon interface, and the right panels show a preview of the current selection and all the properties for the selected item.

To add a command to the current Ribbon panel, simply locate the command in the Action Items panel of the Customize Ribbon dialog box, drag it to the Existing UI panel, and drop it in the location where you want the command to appear. If the command has an icon button associated with it, the button will appear in the specified location. The Preview Window shows exactly how it will appear in the Ribbon panel.

If the command that you drag into the Existing UI panel doesn't have an icon button, you can use the Standard Icon property to load in a custom icon. Other properties let you show or hide the icon or text, enter a new tool label, or specify the height and width of the tool. With the Description property you can enter text that appears in the tooltip.

Making dynamic tools and panels

Some properties, such as the Visible property for standard commands or the Available property for panels, let you choose values of True, False, and Conditional. The Conditional option opens a dialog box, shown in Figure 49.14, where you can specify the conditions required for the command to be visible or available. For example, you can have a button appear only when an Editable Poly object is selected or when Slice Mode is enabled. This gives you the power to make the custom Ribbon dynamic.

FIGURE 49.14

The Conditions dialog box lets you specify the required condition for having the tool or panel appear.

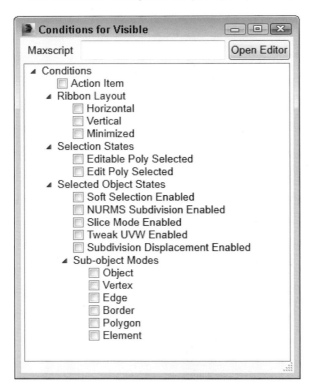

The Conditional dialog box also includes a field for entering MAXScript conditions, such as enabling an object only when multiple objects are selected. The Open Editor button opens a MAXScript Editor window so you can test your conditions.

Saving Ribbon changes

After you're happy with the Ribbon changes, you can press the Save button to immediately apply the changes to the current Ribbon. The Save As button opens a dialog box where you can save the custom Ribbon as a separate file with the .ribbon extension. Saved ribbon files can be reloaded into the Customize Ribbon dialog box using the Load button.

Tutorial: Building a Ribbon panel of primitives

Working with the Ribbon is great, and I typically like to hide the Command Panel when modeling, but occasionally I need to create a primitive object to work with. The Create menu is too cumbersome, so creating a Ribbon panel holding all the primitives is helpful.

To create a Ribbon panel of primitive objects, follow these steps:

1. Click the Graphite Modeling Tools button in the main toolbar to open the Ribbon interface.
2. Right-click the Ribbon title bar and select the Ribbon Configuration→Customize Ribbon menu command.
3. Drag the Tab option in the Ribbon Controls panel and drop it beneath the Object Paint entry in the Existing UI panel. In the Properties panel, enter Primitives for the Name and Title fields.

Note
A new tab cannot be populated until you give it a unique name in the Name field.

4. Drag the Panel item from the Ribbon Controls panel and drop it in the Existing UI panel under the new Primitives tab. Change the Title property to Standard Primitives.
5. In the Action Items panel, set the Group to Main UI and the Category to Objects Primitives, and then locate and drag each of the standard primitives to the new Standard Primitives panel in the Existing UI panel.
6. Drag a Row Break item from the Ribbon Controls panel and drop it beneath the Teapot item in the Existing UI panel to break up the row of icons in the new Standard panel.
7. Click the Save button to apply the changes to the current Ribbon interface.

Figure 49.15 shows the customized Ribbon with the new panel of primitives.

FIGURE 49.15

A custom panel of primitives has been added to the Ribbon interface.

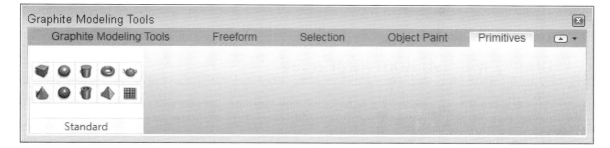

Customizing Modify and Utilities Panel Buttons

 The Modify panel and the Utilities panel in the Command Panel both include a button called Configure Modifier Sets that allows you to configure how the modifiers are grouped and which utility buttons appear in the Utilities panel.

In the Modify panel, the Configure Modifier Sets button is the right-most button directly under the Modifier Stack. This button opens a pop-up menu that lists all the modifier categories. The top pop-up menu command is Configure Modifier Sets, which opens a dialog box, shown in Figure 49.16, when selected. Using this dialog box, you can control which modifiers are grouped with which sets.

FIGURE 49.16

The Configure Modifier Sets dialog box lets you group the modifiers as you want.

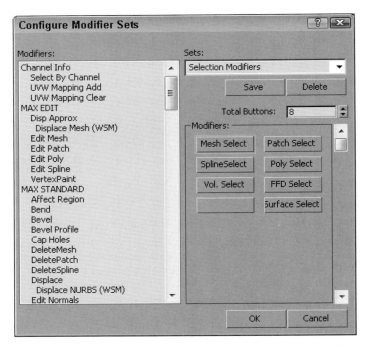

To add a modifier to a set, select the set from the Sets drop-down list and drag the modifier from the list of Modifiers on the left to the button set on the right. To create a new set, simply type a new name into the Sets field. After a set has changed, you need to save it with the Save button.

You can find the same Configure Button Sets button on the Utilities panel. Clicking this button opens a similar dialog box where you can drag from a list of Utilities to a list of buttons on the right. These buttons are then displayed in the Utilities panel. Using this feature, you can add your favorite modifiers and utilities to appear when either panel is opened, saving you from having to locate them.

Working with Custom Interfaces

If you've changed your interface, you'll be happy to know that the Customize menu includes a way for you to save and then reload your custom setup. This feature is especially helpful for users who share a copy of 3ds Max.

Tip

Any custom .ui file can be loaded as the default interface from the command line by adding a –c and the .ui filename after the 3dsmax.exe file (for example, **3dsmax.exe –c my_interface.ui**).

Saving and loading a custom interface

Custom interface schemes are saved with the .ui extension using the Customize→Save Custom UI Scheme menu command. When you save a custom scheme, 3ds Max opens a file dialog box where you can name the .ui file, and then the Custom Scheme dialog box, shown in Figure 49.17. This dialog box lets you choose which customizations to include in the custom scheme. It also lets you select the icon type to use. The options are Light and Dark.

FIGURE 49.17

The Custom Scheme dialog box appears when you're saving a custom interface and lets you select which items to include.

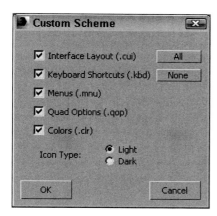

You can load saved user interface schemes with Customize→Load Custom UI Scheme. These custom UI files are saved in the Program Files/Autodesk/3ds Max 2020/UI folder. The default 3ds Max install includes several predefined interface setups located in the UI directory. These standard interfaces are available:

* **DefaultUI:** Default interface that opens when 3ds Max is first installed.

* **Ame-dark:** Displays the standard interface with black windows, backgrounds, and viewports. All the icons and menus are light gray, and many of the icons are different, as shown in Figure 49.18.

* **Ame-light:** Same as the Ame-dark layout, except the icons and menus are black and the backgrounds are all light gray. Many icons are different here, too.

FIGURE 49.18

If you prefer a darker interface, try loading the Ame-dark scheme.

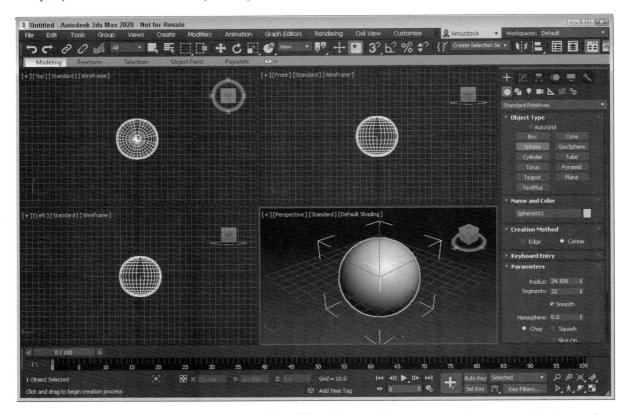

You also can use both the Load Custom UI and Save Custom UI menu commands to save and load any of the custom user interface files types, including these:

* Interface Scheme files (.ui)

* UI files (.cui)

* Menu files (.mnu)

* Color files (.clr)

* Keyboard Shortcut files (.kbd)

* Quad Options files (.qop)

Note

You can set 3ds Max to automatically save your interface changes when exiting. Select the Save UI Configuration on Exit option in the General tab of the Preference Settings dialog box.

Locking the interface

After you're comfortable with your interface changes, locking the interface to prevent accidental changes is a good idea. To lock the current interface, choose Customize→Lock UI Layout (or press the Alt+0 keyboard shortcut). Locking the interface prevents changes by dragging interface elements, but you can still make interface changes using the pop-up menus.

Reverting to the startup interface

When you're first playing around with the software's customization features, really messing things up can be easy. If you get in a bind, you can reload the default startup interface (DefaultUI.ui) with the

Customize→Revert to Startup Layout command. Using the File→Reset menu command does not reset changes to the layout.

Note

If your DefaultUI.ui file gets messed up, you can reinstate the original default interface setup by deleting the DefaultUI.ui file before starting 3ds Max. However, do not overwrite the default UI file because this file is needed to reinstate the default UI.

Switching between default and custom interfaces

The Customize→Custom Defaults Switcher menu command opens an interactive window that presents several options for selecting initial settings and interface schemes, as shown in Figure 49.19. At the top of the window, you can select an option, and then details about the selected option are displayed.

FIGURE 49.19

This window explains the benefits of the different initial settings and scheme choices.

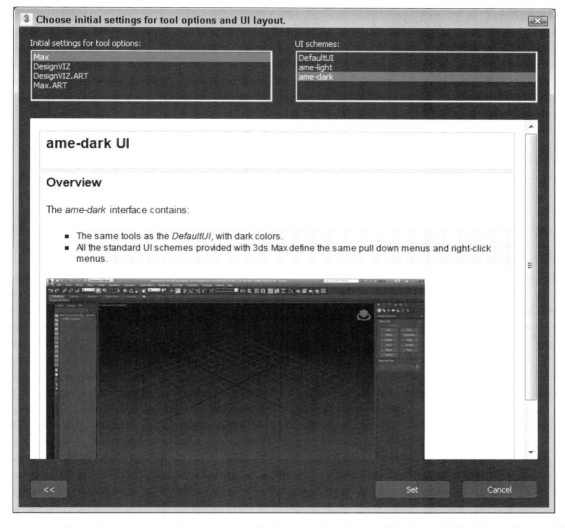

The initial settings for the tool options list include Max, DesignVIZ, DesignVIZ.ART, and Max.ART. These different selections cause the default settings for the various controls to change. For example, the default renderer for 3ds Max is the Scanline renderer, but for the 3ds Max.ART option, ART is the default renderer.

The schemes list includes the same custom interfaces listed earlier, along with any custom interfaces that have been saved.

After selecting the initial settings and scheme to use, click the Set button to commit the selections to the interface. The button with arrows on it in the lower left displays the initial information page again.

Note

If any of the settings within the CurrentDefaults.ini file or if the other files are missing from the new settings folder, the settings and files within the default 3ds Max directory are used.

Summary

You can customize the 3ds Max interface in many ways. Most of these customization options are included under the Customize menu. In this chapter, you learned how to use this menu and its commands to customize many aspects of the interface. Customizing makes the 3ds Max interface more efficient and comfortable for you.

Specifically, this chapter covered the following topics:

* Using the Customize User Interface dialog box to customize keyboard shortcuts, toolbars, quad menus, menus, and colors
* Customizing the Ribbon interface
* Customizing buttons on the Modify and Utilities panels
* Saving and loading custom interfaces

The next chapter shows how you can extend the features of 3ds Max using the Max Creation Graph interface.

Chapter 50
Creating Procedural Content with Max Creation Graphs

IN THIS CHAPTER

Installing Max Creation Graph

Creating Max Creation Graph

Saving MCG Files

Packaging MCG Files

If you get under the hood of Autodesk® 3ds Max® 2020, you'll find that every object, modifier, controller, feature, command and operator has a way to be referenced using MAXScript. This is text command that does the same as clicking its associated button or control in the interface. This gives the software an incredible power because using MAXScript, you can write detailed instructions of exactly you want in your scene and have them run to create the results. This power is accessible by programmers who don't shy away from textual commands.

The problem with MAXScript is that it is made for programmers, which is beyond most artists. Autodesk has created a tool in 3ds Max that gives a visual front end to the MAXScript engine. This front end editing interface is called Max Creation Graph (MCG) and the files it creates are called MCG files. MCG is essentially a visual programming interface that is used to create procedural objects, modifiers and utilities. It works by wiring nodes together in the MCG Editor. MCG files can be saved and shared between users.

All of the MCG commands are located within the Scripting menu. Using this menu, you can install MCG packages, create new MCGs, open existing MCGs and access the MCG editor. The MCG editor works in a similar manner to the Slate Material Editor by connecting adjacent nodes together.

Installing a Max Creation Graph

If you download a Max Creation Graph with the .mcg extension, you can install it using the Scripting→Install Max Creation Graph (.mcg) Package menu command. This opens the dialog box, shown in Figure 50.1. It also lists the creator of the package.

FIGURE 50.1

The Install Max Creation Graph Tool dialog box lets you select and load any .mcg file to install.

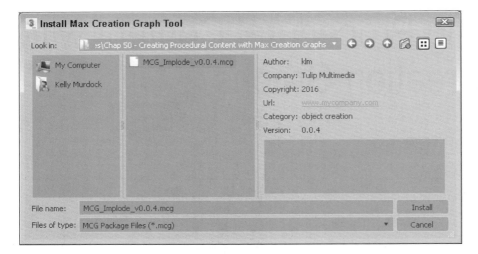

Once installed, the new MCG feature will be integrated into the interface depending on the node that is used. For example, if an Output Modifier node is used, then the new MCG tool will appear in the Modifier List located in the Modify panel. Installed MCG packages are saved locally where 3ds Max is installed in a User Tools→Max Creation Graph→Packages folder.

Note

One of the best ways to learn is to look at existing examples. 3ds Max has several MCG tools built-in to the software such as the animation controllers found in the Animation→MCG Controllers menu. You can load these MCGs directly into the MCG Editor from their installed folder found at Program Files→Autodesk→MaxCreationGraph→Packages.

Creating a Max Creation Graph

All Max Creation Graphs are created using the Max Creation Graph editor. You can access this editor using the Scripting→New Max Creation Graph menu command. The editor opens as a window with several panels, as shown in Figure 50.2. Each new MCG is displayed within a tab located at the top of the interface. To the left is a list or Operator Nodes called the Operator Depot. You can also search for specific nodes using the Search bar located directly above the operator node list.

Note

Be aware that any text in the Search bar will limit the items displayed in the list. You can click the X button to the right of the Search bar to clear the Search text and to display the entire list again.

Tip

Press the X key to immediately access the Search window.

The center panel is the Graph panel where nodes are placed and connected. Double click on an operator to add its node to the Graph panel. Each node is contained with a box and includes input and output connections. Click on a node in the Graph panel to select it. When a node is selected, it is outlined in white and its properties are displayed in the Node Properties panel. This is a helpful place to look at what the node does and how to connect it to other nodes. You can move the selected node by dragging its title in the working area and you can drag over several nodes to select several at once. There is also a View Navigator panel for positioning the current view.

The Message Log displays log information as the graph is being executed. This is helpful for troubleshooting any errors.

Caution

The operators in the Operator Depot are displayed in light gray when the light color interface is used making them hard to read. For this reason, I've switched to the dark color interface when working with MCGs.

FIGURE 50.2

The Max Creation Graph editor interface displays each MCG in a separate tab.

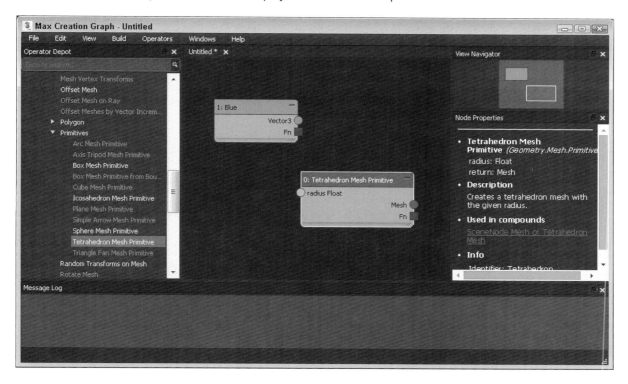

You can pan the graph panel by dragging the middle mouse button and zooming in accomplished by scrubbing the middle mouse button. You can also drag the navigation pane in the upper right corner.

Adding Nodes

Adding nodes to the Graph panel is easy to do. Simply select and drag the node name from the Operator Depot list and drop it in the Graph panel or you can simply double click on the node name. The node appears within a box and its name is at the top of the box. You can move the node around by dragging it within the Graph panel. You can also minimize the box by clicking on its minimize button in the top right corner.

Every new node that is added to the graph is given a number that appears before its name at the top of the node box. This number is helpful for identifying problem nodes when verifying the graph. You can also cut, copy and paste a single node or a series of selected nodes using the commands in the Edit menu.

Nodes can also be deleted using the Delete key.

Connecting Nodes

To the sides of each node box are its inputs and outputs. The inputs are on the left side of the node box and the outputs are on the right side. The output nodes of one node can be connected to the input nodes of another, but only if the connections are compatible. Each of the inputs and outputs represent data that is passed between the nodes. For example, a color node passes color data and cannot be connected to a node input that is expecting a number parameter. You can tell if a set of connections are compatible because they will have the same color.

Note
The MCG Editor does allow you to connect incompatible values, but the connection line will appear red if the values are incompatible.

To connect two nodes together, simply drag from the output or the input of one node to another. This creates a wire that connects the two nodes. Once connected, you can move a node and the connecting wire moves to stay connected. If you hold down the Ctrl and Alt keys together while moving a single node, then all its connected children nodes will move with it. This lets you move several nodes at once.

You can also select and add a connected node by dragging away from an input or output node onto an open area in the Graph panel. This opens a pop-up menu where you can select a new node to add. This pop-up menu also includes a Search bar that you can use to search for a specific node.

Saving the MCG File

When you save a MCG file, the file dialog box gives you the option of saving it to the Max Creation Graph folder where 3ds Max is installed by default. This is also the location that the software first looks when installing an MCG. Saved MCG files have the .maxtools extension. The name of the MCG file is the same name that is used within the 3ds Max interface. For example, if you name the MCG My Cool MCG and it creates a modifier, then that is the name that appears in the modifier list.

Tip
When saving an MCG file, it is a good habit to include MCG in the title of the file in order to make the command easy to locate within the 3ds Max interface.

Although it is not required, you can enter identifying properties for each graph including the graph's author, company, copyright, URL and category using the Graph Properties dialog box, shown in Figure 50.3. The Graph Properties dialog box is opened using the Edit→Edit Graph Properties menu command. It also keeps track of the version number for the graph. All of this properties information is displayed when the MCG file is installed and the Category is used to sort the modifier into different sets within the Command Panel.

FIGURE 50.3

The Graph Properties dialog box lets you enter information about the graph.

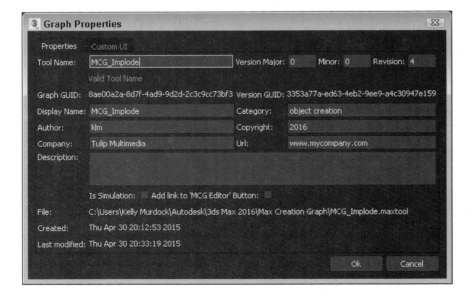

Testing the MCG

Once you have all your nodes connected and the file is saved, you can test it out. To test out a MCG, you'll need to build the graph using the Build→Evaluate menu command. Ctrl+E is the shortcut for this command. After building the file, the file is saved using the .ms extension. This is a MaxScript file that the normal interface can use to make changes to the interface.

Tutorial: Building a Simple MCG modifier

Although the MCG interface can create some amazing features, this tutorial will show some simple functionality using a tool set that is common, the modifier. For this example, we'll start with a simple selected object, perform a unique operation and save it as a modifier. This will let you quickly see the results of your work.

I've always liked the Spherify modifier, but it doesn't accept negative values, so for this graph, we'll do the opposite making an implode modifier.

To create an implode modifier, follow these steps:

1. Select the Scripting→Open Max Creation Graph Editor menu command to open the MCG interface.

2. Within the Operator Depot list locate the Outputs→Tools→Output: Modifier node and double click on it to add it to the graph panel. This node will create the modifier within the 3ds Max interface.

3. Locate and double click on the Geometry→Mesh→Spherify Mesh node to add it to the graph panel. This node performs the Spherify operation.

4. Locate and double click on the Inputs→Tools→Modifier: Mesh node to add it to the graph panel. This node represents the selected object in the scene.

5. Finally, locate and add the Inputs→Tools→Tool Input: Float node. This is a simple floating point number.

6. Connect the Mesh output on the Mesh node to the Mesh input on the Spherify node. Connect the Float output on the Tool Input: Float node to the amount input on the Spherify node. Then, connect the Mesh output on the Spherify node to the Mesh input on the Output node.

7. Within the Float node, name the parameter, Implode. This is the name that will appear within the Command Panel for this parameter. Then set the Min value to -10, the Max value to 0.0 and the Default value to -0.5. This connected graph will look like the one in Figure 50.4.

FIGURE 50.4

The wired graph is ready to save and build.

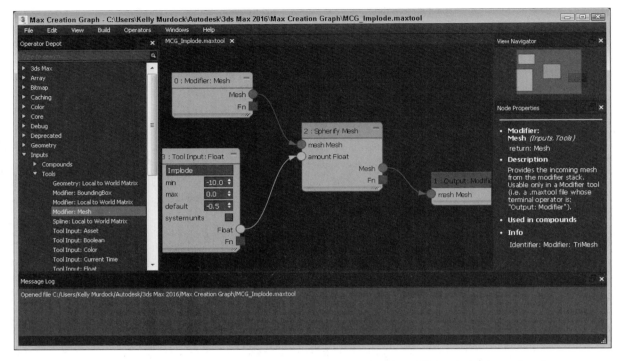

8. Select the File→Save As menu command and name the file MCG_Implode, then select the Build→Evaluate menu command. The build status is displayed in the Message Log pane.

9. Back in the viewport, create a Box object and set its Length, Width and Height Segs to 20 each, then select the new MCG_Implode modifier in the Modify panel and adjust the Implode parameter.

Figure 50.5 shows the box with the negative spherify command applied.

FIGURE 50.5

The inverse Spherify command applied as an MCG modifier.

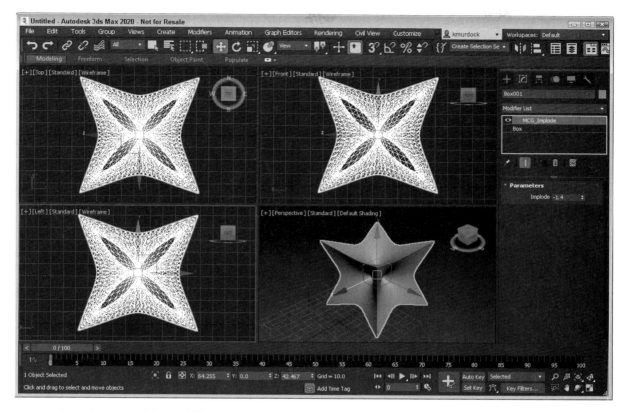

Packaging MCG Files

If you want to share your complete MCG with other users, then you'll want to package it into a single file that you can give to them. The File→Package Tool Graph menu command pulls the .maxtools file along with the .ms file together into a single file with the .mcg file extension. This file type can then be installed on another computer using the Install command, which places all the files in their correct locations.

Installing a MCG file also places it permanently in the interface so that it is available every time that you start 3ds Max. If an MCG is not installed, then you will need to open the MCG Editor and build the graph before you can use it.

Note

After installing an MCG, you typically have to restart the 3ds Max interface before the installed MCG shows up in the interface.

Summary

If you want to add some new features to 3ds Max, but MaxScript seems a little over your head, then you can try using the visual programming features in the Max Content Graph editor. By learning the various nodes and how to connect them, you can create amazing new functionality. This chapter covered the following topics:

* Installing MCG files to extend functionality
* Building MCG graphs using the MCG editor
* Packing MCG graphs for use by other users

The next chapter introduces 3ds Max Interactive, a tool used to port your 3d creations to video game consoles and smart phones.

Chapter 51
Exploring 3ds Max Interactive

IN THIS CHAPTER

Installing 3ds Max Interactive

Learning the 3ds Max Interactive interface

Creating a Live Link to 3ds Max

Loading assets

Exploring the available audio, user interface and level animation tools

Adding interactivity with the Level Flow interface

Deploying a level

So you are really good at creating 3ds Max scenes, but you haven't figured out yet how to create interactive installations like VR or cell-phone games. 3ds Max Interactive is one of those bridge technologies that enable you to take something familiar and common like 3ds Max and produce something unfamiliar and uncommon like a video game.

Keep in mind that 3ds Max Interactive is a complete toolset in itself and this simple little chapter isn't enough to cover it comprehensively. Instead, I hope to introduce this new technology and point you in the right direction for getting started with it.

Installing 3ds Max Interactive

Before you can use 3ds Max Interactive, you'll need to download and install it, you can install it using the Interactive→Get 3ds Max Interactive. This opens a web page with instructions for downloading and installing the software.

Once installed, you'll need to restart 3ds Max. After restarting, the Interactive menu command includes several new commands that let you connect and share assets between the two programs.

Learning the 3ds Max Interactive Interface

If you are familiar with 3ds Max, then the 3ds Max Interactive interface will look like 3ds Max's younger kid brother. Many of the features are different, but the main viewport and navigation tools look and work the same. Figure 51.1 shows the Interactive interface.

FIGURE 51.1

The 3ds Max Interactive interface has several elements that look the same as 3ds Max.

The center viewport shows the scene and the placement of objects. The viewport by default shows only a single view, but you can split it into four views by clicking on the Split Viewport Toggle in the upper right corner of the viewport. To the left of the main viewport is a Toolbar of tools including buttons to Test Level, Run Project, Select, Move, Rotate and Scale Tools and tools to Place objects, Set the Grid and Rotation Steps and finally, to Scatter objects about the scene.

To the right of the main viewport are Explorer and Property Editor panels. Using these panels, you can locate all the objects in the current scene and their relationships and set the properties for the selected object.

At the bottom of the interface are the Asset Browser panel and the Asset Preview panel. This panel lets you select and place all objects loaded into the scene. The Asset Browser also includes a Create Asset button that you can use to create new units for the level including Entity, Materials, Particle Effects, Scatter Brush, Scripts and Terrains.

All panels in the interface can be docked to the sides of the interface or removed as separate dialog boxes as needed. You can locate and open any needed interface dialog box or windows using the Window menu. If you have a specific layout that works for you, you can save and recall it using the Window→Save Layout and Window→Load Layout menu commands.

Creating a Live Link

The easiest way to work with assets in both packages is to establish a Live Link between 3ds Max and 3ds Max Interactive. To establish this link, select the Window→DCC Live Link menu command from within 3ds Max Interactive. This opens a DCC Live Link dialog box, shown in Figure 51.2.

FIGURE 51.2

The DCC Live Link dialog box lets you connect to either 3ds Max or Maya.

Click on the Connect button and the software searches for the correct package and tries to create a link between the two. Then, return to 3ds Max and select the Interactive→Connect menu command. If you have trouble connecting, then check the Port numbers and make sure they are the same. You can change the Port number in 3ds Max using the Interactive→Connect Options menu command.

Building a Level

If you are familiar with 3ds Max, then the 3ds Max Interactive interface will look like a subset of 3ds Max. Scenes in 3ds Max Interactive are called levels and a single project can have multiple levels that are all connected together. The first step in building a level is to place all the environment pieces and the level objects.

Loading assets

Before you can work with levels and objects in 3ds Max Interactive, you'll need to load them into the Asset Browser. To load scenes and objects, you'll first need to save them using the .FBX format. This format can then be loaded into the 3ds Max Interactive using the File→Import menu command. Once loaded, a preview of the scene or object is displayed in the Asset Browser. From there, you can drag and drop the objects into the level.

If both 3ds Max and 3ds Max Interactive are open at the same time and connected using a Live Link, you can use the Interactive→Send All or the Interactive→Send Selection to automatically move assets to the 3ds Max Interactive interface. You'll need to provide a name and folder for the FBX files being sent and on the 3ds Max Interactive side an FBX Import dialog box appears. Once imported the item can be selected in the Asset Browser by the name you gave it. If any changes are made the imported model in 3ds Max, you can update the object with the Interactive→Update menu command in 3ds Max.

Note

You will need to restart 3ds Max after 3ds Max Interactive is installed before you can work with the two packages together.

Placing assets

Objects that appear in the Asset Browser can be dragged and dropped in the viewport. You can then move, rotate and scale the object just as you would in 3ds Max using the transform tools located in the Toolbar.

Placing primitive objects

In addition to the imported objects that can be used in a scene, 3ds Max Interactive also has a small set of primitive objects that you can quickly add to the scene using the Create menu or the Create panel. There is no reason to import a simple sphere or box if you need one when you can use the primitive that are already available.

All the available primitive objects are in the Create menu or you can open the Create panel using the Window→Create menu command. To place an object, simply select it and it appears in the center of the scene. Imported and primitive objects can be selected and transformed using the Transform tools found in the Toolbar to the left of the main viewport. Figure 51.3 shows several primitive objects that has been placed into 3ds Max Interactive level.

FIGURE 51.3

Primitive objects can be easily placed in the level.

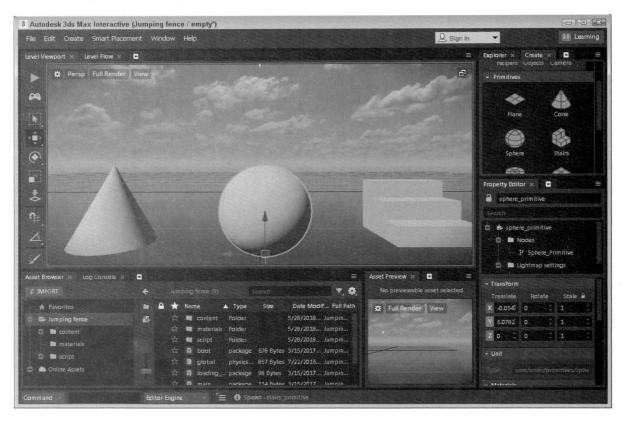

Changing materials

When an object is imported from 3ds Max, its materials are automatically imported along with the object, but you can create new materials and import them also using the Interactive→Maetrial Send Selected menu command in 3ds Max. This imported material appears in the Asset Browser. The current material for each level object is displayed in the Property Editor and using the buttons to the right of the material name, you can change the applied material to any of those available in the project. Figure 51.4 shows the three primitive objects with new materials.

FIGURE 51.4

Level objects can have new materials applied to them.

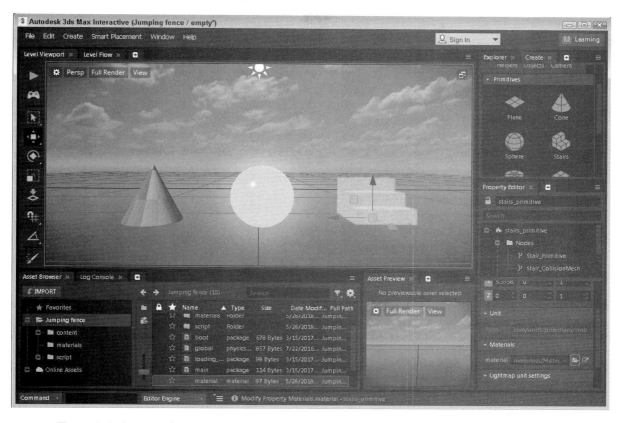

Tutorial: Importing and placing objects in 3ds Max Interactive

In this simple example, we'll move a 3d object from 3ds Max to 3ds Max Interactive and place it in the scene. The quickest way to do this is to create a Live Link between the two packages.

To import and place 3ds Max objects in 3ds Max Interactive, follow these steps:

1. Open 3ds Max Interactive and in the Templates panel of the Projects dialog box, select the Empty template and click the Create button. Name the project, Jumping Fence and set the project folder.

2. Open 3ds Max and load the White picket fence.max file from the Chap 51 directory where the content is downloaded. Then switch to 3ds Max Interactive and choose the Window→DCC Live Link menu command and click the Connect button for 3ds Max.

3. Back within 3ds Max, select the Interactive→Send All menu command. The Select File to Export dialog box opens, give the file the name of White picket fence and click the Save button. Over in 3ds Max Interactive, an Import FBX dialog box appears, click on the Import button and the 3ds Max object will appear in the Asset Browser using the name you entered.

4. Select the imported object from the Asset Browser panel in 3ds Max Interactive and drag and drop it on the main viewport.

Figure 51.5 shows the white picket fence object imported from 3ds Max.

FIGURE 51.5

Objects created in 3ds Max can be easily moved to 3ds Max Interactive.

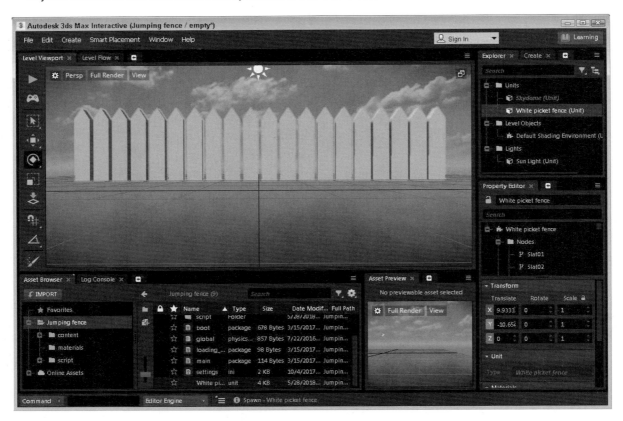

Working with Assets

Interactive scenes combine a lot more than just 3d objects. They often include lots of sound effects, music backgrounds, animation clips, user interface elements and more. 3ds Interactive includes many tools and interface to work with all sorts of assets. Most of these are opened from the Window menu.

Working with Audio

To add audio assets to your interactive scene, you can use the included WWise audio editing tool, shown in Figure 51.6. This audio tool is found in the Window menu and lets you load, record and save audio clips that can be used in scenes. The tool also lets you save multiple audio clips into a single audio bank that can be loaded at the start of the level and played as needed throughout the interactive event.

FIGURE 51.6

The WWise audio tool is a powerful addition to 3ds Max Interactive.

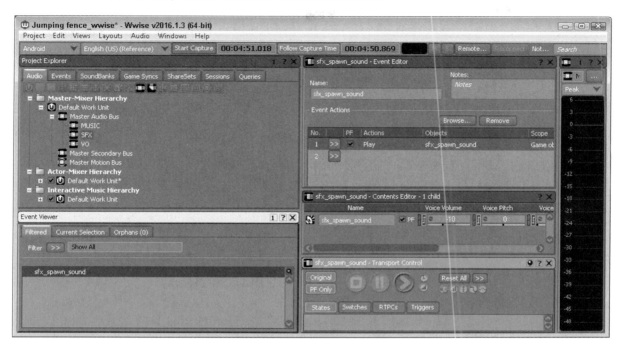

Creating an interface with Scaleform

Another powerful tool included in 3ds Max Interactive is Scaleform, which is used to create user interfaces. With Scaleform Studio, you can add image, text and interactive elements to create advance menus and start-up screens. Scaleform Studio is opened using the Window→Scaleform Studio menu command. Every element can easily be animated and set to respond to user actions. Figure 51.7 shows an example in Scaleform Studio.

FIGURE 51.7

Scaleform Studio is a complete user interface creation tool included in 3ds Max Interactive.

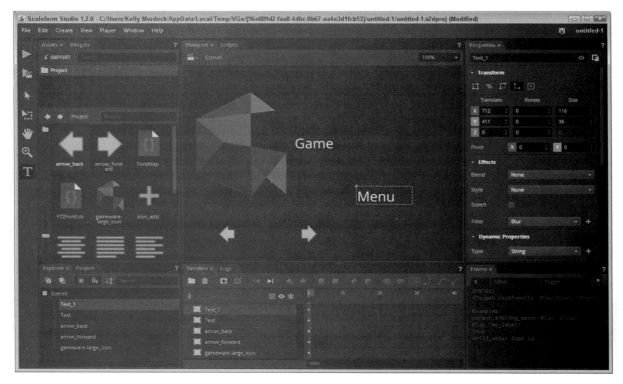

Using the Story Editor

Animation clips created in 3ds Max can be imported into 3ds Max Interactive just like 3d objects are imported using FBX. Imported animation clips can be edited using the Animation Clip Editor. Another way to animate objects in 3ds Max Interactive is directly in the software with the transform tools using the Story Editor. The Story Editor is opened using the Window→Story Editor menu command.

Tutorial: Animating objects with the Story Editor

The Story Editor interface is a little different than animating in 3ds Max, but for quick and easy animation sequences, it works great.

To animate objects directly in 3ds Max Interactive using the Story Editor, follow these steps:

1. Open the Story Editor with the Window→Story Editor menu command, then select the white picket fence object in the viewport and click on the Create a New Story button in the main toolbar of the Story Editor. This adds the selected object to the StoryRoot list.

2. Click on the Live button at the left end of the Story Editor toolbar to begin the animation editing mode.

3. With the fence object selected, press the S key to create a keyframe, then move the Time Slider in the Story Editor to 1 second and drag the fence upward with the Move tool. Then, press the S key to create another keyframe.

4. Continue to move and create new keyframes to make the fence object rise, move left and right and then descend again. The whole animated sequence should only be about 3 seconds.

5. Click on the Live button to exit edit mode. Drag the Time Slider to see the animation.

Figure 51.8 shows the Story Editor with the created keyframes.

FIGURE 51.8

Objects can be animated directly in 3ds Max Interactive with the Story Editor interface.

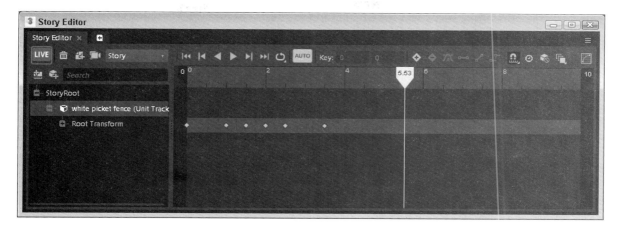

Adding Interactions

Once a scene is placed in 3ds Max Interactive, you can add interactions between the objects in two different ways. 3ds Max Interactive includes a visual scripting feature called Level Flow, found in the Window menu. This interface works in a manner similar to the Slate Material Editor and wiring inputs and outputs of nodes together, as shown in Figure 51.9. If you are a traditional scripter, then you can use the Lua scripting language to program the interactions.

FIGURE 51.9

Interactivity can be added using the Level Flow dialog box.

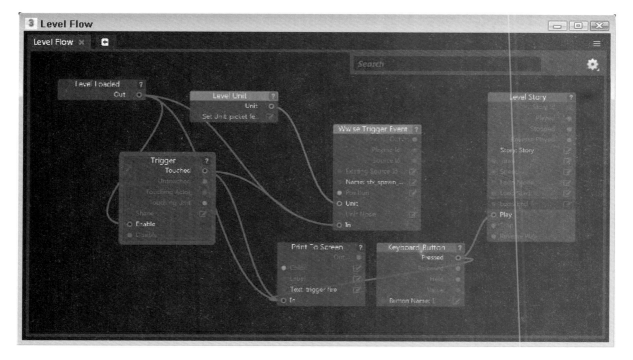

Using the Level Flow interface

Nodes are added to the scene by right clicking on the interface and choosing a node type from the pop-up menu. If you select an object in the viewport, then its node will be listed by name at the top of the right-click pop-up menu. Each node has input and output port marked by colored circles. The color of each circle denotes the type of data that it uses, so you can only connect like colored inputs to like colored outputs. To connect two nodes together, just drag from the input port to the output port or vice versa. A line will connect the two nodes.

Tip

Pressing the tab key opens a search bar where you can type the name of a node and the Level Flow interface will find it.

Within the nodes are also several attributes that you can access by clicking on the checkbox to the right of the attribute name.

Testing interactivity

Once all the nodes are placed and connected, you can test the interactivity using the Text Level button located at the top of the Toolbar or by pressing the F8 key. This loads the level in a interactive window where you can move about the level and test any interactive elements. Press the Escape key will exit the test window.

Tutorial: Using the Level Flow interface to add interactivity

With hundreds of available nodes, it can be tricky to find exactly what you need, but with a little digging the correct node is there. This is a simple example of the kinds of interactions that are available. For this example, we'll display some text when the level loads and then watch for a keypress to display the simple fence animation created in the Story Editor.

To add interactivity to the level using the Level Flow interface, follow these steps:

1. Open the Level Flow interface with the Window→Level Flow menu command or by selecting the Level Flow tab underneath the main menu.

2. Right click on the center of the Level Flow interface and choose the Event→Level Loaded menu to add the Level Load node. Right click and select the Debug→Print to Screen menu command, then connect the Level Load Out port to the Print to Screen In port. Click on the checkbox to the right of the Text attribute and enter the text, 'Press the '1' key make the fence jump' in the text box. This will make the text appear when the level first loads.

3. Right click in the Level Flow window away from the other nodes and select the Input→Keyboard Button. Set the Button Name attribute to 1. Right click and select the Level→Level Story node and set the Story attribute to Story, which is the animated fence story we created earlier in the Story Editor. Then, connect the Pressed Out port for the Keyboard Button node to the Play In port for the Level Story node. Figure 51.10 shows the connected nodes.

FIGURE 51.10

Interactivity is added using the Level Flow interface.

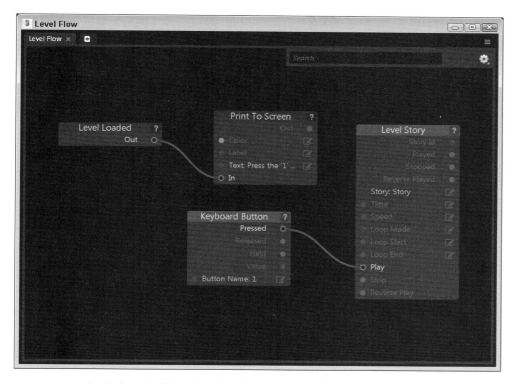

4. Select the File→Save Level menu command to save the level.

5. Select the Level Viewport tab and click on the Test Level button at the top of the Toolbar.

Figure 51.11 shows the Test Level window with the white picket fence.

FIGURE 51.11

The level can be tested in the Test Level window.

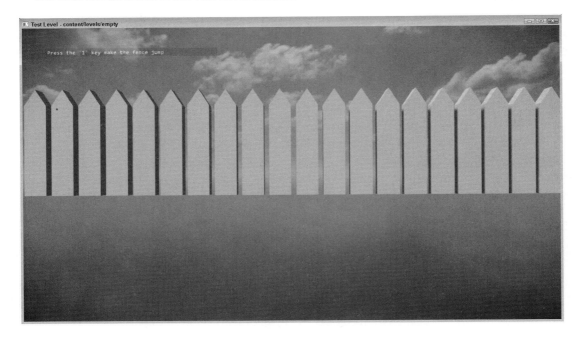

Deploying the Level

After your level is built and tested, you can deploy the finished level using the Deployer dialog box, opened with the Window→Deploy and Connect→Deployer menu command. The Deployer dialog box, shown in Figure 51.12 includes all the settings for porting the level to several different devices.

Note
Before you can deploy levels to the various systems, you'll need to install the necessary development kits.

FIGURE 51.12

The Deployer dialog box is used to send the level to its final destination.

Summary

3ds Max Interactive is a full featured package that enables you to add interactivity to your 3ds Max scenes and publish them on one of the many available interactive systems. The software is advanced and complex and capable of some amazing results, but this chapter just showed the tip of the iceberg.

This chapter covered the following topics:

*   Installing 3ds Max Interactive
*   Learning the 3ds Max Interactive interface
*   Creating a Live Link between 3ds Max and 3ds Max Interactive
*   Loading assets and building a level
*   Exploring the available asset tools for audio, user interfaces, and level animations
*   Adding interactions using the Level Flow interface
*   Deploying the level

The next chapter shows how actions in 3ds Max can be automated and captured using the software's own scripting language, MAXScript.

Chapter 52

Automating with MAXScript

The Autodesk® 3ds Max® 2020 software designers went to great lengths to make sure that you are limited only by your imagination in terms of what you can do in 3ds Max. They've packed in so many different features and so many different ways to use those features that you could use 3ds Max for years and still learn new ways of doing things.

Despite the wide range of capabilities in 3ds Max, there may come a time when you wish for a new 3ds Max feature. With MAXScript, you can actually extend 3ds Max to meet your needs, customize it to work the way you want, and even have it do some of the more monotonous tasks for you.

What Is MAXScript?

In this chapter, you'll look at MAXScript—what it's for and why in the world you would ever want to use it. But before getting into the nitty-gritty details, I'll start with a brief overview.

Simply put, MAXScript is a tool that you can use to expand the functionality of 3ds Max. You can use it to add new features or to customize how 3ds Max behaves so that it's tailored to your needs and style. You also can use MAXScript as a sort of recording device; it can record your actions so you can play them back later, eliminating repetitive tasks.

You can use MAXScript to "talk" to 3ds Max about a scene and tell it what you want to happen, either by having 3ds Max watch what you do or by typing in a list of instructions that you want 3ds Max to execute.

The beauty of MAXScript lies in its flexibility and simplicity: it is easy to use and was designed from the ground up to be an integral part of 3ds Max. But don't let its simplicity fool you; MAXScript as a language is rich enough to let you control just about anything.

In fact, you have already used MAXScript without even knowing it. Some of the buttons and rollouts use bits of MAXScript to carry out your commands. And after you've created a new feature with MAXScript, you can integrate it into 3ds Max transparently and use it just as easily as any other 3ds Max feature.

MAXScript is a fully functional and very powerful computer language, but you don't have to be a computer programmer or even have any previous programming experience to benefit from MAXScript. In the next few sections, you'll look at some simple ways to use MAXScript. For now, just think of a script in 3ds Max as you would a script in a movie or play—it tells what's going to happen, who's going to do what, and when it's going to happen. With your scene acting as the stage, a script directs 3ds Max to put on a performance for you.

One final note before you dive in: MAXScript is so powerful that an entire book could be written about it and every last feature it supports, but that is not the purpose here. This chapter is organized to give you an introduction to the world of MAXScript and to teach you the basic skills you need to get some mileage out of it. What is given here is a foundation that you can build upon according to your own interests and needs.

MAXScript Tools

MAXScript is pervasive and can be found in many different places. This section looks at the MAXScript tools and how different scripts are created and used.

Let's look at some of the tools used in working with MAXScript. 3ds Max has several tools that make creating and using scripts as simple as possible.

The MAXScript menu

The Scripting menu includes commands that you can use to create a new script, open and run scripts, open the MAXScript Listener window (keyboard shortcut, F11) or the MAXScript Editor window, enable the Macro Recorder, open the Visual MAXScript Editor, access the Debugger dialog box, or open the MAXScript Reference.

The New Script command opens a MAXScript Editor window, a text editor in which you write your MAXScript. See the "MAXScript Editor windows" section later in this chapter for more on this editor window. The Open Script command opens a file dialog box that you can use to locate a MAXScript file. When opened, the script file is opened in a MAXScript Editor window, as shown in Figure 52.1. MAXScript files have an .ms or .mrc extension. The Run Script command also opens a file dialog box where you can select a script to be executed.

Note

When you use Run Script, some scripts do something right away, whereas others install themselves as new tools.

FIGURE 52.1

MAXScript is written using standard syntax in a simple text editor window.

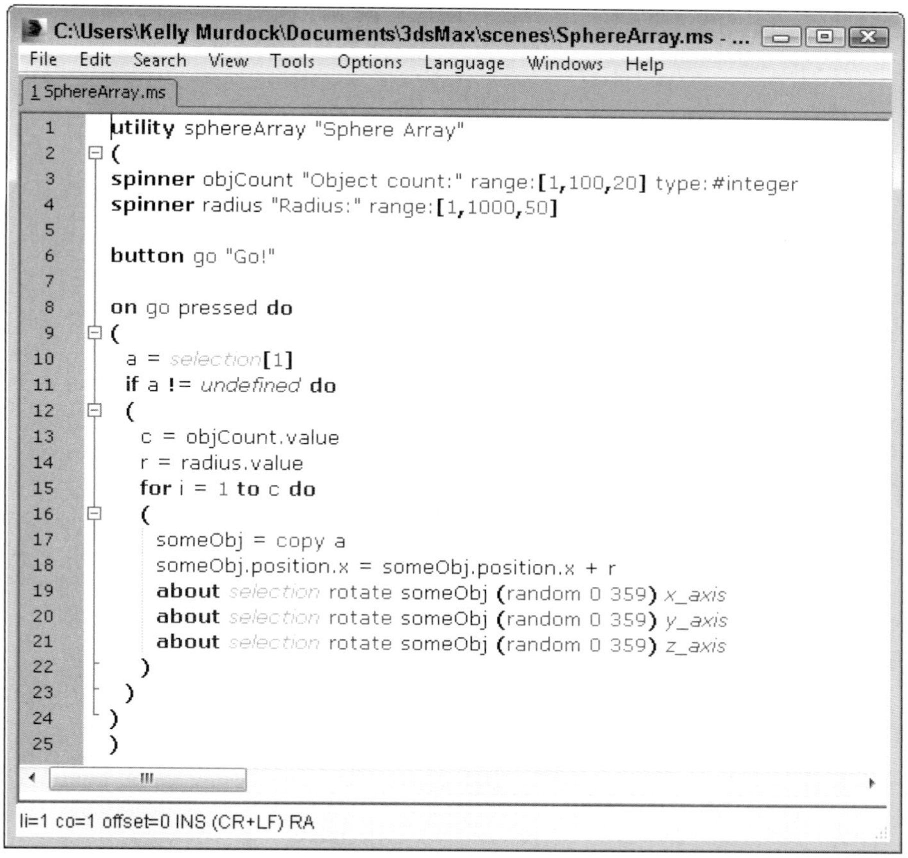

```
C:\Users\Kelly Murdock\Documents\3dsMax\scenes\SphereArray.ms - ...
File  Edit  Search  View  Tools  Options  Language  Windows  Help
1 SphereArray.ms
1    utility sphereArray "Sphere Array"
2    (
3      spinner objCount "Object count:" range:[1,100,20] type:#integer
4      spinner radius "Radius:" range:[1,1000,50]
5
6      button go "Go!"
7
8      on go pressed do
9      (
10       a = selection[1]
11       if a != undefined do
12       (
13         c = objCount.value
14         r = radius.value
15         for i = 1 to c do
16         (
17           someObj = copy a
18           someObj.position.x = someObj.position.x + r
19           about selection rotate someObj (random 0 359) x_axis
20           about selection rotate someObj (random 0 359) y_axis
21           about selection rotate someObj (random 0 359) z_axis
22         )
23       )
24     )
25   )

li=1 co=1 offset=0 INS (CR+LF) RA
```

The MAXScript Listener command opens the MAXScript Listener window. You also can open this window by pressing the F11 keyboard shortcut. The Macro Recorder command starts recording a MAXScript macro. The MAXScript Listener and recording macros are covered later in this chapter.

The MAXScript Utility rollout

You access the MAXScript Utility rollout, shown in Figure 52.2, by opening the Utilities panel in the Command Panel and clicking the MAXScript button. This opens a rollout where you can do many of the same commands as the MAXScript menu.

FIGURE 52.2

The MAXScript rollout on the Utilities panel is a great place to start working with MAXScript.

The MAXScript rollout also includes a Utilities drop-down list, which holds any installed scripted utilities. Each scripted utility acts as a new feature for you to use. The parameters for these utilities are displayed in a new rollout that appears below the MAXScript rollout.

Tutorial: Using the SphereArray script

Here's a chance for you to play around a little and get some experience with MAXScript in the process. In this chapter's folder in the downloadable content is a simple script called SphereArray.ms. It's similar to the Array command found in the Tools menu, except that SphereArray creates copies of an object and randomly positions them in a spherical pattern.

To load and use the SphereArray script, follow these steps:

1. Select Create→Standard Primitives→Box, and drag in the Top viewport to create a Box object that has a Length of **10** and Width and Height values of **1.0**.

2. Select the box object, open the Utilities panel in the Command Panel (the icon is a wrench), and click the MAXScript button.

    The MAXScript rollout appears.

3. Click the Run Script button in the MAXScript rollout to open the Choose Editor file dialog box, locate the SphereArray.ms file from the Chap 52 folder in the downloaded content set, and click Open.

    The SphereArray utility installs and appears in the Utilities drop-down list. (Because SphereArray is a scripted utility, running it only installs it.)

4. Choose SphereArray from the Utilities drop-down list. Make sure that the box object is selected.

5. In the Sphere Array rollout, enter **50** in the Object Count field and **2.0** for the Radius field. Now click the Go! button to run the script.

    The script adds 50 copies of your box to the scene and randomly positions them 2 units away from the box's position.

Figure 52.3 shows the results of the SphereArray MAXScript utility. Notice that the SphereArray script looks much like any other function or tool in 3ds Max.

FIGURE 52.3

The results of the SphereArray MAXScript script

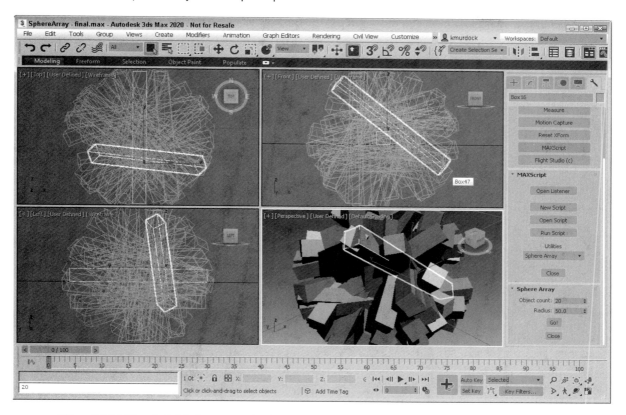

The MAXScript Listener window

Figure 52.4 shows the MAXScript Listener window (keyboard shortcut F11), which lets you work interactively with the part of 3ds Max that interprets MAXScript commands. The top pane of the MAXScript Listener window (the pink area) lets you enter MAXScript commands; the results are reported in the bottom pane (the white area) of the MAXScript Listener window. You also can type MAXScript commands in the bottom pane, but typing them in the top pane keeps the commands separated from the results. If you drag the spacer between the two panes, you can resize both panes as needed.

Note

When you first open the MAXScript Listener, the top pane is not visible, but if you drag on the divider, you can make both panes visible.

FIGURE 52.4

The MAXScript Listener window interprets your commands.

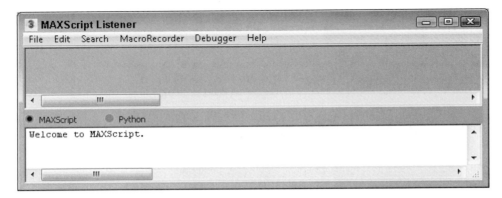

When you type commands into either pane and press Enter, the MAXScript interpreter evaluates the command and responds with the results. For example, if you type a simple mathematical expression such as **2+2** and press Enter, the result of 4 is displayed in blue on the next line. Most results appear in blue, but the results display in red for any errors that occur. For example, if you enter the command **hello there**, a Type error result appears in red because the MAXScript interpreter doesn't understand the command.

Caution

The MAXScript interpreter is very fickle. A misspelling generates an error, but MAXScript is *case-insensitive*, which means that uppercase and lowercase letters are the same as far as 3ds Max is concerned. Thus, you can type **sphere**, **Sphere**, or **SPHERE**, and 3ds Max sees no difference.

The MAXScript Listener window has these menus:

* **File:** You can use this menu to close the window (Ctrl+W), save your work (Ctrl+S), run scripts (Ctrl+R), open a script for editing (Ctrl+O), or create a new script from scratch (Ctrl+N).

* **Edit:** This menu is where you access all the common editing functions you need, such as cutting, pasting, and undoing.

* **Search:** You use this menu for searching through the window to find specific text (Ctrl+F), find the next instance (Ctrl+G), or replace text (Ctrl+H).

* **MacroRecorder:** This menu lets you set various options for the MAXScript Macro Recorder.

* **Debugger:** This menu provides a way to open the Debugger dialog box.

* **Help:** This menu provides access to the MAXScript Reference (F1).

Tutorial: Talking to the MAXScript interpreter

This tutorial gives you a little experience in working with the MAXScript Listener window and a chance to try some basic MAXScript commands.

To start using MAXScript, follow these steps:

1. Choose File→Reset to reset 3ds Max.
2. Choose Scripting→MAXScript Listener (or press F11) to open the MAXScript Listener window.
3. Click anywhere in the bottom pane of the Listener window, type the following, and press Enter:

   `sphere()`

   A sphere object with default parameters is created.
4. Next enter the following in the lower pane, and press the Enter key:

   `torus radius1:50 radius2:5`

3ds Max creates a torus and adds it to your scene. As you specified in your MAXScript, the outer radius (radius1) is 50, and the radius of the torus itself (radius2) is 5. The output tells you that 3ds Max created a new torus at the origin of the coordinate system and gave that torus a name: Torus001.

5. Now use MAXScript to move the torus. In the Listener window, type the following:

$Torus001.position.x = 20

After you press Enter, you see the torus move along the positive X-axis. Each object in 3ds Max has certain properties or attributes that describe it, and what you've done is access one of these properties programmatically instead of by using the rollout or the mouse. In this case, you're telling 3ds Max, "Torus001 has a position property. Set the X-coordinate of that position to 20."

Note

The $ symbol identifies a named object. You can use it to refer to any named object.

6. To see a list of some of the properties specific to a sphere, type the following:

Showproperties $Sphere001

A list of the Sphere001 properties appears in the window.

Figure 52.5 shows the MAXScript Listener window with all the associated commands and results.

FIGURE 52.5

Use the MAXScript Listener window to query 3ds Max about an object's properties.

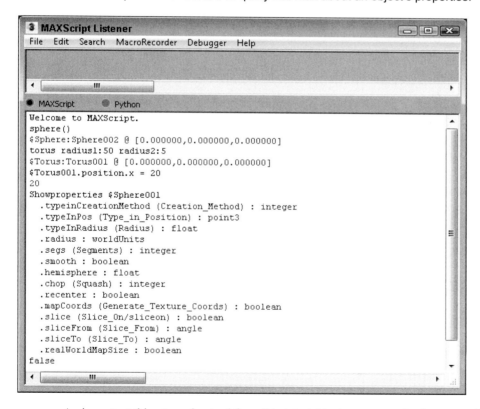

An important thing to understand from this tutorial is that you can do almost anything with MAXScript. Any property of any object that you can access via a rollout is also available via MAXScript. You could go so far as to create entire scenes using just MAXScript, although the real power comes from using MAXScript to do things for you automatically.

Tip

3ds Max remembers the value of the last MAXScript command that it executed, and you can access that value through a special variable: ? (question mark). For example, if you type **5 + 5** in the Listener window, 3ds Max displays the result, 10. You can then use that result in your next MAXScript command by using the question mark variable. For example, you could type **$Torus001.radius2 = ?**, and 3ds Max would internally substitute the question mark with the number 10.

At the left end of the status bar, you can access the MAXScript Mini Listener control by dragging the left edge of the status bar to the right. By right-clicking in this control, you can open a Listener window and view all the current commands recorded by the Listener. Figure 52.6 shows this control with another command along with the objects from the last example.

FIGURE 52.6

The resulting objects created via the MAXScript Listener window

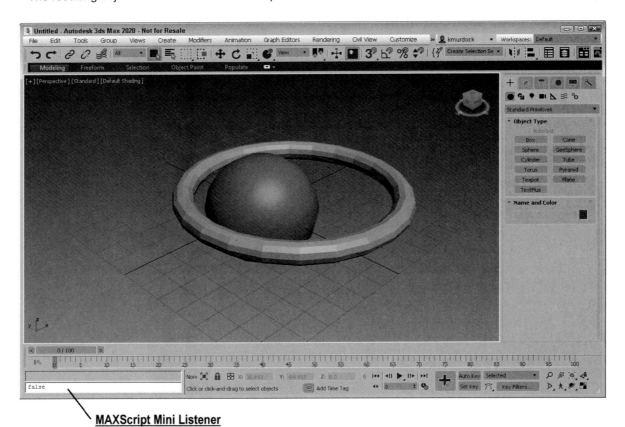

MAXScript Mini Listener

MAXScript Editor windows

The MAXScript Editor window is different from the MAXScript Listener. It enables you to open and edit any type of text file, although its most common use is for editing MAXScript files. Although you can have only one Listener window open, you can open as many editor windows as you want. Each opened script appears in a separate tab at the top of the Editor window.

To open a new MAXScript Editor window, you can choose Scripting→New Script, choose File→New Script from the MAXScript Listener window, or click the New Script button in the MAXScript rollout in the Utilities panel. You also can use MAXScript Editor windows to edit existing scripts.

To create a new script, opening both an Editor window and the MAXScript Listener window is usually best. Then you can try out things in the MAXScript Listener window, and when the pieces of the MAXScript work, you can cut and paste them into the main Editor window. Then you can return to the MAXScript Listener window, work on the next new thing, and continue creating, cutting, and pasting until the script is done.

Tip

The File menu includes several commands for working with script files, including New, Open, Save, Save As, and Revert. Another command exports the opened script to HTML, RTF, PDF, LaTeX, or XML. The Editor also integrates with a Source Control repository. Finally, the Editor's File menu includes the ability to print out the opened script.

The Edit menu includes commands for undoing, redoing, cutting, copying, and pasting script sections between different scripts and to and from the MAXScript Listener window. The Match Brace and Select to Brace commands let you easily create and edit scripts; the Block Comment command allows you to quickly comment out certain areas of script; and the Make Selection Uppercase lets you quickly make variables identifiable.

The Search menu includes Find and Replace features. Another feature finds text in a directory of files. You also can set bookmarks that allow you to quickly move to a specific point in the code. The Views menu includes options for making the editor appear full screen. You also can expand and contract bracketed sections of the script and display editor elements, including a toolbar, status bar, line numbers, and indentation guide.

The Tools menu includes the Evaluate All command. This command is a fast way of having 3ds Max evaluate your entire script. The result is the same as if you had manually selected the entire text, copied it to the MAXScript Listener window, and pressed Enter. An Evaluate Selection option also is available. The Tools menu also includes access to the Visual MAXScript window with the New Rollout and Edit Rollout menu commands. You also have access to the various options and properties files including the User and Global Options, the Abbreviations file, and the MAXScript Properties file, as shown in Figure 52.7.

FIGURE 52.7

The MAXScript Editor can open several script files at once.

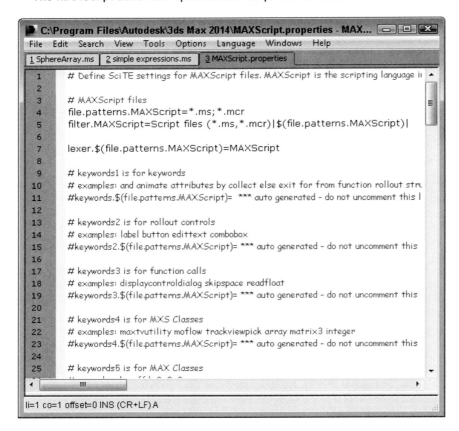

The Macro Recorder

The MAXScript Macro Recorder is a tool that records your actions and creates a MAXScript that can be recalled to duplicate those actions. Using the Macro Recorder is not only a quick and easy way to write entire scripts, but it is also a great way to make a working version of a script that you can then refine. After the Macro Recorder has created a MAXScript from your recorded actions, you can edit the script using a MAXScript Editor window to make any changes you want.

You can turn the Macro Recorder on and off either by choosing Scripting→Macro Recorder or by choosing Macro Recorder→Enable in the MAXScript Listener window. The check mark next to the Macro Recorder command on the MAXScript menu indicates that the Macro Recorder is turned on.

When the Macro Recorder is on, every action is converted to MAXScript and sent to the MAXScript Listener window's top pane. You can then take the MAXScript output and save it to a file or copy it to a MAXScript Editor window for additional editing. The Macro Recorder continues to monitor your actions until you turn it off, which is done in the same way as turning it on.

The MacroRecorder menu in the MAXScript Listener window includes several options for customizing the macro recorder, including

* **Enable:** This option turns the Macro Recorder on or off.

* **Explicit scene object names:** With this option, the Macro Recorder writes the MAXScript using the names of the objects you modify so that the script always modifies those exact same objects, regardless of what object you have selected when you run the script again. For example, if the Macro Recorder watches you move a pyramid named $Pyramid01 in your scene, then the resulting MAXScript will always and only operate on the scene object named $Pyramid01.

* **Selection-relative scene object names:** With this option, the Macro Recorder writes MAXScript that operates on whatever object is currently selected. So if (when you recorded your script) you moved the pyramid named $Pyramid01, you could later select a different object and run your script, and the new object would move instead.

Note

To decide which of these options to use, ask yourself, "Do I want the script to always manipulate this particular object, or do I want the script to manipulate whatever I have selected?"

* **Absolute transform assignments:** This tells the Macro Recorder that any transformations you make are not relative to an object's current position or orientation. For example, if you move a sphere from (0,0,0) to (10,0,0), the Macro Recorder writes MAXScript that says, "Move the object to (10,0,0)."

* **Relative transforms operations:** Use this option to have the Macro Recorder apply transformations relative to an object's current state. For example, if you move a sphere from (0,0,0) to (10,0,0), the Macro Recorder says, "Move the object +10 units in the X-direction from its current location."

* **Explicit subobject sets:** If you choose this option and then record a script that manipulates a set of subobjects, running the script again always manipulates those same subobjects, even if you have other subobjects selected when you run the script again.

* **Selection-relative subobject sets:** This tells the Macro Recorder that you want the script to operate on whatever subobjects are selected when you run the script.

* **Show command panel switchings:** This option tells the Macro Recorder whether or not to write MAXScript for actions that take place on the Command Panel.

* **Show tool selections:** If this option is selected, the Macro Recorder records MAXScript to change to different tools.

* **Show menu item selections:** This option tells the Macro Recorder whether or not you want it to generate MAXScript for menu items you select while recording your script.

Tutorial: Recording a simple script

In this tutorial, you'll create a simple script that squashes whatever object you have selected and turns it purple.

To create a script using the Macro Recorder, follow these steps:

1. Open the Purple pyramid.max file from the Chap 52 folder in the downloaded content set.

   This file includes a simple pyramid object.

2. With the pyramid object selected, choose Scripting→MAXScript Listener (or press F11) to open the MAXScript Listener window.

3. In the MAXScript Listener window, open the MacroRecorder menu and make sure that all the options are set to the relative and not the absolute object settings, thereby telling the script to work on any selected object instead of always modifying the same object.

4. Returning to the MacroRecorder menu, select Enable. The Macro Recorder is now on and ready to start writing MAXScript. Minimize the Macro Recorder window (or at least move it out of the way so you can see the other viewports).

5. Dock the MAXScript Listener window to the Left viewport by clicking the viewport name and choosing Views→Extended Viewports→MAXScript Listener.

   Now you can keep things out of the way while you work.

6. With the Pyramid01 object selected, choose Modifiers→Parametric Deformers→XForm to add an XForm modifier to the object.

7. Select the Non-Uniform Scale tool. Right-click anywhere in the Front viewport to make it active (if it's not already), and then drag the Z-axis gizmo downward to squash the pyramid.

8. In the Modify panel, click the color swatch next to the object name field to open the Object Color dialog box. Pick one of the purple colors, and click OK.

1205

9. The script is done, so in the MAXScript Listener window, choose MacroRecorder→Enable to turn off the Macro Recorder.

10. Now it's time to try out your first MAXScript effort. Add a sphere to your scene. Make sure that it's selected before moving to the next step.

11. In the top pane of the MAXScript Listener window, select all the text, and then hold down the Shift key and press the Enter key to tell 3ds Max to execute the MAXScript.

In Figure 52.8, you can see the script and the sphere that has been squashed and has changed color.

FIGURE 52.8

Running the new squash-and-turn-purple script

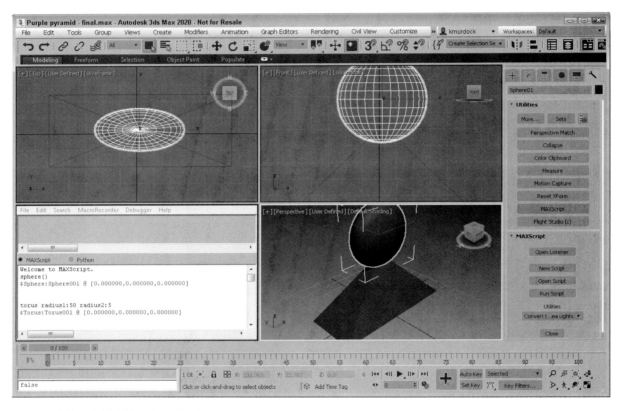

The MAXScript Debugger

The MAXScript Debugger, shown in Figure 52.9, is a separate window that allows you to stop the execution of a script using breaks and look at the variable's values as the script is being run. This information provides valuable information that can help to identify bugs with your script.

FIGURE 52.9

The MAXScript Debugger lets you check the values of variables as the script runs.

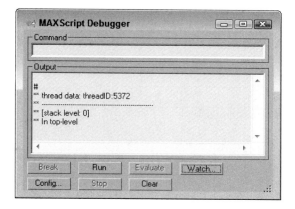

The MAXScript Debugger dialog box is opened using the Debugger menu command in the various MAXScript tool windows or using the Scripting→Debugger Dialog menu command on the main interface. The Command field at the top of the Debugger window lets you input commands directly to the debugger. The Output area displays the results. Several buttons at the bottom of the MAXScript Debugger window let you control how the debugger works:

* **Break:** Causes the current script to break out of its execution
* **Run:** Starts the execution of the current script
* **Evaluate:** Executes the command entered into the Command field
* **Watch:** Opens the Watch Manager window where you can specify distinct variables to watch
* **Config:** Opens the MAXScript Debugger Parameters dialog box where you can configure the debugger
* **Stop:** Halts the execution of the current script
* **Clear:** Clears all the text in the Output field

Watching variables

The Watch button opens the Watch Manager, shown in Figure 52.10. Clicking the Variable column lets you type in a new variable to watch. The Value column displays the value of the listed variables as the script is executed.

FIGURE 52.10

The Watch Manager lets you watch the value of specific variables.

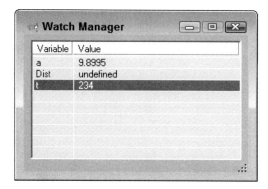

Using debugger commands

Once the execution of a script has been halted by clicking on the Break button, you can enter specific commands in the Command field at the top of the debugger and click the Evaluate button to execute them. Table 52.1 lists the available debugger commands.

TABLE 52.1 MAXScript Debugger Commands

Command	What It Does
threads	Displays a list of all current threads
setThread (thread no.)	Makes the specified thread number the active thread
setFrame (frame no.)	Makes the specified frame number the active frame
getVar (variable)	Gets the value of the specified variable
setVar (variable)	Sets the value of the specified variable
eval (expression)	Evaluates the specified expression
?	Displays a list of debugger commands

Configuring the debugger

The Config button opens the MAXScript Debugger Parameters dialog box, shown in Figure 52.11. This dialog box includes several options for deciding when to break out of the script execution. It also lets you define the time for each break.

FIGURE 52.11

The MAXScript Debugger Parameters dialog box let you set the break cycle time, among other settings.

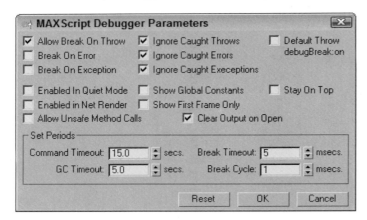

Setting MAXScript Preferences

The Preference Settings dialog box, opened with the Customize→Preferences menu command, includes a panel of MAXScript settings, shown in Figure 52.12. Using these settings, you can set which scripts load automatically, the default settings for the Macro Recorder, and even what font is displayed in the Script Editor window. The Use Fast Node Name Lookup option causes all indexed MAXScript names to be saved in a cache buffer to speed the execution of the script.

FIGURE 52.12

The MAXScript panel in the Preference Settings dialog box includes options for controlling MAXScript.

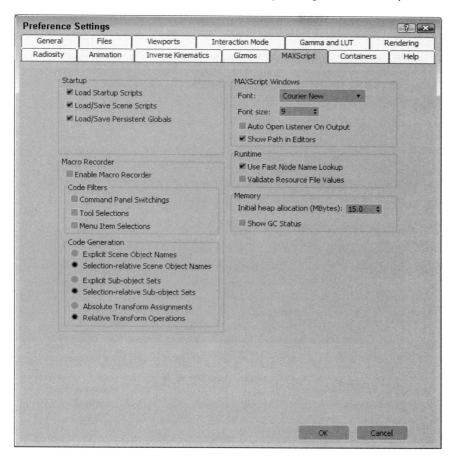

3ds Max includes two directories in its default installation that can be used to automatically load scripts when 3ds Max starts. These directories are scripts and scripts\startup, but you can change them using the Configure System Paths dialog box. The Load Startup Scripts option causes any scripts in these directories to be loaded automatically when 3ds Max starts.

Types of Scripts

All scripts are not created equal, and 3ds Max categorizes different scripts based on how they work. For more information, the MAXScript online help provides exhaustive information on their various options.

The main thing to consider when deciding what type of script to create is the user interface. Ask yourself what the most logical user interface would be for the type of tool you're creating, and this gives you a hint as to which type of script is well suited for the task.

Macro scripts

Macro scripts are scripts created with the Macro Recorder. Any script associated with a toolbar button is considered a Macro script. 3ds Max organizes Macro scripts by their category, which you can change by editing the script file. To call a Macro script from another script, you can use the `macros` command. For example,

```
macros.run "objects" "sphere"
```

runs the "sphere" script in the "objects" category.

Macro scripts generally require no other user input; you just click a button, and the script works its magic.

Scripted utilities

A *scripted utility* is a MAXScript that has its own custom rollout in the Utilities panel, like the SphereArray example. This type of script is particularly useful when your script has parameters that the user needs to enter, such as the radius in the SphereArray script. Scripted utilities are easy to build using the Visual MAXScript Editor.

Scripted right-click menus

When you right-click an object in your scene, 3ds Max opens a pop-up menu of options for you to choose from, much like a quad menu. Scripted right-click menus let you append your own menu items to the right-click menu. If you create a script that modifies some property of an object, making the script available through the right-click menu makes it easily accessible.

Scripted mouse tools

You can use scripted mouse tools to create scripts that handle mouse input in the viewports. These scripts listen for commands from the mouse, such as clicking the mouse buttons and clicking and dragging the cursor. For example, you would use this type of MAXScript if you were making a new primitive object type so that users could create the new objects just like they would a sphere or a box.

Scripted plug-ins

Scripted plug-ins are by far the most complex type of MAXScript available. They mirror the functionality of non–MAXScript plug-ins (which are written in other programming languages such as C++). You can create scripted plug-ins that make new geometry, create new shapes, control lights, act as modifiers, control texture maps and materials, and even produce special rendering effects.

Writing Your Own MAXScripts

This section presents the basics of the MAXScript language and shows you how to use the various parts of MAXScript in your own scripts. You can test any of these scripting commands using the MAXScript Listener window.

Much of the discussion that follows will sound familiar if you've already read the material on expressions found in Chapter 34, "Wiring Parameters." Expressions use many of the same constructs as MAXScript.

Variables and data types

A *variable* in MAXScript is sort of like a variable in algebra. It represents some other value, and when you mention a variable in an equation, you're actually talking about the value that the variable holds. You can think of variables in MAXScript as containers that you can put stuff into and take it out of later. Unlike variables in algebra, however, variables in MAXScript can "hold" other things besides numbers, as you'll soon see.

To put a value into a variable, you use the equal sign. For example, if you type

```
X = 5 * 3
```

in the MAXScript Listener window, 3ds Max evaluates the expression on the right side of the equal sign and stores the result in the variable named X. In this case, 3ds Max would multiply 5 and 3 and would store the result (15) in X. You can then see what is in X by just typing **X** in the MAXScript Listener window and pressing Enter. 3ds Max then displays the value stored in X, or 15.

You can name your variables whatever you want, and naming them something that helps you remember what each variable is for is a good idea. For example, if you want a variable that keeps track of how many objects you're going to manipulate, the name "objCount" would be better than something like "Z."

Note

Variable names can be just about anything you want, but you must start a variable name with a letter. Also, the variable name can't have any special characters in it, like spaces, commas, or quotation marks. You can, however, use the underscore character and any normal alphabetic characters.

Variables also can hold *strings,* which are groups of characters. For example,

```
badDay = "Monday"
```

stores the word "Monday" in the variable `badDay`. You can attach two strings together using the plus sign, like this:

```
grouchy = "My least favorite day is" + badDay
```

Now the variable `grouchy` holds the value "My least favorite day is Monday."

Try this:

```
wontWork = 5 + "cheese"
```

3ds Max prints out an error because it's confused: you're asking it to add a number to a string. The problem is that 5 and "cheese" are two different data types. *Data types* are different classes of values that you can store in variables. You can almost always mix values of the same data type, but values of different types usually don't make sense together.

Note

To see the data type of a variable, use the `classof` command. Using the previous example, you could type `classof grouchy`, and 3ds Max would, in turn, print out `String`.

Another very common data type is Point3, which represents a three-dimensional point. Following are a few examples of using points, with explanatory comments:

```
Pos = [5,3,2]          -- Marks a point at (5,3,2)
Pos.x = 7              -- Change the x coordinate to 7
                      -- Now the point is at (7,3,2)
Pos = Pos + (1,2,5)   -- Take the old value for Pos,
                      -- move it by (1,2,5) to (8,5,7)
                  -- and store the new value in Pos
```

In addition to these basic data types, each object in your scene has its own data type. For example, if you use `classof` on a sphere object, 3ds Max prints out `Sphere`. Data types for scene objects are actually complex data types or structures, which means that they are groups of other data types in a single unit. The pieces of data inside a larger object are called *members* or *properties*. Most scene objects have a member called Name, which is of type String. The Name member tells the specific name of that object. Another common property is Position, a Point3 variable that tells the object's position.

3ds Max has a special built-in variable that represents whatever object is currently selected. This variable is $ (the dollar sign), which is used in the following tutorial.

Tutorial: Using variables

In this tutorial, you learn more about variables in MAXScript by using them to manipulate an object in your scene.

To use variables to manipulate scene objects, follow these steps:

1. Open the Teapot.max file from the Chap 52 folder in the downloaded content set.

    This file has a simple teapot object.

2. Click on the Point-of-View label for the Left viewport, and choose Views→Extended Viewports→MAXScript Listener to open the MAXScript Listener window in the Left viewport.

3. Select the teapot object, then select the lower pane of the MAXScript Listener and type **$**, and press Enter.

    3ds Max displays information about the teapot. (Your numbers will probably be different depending on where you placed your teapot.)

4. Type the following lines one at a time in the top pane to see the property values stored as part of the teapot object:

    ```
    $.position
    $.wirecolor
    $.radius
    $.name
    $.lid
    ```

5. Now type these lines, one at a time, to set the property values of the teapot object:

    ```
    $.lid = false
    $.position.x = -20
    $.segs = 20
    ```

Figure 52.13 shows the commands, their results in the MAXScript Listener window, and the resulting teapot object.

FIGURE 52.13

The script commands entered in the MAXScript Listener affect the objects in the viewports.

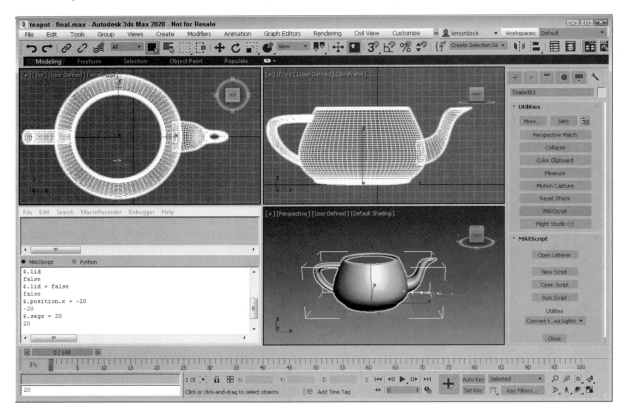

Program flow and comments

In general, when 3ds Max begins executing a script, it starts with the first line of the script, processes it, and then moves on to the next line. Execution of the script continues until no more lines are in the script file. (Later, you'll look at some MAXScript keywords that let you change the flow of script execution.)

3ds Max lets you embed comments or notes in your script file to help explain what is happening. To insert a comment, precede it with two hyphens (--). When 3ds Max encounters the double hyphen, it skips the comment and everything else on that line and moves to the next line of the script. For example, in this line of MAXScript

```
$Torus01.pos = [0,0,0]    -- Move it back to the origin
```

3ds Max processes the first part of the line (and moves the object to the origin) and then moves on to the next line after it reaches the comment.

Using comments in your MAXScript files is very important because as your scripts start to become complex, figuring out what is happening can get difficult. Also, when you come back a few months later to improve your script, comments will refresh your memory and help keep you from repeating the same mistakes you made the first time around.

Note

Because 3ds Max ignores anything after the double hyphen, you can use comments to temporarily remove MAXScript lines from your script. If something isn't working right, you can *comment out* the lines that you want 3ds Max to skip. Later, when you want to add them back in, you don't have to retype them. You can just remove the comment marks, and your script is back to normal.

Expressions

An *expression* is what 3ds Max uses to make decisions. An expression compares two things and draws a simple conclusion based on that comparison.

Simple expressions

The expression

```
1 < 2
```

is a simple expression that asks the question, "Is 1 less than 2?" Expressions always ask yes/no type questions. When you type an expression in the MAXScript Listener window (or inside of a script), 3ds Max evaluates the expression and prints `true` if the expression is valid (like the preceding example) and `false` if it isn't. Try the following expressions in the MAXScript Listener window, and 3ds Max will print the results as shown in Figure 52.14 (you don't have to type in the comments):

```
1 < 2            -- 1 IS less than 2, so expression is true
1 > 2            -- 1 is NOT greater than 2, so false
2 + 2 == 4       -- '==' means "is equal to". 2 + 2 is
                 -- equal to 4, so true
2 + 2 == 5       -- 4 is NOT equal to 5, so false
3 * 3 == 5 + 4   -- 9 IS equal to 9, so true

3 * 3 != 5 + 4   -- '!=' means 'not equal to'. '9 is not
                 -- equal to 9' is a false statement, so
                 -- the expression is false

a = 23           -- store 23 in variable a
b = 14 + 9       -- store 23 in variable b
a == b           -- 23 IS equal to 23, so true
```

FIGURE 52.14

Using the MAXScript Listener to evaluate expressions

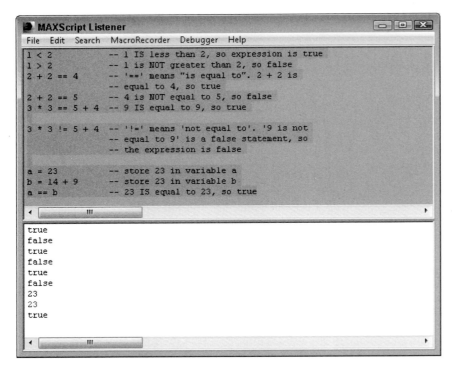

Play around with simple expressions until you're familiar with what they mean and have an intuitive feel for whether or not an expression is going to evaluate to true or false.

Complex expressions

Sometimes you need an expression to decide on more than just two pieces of data. MAXScript has the and, or, and not operators to help you do this.

The and operator combines two expressions and asks the question, "Are both expressions true?" If both are true, the entire expression evaluates to true. But if either is false, or if they are both false, the entire expression is false. You can use parentheses to group expressions, so an expression with the and operator might look something like this:

```
(1 < 2) and (1 < 3)      -- true because (1 < 2) is true AND
                         -- (1 < 3) is true
```

The or operator is similar to and, except that an expression with or is true if either of the expressions is true or if both are true. Here are some examples:

```
(2 > 3) or (2 > 1)       -- even though (2 > 3) is false, the
                         -- entire expression is true because
                         -- (2 > 1) is true
(2 > 3) and (2 > 1)      -- false because both expressions are
                         -- not true
```

Try some of these complex expressions to make sure that you understand how they work:

```
a = 3
b = 2
(a == b) or (a > b)      -- true because a IS greater than b
(a == b) and (b == 2)    -- false because both expressions are
                         -- not true
```

```
    (a > b) or (a < b)        -- true because at least one IS true
   (a != b) and (b == 3)    -- false because b is NOT equal to 3
```

The not operator negates or flips the value of an expression from true to false, or vice versa. For example:

```
   (1 == 2)                  -- false because 1 is NOT equal to 2
   not (1 == 2)              -- true. 'not' flips the false to true
```

Conditions

Conditions are one way in which you can control program flow in a script. Normally, 3ds Max processes each line, no matter what, and then quits; but with *conditions,* 3ds Max executes certain lines only if an expression is true.

For example, suppose you have a script with the following lines:

```
   a = 4
   If (a == 5) then
   (
     b = 2
   )
```

3ds Max would not execute the line b = 2 because the expression (a == 5) evaluates to false. Conditional statements, or "if" statements, basically say, "If this expression evaluates to true, then do the stuff inside the block of parentheses. If the expression evaluates to false, skip those lines of script."

Conditional statements follow this form:

```
   If <expr> then <stuff>
```

where <expr> is an expression to evaluate and <stuff> is some MAXScript to execute if the expression evaluates to true. You also can use the keyword else to specify what happens if the expression evaluates to false, as shown in the following example:

```
   a = 4
   if (a == 5) then
   (
     b = 2
   )
   else
   (
     b = 3
   )
```

After this block of MAXScript, the variable b would have the value of 3 because the expression (a == 5) evaluated to false. Consequently, 3ds Max executed the MAXScript in the else section of the statement.

Collections and arrays

MAXScript has some very useful features to help you manipulate groups of objects. A group of objects is called a *collection.* You can think of a collection as a bag that holds a bunch of objects or variables. The things in the bag are in no particular order; they're just grouped together.

You can use collections to work with groups of a particular type of object. For example, the MAXScript

```
   a = $pokey*
   a.wirecolor = red
```

creates a collection that contains every object in your scene whose name starts with "Pokey" and makes every object in that collection turn red.

MAXScript has several built-in collections that you might find useful, such as cameras and lights, containing all the cameras and lights in your scene. So

```
delete lights
```

removes all the light objects from your scene (which may or may not be a good idea).

An *array* is a type of collection in which all the objects are in a fixed order, and you can access each member of the array by an index. For example

```
a = #()        -- creates an empty array to use
a[1] = 5
a[2] = 10
a[5] = 12
a
```

After the last line, 3ds Max prints out the current value for the array:

```
#(5, 10, undefined, undefined, 12)
```

Notice that 3ds Max makes the array big enough to hold however many elements you want to put in it, and that if you don't put anything in one of the positions, 3ds Max automatically puts in `undefined`, which simply means that array location has no value at all.

One last useful trick is that 3ds Max lets you use the `as` keyword to convert from a collection to an array:

```
LightArray = (lights as array)
```

3ds Max takes the built-in collection of lights, converts it to an array, and names the array `LightArray`.

The members of an array or a collection don't all have to have the same data type, so it's completely valid to have an array with numbers, strings, and objects, like this:

```
A = #(5,"Mr. Nutty",box radius:5)
```

Note

You can use the `as` MAXScript keyword to convert between data types. For example, (5 as string) converts the number 5 to the string "5," and (5 as float) converts the whole number 5 to the floating-point number 5.0.

Loops

A *loop* is a MAXScript construct that lets you override the normal flow of execution. Instead of processing each line in your script once and then quitting, 3ds Max can use loops to do something several times.

For example,

```
j = 0
for i = 1 to 5 do
(
    j = j + i
)
```

This MAXScript uses two variables—i and j—but you can use any variables you want in your loops. The script sets the variable j to 0 and then uses the variable i to count from 1 to 5. 3ds Max repeats the code between the parentheses five times, and each time the variable i is incremented by 1. Inside the loop, 3ds Max adds the current value of i to j. Can you figure out what the value of j is at the end of the script? If you guessed 15, you're right. To see why, look at the value of each variable as the script is running:

```
When                    j    i
---------------------------
First line              0    0
```

```
Start of loop          0    1
After first loop       1    1
Start of second loop   1    2
After second loop      3    2
Start of third loop    3    3
After third loop       6    3
Start of fourth loop   6    4
After fourth loop     10    4
Start of fifth loop   10    5
After fifth loop      15    5
```

A loop is also useful for processing each member of an array or collection. The following MAXScript shows one way to turn every teapot in a scene blue:

```
teapots = $teapot*              -- get the collection of teapots
for singleTeapot in teapots do
(
  singleTeapot.wirecolor = blue
)
```

You can use a `for` loop to create a bunch of objects for you. Try this MAXScript:

```
for I = 1 to 10 collect
  (
  sphere radius:15
  )
```

The `collect` keyword tells 3ds Max to create a collection with the results of the MAXScript in the block of code inside the parentheses. The line

```
sphere radius:15
```

tells 3ds Max to create a sphere with radius of 15, so the entire script created 10 spheres and added them to your scene. Unfortunately, 3ds Max puts them all in the same spot, so move them around a bit so you can see them:

```
i = -50
For s in spheres do
(
  s.position = [i,i,i]
  i = i + 10
)
```

Study this script to make sure that you understand what's going on. You use a `for` loop to process each sphere in your collection of spheres. For each one, you set its position to [i,i,i], and then you change the value of i so that the next sphere is at a different location.

Functions

The last feature of basic MAXScript that you'll look at is the *function*. Functions are small chunks of MAXScript that act like program building blocks. For example, suppose that you need to compute the average of a collection of numbers many times during a script you're writing. The MAXScript to do this might be:

```
Total = 0
Count = 0
For n in numbers do
(
  total = total + n
  count = count + 1
)
average = total / (count as float)
```

Given a collection of numbers called numbers, this MAXScript computes the average. Unfortunately, every time you need to compute the average, you have to type all that MAXScript again. Or you might be smart and just cut and paste it in each time you need it. Still, your script is quickly becoming large and ugly, and you always have to change the script to match the name of your collection you're averaging.

A function solves your problem. At the beginning of your script, you can define an average function like this:

```
Function average numbers =
( -- Function to average the numbers in a collection
 local Total = 0
 local Count = 0
 For n in numbers do
 (
 total = total + n
 count = count + 1
 )
 total / (count as float)
)
```

Now any time you need to average any collection of numbers in your script, you could just use this to take all the numbers in the collection called num and store their average in a variable called Ave:

```
Ave = average num        -- assuming num is a collection
```

Not only does this make your script much shorter if you need to average numbers often, but it makes it much more readable, too. It's very clear to the casual reader that you're going to average some numbers. Also, if you later realize that you wrote the average function incorrectly, you can just fix it at the top of the script. If you weren't using functions, you would have to go through your script and find every case where you averaged numbers and then fix the problem. (What a headache!)

Now take another look at the function definition. The first line

```
Function average numbers =
```

tells 3ds Max that you're creating a new function called average. It also tells 3ds Max that to use this function, you have to pass in one piece of data, and that inside the function you refer to that data using a variable called numbers. It doesn't matter what the name of the actual variable was when the function was called; inside the function, you can simply refer to it as numbers.

Creating functions that use multiple pieces of data is also easy. For example,

```
Function multEm a b c = (a * b * c)
```

creates a function that multiplies three numbers together. To use this function to multiply three numbers and store the result in a variable called B, you would simply enter

```
B = multEm 2 3 4
```

The next two lines

```
local Total = 0
local Count = 0
```

create two variables and set them both to 0. The local keyword tells 3ds Max that the variable belongs to this function. No part of the script outside of the function can see this variable, and if there is a variable outside the function with the same name, changing the variable inside this function won't affect that variable outside the function. That way, you never have to worry about what other variables are in use when someone calls average; even if variables are in use that are named Total or Count, they won't be affected.

The last line

```
total / (count as float)
```

uses the `Total` and `Count` values to compute the average. How does that value get sent back to whoever called the function? 3ds Max evaluates all the MAXScript inside the function and returns the result. Because the last line is the last thing to be evaluated, 3ds Max uses the result of that calculation as the result of the entire function.

Tutorial: Creating a school of fish

Let's look at an example that puts into practice some of the things you've learned in this chapter. In this multipart tutorial, you use MAXScript to create a small school of fish that follows the dummy object around a path.

Part 1: Making the fish follow a path

In this part of the tutorial, you use MAXScript to move one of the fish along a path in the scene. To do this, follow these steps:

1. Open the Fish scene.max file from the Chap 52 folder in the downloaded content set.

   This scene consists of two fish and a dummy object that follows a path. What you need to do is use MAXScript to create a small school of fish that follows the dummy object around the path.

2. Press F11 to open the MAXScript Listener window. In the window, choose File→New Script to open the MAXScript Editor window, and type the following script:

   ```
   pathObj = $Dummy01
   fishObj = $Fish1/FishBody
   relPos = [0,-150,-50]    -- How close the fish is to the path

   animate on
   (
    for t = 1 to 100 do at time t
    (
    fishObj.position = pathObj.position + relPos
    )
   )
   ```

3. Select the Camera01 viewport. Choose Tools→Evaluate All (or press Ctrl+E) to evaluate all the MAXScript in the Editor window, select the Camera01 viewport, and click the Play Animation button.

   The fish rigidly follows the dummy object's path. Figure 52.15 shows one frame of this animation.

FIGURE 52.15

First attempt at making the fish follow a path

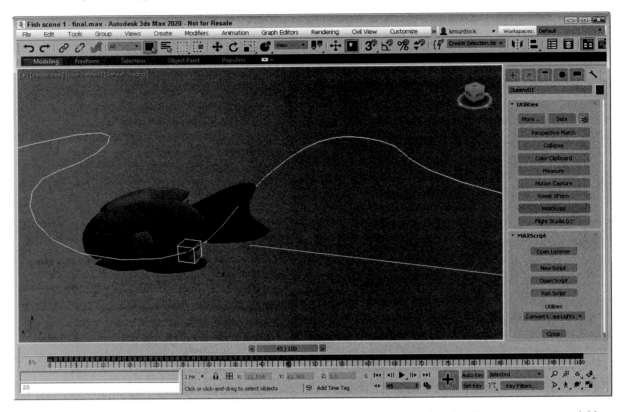

Now you'll explain the MAXScript entered in the previous tutorial. The first few lines create some variables that the rest of the script uses. `pathObj` tells the name of the object that the fish will follow, and `fishObj` is the name of the fish's body. (Notice that you can reference parts of the group hierarchy by using the object name, a forward slash, and then a child part.) Why bother creating a variable for the fish object? After you get this first fish working, you want to apply the same script to another fish. All you have to do is rename `Fish1` as `Fish2`, re-execute the script, and you're finished!

The script also creates a variable called `relPos`, which you use to refer to the relative position of the fish with respect to the dummy object. If you have several fish in the scene, you don't want them all in the exact same spot, so this is an easy way to position each one.

The next block of MAXScript is new: you're using the `animate on` construct. This tells 3ds Max to generate keyframes for your animation. It's the same as if you had pressed the software's Auto Key button, run our script, and then stopped the Animation. So any MAXScript inside the `animate on` parentheses creates animation keyframes. These parentheses define a section of the script you call a block.

Inside the animation block, you have a loop that counts from 1 to 100 (corresponding to each frame of our animation). On the end of the loop line, you have at `time t`, which tells 3ds Max that for each time through the loop, you want all the variables to have whatever values they'll have at that time. For example, if you want the fish to follow the dummy object, you have to know the position of the object at each point in time instead of just at the beginning; so each time through the loop, 3ds Max figures out for you where the dummy object will be.

Inside the loop, you set the fish's object to be that of the dummy object (at that point in time) and then adjust the fish's position by `relPos`.

Part 2: Adding body rotation and tail animation

Next you'll make that fish look a little more lifelike by animating its tail and having it rotate its body to actually follow the path. Also, you'll add a little unpredictability to its motion so that when you add other fish, they aren't exact copies of each other.

To improve the fish's animation, follow these steps:

1. Type the revised version of the script (the new lines are in bold):

```
pathObj = $Dummy01
fishObj = $Fish1/FishBody
fishTail = $Fish1/FishBody/FishTail
relPos = [0,-150,-50]  -- How close the fish is to the path

fishTail.bend.axis = 0 -- 0 is the x-axis
zadd = 4               -- vertical movement at each step
tailFlapOffset = (random 0 100)
tailFlapRate = 25 + (random 0 25)
animate on
(
 for t = 0 to 100 do at time t
 (
  fishObj.position = pathObj.position + relPos
  fishObj.position.z = relPos.z
  relPos.z += zadd

  -- let's say that there's a 10% chance that the fish will
  -- change directions vertically
  if ((random 1 100) > 90) then
  (
   zadd = -zadd
  )

  fishTail.bend.angle = 50 * sin (t * tailFlapRate +
  tailFlapOffset)

  oldRt = fishObj.rotation.z_rotation
  newRt = (in coordsys pathObj pathObj.rotation.z_rotation)

  if ((random 1 100) > 85) then
  (
   fishObj.rotation.z_rotation += (newRt - oldRt) *
   (random 0.5 1.5)
  )
 )
)
```

2. Save your script (File→Save), and then press Ctrl+E to evaluate the script again. This script is saved in the Chap 52 folder as FishPath2.ms. Make the Camera01 viewport active, and click Play Animation. Figure 52.16 shows another frame of the animation. As you can see, the fish is heading in the right direction this time, and the tail is flapping wildly.

FIGURE 52.16

A tail-flapping fish that faces the right direction as it follows the path

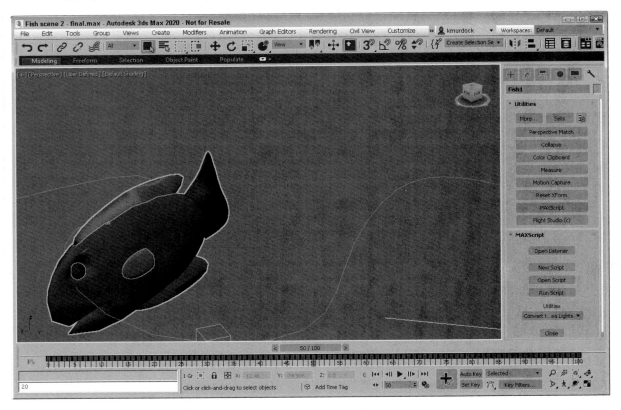

Okay, look at what changed. First, you added a variable to refer to the fish's tail, so that it is easy to change when you add another fish. Also, you accessed the bend modifier of the tail and set its axis to 0, which corresponds to the X-axis. (You can try other values to see that it really does change the axis parameter in the rollout.)

Next, you created some more variables. You use zadd to tell 3ds Max how much to move the fish in the Z-direction at each step. (You don't want our fish to always swim at the same level.) tailFlapOffset and tailFlapRate are two variables used to control the tail flapping. (I explain this when you get to the part of the script that uses them.)

Inside the for loop, notice that you've overridden the fish's Z-position and replaced it with just the relative Z-position, so that each fish swims at its own depth and not the dummy object's depth. Then, at each step, you add zadd to the Z-position so that the fish changes depth slowly. You have to be careful, or your fish will continue to climb out of the scene or run into the ground, so at each step you also choose a random number between 1 and 100 with the function (random 1 100). If the random number that 3ds Max picks is greater than 90, you flip the sign of zadd so that the fish starts moving in the other direction. This is a fancy way of saying, "There's a 90 percent chance that the fish will continue moving in the same direction and a 10 percent chance that it will switch directions."

In the next part, you again access the tail's bend modifier, this time to set the bend angle. To get a nice back-and-forth motion for the tail, you use the sin function. In case you've forgotten all that math from when you were in school, a sine wave oscillates from 1 to –1 to 1 over and over again. By multiplying the function by 50, you get values that oscillate between 50 and –50 (pretty good values to use for your bend angle). You use tailFlapOffset to shift the sine wave so that the tail flapping of additional fish is out of synch slightly with this one (remember, you're trying to get at least a little realism here) and tailFlapRate to make each fish flap its tail at a slightly different speed.

The only thing left for you to do is to make the fish "follow" the path; that is, rotate its body so that it's facing the direction it's moving. The simplest way to do this is to use the following MAXScript (split into two lines to make it easier to read):

```
newRt = (in coordsys pathObj pathObj.rotation.z_rotation)
fishObj.rotation.z_rotation = newRt
```

The `in coordsys` construct tells 3ds Max to give you a value from the point of view of a particular coordinate system. Instead of `pathObj`, you could have asked for the Z-rotation in the world, local, screen, or parent coordinate system, too. In this case, you want to rotate the fish in the same coordinate system as the dummy object. To randomize the direction of the fish a little, you've made the rotation a little more complex:

```
oldRt = fishObj.rotation.z_rotation
newRt = (in coordsys pathObj pathObj.rotation.z_rotation)

if ((random 1 100) > 85) then
(
  fishObj.rotation.z_rotation += (newRt - oldRt) *
                              (random 0.5 1.5)
)
```

First, you save the old Z-rotation in `oldRt`, and then you put the new rotation in `newRt`. Again, you pick a random number to decide whether you'll do something; in this case you're saying, "There's an 85 percent chance I won't change directions at all." If your random number does fall in that other 15 percent, however, you adjust the fish's rotation a little. You take the difference between the new rotation and the old rotation and multiply it by a random number between 0.5 and 1.5, which means you adjust the rotation by anywhere from 50 percent to 150 percent of the difference between the two rotations. So any fish will basically follow the same path, but with a little variation here and there.

Note

3ds Max lets you use shorthand when adjusting the values of variables. Instead of saying `a = a + b`, you can just say `a += b`. Both have the same effect.

Part 3: Animating the second fish

This scene actually has two fish in it (the other one has been sitting patiently off to the side), so for the final part of this tutorial, you get both fish involved in the animation. To animate the second fish alongside the first one, follow these steps:

1. At the top of the script, change these three lines (changes are in bold):
    ```
    pathObj = $Dummy01
    fishObj = $Fish2/FishBody
    fishTail = $Fish2/FishBody/FishTail
    relPos = [50,75,0]    -- How close the fish is to the path
    ```

2. Choose Tools→Evaluate All (or press Ctrl+E) to run the script again, and then animate the fish. Figure 52.17 shows both fish swimming merrily.

FIGURE 52.17

Both fish swimming together

This script generates keyframes for the second fish because you changed the `fishObj` and `fishTail` variables to refer to the second fish. You've also moved the second fish's relative position so that the two don't run into each other.

Learning the Visual MAXScript Editor Interface

Building scripts can be complicated, and piecing together a rollout for a scripted utility can be especially time-consuming and frustrating when done by hand. To help create such custom rollouts, 3ds Max includes the Visual MAXScript Editor. Using this editor, you can drag and drop rollout elements and automatically create a code skeleton for certain events.

Working with textual commands can be time-consuming. In order for the script to work, you need to enter the commands exactly. This can be especially tricky when you're trying to lay out the controls for a rollout. The Visual MAXScript Editor speeds up the creation of rollouts.

To access the Visual MAXScript window, shown in Figure 52.18, select it from the Scripting menu. Another way to access this window is to select Tools→New Rollout or Edit Rollout in the MAXScript Editor window.

FIGURE 52.18

The Visual MAXScript window makes building rollouts easy.

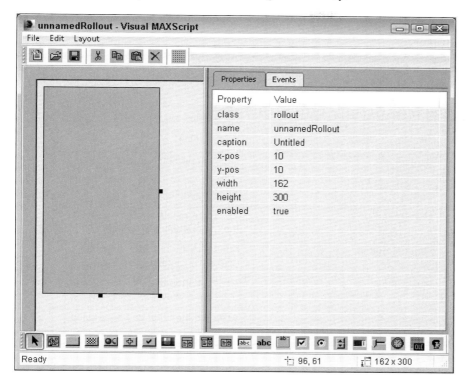

Layouts for a rollout created in the Visual MAXScript Editor can be saved as files with the .vms extension using the File menu. If you access the window from a MAXScript Editor window, the Save menu automatically updates the editor window.

The Editor interface

The window includes two major panes. The left pane is where the various rollout elements are assembled, and the right pane holds the Properties and Event Handlers tabbed panels. The Properties panel lists all the properties and their associated values for the selected element. You can change the property values by clicking on them and entering a new value. For example, if you select a Button element in the left panel, the properties for that control are presented in the Properties panel. If you click the Caption Property, its value becomes highlighted; you can type a new caption, and the new caption appears on the button.

The Events panel lists all the available events that can be associated with the selected element. Clicking the check box to the left of these events can enable the events. For a button element, you can enable the pressed event. With this event enabled, the code includes a function where you can define what happens when this event is fired.

The menus and the main toolbar

At the top of the interface are some menu options and a main toolbar. The File menu also lets you create a new layout (Ctrl+N), save layouts to a file (Ctrl+S), and open saved layouts (Ctrl+O). The Edit menu allows you to cut (Ctrl+X), copy (Ctrl+C), and paste (Ctrl+V) form elements. You can find these same features as buttons on the top toolbar.

The Layout menu includes options for aligning elements left (Ctrl+left arrow), right (Ctrl+right arrow), top (Ctrl+up arrow), bottom (Ctrl+down arrow), vertical center (F9), and horizontal center (Shift+F9); to space elements evenly across (Alt+right arrow) or down (Alt+up arrow); make elements the same size by width, height, or both; center vertically (Ctrl+F9) or horizontally (Ctrl+Shift+F9) in the dialog box; and flip. You can use the Layout→Guide Settings menu command to specify grid snapping and spacing. Grids are enabled using the Toggle Grid/Snap button on the right end of the main toolbar.

You also can access these commands using a right-click pop-up menu when clicking on the left pane.

Toolbar elements

The toolbar along the bottom of the window contains the form elements that you can drop on the form. These buttons and elements include those shown in Table 52.2.

TABLE 52.2 Visual MAXScript Form Elements

Button	Element	What It Does
	Bitmap	Lets you add bitmap images to a rollout
	Button	Adds a simple button
	Map Button	Adds a mapping button that opens the Material/Map Browser
	Material Button	Adds a material button that also opens the Material/Map Browser
	Pick Button	Adds a button that lets you pick an object in a viewport
	Check Button	Adds a button that can be toggled on and off
	Color Picker	Adds a color swatch that opens the Color Picker dialog box when clicked
	Combo Box	Adds a list with several items
	Drop Down List	Adds a list with one item displayed
	List Box	Adds a list with several items displayed
	Edit Box	Adds a text field that can be modified
	Label	Adds a text label
	Group Box	Adds a grouping outline to surround several controls
	Check Box	Adds a check box control that can be toggled on or off
	Radio Buttons	Adds a set of buttons where only one can be selected
	Spinner	Adds an up and down set of arrows that can modify a value field
	Progress Bar	Adds a bar that highlights from left to right as a function is completed
	Slider	Adds a slider control that can move from a minimum to a maximum value
	Timer	Adds a timer that counts time intervals
	ActiveX Control	Adds a generic ActiveX control created by a separate vendor

	Custom	Adds a custom control that can be defined as needed

At the bottom right of the window are two text fields that display the coordinates of the current mouse cursor position and the size of the rollout. The default size of the rollout is 162x300, which is the size needed to fit perfectly in the Command Panel.

Laying Out a Rollout

The rollout space, which appears gray in the left pane, can be selected and resized by dragging the black, square handles at the edges of the form. As you change its size, its dimensions are displayed in the lower-right corner of the interface. With the rollout space correctly sized, you are ready to add elements to the space.

To add one of these elements to the form, click the element button on the toolbar and drag on the form. The element appears and is selected. The selected element is easy to identify by the black handles that surround it. Dragging on these handles resizes the element, and clicking and dragging on the center of the element repositions it within the rollout space.

The Properties and Events panels are automatically updated to show the values and events for the selected element. Properites such as width and x-pos are automatically updated if you drag an element or drag its handles to resize it.

Aligning and spacing elements

Although only one element at a time can be surrounded by black handles, you can actually drag an outline in the rollout space to select multiple elements at once. With several elements selected, you can align them all to the left (Ctlr+left arrow), horizontally centered (Shift+F9), right (Ctrl+right arrow), top (Ctrl+up arrow), vertically centered (F9), or bottom (Ctrl+down arrow).

Multiple elements also can be spaced across (Alt+right arrow) or down (Alt+up arrow). To make several elements the same width, height, or both, use the Layout→Make Same Size menu commands. The Center in Dialog menu aligns elements to the center of the dialog box either vertically (Ctrl+F9) or horizontally (Ctrl+Shift+F9). The Flip command reverses the position of the selected elements.

Figure 52.19 shows a form with several aligned elements added to it.

FIGURE 52.19

You can add control elements to the form in the Visual MAXScript Editor.

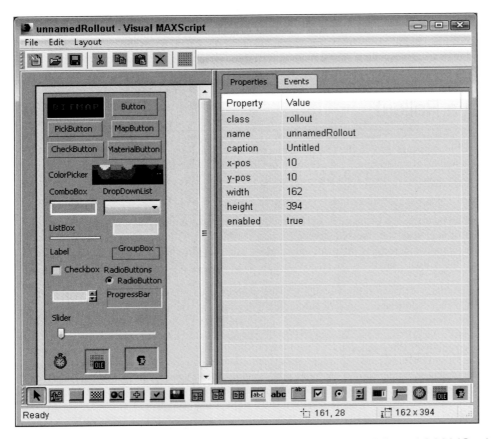

Tutorial: Building a custom rollout with the Visual MAXScript Editor

Now you need some practice using this powerful tool. In this example, you use the Visual MAXScript Editor to lay out a rollout and code the script to make it work.

To create a custom rollout using the Visual MAXScript Editor, follow these steps:

1. Open the BuildCube.max file from the Chap 52 folder in the downloaded content set.

    This file includes a simple sphere object.

2. Choose Scripting→New Script to open the MAXScript Editor window. In the editor window, enter the following:

    ```
    utility buildCube "Build Cube" ( )
    ```

    This line creates a utility named buildCube. The rollout name is Build Cube. Make sure to include a space in between the parentheses.

3. Choose Tools→New Rollout from the window menu. The Visual MAXScript Editor opens. The properties for this rollout are displayed in the Properties panel. Drag the lower-right, black square handle to resize the rollout form.

4. Click the Spinner button on the bottom toolbar, and drag in the rollout form to create a spinner element. In the Properties panel, set the name to **SideNum**, set the caption value to **No. of Side Objects**, set the type to **Integer**, and set the range to **[1,100,5]**. The range values set the lower, upper, and default values for the spinner. Then drag the element handles to resize the element to fit in the form.

5. Click the Spinner button again, and drag in the rollout form to create another spinner element. In the Properties panel, set the name to **length**, set the caption value to **Side Length**, set the type to **Integer**,

and set the range to **[1,1000,50]**. Then drag the element handles to resize the element to fit in the form.

6. Click the Button icon on the bottom toolbar, and drag in the rollout form to create a button below the spinners. In the Properties panel, set the name to **createCube** and the caption value to **Create Cube**. Then drag the element handles to resize the button so the text fits on the button. Open the Events panel, and select the Pressed check box.

7. Drag over the top of both the spinners to select them both, and choose Layout→Align→Right (or press Ctrl+right arrow) to align the spinners. Figure 52.20 shows how the rollout layout looks.

FIGURE 52.20

The rollout laid out in the Visual MAXScript Editor

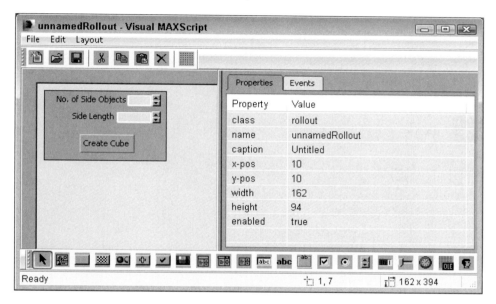

8. With the layout complete, choose File→Save (or press Ctrl+S) to save the layout, and then close the Visual MAXScript window.

   The script code associated with the layout is automatically placed in the editor window.

9. Complete the script by entering the script commands immediately after the open parenthesis that appears on the line following the `on createCube pressed do` event, as shown in Figure 52.21. Then, copy and paste the code from the BuildCube.ms file from the Chap 52 folder in the downloaded content set.

FIGURE 52.21

The MAXScript Editor window is updated with the code from the Visual MAXScript Editor.

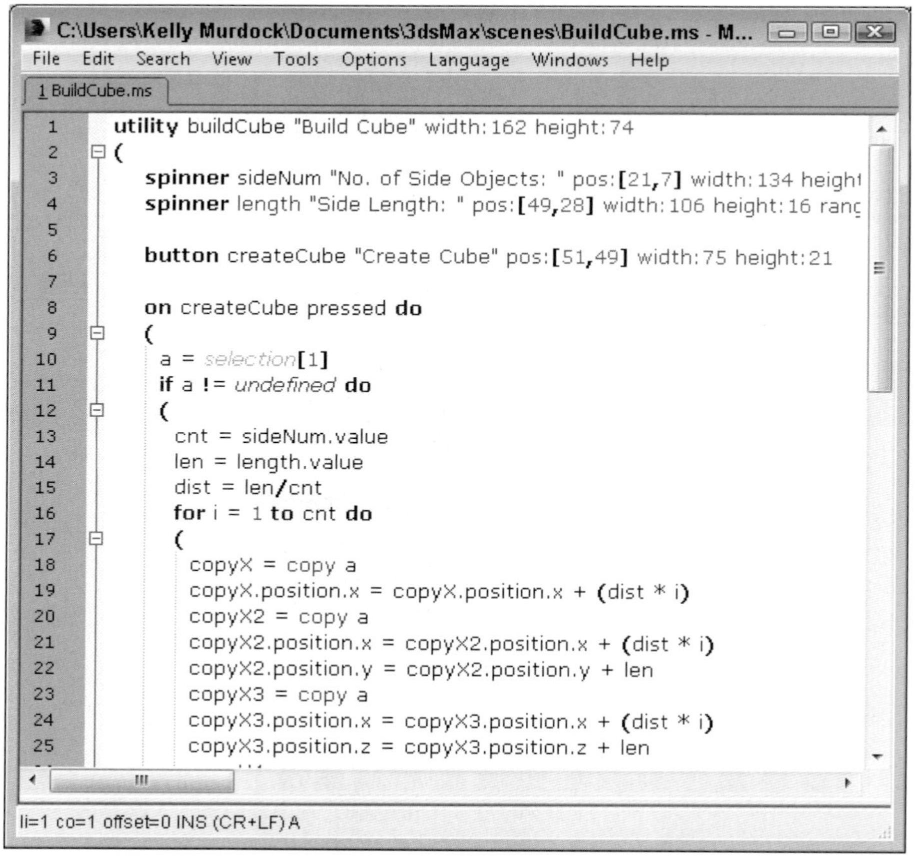

```
C:\Users\Kelly Murdock\Documents\3dsMax\scenes\BuildCube.ms - M...
File   Edit   Search   View   Tools   Options   Language   Windows   Help
1 BuildCube.ms

1      utility buildCube "Build Cube" width:162 height:74
2    ( 
3        spinner sideNum "No. of Side Objects: " pos:[21,7] width:134 heigh
4        spinner length "Side Length: " pos:[49,28] width:106 height:16 ranc
5
6        button createCube "Create Cube" pos:[51,49] width:75 height:21
7
8        on createCube pressed do
9        (
10         a = selection[1]
11         if a != undefined do
12         (
13           cnt = sideNum.value
14           len = length.value
15           dist = len/cnt
16           for i = 1 to cnt do
17           (
18             copyX = copy a
19             copyX.position.x = copyX.position.x + (dist * i)
20             copyX2 = copy a
21             copyX2.position.x = copyX2.position.x + (dist * i)
22             copyX2.position.y = copyX2.position.y + len
23             copyX3 = copy a
24             copyX3.position.x = copyX3.position.x + (dist * i)
25             copyX3.position.z = copyX3.position.z + len

li=1 co=1 offset=0 INS (CR+LF) A
```

10. Open the Utilities panel, and click the MAXScript button. Then click the Run Script button, and select the BuildCube.ms file from the Chap 52 folder in the downloaded content set.

    The utility installs and appears in the Utilities drop-down list in the MAXScript rollout.

11. Select the BuildCube utility from the drop-down list in the MAXScript rollout, and scroll down the Command Panel to see the Build Cube rollout. Select the sphere object, and click the Create Cube button.

    The script executes, and a cube of spheres is created.

Figure 52.22 shows the results of the BuildCube.ms script. You can use this script with any selected object.

FIGURE 52.22

The results of the BuildCube.ms script

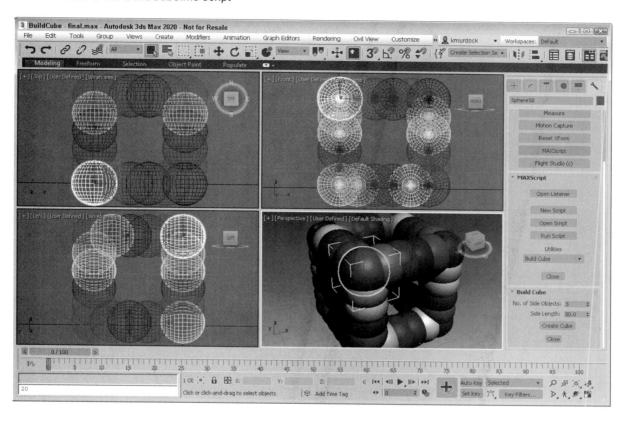

Summary

This chapter gave you a brief introduction to MAXScript, the powerful, built-in scripting language found in 3ds Max. Besides describing the different types of scripts you can create, the chapter covered the following topics:

* The basics of MAXScript
* Using the MAXScript tools such as the MAXScript Editor and MAXScript Listener windows
* Using the Macro Recorder to create scripts
* Using the MAXScript Debugger
* The different script types
* The basics of writing your own scripts
* The Visual MAXScript Editor interface
* The features of each rollout element
* How to create scripted utilities with custom rollouts

Although you can do some amazing things with MAXScript, there is yet another way to extend the software. Using the Software Development Kit (SDK) that comes with 3ds Max, you can program entire new modules as plug-ins that work seamlessly with the software. The next chapter shows how to install and work with third-party plug-ins.

Chapter 53

Expanding 3ds Max with Third-Party Plug-Ins

IN THIS CHAPTER

Understanding plug-ins

Installing, viewing, and managing plug-ins

Looking at plug-in examples

Locating plug-ins

A plug-in is an external program that integrates seamlessly with the Autodesk® 3ds Max® software to provide additional functionality. Autodesk has adopted an architecture for 3ds Max that is open and enables all aspects of the program to be enhanced. 3ds Max ships with a Software Developer's Kit (SDK) that enables users to generate their own plug-ins. Many companies currently produce plug-ins, and other users create and distribute freeware and shareware plug-ins.

The purpose of this chapter isn't to cover all the available plug-ins or to teach you how to create plug-ins, but simply to show you how to install and access plug-ins. At the end of the chapter is a list of web links you can use to locate plug-ins for 3ds Max. The entire architecture of 3ds Max is built around plug-ins, and many of the core components of 3ds Max are implemented as them.

A key feature that allowed 3ds Max to become and remain so popular is that users can download and install plug-ins that extend the software's power and functionality. Plug-ins allow 3ds Max to adapt to the needs of each user as well as keep up with new ideas.

Working with Plug-Ins

After you've located a plug-in that you would like to add to your system, you need to install the plug-in. Most commercial plug-ins come with an executable setup program that automates this for you, but others need to be installed manually, which isn't difficult.

Note

Plug-ins typically don't work from one version of 3ds Max to another. If the plug-in is commercial, the developers usually release an updated version of the plug-in for the new 3ds Max version.

As you begin to add plug-ins to 3ds Max, you may eventually want to see which plug-ins are installed and even disable certain plug-ins. 3ds Max includes tools to view installed plug-ins and to manage your current plug-ins.

Installing plug-ins

For commercial plug-ins that include an installation program, the installation process asks where the 3ds Max root directory is located. From this root directory, the plug-in program files are typically installed in the "plugins" directory, help files for the plug-ins are installed in the "help" directory, and example scenes are installed in the "scenes" directory.

Plug-in program files typically have a .dlc, .dlr, .dlo, .dlu, .dlv, or .dlm extension, depending on the type of plug-in. When 3ds Max loads, it searches the /plugins directory for these files and loads them along with the program files. You can manually install freeware plug-ins simply by copying the plug-in file into the plugins directory and restarting 3ds Max.

You also can place plug-ins in a different directory and load them from this directory. The Configure User and System Paths dialog box, opened with the Customize→Configure User and System Paths menu, includes a panel titled 3rd Party Plug-Ins, shown in Figure 53.1, where you can specify additional plug-in paths.

Find out more about the Configure System Paths dialog box in Chapter 4, "Setting Preferences."

FIGURE 53.1

The Configure System Paths dialog box includes a panel where you can specify where to look for plug-ins.

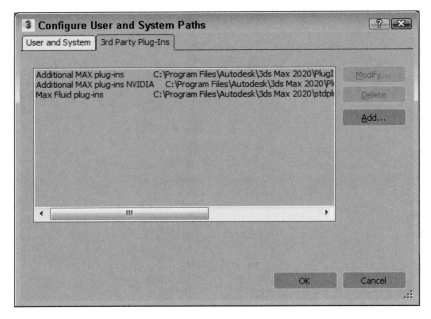

Most commercial plug-ins require that the plug-in be authorized after installation. You must do this before you can use the plug-in. Plug-ins can typically be authorized on-line.

To remove a plug-in, use the uninstall feature that is part of the setup process, or delete the associated program files from the plugins directory.

Plug-ins also can create a help file that explains how to work with the plug-in. These help files are installed in the /help directory where 3ds Max is installed. To view these help files, open the Additional Help dialog box by choosing Help→Plug-In Help. You also can access these help files using the Help button located on the title bar.

Viewing installed plug-ins

To see all the currently installed plug-ins, choose File→Summary Info to open the Summary Info dialog box and click the Plug-In Info button. This opens the Plug-In Info dialog box that lists all installed plug-ins with their details, as shown in Figure 53.2. As you can see, many plug-ins created by Autodesk are installed with just the default installation.

Note

Even if you haven't installed any plug-ins, this dialog box lists many plug-ins. These are core functions in 3ds Max that are implemented as plug-ins.

FIGURE 53.2

The Plug-In Info dialog box includes a list of all the currently loaded plug-ins, both internal and external.

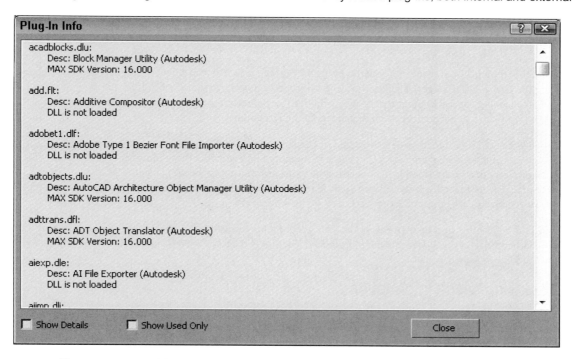

You can manage which installed plug-ins are available using the Plug-in Manager dialog box, shown in Figure 53.3. Open this dialog box by choosing Customize→Plug-in Manager.

FIGURE 53.3

Use the Plug-in Manager dialog box to disable plug-ins.

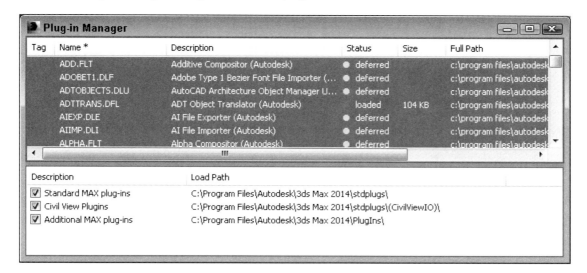

Each column in the Plug-in Manager dialog box includes information about the plug-ins. The columns include Tag, Name, Description, Status, Size, and Full Path. You can sort the list of plug-ins alphabetically by column if you click on the column name. An asterisk appears to the right of the column title that is used to sort.

Each unique directory that is specified within the Configure System Paths dialog box appears in the bottom pane of the Plug-in Manager. Use the check boxes to remove all plug-ins in that directory from the list.

Tip

If you install all your plug-ins into a custom directory, you can use the bottom pane to filter only the plug-ins you've installed.

In the list of plug-ins, you select a specific plug-in by clicking it. You can select multiple plug-ins in the list using the Ctrl and Shift keys. A right-click pop-up menu of options lets you control the selected plug-ins. You also can tag (or mark) certain plug-ins using the Tag Selected option in the right-click pop-up menu. For tagged plug-ins, a white check mark appears in the left column.

You also can choose to load or defer selected or checked plug-ins using the right-click pop-up menu. Plug-ins with a status of loaded are currently loaded in memory and available; these plug-ins are identified with a green circle in the Status column. The deferred plug-ins are waiting in the wings and load when needed; these plug-ins are identified with a yellow circle in the Status column. Plug-ins that are marked Unloaded (with a red circle) are not in memory.

Using the right-click pop-up menu, you also can select Load New Plug-in, which opens the Choose Plug-in File dialog box where you can select a plug-in file. The file is then accessed from this directory and loaded into the Plug-in Manager list.

Displaying scene information

If you like to keep statistics on your files (to see whether you've broken the company record for the model with the greatest number of faces), you'll find the Summary Info dialog box useful. Use the File→Summary Info menu command to open a dialog box that displays all the relevant details about the current scene, such as the number of objects, lights, and cameras; the total number of vertices and faces; and various model settings, as well as a Description field where you can describe the scene. Figure 53.4 shows the Summary Info dialog box.

FIGURE 53.4

The Summary Info dialog box shows all the basic information about the current scene.

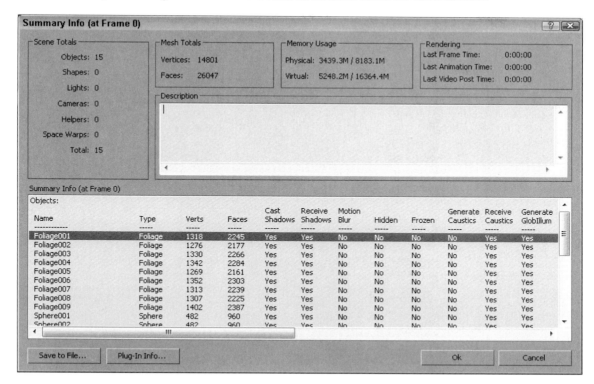

The Plug-In Info button on the Summary Info dialog box displays a list of all the plug-ins currently installed on your system. Even without any external plug-ins installed, the list is fairly long because many of the core features in 3ds Max are implemented as plug-ins. The Summary Info dialog box also includes a Save to File button for saving the scene summary information as a text file.

Tutorial: Installing and using the AfterBurn plug-in demo

If you'd like to try out some plug-ins before purchasing them, visit the developers' websites to download a demo copy of the plug-in.

Caution

These demos are full-featured, but most of them are save-disabled, which prevents you from saving the file that includes the plug-in's features. Look at the readme file as the plug-in is installed to see what has been disabled.

To install a demo plug-in from the 3ds Max install disc or downloaded from a website, follow these steps:

1. Insert the 3ds Max installation disc into your CD-ROM drive. When a menu of options appears, click the Partners and Samples link and then click the Autodesk Certified Animation Plug-Ins link.

   A page of certified plug-in demos appears, including AfterBurn, Absolute Character Tools, DreamScape, and others. Or you could locate a demo plug-in from a website.

2. Click AfterBurn to select it, and click the Install button to launch the installation wizard. Follow the Installation Wizard's instructions, and press Next to complete each step. Click the Finish button when the installation is complete.

3. You need to restart 3ds Max before the plug-ins features become available, so select Start→Program Files→Autodesk→Autodesk 3ds Max 2020→3ds Max 2020 to restart 3ds Max.

4. You can learn to use the AfterBurn plug-in using the help files that were installed; select Help→Plug-In Help, and double-click the AfterBurn reference in the list that appears to open the Help files for the plug-in.

5. After reviewing the help files, select Create→Helpers→Atmospheric→Sphere Gizmo and drag in the Top viewport to create a gizmo.

6. Most of the AfterBurn plug-in features are found in the Environment & Effects dialog box. Select Rendering→Environment (keyboard shortcut, 8), and click the Add button in the Atmosphere rollout. In the Add Atmospheric Effect list, double-click on the AfterBurn Combustion Demo effect. This adds the AfterBurn Combustion effect along with the AfterBurn Renderer to the Effects list and makes several new rollouts appear.

7. In AfterBurn Combustion Parameters rollout, click the Pick Gizmos button and select the Sphere Gizmo icon in the Top viewport.

8. Select Rendering→Render Setup(F10) to open the Render Setup dialog box. Make sure that the Atmospherics option is enabled, and render the Perspective viewport to see the resulting AfterBurn flame effect.

Figure 53.5 shows the resulting fireball created using the AfterBurn demo plug-in.

FIGURE 53.5

This simple fireball was created using the AfterBurn plug-in.

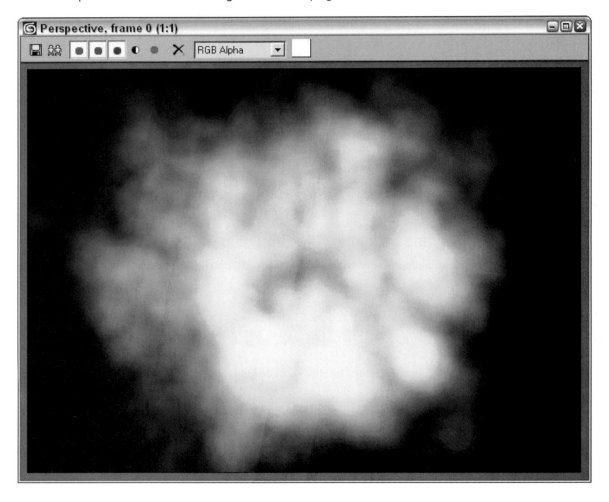

Locating Plug-Ins

Before you can take advantage of plug-ins, you need to locate, acquire, and install them. You can find plug-ins from a variety of sources: commercial, shareware, or freeware.

The first place to look for commercial plug-ins is Digimation. It is not only a plug-in developer, but it also resells many plug-ins for other companies. Another good site to check out is the list at Autodesk. Autodesk evaluates plug-ins as part of the Certified Plug-Ins program to ensure compatibility with 3ds Max.

In addition to the commercially developed plug-ins, many plug-ins are available as freeware or shareware. You can find many of these plug-ins and download them via the Web.

If you're looking for plug-ins, both commercial and shareware, you should visit these sites:

* **Facebook:** facebook.com/Autodesk3dsMax
* **ScriptSpot:** www.scriptspot.com
* **Digimation:** www.digimation.com/
* **BoboLand:** www.scriptspot.com/bobo/
* **Max Plugins.De:** www.maxplugins.de/
* **Highend3d:** www.highend3d.com/3dsmax
* **Turbo Squid:** www.turbosquid.com/

Caution

Plug-ins typically are not compatible between different versions of 3ds Max. For example, a plug-in written for version 2.5 does not work on versions 2018, 2019, or 2020, and vice versa. When downloading and purchasing plug-ins, be sure to get a version that matches your current version of 3ds Max.

Summary

By adding plug-ins, you can increase the functionality of 3ds Max far beyond its default setup. In this chapter, I covered the following topics:

* What plug-ins are and how they can extend 3ds Max
* How to install, view, and manage plug-ins
* Where to find plug-ins

Congratulations, you made it to the last chapter. If you still want more, you can check out the appendixes, which contain some great reference information.

Appendix A
What's New with Autodesk 3ds Max 2020

IN THIS APPENDIX

Finding the new features in 3ds Max

Enjoying the minor enhancements that 3ds Max has to offer

The Autodesk® 3ds Max® 2020 software is as astonishing as its predecessors. With each revision of 3ds Max, I'm always amazed at the new features that are included. 3ds Max is a large and complex piece of software, and just when I think it can't hold anything more, a new revision with a host of new features appears. 3ds Max 2020 is no different.

You can find in-depth coverage of the new features in the various chapters, but this appendix provides a quick overview of these new features, along with references on where to learn more about them. Throughout the book, the New Feature icon identifies the features that are new to 3ds Max 2020.

Note

If 3ds Max needs some improvements that haven't made it into the latest release, you can join Autodesk's Customer Involvement Program using the Help→Feedback→Report a Problem menu command. This program lets you provide feedback and suggestions to the 3ds Max team.

Major Improvements

3ds Max 2020 includes lots of new improvements. Some are considered major because they likely will affect every user's workflow, and others are minor because they are smaller in scope. However, an improvement listed as minor may be the one you've been waiting for.

Note

Within the Introduction section of the 3ds Max Help file is a What's New in Autodesk 3ds Max 2020 page. You also can access the What's New link in the Help menu.

Updated Chamfer modifier

The Chamfer modifier has been updated with several new features including the ability to assign weight values to edges. You can also set the Radius Bias to sharp corners. The Chamfer modifier is presented in Chapter 16, "Deforming Surfaces and Using the Mesh Modifiers."

OSL Maps

Several new Open Shader Language (OSL) maps have been added to the software including Falloff, Simple Gradient, Threads, Waveform and Weave. Using OSL maps is covered in Chapter 18, "Adding Material Details with Maps."

Preview Animation

The Make Preview dialog box has several new features including the ability to name and save preview files. You can also play the preview directly from the dialog box and add custom MAXScript snippets for indicating frame rate. This improved Make Preview dialog box is covered in Chapter 32, "Understanding Animation and Keyframes."

Revit Import

3ds Max now includes support for OSL (Open Shading Language) maps. These open-source maps can be downloaded and loaded within the Slate Material Editor. The Material/Map Browser also includes a large number of preset OSL maps that can be used. OSL maps are presented in Chapter 18, "Adding Material Details with Maps."

Floating Viewports

3ds Max 2020 lets you specify three viewports as floating viewports using the viewport's General label menu. These viewports are fully configurable and can be moved to wherever you need them. Floating viewports are covered in Chapter 2, "Controlling and Configuring the Viewports."

Sharing Views

Using the freely available Autodesk Viewer, you can share views created in 3ds Max with others that don't have the software installed. The viewer runs in a web browser and includes navigation features and the ability to markup the view. Shared views are covered in Chapter 3, "Working with Files, Importing and Exporting."

Hotkey Editor

The new Hotkey Editor provides a single place to manage all your keyboard shortcuts. Using the Hotkey Editor, you can search for specific Actions or Hotkeys to see what is currently assigned. It also recognizes any assigned conflicts. The Hotkey Editor lets you create and manage hotkey sets or quickly switch to the Maya hotkeys. This editor is covered in Chapter 49, "Customizing the Interface."

Minor Improvements

In addition to the major improvements, many minor improvements make working with objects, materials, and other facets of 3ds Max easier. Minor improvements found in version 2020 include the following:

* **Revit Import:** Revit files now have more options when imported making an easier process.
* **Game Exporter in File menu:** The Game Exporter is now available in the File menu.
* **GIF Checker Tool:** Replaces GIF images that are incompatible with Arnold with PNG files.
* **Point Cloud support:** e57 and PLY file formats are supported.
* **Framerate fix:** When splines are rendered, you can fix any twisting problem.
* **Show Expanded Tooltips in Preferences dialog box:** The Help panel of the Preferences dialog box includes a Show Expanded Tooltips option.
* **Transform and Geometry options in Mirror dialog box:** The Transform option in the Mirror panel allows modifiers to also be mirrored.
* **Spline Cap and Twist Correction:** New features in the Rendering panel for Editable Splines.

Appendix B
Installing and Configuring Autodesk 3ds Max 2020

IN THIS APPENDIX

Choosing an operating system

Understanding hardware requirements

Installing 3ds Max 2020

Authorizing the software

Setting the display driver

Before you can enjoy all the great features in the Autodesk® 3ds Max® 2020 software, you must install the software and get your system configured properly. This appendix helps you do just that. Some configuration settings for interfacing with your computer's graphics card can cause trouble if not set up correctly, and these issues are presented also.

Choosing an Operating System

If you're purchasing a new computer to run 3ds Max and have the luxury of customizing your system, you can do several things to make life easier. One of your first decisions is what operating system to use to run 3ds Max. 3ds Max 2020 can run on Windows 7, 8, 8.1, and Windows 10 Professional. You also can run 3ds Max on a Mac using Boot Camp or Parallels Desktop.

Caution
3ds Max 2020 does not run on Windows XP or Vista.

Regardless of the operating system you choose, make sure you have installed the latest Service Pack (which you can download for free from Microsoft's website at www.microsoft.com). Each of these operating systems enables you to run multiple copies of 3ds Max at the same time on a single machine. If you're running the 64-bit version of 3ds Max, you need to be running a 64-bit operating system.

Hardware Requirements

To get good performance from 3ds Max, you need a fairly meaty machine. A good default system to use is a Pentium i7 or i9 or an AMD Athlon-based computer with 8GB or more of RAM (and 8GB of swap space) and a decent-sized hard drive and monitor. If need be, you can get by with as little as 4GB of RAM (with 4GB of swap space), but you may spend lots of time watching your computer churn furiously to keep up. A dual Intel Xeon, a dual AMD Athlon, or a 64-bit Opteron system is recommended.

Note
3ds Max is optimized for any system that has the SSE4.2 extended instruction set.

3ds Max has been compiled to run on 64-bit computers. One element of your system that will probably have the greatest impact on the performance of 3ds Max is the graphics card. Any good graphics card has specialized hardware that will take much of the workload off your computer's CPU, freeing it up to do other tasks. 3ds Max is very graphics-intensive, and a little extra money in the graphics card department goes a long way toward boosting your performance.

The good news is that hardware-accelerated graphics cards are becoming cheaper all the time: You can get great cards for $200–$300. When searching for a graphics card, make sure it can support a resolution of at least 2048 x 1024 at 16-bit color with 3D graphics acceleration, including support for OpenGL and/or DirectX 9.0c and 10. You also want a minimum of 2GB on the graphics card, but look for cards with at least 8GB or more for 3D graphics acceleration. You can use some of the graphics boards built to run computer games; however, be aware that some boards claim to support OpenGL but actually support only a subset of it. Before going out to make your purchase, you can access a list of certified hardware on the Autodesk website using Help→3ds Max Services and Support→Certified Hardware menu.

Tip

A video card is only as good as the available drivers. 3ds Max relies heavily on these drivers. If the driver has trouble, your display flickers, has speed problems, or just plain doesn't work. Both nVidia and ATI video cards have excellent drivers with broad support.

For the complete installation, you need 9GB of hard drive space. You can get by with less if you choose not to install all the components. Another handy piece of hardware to have is a scrollable mouse. A scrollable mouse makes moving through menus and the Command Panel easier, and it gives you a third button, which you can use to navigate the viewports. 3ds Max also supports graphics tablets. Also, you must have an internet connection to validate your license. You can do this in other ways that don't require a connection, but having a connected computer is the quickest way.

Tip

If you can install 3ds Max on a separate physical drive from the drive that Windows runs on, you'll see an increase in speed because 3ds Max doesn't need to compete with Windows for access to the hard drive.

Installing 3ds Max 2020

Installing 3ds Max is straightforward. Here's what you need to do:

1. Download and run the Setup program.

2. When the setup program starts, the 3ds Max 2020 Installation window displays a menu of three options: Install, Create Deployment, and Install Tools and Utilities. The Create Deployment link begins a wizard for installing a network version of 3ds Max, and the Install Tools and Utilities link gives you access to install the supporting tools, including the 3ds Max Software Developer's Kit (SDK). Click the Install link to continue.

Note

The Install Tools and Utilities link includes links to install 3ds Max 2020 SDK and SDK Help for Visual Studio, Network License Manager, Allegorithmic Substance Designer, ArchVision Dashboard, SAMreport-Lite (which requires a license), RPC Plug-in with sample content, EASYnat 2.5 trial, Craft Director Studio, PixelActive CityScape promo, Okino PolyTrans and NuGraf plug-in demos, and TurboSquid Tentacles. You also find info links for learning more about each of these tools and utilities along with links to download a trial of other Autodesk products.

3. In the next screen, the 3ds Max 2020 Setup Wizard shows the Software License Agreement. Choose your country and read the corresponding License Agreement. After you've read the agreement, click the "I Accept" option. Click Next to move on.

Tip

Many of the features of 3ds Max are listed as separate installations on the Products screen, including the Composite 2020, Autodesk Backburner (used for network rendering), and the Autodesk Material Library. If you don't select some of these options, you can save some disk space, but the feature is not installed. You can install them later if you need them.

4. The next screen lets you specify to install a Stand-Alone or Network version of the software. You also can input a serial number and product key or install the software as a 30-day trial. Then click Next.

5. The next screen lists the various components. After reviewing the product list, click the check box for any additional products that you want to install at the same time. Be sure to select the version that matches your operating system.

6. After this, the installation starts and progress bars are updated to show the progress of the installation.

7. When the installation is complete, a screen appears informing you so. The final screen also includes an option to view the Readme file that you can enable.

In the lower-left corner of the installation window are links for Documentation and Support.

Tip

Unlike many Readme files, the software's Readme file includes detailed information about a list of features, including clearer descriptions about the features' idiosyncrasies and a troubleshooting section. If you're having trouble with a specific feature, see the Readme file.

Note

The completed 3ds Max installation does not require Windows to be rebooted.

Registering and Activating the Software

After 3ds Max is installed, you can start 3ds Max using the desktop shortcut or the Start menu, but 3ds Max requires that you register and activate the software through Autodesk for permanent use. The software continues to run for 30 days without activation, but after 30 days it quits working.

An Activation screen appears when you start 3ds Max after installation. Using this screen, you can launch the Activation Wizard or run the software unactivated for 30 days. The Registration Wizard automatically appears when you select the Activate the product option and click the Next button.

The first screen of the Registration Wizard lets you get an activation code or enter an activation code if you have one. To obtain an activation code, you need to enter information such as your name, address, and company name. You also can specify whether this copy is an upgrade. If you're upgrading your 3ds Max version, you need to include the serial number for your previous version. The Serial Number and Request Code are automatically filled in using the numbers you entered during the installation. You can authorize the software using a direct connection to the Web, fax, e-mail, or regular mail.

If you receive an activation code via fax, e-mail, or mail, you can select the "Already have an activation code" option on the first screen that appears when you run 3ds Max and click Next. A screen opens where you can enter the activation code. The wizard registers this number with 3ds Max and completes the registration process.

Caution

The activation code is specific to a particular computer and works only for that computer. If you want to install the software on a separate computer, you need to obtain another activation code.

Setting the Display Driver

When 3ds Max is first run, a Welcome Screen dialog box of Learning Movies appears. This screen shows several short movies that explain the basic features of 3ds Max and are worth previewing. These movies are different for 3ds Max and 3ds Max Design.

As part of the 3ds Max default installation, the Nitrous drivers are installed. Because 3ds Max knows these drivers are available, it uses them to run the software by default. Even though Nitrous is used by default, you can change the display driver to Direct3D or OpenGL if you want.

Tip

If you're using 3ds Max to create game assets, you'll probably want to use the Direct3D drivers. This shows you the same results in the viewports that you will see in the game engine.

If 3ds Max is running, you can change the display driver in 3ds Max by choosing Customize→Preferences and clicking the Choose Driver button in the Viewports panel. This opens the Display Driver Selection dialog box where you can select a different display driver. If you change the graphics driver, you need to restart 3ds Max before the new driver is used.

Note

Most recent graphics cards include support for both OpenGL and DirectX built in, so you may want to try OpenGL or Direct3D to see if they work.

FIGURE B.1

This dialog box lets you choose the display driver to use.

You can choose one of several drivers to use in 3ds Max: Nitrous Direct3D 11, Nitrous Direct3D 9, Nitrous Software, Legacy Direct3D or Legacy OpenGL.

Tip

If you want to change a display driver when starting 3ds Max, select the Start→All Programs→Autodesk→Autodesk 3ds Max 2020→Change Graphics Mode program.

Nitrous Direct3D

The Nitrous Direct3D display drivers are the default drivers and the best ones to use if supported by your video card. The Nitrous drivers are based on Direct3D, but have special features unique to 3ds Max, including the ability to render using non-photorealistic methods. The Nitrous designation means that it has been optimized to work with 3ds Max.

Nitrous Software

The software display driver option is the software's own built-in software graphics driver. Because the software driver does not take advantage of any special graphics hardware that your graphics card supports, your computer's CPU does all the work. The nice thing about this option is that it works on any computer, even if you don't have a good graphics card.

If you're having display problems, try starting 3ds Max using Nitrous software drivers to make sure that everything has installed correctly. Next, try the different graphics drivers to see whether you can move up to something faster.

Legacy Direct3D

The Legacy Direct3D option uses the graphics card's hardware capabilities and simulates anything else it needs in software. This option is available if you have an older Direct3D graphics card. Simulating different features makes Direct3D run on a wide range of computers, but it can be much slower. If your graphics card's drivers support all of Direct3D in hardware, using this driver might give you good performance. If it switches to software mode, however, it will be much slower.

Legacy OpenGL

If your older graphics card supports only OpenGL in hardware, this is a good driver to try out. OpenGL works under all Windows operating systems and is typically present on many older CAD-based graphics cards. In order for 3ds Max to use OpenGL, the drivers must support OpenGL 1.1 or later.

Updating 3ds Max

3ds Max includes a feature for checking for and automatically installing updates. From within 3ds Max, you can select Help→Autodesk Product Information→Check for Updates. This menu command opens a web page where the updates are posted.

You also can schedule 3ds Max to look for updates automatically using the Communication Center button in the InfoCenter toolbar. This button provides access to a dialog box where you can be automatically informed when updates are available.

Moving 3ds Max to Another Computer

A stand-alone license of 3ds Max can exist on only one computer at a time. If you reinstall 3ds Max on a different computer, you need to export the license to the new computer. Exporting a license is accomplished using the License Transfer Utility, shown in Figure B.2. This utility can be found in the Start→All Programs→Autodesk→Autodesk 3ds Max 2020 folder.

Note

If you are running 3ds Max from a network, an alternative to exporting a license is borrowing one using the Help→License Borrowing→Borrow License menu command. The Help→License Borrowing→Return License menu command returns the borrowed license when you're finished.

FIGURE B.2

Use the Portable License Utility to move a 3ds Max license to another computer.

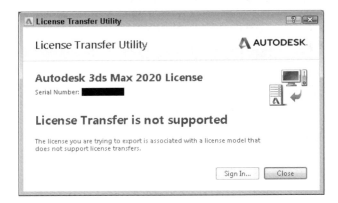

Note

Some licenses, like mine, cannot be transferred. Check with Autodesk to see if a license transfer is possible.

The Portable License Utility requires an internet connection. After you sign in, the utility displays all licenses on the current computer and the identification code for the computer. Select the license to export and click the Export button. This saves the license to a file that you can import using the same utility on the new computer. After you export your license, it continues to work on the original computer for a 24-hour grace period.

Appendix C
Keyboard Shortcuts

IN THIS APPENDIX

Overriding the keyboard shortcuts

Using the Hotkey Map

Main interface shortcuts

Dialog box shortcuts

Character Studio shortcuts

Miscellaneous shortcuts

The key to working efficiently with the Autodesk® 3ds Max® 2020 software is learning the keyboard shortcuts. If you know the keyboard shortcuts, you don't need to spend time moving the mouse cursor all around the interface; you can simply press a keyboard shortcut and get instant access to commands and tools.

Using Keyboard Shortcuts

Most of the major dialog boxes, such as the Material Editor and Track View, have their own set of keyboard shortcuts. When the Keyboard Shortcut Override toggle button on the main toolbar is enabled, keyboard shortcuts work for both the main interface and the separate dialog boxes. If there is a conflict, the dialog box's shortcut takes precedence.

For example, in the main 3ds Max window, the A key toggles the Angle Snap feature on and off, but in the Track View–Curve Editor window, the A key enables Insert Keys mode. If the Curve Editor is open and the Keyboard Shortcut Override toggle is enabled, the Insert Keys mode is enabled. If the Keyboard Shortcut Override toggle is off, the Angle Snap is activated.

If you want to change any of the keyboard shortcuts, the Hotkey Editor dialog box, shown in Figure C.1, provides an interface for viewing and changing keyboard shortcuts. You can open this dialog box using the Customize→Hotkey Editor command. The Hotkey Editor lets you save unique Hotkey Sets that are easy to recall and use. You can even switch to the Autodesk Maya hotkey set using the Hotkey Set drop-down list at the top of the interface.

Chapter 49, "Customizing the Interface," offers more details on creating custom keyboard shortcuts.

New Feature in 2020
The Hotkey Editor is new to 3ds Max 2020.

FIGURE C.1

The Hotkey Editor lets you search by Actions or Hotkey.

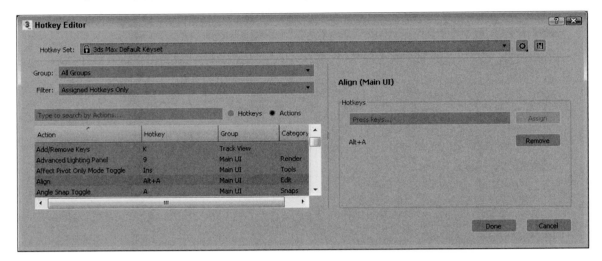

Using the Hotkey Maps

In the Help→3ds Max Resources and Tools menu, you can find the Keyboard Shortcut Map menu command that opens a web browser, shown in Figure C.2, and displays all the current keyboard shortcuts for the main interface.

Several additional images show the hotkey maps for other interfaces including the Track View, Material Editor and UVW Unwrap interfaces.

FIGURE C.2

The Hotkey Maps window displays keyboard shortcuts for multiple interfaces.

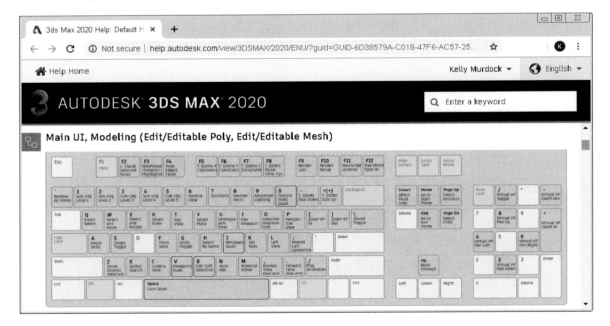

Using the Search Command Field

If you remember the name of a specific command, but you can't find it anywhere, you can use the new Search Command field to locate and execute the command. You can find this feature in the Help menu or you can access it immediately at anytime using the X keyboard shortcut. This opens a small field where you can type the command. As you type, all commands that match the text you enter are presented in a list. You can continue to type or you can select the command from the presented list to execute it. This provides another useful way to get to the commands you need quickly.

Main Interface Shortcuts

Tables C.1 through C.14 present the shortcut keys for the main interface.

TABLE C.1 Menus

Command	Shortcut	Command	Shortcut
File→New	Ctrl+N	Views→Redo View Change	Shift+Y
File→Open...	Ctrl+O	Views→Set Active Viewport→Perspective	P
File→Save	Ctrl+S	Views→Set Active Viewport→Orthographic	U
File→Save As	Shift+Ctrl+S	Views→Set Active Viewport→Front	F
File→References→Asset Tracking Toggle	Shift+T	Views→Set Active Viewport→Top	T
Edit→Undo Scene Operation	Ctrl+Z	Views→Set Active Viewport→Left	L
Edit→Redo Scene Operation	Ctrl+Y	Views→ViewCube→Show/Hide ViewCube	Alt+Ctrl+V
Edit→Hold	Ctrl+H	Views→ViewCube→View Cube Home	Alt+Ctrl+H
Edit→Fetch	Alt+Ctrl+F	Views→SteeringWheels →Toggle SteeringWheels	Shift+W
Edit→Delete	Delete	Views→SteeringWheels →Tour Building Wheel	Shift+Ctrl+J
Edit→Clone	Ctrl+V	Views→Cameras→Create Physical Camera from View	Alt+Shift+C, Ctrl+C
Edit→Move	W	Views→Cameras→Create Standard Camera from View	Alt+Shift+ Ctrl+C
Edit→Rotate	E	Views→xView→Show Statistics	7
Edit→Transform→Transform Type-In	F12	Views→Viewport Background	Alt+B
Edit→Select All	Ctrl+A	Views→Expert Mode	Alt+Ctrl+X
Edit→Select None	Ctrl+D	Graph Editors→Particle	6

		View	
Edit→Select Invert	Ctrl+I	Rendering→Render	Shift+Q, Shift+F9
Edit→Select Similar	Shift+Ctrl+A	Rendering→Render Setup	F10
Edit→Select By Name	H	RenderingEnvironment...	8
Tools→Isolate Selection	Alt+Q	Rendering→Material Editor	M
Tools→Align→Align	Alt+A	Scripting→MAXScript Listener	F11
		Help-> Search 3ds Max Commands...	X
Tools→Align→Quick Align	Shift+A	Edit Menu	Alt+E
Tools→Align→Spacing Tool	Shift+I	Tools Menu	Alt+T
Tools→Align→Normal Align	Alt+N	Group Menu	Alt+G
Tools→Preview - Grab Viewport→Create Preview Animation	Shift+V	Views Menu	Alt+V
Tools→Grids and Snaps→Snaps Toggle	S	Create Menu	Alt+C
Tools→Grids and Snaps→Angle Snap Toggle	A	Modifiers Menu	Alt+M
Tools→Grids and Snaps→Enable Axis Constraints in Snap	Alt+F3	Graph Editors Menu	Alt+D
Views→Undo View Change	Shift+Z	Rendering Menu	Alt+R

TABLE C.2 Main and Floating Toolbars

Command	Shortcut	Command	Shortcut
Undo Scene Operation	Ctrl+Z	Snaps Use Axis Constraints Toggle	Alt+D, Alt+F3
Redo Scene Operation	Ctrl+Y	Snap Type Cycle	Alt+S
Cycle Selection Region	Q	Snap Hit Cycle	Alt+Shift+S
Select by Name	H	Align	Alt+A
Rectangular, Circular, Fence, Lasso Selection, Paint Cycle	Ctrl+F, Q	Normal Align	Alt+N
Window/Crossing Toggle	Shift+O	Place Highlight	Ctrl+H
Select and Move	W	Material Editor	M
Select and Rotate	E	Render Setup	F10

Select and Scale	R	Quick Render	Shift+Q, Shift+F9
Scale Cycle	R, Ctrl+E	Restrict to X	F5
Keyboard Shortcut Override Toggle	Alt+Shift+Ctrl+X	Restrict to Y	F6
Snap Toggle	S	Restrict to Z	F7
Angle Snap Toggle	A	Restrict Plane Cycle	F8

TABLE C.3 Viewports

Command	Shortcut	Command	Shortcut
Maximize Viewport Toggle	Alt+W	Dynamic Resizing	Drag viewport borders
Disable Viewport	Shift+Ctrl+D	Transform Gizmo Size Down	-
Show/Hide Grids	G	Transform Gizmo Size Up	=
Show/Hide ViewCube	Alt+Ctrl+V	Shade Selected Subobject Faces	F2
ViewCube Home	Alt+Ctrl+H	Wireframe/Shaded Toggle	F3
Toggle SteeringWheels	Shift+W	View Edged Faces	F4
Tour Building Wheel	Shift+Ctrl+J	Sound Toggle	\
Show Statistics	7	Default Lighting	Ctrl+L
Create Preview Animation	Shift+V	See-Through Display	Alt+X
Perspective View	P	Redraw All Views	`
Orthographic User View	U	Show/Hide Cameras	Shift+C
Top View	T	Show/Hide Geometry	Shift+G
Front View	F	Show/Hide Helpers	Shift+H
Left View	L	Show/Hide Lights	Shift+L
Camera View	C	Show/Hide Particle Systems	Shift+P
Light View	Shift+4 ($)	Show/Hide Shapes	Shift+S
Show Safeframes	Shift+F	Undo View Change	Shift+Z
High Quality	Shift+F3	Redo View Change	Shift+Y
Configure Viewport Background	Alt+B	Viewports Pop-up Menu	V

TABLE C.4 Key and Time Controls

Command	Shortcut	Command	Shortcut
Selection Lock Toggle	Spacebar	Backup Time One Unit	, (comma)
Auto Key Mode	N	Forward Time One Unit	. (period)
		Go to Start Frame	Home

Set Keys	K	Go to End Frame	End
Play/Stop Animation	/		

TABLE C.5 Viewport Navigation Controls

Command	Shortcut	Command	Shortcut
Zoom Mode	Alt+Z	Zoom Viewport Out	] or scroll wheel backward
Zoom Extents	Alt+Ctrl+Z	Zoom In by Steps	Ctrl+= or [
Zoom Extents All	Shift+Ctrl+Z	Zoom Out by Steps	Ctrl+- or]
Zoom Extents Selected	Z	Pan View	drag with middle button
Zoom Region Mode	Ctrl+W	Interactive Pan	I (held down)
Zoom Viewport In	[or scroll wheel forward	Walk Through mode	Up arrow
Arc Rotate	Ctrl+R, Alt+drag with middle button	Min/ Max Toggle	Alt+W

TABLE C.6 Walk Through Mode

Command	Shortcut	Command	Shortcut
Forward	W, Up Arrow	Decelerate	Z
Back	S, Down Arrow	Increase Step Size	]
Left	A, Left Arrow	Decrease Step Size	[
Right	D, Right Arrow	Reset Step Size	Alt+[
Up	E, Shift+Up Arrow	Level	Shift+Spacebar
Down	C, Shift+Down Arrow	Lock Vertical Rotation	Spacebar
Accelerate	Q	Exit Walkthrough Mode	Esc

TABLE C.7 Quad menus

Command	Shortcut	Command	Shortcut
Viewports Quad menu	V	Viewports Quad menu	V
UI, Editors, Explorers Quad menu	J	MassFX Quad menu	Shift+Alt+right mouse click
Texture Tools Quad menu	0 (zero)	Modeling Quad menu	Ctrl+right mouse click
Lock Selection, Disable and Maximize Viewport	spacebar	Snap Quad menu	Shift+right mouse click

Quad menu			
Animation Quad menu	Alt+right mouse click	Custom 2 Quad menu	Shift+Ctrl+Alt+right mouse click
Lighting/Rendering Quad menu	Ctrl+Alt+right mouse click	Custom 3 Quad menu	Shift+Ctrl+right

TABLE C.8 Subobjects

Command	Shortcut	Command	Shortcut
Subobject mode toggle	Ctrl+B	Subobject Level 4	4
Subobject Level 1	1	Subobject Level 5	5
Subobject Level 2	2	Delete Subobject selection	Delete
Subobject Level 3	3		

TABLE C.9 Hierarchies

Command	Shortcut	Command	Shortcut
Select Ancestor	Page Up		
Select Child	Page Down	Select Entire Hierarchy	Double-click parent

Table C.10 Editable Mesh

Command	Shortcut	Command	Shortcut
Vertex Subobject Mode	1	Bevel Mode	Ctrl+Shift+B
Edge Subobject Mode	2	Chamfer Mode	Ctrl+Shift+C
Face Subobject Mode	3	Extrude Mode	Shift+E
Polygon Subobject Mode	4	Edge Turn	Ctrl+T
Element Subobject Mode	5	Weld Selected	Alt+Shift+W
Attach	Alt+Shift+D	Weld Target Mode	Ctrl+Shift+W
Detach	Ctrl+Alt+D	Ignore Backfacing	Alt+Shift+X
Cut Mode	Alt+C	Break Vertices	Shift+R
Hide Selected	Alt+H	Collapse	Ctrl+Alt+C
Unhide All	Alt+U	Soft Selection	Shift+B

All commands in Table C.11 that are listed in bold can be used as press-and-release keyboard shortcuts. These enable you to access a different command by pressing and holding the keyboard shortcut; when you release the keyboard shortcut, 3ds Max returns to the previous mode. Press-and-release commands are activated only if the Overrides Active option in the Keyboard panel of the Customize User Interface dialog box is enabled.

The following commands can be accessed using press-and-release keyboard shortcuts if you define a keyboard shortcut for them: Bridge, Constrain to None, Constrain to Normal, Paint Deform Push/Pull/Relax, Paint Soft Selection, and Turn.

Note

Don't confuse the keyboard shortcuts that work with the Edit Poly modifier with those for the Editable Poly object. Some of the same keyboard shortcuts work for both, but most don't work for the Editable Poly object.

TABLE C.11 Edit Poly Modifier/Editable Poly

Command	Shortcut	Command	Shortcut
Vertex Subobject Mode	1	**Bevel Mode**	**Shift+Ctrl+ B**
Edge Subobject Mode	2	**Chamfer Mode**	**Shift+Ctrl+ C**
Border Subobject Mode	3	**Extrude Mode**	**Shift+E**
Polygon Subobject Mode	4	Extrude Along Spline Mode	Ctrl+Alt+Shift+E
Element Subobject Mode	5	MeshSmooth	Ctrl+M
Ignore Backfacing in Selections	Alt+Shift+X	**Constrain to Edges**	**Shift+X**
Grow Selection	Ctrl+Page Up	Collapse Poly Object	Ctrl+Alt+C
Shrink Selection	Ctrl+Page Down	Hide	Alt+H
Select Edge Loop	Alt+L	Hide Unselected	Alt+I
Select Edge Ring	Alt+R	Unhide All	Alt+U
Attach	Alt+Shift+D	**Target Weld Mode**	**Shift+Ctrl+ W**
Detach	Ctrl+Alt+D	Make Planar	Alt+M
Break	Shift+R	Cap Poly Object	Alt+P
Connect	Shift+Ctrl+E	Edit Soft Selection Mode	B
Cut	**Alt+C**	Use Soft Selection	Shift+B
Quickslice Mode	**Shift+Ctrl+ Q**	Repeat Last Operation	;

TABLE C.12 Free-Form Deformations

Command	Shortcut	Command	Shortcut
Switch to Control Point Level	Alt+Shift+C	Switch to Set Volume Level	Alt+Shift+S
Switch to Lattice Level	Alt+Shift+L	Switch to Top Level	Alt+Shift+T

TABLE C.13 Edit Normals Modifier

Command	Shortcut	Command	Shortcut
Object Level	5	Paste Normal	Ctrl+V
Normal Level	1	Reset Normals	R
Vertex Level	2	Specify Normals	S
Edge Level	3	Unify Normals	U
Face Level	4	Make Explicit	E
Copy Normal	Ctrl+C	Break Normals	Shift+R

TABLE C.14 Hair Styling

Command	Shortcut	Command	Shortcut
Select Hair by Ends	Ctrl+1	Puff Roots Transformation	Ctrl+P or Shift+Ctrl+3
Select Whole Guide	Ctrl+2	Clump Transformation	Ctrl+M or Shift+Ctrl+4
Select Guide Vertices	Ctrl+3	Rotate Transformation	Shift+Ctrl+5
Select Hair by Roots	Ctrl+4	Scale Transformation	Shift+Ctrl+6
Invert Selection	Ctrl+I	Attenuate Hair	Shift+Ctrl+A
Rotate Selection	Shift+Ctrl+R	Pop Selected	Shift+Ctrl+P
Expand Selection	Ctrl+Up, Ctrl+Page Up	Pop Zero-sized	Shift+Ctrl+Z
Hide Selected Guides	Alt+H	Recomb Hair	Shift+Ctrl+M
Show Hidden Guides	Alt+U	Replace Rest	Shift+Ctrl+T
Hair Brush	Ctrl+B	Toggle Collisions	Shift+Ctrl+C
Hair Cut	Alt+C	Toggle Hairs	Shift+Ctrl+I
Select Hair	Q	Lock Guides	Ctrl+L
Distance Fade	Shift+Ctrl+F	Unlock Guides	Shift+Ctrl+L
Ignore Back Hairs	Alt+Shift+X	Undo Styling	Ctrl+Z
Translate Transformation	Ctrl+T or Shift+Ctrl+1	Split Selected Hair Groups	Shift+Ctrl+-
Stand Transformation	Ctrl+N or Shift+Ctrl+2	Merge Selected Hair Groups	Shift+Ctrl+=

Dialog Box Shortcuts

Tables C.15 through C.22 show shortcut keys that work with the various dialog boxes. The dialog box must be selected for these shortcuts to work. It is possible for modeless dialog boxes to have the dialog box visible but not selected.

TABLE C.15 Slate Material Editor

Command	Shortcut	Command	Shortcut
Assign Material to Selection	A	Zoom Region Tool	Ctrl+W
Delete Selected	Delete	Zoom Extents	Alt+Ctrl+Z
Update Selected Previews	U	Zoom Extents Selected	Z
Auto Update Selected Previews	Shift+U	Show Grid	G
Select Tool	Q	Lay Out All	L
Select All	Ctrl+A	Lay Out Children	C
Select None	Ctrl+D	Hide Unused Nodeslots	H
Select Invert	Ctrl+I	Move Children	Alt+C
Select Children	Shift+Page Down	Material/Map Browser	O
Select Tree	Ctrl+T	Parameter Editor	P
Pan Tool	Ctrl+P	Navigator	N
Pan to Selected	Alt+P	Zoom Tool	Alt+Z

TABLE C.16 Track View

Command	Shortcut	Command	Shortcut
Filters	Alt+F	Nudge Keys Left	Left Arrow
Assign Controller	C	Nudge Keys Right	Right Arrow
Copy Controller	Ctrl+C	Move Highlight Down	Down Arrow
Paste Controller	Ctrl+V	Move Highlight Up	Up Arrow
Make Controller Unique	U	Backup Time One Unit	, (comma)
Add Keys	K	Forward Time One Unit	. (period)
Move Keys	M	Scroll Down	Ctrl+Down Arrow
Snap Frames	S	Scroll Up	Ctrl+Up Arrow
Apply Ease Curve	Ctrl+E	Undo Scene Operation	Ctrl+Z
Apply Multiplier Curve	Ctrl+M	Redo Scene Operation	Ctrl+A
Lock Tangents Toggle	L	Pan	Ctrl+P
Add Parameters	Ctrl+1	Zoom	Alt+Z
Expand Track Toggle	T, Enter	Zoom Region	Ctrl+W
Retime Move Left	Ctrl+Left	Frame Horizontal Extents Keys	Alt+Shift+Ctrl+Z
Retime Move Right	Ctrl+Right	Frame Horizontal Extents	Alt+Ctrl+Z
Scroll Tracks Up	Ctrl+Up	Select All	Ctrl+A

Scroll Track Down	Ctrl+Down	Lock Selection Quad menu	Ctrl+L

TABLE C.17 Schematic View

Command	Shortcut	Command	Shortcut
Connect Tool	C	Refresh View	Ctrl+U
Select Tool	Q	Display Floater	D
Select All Nodes	Ctrl+A	Free Selected	Alt+S
Select None	Ctrl+D	Free All	Alt+A
Select Inverted	Ctrl+I	Hide Selected	Alt+H
Select Children	Shift+Page Down	Shrink Toggle	Shift+S
Pan	Ctrl+P	Move Children	Alt+C
Zoom	Alt+Z	Preferences	Alt+F
Zoom Region	Ctrl+W	Add Bookmark	B
Zoom Extents	Alt+Ctrl+Z	Next Bookmark	Right Arrow
Zoom Extents Selected	Z	Previous Bookmark	Left Arrow
Show Grid Toggle	G	Rename Object	F2

TABLE C.18 Video Post

Command	Shortcut	Command	Shortcut
New Sequence	Ctrl+Shift+N	Zoom Extents	Ctrl+Alt+Z
Pan	Ctrl+P	Zoom Region	Ctrl+W
Execute Sequence	F9		

TABLE C.19 Edit UVWs Dialog Box

Command	Shortcut	Command	Shortcut
Open Edit UVWs interface	Ctrl+E	Snap	S
Move Mode	W	Pan	Ctrl+P
Rotate Mode	E	Zoom	Alt+Z
Scale Mode	R	Zoom Region	Ctrl+W
Texture Vertex Weld Selected	Alt+Shift+W	Zoom Extents	Alt+Ctrl+Z
Texture Vertex Target Weld	Ctrl+Shift+W	Zoom Extents Selected	Z
Break Selected Vertices	Shift+R	Zoom to Gizmo	Shift+Spacebar
Detach Edge Vertices	Ctrl+Alt+D	Update Map	U
Preferences	O	Edit UVWs	Ctrl+E
Hide Selected	Alt+H	Lock Selected Vertices	Ctrl+L

Unhide All	Alt+U	Texture Vertex Contract Selection	Ctrl+Page Down
Filter Selected Faces	Alt+F	Texture Vertex Expand Selection	Ctrl+Page Up
Planar Map Faces/Patches	Enter	Texture Vertex Move Mode	Q
Show Seams in Viewports	Alt+E	Texture Vertex Rotate Mode	E
Freeze Selected	Ctrl+F	Zoom Selected Elements	Alt+Shift+Ctrl+Z
Unfreeze All	Shift+Ctrl+F	Pan	Ctrl+P
Relax Brush cycle	Alt+R	Ignore Back Polygons	Alt+Shift+X
Rotate +90	Ctrl+right	Soft Selection	Shift+B
Rotate -90	Ctrl+left		

TABLE C.20 ActiveShade

Command	Shortcut	Command	Shortcut
Close	Q	Select Object	Q
Draw Region	D	Update	U
Render	F9	Initialize	P

TABLE C.21 Particle Flow

Command	Shortcut	Command	Shortcut
Particle View window	6	Synchronize Particle Flow Layers	Alt+Ctrl+L
Particle Emission Toggle	;	Select All	Ctrl+A
Selected Particle Emission Toggle	Shift+;	Copy Selected	Ctrl+C
Clean Up Particle Flow	Alt+Ctrl+P	Paste Selected	Ctrl+V
Repair Particle Flow Cache System	Alt+Ctrl+C	Open Particle Flow Preset Manager	Alt+Ctrl+M
Reset Particle View	Alt+Ctrl+R		

TABLE C.22 MAXScript Editor

Command	Shortcut	Command	Shortcut
File→New	Ctrl+N	Edit→Box Comment	Ctrl+Shift+C
File→Open	Ctrl+O	Edit→Stream Comment	Ctrl+Shift+Q
File→Open Selected File	Ctrl+Shift+O	Edit→Make Selection Uppercase	Ctrl+Shift+U
File→Close	Ctrl+W	Edit→Make Selection Lowercase	Ctrl+U
File→Save	Ctrl+S	Search→Find	Ctrl+F
File→Save As	Ctrl+Shift+S	Search→Find Next	F3
File→Save a Copy	Ctrl+Shift+P	Search→Find Previous	Shift+F3
File→Revert	Ctrl+R	Search→Replace	Ctrl+H
File→Source Control→Add File	Ctrl+F5	Search→Find in Files	Ctrl+Shift+F
File→Source Control→Remove File	Ctrl+Shift+F5	Search→Go to	Ctrl+G
File→Source Control→Check In	Shift+F5	Search→Next Bookmark	F2
File→Source Control→Check Out	F5	Search→Previous Bookmark	Shift+F2
File→Print	Ctrl+P	Search→Toggle Bookmark	Ctrl+F2
Edit→Undo	Ctrl+Z	View→Full Screen	F11
Edit→Redo	Ctrl+Y	View→Whitespace	Ctrl+Shift+8
Edit→Cut	Ctrl+X	View→End of Line	Ctrl+Shift+9
Edit→Copy	Ctrl+C	Tools→Evaluate All	Ctrl+E
Edit→Paste	Ctrl+V	Tools→Evaluate Line/Selection	Shift+Enter
Edit→Select All	Ctrl+A	Tools→Next Message	F4
Edit→Duplicate	Ctrl+Shift+D	Tools→Previous Message	Shift+F4
Edit→Delete	Delete	Tools→Clear Output	Shift+F5
Edit→Delete All	Ctrl+Alt+D	Tools→Switch Pane	Ctrl+F6
Edit→Match Brace	Ctrl+B	Options→Change Indentation Settings	Ctrl+Shift+I
Edit→Select to Brace	Ctrl+Shift+B	Options→Use Monospaced Font	Ctrl+F11
Edit→Complete Word	Ctrl+Enter	Language→Text	Shift+F11
Edit→Expand Abbreviation	Ctrl+Shift+A	Windows→Previous	Shift+F6
Edit→Insert Abbreviation	Ctrl+Shift+R	Windows→Next	F6
Edit→Block Comment or Uncomment	Ctrl+Q	Help→Help	F1

Miscellaneous Shortcuts

In addition to specific shortcuts for the main interface and the dialog boxes, 3ds Max provides several general shortcuts that can be used in many different places, as listed in Table C.23.

TABLE C.23 General Hotkeys

Command	Shortcut	Command	Shortcut
Numeric Expression Evaluator	Ctrl+N when a spinner field is selected	Highlight Any Text Field	Double-click current value
Cut Value	Ctrl+X	Nudge Selection	Arrow keys
Copy Value	Ctrl+C	Make Viewport Active	Right-click in an inactive viewport
Paste Value	Ctrl+V	Display Quad menus	Right-click in an active viewport
Apply Settings	Enter	Delete Object or Node	Delete
Highlight Next Text Field	Tab	Help	F1
Highlight Previous Text Field	Shift+Tab		

Index

F

G

S

W

X

Z